REVISED EDITION

WORLD GEOGRAPHY TODAY

ANNOTATED TEACHER'S EDITION

HOLT, RINEHART AND WINSTON

Harcourt Brace & Company

Austin • New York • Orlando • Atlanta • San Francisco • Boston • Dallas • Toronto • London

The Authors

Dr. Robert J. Sager is Professor and Chair of Earth Sciences at Pierce College in Tacoma, Washington. Dr. Sager received his BS in geology and geography and MS in geography from the University of Wisconsin and holds a JD in international law from Western State University College in Law. He is the coauthor of several geography textbooks and has written many articles and educational media programs on the geography of the Pacific. Dr. Sager has received several National Science Foundation study grants and was a recipient of the University of Texas NISOD National Teaching Excellence Award. A founding member of the Southern California Geographic Alliance, Dr. Sager is vice president of the Association of Washington Geographers and a member of the Washington Geographic Alliance.

Dr. David M. Helgren is Director of the Center for Geographic Education at San Jose State University in California, where he also teaches in the Department of Geography and Environmental Studies. Dr. Helgren received his PhD in geography from the University of Chicago. He is the coauthor of several geography textbooks and has written many articles on the geography of Africa. Awards from the National Geographic Society, the National Science Foundation, and the L.S.B. Leakey Foundation have supported his many field research projects. Dr. Helgren is a former president of the California Geographical Society and a founder of the Northern California Geographic Alliance.

Requests for permission to make copies of any part of the work should be mailed to: Permissions Department, Holt, Rinehart and Winston, Inc., 6277 Sea Harbor Drive, Orlando, Florida 32887-6777.

Some material in this work previously appeared in WORLD GEOGRAPHY TODAY copyright © 1995, 1992, 1989, 1985, 1980, 1976, 1971, 1966, 1962, 1960 by Holt, Rinehart and Winston, Inc. Copyright renewed 1994 by Lee Durand, Loyal Durand, Philip Durand, Kemper B. Durand, Dorothy Durand, Melitta E. Roemer and Meta Willis. Copyright renewed 1988 by Saul Israel and Melitta E. Roemer. All rights reserved.

For permission to reprint copyrighted material, grateful acknowledgment is made to the following sources: **Amotz Asa-El:** From "Crescent Comeback?" by Amotz Asa-El from *World Monitor*, December 1992. Copyright © 1992 by Amotz Asa-El. **The Dallas Morning News:** From "Crowned in Controversy" by Colleen O'Connor from *The Dallas Morning News*, September 28, 1993. Copyright © 1993 by The Dallas Morning News. **Navajivan Trust:** From *Hind Swaraja* or *Home Rule* by Mohandas K. Gandhi. Copyright 1946 by Navajivan Trust, Ahmedabad, India. **Prentice-Hall/a division of Simon & Schuster, Inc.:** From *The Analects of Confucius*, translated by Arthur Waley. Copyright © 1939 by Macmillan College Publishing Company, Inc. **Franklin Watts, Inc.:** From *The Incas* by Barbara L. Beck, revised by Lorna Greenberg. Copyright © 1966, 1983 by Franklin Watts, Inc. **World Press Review:** From "Poverty Dooms the Planet" by Paulo Nogueira Neto from *World Press Review*, June 1992. Copyright © 1992 by World Press Review. Originally appeared in *Visão*, São Paulo, Brazil.

REFERENCES:

From "Communalism and Community in the African Heritage" by Kofi Asare Opoku from *Dialogue in Community: Essays in Honour of S. J. Samartha*, edited by C. D. Jathanna. Copyright © 1982 by The Karnataka Theological Research Institute.

PHOTO CREDITS:

Page T4–T13(bdr), Dennis Hallinan/FPG; T4, T5(bckgd), Kathleen Campbell/Allstock; T4, T5(t), (b), NOAA; T6, T7(bckgd), Steven L. Hilty/Bruce Coleman, Inc.; T6, T7(t), (b), NOAA; T8, T9(bckgd), Gerald Cubitt; T8, T9(t), (b), NOAA; T10–T11(bckgd), David Ball/The Stock Market; T10, T11, NOAA; T12, T13(bckgd), Mark Snyder/The Stock Market.

Printed in the United States of America
ISBN 0-03-095648-X

3 4 5 6 7   032   98

Consultants and Reviewers

Robert E. Gabler
Professor of Geography
Western Illinois University
Macomb, Illinois

Truman Hartshorn
Professor of Geography
Georgia State University
Atlanta, Georgia

Mohammad Hemmasi
Professor of Geography
University of North Dakota
Grand Forks, North Dakota

Robert B. Kent
Professor of Geography and
    Planning
University of Akron
Akron, Ohio

Christopher T. King
Social Studies Department Chairman
Walton High School
DeFuniak Springs, Florida

Michael Kort
Professor of Social Science
Boston University
Boston, Massachusetts

Jane Young Leatherman
Social Studies Teacher
Durant High School
Durant, Oklahoma

Nancy Lehmann-Carssow
Social Studies Teacher
Lanier High School
Austin, Texas

James F. Marran
Chair Social Studies Department,
    Emeritus
New Trier High School
Winnetka, Illinois

Todd O'Banion
Social Studies Teacher
Scott County High School
Georgetown, Kentucky

Roger Sass
Social Studies Teacher
Kelvyn Park High School
Chicago, Illinois

Ron Scholten
Social Studies Department Chair
Providence High School
Burbank, California

Earl P. Scott
Professor of Geography
University of Minnesota
Minneapolis, Minnesota

Catharine Stoner
Social Studies Teacher
Wentzville High School
Wentzville, Missouri

Bill Takizawa
Professor of Geography
San Jose State University
San Jose, California

Kenneth Wood
Social Studies Teacher
Appleton West High School
Appleton, Wisconsin

TABLE OF CONTENTS

National Standards *T14*

Introduction *T15*

Meeting Individual Needs *T17*

Portfolio Assessment *T18*

Multicultural Awareness *T19*

The World Geography Today *Program* *T20*

To the Student .... *xvi*

Student Atlas *A1*

Skills Handbook *S1*

UNIT 1

Geography: A View of Our World *1*

Interleaf 1A

Chapter 1: The Geographer's World 2

   Chapter 1 Check **9**

Interleaf 9A

Chapter 2: The Earth In Space **10**

   Chapter 2 Check **17**

Interleaf 17A

Chapter 3: The Earth's Atmosphere and Climates **18**

   ◆ **Planet Watch: Watching Over Earth** **22**

   Chapter 3 Check **33**

Interleaf 33A

Chapter 4: The Water Planet **34**

   Chapter 4 Check **41**

Interleaf 41A

Chapter 5: Global Landforms **42**

   Chapter 5 Check **51**

Interleaf 51A

Chapter 6: Patterns of Life **52**

   Chapter 6 Check **61**

Unit 1 Review **62**

UNIT 2

People and Geography 64

Interleaf 65A

Chapter 7: Global Cultures 66

◆ Global Connections: Building a World Culture 70

Chapter 7 Check 77

Interleaf 77A

Chapter 8: Global Economics and Global Population 78

Chapter 8 Check 85

Interleaf 85A

Chapter 9: Resources and Environmental Change 86

◆ Planet Watch: What *You* Do? 90

Chapter 9 Check 97

Unit 2 Review 98

Regions in Transition: What's in a Region? ...100

UNIT 3

The United States and Canada 102

◆ One World, Many Voices 104

Interleaf 105A

Chapter 10 Introducing the United States 106

◆ Mapping the Region 107

Chapter 10 Check 119

Interleaf 119A

Chapter 11: The Northeastern United States 120

◆ Planet Watch: Racing with the Sun 128

Chapter 11 Check 131

Interleaf 131A

Chapter 12: The Southern United States 132

Chapter 12 Check 139

Interleaf 139A

Chapter 13: The Midwestern United States 140

Chapter 13 Check 149

Interleaf 149A

Chapter 14: The Interior West 150

Chapter 14 Check 157

Interleaf 157A

Chapter 15: The Pacific States 158

◆ Cities of the World: Los Angeles 162

◆ Global Connections: The Pacific Rim 166

Chapter 15 Check 171

Interleaf 171A

Chapter 16: Canada 172

◆ News from Around the World: Canada 178

Chapter 16 Check 181

Unit 3 Review 182

UNIT 4

Middle and South America 184

◆ One World, Many Voices 186

Interleaf 187A

Chapter 17: Introducing Middle and South America 188

◆ Mapping the Region 189

Chapter 17 Check 201

Interleaf 201A

Chapter 18: Mexico 202

◆ Themes in Geography: Mexico's Northern
Border Region 206

◆ Cities of the World: Mexico City 210

Chapter 18 Check 213

Interleaf 213A

Chapter 19: Central America and the Caribbean Islands 214

◆ News from Around the World: Costa Rica..............218

◆ Planet Watch: Case Study: Saving the Rain Forests 220

Chapter 19 Check 225

Interleaf 225A

Chapter 20: Caribbean South America 226

Chapter 20 Check 233

Interleaf 233A

Chapter 21: Atlantic South America 234

Chapter 21 Check 243

Interleaf 243A

Chapter 22: Pacific South America 244

Chapter 22 Check 251

Unit 4 Review..252

UNIT 5

Europe 254

◆ One World, Many Voices................................256

Interleaf 257A

Chapter 23: Introducing Europe 258

◆ Mapping the Region 259

◆ Global Connections: Tourism 268

Chapter 23 Check 271

Interleaf 271A

Chapter 24: Northern Europe 272

Chapter 24 Check 281

Interleaf 281A

Chapter 25: West Central Europe 282

◆ Cities of the World: Berlin............................288

◆ Themes in Geography: Polders of the Netherlands..........292

Chapter 25 Check 295

Interleaf 295A

Chapter 26: Southern Europe 296

◆ Planet Watch: Saving Venice........................302

◆ News from Around the World: Europe 305

Chapter 26 Check 307

Interleaf 307A

Chapter 27: Eastern Europe 308

Chapter 27 Check 315

Unit 5 Review..316

Regions in Transition:
Redrawing the European Map 318

UNIT 6

Russia and Northern Eurasia 320

◆ One World, Many Voices................................322

Interleaf 323A

Chapter 28: Introducing Russia and Northern Eurasia..........324

Mapping the Region 325

Chapter 28 Check..333

Interleaf 333A

Chapter 29: Russia 334

◆ Planet Watch: Lake Baykal **338**

◆ Cities of the World: St. Petersburg **344**

◆ News from Around the World: Russia **348**

Chapter 29 Check **349**

Interleaf **349A**

Chapter 30: Ukraine, Belarus, and the Caucasus **350**

Chapter 30 Check **357**

Interleaf **357A**

Chapter 31: Central Asia **358**

◆ Global Connections: The Silk Road **362**

Chapter 31 Check **365**

Unit 6 Review **366**

Regions in Transition:
Facing the Past and Present *368*

UNIT 7

Southwest Asia *370*

◆ One World, Many Voices **372**

Interleaf **373A**

Chapter 32: Introducing Southwest Asia **374**

◆ Mapping the Region **375**

◆ Planet Watch: The Politics of Water **382**

Chapter 32 Check **385**

Interleaf **385A**

Chapter 33: The Eastern Mediterranean **386**

◆ News from Around the World: Israel **391**

◆ Cities of the World: Jerusalem **392**

Chapter 33 Check **397**

Interleaf **397A**

Chapter 34: Interior Southwest Asia **398**

Chapter 34 Check **407**

Unit 7 Review **408**

Regions in Transition:
Differences and Connections *410*

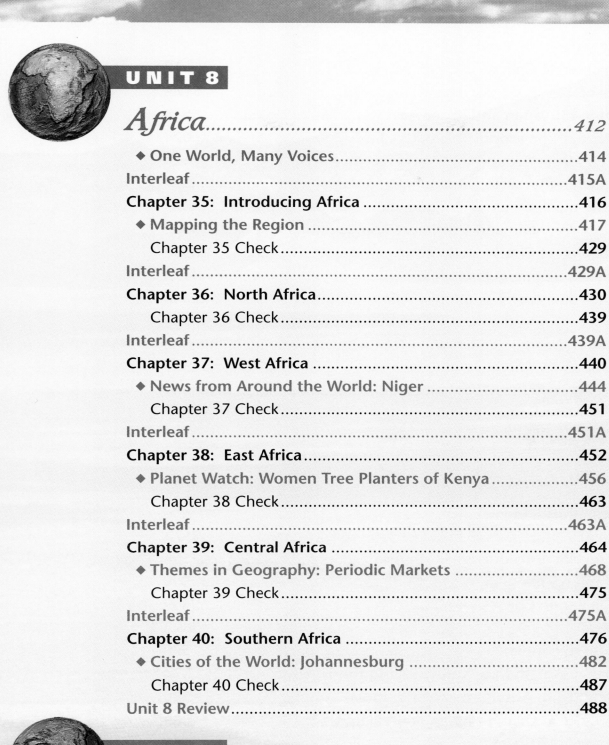

UNIT 8

Africa...412

◆ One World, Many Voices.................................414
Interleaf...415A
Chapter 35: Introducing Africa 416
◆ Mapping the Region 417
Chapter 35 Check.................................429
Interleaf...429A
Chapter 36: North Africa.................................430
Chapter 36 Check.................................439
Interleaf...439A
Chapter 37: West Africa 440
◆ News from Around the World: Niger 444
Chapter 37 Check.................................451
Interleaf...451A
Chapter 38: East Africa 452
◆ Planet Watch: Women Tree Planters of Kenya.................................456
Chapter 38 Check.................................463
Interleaf...463A
Chapter 39: Central Africa 464
◆ Themes in Geography: Periodic Markets 468
Chapter 39 Check.................................475
Interleaf...475A
Chapter 40: Southern Africa 476
◆ Cities of the World: Johannesburg 482
Chapter 40 Check.................................487
Unit 8 Review.................................488

UNIT 9

East and Southeast Asia.................................490

◆ One World, Many Voices.................................492
Interleaf...493A
Chapter 41: Introducing East and Southeast Asia.................................494
◆ Mapping the Region 495
Chapter 41 Check.................................505
Interleaf...505A
Chapter 42: China, Taiwan, and Mongolia 506

Chapter 42 Check...517

Interleaf 517A

Chapter 43: Japan and Korea...............................**518**
   ◆ Cities of the World: Tokyo................................524
   ◆ News from Around the World: South Korea 528
   Chapter 43 Check...529

Interleaf 529A

Chapter 44: Mainland Southeast Asia.........................**530**
   Chapter 44 Check...539

Interleaf 539A

Chapter 45 : Island Southeast Asia **540**
   ◆ Global Connections: Singapore 544
   ◆ Planet Watch: Rice Growers Take a Stand 548
   Chapter 45 Check...551

Unit 9 Review..**552**

UNIT 10

South Asia...554
   ◆ One World, Many Voices...................................556

Interleaf 557A

Chapter 46: Introducing South Asia **558**
   ◆ Mapping the Region 559
   Chapter 46 Check...**569**

Interleaf 569A

Chapter 47: India...**570**
   ◆ Planet Watch: Case Study: Saving the Tiger...............574
   ◆ News from Around the World: India 579
   ◆ Themes in Geography: Indian Villages.....................580
   Chapter 47 Check...**583**

Interleaf 583A

Chapter 48: The Indian Perimeter **584**
   ◆ Global Connections: The Movement of People 588
   Chapter 48 Check...**593**

Unit 10 Review...**594**

UNIT 11

The Pacific World and Antarctica *596*

◆ One World, Many Voices...598

Interleaf 599A

Chapter 49: Australia..**600**

◆ Mapping the Region 601

◆ Cities of the World: Sydney 608

Chapter 49 Check **611**

Interleaf 611A

Chapter 50: The Pacific Islands and Antarctica **612**

◆ News from Around the World: New Zealand....................616

◆ Planet Watch: Preserving Antarctica 618

Chapter 50 Check **621**

Unit 11 Review...**622**

Appendix *624*

Glossary *637*

Index *645*

Acknowledgments *660*

List of Maps

The World: Physical.....................................A2–A3
The World: Political..A4–A5
The United States of America: Physical.......A6–A7
The United States of America: Political A8–A9
North America: Physical...................................A10
North America: Political....................................A11
South America: Physical...................................A12
South America: Political A13
Europe: Physical..A14
Europe: Political...A15
Asia: Physical A16
Asia: Political...A17
Africa: Physical..A18
Africa: Political A19
Australia and New Zealand A20
Pacific Islands: Political....................................A21
The Poles...A22
World Time Zones...S14
Absolute location: Niamey, Niger.........................3
Relative location: Yosemite, California 3
Reading a Weather Map 24
World Climate Regions..............................28–29
Plate Tectonics..47
Continental Drift...48
Elevation Map: Guadalcanal...............................50
Contour Map: Guadalcanal.................................50
Elevation Profile: Guadalcanal 50
The World's Basic Biomes..................................57
Early World Culture Hearths 73
World Religions...74
Regions of the World 101
The United States and Canada:
        Physical–Political 107
The United States and Canada: Climate.........108
The United States and Canada: Population 109
The United States and Canada: Economy.......110
Physiographic Regions of the United
        States and Canada 112
Northeastern United States:
        Physical–Political 121

Southern United States: Physical–Political 133
Natural Regions of Texas.................................134
Midwestern United States:
        Physical–Political 141
The Great Lakes and St. Lawrence Seaway 142
Interior West: Physical–Political 151
Pacific States: Physical–Political 159
Los Angeles...163
Pacific Rim Exporters.......................................167
Canada: Physical–Political 173
Middle and South America:
        Physical–Political 189
Middle and South America: Climate..............190
Middle and South America: Population..........191
Middle and South America: Economy 192
Early Civilizations of Middle and
        South America 195
Mexico: Physical–Political................................203
U.S.–Mexico Border Region 207
Mexico City 211
Central America and the Caribbean:
        Physical–Political 215
Panama and the Kuna......................................221
Caribbean South America: Physical–Political...227
Atlantic South America: Physical–Political.......235
Pacific South America: Physical–Political.........245
Europe: Physical–Political 259
Europe: Climate...260
Europe: Population 260
Europe: Economy 261
Northern Europe: Physical–Political 273
West Central Europe: Physical–Political..........283
Berlin...289
Polders of the Netherlands..............................293
Languages of Switzerland 294
Southern Europe: Physical–Political 297
Venice..302
Eastern Europe: Physical–Political 309
The Baltic Countries and Moldova:
        Physical–Political 319

Russia and Northern Eurasia:
   Physical–Political 325
Russia and Northern Eurasia: Climate.............326
Russia and Northern Eurasia: Population 326
Russia and Northern Eurasia: Economy 327
History of Russian Expansion 328
Russia: Physical–Political..............................335
Lake Baykal 339
St. Petersburg 345
Ethnic Groups of the Former
   Soviet Republics 347
Ukraine, Belarus, and the Caucasus:
   Physical–Political 351
Central Asia: Physical–Political.....................359
The Silk Road 363
Language Groups of Southwest
   and Central Asia....................................369
Southwest Asia: Physical–Political..................375
Southwest Asia: Climate...............................376
Southwest Asia: Population 376
Southwest Asia: Economy 377
The Eastern Mediterranean:
   Physical–Political 387
Central Jerusalem.......................................393
Interior Southwest Asia: Physical–Political.......399
Regional Links...411
Africa: Physical–Political 417
Africa: Climate 418
Africa: Population 419
Africa: Economy...420
Early African Kingdoms 423
Colonialism and Independence in Africa 425
North Africa: Physical–Political 431
West Africa: Physical–Political.......................441

East Africa: Physical–Political 453
Central Africa: Physical–Political 465
Kasindi...469
Southern Africa: Physical–Political 477
Johannesburg 482
Southern Africa: Transportation Network 485
East and Southeast Asia: Physical–Political......495
East and Southeast Asia: Climate...................496
East and Southeast Asia: Population 496
East and Southeast Asia: Economy 497
China, Taiwan, and Mongolia:
   Physical–Political 507
Japan and Korea: Physical–Political................519
Tokyo 525
Mainland Southeast Asia:
   Physical–Political 531
Island Southeast Asia: Physical–Political..........541
Major World Shipping Routes 545
South Asia: Physical–Political........................559
South Asia: Climate.....................................560
South Asia: Population 560
South Asia: Economy 561
India: Physical–Political 571
Realm of the Tiger 575
The Indian Perimeter: Physical–Political..........585
The Pacific World and Antarctica:
   Physical–Political 601
The Pacific World and Antarctica: Climate......602
The Pacific World and Antarctica:
   Population 602
The Pacific World and Antarctica:
   Economy..603
Sydney 609
Antarctica 619

Dear Colleague:

There's a lot of talk these days about national standards. Standards are statements that identify what students should know and be able to do in eight core subject areas (i.e., English, math, science, geography, civics, history, foreign language, and the fine arts). The educational policymakers at the federal and state levels who are promoting standards-based education see them as a way to improve schooling by making the curriculum more rigorous and by raising learning expectations for students. Standards are presented as one of the pieces in the mosaic of the reform movement presently under way in education that is directed at restructuring schools, empowering teachers, recasting the curriculum, and modifying pedagogical practices so that the schools will be more successful in meeting societal needs.

At first glance, standards can be threatening not only because there are so many of them, but also because they were designed to be change agents. And few of us relish change. Doing what we have always done is so much more comfortable than redesigning lesson content and altering teaching styles. Despite their number and forbidding nature, however, when teachers work with standards, they come to value them. The strength of the standards is that they make learning more meaningful because they make it practical. They also make teaching more enjoyable because the standards encourage teachers to make students active learners.

The 18 geography standards are a case in point. They identify the essence of geography as a school subject by indicating to teachers what their students should learn by the end of grades 4, 8, and 12, and what they should be able to do with what they have learned. That means asking appropriate geographic questions; finding, organizing and recording geographic information; presenting that information in a meaningful way; hypothesizing and generalizing about issues in geography; and solving problems that have spatial and environmental elements.

As one of those who has had the responsibility for conceptualizing and writing the geography standards, here is what has consistently guided my thinking and that of the other standards developers as well.

- The standards are voluntary. They are not prescriptive. Neither are they a list of objectives. Rather, they are a composite view of what students should know and be able to do, and not something to go

through and check off. As such, they serve only as frames of reference by which teachers and curriculum committees assess their own curriculum and make informed judgments about what needs to be changed and how change can be effectively implemented.

- The national standards are not curriculum. Their purpose is to help schools set a direction for including geography as both a separate subject and also representing it more fully across all grade levels and in all subject areas. The standards establish that geography is more an integrative discipline than a thematic one.

- The national standards provide specificity about what students should know and understand in geography as well as about how they ought to express that knowledge and understanding through performance. They are resources to aid teachers in organizing their curriculum, preparing assessments, presenting content, and suggesting learning opportunities for students.

As teachers, we know that classroom life is complicated and ever-changing. We also know that there is no single right way to teach. The geography standards were developed with both those precepts in mind. Beyond that, the standards writers also recognized that teachers are decision makers. It is in that role they will use the standards so that they can strengthen what they already teach about geography. By applying the standards, students will use their knowledge of spatial relationships, systematic physical and human processes, and the interaction between people and their environments to make intelligent decisions as adult citizens. Calling the national standards document Geography for Life reinforces the utilitarian value of geography.

This text is reflective of the 18 national standards. The authors and editors have used them as guidelines to structure the content and the skills in each of the units as well as in the ancillary materials. They also serve as benchmarks for assessment.

Learning to use and apply the national geography standards provides you with a chance to grow professionally. Accept the opportunity. Enjoy the challenge. *Carpe diem!* (Or in the parlance of the geographer, *Carpe mundo!*) Have a great adventure.

James F. Marran
Geography Education
Standards Project

INTRODUCTION

World Geography Today is designed to help students reach three important curriculum goals. The first goal is to provide students with knowledge about the physical and human geography of our world. The second is to enable students to enhance their understanding of the interrelatedness of the world's regions and cultures. The third goal is to provide beneficial instruction and practice to students in developing their basic map skills, critical thinking skills, writing skills, and other general social sciences skills. These goals are supported throughout the entire *World Geography Today* program.

The Five Themes of Geography—location, place, human-environment interaction, movement, and region—are emphasized in *World Geography Today*. Students will be alerted to specific themes as they read by the labels that appear next to subsection titles in the Pupil's Edition text. Each Chapter Check review page provides students with opportunities to apply the themes to the chapter content and to field experiences. In addition, the Annotated Teacher's Edition includes numerous Using Geography Themes teaching strategies.

World Geography Today also is fully correlated to *Geography for Life: National Geography Standards*. Your sales representative can provide a document that cross-references the skills in the 1994 national standards and representative page numbers where those skills are taught in *World Geography Today*. Along with the Five Themes of Geography, these national standards offer a valuable instructional guide for geography teachers.

PROGRAM COMPONENTS
THE PUPIL'S EDITION

World Geography Today is organized into 50 chapters in 11 units. The text's clear and engaging narrative is supported by an extensive map program, colorful photographs, and special features that emphasize global awareness, environmental concerns, and the Five Themes of Geography. A special feature of this edition is the illustrated **Skills Handbook** at the beginning of the text, which serves as a continuing reference for students throughout their study of world geography (pages S1–S17).

Components of the Pupil's Edition include an **Atlas** of physical and political maps of the continents, and **Mapping the Region** maps that invite comparisons between the physical features, climate, population, and economic activities and resources within a region. Students will be engaged by the lively subjects covered in **Planet Watch**, **Global Connections**, **Cities of the World**, **Themes in Geography**, **News from Around the World**, **Regions in Transition**, and **One World, Many Voices** features. The **Glossary** provides definitions of all the boldface vocabulary terms in the text and includes their page references in the Pupil's Edition. The **Appendix** includes a metric conversion chart, an illustrated statistical chart of the world's countries, and a special state chart of the United States.

THE ANNOTATED TEACHER'S EDITION

The Annotated Teacher's Edition of *World Geography Today* includes all the Pupil's Edition pages, slightly reduced in size to create top and side margins. This allows for convenient point-of-use teaching strategies and annotations designed to offer full instructional support for all elements in the Pupil's Edition.

Each chapter begins with a two-page Chapter Interleaf that provides several important teaching resources. The interleaf includes a planning guide box correlating all the components in the *World Geography Today* program to the text, suggestions for print and multimedia resources for teachers and students, alternative teaching strategies to meet individual learning styles, and a special multicultural feature emphasizing cultural characteristics of a region, country, or group relevant to the chapter.

Unit Overview provides brief summary statements that describe the focus of each chapter.

Unit Objectives focus on broad geographic concepts covered in each unit.

Using the Illustrations enables students to interpret visual clues in photographs to broaden their geographic knowledge.

Teaching with Sketch Maps at the beginning and end of each unit provides students with an opportunity to draw world regional maps from memory. Such maps have proven invaluable as tools to help students expand their understanding of location, place, and spatial relationships. The Unit 1 Review on page 62 prepares students for how to draw and use sketch maps as they study *World Geography Today*.

Using the Geography Dictionary strategies help students learn key terms listed at the start of each chapter in boldface type in the Pupil's Edition.

Using the Quotation links the introductory quotation at the beginning of each chapter with geographic concepts and details.

Mapping the Region activities enhance students' map skills and enable them to make important connections between a region's physical characteristics, climate, population, and economic activities and resources.

THE LESSON CYCLE

The lesson for each section is based on the following instruction model:

Objectives state the main concepts covered in each section and reflect the Focus questions that guide students' reading at the beginning of each section in the Pupil's Edition.

Motivator introduces the lesson with a high interest activity that engages students in the topic they are about to study. Varied methods using class discussion, teacher modeling, and interpretation of visual materials help students relate their own experiences and prior knowledge to geographic concepts.

Building a Vocabulary refines the definitions of key terms from the chapter Geography Dictionary.

Teaching Strategies use varied techniques to help students develop skills and understand main concepts.

Guided Practice activities and discussions led by the teacher are suitable for the entire class.

Independent Practice activities allow students of all ability levels to work independently or with partners.

Closure synthesizes the main concepts.

Assessment checks students' comprehension of the main ideas in each section.

Reteaching employs varied Reteaching activities in *Teaching Resources* to help students who are having difficulty mastering the content.

Extension activities extend the content of each section and help students to develop research skills.

SIDE MARGIN ANNOTATIONS

The side columns include **Global Perspectives**, **Multicultural Perspectives**, and **Cooperative Learning** activities that support important geographic concepts. **Geographical Sidelights** annotations provide useful background information to extend the text and link geography to other disciplines such as history, economics, and literature. Suggested responses to the Section Review questions, and map, chart, and photo caption questions also are provided.

THE TEACHING RESOURCES

A variety of resources provides many opportunities to reinforce, reteach, and assess students' progress.

Teaching Resources Binder

Blackline masters, organized by unit, include Graphic Organizers, Skills Worksheets, Critical Thinking Worksheets, Challenge/Enrichment Worksheets, Reteaching Worksheets, Chapter Tests, Chapter Portfolio Assessments, Unit Tests, and Unit Portfolio Assessment Worksheets. World and Regional Outline Maps and Alternative Assessment Forms are also included in the binder. The binder also includes *The World Today Gazetteer with Regional Location Maps.* The gazetteer identifies some 2,000 place-names found in *World Geography Today.*

Additional Resources

A variety of other resources are available for you to use with *World Geography Today*:

Regions of the World Map Transparencies and Teacher's Notes include full-color transparencies of maps and graphics in *World Geography Today,* as well as suggested teaching strategies.

Creative Strategies for Teaching World Geography offer expanded strategies using case studies, simulations, field work, and teaching transparencies.

Lecture Notes outline key concepts in the Pupil's Edition, include additional background information, and provide suggestions for using the Outline Maps and teaching transparencies. Barcodes correlate *The World Today* Videodisc Program to the Pupil's Edition.

Study Guide for Spanish Speakers includes Chapter summaries, Section Reviews, and Glossary in Spanish.

World Geography Today **Test Generator** allows for customer-designed tests on Macintosh or IBM-compatible hardware.

The World Today **Videodisc Program** includes 25 full motion segments in English and Spanish, an image bank of more than 500 photographs, a Teacher's Guide, and Student Activity Sheets. Bar codes and brief narrative summaries for these video segments are also located in the side margin of the Annotated Teacher's Edition. Videodisc side (for example, 2A), frame number, and running time in minutes and seconds accompany each barcode.

Global Skill Builder **CD-ROM** is an interactive teaching software tool covering basic social studies skills in 13 skills segments. References in the side margins of the Annotated Teacher's Edition point teachers to segments in *Global Skill Builder* relevant to skills taught throughout the *World Geography Today* program. Your sales representative can provide more information about the *Global Skill Builder* CD-ROM and how it can be integrated into today's classrooms.

USING FIELD WORK and SIMULATIONS

Field work experiences and simulations have proven to be very effective teaching tools for geography. Most students enthusiastically participate in opportunities to examine and solve real life problems afforded by field work and simulations.

Field work generally involves excursions outside the classroom and can include field teaching and research. One reason field work is emphasized more and more today is because it develops students' abilities to observe details, collect data, record data, and reach conclusions.

Simulations also encourage students to apply abstract concepts to practical situations. Using methods such as games, panel discussions, debate, and role playing, students confront many topics relevant to geography and learn to think like geographers, urban planners, entrepreneurs, diplomats, farmers, civic leaders, etc.

In developing field work and simulation experiences, the general model shown below will be a useful guide to you or your students in structuring the activities appropriately:

PLANNING
objectives
background information
special permissions required
time/cost requirements
division of labor

→

IMPLEMENTING
obtain materials needed
spell out procedures

→

DEBRIEFING
tests
class discussion
apply concepts

MEETING INDIVIDUAL NEEDS

World Geography Today recognizes that most classrooms include students from a variety of backgrounds and with a variety of learning styles. Many of the activities in both the Pupil's Edition and the Annotated Teacher's Edition are suited to such students. Activities in the Chapter Interleaf, located at the beginning of each chapter of the Teacher's Edition, focus explicitly on meeting the individual needs of the following categories of students:

Limited English Proficiency (LEP)

Instructional strategies in *World Geography Today* are designed to limit the stress and anxiety for students who face difficulty in communicating in the classroom or in understanding instructions. Students who do not speak English as their first language may display conversational ability in English and understand a major part of a lesson. However, it is important to realize that such students may not yet be able to express themselves in writing or orally to indicate their level of understanding. It is also important to remember that LEP students do not fall within the same ability range, and that many of these students might otherwise be considered gifted or have attained high academic skills in their own language. LEP students also include students whose first language is English but who display insufficient language skills. Because of these diverse characteristics of LEP students, they benefit from a variety of instructional techniques. These include oral introduction to assignments and modeling by the teacher, vocabulary activities that emphasize context, opportunities to engage in role playing and reader's theater, working in small groups or with partners, and responding to visuals.

You might consider the following guidelines for effective classroom interaction with LEP students:

Simplify Content
Speak slowly
Pronounce words clearly
Use a controlled vocabulary
Minimize use of idioms
Context Aids
Use gestures
Use animated facial expressions
Act out meaning of words or concepts
Use varied visual materials
Checking Understanding
Do frequent checks for comprehension
Repeat important concepts
Expand meaning
Use varied questions and techniques

Last, remember that LEP students may be new immigrants to the United States—possibly from war torn countries that are in economic and political turmoil. Students' difficulty in adapting to a new culture could prevent them from taking part in class discussions. Avoid querying them about life in their former country unless they are willing to share their experiences. When developing cooperative projects, try to use LEP students' knowledge, experiences, and cultural traditions as resources in ways that will promote students' self esteem.

Learners Having Difficulty

To assist students who need repetition, have a short attention span, lack self-esteem, and learn at a slower rate than other classmates, activities provide the opportunity for review and reteaching of basic information in the chapter. Some activities also focus students on their own interests and experiences in order to provide additional motivation and to encourage students to make connections. Particular methods employed include the use of audio and visual materials, teacher demonstrations and modeling, interactive activities, extended explanations, and summarizing exercises.

Auditory Learners

Strategies targeting students who respond particularly well to auditory stimuli include lectures (both by teachers and guest speakers), audiotapes (either listened to or created by students), class discussion, role playing, debate, reading aloud, and oral presentation.

Visual Learners

For students who are oriented toward non-verbal activities, strategies suggest the examination, analysis, or interpretation of materials such as photographs, maps, graphs, charts, films, and videocassettes. Activities that make use of chalkboards, bulletin boards, overhead projectors, wall maps, and computers also are recommended. Students might be asked to create graphic organizers, posters, maps, or other visual displays.

Tactile/Kinesthetic Learners

Tactile/kinesthetic learners benefit from interactive activities, hands-on activities, and activities that utilize movement and action. Simulations, field trips, role playing, conducting interviews, using newspaper and magazine articles, constructing models, and creating maps, posters, time lines, and illustrations are all effective strategies.

Gifted Students

Activities requiring research beyond the textbook help meet the needs of gifted learners. Often such activities focus on real-life issues and problem solving through the use of investigation, debate, panel discussions, and other forums for the exchange of ideas. As almost every student is gifted in at least one way or another, teachers are encouraged to use these types of activities with as many students as possible.

PORTFOLIO ASSESSMENT

Many of you may be familiar with the concept of Portfolio Assessment. A student's portfolio—a collection of a student's work in a folder, envelope, box, or other container—serves as a tangible record of his or her progress and performance over a period of time. The variety of portfolio projects and materials provide a multidimensional view of a student's mastery of skills, understanding of information, and ability to apply those skills and knowledge. Portfolios help students, teachers, and parents monitor the learning and performance progress of an individual student.

Accordingly, Portfolio Assessment is an important part of *World Geography Today*. Activities throughout both the Pupil's Edition and the Annotated Teacher's Edition provide students with the opportunity to create a wide range of materials for their portfolios. Many activities suggest that students include their materials in their portfolios. Of course, these suggestions are only guidelines; students should feel free to include any additional materials they consider representative of their progress and understanding.

The emphasis of *World Geography Today's* Portfolio Assessment program lies in the "Building Your Portfolio" activities found in the Unit Reviews at the end of every unit in the book. These activities serve several purposes, as outlined below.

Authentic Assessment

The "Building Your Portfolio" activities are "real world" projects, and they aim to be authentic in both content and structure. Projects might have students analyze how communities and regions are changing, design plans of urban renewal, explain different methods of land usage, compare world economic systems and investigate reasons for economic success, and find solutions to environmental dilemmas.

At the heart of authentic assessment lies the concept that students' most important learning takes place when students create the learning themselves. Rather than asking students to simply memorize information that may be forgotten at the end of the course, the unit projects involve students as the main participant in the learning process. The activities help students acquire skills they will need to know and use in their lives again and again—for jobs, for volunteer organizations, and for informed and active citizenship. As they work on the projects, students will sharpen their research skills, learning what types of information to look for and where to find the information they need. Students will also learn how to present information in a variety of forms, including writing; visual displays such as maps, charts, graphs, and illustrations; and oral presentation. Students' organizational skills also will improve and expand as students complete their research and present their findings.

As students complete their projects, they will truly have the sense that they have created something of their own rather than having copied it. The projects do not have "right" and "wrong" answers: students must come to their own conclusions.

Alternative Assessment

The "Building Your Portfolio" projects provide a particularly valuable opportunity to students who are less comfortable with traditional assessment such as in-class, timed exams. Often such tests are not reliable indicators of a student's true knowledge or abilities, particularly for students who have difficulty expressing themselves in certain formats. The unit projects allow students to work with a degree of flexibility in terms of both content and structure.

The Portfolio Assessment Worksheets in *Teaching Resources* support the unit project. They break down the steps of each project into smaller, more manageable pieces to help guide students through the activity. Each worksheet provides students with a checklist to use in tracking what they have accomplished and what they have yet to do.

The worksheets also assist students in planning their time by helping them pace their work over the entire unit, chapter by chapter, rather than lumping it together at the end of the unit. Students must obtain their teacher's initials as they complete each step, so that teachers can ensure that students are on-target as far as time and content are concerned. Furthermore, checklists require students to make rough drafts of most materials so that students can revise and polish their work.

Choice

Students always have a choice between two projects in each unit. In this way, students have the opportunity to explore their own interests. Students might also be encouraged to suggest their own projects, although such original projects should be discussed with teachers beforehand to ensure appropriateness of content, level, and depth. Also, teachers might suggest to students that they share their completed projects not only with their classmates, but with others in their home or school communities.

Cooperative Learning

The "Building Your Portfolio" activities are yet another opportunity for Cooperative Learning in *World Geography Today*. At the teacher's discretion, students may work cooperatively on the projects rather than individually. Each project is divided into specific steps, allowing group members to take responsibility for one of the steps. Conversely, members might all work on each step together with individual roles for researcher, writer, illustrator, and oral presenter.

MULTICULTURAL AWARENESS

Our society is increasingly influenced by the multicultural nature of our population. The cultural diversity that characterizes and enhances life in the United States is evident in many classrooms today. Also, many of your students may be new arrivals to the United States from regions all over the world. Such students face many challenges in adapting to a new environment. As immigration to the United States increases, it becomes more and more important to foster a spirit of respect, tolerance, and appreciation within the classroom of cultural differences.

Multicultural Education

Educators in recent years have devoted considerable attention to the necessity of and benefits gained by infusing multicultural awareness into the curriculum of all subjects, but especially the social sciences. Useful guidelines have emerged in educational literature to help teachers achieve goals of expanding students' global outlook, celebrating diversity, and enabling students to learn from the experiences of others by sharing differences.

In developing multicultural awareness among your students, they should be provided not only with opportunities to study specific cultural characteristics related to certain groups, but also to analyze events and beliefs from the perspectives of different cultural and ethnic groups.

Creative approaches within the classroom to promote multicultural awareness might be reinforced by using folktales, oral histories, literature selections, music, guest speakers, primary sources, and videocassettes. Try to draw on the resources of your students.

Multiculturalism and World Geography Today

Geography is a subject that is intrinsically linked to an appreciation of cultural diversity and an understanding of how culture influences people's daily lives, attitudes, behaviors, and interaction with their environment.

World Geography Today provides ample opportunities to expand the multicultural awareness of your students.

- **One World, Many Voices** features that open each unit in the Pupil's Edition draw on a wealth of literature excerpts and primary sources that emphasize cultural diversity.
- **News from Around the World** features in each unit focus on aspects of daily life of young people globally. Students are always eager to learn about how people their own age live in other parts of the world.
- **Multicultural Perspectives** strategies in the side margins of the Annotated Teacher's Edition highlight interesting cultural details about a region and/or group. They also include suggestions for thought-provoking class discussions that your students will enjoy.
- **Multicultural Focus** teaching activities in each Chapter Interleaf are ideal multicultural introductions to a specific region or country.

Other resources available with *World Geography Today* also emphasize the multicultural nature of geography. The extended teaching activities in **Creative Strategies for Teaching World Geography** engage your students in projects that explore cultural traits. They will learn how to use a standard framework from which to study different cultures, how to recognize common characteristics of all societies, and how to recognize cultural bias and stereotyping. Students will become adept at applying this model to many enjoyable activities such as "Did You Know You Were Speaking Arabic?", "Japanese Arts: Origami," "How Much Do You Know About Africa?" and "Australian Slang."

You might also find the artistic patterns framing the Multicultural Focus and Multicultural Perspectives activities in the Annotated Teacher's Edition of interest. Each design was carefully researched to reflect an authentic cultural influence in a region. The patterns were adapted from a variety of sources such as jewelry from Africa, American quilts, Islamic art motifs on carpets and tile, fabric patterns from Middle and South America, and wood carvings from the South Pacific region, to name but a few.

SUGGESTED RESOURCES

BOOKS:
Banks, James A. *Teaching Strategies for Ethnic Studies.* 5th ed. Boston: Allyn and Bacon, 1991.

Banks, James A., and Cherry A. McGee Banks, eds. *Multicultural Education: Issues and Perspectives.* Boston: Allyn and Bacon, 1989.

Tiedt, Pamela L., and Iris M. Tiedt. *Multicultural Teaching: A Handbook of Activities, Information and Resources.* 3rd ed. Boston: Allyn and Bacon, 1990.

ARTICLES:
Hilliard, Asa G. "Why We Must Pluralize the Curriculum." *Educational Leadership* (1991,1992 December/January).

LeSourd, Sandra J. "A Review of Methodologies for Cross-Cultural Education." *Social Studies* 83 (1992 January/February).

Weaver, V. "Education that is Multicultural and Global: An Imperative for Economic and Political Survival." *Social Studies* 79 (1988 May/June).

Pointing *ALL* Your Students Toward Greater Comprehension

Pointing Your Students In the Right Direction!

Your students experience *real* geography when you guide them through *World Geography Today*—a program that presents complete, balanced world coverage of both human and physical geography. And *World Geography Today's* in-depth content meets *all* of the National Geography Standards!

Yet there's so much more. . . . A strong tie between geography and students' lives. Renewed emphasis on the five themes. Multicultural perspectives integrated throughout the text. Environmental case studies. An exceptional map program. Powerful features that develop students' civic competence by helping them understand their personal responsibilities in our rapidly changing world. Plus, *World Geography Today* unifies the discussion of Africa into a single regional unit.

All students can enjoy and understand *World Geography Today's* high-interest content. Each chapter is divided into manageable lessons that build students' feelings of success. A strong systematic approach presents every aspect of a geographical region—including culture, natural resources, economics, politics, and history—in a consistent, easy-to-follow format that puts students at ease, enhances understanding, and increases content retention.

THEMES IN GEOGRAPHY features provide timely examples of real-world situations to show students relationships among the five geography themes.

Shown 33% of actual size

THEMES IN GEOGRAPHY

Mexico's Northern Border Region

*B*ordering the United States, northern Mexico lies hundreds of miles north of Mexico's heartland.

Northern Mexico is a vast territory that stretches for almost 2,000 miles (3,218 km) from the Pacific Ocean to the Gulf of Mexico. It is up to 300 miles (483 km) deep and includes

Farther west, the Gulf of California separates the main part of northern Mexico from Baja California, a long, slender peninsula.

To tie the border region more closely to the rest of the country, the Mexican government has tried to increase the region's population and improve its economic conditions. Mexico's most ambitious and successful program to develop the northern border region

U.S.-Mexico Border Region

Although northern Mexico is part of Mexico, its economy depends heavily on the United States, and its people are influenced by U.S. culture. Northern Mexicans in turn influence Americans across the border. **How does a border area such as northern Mexico differ from regions in the central part of a country?**

ONE WORLD, MANY VOICES

LITERATURE

"The Donetz Railroad. A cheerless station, quiet and lonely, gleaming white on the steppe, with walls hot from the sun, with not a speck of shade and, it appears, with not a single human being. The train which brought you here has left; the sound of it is scarcely audible and at last dies away. The neighborhood of the station is deserted, and there are no carriages but your own. You get into it—this is so pleasant after the train—and you roll along the road through the steppe, and by degrees, a landscape unfolds such as one does not see near Moscow—immense, endless, fascinating in its monotony. The steppe, the steppe, and nothing else; in the distance an ancient grave-mound or a windmill; oxcarts laden with coal file by. Birds fly singly low over the plain and the monotonous beat of their wings induces a drowsiness. It is hot. An hour or two passes, and still the steppe, the steppe, and still in the distance the grave-mound. The driver rambles on telling you some long-drawn-out, irrelevant tale, frequently pointing at something with his whip, and tranquillity takes possession of your soul, you are loath to think of the past."

Russian Railroad

Interpreting Literature

1. What type of climate and vegetation region does Chekhov describe?
2. What words does the narrator use to indicate that the scenery changes very little throughout the carriage ride?

Anton Chekhov (1860–1904) is one of Russia's most important literary artists. Ill health from tuberculosis, along with a love of writing, caused him to give up a career in medicine. He wrote many plays and hundreds of short stories. Most of his work concerns the everyday lives of people from many classes of pre-revolutionary Russian society in the late 1800s. Chekhov's tales reveal his understanding of human nature and also express his affection for the Russian physical landscape. The selection on this page is from a story entitled "At Home."

FOR THE RECORD

Laurens van der Post (1906–) is a noted travel writer and novelist. He has said that "one never really knows another country" unless one observes it firsthand through the lives of its citizens. Many of his books describe Africa, where he was born, and his travels to England, France, and Japan. Van der Post was curious about Russia because he was not satisfied with the images of the country he received through newspapers, magazines, and television. In 1961, he began the first of four lengthy trips across the Soviet Union. This excerpt is from *Journey into Russia*, published three years later.

"**T**his Russian now told me that it was right to begin with a feeling of the immensity of the land. I must build on that as perhaps the single greatest fact for understanding his people and his native country. But he feared that it would take time before a stranger could realize how immense the land was. . . .

So wide apart are the two extremities of land that it is day in one when it is still night in another. He himself had once flown in a fast jet aircraft from Baikal [Baykal] to Moscow, made four landings on the way and each time as they landed the sun had set and each time they took to the air the sun had risen again. . . . Already in Turkmenia [Turkmenistan], ploughing and sowing would be done and the spring nearly over, but in Uzbekistan where we would land in a few hours we would find the earth aglow with purple heliotrope, scarlet poppies and wild red tulips. Yet the apple trees around Moscow would not blossom until late in May and in the far north of Siberia the tundra rivers would not be free of ice until the end of June. There by August the ice and frost would start their march in the south. In the extreme south the bamboo shot up at a rate of a yard a day. On the northern Siberian shore it took the straggling larches a century to grow the thickness of a man's middle finger. In the south they would already be harvesting the grain while in Kamchatka the snow was still falling. That, measured in the seasons, was the physical frame of the Soviet Union."

Analyzing Primary Sources

1. What does this selection suggest is most significant about Russia's physical geography?
2. How does Van der Post illustrate Russia's vast size?

ONE WORLD, MANY VOICES in each regional unit includes high-interest literary pieces and primary source readings to promote students' reading skills and personalize geography.

SOUTHERN EUROPE: PHYSICAL–POLITICAL

CHAPTER 26

Southern Europe

"This is still a country of local specialties: wines vary widely from town to town, and so do foods. In San Sebastian they make small pastries designed to look like ham and eggs. In Toledo they make marzipan. Oviedo is famous for its stews, richly compounded of vegetable-broth and black pudding, and Vigo for its eel pies. . . . Uniformity has not yet fallen upon Spain . . . this one [town] makes paper, this one lace, this one swords, this one cars, this one wooden figures of Don Quizote, and the little town of Jijona, near Alicante, makes nothing else but nougat. "

Jan Morris

former Roman villa in **Spain**

The countries of Southern Europe, also known as Mediterranean Europe, are Spain, Portugal, Italy, Greece, and several microstates. The nations occupy peninsulas on Europe

The Mediterranean Sea stretches (3,733 km) from east to west. *Mediter* from a Latin word meaning "middle Indeed, in ancient times, the Mediter considered the center of the Western surrounded by Europe, Africa, and As Gibraltar (juh-BRAWL-tuhr) links the Mediterranean to Atlantic Ocean.

Small waves, weak currents, and low tides charact Mediterranean. As a result, its poorly circulated water polluted. A growing population, coastal industrializati of tourists continue to stress the Mediterranean envi

Mijas, southern **Spain**

296 Unit 5

GEOGRAPHY DICTIONARY

Meseta
plaza
cork
dialect
autonomy
Renaissance
coalition government
balance of trade

> **GEOGRAPHY DICTIONARY** strengthens students' vocabulary skills and helps them read with success by alerting them to important geography terms.

LEGEND ELEVATION

FEET	METERS
13,120	4,000
6,560	2,000
1,640	500
656	200
(Sea level) 0	0 (Sea level)
Below sea level	Below sea level

Size comparison of Southern Europe to the contiguous United States

SPAIN AND PORTUGAL

FOCUS

● *What landforms and climates are found Spain? in Portugal?*

● *What economic activities are in Spain and Portugal? W country's major cities?*

● *What are the historical, pol cultural characteristics of Spain Portugal?*

Physical Geography Spain and Portugal lie on the Iberian (ie-BIR-ee-uhn) Peninsula. Spain covers over 80 percent of the peninsula, while Portugal occupies the remaining western portion. The peninsula is bounded on the north and west by the North Atlantic Ocean and on the south and east by the Mediterranean Sea. (See the map above.) Spain also includes the hilly Balearic (bal-ee-AR-ik) Islands in the Mediterranean and the volcanic Canary Islands in the North Atlantic Ocean.

> **FOCUS QUESTIONS** emphasize key ideas in each section to ensure greater comprehension and guide students in their reading. These questions are reinforced in the section reviews.

Shown 45% of actual size

MAPPING THE REGION

EUROPE

PHYSICAL–POLITICAL

LEGEND ELEVATION

FEET	METERS
13,120	4,000
6,560	2,000
1,640	500
656	200
(Sea level) 0	0 (Sea level)
Below sea level	Below sea level

Ice caps
National capitals
Other cities

Chapter 23 **259**

> **MAPPING THE REGION** includes large, clear maps that enable students to quickly grasp a wealth of geographic information and make comparisons between and within regions.

Shown 40% of actual size

Pointing Your Students
Toward a World View

World Geography Today motivates students
toward a lifelong understanding of world geography
as they begin to think and act like global citizens.

Shown 50% of actual size

NEWS FROM AROUND THE WORLD
looks at teenagers' lives throughout the
world, encouraging students to explore the
cultural geography of other countries as well
as their own.

FACTS IN BRIEF charts bring
relevance to geography as students
compare the population, life
expectancy, literacy rate, and per
capita GDP of the United States to
those of other countries.

FACTS IN BRIEF — The Russian Federation

COUNTRY POPULATION (1994)	LIFE EXPECTANCY (1994)	LITERACY RATE	PER CAPITA GDP (1993)
RUSSIA 149,608,953	64, male 74, female	99% (1993)	$5,190
UNITED STATES 260,713,585	73, male 79, female	97% (1991)	$24,700

Sources: *The World Factbook 1994*, Central Intelligence Agency;
The World Almanac and Book of Facts 1995

Shown 33% of actual size

GLOBAL CONNECTIONS
features show students the
various ways—politically,
economically, culturally,
and environmentally—
in which different regions
are connected and how they
interact with each other.

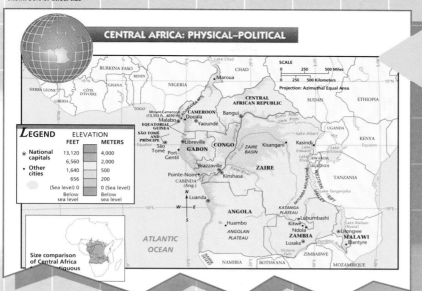

Shown 50% of actual size

CENTRAL AFRICA: PHYSICAL–POLITICAL

SCALE
0   250   500 Miles
0   250   500 Kilometers
Projection: Azimuthal Equal Area

LEGEND
ELEVATION
FEET / METERS
◉ National capitals
• Other cities
13,120 / 4,000
6,560 / 2,000
1,640 / 500
656 / 200
(Sea level) 0 / 0 (Sea level)
Below sea level / Below sea level

Size comparison of Central Africa with the contiguous

Maps throughout the text are accompanied by a **SIZE COMPARISON MAP** of the United States, enabling students to visually compare the size of the United States to that of other countries. **LOCATOR GLOBES** place the region in its global-geographical context.

CITIES OF THE WORLD helps students see how history has shaped the urban geographies of cities throughout the world. Accompanying maps enable students to make comparisons between these cities and their own.

Johannesburg

"*G*old! Gold! Gold!" The cry of "gold!" marked the beginning of Johannesburg.

Unlike many other major cities around the world, Johannesburg, South Africa, owes its existence and growth largely to gold. In 1886, a wandering prospector came across gold-bearing rocks on a farm near a range of low hills known as the Witwatersrand, or "ridge of white waters." The news of the gold strike spread far and wide. Thousands of fortune seekers from around the world made their way to the area. People came mainly from Europe, America, Australia and other parts of southern Africa. A dusty mining camp soon grew up, with a jumble of languages and more.

Gold and businesses related to gold are still important to the city's economy, and the Witwatersrand region is a major source of gold. In fact, nearly half of all the world's gold mined during the past 100 years has come from this region. Although most of the mines have closed, miners continue to work as much as two miles underground to haul out gold-bearing ore east and west of the city in the 62-mile-long (100-km) Witwatersrand.

A Changing City
Today, Johannesburg is the financial and commercial capital of South Africa. It is a city of modern skyscrapers, busy industrial districts, and sprawling suburbs. Most of Johannesburg's residents now work in jobs that have little to do with mining.

Although most blacks still live in vast townships, housing projects, and slums on the outskirts of the city, many have moved to the city center. In contrast, most whites live in comfortable homes in pleasant suburbs of the city. But some blacks, South Africans of mixed race (called coloureds), and those of South Asian descent also can now afford to live in suburbs that once were all white. In addition, schools have become more integrated.

With the end of apartheid, schools in **South Africa**, such as the one attended by these students outside **Johannesburg**, are open to South Africans of all races.

As more blacks move into positions of authority in government and employment, the townships also have begun to change. The more than 2 million residents of Soweto, a large township southwest of downtown Johannesburg, once had to travel to the main city to do even basic shopping. Today, Soweto's homes have better utility services, such as electricity, and many stores and other businesses have opened in the township. Downtown Johannesburg, once

Johannesburg is at the heart of a bustling urban area that includes other modern cities and black townships. Many residents from these cities and townships commute to Johannesburg.

Shown 50% of actual size

Shown 33% of actual size

CHAPTER **39** CHECK

Cameroon rock band

Reviewing the Main Ideas
1. Central Africa is a plains region covered with tropical forests and savannas. Most of the region has a tropical-savanna climate.
2. Central Africa imports many manufactured goods from developed countries. International aid and exports of minerals and cash crops provide income to pay for these goods. Most of the countries in the region must import food.
3. Many people in Central Africa are migrating to the cities in search of work. Cities are overcrowded, and many people live in slums surrounding the cities.
4. Most Central African countries have rich natural resources. Tourist industries can be developed around the beautiful lands and lakes. The development of these resources and the move toward democracy provide Central Africans with hope for the future.

Recalling and Reviewing
1. What are the two important river systems of Central Africa?
2. What political problems have caused Zaire's economy to suffer?
3. Why is Cameroon's economic situation better than that of most of the other countries in Central Africa?
4. What challenges face Angola?

Thinking Critically
1. Using examples from the chapter, explain how the quality and availability of transportation is related to the economic development of an area.
2. Predict which country in Central Africa will enjoy the most economic success by the year 2000. Give reasons for your prediction.

Learning About Your Local Geography

Cooperative Project
Plan a periodic market. In your group, make a list of items grown or produced locally that could be sold or bartered at the market. Your group should decide the best time and location in your community for the market and provide reasons for your choices. Share your group's list of items for sale or barter and the choice for time and location for the market with the whole class. Work with other groups to make a final decision for the time and place to hold the market.

Using the Five Themes of Geography
Make a list of the named countries discussed

LEARNING ABOUT YOUR LOCAL GEOGRAPHY in each Chapter Check increases students' civic competence by encouraging them to investigate local geography and become involved in their own communities.

T23

Pointing Your Students Toward Environmental Awareness

World Geography Today draws students' attention to pressing environmental issues throughout the world.

PLANET WATCH

Case Study: Saving the Rain Forests

Along Panama's Caribbean coast, a great rain forest reaches down to the sea. Giant anteaters, nocturnal tapirs, and armadillos move along the forest floor, while eagles and migrating birds perch high in the forest's canopy. Thousands of species of plants thrive in this humid-tropical environment.

The Kuna have practiced conservation of their rain forest environment for generations.

This rain forest is also home to some 30,000 Kuna, whose villages occupy the small offshore islands.

The Kuna reservation, which encompasses more than 1,100 square miles (2,849 sq. km), stretches some 100 miles (161 km) along the coast of Panama. The coastal waters provide the Kuna with fish, while the fertile mainland plains allow them to grow bananas, sweet manioc, yams, and maize (corn).

Caretakers of the Land

In 1980, a Panama official met with the Kuna. "Why do you Kuna need so much land?" he demanded. "You don't do anything with it. You don't use it. And if anyone else cuts down so much as a single tree, you shout and scream." Rafael Harris, a Kuna leader, responded. "We Kuna need the forest, and we use it and take much from it. But we take only what we need, without having to destroy everything as your people do."

In this area secluded by the rugged San Blas Mountains, the Kuna have managed to head off development and clearing of their precious rain forest. For the Kuna, the plants of the rain forest are a source of sacred medicine. The rain forest also protects their farmlands and fishing areas. Without the trees and plants that absorb rain and hold the soil in place, the land would erode and eventually destroy the fishing grounds on which the Kuna depend.

To protect the rain forest, the Kuna have set up a 230-square-mile (596-sq.-km) pro-

tected reserve on the edge of the ridge that separates the Kuna reservation from the rest of Panama. To help finance their plans, the Kuna called on international environmental groups. In addition, after building a dormitory near the rain forest, they welcomed paying visitors, including scientists, bird watchers, and the Panamanian Boy Scouts. The Kuna collect fees from tourists and sell research rights to scientists visiting the reserve. The Kuna have worked to have their reservation declared a Biosphere Reserve and a World Heritage Site. These United

A Kuna woman stitches a *mola*, a colorful, needlework garment. The Kuna, who live on coral islands off the northern coast of **Panama**, have maintained control over their own cultural and political ways of life.

Nations classifications can help the Kuna in their struggle for international recognition as an independent people.

The Kuna have not relied on outside sources alone, using $100,000 of their own funds. "The main thing about the project is the incredible initiative that has come from the Kuna," said R. Michael Wright of the World Wildlife Fund. "They led us, we didn't lead them. Conservationists come and go, but the Kuna are going to stay." Geographers hope that the Kuna's experiment will provide a model for other peoples living in the world's rain forests.

Panama City and **Colón** offer job opportunities for many Kuna, a well-educated people. What is the distance from the Kuna reservation to Colón?

220 Unit 4

YOUR TURN

Many geographers wish to stop the destruction of the world's rain forests. How might you enlist the help of the people who live in, or earn a living from, the rain forests in this effort? **Plan** a conservation training workshop for a rain forest community. **Outline** the topics you will cover in your workshop, and provide a brief description of the information you will discuss for each

---

PLANET WATCH highlights positive environmental actions in various regions and shows students just what is possible when people work together.

YOUR TURN promotes students' active involvement in finding lasting solutions to real environmental problems.

Pointing Your Students Toward a Geographical Viewpoint

World Geography Today involves students in all aspects of geography and is the first program with a Skills Handbook bound right into the text.

T24

REGIONS IN TRANSITION

Facing the Past and Present

Patterns of trade and culture can change quickly in our modern world. For example, until just a few decades ago, the United States looked primarily to Europe for trade. Many innovations in culture and technology and most of America's immigrants came from Europe. Today, the American focus on Europe has faded. U.S. trade with Japan and other Pacific Rim nations now far exceeds U.S. trade with Europe. New ideas, technology, and immigrants to the United States now come from all around the globe.

Central Asia

With the collapse of the Soviet Union, similar changes are taking place in Central Asia. In the past, many ties linked Central Asia to the Soviet Union. For instance, the economies of the two regions were integrated. Central Asia exported cotton and oil to Russia and the countries of Eastern Europe. In exchange, Central Asia received a variety of manufactured goods from these areas. The Soviet Union also heavily influenced the culture of Central Asia, and many Central Asians learned to speak Russian.

Looking South

Today, Central Asia's links to the former Soviet Union have weakened. At the same time, its ancient ties to Southwest Asia have grown

stronger. The Silk Road, and the travelers that came before it, formed valuable trade links between the oasis cities of Central Asia and the busy Southwest Asian ports on the Mediterranean. Now the peoples of Central Asia are looking southward once again. For example, links are forming between the region's new republics and Turkey. Many people in Central Asia are traditionally Turkic in culture and language. As a result, Turkey's business leaders are working to expand their industries in Central Asia. Regular air travel from Turkey to the capitals of Central Asia now is possible as well.

Religion provides another firm link between Central Asia and Southwest Asia. Islam first arrived in Central Asia in the eighth century, but its role diminished during the Soviet era. Missionaries from the Arab countries and Iran are now committed to reestablishing this connection. In fact, Iran is spending millions of dollars to build roads and rail lines to Central Asia.

Central Asia and Southwest Asia also share a similar climate, environment, and way of life. In both of these dry regions, water conservation and irrigation are basic to daily life. Many of both regions' people grow cotton and herd animals, though this way of life may soon change. Both Central Asia and Southwest Asia are experiencing an invasion of modern Western culture and are struggling to deal with its impact. Many people believe that the ideas and trends that arrive daily through compact discs, videotapes, and satellite television are a threat to their traditional beliefs and ways of life. Such shared fears of cultural loss are bringing Central Asia and Southwest Asia even closer together.

These children are learning about Islam in **Dushanbe, Tajikistan**. Although the former Communist government discouraged the practice of religion, today Islam flourishes in the independent Central Asian republics.

Language Groups of Southwest and Central Asia

LEGEND
Dominant languages:
- Turkic
- Iranic
- Semitic
- Greek
- Other
- Sparsely populated

MAP STUDY This map uses colors to show the major language groups that link ethnic peoples throughout Southwest and Central Asia. Turkic languages, such as Azerbaijani and Uzbek, and Iranic languages, such as Kurdish and Persian, are spoken by ethnic peoples

throughout Central Asia and northern Southwest Asia. Jews and Arabs also are linked by language: both Hebrew and Arabic are Semitic languages. Very often, however, the links between ethnic groups are overshadowed by the differences in culture and history.

Defining the Region

For all these reasons, geographers are constantly reexamining this and other regions of the world. Will they decide to include the new countries of Central Asia in the region of Southwest Asia? Or will Russia regain control of Central Asia and once again push aside the region's ancient economic and cultural links to Southwest Asia? The geographers are watching and waiting.

In the tradition of many Turkic communities, these community elders are meeting to discuss local schools in **Tajikistan**. The Central Asian republics share many traditions with both northern Eurasia and Southwest Asia.

REGIONS IN TRANSITION gives students a fresh perspective on world geography by showing them how geographers organize the world in many different ways.

368   369

Pointing YOU In the Right Direction!

Every tool you need to get the point across is in the *Annotated Teacher's Edition*— including planning guides and teaching strategies for meeting individual needs.

Point-of-use **LESSON PLANS** include global connections, multicultural perspectives, vocabulary development, cooperative-learning activities, cross-curricular connections, and much more!

Shown 37.5% of actual size

PLANNING GUIDES in the chapter interleaf pages reduce your preparation time and provide flexibility in structuring your lessons. Teaching resources— from both the *Annotated Teacher's Edition* and the ancillary package—are clearly organized by section.

THE WORLD TODAY VIDEODISC PROGRAM

You might use the profile of Russia to extend the lesson in Section 4.

ICONS throughout the text indicate where the dynamic resource—*The World Today Videodisc Program*—may be used to enhance your lesson.

Textbook page spread (shown at reduced size)

Reteaching
Assign Section 3 of Reteaching Worksheet 29.

Extension
Have interested students research and write a report on the process of privatization of Russian industry. They should focus on why foreign companies or individuals might or might not want to invest capital in Russian enterprises.

Section 3 Review
ANSWERS
1. Industries are diversifying and are producing more consumer goods. Russians will have a greater selection of consumer goods.
2. Industrial regions: Moscow, St. Petersburg, Volga, and Urals regions; regions with mineral resources: Urals and Far East regions and Siberia; trade: St. Petersburg and Far East regions
3. Possible answer: all regions will be affected, but the Moscow and St. Petersburg regions are likely to be most affected because they are the most populous regions.

SECTION 4
OBJECTIVES
The objectives reflect the Focus questions on this page and are assessed in the review on page 348.
1. Examine political and cultural characteristics of the Russian Federation.
2. Discuss current and future issues facing Russia.

TEACHING SECTION 4
Motivator
Call on students to tell what they know from newspaper articles or television about Russia or the Russian people and their way of life. Write their observations on the board. Ask students for reasons why they think that the Russian people are better or worse off than they were 10 years ago. Ask them to speculate about the Russian people 10 years from now.

Creating a Culture Map
Have students create a large map of Russia and mount it on a bulletin board. Assign students particular ethnic regions to research. Ask them to illustrate the culture of the region using photos or their own drawings. Post the illustrations on the appropriate parts of the map.

Guided Practice
Russia and the United States are both multicultural and multiethnic societies. Point out that American cultural diversity stems largely from immigration, whereas in Russia it results mainly from expansion and migration. Lead a discussion about ways that ethnic diversity affects a society.

Independent Practice
Refer students to the map on this page. Have them analyze the pie graphs to determine which countries appear to have the most ethnically diverse populations. Then have students rank the countries according to the percentage of other ethnic groups within their borders, from the country with the largest percentage (*Tajikistan*) to those with the smallest (*Ukraine and Armenia*).

A tram takes people up one of the steep hills surrounding Golden Horn Bay in **Vladivostok.** Icebreakers keep the harbor open when it freezes during the winter.

in the Sea of Okhotsk. The Kuril (KYUR-eel) Islands, important for commercial fishing, are located east of Sakhalin. Russia took the islands from Japan at the end of World War II, and Japan claims it still owns some of the southern Kurils.

 SECTION REVIEW
1. How are Russia's industries changing? How will these changes affect the daily lives of the Russian people?
2. Which of Russia's regions are industrial? Which have mineral resources? Which are well situated for trade?
3. **Critical Thinking** **Predict** which regions are likely to be most affected by the country's economic changes.

POLITICAL GEOGRAPHY

FOCUS
• What are the political and cultural characteristics of the Russian Federation?
• What current and future issues does Russia face?

The ancient heart of Russia is the upper Volga River basin. The Russian Orthodox Church fled here after Kyiv (formerly Kiev) fell to the Mongols in 1240. By collecting the Mongol ruler's taxes, the princes of Moscow became the leading power in the region. When the Mongols

finally were defeated in the late 1400s, Russian authority spread outward. Military expeditions and land-hungry adventurers increased Moscow's territory in all directions and brought Russian language, laws, and customs to neighboring peoples. The church often followed close behind. The expansion of the Russian state was much like the colonial developments of England and France. Russia, however, invested little to expand its empire overseas.

PLACE **The Russian Federation** The Russian Federation's system of government includes a strong president and a parliament, called the Federal Assembly. The president and members of parliament are elected by Russian citizens. The existence of many political groups in the parliament often has slowed efforts to move Russia toward a market economy. Some legislators think reform has occurred too fast. In addition, many oppose granting more freedoms to some of Russia's numerous ethnic groups.

During the Soviet era, many minority ethnic groups in Russia lived within their own territorial subdivisions. Although these territories had some local power, they were in fact tightly controlled from Moscow. In the aftermath of the fall of the Soviet Union, some of these territories became part of partially self-governing republics within Russia. Twenty-one such republics now make up about 15 percent of Russia's land area. Each republic has its own legislature and leaders, but its degree of authority is much debated.

Republics of the Russian Federation

Adygea	Karelia
Bashkortostan	Khakassia
Buryatia	Komi
Chechnya	Mari El
Ingushetia	Mordvinia
Chuvashia	North Ossetia
Dagestan	Tatarstan
Gorno-Altay	Tuva
Kabardino-Balkaria	Udmurtia
Kalmykia	Yakutia
Karachay-Cherkessia	

These republics of the Russian Federation have large non-Russian populations. The central government in Moscow has given the republics some control over local affairs. Even so, some people in the republics are calling for more freedom and even independence.

Ethnic Groups of the Former Soviet Republics

SKILL STUDY The pie graphs above show the ethnic population breakdown of the former Soviet republics. Centuries of Russian domination and decades of Soviet policies resulted in large ethnic Russian populations in many of the republics. Conflicts between ethnic Russian populations and large minority populations have created problems within the now-independent republics and within the Russian Federation itself. Efforts to ease ethnic tensions must respect minority rights. After Russia, which republic's population has the largest percentage of ethnic Russians?

Although about 82 percent of its current population is ethnically Russian, Russia remains a diverse country with dozens of ethnic groups. All of the Russian republics supposedly were formed to have a dominant minority ethnic group. In many of these republics, however, ethnic Russians form a majority of the population, just as they did in the former Soviet republics. (See the map and graphs above.)

The most populous republic in the federation is Bashkortostan, southwest of the Urals. Yakutia in Siberia has the largest land area. Some republics, such as Chechnya in the south, have used the instability in Russia to push for independence. Fighting between the Russian army and Chechen rebels increased sharply in 1994, claiming thousands of lives. Other ethnic groups, such

as the Tartars along the Volga River, also have sought independence.

Many Russians fear that such desires for independence threaten the future of the federation. Some leaders, however, believe they can bring better government and prosperity to their regions.

Issues Despite the high hopes that followed the breakup of the Soviet Union, bringing freedom and prosperity to the peoples of Russia has been difficult. The reliable certainties of the past have been replaced with an exciting but uncertain present. The concepts of free-market economics, individual liberty, and personal profit are still viewed with suspicion by many people. Nevertheless, progress has been made as increasing numbers of Russians learn about capitalism and

MULTICULTURAL PERSPECTIVES

The novelist Leo Tolstoy once said, "The Russians harness their horses very slowly, but they ride with great speed." Call on students to consider the meaning of this metaphor, and write their responses on the board. Ask students how the saying would apply to the political and economic changes taking place in Russia now. (*Changes may have been slow in coming, but the Russian people will embrace them and move forward quickly to change their society.*)

Skill Study
Kazakhstan

THE WORLD TODAY VIDEODISC PROGRAM

Russia: A New Russian Business
1B 41154 4:29

Play   Pause

346 Unit 6

Chapter 29 347

Extensive Support
at Every Point!

Turn to the complete selection of *World Geography Today* ancillary materials. These resources are organized so they're right **where** you need them **when** you need them. And each high-quality resource is specifically integrated with the textbook, enabling you to reach **all** your students.

TEACHING RESOURCES is a conveniently organized selection of materials that meets all your classroom needs.

- **Unit and Chapter Worksheets and Tests**
    Graphic Organizers
    Skills Worksheets
    Critical Thinking Worksheets
    Challenge/Enrichment Worksheets
    Reteaching Worksheets
    Chapter and Unit Tests
    Chapter and Unit Portfolio Assessments
- **Outline Maps**
- **Alternative Assessment Forms**
- **Answer Key for Worksheets and Tests**

CREATIVE STRATEGIES FOR TEACHING WORLD GEOGRAPHY are dynamic, innovative teaching strategies that can be modified and used at various points in your lesson. The wide range of cooperative-learning activities—including case studies and simulations—motivates students and helps them develop critical-thinking skills.

REGIONS OF THE WORLD: MAP TRANSPARENCIES AND TEACHER'S NOTES provide full-color transparencies to help your students understand key concepts, compare regions, and develop geography skills.

LECTURE NOTES offer convenient chapter outlines, interesting Geographical Sidelights to extend the outlines, correlations to the teaching transparencies, and barcode correlations to *The World Today Videodisc Program.*

ALTERNATIVE ASSESSMENT FORMS help you accurately assess students' progress with blackline-master portfolio, project, and performance assessment forms.

STUDY GUIDE FOR SPANISH SPEAKERS provides chapter summaries, section reviews, and more for your Spanish-speaking students.

THE WORLD TODAY GAZETTEER WITH REGIONAL LOCATION MAPS includes a list of place-names with short descriptions and latitude and longitude information, as well as nine blackline-master location maps that match the regional organization of *World Geography Today.*

The **GLOBAL SKILL BUILDER CD-ROM** contains 13 skill lessons, requiring students to engage in a variety of interactive responses.

TEST GENERATOR software for IBM® PCs and Compatibles and Macintosh® computers provides a quick, easy-to-use tool for creating your own tests.

THE WORLD TODAY VIDEODISC PROGRAM offers you thought-provoking video lessons and short segments.

Shown 20% of actual size

Turn to *The World Today* Videodisc Program

The World Today Videodisc Program is a flexible, curriculum-based resource that's fully integrated with *World Geography Today.* Two videodiscs with English/Spanish soundtracks bring contemporary world cultures to life, helping you motivate your students as never before.

Compelling video segments, detailed maps, and hundreds of still images from every major region reinforce the themes of geography. As your students compare and contrast education, religion, family life, and the environment in 25 countries— including Russia, Kenya, China, Germany, South Africa, and Mexico—they will begin to understand and appreciate cultural similarities and differences.

- The *World Geography Today* Annotated Teacher's Edition includes point-of-use icons so you can easily incorporate the segments into your lessons. Barcode correlations are included in the *Lecture Notes*.

- Use *The World Today* Teacher's Guide to access images in three ways—regionally, thematically, and alphabetically—for maximum instructional flexibility. Complete lesson plans and activity worksheets ensure effective and exciting lessons.

Global Skill Builder CD-ROM

Ignite your students' imaginations and keep them coming back for more with Holt, Rinehart and Winston's latest innovation in instruction—the *Global Skill Builder CD-ROM.* This CD-ROM contains 13 high-interest, lively skill lessons— including Navigating the Internet, Mapping the Earth, Determining Absolute and Relative Location, Using Time Lines, and Creating Graphic Organizers— requiring students to engage in a variety of interactive responses. Designed for both Macintosh® and Windows®, it contains English and Spanish instruction and was specifically developed for your classroom. The program presents context examples from geography and history in a visually based learning environment, ensuring all your students comprehend core social studies skills.

The *Global Skill Builder CD-ROM* also includes a 96-page *User's Guide and Teacher's Manual,* complete with student project sheets. Students have the option of completing each project by using the Internet.

TO THE STUDENT

*W***orld Geography Today** is a guide to learning about the constantly changing world in which you live. Through the chapter text and special features, this textbook is meant to help you understand the study of geography and make sense of our complex world. Hundreds of photographs of people and places will transport you around the world on your journey to learning about other cultures. Full-color diagrams will take you deep beneath the earth's crust, through the canals of a working lock, and to the top of some of the world's highest mountains. Locator maps and globes put the area under study in perspective for you. A geography dictionary in each chapter introduces new words and ideas to help you express your geographic knowledge. And focus and review sections will guide you in your study.

While the emphasis of **World Geography Today** is global, the concepts of geography found in it apply to the place where you live as well. The text seeks to make geography relevant to you by making the connection between regional and global issues and your own community.

ORGANIZATION OF *WORLD GEOGRAPHY TODAY*

It is helpful to think of this book as having three parts. The first part is the front matter. The pages you are reading now are part of the front matter, as are the Table of Contents and the Student Atlas, a collection of handy reference maps of the world and its continents.

The second part of **World Geography Today** is devoted to text describing the study of geography and

the world's major regions. The text is divided into 11 units and 50 chapters.

The third part of your textbook is the back matter. It includes an Appendix with statistical information on and flags of the world's countries. The back matter also includes a Glossary, which lists all the vocabulary words that appear in boldface type throughout the book.

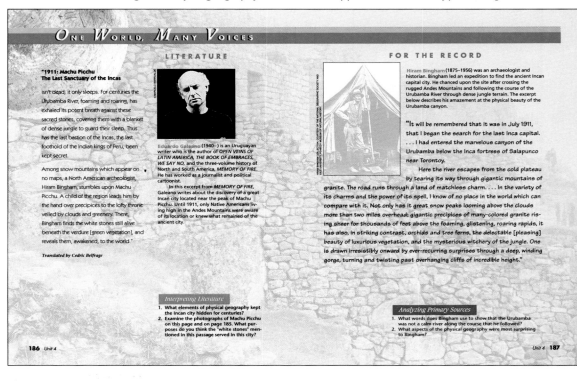

ONE WORLD, MANY VOICES

LITERATURE

**"1911: Machu Picchu
The Last Sanctuary of the Incas**

isn't dead; it only sleeps. For centuries the Urubamba River, foaming and roaring, has exhaled its potent breath against these sacred stones, covering them with a blanket of dense jungle to guard their sleep. Thus has the last bastion of the Incas, the last foothold of the Indian kings of Peru, been kept secret.

Among snow mountains which appear on no maps, a North American archeologist, Hiram Bingham, stumbles upon Machu Picchu. A child of the region leads him by the hand over precipices to the lofty throne veiled by clouds and greenery. There, Bingham finds the white stones still alive beneath the verdure [green vegetation], and reveals them, awakened, to the world."

Translated by Cedric Belfrage

Eduardo Galeano (1940–) is an Uruguayan writer who is the author of *OPEN VEINS OF LATIN AMERICA, THE BOOK OF EMBRACES, WE SAY NO,* and the three-volume history of North and South America, *MEMORY OF FIRE.* He has worked as a journalist and political cartoonist.

In this excerpt from *MEMORY OF FIRE,* Galeano writes about the discovery of a great Incan city located near the peak of Machu Picchu. Until 1911, only Native Americans living high in the Andes Mountains were aware of its location or knew what remained of the ancient city.

FOR THE RECORD

Hiram Bingham (1875–1956) was an archaeologist and historian. Bingham led an expedition to find the ancient Incan capital city. He chanced upon the site after crossing the rugged Andes Mountains and following the course of the Urubamba River through dense jungle terrain. The excerpt below describes his amazement at the physical beauty of the Urubamba canyon.

"It will be remembered that it was in July 1911, that I began the search for the last Inca capital. . . . I had entered the marvelous canyon of the Urubamba below the Inca fortress of Salapunco near Torontoy.

Here the river escapes from the cold plateau by tearing its way through gigantic mountains of granite. The road runs through a land of matchless charm. . . . In the variety of its charms and the power of its spell, I know of no place in the world which can compare with it. Not only has it great snow peaks looming above the clouds more than two miles overhead; gigantic precipices of many-colored granite rising sheer for thousands of feet above the foaming, glistening, roaring rapids, it has also, in striking contrast, orchids and tree ferns, the delectable [pleasing] beauty of luxurious vegetation, and the mysterious witchery of the jungle. One is drawn irresistibly onward by ever-recurring surprises through a deep, winding gorge, turning and twisting past overhanging cliffs of incredible height."

Interpreting Literature

1. What elements of physical geography kept the Incan city hidden for centuries?
2. Examine the photographs of Machu Picchu on this page and on page 185. What purposes do you think the "white stones" mentioned in this passage served in this city?

Analyzing Primary Sources

1. What words does Bingham use to show that the Urubamba was not a calm river along the course that he followed?
2. What aspects of the physical geography were most surprising to Bingham?

Last, the back matter includes the Index. The Index is your guide to the hundreds of topics discussed in the book, with page references for each topic.

SPECIAL FEATURES

You must have certain skills to understand the changes that are affecting our world. In the *Skills Handbook* starting on page S1, you will learn important skills that will help you understand the impact of these changes. In addition, skill lessons throughout the book will give you additional tools for your study as well as opportunities to practice the skills you have acquired.

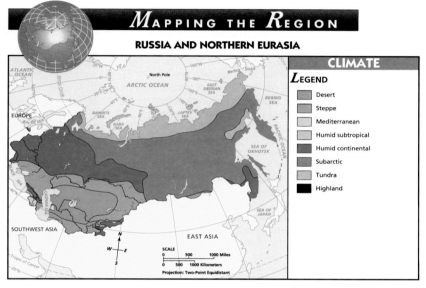

Literature can help you in your study of geography by painting vivid pictures of faraway places. In the beginning of each unit of the text, you will find a feature called *One World, Many Voices*. These literature passages and firsthand accounts describe the geography of various places around the world. You might be familiar with some of the authors of these passages; other authors will be new to you. All will help you understand how different people see the world around them.

To help you learn about all aspects of the world's regions, each unit has a series of physical–political, climate, population, and economy maps called *Mapping the Region*. These full-color maps will help you compare characteristics within and between world regions. As you will learn, however, regions are not fixed areas. Just as the world is constantly changing, so are the ways in which it is divided. Special *Regions in Transition* features highlight areas of the world that are undergoing significant changes.

You also will learn about the themes that tie the study of geography together. These themes are movement, human-environment interaction, location, place, and region. These Five Themes of Geography help us answer the questions of where, why, and how in geography. All of the themes are evident in the changes taking place in the world and are highlighted throughout the text, in special *Themes in Geography* features, and in the review activities.

People around the world share a common environment—the earth. Changes in our environment can affect people around the globe. Each unit's *Planet Watch* feature lets you investigate case studies showing how different people are working in different ways toward a common goal—protecting the environment.

Sometimes the best way to understand change and different regions of the world is to learn about the daily lives of people in other cultures. The *News from Around the World* features are your window to the lives of teenagers in other countries. You will also walk the streets of Sydney, Tokyo, Jerusalem, and other major cities in the *Cities of the World* features.

All of these features and the text are your passport to understanding the world around you. In fact, **World Geography Today** will help you long after you have read the last chapter. The skills and geographic knowledge you are about to learn will help you understand and adapt to future changes in our world.

NEWS FROM

Russia

Teenagers in Russia Have More Free Time

When Russia was part of the Soviet Union, teenagers spent much of their free time in government-sponsored groups. Almost all children joined the Young Pioneers when they reached nine years of age. At age 14, those who did not join Komsomol (the Young Communist League) were considered disloyal to the government and could not expect to attend a university or receive good jobs as adults.

Now that Russia is no longer under communism, few teenagers join political groups. Some Young Pioneer groups still exist, although the organization suffers from lack of funds. Those attending the summer camps run by the Young Pioneers have noticed big changes. No longer intent on promoting communism, Young Pioneer groups

have become far less rigid. Boys and girls used to resemble young soldiers as they saluted and marched with red scarves secured around their necks. "There are no songs, no marching, no red ties anymore," said Artyom V. Chibisov, a 13-year-old boy from Moscow. One young woman had a practical reason for not missing the scarves. "It means less ironing," she said.

As groups such as the Young Pioneers and Komsomol weaken or disappear, young Russians are left with more free time and are seeking new ways to spend it. Although Russian teenagers rarely worked at after-school jobs, Russia's economic crisis offers them even less chance to do so now. Many spend their free time socializing with friends.

Moscow teenagers who enjoy playing rollerball, like those pictured above, arrive each day in Gorky Park to meet friends and practice their skills. Other teens have become swept up in Western music, spending so much time in music-related activities that they are called *rokry* (rockers) or *metallisty* (heavy metal fans).

Many Russian teens, however, have other interests besides sports and music. Some join large groups that work to clean up or conserve the environment. Others remain in the International Clubs that schools organized as the old Soviet Union became more open. In these clubs, teenagers plan parties and meetings and correspond with pen pals in other nations.

progress has been slow. Stable political relationships have yet to be created.

Russia's greatest resource is its people, who are well educated. They have survived the challenges of living in a country characterized by isolation and a harsh environment. During the Soviet era, government officials controlled the lives of the people. Now the people can make their own decisions and determine their own futures. The Russian people have much to offer themselves and the world.

SECTION REVIEW

1. What is significant about the Tatarstan Republic?
2. Name two political challenges facing the Russian Federation.
3. **Critical Thinking List** ways in which a free-market economy and a democratic government might benefit the Russian Federation.

PLANET WATCH

Women Tree Planters of Kenya

Avocado, mango, and acacia trees thrive in the Kenyan highlands, where not long ago farm workers could not even find shade.

Ten million young trees scattered throughout the country offer new hope to Kenyan farmers and families. The trees were planted not by foresters or government workers but by the many women who farm the lands. "We are planting trees to ensure our own survival," says Wangari Maathai, founder of Kenya's Green Belt Movement. The organization encourages tree planting in Kenya's rural communities as a way of fighting deforestation and desertification.

Deforestation

Like many other developing nations, Kenya is a country mainly of farmers and herders. The most productive farmlands are in the highlands, an area once green with trees and lush vegetation. Over the past century, however, much of the land has been stripped. Today, only about 10 percent of the original forests remain. Many trees were cut for firewood, which farm families use in open-hearth cooking.

One by one, the trees began to disappear from the highlands. Without tree roots to hold the soil in place, the land eroded. Erosion of banks along streams and rivers then threatened the water supplies.

The land was losing its ability to sustain the region's growing population as well as its fertility. Farmers moved down to the savannas in search of better land. These plains are home to nomadic peoples and much of Kenya's wildlife.

Wangari Maathai recognized what was happening to her country. The daughter of a farm worker, she is a biologist and the first woman in Kenya to receive a PhD. "When I would visit the village where I was born," she says, "I saw whole forests had been cleared for cultivation and timber. People were moving onto hilly slopes and riverbeds and marginal areas that were only bush when I was a child. Springs were drying up." Maathai was shocked to find children suffering from malnutrition. "[My community was supposed to be a rich, coffee-growing area." Instead of eating nutritious, traditional foods, such as beans and corn, she explains, people were relying on refined foods such as rice because they need less cooking—and thus less firewood.

The Green Belt Movement

On June 5, 1977, in honor of World Environment Day, Maathai and a few supporters planted seven trees in Nairobi, Kenya's capital. This small tree planting began the Green Belt

Many of the trees in the highlands of **Kenya** *have been cleared for crops.*

CITIES OF THE WORLD

Jerusalem

The capital of ancient Israel 3,000 years ago, Jerusalem is the capital of modern Israel today. The city's past and present merge in its skyline of ancient ruins, medieval towers, and modern high-rise buildings.

Jerusalem is of central importance to people of the Jewish faith. For centuries, Jews living outside the city have turned toward Jerusalem in prayer three times each day. Hundreds of millions of non-Jews also care deeply about the city. Christians cherish it as the place where Jesus lived, preached, and was crucified. Muslims honor the site in Jerusalem where they believe the prophet Muhammad rose to heaven.

An Ancient History

Because it is sacred to so many, several groups have fought for control of Jerusalem in its 5,000-year history. It became the center of Jewish life in 1003 B.C., when King David made it his capi-

Perhaps more than in any other city, the threads of human history wind through Jerusalem's streets.

tal. Thirty-three years later, David's son Solomon built his magnificent Temple there. Although the Babylonians conquered Jerusalem and destroyed the Temple in 586 B.C., the Israelites recovered the city and built a second Temple.

Jerusalem was under Roman control when Jesus lived and preached there. A generation later, in A.D. 66, the Jews revolted against Roman rule. The Romans crushed the revolt, however, and in A.D. 70 destroyed the second Temple. The Temple's only remnant, known today as the Western Wall, remains Judaism's most sacred religious shrine.

After the Romans were finally driven out, Jerusalem came under the control of a series of groups, including the Byzantines, Arabs, Christian Crusaders, Egyptians, and Turks. In the late seventh century, the Arabs built the Dome of the Rock and the al-Aqsa Mosque where the Jewish Temple once stood. Later, during the sixteenth century, the Turks constructed the high stone walls that still surround what today is called the Old City.

Modern Jerusalem

The modern history of Jerusalem begins in the 1860s, when the city's growing Jewish population built the first neighborhoods outside the city's walls. Jerusalem continued to spread outward, especially after the British wrestled the city from the Turks in 1917.

When modern Israel formed in 1948, another round in the struggle for Jerusalem began. Neighboring Arab states attacked Israel, and Jordan occupied the western part of the city in 1949 and made it the country's capital. The eastern part, including the Old City, remained under Jordanian control.

Jerusalem was reunified when Israel defeated Jordan, Egypt, and Syria in the Six-Day War in 1967. Today, the Old City is part of a metropolis of more than 500,000 people. Within

Visitors view the Children's Memorial, part of the Holocaust Memorial, or Yad Vashem, in **Jerusalem**. *Israel was founded in 1948, three years after the end of World War II and the defeat of Nazi Germany.*

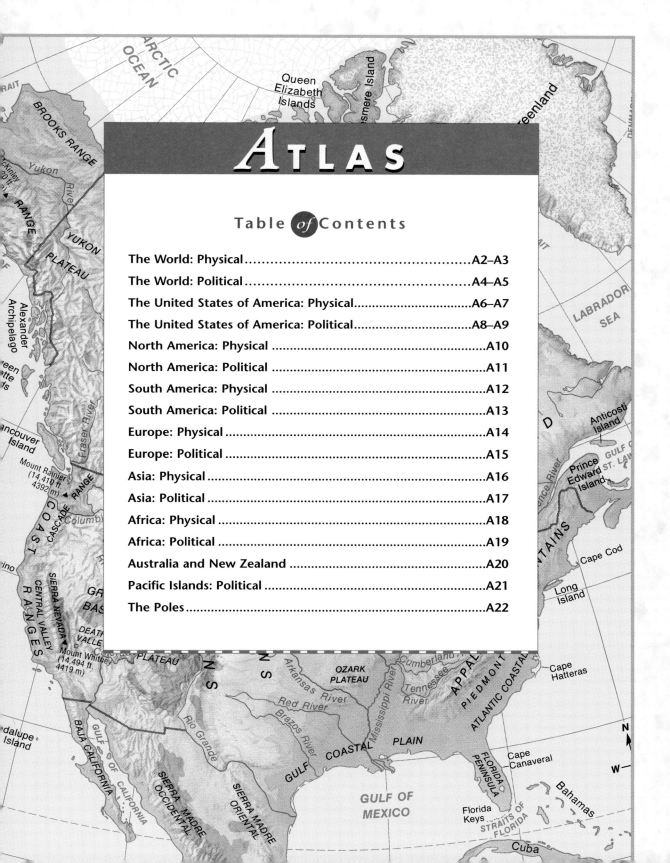

ATLAS

Table *of* Contents

The World: Physical A2–A3

The World: Political A4–A5

The United States of America: Physical............................A6–A7

The United States of America: Political.............................A8–A9

North America: Physical A10

North America: Political A11

South America: Physical A12

South America: Political A13

Europe: Physical A14

Europe: Political A15

Asia: Physical A16

Asia: Political A17

Africa: Physical A18

Africa: Political A19

Australia and New Zealand A20

Pacific Islands: Political A21

The Poles..A22

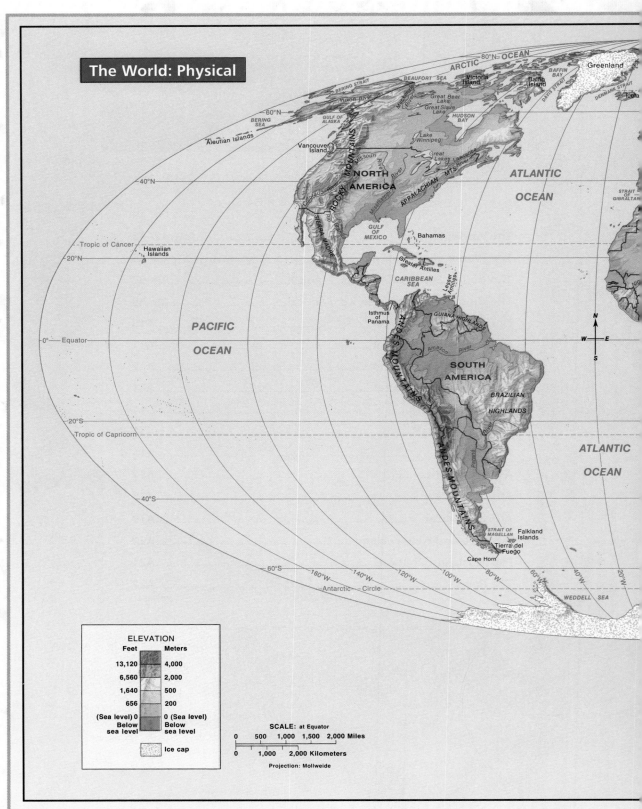

The World: Physical

ELEVATION

Feet	Meters
13,120	4,000
6,560	2,000
1,640	500
656	200
(Sea level) 0	0 (Sea level)
Below sea level	Below sea level

Ice cap

SCALE: at Equator

0   500   1,000   1,500   2,000 Miles

0   1,000   2,000 Kilometers

Projection: Mollweide

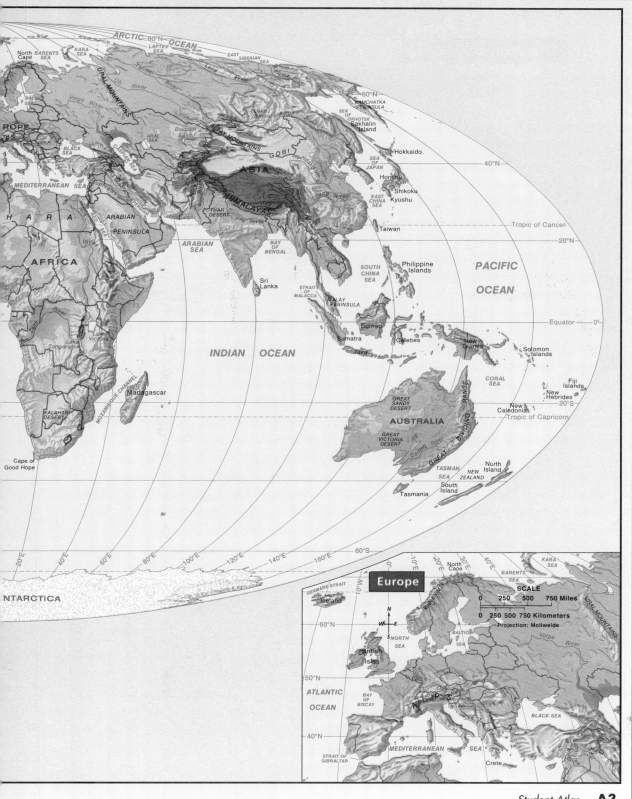

ARCTIC 80°N OCEAN
North BARENTS KARA LAPTEV EAST SIBERIAN SEA
Cape SEA SEA SEA

BALTIC URAL MOUNTAINS Ob River Yenisei River Lena River Kolyma River 60°N
SEA Volga River KAMCHATKA PENINSULA
ROPE ARAL Lake Amur River SEA OF OKHOTSK
CASPIAN SEA Balqash SEA Baikal OKHOTSK Sakhalin Island
BLACK Lake SEA ALTAI MOUNTAINS GOBI Hokkaido
SEA ASIA Huang He SEA OF JAPAN 40°N
MEDITERRANEAN SEA Euphrates River THAR HIMALAYAS Yangtze River Honshu
H A R A ARABIAN DESERT Ganges River Shikoku
PENINSULA Indus River EAST CHINA SEA Kyushu
AFRICA ARABIAN BAY OF Tropic of Cancer
Nile River SEA BENGAL Taiwan
Sri SOUTH Philippine 20°N
Lanka STRAIT CHINA Islands PACIFIC
OF SEA
MALACCA MALAY OCEAN
Zaire River PENINSULA
Lake Borneo
Victoria Sumatra Celebes New Equator 0°
Lake INDIAN OCEAN Java Guinea Solomon
Tanganyika Islands
MOZAMBIQUE CHANNEL CORAL Fiji
Madagascar SEA Islands
GREAT New
SANDY Hebrides
KALAHARI DESERT New 20°S
DESERT GREAT Caledonia
AUSTRALIA DIVIDING RANGE Tropic of Capricorn
GREAT
Cape of VICTORIA Darling River
Good Hope DESERT North
TASMAN NEW Island
SEA ZEALAND
Tasmania South
Island

60°S

NTARCTICA

20°E 40°E 60°E 80°E 100°E 120°E 140°E 160°E

DENMARK STRAIT 0° 10°W 10°E 20°E 30°E 40°E KARA
North BARENTS SEA
Iceland Cape SEA

Europe SCALE
0 250 500 750 Miles
0 250 500 750 Kilometers
Projection: Mollweide

60°N NORTH URAL MOUNTAINS
W E SEA BALTIC Volga River
S SEA
British
Isles
50°N
ATLANTIC BAY
OCEAN OF ALPS
BISCAY BLACK SEA
40°N
STRAIT OF MEDITERRANEAN SEA
GIBRALTAR
Crete

A3

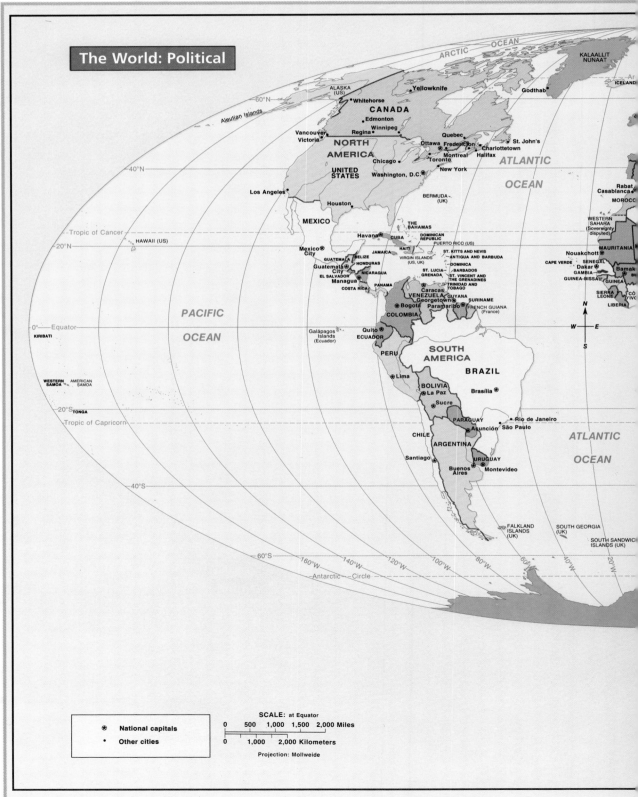

The World: Political

ARCTIC OCEAN

KALAALLIT NUNAAT

ICELAND

ALASKA (US)
Yellowknife
Godthab

Whitehorse

CANADA
Edmonton
Winnipeg
Regina
Vancouver
Victoria
Quebec
Ottawa Fredericton
St. John's
NORTH
AMERICA
Chicago
Montreal Halifax
Toronto
Charlottetown

UNITED
STATES
Washington, D.C.
New York

ATLANTIC

Los Angeles
OCEAN

Rabat
Casablanca
MOROCCO

Houston
BERMUDA
(UK)

WESTERN
SAHARA
(Sovereignty
disputed)

Tropic of Cancer

MEXICO
THE
BAHAMAS

HAWAII (US)
Havana CUBA
DOMINICAN
REPUBLIC
MAURITANIA
Nouakchott

Mexico
City
JAMAICA HAITI
PUERTO RICO (US)
ST. KITTS AND NEVIS
CAPE VERDE
SENEGAL
Dakar Bamak
GUATEMALA BELIZE
VIRGIN ISLANDS
(US, UK)
ANTIGUA AND BARBUDA
GAMBIA
GUINEA-BISSAU GUINEA
Guatemala
City
HONDURAS
DOMINICA
EL SALVADOR
NICARAGUA
BARBADOS
ST. LUCIA
SIERRA
CÔ
Managua
GRENADA
ST. VINCENT AND
THE GRENADINES
LEONE
D'IVO
COSTA RICA
PANAMA
TRINIDAD AND
TOBAGO
LIBERIA
Caracas
VENEZUELA GUYANA
Georgetown SURINAME

Bogotá Paramaribo
FRENCH GUIANA
(France)

COLOMBIA

Galápagos
Islands
(Ecuador)
Quito
ECUADOR

N

PACIFIC
KIRIBATI
Equator

PERU
SOUTH
AMERICA
W
E

OCEAN
BRAZIL
S

Lima

WESTERN
SAMOA
AMERICAN
SAMOA
BOLIVIA
La Paz
Brasília

Sucre

TONGA
Tropic of Capricorn
PARAGUAY
Rio de Janeiro
Asunción São Paulo

CHILE
ATLANTIC

ARGENTINA
Santiago
URUGUAY
OCEAN
Buenos
Aires
Montevideo

FALKLAND
ISLANDS
(UK)
SOUTH GEORGIA
(UK)

SOUTH SANDWICH
ISLANDS (UK)

160°W 140°W 120°W 100°W 80°W 60° 40° 20°

Antarctic Circle

⊛ National capitals
• Other cities

SCALE: at Equator

0   500   1,000   1,500   2,000 Miles

0   1,000   2,000 Kilometers

Projection: Mollweide

COUNTRY	CAPITAL
1 Czech Republic	Prague
2 Slovakia	Bratislava
3 Slovenia	Ljubljana
4 Croatia	Zagreb
5 Bosnia and Herzegovina	Sarajevo
6 Macedonia	Skopje
7 Yugoslavia (Serbia and Montenegro)	Belgrade
8 Lithuania	Vilnius
9 Latvia	Riga
10 Estonia	Tallinn

United States of America: Physical

PACIFIC OCEAN

STRAIT OF JUAN DE FUCA

PUGET SOUND

Mount Rainier 14,410 ft. (4392 m)

Columbia River

COAST RANGES

CASCADE RANGE

BITTERROOT RANGE

COLUMBIA PLATEAU

Franklin D. Roosevelt Lake

Pend Oreille Lake

Flathead Lake

R O C K Y

Milk River

Missouri River

Lake Sakakawea

Red River

Fort Peck Lake

Yellowstone River

Lake Oahe

M O U N T A I N S

SALMON RIVER MTS.

Salmon River

Sawtooth

Snake River

Yellowstone Lake

CONTINENTAL DIVIDE

Bighorn River

Powder River

BLACK HILLS

Cheyenne River

White River

James River

Niobrara River

Klamath River

Goose Lake

Great Salt Lake

Utah Lake

WASATCH RANGE

UINTA MTS.

Gannett Peak 13,804 ft. (4209 m)

FRONT RANGE

North Platte River

South Platte River

Platte River

Republican River

Cape Mendocino

Shasta Lake

Pyramid Lake

Lake Tahoe

GREAT BASIN

SIERRA NEVADA

CENTRAL VALLEY

Sacramento River

San Joaquin River

SAN FRANCISCO BAY

MONTEREY BAY

COAST RANGES

Mount Whitney 14,495 ft. (4418 m)

MOJAVE DESERT

Channel Islands

Colorado River

Lake Mead

GRAND CANYON

Lake Powell

COLORADO PLATEAU

Green River

Colorado River

Gunnison River

Mount Elbert 14,433 ft. (4399 m)

Pikes Peak 14,110 ft. (4301 m)

DIVIDE

SANGRE DE CRISTO MTS.

VALLEY

CONTINENTAL

Smoky Hill River

Kansas River

Arkansas River

Canadian River

Keystone Lake

Eufaula Lake

Lake Texoma

G R E A T   I N T E R I O R   P L A I N S

Salton Sea IMPERIAL VALLEY

PAINTED DESERT

Gila River

SONORA DESERT

GULF OF CALIFORNIA

Pecos River

Colorado River

Brazos River

Rio Grande

Amistad Reservoir

Nueces River

MEXICO

Falcon Lake

Padre Island

GULF

To understand the relative locations of Alaska and Hawaii as well as the vast distances separating them from the rest of the United States, see the world map.

Kauai

Niihau

Oahu

Molokai

Lanai

Kahoolawee

Maui

Mauna Kea (13,796 ft. 4205 m)

Hawaii

PACIFIC OCEAN SCALE

0   75   150 Miles

0   75   150 Kilometers

ARCTIC OCEAN

Arctic Circle

RUSSIA

BERING STRAIT

St. Lawrence Island

St. Matthew Island

Nunivak Island

BROOKS RANGE

Yukon River

Tanana River

Kuskokwim River

ALASKA RANGE

Mount McKinley 20,320 ft. 6194 m)

CANADA

Kodiak Island

GULF OF ALASKA

Alexander Archipelago

BERING SEA

Attu Island

A l e u t i a n   I s l a n d s

SCALE

0   250   500 Miles

0   250   500 Kilometers

Projection: Albers Equal Area

PACIFIC OCEAN

CANADA

MESABI RANGE

Isle Royale

Lake Superior

Lake Michigan

Lake Huron

Lake Ontario

Lake Erie

Mississippi River

Wisconsin River

Des Moines River

Illinois River

Wabash River

Scioto River

Ohio River

Lake of the Ozarks

Lake Barkley

Kentucky Lake

White River

OZARK PLATEAU

OUACHITA MTS.

COASTAL PLAIN

P L A I N S

ALLEGHENY PLATEAU

APPALACHIAN MOUNTAINS

CUMBERLAND PLATEAU

GREAT SMOKY MTS.

BLUE RIDGE MOUNTAINS

P I E D M O N T

ATLANTIC COASTAL PLAIN

Tennessee River

Cumberland River

Kanawha River

Tombigbee River

Alabama River

Coosa River

Chattahoochee River

Pearl River

Savannah River

Santee River

Roanoke River

James River

Potomac River

Susquehanna River

Allegheny River

Finger Lakes

CATSKILL MTS.

ADIRONDACK MTS.

St. Lawrence River

St. Lawrence Seaway

GREEN MTS.

WHITE MTS.

LONGFELLOW MTS.

St. John River

Cape Cod

LONG ISLAND SOUND

Long Island

DELAWARE BAY

CHESAPEAKE BAY

PAMLICO SOUND

Cape Hatteras

Sea Islands

Altamaha River

Okefenokee Swamp

Chandeleur Islands

Mississippi Delta

GULF OF MEXICO

FLORIDA PENINSULA

Cape Canaveral

Lake Okeechobee

The Everglades

Cape Sable

Florida Keys

STRAITS OF FLORIDA

THE BAHAMAS

CUBA

ATLANTIC OCEAN

N
W E
S

SCALE

0	250	500 Miles
0	250	500 Kilometers

Projection: Albers Equal Area

ELEVATION

Feet		Meters
13,120		4,000
6,560		2,000
1,640		500
656		200
(Sea level) 0		0 (Sea level)
Below sea level		Below sea level

Ice cap

90°W 85°W 80°W 75°W 70°W 65°W 60°W 55°W 50°W

50°N
45°N
40°N
35°N
30°N
25°N

United States of America: Political

To understand the relative locations of Alaska and Hawaii as well as the vast distances separating them from the rest of the United States, see the world map.

HAWAII

PACIFIC OCEAN
SCALE
0   75   150 Miles
0   75   150 Kilometers

ALASKA

SCALE
0   250   500 Miles
0   250   500 Kilometers
Projection: Albers Equal Area

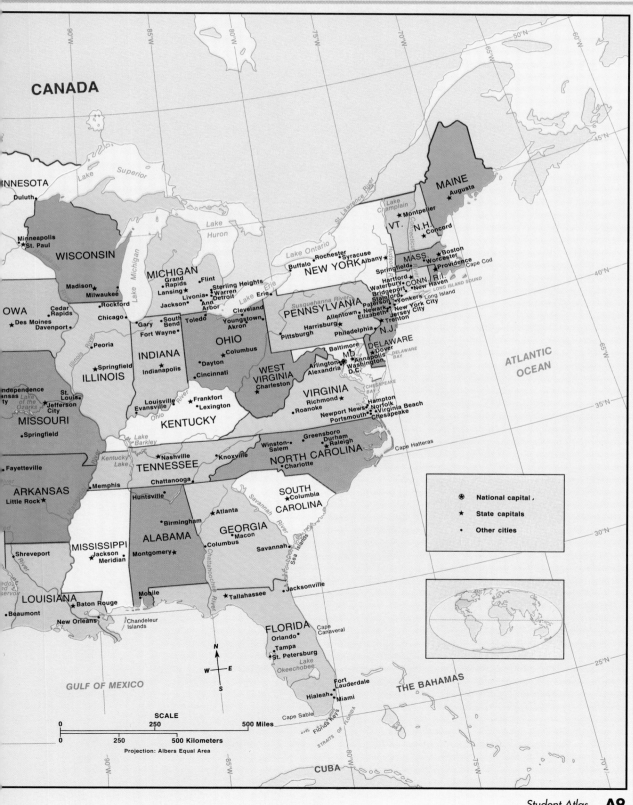

CANADA

MINNESOTA
Duluth

Minneapolis
★ St. Paul

WISCONSIN

Madison ★
Milwaukee

IOWA
Cedar
Rapids
Des Moines
Davenport

ILLINOIS
Springfield ★

MICHIGAN
Grand
Rapids
Lansing ★
Jackson
Flint
Sterling Heights
Warren
Detroit
Livonia
Ann
Arbor

Rockford
Chicago
Gary
South
Bend
Fort Wayne

Peoria

INDIANA
Indianapolis ★

OHIO
Columbus

Dayton
Cincinnati

Toledo

Cleveland
Youngstown
Akron

Lake Superior
Lake Michigan
Lake Huron
Lake Ontario
Lake Erie

NEW YORK
Buffalo
Rochester
Syracuse
Albany

PENNSYLVANIA
Harrisburg ★
Pittsburgh

Susquehanna River

MAINE
★ Augusta

Montpelier
VT.
N.H.
★ Concord

MASS.
Springfield
Worcester ★ Boston
Providence
Hartford
Waterbury
Bridgeport
Stamford
CONN. R.I.
New Haven
Cape Cod

Yonkers Long Island
Paterson
Newark
New York City
Elizabeth
Jersey City
Allentown
Trenton
N.J

LONG ISLAND SOUND

Philadelphia

Baltimore
MD
DELAWARE
Dover
DELAWARE
BAY

Arlington
Alexandria
Annapolis
Washington,
D.C.

ATLANTIC
OCEAN

Independence
Kansas
City
St.
Louis
Jefferson
City

MISSOURI
Springfield

Fayetteville

ARKANSAS
Little Rock ★

Lake
of the
Ozarks

Lake
Barkley

Louisville
Evansville

WEST
VIRGINIA
Charleston

Frankfort ★
Lexington

KENTUCKY

Ohio River

Kentucky
Lake

Nashville ★
Knoxville

TENNESSEE

Chattanooga

Memphis

Huntsville

VIRGINIA
Richmond ★
Roanoke

Hampton
Norfolk
Virginia Beach
Newport News
Portsmouth
Chesapeake

CHESAPEAKE
BAY

Cape Hatteras

Winston-
Salem

Greensboro
Durham
★ Raleigh

NORTH CAROLINA
Charlotte

SOUTH
CAROLINA
Columbia ★

Birmingham

ALABAMA
Montgomery ★

Atlanta ★

GEORGIA
Macon
Columbus

Savannah River

Savannah
Sea Islands

Chattahoochee River

⊛   National capital ,

★   State capitals

•   Other cities

Shreveport

MISSISSIPPI
Jackson ★
Meridian

Mobile

Tallahassee ★

Jacksonville

LOUISIANA
Baton Rouge ★

Beaumont
New Orleans
Chandeleur
Islands

FLORIDA
Orlando
Tampa
St. Petersburg
Lake
Okeechobee

Cape
Canaveral

N
W   E
S

GULF OF MEXICO

Fort
Lauderdale
Hialeah   Miami

Cape Sable

THE BAHAMAS

SCALE
0   250   500 Miles

0   250   500 Kilometers
Projection: Albers Equal Area

Florida Keys
STRAITS OF FLORIDA

CUBA

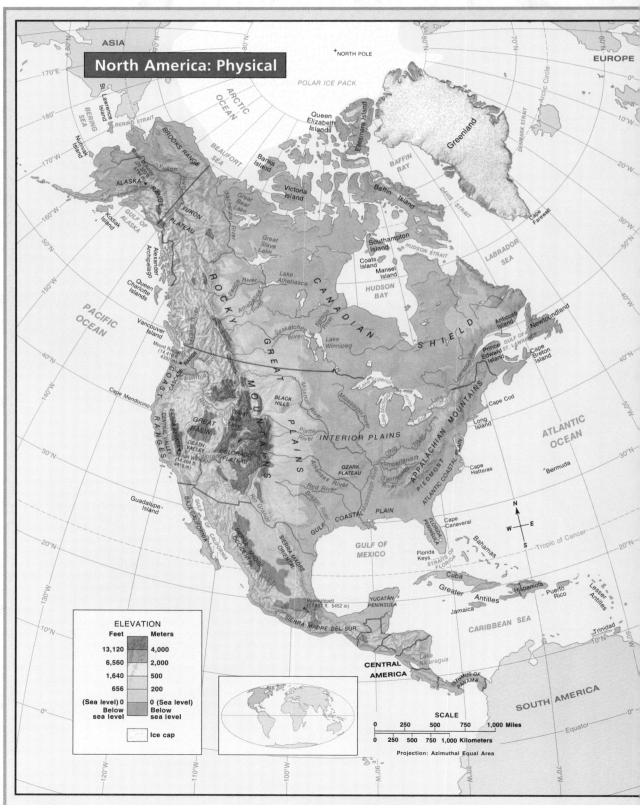

North America: Physical

ASIA

EUROPE

NORTH POLE

POLAR ICE PACK

ARCTIC OCEAN

170°E
180°
170°W
160°W
150°W

BERING SEA

St. Lawrence Island

Nunivak Island

BERING STRAIT

BROOKS RANGE

ALASKA RANGE

Mount McKinley (20,320 ft. 6194 m)

Yukon

YUKON

Kodiak Island

GULF OF ALASKA

PLATEAU

Alexander Archipelago

Queen Charlotte Islands

Vancouver Island

BEAUFORT SEA

Banks Island

Great Bear Lake

Mackenzie River

Victoria Island

Queen Elizabeth Islands

Ellesmere Island

Greenland

DENMARK STRAIT

BAFFIN BAY

Baffin Island

DAVIS STRAIT

Cape Farewell

HUDSON STRAIT

Southampton Island

Coats Island

Mansel Island

HUDSON BAY

LABRADOR SEA

Arctic Circle

10°W
20°W
30°W
40°W

80°N
70°N
60°N
50°N
40°N

Great Slave Lake

Peace River

Athabasca River

Lake Athabasca

ROCKY

CANADIAN

SHIELD

Anticosti Island

Newfoundland

PACIFIC OCEAN

Cape Mendocino

Mount Rainier (14,410 ft. 4392 m)

Fraser River

COAST RANGE

CASCADE RANGE

Columbia River

Snake River

SIERRA NEVADA

CENTRAL VALLEY

DEATH VALLEY

Mount Whitney (14,494 ft. 4419 m)

GREAT BASIN

Great Salt Lake

COLORADO PLATEAU

Colorado River

GREAT

Saskatchewan River

Lake Winnipeg

Nelson River

BLACK HILLS

Missouri River

Platte River

INTERIOR PLAINS

PLAINS

MOUNTAINS

OZARK PLATEAU

Arkansas River

Red River

Ohio R.

Cumberland R.

Tennessee River

Mississippi River

Lake Superior

Lake Michigan

Lake Huron

Lake Erie

Lake Ontario

St. Lawrence River

Prince Edward Island

GULF OF ST. LAWRENCE

Cape Breton Island

Cape Cod

Long Island

APPALACHIAN MOUNTAINS

PIEDMONT

ATLANTIC COASTAL PLAIN

Cape Hatteras

ATLANTIC OCEAN

Bermuda

50°W
60°W

60°N
50°N
40°N
30°N

140°W
130°W
120°W
110°W

Guadalupe Island

BAJA CALIFORNIA

GULF OF CALIFORNIA

SIERRA MADRE OCCIDENTAL

Rio Grande

Brazos River

GULF COASTAL PLAIN

FLORIDA PENINSULA

Cape Canaveral

GULF OF MEXICO

Florida Keys

STRAITS OF FLORIDA

Bahamas

Cuba

Greater Antilles

Jamaica

Hispaniola

Puerto Rico

Lesser Antilles

CARIBBEAN SEA

Trinidad

Tropic of Cancer

Popocatépetl (17,803 ft. 5452 m)

YUCATÁN PENINSULA

SIERRA MADRE DEL SUR

SIERRA MADRE ORIENTAL

CENTRAL AMERICA

Lake Nicaragua

ISTHMUS OF PANAMA

SOUTH AMERICA

Equator

20°N
10°N
0°

N
W   E
S

ELEVATION

Feet	Meters
13,120	4,000
6,560	2,000
1,640	500
656	200
(Sea level) 0	0 (Sea level)
Below sea level	Below sea level

Ice cap

SCALE

0   250   500   750   1,000 Miles

0   250   500   750   1,000 Kilometers

Projection: Azimuthal Equal Area

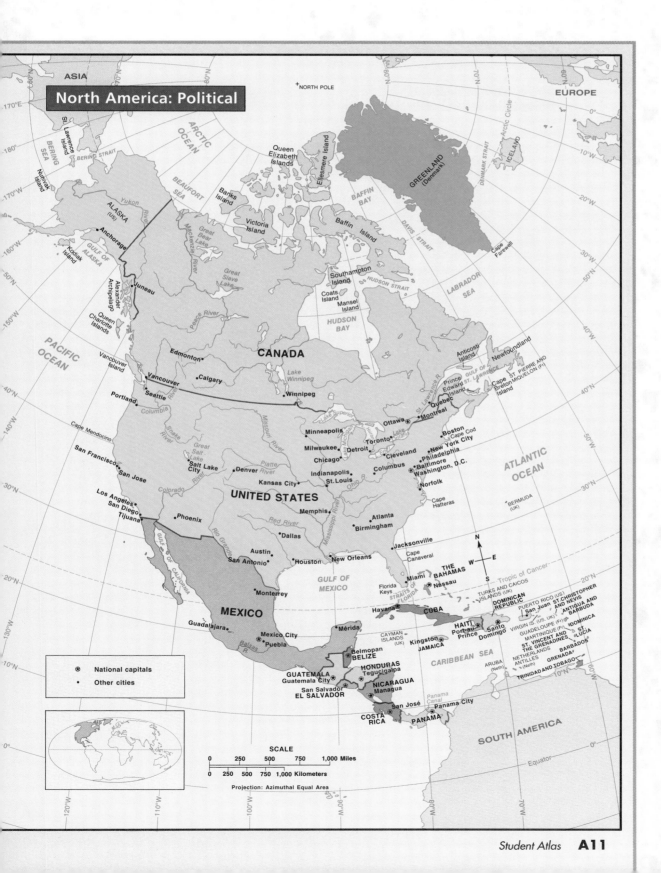

North America: Political

ASIA

EUROPE

+NORTH POLE

ARCTIC OCEAN

St. Lawrence Island

BERING STRAIT

Nunivak Island

BEAUFORT SEA

Banks Island

Queen Elizabeth Islands

Ellesmere Island

ICELAND

GREENLAND (Denmark)

DENMARK STRAIT

Arctic Circle

ALASKA (US)

Yukon River

Anchorage

GULF OF ALASKA

Kodiak Island

Victoria Island

Great Bear Lake

Mackenzie River

Baffin Island

BAFFIN BAY

DAVIS STRAIT

Cape Farewell

Alexander Archipelago

Juneau

Great Slave Lake

Southampton Island

HUDSON STRAIT

LABRADOR SEA

Queen Charlotte Islands

PACIFIC OCEAN

Peace River

Coats Island

Mansel Island

Vancouver Island

Edmonton

CANADA

HUDSON BAY

Anticosti Island

Newfoundland

Vancouver

Calgary

Lake Winnipeg

Prince Edward Island

GULF OF ST. LAWRENCE

Cape ST. PIERRE AND Breton MIQUELON (Fr) Island

Portland

Seattle

Columbia River

Winnipeg

St. Lawrence R.

Cape Mendocino

Snake River

Great Salt Lake

Missouri River

Minneapolis

Milwaukee

Chicago

Detroit

Lake Superior

Lake Michigan

Lake Huron

Toronto

Ottawa

Montreal

Quebec

Lake Erie

Lake Ontario

Cleveland

Boston

Cape Cod

New York City

Philadelphia

ATLANTIC OCEAN

San Francisco

San Jose

Salt Lake City

Denver

Platte River

Kansas City

Indianapolis

St. Louis

Columbus

Baltimore

Washington, D.C.

Los Angeles

San Diego

Tijuana

Colorado River

UNITED STATES

Ohio R.

Norfolk

Cape Hatteras

BERMUDA (UK)

Phoenix

Memphis

Atlanta

Dallas

Red River

Birmingham

Rio Grande

Jacksonville

Cape Canaveral

Austin

San Antonio

Houston

New Orleans

N

W E

S

Monterrey

GULF OF MEXICO

Florida Keys

STRAITS OF FLORIDA

Miami

THE BAHAMAS

Nassau

Tropic of Cancer

TURKS AND CAICOS ISLANDS (UK)

MEXICO

Guadalajara

Mexico City

Puebla

Balsas R.

Mérida

Havana

CUBA

CAYMAN ISLANDS (UK)

Kingston

JAMAICA

HAITI

Port-au-Prince

DOMINICAN REPUBLIC

Santo Domingo

PUERTO RICO (US)

San Juan

VIRGIN IS. (US, UK)

GUADELOUPE (Fr)

MARTINIQUE (Fr)

ST. VINCENT AND THE GRENADINES

NETHERLANDS ANTILLES (Neth)

ST. CHRISTOPHER AND NEVIS

ANTIGUA AND BARBUDA

DOMINICA

ST. LUCIA

BARBADOS

GRENADA

Belmopan

BELIZE

HONDURAS

Tegucigalpa

CARIBBEAN SEA

ARUBA (Neth)

TRINIDAD AND TOBAGO

GUATEMALA

Guatemala City

San Salvador

EL SALVADOR

NICARAGUA

Managua

Panama Canal

COSTA RICA

San José

PANAMA

Panama City

SOUTH AMERICA

Equator

⊛ National capitals

• Other cities

SCALE

0   250   500   750   1,000 Miles

0   250   500   750   1,000 Kilometers

Projection: Azimuthal Equal Area

South America: Physical

CENTRAL
AMERICA

CARIBBEAN SEA

Panama
Canal

GULF
OF
PANAMA

Lake
Maracaibo

Margarita
Island   Tobago
Trinidad

Orinoco River
Delta

ATLANTIC
OCEAN

LLANOS

Meta
River

Orinoco River

Angel Falls

GUIANA   HIGHLANDS

Devil's Island
Cape Orange

Amazon
River
Delta

Malpelo Island

Mount Tolima
(18,425 ft. 5618 m)

M
O
U
N
T
A
I
N
S

Rio   Negro

Galápagos
Islands

Equator

Mount Chimborazo
(20,561 ft. 6267 m)

GULF OF
GUAYAQUIL

Japurá
River

Amazon   River

AMAZON

BASIN

Amazon   River

Tocantins   River

Equator

A
N
D
E
S

Jurua   River

Purus   River

Madeira   River

Tapajos   River

Xingu   River

Tocantins   River

Paranaiba   River

BRAZILIAN

HIGHLANDS

PACIFIC

OCEAN

Mount Huascarán
(22,205 ft. 6768 m)

Ancohuma Peak
(20,956 ft. 6388 m)

MATO GROSSO
PLATEAU

Lake
Titicaca

ATACAMA DESERT

A
N
D
E
S

C H A C O

BRAZILIAN
PLATEAU

Tropic of Capricorn

San Ambrosio
Island

San Félix Island

Tropic of Capricorn

Mount Aconcagua
(22,834 ft. 6960 m)

M
O
U
N
T
A
I
N
S

Uruguay River

Juan Fernández
Islands

PAMPAS

RÍO DE LA PLATA

ATLANTIC

OCEAN

N
W   E
S

ELEVATION
Feet	Meters
13,120	4,000
6,560	2,000
1,640	500
656	200
(Sea level) 0	0 (Sea level)
Below sea level	Below sea level

Chiloé
Island

Chonos
Archipelago

PATAGONIA

GULF OF
SAN MATÍAS

GULF OF
SAN JORGE

Cape Tres Puntas

SCALE
0   250   500   750   1,000 Miles

0   250   500   750   1,000 Kilometers

Projection: Azimuthal Equal Area

BAHÍA
GRANDE

STRAIT OF
MAGELLAN

Tierra del
Fuego

Cape Horn

Falkland
Islands

South
Georgia
Islands

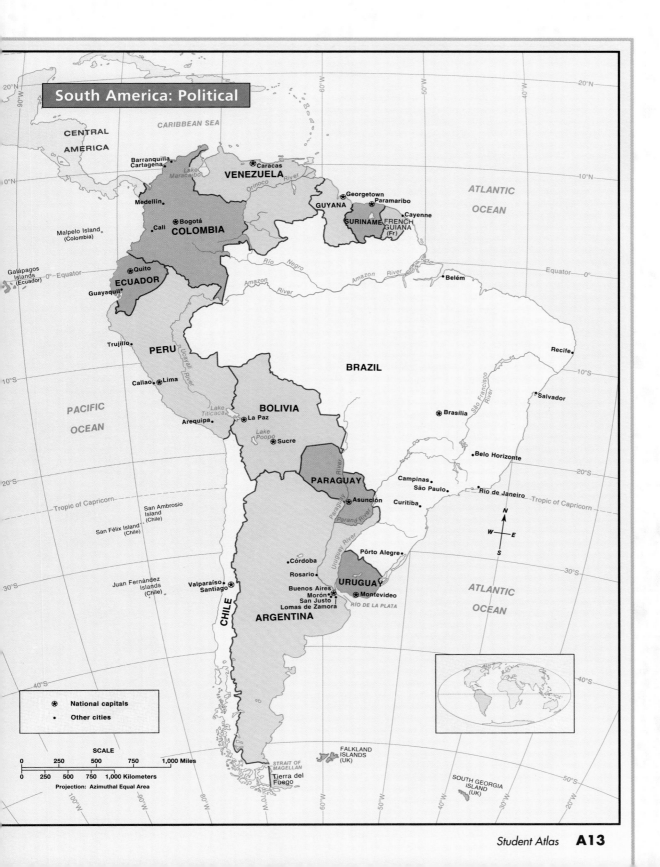

South America: Political

CENTRAL
AMERICA

CARIBBEAN SEA

Barranquilla
Cartagena
*Caracas
VENEZUELA
Lake
Maracaibo
Orinoco River

Medellín
GUYANA
⊛Georgetown
Paramaribo⊛
Cayenne
SURINAME FRENCH
GUIANA
(Fr.)

Malpelo Island
(Colombia)
⊛Bogotá
Cali
COLOMBIA

Galápagos
Islands
(Ecuador)
0°–Equator

⊛Quito
ECUADOR
Guayaquil

Rio Negro
Amazon River
Equator 0°
Belém

Amazon River

Trujillo
Recife

PERU
Ucayali River

Callao Lima

BRAZIL

São Francisco River

10°S

PACIFIC
OCEAN

Lake
Titicaca
Arequipa La Paz
BOLIVIA
Lake
Poopó
Sucre

⊛Brasília

Salvador

Belo Horizonte

Campinas
São Paulo
PARAGUAY
Asunción⊛
Paraguay River
Paraná River

Rio de Janeiro Tropic of Capricorn
Curitiba

N
W E
S

Tropic of Capricorn

San Ambrosio
Island
(Chile)

San Félix Island
(Chile)

Juan Fernández
Islands
(Chile)

Valparaíso
Santiago⊛

CHILE

Córdoba
Rosario
Buenos Aires
Morón
San Justo
Lomas de Zamora
ARGENTINA

Pôrto Alegre

URUGUAY
⊛Montevideo
RÍO DE LA PLATA

Uruguay River

ATLANTIC
OCEAN

ATLANTIC
OCEAN

⊛ National capitals
• Other cities

FALKLAND
ISLANDS
(UK)

SCALE

0   250   500   750   1,000 Miles
0 250 500 750 1,000 Kilometers
Projection: Azimuthal Equal Area

STRAIT OF
MAGELLAN
Tierra del
Fuego

SOUTH GEORGIA
ISLAND
(UK)

Europe: Physical

ELEVATION

Feet	Meters
13,120	4,000
6,560	2,000
1,640	500
656	200
(Sea level) 0	0 (Sea level)
Below sea level	Below sea level

Ice cap

SCALE

0   250   500 Miles

0   250   500 Kilometers

Projection: Azimuthal Equal Area

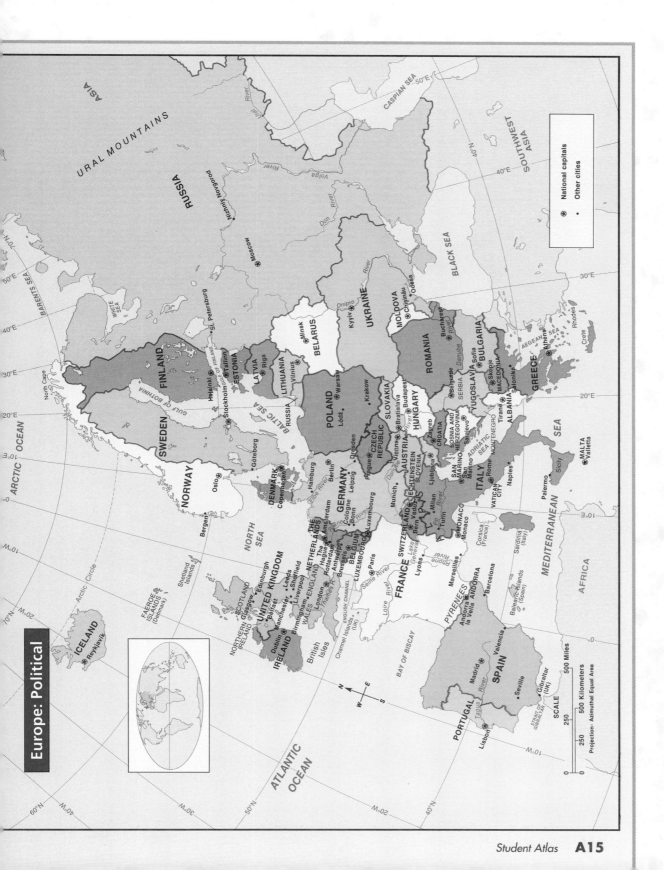

Europe: Political

Legend
⊛ National capitals
• Other cities

ASIA

URAL MOUNTAINS

RUSSIA

Nizhny Novgorod

Moscow

Ural River

Volga River

CASPIAN SEA

SOUTHWEST ASIA

BARENTS SEA

WHITE SEA

St. Petersburg

BLACK SEA

ARCTIC OCEAN

North Cape

FINLAND

Helsinki

GULF OF FINLAND

Tallinn

ESTONIA

Riga

LATVIA

LITHUANIA

Vilnius

BELARUS

Minsk

Dnipro

Kyyiv

UKRAINE

MOLDOVA

Chişinău

Odesa

Danube River

Bucharest

ROMANIA

BULGARIA

Sofia

AEGEAN SEA

Rhodes

Crete

SWEDEN

Stockholm

GULF OF BOTHNIA

RUSSIA

BALTIC SEA

Göteborg

POLAND

Warsaw

Łódź

Kraków

SLOVAKIA

Bratislava

HUNGARY

Budapest

Belgrade

SERBIA

YUGOSLAVIA

MONTENEGRO

Skopje

MACEDONIA

Salonika

GREECE

Athens

ALBANIA

Tirana

NORWAY

Oslo

Bergen

DENMARK

Copenhagen

Hamburg

Berlin

Elbe River

GERMANY

Leipzig

Dresden

Prague

CZECH REPUBLIC

Vienna

AUSTRIA

LIECHTENSTEIN

Danube River

Munich

Cologne

Bonn

Luxembourg

SLOVENIA

Ljubljana

Zagreb

CROATIA

BOSNIA AND HERZEGOVINA

Sarajevo

SAN MARINO

ADRIATIC SEA

Rome

VATICAN CITY

ITALY

Naples

Palermo

Sicily

MALTA

Valletta

MEDITERRANEAN SEA

AFRICA

NORTH SEA

THE NETHERLANDS

Amsterdam

The Hague

Rotterdam

Antwerp

BELGIUM

Brussels

LUXEMBOURG

FRANCE

SWITZERLAND

Bern

Lake Geneva

Vaduz

Milan

Po River

Turin

MONACO

Monaco

Corsica (France)

Sardinia (Italy)

Paris

Seine River

Lyons

Rhône River

Marseilles

Loire River

ICELAND

Reykjavik

FAEROE ISLANDS (Denmark)

Shetland Islands

SCOTLAND

Glasgow

Edinburgh

UNITED KINGDOM

Belfast

NORTHERN IRELAND

IRELAND

Dublin

Manchester

Liverpool

Leeds

Sheffield

Birmingham

WALES

ENGLAND

London

Thames R.

ENGLISH CHANNEL

Channel Islands (UK)

British Isles

BAY OF BISCAY

PYRENEES

ANDORRA

Andorra la Vella

Barcelona

Valencia

Balearic Islands (Spain)

SPAIN

Madrid

Tagus River

Seville

Gibraltar (UK)

STRAIT OF GIBRALTAR

PORTUGAL

Lisbon

ATLANTIC OCEAN

Arctic Circle

60°N

50°N

40°N

30°N

70°N

50°E

40°E

30°E

20°E

10°E

0°

10°W

20°W

30°W

40°W

50°N

40°N

SCALE

0 250 500 Miles

0 250 500 Kilometers

Projection: Azimuthal Equal Area

N E S W

Asia: Physical

ELEVATION

Feet	Meters
13,120	4,000
6,560	2,000
1,640	500
656	200
(Sea level) 0	0 (Sea level)
Below sea level	Below sea level

Ice cap

SCALE

0   500   1,000 Miles
0   500   1,000 Kilometers

Projection: Modified Oblique Conic

Asia: Political

National capitals
Other cities

A17

Africa: Physical

EUROPE

CENTRAL
ASIA

SOUTHWEST
ASIA

STRAIT OF
GIBRALTAR

MEDITERRANEAN   SEA

GULF OF
SIDRA

SUEZ
CANAL

Azores

Madeira
Islands

ATLAS   MOUNTAINS

Canary Islands

S   A   H   A   R   A

LIBYAN DESERT

QATTARA
DEPRESSION

PERSIAN GULF

Tropic of Cancer

Cape
Blanc

EL DJOUF

AHAGGAR
MOUNTAINS

TIBESTI
MOUNTAINS

Lake
Nasser

NUBIAN
DESERT

RED SEA

Cape Verde
Islands

S   A   H   E   L

AIR MTS.

Cape
Verde

Senegal

Niger River

S   U   D   A   N

Lake
Chad

CHAD
BASIN

Nile River

White Nile

Blue Nile

GULF OF ADEN

FOUTA
DJALLON

White Volta R.

Black Volta R.

Benue River

SUDAN
BASIN

ETHIOPIAN
HIGHLANDS

HORN OF AFRICA

SOMALI
PENINSULA

Cape
Palmas

Lake
Volta

ADAMAWA
MTS.

GULF OF
GUINEA

Ubangi
River

Zaire   River

Lake
Albert

Lake
Rudolf

RIFT   VALLEY

Cape
Lopez

Equator

ZAIRE
BASIN

Lake
Edward

Mount Kenya
17,058 ft
(5199 m)

Kasai River

Lake
Kivu

SERENGETI
PLAIN

Mount Kilimanjaro
(19,340 ft)
(5895 m)

INDIAN
OCEAN

EASTERN

Lake
Tanganyika

MASAI
STEPPE

Zanzibar

Ascension
Island

ATLANTIC
OCEAN

Cuanza

MITUMBA   MOUNTAINS

WESTERN

Lake Ruwa

Cape Delgado

Seychelles

River

Lake
Mweru

VALLEY

Comoro
Islands

Lake
Malawi
(Nyasa)

MOZAMBIQUE CHANNEL

Mauritius

Réunion

Lake
Kariba

Zambezi River

Madagascar

ELEVATION

Feet	Meters
13,120	4,000
6,560	2,000
1,640	500
656	200
(Sea level) 0	0 (Sea level)
Below sea level	Below sea level

Okavango
Delta
KALAHARI BASIN

Victoria
Falls

Limpopo River

Tropic of Capricorn

NAMIB   DESERT

KALAHARI
DESERT

Orange   River

Vaal River

DRAKENSBERG
ESCARPMENT

GREAT
KARROO

Cape of
Good Hope

SCALE

0   500   1,000 Miles

0   500   1,000 Kilometers

Projection: Azimuthal Equal Area

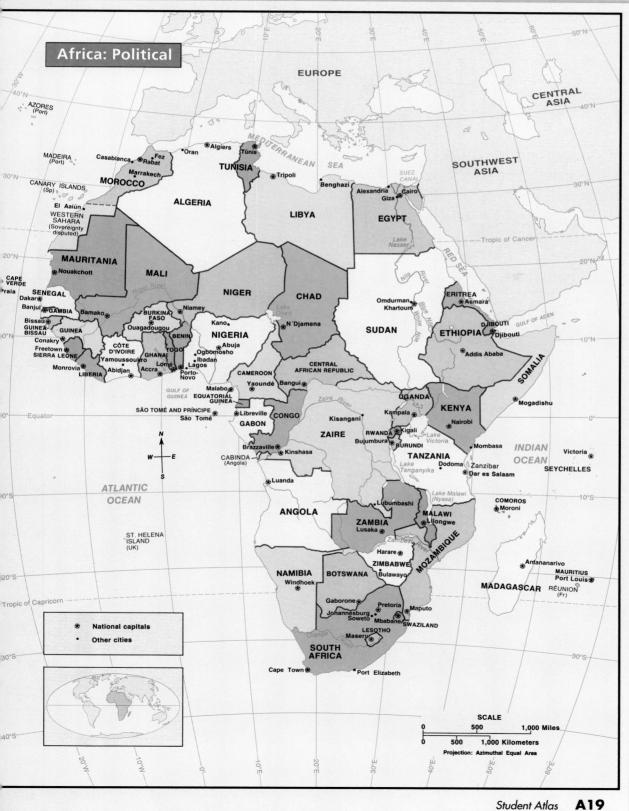

Africa: Political

EUROPE

CENTRAL ASIA

SOUTHWEST ASIA

MEDITERRANEAN SEA

AZORES (Port)

MADEIRA (Port)

CANARY ISLANDS (Sp)

Casablanca • Fez
•Oran
⊛Rabat
Marrakech•

⊛Algiers
⊛Tunis

TUNISIA

•Tripoli

Benghazi•

MOROCCO

El Aaiún•

WESTERN SAHARA (Sovereignty disputed)

ALGERIA

LIBYA

Alexandria• ⊛Cairo
Giza•

EGYPT

Lake Nasser

SUEZ CANAL

Tropic of Cancer

CAPE VERDE
•raia

MAURITANIA

⊛Nouakchott

MALI

NIGER

CHAD

Omdurman• ⊛Khartoum

ERITREA
⊛Asmará

DJIBOUTI

GULF OF ADEN

SENEGAL
Dakar⊛
Banjul⊛ GAMBIA
Bissau⊛ Bamako⊛
GUINEA- BISSAU GUINEA
Conakry⊛
Freetown⊛
SIERRA LEONE
Monrovia⊛
LIBERIA

BURKINA FASO
Ouagadougou⊛
Niamey•
BENIN
CÔTE D'IVOIRE GHANA TOGO
Yamoussoukro• Lomé•
Abidjan⊛ Accra⊛
Porto-Novo

Kano•

N'Djamena•

NIGERIA

Abuja⊛
Ogbomosho•
•Ibadan
Lagos•

SUDAN

CENTRAL AFRICAN REPUBLIC

ETHIOPIA

⊛Djibouti

⊛Addis Ababa

SOMALIA

Blue Nile
White Nile

RED SEA

Nile River

CAMEROON
Yaoundé⊛
Bangui⊛

Malabo⊛ EQUATORIAL GUINEA

GULF OF GUINEA

SÃO TOMÉ AND PRÍNCIPE
São Tomé•

Equator

⊛Libreville
GABON

CONGO

Kisangani•

UGANDA

Kampala⊛

⊛Mogadishu

KENYA

Nairobi⊛

Brazzaville⊛
CABINDA (Angola)

ZAIRE

•Kinshasa

RWANDA ⊛Kigali
Bujumbura⊛ BURUNDI

Lake Victoria

•Mombasa

INDIAN OCEAN

Victoria•

•Luanda

TANZANIA

Dodoma•
•Zanzibar
⊛Dar es Salaam

SEYCHELLES

ATLANTIC OCEAN

Lake Tanganyika

Lake Malawi (Nyasa)

COMOROS
⊛Moroni

ST. HELENA ISLAND (UK)

•Lubumbashi

ANGOLA

ZAMBIA
Lusaka ⊛

MALAWI
Lilongwe⊛

Zambezi River

MOZAMBIQUE

•Antananarivo

MAURITIUS
Port Louis⊛

NAMIBIA

BOTSWANA

Harare ⊛
ZIMBABWE
•Bulawayo

MADAGASCAR

RÉUNION (Fr)

Tropic of Capricorn

Windhoek⊛

Gaborone⊛
Johannesburg•
Soweto•
Pretoria•
Mbabane⊛ ⊛Maputo
SWAZILAND

Orange River

LESOTHO
Maseru⊛

SOUTH AFRICA

Cape Town⊛

•Port Elizabeth

⊛ National capitals
• Other cities

SCALE
0   500   1,000 Miles
0   500   1,000 Kilometers
Projection: Azimuthal Equal Area

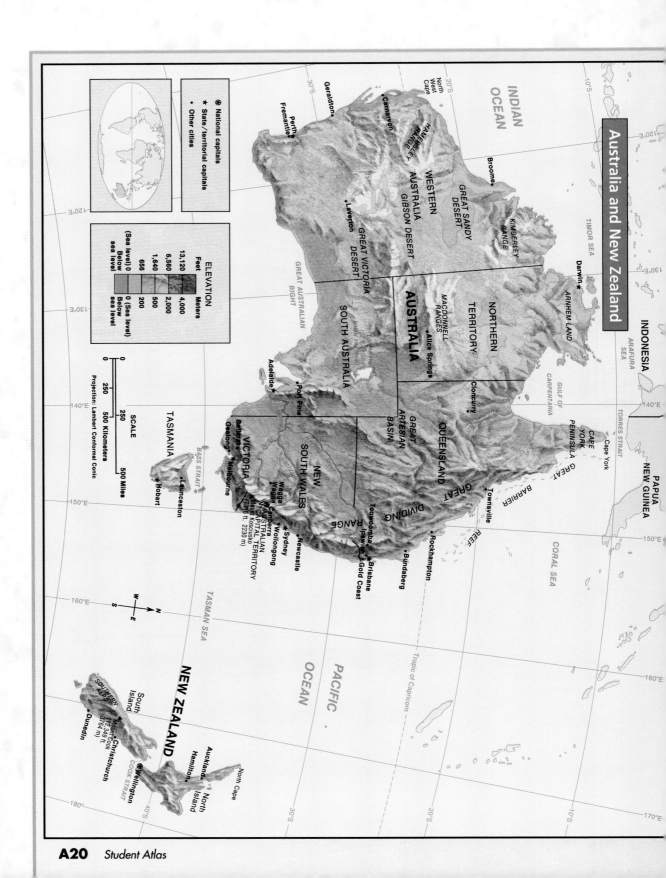

Australia and New Zealand

Pacific Islands: Political

ASIA

NORTH AMERICA

NORTH PACIFIC OCEAN

SOUTH PACIFIC OCEAN

SOUTH CHINA SEA

PHILIPPINE SEA

AUSTRALIA

NEW ZEALAND

TASMAN SEA

CORAL SEA

ARAFURA SEA

TIMOR SEA

INDIAN OCEAN

BONIN ISLANDS (Japan)
VOLCANO ISLANDS (Japan)
NORTHERN MARIANAS (US)
GUAM (US)
⊛ Agana
PALAU
⊛ Koror

MICRONESIA

Truk Is.
FEDERATED STATES OF MICRONESIA
⊛ Palikir

MARSHALL ISLANDS
Eniwetok I.
Kwajalein Island
⊛ Majuro

WAKE ISLAND (US)

MIDWAY ISLANDS (US)

Hawaiian Islands (US)
HAWAII (US)

JOHNSTON ATOLL (US)

KINGMAN REEF (US)
Washington Island
PALMYRA ATOLL (US)
Fanning Island

Starbuck Island

KIRIBATI

HOWLAND ISLAND (US)
BAKER ISLAND (US)

Phoenix Islands
McKean I.
Gardner I.

Gilbert Islands
⊛ Tarawa

NAURU
⊛ Yaren District

TUVALU
⊛ Funafuti

TOKELAU(NZ)

Manihiki Island

Society Islands (Fr)

COOK ISLANDS (NZ)
Rarotonga Island

POLYNESIA

FRENCH POLYNESIA

Marquesas Islands (Fr)

Tuamotu Archipelago (Fr)

⊛ Papeete
Tahiti (Fr)

Tubuai Islands (Fr)

Rapa Island (Fr)

PITCAIRN (UK)
Pitcairn Island
Ducie Island

Easter Island (Chile)

WALLIS AND FUTUNA(Fr)

WESTERN SAMOA
⊛ Apia

AMERICAN SAMOA
Pago Pago

NIUE (NZ)

TONGA
⊛ Nuku'alofa

⊛ Suva
FIJI

SOLOMON ISLANDS
⊛ Honiara
Guadalcanal Island

PAPUA NEW GUINEA
Bismarck Archipelago
New Guinea
⊛ Port Moresby

VANUATU
Espiritu Santo I.
Malekula I.
⊛ Port Vila

NEW CALEDONIA (Fr)
Loyalty Islands(Fr)
⊛ Nouméa

MELANESIA

NORFOLK ISLAND (Aust)

Kermadec Islands (NZ)

Chatham Islands (NZ)

Bounty Islands (NZ)

Auckland Islands (NZ)

CHRISTMAS ISLAND (Aust)

30°N
15°N
Tropic of Cancer
Equator – 0°
Tropic of Capricorn
15°S
30°S
45°S

120°E 135°E 150°E 165°E 180° 165°W 150°W 135°W 120°W

International Date Line

SCALE: At Equator
0   500   1,000 Miles
0   500   1,000 Kilometers
Projection: Mercator

N
W   E
S

⊛ National capitals
• Other cities

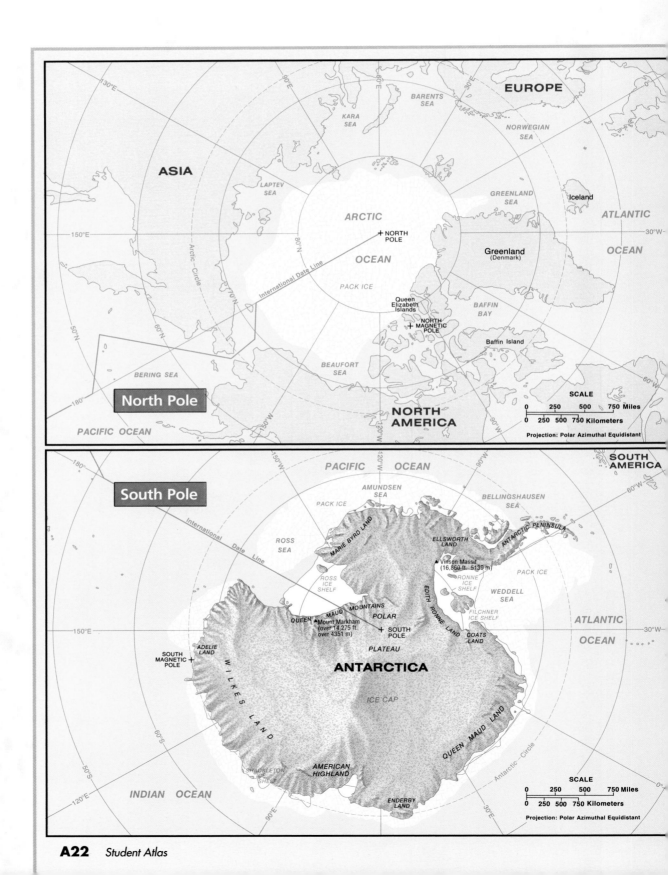

North Pole

EUROPE
BARENTS SEA
KARA SEA
NORWEGIAN SEA
ASIA
LAPTEV SEA
GREENLAND SEA
Iceland
ARCTIC
ATLANTIC
+ NORTH POLE
OCEAN
Greenland (Denmark)
OCEAN
PACK ICE
Queen Elizabeth Islands
BAFFIN BAY
+ NORTH MAGNETIC POLE
Baffin Island
BEAUFORT SEA
BERING SEA
SCALE
0   250   500   750 Miles
0 250 500 750 Kilometers
PACIFIC OCEAN
NORTH AMERICA
Projection: Polar Azimuthal Equidistant

South Pole

SOUTH AMERICA
PACIFIC OCEAN
AMUNDSEN SEA
BELLINGSHAUSEN SEA
PACK ICE
ROSS SEA
MARIE BYRD LAND
ELLSWORTH LAND
ANTARCTIC PENINSULA
International Date Line
Vinson Massif (16,860 ft. 5139 m)
ROSS ICE SHELF
RONNE ICE SHELF
PACK ICE
QUEEN MAUD MOUNTAINS
WEDDELL SEA
POLAR
Mount Markham (over 14,275 ft. over 4351 m)
+ SOUTH POLE
FILCHNER ICE SHELF
ATLANTIC
ADELIE LAND
PLATEAU
COATS LAND
OCEAN
SOUTH MAGNETIC POLE +
ANTARCTICA
ICE CAP
QUEEN MAUD LAND
SHACKLETON ICE SHELF
AMERICAN HIGHLAND
Antarctic Circle
INDIAN OCEAN
ENDERBY LAND
SCALE
0   250   500   750 Miles
0 250 500 750 Kilometers
Projection: Polar Azimuthal Equidistant

A22

Skills Handbook

This Skills Handbook introduces students to basic social studies and geography skills. Concepts covered include the construction, interpretation, and analysis of maps, graphs, and other visual displays of information. Students will also learn about thinking critically and writing. Opportunities to practice these skills are provided throughout the book in exercises marked as "Skill Study" as well as in a variety of other activities. In addition, each Unit Review includes a one-page skill lesson. These lessons focus on higher-order critical-thinking skills and require students to think like geographers. For example, students will learn how to use the geographic method, how to analyze opposing viewpoints, and how to detect cultural bias. They will also find out about links between geography and other fields of study such as economics and history.

VENEZUELA

Georgetown

ATLANTIC OCEAN

SKILLS HANDBOOK

CONTENTS

Mapping the Earth	*S2*
Mapmaking	*S4*
Map Essentials	*S6*
Working with Maps	*S8*
Using Graphs and Diagrams	*S10*
Using Charts and Tables	*S12*
Reading a Time-Zone Map	*S14*
Thinking Critically and Writing	*S16*

GEOGRAPHY DICTIONARY

globe
continent
island
ocean
grid
latitude
longitude
equator
parallel
prime meridian
meridian
degree
hemisphere
map
atlas
map projection
great-circle route
directional indicator
scale
legend
compass rose
mental map

Studying geography requires the ability to understand and use various tools. This Skills Handbook explains how to use maps, charts, and other graphics to help you learn about geography and the various regions of the world. Throughout this textbook, you will have the opportunity to improve these skills and build upon them.

VENEZUELA

IC OCEAN

ILLUSTRATING LATITUDE AND LONGITUDE

On the chalkboard, draw a grid to represent your school grounds. Point out that locations on the grounds can be identified by naming the intersection of horizontal and vertical lines. Point out the similarities between the grounds grid and a map that uses latitude and longitude. Have students indicate the coordinates at which to draw in key buildings, rooms, and courts. Provide students with a series of questions to answer using the completed grid. (Examples: At what intersection is the library located? At what intersection is the auditorium? In terms of the grid, how would you reach the main office from your classroom?)

GLOBAL SKILL BUILDER

You might wish to use **Mapping the Earth** from the interactive *Global Skill Builder* CD-ROM to reinforce students' understanding of maps and globes.

MAPPING THE EARTH

The Globe

A **globe** is a scale model of the earth. It is useful for looking at the entire earth or at large areas of the earth's surface. The earth's land surface is organized into seven large landmasses, called **continents**, which are pictured in the four maps in **Figure 1**. Landmasses smaller than continents and completely surrounded by water are called **islands**. Geographers also organize the earth's water surface into parts, the largest of which is the world **ocean**. Geographers divide the world ocean into four oceans: the Pacific Ocean, the Atlantic Ocean, the Indian Ocean, and the Arctic Ocean. Lakes and seas are smaller bodies of water.

Figure 2 is a diagram of a globe. The pattern of lines that circle the earth in east-west and north-south directions is called a **grid**. The intersection of these imaginary lines helps us find the location of places on the earth. Some mapmakers label the lines with letters and numbers. The grid on many maps and globes, however, is made up of lines of **latitude** and **longitude**.

Lines of latitude are drawn in an east-west direction and measure distance north and south of the **equator**. The equator is an imaginary line that circles the globe halfway between the North Pole and the South Pole. Lines of latitude are called **parallels** because they are

NORTHERN HEMISPHERE

SOUTHERN HEMISPHERE

WESTERN HEMISPHERE

EASTERN HEMISPHERE

▲ **Figure 1: The hemispheres**

USING MAPS IN EVERYDAY LIFE

Ask students if they or people they know use latitude and longitude in daily conversations to indicate the location of places. Point out that it is difficult to keep a mental picture of the latitude-longitude grid system in one's head. Instead, people often use more familiar and informal map systems to describe the location of various places. For example, many maps, such as those found in world and road atlases, have letters and numbers along the sides of the maps to help readers find the location of places.

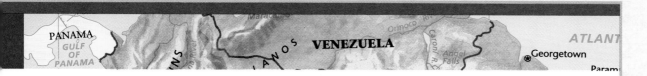

always parallel to the equator. Parallels north of the equator are labeled with an *N*, and those south are labeled with an *S*.

Lines of longitude are drawn in a north-south direction and measure distance east and west of the **prime meridian**. The prime meridian is an imaginary line that runs through Greenwich, England, from the North Pole to the South Pole. Lines of longitude are called **meridians**.

▼ Figure 2: Globe

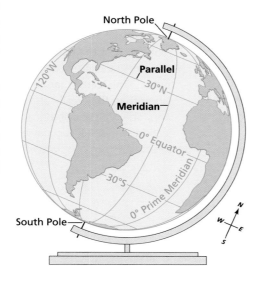

Parallels measure distance from the equator, and meridians from the prime meridian, in **degrees**. The symbol for degrees is °. Degrees are further divided into minutes, for which the symbol is '. There are 60 minutes in a degree.

Lines of latitude range from 0°, for locations on the equator, to 90°N or 90°S, for locations at the North Pole or South Pole. Lines of longitude range from 0° on the prime meridian to 180° on a meridian in the mid-Pacific Ocean. Meridians west of the prime meridian to 180° are labeled with a *W*. Those east of the prime meridian to 180° are labeled with an *E*.

Looking at the globe, you can see that the equator divides the globe into two halves, or **hemispheres**. See **Figure 1**. The half north of the equator is the Northern Hemisphere. The southern half is the Southern Hemisphere.

The prime meridian and the 180° meridian divide the world into the Eastern Hemisphere and the Western Hemisphere. Because the prime meridian separates parts of Europe and Africa into two different hemispheres, some mapmakers divide the Eastern and Western hemispheres at 20° W. This places all of Europe and Africa in the Eastern Hemisphere.

▼ YOUR TURN

1. Look at the Student Atlas map on page A4. What islands are located near the intersection of latitude 20° N and longitude 160° W?

2. Name the four hemispheres. In which hemispheres is the United States located?

3. Name the continents of the world.

4. Name the oceans of the world.

Skill Sidelight

Point out to students that phrases such as "Western influences," "Western World," and "Western society" do not refer to the Western Hemisphere. Rather, these phrases refer to ideas and ways of life associated with western Europe and North America.

Answers to *Your Turn*

1. The Hawaiian islands
2. The United States is located in the Northern and Western hemispheres. The other two hemispheres are the Southern and Eastern.
3. North America, South America, Africa, Europe, Asia, Australia, Antarctica
4. Pacific, Atlantic, Indian, Arctic

S3

Mapmaking
EVALUATING MAP PROJECTIONS

Display for students the three types of map projections described in the textbook. Draw a chart on the board with the headings Advantages and Disadvantages. List each type of projection down the side of the chart. Have students help fill in the chart.

Tell students that many of the world maps used in this textbook are Robinson projections. These maps include the maps of the World Climate Regions (pages 28–29), World Biomes (page 57), Early World Culture Hearths (page 73), World Religions (page 74), and Regions of the World (page 101).

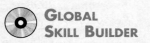
MAP-MAKING

A map is a flat diagram of all or part of the earth's surface. An **atlas** is an organized collection of maps in one book. **M**apmakers have different ways of presenting a round earth on flat maps. These different ways are called **map projections**. Because the earth is round, all flat maps have some distortion. Some flat maps distort size, especially at high latitudes. Those maps, however, might be useful because they show true direction and shape. Some maps, called equal-area maps, show size in true proportions but distort shapes. **M**apmakers must choose the type of map projection that is best for their purposes. Many map projections are one of three kinds: cylindrical, conic, or flat-plane.

▶ **Figure 3a: Paper cylinder**

Cylindrical projections are designed from a cylinder wrapped around the globe. See **Figure 3a**. The cylinder touches the globe only at the equator. The meridians are pulled apart and are parallel to each other instead of meeting at the poles. This causes landmasses near the poles to appear larger than they really are. **Figure 3b** is a Mercator projection, one type of cylindrical projection. The Mercator projection is useful for navigators because it shows true direction and shape. The Mercator projection for world maps, however, emphasizes the Northern Hemisphere. Africa and South America are shown to be smaller than they really are.

▲ Figure 3b:
Mercator projection

▼ **Figure 4a: Paper cone**

Conic projections are designed from a cone placed over the globe. See **Figure 4a**. A conic projection is most accurate along the lines of latitude where it touches the globe. It retains almost true shape and size. Conic projections are most useful for areas that have long east-west dimensions, such as the United States. See the map in **Figure 4b**.

◀ Figure 4b: Conic projection

UNDERSTANDING GREAT-CIRCLE ROUTES

To illustrate the concept of a great-circle route, have students use a globe to identify the great-circle route between Seward, Alaska, and St. Petersburg (Leningrad), Russia. *(The route follows the 150°W/30°E meridian, directly over the North Pole.)* Ask: Along this route, how many degrees of distance separate the two cities? *(60)* Then have students identify the east-west route between the two cities, either on the globe or on a flat map. (Both lie at 60°N.) Point out that when traveling this way, however, one would have to travel 180 degrees of distance.

▲ **Figure 5a: Flat plane**

◄ **Figure 5b: Flat-plane projection**

Flat-plane projections are designed from a plane touching the globe at one point, such as at the North Pole or South Pole. See **Figures 5a** and **5b**. A flat-plane projection is useful for showing true direction for airplane pilots and ship navigators. It also shows true area, but it distorts true shape.

The Robinson projection is a compromise between size and shape distortions. It often is used for world maps, such as the map on pages 28–29. The minor distortions in size at high latitudes on Robinson projections are balanced by realistic shapes at the middle and low latitudes.

▼ **Figure 6a: Great-circle route**

Drawing a straight line on a flat map will not show the shortest route between two locations. Remember, maps represent a round world on a flat plane. The shortest route between any two points on the earth is a **great-circle route**. See **Figures 6a** and **6b**.

Any imaginary line that divides the earth into equal parts is a great circle. The equator is a great circle. Airplanes and ships navigate along great-circle routes.

◄ **Figure 6b: Great-circle route**

Answers to
Your Turn

1. Cylindrical, conic, flat-plane
2. The minor distortions in size at high latitudes are balanced by realistic shapes at the middle and low latitudes.
3. Cylindrical

Map Essentials

DETERMINING DISTANCES

Refer students to the map "East Africa: Physical–Political" on page 453. Ask students to use the scale to determine the following distances. Remind students that using a scale is not an exact process and that distances found will be approximate.

1. The distance between Khartoum, Sudan, and Asmara, Eritrea *(about 375 miles [600 km])*

2. The distance between Kismayu and Mogadishu, Somalia *(about 250 miles [400 km])*

3. The distance between Mount Kenya and Kilimanjaro *(just over 200 miles [320 km])*

Skill Sidelight

Most compass roses in this textbook are placed along lines of longitude. Directions that are true north and true south follow the longitude lines, which are curved on many map projections.

GLOBAL SKILL BUILDER

You might wish to use **Map Legends and Symbols** from the interactive *Global Skill Builder* CD-ROM to reinforce students' understanding of map essentials.

MAP ESSENTIALS

In some ways, maps are like messages sent out in code. Mapmakers provide certain elements that help us translate these codes to understand the information, or message, they are presenting about a particular part of the world. Almost all maps have several common elements: **directional indicators**, **scales**, and **legends**, or keys. **Figure 7**, a map of East Asia, has all three elements.

A directional indicator shows which directions are north, south, east, and west. Some mapmakers use a "north arrow," which points toward the North Pole. Remember, "north" is not always at the top of a map. The way a map is drawn and the location of directions on that map depend on the perspective of the mapmaker. Maps in this textbook indicate direction by using a **compass rose ❶**. A compass rose has arrows that point to all four principal directions, as shown in **Figure 7**.

▲ **Figure 7: East Asia—Physical–Political**

Mapmakers use scales to represent distances between points on a map. Scales may appear on maps in several different forms. The maps in this textbook provide a line scale **❷**. The scales give distances in miles and kilometers (km).

To find the distance between two points on the map in **Figure 7**, place a piece of paper so that the edge connects the two points. Mark the location of each point on the paper with a line or dot. Then, compare the distance between the two dots with the map's line scale. The number on the top of the scale gives the distance in miles. The number on the bottom gives the distance in kilometers. Because the distances are given in intervals, you will have to approximate the actual distance on the scale.

WORKING WITH INSET MAPS AND LEGENDS

Have students examine the inset map on page 221, which shows the lands occupied by the Kuna in Panama. The inset map shows the location of the Kuna lands within the whole of Panama. Have students work in pairs to draw a rough map of the school grounds or a floor of the school building. Pairs should include a legend and an inset map showing their classroom's detailed location within the grounds or, if the main map is of a school building floor, an inset map showing their classroom's internal layout. Display completed maps in the classroom.

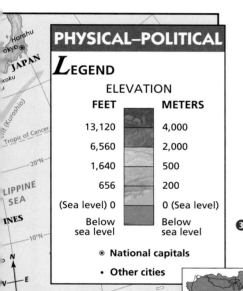

The legend ❸, or key, explains what the symbols on the map represent. Point symbols are used to specify the location of things, such as cities, that do not take up much space on a large-scale map. Some legends, such as the one in **Figure 7**, show which colors represent certain elevations. Other maps might have legends with symbols or colors that represent things such as roads, economic resources, land use, population density, and climate.

Physical–political maps at the beginning of each regional chapter have size comparison maps ❹. An outline of the mainland 48 U.S. states (not including Alaska and Hawaii) is compared to the area under study in that chapter. These size comparison maps help you understand the size of the areas you are studying in relation to the size of the United States.

Inset maps are sometimes used to show a small part of a larger map. Mapmakers also use inset maps to show areas that are far away from the areas shown on the main map. Maps of the United States, for example, often include inset maps of Alaska and Hawaii. (See the map on page 107.) Those two states are too far from the other 48 states to accurately represent the true distance on the main map. Subject areas in inset maps can be drawn to a scale different from the scale used on the main map.

▼ YOUR TURN

Look at the Student Atlas map on pages A4 and A5.

1. Locate the compass rose. What country is directly west of Madagascar in Africa?

2. What island country is located southeast of India?

3. Locate the distance scale. Using the inset map, find the approximate distance in miles and kilometers from Oslo, Norway, to Stockholm, Sweden.

4. Identify the capital of Brazil. Identify the other cities shown in Brazil.

Answers to
Your Turn

1. Mozambique
2. Sri Lanka
3. Less than 500 miles (800 km)
4. Brasília is the capital. Rio de Janeiro and São Paulo also are shown.

Working with Maps
USING PHYSICAL–POLITICAL MAPS

Refer students to the United States and Canada physical–political map on page 107. Have students make one list of the physical features shown on the map and one list of the political features. Ask students to suggest some advantages of having physical and political information on the same map. *(Physical features can be used as points of reference for locating political features and vice versa. Physical features can help explain the location of some political features, such as international boundaries that are marked along rivers, and cities that are located along major rivers or coastlines.)*

GLOBAL SKILL BUILDER

You might wish to use **Using Different Types of Maps** and **Comparing Maps** from the interactive *Global Skill Builder* CD-ROM to reinforce students' study of different kinds of maps.

WORKING WITH MAPS

The Atlas at the front of this textbook includes two kinds of maps: physical and political. At the beginning of most units in this textbook, you will find four kinds of maps. These physical–political, climate, population, and economic maps provide different kinds of information about the region you will study in that unit. These maps are accompanied by questions. Some questions ask you to show how the information on each of the maps might be related.

Mapmakers often combine physical and political features into one map. Physical–political maps, such as the one in **Figure 7** on pages S6 and S7, show important physical features in a region, including major mountains and mountain ranges, rivers, oceans and other bodies of water, deserts, and plains. Physical–political maps also show important political features, such as national borders, state and provincial boundaries, and capitals and other main cities.

▼ Figure 8:
East Asia — Climate

Mapmakers use climate maps to show dominant weather patterns in certain areas. Climate maps throughout this textbook use color to show the various climate regions of the world. See **Figure 8**. Colors that identify climate types are found in a legend that accompanies each map. Boundaries between climate regions do not indicate an abrupt change in dominant weather conditions between two climate regions. Instead, boundaries approximate areas of gradual change between two climate regions.

UNDERSTANDING MAP LIMITATIONS

Use the following questions to elicit a class discussion on the limitations of different kinds of maps: What might be the limitation of physical–political maps and economic maps? *(Both types of maps can show only a limited amount of detail. For example, a physical–political map may show only major rivers in a region. Likewise, an economic map may show only some manufacturing and trade centers or nuclear power plants.)* What might be the limitations of climate and population maps? *(Climate bands and population density shadings mark approximate, not absolute, boundaries. Population maps of large areas cannot show all cities and towns.)*

▲ Figure 9:
East Asia—Population

Population maps show where people live in a particular region and how crowded, or densely populated, regions are. Population maps throughout this textbook use color to show population density. See **Figure 9**. Each color represents a certain number of people living within a square mile or square kilometer. The population maps also use symbols to show metropolitan areas with populations of a particular size. These symbols and the color categories are identified in a legend.

Economic maps show the important resources of a region. See **Figure 10**. Symbols and colors are used to show information about economic development, such as where industry is located or where farming is most common. The meanings of each symbol and color are shown in a legend.

◀ Figure 10: East Asia—Economy

Some kinds of maps are not found in a textbook. **Mental maps** are those maps that we see in our minds. We use mental maps to help us make sense of the world around us. A person can, for example, visualize the community he or she lives in: the location and kinds of buildings there, the activities that take place there, and the surrounding environment. Similarly, mental maps help people structure what they know about any other place: the location and relative size of countries and continents, for example. As people learn more about places, they add more details to their mental maps of these places. Throughout this textbook, you will learn more about the details people use in constructing and using their mental maps.

▼ YOUR TURN

1. How many climate types are shown on the climate map on page 108?
2. In the economy map on page 110, what color represents commercial farming?
3. Use your mental map of your school's neighborhood to write a short paragraph describing the route you take to get to school. Then compare your paragraph with a partner's, and discuss how your mental map differs from that of your partner.

Using Graphs and Diagrams

MAKING BAR AND PIE GRAPHS

On the chalkboard, draw a vertical line and a horizontal line connected at one end to form a 90° angle. Have the class count the number of female students and male students in the class. Work with the class to complete the graph on the board by drawing two bars on the graph: one that shows the number of female students and one that shows the number of male students. Students should suggest a title for the graph and label the axes.

Next, have students divide the number of female students by the total number of students in the class to obtain the percentage of female students in the class. Follow the same procedure for the figure for male students. Work with the class to present this information in a pie graph on the board. Again, students should suggest an appropriate title for the pie graph. Students may want to copy the bar graph and the pie graph and include them in their individual portfolios.

Skill Sidelight

The bar and pie graphs shown here are also found on page 404. The line graph is also found on page 568.

GLOBAL SKILL BUILDER

You might wish to use **Reading Graphs** and **Presenting Data Graphically** from the interactive *Global Skill Builder* CD-ROM to reinforce students' understanding of graphics and their uses.

USING GRAPHS AND DIAGRAMS

Bar graphs are a visual way to present information. The bar graph in **Figure 11** compares the amount of oil reserves in the largest oil-producing countries. Various amounts of oil reserves in billions of barrels are listed on the left side of the graph. Along the bottom of the graph are the names of the 12 countries with the most oil reserves. Above each country is a vertical bar. The top of the bar corresponds to a number along the left side of the graph. For example, Iraq has 100 billion barrels of oil, second to Saudi Arabia's nearly 260 billion barrels.

A pie graph shows how a whole is divided into parts. In this kind of graph, a circle represents the whole, and wedges represent the parts. Bigger wedges represent larger parts of the whole. The pie graph in **Figure 11** shows the percentages of the world's known oil reserves estimated for various groups of countries. The five countries with the most oil have 64.4 percent of the world's known reserves. The other seven countries of the Top 12 have 26.1 percent.

▲ Figure 11: Reading bar and pie graphs

Often, line graphs are used to show such things as trends, comparisons, and size. The line graph in **Figure 12** shows the population growth of India and China over time. The information on the left shows the number of people in millions. The years being studied are listed along the bottom. Lines connect points that show the population in millions at each year under study. This line graph projects population growth far into the future.

Graphing Population Growth in India and China

Source: *Long-range World Population Projections: Two Centuries of Population Growth, 1950–2150*, United Nations

▲ Figure 12: Reading a line graph

CREATING A POPULATION PYRAMID

Obtain from your school administration the number of students in each grade of your school, and, within each grade, how many students are male and how many are female. Distribute these figures to the students, and have them work in small groups to construct a population pyramid showing the school's student population by grade. After students have completed their population pyramids, ask what trends the pyramids show about the school's population. *(For example, if the lower grades have more students, the school's population might be increasing.)*

A climograph compares temperatures and precipitation amounts in different latitudes and locations. See **Figure 13**. Along the left side of the graph is a range of average temperatures. Letters representing each month of the year are listed along the bottom of the graph. Along the right side is a range of average precipitation totals. The climograph uses elements of line and bar graphs. A red line shows the average high temperature in a particular month. The blue line shows the average low temperature each month. The vertical bars along the bottom indicate the average precipitation amount each month.

▲ **Figure 13: Reading a climograph**

▼ **Figure 14: Reading a population pyramid**

Population pyramids show the percentages of males and females by age group in a particular country or region. Population pyramids are split into two sides, one for male and one for female. Along the bottom are numbers that show the percentage of the male or female population that is represented by the age groups. The age groups are listed on the side of the population pyramid. The wider the base of a country's pyramid, the younger the population of that country.

Population pyramids help us understand economic, cultural, and historical patterns of a place or region. Look at the pyramid in **Figure 14**. The wide pyramid base for Kenya indicates a young, rapidly growing population. The pyramid shows that a little over 19 percent of Kenya's female population is made up of females under the age of 4. Nearly 20 percent of the country's male population is made up of males under the age of 4.

YOUR TURN

1. Look at the bar graph in Figure 11. Which countries shown in this graph have less than 50 billion barrels of oil reserves?

2. Study the pie graph in Figure 11. What percentage of the world's proven oil reserves is found in the Top 5 countries?

3. Refer to the line graph in Figure 12. What is the projected population of India by 2150?

4. Look at the climograph in Figure 13. When are average high and average low temperatures coolest in Puerto Montt?

Point out that some climographs show only one average of the temperatures each month for a particular location. Climographs in this textbook show both the average high and the average low temperatures for each month.

This climograph is one of three found on page 194. The population pyramid is one of three found on page 83.

Skill Sidelight

Answers to *Your Turn*

1. The United States, China, Libya, Nigeria
2. 64.4 percent
3. Nearly 2 billion
4. July

Using Charts and Tables

WORKING WITH TABLES

Refer students to the charts in the Appendix at the back of the book (pages 624–636). Have students compare population and land area figures for countries in one of the regional tables (such as the table for Middle and South America). Ask students to identify the most populous country and the largest country in the region. Then have students name the predominant languages in the region.

Skill Sidelight

Point out that the statistical "Facts in Brief" charts in each regional unit include information for the United States. The U.S. statistical information is provided for comparison.

Skill Sidelight

The chart for European languages can also be found on page 267. The time line for China is located on page 508.

GLOBAL SKILL BUILDER

You might wish to use **Using Time Lines, Reading Graphs, Presenting Data Graphically,** and **Creating Graphic Organizers** from the interactive *Global Skill Builder* CD-ROM to reinforce students' understanding of charts, tables, and diagrams.

USING CHARTS AND TABLES

In each chapter on the various regions of the world, you will find tables that provide basic information about the countries under study. In addition, at the back of this textbook are tables with more information about each country of the world.

The countries of Atlantic South America are listed on the left in the table in **Figure 15**. You can match statistical information on the right with the name of each country listed on the left. The categories of information are listed across the top of the table.

FACTS IN BRIEF	Atlantic South America		
COUNTRY POPULATION (1994)	LIFE EXPECTANCY (1994)	LITERACY RATE (1990)	PER CAPITA GDP (1993)
ARGENTINA 33,912,994	68, male 75, female	95%	$5,500
BRAZIL 158,739,257	57, male 67, female	81%	$5,000
PARAGUAY 5,213,772	72, male 75, female	90%	$3,000
URUGUAY 3,198,910	71, male 77, female	96%	$6,000
UNITED STATES 260,713,585	73, male 79, female	97% (1991)	$24,700

Source: *The World Factbook 1994*, Central Intelligence Agency

▲ **Figure 15: Reading a table**

◀ **Figure 16: Reading a chart**

Major Languages of Europe

Indo-European ▶	Germanic ▶	▶	English German Dutch-Flemish Swedish Norwegian Danish Icelandic
	Romance ▶ (Latin or Italic)	▶	Portuguese Spanish French Italian Romanian
	Celtic ▶	▶	Irish Gaelic Scottish Gaelic
	Baltic ▶	▶	Lithuanian Latvian
	Slavic ▶	▶	Russian Ukrainian Belorussian Polish Czech Slovak Serbo-Croatian Slovene Macedonian Bulgarian
	▶ Greek ▶ Albanian		
Uralic ▶	Finno-Ugric ▶	▶	Estonian Finnish Hungarian

Not all charts take the form of tables. The illustration in **Figure 16** shows major languages in Europe. Related languages are grouped together as part of two language families: Indo-European and Uralic. English, Swedish, and Danish are all Germanic languages. Russian, Polish, and Czech are all Slavic languages. Germanic and Slavic languages are all part of the Indo-European language family.

Figure 17: ▶ **Reading a time line**

Time lines provide highlights of important events over a period of time. The time line in **Figure 17** begins at the top with 5000 B.C., when rice was first cultivated by people in present-day China. The time line highlights important events that have shaped the human and political geography of China.

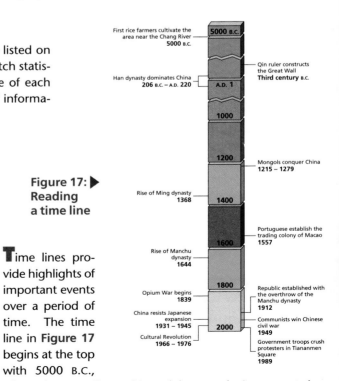

First rice farmers cultivate the area near the Chang River **5000** B.C.

5000 B.C.

Han dynasty dominates China **206** B.C. – A.D. **220**

Qin ruler constructs the Great Wall **Third century** B.C.

A.D. 1

1000

1200

Mongols conquer China **1215 – 1279**

Rise of Ming dynasty **1368**

1400

Portuguese establish the trading colony of Macao **1557**

1600

Rise of Manchu dynasty **1644**

Opium War begins **1839**

1800

Republic established with the overthrow of the Manchu dynasty **1912**

China resists Japanese expansion **1931 – 1945**

Communists win Chinese civil war **1949**

Cultural Revolution **1966 – 1976**

2000

Government troops crush protesters in Tiananmen Square **1989**

CHARTING THE TIME OF YOUR LIFE

Have students create time lines of their own lives or the life of a famous person they admire (a present-day or a historical figure). Time lines should have an appropriate scale and could include such events as birth dates, graduations, significant birthdays, summer vacations, and achievements. Students may wish to include their time lines in their portfolios.

Flowcharts are visual guides that explain different processes. They lead the reader from one step to the next, sometimes providing both illustrations and text. The flowchart in **Figure 18** shows the different steps involved in harvesting cacao and preparing it for use by consumers. Among other things, chocolate and cocoa are made from cacao. The flowchart takes you through the steps of harvesting and processing cacao. Captions help guide you through the four such flowcharts in this textbook.

▲ **Figure 18: Reading a flowchart**

Graphic organizers can help you understand certain ideas and concepts. For example, the word web in **Figure 19** helps you think about the sources of water in the hydrosphere, which is discussed in Chapter 2, "The Earth in Space." In this word web, one source goes on each line. Depending on the subject, a word web can continue to branch out as you gather more and more details.

▲ **Figure 19: Word web**

Skill Sidelight

The cacao illustration is also found on pages 448–449. Three other such crop illustrations appear on pages 144–145, 352–353, and 512–513.

▼ YOUR TURN

1. Look at the statistical table for Atlantic South American countries in Figure 15. Identify the Atlantic South American country with the highest literacy rate.
2. Look at the language chart in Figure 16. Name the two languages that are part of the Celtic language group.
3. Look at the China time line in Figure 17. Identify two important events in China's history between 1200 and 1400.

Answers to *Your Turn*

1. Uruguay
2. Irish Gaelic and Scottish Gaelic
3. Mongol conquest of China (1215–1279), rise of the Ming dynasty in 1368

S13

Reading a Time-Zone Map

CALCULATING TIME DIFFERENCES

Refer students to the time-zone map. Provide students with a series of questions that can be answered by using the time-zone map. Students should assume that no countries are on daylight savings time. (Examples: If it is 1 A.M. in São Paulo, what time is it in Lagos? *[5 A.M.]* If it is 9 P.M. in Sydney, what time is it in Moscow? *[2 P.M.]* If you live in Houston and want to watch a British royal wedding at 10 A.M., London time, how early will you have to be awake? *[4 A.M.]*)

READING A TIME-ZONE MAP

Since the sun is not directly overhead everywhere on Earth at the same time, clocks are adjusted to reflect the difference in the sun's position. The earth rotates on its axis once every 24 hours, so in one hour, the earth makes one twenty-fourth of a complete revolution. Since there are 360 degrees in a circle, we know that the earth turns 15 degrees of longitude each hour ($360° \div 24 = 15°$) in a west-to-east direction. Therefore, if a place on Earth has the sun directly overhead at this moment (noon), then a place 15 degrees to the west will have the sun directly overhead one hour from now. The earth will have rotated 15 degrees during that hour. As a result, the earth is divided into 24 time zones. Thus, time is an hour earlier for each 15 degrees you move westward. Time is an hour later for each 15 degrees you move eastward.

By international agreement, longitude is measured from the prime meridian, which passes through the Royal Observatory in Greenwich, England. Time also is measured from Greenwich and is called Greenwich mean time (GMT). For each time zone east of the prime meridian, clocks must be set one hour ahead of GMT. For each time zone west of Green-

wich, clocks are set back one hour from GMT. When it is noon in London, it is 1:00 P.M. in Oslo, Norway, one time zone east, and 7 A.M. in New York City, five time zones west.

As you can see by looking at the map above, time zones do not follow meridians exactly. Political boundaries often determine time-zone lines. In Europe and Africa, for example, time zones follow national boundaries. The mainland 48 U.S. states,

WORLD TIME ZONES

TRAVELING ACROSS TIME ZONES

Point out to students that many people suffer jet lag when traveling long distances. Because they cross many time zones, their sleep patterns are disrupted. For example, if a traveler arrives from Los Angeles at lunchtime in Tokyo, back home in California it is the middle of the night—when the traveler would most likely be asleep.

Have students determine how many time zones they would have to cross traveling west to east for the following journeys: London to Bombay *(5)*, New York to Cairo *(7)*, Chicago to Jakarta *(13)*.

meanwhile, are divided into four major time zones: Eastern, Central, Mountain, and Pacific. Alaska and Hawaii are in separate time zones to the west.

Some countries have made changes in their time zones. For example, most of the United States has daylight savings time in the summer in order to have more evening hours of daylight.

The international date line is a north-south line that runs through the Pacific Ocean. It is located

at 180°, although it sometimes varies from that meridian to avoid dividing countries. At 180°, the time is 12 hours from Greenwich time. There is a time difference of 24 hours between the two sides of the 180° meridian. The 180° meridian is called the international date line because when you cross it, the date and day change. As you cross the date line from the west to the east, you gain a day. If you travel from east to west, you lose a day.

▼ YOUR TURN

1. In which time zone do you live? Check your time now. What time is it in New York?
2. How many hours behind New York is Anchorage, Alaska?
3. How many time zones are there in Africa?
4. If it is 9 A.M. in the middle of Greenland, what time is it in São Paulo?

Answers to
Your Turn

1. Answers will vary. If students are in the Eastern time zone, the time is the same as New York. The time is one hour behind New York for each time zone moving westward.
2. Five
3. Five (includes Seychelles and Mauritius)
4. 9 A.M. The two locations are in the same time zone. The wide north-south distance between the locations makes no difference in time.

THINKING CRITICALLY AND WRITING

The study of geography requires more than analyzing and understanding information provided by tools like maps and graphs. Throughout this textbook, you will be given opportunities to think critically and write about information you have studied.

THINKING CRITICALLY

At the end of each section and each chapter in this textbook, you will be asked to think critically about some of the information you have just studied. Critical thinking questions might ask you to demonstrate your knowledge or understanding about something in the text. Such questions might ask you to tackle such tasks as evaluating, analyzing, or interpreting information from the text.

Some critical thinking skills include applying what you have learned to solve problems or perform certain tasks. For example, some questions might ask you to examine relationships between pieces of information and to seek possible solutions to problems.

Other critical thinking questions ask you to analyze—to break down information into parts so that it is easier to understand. For example, you might be asked to show how some things discussed in the text are related. On the other hand, some critical thinking questions might ask you to synthesize, or pull information together, to form new ideas, or to develop additional information.

Finally, thinking critically also means evaluating or judging information you have studied. Some questions might ask you to draw conclusions from pieces of information, to give an opinion about something you have read, or to challenge or verify the truth of information you read.

PURPOSES FOR WRITING

Writers have many different reasons for writing. In your study of geography, you might write to accomplish many different tasks.

Sometimes writers *express* their own personal feelings or thoughts about a topic or event. At the beginning of most units in this textbook, you will find passages that tell about some aspect of the regions to be studied. These are expressive passages, some prose and some poetry, written from personal experiences.

Some people write to *inform* the reader about an event, person, or thing. One technique in informative writing is to show cause and effect between events and things. Cause is an action, person, or condition that brings about some result, or effect. Informative writing might also include describing or summarizing events and processes.

In aiming to inform readers, some authors also define or *classify* things or information. Writers might use several techniques, including comparing and contrasting characteristics or advantages and disadvantages of different items under study.

Sometimes one writes to *persuade*—to convince readers to agree with a certain statement or to act in a particular way. Persuasion relies on the use of facts and opinions. Facts are statements or information that can be proven true. Opinions are statements or information that, while perhaps being true, cannot be proven to be true.

WRITING PARAGRAPHS

You will find various kinds of questions at the end of each section, chapter, and unit throughout this textbook. Some questions will require in-depth answers. The following guidelines for writing will help

WRITING PARAGRAPHS

Have students write a paragraph describing an aspect of the neighborhood in which they live. The students might concentrate on any of various aspects of geography: the physical layout of their neighborhood or the people who live there, for example. The paragraph can be expressive, informative, or persuasive. Remind students to use the writing process on this page as a guide. Call on volunteers to read their paragraphs aloud, and have the class decide whether the paragraph is expressive, informative, or persuasive.

The Writing Process

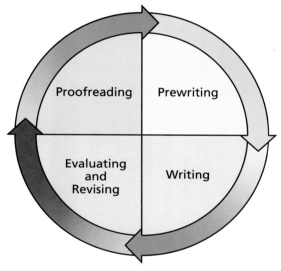

you structure your answers so that they clearly express your thoughts.

Prewriting

Prewriting is the process of thinking about and planning what to write. Prewriting includes gathering and organizing information into a clear plan. Writers also use the prewriting stage to identify the audience and purpose for what is to be written.

Often, writers must conduct research to get the information they need. Research can include finding primary and secondary sources. Primary sources are original information provided by firsthand accounts. Secondary sources pull information from other works. Secondary sources include encyclopedias, newspaper and magazine articles, and biographies.

Writing a Draft

After you have gathered and arranged your information, you are ready to begin writing. Many paragraphs are structured in the following way:

The Topic Sentence The topic sentence states the main idea of the paragraph. Putting the main idea into the form of a topic sentence helps keep the paragraph focused.

The Body The body of a paragraph develops and supports the main idea. Writers use a variety of information, including facts, opinions, and examples, to support the main idea.

The Conclusion The conclusion summarizes the writer's main points or restates the main idea.

Evaluating and Revising

Read over your paragraph and make sure you have clearly expressed what you want to say. Sometimes it helps to read your paragraph aloud or ask someone else to read it. Revise parts of your paragraph that are not clear or that stray from your main idea. You might want to add, cut, replace, or reorder sentences to make your paragraph clearer.

Proofreading

Before you write your final draft, read over your paragraph and correct any spelling, grammar, and punctuation errors.

Writing Your Final Version

After you have revised and corrected your draft, neatly rewrite your paragraph. Make sure your final version is clean and free of mistakes. The appearance of your final version can affect how your writing is perceived and understood.

▼ YOUR TURN

1. What are the steps in the writing process?
2. What elements are found in many paragraphs?

Answers to
Your Turn

1. Prewriting, writing a draft, evaluating and revising, proofreading, writing the final version
2. The topic sentence or main idea, the body, the conclusion

S17

UNIT 1 A VIEW OF OUR WORLD CHAPTERS 1–6

OVERVIEW

In Unit 1, students will learn to identify the main features of the earth's physical geography.

Chapter 1 • THE GEOGRAPHER'S WORLD introduces the study of geography and describes its two main branches: human geography and physical geography.

Chapter 2 • THE EARTH IN SPACE describes Earth as a planet and its relationship to the sun.

Chapter 3 • THE EARTH'S ATMOSPHERE AND CLIMATES explains how the sun and other factors influence weather.

Chapter 4 • THE WATER PLANET describes the importance of water to life on Earth and identifies available types of water resources.

Chapter 5 • GLOBAL LANDFORMS discusses the earth's landforms and the geological forces that shape them.

Chapter 6 • PATTERNS OF LIFE explains the formation of ecosystems and describes world biomes.

OBJECTIVES

1. Learn how professional geographers use the Five Themes of Geography to analyze data and solve problems.

2. Analyze Earth's relationship to the other planets in the solar system.

3. Understand how the movement of the earth affects time, climate, and seasons.

4. Explain how weather and climate are affected by the sun's energy, air pressure, and global circulation systems.

5. Identify the major types of climate.

6. Describe how water has shaped our planet and understand its significance as a natural resource.

7. Identify global landforms and understand how they are formed.

8. Analyze the interrelationships of plants, animals, soil, water, and climate.

9. Learn to draw sketch maps and use them as geographic tools.

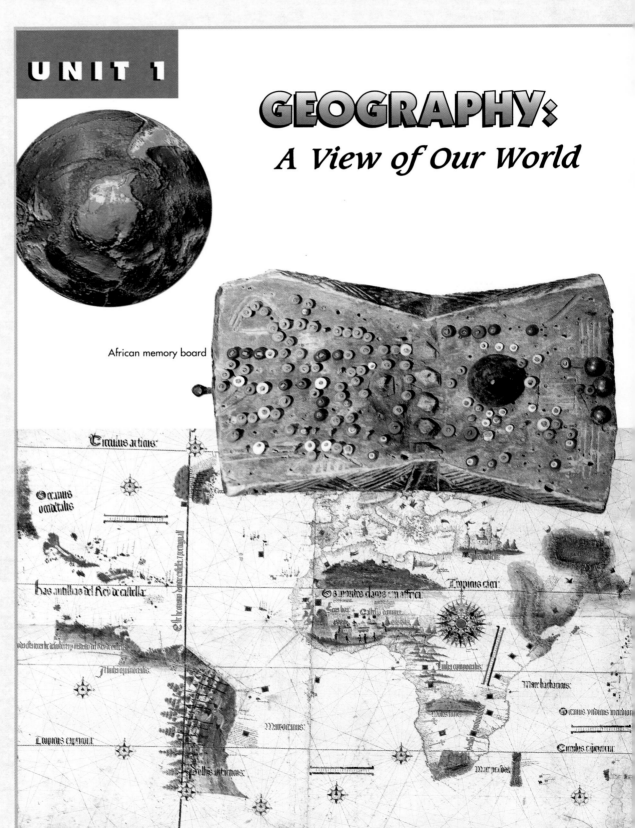

UNIT 1

GEOGRAPHY:
A View of Our World

African memory board

TEACHING UNIT 1

Using the Illustrations

Have students look at the maps shown on these pages without reading the captions. Call on students to tell what maps look old or historical, what map appears modern, and what map(s) may not conform to their own ideas about how maps should look. Then ask students to read the captions and to think of all the ways that the maps differ. *(dates drawn, methods used, areas represented, nationality of mapmakers)* Have students compare the Cantino map with a modern world map or globe. Ask: What continents are most accurately drawn on the Cantino map? *(Africa and Europe)* How do the Cantino and Korean maps reveal something about the extent of navigation or exploration by the early 1500s? *(Many parts of the world are not included on these maps or are drawn inaccurately. Much of the world was uncharted by mapmakers or else was unknown to them at this time.)*

Teaching with Sketch Maps

Tell students that they will learn how to draw sketch maps in Unit 1. They will use them throughout their study of world geography. Explain that sketch maps are freehand maps drawn from memory. Point out that sketch maps will enable them to see their own progress in understanding geographic concepts.

How people view their world is reflected in how they represent it on maps. Some nineteenth-century Africans used a memory board (top, opposite) to locate important sites and record information about their society. The Cantino map (bottom, opposite) shows how little the Europeans in 1502 knew of what they called the New World. The fifteenth-century Korean map (bottom of this page), with an oversized China and Korea, reflected an East Asian view of the world. Today, we can see our world through computers and in great detail, as in this three-dimensional view of the Sphinx in Egypt (below).

UNIT 1 CHAPTERS

1 The Geographer's World
2 The Earth in Space
3 The Earth's Atmosphere and Climates
4 The Water Planet
5 Global Landforms
6 Patterns of Life

PORTFOLIO ASSESSMENT

"Building your Portfolio" on page 63 provides a choice of two unit projects for students to work on as they study Unit 1. The unit projects are supported by Chapter Portfolio Assessment worksheets in the *Teaching Resources.*

Computer-generated image of the Sphinx

Fifteenth-century Korean map

Cantino world map

CHAPTER 1 The Geographer's World
Planning Guide

TEXTBOOK	TEACHING RESOURCES	OTHER AVAILABLE MATERIALS
Section		

1. Geographers Look at the World
Pages 3–7
	Reteaching Worksheet 1.1
	Skills Worksheet 1
	Challenge/Enrichment Worksheet 1
	Critical Thinking Worksheet 1
		Creative Strategies for Teaching World Geography
		  LESSON 1 Understanding Absolute Location
		  LESSON 2 Absolute and Relative Location
		  LESSON 3 Place, Human-Environment
		  Interaction, and Movement
		Study Guide for Spanish Speakers
		Lecture Notes, Chapter 1

2. The Wide World of Geography Pages 7–8	Reteaching Worksheet 1.2 Graphic Organizer 1 Challenge/Enrichment Worksheet 1	*Study Guide for Spanish Speakers* *Lecture Notes,* Chapter 1
Chapter Review Page 9	Chapter 1 Test Chapter 1 Portfolio Assessment See also *Alternative Assessment Forms*	*Test Generator,* Chapter 1

CHAPTER RESOURCES

Lecture Notes, Chapter 1
Teaching Resources
  Chapter 1 Graphic Organizer
  Chapter 1 Skills Worksheet
  Chapter 1 Critical Thinking Worksheet
  Chapter 1 Challenge/Enrichment Worksheet

REVIEW AND ASSESSMENT

Chapter 1 Check, page 9
Chapter 1 Test, *Teaching Resources*
Chapter 1 Portfolio Assessment, *Teaching Resources*
Test Generator, Chapter 1

ADDITIONAL RESOURCES

Books for Teachers

Allen, Russell, *et al. The Geography Learning of High-School Seniors.* Princeton, NJ: National Assessment of Educational Progress, 1990.

The Book of Where; or How to Be Naturally Geographic. Boston: Little, 1982. Available through Global Graphics & Resources, Hillsborough, CA, 1986.

Boyce, Ronald R. *Geographic Perspective on Global Problems: An Introduction to Geography.* New York: John Wiley & Sons, 1982.

Books for Students

The Complete School Atlas. Austin: Holt, Rinehart and Winston, 1995.

Davis, Kenneth. *Don't Know Much About Geography.* New York: Morrow, 1992.

Exploring Your World: The Adventure of Geography. Washington, DC: National Geographic Society, 1993.

Geography on File. New York: Facts on File, 1991.

McClintock, Jack, and David Helgren. *Everything Is Somewhere: The Geography Quiz Book.* New York: Morrow, 1986.

Pratt, Paula Bryant. *Maps: Plotting Places on the Globe.* San Diego: Greenhaven Press, 1994.

The World Today *Videodisc Program*

A variety of still images, short videos, and activities are available for you to use as you teach this chapter. *Lecture Notes* include bar codes for specific photographs.

Multimedia Resources

Geography. (software, IBM, Tandy) Wings for Learning, 1600 Green Hills Road, P.O. Box 66002, Scotts Valley, CA

The Geography Tutor: Map Skills. (VHS, 15 minutes) Tell Me Why, 730 Washington Street, Marina Del Rey, CA 90292.

Picture Atlas of the World. (interactive maps, photo bank, animation, audio on CD-ROM) National Geographic Society, Washington, DC.

World Geograph II—Multimedia Edition. (Multimedia CD-ROM atlas) MECC, 6160 Summit Drive North, Minneapolis, MN 55430-4003.

MEETING INDIVIDUAL NEEDS

Limited English Proficiency (LEP)

Pair LEP students with students who are proficient in English. Instruct the pairs to write down and define the Five Themes of Geography with sketches or short phrases. Then ask them to consider their own community in terms of these Five Themes. Have the pairs make five sketches or drawings that illustrate their community through the Five Themes. Display the illustrations in the classroom.

Learners Having Difficulty

Explain to students that the Focus questions at the beginning of each section will help them to identify the main ideas in the chapter. Have students work in pairs to answer the Focus questions on pages 3 and 7.

Auditory Learners

If possible, arrange for guest speakers who are specially trained geographers or who use geography in their profession to speak to students about their careers. Ask the guest speakers to focus on the role that geography plays as they perform their work. Before the speakers come, have students prepare questions to ask them based on the Five Themes of Geography.

Visual Learners

Have students work in groups to collect pictures that illustrate the Five Themes of Geography. The groups will use the illustrations and write captions to create a set of charts similar to those on pages 3–7 of their textbooks. Allow time for groups to exchange charts.

Tactile/Kinesthetic Learners

Assign the following professions to students: grocery store manager, travel agent, international reporter, truck driver, air-traffic controller, meteorologist, police officer, bus driver. Have them present oral reports that explain the importance of understanding geography for each job. Students should identify the geography themes that are used in these professions.

Gifted Learners

Ask students to imagine that they have been hired to submit a plan for building a recreation center in their community. Students' plans should include an explanation of how the Five Themes of Geography were used to determine the location and construction features of the facility. Their plans may also include a drawing of the building, showing its location in the community.

MULTICULTURAL FOCUS

Read the following passage aloud:

I left Tangier, my birthplace, the 13th of June 1325 with the intention of making the Pilgrimage to [Mecca]. . . . to leave all my friends . . ., to abandon my home as birds abandon their nests.

So begins the memoir of Ibn Battuta, one of the great Muslim geographers of the Middle Ages. Ibn Battuta was only 21 when he set out on his pilgrimage to Mecca. The experiences of this journey fueled a passion for exploration that would lead Ibn Battuta to traverse some 75,000 miles over the next 29 years.

Ibn Battuta traveled throughout Asia, North and West Africa, and into Europe. (You may wish to refer students to the map of his routes in *National Geographic*, December 1991.) Several factors motivated his journeys. His devout faith steered him to most of the world's Islamic lands. He also called on many foreign sultans and emperors, often earning appointments as judge or emissary. Yet perhaps the impulse that drove Ibn Battuta the most was curiosity—the desire to visit new lands and learn about their peoples. His main rule for travel was to try never to follow the same road twice.

Ask students to suggest the benefits of Ibn Battuta's travels to the people of his time and to people today. Point out that a great deal of geographical knowledge was and is gained by the chronicles of explorers such as Ibn Battuta. You might suggest to students that they keep a diary the next time they travel.

OVERVIEW

In Chapter 1, students will learn about the study of geography. Modern geography is about space on the earth, the use of space, and how people interact with their surroundings.

Section 1 describes topics geographers study and the Five Themes of Geography: location, place, human-environment interaction, movement, and region.

Section 2 focuses on the two major branches of geography: human geography and physical geography. Students will learn that human geography deals with historical, political, economic, and other human activity, whereas physical geography is the study of the earth's natural features and their variations. The section also discusses geography as a profession.

About the Author

N. Scott Momaday (1934–) writes about the experiences and cultures of Native Americans. The quotation on this page is from his Pulitzer Prize–winning novel, *House Made of Dawn.* Momaday, of Kiowa and Cherokee heritage, grew up in the canyon lands of New Mexico.

Primary Source

Description of Change: excerpted and bracketed
Rationale: excerpted to focus on main idea; bracketed to clarify meaning

CHAPTER 1

The Geographer's World

> "O̲nce in his life a man [person] ought to concentrate his mind upon the remembered earth, I believe. He ought to give himself up to a particular landscape in his experience, to look at it from as many angles as he can, to wonder about it, to dwell upon it. "
>
> *N. Scott Momaday*

GEOGRAPHY DICTIONARY

site
absolute location
situation
relative location
drought
spatial interaction
cartography
topographic
meteorology

W̲hat does the location of your school or city have in common with an African farmer's method of raising crops or herding animals? What does the location and design of a shopping mall have in common with national parks and tourism? The answer is space. The use of space and how people interact with their surroundings is the focus of geographers, and this focus gives geographers a special way of looking at the world.

In its broadest sense, geography is about discovery of the world around us, because people everywhere make use of space and interact with their environment and with each other. Thus, we use geography every day in finding our way to school, going on trips, or reading about how people live in other countries. We think geographically every time we decide what to do, where to go, how to dress, and even what to eat.

Today, all of the world's peoples are linked. Events far away influence us all, and we share the world's resources, products, and ideas. Geographers study places where products were designed and made. They study how and where products and resources are transported. And they study the people around the world who are involved in these and countless other activities.

Sunrise near Blue Ridge Parkway, **North Carolina**

2

Using the Quotation

Have a student read aloud the quote by N. Scott Momaday. Ask students what they think Momaday means by a person giving "himself up to a particular landscape in his experience." Ask volunteers to describe landscapes that they have found particularly memorable. What special characteristics make the places unforgettable?

Using the Geography Dictionary

Tell students that the first four terms in the Geography Dictionary are all different ways of describing "place" but their meanings are different. Have students use a dictionary to explore the meanings of the words **site**, **location**, and **situation**. Discuss the meanings that apply to the study of geography.

SECTION 1
OBJECTIVES

The objectives reflect the Focus questions on this page and are assessed in the review on page 7.

1. Identify and explain major topics for geographic study.
2. List and explain the Five Themes of Geography.

GEOGRAPHERS LOOK AT THE WORLD

FOCUS

- *What are some topics that geographers study?*
- *What are the Five Themes of Geography?*

Understanding the World Nearby Shopping malls make fascinating geography. For many Americans, a mall is much more than a place to shop. It is a place to socialize with friends and to share news. But for geographers, the building of a mall is a serious study in land use.

Location is a key characteristic of a shopping mall. A mall needs a physical **site**—an exact,

absolute location. So that the foundations of the buildings and the parking lots will function well, the land should be stable and well drained. A site also must have a good **situation**, or **relative location**. Shoppers should not have to travel too far from their homes, and the route to the mall should be fairly simple. For these reasons, malls often are located at exits along highways. This situation means that customers and merchandise can get to the mall easily.

Place also is a key characteristic of a shopping mall. The shopping mall must be an interesting and pleasant place. Much time and money is spent deciding how the mall should look and feel.

FIVE THEMES OF GEOGRAPHY

- LOCATION
- PLACE
- HUMAN-ENVIRONMENT INTERACTION
- MOVEMENT
- REGION

CREATIVE STRATEGIES FOR TEACHING WORLD GEOGRAPHY

LESSON 1 Understanding Absolute Location: Spy Tracking

GLOBAL SKILL BUILDER

You might wish to use **Determining Absolute and Relative Location** from the interactive *Global Skill Builder* CD-ROM to reinforce students' understanding of absolute location and relative location.

FIVE THEMES OF GEOGRAPHY

LOCATION

Every place on Earth has a location. Location is defined in terms of absolute and relative location.

Absolute location: *the exact spot on Earth where something is found*
Example: Niamey, the capital of Niger, is located at 13° 31′ north latitude and 2° 07′ east longitude.

Relative location: *the position of a place in relation to other places*
Example: Yosemite National Park is located north of Los Angeles, California, and east of San Francisco, California.

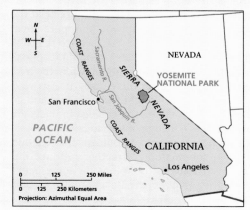

Your Turn
1. Use an atlas to find the absolute location of your city or town.
2. Write a sentence describing the relative location of your home.

Your Turn
1. Locations should indicate longitude and latitude.
2. Sentences might mention local landmarks, buildings, or other cities or states.

Motivator

Focusing on the themes of location and place, have students choose as a topic of discussion a local mall, shopping center, cinema complex, or their school and explain how its location might have been chosen. What advantages and disadvantages might the location have? How does the location influence the site's place characteristics?

Building a Vocabulary

Write the term **spatial interaction** on the board. Circle *spatial*, and ask what noun it comes from. *(space)* Point out that the concept covers all movement of and by humans in their environment. Ask students to suggest ways that spatial interaction has changed over the past century. *(Answers will vary; answers may include changes in transportation and communication technologies.)*

Using Geography Themes

Write the names of the following cities on the board: Columbus, Ohio; London, England; Canberra, Australia; Libreville, Gabon; São Paulo, Brazil. Have students consult classroom resources, atlases, or their textbook to give the absolute location of each city and then describe its relative location. Have students also classify each city by region.

CREATIVE STRATEGIES FOR TEACHING WORLD GEOGRAPHY

LESSON 2 Absolute and Relative Location: Where in the World?

Your Turn

Physical characteristics might include size, building materials, or floor plan of the school. Human characteristics might refer to the type of school or describe the student population.

Geographical Sidelight

A step above regional malls are the huge "megamalls," such as the West Edmonton Mall in Alberta, Canada, and the Mall of America near Minneapolis. These malls combine under one roof hundreds of stores, full-scale amusement parks, and other attractions. The owners promote them as tourist destinations in themselves and count on drawing people from more than 100 miles (161 km) away and even from Asia and Europe.

4

PLACE

Every place on Earth has special characteristics that make it different from every other place.

Every place can be described in terms of its physical characteristics, including weather, land features, plants, and animals.

Example: The Sahel in Africa is hot and dry. Overgrazing by animals and the search for firewood by humans has stripped the Sahel of vegetation.

Every place can be described in terms of its human features. These features include a place's peoples, their cultures, and their ideas.

Example: Yosemite National Park has several hotels and campgrounds to accommodate tourists.

Your Turn

Describe the physical and human characteristics of your school.

This study includes selecting the positions of individual stores. For example, the biggest stores are usually situated at the mall's corners or at the ends of walkways. In this way, shoppers must walk by many little stores on their way to the well-known big stores. The mall also should be appealing to shoppers. If walkways are too wide, the mall will seem empty even on busy days. If walkways are too narrow, people might feel crowded and go elsewhere to shop. Often, the ceilings in malls are very high to lend a feeling of spaciousness.

A mall becomes an environment that encourages human interaction. Attractive product displays draw customers into shops. In some malls, trees and other greenery create a park-like setting. When the weather outside is hot or cold, the mall's pleasant, controlled environment attracts people.

The movement of people and merchandise is also part of the geography of the shopping mall. Trucks, railways, airplanes, and ships move goods from distant places to mall stores. Nearby roads and highways lead to and from shoppers' homes.

The transportation routes to the mall will help define the surrounding area, or region, from which shoppers will come. Shop owners in the mall want as many people as possible to shop there. The mall's operator will want to know how far the mall's region stretches so that he or she can plan advertising strategies that will draw more people to the mall. In many cities and suburbs, developers are building huge "regional malls," which attract people from 50 or more miles (80.5 km) away.

Helping Solve World Problems Since the 1970s, the Sahara, Africa's great desert, has spread southward. This expansion has been caused by **drought** (a long, dry period with no rain) and the overgrazing of land by livestock. The countries of Mauritania, Mali, and Niger have been severely affected, and the place characteristics of the Sahel have changed. Lands once

Guided Practice

Have students draw maps of their neighborhoods, including homes, stores, and other landmarks. Then have them locate their neighborhoods on a map of their town or city. Point out that each one has a slightly different absolute location. Have students also circle items on their maps that show human-environment interaction.

Independent Practice

Have students choose a photograph of a place from a book or magazine. Students should look for a photo that reveals something about three or more of the Five Themes of Geography. Ask them to write a sentence describing how each theme applies to the photo. You may wish to have them present the photo and their summaries to the class.

Closure

Recall for students some of the topics studied by geographers: land use, economic activities, transportation needs, and how human behavior affects the environment. Have them relate each topic to one or more of the Five Themes of Geography.

FIVE THEMES OF GEOGRAPHY

HUMAN-ENVIRONMENT INTERACTION

People are constantly interacting with their surroundings. This geographic theme is called human-environment interaction.

People adapt to their environment.
Example: People who live in the Sahel wrap their heads with heavy material to keep the hot sun and sand out of their faces.

People change their environment.
Example: The developer of a shopping mall chooses to bulldoze the land to level it for buildings and parking lots.

Your Turn
Give one example of how you have adapted to your environment and one example of how you have changed your environment.

---

covered with brush and grass are now bare. The movement of people and animals away from the desert has created overcrowded conditions in nearby environments. Geographers are working with governments and international agencies to help solve this problem.

Geographers may use satellite images and aerial photos to locate the exact boundaries of Africa's overgrazed regions. Teams of researchers, including geographers, then travel to these places for on-site surveys. They study the human-environment interaction in the region. Some investigate how the local economy has declined. Others research the crisis' impact on families and the extent of hunger. Still others check on the health of livestock and soil and water quality. Deterioration of the natural environment is then studied in detail.

Meanwhile, other geographers travel to regions where the drought has been less severe and where lands have not been overgrazed. The

drought and non-drought regions can be compared in order to make recommendations for helping the victims of the drought. Perhaps roads need improvements so that emergency supplies can be moved quickly to all parts of the region.

Solving the world's problems depends increasingly on understanding local and regional geography. Solutions to problems in richer countries may not always work in poorer countries.

Preserving Natural Environments One of the most rapidly growing industries in the world is tourism. In the United States, where national parks are popular destinations, the National Park Service is struggling to maintain the natural beauty of the parks as the number of visitors continues to increase.

Yosemite in California's Sierra Nevada is one of the country's most popular national parks. Yosemite's relative location to San Francisco and Los Angeles is convenient for visitors. Tourists

ASSESSMENT

Assign the Section 1 Review questions on page 7.

Reteaching

Assign Section 1 of Reteaching Worksheet 1.

Extension

Have interested students prepare a short report on one of the following U.S. cities: Atlanta, Las Vegas, Philadelphia, Salt Lake City, St. Louis. The report should focus on why the city is located where it is, how it was founded, what caused it to grow and flourish, and how it fits into its region today.

SECTION 2

OBJECTIVES

The objectives reflect the Focus questions on page 7 and are assessed in the review on page 8.

1. Differentiate between human geography and physical geography.

2. List various geographic professions.

CREATIVE STRATEGIES FOR TEACHING WORLD GEOGRAPHY

LESSON 3 Place, Human-Environment Interaction, and Movement

Your Turn

One example might be to remove trees or shrubs to create wider streets or highways.

COOPERATIVE LEARNING

Organize the class into groups of four or five. Students will apply the geographic theme of movement to transportation routes. Assign each group a U.S. city. Individual members will research air, land, or water routes. One member will draw a map locating the routes. Another will write a statement about how goods, people, or ideas move over these routes.

FIVE THEMES OF GEOGRAPHY

MOVEMENT

People, goods, and ideas move continuously. The movement of people, goods, and ideas is called **spatial interaction**.

People meet their needs either by traveling to other places or by trading with people in other places.

Example: Open markets in the Sahel provide centers where herders and farmers can trade their goods.

As the amount of movement and kinds of transportation change, so do other geographic features.

Example: The movement of tourists into Yosemite National Park has created a need for a shuttlebus service.

Your Turn
Give one example of how the amount of movement has changed geographic features in your community.

---

traveling to California's coastal cities often add Yosemite to their travel plans.

Yosemite's place characteristics include beautiful mountain scenery. Tourists can see breathtaking vistas of steep mountains, green forests, and waterfalls. Hiking trails wind throughout the park. Yosemite Valley also has lovely hotels and camping sites. The park draws visitors from throughout the United States and many foreign countries. Many signs at park facilities are printed in several languages, including Spanish, French, German, and Japanese.

During the 1960s and 1970s, thousands of tourists crowded into the park in a day. On holidays, park employees could expect 40,000 people to visit. The interaction between Yosemite's natural environment and its many visitors created problems. Many of the hiking trails had to be paved. The hotels and campgrounds were over-

flowing. Air pollution from the movement of automobiles in and out of the park clouded the air and limited the views. A steady procession of garbage trucks carried the valley's waste away. Rather than guide visitors through Yosemite's natural wonders, park rangers directed traffic.

Solving the park's transportation problems was important to the future of Yosemite. Geographers and other planners had to make difficult decisions about the building of more roads and hotels. Automobile access to the park has been limited, and shuttlebuses now take tourists into and out of Yosemite Valley. As a result, air pollution has decreased. Preserving the beauty and wildlife of Yosemite and managing tourism in the park are ongoing challenges. The effort to protect Yosemite is an excellent example of geographers working with other professionals to solve environmental problems.

TEACHING SECTION 2

Motivator
Explain that one of the basic tools of geography is the map. Ask students to name the kinds of maps with which they are familiar. *(road maps, city and county maps, world maps, etc.)* Have them explore the difficulties involved in drawing maps. Help them understand that maps are rarely 100 percent accurate because the world is constantly changing.

Building a Vocabulary
The words **cartography** and **meteorology** end in common suffixes: *-graphy*, "writing or representation," and *-logy*, "study or science of." Have students list other words they know that end in these suffixes. Then have students speculate why *meteorology* refers to weather and forecasting. For a clue, have a student read the meanings of the word *meteor* from a dictionary.

Classifying
Human and Physical Geography
Have students work in groups. Give each group a set of index cards labeled with the following phrases, one per card: latitude and longitude, styles of buildings in a city, weather of an area, nations in a common market, monuments and landmarks, people changing the environment, people adapting resources, environment
Continues on page 8

FIVE THEMES OF GEOGRAPHY

REGION

A region is an area defined by common characteristics.

Regions can be defined by more than one characteristic.

Example: The Sahel can be defined as a dry grassland region in Africa because of its physical characteristics. Many people in the Sahel speak Arabic, so it also can be defined as a region by language. And the Sahel can be defined as a region by religion because many people are Muslims, or members of the Islamic faith. Shown at right is a mud mosque in West Africa.

To learn more about how geographers classify regions, read the special feature "What's in a Region?" on pages 100–101.

Your Turn
Name three different regions in which you live.

Your Turn
Answers might refer to climate, language, nationality, or continents.

Section 1 Review
ANSWERS

1. By gathering and providing data about land use, resources, transportation, and human impact on environmentally sensitive areas; possible example: the drought in Africa

2. They are location, place, human-environment interaction, movement, and region. Examples will vary.

3. Location: countries of Mauritania, Mali, and Niger; place: lands once covered are now bare; human-environment interaction: overgrazing has made the situation worse; movement: people are moving away from the desert; region: comparing drought and non-drought regions

The Five Themes of Geography There are a number of common geographic themes in the case studies you have just read about. These common themes are location, place, human-environment interaction, movement, and region. Together these themes are known as the Five Themes of Geography.

One or more of the themes can be found in the study of any topic in geography. As you study the major world regions in the other parts of this textbook, it will be helpful to use the Five Themes of Geography to organize your thinking. Refer to the charts in this section throughout your study of *World Geography Today*.

SECTION REVIEW

1. How can geographers help solve local and regional problems? Use examples from this chapter.
2. Name the Five Themes of Geography and give an example of each.
3. **Critical Thinking** **Apply** the Five Themes of Geography to the drought problem in Africa.

2 THE WIDE WORLD OF GEOGRAPHY

FOCUS

- *What is the difference between human geography and physical geography?*
- *What are some professions that involve geography?*

Branches of Geography By now you know that geography covers a wide range of topics. Often, it is divided into different branches. The two main branches of geography are human and physical geography, and each of these branches can be further divided into many subjects.

Human geography is the study of how people and their activities vary from place to place. It includes political, economic, and cultural factors such as cities and countries, farms and factories, and ways of life. Geographers are unique social scientists in that they borrow elements of many subject areas to conduct their studies. Geographers who work in human geography might specialize in historical, political, economic, medical, population, or urban geography.

affecting people, transportation and communication, absolute location, relative location, vegetation of an area, the Sunbelt states, the Great Plains, spatial interaction, pollution, exports, and imports. Have students separate the cards into two stacks, one for physical geography and the other for human geography.

Guided Practice

List on the board some professions that require specialized geographic knowledge. *(meteorologist, cartographer, urban planner, transportation analyst, environmental consultant, geography teacher)* Have students select one profession from the list and write a paragraph that describes specific activities the geographer might accomplish during one day.

Closure

Relate different professions requiring geographic knowledge to physical or human geography.

ASSESSMENT

Assign the Section 2 Review questions on this page.

Reteaching

Assign Section 2 of Reteaching Worksheet 1.

MULTICULTURAL PERSPECTIVES

Place-names on maps can become a political issue. For example, most Israeli maps call the lands taken from Jordan in the 1967 Six-Day War by the ancient names of Judaea and Samaria. The Palestinian Arabs, who consider these areas their homeland, refer to them as the occupied territories. Another, more neutral, name is the West Bank (geographical location on the Jordan River). Students might research controversies in their communities over changing the place-names of streets or landmarks.

Section 2 Review
ANSWERS

1. Answers include historical, political, economic, medical, population, and urban geography.
2. Answers include studies of water resources, of the distribution of plants and animals, and of the earth's surface.
3. Most professions require an understanding of geography, and geographic knowledge is necessary for good citizenship.

Physical geography is the study of how the earth's natural features vary from place to place. It includes the study of plains and mountains, weather and climates, and plants and animals. Specialty areas of physical geography include the study of water resources, the study of the distribution of plants and animals, and the study of the earth's surface. The interactions of people with water, the weather, soil and minerals, and plants and animals are links between human geography and physical geography.

In this textbook, both human and physical geography are used to study the world's regions. Geographers use regional geography to divide our world into smaller topics for easier study. In this way, different parts of the world can be compared and analyzed. The boundaries of regions might be rivers, mountain ranges, or oceans. Or they might be highways or lines on a map defined by a government. The Five Themes of Geography will serve as a guide to your study of geography and the connections between the earth and its people.

Geography as a Profession Most professions require an understanding of geography. For example, a grocery store manager must know the region of his or her store and what types of products to stock. A physician must understand the physical and social environments of her or his patients. A news reporter may travel all over the world writing articles about people and events. There are, however, several professions that rely on specially trained geographers.

Cartographers are people who make maps. **Cartography** is the branch of geography that studies maps and mapmaking. Although many maps are drawn by hand, computers are revolutionizing map production. Computers store data from old maps as well as satellite images, photographs, and data from surveys and reports. The cartographer then "draws" the map on a computer screen and sends it to a printer that produces a copy almost instantly. Cartographers work for companies that publish maps, atlases, newspapers, magazines, and books. They also are employed by nearly all government agencies. The largest employer of cartographers in the United States has been the United States Geological Survey (USGS), which produces all the **topographic** (surface feature) maps for the United States.

You see geographers at work whenever you watch the nightly news. The person providing information about the weather is called a meteorologist. **Meteorology** is the field of geography that specializes in weather and weather forecasting.

Geographers are taking on an increasingly important role in studying the environment. They are able to research, map, and analyze environmental data. They may gather information about pollution, water quality, hazardous waste, endangered species of plants and animals, the loss of rain forests, and many other environmental issues.

Geographers also make decisions about managing places, guiding the transport of products, and encouraging regional development. Many geographers who work in environmental management investigate land-use practices and advise how lands should be used.

Businesses hire geographers, too. They decide where new stores should be located, schedule routes for airlines and trucking companies, and help identify new markets for businesses. Geographers also work at United Nations offices all around the world. They help organizations such as the Red Cross and the World Bank. Geographers also become experts on different peoples, cultures, and regions of the world.

Geography teachers help develop an informed public. Many people seem to know little about their own country and even less about places far away. In a closely linked world, we must all learn as much as we can about locations, places, human-environment interaction, movement, and regions.

Geographic knowledge also is necessary for good citizenship. People need to be a positive influence on their government's policies. Should we allow suburbs to be built over farmland? Where should garbage and poisonous chemicals be disposed of? How can we encourage economic progress in a poor region? These kinds of questions are geographic. Helping the government and its citizens find answers to these questions is the job of geographers.

2 SECTION REVIEW

1. List three specialty areas of human geography.
2. List three specialty areas of physical geography.
3. **Critical Thinking** **Explain** how geography is useful in everyday life.

CHAPTER 1 CHECK
ANSWERS

Building a Vocabulary

1. Site is a precise, exact location; situation is a location in relation to other places. Absolute location is the exact spot where something is found; relative location indicates position in relation to other places. All describe location.
2. A long rainless period
3. The movement of people, goods, and ideas; movement

Recalling and Reviewing

1. Answers might refer to the Sahel drought.
2. Probably all five themes: location and place to decide where to go; human-environment interaction to plan activities; movement to decide on forms of transportation; and region to know what the area is like
3. Human geography focuses on people and their activities; physical geography focuses on the earth's natural features. Possible answers: human—urban geography; physical—meteorology; both—environmental management

---

CHAPTER 1 CHECK

Cairo, Egypt

Reviewing the Main Ideas

1. Geography is the study of the world, both nearby and far away. Its focus is spatial. Geography can help solve almost any problem that has to do with space and how people use space.
2. There are five fundamental themes that provide an organizational framework for geography: location, place, human-environment interaction, movement, and region.
3. Human geography and physical geography are the two main branches of geography. An understanding of geography is important for many professions and for good citizenship.

Building a Vocabulary

1. Discuss the difference between a site and a situation, and between an absolute location and a relative location. How are the terms related?
2. What is a drought?
3. Define *spatial interaction*. To which geographic theme is it related?

Recalling and Reviewing

1. Give an example of a world problem that geography can help solve.

2. Which of the Five Themes of Geography might you use in planning a trip to a destination of your choice, and how would you use them?
3. What is the difference between human geography and physical geography? Name one profession that uses mostly human geography, one profession that uses mostly physical geography, and one profession that uses both.

Thinking Critically

1. Think of a major news event you have heard about recently in the newspaper or on television. How is geography involved in that event?
2. What are some professions that involve geography that are not mentioned in this chapter? Consider the jobs of your friends and family, and discuss how geography is a part of them.

Learning About Your Local Geography

Individual Project
Many of the review exercises in this textbook will ask you about your "community," and it will be helpful for you to decide in advance what your community is, geographically. Is your community the area around your home, your school, neither, or both? Is your community a street block or rural route, a neighborhood, a whole town? What are the boundaries of your community? Who lives in your community? Write two or three sentences defining your community. You may want to include your definition in your individual portfolio.

Using the Five Themes of Geography

In a list or chart, apply the Five Themes of Geography to a movie you have seen recently— any movie of any kind. In what **locations** did the movie take place? What characteristics did the **places** in the movie have? What evidence of **human-environment interaction** did the movie give? What kinds of **movement** were shown? And finally, how did the movie apply the concept of **region**?

Thinking Critically

1. Answers will vary, depending on the news event students select, but might include that geography plays a role in causing major news events (for example, wars or disputes over land or boundaries) and in determining how problems are solved (for example, economic measures or military intervention).
2. Answers will vary but should indicate an awareness that almost every profession or job involves geography in some way.

Learning About Your Local Geography

Answers will vary. Students should arrive at a solid, working definition of their own community for use in future exercises.

Using the Five Themes of Geography

Answers will vary. Almost every movie involves each of the Five Themes of Geography in some way.

CHAPTER 2 The Earth in Space
Planning Guide

TEXTBOOK	TEACHING RESOURCES	OTHER AVAILABLE MATERIALS
Section		
1. The Sun, the Earth, and the Moon Pages 11–13	Reteaching Worksheet 2.1 Challenge/Enrichment Worksheet 1	*REGIONS OF THE WORLD Map Transparencies:* Angle of the Sun's Rays Hitting Earth *Study Guide for Spanish Speakers* *Lecture Notes,* Chapter 2
2. Solar Energy and the Seasons Pages 14–15	Reteaching Worksheet 2.2 Skills Worksheet 2 Critical Thinking Worksheet 2	*REGIONS OF THE WORLD Map Transparencies:* The Seasons *Study Guide for Spanish Speakers* *Lecture Notes,* Chapter 2
3. The Earth System Page 16	Reteaching Worksheet 2.3 Graphic Organizer 2	*Study Guide for Spanish Speakers* *Lecture Notes,* Chapter 2
Chapter Review Page 17	Chapter 2 Test Chapter 2 Portfolio Assessment See also *Alternative Assessment Forms*	*Test Generator,* Chapter 2

CHAPTER RESOURCES

Lecture Notes, Chapter 2
Teaching Resources
    Chapter 2 Graphic Organizer
    Chapter 2 Skills Worksheet
    Chapter 2 Critical Thinking Worksheet
    Chapter 2 Challenge/Enrichment Worksheet

REVIEW AND ASSESSMENT

Chapter 2 Check, page 17
Chapter 2 Test, *Teaching Resources*
Chapter 2 Portfolio Assessment, *Teaching Resources*
Test Generator, Chapter 2

ADDITIONAL RESOURCES

Books for Teachers

Durham, Frank. *Frame of the Universe.* New York: Columbia University Press, 1985.

Lewis, Richard S. *The Illustrated Encyclopedia of the Universe: Understanding and Exploring the Cosmos.* New York: Crown, 1985.

Sagan, Carl, and Lee Gentry. *Cosmos.* New York: McGraw-Hill, 1980.

Books for Students

Allen, Oliver. *Atmosphere.* Morristown, NJ: Silver Burdett, 1983.

Asimov, Isaac. *How Was the Universe Born?* Milwaukee, WI: Gareth Stevens Publishing, 1989.

Kals, W.S. *Stars and Planets: The Sierra Club Guide to Sky Watching and Direction Finding,* San Francisco, CA: Sierra Club Books, 1990.

Lewis, Richard S. *The Illustrated Encyclopedia of the Universe: Understanding and Exploring the Cosmos.* New York: Crown, 1985

McPhillips, Martin, ed. *The Solar Energy Almanac.* Nashville, TN: Thomas Nelson, 1983.

Wentzel, Donat G. *The Restless Sun.* Washington, D.C.: Smithsonian Institution Press, 1989.

The World Today *Videodisc Program*

A variety of still images, short videos, and activities are available for you to use as you teach this chapter. *Lecture Notes* include bar codes for specific photographs.

Multimedia Resources

Picture Atlas of the World. (interactive maps, photo bank, animation, audio on CD-ROM) National Geographic Society, Washington, DC.

Seasons. (16mm film/VHS, 15 minutes) Lucerne Media, 37 Ground Pine Road, Morris Plains, NJ 07950.

SimEarth: The Living Planet. (Interactive software, simulation game) Broderbund Software, 500 Redwood Boulevard, P.O. Box 6121, Novato, CA 94948.

MEETING INDIVIDUAL NEEDS

Limited English Proficiency (LEP)

Pair LEP students with students proficient in English. Provide each pair with Outline Map 1. Have the teams draw and label the following: Equator, Tropic of Capricorn, Antarctic Circle, Arctic Circle, Tropic of Cancer, North Pole, South Pole. Tell them to label and color the low-latitude areas red, the middle-latitude areas orange, and the high-latitude areas yellow. Pairs should also identify continents, oceans, countries, and cities located in each latitude and discuss the amount of solar energy each would receive.

Learners Having Difficulty

Tell students that the terms listed in the Geography Dictionary at the beginning of each chapter also appear bold-faced in the text. Have students find each boldface term and its definition in Chapter 2. Tell them to find examples of as many terms as possible in the chapter illustrations.

Auditory Learners

Have students call out characteristics of the seasons as you classify them on the board. Then begin a class discussion about what causes the seasons.

Visual Learners

In a darkened room, shine a spotlight on a globe to demonstrate the following: Earth's rotation on its axis, Earth's revolution around the sun, Earth's tilt, and seasons on Earth. Ask the following questions: What is the difference between rotation and revolution as applied to Earth? What factors affect the amount of solar energy that Earth receives? Is the amount of solar energy that Earth receives greater in the low latitudes or in the high latitudes? Why? What causes the seasons? Encourage students to ask their own questions during the demonstrations.

Tactile/Kinesthetic Learners

Have students work in small groups to make three-dimensional models of the solar system using materials such as foam balls or clay. Encourage students to research the distance of the planets from the sun and to make their models to scale. Students should label the parts of the solar system. Display the models in the classroom.

Gifted Learners

Have interested students research the expeditions of a polar explorer such as Roald Amundsen, Richard E. Byrd, Robert Peary, or Matthew Henson.

MULTICULTURAL FOCUS

Point out to students that the mythologies of many cultures have included worship of sun deities. The ancient Greeks, for example, made offerings to Helios, the sun god with an all-seeing eye. Worship of the sun was also central to the culture of the Inca, whose civilization in the Andes Mountains in South America flourished from the thirteenth to the sixteenth century. Read aloud the following Incan legend:

Inti, the sun god, created four brothers and four sisters. They emerged from a window in the earth—a cave. Ten families of followers emerged from two caves alongside. The children of the sun god and their followers set out to find a homeland. When one brother proved troublesome, the other three walled him up in a cave. Later, two other brothers were turned to stone. Only one brother, Manco Capac, was left—to become the founder of the Inca nation.

The legend reveals how Incan emperors were thought to be descended from the sun. A temple was built on the spot where Manco Capac was believed to have first settled. This Temple of the Sun housed a large golden disk on which the rays of the morning sun fell and reflected throughout the shrine. Many Incan ceremonies were held in honor of the sun.

Elicit a class discussion about how the sun affects everyday life in society today. Ask students if they notice any contemporary evidence of "sun worship." *(sunbathing, acquiring a suntan)*

THE EARTH IN SPACE

Section 1: The Sun, the Earth, and the Moon
Section 2: Solar Energy and the Seasons
Section 3: The Earth System

OVERVIEW

Chapter 2 examines Earth as a planet.

Section 1 discusses the solar system and Earth's place in it. It also explains solar energy and Earth's rotation, revolution, and tilt.

Section 2 explores the relationship between latitude and the amount of solar energy that Earth receives. It explains how Earth's revolution and tilt produce the seasons. It identifies the Tropics of Cancer and Capricorn and the Arctic and Antarctic circles. Students will also learn the significance of the solstices and equinoxes.

Section 3 describes the earth system, which includes the atmosphere, lithosphere, hydrosphere, and biosphere.

About the Author

Vincent Willem van Gogh (1853–1890) of the Netherlands was among the most influential artists of modern times. However, his artistic life was marred by insanity and commercial failure. He sold but one painting during his brief life. The quotation on this page is from a letter to his brother Theo and expresses van Gogh's appreciation of the night sky. One of his most famous paintings is *The Starry Night*, which he painted one year before his death by suicide.

Primary Source

Description of Change: excerpted
Rationale: excerpted to focus on main idea

CHAPTER 2

The Earth in Space

❝ The sight of stars always sets me dreaming just as naïvely as those black dots on a map set me dreaming of towns and villages. Why should those points of light in the firmament, I wonder, be less accessible than the dark ones on the map of France? ❞

Vincent van Gogh

GEOGRAPHY DICTIONARY

solar system
planet
satellite
rotation
revolution
tropics
polar region
solstice
Tropic of Capricorn
Antarctic Circle
Arctic Circle
Tropic of Cancer
equinox
earth system
atmosphere
lithosphere
hydrosphere
biosphere

If you look at the sky on a clear night, you can see thousands of stars. With a telescope, you can see millions more. Beyond the telescope's view are billions more. All these stars are part of the universe, made up of all existing things, including space and Earth. Scientists estimate that the universe is 15 to 20 billion years old. Since its birth, according to scientists, the universe has been expanding continuously and is unimaginable in size.

The largest objects in space are stars, which generally are grouped together in huge, rotating collections called galaxies. Many objects that look like individual stars to the naked eye are really billions of stars in a galaxy far, far away. One galaxy, the Milky Way, appears as a bright milky streak across the night sky. Many individual stars in the Milky Way are visible because it is the galaxy in which we live. The sun is actually a medium-size star near the edge of the Milky Way.

When you look up at the night sky, do you ever wonder if there are places with life besides Earth? Scientists wonder this and continue to search for clues. The vast universe holds unlimited possibilities for future discovery.

Earth and Moon

Using the Quotation

Call on a volunteer to paraphrase van Gogh's remarks. *(Dots on a map may be abstract but they represent real ["accessible"] places; he wishes the same were true for the stars.)* Ask students to imagine they see the night sky for the first time and that they know nothing about the distance between Earth and the stars. Do objects in the solar system appear close to Earth?

Using the Geography Dictionary

Write the terms in the Geography Dictionary on the board, out of order. Have students group the terms by using information they already know, or from clues in the words themselves. *(Students are likely to group terms relating to the solar system, proper nouns that relate to place, and terms that have a common suffix: -sphere.)*

SECTION 1

OBJECTIVES

The objectives reflect the Focus questions on this page and are assessed in the review on page 13.

1. Identify the solar system, planets, sun, moon, and Earth.

2. Describe the effects of Earth's rotation, revolution, and tilt.

The Solar System

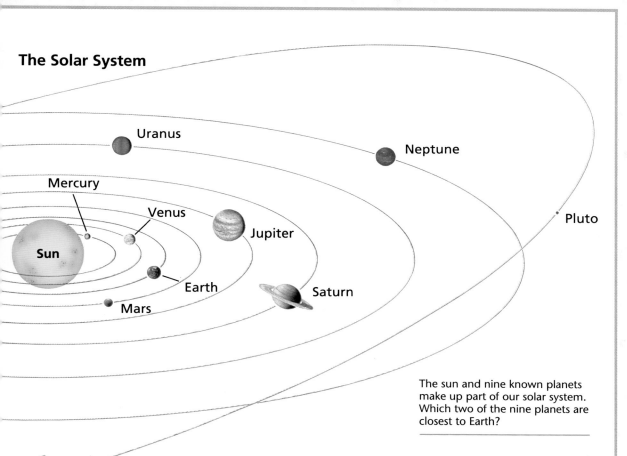

The sun and nine known planets make up part of our solar system. Which two of the nine planets are closest to Earth?

THE SUN, THE EARTH, AND THE MOON

FOCUS

● *What is the solar system?*

● *What factors determine the amount of solar energy that the earth receives?*

The Solar System The sun is the star that is the center of our **solar system**. The nine **planets**, including Earth, that move around the sun are part of the solar system. Planets are spheres, or ball-like objects, that orbit around a star. Jupiter is the largest planet, followed by ringed Saturn. Earth is one of the smaller planets and is fifth in size. It is the third nearest planet to the sun. (See the diagram on this page.)

The Moon Moons also are part of the solar system. A moon orbits around a planet. A body that orbits a larger body is called a **satellite**. Moons are natural satellites. Some planets have many moons—Saturn has 17. Others, such as Venus, have none. Earth has one large moon whose diameter is a little more than one-fourth that of Earth. On a clear night, we often can see the barren volcanic surface of the moon.

The moon, the earth, and the sun exert forces on each other that influence physical processes on Earth such as the movement of the ocean known as tides. Tides are complex, but, generally, tides occur because of the gravitational pull of the moon and sun, and the force produced by the rotation of the earth and moon together. (See the diagram on page 12.)

The Sun The sun is small compared to some other stars. Yet, when compared to Earth, the sun is enormous. For instance, the diameter of the earth is about 8,000 miles (12,872 km). The diameter of the sun is about 865,000 miles (1,391,785 km). The sun is more than 100 times the diameter of Earth and more than 400 times the diameter of

TEACHING SECTION 1

Motivator

Ask students to think of products (such as food, household products, automobiles, and electronic equipment) named after objects in the solar system and in space. List their responses on the board. Ask them to speculate about the manufacturers' reasons for choosing such names for their products.

Building a Vocabulary

Students might know the term **satellite** but associate it only with a human-made object. Point out its connection to **planet** and **solar system**. Then have volunteers read aloud several definitions of *satellite* from the dictionary, all of which specify the relationship between something smaller and something larger.

Cause and Effect

Earth's Rotation, Revolution, and Tilt Have students choose a part of the world that receives much more or much less sunlight (and therefore solar energy) than their own. Have them investigate how the amount of sunlight affects the way of life in the place they have chosen. Some students may wish to illustrate their findings.

Caption Answer

High tides permit some fishing vessels to leave or enter a harbor. Larger ships need high tides to pass shallow harbor entrances. Tidal currents sweep out main channels to keep them deep, and they help clean harbors and beaches. Some recreational activities are regulated by tides.

Geographical Sidelight

Leap year was introduced by Julius Caesar in 46 B.C. Caesar asked Egyptian astronomer Sosigenes to help correct errors in the Roman calendar. Sosigenes developed the Julian calendar, based on a $365\frac{1}{4}$-day year. One extra day was added every fourth year, the leap year.

Sixteenth-century astronomers calculated the solar year to be about 11 minutes longer than $365\frac{1}{4}$ days. To correct this error, the Gregorian calendar was adopted in 1582. Leap years were preserved only in century years divisible by 400. Hence, 1900 was not a leap year; 2000 will be.

Effects of the Moon and Sun on Tides

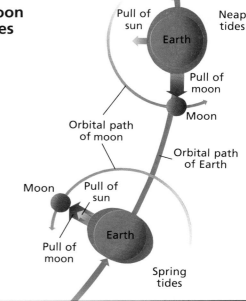

The combined gravitational pull of the moon and the sun produces high spring tides, which occur twice a month. High tides are lower during neap tides, when the pull of the sun is at right angles from the pull of the moon. Why is knowing the schedule of daily high and low tides important to people in coastal areas?

The photo at left shows low tide at **Mont Saint Michel, France**. At right, high tide moves in around the ancient abbey and town.

the moon. The sun looks much larger and brighter than other stars because it is so much closer to the earth. Even though the earth averages 93 million miles (150 million km) from the sun, the next nearest star, Proxima Centauri, is about 25 trillion miles (40 trillion km) from Earth.

Solar Energy Almost all of the earth's energy comes from the rays of the sun. This energy makes life as we know it possible. The amount of solar energy received plays a major role in determining weather, vegetation, animal life, and the ways of life of people around the world. It influences the clothes people wear, the types of homes they live in, the food they grow and eat, and even the types of sports they play.

Three factors control the amount of solar energy that falls on different parts of the earth. These are the earth's rotation, revolution, and tilt.

MOVEMENT **The Earth's Rotation** Think of Earth as having an imaginary rod running through it from the North Pole to the South Pole. This rod represents the earth's axis, and the planet spins around on it. One complete spin of Earth on its axis is one **rotation**, which takes 24 hours. We see the effects of the earth's eastward rotation as the sun "rising" in the east and "setting" in the west. To us, it appears that the sun is moving, while actually it is Earth rotating on its axis.

Solar energy strikes only the half of the earth that is facing the sun. If Earth did not rotate

Guided Practice

Have students imagine that Earth's rotation takes 365 days instead of 24 hours and that, therefore, the same side always faces the sun. Ask them to speculate about whether they and other humans could live on the planet under this condition and, if so, where. Ask what would probably happen to lakes and oceans.

Independent Practice

Have students inquire about the use of solar energy in their communities. Ask them to write a statement that explains why using solar energy in their area is practical or impractical.

Closure

Have students describe Earth's position and movement in relation to the other objects in our solar system. Then have students list specific ways that Earth's tilt affects their lives.

ASSESSMENT

Assign the Section 1 Review questions on this page.

Reteaching

Assign Section 1 of Reteaching Worksheet 2.

Angle of Sun's Rays Hitting Earth

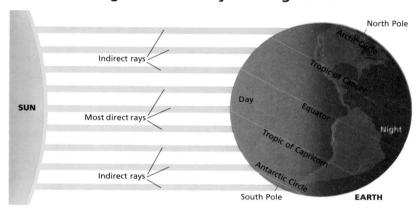

The tilt of the earth's axis and the position of the earth on its orbit determine where the sun's rays most directly strike the planet. Why is the area along the equator always warmer than the polar regions?

on its axis, only the half facing the sun would receive the sun's energy. That side of the earth would be extremely hot. The half of the earth facing away from the sun would always be dark and extremely cold. The earth's rotation makes it possible for the earth's surface to be exposed for a time to the warming effects of daylight and to the cooling effects of darkness. The earth's rotation creates day and night.

The Earth's Revolution In addition to rotating on its axis, the earth revolves around the sun. The earth makes one nearly circular orbit, or one **revolution**, every 365¼ days (one Earth year). Each time you celebrate your birthday, you have just completed another orbit around the sun. For convenience, our calendars have 365 days in one year. To account for the one-fourth day gained each year, an extra day (February 29) is added to the calendar every fourth year. This year, one day longer than the previous three, is called a leap year.

The Earth's Tilt The earth's axis is always pointed toward the same spot in the sky. The north polar axis always points to a star known as the North Star. The position of the axis is fixed in respect to the North Star, but not to our sun. The earth's orbit lies on a plane that runs from the center of the sun to the center of the earth. The earth's axis is tilted 23½° from the perpendicular (90°) to the plane of its orbit. Consequently, as the earth revolves around the sun,

the North Pole points at times toward the sun and at times away from the sun.

When the North Pole leans toward the sun, the North Pole is in constant sunlight, and the Northern Hemisphere receives more of the sun's energy. At this time, the South Pole is in complete darkness, and the Southern Hemisphere receives less of the sun's energy. When the North Pole leans away from the sun and is in complete darkness, the region around the South Pole is bathed in constant sunlight.

What happens when the poles are not leaning toward or away from the sun? If the earth's axis pointed straight up and down all the time in relation to the sun, every day would be the same length. Each day would consist of 12 hours of daylight and 12 hours of darkness. This would be true all over the earth throughout the year. This is not the case, however, since the earth's axis is tilted. During the year, the tilt causes some places to be exposed longer to daylight while others are exposed longer to darkness.

SECTION REVIEW

1. What objects are included in the solar system?
2. What is the difference between rotation and revolution, as applied to Earth?
3. **Critical Thinking** **Analyze** the factors affecting the amount of solar energy that Earth receives.

13

OBJECTIVES

The objectives reflect the Focus questions on this page and are assessed in the review on page 15.

1. Identify the relationship between solar energy and latitude.
2. Explain what causes solstices, equinoxes, and the seasons.

TEACHING SECTION 2

Motivator

Ask students who have ever flown a great distance north or south in the winter or summer to describe the differences in their surroundings when they got off the plane. Tell them that the changes they felt are the result of differences in latitude from the equator. Explain how the tilt of the earth affected their experience.

Building a Vocabulary

Ask volunteers to research in an encyclopedia and report to the class on why the Tropics of Capricorn and Cancer are so named. *(They are named after the constellations Capricornus and Cancer, in whose boundaries the sun appeared 2,000 years ago during the winter and summer solstices, respectively. The names are still used although the constellations have shifted in the sky.)*

MULTICULTURAL PERSPECTIVES

In temperate regions, the four seasons mark Earth's progression around the sun. In the tropics, seasonal differences are so slight that other factors determine the concept of season. In southern Asia and parts of Africa, for example, there is a wet season and a dry season. Ask students what other activities or geographical conditions people might use to differentiate seasons. *(planting/ harvesting, hot/cold)*

GLOBAL PERSPECTIVES

To help students understand the seasonal differences between the hemispheres, have them imagine they will visit Australia in winter. Ask them to write a list of clothing items they will need to pack.

Caption Answer

September equinox and March equinox

2 SOLAR ENERGY AND THE SEASONS

FOCUS

● *What is the relationship between latitude and the amount of solar energy that the earth receives?*

● *What causes the seasons?*

Solar Energy and Latitude As you know, different parts of the earth receive different amounts of solar energy. The areas near the equator receive great amounts of solar energy year-round. These areas are warm all the time. We call these warm, low-latitude areas near the equator the **tropics**. Other areas of the world get very little solar energy. These areas are at high latitudes and are cold most of the time. Because they surround the North and South poles, we call these areas the **polar regions**. In some areas, the amounts of solar energy vary greatly during the year, and they may be warm or cool depending on the time of year. These areas lie between the low-latitude tropics and the high-latitude polar regions and are called the middle-latitude regions.

The amount of energy that a place receives depends on the angle at which the sun's rays strike the earth. Direct, vertical solar rays are much stronger than angled rays in heating the earth's surface. This is because the amount of energy in a direct ray is concentrated on a small area. The same amount of energy in an angled ray is spread over a larger area. (See the diagram on page 13.)

When the North Pole is tilted toward the sun, the sun's most direct rays strike in the Northern Hemisphere. Thus, this hemisphere receives more concentrated solar energy, and temperatures here are warmer. The Southern Hemisphere at this time receives more angled rays with less concentrated energy and is cooler.

When the North Pole tilts away from the sun, the more direct rays strike in the Southern Hemisphere. Now it is the Southern Hemisphere that receives more direct solar energy and has warmer temperatures. At this time, the Northern Hemisphere receives less energy, and its temperatures are cooler.

The Seasons We know these times of greater or lesser heat as the seasons. In each hemisphere, the days are longer, and the sun's energy is stronger, during the summer. In the winter, the days are shorter, and the sun's energy is weaker. During spring and fall, when daylight and darkness are of equal length in both hemispheres, the sun's energy is more evenly distributed. It is the tilt of the earth's axis that causes the Northern and Southern hemispheres to have opposite seasons on the same dates.

The Solstices. At two moments during the year, the earth's poles point toward or away from the sun more than at any other time. The time that the earth's poles point at their greatest angle toward or away from the sun is called a **solstice**. Solstices occur around December 21 and around June 21.

The Seasons

North Pole
September 22 Equinox
December 21 Solstice
June 21 Solstice
23½°
South Pole
Direct rays
Sun
Direct rays
Arctic Circle
Tropic of Cancer
Equator
Earth's yearly path around the sun
Tropic of Capricorn
Antarctic Circle
March 21 Equinox

As the earth revolves around the sun, the tilt of the poles toward and away from the sun brings about seasonal change. When are both the North and the South poles at a 90° angle from the sun?

Because surface temperatures are cooler there, areas represented by blue and green colors radiate less heat than do red areas in these infrared images of Earth. The image at the left, taken in July, shows blue and green colors limited mostly to the Southern Hemisphere, where it is winter. The image at the right, taken in January, shows much of the Northern Hemisphere deep in winter.

In the Northern Hemisphere, the day in December on which the solstice occurs is the shortest day of the year and the first day of winter. In the Southern Hemisphere on that same day, it is the longest day of the year and the first day of summer.

During the December solstice, the sun's most direct rays of energy strike the earth in the Southern Hemisphere, along a parallel 23½° south of the equator. This parallel is called the **Tropic of Capricorn**. The South Pole is tilted toward the sun and receives constant sunlight. All areas located south of the **Antarctic Circle** have 24 hours of daylight. The Antarctic Circle is located at 66½° south of the equator. On this same date, the area around the North Pole experiences constant darkness and is very cold. The parallel beyond which no sunlight shines on this day is known as the **Arctic Circle**. It is located at 66½° north of the equator.

During the June solstice, the Northern Hemisphere experiences the longest day of the year, and it is the first day of summer. On this day, the North Pole is pointed at its greatest angle toward the sun. The direct rays of the sun are at their most northerly position, striking the earth most directly at a line 23½° north of the equator. This line is called the **Tropic of Cancer**. If you were to travel to Australia on the June solstice, it would be the first day of winter and the shortest day of the year.

During the June solstice in the Northern Hemisphere, the sun never sets north of the Arctic Circle. Daylight lasts 24 hours. The opposite occurs during this time south of the Antarctic Circle, where darkness lasts 24 hours.

The Equinoxes. An **equinox**, which means "equal night" in Latin, occurs twice a year when the earth's poles are not pointed toward or away from the sun. At this time, the direct rays of the sun are striking the equator, and both poles are at a 90° angle from the sun. Both hemispheres receive an equal amount of sunlight—exactly 12 hours each. If you were to travel anywhere in the world during an equinox, daylight and darkness would last the same length of time.

Equinoxes occur around March 21 and around September 22. In the Northern Hemisphere, the March equinox signals the beginning of spring. To people living in the Southern Hemisphere, however, the March equinox signals the beginning of fall. The opposite situation occurs around September 22.

Days between the solstices and equinoxes gradually become longer or shorter and warmer or cooler, depending on the hemisphere in which you live. This cycle is repeated each year, creating the four seasons.

2 SECTION REVIEW

1. Is the amount of solar energy that the earth receives greater in the low latitudes or in the high latitudes? Why?
2. What is the difference between a solstice and an equinox?
3. **Critical Thinking** **Estimate** the chances that Australia will get a lot of snow next January.

15

16

A = Atmosphere
B = Biosphere
L = Lithosphere
H = Hydrosphere

The Earth System

The interactions of the atmosphere, the lithosphere, the hydrosphere, and the biosphere make up the earth system. Which elements of this illustration are part of the hydrosphere?

THE EARTH SYSTEM

FOCUS

● *What are the four parts of the earth system?*

● *Why is Earth the only planet in our solar system to support life as we know it?*

Earth is a complex planet. Of all of our solar system's planets, Earth is unique in that it is the only planet that supports life as we know it. Geographers call the interactions of elements on and around the earth the **earth system**. The earth system can be divided into four physical systems that work together to create our physical environment: the atmosphere, the lithosphere, the hydrosphere, and the biosphere.

The **atmosphere** is the layer of gases that surrounds the earth. This mixture of moving gases provides the air you breathe and protects the earth from the sun's intense harmful radiation. The **lithosphere** is the surface of the planet that forms the continents and the ocean floor. All the water of the planet—that in the oceans, on the land, and in the atmosphere—form the **hydrosphere**. (The prefix *hydro-* means "water.") The part of the earth in which all of the planet's plant and animal life exists is called the **biosphere**.

These systems are interrelated, so we cannot draw strict dividing lines between them. The hydrosphere supplies humans with water and also serves as a home to animal and plant life. The hydrosphere affects the lithosphere when, for example, a stream flows over rock, causing the rock to be worn down. Soil can be examined as part of the lithosphere, biosphere, or hydrosphere.

As far as scientists have determined, no other planet in the solar system has four physical systems. Earth's nearest neighbors, Venus and Mars, each have an atmosphere and a lithosphere. But because of extreme heat or cold, these planets lack supplies of water. Without water, there is no life or biosphere. In our solar system, only Earth has the necessary four physical systems that work together to support life.

Is there another planet in a faraway galaxy with just the right temperatures to allow for the formation of water and the existence of life? We have yet to discover it.

SECTION REVIEW

1. What do we call the part of the earth system that forms the surface of the earth?
2. Why is it that Venus and Mars cannot support life as we know it?
3. **Critical Thinking**   **Agree or disagree** with the following statement: "The biosphere is Earth's most important physical system."

Building a Vocabulary

1. A body that orbits a larger body; the moon

2. Polar regions are areas of the earth at high latitudes; tropics are areas of the earth at low latitudes. The middle-latitude regions lie between them.

3. Solstices occur when the poles are pointed at their greatest angles toward or away from the sun. The spring and fall equinoxes occur when both poles are at a 90° angle from the sun.

4. It is the interactions of all the objects on and around the earth.

Recalling and Reviewing

1. Rotation, revolution, and tilt

2. The North Pole is tilted away from the sun. The Northern Hemisphere is receiving less direct solar energy, and the Southern Hemisphere is receiving more of the sun's direct rays.

3. Atmosphere, lithosphere, hydrosphere, and biosphere

CHAPTER **2** CHECK

Rain forest in **Borneo**

Reviewing the Main Ideas

1. The rotation of the earth on its axis makes it possible for the earth's surface to be exposed for a time to the sun's energy. This creates day and night.

2. The sun is a star at the center of our solar system. The earth makes one revolution around the sun every 365¼ days, or one year.

3. The earth's axis points toward the North Star at all times. Therefore, as the earth revolves, it leans toward or away from the sun at different positions in its orbit, causing the seasons.

4. The earth is a unique and complex planet. The physical environment of the earth is made up of four interrelated physical systems.

Building a Vocabulary

1. What is a satellite? Give an example of a natural satellite.

2. Define *polar regions* and *tropics*. What do we call the regions that lie between them?

3. Define *solstice*. How do the winter and summer solstices differ from the fall and spring equinoxes?

4. What is the earth system?

Recalling and Reviewing

1. What factors determine the amount of solar energy that the earth receives from the sun?

2. Describe the position of the earth in relation to the sun during winter in the Northern Hemisphere.

3. What are the four physical systems that make up the physical environment on Earth?

Thinking Critically

1. Would life on Earth be possible if the earth did not rotate on its axis? Why or why not?

2. If you lived in an area near the equator, would you experience all of the seasons? Explain.

Learning About Your Local Geography

Cooperative Project

With your group, take a walk around your immediate school area. Find examples of the four physical systems that make up the physical environment. Present your information in an illustration or chart.

Using the Five Themes of Geography

Carefully examine the photographs on this page. Using the photographs as examples, write two or three sentences that explain how the geographic theme of **location** relates to the amount of solar energy that a place receives.

Antarctica

Thinking Critically

1. No. The half of the earth turned away from the sun would be extremely cold, and the half of the earth turned toward the sun would be extremely hot.

2. No. The equator receives the sun's direct rays year-round; therefore, it is summer at the equator year-round.

Learning About Your Local Geography

Groups' illustrations and charts should be labeled with the four physical systems and should provide examples for each physical system.

Using the Five Themes of Geography

Students should note that one photograph shows a place located in the high latitudes and the other photograph shows a place located in the low latitudes. The location of each place determines the amount of solar energy that place receives. The place in the tropics, or low latitudes, receives direct solar energy year-round. The place in the high latitudes, or polar region, receives less of the sun's direct rays.

CHAPTER 3 The Earth's Atmosphere and Climates

Planning Guide

TEXTBOOK	TEACHING RESOURCES	OTHER AVAILABLE MATERIALS
Section		
1. Global Energy Systems Pages 19–24	Reteaching Worksheet 3.1 Challenge/Enrichment Worksheet 3	*REGIONS OF THE WORLD Map Transparencies:* Sea and Land Breezes Reading a Weather Map *Study Guide for Spanish Speakers* *Lecture Notes,* Chapter 3
2. Atmospheric Effects Pages 25–27	Reteaching Worksheet 3.2 Skills Worksheet 3	*REGIONS OF THE WORLD Map Transparencies:* Orographic Effect of California's Mountains *Study Guide for Spanish Speakers* *Lecture Notes,* Chapter 3
3. Global Climates Pages 27–32	Reteaching Worksheet 3.3 Graphic Organizer 3 Critical Thinking Worksheet 3	*REGIONS OF THE WORLD Map Transparencies:* The World's Climate Regions *Study Guide for Spanish Speakers* *Lecture Notes,* Chapter 3
Chapter Review Page 33	Chapter 3 Test Chapter 3 Portfolio Assessment See also *Alternative Assessment Forms*	*Test Generator,* Chapter 3

CHAPTER RESOURCES

Lecture Notes, Chapter 3
Teaching Resources
    Chapter 3 Graphic Organizer
    Chapter 3 Skills Worksheet
    Chapter 3 Critical Thinking Worksheet
    Chapter 3 Challenge/Enrichment Worksheet

REVIEW AND ASSESSMENT

Chapter 3 Check, page 33
Chapter 3 Test, *Teaching Resources*
Chapter 3 Portfolio Assessment, *Teaching Resources*
Test Generator, Chapter 3

ADDITIONAL RESOURCES

Books for Teachers

Ahrens, C. Donald. *Essentials of Meteorology: An Invitation to the Atmosphere.* New York: West Publishing, 1993.

Critchfield, Howard J. *General Climatology.* New York: Simon & Schuster, 1983.

Jager, Jill. *Climate and Energy Systems: A Review of Their Interactions.* New York: John Wiley & Sons, 1985.

Lockhart, Gary. *The Weather Companion: An Album of Meteorological History, Science, Legend, & Folklore.* New York: John Wiley & Sons, 1988.

Lydolph, Paul E. *Weather and Climate.* Lanham, MD: Rowman and Allanheld, 1985.

Books for Students

Asimov, Isaac. *How Did We Find Out About the Atmosphere?* New York: Walker Publishing Co., 1985.

Battan, Louis J. *Weather in Your Life.* East Lansing, MI: Freeman, 1983.

DeMillo, Rob. *How Weather Works.* Emeryville, CA: Ziff-Davis Press, 1994.

Gribbin, John, ed. *The Breathing Planet.* Cambridge, MA: Basil Blackwell, 1986.

Ramsey, Dan. *Weather Forecasting: A Young Meteorologist's Guide.* Blue Ridge Summit, PA: TAB Books, 1990.

The World Today *Videodisc Program*

A variety of still images, short videos, and activities are available for you to use as you teach this chapter. *Lecture Notes* include bar codes for specific photographs.

Multimedia Resources

The Geography Tutor: Weather and Climate. (VHS, 15 minutes) Tell Me Why, 730 Washington Street, Marina Del Rey, CA 90292.

Picture Atlas of the World. (interactive maps, photo bank, animation, audio on CD-ROM) National Geographic Society, Washington, DC.

World Climate and Landscape Regions. (VHS, 129 minutes) Educational Video Network. Available from Social Studies School Service, P.O. Box 802, Culver City, CA 90232-0802.

MEETING INDIVIDUAL NEEDS

Limited English Proficiency (LEP)

Outline Section 1 on the board as volunteers provide the main ideas. Then pair LEP students with students who are proficient in English. Have pairs outline Sections 2 and 3, using the outline on the board as a model.

Learners Having Difficulty

Point out to students that the Section Review questions in each chapter reflect the main ideas of the sections. At the conclusion of each section in Chapter 3, have students write brief answers to the Section Review questions (pages 24, 27, and 32) and then write two or three sentences that summarize the section.

Auditory Learners

Have students listen to or watch the weather forecasts on the radio or television for one week to learn the format of a weather forecast. Tell them to refer to the terms in the Geography Dictionary on page 18 in their textbooks as they listen. Have them record the terms in the list that were used by the meteorologists and explain how the terms were used.

Visual Learners

Have students locate photographs of places around the world that visually document the climate region of each place. Direct students to categorize the photographs based on the global climate regions discussed in Chapter 3 and shown on the world climate map on pages 28–29.

Tactile/Kinesthetic Learners

Have students collect newspaper or magazine articles describing various storms around the world, such as hurricanes, tornadoes, thunderstorms, monsoons, and typhoons. Then place a large climate map of the world on a bulletin board, and have students use thumbtacks to show the approximate locations of the places mentioned in the newspaper articles. Have students use pieces of string or yarn to connect the thumbtacks to the appropriate articles. Then have them chart the storms and the climate regions in which they appear.

Gifted Learners

Have students identify a major city in one of the climate regions discussed in Chapter 3. They might research ways that climate has influenced construction, settlement patterns, and leisure in the city.

MULTICULTURAL FOCUS

Point out to students that climate influences the kinds of clothes people wear, where they live, and the foods they eat. It can also influence the kind of transportation they use. For example, people who live in western Central Africa, located near the equator, travel mainly by boat along the rivers. This low-latitude climate causes plants and trees to grow quickly. The abundant rainfall and thick forests make it difficult to construct and maintain roads and railroads.

In dry, desert climates, camels are a practical means of transportation for people of nomadic cultures. Camels can travel long distances across hot, dry deserts with little food or water, and they can traverse the soft sand. Camels thus have advantages over trucks and four-wheel-drive vehicles, which could get stuck in the sand or break down from the extreme heat.

People who live in colder climates have developed countless ways of traveling on snow and ice, including skis, snowshoes, sleighs, snowmobiles, and skates. Sled dogs are bred and raised to provide transportation in Arctic climates.

Ask students to consider other ways in which climate affects the kinds of transportation used by people in particular climate regions. Elicit a class discussion about how climate affects the kinds of transportation students themselves use, or other aspects of their daily life.

OVERVIEW

In Chapter 3, students will learn about the earth's atmosphere, weather, and climates.

Section 1 explains how the earth maintains an energy balance and how air and sea currents help exchange heat energy between latitudes. It defines air pressure and describes the four major air pressure zones; global wind belts and ocean currents are also described.

Section 2 discusses the factors that influence weather. It explains humidity and precipitation, storms, and the effect of elevation on weather and climate.

Section 3 describes the major types of climates and how they are classified.

About the Author

Homer (9th–8th century B.C.?) is known for writing adventure epics about ancient Greece. Little has been documented about Homer's life; some scholars doubt that he ever existed. The quotation on this page is from the *Odyssey*, which chronicles a ship's strange and fateful journey.

Primary Source

Description of Change: excerpted
Rationale: excerpted to focus on main idea

CHAPTER 3

The Earth's Atmosphere and Climates

“**T**hen Poseidon drove together the clouds and stirred up the sea. In his hands the Trident; he excited violent squalls, among all the winds. With dense clouds he covered earth and sea; and night advanced from the sky. Together broke the east wind, the south wind and the stormy west wind. And the north winds from cold lands came, driving huge waves. ”

From Homer's *Odyssey*

The earth's atmosphere, a thin layer of gases, is made up mostly of nitrogen and oxygen. The atmosphere is necessary for human life because it contains the oxygen we breathe. The condition of the atmosphere varies constantly from day to day, season to season, and year to year. **Weather** is the condition of the atmosphere at a given place and time.

Weather conditions in an area averaged over a long period of time are called **climate**. Climate includes the expected, or average, weather events, as well as rarer events such as devastating storms. Climate also includes long-term global changes such as ice ages.

In this chapter, you will learn how weather and climate are affected by the sun's energy, air pressure, and the global circulation systems. Both weather and climate also are influenced by physical features such as oceans and mountains. Today, there is concern that human activities also may produce changes in the earth's weather and climate.

GEOGRAPHY DICTIONARY

weather
climate
temperature
greenhouse effect
air pressure
prevailing wind
trade wind
doldrums
westerly
polar wind
front
remote sensor
Landsat
evaporation
humidity
condensation
precipitation
hurricane
typhoon
tornado
elevation
orographic effect
rain-shadow effect
monsoon
steppe
permafrost

Lightning storm in **Washington** state

Using the Quotation

Tell students that the ancient Greeks believed in several gods who controlled all aspects of the physical world. Poseidon was their god of the sea. Elicit student response about ways that Homer's description of a storm differs from that presented by a television meteorologist. Point out that they will learn in Chapter 3 how wind patterns influence the weather.

Using the Geography Dictionary

Point out that wind and water are key elements of both **weather** and **climate.** Write the headings Wind and Water on the board. Have students call out the terms in the Geography Dictionary that relate to each, and write them under the appropriate heading. Point out that some terms can be listed under both.

SECTION 1
OBJECTIVES

The objectives reflect the Focus questions on this page and are assessed in the review on page 24.

1. Describe the cause and effect of air-pressure differences.

2. Explain how the earth maintains an energy balance.

GLOBAL ENERGY SYSTEMS

FOCUS

- *What causes high- and low-pressure areas?*
- *How do winds and ocean currents help the earth maintain an energy balance?*

Every day, a small portion of the sun's total energy reaches the earth. Some of that energy is reflected directly back to space. A small amount stays in the atmosphere, where it is absorbed by atmospheric gases. Much of the sun's energy, however, is absorbed by the earth's surface—its oceans and its continents.

Once the sunlight is absorbed, it is changed into heat energy. For example, if you stand in direct sunlight, you become warm. What you feel is the sunlight changing into heat energy. Just as you warm up by absorbing the sunlight, the earth's surface and the atmosphere also warm up when sunlight is changed into heat. The measurement of this heat in the atmosphere is called **temperature**.

The atmosphere traps energy in much the same way that a greenhouse does. The glass of the greenhouse allows the sunlight to pass through it. But it traps and delays the escape of energy after the sunlight has been changed into heat. Like the greenhouse, the earth's atmosphere allows sunlight to pass through it. When the sunlight is changed into heat energy by the earth's surface, it is trapped by the atmosphere, and the

The Greenhouse Effect

Light from the sun passes through the atmosphere and is changed into heat energy by the earth's surface. Most heat energy later escapes into space. What would happen if too much heat energy remained trapped and could not escape into space?

Caption Answer

The earth would become too warm, which could cause destructive climate changes.

earth is kept warm. This process is called the **greenhouse effect**.

Why doesn't the earth get warmer and warmer, until it is so hot that we cannot survive? It is because all heat energy trapped by the atmosphere eventually escapes back into space. Most of the heat is lost to space at night and during the winter season. This is how Earth maintains an energy balance.

As you have learned, the amount of solar energy received by the earth varies from place to place. The equator receives much more solar energy than do the North and South poles. In the tropics, more solar energy is received than is lost to space. In contrast, the polar regions receive less energy than they lose to space. An imbalance appears to exist. We might expect the tropics to reach boiling temperatures and the polar regions to become colder and colder. Yet, this does not happen.

The areas near the equator do not continuously get warmer and the poles get colder because major exchanges of heat energy take place between them. Excess heat energy is moved from the tropics to the polar areas via the atmospheric winds and ocean currents, and cold air and water are moved from the polar areas to the tropics to be reheated. These systems of global-energy exchange balance temperature extremes between latitudes.

This mirage in **Colombia** is created when sunlight heats the ground and the air above it. Rays of sunlight are bent as they pass through the layers of hot air rising from the desert ground. From a distance, you see what appears to be a pool of water.

Rapidly rising warm air creates unstable weather conditions over the Juruá River near the equator in **Brazil**.

Air Pressure and Wind The measurement of the force exerted by air is called **air pressure**. Air pressure is important in understanding climates because it creates the winds and ocean currents that make up the systems of global-energy exchange. Unequal heating of the earth's surface causes air-pressure differences. When air is warmed, it expands, becomes lighter, and rises. This creates low-pressure areas, which tend to bring unstable conditions. That is, the air rises and cools. As it cools, the water vapor in the air forms clouds. These clouds may bring rain or storms. Persistent low-pressure areas are found along the equator and at about 60° north and south latitudes.

Cold air is dense and heavy and tends to sink. This action produces high-pressure areas, which tend to be stable. As the cold, heavy air sinks, it heats up. This generally causes clear, dry weather. Persistent high-pressure areas are found at the poles and at about 30° north and south latitudes.

Wind is the horizontal motion of air between areas of different pressure. Wind always flows from high- to low-pressure areas. For example,

when air is released from a tire, it flows from inside the tire (high-pressure area) to outside (low-pressure area). In the atmosphere, the seashore illustrates this point. On a sunny day, as the land heats more quickly than the ocean, warm air rises over the land and creates a low-pressure area. Cooler air from over the ocean (high-pressure area) then blows onshore to replace the rising air over the land (low-pressure area). This cooler sea breeze makes the beach very comfortable, even on a warm summer day. During the night, the land cools more rapidly than the ocean. The air then moves in the opposite direction. Then, the cooler land breeze flows offshore from the land (high-pressure area) to replace the warmer, rising air over the ocean surface (low-pressure area). (See the illustration on this page.)

On a global scale, there are four major air-pressure zones: (1) equatorial low pressure, (2) subtropical high pressure, (3) subpolar low pressure, and (4) polar high pressure. (See the diagram on page 21.) At the equator, intense heating by the direct rays of the sun causes warm air to rise, forming the equatorial low-pressure zone. This rising air then flows toward the poles in the upper atmosphere. At about 30° north and south latitudes, the warm air that was rising from the equator now cools and sinks back to the surface. The air piles up, causing the very stable subtropical high-pressure zones. At the poles, the intense cooling causes a continuous sinking of very cold air. This cold, sinking air forms the polar high-

Sea and Land Breezes

Breezes from the ocean cool the warmer land surface during the day. At night, cooler land surface breezes blow toward the ocean. Why does the wind usually flow from cool to warmer areas?

Guided Practice

Have students examine a newspaper weather map of the United States. Point out the high- and low-pressure areas to them. Ask them what type of weather they would expect in cities in those areas. Then ask them to find cities in high- and low-pressure areas and read the short weather descriptions to determine if their predictions were correct.

Independent Practice

Provide each student with Outline Map 1. Ask students to plan an ocean voyage on a sailing ship from Spain to the Caribbean Islands and back. Have them use what they know about wind systems and the diagram on this page to plot a suitable course that would best take advantage of the prevailing winds.

Closure

Ask students to locate on a world map the latitude lines for 60° N and 60° S. Ask if they expect either fluctuating or relatively stable weather conditions at these latitudes. *(fluctuating—they lie within low-pressure zones and are along the polar fronts)*

pressure zones. At about 60° north and south latitudes are the subpolar low-pressure zones. They are formed between the high-pressure areas of the polar zones and the subtropical zones. Air flows from these high-pressure zones toward the subpolar low-pressure zones.

MOVEMENT **Global Wind Belts** Winds move heat and cold across the earth between the high- and low-pressure areas. This movement of heat and cold helps the earth maintain an energy balance. Some parts of the world have winds that usually blow from the same direction. These persistent winds are known as the **prevailing winds**, or global wind belts. These winds blow from areas of high pressure toward areas of low pressure.

The **trade winds** blow from the subtropical high-pressure zone toward the equatorial low-pressure zone. European explorers used the northeast trade winds to sail from Europe to America. Have you ever wondered why Columbus landed in the West Indies instead of along the East Coast of the United States? Look at the arrows showing the northeast trade winds on the diagram on this page. Then examine the world map in the Atlas on pages A4–A5. Notice how these winds would guide a sailing ship from Spain across the Atlantic.

Not all areas of the world are in wind belts. The calm areas with no prevailing winds along the equator are called the **doldrums**. Long ago, sailing ships were sometimes caught in the doldrums for long periods of time.

The middle latitudes are dominated by the **westerlies**. After Europeans settled in North America, mariners sailed with the northeast trade winds to America and returned to Europe with the westerlies. Most of the United States is located in the path of the westerlies. These winds carry weather patterns and storms across the nation from west to east.

In the high latitudes, the winds are more variable but mainly come from the east. These are the **polar winds**. Here, cold masses of air flow out of the high-pressure zones of the frozen Arctic and Antarctic regions. They bring cold conditions into the middle latitudes. Between 40° and 60° latitude, the warm westerlies meet the cold polar winds. This meeting zone of warmer subpolar air and cold polar air is called a polar **front**. A front occurs when two air masses with very different temperatures and amounts of moisture meet. Fronts usually cause stormy weather.

There also are prevailing winds that blow in the upper atmosphere, high above the surface of the earth. Though we do not feel them directly,

Text continues on page 24.

THEME
Place

The horse latitudes and the doldrums are places on the earth's surface where there is very little wind. The horse latitudes lie over the oceans at about 30° north and south latitudes. The term is believed to have originated with ships sailing across the Atlantic to the Caribbean Islands. A voyage prolonged in these latitudes often resulted in a shortage of water. Horses being transported on the ships would die or be thrown overboard. The doldrums is an area along the equator. Sailors used the term to describe this region because of the way passing ships became becalmed there.

Global Pressure and Wind Systems

Winds between the earth's high- and low-pressure areas help regulate the globe's energy balance. What happens when warm westerlies come into contact with cold polar winds?

Polar High Pressure
North Pole
Subpolar Low Pressure
Polar Winds
60°N
Subtropical High Pressure
Westerlies
30°N
Northeast Trade Winds
Equatorial Low Pressure
Doldrums
0°
Southeast Trade Winds
Subtropical High Pressure
30°S
Westerlies
Subpolar Low Pressure
60°S
Polar Winds
South Pole
Polar High Pressure

Caption Answer

A polar front is created. This occurs between the middle and high latitudes.

Predicting

Point out that although geographers, meteorologists, and environmental scientists interpret satellite images in highly technical ways, it is possible to obtain some practical information from them easily. Use the photograph on this page as an example. Ask: What detail do you see in the photograph of Hurricane Andrew on this page that indicates the intensity of the storm? *(The circular pattern shows the relative size of the hurricane.)* How could an image such as this help people in a community to prepare for a storm? *(Knowing the size and path of a hurricane in advance helps individuals and local officials plan for emergencies and make decisions about the necessity of evacuating an area.)*

Geographical Sidelight

Nations sometimes borrow satellites from one another and move them to different orbits. A U.S. weather satellite over the Pacific coast (GOES-6) failed in 1989, and its replacement was not ready for launch for at least five years. Rather than lose weather tracking capability on the West Coast, the United States asked for a loan of Meteosat-3, a European Space Agency satellite then in orbit over the Atlantic Ocean near South America. In February 1993, Meteosat-3 began reading data from its new orbit.

PLANET WATCH

Watching Over Earth

A color-enhanced image from a weather satellite catches Hurricane Andrew as it bears down on **Florida** in 1992.

As you read in Chapter 2, a satellite is a body that orbits a larger body. Some satellites, such as the moon, are natural satellites. There are also human-made satellites, which are becoming more important to the study of geography. Satellites collect and send back information about the earth.

Satellites use different types of **remote sensors** to collect information. Remote sensors are devices attached to satellites that act as "eyes in the sky." These instruments allow scientists to keep track of physical changes that are occurring almost anytime and anywhere on the earth.

A common type of remote sensor is a camera that records light. Satellite cameras take photographs that are broken down into information to be sent back to Earth. Computers on Earth then arrange the information back into images. Satellite cameras mainly record visible light and sometimes infrared light, which is not visible to the human eye. Infrared sensors detect heat given off by clouds, water, and land.

Satellites depend primarily on the earth's gravity to power their movement once they are in orbit.

Tracking the Weather

For centuries, people have looked to the skies for information about the weather. Only in recent times, though, has technology allowed scientists to make accurate forecasts. Part of this new technology is the weather satellite. Today, meteorologists monitor the atmosphere with information gathered from observation stations on Earth and from satellites.

Images collected by the satellites' sensors may show, for example, cloud swirls that warn of a hurricane or typhoon. Or they may reveal dense cloud masses, indicating heavy rain or snow. Weather satellites supplied the data that allowed meteorologists to track the path of Hurricane Andrew, whose 135-mile-per-hour (217-kph) winds battered southern Florida and Louisiana in 1992. Meteorologists knew of the approaching hurricane's power and warned people to leave the area or find shelter. Although the storm caused massive destruction of homes and crops, the advanced warnings saved many lives.

Scientists have depended on weather satellites since 1960, when the United States launched the first TIROS (Television InfraRed Observation Satellite). TIROS 1 was followed by other satellites, all of which collected information about the weather. These early satellites, however, did not remain over one area long enough to detect hourly changes in weather patterns. In 1975, the United States launched GOES (Geostationary Operational Environmental

22

Observing

Help students identify satellite images used on a local or national weather broadcast. Arrange either to bring a portable television to class or to show a videocassette recording of a broadcast to the class. Point out that geostationary weather satellites can deliver a photograph about every 20 minutes. Tell students that weather stations show these pictures in sequence so that they appear as short moving pictures. (such as images that show cloud patterns or storm cells moving across the weather map) Call on students to identify the images in the broadcast they think originated from a weather satellite.

Satellite), the first geostationary satellite. A geostationary satellite orbits at about 22,300 miles (35,890 km) over the equator. At that height, the earth's gravity has exactly enough force to make the satellite move at a speed that matches the earth's rotation. Thus the satellite remains over the same portion of the earth and can gather information continuously from the same area.

Eyeing the Environment

While weather satellites send scientists information about clouds, storms, and oceans, **Landsat** satellites reveal the condition of the environment. Landsat satellites usually orbit around the North and South poles and carry sensors that scan the same area about every 18 days. They detect changes in the earth's surface, such as changes in vegetation and moisture. These satellites view Earth in both visible and infrared light.

After the information gathered by Landsat satellites is processed, each color of the Landsat pictures show a different type of surface feature of the earth. Plants, for example, might appear as bright red. Blue might represent bare soil. Because Landsat photographs do not show true color, they are often called false-color images.

Environmental scientists can use Landsat images to gather information about air pollution, pests that threaten forests and farmlands, or algae that might pollute lakes. They can also monitor human activities such as urban growth or the cutting and burning of rain forests. In

This illustration shows a Landsat satellite in orbit. A new Landsat satellite, built to replace two older satellites, provides even more information about the size and shape of surface features.

addition, Landsat satellites have been used to track forest fires, crop damage, the paths of oil spills, the movement of herds of animals, and ground movement caused by earthquakes.

Geographic Information Systems

Photos from satellites also are used in mapmaking. In fact, some maps are not drawn but are made up of pieced-together satellite photos. Such maps are often used in geographic information systems, or GISs, which use computers to combine the maps with other data. These systems allow people to observe the spatial patterns of a wide variety of phenomena. For example, a GIS might show what roads are subject to flooding.

Weather satellites, Landsat satellites, and GISs have revolutionized the way scientists and geographers view our world. Satellites can help save lives, money, and the environment.

The colors in this satellite image of **New York City** are used to distinguish between different kinds of surface features. To the right of the Hudson River (the dark band in the middle of the photograph) is Manhattan. Central Park is the orange rectangle in the middle of the island.

▼ YOUR TURN

Think of ways in which nations and individuals can make use of Landsat images to protect the environment. **Write** a paragraph describing how the images might be used to highlight an environmental issue that is of concern to you. **Explain** how you and others might use the information to provide solutions to the problem.

Your Turn

Nations and individuals might use Landsat images to monitor drought, water and air pollution, and damage from environmental accidents. Students' paragraphs might suggest that Landsat images be used to inform the public about the nature and severity of a problem. Solutions should be realistic responses to a specific environmental problem.

ASSESSMENT

Assign the Section 1 Review questions on this page.

Reteaching

Assign Section 1 of Reteaching Worksheet 3.

SECTION 2

OBJECTIVES

The objectives reflect the Focus questions on page 25 and are assessed in the review on page 27.

1. Identify the causes of precipitation.
2. Describe how elevation and the orographic effect are related to weather and climate.

GLOBAL PERSPECTIVES

The Gulf Stream, the warm ocean current flowing from Caribbean waters to the North Atlantic, greatly affects life in northern Europe as well as along the eastern seaboard of the United States. Have students research and evaluate its economic and social effects in Iceland and Great Britain.

MAP TRANSPARENCIES

Reading a Weather Map: *REGIONS OF THE WORLD Map Transparencies.*

Section 1 Review

ANSWERS

1. Unequal heating of the earth's surface causes differences in air pressure. Low-pressure areas are unstable, and air rises and cools, causing storms. High-pressure areas are stable, and air sinks and heats, creating dry weather.

2. They distribute heat and cold between the warm tropical regions and the cold polar regions.

3. Without the atmosphere, Earth would not be able to retain any of the sun's energy.

24

MAP STUDY

Weather maps show climate conditions as they currently exist or as they are forecast for a particular time period. Most weather maps have legends that explain what the symbols on the map mean. This map shows a cold front sweeping through the central United States. A low-pressure system is at the center of a storm bringing rain and snow to the Midwest. Notice that temperatures behind the cold front are considerably cooler than those ahead of the front.

Reading a Weather Map

---

these strong winds move energy, storms, and major weather patterns. These high-speed, westerly moving winds are called jet streams. When an airplane experiences turbulence at high altitude, it has crossed the part of a jet stream's path where there are sudden changes in wind speed and direction.

MOVEMENT **Ocean Circulation** The oceans are the most important solar-energy collectors. Most of the sun's energy that reaches the earth's surface is absorbed by seawater. Water heats and cools much more slowly than land. Therefore, areas near oceans, such as islands and coastlines, do not experience great temperature extremes. Large land areas, especially the centers of large continents, have the most extreme temperature ranges.

As the prevailing winds blow across the surface of the ocean, they set in motion the ocean currents. Ocean currents are giant rivers of seawater flowing at the surface of the oceans. The ocean currents generally flow in circular paths. (See the map on pages 28–29.) Warm currents carry heated water from the low latitudes into the cooler high latitudes. Cool currents return cooled water from the higher latitudes to be rewarmed in

the lower latitudes. The ocean currents are one of the major ways in which heat is moved between the warm tropical regions and the cold polar regions. This movement of heat helps the earth maintain its energy balance.

The oceans do much more than move heat. They are a major source of oxygen for the air. Oxygen is produced by tiny ocean plants, which release it into the atmosphere. The oceans also are the major source of moisture for the atmosphere. You will learn more about the oceans in Chapter 4.

SECTION REVIEW

1. What causes differences in air pressure? How does air behave differently in a low-pressure area than in a high-pressure area?

2. How do global wind patterns and global ocean-current patterns help the earth maintain an energy balance?

3. **Critical Thinking** **Evaluate** the importance of the atmosphere to Earth's energy systems.

Have students talk about violent storms that they have seen or that have affected them. Ask them whether they know how the storms were caused. Write their answers on the board. Tell them that in this section they will learn more about the causes of storms.

Building a Vocabulary

Write the word pairs **evaporation/ humidity** and **condensation/precipitation** on the board. Ask students how the words in each pair relate to each other. Point out that they have a cause-and-effect relationship. Without evaporation, there would be no humidity in the air. Rain and snow would not occur if moisture did not condense.

Graphing Precipitation

Have students work in pairs to obtain local monthly precipitation figures for one year from a newspaper office or the weather service. Have them graph the data by type (rain and snow) and total amount. Then have them write a statement about factors that affect local precipitation.

ATMOSPHERIC EFFECTS

FOCUS

- *What causes precipitation?*
- *How is elevation related to weather and climate?*

Water and the Atmosphere Water vapor is an important gas in our atmosphere. Most water vapor in the air is evaporated from the oceans. **Evaporation** is the process by which water is changed from a liquid to a gas. The remainder of the water vapor in the air comes from lakes, plants, and the soil. Without water vapor, there could be no clouds, rain, or storms. The invisible water vapor stores the heat energy necessary to bring about changes in the atmosphere.

The amount of water vapor in the air is called **humidity**. You feel warmer on a humid, muggy day than on a dry day with the same temperature. As perspiration evaporates, it cools your skin. Perspiration will evaporate much more slowly on a humid day, however, because if much moisture is already in the air, there is less room for the air to absorb the moisture evaporating from your skin.

The higher the temperature, the more water vapor the air can hold. When air cools, it will reach a temperature at which it can hold no more water vapor. At this point, **condensation** will occur. Condensation is the process by which water vapor changes from a gas into liquid droplets. You most often see condensation as the formation of clouds, fog, and dew. If the condensed droplets become large enough, they fall as **precipitation**. Rain, snow, sleet, and hail are the four common forms of precipitation. Precipitation is not evenly distributed throughout the year. It may also vary from year to year. Too much precipitation in a short time period may cause floods, and too little precipitation may result in a drought.

Storms The most sudden and violent weather event is a storm, which occurs when energy stored in the water vapor of the atmosphere is released. The large amount of energy released from a storm is shown by the violence and damage that sometimes results.

Storms occur under low-pressure conditions during which rising, unstable air is present. In the middle latitudes, these storms travel with the upper-level, westerly flow and are called middle-latitude storms, or extratropical storms. They can be as large as 1,000 miles (1,609 km) in diameter and travel across oceans and continents during their life cycle. The middle-latitude storms occur along the polar front that forms a boundary line between warm, wet tropical air and cold, dry polar air. Moving fronts occur within the middle-latitude storms as warm and cold air clash.

Storms in the low latitudes differ from those of the middle latitudes. Tropical storms are smaller and, because only warm air is present, they do not produce fronts. They move from the east with the trade winds, bringing rain and wind. (See the photographs on this page.)

Severe tropical storms sometimes form over warm ocean waters. Called **hurricanes** or **typhoons**, these revolving storms bring violent winds, heavy rain, and dangerously high seas. Hurricanes are most destructive to islands and coastal areas. Sometimes they follow warm ocean currents into the middle latitudes, such as along the Gulf and Atlantic coasts of the United States.

Hurricane Bob (left), churning toward the Northeastern **United States** in 1991, and the "Blizzard of 1993" (right) are examples of tropical and extratropical storms that affect weather in the United States.

Guided Practice

As a way of illustrating the principles of condensation and evaporation, have students exhale against a cool mirror or windowpane. Point out that the water vapor in their breath condenses onto the glass. Ask them where they think the water goes as the condensed area shrinks. *(It evaporates into the air.)*

Independent Practice

Have students draw simple illustrations that depict elevation and the orographic effect. Illustrations should include captions explaining how weather and climate are affected in the different areas of the illustration by elevation and the orographic effect.

Closure

Tell students that weather is a complex phenomenon that depends on the interplay of numerous factors. Ask students to name some of the factors discussed so far in the chapter. *(Answers may include heat energy, air pressure, wind belts, ocean currents, fronts, humidity, elevation, orographic effect, rain-shadow effect.)* Then ask which terms describe how high mountains affect climate. *(elevation, orographic effect, rain-shadow effect)*

Geographical Sidelight

Most heat from solar energy does not come directly from the sun's rays. Sunlight passing through the air heats the ground and water, which in turn warm the air above them. So, air is warmest near the earth's surface. Areas at higher elevations are farther away from this source of heat and thus have lower temperatures.

MAP TRANSPARENCIES

An overhead transparency of the illustration on this page is available in *REGIONS OF THE WORLD Map Transparencies.*

Orographic Effect of California's Mountains

As moist air from the ocean moves up the windward side of a mountain, it cools. The water vapor condenses and falls in the form of rain or snow. The descending drier air that moves down the leeward side brings almost no precipitation to lands in the rain shadow of the mountains. This illustration shows that the air that has traveled over two mountain ranges and the Central Valley does not bring much precipitation to the Great Basin desert of the western United States.

Tornadoes are the smallest but most violent of storms. These small, twisting spirals of air can destroy almost anything in their path. They generally form along fronts in middle-latitude storms and are most common in the southeastern and central United States. In fact, the United States has more tornadoes than any other nation.

A thunderstorm, the most common type of storm, is any storm with lightning and thunder present. At any moment, about 2,000 thunderstorms are taking place on Earth. Thunderstorms form most often in the hot, unstable equatorial regions and along fronts in middle-latitude storms. Lightning, which carries a dangerous electrical charge, kills or injures hundreds of people each year. Thunderstorms may drop great amounts of rainfall in a short period of time, producing flash floods.

LOCATION **Elevation and Temperature**
Landforms, particularly high mountains, have a major effect on weather. **Elevation**, or height on the earth's surface above or below sea level, has the greatest impact. An increase in elevation causes a lowering of the temperature. For every 1,000 feet (305 m) up the side of the mountain, the air temperature cools about 3.6°F (2°C). The 19,340-foot (5,895-m) elevation of Kilimanjaro (kil-uh-muhn-JAHR-oh), in Africa, shows us this effect. Even though the mountain is on the equator, its twin peaks are always covered with ice and snow.

Orographic Effect Another effect of elevation on weather and climate is the **orographic** (ohr-uh-GRAF-ik) **effect**. When moist air flowing from the ocean meets a mountain barrier, the air is forced to rise. The higher the air rises, the more it cools. As the rising air cools and condensation begins, clouds form and precipitation occurs in the form of rain or snow. As a result, the side of the mountain facing the wind receives a great deal of moisture. This side of the mountain is called the windward side.

The side of the mountain facing away from the wind is called the leeward side. As the air moves down the leeward side, it warms and

SECTION 3
OBJECTIVES

The objectives reflect the Focus questions on this page and are assessed in the review on page 32.

1. Identify major factors that influence climate.
2. Identify the major climate regions of the world.

---

becomes drier. Deserts commonly form on the leeward side of mountains. This is called the **rain-shadow effect** because areas in a rain shadow do not receive much rain, just as areas in the shade do not receive direct sun. For example, because of the orographic effect of the mountains in the western United States, rain-shadow deserts are located on the leeward side of these mountains. (See the diagram on page 26.)

 SECTION REVIEW

1. What is the relationship between evaporation, condensation, and precipitation?
2. How do elevation and the orographic effect influence land and climate?
3. **Critical Thinking** **Rate** the danger of different types of storms.

 GLOBAL CLIMATES

FOCUS

- *What are the major influences on climate?*
- *Where are the major global climate regions located?*

People living in advanced industrial nations such as the United States have used technology to control the effects of climate on their daily lives. Such societies have invested billions of dollars to cool and heat homes and offices, to build water delivery systems for farms and cities, and to construct transportation systems that can operate in good weather and bad. But for four out of every five people on Earth, climate continues to have an immediate impact on every aspect of daily existence. The clothes people wear, the places they live, and the foods they eat all are influenced by climate.

The major global climate types are classified mainly by temperature and precipitation differences, both of which are influenced by latitude. That is, latitude has the most control over which regions are hot or cold, wet or dry. To a slightly lesser degree, a region's location on land and its elevation also affect climate.

Because the vegetation found in a place is a response to its climate, vegetation is sometimes used to classify climate. Some climate types are named for the dominant vegetation of the region,

which you will learn more about in Chapter 6. Refer to the climate map and chart on pages 28–29 as you read the rest of this chapter.

LOCATION **Low-Latitude Climates** The areas close to the equator mostly have warm temperatures and a large amount of rainfall year-round. These wet, hot areas have a humid-tropical climate. People who live in the humid tropics never experience winter or even cool weather.

Because the equator is constantly being heated by the sun's rays, warm air is always rising in the humid tropics. This continuous rising of warm, unstable air brings almost daily thunderstorms and heavy rainfall. The combination of continuous warm temperatures and heavy rainfall creates ideal conditions for plant growth, and dense tropical rain forests thrive here.

In some tropical areas, especially India and Southeast Asia, the rain is concentrated in one very wet season. During the summer months, moist air flows into these parts of Asia from the warm ocean (high-pressure area) to the hotter land (low-pressure area), bringing heavy rains. During the winter, dry air flows off the cooling continent (high-pressure area) to the warm oceans (low-pressure area), bringing dry conditions to the area. The wind that blows from the same direction for months at a definite season of the year is called a **monsoon**.

Text continues on page 30.

Residents of **Bombay, India**, wade through the high waters caused by the wet monsoon. From where do monsoon winds blow during the wet monsoon season in India?

27

Using the World Climate Map

Identifying Climate Regions

Tell students that colors on maps often suggest certain types of information. Have students predict general characteristics of global climate patterns by studying only the colors on the World Climate map. Tell them to cover the information in the chart below the map with a sheet of paper. Then ask them to select the colors on the map that are likely to depict a hot and dry desert climate *(red)*, a tropical forest climate *(dark green)*, or colder climates *(blues and purple)*. Contrast for students the location and direction of the major ocean currents.

Geographical Sidelight

Geographers are continuing to learn more about global weather and climate patterns. A 1990 article in Science *magazine describes a strong statistical association between rainfall in the West African Sahel region and intense hurricanes in the United States. Prolonged wet periods in the usually dry Sahel may be linked to the landfall of major hurricanes along the U.S. coast. These periods reflect large-scale circulation patterns in the Atlantic, which apparently shift in slow cycles. In 1988–1989, an 18-year drought was broken in the Sahel. As long as the wet period continues, the author predicts, strong hurricanes will occur near the Atlantic coast of the United States.*

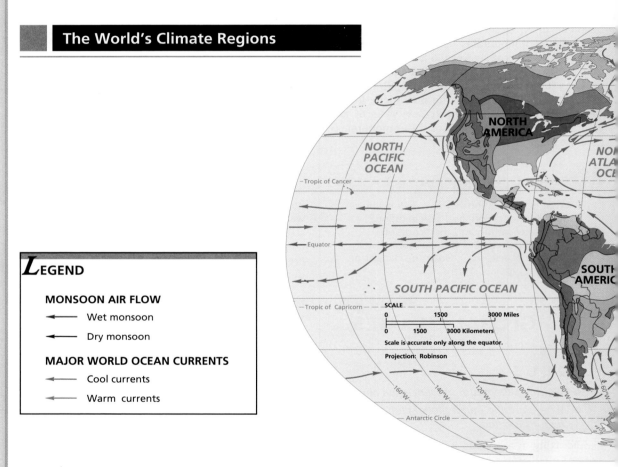

The World's Climate Regions

LEGEND

MONSOON AIR FLOW
- Wet monsoon
- Dry monsoon

MAJOR WORLD OCEAN CURRENTS
- Cool currents
- Warm currents

SCALE
0   1500   3000 Miles
0   1500   3000 Kilometers
Scale is accurate only along the equator.
Projection: Robinson

	Climate	Geographic Distribution	Major Weather Patterns	Vegetation
Low Latitudes	**HUMID TROPICAL**	along Equator; particularly equatorial South America, Zaire Basin in Africa, Southeast Asia	warm and rainy year-round, with rain totaling anywhere from 65 to more than 450 in. (165–1,143 cm) annually; typical temperatures are 90°–95°F (32°–35°C) during the day and 65°–70°F (18°–21°C) at night	tropical rain forest
	TROPICAL SAVANNA	between humid tropics and deserts; tropical regions of Africa, South and Central America, southern and Southeast Asia, Australia	warm all year; distinct rainy and dry seasons; precipitation during the summer of at least 20 in. (51 cm) and in some locations exceeding 150 in. (380 cm); summer temperatures average 90°F (32°C) during the day and 70°F (21°C) at night; typical winter temperatures are 75°–80°F (24°–27°C) during the day and 55°–60°F (13°–16°C) at night	tropical grassland with scattered trees
Dry/Semiarid	**DESERT**	centered along 30° latitude; some middle-latitude deserts in interior of large continents and along western coasts; particularly Saharan Africa, southwest Asia, central and western Australia, southwestern North America	arid; precipitation of less than 10 in. (25 cm) annually; sunny and hot in the tropics and sunny with great temperature ranges in middle latitudes; typical summer temperatures for lower-latitude deserts are 110°–115°F (43°–46°C) during the day and 60°–65°F (16°–18°C) at night, while winter temperatures average 80°F (27°C) during the day and 45°F (7°C) at night; in middle latitudes, the hottest month averages 70°F (24°C)	sparse drought-resistant plants; many barren, rocky, or sandy areas
	STEPPE	generally bordering deserts and interiors of large continents; particularly northern and southern Africa, interior western North America, central and interior Asia and Australia, southern South America	semiarid; about 10–20 in. (25–51 cm) of precipitation annually; hot summers and cooler winters with wide temperature ranges similar to desert temperatures	grassland; few trees
Middle Latitudes	**MEDITERRANEAN**	west coasts in middle latitudes; particularly southern Europe, part of southwest Asia, northwestern Africa, California, southwestern Australia, central Chile, southwestern South Africa	dry, sunny, warm summers and mild, wetter winters; precipitation averages 15–20 in. (38–51 cm) annually; typical temperatures are 75°–80°F (24°–27°C) on summer days; the average winter temperature is 50°F (10°C)	scrub woodland and grassland
	HUMID SUBTROPICAL	east coasts in middle latitudes; particularly southeastern United States, eastern Asia, central southern Europe, southeastern parts of South America, South Africa, and Australia	hot, humid summers and mild, humid winters; precipitation year-round; coastal areas are in the paths of hurricanes and typhoons; precipitation averages 40 in. (102 cm) annually; typical temperatures are 75°–90°F (24°–32°C) in summer and 45°–50°F (7°–10°C) in winter	mixed forest

Locating Climate Regions

Tell students to locate on the World Climate map the region where they live. Have them refer to the chart below the map to identify the climate region for their state. Tell them to read the description of the weather patterns in their climate region. Ask if the description generally conforms to what they observe about weather patterns in their locality.

MAP TRANSPARENCIES

An overhead transparency of the map on these pages is available in *REGIONS OF THE WORLD Map Transparencies*.

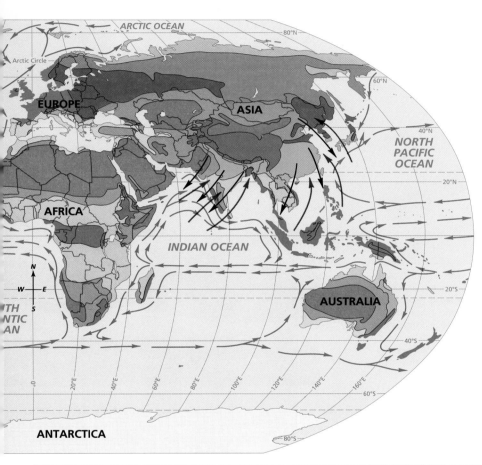

	Climate	Geographic Distribution	Major Weather Patterns	Vegetation
Middle Latitudes	**MARINE WEST COAST**	west coasts in upper-middle latitudes; particularly northwestern Europe and North America, southwestern South America, central southern South Africa, southeastern Australia, New Zealand	cloudy, mild summers and cool, rainy winters; strong ocean influence; precipitation averages 20–60 in. (51–152 cm) annually, with some coastal mountains receiving more than 200 in. (508 cm); average temperature in hottest month usually is between 60°F and 70°F (16°–21°C); average temperature in coolest month usually is above 32°F (0°C)	temperate evergreen forest
	HUMID CONTINENTAL	east coasts and interiors of upper-middle-latitude continents; particularly northeastern North America, northern and eastern Europe, northeastern Asia	four distinct seasons; long, cold winters and short, warm summers; precipitation amounts vary, usually 20–50 in. or more (51–127 cm) annually; average summer temperature is 75°F (24°C); average winter temperature is below freezing	mixed forest
High Latitudes	**SUBARCTIC**	higher latitudes of interior and east coasts of continents; particularly northern parts of North America, Europe, and Asia	extremes of temperature; long, cold winters and short, warm summers; low precipitation amounts all year; precipitation averages 5–15 in. (13–38 cm) in summer; temperatures in warmest month average 60°F (16°C), but can warm to 90°F (32°C); winter temperatures average below 0°F (–18°C)	northern evergreen forest
	TUNDRA	high-latitude coasts; particularly far northern parts of North America, Europe, and Asia, Antarctic Peninsula, subantarctic islands	cold all year; very long, cold winters and very short, cool summers; low precipitation amounts; precipitation average is 5–15 in. (13–38 cm) annually; warmest month averages 40°F (4°C); coolest month averages a little below 0°F (–18°C)	moss, lichens, low shrubs; permafrost bogs in summer
	ICE CAP	polar regions; particularly Antarctica, Greenland, Arctic Basin islands	freezing cold; snow and ice year-round; precipitation averages less than 10 in. (25 cm) annually; average temperatures in warmest month are not higher than freezing	no vegetation
	HIGHLAND	high mountain regions, particularly western parts of North and South America, eastern parts of Asia and Africa, southern and central Europe and Asia	greatly varied temperatures and precipitation amounts over short distances as elevation changes	forest to tundra vegetation, depending on elevation

Ask students to name places in the world they would like to visit. List these places on the board. Ask them what type of weather they expect to find in each, and ask how they know this. Discuss different sources they could use to locate the climate of these places.

Building a Vocabulary

Write the word **monsoon** on the board. Tell students that the word actually has two meanings. It refers to the prolonged, periodic winds that occur in tropical areas during both the wet and dry seasons. However, the word is commonly used to describe the heavy rains in India and Southeast Asia that are brought in from the ocean in the summer months.

Identifying Climate Regions

Have students work in pairs to compare the descriptions of climates on pages 27–32 with the World Climate map on pages 28–29. Write on the board a list of 10 to 20 major world cities, and ask students to identify the climate region in which each is located.

Geographical Sidelight

Climate in an area may vary greatly over time. There are several methods being used to build an indirect picture of climate change. For example, changes in the growth rings of trees accurately chronicle hot and cool periods. In places such as Lapland, where dead trees are often well preserved, the record can extend back further than any living tree. Another method involves boring ice cores from the ice caps of Greenland and Antarctica. Each layer of annual snowfall reveals many clues about Earth's past climate, temperature changes, and even volcanic eruptions. Scientists working in Greenland have taken samples from the ice representing the past 250,000 years.

Climates have shaped the world's landscapes and created regions with distinct characteristics. Much of bitterly cold **Greenland** (above) is covered by an ice cap. A humid-tropical climate nurtures this rain forest in **Suriname, South America** (upper right). Across the same continent, the Andes Mountains of **Peru** (right) provide a climate with conditions that vary by elevation.

Just to the north and south of the humid-tropical climate is the tropical-savanna climate. This wet- and dry-tropical climate is produced by the seasonal change due to the way that the sun's rays strike the areas north and south of the equator. For example, during high-sun season (summer), the sun's rays strike most directly. This high sun increases the temperature and causes low pressure and unstable, rising air that produces heavy rainfall. During the low-sun season (winter), the opposite occurs. As the direct solar rays move to the opposite hemisphere, the subtropical high-pressure zone moves into the area. This causes stable, cool, sinking air and a dry season.

LOCATION **Dry Climates** Though their temperatures may vary greatly, all dry climate regions share aridity, or low annual rainfall. The two types of dry climates are the arid desert climate and the semiarid steppe climate.

Most of the desert climate areas are centered at about 30° north and south of the equator. The dryness is caused by the subtropical high-pressure zone, which brings stable, sinking, dry air all year. Very little rain is produced, and few plants can survive. The largest desert is the Sahara, which stretches across all of northern Africa.

As you read earlier, large deserts also form in the rain shadow to the leeward side of mountains. An example is the Great Basin of the United States. Other deserts are far into the interior of continents, away from moisture-bearing winds. These deserts include the Gobi and Taklimakan deserts of Asia, and the deserts of the dry interior of Australia. These desert regions are blocked from rain by mountain ranges and great distances across large continents. They experience temperature extremes ranging from extremely cold temperatures during winter to hot, scorching temperatures during summer.

Small and very dry deserts are found along the west coasts of continents. Here, the cool ocean currents cause stable conditions, and it may not rain for many years. Examples of dry coastal deserts are found along the west coasts of South America, southwestern Africa, and Mexico.

The semiarid **steppe** climate is a transition area between the arid deserts and the more humid climates. The term *steppe* refers to short-grass vegetation. Areas with steppe climates

Guided Practice

Have students look closely at the World Climate Map on pages 28–29, and help them find desert areas on different continents. Ask them to list factors (such as latitude, interior of continent, orographic effect) that cause these deserts.

Independent Practice

Ask students to write a short description of the climate in their region. Have them use the information available in the textbook to describe the effects that latitude, landforms, elevation, bodies of water, ocean currents, and/or prevailing winds might have on the climate.

Closure

Point out that human beings live in almost all climates of the world, including some of the most extreme. Have them name extreme climates and explain what sorts of adaptations people have had to make to live in these climates.

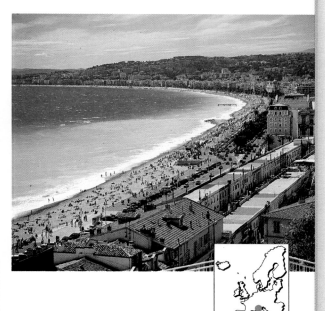

Dunes and hardy vegetation mark a desert-climate region in **West Texas** (below). Sunbathers take advantage of a Mediterranean climate in **Nice, France** (right).

Geographical Sidelight

Traditional housing designs reflect the different climates in which people live. For example, in regions that have much snow and rain, roofs are usually pitched so that rain can run off and snow will not accumulate. In cold areas, much thought is given to insulation and heating. The traditional Inuit igloo uses ice as an effective shelter against the cold. Houses in the tropics are often white, as this color reflects light and absorbs less heat energy from the sun.

receive more moisture than the deserts but less than the more humid areas. Steppe climates generally support grasslands, and trees are rare except along river banks. Today, the natural grasslands in many places have been replaced with fields of food grains. Poor farming practices and overgrazing have turned some natural grasslands into human-made deserts.

LOCATION Middle-Latitude Climates As is the case with low-latitude climates, there are several types of middle-latitude climates. Located between 30° and 40° latitude, the Mediterranean climate is confined mainly to the coastal areas of southern Europe and the west coasts of continents with cool ocean currents. Mediterranean climates usually do not extend far inland beyond mountain ranges. The stable, sinking air of the subtropical high-pressure zone causes long, sunny, dry summers. During the mild winter, however, cool middle-latitude storms enter the region with the westerlies and bring needed rains.

The humid-subtropical climate is much more widespread than the Mediterranean climate. This climate is found on the eastern side of continents with warm ocean currents. The humid-subtropical climate is greatly affected by moist air flowing off the warm ocean waters. Summers are hot and humid. Winters are mild, with occasional frost and some snow. The warm ocean currents cause areas with the humid-subtropical climate to be struck by occasional hurricanes or typhoons.

A climate type influenced mainly by oceans is called the marine-west-coast climate. This climate is generally found on the west coasts of continents in the upper middle latitudes. Temperatures are mild all year. Middle-latitude storms traveling across the oceans in the westerlies bring most of the rainfall to areas with this climate. Winters are foggy, cloudy, and rainy, but summers can be warm and sunny. This climate supports dense evergreen forests in the Northwest United States. The marine-west-coast climate is most widespread in northwestern Europe where the absence of mountain ranges along the coast allows the cool, moist ocean air to spread far into the interior of the continent.

The humid-continental climate is found in latitudes that are subject to periodic invasions of both warm and cold air. This climate has the most changeable weather conditions and experiences four distinct seasons. Its midcontinental locations are responsible for extreme differences between summer and winter temperatures. Because this climate type is situated along the polar front, middle-latitude storms bring rain throughout much of the year and bring snow in winter. Precipitation is heavy enough to support forests.

31

Extension

Give students a choice of writing a short essay or drawing a picture describing or depicting what they consider an ideal climate. Encourage them to include details in their work that make connections to leisure activities, housing, or clothing. Have volunteers read their essays or display their artwork, and lead a discussion about why people like different types of climate.

Section 3 Review
ANSWERS

1. Latitude determines if a place is hot or cold and wet or dry. Places at low latitudes are generally warm with much rainfall. Places at high latitudes are generally cold with much snowfall.

2. Examples should be drawn from the following: low latitude—humid tropical and tropical savanna; dry—desert and steppe; middle latitude—Mediterranean, humid subtropical, marine west coast, and humid continental; high latitude—subarctic, tundra, and polar ice cap.

3. Both are severe and can support little vegetation. Desert climates are very dry and experience extreme temperatures, whereas tundra climates are usually quite cold and have wet areas.

32

Ireland is often called the Emerald Isle for its green countryside. The country is dominated by a marine-west-coast climate.

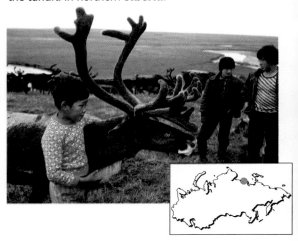

Reindeer and humans have adapted to the cold of the tundra in northern **Siberia**.

LOCATION **High-Latitude Climates** Located in the high latitudes, the subarctic climate is centered above 50° north latitude. This climate has long, dark, cold winters, with temperatures staying well below freezing for half of the year. During the short summers, however, very warm temperatures can occur. The subarctic climate has the greatest annual temperature ranges in the world. Though severe, the subarctic climate supports vast evergreen forests.

The subarctic climate region is very large, extending across northern North America, Europe, and Asia. The Southern Hemisphere lacks this climate type because there is no land at these latitudes.

Another climate with a long winter is the tundra climate. Temperatures are above freezing only during the short summers. The tundra climate takes its name from the only vegetation that can survive there. Tundra vegetation is made up of small, hardy plants such as mosses, lichens (LIE-kuhnz), herbs, and low shrubs. The climate is so severe that no trees grow.

Water below the tundra surface remains frozen throughout the year. This condition is called **permafrost**. Although the melting winter snow creates swamps and bogs on the surface during the summer, permafrost makes it difficult for the water to seep into the ground. These wet areas do, however, support great numbers of insects and birds during the short summer.

The polar ice-cap climate has cold temperatures all year, and snowfall is likely year-round.

Life is almost impossible in polar ice-cap climates. Only animals and plants that can live in the icy polar seas survive here.

LOCATION **Highland Climates** As you learned earlier, an increase in elevation causes a decrease in temperature. Highland climates are determined by elevation rather than by latitude. Mountain areas of the world can have a variety of climate types in a very small area.

The lowest elevations of a mountain will have a climate and vegetation similar to that of the surrounding area. As you climb higher up the mountain, the climate conditions change. Temperatures and air pressure are lower. The cooler temperatures limit the vegetation that can grow, especially trees. At the highest elevations, climate conditions are similar to those of the polar ice-cap climate. Here, temperatures remain below freezing, and ice and snow are always present.

SECTION REVIEW

1. How does the latitude of a place influence its climate?
2. Give an example of a climate type found in each of the following climate regions: low-latitude, dry, middle-latitude, and high-latitude.
3. **Critical Thinking** **Compare and contrast** desert climates and tundra climates.

CHAPTER 3 CHECK
ANSWERS

Building a Vocabulary

1. Weather is the condition of the atmosphere at a given place and time; climate is an average of the weather conditions of an area over a long period.
2. It is the measurement of the force exerted by air. They create the winds and the ocean currents.
3. Trade winds, westerlies, polar winds; calm areas along the equator with no prevailing winds
4. Frozen water below the surface; in tundra regions

Recalling and Reviewing

1. They help distribute heat and cold across the earth.
2. Most water vapor comes from the oceans. As temperature rises, the air can hold more water vapor. When the temperature falls, water vapor condenses and can be seen in the clouds, fog, and precipitation.
3. Low latitude: humid tropical and tropical savanna; middle latitude: Mediterranean, humid subtropical, marine west coast, and humid continental; high latitude: subarctic, tundra, and polar ice cap
4. They are about 30° north and south of the equator. The subtropical high-pressure zone brings stable, sinking, dry air year-round.

CHAPTER **3** CHECK

Space shuttle photo of large thunderstorm east of **Madagascar**

Reviewing the Main Ideas

1. The earth maintains an energy balance by absorbing and releasing heat from the sun. Differences in the heating of the earth's surface create differences in air pressure. These differences create the global wind belts and global ocean currents that move heat and cold across the earth, helping the earth maintain its energy balance.
2. Water vapor is an important gas in the atmosphere. Condensation occurs when water vapor changes from a gas to a liquid. If the condensed droplets become large enough, they fall to the earth as precipitation. Storms occur when the energy stored in water vapor is released.
3. Generally, the global climate regions are classified by differences in temperature and precipitation. Major factors that influence climate include latitude, elevation, and location on land.

Building a Vocabulary

1. What is the difference between weather and climate?
2. Define *air pressure*. How do differences in air pressure affect climate?

3. What are the prevailing winds called at the low, middle, and high latitudes? What are the doldrums?
4. What is permafrost, and where is it found?

Recalling and Reviewing

1. What important role do the oceans play in the global heat or energy exchange system?
2. Where does water vapor come from? How does its form change when the air temperature rises or falls?
3. What climate types are found in the low, middle, and high latitudes?
4. Where are most of the desert climate areas located? Why?

Thinking Critically

1. Imagine that the year is 2050 and you live in a highly technological society. What technologies might be in place to protect you from the climate?
2. Why do you think that the humid-continental climate is not found in the Southern Hemisphere?

Learning About Your Local Geography

Cooperative Project

Watch a local television news show. Take careful notes of the type of information presented by the weatherperson. Then, track the weather in your area for one week. With your group, prepare a weather report using daily figures, charts, and graphs. Present your weather report to the rest of the class.

Using the Five Themes of Geography

An aspect of **human-environment interaction** is the way in which people are affected by the climate in which they live. Make a list of as many factors as possible that describe how the climate affects your daily life. Next to each factor, describe how the use of technology has changed the impact of climate on your life.

Thinking Critically

1. Answers will vary but might include underground cities, totally self-contained environments that utilize connecting walkways, and space suits that do not permit climate conditions to penetrate.
2. It requires a continental location in the middle latitudes, and there are no large land masses in the Southern Hemisphere that meet these requirements.

Learning About Your Local Geography

Weather reports will vary but should include details of the area's weather and should utilize charts and graphs.

Using the Five Themes of Geography

Answers will vary. An example might be heat and humidity, with air-conditioning written next to that factor.

33

CHAPTER 4 The Water Planet
Planning Guide

TEXTBOOK	TEACHING RESOURCES	OTHER AVAILABLE MATERIALS
Section		
1. The Geography of Water Pages 35–37	Reteaching Worksheet 4.1 Challenge/Enrichment Worksheet 4	*REGIONS OF THE WORLD Map Transparencies:* The Hydrologic Cycle *Study Guide for Spanish Speakers* *Lecture Notes*, Chapter 4
2. Water Resources Pages 37–40	Reteaching Worksheet 4.2 Graphic Organizer 4 Skills Worksheet 4 Critical Thinking Worksheet 4	*Study Guide for Spanish Speakers* *Lecture Notes*, Chapter 4
Chapter Review Page 41	Chapter 4 Test Chapter 4 Portfolio Assessment See also *Alternative Assessment Forms*	*Test Generator,* Chapter 4

CHAPTER RESOURCES

Lecture Notes, Chapter 4
Teaching Resources
   Chapter 4 Graphic Organizer
   Chapter 4 Skills Worksheet
   Chapter 4 Critical Thinking Worksheet
   Chapter 4 Challenge/Enrichment Worksheet

REVIEW AND ASSESSMENT

Chapter 4 Check, page 41
Chapter 4 Test, *Teaching Resources*
Chapter 4 Portfolio Assessment, *Teaching Resources*
Test Generator, Chapter 4

ADDITIONAL RESOURCES

Books for Teachers

Bascom, Willard. *Waves and Beaches: The Dynamics of the Ocean Surface.* Arlington, TX: Anchor Books, 1980.

Carson, Rachel. *The Sea Around Us.* New York: Oxford University Press, 1961.

Goldin, Augusta. *Water: Too Much, Too Little, Too Polluted.* Orlando, FL: Harcourt Brace Jovanovich, 1983.

Hundley, Norris. *The Great Thirst: Californians and Water.* Berkeley: University of California Press, 1992.

Books for Students

Dorros, Arthur. *Follow the Water from Brook to Ocean.* New York: HarperCollins, 1991.

Realms of the Sea. Washington, DC: National Geographic Society, 1991.

Simon, Seymour. *Oceans.* New York: Morrow Publishing Co., 1990.

Whipple, A.B.C. *Restless Oceans.* Morristown, NJ: Silver Burdett, 1983.

The World Today *Videodisc Program*

A variety of still images, short videos, and activities are available for you to use as you teach this chapter. *Lecture Notes* include bar codes for specific photographs.

Multimedia Resources

The Geography Tutor: Earth's Physical Features. (VHS, 18 minutes) Tell Me Why, 730 Washington Street, Marina Del Ray, CA 90292

Lagoons. (VHS, 28 minutes) Films for the Humanities & Sciences, P.O. Box 2053, Princeton, NJ 08543-2053

Picture Atlas of the World. (interactive maps, photo bank, animation, audio on CD-ROM) National Geographic Society, Washington, DC.

Planet Earth: The Blue Planet. (Macintosh, IBM-PC software, laserdisk) Coronet MTI, 108 Wilmot Road, Deerfield, IL 60015

Water Power. (16 mm/VHS, 16 minutes) Lucerne Media, 37 Ground Pine Road, Morris Plains, NJ 07950.

MEETING INDIVIDUAL NEEDS

Limited English Proficiency (LEP)

Pair LEP students with students who are proficient in English. Have the pairs make a word bank and fill-in-the-blank sentences using the words and terms in the Chapter 4 Geography Dictionary on page 34. Explain to students that the sentences should contain the definitions of the words or terms. Have the pairs exchange and complete the fill-in-the-blank sentences.

Learners Having Difficulty

Help students create an outline of Chapter 4. Point out that the title of the chapter can be used for the title of the outline; section titles can be used for items designated with Roman numerals; and subsection titles can be used for items designated with uppercase letters. Likewise, explain that the topic sentences in the subsection paragraphs can be used for items designated with Arabic numerals; and details supporting the topic sentence can be used for items designated with lowercase letters.

Auditory Learners

Invite a guest speaker who is knowledgeable about water resources and conservation issues in your community. Have students prepare questions.

Visual Learners

Have students work in small groups to create a drawing or poster showing how people use water. Then have students develop questions based on the information in their illustrations, and exchange the illustrations and questions with another group. Groups will then answer the questions about the illustration they received in the exchange. Display the drawings or posters in the classroom.

Tactile/Kinesthetic Learners

Have students research a major river system in the world—such as the Mississippi, the Amazon, or the Ganges—to create a map. Students' maps should plot and label the following parts of the river system: principal river, tributaries, headwaters, watershed, and estuary. Display students' maps.

Gifted Learners

Have interested students research a specific wetlands area, such as the Everglades in Florida. Ask them to draw a map showing its location. Then have them write a report about the wetlands area they studied, explaining its importance and describing the kinds of wildlife that live there. Students may wish to illustrate their reports.

MULTICULTURAL FOCUS

Point out to students that an aspect of culture is related to human-environment interaction. The farming way of life of the central United States relies heavily on the availability of water. The High Plains region of western Texas, Oklahoma, Kansas, and Nebraska, which suffered from one of the most severe droughts in the nation's history during the 1930s, is now one of America's most productive agricultural areas, largely because of irrigation.

In the 1950s, farmers began tapping into the Ogallala Aquifer, an underground water reserve the size of Lake Huron lying beneath the Great Plains. Water from the aquifer is brought to the surface by pumps and distributed over the fields by sprinkler systems.

Although the Ogallala is estimated to hold a quadrillion gallons of water, the heavy pumping has started to deplete the aquifer in places. Recent efforts to measure the amount of water that individual pumps remove from the aquifer and to limit the number of wells that can be drilled have touched a nerve in local farmers, who depend on the Ogallala for their livelihood. Some believe that their property rights include unrestricted use of the water below their land, although the aquifer is a shared resource.

Point out to students that every time they turn on a tap, they are using a shared resource. Ask students to imagine a situation in which individual water use is rationed all over the world.

CHAPTER 4
THE WATER PLANET
Section 1: The Geography of Water
Section 2: Water Resources

OVERVIEW

In Chapter 4, students will learn about water, a natural resource essential for life on Earth.

Section 1 focuses on the importance of water to life and the necessity of preserving the world's water resources. It describes the ways in which people put water to use, its three important characteristics, and its uneven geographic distribution. Students will learn that available freshwater resources make up less than one percent of Earth's water. The section also explains the process of the hydrologic cycle.

Section 2 discusses the earth's water resources and describes the various types of surface water, the importance of groundwater, and the characteristics of seawater.

About the Author

Ruskin Bond (1934–) is the author of numerous books for children and young adults. His novels and poetry reflect affection for his home country of India and express concern about environmental changes occurring there.

Primary Source

Description of Change: excerpted
Rationale: excerpted to focus on main idea

CHAPTER 4

The Water Planet

66 I never cease to wonder at the tenacity of water—its ability to make its way through various strata of rock, zigzagging, backtracking, finding space, cunningly discovering faults and fissures in the mountain, and sometimes traveling underground for great distances before emerging into the open. Of course, there's no stopping water. For no matter how tiny that little trickle, it has to go somewhere! 99

Ruskin Bond

If you looked at Earth from a spaceship, you might wonder why we do not call our planet *Water*. Almost three-fourths of the earth's surface is covered with water. Earth is the only planet in the solar system that has water, and it is sometimes called the "water planet." The oceans, water frozen as ice, lakes and rivers and other surface water, water found underground, water vapor in the atmosphere, and water in all living things make up the hydrosphere.

Water is the most precious of Earth's natural resources. It is essential for all known life, especially for cell growth. All living things, plant and animal, are made up mainly of water. In fact, your body is about 70 percent water.

Though plentiful on a global scale, fresh water is a scarce resource throughout much of the world. Lack of water limits a country's ability to develop economically and to produce enough food to feed its people. In poor countries, many people have access only to unclean, unsafe water. About 30 percent of the world's population lives in areas with severe water problems.

River in the Carson National Forest, **New Mexico**

GEOGRAPHY DICTIONARY

irrigation
industrialization
hydroelectricity
hydrologic cycle
transpiration
evapotranspiration
headwaters
tributary
watershed
estuary
wetlands
groundwater
water table
aquifer
continental shelf

Using the Quotation

Ask students if they know what this saying means: "Water always finds its level." *(As long as there is an elevation difference and a path, water will flow.)* Ask how this relates to Ruskin Bond's quote, and discuss the idea that water "has to go somewhere."

Using the Geography Dictionary

Write the root word *hydro-* on the board. Tell students that it comes from the Greek word for "water." Point out that *all* of the words in the Geography Dictionary have something to do with water. Challenge students to suggest the link between **industrialization** and water. *(Water is a necessity for most industry.)*

SECTION 1
OBJECTIVES

The objectives reflect the Focus questions on this page and are assessed in the review on page 37.

1. Explain the necessity of preserving our water resources.
2. Describe the three characteristics of water that make it unique.
3. Explain the process of the hydrologic cycle.

The Grand Coulee Dam towers above the town of **Coulee Dam, Washington**. This dam is part of a system of U.S. and Canadian dams on the Columbia River that provides water for hydroelectric power and irrigation. The Grand Coulee Dam produces more electricity than any other hydroelectric dam in the United States.

Geography and History

Irrigation is not a new practice. Farmers in India, Iran, and Afghanistan have irrigated their fields for thousands of years. Iran, for example, has an irrigation system that is more than 2,500 years old. The system, which is still used today, consists of a network of underground tunnels called qanats. *The tunnels carry water from wells at the bases of mountains to fields two or three miles away.*

THE GEOGRAPHY OF WATER

FOCUS

● *Why is water essential to life on Earth?*

● *What characteristics make water a unique substance?*

● *How does water circulate among the parts of the hydrosphere?*

People and Water Water is one of our most important resources. Humans cannot survive for more than about a week without it. How much water do you use each day in your home? In modern society, we often take our water supply for granted. While you may simply turn a faucet to get a glass of water, people in other parts of the world may spend hours each day walking to a community well or stream just for a pail of water.

Water is essential for agriculture, and in dry climates, **irrigation** is necessary for growing crops. Irrigation is the watering of land through pipes, ditches, or canals. In the United States, few crops could be grown in areas of Texas, California, and Arizona if farmers could not get irrigation water. Yet, these states are among the nation's biggest agricultural producers. Millions of dollars are spent each year moving water to these farming regions. In countries with little technological development, people are dependent on rains and careful water conservation to irrigate their fields.

Industry also could not function without water. Water is an ingredient in the manufacture of many products and serves as a coolant and lubricant for many machines. Countries that are beginning to industrialize can do so only where water is available. **Industrialization** is the large-scale manufacturing of goods using machine power.

People and industry also rely on water as a valuable power source. **Hydroelectricity**, a major power source, is produced when water stored behind dams drives engines to produce electric power. Our oceans, rivers, and lakes also provide a world transportation network. To transport bulk cargo such as oil, grain, or automobiles, shipping by water is a preferred method because of low cost.

As the world becomes more densely populated, our water resources become more vulnerable. Pollution from city, farm, and industrial wastes is dumped into our bodies of water. This affects plants and animals, and our sources of fish and other seafood are threatened. In some areas, we have overused our water supplies, and water is becoming scarce. It is important to remember that our quality of life depends upon water, and we must learn to conserve and keep clean our most vital resource.

Characteristics of Water Water is a unique substance with three important characteristics. First, it is the only substance that can exist as a liquid, solid, or gas within the earth's temperature range. This means that water can be

Divide the class into small groups, and ask them to brainstorm a list of all the places where water is found on Earth. One student in each group should list the group's ideas on a sheet of paper. Write the ideas on the board as each recorder reads from the list. Point out any water on Earth that students fail to mention. Ask why water is important to our planet.

Creating a Flowchart

Hydrologic Cycle Check students' understanding of the terms **hydrologic cycle, transpiration,** and **evapotranspiration**. Call on students to describe the hydrosphere, which was defined in Chapter 2. Then have students describe the steps of the hydrologic cycle while you illustrate them on the board using a flowchart.

Guided Practice

As you write the three characteristics of water on the board, have students review the subsection "Geographic Distribution." Then call on volunteers to identify the characteristic that best relates to information presented in the subsection. (*Water exists as a liquid, a solid, and a gas.*)

MAP TRANSPARENCIES

An overhead transparency of the illustration on this page is available in *REGIONS OF THE WORLD Map Transparencies.*

transported and stored nearly anywhere on Earth. Second, water has the ability to dissolve almost anything over time, including the hardest rocks. This allows mineral matter to be dissolved and moved by water in the oceans, rivers and streams, and underground.

Third, water heats and cools very slowly compared to other Earth materials. This characteristic enables water to moderate Earth's air and surface temperatures. Without water, Earth's temperatures probably would be too extreme to support life.

Geographic Distribution

Although water is abundant on Earth, it is not evenly distributed. Because we depend on water for survival, all people must live in a place where they have some access to water. As a result, access to and control of water has been the cause of many conflicts throughout world history.

The majority of the hydrosphere is in the oceans, which contain roughly 97 percent of Earth's water. Of the remaining 3 percent, more than 2 percent is found frozen in the polar and mountain ice sheets. Thus, more than 99 percent of the earth's hydrosphere is either salty or frozen.

As a result, less than one percent of Earth's water is available as a freshwater resource. This very small amount is found underground or in lakes, rivers, and streams. Even smaller, but very essential, amounts are found in the atmosphere and biosphere.

MOVEMENT The Hydrologic Cycle The amount of water on Earth is constant, yet it is always changing form and location. The circulation of water among parts of the hydrosphere is called the **hydrologic cycle**. Water is cycled endlessly through the oceans, ice sheets, lithosphere, atmosphere, and all living things in the biosphere. (See the diagram on this page.)

The hydrologic cycle is driven by solar energy. Most water enters the atmosphere after being heated by the sun's energy and evaporating from the ocean. It becomes water vapor and rises into the air. As it rises, it cools and condenses, changing from a gas to tiny liquid droplets. These droplets join to form clouds, and if the droplets become heavy enough, they will fall back onto the ocean as precipitation. This completes the cycle of evaporation, condensation, and precipitation.

The hydrologic cycle becomes more complex, however, when precipitation falls on the continents and islands. Some precipitation falls as snow on high mountains or polar areas and accumulates as ice. Some flows off the land through rivers and streams, possibly to be stored in lakes, and eventually returns to the ocean.

The Hydrologic Cycle

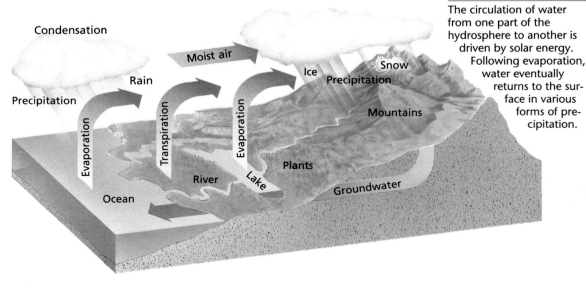

The circulation of water from one part of the hydrosphere to another is driven by solar energy. Following evaporation, water eventually returns to the surface in various forms of precipitation.

36

Independent Practice

Have students create an individual water-use log for one day. They should note the time of day and the way they use water. After completing the log, students should write a paragraph with the following topic sentence: Water is one of the most valuable resources in our daily lives.

Closure

Have students answer the Focus questions on page 35.

ASSESSMENT

Assign the Section 1 Review questions on this page.

Reteaching

Assign Section 1 of Reteaching Worksheet 4.

SECTION 2
OBJECTIVES

The objectives reflect the Focus questions on this page and are assessed in the review on page 40.

1. Distinguish among the earth's various bodies of water.

2. Discuss the importance of groundwater, and identify its source.

3. Identify the major oceans and characteristics of seawater.

The ground absorbs some water by taking it into the air spaces in the soil and rock, where it can be used by plants. Here, the water moves underground to the ocean or is returned to the atmosphere by evaporation from the land or by a process called **transpiration**. Transpiration occurs when plants give off water vapor through their leaves. The evaporation of water from the ground combined with the transpiration of water by the plants is called **evapotranspiration**. This is a measure of the total loss of water from the land to the atmosphere.

Although water is constantly being renewed within the hydrologic cycle, its rate of renewal is very slow. The best technology in the world cannot increase the rate at which water is renewed in the global water cycle.

SECTION REVIEW

1. How is water essential to agriculture and industry?

2. What is the importance of each of the three characteristics of water?

3. Critical Thinking **Analyze** cause-and-effect relationships in the hydrologic cycle.

WATER RESOURCES

FOCUS

● *What are the various types of surface water found on Earth?*

● *What is groundwater, and why is it important?*

● *What are the characteristics of seawater?*

Surface Water As precipitation falls on the continents and islands, it flows down the slopes of hills and mountains toward the lowlands and coasts. The first and smallest streams to form from this runoff are called **headwaters**. As these headwaters join, they form larger streams, and farther downstream, they eventually form rivers. Any smaller stream or river that flows into a larger stream or river is a **tributary**. In the United States, major rivers such as the Ohio and Missouri are tributaries of the Mississippi River.

The **Delmarva Peninsula**, on which **Delaware** and part of **Maryland** and **Virginia** sit, is drained by a number of rivers, including the Choptank River (center) and the Nanticoke River (bottom center). Both rivers empty into Chesapeake Bay, to the left of the peninsula. Delaware Bay is to the upper right of the peninsula.

An area of land that is drained by a river and its many tributaries is a **watershed**, or drainage basin. Very large rivers, such as the Amazon in South America or the Mississippi in the United States, may flow for thousands of miles and be joined by hundreds of tributaries. Their drainage basins might extend over 1 million square miles (2.6 million sq. km) and drain a major portion of a continent. Rivers are a valuable water resource, providing water for agriculture, electricity, transportation, and cities. When a river is polluted from the clearing of land or by the disposal of waste materials, the effects may be felt along the entire river system.

Where rivers meet an inlet of the sea, an **estuary** may form. In this semi-enclosed coastal body of water, seawater and fresh water mix. Estuaries are particularly important because they are rich in fish and shellfish. Good examples of large estuaries are Chesapeake Bay on the Atlantic coast of the United States and Puget Sound on the Pacific coast. As rivers carry fresh water to the coasts, they also bring essential minerals that were removed from the land. These minerals contain nutrients necessary to ocean life.

Section 1 Review
ANSWERS

1. Agriculture: for growing crops; industry: as a coolant and lubricant for machines, as a source of power, for shipping

2. It can be transported and stored almost anywhere on Earth, can dissolve rock and mineral matter, and can moderate temperatures.

3. Evaporation, with cooling, causes condensation, which in turn causes precipitation. Then precipitation ultimately evaporates and starts the cycle over again.

Motivator

Ask students if they are familiar with the phrase "the seven seas." Explain that it refers generally to the navigable waters of the world. Use a globe to point out the continuous world ocean, noting the oceans and seas that are part of it. Ask students if they can name seven "seas." *(North and South Atlantic, Indian, North and South Pacific, Arctic, Antarctic)*

Building a Vocabulary

Write the terms **headwaters, watershed, wetlands, groundwater,** and **water table** on the board. Most of the terms are straightforward. Point out that *-shed* in *watershed* comes from an old sense of the word meaning "divide," as in water that divides the land. Call on students to explain how the name of each term has an obvious connection to its meaning.

Cause and Effect

Remind students that oceans hold more than 97 percent of the earth's water. Have them identify the main characteristic of seawater that prevents the oceans from being a resource for drinking water or for agriculture. *(salt content)* Point out that the uneven distribution of usable water worldwide has social, political, and economic significance.

Geography and the Environment

In the United States, over 300,000 acres (121,500 ha) of wetlands disappear each year, becoming drained and filled for agriculture, industry, and other development. The U.S. Fish and Wildlife Service reported that between the mid-1970s and the mid-1980s, 2.6 million acres (1,053,000 ha) were lost. Put together, this total area would be twice the size of the state of Delaware.

Geographical Sidelight

Some groundwater comes to the surface in the form of springs. Springwater rich in dissolved minerals has always been highly prized by people. Many famous spas have arisen around mineral springs, which have long been credited with curative powers. Bottled mineral waters and springwaters, a mainstay in Europe for a long time, have become popular in the United States in recent years.

These caribou are among the many species that move through and live in the wetlands of **Alaska**. The advantages and disadvantages of economic development in the wetlands of Alaska, as elsewhere, are sharply debated.

When runoff is slowed or stopped inland and water fills a depression on the land surface, a lake forms. Most lakes are freshwater, but some, such as Great Salt Lake in Utah, are salty. Some salty lakes were incorrectly named seas, such as the Caspian Sea between Europe and Asia and the Dead Sea in Southwest Asia. Unlike seas, however, lakes are totally surrounded by land, are usually not at sea level, and do not exchange water with the oceans.

Many lakes, such as the Great Lakes along the United States–Canadian border, were formed by glaciers. The Great Lakes is the largest freshwater lake system in the world. Like the United States and Canada, Russia, Sweden, and Finland have thousands of lakes left from the last ice age. Other lakes, such as Crater Lake in Oregon, have formed in erupted volcanoes. Lakes are an important resource for our water supply, fish, and recreation. In addition, large lakes have a moderating effect on the local climate, allowing for a long fruit-growing season in some areas.

Other important surface waters are **wetlands**. These are land areas that become flooded for at least part of the year. Some wetlands are continuously flooded, while others change with the seasons. Wetlands contain some of the earth's most productive land. Wetlands include coastal marshes, wooded swamps, tundra bogs, and river bottomlands. Wetlands support huge populations of fish, shellfish, and birds, and also filter pollutants from water. One of the best-known wetlands in the United States is the Everglades in southern Florida. Almost half of Alaska is wetlands, and that state holds about 70 percent of the wetlands remaining in the United States. More than half of U.S. wetlands have been destroyed by draining, paving, and filling for farmland, housing, ports, and industry.

Groundwater While the water resources in rivers, lakes, and wetlands are visible to us, there

Water sometimes exists underground in regions where surface water is scarce. How do people gain access to groundwater?

Groundwater

Precipitation

Runoff

Well

Water table

Saturated zone

Groundwater

Aquifer

Guided Practice

Using a wall map, point out the major bodies of water included in the lesson: headwaters, rivers, tributaries, estuaries, lakes, seas, and oceans. Ask small groups of students to use an appropriate map to identify these bodies of water in or near an assigned state, country, or continent.

Independent Practice

Have students in teams create a chart about sources of groundwater contamination. Categories in the chart might include commercial sources (underground gasoline tanks, industrial wastes, landfills), household sources (lawn treatments, garden pesticides, used motor oil), and consequences of each.

Closure

Review with students the differences between surface water, groundwater, and oceans. Call on volunteers to explain the importance of each type of water source.

is a large freshwater resource beneath the earth's surface. This **groundwater** is found in tiny spaces between soil and rock grains. Groundwater is an important resource, especially in dry climates where surface water is scarce.

The major source of groundwater is precipitation. Water from rain or melting snow sinks into the ground and slowly seeps downward. The water stops when there is no more space between the rock grains or when all the spaces are saturated with water. The top of this saturated zone is called the **water table**. To tap this groundwater, we would have to drill a well to a depth below the water table.

The depth of the water table varies. In areas with a wet climate, the water table will be near the surface. In desert areas, it may be hundreds of feet below the surface or it may not even exist. The water table follows the land and will rise under hills and drop under valleys. It will vary from year to year as rainfall patterns change. If a region suffers a drought, the water table falls. Too many wells in a region can also lower the water table.

Like rivers, groundwater can travel long distances but at a much slower rate. Generally, groundwater flows through a rock layer called an **aquifer**. In an aquifer, water can be stored in and carried through porous rock and then released for use. Aquifers may be quite large and may measure hundreds of miles across. The Ogallala Aquifer stretches from Texas to Nebraska and is a major source of water for the irrigation of farms in the Great Plains subregion.

The Oceans As you have learned, most of the earth's water is not found on land. Oceans, the largest bodies of salt water, hold more than 97 percent of the earth's water and cover about 71 percent of the planet's surface. Actually, oceans are one continuous global body of water surrounding all the continents. For geographic identification, this global ocean is divided into four separate oceans: the Pacific, the Atlantic, the Indian, and the Arctic.

The Pacific Ocean is the largest geographic feature on Earth and covers an area larger than all the continents combined. The Atlantic and Indian oceans are each about half the size of the

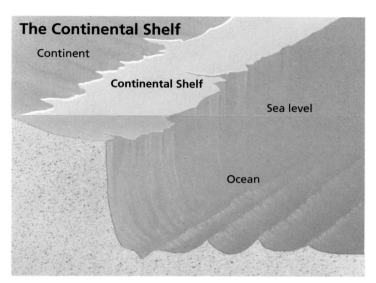

The Continental Shelf

Continent

Continental Shelf

Sea level

Ocean

The continental shelf is the shallowest part of the ocean and slopes gently away from the continents. Much of the mud and sand from rivers is deposited onto the continental shelf.

Pacific Ocean. The Arctic Ocean is much smaller than the other three oceans and is located near the North Pole.

Seas are smaller saltwater bodies connected to the oceans. Examples include the Mediterranean Sea, the Coral Sea, and the Caribbean Sea. Saltwater bodies near coasts that extend into land are sometimes called gulfs, such as the Gulf of Mexico, the Gulf of Alaska, and the Persian Gulf in Southwest Asia. A few large saltwater bays, such as Asia's Bay of Bengal and Canada's Hudson Bay, are actually seas, not bays. Bays are inlets of seas.

The oceans average about 13,000 feet (3,962 m) in depth. There is a variety of features at the ocean bottom, and the depth varies greatly. Flat plains are found in the deep ocean basins, while ridges rise high above them. The greatest ocean depth is in the Mariana Trench, located in the North Pacific Ocean. Its dark, cold waters reach more than 36,000 feet (10,973 m) below sea level.

The most shallow part of the ocean is the **continental shelf**, which slopes gently from the continents. Some parts of the continental shelf were actually land thousands of years ago during the last ice age. As the ice melted, the sea level rose, flooding the continental shelf. If the massive ice sheets of Antarctica and Greenland were to completely melt, low areas far inland would flood.

COOPERATIVE LEARNING

Organize the class into groups of four or five. Have each group design a poster that encourages people to conserve water, with a focus on the concerns of their particular geographical area. The group should brainstorm conservation ideas and use resources such as *50 Simple Things You Can Do to Save the Earth*. Individual members can do the research, design the poster, and collect or draw the illustrations.

GLOBAL PERSPECTIVES

Major oil spills have affected oceans and coastal areas worldwide. Two examples are the tanker *Aegean Sea*, which ran aground near La Coruña, Spain, in 1992, and the *Braer*, which hit the coast of Britain's Shetland Islands in 1993. Have students work in groups to investigate the damage caused by a particular spill, focusing on the international cleanup efforts.

39

Extension

Have interested students research and report to the class on how the oil spill in the Persian Gulf during the 1991 Gulf War threatened the desalinization plants in the area. Then have them discuss the problems and challenges associated with desalinization of water on a large scale.

At the left is a picture of salt flats in **Death Valley**. All salt comes from seas, salt lakes, and other bodies of water composed of brine (salty water). The salt in Death Valley is the remains of an evaporated salt lake. Salt can be removed from sea water at desalinization plants, such as the one in **Saudi Arabia** shown below. Desalinization facilities have been built in many parts of the world, especially in areas where only sea water is available.

Seawater and Its Characteristics

Because water has the unique ability to dissolve almost anything, it contains the materials it dissolves. Thus, seawater contains every element known on Earth, even gold. Most of these dissolved materials are in such small quantities that it is not practical to "mine" seawater. You may have wondered why you taste salt when you swim in the ocean. Salts dissolve easily and are the most common material found in seawater. In fact, about 3.5 percent of seawater is dissolved salts. Because of the salt content, we cannot use the oceans as a resource for drinking water or for agriculture. It is possible to remove salt from seawater, but the process is expensive.

Because water is slow to heat up and cool down, the oceans do not have the temperature extremes of land. On a daily basis, the temperature of ocean water changes less than one degree, while the temperature of land between day and night will vary greatly. The seasonal changes in the ocean also are less than the land's seasonal changes, and the oceans help prevent the climates along the coasts from being too warm or too cool. Because sunlight does not penetrate very deeply into the oceans, only the upper portion of the water is heated.

Like the atmosphere, the ocean is always moving and circulating. The winds push the ocean currents and cause waves, and the waves and currents mix the surface waters. The gravitational pull of the moon and sun causes the tides to rise and fall, which in turn causes many currents, especially in the coastal areas. Even the deepest ocean waters move slowly as a result of varying temperatures or the amount of salt in the water. Though the oceans appear huge, their life-producing regions are limited mainly to the shallow near-shore areas, as most seawater is deep, dark, and cold. Much of the world ocean remains unexplored.

SECTION REVIEW

1. What environmental benefits are provided by rivers, estuaries, lakes, and wetlands?
2. In what climate type is groundwater especially important? Why?
3. **Critical Thinking** **Explain** why the Arctic Ocean freezes but the Pacific Ocean does not.

CHAPTER 4 CHECK
ANSWERS

Building a Vocabulary

1. The watering of land through pipes, ditches, or canals; Arizona, California, and Texas
2. Water stored behind dams drives engines that produce electricity.
3. Hydrologic cycle
4. Headwaters: first streams to form from precipitation on hills and mountains; tributaries: small streams that join larger streams; estuaries: where rivers meet seawater; wetlands: areas that are flooded at least part of the year
5. A drainage basin or an area of land that is drained by a river and its many tributaries

Recalling and Reviewing

1. Water is used for irrigation in agriculture, as a source of power, and as a means of shipping in industry.
2. They carry mineral nutrients that are essential to marine life.
3. Precipitation
4. Pacific, Atlantic, Indian, Arctic

CHAPTER **4** CHECK

Iceberg near **Antarctica**

Reviewing the Main Ideas

1. Water is one of the most important natural resources on Earth. Life as we know it depends on water. Besides its personal uses, water is used for irrigation, as a power source, and in industry.
2. Water has three important characteristics that make it a unique substance: it can exist as a solid, liquid, or gas; it can dissolve almost anything; and it heats and cools very slowly.
3. Water resources on Earth include surface water, such as rivers, lakes, and wetlands; groundwater found beneath the surface of the earth; and seawater.

Building a Vocabulary

1. Define *irrigation*. Name three states in the United States that depend upon irrigation.
2. How is hydroelectricity produced?
3. What is the term used to describe the circulation of water among parts of the hydrosphere?
4. What are the differences among headwaters, tributaries, estuaries, and wetlands?
5. What is a watershed?

Recalling and Reviewing

1. Explain the importance of water to agriculture and to industry.
2. Why are rivers and estuaries important to ocean life?
3. What is the major source of groundwater?
4. Into what four major oceans do geographers divide the global ocean?

Thinking Critically

1. Describe three ways in which life would change if the world's water resources were reduced.
2. Design a poster promoting the protection of wetlands.

Learning About Your Local Geography

Individual Project

Conduct research to find out about water resources in your area. Where does the water in your home come from? Who monitors the water quality? What is the depth of the water table in your area? Is irrigation important? Is there a hydroelectric plant nearby? Organize your information in a written report or in a visual presentation. You may want to include your work in your individual portfolio.

Using the Five Themes of Geography

Write a paragraph about how the geographic themes of **place, human-environment interaction,** and **movement** apply to irrigation. What place characteristics make irrigation necessary, and how does irrigation change places? How do humans change the environment with irrigation? How does water move in irrigation? From where does it come, and to where does it go?

Depoe Bay, Oregon

Thinking Critically

1. Possible answers include reduced food production, limited electrical power, health and pollution problems, and atmospheric changes.
2. Posters will vary but should point out that wetlands contain some of the most productive land on the earth and also provide a home to wildlife. Posters might include figures on the rate at which wetlands are being destroyed.

Learning About Your Local Geography

Reports or illustrations should provide information about water resources in the community.

Using the Five Themes of Geography

Arid places require irrigation for crops to grow. Irrigation can make a dry area agriculturally productive. Humans change the environment by installing pipes and creating ditches and canals to channel water away from its natural course. Irrigation moves water from its source—surface water or groundwater—to the land.

41

CHAPTER 5 Global Landforms
Planning Guide

TEXTBOOK	TEACHING RESOURCES	OTHER AVAILABLE MATERIALS
Section		
1. Forces Shaping the Land Pages 43–46	Reteaching Worksheet 5.1 Critical Thinking Worksheet 5	*Study Guide for Spanish Speakers* *Lecture Notes,* Chapter 5
2. Large Landforms and Plate Tectonics Pages 46–48	Reteaching Worksheet 5.2 Skills Worksheet 5 Challenge/Enrichment Worksheet 5	*REGIONS OF THE WORLD Map Transparencies:* The Drifting Continents *Study Guide for Spanish Speakers* *Lecture Notes,* Chapter 5
3. Landform Development Pages 49–50	Reteaching Worksheet 5.3 Graphic Organizer 5	*Study Guide for Spanish Speakers* *Lecture Notes,* Chapter 5
Chapter Review Page 51	Chapter 5 Test Chapter 5 Portfolio Assessment See also *Alternative Assessment Forms*	*Test Generator,* Chapter 5

CHAPTER RESOURCES

Lecture Notes Chapter 5
Teaching Resources
    Chapter 5 Graphic Organizer
    Chapter 5 Skills Worksheet
    Chapter 5 Critical Thinking Worksheet
    Chapter 5 Challenge/Enrichment Worksheet

REVIEW AND ASSESSMENT

Chapter 5 Check, page 51
Chapter 5 Test, *Teaching Resources*
Chapter 5 Portfolio Assessment, *Teaching Resources*
Test Generator, Chapter 5

ADDITIONAL RESOURCES

Books for Teachers

Eicher, Don L., A. Lee McAlester, and Marcia L. Rottman. *The History of the Earth's Crust.* Englewood Cliffs, NJ: Prentice Hall, 1984.

Menard, Henry. *The Ocean of Truth: A Personal History of Global Tectonics.* Princeton NJ: Princeton University Press, 1986.

Press, Frank, and Raymond Siever. *Understanding Earth.* New York: W.H. Freeman, 1994.

Books for Students

Ballard, Robert D. *Exploring Our Living Planet.* Washington, DC: National Geographic Society, 1983.

Carson, Rob. *Mount St. Helens: The Eruption and Recovery of a Volcano.* Seattle, WA: Sasquatch Books, 1990.

Decker, Robert Wayne, and Barbara Decker. *Volcanoes.* New York: W.H. Freeman & Co., 1989.

Miller, Russell. *Continents in Collision.* Alexandria, VA: Time-Life Books, 1983.

The World Today *Videodisc Program*

A variety of still images, short videos, and activities are available for you to use as you teach this chapter. *Lecture Notes* include bar codes for specific photographs.

Multimedia Resources

The Changing Earth. (transparencies, and duplicating pages) Milliken Publishing Co., 1100 Research Blvd., St. Louis, MO 63132-0579

Earth: The Geology of an Ever-Changing Planet. (VHS, 43 minutes) United Learning, P.O. Box 303, Wilsonville, OR 97070-0303.

The Geography Tutor: Earth's Physical Features. (VHS, 15 minutes) Tell Me Why, 730 Washington Street, Marina Del Rey, CA 90292.

Glaciers. (VHS, 18 minutes) EME, P.O. Box 2805, Danbury, CT 06813-2805.

Picture Atlas of the World. (interactive maps, photo bank, animation, audio on CD-ROM) National Geographic Society, Washington, DC.

MEETING INDIVIDUAL NEEDS

Limited English Proficiency (LEP)

Have students study the photographs in Chapter 5 as a means of previewing the chapter. Then have them work with partners to match the photographs with appropriate terms in the Geography Dictionary.

Learners Having Difficulty

Explain to students how to skim a reading passage to find key information. Then ask them to skim Chapter 5 to find the bold-faced terms. On a set of index cards, have students write each term on one side. On the other side, have them draw a picture to illustrate the term and write a phrase that defines it. Have students use the cards as flash cards until they understand each term.

Auditory Learners

Divide students into three groups. Assign each group a section from the chapter. Have students in each group discuss the answers to the Focus questions and the Section Review questions for their assigned section. Each group should appoint a recorder who writes down the group's answers. When all groups have finished, have each recorder read the questions and answers to the class.

Visual Learners

Provide groups of students with a contour map and a set of questions for the group to answer based on the map (examples: What is the elevation of Point X? Which slope is steeper—Y or Z?).

Tactile/Kinesthetic Learners

Have students work in small groups to make models of landforms using materials such as clay, sand, or plaster. Students' models should include the following: mountains, hills, glaciers, plains, plateaus, islands, valleys, sand dunes, floodplains, alluvial fans, and deltas. Students should label each landform. Then have students identify the landforms found in their community or region. Use the models to begin a class discussion about the forces that shape landforms.

Gifted Learners

Have interested students investigate the Ring of Fire. Students should draw a map of the zone, plotting and labeling the major volcanoes that make up the Ring of Fire. To accompany their maps, have students complete a chart with the following headings: Volcano Name, Location, Height Above Sea Level, Important Facts. Display their work.

MULTICULTURAL FOCUS

Write on the board all the global landforms discussed in Chapter 5. You might then discuss how landforms have affected human geography since ancient times.

Point out that the Western Hemisphere was first settled because of a landform. Indicate the locations of Alaska and Siberia on a wall map. Explain that hunters from Asia walked across what is now the Bering Sea on a land bridge that existed thousands of years ago. Native American groups of North and South America are their descendants.

People also adapted their economic activities to the varied physical features of the earth. For example, the terraced agricultural fields on the sides of mountains in the Andes or the Himalayas are in sharp contrast to the flat, geometrically aligned fields in the American Midwest.

Even language development is influenced by a region's landforms. Peoples living in desert regions may speak a language that has many words for sand, heat, aridity, etc., but few or no words for landforms such as glaciers or mountains.

Some landforms influence human geography in other ways. Mount Fuji, for example, is an imposing and beautiful mountain dominating the landscape near Tokyo. It is considered a sacred place to many Japanese people and is also a national symbol.

Have students brainstorm about ways that the earth's physical features have influenced cultural characteristics.

CHAPTER 5
GLOBAL LANDFORMS

Section 1: Forces Shaping the Land
Section 2: Large Landforms and
Plate Tectonics
Section 3: Landform Development

OVERVIEW

In Chapter 5, students will learn about the geological forces that create land-forms and that wear them down.

Section 1 discusses internal and external forces that change the earth's surface.

Section 2 focuses on large landforms, plate tectonics, and the geological events occurring at plate boundaries.

Section 3 defines primary and secondary landforms and examines land-form development in terms of erosion and tectonic activity.

About the Author

David Attenborough (1926–) is a film pro-ducer and a prolific writer of natural history books. The quotation on this page is from *Life on Earth*. Attenborough, a British citizen, has made many zoological and ethnographic filming expeditions to such loca-tions as Sierra Leone, New Guinea, Zambezi, Borneo, and the Solomon Islands.

Primary Source

Description of Change: excerpted
Rationale: excerpted to focus on main idea

CHAPTER 5

Global Landforms

❝There are few more barren places on earth than the plains surrounding a volcano in the aftermath of its eruption. Black tides of lava lie spilt over its flanks like slag from a furnace. Their momentum has gone but they still creak and boulders still tumble as the flow settles. Steam hisses between the blocks of lava, caking the mouths of the vents with yellow sulphur. Pools of liquid mud, grey, yellow or blue, boiled by the subsiding heat from far below, bubble creamily. Otherwise all is still. ❞

David Attenborough

During a heavy rain, water running in street gutters carries leaves, twigs, litter, and small rocks. If you were to strain the water, you would discover small particles of soil as well. This running rainwater is wearing away the land instead of soaking into it. Wearing away of land is called **erosion.** Water, as well as wind, waves, and ice, causes erosion. Like sandpaper, these agents of erosion smooth the earth. They also transport rock and soil, depositing these particles in new locations.

Other forces deep below the surface of the earth also are at work. These forces sometimes raise masses of rock, building mountains. They may even bend and break the continents. Usually, the forces shaping the earth's surface work only gradually. Sometimes, however, due to a great flood or a volcanic eruption, sudden changes occur. Together, the forces working below and above the earth's surface are constantly changing the world around us.

Hawaii Volcanoes National Park, **Hawaii**

GEOGRAPHY DICTIONARY

erosion
rock weathering
sediment
glacier
sand dune
lava
volcano
fold
fault
earthquake
landform
plain
plateau
plate tectonics
mid-ocean ridge
abyssal plain
subduction
trench
relief
floodplain
alluvial fan
delta

42

Using the Quotation

Ask students to describe volcanic eruptions they may have witnessed, seen on television, or read about. Ask why the area surrounding a volcano might be, as Attenborough describes it, one of the more "barren places on earth." (*destruction of vegetation, livestock, dwellings; layers of lava, ashes, or mud*)

Using the Geography Dictionary

Have students refer to the Geography Dictionary and indicate by a show of hands which terms they know. They are likely to be most familiar with **earthquake** and **volcano**—which are both related to sudden changes to the earth's surface. Point out that most of the terms concern very gradual changes.

SECTION 1
OBJECTIVES

The objectives **reflect the Focus** questions on this page and are assessed in the **review on page 46**.

1. Explain how rock weathering, water, glaciers, and wind change the earth's surface.

2. Describe forces within the earth that change the earth's surface.

FORCES SHAPING THE LAND

FOCUS

- *What external forces cause changes in the earth's surface?*
- *What internal forces cause changes in the earth's surface?*

Rock Weathering At the earth's surface, rocks break and decay. The process that breaks up rocks and causes them to decay is called **rock weathering**. Weathering often is slow and almost impossible to detect, but even the hardest rocks eventually will break into smaller fragments. Given enough time, rocks can dissolve like salt in a pan of water. Some rock weathering is like tooth decay. Chemicals in the air and water pass through rock, slowly eating away at it. Other rock weathering is more like chipping or breaking a tooth. When a forest fire moves over the land, heat can cause rocks to crack. Rocks also can break when water passing through freezes into ice. Similarly, a tree root can pry a rock apart. Rock weathering also is the first step in soil formation.

Weathering breaks rock into smaller particles of mud, sand, or gravel called **sediment**. The weathered sediment is subject to movement due to gravity or removal by erosion. For example, sediment may simply roll or slide downhill in a rockfall or landslide. Sediment also may be eroded and carried away by water, wind, or even moving ice.

Water Whether running in streams and rivers or pounding the beach, water is the most powerful force of erosion and transportation. Eventually, water will carve through the hardest rock. Rivers sometimes carry sediment long distances. Erosion on a hillside can begin as a tiny channel narrower than a pencil. As the water flows farther downhill, the channel might grow wider and deeper, forming a gully—a scar on the land that shows where erosion is particularly active. Rivers also erode and shape the valleys into which they flow. When a river cuts downward for a long period of time, a steep-sided canyon can result. Even in desert regions, rains can erode the bare surfaces between the sparse plants.

Glaciers Thick masses of ice that glide slowly across the earth's surface are called **glaciers**. There are two kinds of glaciers—sheet glaciers and mountain glaciers. Sheet glaciers, or ice sheets, cover large areas. Due to the force of gravity, they flow outward from great domes of ice where snow accumulates. The thick ice in the center of the ice sheet presses downward with great force. The edges of the glacier, which are usually far away from the center, creep outward. Glaciers have the power to level anything in their path. They can move rocks bigger than houses long distances. Ice sheets also can act like

Water and ice are powerful shapers of land. At the left is Utah's **Bryce Canyon**, whose colorful and unusual rock formations were created by centuries of weathering by water and ice. Above, a glacier has carved a path through the Rocky Mountains of **Alberta, Canada**.

Geographical Sidelight

Sand is rock that has been weathered into small grains by the erosive action of cold, heat, rain, ice, and ocean waves. One-fifth of the earth's land area is covered by sand. Its texture ranges from very coarse to very fine. While most sand is silica sand created from parts of rocks that have crumbled away, there are also calcium-rich sands created by the grinding up of seashells, coral, and limestone beds.

43

TEACHING SECTION 1
Motivator

Ask students if they are familiar with such phrases as "terra firma," "hard as a rock," "rock-solid," and "like the Rock of Gibraltar." What qualities of earth and rock do these phrases suggest? *(solidity, permanence, immutability)* Tell students that they will learn how such seemingly permanent features of the earth are changed by wind and water.

Building a Vocabulary

Write the headings Internal Forces Shaping the Earth and External Forces Shaping the Earth on the board. Call on a volunteer to list under the appropriate heading the vocabulary terms in Section 1. Then ask students to study the photographs in Chapter 5 and to identify each as either an example of or the result of an external or an internal force.

Observing

Erosion's Effects   Have students work in teams to explore the school grounds or their neighborhoods and list where they notice the effects of erosion. Ask them to collect rocks of many different sizes, shapes, and compositions. Have them compare the surfaces of the stones to determine the effects that erosion has had on them.

THEME
Interaction

The root systems of trees and other vegetation hold soil in place. They help keep wind and water from eroding fertile topsoil, beach dunes, hillsides, and other areas. Development, overforesting, farming, and other human activities have caused erosion problems such as farmland loss, beach loss, and mud slides in many areas.

Erosion by the Colorado River has created **Marble Canyon** (left), part of Grand Canyon National Park in Arizona. South of Marble Canyon, the Colorado River has cut a gorge as much as one mile (1.6 km) deep and 18 miles (29 km) wide in the Grand Canyon.

blankets, however, protecting the ground underneath from severe cold and other eroding forces.

Today, ice sheets up to 10,000 feet (3,048 m) thick cover most of Antarctica and Greenland. Geographers have discovered that during the last 2 million years there have been many periods when great ice sheets spread across North America and Europe. During these ice ages, glaciers spread as far south as St. Louis, New York City, and Moscow. The last ice age ended about 10,000 years ago. During each ice age, climates changed all across the world. Some areas that are forests today were deserts during the ice ages.

Mountain glaciers are smaller and much more common than ice sheets. Flowing downhill like slow rivers of ice, mountain glaciers are found in high mountain valleys all around the world. Today, there are more than 1,000 mountain glaciers in the Rocky Mountains, the Cascade Mountains, and the Sierra Nevada in the western United States. The ice of mountain glaciers forms from surplus winter snow that does not melt during the summer. This snow is the result of low temperatures caused by high elevation. Mountain glaciers create sharp peaks and U-shaped valleys. During the ice ages, mountain glaciers were more common and much larger than they are today. Usually, they extended farther down the mountainsides, creating much of today's spectacular mountain scenery.

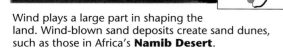

Wind plays a large part in shaping the land. Wind-blown sand deposits create sand dunes, such as those in Africa's **Namib Desert**.

MOVEMENT   **Wind**   Another force that erodes and transports sediment is wind. Because plants protect land surfaces from erosion, wind erosion is greatest where there are the fewest plants—in deserts, along beaches, and on dry, bare ground where people and animals have removed the plants. Wind erosion works in two ways. First, wind wears down hard rocks by blowing sand and other particles against them, often leaving smoothed and polished surfaces. Wind also changes the land by blowing sand and dust from one place to another. Hills of wind-deposited sand are called **sand dunes**. Often, sand dunes are near where the sand was eroded, such as along an ocean beach or beside a

44

Guided Practice

Have the class work in pairs to create a three-column chart with the following headings: External Erosion Forces, Process, and Impact. Have students complete the chart by listing as many forces as they can (*rock weathering, water, glaciers, wind*), along with the process and impact of each. Have them identify forces that are the most evident in the local landscape.

Independent Practice

Have the students answer the following questions based on the charts that they completed in the Guided Practice activity: 1. Which form of external erosion is the most powerful? 2. Which is the quickest? 3. Which causes the most change? Have students explain their answers.

Closure

Review with students the external and internal forces that build up and wear down the land, focusing on these two processes as part of the natural geologic cycle of the planet.

ASSESSMENT

Assign the Section 1 Review questions on page 46.

The Interior of the Earth

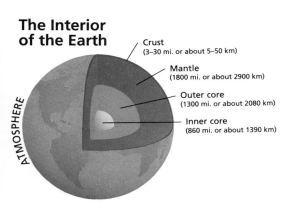

Crust
(3–30 mi. or about 5–50 km)

Mantle
(1800 mi. or about 2900 km)

Outer core
(1300 mi. or about 2080 km)

Inner core
(860 mi. or about 1390 km)

ATMOSPHERE

Temperatures within the earth increase with depth. In the inner core, temperatures may reach as high as 12,000°F (6900°C). How far below the surface is the earth's inner core?

dry stream bed. Sometimes dust particles in deserts are lifted high into the atmosphere and carried great distances. Dust from the Sahara, the great desert in northern Africa, often is transported across the Atlantic to islands in the Caribbean.

Inside the Earth If erosion were the only force shaping the earth's surface, the world around us would have been worn smooth long ago. But as erosion wears down the land, forces inside the earth continue to push up the surface and break it apart.

Scientists have found that our planet is composed of four distinct layers. (See the diagram above.) In the center is the solid inner core, and the next layer is the liquid outer core. The minerals of the core areas are rich in iron. Surrounding the two cores is the mantle, a mostly solid layer. A thin,

rigid crust covers the mantle. If the earth were an apple, the crust would be as thin as the apple's skin. The gases of the atmosphere surround the four layers of the solid earth. The world of people and all other living things is located in a narrow zone at the top of the crust and the bottom of the atmosphere.

Forces Beneath the Crust The earth's surface is bent and broken by heat and by rock movements in the crust and upper mantle. Heat currents travel slowly upward from the core to the crust. These heat currents partially melt the upper parts of the mantle. Nearby parts of the lower crust may be melted as well. The melted, liquid rock, or magma, from within the earth that spills out on the earth's surface is called **lava**.

The crust above the rising heat currents usually rises. Lava may break through and flow over the surface, filling valleys and building the land upward. The opening in the earth's crust through which lava flows is called a **volcano**. (The term *volcano* also is used to refer to the surface feature created by the molten lava.) On the other hand, the crust above the cooler parts of the mantle may sink, creating broad depressions, or low spots, on the earth's surface.

Forces working within the earth not only create lava flows and volcanoes but also bend and break rock. When rock layers are bent, the result is a **fold**. When rock layers break and move, the result is a **fault**. (See the diagram below.) The shock waves, or vibrations, caused by movement along a fault are called **earthquakes**. Whether rocks fold or fault depends on how hard the rocks are and how sudden and strong the forces are. Many of the world's great mountain ranges show much evidence of folding and faulting.

MULTICULTURAL PERSPECTIVES

Many ancient myths and legends explain natural phenomena. In Japanese folklore, a *namazu*, or giant catfish, was believed to cause earthquakes by shaking its body. Ancient Hindus believed that the land was supported by eight mighty elephants. When one shook its head, the earth would move. The Romans believed that earthquakes were caused by Vulcan, the god of fire. Ask students what other natural phenomena have been explained by myths and legends. (*tornadoes, volcanoes, meteors, sunrise*)

Folds and Faults

Mountains formed by faults

Mountains formed by folds

Many large mountain ranges are formed by folding, which occurs when two plates collide and their edges wrinkle, as shown above. Fault-block mountains are formed when parts of the earth's crust are pushed up along a fracture line, or fault.

The weight and hardness of rocks also play a role in shaping the earth's crust. Hard rocks resist erosion and often remain as high ground. In contrast, the erosion of weaker rocks usually produces lowlands. Over the long history of the earth, the lightest rocks have risen to the crust. They now form the continents. If an area of a continent is undergoing rapid erosion, its weight becomes less and less. As a result, the eroded area will steadily rise. Similarly, areas that are being continually buried by the deposit of sediment will tend to sink. Scientists believe these forces that change the surface of the earth have been at work for billions of years.

SECTION REVIEW

1. What processes erode land?
2. What occurs when rock is bent or broken?
3. **Critical Thinking** **Assess** the extent to which human activities cause changes in the earth's surface.

LARGE LANDFORMS AND PLATE TECTONICS

FOCUS

- *What is the theory of plate tectonics?*
- *What are the three types of plate boundaries?*

Plates The shapes on the earth's surface are called **landforms**. Examples of landforms include hills, valleys, nearly flat areas called **plains**, and elevated flatlands called **plateaus**. Landforms vary in size from the smallest indentation made by a raindrop to the largest mountain range. (See the illustration on this page.)

The theory of **plate tectonics** helps explain how the world's large landforms were formed. This theory views the earth's crust as divided into more than a dozen rigid, slow-moving plates. (See the map on page 47.) The plates shift only a few inches each year. Scientists have determined that the processes of plate tectonics usually take millions of years to cause major changes. The plates move like giant rafts cruising slowly over the upper mantle. The less heavy rocks of the continents ride along like passengers on the plate surfaces. Some plates are huge. For example, the North American plate begins under the middle of the Atlantic Ocean and crosses Arctic Canada before reaching its distant edge in northern Japan. Other plates are much smaller. One plate lies totally beneath the Caribbean Sea. The thickness of the plates also varies. They can be more than 60 miles (97 km) thick under the mountains of continents. But they are only about 6 miles (10 km) thick at sites below the oceans where the plates are moving apart.

MOVEMENT **Plate Boundaries** The processes of folding and faulting along plate boundaries build mountains and long valleys. Volcanoes are common, too. In fact, the area bordering the Pacific plate (those areas that lie along the Pacific Ocean) is often called the "Ring of Fire" because it experiences frequent earthquake and volcanic activity. These active regions along the Ring of Fire contrast sharply with the usually inactive interiors of plates. Where plate interiors also are continent interiors, erosion steadily grinds down the land, resulting in wide plains.

Landforms

Many of the landforms shown here can be found together in various regions of the earth. Which landforms can you identify where you live?

Glacier
Mountains
Volcano
Alluvial fan
Canyon
Plateau
Lake
Valley
Plain
Isthmus
River
Hills
Desert
Peninsula
Sandy beach
Floodplain
Delta
Coastline
Strait
Island

46

Hypothesizing

Plate Tectonics   Have students study the map of the lithospheric plates on this page. Point out that part of the California coast lies on the Pacific plate. Based on the map, ask students to predict how the world could look millions of years from now. *(possible responses: coastal California breaks off and drifts; the Atlantic Ocean widens; Australia moves north towards Asia)*

Guided Practice

Have students compare the map of Pangaea on page 48 to a modern globe. Point out the visible clues that indicate that the continents were once connected. *(coastal shapes of the land masses)* Then ask what other kinds of evidence would help verify that now-distant continents were once joined. *(possible answers: fossils of similar plants or animals on separate continents; corresponding landforms or mineral deposits)*

Independent Practice

Have students work in pairs to write short definitions of the following terms related to plates and plate boundaries: **plain, plateau, mid-ocean ridge, abyssal plain, trench**. Students should describe the basic characteristics of each feature type. Students should consult an atlas or a globe to locate and name two specific examples of each feature, noting their absolute and relative locations.

Plate Tectonics

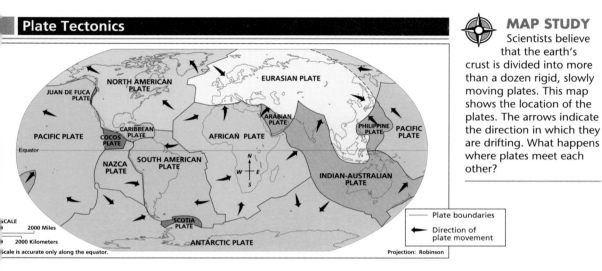

Equator

SCALE
2000 Miles
2000 Kilometers
Scale is accurate only along the equator.

Plate boundaries
Direction of plate movement

Projection: Robinson

MAP STUDY
Scientists believe that the earth's crust is divided into more than a dozen rigid, slowly moving plates. This map shows the location of the plates. The arrows indicate the direction in which they are drifting. What happens where plates meet each other?

Map Study
Volcanic activity and earthquakes are common near plate boundaries. Plate boundaries often also are marked by trenches, mountains, and faults.

COOPERATIVE LEARNING

Organize the class into groups of four or five. Have the students construct rough models of continental plates out of pieces of cardboard. Have them push the pieces together to demonstrate how some mountains or trenches can be formed. Have them envision how the mid-ocean ridge is created as the cardboard plates are moved apart.

There are three types of plate boundaries: those where the plates are moving away from each other, those where plates are pushing against each other, and those where the plates are sliding past each other. Because the crust is thickest and hardest to break where there are continents, most plate boundaries are under the ocean floors. (See the illustration on this page.)

Along the boundaries where plates are moving away from each other, the seafloor spreads, and hot, molten rock moves upward from below to fill the opening. This movement happens because the upper mantle beneath this area of the crust is above heat currents rising from the interior. As a result, the crust is lifted upward, and on the ocean floor a chain of mountains, called a **mid-ocean ridge**, is formed. The mid-ocean ridge system reaches into every ocean floor, extending more than 40,000 miles (64,360 km) in total. (See

the map on page 48.) With no wind and rain to erode them, the mid-ocean ridges remain steep and jagged. Occasionally, a mid-ocean ridge sticks up above the ocean, forming an island. Iceland is the largest of these. Away from the ridges, the ocean floors have large, flat areas. These **abyssal** (uh-BIS-al) **plains** are the earth's flattest and smoothest features.

Along the boundaries where plates collide, a heavier plate may dive under the other plate and sink into the upper mantle. This process occurs above the cooler parts of the mantle and is called **subduction**. This sinking creates a **trench**, a long valley on the ocean floor. Trenches can be more than 600 miles long, 60 miles wide, and 7 miles deep (965 km long, 97 km wide, 11 km deep). They are the oceans' deepest places. Trenches often have a row of volcanic islands on one side. When one of the colliding plates has a continent on it,

Movement at Plate Boundaries

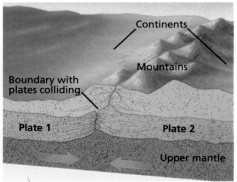

The movement of tectonic plates and the plate boundaries between them create many of the large landforms on the earth. How does the study of plate boundaries explain how some mountains are formed?

Caption Answer

When one plate slips under another or when two plates collide, continental rocks often crumple to form mountains. Some plates slide past each other, creating long creases (faults) in the surface and, sometimes, low mountains.

Geographical Sidelight

German meteorologist and explorer Alfred Wegener introduced the theory of continental drift in 1912. Wegener named Pangaea, meaning "all lands," and Panthalassa, meaning "universal sea."

MAP TRANSPARENCIES

The Drifting Continents: *REGIONS OF THE WORLD Map Transparencies.*

48

This map shows examples of some of the landforms on the floor of the Atlantic Ocean. The Mid-Atlantic Ridge runs the length of the ocean and continues into the Indian Ocean.

World Ocean Floor, Bruce C. Heezen and Marie Tharp, 1977. Copyright by Marie Tharp, 1977. Reproduced by permission of Marie Tharp, 1 Washington Ave., South Nyack, NY 10960.

Continental Drift

Pangaea, a super-continent about 200 million years ago, was surrounded by the earth's single ocean, Panthalassa. New oceanic crust formed as Pangaea broke apart and continental plates drifted away from each other.

the lighter rocks of the continent do not sink. Instead, they crumple and form a mountain range. (See the diagram on page 47.) An example is the Andes Mountains of South America. Sometimes two plates, each carrying a continent, collide. When this happens, no trench forms as the lighter rocks of the continents move together, building huge mountains. The Himalayas between India and China are an example of this.

When plates slide past each other, long creases are formed in the earth's surface, and low mountains may result. Most of these plate boundaries are under the oceans, but the most famous example on land is the San Andreas Fault system in California. There, the Pacific plate slips and jerks northward against the western side of the North American plate, rocking cities such as Los Angeles and San Francisco with earthquakes.

Scientists use the theory of plate tectonics to explain the long history of the earth's surface. They have determined that about 200 million years ago, today's continents were linked as one

supercontinent called Pangaea (pan-JEE-uh). (See the diagram on this page.) Pangaea then broke into two smaller supercontinents named Laurasia and Gondwanaland. These then split further into the modern continents, which continue to drift slowly across the world. The theory of continental drift helps explain the apparent fit in coastal shape and rock types between Africa and South America.

SECTION REVIEW

1. How does the theory of plate tectonics explain the formation of the earth's large landforms?

2. What landforms are formed at each of the three types of plate boundaries?

3. Critical Thinking Determine which type of plate boundary is likely to have the greatest effect on humans.

SECTION 3
OBJECTIVES

The objectives reflect the Focus questions on this page and are assessed in the review on page 50.

1. Differentiate between primary and secondary landforms.

2. Explain the three categories of landforms.

TEACHING SECTION 3
Motivator

Ask students to describe prominent or unusual geological features near where they live, and have them suggest how the landforms might have been formed. Have them consider what they have learned so far about the internal and external forces that continually shape the land. Then discuss how the features affect or have been affected by human habitation and activity.

Building a Vocabulary

Bring to class a jar containing some sediment from a local stream or river. Ask students to explain how sediment is involved in the creation of a **floodplain**, an **alluvial fan,** and a **delta**. Explain that alluvium is the sediment brought down by a river or stream. Finally, have students name any river deltas they know.

LANDFORM DEVELOPMENT

FOCUS

- *What are primary and secondary landforms?*
- *How are landforms classified?*

Primary and Secondary Landforms

Landforms are the result of the struggle between the forces working above and below the earth's surface. Primary landforms are large masses of rock raised by volcanic eruptions and other forces that uplift the earth's surface. A good example is the island of Surtsey, which appeared in 1963. It is the top of a volcano on the mid-ocean ridge south of Iceland. In just a few months, land existed where only waves had washed before.

Yet, even as a primary landform is being formed, the forces of erosion are at work to produce secondary landforms. As soon as the island of Surtsey appeared, the forces of erosion became evident. Ocean waves battered the growing pile of volcanic ash, and sediment washed into the sea. During its first year, the island shrank and grew as the forces of the volcano and the ocean battled each other. Finally, lava flowing out of the volcano hardened into a tough crust over the pile of ash.

Surtsey still exists, but its form is steadily changing. The rain, wind, and pounding surf have worn its edges smooth. Gulleys and valleys are being cut into the island, and beaches have

appeared. Thus, secondary landforms develop from primary landforms due to erosion.

Rain and wind wear down the surface of a primary landform and remove a little more of it each year. For this reason, older secondary landforms are usually smoother, lower, and more rounded than younger ones. Older landforms usually have less **relief**, or smaller differences in elevation, than younger ones. The older age of a landform is usually evident in hillsides that are not very steep and in wide valleys.

Classifying Landforms

Another way to understand landforms is to divide them into three categories, based on how they were formed. One kind of landform is made of rock and has a thin layer of weathered sediment and soil at the surface. Its surface is slowly being lowered by the forces of erosion. Most valleys, plains, and plateaus were formed this way. Another kind of landform is formed by sediment deposited by water, wind, or ice. A sand dune in a desert is an example of this kind of landform. Another example is a **floodplain**. A floodplain is a landform of level ground built by sediment deposited by a river or stream. The third landform is that caused by tectonic activity, the uplifting of the earth's crust. While erosion and deposition change landforms, tectonic activity creates them. In fact, the word *tectonic* comes from the Greek word *tekton,* meaning "builder." Tectonic activity formed landforms of high elevation, such as the Himalayas, and also deeply depressed landforms, such as Death Valley in California.

The terrain in any area is usually a jigsaw puzzle of these three kinds of landforms. For example, a mountain is formed by tectonic activity. The agents of erosion may then form deep valleys in the mountain. The sediment eroded from the mountain may be deposited by a stream on the plain at the mountain's base. This fan-shaped deposit of sediment is called an **alluvial fan**. Later, mud and sand might erode from the alluvial fan and be carried to a river mouth. If it is deposited there, a **delta** may form. Still later, waves might carry the sediment to the ocean floor. Eventually, the sediment could be pulled into a trench and crushed to form a new mountain range.

Landforms and People

The size, shape, and location of major and minor landforms have influenced human settlement throughout history. Just as important has been the ability of people to

GLOBAL PERSPECTIVES

Scientists have learned a great deal about global climate patterns from volcanoes. The 1883 eruption of Krakatau in Indonesia caused atmospheric effects throughout the Northern Hemisphere. The rapid spread of its ash worldwide was an early clue for scientists about the existence of jet streams. Following the 1991 eruption of the Philippines' Mount Pinatubo, scientists tracked a temporary trend of cooler weather over the globe. Thus the unusually cold weather in North America in 1816 may have been produced by the massive 1815 eruption of Tambora in Indonesia. Have students locate these volcanoes on a map or a globe, noting their relative location to the Northern Hemisphere.

Volcanic forces created the island of **Surtsey** near **Iceland**. Wind and water are continuously eroding the face of the island.

Illustrating Ideas

Landforms Have the class work in small groups to create an illustrated glossary of landform terms, including a definition, a picture or drawing, and an example for each term. Tell them to include every type of landform mentioned in the Geography Dictionary on page 42. Students may use atlases, maps, and their own experience to include additional terms.

Guided Practice

Display illustrations of landforms from the glossary students have created, and ask students to classify them as primary or secondary landforms. Review the characteristics discussed in the text for determining primary and secondary landforms. Remind students that as soon as a landform appears, forces such as erosion begin to change it; hence, there are very few primary landforms.

Closure

Review the three categories of landforms described on page 49, and have students give examples for each.

ASSESSMENT

Assign the Section 3 Review questions on this page.

Reteaching

Assign Section 3 of Reteaching Worksheet 5.

Boston, Massachusetts, was founded in 1630 on the Shawmut Peninsula, a small landform left by glaciers from the last ice age. As the town grew in size, the peninsula became crowded, and the citizens began to level glacial hills and fill in bays and inlets. Today, not an inch of the original shoreline remains.

Section 3 Review
ANSWERS

1. Primary landforms, such as volcanoes, are large masses of rock created by the earth's eruptions. Secondary landforms, such as beaches, are created from primary landforms by erosion.

2. Surtsey was formed by a volcano (becoming a primary landform) and then began to shrink and change due to the forces of erosion (becoming a secondary landform).

3. Landforms made of rock, landforms formed by the deposit of sediments, and landforms created by tectonic activity

50

Elevation Map: Guadalcanal

Contour Map: Guadalcanal

MAP STUDY

Elevation is the height of the land above sea level. An elevation map, such as the one for Guadalcanal (above, left) uses color to show how high landforms are. The illustration to the right is a cross-sectional diagram of Guadalcanal. This kind of diagram shows a region's landscape even more clearly because it is a side view, or profile, of a specific region. The diagram here is a profile of Guadalcanal along a line from Point A to Point B in the elevation map.

Vertical and horizontal distances are calculated differently on cross-sectional diagrams. The vertical distance (the height of Mount Makarakomburu, for example) is exaggerated when compared to the horizontal distance between Point A and Point B. This technique

Elevation Profile: Guadalcanal

is called vertical exaggeration. If the vertical scale were not exaggerated, even tall mountains would appear as small bumps on a cross-sectional diagram.

Contour maps, such as the one above right, also show elevation. Instead of colors, map makers use lines that connect points of equal

elevation. Elevation levels are written along the lines. The closer together the lines are, the steeper the terrain in that area. For example, on this contour map, the terrain to the south of Mount Makarakomburu is steeper than the terrain on the northwestern end of the island.

---

adapt to living on, and changing, the earth's landforms. Such adaptation may be as simple as a farmer plowing around the side of a hill to prevent erosion. Or human adaptation may involve technology, such as explosives, that makes it possible to build dams across rivers or to carve highways on the sides of mountains.

Theories about how the continents formed may seem to have little to do with our daily lives. But stories in the news about mud slides destroying roads or villages, or lava from a volcano threatening a nearby town, remind us that the earth is constantly changing. Today, no less than

thousands of years ago, people face the challenge every day of adapting to those changes.

SECTION REVIEW

1. How do primary and secondary landforms differ? Give one example of each.

2. How is Surtsey a good example of a landform that is both primary and secondary?

3. Critical Thinking **Classify** landforms into three categories other than primary and secondary.

CHAPTER 5 CHECK

Coastal plain of **Bangladesh**

Reviewing the Main Ideas

1. The earth's form is constantly being changed by forces inside the earth and on the earth's surface.
2. The forces of running water, waves, glaciers, and wind erode the land. Other influences, such as heat and plate movement, are constantly building new landforms.
3. Primary landforms are those made of large masses of rock that formed as a result of volcanic eruptions and other uplifting forces. Secondary landforms are the result of erosive forces working on primary landforms.

Building a Vocabulary

1. What is the difference between folding and faulting?
2. What is a mid-ocean ridge? How is it formed?
3. What is the term used to describe the differences in elevation of a landform?

Recalling and Reviewing

1. In what ways does erosion of the earth's surface occur? Explain.
2. Explain how the forces within the earth prevent the earth's surface from becoming totally smooth due to erosion.
3. Describe the theory of plate tectonics, and explain how the continents and major continental landforms were made.

Thinking Critically

1. According to the theory of plate tectonics, the great mountain ranges of the world are still forming, although some are more stable than others. What might happen to the Alps as the African plate continues to move under the Eurasian plate?
2. What could you do to prepare for a predicted earthquake? Make a list of items for a survival kit.

Learning About Your Local Geography

Individual Project

Draw a cross-sectional diagram of the landforms found in your area. (Refer to the diagram on page 46.) Label the diagram, and include brief explanations of how each landform might have been formed. You may want to include your diagram in your individual portfolio.

Using the Five Themes of Geography

People changing and adapting to the environment falls under the geographic theme of **human-environment interaction**. Make two lists to show this interaction in your area. In one list, write descriptions of local landforms and the forces of change taking place, such as erosion, weathering, earthquakes, and so on. In the other list, write ways in which people have made changes in the local environment to better suit their needs or to adapt to natural forces of change.

Mitre Peak near **Milford Sound, New Zealand**

51

CHAPTER 6 Patterns of Life
Planning Guide

TEXTBOOK	TEACHING RESOURCES	OTHER AVAILABLE MATERIALS
Section		
1. The World of Plants and Soil Pages 53–56	Reteaching Worksheet 6.1 Challenge/Enrichment Worksheet 6	*REGIONS OF THE WORLD Map Transparencies:* Forest Succession After a Fire *Study Guide for Spanish Speakers* *Lecture Notes*, Chapter 6
2. Biomes: World Plant Regions Pages 58–60	Reteaching Worksheet 6.2 Graphic Organizer 6 Skills Worksheet 6 Critical Thinking Worksheet 6	*REGIONS OF THE WORLD Map Transparencies:* The World's Basic Biomes The Changing Environment *Study Guide for Spanish Speakers* *Lecture Notes*, Chapter 6
Chapter Review Page 61	Chapter 6 Test Chapter 6 Portfolio Assessment See also *Alternative Assessment Forms*	*Test Generator,* Chapter 6
Unit Review Pages 62–63	Unit 1 Test Unit 1 Portfolio Assessment See also *Alternative Assessment Forms*	*Test Generator,* Unit 1

CHAPTER RESOURCES

Lecture Notes, Chapter 6
Teaching Resources
    Chapter 6 Graphic Organizer
    Chapter 6 Skills Worksheet
    Chapter 6 Critical Thinking Worksheet
    Chapter 6 Challenge/Enrichment Worksheet

REVIEW AND ASSESSMENT

Chapter 6 Check, page 61
Chapter 6 Test, *Teaching Resources*
Chapter 6 Portfolio Assessment, *Teaching Resources*
Test Generator, Chapter 6

ADDITIONAL RESOURCES

Books for Teachers

Attenborough, David. *Life on Earth: A Natural History.* Boston: Little, Brown and Company, 1979.

Benarde, Melvin. *Our Precarious Habitat: Fifteen Years Later.* 3d. ed. New York: John Wiley & Sons, 1989.

Goudie, Andrew. *The Human Impact on the Natural Environment.* New York: Cambridge University Press, 1994.

Lovelock, J. *Gaia: A New Look at Life on Earth.* New York: Oxford University Press, 1982.

Sabin, Francene. *Ecosystems and Food Chains.* Mahwah, NJ: Troll Associates, 1985.

Books for Students

Attenborough, David. *The Living Planet: A Portrait of the Earth.* Boston: Little, Brown and Company, 1986.

Brown, Vinson, *Investigating Nature Through Outdoor Projects: 36 Strategies for Turning the Natural Environment Into Your Own Laboratory.* Mechanicsburg, PA: Stackpole Books, 1983.

The Curious Naturalist. Washington, DC: National Geographic Society, 1991.

Gralla, Preston. *How the Environment Works.* Emeryville, CA: Ziff-Davis Press, 1994.

MacMahon, James. *Deserts.* New York: Alfred A. Knopf, 1985.

Maser, Chris. *Forest Primeval: The Natural History of an Ancient Forest.* San Francisco, CA. Sierra Club Books, 1989.

The World Today *Videodisc Program*

A variety of still images, short videos, and activities are available for you to use as you teach this chapter. *Lecture Notes* include bar codes for specific photographs.

Multimedia Resources

The Diversity of Life. (VHS, 25 minutes) National Geographic Society, Washington, DC.

GTV: Planetary Manager. (Multimedia CD-ROM, global environmental issues) National Geographic Society, Washington, DC.

Picture Atlas of the World. (interactive maps, photo bank, animation, audio on CD-ROM) National Geographic Society, Washington, DC.

MEETING INDIVIDUAL NEEDS

Limited English Proficiency (LEP)

As students review the food web illustration on page 53, sketch the following flowchart of a food chain on the board: plant, grasshopper, frog, snake, hawk. Then ask students to follow this model to create and label a food chain on their own. Compare and contrast students' food chains, and discuss the importance of plants as the basis of food chains. Have students identify the biomes in which their food chains would exist.

Learners Having Difficulty

Help students make graphic organizers on the board that illustrate the main ideas of the chapter. The following concepts could be used to make the graphic organizers: ecosystems, plant succession, soil horizons, biomes. Have students copy the graphic organizers, and encourage students to use them to study the chapter.

Auditory Learners

Assign the following topic for debate: "Spending money and energy to end the destruction of the tropical rain forests is a waste because their destruction will not affect my way of life." Divide students into two groups, one group to defend the topic statement and the other to argue against it.

Visual Learners

Have students create posters or collages that illustrate the five basic biome types in the world. Display the completed posters or collages in the classroom.

Tactile/Kinesthetic Learners

You may wish to have students work in groups or as a class to create a small ecosystem that illustrates the interdependence of living things with the physical environment. For example, students might make a marine ecosystem that includes gravel, water, plants, and fish. Another example might include soil, plants, and a frog in a terrarium. Display the ecosystems in the classroom.

Gifted Learners

Direct interested students to research ecosystems in the world that are in danger of being destroyed, such as the rain forests, scrub forests, or the Great Barrier Reef. Have students present an oral report about the ecosystems based on the Five Themes of Geography. Students should conclude their reports with possible solutions to the problems of destruction.

MULTICULTURAL FOCUS

In September 1993, eight volunteers—four men and four women—emerged from two years of near isolation in an enclosed artificial world in the middle of the Arizona desert. The enclosure, called Biosphere 2, is a 3.15-acre (1.27 hectares) sealed model of Earth's environment that was constructed to gain experimental knowledge that could be applied to solving Earth's environmental problems. The participants lived among 3,800 species of plants and animals, eating only food that they grew themselves. This was the first test of what project leaders envision as a 100-year-long experiment.

Biosphere 2 faced criticism that the expensive project was more show than scientific substance. It also encountered several difficulties. Among the most serious was a miscalculation in the amount of organic matter in Biosphere 2's soil, which resulted in an oxygen deficit in the structure. Twenty-three tons of oxygen twice had to be pumped into the dome to maintain its livability. This only fueled controversy over the project's scientific validity. Despite the problems and criticism, "Biospherians" have claimed that much important information was learned from the structure's first test.

Ask students to imagine themselves living in an enclosed environment like Biosphere 2 or a space station. Discuss what daily life might involve. Have students brainstorm the kind of culture—hierarchy or government, modes of communication, arts, recreation, and customs—that might develop.

OVERVIEW

In Chapter 6, students will learn about how plants and animals form ecosystems.

Section 1 describes plant communities, types of soil, and the concept of the food chain. It explains photosynthesis, the connection between plant succession and climax communities, soil horizons, and marine ecosystems.

Section 2 identifies and locates the five basic world biomes—forest, savanna, grassland, desert, and tundra. Students will also learn the types of vegetation found in each biome and understand the interrelationship of all living things in the biosphere.

About the Author

David Attenborough (1926–) is a well-known British documentary filmmaker. His books and films, based on travels to Africa, Asia, and the South Pacific islands, focus on natural history. The quotation on this page about forest ecosystems is from *Life on Earth*.

Primary Source

Description of Change: excerpted

Rationale: excerpted to focus on main idea

CHAPTER 6

Patterns of Life

❝The forests of today are very much the same, in essence, as those that developed soon after the appearance of the flowering plants, fifty million years ago. Then, as now, there were jungles in Asia, dank rain forests in Africa and South America, and cool verdant woods in Europe. . . . Everywhere, leaves sprouted; season after season, century after century, they offered an ever-renewing, inexhaustible supply of food for any animal able to gather and digest them. ❞

David Attenborough

As you learned in Chapter 2, the biosphere contains all living things on Earth. Plants make up most of the biosphere and are the basis for almost all other life forms. In one way or another, all animals depend upon plants for their food supply. Some animals eat only plants, while others eat the animals that eat the plants. Still others, including humans, eat both plants and animals. Plants also are a major source of oxygen for the earth's atmosphere.

Soil provides the nutrients, or nourishment, that plants need to grow and to survive. Soil also supports the food crops necessary to feed the human population. This chapter will help you understand the interrelationships of plants, animals, soil, water, and climate.

GEOGRAPHY DICTIONARY

photosynthesis

food chain

plant community

ecosystem

plant succession

climax community

humus

soil horizon

leaching

coral reef

biome

deciduous forest

coniferous forest

mixed forest

savanna

prairie

extinct

Rain forest of Olympic National Park, **Washington** state

52

Using the Quotation

Attenborough talks about similarities between the forests of fifty million years ago and the forests of today. Have students think about the environment in which they live. In what ways has it changed in their lifetime? In what ways has it stayed the same?

Using the Geography Dictionary

Have students consider how the words in the Geography Dictionary and the chapter title, "Patterns of Life," are connected. Point out that most of the terms concern soil or plants—integral parts of the world's biosphere.

SECTION 1
OBJECTIVES

The objectives reflect the Focus questions on this page and are assessed in the review on page 56.

1. Recognize the importance of plants in sustaining life on Earth.
2. Identify the importance of climate in determining soil type.
3. Describe the stages of a food chain.

THE WORLD OF PLANTS AND SOIL

FOCUS

- *What is a plant community?*
- *What determines the type of soil a location will have?*
- *What is the marine food chain?*

Cacti and other hardy plants make up this desert plant community. Why do you think a desert plant community is not very complex?

Primary Production Every plant is a small factory helping to support life on Earth. These factories convert sunlight into chemical energy through a process called **photosynthesis**. Plants use the sun's energy as fuel to change carbon dioxide from the air, and water and nutrients from the soil, into food. Plants' roots obtain the nutrients from the water, minerals, and gases in the soil. Oxygen is produced as a result of this process and is released, along with water vapor, into the atmosphere. Some of the food produced is used to keep the plant alive, while the rest is used for growth.

Plants are at the bottom of the **food chain** and support animal life. A food chain is a series of organisms in which energy is passed along through living things. Some animals eat plants, and these animals are eaten by other, usually larger, animals. At each stage of the food chain, there are fewer living things. Only a few animals, such as lions, hawks, and humans, are at the top of the food chain.

Plant Growth Generally, plants do not live alone, since they are not independent. Just as you live in a society, surrounded by friends, family, and other people, plants live in groups. Such a group is called a **plant community**. In the humid-tropical climates, these communities are quite large, and thousands of different kinds of plants are found here. The many lush tropical plants in such a community support great numbers of animals, from millions of insects that inhabit the forest floor to birds and monkeys in the treetops. In the desert climate, the plant community is less complex. Here,

Food Web

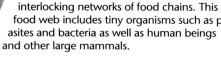

Food chains rarely occur alone in nature. More common are food webs, which are interlocking networks of food chains. This food web includes tiny organisms such as parasites and bacteria as well as human beings and other large mammals.

As you write their responses on the board, call on students to name types of plants or animals that are native to their local environment. Then ask them to think of aspects of their region's climate, soil, or vegetation that might influence what plants or animals are found in their locale.

Building a Vocabulary

Write the terms **plant community**, **plant succession**, and **climax community** on the board. Use an example from your area, such as a nature preserve or a state park, to illustrate the meanings of these terms and the interrelationships between them.

Hypothesizing

Food Chain Ask students to suggest reasons why there are fewer species at the top of the food chain than at the bottom. Point out that only a small percentage of food energy at each step of the chain is passed along to the next. *(Energy loss means that organisms higher on the chain must consume much more to meet their energy needs.)*

Caption Answer
Grass and weeds

MAP TRANSPARENCIES

An overhead transparency of the illustration on this page is available in *REGIONS OF THE WORLD Map Transparencies.*

Geography and Biology

Organic activity has a great effect on soil formation. Leaves, bark, and other parts of dead plants rot into the soil, adding structure and nutrients. Living plants help keep minerals and water in the soil by drawing them up through their roots. Bacteria and other microorganisms feed on organic matter in soil and help break it down. Earthworms work the soil by burrowing into it and passing it through their intestines, altering its chemical composition. Burrowing insects such as ants and large burrowing animals such as rabbits and moles also stir up the soil, mixing material from the upper and lower layers.

54

Difficult conditions after a forest fire mean that the first plants to grow back must be very hardy. In this illustration, which plants are the first to grow back?

there may be only a few highly specialized plants, such as cacti, and a few specialized animals, such as lizards. Specialized plants and animals are those that have adapted to a particular environment.

Each plant evolves in such a way that it plays a special role in its community. Different types of plants require different amounts of sunlight, water, and nutrients from the soil. Some plants, such as grass, have shallow roots that capture water and nutrients close to the soil's surface. Other plants, such as trees, send their roots deep into the ground. Some plants survive in soil that is poor in nutrients, while others require rich soil.

Large trees need a great deal of direct sunlight in order to make food. Ferns, however, can survive with very little sunlight. Only the plants whose requirements for growth are fulfilled in a community will survive there. Variations in sunlight, temperature, water, and soil help explain the geographic distribution of forests or grasslands in different regions of the world.

Nearly all plants depend on other plants to grow. For example, trees provide the shade that keeps ferns growing on the forest floor from receiving too much sunlight. Trees also provide support for clinging vines so that the vines can capture needed sunlight at the top of the trees. Individual types of plants work to aid their own growth and, often, the growth of other plants in their community.

This interdependence of living things combined with the physical environment these organisms inhabit forms an **ecosystem**. Thus, an

Forest Succession After a Fire

A. Forest Fire in Progress

B. Early Plant Growth

C. Middle Stage

D. Forest Recovered

ecosystem is a community of plants and animals that functions as a unit with the climate, soil, and water of the region.

Plant Succession The cluster of plants that constitutes a community does not develop by accident. Rather, the plant community is the result of orderly change in a geographic area over a long period of time.

Imagine an area just after a forest fire. The first plants to grow will be very hardy, and because there are no trees to give shade, the plants must be able to endure direct sunlight. They will grow rapidly when rain falls on the ash and soil. These plants will help prevent erosion by holding the soil in place, and they also will provide shelter for the seeds of other plants.

Grass and weeds may cover the ground for some time, but soon they will be replaced by

Surveying

Plants and Soil To help students understand the differing needs of various plants, have them survey people they know who are successful gardeners. They should ask about soil types, climates, plants that need sunshine, and plants that thrive in shade. Have them compare their findings with those of their classmates.

Guided Practice

Have students refer to the forest succession diagram on page 54. Ask for volunteers to describe each panel of the diagram, mentioning the types of plants that appear at each stage and why they are able to grow. Write these descriptions of the plant succession process for a forest fire on the board, using a flowchart format.

Independent Practice

Have students research and give oral reports about plant succession surrounding Mount St. Helens in Washington state during the first years after its 1980 eruption.

taller plants. Often, these newer plants are shrubs and small trees. Their seeds may have blown in from a distance by the wind or have been carried by animals such as squirrels, bats, and birds. When these seeds are dropped, a new tree may sprout. When the new tree is tall enough and has enough leaves, it will shade the ground. This shade will then make it difficult for the earlier sun-loving plants to survive. Shade-loving plants, such as ferns, will then cover the forest floor. After many years, the area may again look as it did before the fire. (See the diagram on page 54.)

The process by which one group of plants replaces another is called **plant succession**. Each group of plants grows best in one set of conditions. When these conditions change, the growth of a new plant community begins.

The succession process can be controlled by people, but this is not easy to do. If you decide that you want only grass to grow in your yard, you can prevent other plant communities from developing. The grass must be cut frequently to keep tree seedlings from sprouting. Some trees might have to be trimmed so they do not block needed sunlight from the grass. If insects or a lack of fertilizer causes bare spots to appear on your lawn, weeds quickly will fill the open spaces. Time, effort, and money must be spent to prevent succession from changing your grassy lawn into a weed patch. Without your involvement, a new plant community, similar to the original plant community, may in time grow back.

INTERACTION **Climax Community** For each set of climate and soil conditions in a geographic area and for any given time, there is a plant community that is best suited to use the resources there. Only after a long process of plant succession does a stable plant community develop. This last stage of succession is called the **climax community**.

Changes in the environment can alter a climax community. It might be destroyed by natural events such as forest fires, storms, volcanic eruptions, and plant disease. If any of these occur, the succession cycle will start over again. Humans, too, keep climax communities from continuing by burning and clearing the land for farming, by logging, by grazing their animals, and by introducing foreign plants. For these reasons, true climax communities are seldom found today in many areas of the world.

Factors of Soil Formation As you have learned, soil supplies the nutrients plants need to grow and survive, which helps provide the food we need to live. But what is soil? If you dig a shovel into the ground, what do you bring up? Soil is made up primarily of minerals from the weathering of rocks. Most of these minerals are from the rocks found just below the soil layer. Sometimes, weathered rock material is brought to an area by rivers, wind, or glaciers.

Decayed plant or animal matter, called **humus**, also is an important ingredient of soil. Bacteria and insects help break down the humus, and they produce necessary space between the soil particles for gases and water. Weathered rock material, organic matter such as humus, gases, and water must exist together in order for soil to provide plants with the necessities of life.

Soil differs from place to place, and there are hundreds of soil types. Climate is the major factor that determines the type of soil a location will have. A region's climate type controls the amount of sunlight, moisture, and natural vegetation that influence the soil's formation. The second major control is the rock type that provides minerals for the soil. One region may have dark, thick soil with much humus, while another has only a thin layer of sandy soil. Other factors, such as the slope of the land, human activities, and time for soil development, also are important.

Soil Horizons Soil develops very slowly, taking hundreds or even thousands of years to form. As it develops, it forms distinct layers, called **soil horizons**. Most soil has three layers, called the A, B, and C horizons. (See the diagram on page 56.)

The top layer, or A horizon, is called topsoil. This is where decaying plant or animal material accumulates and forms humus. This horizon is very active and has much bacterial and insect activity as well as most of the plant roots. The middle layer, or B horizon, is called subsoil. Only deep plant roots penetrate the B horizon. The C horizon is made up of the broken rock pieces that form the parent material for the soil.

Because climate is the major control of soil formation, there are similarities among soils in the same climate region. For instance, in the humid-tropical climate, warm temperatures and heavy rainfall support lush forests. These tropical rain forests, in turn, supply great amounts of humus to the soil below.

GLOBAL PERSPECTIVES

There are literally thousands of different types of soil around the world. Soil type is determined by the rock that lies beneath it, by temperature, by precipitation, by vegetation, and by many other variables. Soil scientists use a system known as the Comprehensive Soil Classification System. This system groups the world's soils into 10 orders, which are in turn subdivided into 47 suborders, 185 great groups, about 1,000 subgroups, and numerous sub-subgroups. Ask students what characteristics they would expect soil samples from the Sahara, the Nile River delta, the Brazilian rain forest, and the Arctic tundra to have. *(Sahara—dry and sandy; Nile—rich in minerals and moist; rain forest—spongy with much humus; tundra—frozen with little plant life)*

55

Write the following terms around a large circle on the board: *photosynthesis, food chain, plant community, ecosystem, plant succession, climax community, humus, soil horizons, leaching, coral reefs, land ecosystems, marine ecosystems*. Point to any two terms on the circle, and ask a student to explain how they relate to each other. Continue selecting pairs of terms for students to associate until the main concepts of Section 1 have been covered.

ASSESSMENT

Assign the Section 1 Review questions on this page.

Reteaching

Assign Section 1 of Reteaching Worksheet 6.

Extension

Have students research a specific geographical land area or a particular undersea region to determine what species of plants and animals constitute a typical food web there. Using drawings or photographs, students should make a poster that illustrates their findings.

Caption Answer

The many kinds of small marine animals around coral reefs, including living coral-forming animals, are an abundant source of food for many larger species of marine life.

Section 1 Review
ANSWERS

1. Plants help both their own growth and the growth of other plants in the community.
2. Climate
3. On land, plants are eaten by animals, which in turn are eaten by other, usually larger, animals. A few animals, including humans, are at the top of the food chain. Similarly, in the marine food chain, small drifting plants are eaten by small marine animals. These in turn are eaten by larger marine creatures. At the top of the marine food chain are fish, marine birds, and marine mammals.

A
Leaching
B
C
Soil Horizons

Topsoil (Humus)

Subsoil

Broken rock

Solid rock

Most soils have three layers, as shown to the left. Only deep plant roots penetrate the B horizon. Leaching occurs when nutrients are washed downward by heavy rainfall.

Fish are found in abundance near this coral reef in the South Pacific Ocean. Why do you think coral reefs have such rich marine ecosystems?

At the same time, humid-tropical soil retains little of this humus due to a process called **leaching**. Leaching occurs when the nutrients necessary for plant growth are washed downward out of the topsoil by rainfall. The nutrients then cannot be reached by the plant roots. Fortunately, the tropical rain forest continues to produce humus as fast as the rain washes the nutrients away. An area that supports such a lush forest would seem ideal to clear for farming. Just a few years after the trees are removed, however, the soil will lose its nutrients because of the constant leaching, and only scrub plants will grow. Thus, the plants and soil of an ecosystem are interrelated.

Marine Ecosystems Like ecosystems on land, marine (sea) ecosystems depend on sunlight for life. Drifting marine plants convert sunlight and nutrients from seawater into growth. They also produce oxygen for the atmosphere. These sea plants form the bottom of the marine food chain. Small drifting sea animals eat the sea plants, and these small sea animals are then eaten by larger sea animals. At the top of the marine food chain are fish, marine birds, and marine mammals such as seals and whales.

Most marine life is concentrated near the water's surface where sunlight is available to support the food chain, and near coastlines where nutrients are brought to the sea by rivers. The richest marine ecosystems, with many kinds of animals and plants, are coastal marshes and **coral reefs**. A coral reef is a ridge found in warm, tropical, shallow water close to shore. It is made of coral, which is a rocky limestone material formed of the hard skeletons of the coral polyp. Coral polyps are tiny marine animals, usually smaller than the tip of a ballpoint pen. They live in colonies that are anchored to the shallow ocean floor. As old coral polyps die, their limestone skeletons serve as a platform on which new corals grow.

Not all of the ocean is rich in marine life. In fact, much of the ocean is a biological desert—the open ocean far from coastlines as well as the cold, dark depths.

SECTION REVIEW

1. What roles do plants play in a plant community?
2. What is the major factor in soil formation?
3. **Critical Thinking** **Compare** the food chain on land to the marine food chain.

Using the World's Basic Biomes Map

Locating Biomes

Tell students to locate their own state or community on the World's Basic Biomes map and then to identify the biome. Then have them locate and list other places in the world with the same biome. (They can use the World Political map in the Atlas on pages A4–A5 to identify these countries.) Ask students to compare their biome with the climate region of their community (see the World Climate map on pages 28–29). Ask if the biome and climate region of their state or community cover approximately the same area. (*Students are likely to say that they do.*)

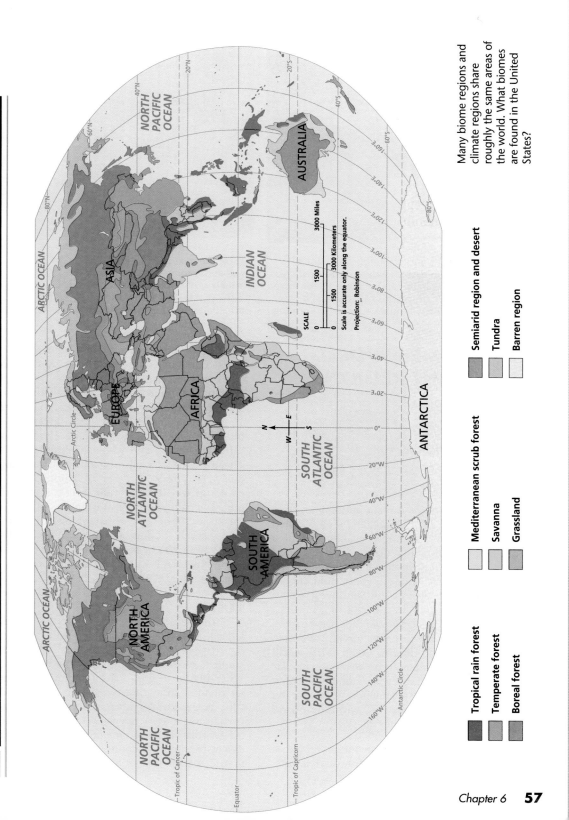

Many biome regions and climate regions share roughly the same areas of the world. What biomes are found in the United States?

Tropical rain forest

Temperate forest

Boreal forest

Mediterranean scrub forest

Savanna

Grassland

Semiarid region and desert

Tundra

Barren region

Map Caption Answer

Tundra, boreal forest, temperate forest, grassland, desert

SECTION 2
OBJECTIVES

The objectives reflect the Focus questions on this page and are assessed in the review on page 60.

1. Identify the five basic world biomes and their locations.
2. Describe characteristics of the five basic world biomes.

TEACHING SECTION 2
Motivator

Ask students how they would characterize the land around where they live. Write their answers on the board. Tell them that in this section they will learn about the five types of plant and animal communities that geographers use to categorize the earth's land areas. As you teach the lesson, refer to the descriptions the students suggested.

Building a Vocabulary

Write the terms **deciduous forest** and **coniferous forest** on the board, and have a volunteer read aloud the Glossary definitions. Provide students with a list of specific types of trees that grow within a temperate region. Have them consult encyclopedias to classify each type as deciduous or coniferous.

Deciduous and coniferous trees color autumn in eastern **North America** (above). Giraffes and scattered trees tower above the grasslands of a savanna biome in **Africa** (above, right). Clockwise from top left on the next page is an example of a grasslands biome in central **North America**, of summer vegetation in the tundra of **Alaska**, and of the desert biome of the Patagonia region in **Argentina**.

Geography and the Environment

The controversy over the northern spotted owl has put the spotlight on old-growth forests. However, no single definition of an old-growth forest is possible, because forests vary from area to area. Foresters and ecologists have offered the following requirements for defining them: two or more tree species with a wide range of ages and sizes; six to eight Douglas firs or other conifers per acre that are at least 200 years old; a multilayered forest canopy; a number of large, standing dead trees at least 20 inches in diameter; and at least ten tons of large fallen logs per acre.

58

2 BIOMES: WORLD PLANT REGIONS

FOCUS

- *What are the five basic biomes, and where are they located?*
- *What types of vegetation are found in each biome?*

What Is a Biome? A plant and animal community that covers a very large land area is called a **biome**. Plants are the most visible part of a biome. If you could look down on the United States from space, you would see the biomes. The forests of the eastern United States would appear green, while the deserts of the Southwest would be light brown.

There are five basic biomes that can be identified throughout the world: forest, savanna, grassland, desert, and tundra. Areas of the world that share the same biome type also have similar climates and soil types. In fact, some biomes and climate types share the same names. Compare the World Biome map on page 57 with the World Climate Regions map on pages 28–29.

PLACE **Forest Biome** The forest biome is tree covered. Within this biome, there are a num-

ber of different climax forest communities. Each forest community is determined by the types of trees growing there. In addition to trees, the forest biome supports a great variety of living things from the treetops to beneath the soil surface.

Tropical rain forests are located in the humid-tropical climate regions. Here, temperatures are warm, and water is abundant all year. These conditions allow plants to grow continuously. Hundreds of different kinds of tall trees make up the tropical rain forest, and the thick umbrella of treetops creates continuous shade on the forest floor. The constant plant growth provides much humus to replace that which is removed by the heavy rains from the tropical soil below the forest. The tropical rain forests are the most complex ecosystem in the world, with thousands of kinds of plants and animals, especially insects.

Temperate forests are located in the middle-latitude climates. Generally, these forests are less dense than tropical rain forests. Thus, enough sunlight reaches the ground to allow shrubs and other plants to grow. The temperate forests can be divided into different types. The **deciduous** (di-SIJ-uh-wuhs) **forests** lose their leaves during a certain season each year. The **coniferous** (koh-NIF-uh-ruhs) **forests** remain green year-round. The densest temperate coniferous forests are located in the marine-west-coast climate regions. Some

Analyzing Biomes

Have students study the biome map on page 57 and determine which biome they live in. Then have them offer examples of natural vegetation in the surrounding area that support the definition of that biome.

Guided Practice

Have students refer to the biome map on page 57 and the photo collage on pages 58–59. Have them match the pictures with a corresponding biome on the map. Ask them whether they think that all areas in a single biome are identical. *(no)* Remind them that a biome is a broad type of plant and animal community.

Independent Practice

Provide each student with a copy of Outline Map 1. Have them label the continents and biomes on their copy of the map.

areas are **mixed forests** where deciduous and coniferous forests blend. These are located primarily in the humid-subtropical and humid-continental climates of the middle latitudes.

Boreal (BOHR-ee-uhl) forests consist mainly of coniferous trees. Because these needle-bearing trees can live under very cold conditions, boreal forests are located in the subarctic climate regions. The floors of these forests build up a layer of tree needles, and due to the long winters, decay is slow. Boreal forests stretch in a broad band across subarctic North America, Europe, and Asia. *Boreal* means "northern," and these forests are not found in the Southern Hemisphere.

Like the Mediterranean climate region, the Mediterranean scrub forests cover a small percentage of the earth's land surface along middle-latitude coasts. Because of the long, sunny, dry summers, most of the vegetation of this forest type is made up of short trees and shrubs. This drought-resistant vegetation has leathery leaves and thick bark that help the plants survive the long summers. Forest and brush fires are frequent during the dry season. Most of the original Mediterranean scrub forests have been lost to clearing for farming, the growth of cities, and overgrazing by herds of domestic animals.

PLACE **Savanna Biome** The **savanna** biome is a tropical grassland with scattered trees

and shrubs. Usually, the savanna is located in areas with distinct wet and dry seasons. The savanna biome occupies tropical regions between the wet, tropical rain forests and the arid deserts. This biome supports great herds of grazing animals and many predators, such as those in East Africa. In Asia and Africa, the savanna has increased in area at the expense of forests. Clearing forests for logging and farming and overgrazing have caused the savanna to spread into what once were forested regions.

PLACE **Grassland Biome** Middle-latitude grasslands are located between the temperate forests and the desert biomes. Because of the semiarid climate, these grasslands usually do not have tree cover, except along rivers. There are two types of middle-latitude grasslands. Those parts that are wetter and closer to the forest biome support a tall grassland cover called **prairie**. The deep fertile soil of the prairie is particularly well suited for growing grains, and the world's largest wheat- and corn-growing areas are found in these prairie regions. Because of agricultural expansion, little original prairie grassland remains.

The drier desert margins of the middle-latitude grasslands support the short-grass vegetation called **steppe**. Though the steppe soil is rich, lower rainfall amounts make this grassland less productive than prairie. Great herds of animals

Geographical Sidelight

Savannas are found in more than 20 countries on four continents. There is a great deal of variety in the physical and biological characteristics of these savannas. For example, in addition to their herds of grazing animals and many predators, African savannas are known for a great diversity of large mammals, such as the rhinoceros, giraffe, elephant, and lion. The savannas of South America are rich in bird life and have a wide range of woody tree species. The savanna biome of Eastern Australia is known for its evergreen species of eucalyptus.

MULTICULTURAL PERSPECTIVES

List on the board the major world biomes described in this section. Call on students to explain how a biome's characteristics can influence how people live. Have students work with a partner to brainstorm about the types of housing, clothing, foods, or recreational activities that might be associated with each biome.

Closure
To help students differentiate among the five basic biomes, write the names of the biomes on the board, and ask students to suggest words and terms that describe each biome. List their suggestions under the appropriate heading on the board. Then have volunteers locate the biomes on a wall map.

ASSESSMENT
Assign the Section 2 Review questions on this page.

Reteaching
Assign Section 2 of Reteaching Worksheet 6.

Extension
Have students research and write a report on conservation efforts within a particular biome, such as a forest or desert. Have them focus on the balance between plant and animal life as well as the effects of human activities.

MAP TRANSPARENCIES
You might wish to show *The Changing Environment* transparencies in *REGIONS OF THE WORLD Map Transparencies.*

Section 2 Review
ANSWERS

1. Forest biome: tropical rain forests near the equator, temperate forests in middle latitudes, boreal forests in high, northern latitudes; savanna biome: tropical regions between the rain forests and the deserts; grassland biome: middle latitudes between the temperate forests and the deserts; desert biome: low and middle latitudes in desert climate regions; tundra biome: arctic regions, in the Antarctic, in high mountain elevations

2. Desert and tundra biomes

3. Many climates can support forests, but different kinds of forests are suited to different climates.

60

once roamed the steppes of the world. But today, for example, the vast grasslands and buffalo herds of the North American Great Plains have been replaced with wheat fields and cattle.

PLACE **Desert Biome** Desert biomes generally are found in the low and middle latitudes in the desert climate regions. Desert biome plants survive by using very little water or by storing water. Cacti are examples of desert plants that store water. Other desert plants, such as sagebrush, have tiny leaves with waxy surfaces and large root systems. They can withstand hot sunny days without letting very much water escape into the air.

Some desert plants actually avoid the dry conditions. These plants have short growing cycles and seeds that can survive underground for several years. When the rare rains come, the seeds sprout and the desert is covered with flowers. These plants die within a few weeks, but not before spreading seeds for the next rains. Because there is so little moisture in the desert biome, plants grow far apart, produce almost no humus for soil formation, and produce limited food for animals.

PLACE **Tundra Biome** Like the desert biome, the tundra biome supports little plant growth and has no trees. Only a few small specialized plants can grow in the cold climate conditions. The permafrost keeps plant roots from penetrating the soil. During the long winter, the plant life is frozen, and only a few strong animals survive. During the short summer, however, the tundra thaws and comes alive with mosses, lichens, low shrubs, and small flowering plants. Animals from other regions arrive to feed on the fresh plant growth and millions of insects. Birds migrate thousands of miles to feed and nest during this brief growing season. The largest areas of tundra are found in the Arctic regions of North America, Europe, and Asia. Smaller areas are located in the Antarctic and at high mountain elevations.

PLACE **Barren Regions** In the far polar regions, ice and snow remain all year, and it is impossible for plants to grow. These areas with no plant life or soil are called barren regions. The largest barren regions are the ice-cap climate regions of Antarctica and Greenland. Here, any animal life must depend upon the ocean for food, as the land provides none. Many high mountain

peaks in other areas of the world also are barren of plant and animal life. Here, elevation causes ice-cap climate conditions outside the polar regions. The cold temperatures, severe winds, snow, and glaciers make it impossible for plants and most animals to survive.

INTERACTION **People and the Biosphere** All living things—plants, animals, and humans—are interrelated within the biosphere. That is, events in one place in the biosphere can affect plants and animals in another part far away. Natural changes in the biosphere take place all the time. For example, a volcanic eruption may destroy a whole ecosystem around a mountain. Then, plants and animals slowly return to the area and repopulate it.

Changes that last for a long period of time may have a greater impact. When an ice age replaces warmer conditions, many plants and animals die. This long-term destruction may cause some plants and animals to become **extinct**, which means that they cease to exist.

Human actions also can have an effect on all parts of the biosphere. As you have learned, some of the biomes have changed size as a result of human activities. The tropical rain forests, for example, are becoming smaller as humans cut the trees for timber and clear the land for farms. Because of the loss of the forest, the soil loses its nutrients. Although the farms may be abandoned and the loggers might move elsewhere, the forest cannot return because the necessary soil was changed. Many kinds of rain-forest plants and animals, some not yet discovered, may become extinct.

While the forest biomes are becoming smaller, other biomes are expanding due to human activities, especially the savanna and desert biomes. If our biosphere is completely changed, what will life be like in the future? Humans have a special responsibility to handle carefully the resources on which all life depends.

SECTION REVIEW

1. Where is each of the five world biomes found?
2. Which biomes support little plant life?
3. **Critical Thinking** **Explain** why there are many types of forests.

CHAPTER 6 CHECK
ANSWERS

Building a Vocabulary

1. A plant community is just one part of an ecosystem. All of a region's plants, animals, climate, soil, and water function together as an ecosystem.
2. An orderly process that occurs over a long period of time as one group of plants replaces another
3. When old coral polyps die, their limestone skeletons serve as a platform on which new corals grow.
4. No longer existing; long-term destruction

Recalling and Reviewing

1. No. Animals need the oxygen and food provided by plants.
2. Land food chain: a plant is eaten by an animal, which is eaten by another, usually larger, animal; marine food chain: a drifting marine plant is eaten by a marine animal, which is eaten by a larger marine animal
3. Forest biome: rain, temperate, and boreal forests; savanna biome: grassland with scattered trees and shrubs; grassland biome: tall and short grasses; desert biome: cacti, sagebrush; tundra biome: mosses, lichens, shrubs, small flowering plants

---

CHAPTER 6 CHECK

Yellowstone National Park one year after the 1988 fire

Reviewing the Main Ideas

1. Life on Earth depends on the earth's plant, animal, and soil resources. Plants are at the bottom of the food chain and support animal life.
2. Soil provides the nutrients plants need to grow and survive. It also supports the food humans grow for survival. Soil is composed primarily of minerals from the weathering of rocks.
3. A plant and animal community that covers a large area of the earth's surface is a biome. There are five basic world biomes.

Building a Vocabulary

1. How is a plant community related to an ecosystem?
2. Describe the process of plant succession.
3. How is a coral reef formed?
4. Define the term *extinct*. What causes a living thing to become extinct?

Recalling and Reviewing

1. Would animals be able to survive on Earth if there were no plants? Why or why not?

2. Describe the stages of either a food chain on land or a marine food chain.
3. What are the five biomes, and what kinds of vegetation would you find in each?

Thinking Critically

1. In what biome type do you live? Give two examples of other places in the world that have the same type of biome as the one in which you live.
2. Predict what will happen to the biosphere if the forest biomes continue to get smaller.

Learning About Your Local Geography

Cooperative Project
With your group, investigate the soil in or near your schoolyard. Is it rocky, sandy, thick with humus, or hard like clay? Draw a cross-sectional diagram of what you think the soil horizon near your school might look like. On the diagram, label the climate type that determines the soil in your region.

Using the Five Themes of Geography

In a paragraph, explain how **human-environment interaction** has resulted in changes to the size of biomes.

Coral reef of **Mili Atoll, Marshall Islands**

Thinking Critically

1. Students should name the biome in which they live and cite two other locations with the same biome.
2. Answers will vary. Students should note that animals are dependent on the oxygen that plants give to the atmosphere. As rain forests are cleared, many plants and animals of the rain forest biome may become extinct.

Learning About Your Local Geography

Students' diagrams may vary but should be clearly labeled and demonstrate that groups performed the activity.

Using the Five Themes of Geography

Possible answer: Humans clear land for farming, cut trees for lumber, expand cities, and overgraze animals. These activities shrink the forest biome. As people and animals are forced onto savanna biomes, these biomes expand.

Unit 1 Review

SKILLS

Point out that sketch maps will help students learn to organize spatial information and understand the basic geographic concepts of location and place. At the beginning of each regional unit in *World Geography Today*, have students draw freehand maps of the region they are about to study. As you bring the unit to a close, ask them to draw another sketch map of the same region. By comparing the maps, students can assess their own progress in acquiring geographic knowledge. Have students include the maps in their individual portfolios.

GLOBAL SKILL BUILDER

You might wish to use **Mapping the Earth** and **Map Legends and Symbols** from the interactive *Global Skill Builder* CD-ROM to reinforce students' understanding of maps and mapmaking.

Practicing the Skill
ANSWERS

1. Student maps should include significant features such as major political boundaries, landforms, and should show awareness of scale, direction, and shape.

2. Student maps should show relative size and placement of classroom objects and layout.

3. A major street or landmark, or a "You Are Here" reference

62

UNIT 1
REVIEW

 SKILLS

Using Sketch Maps

As you have read, maps are an important tool of geographers. But you may occasionally find it useful to make your own maps. Sketching your own map can be a valuable study aid, giving you a better grasp of where places are located. You can study a map in a book or atlas for hours and still not know where things are. If you draw your own map based on what you see in the book, however, you will probably have a better understanding of the subject of the map. You will also be able to remember it better.

Of course, you will not always have access to other people's maps, and sometimes you will have to draw your own map truly "from scratch." You might be having a discussion with a friend about a geographical area during which it will be helpful for you to sketch a map quickly. Drawing your own maps might help you to understand events you read about in the newspaper or learn about on television. Sketch maps also have immediate, practical uses. For instance, someone might ask you for directions to a nearby shopping mall, football field, or museum. You could give directions orally, by saying "go down Main Street and make a left at the first traffic light," but it would probably be more helpful to the person if you drew a map.

How to Draw a Sketch Map

1. Decide what region your map will show. Choose boundaries so that you do not sketch more than you need to.
2. Determine how much space you will need for your map. Things that are the same size as each other in reality should be about the same size as each other on your map.
3. Decide on and note the orientation of your map. Most maps use a directional indicator. On most maps, north is "up."
4. Select reference points so that viewers of your map can quickly and easily figure out what they are looking at. For a sketch map of the world, reference points might be the equator and prime meridian. For a map of your community, a major street or river might be your reference point, or you might include a grid labeled by numbers and letters. Even maps of shopping malls indicate reference points—usually by saying "You Are Here" next to a dot or an asterisk.
5. Decide how much detail your map will show. The larger the area you want to represent, the less detail you will need. For example, a map of the world will not need names of streets, but a map of your community will.
6. You are ready to begin sketching. First, sketch general shapes, such as the continents if you are drawing a world map. If you do not know or cannot remember exact shapes, you can use circles, rectangles, and triangles.
7. Now, fill in more details as they occur to you—names of places, major land features, and so on.
8. Do not spend more than an hour working on your map, and do not try to make it perfect or overly detailed. Many useful maps can be sketched in just a few minutes.

PRACTICING THE SKILL

1. Select any map in this book and copy the main features of the area onto a piece of paper. Do not trace the map.
2. Make a map of your classroom, indicating the location of pieces of furniture, windows, closets, doors, and so on. Use a grid system to divide the map into sections.
3. Suppose a visitor to your community approached you right now and asked you for directions to a nearby library. Sketch a map for the visitor, indicating how to get from your present location to the library. What reference points will be most helpful?

BUILDING YOUR PORTFOLIO

Decide whether students will work individually or in groups on the unit projects described below. Students should then choose which project they will work on. Provide students with copies of the Portfolio Assessment worksheets in *Teaching Resources*. Chapters 1–6 worksheets and the Unit Portfolio Assessment worksheet will guide students step-by-step to complete either project.

BUILDING YOUR PORTFOLIO

Individually or in a group, complete one of the following projects to show your understanding of the geography concepts involved.

A Comparing Earth and Kax

You are a space explorer from the planet Kax across the galaxy. Your spaceship has landed on Earth. Prepare a presentation about what you find on Earth, and how it compares to Kax. For information about Earth, use Unit 1; for information about Kax, use your own imagination. Your presentation should consist of a written report, supported by visual materials, comparing and contrasting Earth with Kax. To prepare your report for Kax, you will need to do the following.

1. Compare the relative locations of Earth and Kax. How far away is each planet from stars, moons, and other planets? How are days and years defined on each planet? What are the seasons like? Diagrams or models may be helpful for presenting this information.
2. Describe the atmosphere and climates of Earth and Kax. What factors influence the climates on each planet? You may want to present this information using charts and maps.
3. Write a paragraph explaining the properties and importance of water on Earth. How important is water to Kax? How do the uses of water on Earth compare to those on Kax?
4. Identify the major landforms on Earth and on Kax. What forces have shaped the landforms on each planet? Create illustrations showing the landforms on both planets. Write captions for your illustrations.
5. Compare the biospheres of Earth and Kax. How does life survive on each planet? What are the biomes on each planet like? Flowcharts will help explain the food chains found on each planet.

Organize the information you have gathered, and write a summary. Present your information to Kaxians (the rest of the class).

B The Sun Is Dying

Scientists estimate that in several billion years, the earth's sun will die out. Suppose, however, that the sun were going to gradually die out over the next 50 years. What implications do you think this would have for Earth? Prepare a presentation to Congress, discussing these implications and suggesting solutions to any problems that might result from the sun's dying. Your presentation should consist of a written report supported by visual materials. You will need to consider the following questions as you prepare your report.

1. How does the earth move in relation to the sun? Without the sun, how might we define days and years? Would we continue to have seasons? Draw a diagram to illustrate what Earth's movement would be with a dying sun.
2. What would the earth's atmosphere and climates be like with a dying sun? Use a climate map to present your theory as to what would happen to Earth's climates.
3. What implications would the sun's dying have for water on Earth? Would the hydrologic cycle be affected? Would water still be able to exist in different forms? Write a paragraph discussing how the earth's water would be affected.
4. Would the sun's dying change the landforms of the earth at all? In a chart, compare the landforms of present-day Earth to those that might exist if the sun were dying.
5. How would life be affected by the sun's dying? What might be necessary to sustain life? How could humans best prepare for the sun's dying? Make a list of specific actions the government might take to prepare for the catastrophe.

Organize the information you have gathered, and write a summary. Present your information to Congress (the rest of the class).

Project A:
Descriptions of Kax will vary, but should show a logical comparison between the imaginary planet and Earth. Students should accurately record the relative location, atmosphere, climates, hydrosphere, landforms, and biosphere of Earth as they are described in Unit 1.

Project B:
Student projects should emphasize how the sun is vital to life on Earth.

UNIT 2 PEOPLE AND GEOGRAPHY *CHAPTERS 7–9*

OVERVIEW

In Unit 2, students will learn about the ways that humans interact with their environment.

Chapter 7 • GLOBAL CULTURES introduces human geography and the elements of world cultures.

Chapter 8 • GLOBAL ECONOMICS AND GLOBAL POPULATION defines economic geography and population geography and explores the relationship between economics and population in the world's nations.

Chapter 9 • RESOURCES AND ENVIRONMENTAL CHANGE describes the distribution and use of Earth's soil, forest, water, air, and fuel resources.

OBJECTIVES

1. Understand how culture traits and culture regions can influence the development of societies and create conflicts among nations.
2. Explain how people acquire and use the resources necessary to modify their environments.
3. Identify aspects of the information age.
4. Describe characteristics of developed and developing nations and how they affect economic, political, and environmental issues.
5. Analyze global population patterns.
6. Identify different types of energy sources, and explain the benefits and consequences associated with them.
7. Understand the effects of uneven distribution of resources.
8. Analyze different points of view and interpretations by geographers.

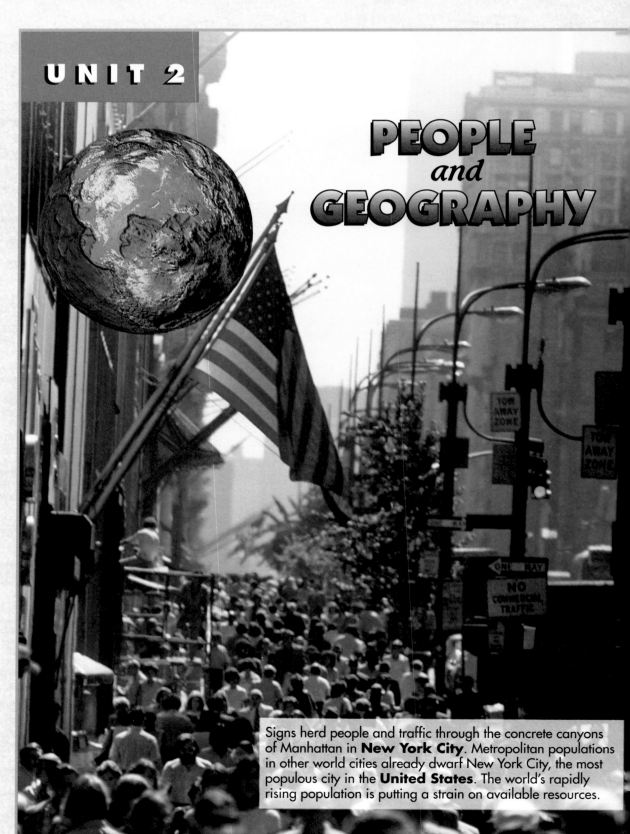

UNIT 2

PEOPLE and GEOGRAPHY

Signs herd people and traffic through the concrete canyons of Manhattan in **New York City**. Metropolitan populations in other world cities already dwarf New York City, the most populous city in the **United States**. The world's rapidly rising population is putting a strain on available resources.

TEACHING UNIT 2

Using the Illustrations

List on the board the world's 10 most populated cities—Tokyo, Mexico City, São Paulo, Seoul, New York, Osaka-Kobe-Kyoto, Bombay, Calcutta, Rio de Janeiro, Buenos Aires—and locate them on a wall map. Then have students look at the photograph on page 64 as you read the caption aloud. Ask students to speculate about what the people shown in the photograph of New York have in common with people living in the other cities listed on the board. *(They might share problems related to housing, employment, and transportation.)* Point out that Unit 2 explains how diverse groups of people globally confront some of the same problems in meeting basic needs for food, water, shelter, and earning a living.

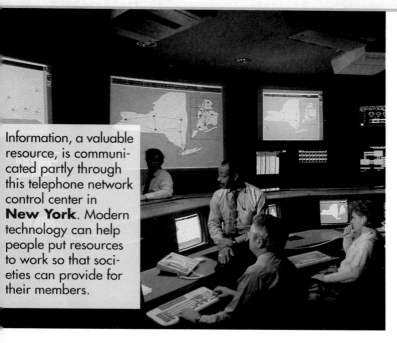

Information, a valuable resource, is communicated partly through this telephone network control center in **New York**. Modern technology can help people put resources to work so that societies can provide for their members.

UNIT 2 CHAPTERS

7 *Global Cultures*

8 *Global Economics and Global Population*

9 *Resources and Environmental Change*

PORTFOLIO ASSESSMENT

"Building Your Portfolio," on page 99, provides a choice of two unit projects for students to work on as they study Unit 2. The unit projects are supported by Chapter Portfolio Assessment worksheets in *Teaching Resources.*

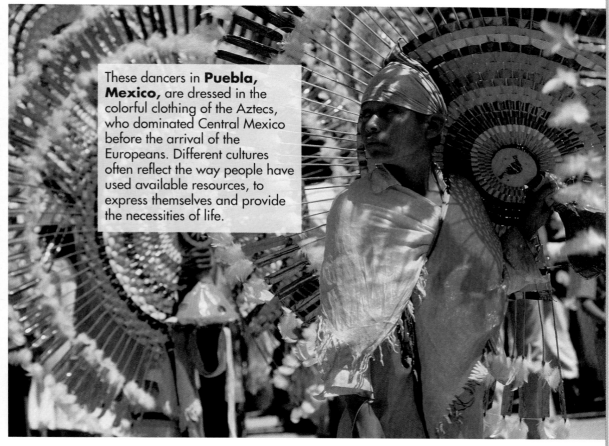

These dancers in **Puebla, Mexico,** are dressed in the colorful clothing of the Aztecs, who dominated Central Mexico before the arrival of the Europeans. Different cultures often reflect the way people have used available resources, to express themselves and provide the necessities of life.

CHAPTER 7 Global Cultures
Planning Guide

TEXTBOOK	TEACHING RESOURCES	OTHER AVAILABLE MATERIALS
Section		
1. Human Geography Pages 67–69	Reteaching Worksheet 7.1 Critical Thinking Worksheet 7 Challenge/Enrichment Worksheet 7	*Creative Strategies for Teaching World Geography* LESSON 4 The ABCs of Culture *Study Guide for Spanish Speakers* *Lecture Notes*, Chapter 7
2. People Shape the Land Pages 69–73	Reteaching Worksheet 7.2 Skills Worksheet 7	*REGIONS OF THE WORLD Map Transparencies:* The Changing Environment *Study Guide for Spanish Speakers* *Lecture Notes*, Chapter 7
3. Culture and World Events Pages 74–76	Reteaching Worksheet 7.3 Graphic Organizer 7	*REGIONS OF THE WORLD Map Transparencies:* World Religions *Creative Strategies for Teaching World Geography* LESSON 5 Using the Mural Method to Study Culture *Study Guide for Spanish Speakers* *Lecture Notes*, Chapter 7
Chapter Review Page 77	Chapter 7 Test Chapter 7 Portfolio Assessment See also *Alternative Assessment Forms*	*Test Generator*, Chapter 7

CHAPTER RESOURCES

Lecture Notes, Chapter 7
Teaching Resources
    Chapter 7 Graphic Organizer
    Chapter 7 Skills Worksheet
    Chapter 7 Critical Thinking Worksheet
    Chapter 7 Challenge/Enrichment Worksheet

REVIEW AND ASSESSMENT

Chapter 7 Check, page 77
Chapter 7 Test, *Teaching Resources*
Chapter 7 Portfolio Assessment, *Teaching Resources*
Test Generator, Chapter 7

ADDITIONAL RESOURCES

Books for Teachers

Brislin, Richard. *Understanding Culture's Influence on Behavior.* Orlando, FL: Harcourt Brace Jovanovich, 1993.

Foote, Kenneth E., ed., *Re-reading Cultural Geography.* Austin: University of Texas Press, 1994.

George, Luvenia. *Teaching the Music of Six Different Cultures.* Danbury, CT: World Music Press, 1987.

Books for Students

Burger, Julian. *The Gaia Atlas of First Peoples: A Future for the Indigenous World.* New York: Anchor Books, 1990.

Harpur, James. *The Atlas of Sacred Places: Meeting Points of Heaven and Earth.* New York: Henry Holt, 1994.

Reese, Lyn, and Jean Wilkinson, eds. *Women in the World: Annotated History Resources for the Secondary Student.* Metuchen, NJ: Scarecrow Press, 1987.

The World Today *Videodisc Program*

A variety of still images, short videos, and activities are available for you to use as you teach this chapter. *Lecture Notes* include bar codes for specific photographs.

Multimedia Resources

Cross Cultural Studies. (VHS, Parts 1–4; 30–49 minutes each) Educational Design. Available through Social Studies School Service, 10200 Jefferson Boulevard, P.O. Box 802, Culver City, CA 90232–0802.

Material World: A Global Family Portrait. (Multimedia CD-ROM, average families from diverse nations) Star Press Multimedia, P.O. Box 629000, El Dorado Hills, CA 95762.

Picture Atlas of the World. (interactive maps, photo bank, animation, audio on CD-ROM) Washington, DC: National Geographic Society.

MEETING INDIVIDUAL NEEDS

Limited English Proficiency (LEP)

To help reinforce understanding of the features that make up a culture, create a graphic organizer on the board. Classify elements of culture such as language, religion, government, economics, food, clothing, architecture, and family life. Have students name examples of each to describe culture in the United States, and write their responses on the board.

Learners Having Difficulty

On a world map, have volunteers identify the culture region that they or their parents, grandparents, or ancestors came from. Ask them to name features of their culture such as language, religion, food, clothing, family life, or occupation that were passed down from generation to generation through teaching or example.

Auditory Learners

Point out to students that language often reflects the infusion of new ideas or traits into a culture. Have students try to identify the following terms and the cultures from which they originated: fiesta *(Spain and Latin America)*, crepe *(France)*, taco *(Mexico)*, judo *(Japan)*, Nintendo *(Japan)*, frankfurter *(Germany)*, samba *(Brazil)*.

Visual Learners

Help reinforce the concept of culture by having students collect photographs that illustrate cultural traits. Students might refer to them for help in designing travel brochures about a specific location. Display the brochures in the classroom.

Tactile/Kinesthetic Learners

Have students use newspapers and magazines to research contemporary world conflicts and their causes. Then have students plot and label the locations of the conflicts on Outline Map 1 and identify causes. Use the maps to elicit a class discussion about how cultural differences may have contributed to the conflicts. Display the maps on a bulletin board.

Gifted Learners

Invite interested students to do research to find examples of current innovations in the United States. Have students give an oral report that describes the innovations and identifies the regions where they are being put to use. Students should offer reasons why the innovations may or may not become cultural traits in the United States.

MULTICULTURAL FOCUS

Point out to students that the United States is a leading exporter of popular entertainment to other countries. People around the world can regularly view American movies, music, television serials and crime shows, and sporting events. As American mass entertainment travels around the globe, it carries messages about our culture, our beliefs, and our attitudes. Ask students: What kinds of information about a society come through entertainment?

For the United States, global trade in the "entertainment age" has largely been one-way. Although information about other countries is available from many sources, Americans have limited direct access to foreign popular culture. Ask: If we received entertainment from other countries, what kinds of things would we learn about them that we do not already know? Ask if students have ever seen a foreign movie. Ask: What did they learn from it about the culture of the country?

Many Americans already have access to 40 or more channels on cable television. Fiber-optic technology has prepared the way for a new explosion of home entertainment choices with the promise of some 500 cable channels. Have students imagine that 10 percent of these channels will be filled with foreign-produced programming. Ask: What types of programs would students most like to see? What types of information would be most important to learn about other countries? Have students suggest formats for programs that promote understanding of other cultures.

Section 1: Human Geography
Section 2: People Shape the Land
Section 3: Culture and World Events

OVERVIEW

In Chapter 7, students will learn about cultural, or human, geography, including the elements that define cultures.

Section 1 explains the aspects that link or separate global cultures. It analyzes the connections between technological innovation and cultural change.

Section 2 investigates the use of natural resources for subsistence and commerce. It examines the relationship between urbanization and industrialization.

Section 3 describes how elements of culture such as nationalism, religion, politics, and economics can cause conflict. It also defines the information age and explains how new technologies will affect people in the future.

About the Author

Mongi Bousnina is a government official in Tunisia. As head of the nation's culture ministry, he works to preserve and emphasize the North African country's national identity and cultural traditions.

CREATIVE STRATEGIES FOR TEACHING WORLD GEOGRAPHY

LESSON 4 The ABCs of Culture: A Framework for Understanding Culture

Primary Source

Description of Change: excerpted
Rationale: excerpted to focus on main idea

CHAPTER 7

Global Cultures

❝O̲ur most important task is to develop a spirit of democratic pluralism, a culture that respects the other person, whether he's a neighbor or the people across the water. But to have the strength for that kind of respect, one must be confident in one's own heritage and national personality. ❞

Mongi Bousnina

I̲n Unit 1, you learned about the physical geography of our world and the forces at work that continue to shape the earth's surface. Landforms, climate, water resources, and vegetation are key features of our environment. Much of the world around us, however, has been formed by people. Cities and villages, farm fields and grocery stores, and highways and airports have been created to meet our ever-changing needs.

People have divided the earth into nearly 200 countries, which are home to people of varying languages, religions, economic systems, and political views. The study of the world's people—how they live and how their activities vary from place to place—is called cultural or human geography.

1993 State of the Union Address before the U.S. Congress

GEOGRAPHY DICTIONARY

culture

culture trait

culture region

ethnic group

innovation

diffusion

acculturation

domestication

subsistence agriculture

commercial agriculture

urbanization

culture hearth

nationalism

fundamentalism

totalitarian government

democratic government

tariff

quota

66

Using the Quotation

Call on volunteers to write three or four dictionary definitions of *pluralism* on the board. Point out that *plural* means "more than one." Next, read aloud the quotation on page 66, and ask students to consider ways that societies can be pluralistic. Have volunteers restate the quotation in their own words.

Using the Geography Dictionary

Write four headings on the board: Elements of Culture, Beliefs, Meeting Needs, Government. Then have students classify the chapter's vocabulary terms under the appropriate heading or headings. Point out that some of the terms may fit under more than one heading.

SECTION 1
OBJECTIVES

The objectives reflect the Focus questions on this page and are assessed in the review on page 69.

1. Define and explain key concepts related to culture.
2. Discuss culture regions and how cultures change.

HUMAN GEOGRAPHY

FOCUS

- *What elements make up a culture?*
- *What is a culture region? Why do cultures change?*

What Is Culture? The key term in understanding human geography is **culture**. Culture includes all the features of a society's way of life. Culture is learned, passed down from generation to generation through teaching, example, and imitation. Culture includes language, religion, government, economics, food, clothing, architecture, and family life. It includes all of a society's shared values, beliefs, institutions, and technologies—or skills, tools, and abilities. Culture gives roles to individuals and establishes relationships within groups.

To better understand culture, we can divide it into activities and behaviors that people repeatedly practice. These activities and behaviors are called **culture traits**. There are many kinds of culture traits. Some, such as learning to read and do arithmetic, are much the same around the world even though various cultures use different alphabets or symbols. Many culture traits, however, vary from place to place. For example, a typical U.S. teenager usually eats with a knife, fork, and spoon. Chinese teenagers, however, eat with chopsticks, and Malaysian teenagers have been taught to eat with their fingers. Each of these learned culture traits is considered the best method in its own culture.

Culture traits often are linked together. For example, Islam and Judaism are two religions that do not allow the eating of pork. Therefore, in Southwest Asia, pigs usually are not raised by either Islamic or Jewish farmers. Islam also does not allow alcoholic beverages, while Judaism permits wine. As a result, grapes grown on an Islamic farm are eaten fresh or are dried into raisins, while the grapes grown on a Jewish farm may be made into wine. These examples show that the culture traits of religion and farming are linked.

Culture traits also change through time. Sometimes, these changes are as simple as new clothing styles. For example, ask your parents or other older adults what clothing was fashionable when they were teenagers. Often, changes in culture traits through time are more complex. In the United States, for example, changes in transportation have changed shopping habits. People once rode horse carts or streetcars to downtown or central areas to do their shopping. Over time, these modes of transportation were replaced by buses and automobiles. Today, highway networks and mass transit systems take shoppers away from city centers to huge shopping malls in the suburbs.

REGION **Culture Regions** An area that has many shared culture traits is called a **culture region**. One country may be a single culture region. France and Japan are two examples.

COOPERATIVE LEARNING

Organize the class into groups of four or five. Assign each group a different topic such as food, shopping, dance, clothing, sports, or music. Then ask students to identify culture traits for their topic that are present today in their communities. Each group should poll older people (parents, neighbors, grandparents) to find out how the traits may have changed over time. Direct one member of each group to compile the information and one or more to illustrate their findings. A spokesperson from each group can present the ideas to the class.

Culture traits help determine how a society marks important events in people's lives. Traditional clothing and rituals mark the wedding days of these people in **Slovakia** (far left), **Indonesia** (middle), and **Tanzania** (above).

67

TEACHING SECTION 1
Motivator

Ask students to name traditional events that occur regularly in their communities, and write their responses on the board. *(parades, arts and crafts shows, garage sales, sports tournaments, concerts)* Point out that these events are a part of a community's culture.

Building a Vocabulary

Draw a graphic organizer on the board that suggests the hierarchical arrangement of **culture** (the broadest term), **culture region** (what delineates a culture), and **culture trait** (specific characteristics of a culture or culture region). Point out that **diffusion** relates to *diffuse*—"to spread out."

Guided Practice

Remind students that a culture region is not always defined by political boundaries; rather, a country may contain several culture regions. Ask students to name some culture regions within the United States and to explain why they may be less distinct today than 100 years ago. *(influence of television, mobility of population, standardized education, business franchises)*

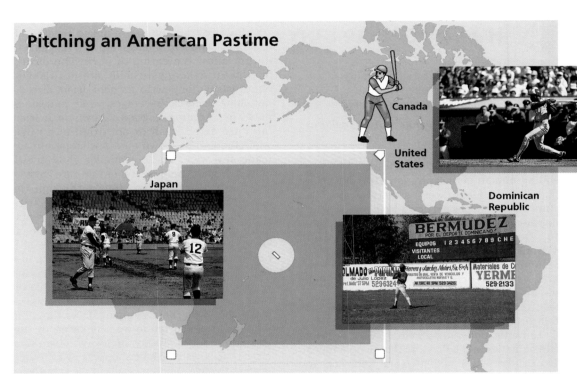

Pitching an American Pastime

Canada · United States · Japan · Dominican Republic

Baseball might be as American as apple pie, but fans around the world are evidence of its wide diffusion from the Yankee bullpen. About 16 percent of players on Major League Baseball 1993 spring rosters were born outside the United States. Many of those players were from Middle and South America.

Sometimes, though, political boundaries cut across culture regions. As a result, some countries contain several culture regions. This is particularly true in the countries of Africa. In countries such as Nigeria and Zaire, many **ethnic groups** representing different culture regions live within the boundaries of one country. An ethnic group is a population that shares a common cultural background.

A culture region also can be made up of several countries. An example is Central America, which includes Guatemala, Honduras, Nicaragua, El Salvador, Costa Rica, and Panama. These countries share similar languages, religions, and historical traditions.

MOVEMENT **Culture Change** Cultures develop slowly, but they are changing constantly. New culture traits are added as older traits fade from memory. Two concepts are important in understanding how cultures change. The first is **innovation**. People are always thinking of new ways to do things. Only those new ideas that are useful and valuable will last, however. New ideas that are accepted into a culture are called innovations. While some innovations only happen once and then spread throughout the world, other innovations happen in more than one place and at different times. Boat building, for example, was developed independently by people all around the world. Though the boat designs differ in detail, boats appear in the artwork of all ancient civilizations.

The second concept necessary for understanding culture change is **diffusion**. Cultural diffusion happens when an innovation or other culture trait spreads into another culture region, which then adopts that element into its own culture. For example, while jazz music partly has its roots in African music, it took hold as an art form in New Orleans and now has spread throughout the world. Video games and compact discs are two examples of innovations that have diffused into other culture regions during your lifetime.

Independent Practice
Have students label on Outline Map 1
the regions of the world that share
languages, religions, and political or
economic systems. They may use their
textbook or atlases as references.
Students should write a statement at
the bottom of their map that suggests
ways culture traits spread throughout the world. (*tourism, music, art,
books, radio, television, movies*)

Closure
Have students name examples of culture traits and match them with culture regions.

ASSESSMENT

Assign the Section 1 Review questions
on this page.

Reteaching
Assign Section 1 of Reteaching
Worksheet 7.

SECTION 2
OBJECTIVES

The objectives reflect the Focus
questions on this page and are
assessed in the review on page 73.
1. Identify different ways people
acquire necessary resources.
2. Analyze reasons for urbanization
and industrialization.

Culture traits diffuse to new regions in two
ways. First, an innovation might gradually spread
from its source. Imagine a ripple of water spreading outward in a pond when a rock is thrown into
the water. In the United States, fashions spread in
this way from a center of innovation such as Los
Angeles, California. Likewise, country music diffuses from Nashville, Tennessee, to the rest of the
country. Second, culture traits spread when people
move from a region and carry the innovation with
them to a new area. America's 500-year period
of settlement by Europe, Africa, and other areas
of the world had a strong impact on American
culture. For example, European colonists brought
the Christian religion to the Americas, and
Europeans and Africans brought new foods and
cooking methods.

When one culture changes a great deal
through its meeting with another culture, the
process is called **acculturation**. The adoption of
U.S. blue jeans and fast foods by teenagers
throughout the world is an example of acculturation. Acculturation is usually a two-way process
in which each culture changes the other. For
example, while U.S. fast-food restaurants have
spread hamburgers and fried chicken worldwide,
Chinese, Mexican, Japanese, and Italian foods
have spread across the United States.

One of the best examples of cultural diffusion is the spread of languages. Geographers
study the spread of languages to trace patterns of
political, economic, and religious expansion of
cultural groups over time. Language is probably
the most important culture trait. People who cannot talk with each other have a difficult time sharing ideas. Often, the spread of innovations is
accompanied by the language used to describe
them. As U.S. popular culture diffuses around the
world, so does the English language.

SECTION REVIEW

1. List eight parts of society that are
influenced by culture.
2. What happens when different culture
regions come into contact with each other?
3. Critical Thinking   **Assess** whether an
individual can belong to more than one
cultural group.

2 PEOPLE SHAPE THE LAND

FOCUS

● *In what ways have people acquired the
resources they need to live?*

● *Why do cities develop and grow? What
purposes does industrialization serve, and
what problems does it create?*

All people have the same basic needs: food, clothing, and shelter. Food, water, and the materials
needed for daily life are basic resources. To live,
people must gather these resources from their
environment. All of the resources we need cannot
be found in one place, however. People must
travel and trade to obtain resources from other
regions. How people have used the land to
acquire their resources has greatly affected
Earth's human geography. Throughout history,
new discoveries and their diffusion have altered
places dramatically.

MOVEMENT   **Hunting and Gathering**
Hunting and gathering was the main way of life
for most of human history. Hunter-gatherers live
by foraging, or collecting foods from the surrounding land. To be successful, hunter-gatherers
must know their environments very well.

In early times, hunter-gatherers often
roamed across large areas. They moved their
camps with the seasons, in search of different
plants and animals throughout the year. There are
few hunter-gatherers left today. They survive only
in isolated areas where farming is difficult
because of harsh climates or poor soils. Among
these survivors are the San (Bushmen) of the
Kalahari Desert in Africa and the Aborigines of
the Australian deserts. Small groups of hunter-
gatherers also live in the humid-tropical forests of
South America. The Inuit (Eskimo) who hunt and
fish along the Arctic shores of North America and
Siberia are hunter-gatherers as well.

Even though hunter-gatherers often have
simple tools and limited technology, they all have
complex social customs and religious beliefs.
Most hunter-gatherers live without the modern
products other cultures have come to rely on.

Text continues on page 72.

**Geographical
Sidelight**
*The San (Bushmen) peo-
ple of Namibia have lived
in the Kalahari Desert for
40,000 years. Survival of
their culture was threat-
ened by the South African
government, which took
away most of their land.
The Namibian govern-
ment, newly independent
from South Africa, is
encouraging the San to
revert to their traditional
way of life by returning
their land to them.*

Section 1 Review
ANSWERS

1. Possible answers:
language, religion,
government, econom-
ics, foods, clothing,
architecture, family
life, values, beliefs,
institutions, tech-
nologies
2. They influence each
other, each adopting
traits of the other(s).
3. Many individuals
belong to more than
one cultural group,
such as an ethnic group
and a religious group.

69

Global Connections

Identifying Cultural Influences

Point out that, throughout history, cultural information has traveled from one part of the world to another primarily because of trade, exploration, and migration. Have students think of additional means of spreading cultural information in modern times. *(Responses may include advertising, textbooks, literature, newspapers and magazines, radio, television,* movies, computers, personal relationships.) Ask them to write a paragraph that describes how *they* have obtained information about other cultures. Have students specify the method that has provided the most information to them, and have them include in their paragraphs examples of what they learned through this method. Call on volunteers to read their paragraphs aloud.

GLOBAL CONNECTIONS

Building a World Culture

Thousands of Russians jam this stadium in **Moscow** for a music festival. American rock music is increasingly popular in the republics of the former Soviet Union.

In the past few decades, modern technology has brought different peoples into closer contact than ever before. Jet airplanes and high-speed trains carry immigrants, tourists, and travelers quickly from nation to nation. Television, radio, and motion pictures have turned the world into one giant theater. Music television is available in more than 70 countries, bringing together performers and audiences from around the world.

An Emerging World Culture

Throughout the world, cultures are mixing and mingling to form a new "world culture." Although they may live oceans apart, people are exposed to common music, food, clothing, and ideas. In this new world culture, music plays a central role. Boris Grebenshikov, a well-known popular musician in Russia, recalls that he first heard rock music on the Voice of America, a U.S. radio station that broadcast into the Soviet Union. The freedom and energy expressed in the music, he says, "changed my life, and the lives of my friends, forever." For Grebenshikov and young people like him, rock music became a means of expressing their individuality and of defying authority—even the authority of the Communist government.

Blue jeans are popular with consumers around the world, including these young shoppers in **Tokyo, Japan**. U.S. companies exported more than 90 million pairs of pants made from blue denim to other countries in 1994.

Modern transportation and communication have bridged the geographical obstacles that once separated countries and cultures.

Maulana Karenga teaches these African American children in **Los Angeles** about *Kwanzaa*, an annual celebration of black culture. Karenga developed *Kwanzaa* in the United States in 1966. The word is part of a phrase that means "first fruits" in the African language of Swahili. Based on a traditional African festival, *Kwanzaa* is an example of how cultures from around the world help shape U.S. culture.

In the same way that rock music represents the music of the new world culture, blue jeans represent the clothing. Blue jeans are so sought after in Russia, for example, that jean-clad tourists often are asked if they would be willing to sell the pants they are wearing. In some countries, even used jeans sell at high prices.

The food of the new world culture, like the clothing, is casual and convenient. Fast-food outlets have sprung up around the world. In Moscow, people form lines stretching entire city blocks to sample the menu at a new hamburger franchise. In Ho Chi Minh City, Vietnam, people flock to the *Ca-li-pho-nia Ham-bu-gó*, the "California Hamburger."

The American Influence

Rock music, blue jeans, and fast food are all products of American popular culture. People in less prosperous nations view these products as symbols of a wealthy nation, able to produce many consumer goods. Yet some critics question this emphasis on consumer items. They worry that people may value wealth and material possessions more than tradition, family, community, and spirituality.

In contrast, those who defend American popular culture say that it reflects important ideals such as democracy. They point out that rock music, blue jeans, and fast food are available to many people rather than to a privileged few. Others say that American culture reflects the important ideal of diversity. American culture is molded by a variety of ethnic groups and appeals to different peoples around the world.

Although American popular culture is important in shaping the new global culture, cultural exchange is not a one-way process. American musician Paul Simon, for example, found inspiration in the music of South Africa. Cultural exchanges take place between many other societies as well. The English band 3 Mustaphas 3 combines music styles from Latin America, Africa, and the Balkans. The Japanese group Shang Shang Typhoon mixes traditional Japanese styles with reggae, funk, and salsa.

A World of Bridges, Not of Walls

Around the world, cultural ideas are crossing borders and barriers that once separated people. One symbol from the past is the Berlin Wall, which divided Germany and separated Communist and Western societies. Today, a new symbol is needed for the bonds of shared tastes and ideas being forged by the emerging world culture.

One such symbol of world culture was offered by Gianni de Michelis, chairman of the World Arts Council. "Walls," de Michelis said, "are the architectural emblem of the age we have just left; bridges of the age ahead."

Country Music, Japanese-Style

Shogi Tabuchi of Japan studied classical violin, but he also admired many forms of American music. A chance meeting with country music legend Roy Acuff sparked friendship and encouragement from the entertainer, and Tabuchi soon began a musical career in the United States. Now he performs at his own theater in Branson, Missouri, and delights audiences by playing bluegrass and Cajun fiddle music, songs from Broadway musicals, and American pop tunes from the 1940s.

▼ YOUR TURN

Do you agree with critics of a world culture or those who believe a world culture brings people together? **Imagine** you are a delegate to the United Nations who serves on a commission concerned with promoting cultural exchange. **Write** a letter reflecting your opinion about a world culture.

71

Motivator

Have students consider this situation: You have been transported to an uninhabited area. Ask: What problems must you solve in this new environment in order to survive? Have students generalize about people's basic needs. Tell them that they will learn how people throughout history have acquired necessary resources.

Building a Vocabulary

Help students understand **domestication** by identifying ways that the term *domestic* is used. *(living near humans, related to household or family)* Contrast **subsistence agriculture** and **commercial agriculture** by pointing out that subsistence comes from a Latin word meaning "to maintain"; *commercial* relates to *commerce*, referring to the sale of products.

Guided Practice

Have students work in groups to create poster-sized charts showing the progression of human activity from hunting and gathering through industrialization. Suggest that agriculture, village life, and urbanization be included as intermediate stages of development. Ask groups to include under each stage the innovations that were necessary for it to come about.

MULTICULTURAL PERSPECTIVES

Remind students that when there is a movement of peoples within a region or from one region to another, such movement promotes the sharing of ideas and technologies. Have students name ideas and technologies that they might share if they were to move to an area considerably different from where they now live (for example, from urban to rural, or from English-speaking to non-English-speaking).

THEME
Movement

The U.S. population in 1900 was 40 percent urban and 60 percent rural. In 1990, it was 75 percent urban and 25 percent rural. Two great periods of farm-to-city migration came during the economic expansions of the 1920s and the 1940s and 1950s.

CREATIVE STRATEGIES FOR TEACHING WORLD GEOGRAPHY

LESSON 5 Using the Mural Method to Study Culture: A Case Study of the Mbuti

72

This satellite image shows ordered crop patterns near the Salton Sea in **California** contrasting with the irregular farm plots found across the border in **Mexico**. More efficient cultivation techniques make for more productive land on the northern side of the border.

INTERACTION **Agriculture** Discovering how to raise animals and how to plant crops were perhaps the two most important innovations in human history. The development of agriculture (farming) changed how people interacted with their environments. Agriculture appeared when hunter-gatherers studied a plant or animal so closely that they found out how to grow or tame it. This innovation is called **domestication**. Geographers believe agriculture first appeared about 10,000 years ago in several areas of southwestern Asia.

Farming is a more reliable source of food than hunting and gathering. To grow food, however, the earth's natural vegetation had to be altered. As agriculture spread, forests were cleared. Waters were diverted to irrigate drylands. Food now came from fields and pastures, and people did not have to move with the seasons. Agriculture permitted people to grow more food year after year.

Farming also encouraged village life. Because people moved less often, they could live in larger groups. They shared innovations easily, and social life became more complex. People began forming governments to create social order, and religions became more structured. Large areas of land were then devoted to agriculture. More food led to increased populations.

Many people still make their living by agriculture. Two types of agriculture are common today. In traditional or **subsistence agriculture**, people grow food on small farms mostly for their own families. Some surplus food may be traded or sold. Perhaps one or two crops are raised to sell for cash at a nearby market. In subsistence agriculture, tools are basic. The farm's family members and the farm animals are the main source of labor. Subsistence agriculture is common today in the world's poorer countries, where it may be a town or village's main economic activity.

In **commercial agriculture**— the growing of crops for sale—farms are large and use modern technology. High-yield seeds, industrial fertilizers, and powerful machinery are typical. Farm animals are raised using scientific methods. Commercial agriculture requires fewer people, yet the yields far exceed the needs of the farmers. In the United States, only a few million farmers produce the food on commercial farms for more than 260 million people. A large amount remains to sell to other countries. Commercial agriculture is practiced throughout the world's wealthier countries.

City Life
Hunter-gatherers lived in small camps, and the first farmers lived in villages surrounded by fields. Once food was easier to grow, people had more time to work at other things. Some workers specialized in making tools or clothing. Others worked in government and religion or became traders and merchants. These types of work are key characteristics of cities. Cities first appeared more than 5,000 years ago. Growth in the proportion of people living in towns and cities is called **urbanization**.

The development of cities signaled a change in human geography. People in cities lived close together, and communication was easier. As a result, new ideas and technologies could be shared and tested quickly. The first cities became centers of innovations. Regions of important new ideas and developments are called **culture hearths**. For example, Ancient Mesopotamia (mehs-uh-puh-TAY-mee-uh), where writing developed, is a culture hearth. Culture hearths developed in many areas of the world and at different times. Cultural diffusion became more rapid as trade carried goods and new ideas from city to city.

Cities continue to play a major role in world geography as the source of most innovations.

Early World Culture Hearths

Early culture hearths were scattered across the world and diffused ideas and inventions to other regions. To what region did some ideas and inventions from Mesoamerica diffuse?

Geography and History

In the early 1800s, as the textile industry in England was becoming more mechanized, groups of hand weavers, knitters, and other craftspeople rebelled against the new technology that threatened their livelihoods by destroying as many machines as they could. The rebels came to be called "Luddites" for the legendary Ned Ludd, an apprentice who lost his temper and smashed his knitting frame.

Map Caption Answer

Southwestern and southern United States

Cities are centers of manufacturing, communications, education, government services, and the arts. Television, radio, and computer networks, along with transportation facilities, link major cities throughout the world.

Industrialization In the early cities, workers used simple tools to make clothing and other goods by hand. Such goods were custom-made and carefully crafted, but they were scarce and costly. Industrialization allowed large amounts of standardized goods to be made at a lower cost. It also dramatically altered the earth's physical and human geography.

Industrialization spread rapidly in western Europe during the eighteenth and nineteenth centuries. In world history, this period of time was known as the Industrial Revolution. Factories were clustered near large rivers where running water provided power. Later, coal and petroleum provided even more efficient sources of power and freed industrial sites from having to be located near water. As a result, most factories sprang up in cities where large pools of labor were available. In turn, industrialization led to increased urbaniza-

tion as people from the countryside moved into the cities to find jobs in factories.

Industrialization continues to spread around the world. New factories are being built in countries where only a few years ago subsistence farming was the main economic activity. These countries hope that industrialization will improve their economies. Changing to an industrial economy can be expensive, however, and requires resources such as water, fuel, and low-cost labor. In addition, industrialization creates pollution and industrial waste and depletes natural resources.

SECTION REVIEW

1. Why was the development of agriculture an important innovation?
2. Why do urbanization and industrialization often accompany each other?
3. **Critical Thinking** **Imagine** that a previously isolated hunter-gatherer group suddenly came into contact with an industrial society. What might be the consequences of this meeting?

Section 2 Review
ANSWERS

1. It provided a more reliable source of food than hunting and gathering, encouraged village life, and allowed populations to grow.
2. People living in cities generate ideas and labor to fuel industrialization; industrialization encourages people to come to cities for jobs.
3. Possible answer: The hunter-gatherers' way of life would probably be changed by the industrial society. Both societies could learn from each other.

SECTION 3
OBJECTIVES

The objectives reflect the Focus questions on this page and are assessed in the review on page 76.

1. Analyze how nationalism, religion, value differences, political ideologies, and uneven distribution of resources can cause conflict.
2. Discuss possible consequences of the advent of the information age.

TEACHING SECTION 3
Motivator

Provide students with copies of a major newspaper. Working in groups, have students circle the headlines of stories about conflicts at local, national, and international levels. A student in each group should list the source or underlying cause of each conflict. Have the groups share their data. Record it on the board and save it for the Analyzing activity.

Building a Vocabulary

Write **nationalism** and **fundamentalism** on the board. Have volunteers define the root words. Point out that the suffix *-ism* denotes a theory or doctrine and can apply to such things as feelings for one's country, religion, or politics and economics. Next ask the class to give examples and define other "isms." (*examples: Judaism, Catholicism, Buddhism, capitalism, patriotism*)

GLOBAL PERSPECTIVES

Have students locate the former Soviet Union and Yugoslavia on a pre-1990 map or globe. Note that in the early 1990s both of these countries split up into smaller national units with new forms of government. Point out that cultural identities, cultural traits, and nationalistic feelings led to both restructurings. Ask students to recall what they may have seen on television or read in newspapers about cultural conflicts over religion, values, economics, or politics in these places.

MAP TRANSPARENCIES

An overhead transparency of the map on this page is available in *REGIONS OF THE WORLD Map Transparencies.*

Map Study
Answers will vary.

CULTURE AND WORLD EVENTS

FOCUS

- *What are the major causes of conflicts between nations or cultures?*
- *What is the information age?*

Sometimes, disagreements among people and among countries can be understood by studying the countries' different culture traits. This section examines the main causes of cultural conflict.

Nationalism Most people feel proud of their culture and country. Feelings of pride and loyalty for one's country are called **nationalism**. Nationalism often is expressed in patriotic songs, symbols, and writings. Unfortunately, one group's pride can conflict with that of another group. When one group believes that it is better than others, discrimination, or unfairness, toward different groups results. Competing nationalisms can sometimes lead to war. For example, the former country of Yugoslavia dissolved violently into separate ethnic territories during the early 1990s. And conflicts among ethnic groups in India have caused the country to redefine its internal boundaries many times.

Religion Religion is a key culture trait in many societies. Religions advise people on personal behavior, sacred times, and sacred places.

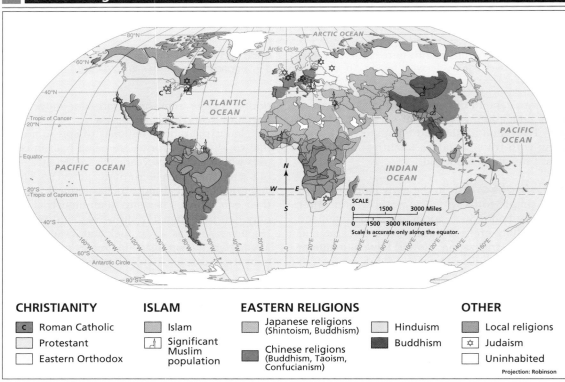

World Religions

CHRISTIANITY
- C Roman Catholic
- Protestant
- Eastern Orthodox

ISLAM
- Islam
- Significant Muslim population

EASTERN RELIGIONS
- Japanese religions (Shintoism, Buddhism)
- Chinese religions (Buddhism, Taoism, Confucianism)
- Hinduism
- Buddhism

OTHER
- Local religions
- Judaism
- Uninhabited

Projection: Robinson

MAP STUDY This map shows which religions are dominant in various parts of the world. The followers of most major religions are found in more than one region of the world. Significant numbers of followers of Judaism, for example, can be found in Southwest Asia, Europe, and North America. Which religions have places of worship in or near your community?

Analyzing

Causes of Conflicts Have students review the class list of conflicts developed in the Motivator activity and suggest categories for classifying the sources of conflict. Ask students whether or not there is a relationship between conflict and culture traits.

Guided Practice

Direct students to the map of world religions on page 74. Have them work in pairs to find out where these religions began and the current number of followers of each. Combine the findings into a chart. Have students develop generalizations about the distribution of each religion. Then have them hypothesize about how religions spread from their region of origin.

Independent Practice

Have students use encyclopedias or other library resources to identify professions that will be important in the twenty-first century. Ask students to select 10 professions and then list the job skills that will be necessary for success in these fields. Have them label the professions that are related to the "information age."

People also may receive a sense of personal fulfillment from their religion.

Conflicts emerge when people of one religion believe that theirs is the one true religion. When people do not tolerate other religions, disputes can lead to unrest and even war. Because each side believes its cause is holy, religious conflicts can be bitter and long lasting.

There are many modern examples of religious conflict. In Sri Lanka, the Buddhist Sinhalese have warred with the Hindu Tamils. In Northern Ireland, Roman Catholics and Protestants have battled each other. In Southwest Asia, religious differences among Jews, Muslims, and Christians have caused bitter conflicts.

Traditional Versus Modern Values

Some cultures adapt to change easily, while other cultures do not. Changes that appear to threaten traditional religious practices or the traditional roles of men and women may be resisted. In some cases, religious **fundamentalism** has been a direct reaction to the spread of Western culture and values throughout the world. Fundamentalism is a movement that stresses the strict following of basic traditional principles. An example is Islamic fundamentalism now spreading in Southwest Asia and North Africa.

Conflicts over values arise from several causes. Often, young people accept innovations that older generations reject, such as popular music and new styles of clothes. People who live in rural areas sometimes resist new technology, while city dwellers welcome it.

Politics

How a group of people should be governed has been a source of conflict throughout world history. Some countries are governed by one person and a few advisers. In these **totalitarian governments**, only a few people decide what is best for everyone. In other countries, everyone has a voice in their govern-

ment. These are **democratic governments**. Human rights often are violated by totalitarian governments. In contrast, democratic governments promote human rights and value individual freedom.

In countries with many political parties, conflicts may arise over which party should govern the country. Even when democratic elections are held, sometimes the losing party or parties may not honor the election. Conflicts also appear over different ideas about the basis for a country's laws. For example, should laws be based on religious rules or on the ideas of political philosophers or elected officials?

The United Nations is a worldwide organization that tries to settle problems among and within countries. Representatives from countries can work with the United Nations to solve problems without violence. UN peacekeeping forces are on duty in many places. In the early 1990s, UN forces aided the United States in its effort to stop civil war and feed starving people in Somalia. UN forces are not always successful. In the former Yugoslavia, UN forces at times were blockaded from providing food and medical relief.

Economics The world's resources are not evenly distributed. Some countries have many raw materials and successful industries. Others have only a few. These inequalities may encourage a poor but well-armed country to invade its richer neighbor. In the past, wealthy countries have colonized other lands to acquire valuable resources.

Sometimes, economic conflicts between countries are nonviolent struggles. The conflicts may appear

Minimum Voting Ages (in selected countries)

Country	Minimum Voting Age
United States	18
Germany	18
Japan	20
United Kingdom	18
Israel	18
India	18
Russia	18
Iran	15
New Zealand	18
Malaysia	21
Brazil	16
Mexico	18
Kenya	18
Egypt	18
Sweden	18

Source: *The World Factbook 1994,* Central Intelligence Agency

This chart provides a random sample of legal voting ages in various countries. As of 1994, Iran had one of the lowest legal voting ages in the world. Teenagers in the United States can legally vote at age 18.

GLOBAL PERSPECTIVES

Modern telecommunication systems, especially satellite television, have changed the speed at which the world becomes informed about international events. The term *global village* is sometimes used to describe the way the world has "shrunk" because billions of people share the same experiences through television and radio broadcasts. Ask students to name some of the international events in recent years that they have become aware of because of television. *(Students might mention the Persian Gulf War, the conflicts in Somalia and Bosnia, or the collapse of Soviet communism and the removal of the Berlin Wall.)*

75

Closure

Review with students sources of conflict among peoples. Elicit a discussion about how these conflicts could be resolved. What role might technology play? Then ask: Can you imagine one global nation within 50 years? Why or why not?

ASSESSMENT

Assign the Section 3 Review questions on this page.

Reteaching

Assign Section 3 of Reteaching Worksheet 7.

Extension

Have interested students use an atlas to research the distribution patterns of world resources such as coal, oil, iron, fresh water, and arable land. Students should report their findings in class and suggest global problems that could result from unequal distribution of resources.

Section 3 Review
ANSWERS

1. Possible answers include nationalism, religion, differences in values, ideas about government and laws, and the uneven distribution of resources.

2. Possible answers include talking on the telephone, watching television, using computers and video technology, reading, and attending school.

3. Answers will vary. With improved communications, people might have fewer differences that could cause conflicts. Dramatic changes brought by the information age, however, might increase conflicts, such as between people who have the new technologies and those who do not.

76

The technological revolution in the last half of this century has dramatically changed the way we exchange information. The telecommunications satellite (above), telephone bank (above, right), and high school computer lab (right) all are tools of the information age.

over trading rules. For example, countries often establish **tariffs**, which are taxes placed on imports and exports. A similar tool is a **quota**, which limits the amount of a particular good that can be imported. Tariffs and quotas protect a country's industries from foreign competition. They can, however, cause ill will among the countries in competition with each other.

Issues in Human Geography In spite of cultural conflicts throughout the world, the earth's people are linked closely. Innovations in high technology, particularly computers and satellites, are permitting rapid and easy communications. Information now travels great distances almost instantly. Many people around the world watch the same television programs, eat the same foods, and speak the same language.

In this "information age," our definitions of resources and what is of economic value are changing. For example, fewer raw materials are needed in the information age. Fuels such as oil, coal, and natural gas are used little in moving information. Instead, computer chips are made from silicon, an element common in sand.

As a result of changes brought about by the information age, jobs in traditional industries are disappearing. Workers must retrain for new jobs in the information industry. Today, the world's richest countries are those with educated work forces. Clearly, the skills you are learning in school are more important than ever before.

 SECTION REVIEW

1. List and explain four sources of world conflicts.
2. How does your daily life reflect the information age? Give examples.
3. **Critical Thinking** **Decide** whether the arrival of the information age will increase or decrease the number of world conflicts. Explain your answer.

CHAPTER 7 CHECK
ANSWERS

Building a Vocabulary

1. A population that shares a common cultural background. Possible answers: African Americans, Asian Americans, European Americans, Hispanic Americans, and Native Americans

2. Acculturation

3. In subsistence agriculture, people grow food mostly for their own families, whereas in commercial agriculture, people grow enough to sell to others.

4. In totalitarian governments, one person or a few make all the decisions affecting everyone; in democratic governments, all citizens have a voice.

5. Tariffs are taxes placed on imports and exports. Quotas limit how much of a particular good can be imported. Both ensure that industries do not face excessive foreign competition.

Recalling and Reviewing

1. Innovation: as people think of new ways to do things, these ways can become part of the culture; diffusion: culture traits spread from region to region.

2. Agriculture provided a more reliable source of food and enabled people to live in larger groups and to form villages.

3. In the new era, information can travel long distances almost instantly.

Thinking Critically

1. Culture includes all aspects of life, so no one is "without culture." Everyone is part of one or more cultures.

2. Answers will vary but might include nationalism, religion, differences in values, politics, and economics.

Learning About Your Local Geography

Discussions should focus on the idea that people define themselves in different ways and that these inevitable differences should be respected.

Using the Five Themes of Geography

Students' collages or drawings should indicate an awareness of how the themes of place, movement, and region apply to cultures.

CHAPTER 7 CHECK

Modern farming in **Kansas**

Reviewing the Main Ideas

1. Culture includes all aspects of a society's way of life. A culture is made up of many culture traits, and a culture region is an area where many culture traits are shared. Cultures change through innovation and diffusion.

2. All people have the same basic needs for food, shelter, and clothing. Throughout history, the innovations of agriculture, urbanization, and industrialization have affected how our basic needs are met.

3. World conflicts are caused by many factors, including nationalism, differences in religious and political beliefs, different values, and the uneven distribution of resources. The current era, in which geographical distance no longer slows world communication, is called the information age.

Building a Vocabulary

1. What is an ethnic group? Name three ethnic groups who live in the United States.

2. What is the term used to describe the change of one culture through its meeting with another culture?

3. What is the difference between subsistence agriculture and commercial agriculture?

4. Compare and contrast totalitarian governments and democratic governments.

5. What are tariffs? What is a quota? Why are they used?

Recalling and Reviewing

1. How do innovation and diffusion cause cultures to change?

2. How was the development of agriculture related to the growth of cities?

3. Why is the present era called the information age?

Thinking Critically

1. Some people believe they are not part of any culture. Do you think it is possible for someone not to be part of any culture? Explain.

2. In your opinion, what one factor is most likely to cause conflicts between countries or cultures? Support your answer with examples.

Learning About Your Local Geography

Cooperative Project

On a sheet of paper, write down at least two cultural, ethnic, or other types of groups of which you, as an individual, consider yourself to be a part. As a class, make a master list of all the groups mentioned on the individual lists. Discuss your findings. Are many different groups listed, or just a few? What are some of the various ways people define themselves? What can you learn from people who define themselves differently than you do?

Using the Five Themes of Geography

From pictures in magazines, newspapers, and other materials, make a collage showing human **place** characteristics of various culture **regions**. Show evidence of **movement** between various culture regions. Make sure you have permission before cutting up any materials. If you do not have access to materials from which to cut pictures, draw your own illustrations.

77

CHAPTER 8 Global Economics and Global Population

Planning Guide

TEXTBOOK	TEACHING RESOURCES	OTHER AVAILABLE MATERIALS
Section		
1. Global Economics Pages 79–80	Reteaching Worksheet 8.1	*Study Guide for Spanish Speakers* Lecture Notes, Chapter 8
2. Developed and Developing Countries Pages 80–82	Reteaching Worksheet 8.2 Graphic Organizer 8 Critical Thinking Worksheet 8	*Study Guide for Spanish Speakers* Lecture Notes, Chapter 8
3. Population Geography Pages 82–84	Reteaching Worksheet 8.3 Skills Worksheet 8 Challenge/Enrichment Worksheet 8	*REGIONS OF THE WORLD Map Transparencies:* Population Pyramids in 1995 *Study Guide for Spanish Speakers* Lecture Notes, Chapter 8
Chapter Review Page 85	Chapter 8 Test Chapter 8 Portfolio Assessment See also *Alternative Assessment Forms*	*Test Generator,* Chapter 8

CHAPTER RESOURCES

Lecture Notes, Chapter 8
Teaching Resources
    Chapter 8 Graphic Organizer
    Chapter 8 Skills Worksheet
    Chapter 8 Critical Thinking Worksheet
    Chapter 8 Challenge/Enrichment Worksheet

REVIEW AND ASSESSMENT

Chapter 8 Check, page 85
Chapter 8 Test, *Teaching Resources*
Chapter 8 Alternative Assessment, *Teaching Resources*
Test Generator, Chapter 8

ADDITIONAL RESOURCES

Books for Teachers

Ehrlich, Paul R., and Anne H. Ehrlich. *The Population Explosion.* New York: Simon & Schuster, 1990.

Parisi, Lynn S., and Robert D. LaRue. *Global/International Issues and Problems: A Resource Book for Secondary Schools.* Santa Barbara, CA: ABC-CLIO, 1989.

Winckler, Suzanne, and Mary M. Rodgers. *Our Endangered Planet: Population Growth.* Minneapolis, MN: Lerner Publications Co., 1991.

Books for Students

Ballard, John. *Monsoon: A Novel to End World Hunger.* Sarasota, FL: New Horizons Press, 1986.

Gallant, Roy A. *The Peopling of Planet Earth: Human Population Growth Through the Ages.* New York: Macmillan Publishing Co., 1990.

Hanmer, Trudy. *The Growth of Cities.* New York: Franklin Watts, 1985.

Leone, Bruno, series ed. *Population: Opposing Viewpoints.* San Diego: Greenhaven Press, 1995.

Newton, David E. *Population: Too Many People?* Hillside, NJ: Enslow Publishers, 1992.

Polesetsky, Mathew, ed. *Global Resources: Opposing Viewpoints.* San Diego, CA: Greenhaven Press, 1991.

The World Today *Videodisc Program*

A variety of still images, short videos, and activities are available for you to use as you teach this chapter. *Lecture Notes* include bar codes for specific photographs.

Multimedia Resources

The Geography Tutor: Global Problems and Issues. (VHS, 15 minutes) Tell Me Why, 730 Washington Street, Marina Del Rey, CA 90292.

Global Links. (six videocassettes in series, 28 minutes each) WETA/World Bank. Available through Social Studies School Service, 10200 Jefferson Boulevard, P.O. Box 802, Culver City, CA 90232–0802.

Picture Atlas of the World. (interactive maps, photo bank, animation, audio on CD-ROM) National Geographic Society, Washington, DC.

MEETING INDIVIDUAL NEEDS

Limited English Proficiency (LEP)

Help students create graphic organizers on the board that demonstrate the main ideas of Chapter 8: four categories of economic geography, economic indicators, developed and developing countries, capitalist and command economies, three stages of population growth. Have students copy the graphic organizers. Encourage them to refer to the organizers as they study Chapter 8.

Learners Having Difficulty

Have students work in pairs to answer each section's Focus and Section Review questions. Then have them work together to write short summaries of each section.

Auditory Learners

Lead a class discussion on ways that the four categories of economic activities—primary, secondary, tertiary, and quaternary—can be seen in your school. Begin with items in the classroom such as shoes, desks, pens, and lighting fixtures, and discuss the raw materials used to make them and how they were likely produced. Also have students consider the activities related to the school cafeteria, store, library, and gym, as well as to administration, maintenance, and transportation.

Visual Learners

Help students create a chart on the board that shows the characteristics of developed, developing, and middle-income countries. Then provide students with photographs of scenes in different countries. Help them categorize the pictured countries based on the chart.

Tactile/Kinesthetic Learners

Have students investigate how their town or city has changed over a period of 30 years in population size and density, types of economic activities, and environment. Students might use local library resources or interview local residents.

Gifted Learners

Have students make a demographic chart of one developed, one developing, and one middle-income country. Students might include the following data: population, birth and death rates, percent of annual population increase, infant mortality rate, life expectancy, urban population, literacy rate, per capita GDP. Possible references include *Statesman's Yearbook*, *World Population Data Sheet*, and *United Nations Demographic Yearbook*.

MULTICULTURAL FOCUS

Explain to students that, in spite of shortages of housing, food, jobs, and land, some people living in heavily populated countries may resist family planning. In places where subsistence farming is the norm, people need many children to help farm the land. Children are seen as a guarantee of survival. According to a folk story that is told in Brazil, a man explains why he has nine children as follows: "Because three die when they are little; three migrate to São Paulo, Rio de Janeiro, or Brasília; and three stay here to take care of us when we get old."

Overpopulation of a country forces leaders to make difficult decisions regarding such things as land use. Have students decide the best use of a 25-acre (10-hectare) parcel of forested land at the edge of a city in a densely populated developing country. Divide the class into four groups, and assign each group to represent one of the following: a farmer who wants to grow soybeans, a city planner who wants to build housing for the homeless, a conservationist who wants to protect the forest, and a business owner who wants to build a factory that will supply work for 500 people. Allow the groups 10 minutes to discuss why the person they represent should be allowed to have use of the land. Then have students appoint one student in each group to give a persuasive speech for the group's client. Finally, have the class vote on how to use the land.

CHAPTER 8

GLOBAL ECONOMICS AND GLOBAL POPULATION

Section 1: Global Economics
Section 2: Developed and Developing Countries
Section 3: Population Geography

OVERVIEW

In Chapter 8, students will learn about economic geography and population geography and the relationship between economics and population in developed and developing countries.

Section 1 defines the four categories of economic activity. It explains the indicators that are used to measure a country's economic level.

Section 2 focuses on economic development, examining the characteristics of countries at various stages of development and the political and economic factors that influence development.

Section 3 explains demography and identifies population issues. It explains the stages of population growth and the factors that affect growth.

About the Author

Nafis Sadik is from Pakistan. She is executive director of the United Nations Population Fund.

Primary Source

Description of Change: excerpted
Rationale: excerpted to focus on main idea

CHAPTER 8

Global Economics and Global Population

❝Whether we look at population, environment, or development, the next 10 years will be critical for our future. The decisions we make or don't take will widen or narrow our options for a century to come. They could decide the fate of the earth as a home for human beings. ❞

Nafis Sadik

In the past hour, about 10,000 babies were born. Indeed, the world's population is fast approaching 6 billion people. In fact, we are living during the period of greatest population increase in human history. Can our planet continue to support this ever-increasing global population? The question is an urgent one for the more than 1 billion people around the world who will go to bed hungry tonight.

Economic progress varies among the world's countries and from place to place within individual countries. A country's level of economic development also changes over time. For example, in its earliest years, the United States was a poor country of farmers. Today, it is the richest country in the world. Yet pockets of poverty still exist within its borders. All countries, regardless of their level of economic development, wish to improve their economies. This chapter explores the important link between global economics and global population.

New York City street festival

GEOGRAPHY DICTIONARY

primary industry
secondary industry
tertiary industry
quaternary industry
gross national product (GNP)
gross domestic product (GDP)
literacy
infrastructure
telecommunications
developed country
developing country
foreign aid
multinational company
free enterprise
capitalism
market economy
command economy
communism
demography
birthrate
death rate
emigrant
immigrant

GLOBAL ECONOMICS

FOCUS

- *What are the categories of economic activities?*
- *How do geographers measure varying levels of economic development?*

Economic Geography The branch of geography concerned with how people use Earth's resources, how they earn a living, and how products are distributed is called economic geography. People's economic activities can be divided into four categories—primary (first order), secondary (second order), tertiary (third order), and quaternary (fourth order) industries.

Primary industries include agriculture, forestry, and mining. These activities directly use natural resources and raw materials and are found at or near sources of those materials. Agriculture and forestry are found in regions with suitable soils and climates. Mines are located where rich mineral deposits have been found.

Secondary industries take goods made in primary industries and change the goods into products that are useful to consumers. Secondary industries involve manufacturing. For example, a dairy takes a raw material (raw milk) and processes it into products such as ice cream and butter. Often, a product will go through several stages before it becomes a finished good. Many factors determine the locations of manufacturing industries. Some industries, such as dairies and bakeries, are located near markets because their products may spoil if they must be transported too far. Modern refrigeration and improved roads, however, have made longer hauls possible. Other industries are located where special manufacturing needs can be met easily. These may be needs for cheap power sources, skilled workers, or specialized transportation facilities.

Tertiary industries are service industries. These industries provide services to primary and secondary industries, to communities, and to individual consumers. Service industries include stores, restaurants, banking, insurance, transportation, and government agencies. Services usually are located close to where people live.

These photographs of the dairy business show four industries at work. Dairy farming (top), a primary industry, provides the milk needed for processing into cheese (middle, left), a secondary industry. A grocery store, a tertiary industry, offers a customer a selection from the dairy case (middle, right). A specialist, part of a quaternary industry, conducts research to find ways to improve dairy production and quality (bottom).

The use of computers and electronic communications is changing some service industries. Today, consumers can buy a wide variety of catalog items over the telephone. Computers allow some people to work at home, which gives them more flexibility in deciding where they want to live. A worker living in Jackson Hole, Wyoming, for example, may work for a firm in New York City.

Some geographers include this movement and processing of information in a fourth category of economic activity. **Quaternary industries**, a subdivision of tertiary industries, are performed by professionals having specialized skills or knowledge. Information research, management, and administration are quaternary industries. Your school principal is an example of a worker involved in a quaternary industry.

Economic Indicators As economic activities vary, so do economic conditions. Accurately measuring levels of economic development is difficult, but geographers can use several means to compare data and make generalizations. It is important

Building a Vocabulary

Write **primary, secondary, tertiary,** and **quaternary industries** on the board, and have students read the examples of each given on page 79. Explain that secondary industries depend on primary ones for raw materials, tertiary industries rely on secondary ones for products and customers, and quaternary industries depend on tertiary ones that need help in providing services.

Guided Practice

Provide students with a list of 20 or 25 countries. Have small groups divide the list and determine the GNP, GDP, and per capita GDP of those countries, using almanacs and encyclopedias. Put all the data into a chart on the board. Then have students rank the countries by each of the three measures. Discuss variations in the rankings and what they signify.

Closure

Have students list the four types of economic activity and suggest a condition necessary for each.

ASSESSMENT

Assign the Section 1 Review questions on this page.

Reteaching

Assign Section 1 of Reteaching Worksheet 8.

Geography and Economics

GNP may be the more familiar measure of a country's economic output, but GDP is now used more often to evaluate the true size of a country's economy.

Section 1 Review
ANSWERS

1. They have different geographical needs. Primary industries must be near natural resources or raw materials, whereas secondary industries need to be near markets or where manufacturing needs can be met; tertiary industries are often near where people live.

2. Possible answers: GNP, GDP, per capita GDP, level of industrialization, literacy rates, infrastructure, telecommunications

3. Literacy encourages economic development because people who can read and write are more likely to push for and bring about development.

This car manufacturing plant is evidence of the great improvements in the economy of **South Korea** since the Korean War. Manufacturing, a secondary industry, is an important part of South Korea's economic growth. Automobiles are among South Korea's most important exports.

to remember that a country's economic level says little about its cultural values and contributions.

One measure of a country's economic development is **gross national product** or **GNP**. Gross national product is the total value of goods and services produced by a country in a year, both inside and outside the country. Some geographers use **gross domestic product (GDP)** instead of gross national product. Gross domestic product includes only those goods and services produced within a country and does not include income earned outside that country. This number becomes more useful when it is divided by the number of people in that country. This figure tells us the per capita GDP, which then can be used to compare the economic level among countries.

Another way to describe levels of economic development is to identify the extent to which a country has industrialized. The world's poorest countries have no industrial activity. The next poorest countries are industrializing, which means they are in the process of developing manufacturing and industry. Some countries are beginning to move out of the industrial stage into a postindustrial stage. In a postindustrial country, more people are employed in tertiary and quaternary industries than in primary and secondary industries. This trend, however, has resulted in industrial decline in some regions, as experienced in the Northeastern and Midwestern United States. The full impact of this trend will not be known for some time.

Other measures of economic development are literacy, infrastructure, and telecommunications.

Literacy is the ability to read or write. In general, the higher a country's literacy rate, or the higher the number of people who can read or write, the more economically developed the country. **Infrastructure** includes roads, bridges, transportation facilities, power plants, water supply, sewage treatment facilities, and government buildings that are necessary to build industries and to move goods in and out of a country quickly and easily. **Telecommunications** is electronically transmitted communication. Today, a rapid, efficient telecommunications network is important for developing high-technology industries.

SECTION REVIEW

1. Why do different kinds of economic activities take place in different areas?
2. List three different ways of measuring a country's economic development.
3. **Critical Thinking**   **Explain** the relationship between literacy rates and economic development.

2   DEVELOPED AND DEVELOPING COUNTRIES

FOCUS

● *How do developed countries and developing countries differ?*

● *What are the various types of economic and political systems?*

Categorizing the world's countries by economic level identifies the world's richest and poorest countries and those that lie in between. The world's wealthy countries are called **developed countries**. These include many of the countries in Europe, the United States, Canada, Japan, Australia, Singapore, and others.

Developed countries share a number of common characteristics. Among these are a good educational system, widely available health care, and many manufacturing and service industries. Developed countries are industrialized and participate in international trade. People mostly live in cities. The few people who work as farmers use

SECTION 2
OBJECTIVES

The objectives reflect the Focus questions on page 80 and are assessed in the review on page 82.

1. Compare and contrast developed and developing countries.
2. Identify different economic and political systems.

TEACHING SECTION 2
Building a Vocabulary

Write **free enterprise, capitalism,** and **market economy** on the board, and review the concepts. Have students look up the word *capital* in the dictionary. Explain how *capital* (value of accumulated goods) relates to *capitalism.* (Remind them that the suffix *-ism* means a theory or doctrine.) Point out why free enterprise must be the basis of a market economy.

Classifying
Economic Development Review the characteristics of developed countries. Then return to the list of countries chosen for Section 1 Guided Practice. Have students find in an almanac the following data for each: urban/rural population, literacy rate, life expectancy, labor force characteristics. Have them use the data to classify the countries as developed, middle-income, or developing.

---

modern technology. In developed countries, most people have access to a telephone and are part of the global network formed by modern telecommunications. The wealthiest of the developed nations have per capita GDPs above $19,000.

About three-quarters of the world's people live in the poorer countries, which often are called **developing countries**. Many of the people in developing countries live by subsistence farming in rural areas. These countries have few manufacturing and service industries to provide jobs. Though capital cities and other major cities may be large, poverty and unemployment are widespread. Many people migrate, or move, from rural areas to cities. Many city dwellers are underemployed, working only part time.

In developing countries, health services are limited and schools often are crowded. The literacy rate is usually low. Modern telecommunications are not usually found outside of major cities. Exporting agricultural products and minerals to developed countries often is an important source of income for developing countries. Some of the world's poorest countries have per capita GDPs of less than $600.

Between the world's richest and poorest countries are middle-income countries such as Argentina, Mexico, Venezuela, Brazil, Hungary, South Korea, and Malaysia. These countries combine characteristics of developed and developing countries. Some middle-income countries, such as South Korea and Taiwan, are newly industrialized. Their per capita incomes are growing rapidly and soon will be in the developed category. Other countries, such as Argentina, have remained in the middle category for decades.

Choices in Economic Development
How can a poor country improve its economy? If it has many natural resources such as oil fields, mineral deposits, or valuable crops, then development of these resources can lead to economic progress. Income from economic activity can be invested in education and infrastructure.

For a developing country with few natural resources, the economic choices often are very limited. Sometimes the governments of developed countries help developing countries with loans and gifts of money. Some of this **foreign aid** goes to ease human suffering.

Developed countries often debate how to help developing countries. Some wealthy countries provide little help. They argue that there are too few resources to share and that by helping others their own economic growth will suffer. Other developed countries give aid only to countries with rich natural resources or a low rate of population growth. Still other countries give aid only to those countries with which they are on close political terms.

Multinational companies build factories in many poor countries around the world. These companies are businesses with activities in many countries. Multinational firms seek raw materials and labor at low cost. Multinational companies provide jobs and often transfer technology from the developed countries to developing countries. Multinational companies sometimes are criticized, however, for harming the natural environment or changing the cultures of developing countries.

Politics and Economics
Economic development is of concern to all governments. What kinds of economic and political ideas lead to economic progress?

The world's economically developed countries share two important characteristics. The first is an economic system guided by **free enterprise**. Under free enterprise, prices are determined mostly through competition in which buyers and sellers are free to choose what and when to sell and buy. As a result, goods that people want are produced in large amounts and are sold at fair prices. Under free enterprise, people are free to move in search of higher wages and better working conditions.

Free enterprise is the basis of **capitalism**, an economic system in which resources, industries, and businesses are owned by private individuals. Capitalism is also called a **market economy**. In a market economy, consumers determine what is to be bought or sold by buying or not buying certain goods and services. The term *market* refers to a system in which goods and services are exchanged freely.

The second important characteristic of the developed countries is democracy. People must have personal freedom to make free enterprise work. During the past few years, all of the countries whose economies have improved also have expanded free enterprise and democratic ideas.

In contrast to economies based on free enterprise are **command economies**. In a command economy, the government determines wages, the kinds and amounts of goods produced, and the prices of

COOPERATIVE LEARNING

Organize the class into groups of four or five. Tell students to imagine that an automobile company wants to build a factory in a developing country. Have some members of each group play the role of automobile executives, and have others be officials of the developing country. Assign one member of each group to research (or invent) characteristics that describe the country (location, labor force, resources, environment). Each side should then present the benefits to the other of locating the factory there. They should also include questions or concerns.

Guided Practice

Have students find an example in Section 2 of a country that has a democratic government but a mostly government-controlled economy. *(India)* Then ask them to find an example of a nondemocratic government that allows some degree of free enterprise. *(China)* Call on students to suggest reasons why democracy and free enterprise seem to go hand in hand to build successful economies.

Closure

Review with students the characteristics of developed and developing countries and of market and command economies. Then complete a word web on the board that illustrates these concepts.

ASSESSMENT

Assign the Section 2 Review questions on this page.

Reteaching

Assign Section 2 of Reteaching Worksheet 8.

Geographical Sidelight

The UN Population Fund estimates that 95 percent of the world's population growth in the coming decades will occur in developing countries.

Section 2 Review
ANSWERS

1. Developed countries have high GDPs, good educational systems, available health care, industrialization, and telecommunications. Developing countries have low GDPs, little industry, much poverty and unemployment, and limited health services, education, and telecommunications.

2. In capitalism, private individuals own resources, industries, and businesses. In communism, the government controls all economic and political activity.

3. Systems will vary. Students should include a type of government and mention how goods are produced and wages are determined. Some students may try to balance political freedom and economic equality.

82

---

goods. Communist countries have command economies. **Communism** is an economic and political system in which the government owns or controls almost all of the means of production. These definitions are generalizations in today's world, however. Since the breakup of the Soviet Union, the once-Communist countries of Eastern Europe and the former Soviet Union have been struggling to reform and improve their economic systems.

The world's developing economies often are a mix of command economies and free enterprise. India, for example, is a democracy, but much of its economy is controlled by the government. Though China has greatly expanded free enterprise, democracy in the country has been put down, sometimes violently.

Many developing countries' economies are hindered by poor government. In some countries, few people are trained in government and business. In other countries, much of the wealth is controlled by a small number of people who pressure the government to pass laws favorable to them. Such governments are often unstable. As the developing countries turn toward free enterprise and democracy, their economic conditions can be expected to improve.

2 SECTION REVIEW

1. What determines whether a country is considered developed or developing?
2. What is the main difference between capitalism and communism?
3. **Critical Thinking** **Formulate** your own ideal political and economic system.

POPULATION GEOGRAPHY

FOCUS

- *How is the world's population distributed, and why? Why has the population grown over time?*

- *What factors will affect the size of future populations?*

 PLACE **Global Population Density** The geography of the human population is uneven.

Some regions are densely populated, while others have many fewer people. Some regions have aging populations, while others have mostly children. Population geography is the study of the variations and changes in a population's makeup, distribution, movement, and relationship to its environment. **Demography** is the study of the human population that emphasizes statistical characteristics, or facts and figures. It is used to compare population distribution, population density, trends in population, and so on.

It is easy to understand why some parts of the world have fewer people. These regions usually have harsh climates, and farming is difficult. In contrast, understanding patterns of dense population settlement is more complex.

There are four large regions of dense human settlement. These are eastern Asia, particularly eastern China; south Asia, particularly India and Bangladesh; Western Europe; and eastern North America. A few smaller areas of dense population include southeastern Brazil, Nigeria, and southern California. Clearly, regions with environments favorable to farming usually are densely settled. The fertile coastal lowlands and river valleys of China, India, and Egypt have been heavily populated for more than 2,000 years. Other factors also lead to population growth and settlement in regions, however. The populations of Western European and eastern North American cities increased dramatically when these regions industrialized.

Population Growth Archaeologists believe the world's population at the end of the last ice age was less than 10 million. With the development of agriculture, populations grew rapidly. Two thousand years ago, the world's population numbered about 250 million, growing only gradually during the next centuries. In periods of prosperity, populations rose, only to decline during periods of war or widespread outbreak of disease. By about A.D. 1650, world population finally had doubled to 500 million. Soon after, populations began to increase steadily. With modern health practices, people began to live longer. More babies lived to become adults. Food supplies increased, and cities grew rapidly.

By 1850, the world's population had doubled to 1 billion. Since then, we have witnessed explosive population growth. World population passed the 2 billion mark around 1940, doubling

SECTION 3
OBJECTIVES

The objectives reflect the Focus
questions on page 82 and are
assessed in the review on page 84.

1. Identify and explain factors determining population distribution
and size over time.

2. Discuss issues affecting future
population patterns.

TEACHING SECTION 3
Motivator

Ask students to characterize their
town or state as densely or sparsely
populated. You may wish to refer
them to the U.S. population map on
page 109. Do they believe that their
region is growing? How fast? Point
out that demographers keep track of
such information. Ask students to
suggest ways that population data is
obtained. (*census, birth and death
records*)

Building a Vocabulary

Write **emigrants** and **immigrants** on
the board. Point out that these words
are sometimes confused. Have volunteers write sentences on the board to
illustrate the meaning and usage of
each word. (*For example: The Cuban
emigrants left Havana to go to Miami.
Many immigrants have moved to
Florida from the Caribbean area.*)

---

to 4 billion in 1975. In the year 2000, the population
of the world will be 6 to 7 billion. Demographers
predict that if this trend continues, world population will surpass 10 billion in the next century.

Stages of Population Growth

What fac-
tors lead to rapid population growth? Demographers explain that a country experiences three
stages of population growth as its economy develops. As a model, geographers use Western Europe,
which went through industrialization and urbanization beginning in the mid-1700s. Of course, not
all countries follow this model, but it is useful to
study how and why population growth in a region
changes in relation to economic development.

Geographers use birthrates and death rates
to explain the three stages of population growth.
Birthrate is the number of births per 1,000 people
in a given year. **Death rate** is the number of
deaths per 1,000 people in a given year. If a country's birthrate exceeds its death rate, the population will increase. If the number of deaths is more
than the number of births, the country's population will decrease.

During the first stage of population growth of
a country, the birthrate and the death rate are
high but equal. Though there are many births,
widespread disease takes its toll on the population. As health care and safe water become more
widespread, the death rate declines steadily while
the birthrate remains high. In this second stage,

World Population Growth

Source: Population Reference Bureau, Inc.

The world's population has increased rapidly since the
late 1700s. This sign reminds residents of **New Delhi,
India,** of their country's birthrate and population.

---

family sizes increase, and more children than
before survive to adulthood. Population growth is
rapid. In the third stage, the birthrate drops, and
most children live to adulthood. Family sizes
decrease as more people decide to have fewer
children. The birthrate and death rate again
become about the same, though now at lower levels. The country's rate of population growth again
becomes low. The population now is much higher
than in the first stage, however.

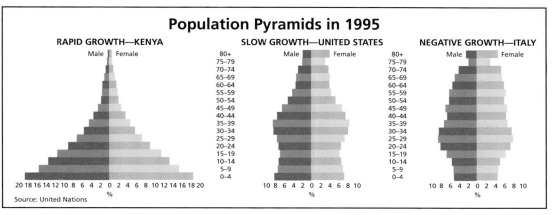

These population pyramids compare three
countries' population distributions by age
and sex. Italy, with a relatively high percentage of older residents, is an example
of a negative-growth country. The population of Kenya is rapidly growing, leaving

the country with a larger percentage of
younger residents. The bulge in the 20–40
age range in the United States pyramid is a
result of the post–World War II baby
boom. Can you predict what the U.S.
pyramid will look like in the year 2020?

MAP TRANSPARENCIES

Population Pyramids:
*REGIONS OF THE
WORLD Map
Transparencies*

Chart Caption Answer

The 2020 pyramid will
show a larger percent-
age of the population
aged 50 to 70 years old.

83

Have students locate the four large regions of dense human settlement on the World Political Map in the Student Atlas on pages A4–A5. Have students also identify the other densely populated areas noted in Section 3. Then ask them to locate all of these regions on the World Physical Map in the Atlas. Have them suggest some of the reasons why these regions are heavily populated.

Closure

Ask students if they think that developing countries have more emigration or immigration. Why?

ASSESSMENT

Assign the Section 3 Review questions on this page.

Reteaching

Assign Section 3 of Reteaching Worksheet 8.

Extension

Have interested students research a historical disease epidemic—such as the Black Death, the devastation of Native Americans by smallpox, a worldwide flu epidemic, or the present-day pandemic of AIDS. Students should consider how the epidemic related to and affected (or relates to and might affect) population growth.

Section 3 Review
ANSWERS

1. Some are more densely populated than others because their environments are better suited to agriculture or because cities and industry attract people. Factors explaining changing population size include birthrates and death rates, and numbers of emigrants and immigrants.

2. Providing food for an increasing population and dealing with the spread of disease

3. No. Starvation and disease caused by inadequate diets would raise the death rate, preventing further growth.

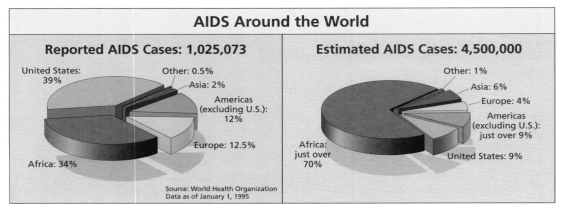

AIDS Around the World

Reported AIDS Cases: 1,025,073

United States: 39%
Other: 0.5%
Asia: 2%
Americas (excluding U.S.): 12%
Europe: 12.5%
Africa: 34%

Estimated AIDS Cases: 4,500,000

Other: 1%
Asia: 6%
Europe: 4%
Americas (excluding U.S.): just over 9%
United States: 9%
Africa: just over 70%

Source: World Health Organization
Data as of January 1, 1995

Some demographers believe projections for population growth and life expectancy rates must be revised for some parts of the world because of the toll taken by the disease AIDS. These pie graphs show the worldwide distribution of reported and estimated AIDS cases from the beginning of the pandemic in the late 1970s or early 1980s to the end of 1994. (Diseases become pandemic when they are wide-spread in many locations at the same time.) The World Health Organization estimated that by the end of 1994, about 19.5 million people, including 1.5 million children, had become infected with the virus that causes AIDS. By the year 2000, the total number of infected individuals has been projected at between 30 and 40 million, with the vast majority of infections taking place in developing countries.

A country's population growth or decline also depends upon the number of its emigrants and immigrants. An **emigrant** is someone who moves out of a country. An **immigrant** is someone who moves into a country. If a country has more immigrants than emigrants, its population will increase.

Population change and economic success are related. In all the world's developed countries, the period of rapid population growth has ended. These countries are experiencing low death rates and low birthrates. People in these countries live mostly in cities and die at an old age. Women have the opportunity to work outside the home, and many choose to have smaller families. The populations of these countries are changing little.

In contrast, the world's developing countries have not completed the three stages of population growth. Often, with aid from the developed countries to improve public health, death rates have dropped dramatically among developing countries. Though birthrates have begun to decline, they still exceed the death rates. In these countries, large families are often important to the family economically. Children are needed to work around the home or farm. International organizations have begun providing education and information to these countries about controlling population growth.

Future Growth Because the world's developing countries are home to a high number of children who will themselves have children, rapid population growth will continue. Providing for this still-expanding population will mean increases in food production, industrial development, and education within the developing regions.

An issue with an uncertain effect on future population growth or decline is the spread of disease. Of particular concern is AIDS (Acquired Immune Deficiency Syndrome) and the virus that causes it (HIV). According to the UN World Health Organization Global Programme on AIDS, "With every day that passes, 5000 more people worldwide become infected with the virus." Many more cases go unreported or undiagnosed.

SECTION REVIEW

1. Why are some areas more densely populated than others? What factors explain why populations change in size?
2. What are some of the challenges future populations will face?
3. **Critical Thinking** **Predict** what would happen if world population exceeded world food production. Would growth still continue?

CHAPTER 8 CHECK
ANSWERS

Building a Vocabulary

1. GNP: total value of goods/services produced by a country in a year, both inside and outside a country; GDP: total value of goods/services produced within a country
2. It allows goods to be produced/transported easily.
3. Electronically transmitted communication
4. Demography
5. An emigrant leaves a country; an immigrant enters.

Recalling and Reviewing

1. They are primary, secondary, tertiary, and quaternary. Examples will vary but might include farmer, factory worker, salesperson, and school principal.
2. Better educational systems and health care, more technology and industries, greater access to telecommunications, less poverty and unemployment
3. Answers may include limited resources, failure of multinational companies to consider needs of host countries, and rapidly increasing populations.
4. First stage: birthrate and death rate are high but equal, so population size remains constant. Second stage: birthrate remains high, but health care improves, leading to a lower death rate, so population grows. Third stage: birthrate drops as people decide to have fewer children, and population growth rate drops.

Thinking Critically

1. Per capita GDP is only an average figure of value, indicating nothing about individual wealth. A country can have a very high per capita GDP and much poverty.
2. Answers will vary. Students may say that developed countries should help developing countries out of a sense of moral obligation, or to help reduce world conflicts.

CHAPTER **8** CHECK

Mall of America in **Bloomington, Minnesota**

Reviewing the Main Ideas

1. Economic activities can be divided into four categories: primary, secondary, tertiary, and quaternary. Geographers have several different ways of measuring economic development.
2. The countries of the world can be grouped according to their economic conditions. Most of the world's people live in developing countries. Developed countries usually have an economic system based on free enterprise, whereas the economic systems of developing countries often are a mix of command and market economies.
3. Some regions of the world are densely populated, while other regions are sparsely populated. The world's population has increased dramatically over the past few centuries.

Building a Vocabulary

1. What is the difference between gross national product and gross domestic product?
2. Why is infrastructure necessary for industrialization?
3. Define *telecommunications*.
4. What do we call the study of the statistical characteristics of the human population?
5. What is the difference between an emigrant and an immigrant?

Recalling and Reviewing

1. List the four categories of economic activities, and give an example of a job or profession in each category.
2. What advantages do people in developed countries often have that many people in developing countries do not have?

3. What are some of the problems faced by developing countries in their efforts to improve their economies?
4. Describe the three stages of population growth.

Thinking Critically

1. If a country has a very high per capita GDP, what can you tell about how much money any given individual in that country might have?
2. Do you think that developed countries should try to help developing countries? Explain.

Learning About Your Local Geography

Individual Project
Collect information from an almanac or statistical abstract on how your state's population is distributed. What percentage of your state's population lives in urban or metropolitan areas, and what percentage lives in rural or nonmetropolitan areas? Construct a pie graph to present your information. You may want to include your graph in your individual portfolio.

Using the Five Themes of Geography

In this chapter, you have learned that the world can be divided into developed, developing, and middle-income **regions**. Make a chart with three columns, one for each of these regions. Fill in the chart with the human **place** characteristics that often are associated with each of these regions.

Learning About Your Local Geography
Graphs should be clearly labeled and accurate, indicating thorough research.

Using the Five Themes of Geography
Charts should include information about wealth, economic activities, health care, education and literacy, infrastructure, and telecommunications.

CHAPTER 9 Resources and Environmental Change

Planning Guide

TEXTBOOK	TEACHING RESOURCES	OTHER AVAILABLE MATERIALS
Section		
1. Soil and Forests Pages 87–88	Reteaching Worksheet 9.1 Challenge/Enrichment Worksheet 9	*Creative Strategies for Teaching World Geography* LESSON 8 Our Disappearing Rain Forests *Study Guide for Spanish Speakers* *Lecture Notes*, Chapter 9
2. Water and Air Pages 89–93	Reteaching Worksheet 9.2 Skills Worksheet 9	*Study Guide for Spanish Speakers* *Lecture Notes*, Chapter 9
3. Energy Resources Pages 93–95	Reteaching Worksheet 9.3 Graphic Organizer 9 Critical Thinking Worksheet 9	*Study Guide for Spanish Speakers* *Lecture Notes*, Chapter 9
4. Population and Resources Page 96	Reteaching Worksheet 9.4	*Study Guide for Spanish Speakers* *Lecture Notes*, Chapter 9
Chapter Review Page 97	Chapter 9 Test Chapter 9 Portfolio Assessment See also *Alternative Assessment Forms*	*Test Generator,* Chapter 9
Unit Review Pages 98–99	Unit 2 Test Unit 2 Portfolio Assessment See also *Alternative Assessment Forms*	*Test Generator,* Unit 2

CHAPTER RESOURCES

Lecture Notes, Chapter 9
Teaching Resources
   Chapter 9 Graphic Organizer
   Chapter 9 Skills Worksheet
   Chapter 9 Critical Thinking Worksheet
   Chapter 9 Challenge/Enrichment Worksheet

REVIEW AND ASSESSMENT

Chapter 9 Check, page 97
Chapter 9 Test, *Teaching Resources*
Chapter 9 Portfolio Assessment, *Teaching Resources*
Test Generator, Chapter 9

ADDITIONAL RESOURCES

Books for Teachers

Benards, Neal, ed. *The Environment in Crisis, Opposing Viewpoints.* San Diego, CA: Greenhaven Press, 1991.

Brower, Michael. *Cool Energy: The Renewable Solution to Global Warming.* Cambridge, MA: Union of Concerned Scientists, 1990.

Clark, John. *The Political Economy of World Energy: A Twentieth Century Perspective.* New York: Harvester Wheatsheaf, 1990.

Leggett, Jeremy, ed. *Global Warming: The Greenpeace Report.* New York: Oxford University Press, 1990.

Books for Students

The Earthworks Group. *50 Simple Things You Can Do to Save the Earth.* Berkeley, CA: Greenleaf Pubs., 1990.

Elkington, John, and Julia Hailes, Douglas Hill, and Joel Makower. *Going Green: A Kid's Handbook to Saving the Planet.* New York: Penguin Books, 1990.

Miller, Christina G., and Louise A. Berry. *Jungle Rescue: Saving the New World Tropical Rain Forests.* New York: Atheneum Macmillan, 1991.

The World Today *Videodisc Program*

A variety of still images, short videos, and activities are available for you to use as you teach this chapter. *Lecture Notes* include bar codes for specific photographs.

Multimedia Resources

Decisions, Decisions Series: The Environment. (Interactive software, simulation game on local environmental issues) Tom Snyder Productions, 80 Coolidge Hill Road, Watertown, MA 02172-2817.

Protecting Our Environment Series. (film/VHS/videodisc, 13–16 minutes each program) AIMS Media, 9710 DeSoto Avenue, Chatsworth, CA 91311-4409.

MEETING INDIVIDUAL NEEDS

Limited English Proficiency (LEP)

Write the terms in the Chapter 9 Geography Dictionary on the board. Call out the terms that refer to processes, such as soil salinization, desertification, and reforestation. Then point out the terms in the list that have connections to each other, such as nonrenewable resource and renewable resource, reforestation and deforestation, and photochemical smog and fossil fuel.

Learners Having Difficulty

Help students create cause-events-effects concept maps on the board to reteach the main ideas of the chapter. Map the following topics from the chapter: topsoil erosion, deforestation, desalination, air pollution, the Chernobyl disaster, uneven distribution of resources. Have students copy the concept maps and use them to study the chapter.

Auditory Learners

If possible, arrange for guest speakers who are environmentalists to speak to students about local and worldwide environmental concerns. Ask the speakers to focus on ways in which students can help protect the environment. Have students prepare questions in advance to ask the speakers.

Visual Learners

Have students create posters illustrating an environmental concern that was addressed in the chapter. The posters should contain rules or slogans such as "Return to the soil what you get from it." Display the posters in the classroom.

Tactile/Kinesthetic Learners

Have students create charts showing energy sources and their advantages and disadvantages. Then have them rank the energy sources in the order of the most advantageous to the least advantageous. Display the charts in the classroom.

Gifted Learners

Have interested students do research to identify the countries with reserves of nonrenewable resources, such as oil, coal, and minerals. Then have students make pie graphs showing the percentage of reserves that countries have for each nonrenewable resource. Students might use reference books such as *Goode's World Atlas* to find this information. Students who are interested in microcomputers may wish to use a spreadsheet program to create their graphs. Display the graphs in the classroom.

MULTICULTURAL FOCUS

In indigenous societies around the world, cultural patterns have evolved that naturally help preserve the local ecosystem. For example, the Wet'suwet'en and Gitksan peoples of the Canadian Pacific protect salmon supplies, traditionally their staple food, through religious customs. They believe that salmon spirits offer their bodies to humans for food and that these spirits will punish humans who waste or misuse this gift or disrupt salmon habitats. A reverence for animal life also is felt by the Haida of western Canada. Their view of animals as special beings with greater intelligence than humans tends to discourage overkill and waste.

The Kuna of Panama have strict tribal rules governing the harvesting of lobsters, turtles, and game. The Kuna are also conscientious managers of the forests they inhabit, with regulations on how trees can be cut. Other peoples seek to preserve forests or particular trees they deem sacred. The Gorowa and the Gabra of East Africa, for example, set aside ancient groves for ceremonies and burials. Local cultures in India worship certain trees and thus protect them.

Ask students to give examples of ecosystems, natural resources, and endangered species that are protected by laws, customs, and traditions in the United States or in their particular state. *(the bald eagle, state flowers and birds, wetlands and old-growth forests, state and national parks and monuments, wilderness conservation areas)*

Section 1: Soil and Forests
Section 2: Water and Air
Section 3: Energy Resources
Section 4: Population and Resources

OVERVIEW

In Chapter 9, students will learn about nonrenewable and renewable resources and how they are distributed around the world.

Section 1 identifies soil and forest resources and describes the causes and effects of deforestation.

Section 2 focuses on water resources and air quality and the effects on both of population growth and human activity.

Section 3 examines nonrenewable and renewable energy sources and describes how they are used and obtained.

Section 4 considers the global implications of population growth and the uneven distribution of resources.

About the Author

Dalai Lama is the title of Tibet's traditional ruler and Buddhist spiritual leader. The present Dalai Lama (1935–   ) was officially selected in 1940 when he was just a boy. He has lived in exile in India since 1959 because of efforts by the Chinese government to repress Buddhism in Tibet.

Primary Source

Description of Change: excerpted
Rationale: excerpted to focus on main idea

CHAPTER 9

Resources and Environmental Change

❝Now it seems as though our mother planet is telling us, "My children, my dear children, behave in a more harmonious way. My children, please take care of me."❞

The Dalai Lama

The world's almost 6 billion people draw from a common pool of Earth's natural resources for survival. Yet, resource depletion and pollution do not know geographic boundaries. An oil spill in the ocean can destroy fisheries along a coast many miles away. Air pollution, carried by winds, can destroy forests and lakes hundreds of miles away. The destruction of forests in one nation may cause severe flooding and soil erosion to a nation downstream. Soil erosion may decrease crop yields and increase costs of food, or even lead to starvation.

The earth's resources are divided into two major categories: nonrenewable and renewable. **Nonrenewable resources** are those that are not replaced by natural processes or are replaced at extremely slow rates. Examples of nonrenewable resources include oil, coal, and minerals. **Renewable resources** are those that can be replaced by the earth's natural processes. These include solar energy, soil, water, and forests. Even renewable resources, however, can be lost if we are careless or wasteful.

The population of the United States makes up only 5 percent of the world's population, yet we use 25 percent of its resources and produce 50 percent of its solid waste. We have a special responsibility to look after our planet's natural resources.

GEOGRAPHY DICTIONARY

nonrenewable resource
renewable resource
crop rotation
soil salinization
desertification
reforestation
deforestation
sustained yield use
desalinization
photochemical smog
acid rain
ozone
fossil fuel
petrochemical
geothermal energy

Hikers in the rain forest of **Borneo**

SOIL AND FORESTS

FOCUS

- *Why is soil conservation important to agriculture?*
- *What are the causes and effects of deforestation?*

Soil Throughout history, people have developed ways to conserve and enrich the soil. "Return to the soil what you get from it" has been the rule for thousands of years in China. The world's farmers know that the soil is like a bank. It is impossible to continue making withdrawals without making deposits.

The terraced rice paddy, developed in Asia thousands of years ago, is an example of how farmers learned to preserve the soils and store water even on the steepest slopes. Many of today's farmers also are aware of the importance of soil conservation. They use contour plowing, which involves plowing and planting crops in rows across the slope of the land to prevent runoff from eroding topsoil. Farmers also reduce wind erosion by planting rows of trees along the edges of fields to block the wind.

The fertility of the soil can be preserved in several ways. Use of natural or chemical fertilizers is one; **crop rotation** is another. In crop rotation, a field planted one year with corn, a soil-depleting crop, may be planted the next year with alfalfa or some other soil-enriching crop.

Many dry regions that depend upon irrigation are suffering from **soil salinization**, or salt buildup in the soil. The evaporation of irrigation water leaves behind salt. High salt content becomes destructive to crops and eventually destroys the soil's productivity. Soil salinization has reached problem levels in areas of the southwestern United States, Egypt, and India.

Many nations have lost their precious soil to **desertification**, a process by which the loss of plant cover and soil eventually leads to desert-like conditions. In developing countries, desertification is primarily the result of overgrazing and the removal of trees over many years. Climatic change also is involved. The condition spreads as people move their herds from overgrazed areas to vegetated areas, and the process repeats itself. Thousands of

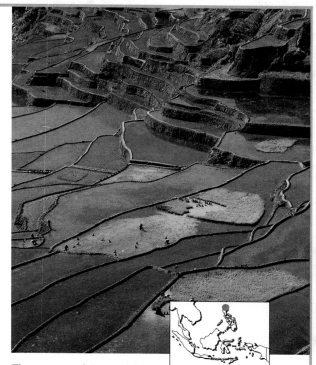
These terraced rice paddies in the **Philippines** help prevent soil erosion in the hilly terrain. The terraces keep rainwater from carrying soil down the hillsides.

people have died from starvation in regions of Africa as fertile grasslands have turned to desert.

In developed countries, such as the United States, loss of farmland is mainly the result of suburban and urban expansion across some of the best farmland. Today, because the world's growing population must be fed, our soil resource is more valuable than ever.

Forests Forests, if properly used and cared for, are valuable renewable resources. The first settlers in the United States found a land of seemingly endless forests. As the nation's population grew following European settlement, many forests were cut to provide timber for fuel, homes, and industry. New trees were not planted to replace those that were destroyed.

It was not until the early 1900s that the federal government took steps to conserve our forests. The government declared certain areas national forests to control the cutting of the trees and to preserve wildlife and recreation in those areas. In national forests that are logged, or cut for lumber, **reforestation** is required.

GLOBAL PERSPECTIVES

Centuries of rapid population growth and deforestation in India's Western Ghats have left thousands of people destitute. In 1975, in the village of Ralegan Sidhi, local leader Kishan Baburao Hazare began implementing watershed development practices to slow erosion and build up groundwater levels. Villagers dug ponds to catch water and restore the water table, terraced the hillsides, and planted trees to anchor the soil. Because of the success of the village's program, the annual per capita income there has increased eightfold since 1976. Ask students if they know of any ways that their own communities have responded to environmental problems.

Building a Vocabulary

Write these vocabulary terms on the board: **crop rotation, soil salinization, desertification, reforestation, deforestation,** and **sustained yield use**. Point out that these terms fit into two categories: environmental problems and solutions to environmental problems. Have students categorize the terms and explain any connections between them.

Guided Practice

List on the board major soil conservation techniques (terracing, contour plowing, crop rotation) and the effects of deforestation (loss of wildlife habitat, recreation areas, atmospheric oxygen, and potential sources for new medicines). Have students create drawings that depict either topic. Summary captions should explain the importance of soil conservation or why deforestation is a serious global problem.

Closure

Have students suggest actions that could help prevent the destruction of forests and the loss of soil.

ASSESSMENT

Assign the Section 1 Review questions on this page.

Reteaching

Assign Section 1 of Reteaching Worksheet 9.

CREATIVE STRATEGIES FOR TEACHING WORLD GEOGRAPHY

LESSON 8 Our Disappearing Rain Forests: What Shall We Save?

The blue and white areas in this satellite image show large paths that have been cleared through the rain forest of **Brazil.** The thick forest vegetation shows as red. Deforestation is a danger to the world's overall environment, to local wildlife, and to the native peoples whose livelihoods depend on the rain forests.

Section 1 Review
ANSWERS

1. Farmers can practice soil conservation by using contour farming and by planting rows of trees along the edges of fields. Fertility can be preserved through the use of fertilizers or through crop rotation.

2. No. Original old-growth forests are lost forever when they are cut.

3. Advantages: trees survive, wildlife is preserved, and forests continue to produce oxygen and absorb carbon dioxide; disadvantages: industries based on lumber are hurt, and jobs are lost.

88

Reforestation is the process by which forests are renewed through the planting of seeds or young trees. Other nations, such as Canada, Sweden, Germany, Finland, and New Zealand, also practice reforestation. In many areas, the original-growth forests have been lost and replaced with tree farms.

In the tropical rain forests of Asia, Africa, and Latin America, **deforestation**, or the clearing of forests, is occurring at a rapid rate. In some nations, including Bangladesh and Haiti, all the original rain forests have been destroyed. Other nations, such as Costa Rica and Australia, are trying hard to preserve what is left of their original rain forests.

One-third of the earth's remaining rain forests are located in Brazil, Zaire, and Indonesia. Before the rain forests vanish completely, it is hoped that these countries will realize the consequences of losing the trees as well as the soil, wildlife, and medical benefits of forest plants.

The loss of the rain forests is mainly due to the logging of tropical hardwoods, clearing for farmland and pastures, and cutting for fuel. Charcoal from wood is the main source of energy for cooking and heating in most developing coun-

tries. In fact, more than 2 billion people rely on wood as their main source of fuel. Logging companies in these countries also continue to cut vast stretches of rain forest for lumber, offering no plans for reforestation. Government officials in developing countries defend these actions by saying that cutting the rain forests provides jobs and income to invest in their economies.

In the middle latitudes, the native temperate forests also are being logged at a rapid rate. This is especially true in northwestern North America, where U.S. and Canadian lumber companies clearcut, or remove all trees from, large areas of the forests for domestic lumber and export to Japan and other countries. Though much of the logging is on replanted tree farms, there is continued cutting of original old-growth forest, a resource that can never be replaced.

We must recognize the importance of preserving original-growth forests. By practicing **sustained yield use**, countries can take products from the forest such as rubber, fruits, nuts, and medicinal plants without destroying the trees. Sustained yield use of a resource ensures that the resource is well managed and remains renewable. When the forests are logged or cut for fuel, reforestation must be practiced. Forests are an extremely important part of the biosphere because they produce oxygen and absorb carbon dioxide, helping maintain the proper ratio of gases in the lower atmosphere.

SECTION REVIEW

1. How can farmers practice soil conservation and preserve soil fertility?
2. Are forests always considered renewable resources? Why or why not?
3. **Critical Thinking** **Discuss** advantages and disadvantages of practicing sustained yield use in forests.

Using the Quotation
Read aloud the quotation on page 86, and explain who the Dalai Lama is. Ask students to suggest ways that the planet is "telling us" that certain human activities are threatening the earth's ability to provide needed resources. Then ask students which of these activities could be altered to help reverse damaging trends.

Using the Geography Dictionary
Preview the technical terms by breaking them down into root or common words. *(For example, **desert** and -**ification**—the process of becoming a desert)* Then write four headings on the board: Soil, Forests, Water, and Air. Have students identify the terms that relate to each of the headings. Point out that some terms may apply to more than one heading.

SECTION 1
OBJECTIVES

The objectives reflect the Focus questions on this page and are assessed in the review on page 88.

1. Explain the importance of soil conservation.

2. Discuss the causes and effects of deforestation.

SOIL AND FORESTS

FOCUS

- *Why is soil conservation important to agriculture?*
- *What are the causes and effects of deforestation?*

Soil Throughout history, people have developed ways to conserve and enrich the soil. "Return to the soil what you get from it" has been the rule for thousands of years in China. The world's farmers know that the soil is like a bank. It is impossible to continue making withdrawals without making deposits.

The terraced rice paddy, developed in Asia thousands of years ago, is an example of how farmers learned to preserve the soils and store water even on the steepest slopes. Many of today's farmers also are aware of the importance of soil conservation. They use contour plowing, which involves plowing and planting crops in rows across the slope of the land to prevent runoff from eroding topsoil. Farmers also reduce wind erosion by planting rows of trees along the edges of fields to block the wind.

The fertility of the soil can be preserved in several ways. Use of natural or chemical fertilizers is one; **crop rotation** is another. In crop rotation, a field planted one year with corn, a soil-depleting crop, may be planted the next year with alfalfa or some other soil-enriching crop.

Many dry regions that depend upon irrigation are suffering from **soil salinization**, or salt buildup in the soil. The evaporation of irrigation water leaves behind salt. High salt content becomes destructive to crops and eventually destroys the soil's productivity. Soil salinization has reached problem levels in areas of the southwestern United States, Egypt, and India.

Many nations have lost their precious soil to **desertification**, a process by which the loss of plant cover and soil eventually leads to desert-like conditions. In developing countries, desertification is primarily the result of overgrazing and the removal of trees over many years. Climatic change also is involved. The condition spreads as people move their herds from overgrazed areas to vegetated areas, and the process repeats itself. Thousands of

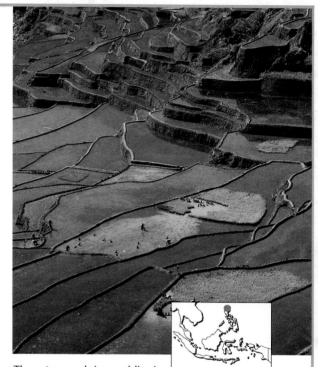

These terraced rice paddies in the **Philippines** help prevent soil erosion in the hilly terrain. The terraces keep rainwater from carrying soil down the hillsides.

people have died from starvation in regions of Africa as fertile grasslands have turned to desert.

In developed countries, such as the United States, loss of farmland is mainly the result of suburban and urban expansion across some of the best farmland. Today, because the world's growing population must be fed, our soil resource is more valuable than ever.

Forests Forests, if properly used and cared for, are valuable renewable resources. The first settlers in the United States found a land of seemingly endless forests. As the nation's population grew following European settlement, many forests were cut to provide timber for fuel, homes, and industry. New trees were not planted to replace those that were destroyed.

It was not until the early 1900s that the federal government took steps to conserve our forests. The government declared certain areas national forests to control the cutting of the trees and to preserve wildlife and recreation in those areas. In national forests that are logged, or cut for lumber, **reforestation** is required.

GLOBAL PERSPECTIVES

Centuries of rapid population growth and deforestation in India's Western Ghats have left thousands of people destitute. In 1975, in the village of Ralegan Sidhi, local leader Kishan Baburao Hazare began implementing watershed development practices to slow erosion and build up groundwater levels. Villagers dug ponds to catch water and restore the water table, terraced the hillsides, and planted trees to anchor the soil. Because of the success of the village's program, the annual per capita income there has increased eightfold since 1976. Ask students if they know of any ways that their own communities have responded to environmental problems.

TEACHING SECTION 1

Building a Vocabulary

Write these vocabulary terms on the board: **crop rotation, soil salinization, desertification, reforestation, deforestation,** and **sustained yield use.** Point out that these terms fit into two categories: environmental problems and solutions to environmental problems. Have students categorize the terms and explain any connections between them.

Guided Practice

List on the board major soil conservation techniques (terracing, contour plowing, crop rotation) and the effects of deforestation (loss of wildlife habitat, recreation areas, atmospheric oxygen, and potential sources for new medicines). Have students create drawings that depict either topic. Summary captions should explain the importance of soil conservation or why deforestation is a serious global problem.

Closure

Have students suggest actions that could help prevent the destruction of forests and the loss of soil.

ASSESSMENT

Assign the Section 1 Review questions on this page.

Reteaching

Assign Section 1 of Reteaching Worksheet 9.

CREATIVE STRATEGIES FOR TEACHING WORLD GEOGRAPHY

LESSON 8 Our Disappearing Rain Forests: What Shall We Save?

Section 1 Review

ANSWERS

1. Farmers can practice soil conservation by using contour farming and by planting rows of trees along the edges of fields. Fertility can be preserved through the use of fertilizers or through crop rotation.

2. No. Original old-growth forests are lost forever when they are cut.

3. Advantages: trees survive, wildlife is preserved, and forests continue to produce oxygen and absorb carbon dioxide; disadvantages: industries based on lumber are hurt, and jobs are lost.

88

The blue and white areas in this satellite image show large paths that have been cleared through the rain forest of **Brazil.** The thick forest vegetation shows as red. Deforestation is a danger to the world's overall environment, to local wildlife, and to the native peoples whose livelihoods depend on the rain forests.

Reforestation is the process by which forests are renewed through the planting of seeds or young trees. Other nations, such as Canada, Sweden, Germany, Finland, and New Zealand, also practice reforestation. In many areas, the original-growth forests have been lost and replaced with tree farms.

In the tropical rain forests of Asia, Africa, and Latin America, **deforestation**, or the clearing of forests, is occurring at a rapid rate. In some nations, including Bangladesh and Haiti, all the original rain forests have been destroyed. Other nations, such as Costa Rica and Australia, are trying hard to preserve what is left of their original rain forests.

One-third of the earth's remaining rain forests are located in Brazil, Zaire, and Indonesia. Before the rain forests vanish completely, it is hoped that these countries will realize the consequences of losing the trees as well as the soil, wildlife, and medical benefits of forest plants.

The loss of the rain forests is mainly due to the logging of tropical hardwoods, clearing for farmland and pastures, and cutting for fuel. Charcoal from wood is the main source of energy for cooking and heating in most developing coun-

tries. In fact, more than 2 billion people rely on wood as their main source of fuel. Logging companies in these countries also continue to cut vast stretches of rain forest for lumber, offering no plans for reforestation. Government officials in developing countries defend these actions by saying that cutting the rain forests provides jobs and income to invest in their economies.

In the middle latitudes, the native temperate forests also are being logged at a rapid rate. This is especially true in northwestern North America, where U.S. and Canadian lumber companies clear-cut, or remove all trees from, large areas of the forests for domestic lumber and export to Japan and other countries. Though much of the logging is on replanted tree farms, there is continued cutting of original old-growth forest, a resource that can never be replaced.

We must recognize the importance of preserving original-growth forests. By practicing **sustained yield use**, countries can take products from the forest such as rubber, fruits, nuts, and medicinal plants without destroying the trees. Sustained yield use of a resource ensures that the resource is well managed and remains renewable. When the forests are logged or cut for fuel, reforestation must be practiced. Forests are an extremely important part of the biosphere because they produce oxygen and absorb carbon dioxide, helping maintain the proper ratio of gases in the lower atmosphere.

SECTION REVIEW

1. How can farmers practice soil conservation and preserve soil fertility?

2. Are forests always considered renewable resources? Why or why not?

3. Critical Thinking   Discuss advantages and disadvantages of practicing sustained yield use in forests.

SECTION 2
OBJECTIVES

The objectives reflect the Focus questions on this page and are assessed in the review on page 93.

1. Explain how the growing world population affects water quantity and quality.
2. Discuss the greenhouse effect and the consequences of a possible global warming.

TEACHING SECTION 2
Motivator

Point out that the average American uses 168 gallons (636 l) of water daily, which amounts to about 107,000 gallons (404,995 l) per family a year. Ask students why the efforts of individuals to conserve water are important. Have them name some household water-saving practices. *(turning off tap while brushing teeth, quick showers, water-saving shower-heads and toilets)*

Building a Vocabulary

Write **desalinization** and **photochemical smog** on the board. Review the meanings of the roots *salin-*, from the Latin word for "salt," and *photo-*, meaning "light." *Photochemical* refers to chemical reactions caused by or involving sunlight. Ask students to name other words or terms with *salin-* and *photo-*. *(soil salinization, saline; photosynthesis, photography)*

2 WATER AND AIR

FOCUS

- *What effect does a growing world population have on water quality?*
- *What might be changing the earth's climate?*

INTERACTION **Water Quantity** With the rapidly growing world population, maintaining water quantity and quality are serious challenges in many areas of the world. In the western United States, Australia, and Russia, expensive water transfer systems with dams, reservoirs, and canals transport water long distances. For example, most of California's cities and agriculture are dependent on water that is transferred hundreds of miles across desert and mountains.

In many arid parts of the world, deep wells are the only source of water. As water demands increase, the water table drops, forcing the drilling of deeper wells and further depleting the aquifers.

To solve the problem of inadequate water supplies, some countries rely on **desalinization** plants, which change salty seawater into fresh water. But desalinization is costly, especially for the energy that must be generated to produce fresh water. Presently, two-thirds of the desalinization plants are in Southwest Asia, where

A laboratory technician checks the water quality at a water treatment plant in **Michigan**. Most people in the United States get their water from public water supply systems, which use treatment plants to purify the water.

water is expensive and energy is cheap. Desalinization also is used where there is no alternative, such as in the Florida Keys and in the islands of the Caribbean.

Drought has caused water shortages in Africa, Australia, and parts of the United States. To survive droughts and meet the continued demand for more water, people need to conserve water. Water conservation must be practiced on farms, in factories, in cities, and in our homes.

Flood Control Although water shortages affect many countries, floods also create hazards. As natural disasters, floods rank high in the taking of human life. In fact, floods have killed millions of people over the centuries. The Huang He (Yellow River) of northern China is known as "China's sorrow." The Huang He has drowned 1 million people in a single flood.

Dams are built to control flood waters. After the building of the Aswān High Dam in Egypt, floods from the Nile River were minimized. As a result, however, the agricultural fields along the riverbanks were no longer enriched by fertile sediments deposited by the floodwaters. Egyptian farmers now must use chemical fertilizers to replenish their soils.

INTERACTION **Water Quality** As the population of the world continues to grow along with the expansion of agriculture and industry, the by-products of human activity also increase. As a result, impure water supplies can endanger human life. This is particularly true in developing nations that are densely populated. There, when sewage comes into contact with the drinking-water supply, diseases may spread.

Sewage is not the major source of water pollution in industrialized nations, which have sewage-treatment plants. In these countries, water is contaminated by industrial chemicals, fertilizers, pesticides, metals, and oil. Farms also pollute rivers and streams with the runoff of pesticides and chemical fertilizers, as do homes with the runoff from lawns.

The oceans are the final dumping ground for some human wastes. Some pollution occurs by accident, as when an oil tanker sinks. In other cases, pollution is caused by the direct dumping of dangerous toxic-chemical wastes and radioactive materials. Most marine pollution comes from rivers, which wash all types of human waste to the

Text continues on page 92.

MULTICULTURAL PERSPECTIVES

Point out that societies throughout history have linked water to their spiritual beliefs. For example, many Native American groups revered rain gods or goddesses, and the ancient Greeks and Romans identified the god of the sea as among their most powerful deities. In Christianity, baptism symbolizes spiritual rebirth. People of the Hindu religion in India consider the water of the Ganges River to be a sacred, healing force. Lead a class discussion about why water has been almost universally linked with religious beliefs. *(It is vital to survival of humankind.)*

Planet Watch

Summarizing

Help students understand the main ideas expressed in this feature. Write "Think Globally, Act Locally" on the board. Call on volunteers to provide their own interpretation of this saying. Ask volunteers to describe how they may have altered their own behavior on a daily basis as a result of their awareness of environmental problems. List their responses on the board.

Photo Caption Answer

Answers will vary. The question's purpose is to encourage students to think about their own habits.

90

PLANET WATCH

What Can You Do?

Persuading Industry

A ninth-grade biology student in Maine learns that tuna nets have been needlessly capturing and killing hundreds of thousands of dolphins each year. He and 75 other students write to tuna company executives, pleading that they stop the practice. As a result of public pressure such as this, within months the nation's three leading tuna companies pledge to stop killing dolphins.

Persuading Government

A group of New Hampshire tenth graders become concerned that their lunch trays are made of styrofoam. The students write and distribute a pamphlet discussing the environmental problems that result from making, using, and disposing of styrofoam. Shortly afterwards, the local school board bans styrofoam trays, and the town council decides to make the government offices and schools styrofoam free.

People around the world are choosing to tackle the earth's environmental problems. As the examples above show, kids play a major role in this effort. When you take positive steps to protect the environment and Earth's resources, your actions benefit the entire planet. Environmentalists call this "thinking globally, acting locally." Many of the most important accomplishments are a result of people adjusting everyday behaviors in simple, small ways.

Reduce, Reuse, Recycle

It is estimated that Americans generate enough trash each year to fill a line of trash trucks that would stack halfway to the moon. You yourself may have thrown away 60 pounds (27 kg) of plastic last year alone.

Have you ever thought about what happens to all that trash? While putting a piece of trash in a garbage can may seem like getting rid

This **New York** landfill provides a feast for hungry seagulls. For humans, the landfill is a warning about the burdens on our environment. How much trash do you think you throw away in one week?

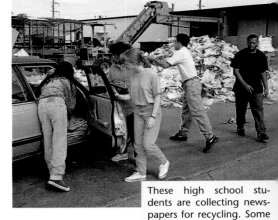

These high school students are collecting newspapers for recycling. Some cities have residential collection programs for used paper, glass, and plastic.

of the item once and for all, really it is just being moved to another place. An astounding 80 percent of all trash goes straight to landfills where waste is buried between layers of earth. Much of the trash will sit there for hundreds of years.

The earth is running out of space for all this garbage. To cut down on the amount of trash piling up around us, one major thing you can do is follow a simple strategy called "Reduce, reuse, recycle."

1. *Reduce* the amounts of disposable goods, water, and energy you use. Not only will this help the environment, it will also save money. For example, write on both sides of notebook paper, so that you will use only half as

Ask several volunteers to bring local phone directories to class. Have students work with partners to search the directories to identify private environmental organizations (similar to the ones listed below) or departments within city or state government that are related to environmental concerns. After they have searched the directories, discuss whether students believe that their community emphasizes environmental concerns.

Point out that their community might show a strong commitment to its environment through volunteer efforts alone. Call on students to describe any actions that have been taken in their community to address environmental problems.

much. If you go shopping, bring a cloth bag in which to carry goods home. While brushing your teeth or washing dishes, turn the water off while you are not actually using it. Turn the lights off when you leave a room.

2. **Reuse** products you already have instead of throwing them out and buying new ones. For instance, if you take your lunch to school in a brown bag, fold up the bag and use it again the next day. If you go on a picnic and use plastic forks and spoons, do not throw them out. Take them home, wash them, and use them again for your next picnic.

3. **Recycle** items that you can no longer reuse, and use recycled products. Most areas now have recycling centers or even residential pick-up for items such as cans, glass bottles and jars, newspapers, cardboard, and plastic. Set aside a place in your home and school to put goods to be recycled. Even if everyone recycled all the time, however, it would do little good if no one ever used products made from recycled materials! Today, many such products are available. Even some notebook paper is made from recycled paper.

Should you wish to take your efforts a little further, you can become involved with environmental groups that share your concerns. Such organizations can make your voice heard and can bring people together to make an even greater difference.

Here are some of the many organizations committed to protecting Earth and its resources:

America the Beautiful Fund
Free Seeds for a Green Earth Program
219 Shoreham Bldg.
Washington, DC 20005

American Forests Association
Global Releaf Program
P.O. Box 2000
Washington, DC 20013

The Children's Rainforest, U.S.
P.O. Box 936
Lewiston, ME 04243–0936

Citizens Clearinghouse for Hazardous Waste
P.O. Box 6806
Falls Church, VA 22040

Earth Day USA
P.O. Box 470
Peterborough, NH 03458

EarthSavers
Dept. ESC
1400 16th St., N.W.
Washington, DC 20036

Educators for Social Responsibility
23 Garden St.
Cambridge, MA 02138

Greenpeace
1436 U St., N.W.
Washington, DC 20009

Kids Against Pollution
(KAP)
Tenakill School
275 High St.
Closter, NJ 07624

Kids for a Clean Environment
(Kids FACE)
P.O. Box 158254
Nashville, TN 37215

National Recycling Coalition
1727 King St., Suite 105
Alexandria, VA 22314–2720

National Wildlife Federation
1400 16th St.
Washington, DC 20036–2266

Renew America
1400 16th St., N.W., Suite 710
Washington, DC 20036

Sierra Club
730 Polk St.
San Francisco, CA 94109

Student Environmental Action Coalition
P.O. Box 1168
Chapel Hill, NC 27514

Toxic Avengers
c/o El Puente
211 South 4th St.
Brooklyn, NY 11211

World Resources Institute
1709 New York Ave., N.W.
Washington, DC 20006

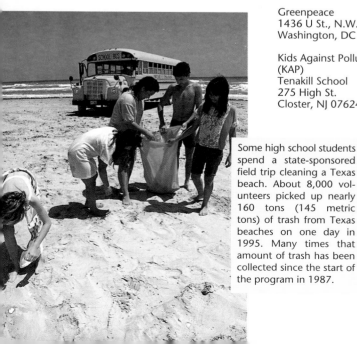

Some high school students spend a state-sponsored field trip cleaning a Texas beach. About 8,000 volunteers picked up nearly 160 tons (145 metric tons) of trash from Texas beaches on one day in 1995. Many times that amount of trash has been collected since the start of the program in 1987.

▼ YOUR TURN

Organize a clean-up of your school grounds. With several classmates, pick up any trash you find and put it in the garbage. Make a list of five items you pick up. **Explain** how each item on your list might have been reduced (not used at all), reused, or recycled.

Your Turn

Advise cleanup teams to wear gloves and to avoid picking up hazardous materials such as broken glass. (Unbroken glass objects such as jars or bottles could be included in the project.) Students' lists might include items such as newspaper, Styrofoam containers, glass, paper, or plastic.

Summarizing

Have students review the discussion on page 93 about the greenhouse effect and climate change. Ask them to name some of the natural conditions and human activities that may contribute to either of these conditions. Then have them write a paragraph summarizing the role of carbon dioxide in global warming.

Guided Practice

Ask students to name natural phenomena that link global ecosystems. *(winds, rain, waterways, animal migration)* Then have them explain ways that pollution created in one state or country can contaminate the air, water, or land of another. Discuss the types of problems that can arise among states or countries that "share" pollution and how these problems can be solved.

Independent Practice

Have students create a chart with these three headings: Pollution Sources, Regulated Solutions, Behavioral Solutions. Ask students to complete the chart, listing as many sources of pollution as they can in the first column *(for example, automobiles)*, a regulated solution for each problem in the second column *(clean air laws)*, and a solution that depends on people altering their behavior in the third column *(carpooling)*.

Geographical Sidelight

Halon 1301 and Halon 1211 are two other gases that are destroying the ozone layer. Halons are synthetic compounds used in fire extinguishers. Chlorofluorocarbons are more concentrated in the atmosphere, but halons destroy ozone 10 to 40 times faster. In 1987, the United States and 22 other nations agreed to stop producing halons by the year 2000.

sea. Such pollution can contaminate fish and shellfish and can make people sick or even poison them.

Some water, such as that of rapidly flowing rivers, recovers quickly from pollution unless chemicals are present. Water that circulates slowly, in deep lakes or groundwater, may stay polluted for centuries.

Air Pollution Though it is often taken for granted, air also is a natural resource. It is an essential gas that we breathe, it protects us from harmful radiation, and it helps maintain the temperature balance of the planet. Global winds also link ecosystems in remote regions of the planet. For example, dust from Africa travels to the Amazonian rain forests, where it may help replace nutrients in the soil. Although we may have an endless supply of air, our atmosphere is subject to change from human pollution.

Air pollution, or smog, is a serious problem facing many large urban and industrial areas. *Smog* originally meant "smoke plus fog," but it now refers to both industrial air pollution and **photochemical smog**. Photochemical smog is produced when sunlight interacts with exhaust gases to produce visible and dangerous air pollution.

Automobiles release exhaust fumes, and factories pour smoke and fumes into the atmosphere. Some cities, such as Mexico City and Los Angeles, have become infamous for their smog. In many developed countries, clean air laws and modern technology have slowed or reduced air pollution. In their rush to industrialize, however, most developing nations face increasing smog problems in their cities.

Acids released by industrial smokestacks combine with water vapor in the atmosphere, causing **acid rain**. These acid droplets can harm lakes, forests, and human health. Often, the industries of one nation produce the pollutants that cause acid rain in another nation hundreds of miles downwind. Industries in Germany and in Poland may harm lakes and forests in Sweden and in Finland.

Global Air Pollution Air pollution is a global problem that knows no political boundaries and has the potential to change the earth's

Russian factories belch pollution into the air around **Murmansk** (above). Scientists are worried that some pollutants being pumped into the atmosphere are damaging the earth's ozone layer. In the satellite images at the right, blue shades show areas with lower ozone levels than areas with brown and yellow shades. The image at the bottom shows much less ozone over Antarctica in 1987 than there was eight years earlier (top).

climate. In the upper atmosphere is a region known as the **ozone** layer. Ozone (O_3) is a gas formed from an interaction between oxygen and sunlight. The ozone layer protects life beneath it by filtering out dangerous ultraviolet solar radiation. Recent evidence collected from satellites, high-flying research aircraft, and weather balloons indicates that the ozone layer has thinned over the Antarctic. Here, there is a "hole" in the ozone layer. As this ozone hole has grown, skin cancer rates in the Southern Hemisphere have risen.

Scientists also are studying the upper atmosphere over the Arctic region for evidence of an ozone hole. Should an Arctic ozone hole exist, skin cancer rates in North America and Europe are likely to increase. Scientists also are very concerned about the effect of the loss of ozone protection on the rich Arctic and Antarctic waters, where the food chain may be damaged by increased ultra-

Closure

Review the main topics in the section—water quantity and world population, flood control, water quality, air pollution, and global air changes. Have students discuss ways in which the issues and challenges relate to their own lives.

ASSESSMENT

Assign the Section 2 Review questions on this page.

Reteaching

Assign Section 2 of Reteaching Worksheet 9.

Extension

Interested students can work together to identify cities of the world that have serious smog problems. They should label the cities and their countries on Outline Map 1 and color code the countries as developed, middle-income, and developing.

SECTION 3

OBJECTIVES

The objectives reflect the Focus questions on this page and are assessed in the review on page 95.

1. Identify fossil fuels and their uses.

2. Discuss the pros and cons of nuclear energy as a power source.

3. Identify renewable energy sources and their benefits.

violet radiation. Increased ultraviolet radiation may damage or kill plants and bacteria. Plants and bacteria are essential to continued life on Earth.

A group of chemicals called CFCs (chlorofluorocarbons) apparently rise in the atmosphere and set off chemical reactions that destroy the ozone layer. CFCs are released from air conditioners, refrigerators, liquid cleaners, and plastic foams. Concerned nations, including the United States, Canada, and some European countries, have agreed to decrease the manufacture of products that release CFCs. Unfortunately, some developing nations find that substitutes for CFCs are too expensive to produce.

Global warming caused by the greenhouse effect may be a further threat to the atmosphere. The greenhouse effect is the result of an apparent buildup of carbon dioxide (CO_2) in the lower atmosphere. This buildup traps heat and causes a long-term warming of the planet. Many scientists have noted that massive coal burning causes increased heat to be trapped in the atmosphere.

When the earth has natural fluctuations in carbon dioxide, the climate changes. Ice ages occurred when carbon dioxide content was low. Warmer periods of the earth's history occurred when carbon dioxide had increased in the atmosphere. Natural conditions such as volcanic activity and forest fires release great amounts of carbon dioxide. Humans also produce carbon dioxide by burning fuels. An increase in carbon dioxide in the atmosphere over the past two centuries coincides with industrial development.

Although we cannot be sure of the effects of global warming and climate change, some scientists warn that sea levels will rise, some types of vegetation will die, and entire ecosystems will be destroyed. Scientists say two ways to protect ourselves from the greenhouse effect are to decrease the burning of fuels that pollute and to stop clearing the world's forests.

2 SECTION REVIEW

1. How are water quantity and water quality affected by a growing world population?
2. Why is there concern that the earth's climate might be changing?
3. **Critical Thinking** **Decide** whether clean air is a renewable or nonrenewable resource, and explain your answer.

ENERGY RESOURCES

FOCUS

● *What fossil fuels are used as major energy sources?*

● *Why is the use of nuclear energy as a power source declining?*

● *What are some renewable energy sources?*

Fossil Fuels Coal, petroleum, and natural gas are known as **fossil fuels**. Fossil fuels are believed to have been formed slowly from the buried remains of prehistoric plants and animals. Coal originates from ancient swamps and bogs; petroleum and natural gas were formed from ancient marine plants and animals.

Since the Industrial Revolution, great amounts of fossil-fuel energy have been needed for factories, transportation, and farms. During the previous two centuries, coal was the major resource. In this century, petroleum and natural gas have replaced coal as the preferred energy sources. In the next century, as petroleum and gas supplies decline, what will be the major energy resource?

Coal is still an important source of energy. Most of it is mined and burned to produce electricity. Along with iron ore, it is also used to make steel, a basic material for all industrial nations. Coal is the source of many chemical products as well, such as synthetic rubber, synthetic fabrics, plastics, and paints.

Because coal is a solid fuel, it has certain disadvantages, and burning it causes acid rain and air pollution. Through expensive liquification and gasification processes, coal can be changed into a cleaner-burning liquid fuel or gas. The nations with the greatest coal reserves include the United States, Russia, China, and Australia.

The main use of petroleum is for liquid fuels such as gasoline, diesel, and jet fuel. Petroleum also provides fuel oils for heat and electrical generation. Many chemicals also are produced from petroleum. These **petrochemicals** include chemical fertilizers, pesticides, food additives, explosives, and medicines. No nation can be an industrial power without a dependable source of petroleum and a petrochemical industry. Petroleum is a

Section 2 Review
ANSWERS

1. Demand for water increases as population increases. Deep wells will need to be drilled, and aquifers may dry up. In developing countries, population growth will affect water quality by adding pollutants such as sewage to water supplies. In industrialized nations, increased industrial and agricultural activity will pollute water with chemicals, metals, and oil.

2. The greenhouse effect, caused by the trapping of gases in the earth's atmosphere, may result in global warming. Effects of global warming might include a rise in sea levels, the loss of some vegetation, and the destruction of ecosystems.

3. Possible answer: nonrenewable because it is difficult to clean polluted air

Motivator

Ask students to name the sources of energy their families use every day in their homes and for transportation and recreation. List them on the board as they are given. Remind students to include fuels used to produce household power. *(coal, nuclear energy)* Then ask them how their lives might be affected if certain energy sources became limited or unavailable.

Evaluating Energy Sources

Have students name the energy sources listed in the section *(coal, petroleum, natural gas, nuclear, hydroelectric, geothermal, solar, wind)*, as well as any others they can think of. *(wood, blubber)* Have them evaluate the pros and cons of each source. *(For example, petroleum has many practical uses but is nonrenewable and causes pollution.)*

Guided Practice

Write *petroleum, coal, natural gas,* and *electricity* on the board. Ask volunteers to list the ways these fuels are transported from their source to the consumer. *(pipelines, oceangoing tankers, trucks, river barges, railroads, wires)* Then discuss which of these methods may pose the greatest threat to the environment.

THEME
Movement

Point out that, during the period following World War II in the United States, many urban mass transit systems were cut back or eliminated in favor of highway building. Since the 1960s, problems with pollution, energy consumption, and traffic have led many cities—including San Francisco, Washington, D.C., and Atlanta—to develop new rapid transit systems.

COOPERATIVE LEARNING

Organize the class into groups of four or five. Assign each group one of the renewable energy sources described in Section 3. Have all groups design an advertisement promoting their energy source, and have them illustrate the ad with examples of its use. Specify different group roles such as researcher, art director, and copywriter. Each group should select an account executive to make the presentation of the ad to the class.

Coal mine

Nuclear power plant

Natural gas flame

Offshore oil rig

nonrenewable resource, however, and the world's supply may last only another 50 years or so. About one-fourth of the world's petroleum comes from beneath coastal waters, and this percentage may increase as the reserves on land are depleted.

The largest petroleum reserves are found in Saudi Arabia and in the neighboring Persian Gulf countries. In the United States, the largest oil reserves are in Alaska, Texas, and Louisiana. Oil is shipped mainly by sea in large tankers, and it is the leading commodity traded in the world today. Oil and oil products also are a major cause of air and marine pollution.

The use of natural gas is growing rapidly. Natural gas often is found in the same wells with oil, but it may also be found separately. The largest reserves of natural gas are in Russia and the Persian Gulf nations.

Natural gas is the cleanest-burning fossil fuel and can be shipped great distances through pipelines. In fact, there are more than 1 million miles of natural-gas pipeline in the United States. Natural gas drilled in Texas heats homes in Massachusetts, and people cooking with gas stoves in California receive natural gas from New Mexico. The United States has abundant supplies of natural gas. Like petroleum reserves, however, they are limited and may become more expensive and scarce over the next several decades.

Nuclear Energy The world's first nuclear power plant began producing electricity in Britain in 1956. Today, there are more than 400 nuclear power plants scattered around the world in more than 25 nations. The United States has just over 100 operating nuclear power plants, which produce about 20 percent of the nation's electricity.

In France, nuclear power plants produce about three-fourths of the electricity. In other nations, such as South Korea and Hungary, nuclear power plants produce almost half of the electricity. Japan depends heavily on nuclear energy.

Originally, nuclear power was seen as a clean, inexpensive power source. Today, however, the nuclear power industry faces two serious problems: nuclear accidents and nuclear waste. In 1986, a nuclear accident occurred at Chernobyl, in the former Soviet Union. Many people died from radiation, and thousands of others who were contaminated will have shorter lives. Many square miles of land around Chernobyl were evacuated, and radiation spread over much of Europe. The accident was devastating to agriculture in the region.

All nuclear power plants produce nuclear waste, and exposure to these radioactive substances is dangerous. Radioactive wastes decay slowly, which means that they must be stored and monitored for thousands of years. The best storage method developed so far is to bury the waste in sealed containers deep underground in solid rock. Transporting these waste products to storage facilities involves a risk of accidents.

Since the Chernobyl accident, public opposition to expansion of nuclear power has grown. Some nations, such as Sweden and Germany, are phasing out existing power plants. The United States, meanwhile, foresees almost no growth in this power source. A few nations, including Denmark and New Zealand, have declared themselves "nuclear free" and have no intention of ever building nuclear power plants.

Renewable Energy Sources Flowing water is an important energy resource. Hydroelectric

Geothermal power plant

Hydroelectric dam

Wind turbines

Solar power panels

power, or hydropower, produced by the force of running water is a renewable and relatively pollution-free source of electrical energy.

Development of hydropower requires rivers with a steady flow, drops in elevation, and the construction of dams, which are very expensive. In some areas, such as the Columbia River in the northwestern United States, numerous dams provide great amounts of electricity. Overall, 10 percent of the United States' electricity is produced by hydropower. Russia, Canada, New Zealand, and Norway also produce large amounts of hydroelectricity. Unfortunately, hydropower can change the natural flow of rivers, flooding vast areas under reservoirs and interfering with the migration of fish, such as salmon. It also depletes farmland.

Another source of renewable energy is underground heat, or **geothermal energy**. Geothermal energy originates in the hot interior of the planet and escapes to the surface through hot springs and steam vents. The escaping steam can be used to generate electricity. Today, geothermal plants produce electricity in countries all over the world, including Ethiopia, Kenya, Nicaragua, India, Indonesia, the United States, Japan, New Zealand, Italy, Mexico, and Iceland.

A third renewable energy source is the sun. Solar energy involves capturing the sun's light energy and converting it into heat or electricity. Many homes in sunny climates now have solar panels that collect sunshine and convert it into heat for hot water. An experimental solar electric plant is now operating in the desert of southern California. If this plant is successful, many more may be built in the future.

For centuries, wind power has been used to grind grain and pump water from wells with windmills and to power sailing ships. Wind also can be harnessed to produce electricity by the use of machines called aerogenerators. On a small scale, aerogenerators may be used on tropical islands to generate electricity for a school. On a much larger scale, "wind farms" with hundreds of aerogenerators can produce electricity for much larger areas. California, for example, supplies 1 percent of its electricity needs from wind power.

In the air, in oceans, underground, and in living things, Earth has virtually unlimited energy resources. Many of these energy sources are nonrenewable, and they may be in short supply in the future. We must continue to seek and improve alternative energy sources and use them more efficiently. Only then can we maintain our energy supply and also reduce environmental hazards.

SECTION REVIEW

1. What are the main uses of coal, petroleum, and natural gas?
2. What are the positive and negative aspects of nuclear energy as a power source?
3. **Critical Thinking** **Imagine** what might happen if we do not seek alternative, renewable energy sources.

95

SECTION 4
OBJECTIVES

The objectives reflect the Focus questions on this page and are assessed in the Section Review.

1. Discuss problems associated with rapid population growth.

2. Describe human activities resulting from the unequal distribution of resources.

TEACHING SECTION 4
Motivator

Have students imagine that the number of students in their classroom has tripled and that they must redistribute desks, textbooks, and supplies.

Guided Practice

Have students give examples of conflicting interests involving an environmental or resource issue.

Closure

Ask students whether they feel optimistic or pessimistic about the future of the planet and why.

ASSESSMENT

Assign the Section 4 Review questions on this page.

Reteaching

Assign Section 4 of Reteaching Worksheet 9.

Section 4 Review
ANSWERS

1. To limit population growth, in order to put less strain on limited resources, thereby raising the standard of living

2. World trade, and conflicts that occur when people try to obtain resources that are not in their own territory

3. It depends on the country. For a country that already has a low population growth rate, a high growth rate might be positive. But for developing countries with an already high growth rate, a low growth rate is usually the goal. Worldwide, a high growth rate is of greater concern.

96

POPULATION AND RESOURCES

FOCUS

● *Why can a rapid population growth rate be a problem for a nation?*

● *What are some consequences of uneven resource distribution?*

Population Growth As you have learned, the world's population is not evenly distributed. Extreme climates, rugged terrain, and lack of water and fertile soil prevent many areas from being populated. Most of the productive regions of the world today, however, are populated.

Population growth rates are important to the future of the world. Some nations, such as Kenya and Pakistan, have rapid growth rates. Many industrial nations have slower growth rates. These include the United States, Canada, Russia, Australia, New Zealand, and most European countries. Some nations, such as China and Indonesia, are trying hard to control population growth.

Why would a nation want to control population growth? Control of population growth would contribute to a higher standard of living for its citizens. Nations must provide jobs, education, roads, and medical care for its people. A slower growth rate will put less strain on the limited resources of that nation. The nations with the highest growth rates are generally the poorest nations.

In contrast, some nations, such as Singapore, Thailand, Japan, and several European countries, have been so successful at controlling population that they are now concerned about low birthrates. With low birthrates, there are fewer young people entering the work force. This may lower productivity and consumer spending. At the same time, the percentage of elderly people is growing. This may cause an economic strain because of the increased cost of health care.

Balancing Population and Resources
The present population of the world is more than 5.5 billion. If current growth rates continue, it is estimated that the world's population will be 8 billion by 2020. More than 1 million people are added to the world every five days. Can the earth support this many people?

Natural resources, like people, are not distributed evenly. Some regions have surplus water, food, or energy resources. Other areas have shortages. Uneven distribution explains why we need world trade, but it also leads to world conflict as nations try to secure needed resources. The people of the world must share and use the planet's resources wisely.

Many people are optimistic about the future. They believe that as the earth's population increases, scientists will discover ways to produce new kinds of foods. Farmers will increase crop yields through new methods, more food and water will be taken from the sea, and high-rise buildings will permit more people to live in less land space.

Yet, other people question whether we have already reached the point of limitation for important resources and food production. Will substitutes be found for many of our dwindling natural resources? Can expanding food production keep up with the population growth? Can we continue to pollute the environment without harming our planet, our nation, and ourselves? Will countries continue to trade valuable resources? Will the richer nations share their wealth and resources with the less fortunate nations?

It is important to remember that people sometimes have conflicting interests. Politicians, economists, environmentalists, scientists, wealthy people, and poor people all have different sets of priorities. A better understanding of the world's geography will help us face these challenging issues. Turn to the "Planet Watch" feature on pages 90–91 for information about how you can help protect the environment and conserve our resources.

SECTION REVIEW

1. Why do some nations try to practice population control?

2. What human activities occur due to the unequal distribution of resources?

3. Critical Thinking **Evaluate** which would be of greater concern to a country: a high growth rate or a low growth rate.

CHAPTER 9 CHECK
ANSWERS

Building a Vocabulary
1. Resources that can be replaced by natural processes; resources that cannot be replaced by natural processes
2. Soil salinization is a high salt buildup that will destroy a soil's productivity. Desertification creates desert-like conditions.
3. Reforestation is the action of replanting trees; deforestation is the process of clearing trees.
4. Turning salty seawater into fresh water; expense

5. It is an air pollutant caused by the combination of acid particles with water vapor. It can travel from its source in one country and pollute the air of neighboring countries.

Recalling and Reviewing
1. Contour farming and planting rows of trees along fields; through the use of fertilizers and crop rotation
2. Original old-growth forests can never be replaced.
3. The ozone layer blocks harmful radiation.
4. The burning of fossil fuels produces carbon dioxide. The buildup of CO_2 is a major factor in the greenhouse effect.
5. Controlling the population has left them with fewer young people to enter the work force and fewer consumers.

Thinking Critically
1. Possible answer: In developed nations, industrialization generally causes environmental problems. In developing countries, environmental problems may be caused by survival needs, such as the need to cut wood for heating/cooking. Economically developed countries have more money to combat pollution.
2. Possible answers: higher prices for gasoline, oil spills from transporting the oil, increased use of alternative energy sources

Learning About Your Local Geography
Reports will vary but should describe environmental studies done in students' communities.

Using the Five Themes of Geography
Answers will vary, depending on how students' homes are heated and/or air-conditioned.

CHAPTER 9 CHECK

A nuclear power plant towering over the French countryside

Reviewing the Main Ideas
1. With an ever-growing population, the world faces a serious problem of resource depletion. Air, water, forests, soil, and fossil fuels are some of the natural resources people must work together to preserve and use responsibly.
2. Human activities, such as industrialization, urbanization, and the burning of fossil fuels, have caused serious environmental pollution. Global warming, caused by the trapping of greenhouse gases in the atmosphere, may result in climate change.
3. The unequal and uneven distribution of resources results in world trade but also in world conflict. In an interdependent world, we must all work together to secure the future health of our shared planet.

Building a Vocabulary
1. Define *renewable* and *nonrenewable resources*.
2. How do soil salinization and desertification threaten the world's soil?
3. What is the difference between reforestation and deforestation?
4. Define *desalinization*. Why is this process not always practical?
5. What is acid rain? Why is it of concern to nations other than the one producing it?

Recalling and Reviewing
1. What are two methods farmers use to prevent soil erosion? How do farmers protect the soil's fertility?
2. Forests are renewable resources, but it is still necessary to conserve them. Why?
3. Why is the recent discovery of a hole in the ozone layer a cause for concern?
4. How does the burning of fossil fuels contribute to the greenhouse effect?
5. Why are some nations concerned about the success of their attempts to control population?

Thinking Critically
1. What is the difference between environmental problems in developing and developed nations? How is pollution related to economic development?
2. The United States currently imports 50 percent of its oil. By the year 2010, the United States may be importing as much as 75 percent of its oil. What might be the consequences of this increase?

Learning About Your Local Geography

Individual Project
Find out about any future plans for development in your community. Then, make inquiries about the kinds of environmental studies that are completed before a development plan is approved by city or county officials. Present your information in a brief report accompanied by a flowchart.

Using the Five Themes of Geography
Find out how your home is heated and/or air conditioned, and write a paragraph about how the geographic themes of **region, movement,** and **human-environment interaction** play a role. How does the region you live in affect your home's energy needs and the forms of energy that are available? How does the energy move from its place of origin to your home? What consequences might the forms of energy you use have for the environment? What could you do to reduce the amount of energy you use?

Unit 2 Review

SKILLS

Explain that geographic controversies often involve many layers of disagreements and frequently have more than two "sides." Students should acquire as much information as possible to learn about the various points of view. For example, students might wish to find out more about the international political and economic aspects of the debate over global warming. Some experts believe that the current industrialization of developing countries will increase global warming and thus should be slowed or stopped. Other experts, however, point out that developing countries should have the same opportunities to industrialize as developed countries had in previous decades.

Practicing the Skill
ANSWERS

1. The authors of the first quote think that continued population growth will exhaust Earth's resources. The author of the second quote believes that population growth may be positive. The quotations disagree over whether new ideas and technologies can offset population increases.

2. The second quote's point of view implies that governments should not be concerned about population growth, whereas the first quote's point of view implies that they should be and might want to encourage population control.

3. Answers will vary but might include information on past population increases, on projected future growth, on resources, and on waste. Such information is available in books and articles at the library.

98

UNIT 2 REVIEW SKILLS

Analyzing Points of View

Have you ever had a disagreement with a friend in which you were both convinced that you were right and the other was wrong? Sometimes it is difficult to get the facts on a particular topic, and sometimes the facts seem to contradict each other. As a member of our global society, you will need to learn how to understand the various, and often opposing, points of view held by geographers. Ultimately, you will have to make decisions about which opinions you agree with and what actions you will take to support your opinions.

A topic on which many geographic experts have different points of view is the greenhouse effect. Andrew Revkin, the author of a book on global warming, believes that global warming could have some very serious, negative consequences. He states:

"If levels of carbon dioxide continue to rise at the current rate, there is a significant chance that disruptive climate shifts will occur within the lifetimes of children born today. The list of potential consequences is familiar: coastal flooding, searing droughts, wars over shrinking water supplies, accelerated extinction of species."

In contrast, Robert C. Balling, the author of another book on global warming, believes that if the earth is warming, it is not warming much, and small increases in the earth's temperature might even be good:

"There is a large amount of . . . evidence suggesting that . . . the highly touted greenhouse disaster is most improbable. The empirical evidence suggests that our future will see a rise in temperature of approximately 1.0°C (1.8°F) . . . the earth will probably be wetter, cloudier, with substantial increases in soil moisture; droughts may diminish in frequency, duration, and intensity. . . . In the case of the greenhouse effect, we can find legitimate benefits from our environmental impact."

This difference of opinion poses difficult questions. Is global warming a threat large enough to justify spending a large amount of money trying to control it? If so, who should take the primary responsibility for this control? These debates are far from being resolved.

PRACTICING THE SKILL

Another topic over which geographers differ is population growth. As you know, the earth's human population is growing rapidly. Not everyone, however, agrees on what effects this rapid growth will have. Read the two quotations below, and then answer the questions that follow.

"There are limits to the rates at which [the] human population . . . can use materials and energy, and there are limits to the rates at which wastes can be emitted without harm. . . . Human society is now using resources and producing wastes at rates that are not sustainable. . . . Even with much more efficient institutions and technologies, the limits of the earth's ability to support population . . . are close at hand."

Meadows, Meadows,
and Randers,
Population Researchers

"The effect of higher population density actually seems to be positive. . . . The central benefit of more people in a more developed world is that there are more . . . people to invent new ideas. . . . If population had not increased about 8,000 years ago and made hunting and gathering become less productive, we would be having wild roots, rabbits, and berries for lunch."

Julian Simon, Economist

1. What do the authors of each quotation think about population growth? About what do the two quotations disagree?
2. What are the implications of each point of view for how governments should think about and make decisions on population growth?
3. What additional information might help you decide which quotation you agree with? How might you find this information?

BUILDING YOUR PORTFOLIO

Decide whether students will work individually or in groups on the unit projects described below. Students should then choose which project they will work on. Provide students with copies of the Portfolio Assessment Worksheets in *Teaching Resources*. Chapters 7–9 worksheets and the Unit Portfolio Assessment Worksheet will guide students step-by-step to complete either project.

BUILDING YOUR PORTFOLIO

Individually or in a group, complete one of the following projects to show your understanding of the geography concepts involved. Both projects are aimed at class presentations. Your report may also take the form of a community education project. You will need to speak with your teacher about specific activities.

A When Different Cultures Meet

From the point of view of many Americans, an important example of different cultures coming into contact has been the meeting of Native Americans with Europeans and European Americans. This contact was not a one-time event—it has lasted several hundred years. These cultures are still interacting today.

Research the results of the meeting of one or more groups of Native Americans with one or more groups of European Americans. Prepare an oral report, supported by written and visual materials, about your findings. To complete your report, think about the following questions.

1. How were Native American cultures different from European cultures? Demonstrate how Native American cultures differed from European cultures, and how these cultural differences caused conflicts. Focus on three or four culture traits. Use illustrations, with captions, to present this information.
2. How did Native American populations change in size after contact with Europeans? What factors were responsible for these changes? In a graph or chart, show how the size of Native American populations has changed over time, from the first meeting with Europeans to today.
3. Did Native Americans and Europeans have similar views about the physical environment? How might different views about the use of the environment cause problems for people who want to share the same environment? Summarize your findings in a few statements.

Organize the information you have gathered, and present your information to the rest of the class. Conclude with a few statements giving your reactions to what you have learned. Do you think there is any way that conflict between European and Native American cultures might have been avoided?

B The Geography of Industry

As you learned in this unit, many countries have industrialized or are industrializing. Investigate industrial development and its effects on a local level. Select a city or large town near where you live. For that city or town, identify one neighborhood or community that contains much industry, and one neighborhood or community in which there is very little industry. Prepare an oral report, supported by written and visual materials, comparing the two communities. To complete your report, think about the following questions.

1. Do the two communities have similar ethnic compositions? Make two bar graphs. One should show the proportions of people of various ethnic groups living in the industrial community. The other one should show the proportions of people of various ethnic groups living in the nonindustrial community.
2. What are the economic activities of people in each community? Make two lists, one showing the most common occupations of people who live in the industrial community, and one showing the most common occupations of people who live in the nonindustrial community.
3. How do the environmental problems faced by the people in each community differ? On a poster, illustrate the environmental hazards in the industrial community. Write a few sentences on the poster about how these hazards compare to those in the nonindustrial community.

Organize the information you have gathered, and present your information to the rest of the class. As you do so, suggest some possible reasons for any differences you have discovered between the two communities. Also, discuss the implications of your findings for the city government and for the members of both communities.

Project A:

Students should discuss the meeting of one or more groups of Native Americans with one or more groups of European Americans. In particular, students should focus on cultural differences and how these differences caused conflicts, on the demographic consequences of the meeting, and on cultural views on the environment. Students should give their reactions to what they have learned.

Project B:

Projects should compare and contrast two local communities, one with much industry and one with little industry. Cultural, economic, and environmental aspects should be considered. Students should carefully consider cause-and-effect relationships and should suggest reasons for any differences they have found. They should also discuss the implications of their findings.

Regions in Transition

Identifying Local Regions

Have students list the regions in which they live, go to school, and engage in other daily activities. List the regions on the board as students suggest them, placing the largest regions (continent, country) at the top and placing smaller regions (neighborhoods, blocks) at the bottom. Note the point at which students list different regions in which they live (probably at the county, city, or neighborhood levels). Take all the regions below this point, and have students create a large map showing them. The map legend should explain the characteristics defining each region shown.

THEME
Region

Point out that geographers are constantly revising the ways they divide the world into regions. Physical, political, economic, and cultural changes necessitate the drawing of new regional boundaries.

Physical changes are often slow to occur but are still important. Continents move and break apart, mountains rise up, rivers run dry. Political changes perhaps bring the most obvious regional redefinitions. For example, one large country may break up into several smaller ones, or many small countries may unite. Economic changes to regions come as different resources or economic activities decline or increase in importance. For example, a community once known for its manufacturing industries may now be part of an area that focuses on services. Finally, changing cultural characteristics can reshape regions as people of various language, religious, or ethnic groups move across regional boundaries.

REGIONS IN TRANSITION

What's in a Region?

Think for a moment about where you live, where your school is located, and where you usually shop. These places are all part of your neighborhood. In geographic terms, your neighborhood can be called a **region**. A region is an area that has more common characteristics than differences. Each region has location, area, and unique descriptive elements.

Geographers classify places by region. Dividing the world into manageable parts allows geographers to compare one region to another, or to study aspects of a single region.

Try to list the many regions in which you live. You live on a continent, in a country, in a state, in a county, and in a city or town. Your zip code, your telephone area code, and your school district are regions. All of these regions can be mapped.

Regions often are organized into smaller regions called **subregions**. For example, Africa is an enormous region that can be divided into several subregions: North Africa, West Africa, East Africa, Central Africa, and Southern Africa. Each of these subregions can be further organized into smaller subregions. Your hometown may be organized into several subregions, such as fire, police, and school districts.

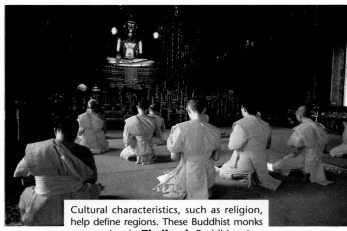

Cultural characteristics, such as religion, help define regions. These Buddhist monks are praying in **Thailand**. Buddhism is a major religion in **Southeast Asia**.

Regional Characteristics

There are limitless characteristics for forming and mapping regions. The four most common types of determining characteristics are physical, political, economic, and cultural. Physical regions are based on the natural features of the earth, such as continents, seas, landforms, and climates. Political regions are based on the nations of the world and their subregions, such as states, provinces, and cities. Economic regions are defined by activities such as agriculture and other industries, or they may center around transportation systems and world trade routes. Cultural regions may classify areas by language, religion, or ethnic groups.

Regional Boundaries

All regions have boundaries, or borders. Boundaries indicate where the characteristics of one region meet the differing characteristics of a neighboring region. The boundaries of some regions are distinct and can be shown as lines on a map. For example, look at the world political map on pages A4–A5. Notice the boundaries that separate countries.

Other boundaries may also appear on maps. For example, turn to the United States physical map on pages A6–A7. Coastlines,

High mountain ranges, such as the Alps here in northern **Italy**, are found in many regions of the world. Ways of life, terrain, and climates in mountainous regions may differ greatly from those in regions with lower elevations.

100

Analyzing Maps

Have students analyze the regional boundaries on the map they created for the **Identifying Local Regions** exercise. Note areas where regions overlap, and ask students how they might describe these areas. *(as transition zones)* Then ask whether any regions shown on the map are uniform or nodal regions.

Using Geography Themes

Point out that location and place characteristics are the major factors in how regions are defined. Then mention that movement and human-environment interaction can also play a role. Ask students what types of human-environment interaction might create regions or change regional boundaries. *(constructing dams that create lakes or change the course of rivers, deforestation)* Ask: How may movement affect defined regions? *(The migration of culture groups can change the characteristics of an area.)*

Geographical Sidelight

Some geographers prefer to call nodal regions "functional regions", in order to emphasize how the center (node) functions in relationship to the areas that surround it and the activities that occur within the region. Also, the terms "uniform region" and "formal region" are used interchangeably.

National Standards in Geography define regions in three basic ways: formal, functional, and perceptual. Formal regions are characterized by the presence of a common human property (e.g., language, religion, nationality, political identity) or common physical property (e.g., climate, landform, vegetation cover). Functional regions are organized around a node or focal point with surrounding areas linked to that node through transportation or communication systems or economic associations. Perceptual regions reflect human feelings or attitudes and are defined by subjective images of the area (e.g., Dixie, Southern California, the upper Midwest).

Regions of the World

mountain ranges, and rivers show regional boundaries. Some regional boundaries are less clear, however. Areas called **transition zones** exist when different characteristics such as language, religion, and economic activity overlap into the same area. For example, when a city's suburbs are expanding into a rural area, a transition zone forms. During this period, it is difficult to determine where the exact boundary between the rural and urban areas lies.

Types of Regions

Geographers identify two types of regions. The first type is a **uniform region**—an area distinguished by one or more common characteristics. The country of Mexico is a uniform region because its people share a common government, language, and culture.

In dividing the world into nine regions, the authors of this textbook took into account physical, cultural, political, and economic factors. As you read this textbook, think about how you would have divided the world into regions.

The second type, a **nodal region**, is an area distinguished by the movement or activities that take place in and around it. A node is the region's center point or point of activity. Paris, France, is at the center of a nodal region defined by the distances that goods, services, and workers travel to and from the city each day. A shopping center, an airport, or a seaport also can be a region's node.

Often, uniform and nodal regions overlap to form complex world regions. In Units 3 through 11 of this textbook, nine regions form the basis of our geographic study of the world. (See the map above.) Each of these major world regions has unifying characteristics based on physical location and on historical, cultural, economic, and political elements.

Economic activities, such as oil production here in **Saudi Arabia**, are important characteristics of the world's regions.

UNIT 3 THE UNITED STATES AND CANADA
CHAPTERS 10–16

OVERVIEW

In Unit 3, students will learn about the tremendous geographical diversity of the United States and Canada.

Chapter 10 • INTRODUCING THE UNITED STATES provides a geographical overview of the United States.

Chapter 11 • THE NORTHEASTERN UNITED STATES describes how population, cultural diversity, and urbanization characterize the Northeastern United States.

Chapter 12 • THE SOUTHERN UNITED STATES introduces the physical, economic, and urban geography of the region.

Chapter 13 • THE MIDWESTERN UNITED STATES highlights the region's agricultural activity and industry.

Chapter 14 • THE INTERIOR WEST describes the Great Plains, Rocky Mountains, and Intermountain West.

Chapter 15 • THE PACIFIC STATES focuses on California, Oregon, Washington, Alaska, and Hawaii.

Chapter 16 • CANADA describes significant characteristics of Canada.

OBJECTIVES

1. Describe the major landform regions and climates found in the United States and Canada.

2. Identify major urban areas in the United States and Canada and the social, political, and economic challenges they face.

3. Understand how the development of natural resources found in North America has benefited both the United States and Canada.

4. Examine the cultural geography of the United States and Canada.

5. Identify environmental challenges that confront industries and communities in the United States and Canada.

6. Interpret special-purpose maps to analyze the relationships among climate, population patterns, and economic resources in the United States and Canada.

7. Learn how models are used to analyze geographic data.

UNIT 3

The UNITED STATES and CANADA

The Grand Canyon in **Arizona** is among North America's most spectacular physical features. The Colorado River carved the mile-deep canyon. Rock layers hundreds of millions of years old are visible in the colorful sides of the canyon. Rocks at the bottom of the canyon are several billion years old.

TEACHING UNIT 3

Using the Illustrations

Direct students' attention to the photograph of the Grand Canyon on page 102. Ask for a show of hands by students who have seen the vast canyon, either in person or in a movie or video or from illustrations in a book. Ask them to share their impressions with the rest of the class. Point out that this natural wonder is one of the major tourist draws in the United States. Ask students to identify another tourist attraction shown on these pages. *(Olympic Stadium in Montreal)*

Have students look at the picture of the Hopi wood carving of the Sun Kachina. Point out that the Hopi are Native Americans who predominantly live in the southwestern part of the United States. Point out that many religious symbols in Native American cultures reflect elements of physical geography such as the sun.

Teaching with Sketch Maps

Have students draw a map of the United States and Canada without consulting any resources. Direct them to form a mental picture of the region and to transfer what they remember to paper. Tell them to include as many physical features, state and provincial boundaries, and cities as they can. They might add this map to their portfolios and compare it with another they will draw after they have completed Unit 3.

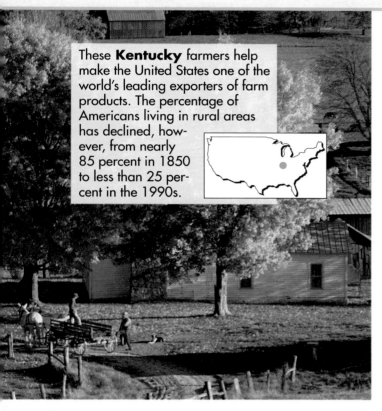

These **Kentucky** farmers help make the United States one of the world's leading exporters of farm products. The percentage of Americans living in rural areas has declined, however, from nearly 85 percent in 1850 to less than 25 percent in the 1990s.

UNIT 3 CHAPTERS

10 Introducing the United States

11 The Northeastern United States

12 The Southern United States

13 The Midwestern United States

14 The Interior West

15 The Pacific States

16 Canada

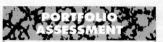

PORTFOLIO ASSESSMENT

"Building Your Portfolio" on page 183 provides a choice of two unit projects for students to work on as they study Unit 3. The unit projects are supported by Chapter Portfolio Assessment worksheets in the *Teaching Resources*.

THE WORLD TODAY VIDEODISC PROGRAM

You might wish to use the profile of Canada as you teach Unit 3.

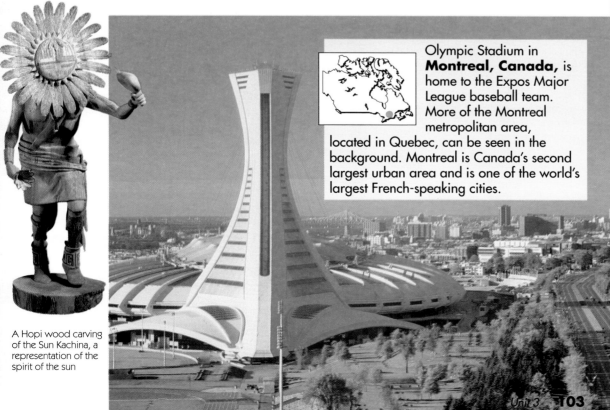

A Hopi wood carving of the Sun Kachina, a representation of the spirit of the sun

Olympic Stadium in **Montreal, Canada,** is home to the Expos Major League baseball team. More of the Montreal metropolitan area, located in Quebec, can be seen in the background. Montreal is Canada's second largest urban area and is one of the world's largest French-speaking cities.

One World, Many Voices

LITERATURE

Introduce the poem on this page by telling students that it describes the eruption of a volcano. First ask a volunteer to read aloud the explanatory note about the title; then have students locate Mount St. Helens on the map on page 107. Explain that the author has given the mountain human characteristics as a way of presenting an environmental message. Point out that many environmentalists object to the logging industry's practice of cutting timber in old-growth forests such as those in the Cascade mountain range where Mount St. Helens is located. Elicit student responses about ways that Loo-Wit reacts to something being done to her.

More About Wendy Rose

Wendy Rose is known primarily as a poet, but she is also an artist. She has drawn original illustrations for several children's books, including her own work. Rose studied anthropology at the University of California at Berkeley. Currently, she teaches at Fresno City College.

Wendy Rose (1948–) was born in California. She began her writing career in 1967. Many of her published books of poetry reflect her Native American heritage and emphasize environmental themes. "Before I learned how to write I expressed myself through making up songs. It has always been a tradition of my people [the Hopi of Arizona] to celebrate everything in life with song. For all occasions, both happy and sad, I sang my feelings."

ONE WORLD, MANY VOICES
LITERATURE

Loo-Wit*

The way they do
this old woman
no longer cares
what we think
but spits
her black tobacco
any which way
stretching
full length
from her bumpy bed.
Finally up
she sprinkles
ash on the snow,
cold buttes
promise nothing
but the walk
of winter.
Centuries of cedar
have bound her
to earth,
huckleberry ropes
lay prickly
on her neck.
Around her
machinery growls,
snarls and ploughs
great patches of her skin

She crouches in the north,
her trembling
the source
of dawn.

Light appears
with the shudder
of her slopes,
the movement
of her arm.
Blackberries unravel,
stones dislodge,
it's not as if
they were not warned.

She was sleeping
but she heard the boot scrape, the creaking floor,
felt the pull of the blanket
from her thin shoulder.
With one hand free she finds her
weapons
and raises them high, clearing the
twigs from her throat
she sings, she sings, shaking the
sky
like a blanket about her
Loo-wit sings and sings and sings!

Loo-Wit is a name that Native Americans of the Cowlitz group gave to Mount St. Helens, a volcano in Washington state. Mount St. Helens erupted violently in 1980 after being inactive for more than 120 years.

Interpreting Literature

Answers

1. Answers will vary. Volcanoes are very old physical landforms that can be inactive for many years. An eruption might be compared to awakening from a long sleep.

2. Her arm movements dislodge stones from the mountain and cause berries to fall from the trees, and she raises her weapons. She clears her throat and sings.

Interpreting Literature

1. Why is this image of an elderly woman awakening from her sleep appropriate for a volcano such as Mount St. Helens?

2. How does Loo-Wit warn of the impending eruption? What action by Loo-Wit represents the eruption itself?

104

FOR THE RECORD

Introduce this excerpt by showing students on a wall map how much territory was added to the United States in 1803 as a result of the Louisiana
Purchase. Call on students to suggest why the president was interested in
details such as climate, natural resources, and the course of rivers. *(He
knew that such information was relevant to the future economic development
of the United States and would be of benefit to people who would move into
this region from the east.)* Point out that Thomas Jefferson was thinking as a
geographer when he issued these detailed instructions to Lewis and Clark.

More About Thomas Jefferson

Few U.S. presidents have been as
admired as Thomas Jefferson. His life
was dedicated to public service and
to the development of the United
States. Jefferson is remembered not
only for his authorship of the Declaration of Independence, but also for
his considerable accomplishments as
a diplomat, an architect, an educator,
and a philosopher.

FOR THE RECORD

Thomas Jefferson, third president of the United
States, played an important role in acquiring the
vast territory known as the Louisiana Purchase.
Bought from France in 1803, this land extended
the boundary of the new nation from the
Mississippi River to present-day Montana.
Jefferson commissioned young army officers
Meriwether Lewis and William Clark to lead an
exploration team across this region. A letter of
instructions to Lewis reveals the president's
keen interest in the geography of the territory.

*To Meriwether Lewis, esquire, Captain of the 1st regiment of
infantry of the United States of America:*

The object of your mission is to explore the Missouri river, & such principal
stream of it, as, by its course & communication with the waters of the Pacific
Ocean, may offer the most direct & practicable water communication across
this continent, for the purposes of commerce.
   Beginning at the mouth of the Missouri, you will take observations of lati-
      tude & longitude, at all remarkable points on the river, & especially
      at the mouths of rivers, at rapids, at islands & other places. . . .
      Other object worthy of notice will be
      the soil & face of the country, its growth & vegetable
         productions . . .
      the animals of the country . . .
      the mineral productions of every kind . . .
      Volcanic appearances.
      climate as characterized by the thermometer, by the propor-
         tion of rainy, cloudy & clear days, by lightning, hail, snow, ice,
         by the access and recess of frost, by the winds prevailing at
         different seasons.

William Clark's journal
observations, March 2, 1806

Analyzing Primary Sources

1. What did President Jefferson consider the most important
   purpose of the Lewis and Clark expedition?
2. What elements of physical geography in the Louisiana
   Purchase territory were of special interest to Jefferson?

*Native American inhabi-
tants of the territory were
instrumental to the suc-
cess of Lewis and Clark's
venture. The Mandan
Indians allowed the expe-
dition team to spend the
winter of 1804–1805 in their
camp on the Missouri
River. And, when the
explorers prepared to cross
the Rocky Mountains, the
Shoshone resupplied the
party and prepared them
for the many dangers they
would encounter travers-
ing the rugged range.*

Primary Source

Description of Change:
excerpted
Rationale: excerpted to
focus on main idea

Analyzing Primary Sources

Answers

1. To explore the
   Missouri River to
   determine its
   potential as a
   means to trans-
   port goods and
   people to the
   Pacific Ocean
2. Navigable quali-
   ties of the river,
   potential for min-
   eral wealth in the
   territory, climate
   patterns

105

CHAPTER 10 Introducing the United States

Planning Guide

TEXTBOOK	TEACHING RESOURCES	OTHER AVAILABLE MATERIALS
Section		
1. Physical Geography Pages 111–115	Graphic Organizer 10 Reteaching Worksheet 10.1	*Creative Strategies for Teaching World Geography* LESSON 6 Colonization of Country "x" *REGIONS OF THE WORLD Map Transparencies:* United States and Canada Origins of Immigrants to the United States *Study Guide for Spanish Speakers* *Lecture Notes,* Chapter 10
2. Human Geography Pages 115–117	Reteaching Worksheet 10.2 Skills Worksheet 10	*REGIONS OF THE WORLD Map Transparencies:* Ethnic Diversity in the United States *Study Guide for Spanish Speakers* *Lecture Notes,* Chapter 10
3. Economic Geography Pages 117–118	Reteaching Worksheet 10.3 Critical Thinking Worksheet 10 Challenge/Enrichment Worksheet 10	*Study Guide for Spanish Speakers* *Lecture Notes,* Chapter 10
Chapter Review Page 119	Chapter 10 Test Chapter 10 Portfolio Assessment See also *Alternative Assessment Forms*	*Test Generator,* Chapter 10

CHAPTER RESOURCES

Lecture Notes, Chapter 10
Teaching Resources
   Chapter 10 Graphic Organizer
   Chapter 10 Skills Worksheet
   Chapter 10 Critical Thinking Worksheet
   Chapter 10 Challenge/Enrichment Worksheet

REVIEW AND ASSESSMENT

Chapter 10 Check, page 119
Chapter 10 Test, *Teaching Resources*
Chapter 10 Portfolio Assessment, *Teaching Resources*
Test Generator, Chapter 10
Alternative Assessment Forms

ADDITIONAL RESOURCES

Books for Teachers

Dilsaver, Lary M., and Craig E. Colten. *The American Environment.* Lanham, MD: Rowman & Littlefield, 1992.

McKnight, Tom L. *Regional Geography of the United States and Canada.* Englewood Cliffs, NJ: Prentice Hall, 1992.

Newcombe, Jack, ed. *Travels in the Americas.* New York: Weidenfeld & Nicholson, 1989.

Books for Students

America's Hidden Wilderness: Lands of Seclusion. Washington, DC: National Geographic Society, 1988.

America's Historic Places: An Illustrated Guide to Our Country's Past. New York: Reader's Digest Association, 1988.

America's Wild and Scenic Rivers. Washington, DC: National Geographic Society, 1983.

Multimedia Resources

America Alive GUIDisc. (Multimedia CD-ROM, interactive tour) MediAlive, 766 San Aleso Avenue, Sunnyvale, CA 94086.

A Day in the Life of America. (16mm film/VHS, 18 min.) Aims Media, 9710 DeSoto Avenue, Chatsworth, CA 91311–4409.

GTV: A Geographic Perspective on American History. (Multimedia CD-ROM) National Geographic Society, Washington, DC.

Picture Atlas of the World. (interactive maps, photo bank, animation, audio on CD-ROM) National Geographic Society, Washington, DC.

USA Geograph II—Multimedia Edition. (Multimedia CD-ROM atlas) MECC, 6160 Summit Drive North, Minneapolis, MN 55430-4003.

MEETING INDIVIDUAL NEEDS

Limited English Proficiency (LEP)

To help emphasize the cultural diversity of the United States, help students make a pie chart showing the ethnic diversity of students in the class. Students might use the pie charts on page 116 as models for their charts.

Learners Having Difficulty

Help students review the main ideas of Chapter 10 by having them use the maps on pages 107–110 to answer the Focus questions and the Section Review questions for each section. Students should skim the section to find the answers for any questions that cannot be answered from the maps.

Auditory Learners

Characteristics that have helped make the United States a world leader are discussed throughout this chapter. Call on students to take turns locating and reading aloud parts of the chapter that describe these attributes, while a volunteer lists them on the board. Have students copy the characteristics on a piece of paper and rank them in order of importance. Hold a class discussion about the rankings.

Visual Learners

Divide the class into three groups. Direct each group to compare one of the following pairs of maps on pages 107–110: Population and Physical–Political, Population and Climate, Population and Economy. Have each student write generalizations based on the maps. Ask a member of each group to present the group's statements to the class.

Tactile/Kinesthetic Learners

To illustrate the growth of the Sunbelt and the western United States in recent decades, have students investigate the relocation and expansion of major league baseball teams since 1940. Note that, until 1953, 16 teams played in the same 10 cities—mainly in the northern and eastern states—for 50 years. On an outline map, have students plot and label the locations of pre-1953 teams in one color and each successive wave of relocation or expansion in different colors, noting the year of each team's establishment in a new city.

Gifted Learners

Ask interested students to write a story or poem that uses "local color"—that describes or weaves in distinctive characteristics of their particular locale.

MULTICULTURAL FOCUS

Students may be familiar with the term "melting pot" as it applies to the cultural history of the United States. Call on volunteers to explain in their own words what this term means. Discuss ways that immigrants were often expected to give up aspects of their ethnic or former national heritages in order to assimilate into a homogeneous "American" culture. Today, many people prefer to characterize the United States as a "tossed salad"—a society that recognizes and affirms the cultural differences that exist among the diverse peoples that make up our population.

People often honor their ethnic or cultural heritage by describing themselves as Italian-American, African-American, etc. However, labels such as these can be misleading. In the 1990 U.S. Census, more than 22 million people identified themselves as of "Spanish/Hispanic" origin. To some, this label is limiting because it lumps people with many cultural, national, language, and racial differences into the same category. Many people identified as "Hispanic" now prefer the term "Latino." Cuban-Americans or Puerto-Rican Americans may find both terms too general. People of African heritage may have Spanish surnames and originally come from Spanish-speaking countries, yet they do not identify themselves as "Hispanic."

CHAPTER 10

INTRODUCING THE UNITED STATES

Section 1: Physical Geography
Section 2: Human Geography
Section 3: Economic Geography

OVERVIEW

Chapter 10 provides students with a geographic overview of the United States.

Section 1 describes the physical features of the United States and highlights landform regions and climates.

Section 2 discusses the people of the United States. It summarizes early settlement activity and explores the cultural diversity of the present population. Students will learn how the distribution of the population has changed over time.

Section 3 examines factors that have enabled the United States to build a strong economy and discusses challenges to the nation's economic well-being.

MAP TRANSPARENCIES

You might wish to show the Origins of Immigrants to the United States transparency, in *REGIONS OF THE WORLD Map Transparencies.*

CREATIVE STRATEGIES FOR TEACHING WORLD GEOGRAPHY

LESSON 6 Colonization of Country "x"

CHAPTER 10

Introducing the United States

Presidential Seal of the
United States

GEOGRAPHY DICTIONARY

piedmont

Fall Line

basin

contiguous

Gulf Stream

seaboard

plantation

edge city

ghetto

trade deficit

The United States and Canada share the majority of the North American continent as well as the world's longest peaceful boundary. The United States shares a southern boundary with Mexico.

The United States consists of 50 states and often is referred to as America or the USA. It features a federal system in which the national government shares power with the states. The United States is a democracy, a form of government in which all citizens have a voice. Regarded as a "land of opportunity" throughout its history, the United States continues to attract millions of immigrants from around the world.

The United States ranks fourth in the world in land area and ranks third in population. The country faces oceans and seas in three directions, making its location ideal for international trade. With the breakup of the Soviet Union in the early 1990s, the United States remains the world's single undisputed military superpower.

This unit explores the diverse and complex geography of the United States and Canada. Chapters 10 through 15 examine the United States; Chapter 16 discusses Canada.

The U.S.–Canadian border between **New York** and **Quebec**

Using the Geography Dictionary

Write the words in the Geography Dictionary on the board, and pronounce them for the class. Call on a volunteer to choose a word and to attempt to define it. If the student's definition is correct, have him or her erase the word from the board. Call on another volunteer to define a word and then to erase it if correct. Continue in this way until the class can no longer define any of the words remaining on the board. Have students look up undefined terms in the Glossary.

Using the Physical–Political Map

Call attention to the physical–political map of the United States and Canada on this page. Have volunteers locate the area where they live and name other areas where they have lived or visited. Then have students compare the physical and political features of the United States and Canada.

MAP TRANSPARENCIES

Overhead transparencies from the maps on pages 107–110 are available in *REGIONS OF THE WORLD Map Transparencies.*

MAPPING THE REGION

THE UNITED STATES AND CANADA

PHYSICAL–POLITICAL

LEGEND

Ice caps

⊛ National capitals

• Other cities

ELEVATION

FEET		METERS
13,120		4,000
6,560		2,000
1,640		500
656		200
(Sea level) 0		0 (Sea level)
Below sea level		Below sea level

SCALE

0   500   1000 Miles

0   500   1000 Kilometers

Projection: Azimuthal Equal Area

MAPPING THE REGION

Using the Climate Map

Direct students' attention to the climate map of the United States and Canada on this page. Note that Canada has one climate type that covers most of its territory (subarctic). Ask students to identify the climate regions of the United States that are not found in Canada. (*humid subtropical, Mediterranean, tropical savanna, desert, humid tropical*) Have students suggest reasons for these climate differences. (*Higher latitudes of Canada produce a colder climate.*)

MAPPING THE REGION
THE UNITED STATES AND CANADA

CLIMATE

◆ Most of Canada has a subarctic or tundra climate.

◆ Most of the eastern United States has a humid-continental or humid-subtropical climate.

◆ A variable highland climate dominates large western areas.

Find your home state on the physical–political map on page 107.

(?) In which climate region do you live?

Compare this map to the population map on page 109.

(?) Which climate zone in Canada is the least populated?

(?) Which metropolitan areas with more than 1 million inhabitants are located in a desert climate?

LEGEND

■	Humid tropical	■	Marine west coast
■	Tropical savanna	■	Humid continental
■	Desert	■	Subarctic
■	Steppe	■	Tundra
■	Mediterranean	□	Ice cap
■	Humid subtropical	■	Highland

CLIMATE

Answers

1. Answers will vary.
2. Tundra
3. Phoenix

MAPPING THE REGION

Using the Population Map

Direct students' attention to the population map of the United States and Canada on this page. Have them compare the overall population density of the two countries. *(Students should notice that the United States is more densely populated than Canada.)* Refer students back to the climate map on the opposite page. Tell them to search for patterns that cover the same areas on both maps. *(One example is the sparsely populated tundra and subarctic climate zones of Canada.)*

Mapping the Region

THE UNITED STATES AND CANADA

POPULATION

◆ The greatest concentrations of population in the United States are along the East Coast and in California.

◆ Large parts of Canada and Alaska are sparsely populated.

◆ Canada's greatest urban region is in the southeastern part of the country.

❓ Which metropolitan areas in Canada and the United States have more than 2 million inhabitants?

❓ Which state has the most metropolitan areas with populations of more than 1 million?

Compare this map to the economic map on page 110.

❓ Name some economic activities found in the more sparsely populated areas of the western United States.

LEGEND

POPULATION DENSITY

Persons per sq. mile	Persons per sq. km
520	200
260	100
130	50
25	10
3	1
0	0

● Metropolitan areas with more than 2 million inhabitants

● Metropolitan areas with 1 million to 2 million inhabitants

POPULATION

Answers

1. Montreal, Toronto, Boston, New York, Philadelphia, Washington, D.C., Detroit, Atlanta, Miami, Dallas, Houston, Chicago, Minneapolis, Phoenix, San Diego, Los Angeles, San Francisco, St. Louis

2. California

3. Livestock raising, mining, commercial farming, forestry

109

MAPPING THE REGION

Using the Economy Map

Have students locate the state in which they live on the map below. Tell them to identify all the economic activities and resources that are found in their state. Then ask them to select the activity or resource that is most prevalent in their state. Ask students to study the map to discern economic patterns in the United States and Canada based on the information on the map. *(Mineral resources are generally located in the western United States. Oil and natural gas are prevalent in coastal areas of Louisiana and Texas. Commercial farming is widespread through the middle region of the United States. Hydro-electric power is generated in eastern and western Canada.)*

MAPPING THE REGION
THE UNITED STATES AND CANADA

SCALE
| 0 | 100 | 200 Miles |
| 0 | 100 | 200 Kilometers |

Projection: Albers Equal Area

To understand the relative location of Hawaii as well as the vast distance separating it from the rest of the United States, see the map on page A4.

SCALE
| 0 | 500 | 1000 Miles |
| 0 | 500 | 1000 Kilometers |

Projection: Azimuthal Equal Area

ECONOMY

Answers

1. These locations are ideal for the transportation of goods to market.

2. Alaska, Texas, Louisiana, states along the Appalachian Mountains, northern and central Interior West, California

3. Steppe, highland, desert

ECONOMY

◆ Large central regions of the United States and Canada are used for farming.

◆ Major manufacturing centers are scattered throughout most regions of the United States.

◆ Timber is an important natural resource in Canada and the United States, especially in the southeastern and northwestern states.

? Why are many manufacturing and trade centers located along rivers, the coasts, and the Great Lakes?

Compare this map to the physical–political map on page 107.

? In which states and provinces are oil and coal production important economic activities?

Compare this map to the climate map on page 108.

? What climates dominate the livestock-raising areas in the western United States?

LEGEND

☷ Livestock raising	☞ Dairying	Hydroelectric power
Commercial farming	Limited economic activity	Ⓐ Gold
Manufacturing and trade	⚒ Coal	Ⓐ Silver
Commercial fishing	⬙ Natural gas	Ⓤ Uranium
Forests	⚑ Oil	Other miner
	✳ Major nuclear power plants	Timber

A blue, misty haze gives the Great Smoky Mountains their name. Shown here in **North Carolina**, the mountain range is part of the Appalachian Mountains.

PHYSICAL GEOGRAPHY

FOCUS

- *What are the major landform regions in the United States?*
- *What different climates are found in the United States?*

REGION **Landforms** The United States features some of the world's most spectacular scenery. Its landforms vary from vast, flat plains to high mountains and plateaus to volcanic islands. (See the map on page 107.)

The landforms of the eastern half of the United States are older than those of the western half. The mountains have eroded, and rolling hills and flatlands dominate most of the region. In contrast, the western United States has a younger landscape. Steep mountains, active volcanoes, deep canyons, high plateaus, and wide plains are common here.

The region known as the Coastal Plain stretches along the Atlantic Ocean and Gulf of Mexico from New York to the border with Mexico.

This low region lies close to sea level and rises only gradually inland. The Coastal Plain is narrowest along the Atlantic coast, widens to the south in the Carolinas and Georgia, and includes all of Florida. Along the Gulf of Mexico, it widens considerably and includes all of the lower Mississippi River valley as far north as southern Illinois.

The Piedmont is a rolling plateau region inland from the Coastal Plain. A **piedmont** is an area at or near the foot of a mountain region. The Piedmont begins in New Jersey and stretches as far south as Alabama. Many of America's first cities, including Philadelphia, Pennsylvania, and Richmond, Virginia, were established along the **Fall Line**, where the Coastal Plain meets the Piedmont. Here, rivers plunge over rapids and waterfalls from the hard rock underlying the Piedmont to the softer rocks underneath the Coastal Plain. These rivers provided waterpower for the early towns and cities.

The Appalachian Mountains rise to the west and north of the Piedmont. The Appalachians were formed more than 300 million years ago when, according to the theory of plate tectonics, eastern North America collided with Africa. The mountains' peaks have been lowered and smoothed by erosion. The Appalachians stretch from Maine to Alabama and are divided into several ranges. Their highest peaks rise to more than 6,000 feet (1,829 m) in North Carolina's Blue Ridge Mountains and New Hampshire's White Mountains. A major portion of the Appalachians is made up of a series of parallel ridges and valleys. Well-known features in the Appalachians are the Blue Ridge, Catskill, and Pocono mountains.

FACTS IN BRIEF	United States		
COUNTRY POPULATION (1994)	LIFE EXPECTANCY (1994)	LITERACY RATE (1991)	PER CAPITA GDP (1993)
UNITED STATES 260,713,585	73, male 79, female	97%	$24,700

Source: *The World Factbook 1994*, Central Intelligence Agency

The United States has a higher gross domestic product (GDP) per capita than Canada. Canadians, however, have a longer life expectancy than Americans. See the chart on page 174.

Locating

Landforms Call on volunteers to locate the following on a wall map of the United States: Alaska Range, Appalachian Mountains, Brooks Range, Cascade Range, Coast Ranges, Rocky Mountains, Sierra Nevada. Ask students questions such as, Which is the longest and highest range? and Which range has the highest peak?

Cause and Effect

Climate Have students complete a chart showing factors that influence the climate in various areas of the United States. *(For example, in the southeastern quarter of the United States, the warm waters of the Gulf of Mexico and the warm Gulf Stream current cause summers to be hot and humid and winters to be mild.)*

Guided Practice

Write the names of these major landform regions on the board: Coastal Plain, Piedmont, Interior Plains, Intermountain, Pacific Coast. Ask students to find words in the text that describe the location of each region. Call on volunteers to read each description aloud as other volunteers trace the region's boundaries on a wall map of the United States.

THEME
Place

Mount McKinley, located in Denali National Park in Alaska, actually has two peaks—the South Peak (20,320 feet, or 6,194 meters) and the North Peak (19,470 feet, or 5,934 meters). Mount McKinley was named in 1896 for William McKinley shortly before he was elected to the U.S. presidency. Native Americans of the region called the peak Denali, which means "the High One." Efforts were made in the mid-1970s to restore the Native American name.

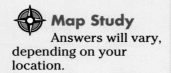 **Map Study**

Answers will vary, depending on your location.

Physiographic Regions of the United States and Canada

 MAP STUDY The colors in this map designate the physiographic regions of the United States and Canada. A physiographic region, or province, has the same general types of landforms throughout. In which physiographic region do you live?

The Ozark Plateau, although separated from the Appalachians, also is an old, eroded upland. It lies in Missouri, Arkansas, and Oklahoma.

There are many fertile river valleys in the Appalachians. Among these are the Connecticut, Hudson, and Shenandoah valleys. In the western part of the highlands, a long plateau slopes toward the plains of the Midwest. This is known as the Allegheny Plateau in the northeast and the Cumberland Plateau farther south.

Between the Appalachians and the Rocky Mountains is a vast area called the Interior Plains. The Mississippi River and its many tributaries drain most of the region. The northern Interior Plains, north of the Ohio and Missouri rivers, were covered by glaciers during the last ice age. The great ice sheets changed the landscape. Today, this area has rolling hills, thousands of lakes (including the five Great Lakes), river systems, and a wide variety of productive soil types.

The Great Plains stretch from Canada to Mexico in a north-south direction and from approximately 100° west longitude to the foot of the Rocky Mountains in an east-west direction. Although the Great Plains are the flattest portion of the Interior Plains, they are not as low in elevation. At the eastern edge of the Rocky Mountains, the Great Plains reach elevations of more than 5,000 feet (1,524 m).

The Rocky Mountains extend north and south from Canada to New Mexico and from the Great

Independent Practice

Have students draw pictures that illustrate the general climate characteristics of the landform region in which they live.

Closure

Ask students to identify the landform region of the places shown in the photographs on pages 111, 113, 114, and 115.

ASSESSMENT

Assign the Section 1 Review questions on page 115.

Reteaching

Assign Section 1 of Reteaching Worksheet 10.

Plains west to the Intermountain region. Many of the highest peaks reach more than 14,000 feet (4,267 m) above sea level. Their jagged peaks and deep valleys are evidence of glacial action. The Rocky Mountains are not a single range but a series of several ranges separated by high plains and valleys.

The Intermountain region is located between the Rocky Mountains on the east and the Cascade and Sierra Nevada ranges on the west. Because rivers in this region (except the Colorado River) never reach the ocean, much of the region is called the Great Basin. A **basin** is a low area of land, generally surrounded by mountains. The rugged landscape of the Intermountain region consists mainly of high plateaus with deep canyons, isolated mountain ranges, and desert basins. The lowest point on the continent of North America, Death Valley in California, lies at the western edge of the Great Basin. Death Valley drops 282 feet (86 m) below sea level.

The Pacific coast region is characterized by two major mountain systems and a series of valleys between the mountains that extend in a north-south direction. On the eastern edges of this region are the Sierra Nevada and Cascade ranges. The Sierra Nevada is a giant faulted mountain range that runs almost the length of California. It is the longest and highest range in the 48 **contiguous** states. Contiguous states are those that border each other as a single unit. Only Alaska and Hawaii are not contiguous states. To the north of the Sierra Nevada, in northern California, Oregon, and Washington, are the Cascade Mountains. This range consists of a series of high volcanoes, such as Mount Rainier, Mount Hood, Mount Shasta, and Mount St. Helens.

Along the Pacific Ocean are the rugged Coast Ranges that stretch from California to Canada. Between the Coast Ranges and the Sierra Nevada and Cascades are three fertile valleys: the Puget Sound Lowland, the Willamette Valley, and the Central Valley of California.

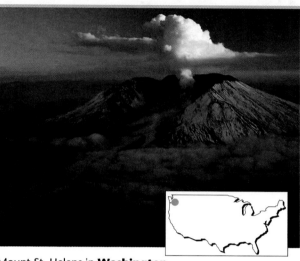

Mount St. Helens in **Washington** state is one of a number of volcanoes in the western United States. What is the cause of the volcanic activity in the western part of North America?

The rugged western United States is part of the Pacific Ring of Fire, with its active volcanoes and earthquake faults. This is where, according to the theory of plate tectonics, the North American plate is colliding with the Pacific plate. All of the continental United States is on the North American plate except for coastal California, which is located on the Pacific plate. This part of California is separated from the North American plate along the San Andreas Fault, where major earthquakes occur.

The two westernmost states, Alaska and Hawaii, also are geologically active. All of the Hawaiian Islands are a result of huge volcanoes that have grown from the seafloor. Southeast and south-central Alaska is very mountainous. North America's highest peak, Mount McKinley in the Alaska Range, is 20,320 feet (6,194 m) in elevation. Except for the Brooks Range, northern and interior Alaska have flat and hilly terrains. The volcanic Aleutian (uh-LOO-shuhn) Islands extend into the Pacific from Alaska.

REGION **Climate** The 48 contiguous states are located in the middle latitudes and have a variety of climates. The climates of the United States are influenced mainly by this middle-latitude location, a large land area, high mountain ranges, westerly winds, and ocean currents. The United States has 11 climate types, the greatest variety of climates of any nation. (See the map on page 108.)

A humid-continental climate covers the northeastern quarter of the nation. It stretches westward from the Atlantic coast to approximately 100° west longitude. This climate region has four distinct seasons, including a warm, humid summer and a cold, snowy winter. The nearness of the Great Lakes and the Atlantic Ocean moderates temperatures slightly and is a source of increased precipitation.

Photo Caption Answer

Some scientists believe that the collision of the North American and Pacific plates has created active volcanoes and earthquake faults.

Geography and Seismology

Seismologists constantly monitor movement along the San Andreas Fault. The fault begins off the coast of northwestern California and extends about 600 miles (965 km) to the southeastern part of the state. Since 1986, a number of moderately strong earthquakes have occurred in southern California. These clusters of quakes have led several scientists to predict that a major earthquake will occur somewhere along the fault before 2018.

113

Extension

Have each student play the role of a meteorologist for a major television station in one of the following cities: Buffalo, New York; Chicago, Illinois; Denver, Colorado; Fairbanks, Alaska; Los Angeles, California; Miami, Florida; Phoenix, Arizona; and Seattle, Washington. Ask students to research the climates of these cities and to write a weather forecast for a typical summer or winter day in the chosen city.

SECTION 2
OBJECTIVES

The objectives reflect the Focus questions on page 115 and are assessed in the review on page 117.

1. Identify groups of people that have inhabited what is now the United States.

2. Describe population composition and distribution patterns in the United States.

TEACHING SECTION 2
Motivator

Explain the meaning of this statement: The United States is sometimes considered a cultural "melting pot." *(It is the idea that although the United States has drawn many diverse peoples, it now has a single uniform culture.)* Ask students whether they agree that the United States is a melting pot. Have them support their opinions.

Geographical Sidelight

Parts of the Great Plains, especially in Kansas and Oklahoma, are known as "tornado alley" because of the many tornadoes that occur there. Underground shelters are common in some towns of tornado alley.

Geography and Literature

Read to the class an American tall tale from an anthology of tall tales. Explain to the students that such tales were a feature of American frontier life. Some tales claim to explain how a landform such as a mountain or canyon was created. Others tell of the exploits of a superhero. Tall tales were a form of entertainment for pioneers, who enjoyed the exaggeration that made the story bigger and better every time it was told. (You might consult Tall Tales of Davey Crockett, *published by the University of Tennessee Press in 1987, edited by Michael A. Lofaro.)*

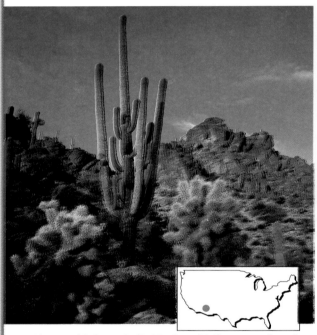

These teddy bear cholla (CHOL-yah) (foreground) and saguaro cacti near **Tucson, Arizona**, are native to the desert climate of the southwestern United States.

Lightning arcs over the Space Needle and downtown **Seattle, Washington**. Seattle is located in a wet, marine-west-coast climate.

A humid-subtropical climate is found in the southeastern quarter of the United States. This region stretches from the Atlantic and Gulf of Mexico coasts to about 100° west longitude. The climate is influenced by the warm waters of the Gulf of Mexico and by the warm **Gulf Stream** current in the Atlantic Ocean. The Gulf Stream is an ocean current that moves warm tropical water northward. Summers are hot and humid, and winters are mild. Rainfall is distributed fairly evenly throughout the year, and thunderstorms are common. Some snow and frost may occur in winter. The very southern tip of Florida extends beyond the humid-subtropical region. This area has a tropical-savanna climate.

West of 100° west longitude is the steppe climate of the Great Plains. This semiarid climate supports vast grasslands but has few trees. Air from the deserts, Arctic Canada, and the Gulf of Mexico invade the region. Summers are hot, while winters can be very cold. The Great Plains region is subject to violent weather, including thunderstorms, hailstorms, tornadoes, and dust storms. Droughts are also a major concern here.

The interior western states have a variety of climates because of the mountainous terrain. The Rocky Mountains have highland climates, with temperatures and precipitation dependent upon elevation and exposure. The Intermountain region, located between the major mountain areas of the west, has mainly desert and steppe climates due to its rain-shadow location.

The Pacific coast region has two climates, marine west coast and Mediterranean. The mild marine-west-coast climate dominates the Pacific Northwest. It has cool, cloudy, wet winters and mild, sunny summers. The Mediterranean climate bathes southern and central California. This climate is known for its long, sunny, dry summers and its mild, wet winters.

Hawaii is the southernmost state and the only one that is fully within the tropics. Because the Hawaiian Islands fall in the easterly trade wind belt, they are wetter on the windward, eastern sides where they have a humid-tropical climate. On the leeward, western slopes there is a drier tropical-savanna climate.

Alaska is the northernmost state and the only one that extends into the subarctic and tundra climates. It actually has three climates: a narrow strip of marine-west-coast climate along the southeast coast, a subarctic climate in the interior, and a tundra climate along the north and west coasts. The subarctic climate area is forested and has severe seasonal changes. Winters are

Building a Vocabulary

Write **seaboard, plantation, edge city,** and **ghetto** on the board. Point out that these terms all refer to types of places. Call on volunteers to give definitions of each term. Write responses on the board, and then have students compare the responses with the Glossary definitions.

Analyzing

Population Distribution Point out to students that more than half the U.S. population lives in areas adjacent to the Atlantic and Pacific oceans, the Gulf of Mexico, and the Great Lakes. Ask students to suggest reasons why coastal areas are the most densely populated. *(Possible answers include location of earliest settlements, access to water and land transportation, industrialized cities, entry point for immigration, natural beauty.)*

Guided Practice

On the board, sketch line graphs representing each of the following U.S. population characteristics over time: amount of cultural diversity, average age, percentage of population in the Sunbelt, percentage of population in suburbs. Point out that these graphs all show increases. Ask students what graphs might show decreases rather than increases. *(Students might mention population in rural areas.)*

cold and long, while summers are short and relatively warm. The tundra climate area is cold all year. It has no trees, and only small plants grow.

SECTION REVIEW

1. Name four major mountain ranges in the United States. Where is each one located?
2. Name and describe the climate region in which you live.
3. **Critical Thinking**   **Explain** why the United States has such a wide variety of landform regions and climate types.

HUMAN GEOGRAPHY

FOCUS

- *Over time, what groups of people have settled in what is now the United States?*
- *What are the characteristics of the population of the United States today, and how is the population distributed?*

MOVEMENT **Early Settlement** The first inhabitants of North America were Native Americans, who probably arrived from Asia between 10,000 and 40,000 years ago. The first Europeans to arrive were Viking explorers about

A.D. 1100. By this time, several million Native Americans lived throughout the North American continent.

The Spanish arrived about 500 years ago, claiming mainly the southern part of the present-day United States from Florida to California. They were soon followed by the French, who explored major river systems and claimed the interior of North America. The British, however, became the major influence during the nation's early history, establishing 13 colonies along the Atlantic **seaboard**. A seaboard is a land area near the ocean. Among the earliest settlers along the eastern seaboard were Africans, who first arrived in Jamestown in 1619.

Beginning in the 1600s, millions of Africans were brought to the American colonies and sold as slaves to work on southern **plantations**. A plantation is a large farm that concentrates on one major crop, such as cotton or tobacco. Although the slave trade with Africa ended in 1808, the tragic practice of slavery did not end completely until after the Civil War.

After gaining independence from Britain, the United States grew as immigrants arrived, mainly from Europe. By the mid-1800s, the United States stretched from the Atlantic to the Pacific coast and had formed its present borders with Canada and Mexico. During this time, as the nation's growing population gradually moved westward, more states were added to the original 13. Diseases and warfare with settlers killed many Native Americans. The surviving populations were driven farther west, where they were eventually forced onto reservations.

Section 1 Review
ANSWERS

1. Possible answers: Appalachians, from Maine to Alabama; Rocky Mountains, from Canada to New Mexico; Sierra Nevada, in California; Cascades, in northern California, Oregon, and Washington; Coast Ranges, along Pacific Coast; Alaska and Brooks ranges, in Alaska
2. Students should name the climate in which they live and describe its seasons and weather patterns.
3. It is large, spanning many degrees of longitude and latitude.

Some archaeologists believe the Great Serpent Mound—constructed near present-day **Cincinnati, Ohio**, by Native Americans—was used as a burial site and for defense.

115

Independent Practice

Have students make time lines showing the arrival of various groups in America, from prehistoric times to the present. If time permits, assign research on groups not specified in the text.

Closure

Call on students to report one fact they learned in this section about the settlement or population of the United States. Use students' responses to elicit a class discussion relating early settlement to current trends in population distribution and diversity.

ASSESSMENT

Assign the Section 2 Review questions on page 117.

Reteaching

Assign Section 2 of Reteaching Worksheet 10.

Extension

Have interested students write a paragraph describing the population composition and distribution of their local community. Have them note ways the community might be changing.

GLOBAL PERSPECTIVES

Point out that war, poverty, and persecution in other parts of the world influence immigration to the United States. Have students look for newspaper articles about world events or conditions that might cause people to seek a better life in this country. Post students' findings on a class bulletin board.

MAP TRANSPARENCIES

An overhead transparency of the graphs on this page is available in *REGIONS OF THE WORLD Map Transparencies.*

Ethnic Diversity in the United States

Racial and Ethnic Distribution of the U.S. Population: 1990

Asian: 2.8%
Native American: .7%
Hispanic: 9%
Black: 11.8%
Non-Hispanic White: 75.7%

Projected Racial and Ethnic Distribution of the U.S. Population: 2050

Asian: 10.1%
Hispanic: 21.1%
Native American: 1.1%
Black: 15%
Non-Hispanic White: 52.7%

Source: U.S. Bureau of the Census

These pie graphs show the racial distribution of the U.S. population in 1990 and the projected distribution for 2050. The percentage of Americans of Hispanic (Latino) origin and of Asian Americans is projected to increase dramatically by 2050. The figures for Hispanic origin include all those who identified themselves as Hispanic in surveys conducted by the U.S. Bureau of the Census.

Population Diversity Today, the United States has the most culturally diverse population in the world. It is about 75 percent European American (also called non-Hispanic white by the U.S. Bureau of the Census), 12 percent African American, 9 percent Hispanic (Latino), 3 percent Asian, and 1 percent Native American (Indian, Inuit, and Hawaiian).

Europeans were the dominant immigrant group in America until the 1960s. Recent immigration, however, has been mostly from Middle and South America and Asia. In the past few decades, millions of Hispanics (Latinos) have immigrated to the United States, especially from Mexico, Puerto Rico, Cuba, and Central America. Early Asian immigrants included Chinese laborers brought to the West Coast between 1849 and 1882. Since the 1960s, Asians have arrived from many countries, including the Philippines, Vietnam, Korea, China, and India. In recent years, many people have immigrated to the United States illegally. Most come to seek jobs or to escape political instability in their home countries.

The United States is predominantly Christian in religion, though nearly every world religion is represented here. The population is about one-half Protestant and one-third Roman Catholic. There also are significant numbers of Jews, Mormons, and Muslims.

The U.S. population, which ranks third in the world after China and India, is only about 5 percent of the total world population. With low birthrates by world standards, the population of the United States will form an even smaller percentage in the future. Low birthrates mean that there are fewer young people, and high health standards mean that the nation's population is living longer. In the early 1990s, half of Americans were older than 33. It is estimated that by 2050 half of Americans will be over 40.

Population Distribution The densest settlement in the early years of the United States was along the eastern seaboard, especially in the northeastern quarter of the country. Today, the greatest concentrations of population are still along the East Coast and in California, the most populous state. As the United States has developed from an industrial society into a technological society over the past few decades, many people have moved from the northern states to the Sunbelt states, particularly California, Texas, Florida, and Arizona. In recent years, some of the most rapid population growth has been in the less populated states of the Intermountain West and Pacific Northwest. (See the map on page 109.)

The United States is a highly urbanized nation. Its three largest metropolitan areas are New York, Los Angeles, and Chicago. Most U.S. metropolitan areas contain large central cities and ever-growing suburbs. Most commuters now drive from suburb to suburb instead of from suburb to city. **Edge cities**, or suburbs with large employment and commercial centers, are common.

This suburbanization has left many of America's inner cities abandoned and in decay. These areas have become pockets of urban poverty with high crime rates, substandard schools, and little economic opportunity. Many areas are **ghettos**, parts of cities where minority

SECTION 3
OBJECTIVES

The objectives reflect the Focus questions on this page and are assessed in the review on page 118.

1. Explain why the United States is a world leader in many respects.
2. Discuss challenges facing the United States.

TEACHING SECTION 3
Motivator

Ask students to think of ways in which the United States is a world leader. Have them suggest characteristics that have made the United States a world leader. Tell students that they will learn in Section 3 about some of the United States' economic advantages.

Building a Vocabulary

Call on students to explain what a **trade deficit** is. Tell them not to confuse "trade deficit" with "budget deficit," a term they may have heard in the news. Tell them what a budget deficit is. Then call attention to the reference to "national debt" on page 118. Have students suggest the meaning. Ask students how the three terms, while having different meanings, might be related. *(All are economic challenges facing the United States.)*

---

groups are concentrated because of economic pressure and social discrimination.

SECTION REVIEW

1. List six major groups of people who have inhabited what is now the United States.
2. What is meant by suburbanization?
3. **Critical Thinking** **Analyze** the changes occurring in the makeup and distribution of the population of the United States.

ECONOMIC GEOGRAPHY

FOCUS

- *How does the United States compare to other nations of the world in resources, industry, communications, and trade?*
- *What political, economic, and domestic challenges does the United States face?*

Resources Rich natural resources and a productive population have made the United States an industrial giant and the world's leading agricultural nation. Even though less than 3 percent of the population farms, the nation easily feeds its people and is a major food exporter. U.S. farmers produce the world's widest variety of agricultural products, including huge supplies of corn, wheat, and soybeans. Much of this success is due to the nation's wide variety of climates and rich soils.

Water also has been important to the nation's economic development. Many rivers, including the Tennessee, Colorado, and Columbia, are a source of irrigation and hydroelectricity. The Great Lakes and rivers such as the Mississippi, Ohio, Hudson, and Columbia are excellent inland water routes. The waters of the nearby continental shelf provide rich ocean resources.

Forests are another valuable natural resource to the country. One-third

of the United States is forested, with the major commercial forests located in the southeastern United States and in the Pacific Northwest. Early logging depleted the forests of the Northeast and the Midwest. Much of today's logging, both for domestic use and export, takes place on private tree farms and in national forests.

The United States is rich in energy and mineral resources. Almost 30 percent of the world's coal reserves are in the United States, which is a major coal exporter. Most of the coal is mined in the Appalachian and Rocky mountains and on the Interior Plains.

The United States also is a major oil producer, although it uses more than it produces. More than one-third of its needed oil must be imported. Most domestic oil production takes place on the Gulf Coast of Texas and Louisiana, and in California and Alaska.

INTERACTION **Industrial Growth** During the nineteenth century, the United States experienced rapid industrialization. America's early steel industry was centered in Midwest cities and was dependent on the region's coal and iron ore. The nation's infrastructure was a major boost to U.S. industrial leadership. The early development of waterways and railroads helped industry spread into the interior of the country.

Free enterprise encouraged innovation, inventions, and competition. Henry Ford's mass automobile production is a good example of American

A dry landscape near **Yuma, Arizona,** has been transformed into productive farmland by irrigation systems. Much of the western and southwestern United States has a climate that is too dry to support agriculture without irrigation.

Section 2 Review
ANSWERS

1. Possible answers: Native Americans, Spanish, French, British, Africans, other Europeans, Hispanics (Mexicans, Puerto Ricans, Cubans, Central Americans), Asians (Chinese, Filipinos, Vietnamese, Koreans, Indians)
2. The movement of large numbers of people from cities to suburbs
3. Makeup: increasingly culturally diverse and older; distribution: moving to Sunbelt and western states, and moving to suburbs

Analyzing U.S. Industry

Have students list factors that contributed to industrial growth in the United States. *(natural resources, infrastructure, free-enterprise system, skilled workers)* Ask them to specify the factor they think was the most important.

Guided Practice

Ask students to name a job they would like to have in the future. Write their responses on the board. As a group, classify the job choices as agriculture, manufacturing, or services. Make the point that most Americans now work in service industries. Discuss reasons for the change in employment patterns (such as the growth of technology).

Closure

Have students create a phrase that sums up the section (*such as, Many Resources, Many Challenges*)

ASSESSMENT

Assign the Section 3 Review questions on this page.

Reteaching

Assign Section 3 of Reteaching Worksheet 10.

Geography and Economics

In late 1993, the U.S. Congress approved the North American Free-Trade Agreement (NAFTA) to increase free trade among the United States, Canada, and Mexico. The free-trade zone created by the agreement has more than 360 million consumers and a combined economic output of almost $7 trillion.

Section 3 Review
ANSWERS

1. Possible answers: rich natural resources, strong infrastructure, free enterprise and innovation, productive and diverse population

2. Possible answers: international—trade deficit and role of the United States in world conflicts and affairs; domestic—high national debt, poverty, rising crime rates, urban decay, environmental pollution

3. Answers will vary; students might say that with its many natural and human resources the United States is well equipped to remain a world leader for the next several decades.

The world's most mobile people help make U. S. airports the world's busiest. In one year, more than 45 million passengers pass through the terminals of **Los Angeles International Airport**, shown here.

innovation. The United States was the first nation to develop this method of production. American industry has also been a leader in industrial research, which led to the development of products such as plastics, nylon, aircraft, computers, and computer software.

Today, American companies are world leaders in fields such as aerospace, computer technology, pharmaceuticals (medicinal drugs), chemicals, and construction engineering. The United States also is the leader in the entertainment industry (movies, television, and music) and the fast-food industry. Service industries, especially banking, education, food services, health services, retail sales, tourism, and communications, are the most rapidly growing area of the U.S. economy.

With the world's largest highway system and more than 135 million automobiles, Americans are the most mobile people in the world. The nation also has a vast aviation network. Many Americans, in homes and offices, have access to the world's most developed telecommunications network. Televisions, computers, fax machines, and telephones are a part of everyday life.

International Trade The United States is a world leader in international trade. For the past 20 years, however, the nation has had a **trade deficit**. A trade deficit occurs when the value of a nation's imports exceeds that of its exports. Although the United States exports aircraft, computers, entertainment products, and food, it imports (in value) more oil, automobiles, and electronic consumer products.

The United States is a major part of the global economy, and imports and exports will

continue to grow. An important challenge for the nation is to reverse the trade deficit and continue innovation in the face of economic competition.

The United States has worked with other countries to lower barriers to world trade. For example, the United States approved the North American Free-Trade Agreement (NAFTA) with Canada and Mexico in 1993. Also, the United States and other countries finished work on the new General Agreement on Tariffs and Trade (GATT) in 1994. These agreements lowered tariffs and other trade barriers.

Issues The United States is a wealthy and democratic nation. Its future lies in solving problems created by a changing political and economic world. Major issues in foreign affairs include trade and the U.S. role in world conflicts and affairs. The United States also faces domestic challenges such as a high national debt, poverty, crime, urban decay, and pollution. With the most productive, innovative, and diverse people in the world, the United States is well equipped to confront the challenges of the twenty-first century.

SECTION REVIEW

1. Describe three characteristics of the United States that have helped it become a world leader.
2. Name two international and two domestic challenges facing the United States.
3. **Critical Thinking**   **Assess** the ability of the United States to remain a world leader over the next 50 years.

CHAPTER 10 CHECK

ANSWERS

Building a Vocabulary

1. Piedmont: area near the foot of a mountain region; Fall Line: where the Coastal Plain meets the Piedmont
2. Alaska and Hawaii
3. An ocean current that moves tropical water northward; the humid-subtropical climate of the southeastern quarter of the United States
4. A large farm that produces one major crop; enslaved peoples from Africa

Recalling and Reviewing

1. The western continental United States and Alaska and Hawaii are the most geologically active because of earthquake faults and active volcanoes.
2. From Middle and South America and Asia; to seek jobs or to escape political instability
3. Because the United States uses more oil than it produces
4. Service industries, especially banking, education, food services, health services, retail sales, tourism, communications

---

CHAPTER 10 CHECK

Loading grain onto a freighter in **Duluth, Minnesota**

Reviewing the Main Ideas

1. The United States has a wide range of landforms, including plains, mountains, plateaus, and volcanic islands. The United States also has many climate types, from humid tropical in Hawaii to desert in the western United States to tundra in Alaska.
2. Many different groups of people have lived in what is now the United States. Today, the population is ethnically and religiously diverse, and it is aging. It is shifting toward southern and western states and out of cities into suburbs as well.
3. The United States is rich in natural and human resources and is a world leader in industrialization and international trade. Even so, the United States faces many challenges.

Building a Vocabulary

1. Define *piedmont* and *Fall Line.*
2. Which two states are not contiguous with the other states?
3. What is the Gulf Stream? Which region's climate is influenced by it?
4. What is a plantation? Who provided the labor for early U.S. plantations?

Recalling and Reviewing

1. Which parts of the United States are the most geologically active? Why?
2. From where do most recent immigrants to the United States come? For what reasons are they immigrating to the United States?
3. If the United States is a major oil producer, why must oil be imported?
4. Which U.S. industries employ the most people?

Thinking Critically

1. Do you think the presence of tariffs would serve to increase or decrease a trade deficit? Explain.
2. Of all of the challenges facing the United States, which one do you think is the most urgent, and why? List some things you might do to help the United States confront this challenge.

Learning About Your Local Geography

Individual Project

How has the average age of your community's population changed in recent years? Request information from a community elementary school about student enrollments over the past 20 years. Has the number of students in the school increased, decreased, or stayed the same? Graph your data. What can you conclude from your data about the average age of your community's population? Write a caption for your graph, suggesting reasons for any shifts. You may want to include your graph in your individual portfolio.

Using the Five Themes of Geography

Create a graphic organizer to show how the geographic themes of **location**, **place**, and **movement** have interacted to make the United States a world leader in many respects. What advantages does the United States have because of its location? What physical and human place characteristics have enabled the United States to make the most of its location? How has movement played a role in helping the United States use both its location and its place characteristics?

A Mexican-style rodeo near **Austin, Texas**

Thinking Critically

1. Tariffs would probably reduce a trade deficit because they make imported goods more expensive, reducing imports.
2. Answers will vary. Students might mention voting and other means of government participation as ways of facing challenges.

Learning About Your Local Geography

Captions should demonstrate an understanding that as elementary school enrollments increase, the average age of the community population is probably falling, whereas if enrollments are decreasing, the average age of the community is probably rising.

Using the Five Themes of Geography

Graphic organizers will vary but should indicate advantages of location (with Pacific and Atlantic coasts, the United States is well situated for foreign trade), place (it is rich in many resources), and movement (it has an infrastructure allowing relatively easy movement of people, goods, and ideas).

CHAPTER 11 The Northeastern United States

Planning Guide

TEXTBOOK	TEACHING RESOURCES	OTHER AVAILABLE MATERIALS
Section		
1. New England Pages 121–123	Reteaching Worksheet 11.1 Skills Worksheet 11 Challenge/Enrichment Worksheet 11	*Study Guide for Spanish Speakers* *Lecture Notes,* Chapter 11
2. The Middle Atlantic States Pages 124–126	Reteaching Worksheet 11.2	*Study Guide for Spanish Speakers* *Lecture Notes,* Chapter 11
3. Megalopolis Pages 127–130	Reteaching Worksheet 11.3 Graphic Organizer 11 Critical Thinking Worksheet 11	*Study Guide for Spanish Speakers* *Lecture Notes,* Chapter 11
Chapter Review Page 131	Chapter 11 Test Chapter 11 Portfolio Assessment See also *Alternative Assessment Forms*	*Test Generator,* Chapter 11

CHAPTER RESOURCES

Lecture Notes, Chapter 11
Teaching Resources
    Chapter 11 Graphic Organizer
    Chapter 11 Skills Worksheet
    Chapter 11 Critical Thinking Worksheet
    Chapter 11 Challenge/Enrichment Worksheet

REVIEW AND ASSESSMENT

Chapter 11 Check, page 131
Chapter 11 Test, *Teaching Resources*
Chapter 11 Portfolio Assessment, *Teaching Resources*
Test Generator, Chapter 11
Alternative Assessment Forms

ADDITIONAL RESOURCES

Books for Teachers

Kircher, John C. *A Field Guide to Eastern Forests: North America.* Boston: Houghton Mifflin, 1988.

O'Brien, Robert. *The Encyclopedia of New England.* New York: Facts on File, 1985.

Yardley, Jonathan. *States of Mind: A Personal Journey Through the Mid-Atlantic.* New York: Villard Books, 1993.

Books for Students

Durham, Michael. *The Mid-Atlantic States.* New York: Stewart, Tabori & Chang, 1989.

Kane, Joseph Nathan. *Nicknames and Sobriquets of U.S. Cities, States, and Counties.* 3d ed. Metuchen, NJ: Scarecrow Press, 1979.

New England: Land of Scenic Splendor. Washington, DC: National Geographic Society, 1989.

Multimedia Resources

Boston and Vicinity: Structural Changes. (VHS, 13 minutes) Encyclopedia Britannica Educational Corporation, 310 South Michigan Avenue, Chicago, IL 60604–9839.

Our Friends in Historic Cities. (Philadelphia, Pennsylvania/Washington, DC) (VHS, 7 minutes) Encyclopedia Britannica Educational Corporation, 310 South Michigan Avenue, Chicago, IL 60604–9839.

Picture Atlas of the World. (interactive maps, photo bank, animation, audio on CD-ROM) National Geographic Society, Washington, DC.

MEETING INDIVIDUAL NEEDS

Limited English Proficiency (LEP)

Pair LEP students with students who are proficient in English. Have the pairs skim the chapter to find the definitions of the terms in the Geography Dictionary. Have the students take turns explaining the meanings of the terms in their own words. Then ask each pair to determine together whether each term helps describe the physical, human, or economic geography of the Northeastern United States.

Learners Having Difficulty

Help students review the information presented in the chapter by creating a table on the board. Use these column headings: States, Landforms, Climate, Agriculture, Natural Resources, Industrial Growth, Issues, Major Cities. Use these row headings: New England, Middle Atlantic. Ask students to skim the chapter to find information to complete the table, while volunteers write it in on the board.

Auditory Learners

Assign students one of the five major cities of Megalopolis. Have students use information in the chapter to present an oral report about the city, including such information as location, main features, and historical significance.

Visual Learners

Have students collect pictures or create drawings of important features and points of interest in the Northeastern United States. Then have students work in small groups to make travel brochures for the region, using the pictures.

Tactile/Kinesthetic Learners

Have students work in pairs to create maps that show New England's landforms and the factors that influence its climate. When students have finished, hold a discussion about the effect of New England's landforms and climate on its ways of life. Ask: What crops are best suited to the short growing season? What adaptations and changes in landscape have people made to live in this climate? How do New England's landforms and climate affect tourism?

Gifted Learners

Have interested students investigate the colleges and universities of the Northeastern United States. Have them present an oral report on a college or university that they might like to attend in this region. Their oral reports might include information such as location, date founded, reasons founded, and academic specialties.

MULTICULTURAL FOCUS

Explain to students that the climate of New England helped shape the type of local government in the region. Harsh winters encouraged townspeople to build their homes near one another for aid and comfort. Close community living fostered the New England town meeting, a legislative body in which every qualified citizen has a voice in local government.

The town meeting is an example of direct democracy, in which people vote on matters individually as citizens rather than through representatives. The town meetings held in colonial times were attended by most white adult townsmen. The taste of self-government that these meetings provided helped encourage colonists to revolt against British rule.

The town meeting is still a part of the New England way of life. In some small towns, townwide gatherings are held as they were in colonial times, to pass laws, discuss improvements, levy taxes, and elect the officers who run the town between meetings. In many larger towns and cities, voters elect representatives to attend the meetings.

The town meeting form of government has also been adopted in other parts of the country. The concept of a town meeting has also been used on a national level—through call-in television and radio programs—as a forum for discussing major issues.

Have students describe their local government or student government and compare it to a town meeting. Elicit a class discussion about the pros and cons of direct democracy and representative democracy.

OVERVIEW

Chapter 11 describes the Northeastern United States. Dense population, cultural diversity, and heavy urbanization characterize this region.

Section 1 discusses how the landforms and climates of New England have affected its economy and how the economy of this region is changing.

Section 2 highlights the major landform regions and climates of the Middle Atlantic states. This section also explains the changing importance of this region's resources and location to its development.

Section 3 describes the major cities of Megalopolis and the challenges that face this nodal region.

About the Author

E. B. White (1899–1985) was an essayist, a poet, and the author of the children's classic *Charlotte's Web*. His columns in *The New Yorker* magazine during the 1930s and 1940s often profiled life in the city.

Primary Source
Description of Change: excerpted
Rationale: excerpted to focus on main idea

CHAPTER 11

The Northeastern United States

❝ There are roughly three New Yorks. There is, first, the New York of the man or woman who was born here, who takes the city for granted and accepts its size and turbulence as natural and inevitable. Second, there is the New York of the commuter—the city that is devoured by locusts each day and spat out each night. Third, there is the New York of the person who was born somewhere else and came to New York in quest of something. ❞

E. B. White

The Northeast is the most populated and urbanized region of the United States. It has the nation's greatest concentration of factories, banks, universities, and transportation centers. The region is home to New York City, the nation's largest city, and Washington, D.C., the nation's capital.

This chapter organizes the Northeastern United States into two subregions that are overlapped by a third subregion. The New England states are located in the far northeastern portion of the country. The Middle Atlantic states stretch eastward from the Appalachians to the Atlantic coast. **Megalopolis** (megh-uh-LAHP-uh-luhs) is a giant urban area that extends along the eastern seaboard. This subregion overlaps the New England and Middle Atlantic states.

Subway tokens for the **New York City** Transit Authority

GEOGRAPHY DICTIONARY

Megalopolis
textile
granite
peninsula
moraine
second-growth forest
biotechnology
truck farm
bituminous coal
coke
anthracite coal
borough
break-of-bulk center

The Statue of Liberty and **New York City**

Using the Quotation

Read the quotation on page 120 aloud. Ask students if this passage could apply to other big cities. Have them explain how the last sentence is related to the photograph on this page. (*Immigrants often see the Statue of Liberty on entering the United States, and most immigrants come "in quest of something."*)

Using the Geography Dictionary

Have students look in the Glossary to find the definitions of the terms in the Geography Dictionary. Then ask students to explain the meanings of the terms in their own words. Ask them to determine whether each term helps describe the physical geography, human geography, or economic geography of the Northeastern United States.

SECTION 1
OBJECTIVES

The objectives reflect the Focus questions on this page and are assessed in the review on page 123.

1. Explain how New England's landforms and climates have influenced its economy.

2. Analyze issues affecting New England's future economic growth.

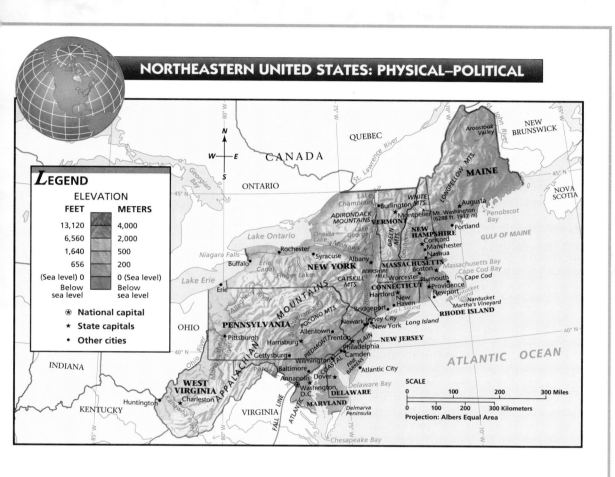

NORTHEASTERN UNITED STATES: PHYSICAL–POLITICAL

Using the Physical–Political Map: Northeastern United States

Ask volunteers to name the states that are part of New England and the Middle Atlantic states. Indicate the area covered by Megalopolis, and name its major cities. Point out how the boundaries of Megalopolis overlap the other two regions.

NEW ENGLAND

FOCUS

- *How have New England's landforms and climates affected the region's economy?*

- *Why is New England's economy changing, and in what ways is it changing?*

The New England states include Maine, New Hampshire, Vermont, Massachusetts, Rhode Island, and Connecticut. The population of this region is about 5 percent of the nation's total.

New England has played a major role in U.S. history. The Pilgrims arrived at Plymouth, on Massachusetts Bay, in 1620. They soon were followed by other colonists, who traded furs, farmed, fished, and logged the forests. In the eighteenth century, New England became the center of the American Revolution.

During the nineteenth century, New England became the first industrialized region in North America. Its oldest industries, including **textiles** (cloth products), timber, and shoes, are being replaced by industries centered around high technology and finance. These new industries are concentrated mainly in southern New England.

REGION **Landforms** The New England states are famous for their scenic beauty and variety of landforms. (See the map above.) The Appalachian Mountains cross much of northern New England. The major ranges are the White Mountains in New Hampshire, the Green Mountains in Vermont, and the Longfellow Mountains in Maine. The highest peak, Mount Washington in the White Mountains, reaches an elevation of 6,288 feet (1,917 m). Because of glacial erosion during the last ice age, northern New England has thousands of lakes and thin, rocky soils.

Southern New England primarily is a glaciated, hilly region. As the great ice sheets pushed

Motivator

Discuss the effect of New England's climate on its ways of life. Ask the following questions to guide the discussion: How long is the growing season? What crops are best suited to this environment? What adaptations must people make to live in this climate? (*food, shelter, clothing*) What changes to the land have people made to live in this environment? (*irrigation, dams, reforestation*)

Building a Vocabulary

Help students understand the meaning of **biotechnology.** Tell them that *bio-* means "life," *techno-* means "craft" or "art," and *-logy* means "study of." *Biotechnology,* therefore, refers to the study of ways to craft, or change, life forms. Point out that *biotechnology* as applied to industry is the basis for future economic growth in New England.

Role Playing

Regional Characteristics Write the following terms on slips of paper: pilgrim, farmer, fisher, miner, forester, textile manufacturer, college professor, biotechnologist. Have each student select a slip. Allow a few minutes for students to prepare roles. Call on volunteers to describe New England from the point of view of the person on their slip.

their way to the Atlantic coast, they deposited rock material, forming hundreds of hills. The Berkshires of western Massachusetts form the highest hill region in southern New England. Bunker Hill, the site of a famous Revolutionary War battle, probably is the best-known glacial hill. The only plains in New England are along sections of the Atlantic coast and in the lower Connecticut River valley.

New England's coastline varies greatly from north to south. In the north, the scenic coast of Maine is made of eroded **granite**. Granite is a speckled, hard, crystalline rock formed deep in the earth's crust. It becomes exposed at the earth's surface after millions of years of uplift and erosion. Some of the world's greatest tides occur along this coast, which is rugged and rocky with many narrow inlets, bays, **peninsulas**, and offshore islands. A peninsula is a landform that is surrounded by water on three sides.

Glacial deposits are responsible for the shoreline features of the southern New England coast. Cape Cod and the islands of Nantucket and Martha's Vineyard are glacial **moraine** materials piled up by the huge ice sheets. A moraine is a ridge of rocks, gravel, and sand deposited along the margins of a glacier or ice sheet.

Climate New England has a humid-continental climate. Each year, autumn's brightly colored, changing leaves attract tourists, while cold, snowy winters are inviting to winter sports enthusiasts. Fog is common along the coast, especially

The Old South Meeting House, built in 1729, is a link between colonial and modern **Boston, Massachusetts**. About 24,000 people lived in Boston around 1800. Today, the metropolitan area is home to nearly 3 million New Englanders.

in summer, when the warm Gulf Stream waters meet the cool waters of the Labrador Current, which flow south from the Arctic. Occasionally during late summer, hurricanes strike coastal southern New England. More common are winter storms from the North Atlantic, called northeasters. These middle-latitude storms bring cold, snowy weather with strong, gusting winds and high ocean waves.

Agriculture Because of the region's rocky terrain and short growing season, agriculture in New England is quite limited. The Connecticut River valley is the leading agricultural area. The swampy lowlands of Cape Cod grow more than half of all the cranberries produced in the United States. The Aroostook (uh-ROOS-tuhk) Valley in northern Maine produces about 10 percent of the country's potatoes.

New England's cool, moist climate and hilly pastureland make dairy farming the region's most important agricultural activity. Poultry farms also can be found. Farming in New England has always been a small-scale operation and has been in decline for more than a century. Many farms have been abandoned, while others have been bought by urban dwellers to use as second homes or retirement retreats.

Resources The shallow waters off New England's coast are among the world's most productive fishing grounds. Where the cool Labrador Current flows over the shallow continental shelf, fish are plentiful. Cod and lobster are the region's

Autumn explodes in color around this **Vermont** town. Look at the biome map on page 57. Which biomes are dominant in the Northeast United States?

Guided Practice

On the board, list features of New England's land and climate. *(thin, rocky soils; hilly; short growing season; swift streams; Atlantic harbors)* Ask students to tell how each feature led to the growth of early industry in the region. Discuss how and why New England's industries have changed over the years.

Closure

Ask students to summarize the ups and downs of industrial growth in New England.

ASSESSMENT

Assign the Section 1 Review questions on this page.

Reteaching

Assign Section 1 of Reteaching Worksheet 11.

Extension

Have students make a travel brochure describing an "autumn leaves" tour of New England. Have them create an itinerary from the map on page 121, research the towns on the route, and draw illustrations of typical scenes of the countryside.

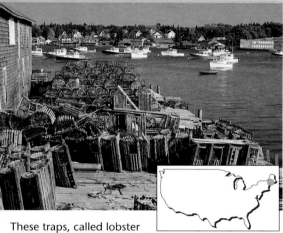

These traps, called lobster pots, are tools of the trade for lobster catchers of **Bernard, Maine**. Lobstering is an important part of the economy of coastal New England.

most valuable seafoods. The most productive waters have been around Georges Bank, an underwater plateau off the Massachusetts coast. After years of overfishing and declining fish stocks, however, the U.S. government ordered restrictions on fishing in some areas around Georges Bank in 1994.

Softwood forests, especially tall white pines, were an important natural resource for New England's early shipbuilding industry. Nearly all forests in New England are now **second-growth forests**. These are the trees that cover an area after the original forest has been removed. New England supports a pulpwood industry in Maine.

New England's few minerals cannot support a mining industry. The region's only major mining resource is building stone, especially Vermont granite and marble.

INTERACTION **Industry** New England became the nation's first industrial area. Swift streams provided power for the mills, and the relatively large local population provided a ready labor source. Boston served as a gateway city, where raw materials were imported and finished products could be shipped out by sea or by land. Many textile-mill towns and shoe factories sprang up along New England's rivers.

By the twentieth century, these industries had declined. Cheaper labor and more efficient factories could be found in the southern states and overseas. To remain competitive, New England has developed new industries and modernized its remaining traditional industries. For example, the

region now makes textile and footwear products geared for outdoor and sporting activities.

New England long has been a national leader in higher education. More than 250 colleges and universities are located here, including Harvard, Yale, and Massachusetts Institute of Technology (MIT). This leadership in education made it possible for the region to become a leader in high-technology industries, such as electronics and computers. With the end of the Cold War, the region's economy has worked to recover from cuts in defense spending, which once provided thousands of jobs.

With some of the nation's leading hospitals and medical schools, New England's future economic growth may also be in **biotechnology**. Biotechnology is the application of biology to industrial processes, such as pharmaceutical and genetic engineering research and the manufacture of products from that research. In addition, New England is the center for numerous service industries. Many banks, investment houses, and insurance companies have their headquarters here.

Tourism also is important to New England's economy. Mountains, beaches, and historic cities and towns attract tourists from around the world. Popular ski resorts are scattered throughout the region's scenic, forested mountains.

Issues New England's future growth must be balanced against the need to protect the region's environment. Increased industrialization and urbanization, especially in southern New England, will place increasing pressure on the surrounding rural areas. Tourism and dense coastal development may also strain the region's resources. Controlling acid rain and other forms of pollution remains a major challenge.

SECTION REVIEW

1. Why was New England better suited for industrial growth than for agriculture?
2. What industries will dominate New England's economy in the near future? Why?
3. **Critical Thinking** **Suggest** possible origins of the name "New England." Do you think the name is appropriate for the region today?

GLOBAL PERSPECTIVES

Hold a class discussion comparing free trade with a quota system. The U.S. textile industry has, for example, supported quotas on imported goods as a means of keeping U.S. industry competitive with overseas manufacturers. Some studies predict thousands of job losses in the textile industry alone if quotas are phased out.

Section 1 Review
ANSWERS

1. Terrain and climate limited agriculture; forests encouraged shipbuilding; streams were a power source for factories.
2. Industries based on electronics, computers, biotechnology, finance, and tourism; traditional industries have declined as cheaper labor and more efficient factories have become available elsewhere.
3. Many of the region's first European settlers were from England. The name may no longer be appropriate since many residents have other cultural roots.

SECTION 2
OBJECTIVES

The objectives reflect the Focus questions on this page and are assessed in the review on page 126.

1. Identify landform regions and climate types of the Middle Atlantic states.
2. Discuss the importance of resources and location to the Middle Atlantic states.

TEACHING SECTION 2
Motivator

Ask students to imagine that a document has just been found indicating that the Middle Atlantic States region is still the property of Great Britain. Discuss with students the immediate and future problems that this situation would create in the United States.

Building a Vocabulary

Write the terms **bituminous coal, coke,** and **anthracite coal** on the board, and explain their differences to students. Bituminous (soft) coal has more moisture, has a lower carbon content, and is less compressed than anthracite (hard) coal. Bituminous coal is used as an industrial fuel and in making coke. Note that anthracite coal, which is nearly pure carbon, is used as a domestic fuel.

THEME
Location

Have students locate the Middle Atlantic states on a wall map of the United States or the world. Help them see the strategic location of the region relative to other areas. Emphasize the advantages of this location for exporting and importing raw materials and manufactured products.

Chart Caption Answer

New York and Pennsylvania

THE MIDDLE ATLANTIC STATES

FOCUS

● *What are the major landform regions and climate types of the Middle Atlantic states?*

● *How were resources and location once important to the Middle Atlantic states, and why are they less important today?*

REGION The Middle Atlantic states include New York, Pennsylvania, New Jersey, Delaware, Maryland, and West Virginia. The nation's capital, Washington, District of Columbia, (or Washington, D.C.) also is included in this region. About one-fifth of the U.S. population lives here, making it one of the world's most industrialized and urbanized areas.

The Middle Atlantic states have played an important role in the economic development of the United States. Vast coal deposits, together with important land and water routes, helped the region emerge as an industrial giant during the nineteenth and early twentieth centuries. This industrialization attracted many immigrants to the region. Today, the Middle Atlantic region is where many of the nation's economic and political decisions are made.

Landforms Three landform regions cross the Middle Atlantic states: the Coastal Plain, the Piedmont, and the Appalachian Mountains. (See the map on page 121.) The Coastal Plain stretches across all of the Middle Atlantic states except West Virginia. This flat plain does not rise much above sea level. Long Island and Chesapeake Bay are both part of the Coastal Plain.

Chesapeake Bay, fed by the Susquehanna River, is the largest and most productive estuary on the Atlantic coast. As the sea level rose after the ice-age glaciers melted, the sea drowned the lower river valleys and formed the bay. Chesapeake Bay is a major supplier of oysters and crabs and has several important seaports.

Inland from the Coastal Plain is the Piedmont region, which gently slopes down from the Appalachians to meet the Coastal Plain. Because of the available waterpower at the Fall Line, many cities developed here.

The northern part of the Appalachian Mountains crosses the Middle Atlantic States region in

FACTS IN BRIEF

States of the Northeastern United States

STATE POPULATION (1993)	POP. RANK (1993)	AREA (1990)	AREA RANK (1990)	PER CAPITA INCOME (1993)
CONNECTICUT 3,277,000	27	5,544 sq. mi. 14,359 sq. km	48	$28,110
DELAWARE 700,000	46	2,397 sq. mi. 6,208 sq. km	49	$21,481
MAINE 1,239,000	39	33,741 sq. mi. 87,389 sq. km	39	$18,895
MARYLAND 4,965,000	19	12,297 sq. mi. 31,489 sq. km	42	$24,044
MASSACHUSETTS 6,012,000	13	9,241 sq. mi. 23,934 sq. km	45	$24,563
NEW HAMPSHIRE 1,125,000	41	9,283 sq. mi. 24,043 sq. km	44	$22,659
NEW JERSEY 7,879,000	9	8,215 sq. mi. 21,277 sq. km	46	$26,967
NEW YORK 18,197,000	2	53,989 sq. mi. 139,832 sq. km	27	$24,623
PENNSYLVANIA 12,048,000	5	45,759 sq. mi. 118,516 sq. km	33	$21,351
RHODE ISLAND 1,000,000	43	1,231 sq. mi. 3,188 sq. km	50	$21,096
VERMONT 576,000	49	9,615 sq. mi. 24,903 sq. km	43	$19,467
WEST VIRGINIA 1,820,000	35	24,232 sq. mi. 62,761 sq. km	41	$16,209

Rankings are for all 50 states.
Source: *Statistical Abstract of the United States 1994*

Which are the two most populous states of the Northeastern United States?

a northeast-southwest direction. Several major rivers cut through some of the Appalachian ranges here on their way to the Atlantic. These include the Potomac, Susquehanna, Delaware, and Hudson rivers.

Climate The Middle Atlantic states have two major climate types: a humid-continental climate in the north and a humid-subtropical climate in the south. The dividing line between the two climates stretches west from the Atlantic coast south of New York City.

Summers in both climate regions can be very hot and humid. Winds passing over the warm Gulf Stream and the Gulf of Mexico bring hot, humid air inland. During winter, cold periods occur when arctic air from the north enters the region. Occasional snowstorms can cause serious transportation problems for the major cities. Farther

Contrasting

Resources and Location Ask students to list factors that enabled the Middle Atlantic States region to become an industrial giant. *(vast coal deposits, access to land and water transportation, a large labor force)* Explain why the region is now facing industrial decline. *(The U.S. economy is not based on manufacturing as much as it once was because of a shift to service and information industries.)*

Guided Practice

Draw a flowchart on the board to convey why some areas in the Middle Atlantic states face a tough economic future. Begin with the closing of a steel processing plant. Show how unemployment can lead to workers' moving elsewhere to seek work. Then call on students to suggest how this loss of population affects the local economy. Add their answers to the flowchart.

Independent Practice

Provide students with copies of Outline Map 9. Have them complete the map to show the states, capital cities, and significant landforms of the Middle Atlantic states. In addition, have them label the two major climate regions and the dividing line between the two.

inland and northward in the Middle Atlantic states, winters are colder and snowfall can be heavy, especially in the Appalachians and in upstate New York. Like the New England coast, the Middle Atlantic coast is subject to hurricanes and northeasters, which can cause severe damage to coastal communities.

Agriculture Agricultural production in the Middle Atlantic states supplies the region's huge population. Fresh fruits and vegetables are produced on farms on Long Island and in New Jersey, Delaware, and Maryland. These farms are sometimes called **truck farms** because their products can be trucked easily to city markets. Some areas of the region specialize in certain crops. Potatoes are grown in New Jersey and on Long Island. Vineyards are found in the Finger Lakes area of New York and on Long Island. Other farms in the region produce milk, poultry, and grains.

During recent years, there has been a decrease in the amount of farmland in the Middle Atlantic states. Part of this loss has resulted from urban expansion into rural areas. Valuable farmland is being bought for housing and industry, while many farms in more remote rural areas

Snow blankets **Westchester County** north of New York City. Winters generally are colder farther inland from coastal areas.

have been abandoned or combined into larger agricultural operations.

Resources In this highly industrialized region, coal has been the key natural resource. One of the largest **bituminous coal**, or soft coal, areas in the world is found in West Virginia, the western counties of Pennsylvania, and bordering areas in Ohio and Kentucky. Bituminous coal is used to produce electricity and chemicals. It also is used in the production of **coke**, or coking coal, which is used in blast furnaces to purify iron ore for making steel. Bituminous coal mining became the basis for industrialization in the Middle Atlantic states.

Deposits of **anthracite coal**, or hard coal, are found in eastern Pennsylvania. Anthracite is a higher grade coal than bituminous, but because it is more expensive to mine, anthracite mining has declined.

The Appalachian coal-producing region has suffered economic hardships over the past several decades. This area has some of the highest unemployment and poverty rates in the nation. Thousands of people have left the region, seeking jobs elsewhere. The coal industry has left behind abandoned mine debris and polluted water.

The Middle Atlantic states have several forests and wilderness areas. The Adirondack and Catskill mountains in New York, the Pine Barrens in New Jersey, and many mountain ridges in Pennsylvania and West Virginia remain forested and thinly populated. These forests are not major wood producers but instead serve as wildlife reserves and as recreation areas for the huge urban populations nearby.

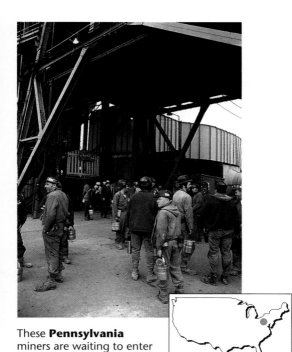

These **Pennsylvania** miners are waiting to enter one of the many coal mines in the Middle Atlantic states.

Geography and the Environment

Both coal mining and the burning of coal for energy have damaging effects on the environment. In recent decades, laws have been passed to reverse some of these effects. For example, owners of strip mines are now required to restore damaged land to its original condition. Special scrubbers have been added to smokestacks to reduce the amount of pollution entering the air.

Geographical Sidelight

The Appalachian Trail is a scenic hiking path that follows the crest of the Appalachian Mountains from Maine to Georgia. The trail, completed in 1937, passes through 14 states.

125

Review with students why and how the Middle Atlantic states have been important to the country as a whole.

ASSESSMENT

Assign the Section 2 Review questions on this page.

Assign Section 2 of Reteaching Worksheet 11.

Extension

Have students draw a line graph showing U.S. coal consumption over the past 50 years, using information in almanacs. Ask students to share their completed graphs and to draw conclusions about the use of coal in the United States.

SECTION 3
OBJECTIVES

The objectives reflect the Focus questions on page 127 and are assessed in the review on page 130.

1. Describe characteristics of each major city of Megalopolis.
2. Examine challenges facing Megalopolis.

MULTICULTURAL PERSPECTIVES

Native Americans often gave names to places that described their features. *Monongahela* is a Native American word meaning "high banks breaking off and falling down in places." Discuss the origins of other place-names in this section.

Section 2 Review
ANSWERS

1. Landform regions: Coastal Plain—flat plain; Piedmont—gentle slope from mountains to plain; Appalachians—mountains, rivers. Climate types: humid continental and humid subtropical—hot, humid summers, cold winters, storms

2. Access to national and world markets enabled export of finished products made using nearby resources. Today, traditional manufacturing industries are less important.

3. Answers will vary. Increased tourism might help the economy but endanger the environment.

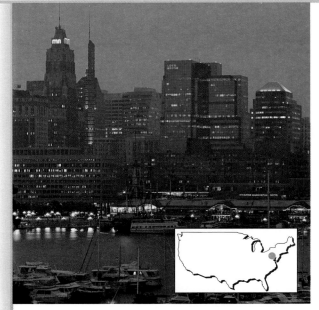

Office towers, hotels, shops, and apartments have replaced the decay that once haunted the Inner Harbor of **Baltimore, Maryland**. Many old industrial cities of the Northeastern United States are faced with the task of urban renewal.

INTERACTION **Industry and Urban Growth**

The Middle Atlantic states have benefited from their transportation access to national and world markets. This access along with a large labor force helped maintain the region's early industrial growth. As canals, roads, and railroads were built, the Middle Atlantic states found it easy to obtain raw materials from almost every part of the country and the world. Today, nearly every kind of manufacturing and service industry can be found in the region.

Industrial activity spread to several areas, such as upstate New York and western Pennsylvania, outside the region's major cities. New York's industries are located chiefly along the Hudson and Mohawk river valleys. This is the route of the Erie Canal, which was the first route to connect trade between the Atlantic Ocean and the Great Lakes. The canal was completed in 1825 and soon attracted industrial development in several cities. Albany, New York's state capital, is located near the junction of the Hudson and Mohawk rivers. Buffalo, the state's second largest city, is an industrial port on Lake Erie.

Pittsburgh, Pennsylvania, the largest industrial city in the Appalachians, also is the nation's

largest inland harbor. It began as the French-built Fort Duquesne in 1754, where the Allegheny and Monongahela (muh-nahn-guh-HEE-luh) rivers join to form the Ohio River. In 1758, the fort was captured by the British and renamed Fort Pitt. Situated in the coal-producing region of western Pennsylvania, Pittsburgh became a major steel producer. As a result, it also became one of the nation's most polluted cities. The Pittsburgh urban area suffered economic problems and population losses with the decline of the steel industry. In recent years, new high-tech industries have helped renew the city, and it now has cleaner air.

The tourism industry is vital to many cities and towns throughout the Middle Atlantic States region. The major beach destinations include eastern Long Island, the Delmarva Peninsula, and the New Jersey shore, especially Atlantic City. Tourists also flock to the scenery of the Blue Ridge, Catskill, Adirondack, and Pocono mountains, and Niagara Falls. Countless historic and battlefield sites, such as Gettysburg, also are popular.

Issues The Middle Atlantic states played a key role in the industrialization of America. The region, however, now faces industrial decline as the nation moves from manufacturing to service and information industries. Since the 1970s, the Middle Atlantic states have experienced unemployment and migration to the Sunbelt. The industrial decline has affected most severely the small communities that depended on traditional industries such as coal mining, textiles, and steel. In contrast, the region's large cities have a greater economic diversity and major ports, which provide employment opportunities and a foundation for future growth.

SECTION REVIEW

1. What are the characteristics of the Middle Atlantic states' various landform regions? their climate types?
2. How did the location of the Middle Atlantic states help the region make the most of its resources? Why does the region now face economic decline?
3. **Critical Thinking** **Analyze** advantages and disadvantages of a plan to make the region's wilderness areas more accessible to tourists.

TEACHING SECTION 3

Motivator

Ask students to consider the relationship between the school office and the rest of the building. How and why is the office a nodal region? (See page 101 for a definition of this term.) Explain that in this section students will study Megalopolis, a nodal region of the United States.

Building a Vocabulary

To help students understand the meaning of **Megalopolis,** tell them that *mega-* or *megalo-* means "huge" or "great" and that *polis* means "city." Thus, the literal meaning of *Megalopolis* is "great city." Ask students to think of other words or place-names with these word parts. (*megabucks, megabyte, Minneapolis, Indianapolis*)

Explaining

Characteristics of Megalopolis

Remind students of the discussion about the school office in the Motivator activity. Ask students to explain why Megalopolis is called a nodal region. Have them locate and read sentences in Section 3 that describe movement in Megalopolis.

MEGALOPOLIS

FOCUS

- *What are the major cities of Megalopolis? What characteristics make each city important?*

- *What challenges do the cities of Megalopolis face?*

REGION The ongoing commercial and industrial growth of the Atlantic seaboard has led to the development of a continuous urban corridor. This corridor, called Megalopolis, was named by a French geographer in the 1960s. Gradually, the five major nodal cities of the Northeastern United States—Boston, New York, Philadelphia, Baltimore, and Washington, D.C.—spread toward each other. (See the map on page 121.) Excluding Washington, D.C., the original growth was due to the cities' good port sites, which linked the nation's interior to world trade routes.

The boundaries of Megalopolis continue to change and expand. Expansion has been most rapid as new highways allow commuters easy access to the cities. Subways, high-speed rails, and airline shuttles offer commuters an alternative to the automobile. In fact, this nodal region is connected by some of the densest railroad and interstate highway networks in the nation. The region has some of the world's busiest seaports and most crowded airports, as well.

Between these major cities are expanding suburbs and edge cities. Some farms and even forested land still exist within Megalopolis, however. Preserving the rural character within the expanding urban sprawl is a major challenge for the region.

Boston Megalopolis is anchored in the north by Boston, the capital of Massachusetts and New England's largest industrial, educational, financial, and commercial center. The city was founded as a British harbor in 1630, and its main commercial activities were connected with the sea. Boston Harbor has been modernized, and the waterfront and marketplace areas have been renovated. Much of Boston's recent growth has centered around highways called "electronics parkways." This term refers to the high-technology industries established there.

MOVEMENT **New York** South of Boston is New York City. Situated at the mouth of the Hudson River, the city has an excellent natural harbor with several bays and islands. It was originally settled by the Dutch, who named it New Amsterdam. The British seized the city in 1664 and renamed it New York. Today, it is the central node of Megalopolis. New York City is one of the world's most important cities and the site of the United Nations headquarters.

New York City is a major seaport and rail center and has three major airports. As the nation's commercial and financial center, the city is the site of stock exchanges, corporate headquarters, major banks, law firms, publishers, and advertising agencies. It is also the heart of the nation's garment industry and is a world fashion center. New York City serves as the nation's center for theater, art, and entertainment as well, and is the headquarters for the country's major television networks.

The city is divided into five **boroughs,** or administrative units. These are Manhattan, Staten Island, Brooklyn, the Bronx, and Queens. The commercial heart of New York City is Manhattan Island. Because New York City always has been a landing point for immigrants to the United States, it is a city of great ethnic diversity.

The importance of movement in a place's economic growth is clearly shown by New York City. The city surpassed its rivals, Philadelphia and Boston, in population and commerce when the Erie Canal connected it to the Great Lakes in 1825. As a result of the canal, New York City experienced tremendous growth and benefited from its role as a **break-of-bulk center.** A break-of-bulk center is a place where shipments of goods are moved from one mode of transportation to another, such as from train to ship.

The city's urban area has spread east across Long Island, north into Connecticut, and west across the Hudson River into New Jersey. The three-state region is connected by interstate highways, toll roads, a large subway and rail system, bridges, several underwater tunnels, and a ferry system.

Philadelphia Philadelphia is the fifth largest urban area in the United States and is an important port and industrial center. Located on the lower Delaware River, it has access to the Atlantic through Delaware Bay. Philadelphia was planned

Text continues on page 130.

127

Planet Watch

Investigating New Products

Have students suggest why electric cars are not widely available today. *(Models are costly to produce; people are not yet convinced that electric cars are practical.)* Point out that new technologies often are expensive or impractical. For example, the first fully electronic computer, developed in 1946, required 1,500 square feet (140 sq. m) of floor space and weighed more than 30 tons (27 metric tons). Have students work in small groups to research the development of consumer items such as microwave ovens, televisions, and calculators. Instruct students to include the cost and practicality of the products at the time they were invented. Encourage students to illustrate their reports and present them to the class.

Racing with the Sun

Each Memorial Day weekend, hundreds of thousands of people gather to watch the Indianapolis 500, a world-famous, high-speed automobile race. Noise shatters the quiet of a lazy holiday, fuel flows, and exhaust fumes pollute the air.

As cars circle the track in Indianapolis, Indiana, each year, another, quieter race is taking place. The drivers are younger; some are high-school students. One year, drivers raced from Albany, New York, to the finish line in Plymouth, Massachusetts. Although the cars reached speeds of almost 70 miles per hour (113 kph), they spent five days traveling the 250 miles (402 km) between the two cities, a snail's pace compared to the "Indy 500."

Solar-Charged Cars

The name *Tour de Sol* offers a clue to what the annual race is about. *Sol* means "sun" in Latin, and the young drivers rely on energy from the sun. Lift the hood of a Tour de Sol car, and you will not find a methanol or gasoline engine. Instead, you might see an electric motor and batteries. The batteries are connected to solar cells that convert sunlight into electric current. One full day of sunlight can store up to 50 miles (80 km) of power. When a Tour de Sol car needs a recharge, the driver makes a pit stop and utilizes the sun, rather than gasoline, to provide energy.

The goal of the creators of the Tour de Sol is to encourage the development and everyday use of solar-charged electric cars.

A winning car in one Tour de Sol, the *Electric Hilltopper* would never have been mistaken for one of the sleek cars in the Indianapolis 500. Owned by a high-school teacher, the car once was destined for the junkyard. But some enterprising students saw a new use for the tired automobile.

The students worked for nine months. They removed the engine and other parts to make the car as light as possible. Then they added an electric motor and 16 batteries. The *Electric Hilltopper* set a Tour de Sol record, covering more than 100 miles (161 km) before needing its batteries recharged. "The car ran like a dream," one proud student beamed. "We never had to take a tool out of the toolbox."

Although fun and competition are all part of the Tour de Sol and similar races, such as Sunrayce '95 from Indianapolis to Colorado, the race has a serious purpose. The creators of these races believe that an alternative to the gasoline engine must be found. Their mission is to encourage the development and everyday use of solar-charged electric cars.

Gasoline engines are responsible for much of the air pollution that threatens our environment. Exhaust fumes help to produce the thick clouds of smog that cause watery eyes, clogged throats, and perhaps even lung cancer. Gases emitted from automobiles also contribute to acid rain and global warming.

Solar car exhibition in **Putnam, Connecticut**

Using Geography Themes

Call on a volunteer to write the five geography themes on the board. Have students select the one that best applies to the development of solar-powered technologies *(human-environment interaction)*, and have them write a paragraph justifying their choice.

Designing a Solar Environment

Have students imagine that they are solar engineers. Ask them to draw a model of their homes, classroom, or school. Instruct them to redesign the structures to allow replacement of electricity or gas with solar power.

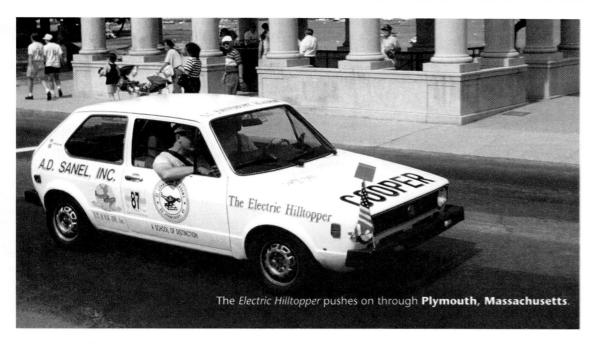

The *Electric Hilltopper* pushes on through **Plymouth, Massachusetts**.

The Solar White House

President Jimmy Carter enthusiastically supported solar power. In 1979, he installed solar panels on the roof of the White House to draw energy for heating water. They were removed several years later, however, during the remodeling of the presidential mansion.

A Renewable Resource

Unlike petroleum products, the sun provides a clean and inexpensive energy source. Better yet, solar energy is renewable—there is an endless supply. In contrast, petroleum is a fossil fuel, formed millions of years ago from microscopic marine plants and animals. Once pumped from the ground, petroleum takes millions of years to replace.

People have always made use of solar energy to meet their needs. In the southwestern United States, for example, many Native Americans built their homes of adobe brick, or sun-dried clay and mud. The adobe kept the homes cool in summer and warm in winter.

Today, people living in sunny climates can place solar panels on roofs in order to heat water for homes and office buildings. Automobile manufacturers, meanwhile, are experimenting with solar and electric cars.

One Tour de Sol winner hopes to find marketable success with his solar-powered car. He claims that his efficient design will allow people to commute to work without using a drop of gasoline. The car would reach a speed of 60 miles per hour (97 kph) and travel 200 miles (322 km) before needing a recharge. Stressing other advantages, one solar-car enthusiast says, "If you're sitting in a traffic jam in the sun, this car isn't using power, it's gaining power. There's no pollution, no noise. That's the most amazing part."

Solar energy can power more than cars. Solar panels on the roof collect sunlight to heat water for this **Vermont** home. Satellites and other spacecraft also use solar panels for power.

▼ YOUR TURN

Write an eyewitness account of a Tour de Sol race. In your account, **describe** one of the cars, and explain the role of the race in protecting the environment. You may also want to **invent** a solar-powered machine. Write a paragraph describing your invention.

Your Turn

Eyewitness accounts will vary. They should include descriptive details about the event, the spectators, and the cars.

Students' paragraphs on solar-powered inventions might include details about size, speed, cost, and use.

Provide students with a supply of newspapers. Have them look for articles with datelines from the major cities of Megalopolis. Ask them to report on articles they find. Use the articles to emphasize the region's influence on other parts of the country, especially in regard to government, finance, and culture. Also note discussions of challenges facing Megalopolis.

Have students construct a graphic organizer for the five major cities of Megalopolis. The organizer should include facts about each city's location, history, and industries. Call on students to answer questions about the cities, using their organizers as references.

Ask students to list common features that enable the five metropolitan areas to be grouped into a region.

ASSESSMENT

Assign the Section 3 Review questions on this page.

Reteaching

Assign Section 3 of Reteaching Worksheet 11.

Section 3 Review
ANSWERS

1. Boston: industry, education, finance, commerce, "electronics parkways"; New York: seaport, rail center, break-of-bulk center, commerce, finance, industry, fashion, art, entertainment, five boroughs, diversity; Philadelphia: port, industry, history, education; Baltimore: port, industry, Johns Hopkins, redevelopment; Washington, D.C.: U.S. capital, politics, tourism

2. Possible answers include substandard housing, crowded schools, a lack of services, traffic jams, air pollution, and disposal of sewage and toxic waste materials.

3. Possible answer: railroads are a major means of transportation in Megalopolis, so a strike of railroad workers would place greater strain on highways and might even temporarily curtail some industries in the region.

by William Penn in the late 1600s and was the nation's most important city until the early 1800s. It is where the Declaration of Independence was signed in 1776.

Today, Philadelphia is a major manufacturing, pharmaceutical, banking, and educational center. Though many older sections of the city have suffered urban decay, downtown rebuilding programs have helped reverse this trend. The Philadelphia urban area has spread west farther into Pennsylvania, east beyond Camden, New Jersey, and south along the Delaware River.

Baltimore Baltimore, Maryland, is a seaport on the western shore of Chesapeake Bay. Founded along the Fall Line in 1729, it grew as a major port because of its rail connections to interior coal mines, steel mills, and farms. Its main industries are large oil refineries and steel mills. The city is also home to Johns Hopkins Hospital and University. A redevelopment program has helped rebuild the inner harbor area as well as some of the poorer neighborhoods.

Washington, D.C. Anchoring Megalopolis in the south is Washington, D.C. The nation's capital occupies a unique place among American cities. It is a federal district, located outside the authority of any state. There has long been an active movement to make the district the 51st state.

The District of Columbia was founded for one reason—to be the seat of the nation's government. The original city was designed in the late 1700s by French engineer and architect Pierre L'Enfant (lahn-FAHNT). Later work on the city was done by Benjamin Banneker, an African American mathematician, astronomer, and surveyor.

Washington, D.C., is a city mainly of politicians, lawyers, journalists, diplomats, military personnel, and tourists. In fact, tourism has become the city's second major source of income after government employment. Each year, millions of tourists are attracted to the city's many sites, including the U.S. Capitol, the White House, the Smithsonian Institution, the Vietnam Veterans Memorial, and hundreds of other historical and political landmarks.

Thousands of people who work in Washington, D.C., commute daily from their homes in suburbs and edge cities outside the boundaries of the district. The expanding Washington, D.C., urban area

has grown so large that it has spilled over into Maryland and the southern state of Virginia. The southern boundary of Megalopolis now is approaching Richmond, Virginia.

Issues Changes in U.S. and world economies over the past 30 years have had a dramatic impact on Megalopolis. As traditional manufacturing has declined, factory jobs have disappeared. In turn, neighborhoods in large cities, which once depended on factory jobs, have declined. Substandard housing, crowded schools, and a lack of services have been problems for all the major urban areas of the region.

Meanwhile, new businesses in service industries and high technology have opened in cities on the edge of large urban centers. These edge cities around Boston, New York, Philadelphia, Baltimore, and Washington, D.C., have experienced explosive growth. As the population has spread out, new problems such as long commutes and traffic jams have become everyday challenges. Farmland and open space are being used for housing developments, business centers, and recreation areas.

Air pollution and disposal of sewage and toxic waste materials are vital concerns for Megalopolis. Indeed, pollution threatens rich ecosystems in the region, such as Chesapeake Bay. Environmental awareness by corporations and by the general public, though, is beginning to improve the situation. For example, the Hudson River, once polluted from the Adirondacks to New York Harbor, is recovering. The river now supports a healthy fish population, and people once again are making their homes nearby. Protecting its precious water resources, especially coastal bays and rivers, is a major challenge for the entire Northeastern United States.

---

3 SECTION REVIEW

1. List the major cities of Megalopolis. Next to each city, write down two of that city's important features.
2. What are two challenges faced by Megalopolis?
3. **Critical Thinking** **Predict** what effects a major strike of railroad workers might have on Megalopolis.

CHAPTER 11 CHECK
ANSWERS

Building a Vocabulary

1. Granite is a type of hard rock. A moraine is a ridge of rocks and other materials deposited along a glacier's edges. Uplift and erosion have made granite common, and glaciers made moraines.
2. The original forests were cut for industry.
3. The application of biology to industrial processes
4. Because their products are trucked to markets
5. Boroughs

Recalling and Reviewing

1. New England is less agriculturally productive than the Middle Atlantic States region because of its rocky terrain and short growing season.
2. New England: autumn colors, winter sports, mountains, beaches, historic cities and towns; Middle Atlantic: beaches, mountains, Niagara Falls, historic sites; Megalopolis: entertainment, arts, landmarks
3. Air pollution from industry, mining, and urbanization; acid rain and other water pollution; sewage disposal, toxic waste

CHAPTER **11** CHECK

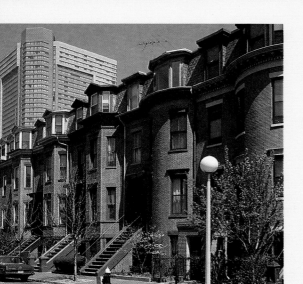

Row houses in **Boston, Massachusetts**

Reviewing the Main Ideas

1. In the nineteenth century, New England was well suited for industrial development because of its landforms, climates, and resources. The region's current industries are based increasingly on high technology.
2. Because of access to major land and water routes and resources such as coal, the Middle Atlantic states played a key role in the industrialization of the United States. Today, many of the nation's economic and political decisions are made in the region.
3. Megalopolis, a region overlapping the New England and Middle Atlantic states, is a continuous urban area along the eastern seaboard. The five major cities of the region are important industrial, commercial, financial, educational, and entertainment centers. They also face many economic and environmental challenges.

Building a Vocabulary

1. What is granite? What is a moraine? What natural forces have made granite and moraine materials common along the New England coastline?
2. Why are most of New England's forests second-growth forests?

3. What is biotechnology?
4. Why are some farms called truck farms?
5. What is the term for New York's five administrative units?

Recalling and Reviewing

1. How does New England's agriculture differ from the agriculture of the Middle Atlantic states? What accounts for the differences?
2. What makes each of the Northeastern United States subregions attractive to tourists? List three features of each subregion that attract tourists.
3. What environmental challenges are faced by New England, the Middle Atlantic states, and the cities of Megalopolis?

Thinking Critically

1. Imagine that you were a visitor to the Northeastern United States in the early nineteenth century. How would the region look in contrast to the way it looks today? How do you think it will look 100 years from today?
2. Evaluate the following statement as true or false: "Regions are defined solely on the basis of location; therefore, any individual place cannot be part of more than one region." Explain your evaluation, using examples from the chapter.

 Learning About Your Local Geography

Cooperative Project

With your group, research the extent to which a nearby city or town has expanded over the past 20 years. Look for the growth of housing developments, office buildings, shopping centers, and malls. Locate or draw two maps, one of the city or town 20 years ago, and one of the city or town today.

Using the Five Themes of Geography

Construct a flowchart linking the two themes of **location** and **region** in the Northeastern United States. How did the location of the region and its cities affect its growth? How, in turn, did the growth of the region influence how some geographers divide the Northeast into subregions?

Thinking Critically

1. In the early 1800s, industry would be developing, but there would be more rural areas than there are today. Answers will vary regarding how the region will look in 100 years.
2. Students should say that the statement is false. Regions are not necessarily defined solely on the basis of location, and a place can be part of more than one region, as are the cities of Megalopolis.

Learning About Your Local Geography

Maps should show how cities or towns have grown to take over surrounding rural areas.

Using the Five Themes of Geography

Flowcharts should indicate that the region's location along the Atlantic coast made it one of the country's first settled regions, and its access to both national and world markets made it well situated for commercial and industrial leadership. Early growth in the region was rapid, which led to the rise of Megalopolis, the overlapping subregion.

131

CHAPTER 12 The Southern United States

Planning Guide

TEXTBOOK	TEACHING RESOURCES	OTHER AVAILABLE MATERIALS
Section		
1. Physical Geography Pages 133–135	Reteaching Worksheet 12.1 Critical Thinking Worksheet 12 Graphic Organizer 12	*Study Guide for Spanish Speakers* *Lecture Notes,* Chapter 12
2. Economic Geography Pages 135–137	Reteaching Worksheet 12.2 Challenge/Enrichment Worksheet 12	*Study Guide for Spanish Speakers* *Lecture Notes,* Chapter 12
3. Urban Geography and Regional Issues Pages 137–138	Reteaching Worksheet 12.3 Skills Worksheet 12	*Study Guide for Spanish Speakers* *Lecture Notes,* Chapter 12
Chapter Review Page 139	Chapter 12 Test Chapter 12 Portfolio Assessment See also *Alternative Assessment Forms*	*Test Generator,* Chapter 12

CHAPTER RESOURCES

Lecture Notes, Chapter 12
Teaching Resources
    Chapter 12 Graphic Organizer
    Chapter 12 Skills Worksheet
    Chapter 12 Critical Thinking Worksheet
    Chapter 12 Challenge/Enrichment Worksheet

REVIEW AND ASSESSMENT

Chapter 12 Check, page 139
Chapter 12 Test, *Teaching Resources*
Chapter 12 Portfolio Assessment, *Teaching Resources*
Test Generator, Chapter 12
Alternative Assessment Forms

ADDITIONAL RESOURCES

Books for Teachers

Dunbar, Anthony P. *Delta Time.* New York: Pantheon Books, 1990.

Mohl, Raymond, ed. *Searching for the Sunbelt.* Knoxville: University of Tennessee Press, 1990.

Raitz, Karl B. *Appalachia, a Regional Geography.* Boulder, CO: Westview Press, 1984.

Books for Students

Heat-Moon, William Least. *Blue Highways: A Journey into America.* Boston: Little, Brown, 1983.

Jenkins, Peter. *A Walk Across America.* New York: Morrow, 1979.

Twain, Mark. *Life on the Mississippi.* New York: Bantam Books, 1985.

Multimedia Resources

American Visions: The Great Smoky Mountains. (VHS, 30 minutes) Lumivision.

The Mississippi System: Waterways of Commerce. (VHS, 16 minutes) Encyclopedia Britannica Educational Corporation, 310 South Michigan Avenue, Chicago, IL 60604–9839.

Picture Atlas of the World. (interactive maps, photo bank, animation, audio on CD-ROM) National Geographic Society, Washington, DC.

MEETING INDIVIDUAL NEEDS

Limited English Proficiency (LEP)

Have students use a physical-political wall map of the United States to locate the Southern United States and to identify the states, major cities, river systems, and prominent landforms in the region.

Learners Having Difficulty

Have students review the important facts presented in Chapter 12 by studying the maps, photographs, and illustrations in each section. Ask students to write one or two questions about each illustration. Organize students into groups, and have them answer each other's questions.

Auditory Learners

Point out to students that the southern region has been the birthplace of many American musical styles, including bluegrass, country, western swing, *conjunto,* gospel, jazz, blues, rock and roll, and zydeco. Representative performers in each category include, respectively, Bill Monroe, Hank Williams, Bob Wills, Flaco Jiménez, Mahalia Jackson, Louis Armstrong, Muddy Waters, Elvis Presley, and Clifton Chenier. Give students a musical tour of the South by playing selections from several of these styles.

Visual Learners

Ask students what they know about the role of the South in the Civil War and the reasons for the conflict. Arrange to show portions of Ken Burns's television series *The Civil War.* Then elicit a class discussion about how the human and physical geography of the North and the South influenced the development and the outcome of the conflict.

Tactile/Kinesthetic Learners

Tell the class that it has been asked by a family to plan a two-week car trip through the Southern United States. Have students prepare a trip itinerary and map showing the route the family should take, and have them use a road atlas to estimate the total mileage of the journey. Have students write a brief paragraph explaining their choices of things for the family to see and do on the trip.

Gifted Learners

Have interested students research the Tennessee Valley Authority to find out why it was formed and if it has fulfilled its original purpose. Ask students to create a map of the region covered by the TVA, showing its dams, reservoirs, and canals.

MULTICULTURAL FOCUS

Write the following list of terms on the board: chili con carne, collard greens, crawfish, fajitas, grits, gumbo, hominy, jambalaya, okra, pralines, shrimp creole, succotash.

Ask students to identify and explain the terms that they know. Students may realize that these are foods that originated in the Southern United States; many of them have become popular throughout the country. Point out to students that the variety of foods in southern cooking reflects the region's ethnic diversity.

Some traditional dishes enjoyed by rural, southern African Americans and rural southern whites, such as collard greens, possibly reflect a Native American influence. Crawfish, rice, and cayenne peppers are some of the basic ingredients in Cajun food. Cajuns are descendants of French Nova Scotians who migrated to Louisiana in the mid-1700s. Shrimp creole is Creole cuisine, which, according to Louisiana chef Paul Prudhomme, is the mixture of the food traditions of the French, Spanish, Italians, Native Americans, Africans, and other ethnic groups that lived in New Orleans at various times over the years. Tex-Mex dishes frequently use cornmeal and hot chilies; as their name implies, they are largely influenced by Mexican cooking.

Have students bring to school a favorite family recipe that expresses some part of their family's ethnic identity. Compile the recipes in a classroom cookbook to illustrate the ethnic diversity within the class.

CHAPTER 12

THE SOUTHERN UNITED STATES

Section 1: Physical Geography
Section 2: Economic Geography
Section 3: Urban Geography and Regional Issues

OVERVIEW

Chapter 12 introduces students to the 12 states that make up the Southern United States.

Section 1 describes the physical geography of the region and how it affects life in the Southern United States.

Section 2 outlines the economic geography of the Southern United States, highlighting changes in the region's agricultural activity and the growth of industry.

Section 3 discusses the region's urban geography, focusing on challenges brought on by a rising population and increased urbanization.

About the Author

Charlotte Forten (1837–1914) was an African American schoolteacher who traveled to the coastal region of South Carolina during the Civil War. She was among many young people from New England who came south to open schools for newly freed slaves. The quotation on this page is from her personal journal.

Primary Source

Description of Change: excerpted and bracketed
Rationale: excerpted to focus on main idea, bracketed to clarify meaning

CHAPTER 12

The Southern United States

66 **N**ever saw anything more beautiful than these trees [live oaks]. . . . They are the first objects that attract one's attention here. They are large, noble trees with small glossy green leaves. Their great beauty consists in the long bearded moss with which every branch is heavily draped. This moss is singularly beautiful, and gives a solemn almost funereal [gloomy] aspect to the trees. 99

Charlotte Forten

The Southern United States stretches in a great arc from Virginia to Texas. The states of this region are Virginia, North Carolina, South Carolina, Georgia, Kentucky, Tennessee, Alabama, Florida, Mississippi, Arkansas, Louisiana, and Texas. The largest state in the region, Texas, shares a long international border with Mexico.

Historically, the region was rural and agricultural. The majority of the population lived on small farms, but enslaved African Americans provided the labor for large cotton, rice, and tobacco plantations in the region. After the Civil War and the end of slavery, many African Americans as well as whites migrated to northern cities in search of factory jobs.

Since the 1960s, there has been a large movement of people back to the South from the North. In addition, thousands of people have moved to the region from the Caribbean, Mexico, and Middle and South America.

Fiddle from **Louisiana**

GEOGRAPHY DICTIONARY

barrier island

bayou

levee

lock

tenant farmer

diversify

bilingual

Louisiana live oak trees draped in Spanish moss

Using the Quotation

Ask students to close their eyes while you read the quotation on page 132 aloud. Have them imagine the large oak trees draped with moss. Ask why the moss is "bearded." *(The moss looks like a long beard hanging down.)* Then have students look at the photograph on page 132. What does moss hanging from trees suggest about the landforms and climate of the South? *(low-lying, damp land; warm, humid climate)*

Using the Geography Dictionary

Write the words in the Geography Dictionary on the board, and pronounce them for the class. Tell students that four of the words have something to do with water. Have students look up all the words in the Glossary to discover the four water words. **(barrier island, bayou, levee, lock)**

The objectives reflect the Focus questions on this page and are assessed in the review on page 135.

1. Identify and describe the landform regions of the Southern United States.
2. Describe the regions' climates.
3. Explain the importance of river systems to the region.

Using the Physical-Political Map: Southern United States

Direct students' attention to the map on this page. Point out physical characteristics by asking questions such as: Which three states are not coastal states? *(Arkansas, Tennessee, and Kentucky)* Which state shares a border with Mexico? *(Texas)* Which state is mostly a peninsula? *(Florida)* Then review the definitions of *piedmont* and *Fall Line* from Chapter 10, and have students locate these on the map.

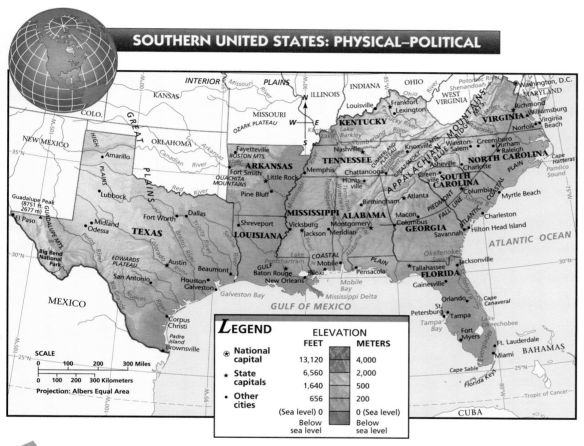

SOUTHERN UNITED STATES: PHYSICAL–POLITICAL

PHYSICAL GEOGRAPHY

FOCUS

- *What landform regions are found in the Southern United States?*
- *What are the characteristics of climates in the southern states?*
- *Why are river systems important to the southern region?*

REGION **Landforms** The dominant landform region of the Southern United States is the Coastal Plain. Miles of sandy beaches mark where the plain meets the waters of the Atlantic and the Gulf of Mexico. Inland from the Coastal Plain are the other major landform regions—the Piedmont, the southern Appalachian and Ozark highlands, and the Interior Plains. West Texas extends into the Great Plains. (See the map above.)

The Coastal Plain stretches inland from the Atlantic Ocean and the Gulf of Mexico. Most of this low area has a shoreline bordered by **barrier islands**. A barrier island is a long, narrow, sandy island separated from the mainland by a shallow lagoon or wetland.

Lying just above sea level, the Coastal Plain holds the nation's largest wetlands. These include the Everglades in Florida, the Okefenokee Swamp in Georgia, the Mississippi Delta in Louisiana, and hundreds of coastal marsh areas behind the barrier islands. The Coastal Plain also contains the Mississippi Delta, the lower Mississippi Valley, and all of the Florida peninsula. The plain then stretches inland to the Fall Line, where it meets the Piedmont. The rolling hills of the Piedmont rise inland in central Virginia, in the Carolinas, and in northern Georgia.

Adding to the beauty of the region are the southern Appalachian Mountains. These include the Blue Ridge and Great Smoky mountains and the Cumberland Plateau. The Appalachians cross the western part of Virginia and the Carolinas, the

Motivator

Show students pictures of scenes such as the following: marshland, white sand beaches, rolling hills, low-lying mountains, plains. Refer to the map of the Southern United States (page 133), and ask where in this region students might see such scenes.

Building a Vocabulary

Have students look up **lock** in a dictionary and find the meaning related to Section 1. Note that **levee** and *levy* sound alike. Help students to distinguish between these two different words. Ask students why they would or would not like to live on a **barrier island.** *(The surroundings would be beautiful, but there might be danger from storms.)*

Guided Practice

On the board, draw a diagram of the river systems in the region. Have each student copy the diagram on a sheet of paper and label the rivers. Call on volunteers to come up and label the rivers on the diagram on the board. Discuss the importance of each river system to the region.

C∞PERATIVE LEARNING

Organize the class into groups of four or five. Direct the groups to research river control techniques and to prepare a written report. Assign one technique to each group member, who should find out its purpose and how it operates. He or she should also find out more about how the technique is used in the river systems of the Southern United States. Groups might provide illustrative diagrams or models.

Geography and the Environment

In recent years, residential development has proceeded rapidly along unleveed floodplains to the north and east of New Orleans. Although many new homeowners want the government to construct flood-control projects, other residents worry that such projects will lead to the loss of more wetlands in the area. The wetlands contain plants such as cypress and tupelo gum trees and wildlife such as black bears, cougars, herons, and eagles.

Natural Regions of Texas

MAP STUDY The large state of Texas has a variety of landforms, climates, and biomes. Texas can be divided into natural regions, which are part of the four major physiographic regions in Texas. See the map on page 112 for the physiographic regions of the United States and Canada. Environmental conditions are similar within each of the state's natural regions.

eastern portions of Kentucky and Tennessee, and northern Georgia and Alabama.

The Ozark Plateau lies mainly in Arkansas but extends into Oklahoma and Missouri. This ancient plateau region has eroded into rugged hill country. The Arkansas River forms a valley through the center of the Ozarks.

Central Kentucky and Tennessee and portions of Texas are part of the Interior Plains, while most of central and western Texas is part of the Great Plains. The northwest part of Texas' Great Plains region is called the High Plains, where elevations reach between 2,000 and 5,000 feet (610 and 1,524 m). South of the High Plains is the Edwards Plateau. This hilly area features many springs where groundwater reaches the surface.

Eastern Texas has the most in common with the rest of the southern states. Most of eastern

Texas lies in the Coastal Plain and is lowland with gently rolling hills. Southwest Texas is a region of mountain ranges and desert basins. Most of the mountains are an extension of the Rocky Mountains.

Climate People living in the Southern United States are accustomed to a humid-subtropical climate with long, hot, and humid summers and mild winters. In the higher elevations of the Appalachians and Ozarks, temperatures are cooler. Because of its large size, Texas has a more varied climate pattern. While East Texas has a humid-subtropical climate, drought-prone West Texas is higher and drier. Here, climates vary among steppe, desert, and highland types.

For most of the southern region, rainfall averages 40 to 60 inches (102 to 152 cm) per year. Thunderstorms, accompanied by dangerous lightning and tornadoes, bring much rain to the region. In the Appalachians, annual rainfall totals may reach more than 100 inches (254 cm), and some areas experience occasional winter snowstorms. Invasions of cold arctic air, called "northers," have caused crop damage as far south as southern Texas and Florida.

During late summer and early fall, hurricanes from the Atlantic Ocean or Gulf of Mexico may strike coastal areas. In 1992, Hurricane Andrew cut a path of destruction across south Florida and Louisiana. The worst natural disaster in the United States was the 1900 hurricane that flattened Galveston, Texas, and claimed more than 6,000 lives. Cities like Galveston that are located on barrier islands are threatened by hurricanes and severe coastal erosion.

INTERACTION **River Systems** The Mississippi is one of the largest river systems in the world. Along with its tributaries, which include the Ohio and Missouri rivers in the north and the Arkansas, Tennessee, and Red rivers in the south,

134

Independent Practice

Give all students a copy of Outline Map 9. Have them complete the map to show the state names, capital cities, landform regions, and major rivers of the Southern United States. Have students also indicate variations in climate within the region.

Closure

Have students list characteristics that make the 12 southern states a region.

SECTION 2
OBJECTIVES

These objectives reflect the Focus questions on this page and are assessed in the review on page 137.

1. Discuss changing patterns of agriculture in the Southern United States.
2. Identify resources and industries important to the economy of the Southern United States.

the Mississippi drains nearly half of the continental United States.

Sediment is carried by the river and deposited at its mouth, forming the large Mississippi Delta. The delta is located in Louisiana, where the river empties into the Gulf of Mexico. The delta's land is low and swampy and cut by small, sluggish streams called **bayous**.

The Mississippi has often flooded its banks and caused much destruction to the areas along its floodplain. Yet, the river serves as a means of transportation, and the floods are a source of fertile new soil. Authorities have tried to lessen flood danger and damage from the Mississippi with better flood forecasting, dams along the river's upper tributaries, and an improved system of **levees** (LEHV-eez). A levee is a ridge of earth along a riverbank that hinders flooding.

The Tennessee River originates in the Appalachians and later joins the Ohio River. In 1933, the federal government formed the Tennessee Valley Authority (TVA). The TVA built dams, reservoirs, and canals in the Tennessee River valley to control flooding, slow soil erosion, allow barge navigation, produce electricity, and increase economic development.

The newest river project is the Tenn-Tom Waterway. Partly as a result of the construction of **locks** on the Tombigbee River, the Tennessee River now connects the Ohio River to the port of Mobile, Alabama, on the Gulf of Mexico. A lock is a part of a waterway enclosed by gates at each end. It is used to raise or lower boats as they pass from one water level to another. These channel and lock systems allow barge transportation to the Gulf ports and provide flood control and hydroelectricity.

The lower Rio Grande forms the international boundary between the United States and Mexico. The irrigation waters from this river are important to the productive farms along this borderland.

SECTION REVIEW

1. Describe three of the landform regions of the Southern United States.
2. What kinds of storms is the southern region likely to experience? What damage can these storms cause?
3. **Critical Thinking** **Assess** the importance of river control techniques to the Southern United States.

2 ECONOMIC GEOGRAPHY

FOCUS

- *Why and how has agriculture been changing in the Southern United States?*
- *What resources and industries are important to the region's economy?*

Agriculture The climate, rainfall, and soil on the Coastal Plain are well suited to growing cotton. Cotton became the leading commercial crop of the South in the years before the Civil War. After the end of slavery, many former slaves and some whites became **tenant farmers**. A tenant farmer is one who rents a plot of land on which to grow crops.

From the 1930s to 1980s, cotton production in the region fell. Pests, such as the boll weevil, foreign competition, expensive mechanization, and synthetic materials all contributed to this decline. Since then, cotton demand has increased, but now most of the nation's cotton is grown on large irrigated and mechanized farms farther west. Today, the leading cotton-producing state is Texas.

Tobacco was introduced to the early settlers by Native Americans. Tobacco is grown mainly on the Coastal Plain and the Piedmont. The farming of tobacco is now mechanized, and the United States remains one of the world's largest producers and exporters of this profitable crop. Health concerns related to tobacco products have forced most tobacco farmers to **diversify**, or to produce a variety of crops.

Today, the Southern United States leads the nation in the production of citrus fruits. Oranges, lemons, and grapefruits are grown in central Florida and in the lower Rio Grande valley of Texas. Florida produces about 75 percent of the nation's oranges and grapefruits.

Many agricultural areas of the Southern United States specialize in certain crops. These crops include winter vegetables from the lower Rio Grande valley; peaches from Georgia; sugarcane from Louisiana, Florida, and Texas; rice from Arkansas and Mississippi; and peanuts and pecans from Georgia and Alabama.

Soybeans have become one of the region's most valuable crops as well, particularly in the

Section 1 Review
ANSWERS

1. Possible answers: Coastal Plain—shoreline, barrier islands, wetlands; Piedmont—rolling hills, inland; Appalachian Mountains—mountain ranges, plateau; Ozark Plateau—hill country; Interior/ Great Plains—High Plains, plateau, hilly, springs
2. Thunderstorms, tornadoes, snowstorms, northers, hurricanes; crop damage, property destruction, deaths
3. Without levees, dams, reservoirs, canals, and locks, the Southern United States would be subject to greater flooding; river transportation would be difficult or impossible; electricity would be scarcer; and farms would be less efficiently irrigated.

135

Motivator

Ask students to name products they use that may have come from the Southern United States. *(orange juice, cotton shirts, peanuts)*

Building a Vocabulary

Point out that **diversify** is similar in meaning to *divide*. Ask how the old saying "Don't put all your eggs in one basket" fits the term *diversify*.

Predicting

Economic Geography   Write the following occupations common in the Southern United States on the board: citrus grower, tobacco farmer, cotton grower, tenant farmer, oil producer, lumber mill operator. Ask students to think of risks or drawbacks for each occupation. *(For example, an unexpected frost could wipe out a citrus crop. Antismoking trends hurt the tobacco industry.)*

Guided Practice

Have students identify some of the fabrics from which their clothes are made (cotton, polyester, nylon, rayon). List responses on the board. Point out that many of the non-cotton fabrics were not developed until the 1930s and 1940s, and that, previously, more textiles were made from cotton. Explain how this change relates to the Southern United States. *(The major cotton-producing region suffered when new fabrics were introduced.)*

Chart Caption Answer

Texas and Florida

GLOBAL PERSPECTIVES

Not only does the southern region hold many aerospace-related industries, it also hosts some of the major U.S. space exploration and launching facilities, including the John F. Kennedy Space Center in Cape Canaveral, Florida, and the Lyndon B. Johnson Space Center in Houston, Texas. The space shuttles are launched from Cape Canaveral. During flight, however, they are controlled by teams in Houston.

The shuttles put satellites into place and maintain them, thus playing an important role in global communications and in monitoring the earth's weather and environment. The shuttles also carry experiments and projects for other countries.

Lead a discussion based on the following question: What benefits might international cooperation in the exploration of space bring to participating nations?

136

FACTS IN BRIEF — States of the Southern United States

STATE POPULATION (1993)	POP. RANK (1993)	AREA (1990)	AREA RANK (1990)	PER CAPITA INCOME (1993)
ALABAMA 4,187,000	22	52,237 sq. mi. 135,294 sq. km	30	$17,234
ARKANSAS 2,424,000	33	53,182 sq. mi. 137,741 sq. km	28	$16,143
FLORIDA 13,679,000	4	59,988 sq. mi. 155,369 sq. km	23	$20,857
GEORGIA 6,917,000	11	58,977 sq. mi. 152,750 sq. km	24	$19,278
KENTUCKY 3,789,000	24	40,411 sq. mi. 104,664 sq. km	37	$17,173
LOUISIANA 4,295,000	21	49,650 sq. mi. 128,594 sq. km	31	$16,667
MISSISSIPPI 2,643,000	31	48,286 sq. mi. 125,061 sq. km	32	$14,894
NORTH CAROLINA 6,945,000	10	52,672 sq. mi. 136,420 sq. km	29	$18,702
SOUTH CAROLINA 3,643,000	25	31,189 sq. mi. 80,780 sq. km	40	$16,923
TENNESSEE 5,099,000	17	42,145 sq. mi. 109,156 sq. km	36	$18,434
TEXAS 18,031,000	3	267,277 sq. mi. 692,247 sq. km	2	$19,189
VIRGINIA 6,491,000	12	42,326 sq. mi. 109,624 sq. km	35	$21,634

Rankings are for all 50 states.
Source: *Statistical Abstract of the United States 1994*

Which two states in the Southern United States are the third and fourth most populous in the country?

lower Mississippi Valley with its rich alluvial soil. Corn, for livestock feed, also has become more important. Florida, Georgia, and Texas are major beef-cattle producers. The poultry industry is important in Arkansas, Georgia, and Mississippi.

Resources   The coastal waters of the Gulf of Mexico and the Atlantic are rich in ocean life. Though all the southern coastal states have fishing industries, Louisiana and Texas are the leaders. The shallow waters around the Mississippi Delta produce great quantities of oysters, shrimp, and other seafood.

Before the region filled with settlers, nearly all of the Southern United States was forested. Today, second-growth timber covers more than half the area. The southern states generate about one-third of the nation's commercial forest production. Southern pine is plentiful on the Coastal

Plain, and hardwoods are common in the Mississippi Valley and on the Piedmont. The softwood forests and tree farms support a paper and pulp industry.

The major mineral and energy resources of the Southern United States include coal, sulfur, salt, phosphates, oil, and natural gas. Coal and iron-ore deposits supplied the steel mills of Birmingham, Alabama, in the late 1800s. A century later, the iron has been mined out and production is down. The city now depends more on service industries, although it still produces metal products.

Texas and Louisiana lead the nation in the production of sulfur and are also important salt-mining states. Florida, meanwhile, has the largest phosphate-mining industry in the nation. Much of the product, which is used to make fertilizer, is exported to Japan and other Asian nations.

Oil is found in West Texas and in coastal Texas and Louisiana. In fact, these two states produce about one-third of the country's crude oil. Much of the production is from offshore oil rigs in the continental shelf waters of the Gulf of Mexico.

With low oil prices on the world market during the 1980s and early 1990s, Texas and Louisiana both experienced high unemployment rates and a loss in government income. The oil and gas industries here suffer from "boom and bust" periods, causing ups and downs in the regional economy.

Industry   The Southern United States now contains about 90 percent of the U.S. textile industry. This industry is heavily concentrated on the Piedmont of Georgia, in the Carolinas, and in Virginia. Most of the large cotton and synthetic-fiber mills are located in cities in North Carolina and Virginia.

Oil refineries are concentrated along the Texas Gulf Coast, producing gasoline, kerosene, and lubricating oils. Associated petrochemical products include plastic, synthetic rubber, paint, tar, fiber, asphalt, explosives, fertilizer, and insecticide. Most petrochemical plants lie in the Houston, Texas, area and along the lower Mississippi in the area around Baton Rouge, Louisiana.

Many new industries related to high technology, publishing, and aerospace have located in the Southern United States. Cities that boast new

Closure

Have students list the 12 states in the Southern United States and write one fact about the economy of each.

SECTION 3
OBJECTIVES

The objectives reflect the Focus questions on this page and are assessed in the review on page 138.

1. List major cities of the Southern United States, and identify features that make each important.
2. Discuss challenges facing the Southern United States.

TEACHING SECTION 3
Motivator

Have students add the major cities of the southern states to Outline Map 9. Ask them to speculate on how each city's location has influenced its growth.

Building a Vocabulary

Help students understand the meaning of **bilingual** by examining its component parts: *bi-* means "two" and *lingual* refers to language.

URBAN GEOGRAPHY AND REGIONAL ISSUES

FOCUS

● *What are the major cities of the Southern United States?*

● *What challenges does the region face?*

The Southern United States is home to some of the nation's oldest cities, including St. Augustine, Florida, founded by Spanish settlers in 1565. Today, the region's metropolitan areas are among the fastest growing in the country as more and more people move to the Sunbelt states.

Atlanta Georgia's largest city, Atlanta is the major transportation hub of the Southern United States. Atlanta's selection as host of the 1996 Olympics is evidence of the city's importance as a major commercial center. The metropolitan area, with nearly 3 million people, is one of the largest urban regions in the southeastern United States.

Miami The rapid growth along the southeast coast of Florida started with tourism and retirement communities. Today, Miami is a major center for transportation, especially with its close connections to the Caribbean, Mexico, and Central America. Electronics, aerospace, and medical technology industries also are established here. Much of Miami's recent population growth is the result of immigration from Cuba, Nicaragua, and other Central American countries.

Tampa and St. Petersburg Located on Florida's Gulf Coast, this Sunbelt metropolitan area grew rapidly during the 1980s. Much of the growth was due to a large migration of retirees from the northern states, which fueled a construction boom. Tampa's port is the export harbor for products from Florida's phosphate mines and citrus farms.

New Orleans Located on the Mississippi River about 100 miles (161 km) from the Gulf of Mexico, New Orleans was founded by the French in 1718. Once a riverboat port, New Orleans now is a major oil-tanker port. Its main industries include petroleum, chemicals, and metal. Its famous jazz

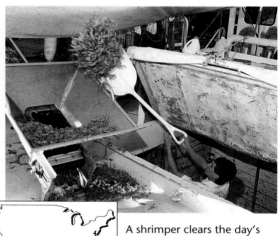

A shrimper clears the day's catch from the hold of a shrimping boat along the **Louisiana** coast. The Gulf Coast is the richest shrimp-producing region in the United States.

industries and research facilities include Austin, Texas; Tampa and Orlando, Florida; Raleigh, North Carolina; and Nashville, Tennessee. The region's low labor costs, transportation access, and relatively inexpensive land have also drawn many foreign companies.

Tourism is a major industry in many southern states, particularly Florida. The state's warm winters attract millions of visitors to beach resorts, the Florida Keys, and Everglades National Park. Coastal resorts such as Padre Island in Texas and Hilton Head and Myrtle Beach in South Carolina attract tourists to their sandy beach locations. Other major destinations in the region include scenic areas of the Ozarks and southern Appalachians, especially Great Smoky Mountains National Park in Tennessee.

2 SECTION REVIEW

1. What crops have been important in the Southern United States?
2. What industries are found in the southern region?
3. **Critical Thinking** **Predict** what might happen to the economies of Texas and Louisiana when their oil resources are depleted. How might these states prepare?

Section 2 Review
ANSWERS

1. Cotton, tobacco, fruits, winter vegetables, rice, nuts, soybeans, corn
2. Paper, pulp, steel, fertilizer, textiles, plastic, synthetic rubber, paint, tar, fibers, asphalt, explosives, insecticides, high technology, publishing, aerospace, tourism
3. Many Texas and Louisiana industries are based on oil; the economies will suffer when the oil is depleted. The states might prepare by further developing industries that are not based on oil.

137

Jazz musicians such as these are popular performers in **New Orleans**. African Americans made New Orleans the birthplace of a brand of music known as Dixieland jazz.

and annual Mardi Gras festivals help make tourism a major source of income as well. Much of New Orleans is below sea level, so serious drainage problems occur during storms. Pollution threatens the city's water supply.

Houston Houston is the nation's fourth largest metropolitan area, its third busiest port, and Texas' largest city. The city was founded in 1836 and is named after Sam Houston, who helped gain Texas' independence from Mexico that same year. Its recently diversified economy includes a large medical complex and service and convention industries. Houston's many refineries and petrochemical and oil-equipment companies reflect its location near the local oil fields.

Dallas-Fort Worth Area About 250 miles (402 km) north of Houston is the Dallas-Fort Worth metropolitan area, known locally as "the Metroplex." This large urban area has grown around two nodes that are 30 miles (48 km) apart. Dallas was founded in 1841 as a railroad shipping stop for cattle and cotton. Today, it is a large cosmopolitan city and center for electronics, banking, insurance, and fashion. To the west is Fort Worth, an early cattle-marketing center. Today, it is a center for petroleum, grain, and aerospace, though the city retains its large cattle stockyards. Between Dallas and Fort Worth are many high-tech industries and recreational facilities, as well as one of the nation's busiest airports.

San Antonio Located in hilly south central Texas, San Antonio has experienced rapid growth. It was founded as a Spanish mission town in 1718 and is one of the region's oldest cities. It is home to food processing, electronics, and aerospace industries, and several military bases. The city is also Texas' major tourist destination. San Antonio is a **bilingual** city with half of its population of Hispanic (mainly Mexican American) origin. A bilingual person has the ability to speak two languages.

INTERACTION **Issues** Although economic progress in the southern Sunbelt cities and coastal areas has been rapid, several southern states suffer the lowest per capita incomes, health standards, and literacy rates in the United States. High unemployment and poverty rates plague Appalachia, part of which is in the Middle Atlantic states, and the lower Mississippi Valley and Delta area.

Rapid growth in urban areas has contributed to traffic congestion, air pollution, ghetto growth, and housing shortages. Newcomers have strained public facilities such as schools, water supplies, and public transportation. Thousands of people have settled on the barrier islands. Such high population densities put a strain on the already-limited roads and bridges. Future hurricanes may endanger the lives and property of the many people who live here.

Pollution and new drainage systems have caused environmental damage to the region's wetlands. The Florida Everglades, a giant "river of grass," has shrunk as natural drainage systems have been altered by increased farming demands. Lower water levels also expose the Everglades and its wildlife to increased dangers from hurricanes and from fires during periods of drought.

SECTION REVIEW

1. Which cities of the Southern United States are transportation centers or ports? To which cities is the electronics industry important?
2. What challenges exist in the Southern United States as a result of growth?
3. **Critical Thinking** **Evaluate** the following statement: "Miami and San Antonio both have large Spanish-speaking populations. Therefore, cultures in the two cities are similar."

CHAPTER 12 CHECK

ANSWERS

Building a Vocabulary

1. A long sandy island separated from the mainland by a shallow lagoon or wetland
2. Bayou
3. A levee helps prevent the river from flooding; a lock allows boats to pass from one water level to another.
4. Farmers might diversify to protect their interests in case demand for any one of their products is reduced.

Recalling and Reviewing

1. Like other southern states, Texas has Coastal Plain, Interior Plains, and mountain landform regions. Unlike other southern states, Texas also has a Great Plains region and desert basins. Texas has more varied landform regions because of its large size and location.
2. Possible answers include the decline of cotton production due to pests, competition, mechanization, and synthetic materials; health concerns related to tobacco; exhaustion of iron-ore deposits; and low oil prices on the world market.
3. People from northern states, Cuba, Nicaragua, Mexico, and other Middle and South American countries

---

CHAPTER 12 CHECK

Boating in the **Florida Everglades**

Reviewing the Main Ideas

1. The Southern United States has a variety of landforms, ranging from plains to mountains. Most of the region has a humid-subtropical climate. The Mississippi River, one of the largest river systems in the world, has been critical in the development of the economy of the Southern United States.
2. Cotton and tobacco traditionally have been important crops in the southern region, but fruits, vegetables, rice, soybeans, and corn have become increasingly important. Ocean, forest, and mineral resources are also significant. Industries in the region are based on textiles, oil, technology, publishing, aerospace, and tourism.
3. Some of the largest urban areas of the Southern United States are Atlanta, Miami, Tampa and St. Petersburg, New Orleans, Houston, Dallas-Fort Worth, and San Antonio. Rapid growth has presented the region with a number of challenges.

Building a Vocabulary

1. What is a barrier island?
2. What is the term for a small stream that cuts through the swampy land of a delta?

3. How are levees and locks used to control rivers and river systems?
4. Why might farmers diversify their crops?

Recalling and Reviewing

1. Compare and contrast the landform regions found in Texas with the landform regions found in other parts of the Southern United States. What accounts for the differences?
2. What are some of the economic challenges that the southern states have faced in recent years?
3. What groups of people are migrating to the southern region?

Thinking Critically

1. What precautions might people who live in the southern region take against hurricanes and floods? Design an evacuation plan for the residents of barrier islands in the case of a severe hurricane.
2. Imagine that you are the president of a major corporation based in the Northeastern United States, and you are considering moving your corporate headquarters to the southern region. What factors should you consider in making your decision?

Learning About Your Local Geography

Individual Project
Investigate the last major storm that caused damage to your community. Interview long-term residents of your community to determine when the storm occurred. Next, go to the library and look through old newspapers from the time of the storm. Create illustrations showing damage done by the storm. Write captions for your illustrations. You may want to include your illustrations in your individual portfolio.

Using the Five Themes of Geography

Write three to five statements discussing the themes of **movement** and **place** in the Southern United States. What place characteristics encourage movement into and within the region? What place characteristics allow such movement to occur?

Thinking Critically

1. Possible precautions include awareness of weather conditions; being prepared to evacuate on short notice; keeping canned food, bottled water, and batteries on hand; identifying a secure place to which to retreat. Evacuation plans will vary but should consider limited capacities of roads and bridges.
2. Possible answers include resource needs, transportation access, and labor and land costs.

Learning About Your Local Geography

Illustrations should show damage done by the area's last major storm. Captions should reflect awareness of storm's effects.

Using the Five Themes of Geography

Place characteristics encouraging movement into the region include a humid-subtropical climate, low labor and land costs, good transportation access, and scenic areas. Storms can necessitate movement within the region. Place characteristics allowing movement include a developed river system and cities serving as transportation hubs.

139

CHAPTER 13 The Midwestern United States

Planning Guide

TEXTBOOK	TEACHING RESOURCES	OTHER AVAILABLE MATERIALS
Section		
1. Physical Geography Pages 141–143	Reteaching Worksheet 13.1 Skills Worksheet 13 Challenge/Enrichment Worksheet 13	*Study Guide for Spanish Speakers* *Lecture Notes,* Chapter 13
2. Agriculture Pages 143–145	Reteaching Worksheet 13.2	*Study Guide for Spanish Speakers* *Lecture Notes,* Chapter 13
3. Resources and Industry Pages 145–147	Reteaching Worksheet 13.3 Graphic Organizer 13	*Study Guide for Spanish Speakers* *Lecture Notes,* Chapter 13
4. Urban Geography Pages 147–148	Reteaching Worksheet 13.4 Critical Thinking Worksheet 13	*Study Guide for Spanish Speakers* *Lecture Notes,* Chapter 13
Chapter Review Page 149	Chapter 13 Test Chapter 13 Portfolio Assessment See also *Alternative Assessment Forms*	*Test Generator,* Chapter 13

CHAPTER RESOURCES

Lecture Notes, Chapter 13
Teaching Resources
    Chapter 13 Graphic Organizer
    Chapter 13 Skills Worksheet
    Chapter 13 Critical Thinking Worksheet
    Chapter 13 Challenge/Enrichment Worksheet

REVIEW AND ASSESSMENT

Chapter 13 Check, page 149
Chapter 13 Test, *Teaching Resources*
Chapter 13 Portfolio Assessment, *Teaching Resources*
Test Generator, Chapter 13
Alternative Assessment Forms

ADDITIONAL RESOURCES

Books for Teachers

Borchert, John R. *America's Northern Heartland.* Minneapolis: University of Minnesota Press, 1987.

Cronon, William. *Nature's Metropolis: Chicago and the Great West.* New York: W.W. Norton, 1991.

Morton, Michael, ed. *A Sense of Place: Essays in Search of the Midwest.* Iowa City: University of Iowa Press, 1988.

Books for Students

Fleming, Alice, ed. *America Is Not All Traffic Lights.* Boston: Little, Brown, 1976.

Siebert, Diane. *Heartland.* New York: Crowell, 1989.

Multimedia Resources

Great Lakes, Bitter Legacy. (VHS, 60 minutes) PBS Video, Public Broadcasting Service, 1320 Braddock Place, Alexandria, VA 22314–1698.

Picture Atlas of the World. (interactive maps, photo bank, animation, audio on CD-ROM) National Geographic Society, Washington, DC.

The St. Lawrence Seaway. (VHS, 46 minutes) Films for the Humanities & Sciences. P.O. Box 2053, Princeton, NJ 08543–2053.

The United States Regions Series: The Midwest. (VHS, 18 minutes) Encyclopedia Britannica Educational Corporation, 310 South Michigan Avenue, Chicago, IL 60604–9839.

MEETING INDIVIDUAL NEEDS

Limited English Proficiency (LEP)

Have students work in small groups to create posters showing the important features of the Midwestern United States as presented in Chapter 13. The posters should highlight the region's physical geography, agricultural products, resources, industries, and urban geography. Display the completed posters in the classroom.

Learners Having Difficulty

Have students work in small groups, with individuals or pairs assigned the Section Review questions for each section. Ask students to skim the sections to find the answers to the questions. Have them record the answers. When they have finished, have students regroup and then share their answers with the other members.

Auditory Learners

Chapter 13 explains several reasons for the global importance of the Midwestern United States. Call on volunteers to explain how the following factors have affected the region's importance today: climate, Great Lakes, upper Mississippi River system, agricultural products, natural resources, industries, interstate highway system, and important cities.

Visual Learners

Provide students with large outline maps of the Midwestern United States. Have students create maps showing the transportation network in this region. The maps should include major waterways, airports, and interstate highways. Next, have students analyze their maps and explain the following statement: "The Midwest serves as the transportation center of the United States." Display the maps in the classroom.

Tactile/Kinesthetic Learners

Have students work in small groups to make matching games to exchange with another group. Provide each group with two sets of index cards. On each card in the first set, students should write the name of a state or city in the Midwest. On each card in the second set, students should name or draw a picture of an important feature for each place named on the first set. Have the groups exchange their cards with another group and play the game.

Gifted Learners

Have students research and construct a scale model diorama of a flood control levee similar to those used along the Mississippi River.

MULTICULTURAL FOCUS

The Midwest is often identified with one of America's most venerated institutions, small-town life. The origins of many small towns lie with the family-owned farms that once typified the region.

Many towns arose in farming areas as local administrative centers. In the eastern half of the United States, the location of county seats—towns containing the county government and courthouse—was determined by the round-trip distance that a farmer from an outlying region could travel in a horse-drawn wagon in one day—about 20 miles. Towns also provided farmers with stores, exchanges, and churches.

Decreasing numbers of farmers, increased mobility, and urbanization have all contributed to a gradual decline in small towns over the last century. In 1910, 17.5 percent of Americans lived in communities of fewer than 10,000 people; that percentage had dropped to 11.2 by 1980.

In spite of these changes, however, Main Street remains to many people a symbol of the American way of life. Small-town living implies security, shared values, and a concern for fellow residents. On the other hand, small-town dwellers can also experience a lack of privacy, limited economic opportunities, and pressure to conform to strict standards of behavior.

Midwestern small-town life has been a popular theme in American literature explored by authors such as Mark Twain and Garrison Keillor. Provide students with a story to read about small-town life. Discuss how the story reflects the geography of towns, the Midwest, and the United States.

CHAPTER 13

THE MIDWESTERN UNITED STATES

Section 1: Physical Geography
Section 2: Agriculture
Section 3: Resources and Industry
Section 4: Urban Geography

OVERVIEW

In this chapter, students will be introduced to the eight states that form the Midwestern United States.

Section 1 describes the physical geography of the region and how it affects life in the Midwest.

Section 2 focuses on the agricultural activity of the Midwest, especially the importance of corn, soybean, and dairy production.

Section 3 discusses the region's industrial growth and outlines challenges facing the region.

Section 4 explains the importance of the region as a transportation center and highlights the major cities.

About the Author

William Cronon (1954–) is a historian. His books emphasize the connection between history and the environment and analyze the ways that people change the land. The quotation on this page is from *Nature's Metropolis: Chicago and the Great West.*

Primary Source

Description of Change: excerpted
Rationale: excerpted to focus on main idea

CHAPTER 13

The Midwestern United States

GEOGRAPHY DICTIONARY

dredge

channel

township and range system

Corn Belt

Dairy Belt

> **❝ I** came to know and care for a landscape that few who are not midwesterners ever call beautiful. Travelers, whether in the air or on the ground, usually see the Middle West less as a destination than as a place to pass through. Only after a long while does one appreciate that the very plainness of the countryside is its beauty. . . . When people speak . . . of the American heartland, this is one of the places they mean. **❞**
>
> *William Cronon*

Butter churn

The Midwest includes the eight states of Ohio, Indiana, Illinois, Michigan, Wisconsin, Minnesota, Iowa, and Missouri. As the nation's agricultural and industrial center, the region has great economic importance. Most of the nation's corn and much of its meat and dairy products come from the Midwest. Access to the Atlantic Ocean through the Great Lakes–St. Lawrence Seaway and to the Gulf of Mexico through the Mississippi River system make the region well situated for international trade.

Prior to European exploration, the Midwest was home to Native Americans who hunted and farmed in the region. By the 1840s, most of the Native American lands were taken by European settlers. As transportation routes from the East Coast developed, thousands more settlers moved to the Midwest, attracted to the promising farmlands and growing cities. In the late 1800s, African Americans, many of whom were former slaves, migrated from the southern states seeking jobs in the new Midwestern cities.

A dairy farm near **Baraboo, Wisconsin**

Using the Quotation
Read the quotation to the class. Ask students to describe both the photograph on page 140 and other Midwestern landscapes to which the author could be referring. Discuss how something plain can be beautiful. Note that the Midwest is called "the American heartland." Ask students to suggest what characteristics might make the Midwestern region comparable to a human heart. *(hard-working, life-sustaining center of activity)*

Using the Geography Dictionary
Pronounce the five terms, and provide brief definitions. Note that **dredge** and **channel** refer to ways of making rivers more navigable. **Township and range system, Corn Belt,** and **Dairy Belt** refer to the agriculture industry in the Midwest. Ask students how the two sets of terms might be related. *(Farmers use rivers to transport products to markets.)*

SECTION 1
OBJECTIVES

The objectives reflect the Focus questions on this page and are assessed in the review on page 143.

1. Describe landforms and climates of the Midwestern United States.
2. Identify lakes and rivers in the Midwest, and explain their importance.

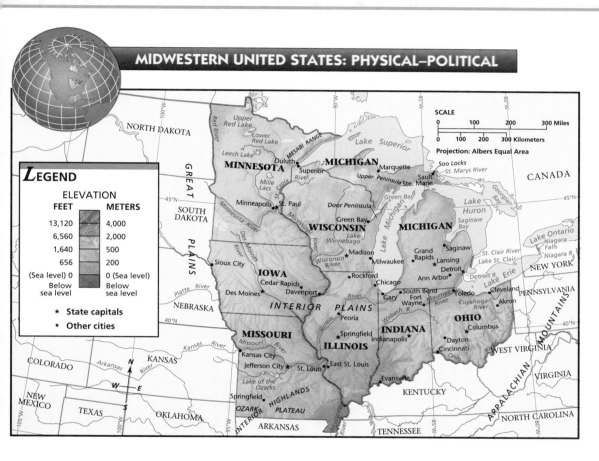

MIDWESTERN UNITED STATES: PHYSICAL–POLITICAL

Using the Physical–Political Map: Midwestern United States

Direct students' attention to the map on this page. Note that although the Midwestern states do not border an ocean, industries are able to use water routes for trade with countries throughout the world. Refer students to the maps on pages 107 and 142. Have them trace the courses of the Ohio and Mississippi rivers to the port of New Orleans, and the route from the Great Lakes through the St. Lawrence Seaway to the Atlantic Ocean.

PHYSICAL GEOGRAPHY

FOCUS

- What are the characteristics of landforms and climates found in the Midwest?
- What lakes and rivers are found in the region, and why are they important?

REGION **Landforms and Climate** The Midwest lies almost entirely within the Interior Plains, a region of flat plains and low hills. More varied landforms are found in the eroded hills of the ancient Canadian Shield, which extends south from Lake Superior. The most rugged areas are in the Ozark Plateau, which stretches into southern Missouri. (See the map above.)

During the last ice age, huge ice sheets covered the Midwest as far south as the Missouri and Ohio rivers. The northern part of the region was eroded glacially, leaving behind ice-scoured rocks, thin soils, and thousands of lakes. Minnesota alone has more than 10,000 lakes. Most of the southern portion of the Midwest has thick, rich soils that were deposited by glaciers.

The entire Midwest region has a humid-continental climate with four distinct seasons. Generally, the northern Midwest states along the Great Lakes have cool, short summers with severe winters. The southern Midwest states along the Ohio and Missouri river valleys have longer summers filled with hot, humid days. While winters tend to be milder to the south, all of the Midwest region experiences cold arctic air and snow in winter. The region receives adequate rainfall for agriculture, though summer droughts occasionally occur. The entire region is subject to severe thunderstorms and tornadoes.

The Great Lakes The Great Lakes are the largest freshwater lake system in the world and contain one-fifth of the world's fresh surface

Motivator

Write the names of the five Great Lakes on the board, and challenge students to make a word using the first letter in the name of each lake. *(HOMES)* They can use this word to help them remember the names of the lakes. Call on a volunteer to locate the Great Lakes on a wall map. Ask questions such as: Which lake is largest? smallest? Which ones border Canada?

Describing

Landforms Have students describe ways glacial action affected landforms in the Midwest. *(For example, glaciers eroded the northern part of the region.)* Then have them describe ways people have changed landforms. *(For example, people have dredged and channeled rivers.)*

Guided Practice

Have students imagine that they are going on a trip to the Midwest. Have them tell what kind of clothes they would pack for summer in Duluth, Minnesota; summer in St. Louis, Missouri; winter in Des Moines, Iowa; and winter in Saginaw, Michigan. Ask students to relate clothing selections to their knowledge of the region's climate.

Geographical Sidelight

The 1993 floods of the upper Mississippi River affected nine states, left approximately 70,000 homeless, and caused property damage estimated at 12 billion dollars. Describing the flood as one of the worst in U.S. history, some analysts criticized the extensive levee system along the Mississippi. They believed that the levees increased and prolonged flooding by preventing the rising river from spreading out and draining into the land.

Section 1 Review
ANSWERS

1. Landform region: Interior Plains; climate region: humid continental
2. The Great Lakes: Superior, Michigan, Huron, Erie, and Ontario; major rivers: St. Marys, St. Clair, Detroit, Niagara, St. Lawrence, Mississippi, Ohio, Missouri, Illinois. Rivers have been modified by the construction of locks and by dredging and channeling.
3. Fewer lakes and therefore fewer rivers; less fertile soils in the region's southern portion

The Great Lakes and St. Lawrence Seaway

The St. Lawrence Seaway permits the passage of ships between ports on the Great Lakes and the Atlantic Ocean. The waterway includes canals and locks that allow ships to move from one water level to another. The eastern end of Lake Erie, at the western end of the waterway, is about 570 feet (174 m) above sea level. Farther east, Montreal lies at about 100 feet (30 m) above sea level. Ships moving from one water level to another enter a lock. The water level inside the lock is raised or lowered to match the water level in the portion of the waterway ahead of the ship.

Moving Through a Canal Lock

water. They include lakes Superior, Michigan, Huron, Erie, and Ontario. All of the Great Lakes were carved by ice sheets during the last ice age. All but Lake Michigan are shared with Canada.

Lake Superior is the largest and deepest of the lakes and has the highest lake-level elevation. It is connected to neighboring lakes Michigan and Huron by the St. Marys River. They, in turn, are connected to Lake Erie via the St. Clair River, Lake St. Clair, and the Detroit River. Lake Erie is connected to the lowest lake, Lake Ontario, by the Niagara River, which flows over Niagara Falls.

Lake Ontario flows into the St. Lawrence River, which empties into the Atlantic Ocean. Locks between the lakes allow for interlake shipping.

The St. Lawrence Seaway, completed in 1959, is one of the world's greatest inland navigation and transportation systems. It allows oceangoing ships from the Atlantic to enter the St. Lawrence River in eastern Canada and travel into the heartland of North America as far as Chicago, Illinois, or Duluth, Minnesota. Ice during the severe winters causes the seaway to close for several months each year.

Closure

Have students list the eight Midwestern states and tell why they form a region.

ASSESSMENT

Assign the Section 1 Review questions on this page.

Reteaching

Assign Section 1 of Reteaching Worksheet 13.

SECTION 2
OBJECTIVES

The objectives reflect the Focus questions on this page and are assessed in the review on page 145.

1. Analyze patterns of farm ownership and land use in the Midwest.
2. List major agricultural products of the Midwest region, and discuss their global significance.

TEACHING SECTION 2
Motivator

Tell students that many Midwestern farmers are pleased if a certain crop is "knee high by the Fourth of July." Have the class guess the crop this familiar saying refers to. *(corn)*

Building a Vocabulary

Draw a grid on the board to represent the township and range system. Discuss the purpose of this system.

AGRICULTURE

FACTS IN BRIEF
States of the Midwestern United States

STATE POPULATION (1993)	POP. RANK (1993)	AREA (1990)	AREA RANK (1990)	PER CAPITA INCOME (1993)
ILLINOIS 11,697,000	6	57,918 sq. mi. 150,008 sq. km	25	$22,582
INDIANA 5,713,000	14	36,420 sq. mi. 94,328 sq. km	38	$19,203
IOWA 2,814,000	30	56,276 sq. mi. 145,755 sq. km	26	$18,315
MICHIGAN 9,478,000	8	96,705 sq. mi. 250,466 sq. km	11	$20,453
MINNESOTA 4,517,000	20	86,943 sq. mi. 225,182 sq. km	12	$21,063
MISSOURI 5,234,000	16	69,709 sq. mi. 180,546 sq. km	22	$19,463
OHIO 11,091,000	7	44,828 sq. mi. 116,105 sq. km	34	$19,688
WISCONSIN 5,038,000	18	80,374 sq. mi. 208,169 sq. km	16	$19,811

Rankings are for all 50 states.
Source: *Statistical Abstract of the United States 1994*

Which state had the highest per capita income in 1993 in the Midwestern United States?

FOCUS

● *What patterns of farm ownership and land use are found in the Midwest?*

● *What agricultural products are grown in the region? How are they important on a global scale?*

Hot, humid summers, fertile soils, level land, and agricultural technology have made the Midwest one of the most productive farming regions in the world. It is the United States' leading producer of dairy products, pork, corn, soybeans, and many other vegetables.

From an airplane, the land of the Midwest appears to be divided into squares. The settlement pattern is one in which fields, roads, and towns all are aligned according to the **township and range system**. The federal government started this survey system in the late 1780s to divide and sell land northwest of the Ohio River. Each township was divided into 36 sections, each one-mile square. During the late 1800s, in order to settle the land, the government offered homesteaders free land if they would farm it for at least five years.

Many Midwestern farms have been family owned for generations and are productive, highly mechanized, and well maintained. Changing economic conditions over the past two decades, however, have led to a decrease in the number of family-owned farms and an increase in the number of large, corporate-owned farms. These changes have affected many of the Midwest's small rural towns that serve farmers and their families. As the number of family farms drops, these towns, especially those located away from major highways, suffer economic decline.

In a world with an ever-growing population, the agricultural yields of the Midwest are among America's most important exports. The Midwest, with its productive soils and efficient farmers, will maintain its global importance.

The Corn Belt Corn originated in Mexico (called "maize" there) and was introduced to the early European settlers by Native Americans. Although corn is grown in every Midwestern state, it is concentrated in the **Corn Belt**, the core of which stretches from Ohio to Iowa. The Corn

Chart Caption Answer

Illinois

COOPERATIVE LEARNING

Organize the class into groups of four. Have students work in pairs within each group. Give each pair two copies of Outline Map 9. One partner should complete a map to show the states in the Corn Belt while the other partner coaches. Then the partners should switch roles to complete a map of the Dairy Belt. Have the partners compare their maps with the maps of the other pair in the group.

MOVEMENT — The Upper Mississippi River

The drainage patterns of the Midwest were formed by the melting of the great ice sheets. Today, the upper Mississippi River system drains the region. The major tributaries include the Ohio, Missouri, and Illinois rivers. These rivers have been **dredged** and **channeled**, which means they have been dug out and straightened. The dredging and channeling allows barge traffic to move between the Midwest and the Gulf of Mexico. In 1993, two months of heavy rainfall swelled the upper Mississippi and its tributaries, flooding much of the Midwest.

SECTION REVIEW

1. Within which landform region is most of the Midwest located? within which climate region?
2. Name the Great Lakes and the major rivers of the Midwest. How have the rivers been modified to allow for shipping?
3. **Critical Thinking** **Imagine** that the last ice age had not occurred. How might the physical geography of the Midwest be different?

Writing to Explain

Agriculture Have students write a letter inviting scientists and government leaders from a foreign country to visit a Midwestern farm. The letter should explain where the farm is located, describe the kind of farm (family farm, corporate farm), and list the crops and farming techniques the visitors will see. Ask volunteers to read their letters.

Guided Practice

Divide the class into three groups: Corn Growers Association, Soybean Growers Association, and Dairy Farmers Association. Have each student write down one or two facts from the text relevant to his or her assigned group. Call on students in each group to state a different fact until all the information has been covered.

Independent Practice

Give students a copy of Outline Map 9. Have them label the map to show the states, capital cities, rivers, lakes, and major landforms of the Midwestern states.

Closure

As a class, list the agricultural products of the Midwest. Have students tell which product they think is most important and why.

THEME:
Interaction

Corn is one of the world's most important crops. It has been used for food for thousands of years. Human efforts have transformed wild corn into a cultigen, or a product of human culture. Today's corn plants, for example, have very high seed yields but reproduce very inefficiently without human intervention.

MULTICULTURAL PERSPECTIVES

Tell students that fish and game conservation laws have caused problems in some parts of the Midwest. Immigrants from Southeast Asia, for example, have found it difficult to understand the laws' restrictions since, in their own culture, hunting and fishing are not regulated sports—they are necessary to maintain life. Hold a discussion on ways these cultural barriers might be overcome. *(community meetings, tours of recreational areas)*

Belt also extends into southern Wisconsin and Minnesota, northern Missouri, and the eastern portions of the Great Plains states of Nebraska and South Dakota. (See the map on page 141.) Iowa and Illinois are the country's two leading corn producers.

The Corn Belt is often called the Feed and Grain Belt because most of the corn grown in the Midwest is used to feed livestock. Many Corn Belt farmers raise beef cattle and hogs. In fact, Iowa and Illinois are the leading pork-producing states. The animals are fed corn to fatten them before they are sent to market. Many western ranches send their cattle to the Midwest to be fattened in feedlots before marketing the cattle.

A small percentage of the corn grown in the Midwest is consumed as fresh sweetcorn, as canned and frozen corn, in breakfast cereals, and in cooking oils. By-products of corn, such as corn oil, cornstarch, and corn syrup, are found in many other store products, including soft drinks, instant coffee, and cake mix. Even some plastic grocery bags are derived from corn.

A significant change in the Corn Belt in recent decades has been the expansion of soybean production. Soybeans originated in China and were introduced into the Midwest after World War II. By 1980, the United States was the world's leading soybean producer, with the Midwest its major growing region. Soybeans are commonly used to produce margarine, vegetable oil, and bean curd (tofu). Almost half of the soybeans produced in the United States are exported, mainly to East Asia.

Soil loses its fertility if fields are planted with the same crop year after year. To prevent this loss, Midwestern farmers practice crop rotation. Crops such as soybeans and clover are substituted for corn in some years. The soybeans and clover help return the necessary mineral balance to the soil. Commercial fertilizers also keep production levels high.

The Dairy Belt The **Dairy Belt** is located north of the Corn Belt, where the summers are cooler and soils become rocky and less fertile. This region covers almost all of Wisconsin and most of Michigan and Minnesota. The Dairy Belt also extends into Iowa and into some Northeastern states. Known as "America's Dairyland," Wisconsin produces more milk, butter, and cheese than any other state. Products from Dairy Belt farms are sold in nearby Midwestern urban areas and are shipped by refrigerated trucks to the rest of the country.

Much of the land in the Dairy Belt is devoted to pasture and to the cultivation of the hay, oats, and corn used to feed dairy cattle. Midwestern dairy farms are efficient and use scientific breeding methods to improve their herds. Strict laws regulate the dairy industry to ensure that dairy products are safe and pure. In recent years, the Midwest's dairy farmers have had problems because of large dairy product surpluses, low prices, and the changing American diet.

Many dairy farms have woodlots, which can be used to earn extra income from the sale of firewood, maple syrup, Christmas trees, and wood pulp. Because of the warming effect of the Great Lakes on nearby land, specialized fruit-growing

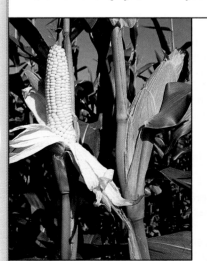

Corn: From Field to Consumer

• Most ears of corn in the United States are about nine inches long. The kernels, arranged in rows on the ear, are edible seeds.

• In the United States, corn usually is harvested anytime from August to late October.

• The corn is dried, stored in bins, and then sold or used on the farm for livestock feed.

Combine harvesting

ASSESSMENT

Assign the Section 2 Review questions on this page.

Reteaching

Assign Section 2 of Reteaching Worksheet 13.

Extension

Have students research the subject of organic farming and report to the class.

SECTION 3
OBJECTIVES

The objectives reflect the Focus questions on this page and are assessed in the review on page 147.

1. Identify primary industries of the Midwest region.
2. Discuss challenges faced by the Midwest and how these challenges have been addressed.

TEACHING SECTION 3
Analyzing Industry

Have students work in pairs to review the industry of the Midwest. Ask one partner to be responsible for Great Lakes industrialization and the other partner to be responsible for Ohio Valley industrialization. Partners should locate key cities in each area, tell why the area became industrialized, and trace the movement of resources and manufactured products.

areas are located around the Great Lakes, especially in western Michigan and the Door Peninsula in Wisconsin. Farmers grow cherries, cranberries, apples, and other fruits.

SECTION REVIEW

1. How have farm ownership patterns been changing in the Midwest? How do farmers keep the soil productive?
2. What products come from the Corn Belt, and where do they go?
3. **Critical Thinking** **Discuss** the relevance of science and technology to agriculture in the Midwest.

RESOURCES AND INDUSTRY

FOCUS

● *What are the primary industries of the Midwest?*

● *What challenges does the region face? How have these challenges been addressed?*

Forests After the major forests of the Northeastern states were logged, the lumberjacks moved into Michigan, Wisconsin, and Minnesota. Today, the region's virgin pine forests are gone,

although many second-growth forests are taking their place. Lumbering is no longer a major industry. Instead, the forests of northern Wisconsin and Minnesota and of Michigan's Upper Peninsula now serve as major recreational areas. The region's forests, lakes, streams, and hills attract millions of tourists, who come to camp, hunt, fish, and ski. The Ozarks of southern Missouri remain largely forested and serve as a tourist center. The forests here also provide valuable wildlife and watershed areas. Recreation and tourism have become vital to the Midwest and join agriculture and industry as a cornerstone of the region's economy.

Industry The Midwest has been an important industrial region for most of the twentieth century. Mining interests grew with discoveries of lead in Wisconsin, copper in Michigan, and iron in Minnesota. The early settlement and growth of the region's many industrial cities were based in part on the cities' locations on or near important transportation routes. Key industries include food processing and the manufacturing of iron, steel, automobiles, and machinery.

Some industrial cities of the Midwest grew as ports on the Great Lakes. Here, they had access to the coal deposits of the Appalachian Mountains, coal mines in Illinois, Indiana, and Ohio, and the iron-ore deposits of upper Michigan and northern Wisconsin and Minnesota. The iron ore was shipped out of the Lake Superior ports of Duluth and Superior to industrial cities such as Gary, Milwaukee, Detroit, Toledo, and Cleveland.

Section 2 Review
ANSWERS

1. There used to be more farms owned by families; now, corporate-owned farms are increasingly common. Crop rotation keeps soil productive.

2. Corn Belt products include corn, beef, pork, soybeans, and clover. Some corn is used to feed livestock and remains in the Midwest; corn also is processed into other products that can be exported out of the region. Soybean products include oils and bean curd and often are exported to East Asia.

3. Because of science and technology, agricultural products can be produced in greater quantities (fertilizer) and with higher quality (breeding methods). They also have increased applications (using corn products in the manufacturing of bags) and accessibility (refrigerated trucks).

● Corn can be processed in a variety of ways. Some corn is cooked and canned, or ground and used for livestock feed. Corn also might be wet-milled or dry-milled, after which various grain parts are used to make different products.

● Corn by-products, such as cornstarch and corn syrup, are used to make breads, breakfast cereals, puddings, and snack foods. Corn oil is used for cooking, and cornstarch is a common ingredient in many medicines.

Cooking
Corn
Other grains
Corn soaking
Wet milling
Dry milling
Steam cleaning
Milling
High-pressure pump
Milling
Sterilization
Coarse grinding
Grain separating
Corn oil
Cornmeal
Filling
Glue
Cornstarch
Canning
Livestock feed
Candy
Tortillas
Corn flour

Guided Practice

Make two columns on the board headed Challenges and Advantages. For the Challenges column, have the class list difficulties facing Midwestern industry, such as foreign competition or depletion of high-grade iron ore. For the Advantages column, have the class give examples of the positive signs for Midwestern industry, such as its experienced labor force and excellent transportation network.

Closure

Have students list past and present natural resources of the Midwest and match them with industries they support.

ASSESSMENT

Assign the Section 3 Review questions on page 147.

Reteaching

Assign Section 3 of Reteaching Worksheet 13.

Extension

Read Carl Sandburg's "Chicago" to the class. The poem captures the spirit of the Midwest's great industrial centers in the early twentieth century. Discuss images a poet might use today to describe industry in the Midwest.

The 125-mile-long (201-km) Lake of the Ozarks in southern **Missouri** is a popular destination for vacationers. The hydroelectric dam that forms the lake on the Osage River provides power for the St. Louis area.

Today, iron is mined only in the Mesabi (muh-SAHB-ee) Range in northern Minnesota and in Michigan and Missouri. After many years of mining, the high-grade ores have become exhausted. The area, however, does have a plentiful supply of low-grade iron ore called taconite (TAK-uh-nite).

In many ways, the Great Lakes are the economic lifeline of the Midwest. They furnish inexpensive transportation for bulk cargo materials. Because of the transport connections provided by the St. Lawrence Seaway, the products of the Midwest can reach and compete more easily in world markets.

The industrialized Ohio River valley stretches across southern Ohio, Indiana, and Illinois. Barges carry coal and other bulk goods to the Ohio Valley cities. The Ohio River empties into the Mississippi River and eventually the Gulf of Mexico.

The Ohio Valley cities produce a variety of goods, including steel, plastics, glass, fertilizers, textiles, and machinery. The largest Midwest industrial cities on the Ohio River are Cincinnati, Ohio, and Evansville, Indiana. Tributaries of the Ohio connect to Dayton and Columbus, Ohio, and to Indianapolis, Indiana.

INTERACTION **Issues** The Midwest's manufacturing industries have experienced the same decline as those in the Northeastern states. Traditional industries based on coal, iron, steel, and automobiles have suffered from foreign competition. Many outdated factories with high labor costs and pollution problems have closed, which has increased unemployment in the region. Some factories, especially in the Ohio Valley, burn local coal, which has caused acid rain in the Northeast and in eastern Canada. The high cost of cleaning up these industries may lead to further closings.

Some midwesterners have migrated to the Sunbelt states in the South and West to seek employment. Even so, the Midwest's central location, excellent transportation network, and experienced labor force have strong potential for industrial redevelopment. An example of this redevelopment is the automobile industry. American automobile manufacturers have renovated factories, and Japanese automakers have built assembly plants in states such as Ohio and Indiana. Other areas have added new industries and now produce high-technology engineering products, such as precision machine tools and automated machines.

Of major concern to the people of the Midwest and to Canada's neighboring provinces is the water quality of the Great Lakes. The people of this region depend on these lakes not only for transportation but also for drinking water, industrial water, fishing, and recreation. All the Great Lakes have suffered from numerous types of pollution, including pesticides, industrial metals, and toxic wastes, and from erosion from construction and dredging.

Lake Erie, however, once considered the most polluted body of water in the world, now sustains a variety of plants and animals. Stricter pollution laws have made many of the rivers in the Midwest, including the industrial Ohio River, cleaner today than they have been in decades.

146

SECTION 4
OBJECTIVES

The objectives reflect the Focus questions on this page and are assessed in the review on page 148.

1. Explain the significance of the interstate highway system to the Midwest.
2. List major cities of the Midwest, and summarize important characteristics of each city.

TEACHING SECTION 4
Motivator

Provide students with copies of a map showing the U.S. interstate highway system. Have them identify the interstate highways that cross the Midwest and the major cities along these routes. Emphasize that interstate highways, along with rivers and lakes, provide transportation routes for the region's products.

Planning a Route

Movement Have students work in small groups to plan the best route for transporting products from the Midwest to various parts of the world. Provide groups with maps and situations such as these:
- Milk from Medford, Wisconsin, to Chicago, Illinois
- Grain from Cresco, Iowa, to Caracas, Venezuela
- Iron ore from Hibbing, Minnesota, to Copenhagen, Denmark

SECTION REVIEW

1. What industries are supported by the Midwest's forest resources? List three other industries of the Midwest.
2. Why has the Midwest faced industrial decline in recent years? Is redevelopment a possibility? Why or why not?
3. **Critical Thinking** **Assess** the ability of industrial corporations to reduce labor costs without cutting workers' salaries and without increasing unemployment.

URBAN GEOGRAPHY

FOCUS

- *How is the interstate highway system significant to the Midwest?*
- *What are the major cities of the Midwestern United States? What are some characteristics of each?*

MOVEMENT **The Interstate Highways**

The United States is crisscrossed by a network of more than 45,000 miles (72,405 km) of multilane interstate highways. This network links major cities from coast to coast. Interstate highways

Many large cities, such as **Chicago,** rely on rail transportation to help ease heavy traffic on clogged highways.

The Sears Tower, at the far left, in downtown **Chicago** is the world's tallest building. Chicago is the core of the third largest metropolitan area in the United States.

have a simple purpose: to move people and goods from city to city.

Before the late 1950s, roads linking major U.S. cities passed through many small towns. The building of the interstate system changed this situation because the new highways bypassed towns and villages. Travelers benefited by saving time. By pulling business away from towns, however, the interstate system caused economic decline in some bypassed towns.

As the transportation center of the United States, the Midwest relies on the interstate system as well as other forms of transportation. The region is supported by an intricate network of railroads, rivers, and highways, which connect the Midwest's major metropolitan areas.

Chicago The largest of the Midwest's industrial and commercial centers and the nation's third largest metropolitan area is Chicago, Illinois. Chicago was originally the site of Fort Dearborn and became a city in 1837. Its location at the southern tip of Lake Michigan and its network of major railroads caused Chicago to grow rapidly and to become the nation's transportation center. In the late 1800s, the city swelled with immigrants who were attracted by jobs in Chicago's steel mills, meat-packing houses, and grain elevators.

Today, Chicago is the busiest port on the Great Lakes and has the world's busiest airport. Its

Section 3 Review
ANSWERS

1. Forests: recreation and tourism; other industries: iron, steel, automobiles, machinery, food processing, plastics, glass, fertilizers, textiles
2. Industrial decline has been caused by foreign competition, outdated factories, high labor costs, pollution problems, and migration to Sunbelt states. Redevelopment is a possibility because of the region's central location, transportation network, and experienced labor force.
3. Answers will vary. Some students may conclude that corporations are unable to reduce labor costs without cutting salaries or increasing unemployment.

Independent Practice

Have students make a table of the major cities of the Midwest. The table should list facts about each city's location, size, date of founding, industries, and attractions. Students can find additional information in encyclopedias and almanacs.

Closure

Have students take turns naming a major Midwestern city and telling a fact about it.

ASSESSMENT

Assign the Section 4 Review questions on this page.

Reteaching

Assign Section 4 of Reteaching Worksheet 13.

Extension

Ask students to imagine that they are citizens of a small Midwestern town. They have just learned that the federal government plans to build an interstate highway near their town. Have students write a paragraph discussing the impact the interstate might have on their lives.

downtown area, called "the Loop," is the site of several of the world's tallest buildings. The Chicago area manufactures steel, electronic equipment, machinery, clothing, and food products. The city is primarily a business center, however. Chicago's universities, theaters, zoos, and museums make the city the cultural capital of the Midwest.

Detroit Located on the Detroit River between Lake Erie and Lake St. Clair, Detroit was established as a French fur-trading fort in 1701. Detroit, Michigan, grew to become the Midwest's second largest city and in 1994 was the nation's seventh largest. Here, Henry Ford built his first automobile factory, which produced its first car in 1908. Since then, Detroit has been the leading automobile producer in the nation. Recent cutbacks in the auto industry have led to high unemployment rates and a population decline in the city.

St. Louis St. Louis, Missouri, is located on the west bank of the Mississippi River, just south of its junction with the Missouri and Illinois rivers. The city began as a French fur-trading post in 1764. It later became the center for pioneers heading west and the nation's leading riverboat port. St. Louis remains a transportation hub, but it is mainly an industrial city. The urban area pro-

duces food products, aircraft, beer, chemicals, machinery, and clothing. The city's waterfront district has undergone urban renewal.

The Twin Cities Minneapolis and St. Paul, Minnesota, called the Twin Cities, are located where the Minnesota River joins the upper Mississippi River. Both cities serve as a major distribution center for the agricultural produce of the Midwest and the Great Plains. They also serve as the retail center of the upper Midwest. An enormous enclosed shopping mall sprawls in the Twin Cities urban area. The cities are leaders in food processing and machinery and electronics manufacturing. St. Paul is the state capital.

Cleveland The major port and manufacturing city on Lake Erie is Cleveland, Ohio. It was established at the mouth of the Cuyahoga River in 1796. In 1834, Cleveland was connected to the busy Ohio River by canal and grew rapidly. Besides being a coal and iron-ore port, Cleveland has a variety of industries, including steel, clothing, paint, and chemical. After years of battling industrial decline and pollution, the city is making a comeback as a high-technology center.

Kansas City The sister cities of Kansas City, Missouri, and Kansas City, Kansas, are located at the junction of the Missouri and Kansas rivers. This site was the starting point of the Oregon Trail and became a pioneer supply town. By 1869, Kansas City, Missouri, was one of the nation's major railroad centers and was the site of grain elevators, stockyards, and meat-packing plants. Today, it is a major financial and retail center. Both cities serve the Midwest and Great Plains agricultural areas.

Section 4 Review
ANSWERS

1. It allowed travelers to bypass some towns, thereby pulling business away from those towns.

2. Answers include Chicago, Detroit, St. Louis, Minneapolis, St. Paul, Cleveland, and Kansas City. Each city is located on a major lake or river or at the junction of rivers, so all serve as transportation hubs or distribution centers.

3. Answers will vary. Students might say no, because the Midwestern cities are separated by greater distances than are the cities of Megalopolis.

The Gateway Arch rises over **St. Louis, Missouri,** to a height of 630 feet (192 m).

SECTION REVIEW

1. How has the interstate system caused decline in some Midwestern towns?
2. Name four cities of the Midwest. How is each city's relative location to bodies of water significant?
3. **Critical Thinking   Determine** whether the major cities of the Midwest form a subregion similar to the Northeast's Megalopolis. Support your answer.

148

CHAPTER 13 CHECK
ANSWERS
Building a Vocabulary

1. It has been dug out (dredged) and straightened (channeled). Dredging and channeling allow rivers to carry barge traffic between specific places.
2. Corn Belt: Ohio, Indiana, Illinois, and Iowa, and stretching into Wisconsin, Minnesota, Missouri, Nebraska, and South Dakota; the Dairy Belt is located north of the Corn Belt, in Wisconsin, Michigan, Minnesota, Iowa, and parts of some Northeastern states.

Recalling and Reviewing

1. Plains with rich soils; usually adequate rainfall
2. Without its water routes, the region might be able to feed itself agriculturally but would not be able to export such large amounts of its products to other areas.
3. Corporations control several of the region's industries and increasingly control farms.
4. Many towns are suffering economic decline because the number of family-owned farms has decreased and because interstate highways have pulled business away. Some cities have experienced much unemployment and population declines but are being redeveloped.

CHAPTER 13 CHECK

The Soo Locks of **Sault Ste. Marie, Michigan**

Reviewing the Main Ideas

1. Landform regions of the Midwest include the Interior Plains, the Canadian Shield, the Appalachian Mountains, and the Ozark Plateau. The Midwest has a humid-continental climate and contains the Great Lakes and many important rivers that are key to the region's economy.
2. The Midwest, with the Corn Belt and the Dairy Belt, is one of the world's most productive agricultural regions. Farms are owned by families and corporations, and farmers use various methods to keep soils fertile.
3. Industries of the Midwest include tourism, food processing, plastics, glass, fertilizers, textiles, and the manufacturing of steel, automobiles, and machinery. Although the region's industries have been undergoing a period of decline, there is potential for redevelopment.
4. Many Midwestern cities, including Chicago, Detroit, St. Louis, the Twin Cities, Cleveland, and Kansas City, are located on key shipping routes and are important industrial and transportation centers.

Building a Vocabulary

1. If a river has been dredged and channeled, what has happened to it? What purposes do dredging and channeling serve?
2. In which states is the Corn Belt located? Where is the Dairy Belt located?

Recalling and Reviewing

1. What landform and climate characteristics make the Midwest suitable for agriculture?
2. Discuss the following statement: "Without its lake and river systems, the Midwest would be neither a major agricultural region nor a major industrial region."
3. How do corporations play a role in the economy of the Midwestern region?
4. How are Midwestern towns and cities changing in response to economic conditions?

Thinking Critically

1. How do product surpluses and low prices present problems for farmers? Suggest some solutions for these problems.
2. Using examples from the chapter, explain how pollution created by one country can affect other countries. Who do you think should bear the responsibility for pollution that crosses international boundaries?

Learning About Your Local Geography

Cooperative Project
With your group, conduct interviews to find out how goods that are locally grown or produced are transported to their destinations and what some of those destinations are. Make a chart showing the major transportation routes in and out of your area.

Using the Five Themes of Geography

Write a paragraph about how the themes of **location** and **region** are evident in the Midwestern United States. How has the region's location affected its development? Into what subregions might you divide the Midwestern region?

Thinking Critically

1. Product surpluses and low prices make it difficult for farmers to make a profit, because they cannot sell enough of their products. Those that they do sell do not earn much. Solutions include paying farmers to grow less and advertising products to increase popularity.
2. Acid rain originating in the Ohio Valley and pollution affecting the Great Lakes have been a source of concern to people in Canada, who need clean water. Answers to the second part of the question will vary.

Learning About Your Local Geography

Charts should show the major transportation routes in and out of the area.

Using the Five Themes of Geography

Paragraphs should state that the central location of the Midwestern United States allowed the region to become a national center of transportation and industry. Possible subregions might be based on agricultural products.

149

CHAPTER 14 The Interior West
Planning Guide

TEXTBOOK	TEACHING RESOURCES	OTHER AVAILABLE MATERIALS
Section		
1. The Great Plains Pages 151–153	Reteaching Worksheet 14.1	*Study Guide for Spanish Speakers* *Lecture Notes,* Chapter 14
2. The Rocky Mountains Page 154	Reteaching Worksheet 14.2 Critical Thinking Worksheet 14	*Study Guide for Spanish Speakers* *Lecture Notes,* Chapter 14
3. Intermountain West Pages 155–156	Reteaching Worksheet 14.3 Graphic Organizer 14 Challenge/Enrichment Worksheet 14	*Creative Strategies for Teaching World Geography* LESSON 7 Studying States by Using Murals *Study Guide for Spanish Speakers* *Lecture Notes,* Chapter 14
Chapter Review Page 157	Chapter 14 Test Chapter 14 Portfolio Assessment See also *Alternative Assessment Forms*	*Test Generator,* Chapter 14

CHAPTER RESOURCES

Lecture Notes, Chapter 14
Teaching Resources
   Chapter 14 Graphic Organizer
   Chapter 14 Skills Worksheet
   Chapter 14 Critical Thinking Worksheet
   Chapter 14 Challenge/Enrichment Worksheet

REVIEW AND ASSESSMENT

Chapter 14 Check, page 157
Chapter 14 Test, *Teaching Resources*
Chapter 14 Portfolio Assessment, *Teaching Resources*
Test Generator, Chapter 14
Alternative Assessment Forms

ADDITIONAL RESOURCES

Books for Teachers

Bruhns, Roger. *Dreams in Dry Places.* Lincoln: University of Nebraska Press, 1990.

Coffey, Marilyn. *Great Plains Patchworks.* Ames: Iowa State University Press, 1989.

Heat-Moon, William Least. *PrairyErth.* Boston: Houghton Mifflin, 1991.

Books for Students

Eberhart, Perry. *Ghosts of the Colorado Plains.* Athens, OH: Swallow Press, 1986.

Fishbein, Seymour L. *Yellowstone Country: The Enduring Wonder.* Washington, DC: National Geographic Society, 1989.

Momaday, N. Scott. *The Way to Rainy Mountain.* Albuquerque: University of New Mexico Press, 1969.

Multimedia Resources

The American Indian: A Multimedia Encyclopedia. (Multimedia CD-ROM) Facts on File, Inc., 460 Park Avenue South, New York, NY 10016.

The New Range Wars. (VHS, 60 minutes; teacher's guide available) PBS Video, Public Broadcasting Service, 1320 Braddock Place, Alexandria, VA 22314–1698.

Rafting Through the Grand Canyon. (VHS, 60 minutes) NOVA-WGBH Collection. Available through Films for the Humanities & Sciences, P.O. Box 2053, Princeton, NJ 08543–2053.

The United States Regions Series: The Southwest. (VHS, 22 minutes) Encyclopedia Britannica Educational Corporation, 310 South Michigan Avenue, Chicago, IL 60604–9839.

MEETING INDIVIDUAL NEEDS

Limited English Proficiency (LEP)

Pair LEP students with students who are proficient in English, and provide the pairs with index cards. On one side of each card, have students write a vocabulary term from the Geography Dictionary. On the other side, have students define each term with sketches or short phrases. Have them use the completed cards as flash cards until they have mastered the terms and their definitions.

Learners Having Difficulty

Help students review Chapter 14 by making an outline on the board of the regions of the Interior West of the United States—the Great Plains, the Rocky Mountains, and the Intermountain West. Call on volunteers to give further details about each region, and write them on the board.

Auditory Learners

Obtain a copy of one of the following books: *The Way to the Western Sea: Lewis and Clark Across the Continent* by David Lavender or *The Journals of the Lewis and Clark Expedition* published by the University of Nebraska Press. Read excerpts from the books that give detailed descriptions of the Great Plains and the Rocky Mountains. Discuss the excerpts with students.

Visual Learners

Divide students into three groups. Have each group create a collage of drawings or pictures showing the physical geography of one of the three regions of the Interior West. Display the completed collages in the classroom.

Tactile/Kinesthetic Learners

Have students work in small groups or in pairs to create relief maps of the Interior West. Ask them to include the Continental Divide, the tree line, and the three Interior West regions.

Gifted Learners

Read the following quotation about the Grand Canyon to students:

> Let this great wonder of nature remain as it now is. Do nothing to mar its grandeur, sublimity and loveliness. You cannot improve upon it. But what you can do is to keep it for your children's children, and all who come after you, as the one great sight which every American should see. . . .
>
> — Theodore Roosevelt, May 6, 1903

Have students research Roosevelt's role in establishing the national park system.

MULTICULTURAL FOCUS

Sixty percent of the residents of the state of Utah are Mormons—members of The Church of Jesus Christ of Latter-day Saints. The religion was founded in western New York State in 1830 by Joseph Smith, known as "The Prophet," who claimed that he had received divine guidance to restore the church instituted by Jesus Christ. Facing prejudice and persecution, his followers eventually migrated west under the leadership of Brigham Young, in search of "a land that nobody wanted." In 1847, they settled in the Salt Lake valley of Utah.

The Mormons worked hard to develop the area agriculturally, planting crops on their first day in the new land. They diverted streams from the nearby Wasatch mountains to irrigate the rock-hard soil and make their desert bloom.

The area around Salt Lake City is still a productive farming region.

During the late 1800s, Mormon settlements spread from the Salt Lake valley throughout the Intermountain West. Mormon relations with the Native Americans of the region, the Ute, were relatively peaceful. Mormons actively sought to convert them to Mormonism.

Have students trace the settlement of their own geographic region, beginning with the Native Americans who originally lived there. Have students focus on the following questions: Who were the original inhabitants? What brought the first settlers there from outside the region? What economic activities did they pursue? What hardships did the original inhabitants and settlers face? How were relations between the two groups?

OVERVIEW

In this chapter, students will learn about the Interior West, the region extending from the middle of the United States to the Sierra Nevada and Cascade ranges in the west. The chapter divides the region into three subregions.

Section 1 describes the landforms, climate, and resources of the Great Plains. It explains the history of settlement and agriculture in the region and discusses challenges the region's residents face.

Section 2 focuses on the physical and economic geography of the Rocky Mountains.

Section 3 discusses the landform regions and fast-growing cities of the Intermountain West.

About the Authors

Peter and Barbara Jenkins are travel writers. The quotation on this page is from *The Walk West: A Walk Across America 2,* which chronicles their experiences on an extensive walking tour of the United States.

CREATIVE STRATEGIES FOR TEACHING WORLD GEOGRAPHY

LESSON 7 Studying States by Using Murals

Primary Source

Description of Change: excerpted
Rationale: excerpted to focus on main idea

CHAPTER 14

The Interior West

❝ **B**eing in the midst of these mountains was different from anything we'd ever known. Inside the Rockies even the colors of things were different. Instead of the orange and red skies that seemed to cover the earth on the plains and in West Texas, the sky was so blue you had to wear sunglasses to look into it. . . . Slashes of white rock cut upward into the blue sky that dared you to try to look to the end of it. ❞

Peter and Barbara Jenkins

GEOGRAPHY DICTIONARY

badlands

chinook

Wheat Belt

shelterbelt

center pivot irrigation

strip mining

land reclamation

Continental Divide

tree line

ghost town

aqueduct

toponym

In books, in movies, and on television, the Interior West of the United States has often been romanticized as the "Old West." To geographers, this region generally encompasses most of the drier and mountainous areas of the western United States.

The Interior West states include North Dakota, South Dakota, Nebraska, Kansas, Oklahoma, Montana, Wyoming, Colorado, Idaho, Utah, Nevada, New Mexico, and Arizona. These states can be grouped into three subregions: the Great Plains, the Rocky Mountains, and the Intermountain West.

Water is the vital resource for the entire Interior West region. Irrigation agriculture is the major consumer of this precious resource. Rapid urban development and the coal-mining industry are making competitive claims to the water, however. In some areas, underground water supplies are dangerously low. Future growth in the Interior West will require that communities balance water supply, preservation of the environment, and economic growth.

A painted cowhide dance shield of the Native American Dakota people

The Rocky Mountains in the White River National Forest of **Colorado**

150

Using the Quotation

Have a volunteer read the quotation aloud to the class. Then call on volunteers who have visited, read about, or seen pictures of mountain regions to comment on the quote. Ask: Did they share the authors' perceptions? What other aspects of the mountain scenery made an impression on them? Did they, for example, notice a tree line or other changes in vegetation?

Using the Geography Dictionary

Write the terms in the Geography Dictionary on the board, and pronounce them for the class. Point out that most of the terms are compound words. Ask students to suggest meanings of the compound words by analyzing their parts. Then have students check their responses by looking up the words in the Glossary.

SECTION 1
OBJECTIVES

The objectives reflect the Focus questions on this page and are assessed in the review on page 153.

1. Describe the landforms and climate of the Great Plains.
2. List past and present economic activities on the Great Plains, and explain their effects.
3. Discuss challenges facing the region.

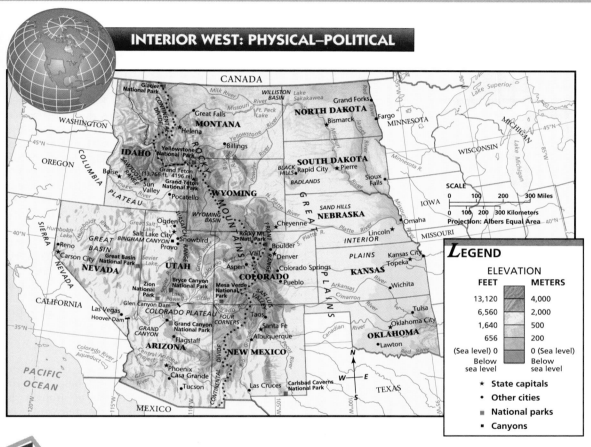

INTERIOR WEST: PHYSICAL–POLITICAL

LEGEND

ELEVATION

FEET	METERS
13,120	4,000
6,560	2,000
1,640	500
656	200
(Sea level) 0	0 (Sea level)
Below sea level	Below sea level

★ State capitals
• Other cities
■ National parks
▪ Canyons

Using the Physical–Political Map: Interior West

Have students describe the features that define the Interior West region. Point out the three subregions (the Great Plains, the Rocky Mountains, and the Intermountain West) discussed in the chapter, and note that some states are part of more than one subregion.

THE GREAT PLAINS

FOCUS

● *What are the characteristics of the Great Plains' landforms and climate?*

● *What economic activities have taken place in the Great Plains?*

● *What challenges do the region's residents face?*

REGION **Landforms** The Great Plains lie in the middle of the United States between the Interior Plains on the east and the Rocky Mountains on the west. The region includes Oklahoma, Kansas, Nebraska, North Dakota, South Dakota, and parts of New Mexico, Colorado, Wyoming, Montana, and Texas. (See the map above.)

The Great Plains were formed by sediment that eroded from mountains and was deposited by rivers. The plains rise gradually from east to west and resemble a huge, tilted table. Although the Great Plains are known for their flat, sweeping horizons, a few areas feature more varied landforms. The Sand Hills of Nebraska are ancient, grass-covered sand dunes. In the Dakotas, rugged areas of soft, sedimentary rock called **badlands** are found. Badlands are lands that have been eroded by wind and water into small gullies and have been left without vegetation and soil.

Climate A steppe climate dominates the region. The Great Plains are semiarid and become drier toward the west. Temperatures can be extreme, with winters in some areas reaching to -40°F (-40°C) and summers reaching above 100°F (38°C). In winter, blizzards occur, bringing cold winds with blowing snow. In summer, hot, moist air from the Gulf of Mexico brings severe thunderstorms and destructive tornadoes.

Droughts are the major climate hazard of the region and occur every few decades. During drought periods, dust storms may cover huge areas. In addition, strong, dry mountain winds,

Motivator

Ask for a show of hands by students who have heard of the Great Plains. Have them suggest images that might describe this region. *(for example, a large, flat area covered by prairie grasses)* Refer them to the map on page 151. Tell them that they should be able to locate this region by studying elevation.

Building a Vocabulary

Write the vocabulary terms for Section 1 on the board. Ask the class which words relate to landforms and climate (*badlands, chinook*), which words relate to agriculture (*Wheat Belt, shelterbelt, center pivot irrigation*), and which words relate to mining (*strip mining, land reclamation*).

Recognizing Cause and Effect

Economic Geography   Ask the class to name past and present economic activities on the Great Plains. List their responses on the board. Then go through the list, and ask students to suggest the impact each activity has had on the environment. Discuss current efforts to protect the environment while carrying on economic activities.

GLOBAL PERSPECTIVES

Point out that in 1991 the United States sent millions of bushels of wheat to Russia and other former Soviet republics, to help alleviate winter food shortages. As a result, bread prices in the United States rose. Ask students to explain the connection. *(The large shipment reduced the wheat supply, making it possible for wholesalers to increase prices.)*

Chart Caption Answer

Six: Montana, New Mexico, Arizona, Nevada, Colorado, Wyoming; Wyoming ranks last in population.

FACTS IN BRIEF   States of the Interior West

STATE POPULATION (1993)	POP. RANK (1993)	AREA (1990)	AREA RANK (1990)	PER CAPITA INCOME (1993)
ARIZONA 3,936,000	23	114,006 sq. mi. 295,276 sq. km	6	$18,121
COLORADO 3,566,000	26	104,100 sq. mi. 269,619 sq. km	8	$21,564
IDAHO 1,099,000	42	83,574 sq. mi. 216,457 sq. km	14	$17,646
KANSAS 2,531,000	32	82,282 sq. mi. 213,110 sq. km	15	$20,139
MONTANA 839,000	44	147,046 sq. mi. 380,849 sq. km	4	$17,322
NEBRASKA 1,607,000	37	77,359 sq. mi. 200,360 sq. km	17	$19,726
NEVADA 1,389,000	38	110,567 sq. mi. 286,369 sq. km	7	$22,729
NEW MEXICO 1,616,000	36	121,598 sq. mi. 314,939 sq. km	5	$16,297
NORTH DAKOTA 635,000	47	70,704 sq. mi. 183,123 sq. km	19	$17,488
OKLAHOMA 3,231,000	28	69,903 sq. mi. 181,049 sq. km	21	$17,020
SOUTH DAKOTA 715,000	45	77,121 sq. mi. 199,743 sq. km	18	$17,666
UTAH 1,860,000	34	84,904 sq. mi. 219,901 sq. km	13	$16,180
WYOMING 470,000	50	97,819 sq. mi. 253,351 sq. km	9	$19,539

Rankings are for all 50 states.
Source: *Statistical Abstract of the United States 1994*

How many states of the Interior West are among the 10 largest states in area in the country? Which state has the smallest population of all 50 states?

---

called **chinooks**, descend from the Rockies and affect the western portions of the Great Plains.

The vegetation of the Great Plains is prairie and steppe grasses, shrubs, and sagebrush. The wetter eastern plains originally were covered with continuous tall-grass prairie. The drier High Plains had shorter steppe grasses. Today, most of the original grasslands have been changed by agriculture and grazing.

The End of the Open Range Because of the lack of moisture, most of the Great Plains has been better suited to grazing than to farming. During the "cowboy era" of the mid-1800s, herds of cattle and flocks of sheep roamed the open range of the Great Plains. They replaced the huge buffalo herds hunted by the Native Americans

prior to the arrival of the railroads and settlers. Within a short time, the buffalo herds had been killed off by hunters, and the Native Americans were forced onto government reservations.

The Homestead Act of 1862 granted 160 acres to settlers willing to farm the land for five years, causing rapid population growth in the region. Eventually, barbed wire for fencing, steel plows that cut through the tough sod, and windmill pumps for wells ended the era of the open range. Fences divided the plains into ranches and farms. Today, livestock raising is still a major economic activity, but it often is combined with wheat farming.

Wheat is one of the most valuable U.S. food crops and an important export for the country. Nearly all Americans eat wheat in one form or another every day. The **Wheat Belt** stretches across the Dakotas, Montana, Nebraska, Oklahoma, Colorado, and Texas. In recent decades, the Wheat Belt has extended westward into drier steppe grasslands. This expansion occurred with the development of large-scale irrigation.

Hailstorms, insects, and plant disease are constant challenges in the Wheat Belt. Soil erosion, particularly during periods of drought, is another major concern. In windy areas, rows of trees, called **shelterbelts**, block the wind and protect the soil.

Irrigation Agriculture The western part of the region generally receives less than 20 inches (50 cm) of rainfall per year, making irrigation nec-

A storm rolls over these lonely structures on the Great Plains of **Wyoming**. To see the area covered by the Great Plains, see the map of the physiographic regions of the United States and Canada on page 112.

Closure

Divide the class into groups. Give each group a piece of paper. Have one group member write a fact about the Great Plains and then pass the paper to the person on the right. That student should write a different fact and pass the paper to the next student. The group should continue in this way until the information in the text has been covered. Using the groups' lists, have one student compile a composite list, and post it in the classroom.

ASSESSMENT

Assign the Section 1 Review questions on this page.

Reteaching

Assign Section 1 of Reteaching Worksheet 14.

Extension

You might recommend Willa Cather or Laura Ingalls Wilder's books about life on the Great Plains.

SECTION 2
OBJECTIVES

The objectives reflect the Focus questions on page 154 and are assessed in the review on page 154.

1. Describe the physical geography of the Rocky Mountains.
2. Summarize important economic activities of the Rocky Mountain region.

essary for crop growth. Underground water supplies are pumped for use in the fields. The largest source of water under the Great Plains is the Ogallala Aquifer, which stretches from South Dakota to Texas. For many years, this aquifer supplied all the farmers' water needs. With increased irrigation and wasteful water practices, however, the water table began to drop. Today, the depletion rate has slowed as a result of more efficient irrigation methods.

A number of rivers cross the Great Plains. In the past, the rivers and tributaries were full in the spring, but in the summer the water slowed to a trickle. Today, major dam-building projects control the flow of these river systems and make farming less risky.

From the air, much of the irrigated Great Plains appears as a series of green circles. This pattern is caused by **center pivot irrigation**, which employs long sprinkler systems mounted on huge wheels that rotate slowly. Center pivot irrigation has been introduced throughout the western United States. Major irrigation crops of the Great Plains include corn, sugar beets, and alfalfa, which is used for cattle feed.

Farming Issues Although the Great Plains region is productive agriculturally, the farmers are challenged continually by climate and world markets. During the 1930s, the combination of a drought and the Great Depression meant disaster for thousands of farmers on the plains.

Today, farmers also are concerned about the price of wheat on world markets—low prices can be as disastrous to them as drought. Political decisions, such as how much the United States will sell to the largest wheat importers, China and Russia, influence prices. During the 1980s, a surplus of wheat caused prices to drop. Thousands of farms were abandoned, and many rural plains towns began shrinking. States such as North Dakota actually lost population. Recently, some of the region's small towns have been undergoing renewal.

Resources and Industry Both coal and oil have been found beneath the Great Plains. Oil production on the plains has been mainly in Oklahoma, Wyoming, Colorado, and eastern New Mexico. The largest coal deposits are in Wyoming. The coal, buried just beneath the surface, can be removed by **strip mining**. In strip mining, soil and rock are stripped away by large machines to get at the coal. The law now requires that mining companies use **land reclamation** methods, so that the land is usable after the mining is finished. To reclaim the land, the topsoil is saved, and the land surface is returned to its original contours. If land reclamation is not practiced, serious erosion occurs, and an ugly pit remains.

Located at the foot of the Rockies, Denver is the dominant city and economic center of the Great Plains and of the adjoining Rocky Mountains region. It is also the capital of Colorado. Unfortunately, Denver's rapid growth has caused severe air pollution.

Workers have dug this pit to mine coal near **Sheridan, Wyoming**.

SECTION REVIEW

1. Describe the dominant landform and dominant climate of the Great Plains region.
2. What economic activities take place in the region, and how has the environment been modified?
3. **Critical Thinking** **Explain** how international concerns play a role in the region.

Section 1 Review
ANSWERS

1. Landform: the Great Plains, which rise gradually from east to west and have flat, sweeping horizons; climate: semiarid steppe with extreme temperatures, cold winters, hot summers, storms, droughts, and winds
2. Economic activities include livestock raising, wheat farming, the raising of other crops, and strip mining. The environment has been modified in that natural grasses have been replaced by wheat, through the construction of fences, plows, pumps for wells, shelterbelts, dams, and irrigation systems.
3. International concerns, such as those affecting how much wheat the United States will export and what its world market price will be, make for economic uncertainty in the region and can even cause farm abandonment and population decreases.

Motivator

Ask students what images come to mind when they think of the Rocky Mountains. Some may say hiking or skiing. Point out that tourism is a valuable industry in the region.

Building a Vocabulary

Discuss reasons why communities become **ghost towns**. Contrast *ghost town* and *boomtown*.

Illustrating Ideas

Physical and Economic Geography
Locate the Rocky Mountains on a wall map. Draw a sketch of the mountains on the board. Ask volunteers to label and explain the Continental Divide, the tree line, and climate zones. Next to the drawing, have students list natural resources found in the Rockies that have contributed to the economic growth of the region. *(minerals, forests)*

Reporting Information

Point out ways that scenic areas contribute to a region's economic growth. Have each student choose one of the national parks located in the Rockies and write a brief report on it. The report should include when the park was established and what attractions it offers. Students should also note steps the government is taking to correct problems caused by overcrowding in national parks. Have students share their reports.

THEME
Location

Yellowstone is the country's oldest national park. Established in 1872, it has about 200 geysers—including Old Faithful—and 10,000 hot springs.

Section 2 Review
ANSWERS

1. The region has several vegetation zones because the different elevations have different climates. Vegetation zones include semiarid grasslands at the foot of the mountains, forested mountain slopes below tree line, and tundra above tree line.

2. The depletion of minerals such as gold and silver; tourism

3. Possible answer: increased development.

154

---

2 THE ROCKY MOUNTAINS

FOCUS

- *What are the physical features of the Rocky Mountains?*
- *What economic activities have been important in the region?*

REGION **Physical Geography** West of the Great Plains are the Rocky Mountains. The Rockies have rich natural resources and great scenic beauty. They stretch from Canada through Idaho, Montana, Wyoming, Utah, Colorado, and New Mexico. The Rockies are actually a series of mountain ranges separated by passes and wide valleys. The crest of the Rocky Mountains forms the **Continental Divide**, which divides the major river systems of North America into those that flow eastward and those that flow westward. Snowcapped peaks more than 12,000 feet (3,658 m) high are numerous in the Rockies.

The highland climates of the Rockies are variable, depending upon elevation. Semiarid grasslands usually are found at the foot of the mountains, while most of the slopes are forested. The forests capture the winter snowfall and form the source for the rivers that flow across the Great Plains and Intermountain West states. The highest mountain peaks reach above the **tree line** to a zone of tundra, rock, and ice. The tree line is the elevation above which trees cannot grow.

Economic Geography Most of the Rocky Mountains region is still wilderness or forest land. Several highways and railroads cross the mountain passes, but no major cities have developed in the Rockies. Most of the residents of the region live in small towns that are dependent on mining, ranching, forestry, or tourism.

Early prospectors struck major veins of gold and silver in Montana, Idaho, Colorado, Utah, and New Mexico. Utah, Montana, and New Mexico are leading copper-producing states. Lead and many other ores also are found in the Rocky Mountains.

Though mining communities are scattered throughout the region, many have become **ghost towns**. Ghost towns are towns that once flourished but are deserted today, usually because of the depletion of a nearby resource such as gold or silver. Oil also has been found on the western slopes of the Rockies. In the northern Rockies of Idaho and Montana, logging supports lumber and pulp mills. Grazing of sheep and cattle is important in some Rocky Mountain areas. Farming is limited, however.

One of the most valuable industries in the Rocky Mountains region is tourism. Popular Rocky Mountain ski areas include Aspen, Vail, Sun Valley, and Taos (TAUS). The federal government has set aside large scenic areas in the region as national parks. Among these are Yellowstone, Grand Teton, Rocky Mountain, and Glacier national parks. Millions of tourists come to these parks to camp, hike, mountain climb, and fish. The largest landowner in the Rocky Mountains is the federal government. Private businesses must acquire permits to mine, log, graze cattle, or build ski resorts on these government lands.

A skier takes advantage of the snow in Taos Ski Valley in **New Mexico**. Each year tourists flock to the slopes of ski resorts that dot the Rocky Mountains.

2 SECTION REVIEW

1. Why do the Rockies have several vegetation zones, and what are they?
2. Why has mining activity decreased in the region? What is one of the region's most valuable economic activities today?
3. **Critical Thinking** **Predict** what might happen in the Rocky Mountains if the federal government sold off its lands here.

INTERMOUNTAIN WEST

FOCUS

- *What is the physical and economic geography of the Intermountain West?*

- *How is the population distributed in the Intermountain West?*

REGION **Physical Geography** Between the Rocky Mountains and the Sierra Nevada and Cascade ranges lies the vast Intermountain West. The Intermountain West includes all of Arizona, Nevada, and Utah, southern Idaho, and the western portions of Colorado and New Mexico. The Intermountain West region can be further divided into three landform regions: the Basin and Range region (also known as the Great Basin), the Colorado Plateau, and the Columbia Plateau.

The Basin and Range region is centered in Nevada. It is a region of desert basins separated by faulted mountains. The central portion of this area is called the Great Basin because its river waters do not reach the sea. Instead, the rivers flow into low basins and evaporate, leaving behind dry lake beds or salt flats.

The Colorado Plateau is a beautiful uplifted area of horizontal rock layers that have been eroded by wind and water. This region is known for its deep canyons, the largest of which is the Grand Canyon of the Colorado River. The Columbia Plateau lies in the northern part of the Intermountain West. It is centered in southern Idaho but extends into eastern Washington and Oregon. The Snake River has cut deep canyons into the region's plateaus.

The climate of the Intermountain West region varies. Generally, the southern part, in Arizona, New Mexico, and Nevada, is desert. The northern part, in Utah and southern Idaho, is mainly steppe. Some of the low regions, especially the Great Basin, experience the rain-shadow effect and receive almost no rain at all.

Agriculture While only a small portion of the Intermountain West is used as farmland, farming is an important part of the economy. In the warmer and drier south, all farming requires irrigation. The Colorado River is the lifeblood of the southwestern states as it flows more than 1,450 miles (2,333 km) from the Rocky Mountains to the Gulf of California. By agreement, several states and Mexico share the Colorado River's valuable waters. Dams, including giant Hoover Dam and Glen Canyon Dam, now control the once wild river.

Today, agriculture in all of these dry states is possible because of the Colorado River basin projects. Arizona and southern California have **aqueduct**, or canal, systems to carry water hundreds of miles from the Colorado River. The desert city of Las Vegas, Nevada, also receives most of its water from the Colorado River. As the most regulated and tapped river in the United States, the Colorado is virtually dry by the time it reaches the Gulf of California.

Other Resources Mining is vital to the economy of the Intermountain West. The most plentiful ores are copper, gold, silver, lead, zinc, and uranium. The Basin and Range area in Arizona and Utah is the richest copper-mining area in the United States. Coal is mined in Utah, Arizona, and New Mexico. Mining development has come under scrutiny, however, because of its impact on water and air quality.

The Intermountain West is a favorite tourist destination, particularly Arizona and Utah. Grand Canyon National Park in northwestern Arizona attracts millions of tourists each year. The canyon is more than one mile (1.6 km) deep, carved by the Colorado River over millions of years. Other heavily visited national parks in the Intermountain West are Zion Canyon and Bryce Canyon in Utah, Mesa Verde in southwestern Colorado, and Great Basin in Nevada.

Urban Geography The Intermountain West began gaining population after World War II when the federal government built dams and water projects, military bases, and major highways here. The widespread use of air conditioning made the region even more attractive. Recent growth here has been the most rapid in the nation.

People and industry are attracted to the area because of the wide-open spaces and warm, sunny climate. Cities have grown rapidly, especially Phoenix, Tucson, and Las Vegas. Many retirement communities have sprung up in the middle of the desert. The growing population requires large quantities of water—a limited resource.

Phoenix, the capital of Arizona, is the largest city in the Intermountain West. It contains more

156

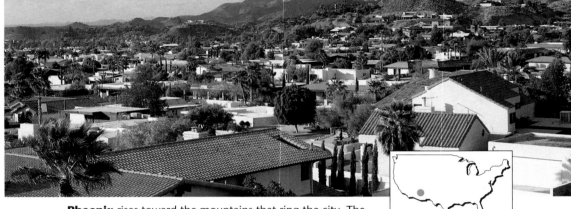

Phoenix rises toward the mountains that ring the city. The population of this rapidly growing city increased by nearly 400,000 from 1970 to 1990.

than half of the population of Arizona in its expanding urban area. Like Denver, Phoenix is experiencing environmental problems related to rapid population growth. Today, the state has an urban-based economy that is dependent on industry, tourism, and retirement communities.

The second largest urban area in the Intermountain West is Salt Lake City, the capital of Utah. The area has recently become a center for computer and communications industries. More than 65 percent of Utah's population lives in Salt Lake City and other nearby cities.

Salt Lake City is the headquarters of The Church of Jesus Christ of Latter-day Saints, whose members are called Mormons. Mormon pioneers, led by Brigham Young, settled here in 1847. They developed a large-scale irrigation system in this arid region. More than 60 percent of Utah's population is Mormon.

Human Geography The earliest inhabitants of the Intermountain West were the Native Americans who arrived in the region more than 12,000 years ago. Agricultural Native American cultures were established by A.D. 700. When the Spanish arrived in the sixteenth century, there were many different Native American groups living in the region.

A large Native American population lives on the Colorado Plateau of Arizona and New Mexico. The Navajo and the Hopi are the largest of these groups. The Navajo reservation, which is located mainly in northeastern Arizona but also extends into Colorado, New Mexico, and Utah, is the largest reservation in the United States. For many years, the Navajo lived as nomadic herders, moving over the plateau with their sheep, goats, and

horses in search of pasture. Today, some Navajo are farmers as well as herders. Though deposits of coal, uranium, and oil have been found on the Navajo reservation, poverty here persists.

The Hopi reservation, also in northeastern Arizona, is surrounded completely by Navajo land. The Hopi are mainly farmers and craftsworkers who live in small farming villages on high plateaus. Some Hopi also graze sheep and goats.

The impact of Spanish settlement is illustrated in the many Spanish **toponyms** that can be found on maps of the southwest United States. Toponyms are names of places and often reflect the history and culture of a region. Examples are Las Vegas (the meadows), Casa Grande (grand house), and Las Cruces (the crossroads).

Spanish architecture and language in the region reflect this cultural influence. Today, the majority of the population of New Mexico speaks Spanish. Many of the farms in the southwestern United States are dependent on migrant labor from Mexico. The continuing migration from south of the border ensures that Hispanic (Latino) culture will remain important in this region.

 SECTION REVIEW

1. Given the region's landforms and climates, how is agriculture possible?
2. What is the cultural background of the region?
3. **Critical Thinking** **Imagine** that aqueduct systems and air conditioning had not been invented. Would the Intermountain West region still be undergoing rapid growth? Explain your answer.

CHAPTER 14 CHECK
ANSWERS

Building a Vocabulary

1. Lands that have been eroded into gullies and have little or no vegetation and soil
2. Chinook
3. A row of trees that blocks wind and protects fields from soil erosion
4. Because of center pivot irrigation, a rotating sprinkler system
5. Strip mining; making the land usable again

Recalling and Reviewing

1. It supplies wheat for much of the United States and for other countries.
2. Mining has declined in the Rocky Mountains but is still very important in the Intermountain West. Farming is limited in the Rocky Mountains, but irrigation allows agriculture in the Intermountain region, and farming is an important part of that region's economy. Tourism is important to the economies of both regions.
3. Droughts and water shortages, the effects of mining, and air pollution and other problems related to urban growth.

Thinking Critically

1. Plains dominate both regions, although the Great Plains are higher. Both are internationally significant because of their agricultural products. The Midwest is more industrialized and urbanized.
2. Answers will vary but might include providing sufficient clean water, adequate health care, and educational opportunities on the reservations, and treating Native Americans with respect and dignity.

Learning About Your Local Geography

Paragraphs should explain the origins and meanings of the names of students' communities.

Using the Five Themes of Geography

Illustrations of the Great Plains might show irrigated farmlands, fenced ranches, or strip mining. Illustrations of the Rocky Mountains might show mountain climbers, mining, ghost towns, logging, or ski resorts. Illustrations of the Interior West might show rivers with dams, aqueduct systems, mining, or cities.

CHAPTER 14 CHECK

Hoover Dam on the Colorado River between **Nevada** and **Arizona**

Reviewing the Main Ideas

1. Flat plains and a steppe climate dominate the Great Plains. The land has been used for grazing, but now wheat farming and strip mining are the major economic activities. Various political, economic, and environmental challenges continue to face the region.
2. The Rocky Mountains, a series of mountain ranges, form the Continental Divide. Climates and vegetation vary according to elevation. Mining activities have been important in the Rocky Mountains region but have decreased with the depletion of some minerals, and tourism is now one of the region's most important industries.
3. Despite dry climates in the Intermountain West region, irrigation has made agriculture possible. Mining and tourism are also important parts of the region's economy. Cities of the region are growing rapidly. Retirees, Mormons, Native Americans, and Spanish-speakers make up a large part of the region's population.

Building a Vocabulary

1. What are badlands?
2. What term is used to describe a strong, dry mountain wind?

3. Define *shelterbelt*.
4. Why do parts of the Great Plains appear to be a series of green circles when viewed from above?
5. What technique is used to mine coal? What does land reclamation accomplish?

Recalling and Reviewing

1. How is the Great Plains region internationally important?
2. Compare and contrast the economic activities found in the Rocky Mountains with those of the Intermountain West.
3. What environmental concerns do the Great Plains and Intermountain West regions share?

Thinking Critically

1. In what respects do the Great Plains resemble the Midwest? How do they differ?
2. Propose a way to reduce the poverty found on Native American reservations without radically altering Native American cultures.

Learning About Your Local Geography

Individual Project
Conduct library research to find out the origin and/or meaning of your community's toponym. Does it reflect the history or culture of your community? Was it the name of an individual or family? Does it describe the physical setting of your community? Write a paragraph explaining the toponym. You may want to include your paragraph in your individual portfolio.

Using the Five Themes of Geography

Illustrate how the themes of **region, place,** and **human-environment interaction** apply to the Interior West by drawing at least one picture of each region discussed in the chapter. Each illustration should show physical and human place characteristics and evidence of human-environment interaction in one of the regions. Do not label your illustrations. Have other class members identify the region shown in each illustration.

157

CHAPTER 15 The Pacific States
Planning Guide

TEXTBOOK	TEACHING RESOURCES	OTHER AVAILABLE MATERIALS
Section		
1. California Pages 159–161	Reteaching Worksheet 15.1 Challenge/Enrichment Worksheet 15	*Study Guide for Spanish Speakers* *Lecture Notes,* Chapter 15
2. Oregon and Washington Pages 164–165	Reteaching Worksheet 15.2	*Study Guide for Spanish Speakers* *Lecture Notes,* Chapter 15
3. Alaska Pages 165–169	Reteaching Worksheet 15.3	*Study Guide for Spanish Speakers* *Lecture Notes,* Chapter 15
4. Hawaii Pages 169–170	Reteaching Worksheet 15.4 Graphic Organizer 15 Skills Worksheet 15 Critical Thinking Worksheet 15	*Study Guide for Spanish Speakers* *Lecture Notes,* Chapter 15
Chapter Review Page 171	Chapter 15 Test Chapter 15 Portfolio Assessment See also *Alternative Assessment Forms*	*Test Generator,* Chapter 15

CHAPTER RESOURCES

Lecture Notes, Chapter 15
Teaching Resources
    Chapter 15 Graphic Organizer
    Chapter 15 Skills Worksheet
    Chapter 15 Critical Thinking Worksheet
    Chapter 15 Challenge/Enrichment Worksheet

REVIEW AND ASSESSMENT

Chapter 15 Check, page 171
Chapter 15 Test, *Teaching Resources*
Chapter 15 Portfolio Assessment, *Teaching Resources*
Test Generator, Chapter 15
Alternative Assessment Forms

ADDITIONAL RESOURCES

Books for Teachers

Bockstoce, John R. *Arctic Passages.* New York: Hearst Marine Books, 1991.

Kanahele, George S. *Ku Kanaha, Stand Tall: A Search for Hawaiian Values.* Honolulu: University of Hawaii Press, 1986.

McPhee, John. *Coming into the Country.* New York: Farrar, Straus & Giroux, 1977.

Muir, John. *The Yosemite.* San Francisco: Sierra Club Books, 1989.

Books for Students

Alastair, Scott. *Tracks Across Alaska: A Dog Sled Journey.* New York: Atlantic Monthly Press, 1990.

America's Seashore Wonderlands. Washington, DC: National Geographic Society, 1985.

Egan, Timothy. *The Good Rain: Time and Terrain in the Pacific Northwest.* New York: Knopf, 1990.

Multimedia Resources

American Visions: Olympic. (VHS, 30 minutes) Lumivision.

Hawaiian Legacy. (VHS, 60 minutes) PBS Video, Public Broadcasting Service, 1320 Braddock Place, Alexandria, VA 22314–1698.

Los Angeles: The Making of a City. (VHS, 29 minutes) Films for the Humanities & Sciences, P.O. Box 2053, Princeton, NJ 08543–2053.

National Parks of America. (Multimedia CD-ROM) Multicom Publishing, Inc., 1100 Olive Way, Suite 1250, Seattle, WA 98101.

MEETING INDIVIDUAL NEEDS

Limited English Proficiency (LEP)

Create a graphic organizer on the board that illustrates the similarities and differences among the Pacific states. Ask students to volunteer information as you write it on the board. Have students copy the organizer and use it to study the chapter.

Learners Having Difficulty

Orally review each of the Pacific states' principal landforms as you indicate their locations on a wall map.

Auditory Learners

Assign pairs of students one of the following topics: Should offshore oil deposits found on California's continental shelf be mined? Should dams that prevent salmon from returning to their upstream spawning grounds be eliminated? Should the lumber industry in Washington and Oregon be allowed to clear-cut forests? Should the federal government allow the oil industry drilling rights in the Arctic National Wildlife Refuge? Should Hawaii encourage industrial expansion? Assign individuals pro and con positions, and have them debate the issues, using their textbook as a resource. Have pairs summarize their discussions for the class.

Visual Learners

Provide students with outline maps of the Pacific states and a list of precipitation data for the region from an almanac. Have students plot the points on the map and shade some areas that have similar levels of precipitation. Then have students analyze their maps and make up true/false questions based on their findings. Have students exchange questions with another student and answer the questions they receive.

Tactile/Kinesthetic Learners

Divide the class into five groups, and assign each group a Pacific state. Have students use local grocery store advertisements to create posters showing the variety of agricultural products grown in their assigned state. Use the posters as the focus of a discussion of the economic geography of the Pacific region.

Gifted Learners

Have students research state building codes that are required because of the frequent earthquakes in California and Alaska. Ask students to construct a scale model of a building or other structure showing the special measures taken to help make it earthquake-proof.

MULTICULTURAL FOCUS

Point out to students that the Pacific states region is perhaps the most culturally diverse region in the United States. People from many parts of the world have moved there in different waves of immigration brought on by a variety of events and motivations. For example, many Chinese moved to California in the mid-1800s to pursue opportunities working in industry and agriculture. Many Mexicans moved north during World War II when California experienced a critical shortage of farm laborers. Explain that, in most cases, the different groups that make up the region's population came there originally because they saw it as a land of opportunity. To help students learn more about the different waves of immigrants to the region, have them prepare bulletin board displays about each group. Start by writing a list on the board of the main culture groups that live in the region, including Mexicans, Chinese, Japanese, Polynesians, Africans, Europeans, Filipinos, Aleut, Inuit, and any other groups you think students would like to research.

Organize the class into groups, and assign each group a different culture group to research. Students should perform library research, write short essays, collect pictures, draw illustrations, and collect poems and other works by people in the assigned culture group, and they should assemble the collection into a presentation.

CHAPTER 15
THE PACIFIC STATES

Section 1: California
Section 2: Oregon and Washington
Section 3: Alaska
Section 4: Hawaii

OVERVIEW

In this chapter, students learn about the Pacific States. Over the years, the Pacific states' beauty and natural resources have attracted many newcomers. Now these states face the challenge of meeting the economic needs of a diverse and growing population while protecting the environment.

Section 1 describes California, the nation's leading state in agriculture and industry.

Section 2 highlights Oregon and Washington, the core of the Pacific Northwest region.

Section 3 focuses on Alaska, the largest state in the Union.

Section 4 describes the land, economy, and population of Hawaii.

About the Author

John Muir (1838–1914) was a famous American naturalist. His travel journals and diaries describing the western United States and Alaska have been widely read. His influence led President Theodore Roosevelt to establish national forest preserves.

Primary Source

Description of Change: excerpted
Rationale: excerpted to focus on main idea

CHAPTER 15

The Pacific States

❝The last of the Coast Range foothills were in near view. . . . Their union with the valley is by curves and slopes of inimitable beauty. They were robed with the greenest grass and richest light I ever beheld, and were colored and shaded with myriads of flowers of every hue, chiefly of purple and golden yellow. ❞

John Muir

At first glance, you might wonder why the Pacific states of California, Oregon, Washington, Alaska, and Hawaii are grouped together as a region. After all, Alaska boasts the record low temperature for the entire United States, while Hawaii is tropical and the only island state. California is home to world-famous tourist attractions such as Hollywood and the Golden Gate Bridge. Washington and Oregon delight residents and visitors with rugged mountains, wild and scenic rivers, craggy coastlines, and green valleys.

Upon closer look, however, the similarities among the five states become more apparent. Physical geographers would note that these states share an active geological environment characterized by mountains, volcanoes, and earthquakes. Cultural, political, and economic geographers might point out the increasing trade, investment, and immigration that connect these states with the Pacific Rim countries.

Furthermore, the Pacific states face a common challenge. They must preserve and protect the fragile wilderness areas, fertile agricultural lands, and valuable natural resources that first attracted people to the region.

GEOGRAPHY DICTIONARY

agribusiness
multicultural region
caldera
panhandle
hot spot

Wheel from a wagon that carried settlers to **California**

The Pacific Coast Highway north of **Half Moon Bay, California**

Using the Quotation

Read aloud the quotation by John Muir. Tell students to locate the landform region mentioned in the quote (Coast Ranges) on the map on this page. Point out that although Muir describes a particular place in California, the entire Pacific region is known for its scenic beauty.

Using the Geography Dictionary

Pronounce the words in the Geography Dictionary. Have students look up the words in the Glossary. Then ask students to identify the three words associated with physical geography. (*caldera, panhandle, hot spot*) Have them connect the remaining words with other branches of geography. (*agribusiness*—*economic, **multicultural region***—*cultural*)

SECTION 1
OBJECTIVES

The objectives reflect the Focus questions on this page and are assessed in the review on page 161.

1. Describe major landform and climate regions of California.

2. Analyze the economic geography and urban geography of California.

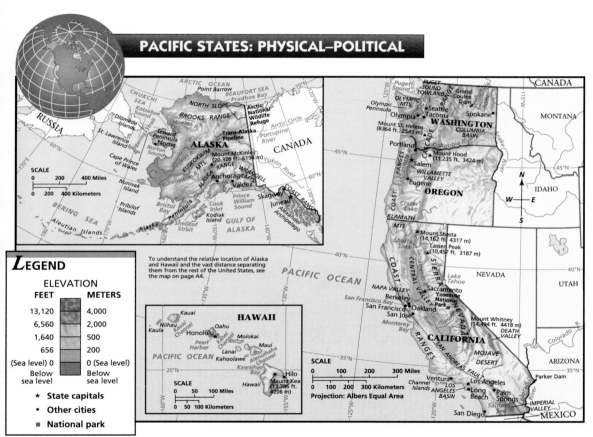

PACIFIC STATES: PHYSICAL–POLITICAL

LEGEND

ELEVATION

FEET	METERS
13,120	4,000
6,560	2,000
1,640	500
656	200
(Sea level) 0	0 (Sea level)
Below sea level	Below sea level

★ State capitals
• Other cities
■ National park

Using the Physical-Political Map: Pacific States

Tell students to examine the map on this page. Call on them to identify some obvious physical characteristics and features of the states they will study in Chapter 15. *(For example, they should notice the range of the San Andreas Fault in California, notice that Hawaii is comprised of many islands, and point out the mountain ranges running parallel to the Pacific coast in California, Oregon, and Washington.)*

CALIFORNIA

FOCUS

● *What are the major landform and climate regions of California?*

● *Where are the three major urban areas of California located?*

REGION **Landforms** California can be divided into four major landform areas: the Coast Ranges, the Sierra Nevada, the Central Valley, and the desert basins and ranges. The Coast Ranges form a rugged coastline along the Pacific. They are made up of parallel mountain ranges and are separated by earthquake-fault systems. There are a few lowland areas between the mountains and the Pacific, the largest of which is the Los Angeles Basin.

The Sierra Nevada range lies inland and east of the Coast Ranges and is one of the longest and

highest mountain ranges in the United States. Mount Whitney, reaching an elevation of 14,494 feet (4,418 m), is the highest peak in the 48 contiguous states.

Between the Sierra Nevada and the Coast Ranges is a narrow plain known as the Central Valley, one of the richest agricultural regions in the United States. Stretching more than 400 miles (644 km), the Central Valley is irrigated by the many rivers that flow from the Sierra Nevada.

To the east of the Sierra Nevada are desert basins and mountain ranges separated by fault zones. At 282 feet (86 m) below sea level, Death Valley, California, is the lowest point in all of North America.

Earthquake activity is common in California due to the geologic structure of the Pacific States region. The San Andreas Fault system, formed where the Pacific and North American plates meet, stretches from offshore north of San Francisco to the southeast through southern California. The Pacific plate is slowly moving northward along the San Andreas Fault, past the

Motivator

Play a recording of a California song, such as "California, Here I Come" or "California Dreaming." Note that the motto of westward pioneers was "California or Bust." Ask students why "the Golden State" has always appealed to so many people. *(People are attracted by the state's warm climate, beautiful scenery, and natural resources.)*

Building a Vocabulary

To help students retain the meanings of **agribusiness** and **multicultural region**, note that *agri-* or *agro-* refers to agriculture, and *multi-* means "many". Have students work in pairs to list other words with these word parts. *(agronomy, agrochemical, multiply, multitude, multimillionaire, multidimensional, multitrack)*

Synthesizing Information

California's Geography Point out that many television game shows are produced in Los Angeles. Divide students into groups, and have them devise a list of questions about California's physical, economic, and urban geography that might be used to quiz contestants on a game show. Have groups exchange lists and answer the questions.

Chart Caption Answer

Alaska, California

Geography and Economics

In 1993, California's economy accounted for 13 percent of the U.S. gross domestic product. This makes California the world's eighth largest economy. Many analysts say that the economic well-being of the nation as a whole depends to a large extent on the economic well-being of California. For example, when the nation experienced an economic recession in the early 1990s, its recovery hinged largely on California's ability to turn around its own economic downturn. California alone accounted for 38 percent of all U.S. job losses between mid-1990 and the end of 1992.

160

FACTS IN BRIEF

The Pacific States of the United States

STATE POPULATION (1993)	POP. RANK (1993)	AREA (1990)	AREA RANK (1990)	PER CAPITA INCOME (1993)
ALASKA 599,000	48	615,230 sq. mi. 1,593,446 sq. km	1	$22,846
CALIFORNIA 31,211,000	1	158,869 sq. mi. 411,471 sq. km	3	$21,821
HAWAII 1,172,000	40	6,459 sq. mi. 16,729 sq. km	47	$23,354
OREGON 3,032,000	29	97,093 sq. mi. 251,471 sq. km	10	$19,443
WASHINGTON 5,255,000	15	70,637 sq. mi. 182,950 sq. km	20	$21,887

Rankings are for all 50 states.
Source: *Statistical Abstract of the United States 1994*

Which of the Pacific states has the largest area of all 50 states? Which state has the largest population of all 50 states?

North American plate. The shock waves caused by moving rock along the fault line can create severe earthquakes, such as the 1994 Los Angeles earthquake. Scientists warn that future severe earthquakes threaten California.

Climate A marine-west-coast climate covers the northern coast of the state, while a Mediterranean climate dominates most of coastal southern and central California. Temperatures are warm all year, even in winter. Dry, hot summer winds heighten the chance of dangerous brushfires.

The western slopes of the Sierra Nevada act as the main watershed for California. The mountains store snow and rainwater in snowfields, in lakes, and behind dams in reservoirs. The water is then distributed to farmlands and cities below.

The Sierra Nevada range blocks Pacific moisture from reaching the basins of eastern California.

The San Andreas Fault scars the **California** landscape between Los Angeles and San Francisco.

Desert and steppe climates exist here. The hottest temperature and lowest rainfall records in North America have been recorded in Death Valley, where summer temperatures frequently soar to more than 120°F (49°C) and annual rainfall averages less than two inches (5 cm).

Agriculture California is the nation's leading agricultural producer. The southern and central parts of the state have a year-round growing season. This enables California to be the major supplier of fresh vegetables and fruits to the rest of the nation throughout the winter. Most farming in California takes place on large modern farms, many of them owned by big corporations. This is called **agribusiness**.

Because of the dry climate, California's agriculture can be productive only with the help of irrigation, which uses 85 percent of California's water supply. The greatest variety of crops in North America, including nut trees and fruit trees, is grown in the flat, irrigated Central Valley. The southern portion of the Central Valley—the San Joaquin Valley—is one of the world's major cotton-growing areas.

The valleys of coastal California grow specialty crops such as artichokes and garlic, while the Napa Valley near San Francisco is world famous for its vineyards. The valleys and basins of southern California supply many citrus fruits, such as oranges, lemons, and limes.

INTERACTION **Resources and Industry**

California is rich in mineral resources. When gold was discovered in the Sacramento Valley in 1849, a gold rush brought thousands of fortune seekers to California. Today, however, California's most valuable mines are those that provide the sand and gravel needed in construction.

California's oil deposits lie offshore on the state's continental shelf. The future development of these resources is a source of controversy. An oil spill could be disastrous to the area's fishing industry, tourism, and rich ocean life. On the other hand, future oil production would benefit the state's economy.

California's most precious natural resource is water. Aqueducts transport water hundreds of miles from the Colorado River and the rivers of northern California and the Sierra Nevada to meet the needs of dry southern and central California. Overuse of underground water for irrigation from

Guided Practice

On the board, begin an outline titled "California's Geography." Write three headings under the title: Landforms, Climate, Urban Areas. Have volunteers take turns at the board writing in appropriate information under each heading until the outline has been correctly filled in. Then direct students to copy the information on a sheet of paper and to use it as a study aid.

Independent Practice

Have students list various occupations that Californians might have, based on the state's agriculture, resources, and industries.

Closure

Have students explain why they would or would not like to live in present-day California.

ASSESSMENT

Assign the Section 1 Review questions on this page.

Reteaching

Assign Section 1 of Reteaching Worksheet 15.

The Golden Gate Bridge links **San Francisco** with communities to the north. The San Francisco–Oakland Bay Bridge can be seen in the background. Bridges, ferries, and light rail aid travel and commerce between the many communities of California's second largest urban area.

wells in the Central Valley has lowered the water table. Droughts also strain the state's water supply.

California is the nation's leading industrial state. Its industries include aerospace, construction, entertainment, computers and microelectronics, food processing, and tourism. The state has many offshore fisheries, and a major lumber industry is found in the rainy northwest corner. Foreign trade, especially with the Pacific Rim nations, plays a vital role in California's economy. (See "Global Connections" on pages 166–167.)

Urban Geography More than 90 percent of Californians live in urban areas and are concentrated in three coastal areas: the Los Angeles Basin, the San Francisco Bay Area, and San Diego County.

The Los Angeles metropolitan area, the second most populated U.S. urban area, is considered the nation's most **multicultural region**. A multicultural region is an area in which people of many cultures have settled. Los Angeles is best known for its role in the entertainment industry, which developed in a section of the city called Hollywood. Today, Los Angeles is the largest manufacturing center in the nation, focusing on aerospace, electronics, furniture, and clothing industries. (See "Cities of the World" on pages 162–163.)

The San Francisco Bay Area is the second largest urban area in California. The city is located on a peninsula that separates San Francisco Bay from the Pacific Ocean and has one of the world's finest natural harbors. Tourism and banking provide most of the city's income.

The San Francisco urban area has expanded to include cities such as Oakland and Berkeley in

the East Bay. South of San Francisco are San Jose and other cities of the Silicon Valley, named for the silicon material used in the production of computer chips. It is the world's largest center for computer firms.

San Diego is California's third largest urban area and one of the nation's fastest-growing cities. It lies just north of the border with Mexico. Like many other cities in the southwestern United States, the rich cultural heritage of Spain and Mexico is reflected in San Diego's architecture, art, and music. San Diego's major industries are centered around its large, natural harbor.

Issues Most of the population growth in California from 1940 to 1970 was caused by people moving from other states. Today, much of the growth is a result of immigration from Middle and South America and Asia. In the Los Angeles area, nearly 100 languages are spoken, and Spanish is the major language in many parts of the state. By the year 2000, there will be no ethnic majority group.

While greater cultural diversity has enriched California in many ways, rapid growth and economic and urban changes have created problems for all residents. High housing costs, loss of farmland, water shortages, environmental hazards such as earthquakes and fires, increased crime, traffic congestion, and air pollution are some of the challenges facing Californians.

 SECTION REVIEW

1. What are the major landform and climate regions of California?
2. What are California's three major urban areas?
3. **Critical Thinking** **Suggest** three ways that individual Californians might conserve water and/or energy.

Text continues on page 164.

Geography and the Environment

California is trying out a new way to resolve disputes between environmentalists and developers. For instance, state and local governments, developers, and environmentalists have agreed to set aside certain areas as preserves for an endangered bird species, while allowing development on the rest of the bird's habitat. It is hoped that this project will protect as many as 50 other plant and animal species.

Section 1 Review
ANSWERS

1. Landform regions: Coast Ranges, Sierra Nevada, Central Valley, desert basins and ranges; climate regions: marine west coast, Mediterranean, desert, and steppe
2. The Los Angeles Basin, the San Francisco Bay Area, and San Diego County
3. Possible answers include installing water-saving devices on faucets, keeping and reusing water (such as for watering gardens), and carpooling or taking mass transportation to work or school.

161

Cities of the World

Researching and Sequencing

Have students work in groups of three or four to research and report on the growth and development of Los Angeles, from its beginnings as a Spanish settlement to the present. Instruct half the groups to present their findings in the form of an illustrated time line and the remaining groups in the form of an illustrated chart. Display the time lines and charts in the classroom.

CITIES OF THE WORLD

Los Angeles

The mild climate and Pacific trade links of Los Angeles continue to attract thousands of newcomers each year.

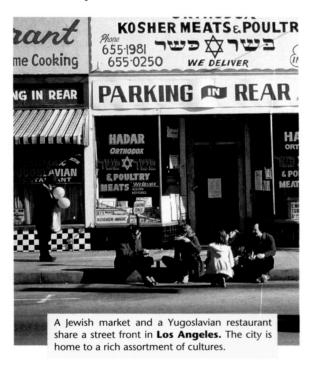

A Jewish market and a Yugoslavian restaurant share a street front in **Los Angeles.** The city is home to a rich assortment of cultures.

Los Angeles was founded in 1781 as El Pueblo de Nuestra Señora la Reina de los Angeles (The Village of Our Lady the Queen of the Angels). This Spanish settlement along the Los Angeles River rested in a fertile basin rimmed by mountains. Belonging first to Spain and later to Mexico, the pueblo came to be called Los Angeles. Even today, Los Angeles is often referred to as the "City of Angels."

The Los Angeles of today bears little resemblance to that original settlement along the river, which is now dry much of the year.

The Los Angeles Basin encompasses 34,000 square miles (88,060 sq. km) and includes Los Angeles, Orange, Riverside, San Bernadino, and Ventura counties. These five counties are home to more than 14 million people. About 3.5 million live in the city of Los Angeles itself.

Los Angeles's pattern of growth has differed from that of many other cities. Until 1956, no buildings could be higher than 140 feet (43 m). As a growing population pushed out beyond the city limits to find more land, manufacturing and business interests followed, creating new economic centers. As a result, the greater Los Angeles urban area has not one node, but a number of them. The city has been described as "a series of constellations" that form the "metropolitan galaxy."

Cultural and Economic Diversity

Los Angeles has the nation's most culturally diverse population. The area is home to more people of Mexican descent than any other metropolitan area except Mexico City. It has the nation's largest concentration of Asians, many of whom have their roots in Japan, China, Korea, the Philippines, Vietnam, Cambodia, and Thailand. African Americans make up around 13 percent of the population.

Los Angeles has become a leading manufacturing area and a center of retailing, finance, trade, and foreign investment. Los Angeles–Long Beach is now the nation's busiest port and a gateway of trade to the Pacific Rim nations. To many people, however, Los Angeles is most noted as the center of the nation's large entertainment industry.

Growing Pains

Like most of the world's major cities, Los Angeles and its infrastructure have suffered growing pains. The region's huge network of highways can no longer accommodate such a sprawling, densely populated area. Commuters, many of whom must travel as much as 70 miles (113 km) or more to work, face miles of traffic jams on freeways that rank as the nation's most congested.

Have students write a brief paragraph explaining which of the Five Themes of Geography best relates to the founding of Los Angeles. *(region— it was a culturally homogeneous Spanish settlement)* Also have students state which themes best relate to Los Angeles today. *(movement— freeway system, commuters, immigration; human-environment interaction—smog created by automobile exhaust, rising population,* *depletion of water supply)* Have them explain their answers. You might have volunteers read their paragraphs aloud.

Have students create a travel poster or a collage about Los Angeles, using the information presented in this feature. Students might draw their own illustrations or use photographs from old magazines. Posters and collages should include details that relate to the physical, cultural, and economic geography of Los Angeles. Display students' completed work.

Geography and the Environment

A new program to reduce Los Angeles smog was approved in 1993. The plan, called the Regional Clean Air Incentive Market (RECLAIM), affects 390 of the Los Angeles area's biggest pollution-producing companies by setting limits on the amount of pollutants they can emit. Companies that reduce emissions more than the required amount can sell "pollution credits" to companies having more difficulty in cutting down on their pollutants. Supporters believe this financial incentive will make the plan effective and could reduce the emissions of some pollutants by more than 50 percent by 2003.

The sprawling **Los Angeles** metropolitan area is located in the most urbanized state in the United States. This map shows only some of the many cities that crowd the Los Angeles Basin.

Los Angeles remains one of the nation's smoggiest cities.

Rapid population growth also has strained the city's water supply. Most of the water for this semiarid region is piped in from northern California and from the Colorado River. Long periods of drought in recent decades, coupled with a population growth of 17.5 percent between 1980 and 1990, have resulted in critical water shortages. Overburdening of the city's waste disposal system has resulted in raw waste being dumped into Santa Monica Bay.

These clogged highways shoulder much of the blame for the city's air pollution. Because the Los Angeles area lies in a basin, pollutants from automobile exhaust are trapped by the surrounding mountains, creating smog. Despite the passage of laws aimed at improving air quality,

Los Angeles is working hard to solve its growth-related problems. In fact, the city's growing diversity may be its greatest resource. As one writer and teacher in Los Angeles notes, "If L.A. is to have a future, all kinds of people need to make linkages to the inner city. We need to know that this city is full of hard-working, concerned people of all races."

Geographical Sidelight

A dry climate and the hot Santa Ana winds that blow from the inland deserts make the Los Angeles area particularly susceptible to brushfires. The strong winds can spread fires quickly and allow flames to jump fire lines. Fires in the Los Angeles area in the fall of 1993 burned some 200,000 acres (81,000 ha) and destroyed more than 1,000 homes, including those of some celebrities.

Commuters travel along a reconstructed part of the Santa Monica Freeway, one of the **Los Angeles** area's lifelines for millions of drivers. The freeway was reopened just three months after an earthquake killed more than 60 people and caused billions of dollars in damage in 1994.

OBJECTIVES

These objectives reflect the Focus questions on this page and are assessed in the review on page 165.

1. Describe major landform and climate regions of Oregon and Washington.
2. Identify the importance of natural resources to Oregon and Washington.

TEACHING SECTION 2

Motivator

Have students review the location and landforms of Oregon and Washington on the map on page 159, and have them suggest economic activities they might expect to find.

Building a Vocabulary

On the board, draw a "before" and "after" picture to illustrate **caldera** (*a mountain, then a depression*). Point out the connection between this word and the word *caldron*. (a large kettle)

Illustrating Ideas

Physical Geography   Ask students to recall the meaning of *orographic effect* (Chapter 3). Ask them to apply this concept to the Cascades in Washington and Oregon. Have them draw a sketch of the Cascades showing how the mountains divide these states into two climate regions. Discuss how climate affects economic activity in each area.

GLOBAL PERSPECTIVES

Note that Washington and Oregon do not favor protectionist trade policies. The economies of these states depend on open trade agreements and friendly relations with Pacific nations, especially Japan. From 1980 to 1990, exports from Washington grew by 166 percent while exports nationwide grew by only 75 percent. Ask students to suggest possible reasons for this growth.

THEME
Region

A new region based on industry rather than political boundaries is developing in the Northwest. Referred to as Cascadia, the region includes Washington, Oregon, Montana, Idaho, and Alaska and the Canadian provinces of Alberta and British Columbia.

2 OREGON AND WASHINGTON

FOCUS

- *What climate type makes Oregon and Washington suitable for agriculture?*
- *What are the major resources in Oregon and Washington, and what issues surround their development?*

REGION **Landforms**   Oregon and Washington form the core of the Pacific Northwest region. Four landform regions dominate: the Coast Ranges, the Puget Sound and Willamette Valley lowlands, the Cascades, and the Columbia Basin. The Coast Ranges form the spectacular Pacific Northwest coastline. (See the map on page 159.)

Between the Coast Ranges and the Cascade Range are two lowland regions. The Puget Sound Lowland is in Washington, and the Willamette Valley lies in Oregon. These rich farmlands contain most of the population of each state.

The Cascades are a volcanic mountain range stretching across Washington and Oregon into northern California. The range includes Oregon's Crater Lake, the deepest lake in the United States. Crater Lake fills a huge **caldera**, a large depression formed after a major eruption and collapse of a volcanic mountain. Because of their location on the Pacific Ring of Fire, western Washington and

Crater Lake is what remains of a volcanic mountain that literally blew its top more than 6,000 years ago. The deep and scenic **Oregon** lake is a popular tourist destination.

Oregon experience earthquakes and volcanic eruptions.

To the east of the Cascades is a region of dry basins and mountains. Much of this area is known as the Columbia Basin and is drained by the Columbia River, the largest river flowing into the Pacific Ocean. The Columbia River is the nation's largest producer of hydroelectricity and provides irrigation waters for a rich agricultural region in eastern Washington.

Climate   The Cascades divide Washington and Oregon into two climate regions. To the west, or windward side, of the Cascades is a marine-west-coast climate. Westerly winds off the Pacific Ocean yield mild temperatures year-round, with cloudy, rainy winters and warm, sunny summers.

The moisture that the westerlies bring from Pacific storms falls on the western slopes of the mountains, supporting dense, coniferous forests. In fact, the Olympic Mountains in Washington receive around 150 inches (381 cm) of rain per year and support a temperate rain forest.

To the east, or leeward side, of the Cascades, it is dry, with desert and steppe climates. Most of the moisture from Pacific storms is blocked, and grassland and shrubs dominate.

Agriculture   Farming is important to the economies of Washington and Oregon. Dairy, fruit, and vegetable farming is concentrated in the Willamette Valley in Oregon and in the Puget Sound Lowland in Washington.

Though Washington is known for its apples and vineyards, the state grows a variety of crops. One of the nation's most productive agricultural regions is located in dry eastern Washington. The rich, irrigated farmlands of the Columbia Basin produce fruits and vegetables with help from immigrant labor, mainly from Mexico.

Resources and Industry   Fish and forests are two of the most important resources in Washington and Oregon. These resources supported numerous groups of Native Americans for thousands of years. Today, fishing fleets catch salmon, halibut, and herring. Concern for spawning salmon has grown with the construction of dams

Provide students with a list of land-forms, climates, places, agricultural products, resources, industrial products, and issues. Have them indicate whether each item refers to Washington, to Oregon, or to both states. For each item, also have them write a sentence based on information in Section 2 that supports their answer.

Closure

Have students work with partners to answers the Focus questions on page 164.

ASSESSMENT

Assign the Section 2 Review questions on this page.

Reteaching

Assign Section 2 of Reteaching Worksheet 15.

SECTION 3
OBJECTIVES

The objectives reflect the Focus questions on this page and are assessed in the review on page 169.

1. Identify Alaska's resources.
2. Describe the cultural geography of Alaska.

Clear-cutting has left irregular patterns in the Olympic National Forest of **Washington** state. Some of the largest tree species in the world can be found in the old-growth forests of the Pacific Northwest. Even under the best conditions, it would take centuries for these trees to be replaced by other trees as large.

on the Columbia and other rivers. The dams prevent some species of salmon from returning to their upstream spawning grounds, greatly decreasing their numbers.

The forests of Oregon and Washington support a major lumber industry. Tree farms and clear-cuts mark the mountains of the two states. Clear-cuts are areas where all the trees have been cut down. Most of the original forests have been cut, and lumbering is now on second-growth timber. These forests also provide erosion protection, water storage, a home for wildlife, and recreation sites. A major issue facing the region is conflict between the lumber industry and those who support environmental protection.

Urban Geography Seattle, Washington, is the largest city in the Pacific Northwest. Seattle is the home of the world's largest aircraft producer and a computer products industry. The Puget Sound urban area is one of the fastest-growing regions in the United States. With their Pacific access to Asia, the Puget Sound ports of Seattle and Tacoma are among the nation's busiest.

Half of Oregon's population lives in the urban area of Portland, the state's largest city. Portland began as a trading post in 1845 at the junction of the Columbia and Willamette rivers. Today, it is an important seaport and an industrial, computer, and food-processing center.

Issues Oregon and Washington are growing at a rapid pace. The region's residents face growth problems similar to those in California. Rapid urbanization here will result in the loss of valuable farmland and forest areas. Other environmental issues, especially those dealing with the use of forest, fish, and water resources, also challenge the Pacific Northwest.

2 SECTION REVIEW

1. Where are the most agriculturally productive areas in Oregon and Washington?
2. What are the pros and cons of resource development in this region?
3. **Critical Thinking** **Propose** a compromise solution to the conflict between the lumber industry and environmental groups.

3 ALASKA

FOCUS

● *What are Alaska's important resources?*

● *Who were the native groups living in Alaska before Europeans arrived?*

LOCATION **Landforms** Alaska, covering an area one-fifth the size of the rest of the United States, is the largest state. Alaska occupies a huge peninsula connected to North America on the state's eastern border with Canada. The Aleutian Islands extend to the southwest, forming an arc of volcanic islands. Southeast Alaska includes hundreds of offshore islands and a narrow, rugged strip of land along the Canadian border, called the Alaskan **panhandle**. A panhandle is a narrow arm of land attached to a larger region. (See the map on page 159.)

The mountain ranges of Alaska have some of the most rugged mountains in the world. The highest peaks are in the Alaska Range, located in the south central part of the state. Mount McKinley, also called Denali (duh-NAHL-ee), reaches a height of 20,320 feet (6,194 m). The Coast Mountains form the border between the Alaskan panhandle and Canada, while the Brooks Range runs across northern Alaska. Alaska has

Text continues on page 168.

Section 2 Review
ANSWERS

1. The Willamette Valley, the Puget Sound Lowland, eastern Washington, and the irrigated Columbia Basin

2. Pro: resource development might help the economy. Con: rapid growth and urbanization could result in loss of farmland and forest. Logging and the damming of rivers may endanger some species, increase erosion, and take away recreation sites.

3. Answers will vary. One possible compromise might involve setting aside some areas for environmental protection and allowing logging in the remaining areas.

Global Connections

Characterizing Regions

Use a globe to contrast the dimensions of the Pacific and Atlantic oceans. Point out that the Pacific stretches across an entire hemisphere; turn the globe to illustrate this concept. Then explain that the terms *Pacific Rim* and *Pacific Basin* often are used together. On a wall map or globe, indicate that Pacific Rim nations are located along the outside borders of the Pacific Ocean, whereas *Pacific Basin* refers to the islands located inside its borders. You might illustrate these terms by using a large bowl. Indicate the location of Pacific Rim nations on the bowl's rim and islands of the Pacific Basin inside the bowl. Have volunteers use a dictionary, and write different definitions of *rim* and *basin* on the board. Call on students to choose the definitions that best apply to the geographic characteristics of the Pacific Rim and the Pacific Basin.

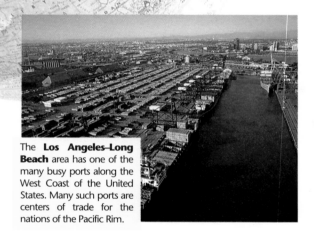

GLOBAL CONNECTIONS

The Pacific Rim

The **Los Angeles–Long Beach** area has one of the many busy ports along the West Coast of the United States. Many such ports are centers of trade for the nations of the Pacific Rim.

Historical, cultural, and economic ties bind the peoples of the Pacific Rim together across thousands of miles.

In 1869, at Promontory Point, Utah, workers drove the last spike into the transcontinental railroad, linking the American West with the rest of the United States. Soon a vast network of rails crisscrossed the nation. Trains moving eastward carried lumber from the forests of Oregon and Washington and agricultural products from the fertile valleys of California. Traveling westward, the trains brought manufactured goods from the industrial centers of the East.

One California entrepreneur, however, believed that future opportunities for trade lay in a different direction. Henry Huntington, a powerful railroad owner, did not look to overland routes east. Instead, he looked out at the great ocean bordering California. "Los Angeles," Huntington said in 1912, "is destined to become the most important city in the country, if not in the world. Its front door opens on to the Pacific, the ocean of the future."

Ocean Trade

Today, Huntington's "ocean of the future" is the bustling seaway of trade he predicted. The nations and U.S. states bordering the Pacific Ocean make up a region called the Pacific Rim. The relationship among these nations and states can no longer be described only in geographic terms. Their location on the Pacific has bound them economically as well.

Indeed, trade among the Pacific Rim nations has surged. Television sets, VCRs, stereos, clothing, and automobiles travel from Asian nations across the Pacific to eager customers. In turn, airplanes crafted in Seattle, timber cut in the forests of the Pacific states, and computer parts made in California regularly arrive in Asian ports. Agricultural products also are an important part of Pacific trade. For example, New Zealand exports kiwi fruit, and Chile exports grapes. Japanese shoppers buy California oranges when they go to market.

A Tapestry of Peoples

Just as trade has forged economic links between the Pacific states and Asia, immigration has woven cultural ties. During the twentieth century, Asians have crossed the Pacific in record numbers, hoping to find a better life and greater opportunities. Immigrants from China, Japan, South Korea, Vietnam, Thailand, and the Philippines all make their homes in the Pacific states.

As a result, the region is a tapestry of cultural diversity. Seattle and Portland now have large Asian American populations. Los Angeles and San Francisco have "Little Saigons" and "Little Manilas" as well as large "Chinatowns" and "Japantowns."

Cultural and economic exchange also characterizes the nations of the Pacific Rim that lie south of the United States. The number of immigrants flowing north from Mexico, Central America, and South America has increased dramatically since the 1960s. Trade also has increased. California, for example, exports a variety of goods to Mexico. Mexico, in turn, ships

Using Geography Themes

To illustrate the theme of movement, ask students to examine products in their own homes (clothing, food, appliances) to determine which ones were imported from Pacific Rim countries. On the board, create a master list from the examples cited by the students. Ask them to categorize the types of products.

Mapping the Pacific Rim

Provide students with Outline Map 24, "The Pacific Rim." Have them label the countries, states, and cities identified in the Global Connections feature. Then have students draw connecting lines from these locations to the destinations of their trade products mentioned in the text.

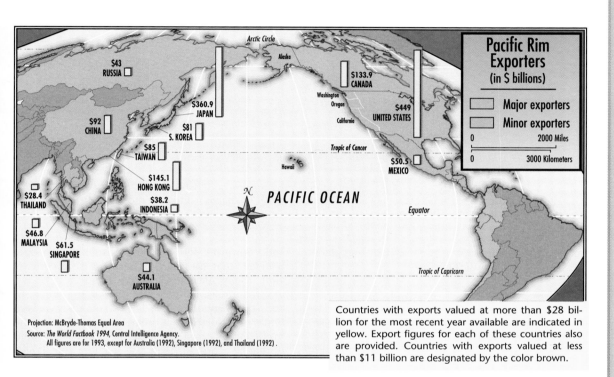

Countries with exports valued at more than $28 billion for the most recent year available are indicated in yellow. Export figures for each of these countries also are provided. Countries with exports valued at less than $11 billion are designated by the color brown.

GLOBAL PERSPECTIVES

Many of the Pacific Rim nations belong to the Asia Pacific Economic Cooperation (APEC) council, a growing organization oriented toward free trade. APEC formed in 1988 and by 1993 accounted for more than one-third of the world's trade. Member nations include Australia, Brunei, Canada, China, Hong Kong, Indonesia, Japan, Malaysia, Mexico, New Zealand, Papua New Guinea, the Philippines, Singapore, South Korea, Taiwan, Thailand, and the United States. Together, these 17 countries have more than 2 billion people and a combined GDP of $12 trillion—half the world's total.

agricultural produce and other goods to the United States. Political and economic problems within Middle and South America, however, have kept these nations from sharing many of the benefits of Pacific Rim trade.

On the Horizon

The exchange of goods and cultures taking place in the Pacific Rim may increase as China, Mexico, and other countries begin to play a more important economic role in the region.

Ties among the Pacific Rim states and nations could weaken, however, if Asian nations join together to increase trade mainly among themselves. The Pacific states, too, may concentrate on expanding trade with other nations in the Americas. Already, the United States, Mexico, and Canada have sought to lower trade barriers and join together to form a vast free-trade zone.

Regardless of what nations decide to do, the Pacific Rim will remain an important economic and cultural region. The links of trade, investment, immigration, and culture that have been forged will not be broken easily.

These Vietnamese-Americans prepare to celebrate Tet, the Vietnamese New Year festival, in **San Francisco**. The Pacific states are home to many immigrants from Pacific Rim countries. Vietnamese immigration to the United States greatly increased during the 1970s, especially after the end of the Vietnam War in 1975.

▼YOUR TURN

Imagine you live in one of the states or nations of the Pacific Rim. **Write** a letter to a friend living in the eastern United States. **Explain** how location and movement tie the Pacific Rim states and nations together as a region. **Describe** a few daily activities in your region that your friend might find interesting.

Your Turn

Letters will vary. Responses should emphasize the movement of trade products and people among Pacific Rim nations and should illustrate the cultural diversity within the Pacific Rim.

167

Ask students to draw a picture of something they associate with Alaska. Then have them share their drawings. Note that many people think of snow and ice when they think of Alaska. Point out that such images represent only one part of this state. Have students identify additional images of Alaska as they study Section 3.

Mapping Economic Resources

Students will create their own economic resources map of Alaska. First, have them use the map on page 159 as a reference for drawing a large freehand map of the state. Then tell them to include on their maps the locations mentioned in the "Resources" subsection on page 168. Instruct them to draw in the Trans-Alaska Pipeline System and symbols for oil, forests, and fish. They can consult the economy map on page 110 for help.

Guided Practice

Use a wall map to show where the following groups settled in Alaska prior to the arrival of Europeans: fishers and nomadic hunters (subarctic interior), Aleuts (Aleutian Islands, Alaskan Peninsula), Inuit (near Bering Sea and Arctic Ocean).

Geographical Sidelight

In 1978, stone tools and lance points related to hunting were discovered on a mesa in northern Alaska. Archaeologists have now determined that these artifacts are 11,700 years old. Ask students what artifacts of contemporary American culture might be discovered thousands of years from now.

Mount McKinley, the highest mountain in North America, dominates the landscape of Denali National Park in **Alaska**. Native Americans named the mountain *Denali*, which means "The Great One."

more than 3 million lakes, and ice fields and glaciers cover 4 percent of the state.

Climate The mountain ranges of Alaska divide the state into three major climate regions: marine west coast, subarctic, and tundra. A marine-west-coast climate with mild temperatures is found mainly along coastal southeast Alaska. Precipitation, mainly in the form of rain, reaches up to 200 inches (508 cm) per year here, supporting a dense temperate forest. The interior of Alaska has a subarctic climate. Summers here are short, while winters are long and severe. Most precipitation is in the form of snow and supports a boreal forest. The North Slope area along the Arctic Ocean coast has a tundra climate.

Resources In 1867, U.S. Secretary of State William H. Seward purchased Alaska from Russia for $7.2 million. Alaska was called "Seward's Folly"

The Trans-Alaska Pipeline snakes across mountain ranges and tundra. Completed in 1977, the pipeline has become a human addition to the environment of **Alaska**.

and "Seward's Ice Box" because many people believed that he had purchased a frozen wasteland. Today, Alaska is known for its wealth of natural resources. Oil generates about 85 percent of the state's income. The largest oil reserves are under the North Slope near Prudhoe Bay on the Arctic Ocean coast. Because the Arctic Ocean and Bering Sea are blocked by sea ice, the Trans-Alaska Pipeline System (TAPS) was built to transport the oil to the ice-free port of Valdez, 800 miles (1,287 km) south.

As Alaska's oil resources begin to decline, however, there is pressure to seek new sources in the Arctic National Wildlife Refuge (ANWR). Proposals to drill in the refuge have sparked debate between people who want increased resource development and those who see such development as a threat to the area's environment. In 1989, the nation's largest oil spill occurred when an oil tanker ran aground in Alaska's Prince William Sound. The spill brought into question the future development of the state's oil resources.

Alaska is the leading fishing state in the United States. Huge catches of cod, king crab, and salmon are made each year. Forestry also is important to the economy of southeast Alaska. Alaska's scenic forests, mountains, and glaciers have made tourism another important source of income for the state.

Human Geography Three major groups of people settled Alaska before the first Europeans arrived: fishers and nomadic herders, the Aleuts, and the Inuit, or Eskimo. Fishers lived along the southeastern coast, and nomadic hunters lived in the subarctic interior of Alaska. The Aleuts

Closure
Have students write a license-plate slogan for Alaska. *("Land of 3 Million Lakes"; "Last Frontier"; "49th State")*

ASSESSMENT

Assign the Section 3 Review questions on this page.

Reteaching
Assign Section 3 of Reteaching Worksheet 15.

SECTION 4
OBJECTIVES

These objectives reflect the Focus questions on this page and are assessed in the review on page 170.

1. Analyze the effect of the physical environment on ways of life in Hawaii.
2. Identify economic and environmental issues concerning Hawaii.

occupied the Alaskan Peninsula and the Aleutian Islands and were engaged mainly in seal hunting and fishing. The largest native Alaskan group was the Inuit (Eskimo). They lived along the tundra coasts of the Arctic Ocean and Bering Sea. Although this region has one of the most severe climates of any inhabited area in the world, the Inuit (Eskimo) were able to meet all their needs by hunting seals, walrus, and whales.

Today, native Alaskans account for about 15 percent of the state's population. During the settlement of Alaska, many native Alaskans lost their homelands. The Alaska Native Land Claims Settlement Act of 1971 helped them regain much of what they had lost. Native-owned corporations now earn income from oil, mining, and forestry on native lands.

The first Europeans to reach Alaska were Russian explorers. Russia ran a fur-trading colony in Alaska from 1743 until 1867, when Russia sold Alaska to the United States. Alaska's population grew after gold was discovered in the late 1800s. Since statehood in 1959, the oil industry has spurred more population growth. Today, more than a half million people live in the state, mostly in urban areas.

Issues The Alaskan economy suffers from a high cost of living and seasonal unemployment in forestry and fishing. With little agriculture and industry, most food and goods must be shipped in from other states.

The federal government owns more than 72 percent of the land, which is protected in national parks, wildlife refuges, national forests, and wilderness areas. Concern for the Alaskan wilderness is a consideration in any future exploitation of the state's rich resources.

SECTION REVIEW

1. Why are Alaska's oil reserves a source of controversy?
2. Identify the three native Alaskan groups. Where did each group live? How did each group survive?
3. **Critical Thinking   Decide** whether William H. Seward was wise or foolish to purchase Alaska for the United States in 1867. Explain your answer.

The setting sun casts a golden glow on the cliffs of Kalalau Valley on **Kauai**. Rugged cliffs such as these make road building impossible in some parts of this Hawaiian island.

HAWAII

FOCUS

● *How does Hawaii's physical environment affect its economy?*

● *What major economic and environmental issues does Hawaii face?*

LOCATION **Island Environment** Hawaii lies in the Pacific Ocean, about 2,500 miles (4,023 km) south of Alaska. The Hawaiian Islands are a chain of eight major islands and more than 100 smaller islands that stretch for over 1,500 miles (2,414 km). The largest and southernmost island is Hawaii. The other large islands include Maui, Kauai, Molokai, Lanai, and Oahu, the most populated island. (See the map on page 159.)

The islands are volcanic mountains rising above the Pacific, though the only island with active volcanoes is Hawaii. Hawaii sits on a volcanic **hot spot** in the Pacific plate. A hot spot is an area where molten material from the earth's interior rises through the crust. The larger islands are mountainous with flat land only along narrow coastal plains and in a few mountain valleys. The coast is rugged and scenic, with many coral reefs that erode into fine, white beach sand.

Section 3 Review
ANSWERS

1. Oil, while generating most of the income for the state, is also a source of controversy because of the chance of oil spills that might damage the state's environment.
2. Fishers and nomadic hunters: southeastern coast and in the subarctic interior, fished and hunted; Aleuts: Alaskan Peninsula and Aleutian Islands, hunted seals; Inuit (Eskimo): tundra coasts of Arctic Ocean and Bering Sea, hunted seals, walrus, and whales
3. Answers will vary. Some students might say that Seward was foolish because, as far as he knew, Alaska was a frozen wasteland. Other students might say that Seward was wise because Alaska has important resources.

169

Guided Practice

Point out ways in which Hawaii's landforms are both positive and negative for the state's economy. *(For example, volcanic mountains provide fertile soil for sugarcane and pineapples and attract tourists, but they limit land space for industry and a more diversified economy.)* Then have students complete the following sentence: "Hawaii's climate has a positive influence on its economy because…"

Independent Practice

Have students write a paragraph explaining how agricultural diversification may help Hawaii both economically and environmentally.

Closure

Ask students to sketch a map of the eight main Hawaiian islands, and then to write on the map a few captions about particular islands or about the state as a whole.

ASSESSMENT

Assign the Section 4 Review questions on this page.

Reteaching

Assign Section 4 of Reteaching Worksheet 15.

THEME
Interaction

In the early 1900s, Hawaiian sugarcane growers became concerned that rats were nibbling at sugar crops and thus at profits. To combat the rats, the growers introduced the mongoose to the islands. The mongooses, however, also preyed on other Hawaiian wildlife. One affected species was the nene, or Hawaiian goose (Hawaii's official state bird). In fact, the nene nearly became extinct. Although breeding efforts have increased the nene population, the bird's future remains uncertain.

Section 4 Review
ANSWERS

1. Fertile volcanic soils and the tropical climate are good for growing certain crops; the beauty of the physical landscape and the tropical climate attract tourists.
2. A high cost of living
3. Possible answer: hurricane damage might make Hawaii's natural areas less accessible or less attractive to tourists, thereby reducing Hawaii's major economic resource. Hurricanes might also destroy crops.

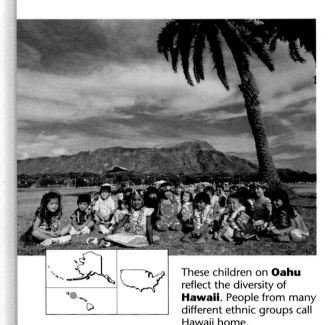

These children on **Oahu** reflect the diversity of **Hawaii**. People from many different ethnic groups call Hawaii home.

Climate Hawaii has little daily or seasonal temperature change. Honolulu's coldest month is January, which averages 72°F (22°C). August, the warmest month, averages 81°F (27°C). Only the high volcanic peaks have cool temperatures.

The northeast trade winds dominate the weather and cause heavy rainfall on the eastern, windward slopes of the islands. This produces a humid-tropical climate. Part of Kauai is considered the wettest place in the world. The western, leeward slopes of the islands generally receive less annual rainfall.

Economic Geography Hawaii's tropical climate and fertile volcanic soils provide excellent growing conditions for sugarcane, pineapples, and other tropical crops. Although less than 3 percent of the population is engaged in agriculture, it is important to the local economy. Tourism now dominates Hawaii's economy. Several million tourists, mostly from other states and Japan, visit the islands each year. More than 30 percent of all jobs are in the tourist industry.

Cultural Geography The original settlers of Hawaii were Polynesians from the South Pacific islands. They arrived as early as the fourth century,

forming chiefdoms on each of the major islands. The early European settlers on the islands were missionaries and whalers. They drastically changed life on the islands. Settlers and traders brought farm animals, plants, and manufactured goods. They also brought diseases that were fatal to many native Hawaiians.

During the nineteenth century, European settlers developed sugarcane and pineapple plantations. Laborers emigrated from China, Japan, the Philippines, Samoa, and Portugal to work these fields. Hawaii became a U.S. territory in 1900, and gained statehood in 1959. Since 1900, there has been a continuous flow of workers and settlers from the mainland United States and the Pacific Rim nations, especially Japan.

Today, Hawaii has more than 1 million people representing a great mixture of cultures. Asian Americans form more than 50 percent of the population. Of these, the greatest percentage is Japanese American, followed by Filipino American and Chinese American. European Americans are the second largest group, accounting for one-third of the total population. Native Hawaiians make up about 13 percent of the population. More than 80 percent of the population lives in and around the capital city of Honolulu on Oahu.

Issues The major concern for Hawaii is its dependence on materials and income from outside sources. Because the islands lack industry and a diversified agriculture, more than 80 percent of all materials, energy, and food must be imported. This economic situation keeps the cost of living in Hawaii high. Like the other states of the Pacific States region, Hawaii must find a balance between development and preservation of its natural environment.

SECTION REVIEW

1. How do Hawaii's landforms and climate affect its economy?
2. What problem results from Hawaii's dependence on outside resources for most of its income and materials?
3. **Critical Thinking** **Analyze** how severe hurricane damage might affect Hawaii's economy.

Building a Vocabulary

1. A geographic area in which people of many cultures have settled; Los Angeles, California
2. A depression formed after a major eruption and collapse of a volcanic mountain; Crater Lake
3. Possible answers: Oklahoma, Texas, Florida, Connecticut, New Mexico
4. Molten material from the earth's interior rising through the crust; Hawaii

Recalling and Reviewing

1. Marine-west-coast climate along the northern coast; Mediterranean climate in coastal southern and central California; desert and steppe climates in eastern California
2. Forests, fish
3. Oil, fish, forests, scenery
4. It is dependent on outside sources for materials and income.

CHAPTER **15** CHECK

Exit Glacier in Kenai Fjords National Park, **Alaska**

Reviewing the Main Ideas

1. The Pacific States region includes California, Oregon, Washington, Alaska, and Hawaii. Although landforms and climate throughout the states vary, the whole region experiences volcanic activity and earthquakes.
2. California is the most urbanized state, with three major urban areas—the Los Angeles Basin, the San Francisco Bay Area, and San Diego County. Oregon and Washington also have urbanized areas, particularly in the Portland and Puget Sound metropolitan areas.
3. Alaska is the largest state, and oil is its most important resource. Hawaii is tropical and is the only island state. Hawaii depends mostly on tourism for income.
4. The Pacific states must balance future development with environmental considerations.

Building a Vocabulary

1. What is a multicultural region? Give an example of a multicultural region from the chapter.
2. Define *caldera*. What lake discussed in this chapter fills the caldera of an extinct volcano?
3. Name two states other than Alaska that have panhandles.
4. What causes a hot spot? Where in the Pacific States region would you find a hot spot?

Recalling and Reviewing

1. What are the major climate regions of California?
2. What are the major resources in Oregon and Washington?

3. What are Alaska's important resources?
4. Why does Hawaii have a high cost of living?

Thinking Critically

1. Imagine that you live in California in the year 2010, when there will be no ethnic majority group. What new issues might this population shift create for California?
2. What do you think everyday life is like for many of the people who live in Alaska? Hawaii? Are there any similarities between the two states?

 ## Learning About Your Local Geography

Cooperative Project

With your group, use the library and other community resources to find out if there is a conflict in or near your area between resource development and environmental protection. Present your information to the rest of the class in an interview, debate, or news report.

Using the Five Themes of Geography

Sketch a map of the Pacific states. On your map, place some arrows to show **movement** into the various parts of the region. From what directions have people arrived into the region? Also, use symbols and/or shading to show some of the challenges to the region brought about by **human-environment interaction**. Make sure your map has a legend.

Harvesting grapes in the San Joaquin Valley, **California**

Thinking Critically

1. Possible answers include issues involving languages spoken and the ways people of different ethnic groups relate to each other.
2. Alaska's harsh climate may be a challange in daily life; Hawaii is usually warm. Both depend on other states for most food and goods, making the cost of living high. Both face development-versus-environment issues.

Learning About Your Local Geography

Projects will vary but should demonstrate careful research.

Using the Five Themes of Geography

Arrows should show the directions from which people have arrived and continue to arrive in each of the Pacific states. Symbols and shading might show oil spills, water shortages, air pollution, and wildlife and forestry concerns.

CHAPTER 16 Canada
Planning Guide

TEXTBOOK	TEACHING RESOURCES	OTHER AVAILABLE MATERIALS
Section		
1. Physical Geography Pages 173–174	Reteaching Worksheet 16.1	*Study Guide for Spanish Speakers* *Lecture Notes,* Chapter 16
2. Historical Geography Pages 175–176	Reteaching Worksheet 16.2 Skills Worksheet 16	*Study Guide for Spanish Speakers* *Lecture Notes,* Chapter 16
3. Canada's Provinces and Arctic Frontier Pages 176–179	Reteaching Worksheet 16.3 Graphic Organizer 16 Critical Thinking Worksheet 16 Challenge/Enrichment Worksheet 16	*The World Today* Videodisc Program *Study Guide for Spanish Speakers* *Lecture Notes,* Chapter 16
4. Canadian Issues Page 180	Reteaching Worksheet 16.4	*Study Guide for Spanish Speakers* *Lecture Notes,* Chapter 16
Chapter Review Page 181	Chapter 16 Test Chapter 16 Portfolio Assessment See also *Alternative Assessment Forms*	*Test Generator,* Chapter 16
Unit Review Pages 182–183	Unit 3 Test Unit 3 Portfolio Assessment See also *Alternative Assessment Forms*	*Test Generator,* Unit 3

CHAPTER RESOURCES

Lecture Notes, Chapter 16
Teaching Resources
   Chapter 16 Graphic Organizer
   Chapter 16 Skills Worksheet
   Chapter 16 Critical Thinking Worksheet
   Chapter 16 Challenge/Enrichment Worksheet

REVIEW AND ASSESSMENT

Chapter 16 Check, page 181
Chapter 16 Test, *Teaching Resources*
Chapter 16 Portfolio Assessment, *Teaching Resources*
Test Generator, Chapter 16
Alternative Assessment Forms

ADDITIONAL RESOURCES

Books for Teachers

Fraser, Marian Botsford. *Walking the Line.* San Francisco: Sierra Club Books, 1989.

Lipset, Seymour Martin. *Continental Divide: The Values and Institutions of the United States and Canada.* New York: Routledge, 1990.

Martin, Desmond. *A Short History of Canada.* Toronto: McClelland & Stewart, 1994.

Books for Students

The Canadian Encyclopedia. 2d ed. Edmonton: Hurtig, 1988. Distributed by Gale Research, Detroit.

Halsey, David. *Magnetic North: A Trek Across Canada.* San Francisco: Sierra Club Books, 1990.

Malcolm, Andrew H. *The Canadians.* New York: Times Books, 1985.

The World Today *Videodisc Program*

A variety of still images, short videos, and activities are available for you to use as you teach this chapter. *Lecture Notes* include bar codes for specific images and videos.

Multimedia Resources

The Canadians: Their Cities. (VHS, 16 minutes) Encyclopedia Britannica Educational Corporation, 310 South Michigan Avenue, Chicago, IL 60604–9839.

The Canadians: Their Land. (VHS, 16 minutes) Encyclopedia Britannica Educational Corporation, 310 South Michigan Avenue, Chicago, IL 60604–9839.

Montreal. (VHS, 40 minutes) Films for the Humanities & Sciences, P.O. Box 2053, Princeton, NJ 08543–2053.

Picture Atlas of the World. (interactive maps, photo bank, animation, audio on CD-ROM) Washington, DC: National Geographic Society.

MEETING INDIVIDUAL NEEDS

Limited English Proficiency (LEP)

Have students invent and draw an international symbol for each of the following situations or places: pedestrian crossing, slippery when wet, watch for falling rocks, no left turn, rest rooms, no dogs allowed, airport, and railroad station. Discuss the importance of such symbols, especially in a bilingual country such as Canada.

Learners Having Difficulty

Have students work in pairs to answer the Chapter 16 Section Review questions. Have students use the answers to write two- or three-sentence summaries of each section. Then have them compile the section summaries to form a chapter summary. Encourage students to use the summaries to help them study the chapter.

Auditory Learners

Have students gather in small groups to discuss how the Five Themes of Geography relate to Canada. Assign one student to lead the discussion and another to take notes. When groups have finished their discussions, have a volunteer from each group use the notes to briefly summarize the discussion for the class.

Visual Learners

Help students create a three-column chart of important features of the Canadian provinces and territories. Write the following column headings on the board: Physical Geography, Human Geography, Economic Geography. Call on students to provide information for each of the provinces and territories, and have a volunteer write the facts under the appropriate heading.

Tactile/Kinesthetic Learners

Have students create on a long strip of newsprint or computer paper a schematic diagram of the route of Canada's transcontinental railroad. Students should plot and label major cities, major landforms, and the provinces through which the railroad passes. The diagram might also include pictures of natural vegetation and animals that may be seen from a train on the route. Display the diagram along a wall in the classroom.

Gifted Learners

Ask interested students to research the terms of the North American Free Trade Agreement among the United States, Mexico, and Canada. Have students report their findings in the form of a news talk show.

MULTICULTURAL FOCUS

The Canadian Radio-television and Telecommunications Commission, a government agency, has established content quotas for Canadian radio and television programming. These quotas require that at least 30 percent of all musical radio programming and 60 percent of all television programming be written or performed by Canadians.

The Canadian government has imposed these quotas to help protect Canadian culture from U.S. domination. U.S. firms control over 90 percent of Canada's film, video, recording, and book publishing industries. As a result, much of the popular culture Canadians receive is American, a fact that disturbs Canadians who want to maintain their own cultural identity.

The quotas also encourage the production of television and radio shows in Canada, which helps bolster the domestic entertainment market and create work for Canadians. Some believe that this has helped stem the flight of Canadian actors, musicians, and other performers to the United States in the promise of more lucrative careers.

Elicit a class discussion about the importance of popular culture (movies, music, television) to a nation. Ask students to imagine how they would feel if only Canadian songs and performers appeared on MTV, or if their favorite magazines focused mainly on Canadian news, sports, and personalities, or if their local theater showed only Canadian films. Ask students if they feel that the promotion of national popular culture is worthy of government support.

CHAPTER 16

CANADA

Section 1: Physical Geography
Section 2: Historical Geography
Section 3: Canada's Provinces and Arctic Frontier
Section 4: Canadian Issues

OVERVIEW

Chapter 16 introduces Canada, the world's second largest country.

Section 1 describes Canada's physical features, climate, and vegetation and examines its wealth of natural resources.

Section 2 explores Canada's history and explains the origin of its English and French cultural influences.

Section 3 summarizes the history, industry, and cultural characteristics of Canada's 10 provinces and three territories.

Section 4 focuses on two issues that will influence Canada's future—regionalism and Canada's economic and cultural ties to the United States.

About the Author

Margaret Atwood (1939–) of Canada is a well-known poet, novelist, and author of children's books. The quotation on this page is from a selection in *True North*, a travel anthology about Canada.

About the Artifact

Some Native Americans use totems (carved and painted vertical logs) as symbols for tribes, clans, or families. Each totem, with one or more animal images, is unique to a particular group.

Primary Source

Description of Change: excerpted
Rationale: excerpted to focus on main idea

CHAPTER 16

Canada

❝ In northern Canada, the roads are civilization, owned by the collective human *we*. Off the road is *other*. Try walking in it, and you'll soon find out why all the early traffic here was by water. "Impenetrable wilderness" is not just verbal. ❞

Margaret Atwood

GEOGRAPHY DICTIONARY

newsprint
portage
deport
dominion
parliament
hinterland
potash
muskeg
regionalism

A Native American totem pole in **British Columbia, Canada**

Canada is the world's second largest country, after Russia. It covers about two-fifths of the North American continent, stretching east to west for over 3,400 miles (5,471 km) and north to south for over 2,000 miles (3,218 km). The population of this large nation numbers only about 27 million—less than that of California. Canada is divided into 10 provinces and three northern territories. The harsh climates of the Canadian north have kept Canada from becoming evenly populated. More than 75 percent of Canadians live along the nation's southern border within 100 miles (161 km) of the United States.

Today, the economies and cultures of Canada and the United States are closely linked. The Canadian–United States border is the longest unguarded boundary in the world. Each country is the other's most important trading partner, and many companies do business in both countries. Cultural exchange between the two countries has helped shape both societies.

Kluane National Park, **Yukon, Canada**

172

CANADA: PHYSICAL–POLITICAL

Size comparison of Canada to the contiguous United States

LEGEND

- Ice caps
- ⊛ National capital
- ★ Provincial capitals
- • Other cities

ELEVATION

FEET	METERS
13,120	4,000
6,560	2,000
1,640	500
656	200
(Sea level) 0	0 (Sea level)
Below sea level	Below sea level

SCALE
0   250   500 Miles
0   250   500 Kilometers
Projection: Azimuthal Equal Area

PHYSICAL GEOGRAPHY

FOCUS

● *What landforms, climates, and vegetation does Canada have?*

● *What are Canada's primary resources?*

REGION **Landforms** Canada can be divided into six major landform regions: the Appalachian Mountains, the St. Lawrence Lowlands and Great Lakes, the Canadian Interior Plains, the Canadian Shield, the mountains of the Canadian west, and the mountains of the eastern Arctic.

Canada's Appalachian Mountains are in southeastern Quebec and in the Maritime Provinces. The St. Lawrence River valley and the Great Lakes were major routes of waters from melting glaciers. Today, these parts of Ontario and Quebec contain some of Canada's most fertile soils. The Canadian Interior Plains stretch east to west across Manitoba, Saskatchewan, and eastern Alberta. The Canadian Shield is a semicircular band of rocky uplands and plateaus surrounding Hudson Bay. The mountains of the Canadian west

Have students review the maps on pages 107–110. Then ask them to write statements regarding Canada's landforms, climate, population, and natural resources. Call on volunteers to read their statements. As students read through Section 1, have them verify the accuracy of their statements.

Discussing Canada's Resources

Tell students that in frontier days in the United States "Go west, young man!" was a challenge to excitement and adventure. Have students explain why "Go north!" might be an appropriate challenge for Canada's young people today. *(Northern Canada is undeveloped)* Discuss Canada's developed resources (forests) and undeveloped resources (minerals, coal).

Guided Practice

Write the names of Canadian towns and cities on slips of paper. Have students select a slip. After a few minutes of preparation, have them locate the town or city on a map and identify the climate, landforms, and resources of the area.

Chart Caption Answer

United States

Photo Caption Answer

Humid continental

Section 1 Review

ANSWERS

1. Landform regions: Appalachian Mountains, St. Lawrence Lowlands and Great Lakes, Canadian Interior Plains, Canadian Shield, Canadian western mountains, eastern Arctic mountains; climates: humid continental, marine west coast, subarctic, tundra

2. Nickel, zinc, uranium, lead, copper, gold, silver in the Canadian Shield; coal in Nova Scotia, Saskatchewan, Alberta, and British Columbia; petroleum and natural gas in Alberta

3. Water resources support a fishing industry and provide hydroelectricity. Forest resources support lumber, paper, and pulpwood industries.

FACTS IN BRIEF Canada			
COUNTRY POPULATION (1994)	LIFE EXPECTANCY (1994)	LITERACY RATE (1991)	PER CAPITA GDP (1993)
CANADA 28,113,997	75, male 82, female	99%	$22,200
UNITED STATES 260,713,585	73, male 79, female	97%	$24,700

Sources: *The World Factbook 1994*, Central Intelligence Agency; *The World Almanac and Book of Facts 1995*

Canada's population is a little more than one-tenth that of the United States. Which country has the highest per capita GDP?

---

connect the United States Rockies and Coast Ranges with the mountains of Alaska.

Climate Canada is frozen more often than it is thawed. The central and eastern parts of southern Canada have a humid-continental climate. Winters are cold and snowy, and the short summers are usually cool and mild. The warmest part of Canada is in southwestern British Columbia with its marine-west-coast climate. The lowlands of British Columbia have rainy winters, while the mountains have heavy snow. The climates of central and northern Canada are severe. Wide areas have a subarctic climate. The coast and islands of the Arctic have a tundra climate. Permafrost underlies about half of Canada.

Resources Canada's Atlantic and Pacific coastal waters are among the world's richest fishing grounds, though overfishing has taken its toll on these waters. The long, indented coastline of the Atlantic seaboard provides many good harbors. On the Pacific coast, salmon fishing is of major importance. The many inland lakes and streams of Canada provide a plentiful supply of freshwater fish and attract many tourists.

A belt of needle-leaf forests stretches all the way from Labrador to the Pacific coast. These softwood trees provide good lumber and pulp for paper. The United States, the United Kingdom, and Japan look to Canada for many of their lumber, **newsprint** (an inexpensive paper used mainly for newspapers), and pulpwood products.

Minerals are the most valuable of Canada's natural resources. The Canadian Shield, once thought to be a wasteland, has proven to be a source of many mineral deposits. Canada is a leading source of the world's nickel, zinc, and ura-

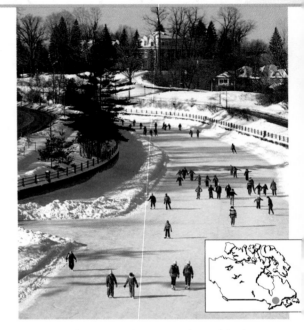

Skating on frozen canals is a popular activity in **Ottawa**, the Canadian capital. Compare the physical–political and climate maps on pages 107 and 108. In which climate region is Ottawa located?

---

nium. It also is a leading producer of lead, copper, gold, and silver.

Experts believe that the coal fields of Canada may be among the largest in the world. The coal deposits are generally very deep and are located in unpopulated areas. Soft coal is mined in Nova Scotia and in the western provinces of Saskatchewan, Alberta, and British Columbia.

Canada is a large-scale producer of petroleum. About two-thirds of Canada's petroleum and about four-fifths of its natural gas come from Alberta. Major deposits have been discovered off the east coast and off the Arctic coast as well. Canada also is one of the world's largest producers of hydroelectricity. Hydroelectric power plants generate about two-thirds of Canada's electricity.

SECTION REVIEW

1. List four landform regions and two climates found in Canada.
2. What mineral resources does Canada have, and where is each one found?
3. **Critical Thinking** **Discuss** the importance of water and forest resources to Canada.

Closure

Have pairs of students quiz each other on Canada's landforms, climates, and resources.

ASSESSMENT

Assign the Section 1 Review questions on page 174.

Reteaching

Assign Section 1 of Reteaching Worksheet 16.

SECTION 2
OBJECTIVES

These objectives reflect the Focus questions on this page and are assessed in the review on page 176.

1. Summarize Canada's political history.
2. Discuss the settlement of various parts of Canada.

TEACHING SECTION 2
Building a Vocabulary

Write **portage** and **deported** on the board. Note that they have a common word root -*port*, meaning "to carry." Thus, *portage* refers to a place across which cargo can be carried, and *deported* means "to carry away from." Ask students to think of other words with this root. (*portable, transport, airport, seaport*)

2 HISTORICAL GEOGRAPHY

FOCUS

- *How have other countries played a role in Canada's history?*
- *How were various parts of Canada settled, and by whom?*

MOVEMENT **New France** The first Europeans to land on Canada's eastern shores were Viking adventurers, who arrived between A.D. 1000 and 1200. They left no permanent settlements, however. The formal exploration of Canada by Europeans began in 1497 when John Cabot landed on Newfoundland and other islands. During the first part of the 1500s, Canada's eastern coast was visited by people from many European nations. Trade between Canada's Native Americans and Europeans developed quickly.

The first great explorer of the St. Lawrence River was Jacques Cartier of France. In 1535, he traveled upstream as far as the site of Montreal. Cartier and the French explorers who followed him had three goals: to search for a northwest passage to China and India, to develop a fur trade, and to convert the Native Americans to Roman Catholicism. The territories claimed by the French during this time were known as New France.

The French realized that the St. Lawrence River and the Great Lakes were an avenue into the interior of North America. The lakes and rivers were connected by easy **portages**. A portage is a low land area across which boats and their cargoes can be carried. French settlers came to farm on the fertile lowlands along the St. Lawrence and in Nova Scotia. Fur trappers and trading posts spread into central North America.

British North America In 1713, England acquired Nova Scotia for its fishing ports and farmlands. The French who were living in Nova Scotia were **deported** to what was then French Louisiana. To deport means to send out of a country. In the early 1700s, England's expanding American colonies and France's interests in North America came into conflict, which, along with other conflicts elsewhere, caused France and England to go to war in 1756. In North America, this war drew to a close in 1759 when the British captured the city

The Canadian Parliament buildings in **Ottawa, Ontario**, reflect the European influence of the country's colonial past. Canada's government is a combination of U.S. and British government systems.

of Quebec. To promote peace, the British recognized French traditions and laws. Both French and English became official languages.

British settlement of Canada increased dramatically during the American Revolution. To remain under British rule, many colonists who remained loyal to Britain left the colonies to live in southern Ontario. Their descendants have been economically and politically powerful throughout Canada's history.

In 1791, Canada was divided into Upper Canada (present-day Ontario) and Lower Canada (present-day Quebec). At this time, Canada's Maritime Provinces along the eastern coast were separate British colonies. In the early 1800s, Canada grew steadily. The English and French communities began arguing over language and immigration policy. The cities of Quebec, Montreal, and Toronto grew into major commercial centers. Farther west, the Hudson's Bay Company and the North West Company extended their fur-trading posts to the Pacific coast and to the subarctic region.

Dominion of Canada In 1867, the British Parliament passed the British North America Act, which created the Dominion of Canada. A **dominion** is a territory or sphere of influence. The dominion joined together the provinces of Ontario, Quebec, Nova Scotia, and New Brunswick to be governed by a **parliament** (a lawmaking body) and a prime minister. The city of Ottawa, in Ontario but on the Quebec border, became the capital. The other present-day provinces joined the dominion

Geographical Sidelight

Canada's once-gigantic fur industry is shrinking. Campaigning by animal rights advocates is largely responsible for a significant drop in the sale of fur coats. This, in turn, has led to the loss of hundreds of jobs for those who trap, dress, make, and sell furs. Even the Hudson's Bay Company—synonymous with Canada's fur trade—has closed fur salons in its department stores.

175

Describing Settlement

Note that Canada developed like a ribbon from east to west along its southern border. Have students describe the different groups that settled each region. Using a wall map of Canada, start at the East Coast with Viking adventurers and work across to the West Coast with fur traders and railroad builders.

Guided Practice

Tell students that July 1 is Dominion Day in Canada. Have them find out the meaning of this day and compare it with July 4 in the United States. *(Dominion Day celebrates the proclamation in 1867 of Canada's self-governing status.)*

Independent Practice

Have students write a chronology of Canada's history. Tell them to use the dates and events in Section 2.

Closure

Discuss how Canada might be different today if Canadians had rebelled against British rule.

ASSESSMENT

Assign the Section 2 Review questions on this page.

Reteaching

Assign Section 2 of Reteaching Worksheet 16.

Section 2 Review
ANSWERS

1. France, Britain; the act that created the Dominion of Canada
2. Native Americans, French, British, African Americans, other European immigrants, East and Southeast Asians, people from the Caribbean
3. Both the French and the British came to Canada for farming. The French also pursued the fur trade, and the British sought fishing ports.

176

---

later. Each Canadian province has its own parliament, headed by a provincial prime minister. Provincial governments have control over taxes, education, and civil rights.

During the 1800s, many African Americans from the United States moved to eastern Canada to escape slavery. Most of Canada's west was settled after the country's transcontinental railroad was finished in 1885. Today, many of the nation's citizens are immigrants from Europe, East and Southeast Asia, and the Caribbean.

Ties between Canada and Britain have remained close. Canada's national government has a parliament like Britain's, and Britain's queen also is Canada's queen.

② SECTION REVIEW

1. What countries have ruled Canada? What was the British North America Act?
2. Over time, what groups of people have settled in Canada?
3. **Critical Thinking** **Compare and contrast** historical reasons for French and British settlement in Canada.

③ CANADA'S PROVINCES AND ARCTIC FRONTIER

FOCUS

- *What are Canada's provinces and territories?*
- *What economic characteristics does each province or territory have?*
- *What cultures are found in Canada, and how do they interact?*

Maritime Provinces and Newfoundland

New Brunswick, Nova Scotia, and Prince Edward Island are called the Maritime Provinces. Newfoundland includes the island of Newfoundland as well as Labrador on the nearby mainland. These provinces are Canada's eastern **hinterland**. A hinterland is a region beyond the center of a country or city.

Life often is hard in eastern Canada. Economic development has depended on the resources of the

---

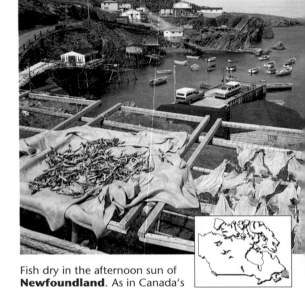

Fish dry in the afternoon sun of **Newfoundland**. As in Canada's other eastern provinces, most residents live near or along the Atlantic coast, where fishing is a major occupation.

sea and forests. The short growing season and thin, rocky soils make farming difficult, although small farms that produce a variety of crops are typical.

Commercial fishing is an important economic activity. The Grand Banks, an area of shallow waters off the eastern shore of Newfoundland, forms one of the world's great fishing grounds. The region's thick fog and winter storms can make fishing dangerous, however.

Many people in the Maritime Provinces are descended from families that originally emigrated from the British Isles. They are separated from the rest of English-speaking Canada by French-speaking Quebec. Many families from Quebec have moved to New Brunswick, however, and New Brunswick claims to be Canada's only truly bilingual province.

Most people in Canada's eastern provinces reside in coastal cities. These cities are manufacturing centers, as well as fishing and freight ports. The region's other population centers are in the few inland areas best suited for farming. Halifax, Nova Scotia, is the region's leading city.

Quebec The chain of cities that extends from Windsor, Ontario, to the city of Quebec is Canada's greatest urban region. The province of Quebec is distinctive in North America because of its strong French-Canadian culture. More than one-fourth of all Canadians live in the province. For

These objectives reflect the Focus questions on page 176 and are assessed in the review on page 179.

1. List the Canadian provinces and territories.
2. Describe economic characteristics of each province/territory.
3. Analyze Canada's cultural geography.

See how many of Canada's provinces and territories the class can name without consulting a map. Ask students to suggest why Canada has fewer political subunits than the United States. *(Since Canada is sparsely populated, especially in the north, fewer governmental divisions are needed.)*

Building a Vocabulary

Have students look up the origins of the terms **hinterland** *(German)* and **muskeg** *(Algonquian)*. Then have a volunteer look up **potash** in an encyclopedia. Note that *potash* literally means "pot ash," because it was originally obtained from the residue of wood ash, which was boiled in large pots. The chemical element potassium takes its name from potash because the metal was first isolated from this mineral.

---

most of them, French is their first language. With nearly four centuries of experience in North America, these French Canadians are proud to call themselves Quebecois (kay-beh-KWAH), rather than French. Many people speak only French, official papers of the provincial government are prepared only in French, and signs along roads and on businesses appear only in French.

Throughout Quebec history, many French Canadians have argued that the province should become an independent country. The issue of Quebec's independence has threatened to break up Canada many times. An independent Quebec would separate the eastern provinces from Ontario. The western provinces might then demand their independence as well.

Quebec, Canada's first city and the provincial capital, is located where the shores of the St. Lawrence estuary pinch together. Many of the city's older sections, which feature narrow streets and stone walls, have been preserved. Quebec is a symbol of Quebecois culture and is a popular tourist destination.

Montreal is the financial and industrial center of Quebec. The city began on an island in the St. Lawrence River. Today, Montreal serves as an important port on the St. Lawrence Seaway. To combat the winter cold, many of the buildings in the city center are connected by underground arcades and overhead glass tunnels.

Montreal was once an island of British-Canadian culture in French Quebec. Now, important jobs are being filled by the Quebecois. After Paris, Montreal is the second largest French-speaking city in the world.

The French colonial past is clearly seen in the architecture of the city of **Quebec**. Where in the United States is French influence most noticeable?

Ontario As in Quebec, most of the people of Ontario live in the far south. The cities of southern Ontario are the fastest growing in Canada. Southeastern Ontario is the chief manufacturing district of the country. Toronto, Hamilton, Kitchener, Windsor, and several smaller cities are grouped together in this region.

Toronto is the provincial capital and the industrial, financial, cultural, and educational center of Canada. It houses Canada's largest stock exchange, and most large Canadian corporations are headquartered here. The nation's finest museums, universities, symphony orchestras, and other cultural institutions are concentrated in the Toronto region. Metropolitan Toronto has a population of about 3 million.

Toronto reflects Canada's British heritage much like the city of Quebec symbolizes French Canada. British customs are still widespread, and French is seldom heard or seen on signs in Toronto. Instead, recent immigrants from Italy, the Caribbean, and China give the city its international flavor.

Ottawa, Canada's capital city, lies in the province of Ontario. Its twin city is Hull, just across the Ottawa River in Quebec. The capital region is fully bilingual. Ottawa has grand government buildings, more than 1,500 acres (607 ha) of parks, and several universities. Manufacturing activities concentrate on forest products.

Prairie Provinces Manitoba, Saskatchewan, and Alberta are called the Prairie Provinces. Much of the southern parts of these provinces was once covered with prairie grasslands. Today, the region is the center of a rich wheat belt. Highly mechanized farms require few people, and the Prairie Provinces remain only thinly settled. Still fewer people live to the north, where the grasslands give way to forests.

Canada's population is quite small in proportion to its production of wheat. Consequently, in most years, Canada has more wheat available for export than any other country. Changes in the foreign wheat market and uncertain weather conditions cause difficulties for the wheat farmers, however. As in the United States, the uncertainty of prices and growing conditions has led Canadian farmers to request aid from the government.

Winnipeg, the capital of Manitoba, is the chief city of the eastern prairies. All east-west rail

THEME
Location

Point out Ottawa on a map. Ask students why the capital was located there. (The location represents a compromise between English- and French-speaking Canada.)

COOPERATIVE LEARNING

Organize the class into groups of six. Assign each group member one of the following areas: Quebec, Ontario, Maritime Provinces, Prairie Provinces, British Columbia, Canadian North. Then have all students assigned Ontario meet together, and all students assigned each other area meet together. Have these subgroups learn about the history, industry, and culture of their respective area. Finally, have students return to their original groups and teach the other members about their province or region.

Photo Caption Answer

Louisiana

Cultural Geography Draw a rough map of Canada on the board. Have each student do the same on a sheet of paper. Ask students to label the English-speaking, French-speaking, Inuit, and bilingual areas. Discuss the historic reasons underlying the formation of these areas. Ask students what might be the advantages and disadvantages of Canada having one official language.

Guided Practice

Write the following occupations on the board: pilot, salmon fisher, prime minister, prospector, wheat farmer, oil driller, naval commander, stockbroker. Ask students to show on a wall map the place in Canada where each of these occupations might be found. Have them explain their choices, relating them to the economic characteristics of the provinces and territories.

Closure

Have students work in pairs to draw illustrated maps of Canada's provinces and territories. Each pair should choose one province or territory to illustrate. Within an outline map of the selected area, each pair should show scenes and symbols of the area's economic and cultural characteristics. Extended captions should explain the map's contents.

MULTICULTURAL PERSPECTIVES

Canada, like the United States, is a nation of immigrants. Many Canadians have traditionally thought of themselves as (for example) of French origin or of British origin rather than as Canadians. In the 1991 census, however, the number of people who indicated their ethnic origin as "Canadian" rose sharply from previous years. Discuss with students the factors that they think influence the way an immigrant group views itself over time.

Geography and Education

The organization of Montreal's public schools reflects religion and language preferences. Types of schools include English-language Roman Catholic, French-language Roman Catholic, and French-language Protestant. Most Jewish students attend private schools of their faith or public Protestant schools.

NEWS FROM
AROUND THE WORLD
Canada

A hockey game in **Quebec Province**

Teenagers in Canada Struggle with Divided Loyalties

Many teenagers in Canada are accustomed to waking before dawn on chilly Saturday mornings. That may be the only time their hockey team can practice at one of the busy 24-hour rinks. When rinks are full, teens put up homemade nets and play hockey on frozen driveways and streets.

Winter sports are a way of life in much of Canada. The winters are cold, snowy, and long. Still, the snow and ice cannot cover up some basic differences that have not only divided Canadians but have threatened to split the country as well.

Imagine that you are a typical student in Ontario, Canada. English is your first language at home and in school, and you study British history in your classes. Perhaps you come home to tea in the afternoon.

In contrast, imagine that you live in Quebec, Canada's only predominantly French-speaking province. Your classes are probably taught in French. You might eat croissants for breakfast and tourtiere, or pork pie, for lunch. Indeed, as a Quebecois, your life might be quite different from that of a student living only a few hundred miles away in the province of Ontario.

Teenagers are keenly aware of the differences that exist between Quebec and other regions of Canada. "Sometimes there seems to be a feeling that you can't be Quebecois and Canadian at the same time," a young man comments.

Still, young Canadians have a great deal in common with each other and with young people around the world. Many Canadian high school students work at part-time jobs to earn money for college. Many also are concerned about their future careers and job opportunities.

Many of them worry, too, about the possible breakup of their homeland. They hope that the Canadian provinces can overcome their differences and remain unified. One young woman says, "I often wonder how I'd feel if Quebec actually separated. I love Montreal, but I'm Canadian."

traffic passes through Winnipeg. As an important collection and shipping point for the region's products, Winnipeg serves Canada much like Chicago serves the United States.

Saskatchewan is predominantly agricultural, although it contains one of the world's largest deposits of **potash**. This mineral is an important raw material for manufacturing fertilizers. Regina, Saskatchewan's capital, is a center of the potash industry as well as of the province's food-processing industries.

Alberta's wealth is based on fossil fuels, particularly oil. The province's oil and natural gas travel in pipelines to eastern and western Canada and to the United States. The spectacular Rocky Mountains of western Alberta are popular with

Extension

Have students look for newspaper and magazine articles about Canada. Ask them to share information from these articles with the class. If possible, have them obtain Canadian newspapers and magazines for a different perspective on U.S. news.

SECTION 4
OBJECTIVES

These objectives reflect the Focus questions on page 180 and are assessed in the review on page 180.

1. Examine the origins and effects of regionalism in Canada.
2. Explain how economic ties have been an issue for Canada and its regions.

tourists from around the world. Alberta has two rapidly growing cities, Calgary and Edmonton. Each is an important oil and agricultural center.

British Columbia The province of British Columbia lies on the Pacific coast of Canada. It is a land of rugged mountains, intermountain plateaus, and fertile river valleys. Large mountain ranges extend the length and width of the province. These mountains have long isolated the province from the rest of Canada.

British Columbia is rich in natural resources. Much of the province is covered with forests of fir, spruce, and cedar trees. Salmon fishing and mining industries also are important to the province's economy. Because of its location on the west coast, British Columbia can trade easily with other ports around the Pacific Rim. Water resources play a large role in British Columbia's economy, and mountain streams provide inexpensive electric power.

Vancouver is Canada's third largest urban region. It grew to become western Canada's greatest city after the first transcontinental railway built its end point at the mouth of the Fraser River. Vancouver has Canada's major ice-free harbor and is Canada's outlet to the Pacific Ocean. Victoria, the capital of British Columbia, is located at the southern tip of Vancouver Island. This government center is the home port for a large fishing fleet.

The Canadian North Canada's vast northern lands are divided into the Yukon, the Northwest Territories, and Nunavut. Nunavut is a new territory that was created for and is administered by the native Inuit (Eskimo) who live there. These three regions occupy more than one-third of Canada's land surface but are home to only about 80,000 people. Dense forests and tundra, mostly underlain by permafrost, separate the isolated towns and villages. Thousands of rivers and millions of lakes characterize the Canadian North. **Muskegs** are other common landforms. A muskeg is a forested swamp that melts only during the mosquito-breeding season in the summer.

Airplanes and satellite communications recently have brought northern Canada into the national sphere. In spite of its severe climate, the region holds important promise for Canada's future. Rich deposits of metals and fossil fuels have been discovered here, and the supplies of

A farmer works his land in the Canadian wheat belt near **Saskatoon, Saskatchewan**. Wheat is the primary crop throughout the Great Plains from the Canadian prairie provinces to Texas.

fresh water and potential for hydroelectricity are enormous.

The people of the Canadian north are a mixture of hardy frontier people. The residents include the Inuit (Eskimo) and Native Americans whose ancestors fished and hunted in the region hundreds of years ago, as well as people who have migrated from Canada's cities in the south.

③ SECTION REVIEW

1. List Canada's provinces and territories.
2. Which Canadian provinces are primarily industrial? agricultural?
3. **Critical Thinking** **Analyze** Canada's cultural geography. How have cultural differences been a challenge at times?

THEME
Location

Vancouver's location on the Pacific Rim has resulted in the city's dramatic growth in size and importance during the twentieth century. One source of growth over the past two decades has been Hong Kong Chinese, who have been investing in and immigrating to the area because of fears of sweeping changes following Hong Kong's transfer to China in 1997.

Section 3 Review
ANSWERS

1. Provinces: New Brunswick, Nova Scotia, Prince Edward Island, Newfoundland, Quebec, Ontario, Manitoba, Saskatchewan, Alberta, British Columbia; territories: Yukon, Northwest Territories, Nunavut
2. Industrial: Quebec, Ontario; agricultural: Manitoba, Saskatchewan
3. Cultures of most provinces are primarily British; Quebec is largely French. Some areas are neither British nor French, such as Nunavut. Some people believe that Quebec should be an independent country due to its French heritage.

Motivator

Have students discuss the advantages and disadvantages of living in a large country such as Canada or the United States. Point out that almost two-thirds of the world's countries—including some of the richest—have fewer than 10 million people, and close to one-third have fewer than one million.

Guided Practice

Call on a volunteer to define the term **regionalism**. Have students cite reasons for regionalism in Canada. *(physical barriers such as the Rocky Mountains, language differences, autonomy of the provinces)* Discuss the ties that link Canada's regions and that link Canada to the United States.

Closure

Have students write an editorial from a Canadian perspective on whether or not Canada should remain united.

ASSESSMENT

Assign the Section 4 Review questions on this page.

Reteaching

Assign Section 4 of Reteaching Worksheet 16.

Section 4 Review
ANSWERS

1. Maritime and Western provinces are suspicious of Ontario's and Quebec's influence in the national government; Ontario and Quebec have differences over language and culture; Alberta is protective of its oil income; British Columbia has voiced detachment due to its location.

2. Poor regions rely on wealthier regions; the Northwest Territories are dependent on British Columbia for stock trading; all depend on Alberta's oil. Internationally, the Canadian and U.S. economies are closely linked.

3. Possible answer: economic ties between regions counteract regionalism, which could divide Canada into several countries.

180

CANADIAN ISSUES

FOCUS

- *What are the origins of regionalism in Canada? What effects has it had?*
- *How have economic ties been an issue for Canada and its regions?*

REGION **Regionalism** When Canadians deal with each other, they often show considerable **regionalism**. Regionalism is the political and emotional support for one's region before support for one's country. The threat to divide Canada into several countries has surfaced repeatedly in Canadian history. Although Canada's national government has international duties similar to those of the U.S. federal government, the Canadian provinces are much more powerful than the states in the United States.

This situation, combined with the relative isolation and separation of Canadian settlements, keeps regionalism alive. People in the Maritime Provinces and in the Canadian west have long been suspicious of events in Ottawa, where Ontario and Quebec seem to dominate. People in Quebec and Ontario have been haggling over language and culture for more than two centuries. Alberta would rather not share its petroleum industry revenues with Ottawa. British Columbia, separated by high mountains and long distances, has often voiced detachment from the rest of Canada.

Even so, most Canadians support a united Canada rather than several small countries. Canadians continue to find strength in their diversity. In doing so, they set a positive example for many other countries where ethnic diversity has led to unrest and war.

Economic Ties The many economic ties between the various provinces are another factor holding Canada together. Poor regions, such as Newfoundland and rural Manitoba, depend on the bankers and industrialists of Toronto and Montreal. Development of resources in the Northwest Territories relies on trading at the Vancouver Stock Exchange. All Canadians depend on the oil resources of Alberta.

Internationally, Canada's main economic ties are with its southern neighbor. Often, the United

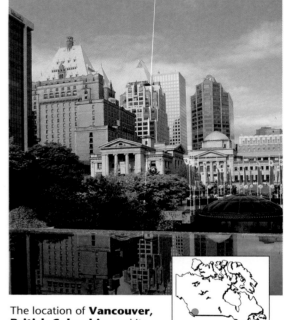

The location of **Vancouver, British Columbia**, and its busy port make the city Canada's main trade link to other Pacific Rim countries.

States buys Canada's natural resources in exchange for manufactured goods. Both Toronto and Montreal do more business with U.S. firms than they do with each other. Many U.S. companies are giants in the Canadian economy.

The United States has nearly 10 times the population of Canada, and U.S. mass culture frequently seems to overwhelm Canadians. Canadians, however, continue the struggle to maintain their own distinct culture. They provide another voice in the varied culture of North America.

SECTION REVIEW

1. Which Canadian provinces have been most affected by regionalism, and in what ways?
2. How are Canada's regions dependent on each other? What economic ties exist between Canada and the United States?
3. **Critical Thinking** **Discuss** the relationship between regionalism and interregional economic ties.

CHAPTER 16 CHECK
ANSWERS
Building a Vocabulary
1. Portages, land areas between bodies of water across which boats and cargoes can be carried, gave the French access to lakes and rivers that led to the interior.
2. A region beyond the center of a country or city; the Maritime Provinces and Newfoundland
3. The manufacturing of fertilizers
4. Muskeg

Recalling and Reviewing
1. Climates are milder in southern Canada; the two countries have many economic and cultural ties.
2. In the tension between French Canadians, whose ancestors were Canada's first European settlers and British Canadians, whose ancestors came later
3. Tensions between people of different cultures, particularly French Canadians and British Canadians, have threatened to break up Canada. Nevertheless, most people support a united Canada, and the nation's diversity provides many traditions.

CHAPTER 16 CHECK

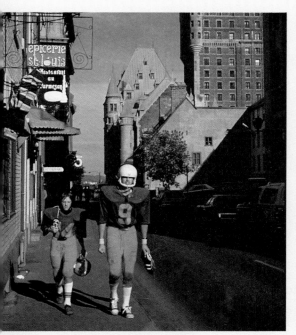

Old city section of **Quebec**

Reviewing the Main Ideas
1. Canada has diverse landforms and mostly cold climates. Resources include fish, forest products, minerals, and hydroelectricity.
2. The French formed the first permanent European settlements in Canada. Over time, Britain gradually gained control over most of Canada. Canada has maintained ties to Britain. Today, immigrants to Canada come from Europe, East and Southeast Asia, and the Caribbean.
3. Canada has 10 mostly self-governing provinces and three northern territories. The provinces and territories have a wide range of economic and cultural characteristics.
4. One major Canadian issue has to do with regionalism, the support for one's region before one's country. Another Canadian issue involves economic ties, on both national and international levels.

Building a Vocabulary
1. How did portages help the French explore and settle the Canadian interior?

2. Define *hinterland.* Which provinces make up Canada's eastern hinterland?
3. What is potash used in?
4. What term describes a cold forested swamp that melts only in the summer?

Recalling and Reviewing
1. Why does a large portion of Canada's population live close to the border with the United States? What effects has this closeness had on Canada?
2. How is Canada's history reflected in the country today?
3. What challenges has Canada's cultural diversity presented? How has this diversity also been a source of strength to the country?

Thinking Critically
1. How might going to school in a country that has two official languages be different from going to school in a country where only one language dominates?
2. Imagine that several Canadian provinces became independent countries. What effects might Canada's breaking up have on the United States?

 ## Learning About Your Local Geography

Individual Project
Research your community's cultural background. From what countries or areas did the first settlers of your community originate? What ethnic backgrounds do current residents have? What languages are spoken in your community? Make a list of the various cultural and ethnic groups that have lived in your community over time. You may want to include your list in your individual portfolio.

Using the Five Themes of Geography
Write a paragraph about the themes of **place** and **location** in Canada. How have physical place characteristics influenced the location of Canada's people and cities? How has location influenced Canada's human place characteristics?

Thinking Critically
1. In a country with two official languages, schools might provide instruction equally in both languages in all subject areas, and many students might be bilingual.
2. If Canada broke up into several smaller countries, trading patterns would likely be disrupted. The United States would have to forge new relations with each Canadian country.

Learning About Your Local Geography
Lists should show the cultural and ethnic groups that have lived in students' communities.

Using the Five Themes of Geography
Possible answer: climates are milder in southern Canada, so most of Canada's population is located there. As such, much of Canada's population lives close to the United States, so the cultures of the two countries tend to be similar.

Unit 3 Review

SKILLS

Point out the differences between models and maps. Whereas a map shows the layout of a particular, unique place (and usually a real, existing place), a model shows the layout of a type of place, such as a city. Models can also show the frameworks of concepts or objects. Point out that cities are just one example of places that can be illustrated by models. Have students draw models of high schools and elementary schools. Models might show the location of classrooms, the cafeteria, the library, the main entrance, and offices. Models should not map out any actual schools, but rather they should represent what students believe to be the layout of a "typical" school.

Practicing the Skill

ANSWERS

1. Possible answer: due to noise, pollution, unattractive scenery, and other factors, property values in areas next to industrial areas might be lower than in areas farther away, and thus such property might be more affordable for people earning low incomes.

2. Such a model might show several individual-city models touching each other at the edges.

3. Students should identify the central business district, the industrial area, and residential areas of a nearby city and should decide which model the city resembles most.

182

UNIT 3
REVIEW   SKILLS

Using Models in Geography

When geographers present ideas, they sometimes find it helpful to use **models**. Here, model does not mean a small structure made of cardboard and glue, like a "model" airplane. Rather, the models used by geographers are more like the plans or patterns of different types of airplanes. You might look at these plans to construct an actual airplane (miniature or otherwise), or you might just study the plans to gain a greater understanding of the parts of different types of airplanes.

Models are particularly useful for understanding how cities are arranged. For example, why are skyscrapers and houses located in different areas? Over the past decades, geographers have suggested several models to answer such questions. The goal of these models is not to explain the layout of an existing city, such as New York or Los Angeles, but to make generalizations that will apply to many cities. Models change over time as geographers gain new information and as cities themselves change.

Most models of cities show three major categories of land use. The first category is the central business district, or CBD—the area people often call "downtown." It contains banks, office buildings, hotels, and many kinds of stores and shops. Because these businesses need to be reached by people from all over the city, most models place the CBD in the middle of the city. The second category of urban land use is the industrial and manufacturing area. The third category consists of the residential areas, where most people live.

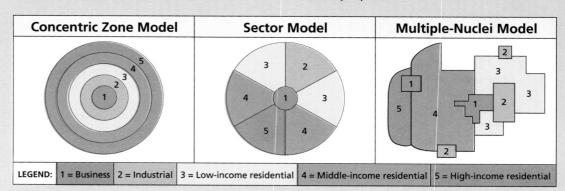

Concentric Zone Model | **Sector Model** | **Multiple-Nuclei Model**

LEGEND: 1 = Business | 2 = Industrial | 3 = Low-income residential | 4 = Middle-income residential | 5 = High-income residential

In this early model, the city looks like a bull's-eye. The center circle is the CBD. The next ring is the manufacturing area. The three outer rings are the residential areas, with low-income areas toward the inside and high-income suburban areas as the outermost ring.

The Sector Model also places the CBD in the middle of a circle, but the rest of the circle looks like a pie rather than a bull's-eye. Transportation routes, such as freeways, rail lines, and rivers, run between the CBD and the different sectors, or pieces of the pie.

In this non-circular model, developed in the 1940s, there are several small business and industrial districts, or nuclei, in some of the outer areas in addition to the more central CBD and industrial areas. This model attempts to show the increased complexity of cities.

PRACTICING THE SKILL

1. Why do you suppose low-income residential areas tend to be next to industrial areas?

2. Recent models of cities attempt to combine elements of the previous models with the idea that, in modern times, a city's surrounding areas tend to overlap with those of other cities and towns. What might such a model look like?

3. Identify the central business district, industrial area, and residential areas of the nearest city. Which model does the city resemble most?

BUILDING YOUR PORTFOLIO

Decide whether students will work individually or in groups on the unit projects described below. Students should then choose which project they will work on. Provide students with copies of the Portfolio Assessment Worksheets in *Teaching Resources*. Chapters 10–16 worksheets and the Unit Portfolio Assessment Worksheet will guide students step-by-step to complete either project.

Teaching with Sketch Maps

Have students sketch a map of the United States and Canada from memory. Tell them to include as many details as they can about place names, landforms, and boundaries of regions they studied in Unit 3. Students should compare this map with the sketch map they completed at the beginning of Unit 3 to note their improvement in identifying geographic features.

BUILDING YOUR PORTFOLIO

Individually or in a group, complete one of the following projects to show your understanding of the geography concepts involved.

A Economic and Demographic Shifts

In this unit, you have been reading about some of the major changes the United States and Canada have been experiencing recently: the shift from primary and secondary industries to service industries, the displacement of traditional industries by high-tech industries, the increase of corporate-owned farms as compared to family-owned farms, and the population shifts to the Sunbelt and to suburbs. Research how these changes have affected your local area, and prepare a report on your findings. To complete your report, you will need to do the following.

1. Make bar or pie graphs showing how the economic activities of your area have changed over the past several years.
2. Write a paragraph discussing how manufacturing industries in your area have changed, and how the changes have affected local workers.
3. In a chart, show the extent to which patterns of farm ownership have been changing in your area.
4. Construct a line graph showing how your area's population has changed in size over time as a result of the national population shift to the Sunbelt.
5. With maps, show whether the demography of a local town or city indicates a widespread move into suburbs.
6. In several illustrations and captions, predict how these changes will affect the future of your area and your own future, as an individual.

Organize your materials, and present your report to the rest of the class.

B Urban Renewal

You are a city planner. For a conference called "The Renewal of American Cities," prepare a proposal for how cities might go about revitalizing themselves. For your proposal, you will need to do the following.

1. Draw a series of illustrations, with captions, that show problems faced by urban areas. What characteristics might make some areas unpleasant for residents and unattractive to visitors?
2. Construct a table identifying three U.S. and/or Canadian cities that have areas in need of revitalization and showing some of the causes of urban decline in each city.
3. Make a list of three cities that have already gone through renewal. Next to each city, discuss when the renewal was finished, what forms it took, and what problems the city faced during the renewal.
4. Graph before-and-after crime rates, income levels, and other information about the revitalized areas of the three cities that have already gone through renewal. With your graphs, include captions discussing what the renewal accomplished and how it changed residents' daily lives.
5. Write a paragraph about how to avoid the problems during revitalization that other cities have faced. Are some problems unavoidable?
6. Create a graphic organizer showing your plan for urban renewal. What problems does your plan address, and how does it address them? Does your plan create any new problems?

Organize your materials, and present your proposal at the conference (in class).

Project A:

Reports should show the extent to which local areas have been affected by the economic and demographic changes being experienced by the United States and Canada. Students should consider economic activities, industries, patterns of ownership, population size, and population distribution. Students should also predict how the various changes will affect the future of the area and their own futures. They might illustrate what they expect to do for a living and where they expect to live when they are adults.

Project B:

Proposals should show problems common to urban areas, identify cities in need of revitalization, discuss cities that have already gone through renewal, address problems involved in renewal, and present a new plan for urban renewal. Students might address how urban renewal can balance the needs of a city as a whole with the needs of individual residents, particularly those who live in the areas in need of revitalization. Plans might include the introduction of community education and recreational programs and facilities in addition to changes in the outward appearance of the community.

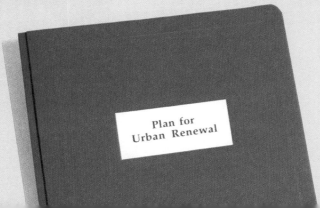

Plan for Urban Renewal

183

UNIT 4 MIDDLE AND SOUTH AMERICA   CHAPTERS 17–22

OVERVIEW

In Unit 4, students will learn about the physical geography and human geography of the countries of Middle and South America.

Chapter 17 • INTRODUCING MIDDLE AND SOUTH AMERICA describes the landforms, climates, history, culture, resources, agriculture, and industry of Middle and South America.

Chapter 18 • MEXICO focuses on the physical geography, economy, and population diversity of Mexico.

Chapter 19 • CENTRAL AMERICA AND THE CARIBBEAN ISLANDS describes the seven countries of Central America and the culture of the Caribbean islands.

Chapter 20 • CARIBBEAN SOUTH AMERICA features Colombia, Venezuela, and the Guianas.

Chapter 21 • ATLANTIC SOUTH AMERICA explores the countries of eastern South America.

Chapter 22 • PACIFIC SOUTH AMERICA describes the landforms, climate regions, economies, and political situations of Ecuador, Peru, Bolivia, and Chile.

OBJECTIVES

1. Describe the landforms and climates of Middle and South America.

2. Analyze how the development of natural resources may affect the economies of Middle and South American nations.

3. Identify the ancient Native American civilizations in Middle and South America.

4. Understand the effects of European colonization in Middle and South America.

5. Describe the political, economic, social, and cultural issues that confront Middle and South America.

6. Interpret special-purpose maps to analyze the relationships among climates, population patterns, and economic resources in Middle and South America.

7. Learn to use steps of the geographic method to collect and analyze geographic data.

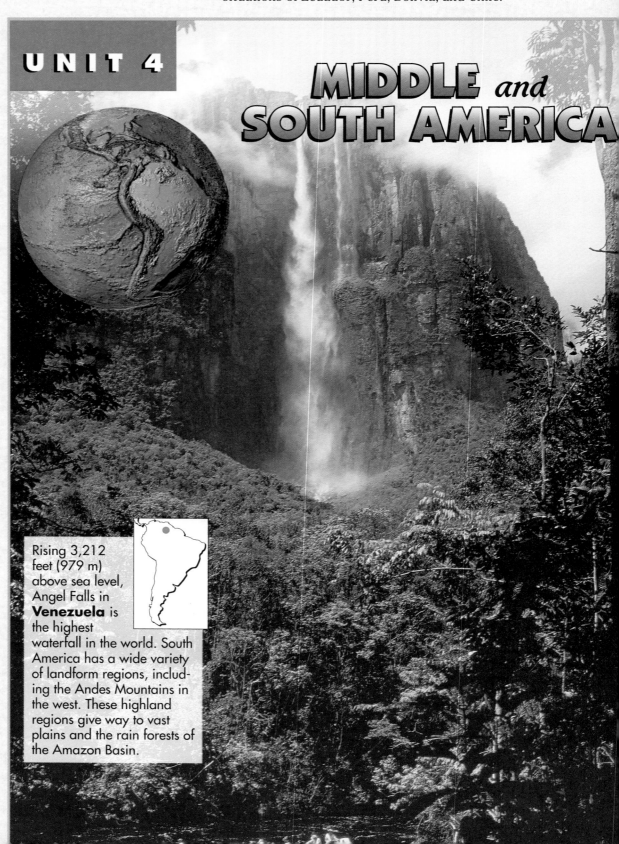

UNIT 4

MIDDLE and SOUTH AMERICA

Rising 3,212 feet (979 m) above sea level, Angel Falls in **Venezuela** is the highest waterfall in the world. South America has a wide variety of landform regions, including the Andes Mountains in the west. These highland regions give way to vast plains and the rain forests of the Amazon Basin.

184

TEACHING UNIT 4

Using the Illustrations

Have students look at the photographs on these pages. Tell them to locate the source of Angel Falls by referring to the physical-political map on page 227. *(Caroni River)* Identify details in the photograph of Machu Picchu that suggest the highly developed culture of the Inca peoples. *(ruins of intricately made stone walls)* Ask: What do you notice in the photograph that suggests why the Spanish explorers never found the Incan city when they conquered Peru in the 1500s? *(its secluded location high in the Andes Mountains)* What geographic themes would best apply to Machu Picchu? *(human-environment interaction; place)* Tell students that they can learn more about this unusual place in *One World, Many Voices* on pages 186–187.

Teaching with Sketch Maps

Have students draw a map of Mexico, the Caribbean region, and South America from memory. Direct them to form a mental picture of the region and to transfer what they remember to paper. Tell them to include as many bodies of water, political boundaries, country names, and cities as they can. They might add this map to their portfolios and compare with another they will draw after they have completed Unit 4.

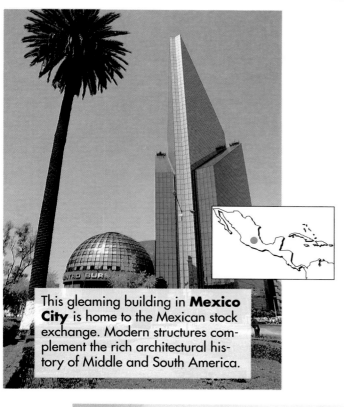

This gleaming building in **Mexico City** is home to the Mexican stock exchange. Modern structures complement the rich architectural history of Middle and South America.

UNIT 4 CHAPTERS

17 Introducing Middle and South America

18 Mexico

19 Central America and the Caribbean Islands

20 Caribbean South America

21 Atlantic South America

22 Pacific South America

Gold Incan medallion from **Bolivia**

PORTFOLIO ASSESSMENT

"Building your Portfolio" on page 253 provides a choice of two unit projects for students to work on as they study Unit 4. The unit projects are supported by Chapter Portfolio Assessment worksheets in the *Teaching Resources*.

THE WORLD TODAY VIDEODISC PROGRAM

The videodisc profiles the following countries in Unit 4: Mexico, Brazil, Ecuador, and Argentina.

...d decoration **Puerto Rico**

Clouds shroud the ancient Incan city of **Machu Picchu** high in the Andes Mountains of **Peru**. Middle and South America was home to many thriving civilizations before the arrival of the first Europeans in the late fifteenth century. Many people in the region can trace their ancestry to the original inhabitants of the region.

One World, Many Voices

LITERATURE

Inform students that the literature excerpt on this page and the primary source reading on the opposite page are related to the same subject: the 1911 "discovery" of the ruins of Machu Picchu, an ancient Incan city in Peru. Provide students with background information about the Incan Empire, and have them locate Machu Picchu on the map on page 195. Point out that one reason Spanish explorers never came across the city during the 16th century was its remote location. Draw attention to the literary style used by the author, Eduardo Galeano, to describe a historical event.

More about Eduardo Galeano

Eduardo Galeano of Uruguay spent nine years writing *Memory of Fire*, from which the excerpt below is taken. He sought to present the complex history of the Americas by focusing on the daily lives of individuals, and on small "moments" in history that became very significant. Galeano has also lived in Spain and Argentina.

The Incan Empire flourished during the 1400s. One characteristic of this advanced civilization was engineering skill. Masons cut the large stones used for the temples, houses, and fortified walls so perfectly that no mortar was necessary for fitting the stones together. The Inca also built complex irrigation networks and roads.

Interpreting Literature

Answers

1. Its location within a remote part of the Andes and dense jungle vegetation

2. Answers will vary, but students should realize by looking at the photographs that the "white stones" are remnants of the buildings in Machu Picchu and the fortified walls that protected the city.

ONE WORLD, MANY VOICES

LITERATURE

"1911: Machu Picchu The Last Sanctuary of the Incas

isn't dead; it only sleeps. For centuries the Urubamba River, foaming and roaring, has exhaled its potent breath against these sacred stones, covering them with a blanket of dense jungle to guard their sleep. Thus has the last bastion of the Incas, the last foothold of the Indian kings of Peru, been kept secret.

Among snow mountains which appear on no maps, a North American archeologist, Hiram Bingham, stumbles upon Machu Picchu. A child of the region leads him by the hand over precipices to the lofty throne veiled by clouds and greenery. There, Bingham finds the white stones still alive beneath the verdure [green vegetation], and reveals them, awakened, to the world."

Translated by Cedric Belfrage

© MARCELO ISARRUALDE

Eduardo Galeano (1940–) is an Uruguayan writer who is the author of *OPEN VEINS OF LATIN AMERICA, THE BOOK OF EMBRACES, WE SAY NO,* and the three-volume history of North and South America, *MEMORY OF FIRE*. He has worked as a journalist and political cartoonist.

In this excerpt from *MEMORY OF FIRE*, Galeano writes about the discovery of a great Incan city located near the peak of Machu Picchu. Until 1911, only Native Americans living high in the Andes Mountains were aware of its location or knew what remained of the ancient city.

Interpreting Literature

1. What elements of physical geography kept the incan city hidden for centuries?
2. Examine the photographs of Machu Picchu on this page and on page 185. What purposes do you think the "white stones" mentioned in this passage served in this city?

186

FOR THE RECORD

Point out that the route Hiram Bingham followed to seek the Incan capital was very treacherous. The expedition team he led through the Andes Mountains climbed rugged peaks and crossed dangerous river rapids. After students have read this primary source, focus on details in the literature feature on the opposite page that also provide details about Bingham's expedition. Tell students that the "child of the region" mentioned in the Galeano selection refers to the Native American child who helped guide Bingham to the site of Machu Picchu.

More about Hiram Bingham

Hiram Bingham grew up in Honolulu. There, he developed an interest in archaeology and mountain climbing while exploring Hawaiian peaks with his father, a missionary. Bingham undertook several archaeological expeditions across South America between 1906 and 1915. He became governor of Connecticut in 1924 and was later elected to the United States Senate.

FOR THE RECORD

HIRAM BINGHAM COLLECTION, COURTESY OF THE NATIONAL GEOGRAPHIC SOCIETY AND YALE UNIVERSITY, PEABODY MUSEUM OF NATURAL HISTORY.

Hiram Bingham (1875–1956) was an archaeologist and historian. Bingham led an expedition to find the ancient Incan capital city. He chanced upon the site after crossing the rugged Andes Mountains and following the course of the Urubamba River through dense jungle terrain. The excerpt below describes his amazement at the physical beauty of the Urubamba canyon.

"It will be remembered that it was in July 1911, that I began the search for the last Inca capital. . . . I had entered the marvelous canyon of the Urubamba below the Inca fortress of Salapunco near Torontoy.

Here the river escapes from the cold plateau by tearing its way through gigantic mountains of granite. The road runs through a land of matchless charm. . . . In the variety of its charms and the power of its spell, I know of no place in the world which can compare with it. Not only has it great snow peaks looming above the clouds more than two miles overhead; gigantic precipices of many-colored granite rising sheer for thousands of feet above the foaming, glistening, roaring rapids, it has also, in striking contrast, orchids and tree ferns, the delectable [pleasing] beauty of luxurious vegetation, and the mysterious witchery of the jungle. One is drawn irresistibly onward by ever-recurring surprises through a deep, winding gorge, turning and twisting past overhanging cliffs of incredible height."

Analyzing Primary Sources

Answers

1. Bingham's description of the river as it "escapes" the plateau, "tearing its way" through the mountains, with "roaring rapids" depicts a powerful and fast-moving river.
2. He was surprised to see rugged mountains topped with snow near a tropical forest or jungle. He had not expected the striking contrasts provided by the physical landscape.

Primary Source
Description of Change: excerpted and bracketed
Rationale: excerpted to focus on the main idea; bracketed to clarify meaning

Analyzing Primary Sources

1. What words does Bingham use to show that the Urubamba was not a calm river along the course that he followed?
2. What aspects of the physical geography were most surprising to Bingham?

CHAPTER 17 Introducing Middle and South America

Planning Guide

TEXTBOOK	TEACHING RESOURCES	OTHER AVAILABLE MATERIALS
Section		
1. Physical Geography Pages 193–194	Reteaching Worksheet 17.1	*REGIONS OF THE WORLD Map Transparencies:* Middle and South America *Creative Strategies for Teaching World Geography* LESSON 9 World Geography Class Action News *Study Guide for Spanish Speakers* *Lecture Notes,* Chapter 17
2. Historical Geography Pages 195–197	Reteaching Worksheet 17.2 Graphic Organizer 17 Challenge/Enrichment Worksheet 17	*Study Guide for Spanish Speakers* *Lecture Notes,* Chapter 17
3. Economic Geography Pages 197–198	Reteaching Worksheet 17.3	*Study Guide for Spanish Speakers* *Lecture Notes,* Chapter 17
4. Regional Issues Pages 199–200	Reteaching Worksheet 17.4 Critical Thinking Worksheet 17 Skills Worksheet 17	*Study Guide for Spanish Speakers* *Lecture Notes,* Chapter 17
Chapter Review Page 201	Chapter 17 Test Chapter 17 Portfolio Assessment See also *Alternative Assessment Forms*	*Test Generator,* Chapter 17

CHAPTER RESOURCES

Lecture Notes, Chapter 17
Teaching Resources
   Chapter 17 Graphic Organizer
   Chapter 17 Skills Worksheet
   Chapter 17 Critical Thinking Worksheet
   Chapter 17 Challenge/Enrichment Worksheet

REVIEW AND ASSESSMENT

Chapter 17 Check, page 201
Chapter 17 Test, *Teaching Resources*
Chapter 17 Portfolio Assessment, *Teaching Resources*
Test Generator, Chapter 17
Alternative Assessment Forms

ADDITIONAL RESOURCES

Books for Teachers

Blakemore, Harold, and Clifford Smith, eds. *Latin America: Geographical Perspectives.* 2d ed. New York: Routledge, 1983.

Bromley, Rosemary, and Ray Bromley. *South American Development: A Geographical Introduction.* 2d ed. New York: Cambridge University Press, 1982.

Garduno, Flor. *Witnesses of Time.* New York: Thames & Hudson, 1992.

McGuire, Stryker. *Streets with No Names: A Journey into Central and South America.* New York: Atlantic Monthly Press, 1991.

Skidmore, Thomas E., and Peter H. Smith. 3d ed. *Modern Latin America.* NY: Oxford, 1992.

Books for Students

Collier, Simon, et al, eds. *The Cambridge Encyclopedia of Latin America and the Caribbean.* New York: Cambridge University Press, 1992.

Dostert, Pierre Etienne. *Latin America.* Washington, DC: Stryker-Post Publications, 1988.

Swanson, Earl H., et al. *The Ancient Americas.* New York: Peter Bedrick Books, 1989.

Multimedia Resources

The Amazon. (VHS, 55 minutes) Films for the Humanities & Sciences, P.O. Box 2053, Princeton, NJ 08543–2053.

Latin America. (16mm film, VHS, 60 minutes) Social Studies School Service, P.O. Box 802, Culver City, CA 90232–0802.

Let's Visit South America. (Multimedia CD-ROM, interactive tour) Queue, 338 Commerce Drive, Fairfield, CT 64305.

Picture Atlas of the World. (interactive maps, photo bank, animation, audio on CD-ROM) National Geographic Society, Washington, DC.

MEETING INDIVIDUAL NEEDS

Limited English Proficiency (LEP)

Write the following words from the Chapter 17 Geography Dictionary on the board: **pampa, llano, mestizo, mulatto, cacao,** and **hacienda.** Pronounce the words for the students, and then have students say them out loud. Point out that all these words are from the Spanish language. Explain the meanings of terms with which students are not familiar.

Learners Having Difficulty

Have students look at the maps on pages 189–192 while you ask the following questions: What are the countries of Middle America? Of South America? What are the major landforms of Middle and South America, and where are they located? What different climate types can be found in this region? What are the major cities of the region, and where are they located? What areas are the most densely populated? What are the major natural resources, agricultural products, and industries in the region?

Auditory Learners

Lead a discussion on how European colonization of Middle and South America changed the way of life of the indigenous peoples there. Have students hypothesize what the culture of the region might be like today if the Europeans had not come.

Visual Learners

Have students create a picture time line of the historical geography of Middle and South America. Each student might illustrate a different date or event. Have students work together to assemble the time line in the classroom. Ask them to think of a question based on the time line. Call on volunteers to ask their question to the class.

Tactile/Kinesthetic Learners

Arrange newspaper and magazine articles about issues affecting Middle and South America on a bulletin board under the headings Population Issues, Political Issues, and Economic Issues. Assign each article to two or three students. Have each student write a question about the story on the front of an index card and write the answer on the back. Tack the cards on the bulletin board next to the appropriate articles.

Gifted Learners

Have interested students research a country in Middle or South America to find places that might be of interest to tourists. Ask students to pretend to be a guide for their chosen country and take the class on a "tour" of the country, using maps, pictures, artifacts, or other visual aids.

MULTICULTURAL FOCUS

Many of the ethnic groups in Middle and South America have in common the Roman Catholic religion. Catholicism was spread throughout the region by missionaries who followed Spanish and Portuguese explorers and colonists.

When a Catholic child is baptized, the parents choose two people to be the child's *padrinos,* or godparents. Godparents take on the responsibility for the religious upbringing of the child in case the parents die or are unable to fulfill this obligation. In Hispanic (Latino) cultures, the *madrina* (godmother) and the *padrino* (godfather), although not necessarily married to each other, are a lot like a second family for the child.

A strong bond of spiritual kinship exists between the *padrinos,* the child, and the child's family. The *madre* (mother) of the child and the *madrina* call each other *comadre,* or co-mother. Likewise, the *padre* (father) of the child and the *padrino* are *compadres,* or co-fathers. Apart from its religious basis, co-parenthood in Latin America serves many social functions by extending a family's ties to other families or to a benefactor of higher social status. The networks of relationships—almost like extended families—created by these bonds provide a support system in communities.

Have students relate the notions of *padrinos* and ritual kinship to their own experience. Ask them if they have godparents, family friends, or relatives who are like *padrinos* to them. Ask: What other types of bonds bring individuals together into family-like networks?

CHAPTER 17

INTRODUCING MIDDLE AND SOUTH AMERICA

Section 1: Physical Geography
Section 2: Historical Geography
Section 3: Economic Geography
Section 4: Regional Issues

OVERVIEW

Chapter 17 gives an overview of South America and Middle America, which includes Mexico, Central America, and the Caribbean islands.

Section 1 describes the diverse landforms and climates in this region, which includes the soaring Andes Mountains, the vast Atacama Desert, and the world's largest tropical rain forest, the Amazon.

Section 2 outlines the history of the region and explains the effects of European colonialism on Middle and South America.

Section 3 discusses the natural resources, agriculture, and industry of Middle and South America.

Section 4 presents economic and political issues facing the region, many of which affect events in the United States as well.

CREATIVE STRATEGIES FOR TEACHING WORLD GEOGRAPHY

LESSON 9 World Geography Class Action News

CHAPTER 17

Introducing Middle and South America

Fifteenth-century plaster carving from a monastery in **Oaxaca, Mexico**

Geographers have used cultural and historical geography to define a region called Middle America. Middle America includes Mexico, the seven countries of Central America, and the many islands of the Caribbean Sea. The region also includes the Isthmus of Panama, which links North America to South America. An **isthmus** is a narrow neck of land that acts as a bridge to connect two larger bodies of land.

Middle America is one of the most culturally diverse regions in the world. Though the Caribbean islands were originally inhabited by **indigenous** (native) peoples and later colonized by European countries, millions of African slaves were transported to the Caribbean to work its sugar plantations. As a result, the major cultural influence is African.

South America extends from the warm and sunny beaches of Point Gallinas in Colombia to the cold and stormy seas around Cape Horn, about 4,500 miles (7,241 km) to the south. Its peoples are as diverse as its physical geography. The continent's many ethnic groups are separated by geographical obstacles, including the Andes Mountains and the forests of the Amazon. The region is united, however, by the Roman Catholic religion and the Spanish language (except for Brazil, where Portuguese is spoken). The region's architecture, laws, language, and religion are lasting reminders of Europe's colonial rule of South America.

GEOGRAPHY DICTIONARY

isthmus

indigenous

pampas

llanos

indigo

mestizo

mulatto

cacao

hacienda

labor intensive

shantytown

land reform

Ancient Mayan pyramid in **Yucatán, Mexico**

Using the Geography Dictionary

Pronounce the words in the Geography Dictionary on page 188. Have student volunteers define any terms they can. Then ask other students to look up and paraphrase the definitions for the remaining terms. Point out that a number of the terms **(pampas, llanos, indigo, mestizo, mulatto, cacao, hacienda)** are Spanish in origin. Note that the two regions discussed in this chapter are dominated by the Spanish language.

MAPPING THE REGION
Using the Physical-Political Map

As students study the map on this page, call their attention to the physical features they will study in this chapter. Ask them to name the dominant landforms. *(Amazon Basin, Andes Mountains)* Review the definition of *basin* (page 113), and have them note the elevation patterns in the huge Amazon Basin.

MAP TRANSPARENCIES

Overhead transparencies from the maps on pages 189–192 are available in *REGIONS OF THE WORLD Map Transparencies.*

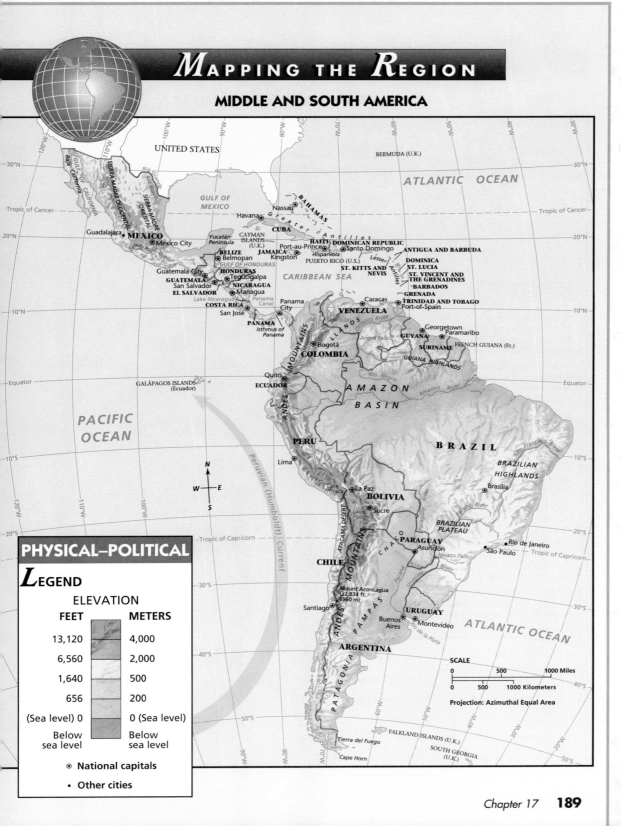

MAPPING THE REGION
MIDDLE AND SOUTH AMERICA

PHYSICAL–POLITICAL

LEGEND

ELEVATION

FEET	METERS
13,120	4,000
6,560	2,000
1,640	500
656	200
(Sea level) 0	0 (Sea level)
Below sea level	Below sea level

⊛ National capitals

• Other cities

MAPPING THE REGION

Using the Climate Map

Direct students' attention to the climate map on this page. Point out the large variety of climate types in the region. Ask students to compare this map to the physical-political map on page 189. You might have them write sentences that describe the relative locations of the main climate areas of South America. *(For example: "The highland climate runs the length of the Andes." "The humid tropical climate covers the Amazon Basin.")* Ask students to study the map legend and then choose the climate that is *not* found on the continent of South America. *(subarctic)*

MAPPING THE REGION
MIDDLE AND SOUTH AMERICA

CLIMATE

◆ Variable highland climates dominate large parts of central Mexico, Central America, and western South America.

◆ Much of eastern Central America and equatorial South America is dominated by a humid-tropical climate.

◆ Only small areas of south-western and southeastern South America have a cool, moist, marine-west-coast climate.

Compare this map to the physical–political map on page 189.

⁇ Which countries have desert climates?

⁇ Which climate regions exist in Chile?

Compare this map to the population map on page 191.

⁇ Which climate types are found in the areas around São Paulo and Rio de Janeiro?

LEGEND

- Humid tropical
- Tropical savanna
- Desert
- Steppe
- Mediterranean
- Humid subtropical
- Marine west coast
- Subarctic
- Highland

SCALE
0   500   1000 Miles
0   500   1000 Kilometers

Projection: Azimuthal Equal Area

CLIMATE

Answers

1. Mexico, Peru, Chile, Argentina, Venezuela
2. Desert, highland, steppe, Mediterranean, marine west coast
3. Humid subtropical, marine west coast

190

MAPPING THE REGION
Using the Population Map

Have students examine the population map on this page. You might have them describe the general pattern of population distribution in South America. *(Most people live near the coast or in highland regions; the interior is very sparsely populated.)* Then ask students to suggest reasons why the population is distributed in this way. *(Coasts: more accessible than the interior; early settlements grew into towns and cities. Highlands: more temperate climate than at lower elevations.)*

MAPPING THE REGION
MIDDLE AND SOUTH AMERICA

POPULATION

◆ Many of Middle and South America's largest cities are near coasts or in highland regions.

◆ Central Mexico, including Mexico City, is the most densely populated region of Mexico.

◆ Most of interior South America has a lower population density than many coastal regions have.

⟨?⟩ Which country has the most metropolitan areas with more than 1 million people?

⟨?⟩ Name the metropolitan areas of Mexico that have more than 2 million people.

Compare this map to the economic map on page 192.

⟨?⟩ Name the major manufacturing and trade centers that have populations of more than 2 million.

LEGEND
POPULATION DENSITY

Persons per sq. mile	Persons per sq. km
520	200
260	100
130	50
25	10
3	1
0	0

● Metropolitan areas with more than 2 million inhabitants

● Metropolitan areas with 1 million to 2 million inhabitants

SCALE
0   500   1000 Miles
0   500   1000 Kilometers
Projection: Azimuthal Equal Area

MAPPING THE REGION

Using the Economy Map

Have students examine the map on this page closely. Call on volunteers to name the countries that have oil as a resource. *(Mexico, Venezuela, Colombia, Ecuador, Bolivia, Chile, Argentina, Brazil)* Ask: In which parts of South America has hydroelectric power been developed? Discuss how the availability of oil and hydroelectric power might affect the economies of individual countries. *(The economies would likely be strengthened if these resources were developed.)*

SECTION 1

OBJECTIVES

The objectives reflect the Focus questions on page 193 and are assessed in the review on page 194.

1. Explain the geologic processes that created the landforms of Middle and South America.

2. Describe the climates and river systems of Middle and South America.

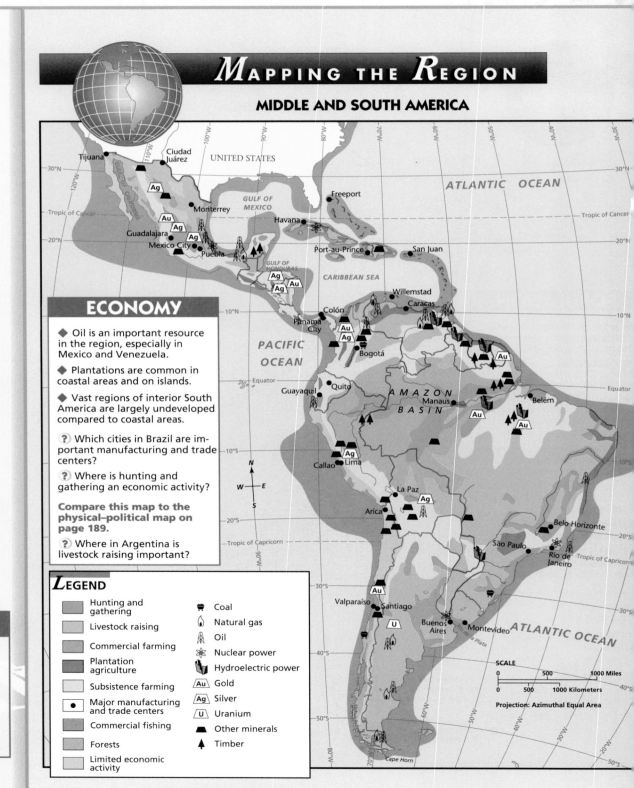

MAPPING THE REGION

MIDDLE AND SOUTH AMERICA

ECONOMY

◆ Oil is an important resource in the region, especially in Mexico and Venezuela.

◆ Plantations are common in coastal areas and on islands.

◆ Vast regions of interior South America are largely undeveloped compared to coastal areas.

(?) Which cities in Brazil are important manufacturing and trade centers?

(?) Where is hunting and gathering an economic activity?

Compare this map to the physical–political map on page 189.

(?) Where in Argentina is livestock raising important?

LEGEND

- Hunting and gathering
- Livestock raising
- Commercial farming
- Plantation agriculture
- Subsistence farming
- • Major manufacturing and trade centers
- Commercial fishing
- Forests
- Limited economic activity

- 🜨 Coal
- 🜁 Natural gas
- 🜔 Oil
- ✳ Nuclear power
- 🏭 Hydroelectric power
- Au Gold
- Ag Silver
- U Uranium
- ▲ Other minerals
- ♣ Timber

SCALE
0 ___ 500 ___ 1000 Miles
0 ___ 500 ___ 1000 Kilometers

Projection: Azimuthal Equal Area

ECONOMY

Answers

1. Manaus, Belém, Belo Horizonte, Rio de Janeiro, São Paulo

2. Amazon Basin

3. Pampas, Patagonia

TEACHING SECTION 1

Motivator

Ask students what images come to mind when they think of the land-forms and climates of Middle and South America. Students may say lush mountain valleys and tropical rain forests. Point out that the region also has a wide variety of other land-forms and climates, including plains and deserts, which students will learn about in Section 1.

Building a Vocabulary

Remind students that they studied the Great Plains region of the United States in Unit 3. Write **pampas** and **llanos** on the board. Tell them that these terms refer to large areas of plains in South America.

Guided Practice

List the climates of Middle and South America on the board. Then call on volunteers to show the location of each climate on a wall map. Ask students to describe characteristics of each type of climate.

PHYSICAL GEOGRAPHY

FOCUS

- *How were the various landforms of Middle and South America created?*

- *What climate types and what rivers are found in Middle and South America?*

REGION **Landforms** According to the theory of plate tectonics, South America and Africa broke apart from Gondwanaland more than 100 million years ago. As South America moved westward, other continental pieces from the Pacific region collided with the west coast. These collisions may have formed the first Andes Mountains. A few million years ago, South America began to ride up over the Nazca plate under the southeastern Pacific Ocean. Since this first collision, the Andes have continued to build as the result of folding, faulting, and volcanic activity. The other mountains of South America are the Guiana (gee-AN-uh) Highlands and the highlands of eastern Brazil. These eroded highlands were formed much earlier than the Andes.

Plains cover most of South America. The largest plains are in the Amazon River basin. The other large plains are the **pampas** of Argentina and the **llanos** (YAHN-ohz) of Colombia and Venezuela. These plains are the result of erosion steadily smoothing the land's surface over millions of years.

The theory of plate tectonics also explains the creation of islands around the Caribbean Sea. As the South American plate drifted westward from Africa, the smaller Caribbean plate cruised eastward. This Caribbean plate carried only a few continental pieces. These pieces became the mountainous islands of the Greater Antilles—Cuba, Jamaica, Hispaniola, and Puerto Rico. At the eastern edge of the Caribbean plate was a chain of volcanic islands—the Lesser Antilles.

Central America and Mexico are believed to have been formed from a jumble of small plates attached to the rim of the North American and Caribbean plates. Mexico consists mostly of a high central plateau with mountains along each side. Central America has many short mountain ranges and huge volcanoes. These mountains and an ocean trench off the Pacific coast of southern Mexico and Central America indicate a continuing plate collision. Scientists believe that Central

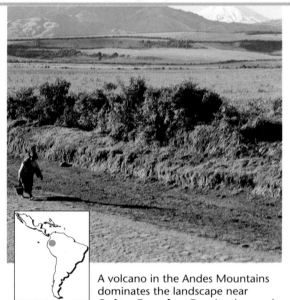

A volcano in the Andes Mountains dominates the landscape near **Quito, Ecuador**. Despite the area's tropical location, many people in the region, such as the woman in the foreground, dress for the cooler highland climates of the Andes Mountains.

America and South America became connected only a few million years ago. Earthquakes and volcanoes are a constant threat throughout Middle America and the western side of South America.

Climate Because the region of Middle and South America extends across nearly 90 degrees of latitude, it includes a full range of middle-latitude as well as tropical climate regions. South America contains the largest humid-tropical region in the world—the huge forested area centered on the Amazon River. Rain falls here almost every day. The region's other areas of humid-tropical climate are along the eastern coasts where the moist trade winds move on shore.

Many areas of Middle and South America experience wet summers and dry winters typical of the tropical-savanna climate. The many areas of dry forests on the Caribbean islands and the Pacific coast of Central America have a similar climate.

Southern South America has a variety of middle-latitude climates. Far southern Chile is influenced by the westerlies. The climate found here is cool and wet, similar to that of the United States' Pacific Northwest. Snow and rainfall are particularly heavy in the southern Andes. East of the Andes in southern Argentina, however, the mountains block moisture from reaching the area. As a result, semiarid steppe and desert climates

Closure

Have a student name a climate, land-form, or major river of Middle and South America. The student should then call on another student to locate the climate, landform, or river on a map.

ASSESSMENT

Assign the Section 1 Review questions on this page.

Reteaching

Assign Section 1 of Reteaching Worksheet 17.

Extension

Have interested students read an account from an article or book about a journey on one of South America's rivers. Ask students to share their impressions of the article or book with the class.

SECTION 2
OBJECTIVES

The objectives reflect the Focus questions on page 195 and are assessed in the review on page 197.

1. List and describe ancient Native American civilizations in Middle and South America.
2. Analyze the effects of European colonization, and of subsequent independence, on Middle and South America.

MAP TRANSPARENCIES

An overhead transparency of the graphs on this page is available in *REGIONS OF THE WORLD Map Transparencies.*

Skill Study

Precipitation amounts are higher during the winter.

Section 1 Review
ANSWERS

1. Andes: originally formed by the collisions of continental pieces; mountainous Greater Antilles: formed by drifting of continental pieces from the Caribbean plate; Lesser Antilles: formed by volcanoes; mountains of Central America and Mexico: formed from a jumble of small plates that continue to collide

2. The region has several climate types because it spans much latitude. The climate types include humid tropical, tropical savanna, steppe, desert, Mediterranean, humid subtropical, and highland. Major river systems are the Amazon, the Paraná, and the Orinoco.

3. Mountainous terrain might hinder the movement of aid to damaged areas.

194

Graphing Climate in South America

SKILL STUDY The first two climographs above show the effects of the Andes Mountains on climates in South America. To learn how to read climographs, see page S11 of the Skills Handbook. Although both are located near the equator, the humid-tropical climate of Manaus, Brazil, is much warmer and moister than the highland climate of Quito, Ecuador. The Ecuadorean capital is located in the higher elevations of the Andes Mountains. Other than having cooler temperatures, what is the weather of Puerto Montt, Chile, like during the winter?

are found. Farther north, the central valley of Chile has a Mediterranean climate, with winter rains and summer droughts. A large region with humid-subtropical climates is found in northeastern Argentina and southern Brazil.

The driest region of Middle and South America includes the Atacama Desert of northern Chile and the nearby coastal deserts of Peru and southern Ecuador. Subtropical high pressure brings dry weather to these areas all year. Because the high Andes block moisture from the east, the western slopes of the mountains are very dry.

The mild climates on mountain slopes between the elevations of 3,000 and 6,000 feet (914 and 1,829 m) are particularly important in tropical Middle and South America. The mountain valleys have fertile soils and water from mountain streams. These pleasant environments are some of the densest areas of settlement in the region.

The pattern of the highland climates of the region is important in understanding population distribution in the region. (See the maps on pages 190 and 191.) This climate pattern can be seen from the mountains of Mexico and Central America through the Andes of South America. You know that temperatures become cooler at higher elevations. For this reason, the region's highland climates have been divided into elevation zones along mountain slopes. Geographers are able to study how population distributions and ways of life relate to these climate patterns. You will learn more about these elevation zones in Chapter 20.

River Systems Three major river systems drain the eastern side of South America: the Amazon, the Paraná, and the Orinoco. Beginning on the eastern side of the Andes, the Amazon River stretches more than 4,000 miles (6,436 km) before it reaches the Atlantic Ocean. Curiously, the Orinoco and Amazon river basins are linked. (See the map on page 189.) The Río Negro, which flows mainly south into the Amazon, has an upper tributary that flows north into the Orinoco during part of the year. Farther south, the Paraná River system empties into the Río de la Plata estuary, located between Argentina and Uruguay. The headwaters of the Paraná River system are located in the highlands of eastern Brazil and in western Argentina along the eastern slopes of the Andes.

SECTION REVIEW

1. What forces created the different mountain ranges of Middle and South America?
2. Why does the region of Middle and South America have several climate types, and what are they? What are the region's major river systems?
3. **Critical Thinking**   **Imagine** a country in Middle America experiences a severe earthquake. How might the physical geography of the region affect that country's ability to respond?

Motivator

Tell students that the ancient Mayan civilization had well-developed systems of mathematics and astronomy. The Mayas had an accurate solar calendar, kept precise tables of the locations of the Moon and Venus, and could predict solar eclipses. Tell students that in this section they will learn about civilizations that existed in Middle and South America before the Europeans arrived.

Building a Vocabulary

Write **mulatto** and **mestizo** on the board. Discuss how *mulatto* and *mestizo* reflect the cultural diversity of Middle and South America.

Guided Practice

Ask students to name some of the effects of European colonization on Middle and South America. List their answers on the board.

---

② HISTORICAL GEOGRAPHY

FOCUS

- *What Native American civilizations existed in Middle and South America before the arrival of Europeans in the region?*

- *How did European colonization affect the region's land and peoples? What effects did independence have?*

Settlement Patterns By the time Christopher Columbus arrived in the Americas, Native Americans had lived in the region for thousands of years. Columbus thought he had reached the islands off the coast of India. He called the indigenous peoples he found living there "Indios," or Indians. Columbus actually had landed on an island in the Bahamas, southeast of Florida. He claimed the island for Spain and named it San Salvador (Holy Savior). Emigration from Spain was encouraged, and by the mid-1500s, there were thousands of Spanish colonists on the larger Caribbean islands.

Most Native Americans were farmers and hunters who lived in small villages. Some groups traded salt, shells, and other goods with their neighbors. There were also, however, great Native American civilizations that built impressive monuments and large cities housing tens of thousands of people.

The ancient Olmecs thrived in eastern and southeastern Mexico from about 1200 B.C. to about 400 B.C. They developed extensive trade and were the first people to build religious pyramids, greatly influencing later Middle American cultures.

The Mayas, considered one of the most advanced Native American cultures, flourished between about A.D. 250 and 900. Ruins of once brilliant Mayan cities now lie in the forests of Guatemala, Honduras, and the Yucatán Peninsula of Mexico and Belize. Sophisticated Native American societies also flourished elsewhere on the mainland of Middle America and in northern South America.

At the time the Spanish arrived on the continents, there existed two monument-building civilizations. The empires of the Aztecs in central Mexico and of the Incas in western South America developed about A.D. 1300 and continued the achievements of earlier civilizations. The Aztec Empire was centered at Tenochtitlán (tay-nawch-tee-TLAHN) on the present site of Mexico City. Tenochtitlán and the Aztec Empire were conquered in 1521 by the Spanish explorer Hernán Cortés (kawr-TEHZ). The Incan Empire extended along the Andes Mountains from Ecuador to Chile.

Many factors contributed to the decline of Native American civilizations. The Europeans, with guns and armor, had a technological advantage in warfare. Furthermore, diseases that the Europeans brought with them, such as smallpox and measles, were devastating to the Native

Early Civilizations of Middle and South America

Tollán
Teotihuacán • Chichén Itzá
Tlacopán Lake Texcoco Mayapán
Tenochtitlán• Cholula
Xochicalco
Monte Albán
Xoconocho• Tikal

CARIBBEAN SEA

PACIFIC OCEAN

SOUTH AMERICA

Equator Quito

ANDES MOUNTAINS

Machu Picchu
Cuzco
Lake Titicaca
Lake Poopó

Tropic of Capricorn

LEGEND

- Olmecs (1200–400 B.C.)
- Mayans (A.D. 250–900)
- Aztec Empire (1519)
- Incan Empire (1532)
- ★ Capitals
- • Other cities and cultural centers

SCALE
0   500   1000 Miles
0   500   1000 Kilometers
Projection: Azimuthal Equal Area

 MAP STUDY Historical maps are important tools for understanding earlier times. Some historical maps identify earlier civilizations, major cities, and important events. This map shows the location and dates of civilizations in Middle and South America before European colonization. Which civilization dominated western South America at the time the Spanish reached the region in the first decades of the sixteenth century?

Independent Practice

Give students Outline Maps 11 and 13, and ask them to label and date the areas where Olmec, Maya, Aztec, and Inca civilizations were located. Then have students label the maps to show the areas that were colonized by the Spanish, Portuguese, French, British, and Dutch. Finally, have students place arrows on the maps to show how the region's trading patterns changed following independence.

Closure

Have students name similarities and differences between the history of Middle and South America and that of North America.

ASSESSMENT

Assign the Section 2 Review questions on page 197.

Reteaching

Assign Section 2 of Reteaching Worksheet 17.

Extension

Have students research archaeological sites of one of the ancient civilizations of Middle and South America. Ask them to find or draw a picture of an artifact (piece of pottery, jewelry, tool) they might discover at the site. Have students explain their pictures to the class.

Geography and Language

At one time, thousands of Native American languages existed in Middle and South America. Scholars estimate that in Brazil alone 1,200 languages were spoken by indigenous peoples in the early 1500s. Today, linguists count 180 languages in that country. The death of millions of Native Americans during the colonial period and the continued Europeanization of native cultures led to the great decline in the number of Native American languages.

Americans. When Cortés arrived in the Valley of Mexico, perhaps three-quarters of the Aztec people were ill. Diseases from Europe eventually killed millions of Native Americans.

European Colonization By the late fifteenth century, Spain and Portugal were rich and powerful. Both countries were eager for the profits possible by having overseas empires. A treaty in 1494 divided the Western Hemisphere into Spanish and Portuguese territories. Portugal received a large triangle of eastern South America, including present-day Brazil.

Spanish and Portuguese colonists quickly settled the region. They extended the Native Americans' gold and silver mines and opened new ones. Ranching and farming were the key economic activities, however. The colonists brought with them European crops, livestock, and farming methods. They also adapted Native American crops, such as maize (corn), beans, and potatoes, to their needs. Spanish colonists were awarded large land grants by the Spanish king. Native Americans living on this land were forced to work the farms and ranches. The idea that land could be owned was new to the indigenous peoples.

By the mid-1600s, the French, British, and Dutch began to acquire lands that the Spanish had not colonized. These included the smaller islands of the Caribbean, as well as Jamaica and Haiti. Their only mainland settlements were British Honduras (now Belize), British Guiana (now Guyana), French Guiana, and Suriname (soohr-uh-NAMH-uh). These colonies were all based on plantation agriculture. Major plantation crops were sugarcane and **indigo**. Indigo is a plant used to make a blue dye. Later, bananas became a key plantation crop. For more than two centuries, African slaves provided most of the labor for the plantations. When slavery ended, the British imported field workers from India, while the Dutch brought in Indonesians.

Because there were few women among the colonists, European settlers often married Native American or African women. Many Middle and South Americans are **mestizos**, or people with both Native American and European ancestors. The descendants of European plantation colonists and Africans are known as **mulattos**. Today, the geography of the peoples, languages, and cultures in the region is mostly a result of the process of colonization.

Independence By 1800, ideas of independence introduced by the American and French revolutions were well known throughout Middle and South America. Between 1808 and 1825, many colonies gained independence. Yet, the economies and societies of the various countries usually changed little. Rich landowners, merchants, and military generals merely replaced the colonial government. The lives of rural peoples remained much the same.

With independence, Middle and South America began trading more with Great Britain and the United States rather than with Spain. Great Britain protected the new Middle and South

---

Languages of Middle and South America

Spanish ▶		▶ Mexico, Argentina, most other countries of Middle and South America
Portuguese ▶		▶ Brazil
French ▶		▶ Haiti, French Guiana, Guadeloupe, Martinique
Dutch ▶		▶ Suriname, Aruba, Netherlands Antilles
English ▶		▶ Belize, Guyana, some Caribbean countries and territories
Native American ▶	Nahuatl	▶ Mexico, El Salvador
	Quiché	▶ Guatemala
	Quechua	▶ Peru, Bolivia, Ecuador, Argentina, Colombia
	Aymara	▶ Bolivia, Peru
	Guaraní	▶ Paraguay, Argentina, Brazil

NOTE: Nahuatl, Quiché, Quechua, Aymara, and Guaraní are only five of the many Native American languages and dialects spoken in Middle and South America, especially in southern Mexico, Central America, and Andean South America.

Spanish and Portuguese are the most widely spoken languages in Middle and South America. Many Native American languages, however, still are spoken throughout the region.

The objectives reflect the Focus questions on this page and are assessed in the review on page 198.

1. Identify natural resources and industries important to the economies of countries in Middle and South America.
2. Categorize agricultural systems in Middle and South America.

TEACHING SECTION 3
Building a Vocabulary

Point out that **cacao, hacienda,** and **labor intensive** are all related to economic activities in Middle and South America. Ask students to write sentences using these terms. Call on volunteers to read their sentences.

Identifying

Natural Resources Recall that gold and silver lured European adventurers to Middle and South America. Have students name the mineral resources that might attract people to the region today. Ask students which mineral they think is most important to the economies of Middle and South America.

ECONOMIC GEOGRAPHY

FOCUS

- *What natural resources and industries are important to the economies of the region's countries?*
- *What forms of agriculture are common in Middle and South America today?*

Natural Resources The countries of Middle and South America have rich mineral deposits, fertile soils, and climates suitable for growing many different crops. Dams on many rivers provide irrigation water and generate electricity. Several countries also have forest and fish resources.

The mineral wealth that attracted the Spanish to the region centuries ago is still being developed. Gold and silver deposits have been found in Brazil, Mexico, Colombia, and Peru. Metals such as copper are being mined in parts of the Andes, in central Brazil, and in Mexico's mountains. Brazil has enormous iron-ore reserves. Bauxite, the ore of aluminum, is widely available in northeastern South America and in Jamaica.

Petroleum is found in many countries of Middle and South America as well. The greatest oil reserves are in Mexico and Venezuela. Oil deposits also have been developed in the upper Amazon Basin. More recent oil discoveries have been made off the coasts of Brazil, Argentina, and Chile.

Agriculture Several types of agriculture are practiced in Middle and South America today. Subsistence agriculture remains common in many remote areas, such as in inland forests and in the mountains. Subsistence farmers often live in small villages. Each day, they walk to their nearby fields.

In contrast, modern commercial agriculture, or agribusiness, uses the latest farming methods and mechanized equipment. Farm products are processed and exported to international markets. Areas with agribusiness are generally prosperous agricultural regions. Commercial farming is well developed in southern Brazil and southward to the pampas of Argentina, in the central valley of Chile, and in northern Mexico. Fruits and vegetables from each of these regions can be bought in stores throughout the United States.

This statue in **Cartagena, Colombia,** honors Simón Bolívar for his role in leading South America's fight for independence. Statues of Bolívar, who is called *El Libertador* (The Liberator) by some South Americans, can be found in cities throughout the continent.

American states with its navy. The United States offered diplomatic protection under the Monroe Doctrine, which declared the countries of Middle and South America to be off-limits to new European colonization. The influence of the United States has remained strong throughout the region.

 SECTION REVIEW

1. What two Native American empires were encountered by Spanish explorers in Middle and South America? Where was each empire located?
2. How did the region's agriculture change as a result of European colonization?
3. **Critical Thinking** **Agree or disagree** with the following statement: "The colonization of Middle and South America by Spain and Portugal had both positive and negative effects." Support your answer.

Section 2 Review
ANSWERS

1. The Aztec Empire, in central Mexico, and the Incan Empire, in western South America
2. Europeans introduced new crops, livestock, farming methods, and the concept of land ownership.
3. Answers will vary. Some students may say that there were both positive and negative effects; other students may say that the effects were mostly positive or mostly negative. Possible positive effects include the introduction of new crops and farming methods to the region; possible negative effects include the disruption of Native American societies and the introduction of certain diseases.

197

Guided Practice

Discuss with students similarities and differences between subsistence agriculture and commercial agriculture. Ask students to describe some of the different types of Middle and South American agricultural systems, including plantations and haciendas.

Independent Practice

Have students make a graphic organizer showing the natural resources, agricultural products, and manufactured products of Middle and South America.

Closure

Have students take turns naming a food or product they use every day that might have had its origin in Middle or South America.

ASSESSMENT

Assign the Section 3 Review questions on this page.

Reteaching

Assign Section 3 of Reteaching Worksheet 17.

GLOBAL PERSPECTIVES

Bananas were eaten mostly in the tropical areas where they were grown until the mid-1800s, when U.S. and European merchants started to market them in other regions. Today, the banana industry is a multi-million dollar enterprise controlled by multinational corporations and the eight-member Union of Banana Exporting Countries, based in Panama. In 1990, the United States imported over 3.4 million tons (3.1 million metric tons) of bananas, mainly from Ecuador, Costa Rica, and Honduras. Lead a discussion about how the introduction of large, foreign-owned plantations might have affected the countries of Middle and South America.

Manufacturing jobs, such as those provided by this steel plant in **Monterrey, Mexico**, attract many people to urban areas throughout Middle and South America. The pace of economic growth must increase to catch up with the region's soaring population growth.

Today, plantation agriculture is a form of modern commercial agriculture. It is practiced on many Caribbean islands and along the region's humid coasts. Most plantation products, including sugar, bananas, and **cacao**, are exported to developed countries. Cacao is a small tree on which the cocoa bean grows. The cocoa bean is used in making chocolate. Modern-day plantations usually are part of large, foreign-owned corporations.

Another form of agriculture common in Middle and South America is the **hacienda** system. Introduced by the Spanish, haciendas are large family-owned estates. Peasants live on these estates and have small plots on which to grow their own crops. In return for these plots and services granted by the landowner, the peasants also must work the owner's land as a form of payment. More symbols of the landowners' social status than profit-making operations, haciendas are often unproductive. Even so, they remain a basic form of land organization in much of the region.

Industry As manufacturing increases in Middle and South America, the number of jobs in urban areas also increases. The largest industrial region is located in an arc of cities that extends from Rio de Janeiro in southeastern Brazil to Buenos Aires (bway-nuhs AR-eez) in eastern Argentina. Most capital cities of Middle and South America have developed into industrial regions.

Most Middle and South American factories produce food items, consumer goods, or building materials for local markets. Automobiles, trucks, and railway cars are produced in the larger countries. Industries that manufacture export products are often **labor intensive**. That is, they require a large work force. Often assembled by hand, export products include clothing, furniture, and small appliances. While these labor-intensive industries provide jobs and the training that may lead to better jobs, factory workers often face a life of hard work with little reward.

Tourism is a growing industry in the region, particularly in the islands of the Caribbean. The income generated from tourism is important to many countries' economies. The benefits of the tourist industry to local island communities are sometimes less than expected, however, because hotel and resort workers generally receive low pay.

SECTION REVIEW

1. List four natural resources and four industries found in Middle and South America.
2. Compare and contrast plantations and haciendas.
3. **Critical Thinking**   **Determine** which forms of agriculture are labor intensive, and explain your answer.

198

SECTION 4
OBJECTIVES

The objectives reflect the Focus questions on this page and are assessed in the review on page 200.

1. Examine the relationships between economics and politics in Middle and South America.
2. Discuss foreign influences on Middle and South America.

TEACHING SECTION 4
Motivator

Show students news items about Middle and South America from newspapers and magazines. Ask why young people in North America should be concerned about events in Middle and South America. Emphasize the need for people in the Western Hemisphere to work together for political and economic stability.

Building a Vocabulary

Write **land reform** and **shantytown** on the board. Ask students to look up the definitions of these two words in the Glossary, and have a volunteer read them aloud. Ask students how these terms might relate to each other. *(Land reform could ease rural economic problems that lead people to migrate to the cities and live in shantytowns.)*

Economic progress, especially in the developing nations of the world, often has proceeded at the expense of the environment. Large parts of the Amazon rain forests, such as here in **Brazil**, have been burned and cleared. Developers want to use the cleared land for farming and mining.

REGIONAL ISSUES

FOCUS

- *How are economics and politics linked in the region?*
- *How do countries outside the region affect countries in the region? How do countries in the region affect countries outside the region?*

REGION **Population Issues** Middle and South America are experiencing rapid population growth. If current growth rates continue, many countries will double their populations in 20 to 30 years. Building schools and roads as well as creating jobs for this increasing number of people will be difficult.

Rapid population growth often leads to increased migration from poor rural areas to cities. Rural migrants come to the cities seeking jobs and a better life. Most find housing in the **shantytowns** that form the large slums surround-ing many Middle and South American cities. A shantytown is a poor settlement of small makeshift shelters. City services such as water and waste disposal are not available, and most roads are unpaved.

Even in the richest countries of Middle and South America, there are many poor people. Middle and South America contain sharp economic and social contrasts. Lack of economic development has contributed to political instability in the region.

Political Issues While much of Middle and South America today is democratic, most countries have histories of repeated changes in government. Typically, different groups within the wealthy families and the military have battled for control of a country's government. Changes in government usually have been minor and have

Section 3 Review
ANSWERS

1. Resources: hydro-electricity, forests, fish, gold, silver, copper, iron, bauxite, petroleum; industries: food items, consumer goods, building materials, automobiles, trucks, railway cars, clothing, furniture, appliances, tourism
2. Plantations are a form of modern commercial agriculture, export crops to developed countries, and are often part of foreign-owned corporations. Haciendas are large family-owned estates worked by peasants who also do some subsistence farming; unlike plantations, haciendas are often unproductive.
3. Subsistence agriculture (on a large scale) and agriculture on haciendas; neither uses much technology, so both require a large work force.

Guided Practice

Organize the class into several small groups. Have students imagine that they belong to a political party in Middle and South America. Ask each group to write a platform, stating the party's position on various issues discussed in Section 4. *(for example, the breaking up of large landholdings, the involvement of foreign countries and companies, or the development of rain forests)* Have groups share their platforms.

Closure

Have students choose the regional issue they think is the most critical in Middle and South America and explain their choice.

ASSESSMENT

Assign the Section 4 Review questions on this page.

Reteaching

Assign Section 4 of Reteaching Worksheet 17.

Extension

Point out that more than half of South America's population is under the age of 30. These young people are leading the way for democratic reforms in the region. Have the class look for examples of reform movements in news articles and discuss their implications.

Section 4 Review
ANSWERS

1. Economic inequality and lack of economic development; through land reform, other policies of economic equalization and development, and increased regional cooperation

2. International forces include the accumulation of debts to foreign countries, the presence of multinational corporations, deforestation, the profitability of illegal drug growing and trading, and the formation of organizations such as the OAS.

3. The countries in the region have difficulty paying money back to foreign lenders. There are no easy solutions to debt crises. Further borrowing should be avoided if possible.

200

done little for most of the country's people. Wealth and land often remain in the hands of the fortunate few.

Political instability may be lessened in Middle and South America by **land reform**. Land reform means breaking up the large landholdings and allowing small farmers to own their own land. For generations, farmers have worked for shares of the crops they produced or for low wages on lands that were part of plantations and haciendas. These large landholdings usually have been owned by powerful people. Though land reform has been slow in many countries, it became widespread in Mexico after the 1910 Mexican Revolution.

Economic Issues Sometimes, citizens become angry at the policies of their national government. Often, money that was supposed to aid development of a poor region goes to meet less important needs in a wealthier capital city. Also, many people in Middle and South America believe that their countries are not truly independent. These people think that there is too much foreign investment and control.

Foreign governments, trying to encourage development, lent a great deal of money to various countries in the region during the economic boom years of the 1970s. These loans, however, have caused almost as many problems as they have solved. The borrowing countries now have difficulty paying back the loans. As a result, many Middle and South American countries have large debts, which make further economic development even more difficult.

Although multinational corporations from the United States, Europe, and Japan provide investments and technology, these companies have been accused of interfering in local politics and of exploiting local resources. Today, most countries in the region have a cautious attitude toward foreign corporations. Tax laws encourage multinational corporations to share ownership and profits with the local communities.

Many of the region's economic troubles echo throughout the world. For example, cutting the rain forests for lumber, or clearing them for farmland, provides an important source of income for many people in Central America and Brazil. This widespread deforestation, however, may contribute to global warming and the extinction of some species. In addition, deforestation may be leaving homeless some indigenous peoples who live in the rain forests.

Another regional issue with global consequences is the growing and trafficking of illegal drugs. Illegal drugs carry a high price. Many Middle and South American farmers find that they can earn more money by growing crops used for drugs than they can from growing other kinds of crops. Likewise, people who are willing to trade these drugs among countries often can make large profits. The large illegal-drug markets in the United States and Europe often are the final destinations for these dangerous products. Political and military conflicts have broken out over the illegal-drug issue, such as the conflict between the United States and Panama in 1989.

REGION **Regional Cooperation** In an effort to promote increased regional cooperation, almost all of the countries of Middle and South America, with the United States, formed the Organization of American States (OAS) in 1948. The goals of the OAS include keeping peace and increasing security in the Western Hemisphere, settling disputes among member nations, and promoting economic, social, and cultural cooperation.

Several of the countries of Middle and South America also have made efforts at regional economic cooperation on a smaller scale by forming organizations such as the Latin-American Integration Association, the Central American Common Market, and the Andean Group. Disagreements among the member countries and their differing goals and levels of development, however, have hindered these organizations' efforts and progress. For example, Ecuador and Peru fought a brief border skirmish in 1995.

4 SECTION REVIEW

1. What factors have caused political instability in some of the countries of Middle and South America? How might greater stability be achieved?
2. Using examples, discuss the international forces at work within the region.
3. **Critical Thinking** **Explain** why many countries in Middle and South America have large debts. Do you think that there is any solution to these "debt crises"? Why or why not?

Building a Vocabulary
1. Isthmus
2. Native; Olmecs, Mayas, Aztecs, Incas
3. Plains; Argentina (pampas) and Colombia and Venezuela (llanos); by erosion smoothing the land's surface over time
4. To make a blue dye
5. To bring about increased equality in land ownership

Recalling and Reviewing
1. Mining is important in several mountain regions; valleys and climates are suited for many types of agriculture; and rivers provide irrigation water and a source of power. Scenic landforms and warm climates make tourism important in some areas.
2. Portuguese, French, British, Dutch, Africans, people from India, Indonesians, and people of mixed ancestry
3. Europeans believed that land could be owned, but this idea was new to Native Americans, and they were forced to work for the Europeans. Land was never evenly distributed, and even today much of the land is owned by a few people.

CHAPTER **17** CHECK

Harvesting sugarcane in **Jamaica**

Reviewing the Main Ideas

1. The theory of plate tectonics helps explain the origins of landforms found in Middle and South America. The region has a wide range of tropical and middle-latitude climates, and three major river systems.
2. Ancient Native American civilizations flourished in Middle and South America. When Europeans colonized the region, many Native Americans died or were forced to work for the Europeans. Cultures in the region today are a mix of Native American, European, and African heritages.
3. The region of Middle and South America has many natural resources. Subsistence and commercial agriculture both are common in the region. Industry is increasing.
4. Many of the countries of Middle and South America have rapidly growing populations and much poverty, which have contributed to political instability. Other regional issues involve relationships with countries outside the region.

Building a Vocabulary

1. What term describes a strip of land that acts as a bridge between two larger bodies of land?
2. Define *indigenous*, and list some of the indigenous groups of Middle and South America.
3. What are pampas and llanos? Where are they found, and how were they created?
4. For what is indigo used?
5. What is the purpose of land reform?

Recalling and Reviewing

1. How have the landforms, climates, and rivers of Middle and South America influenced the region's economic activities?
2. Other than Native Americans and the Spanish, what groups of people have resided in Middle and South America?
3. How did European ideas of land ownership create problems in Middle and South America? How are these ideas still visible in the region today?
4. What interests has the United States had in the countries of Middle and South America, historically and currently?

Thinking Critically

1. The term *revolution* means a complete and total change in the way of life of a majority of a country's people. Given this definition, discuss the following statement: "There have been very few true revolutions in Middle and South America."
2. Which of the following factors do you think is most responsible for the challenges facing Middle and South America today: past colonialism, present foreign involvement, or rapid population growth? Support your answer.

Learning About Your Local Geography

Individual Project
Many of the countries of Middle and South America have experienced much political instability. Investigate politics in your local area. Are elections held regularly? For what purposes are they held? Use library resources to make a chart showing this information. You may want to include your chart in your individual portfolio.

Using the Five Themes of Geography

Create and write the clues for a crossword puzzle about the themes of **region** and **place** in Middle and South America. The words in the puzzle should include subregions and human and place characteristics of the region, as discussed in the chapter.

4. Historically, the United States has had an economic interest in ensuring that the countries remained independent because of trade. Currently, U.S. corporations have invested in the development of the region. Also, the United States is concerned with illegal drug cultivation in the region.

Thinking Critically
1. The statement is valid. Except for the 1910 Revolution in Mexico, most of the political changes in Middle and South America, including the wars for independence and current shifts of power, have not brought much change to the lives of most people.
2. Answers will vary; all three factors are to some extent responsible.

Learning About Your Local Geography
Charts should show information about local elections.

Using the Five Themes of Geography
Puzzles will vary but should include terms that apply to the themes of region and place.

CHAPTER 18 Mexico
Planning Guide

TEXTBOOK	TEACHING RESOURCES	OTHER AVAILABLE MATERIALS
Section		
1. Physical Geography Pages 203–204	Reteaching Worksheet 18.1 Critical Thinking Worksheet 18	*Study Guide for Spanish Speakers* *Lecture Notes,* Chapter 18
2. Economic Geography Pages 204–208	Reteaching Worksheet 18.2 Skills Worksheet 18	*Study Guide for Spanish Speakers* *Lecture Notes,* Chapter 18
3. Mexico's Culture Regions Pages 208–209	Reteaching Worksheet 18.3 Graphic Organizer 18 Challenge/Enrichment Worksheet 18	*The World Today* Videodisc Program *Study Guide for Spanish Speakers* *Lecture Notes,* Chapter 18
4. Mexico's Future Page 212	Reteaching Worksheet 18.4	*Study Guide for Spanish Speakers* *Lecture Notes,* Chapter 18
Chapter Review Page 213	Chapter 18 Test Chapter 18 Portfolio Assessment See also *Alternative Assessment Forms*	*Test Generator,* Chapter 18

CHAPTER RESOURCES

Lecture Notes, Chapter 18
Teaching Resources
    Chapter 18 Graphic Organizer
    Chapter 18 Skills Worksheet
    Chapter 18 Critical Thinking Worksheet
    Chapter 18 Challenge/Enrichment Worksheet

REVIEW AND ASSESSMENT

Chapter 18 Check, page 213
Chapter 18 Test, *Teaching Resources*
Chapter 18 Portfolio Assessment, *Teaching Resources*
Test Generator, Chapter 18
Alternative Assessment Forms

ADDITIONAL RESOURCES

Books for Teachers

Canby, Peter. *The Heart of the Sky: Travels Among the Maya.* New York: Kodansha America, 1994.

Hellman, Judith Adler. *Mexican Lives.* New York: The New Press, 1994.

Sabloff, Jeremy A. *The Cities of Ancient Mexico: Reconstruction of a Lost World.* New York: Thames & Hudson, 1989.

Books for Students

Constable, George, ed. *Mexico.* Alexandria, VA: Time-Life Books, 1985.

Diaz del Castillo, Bernal. *Cortez and the Conquest of Mexico by the Spaniards in 1521.* (abridged) Hamden, CT: Linnet, 1988.

Lewis, Oscar. *The Children of Sanchez.* New York: Random House, 1979.

Meyer, Carolyn, and Charles Gallenkamp. *The Mystery of the Ancient Maya.* New York: Macmillan, 1985.

Riding, Alan. *Distant Neighbors: Portrait of the Mexicans.* New York: Knopf, 1985.

The World Today *Videodisc Program*

A variety of still images, short videos, and activities are available for you to use as you teach this chapter. *Lecture Notes* include bar codes for specific images and videos.

Multimedia Resources

Chapultepec. (VHS, 40 minutes) Films for the Humanities & Sciences, P.O. Box 2053, Princeton, NJ 08543–2053.

Let's Visit Mexico. (Multimedia CD-ROM, interactive tour) Queue, 338 Commerce Drive, Fairfield, CT 64305.

The Mexican Way of Life. (16mm film, VHS, Laser videodisc, 23 minutes) Aims Media, 9710 DeSoto Avenue, Chatsworth, CA 91311–4409.

Mexico: The Land and People. (VHS, 24 minutes) Encyclopedia Britannica Educational Corporation, 310 South Michigan Avenue, Chicago, IL 60604–9839.

Picture Altas of the World. (interactive maps, photo bank, animation, audio on CD-ROM) National Geographic Society, Washington, DC.

MEETING INDIVIDUAL NEEDS

Limited English Proficiency (LEP)

Organize students into small groups or pairs. Have students make drawings or clip pictures from magazines or travel brochures to illustrate how each of the Five Themes of Geography applies to Mexico. Have students create posters using the sketches or pictures, labeling each theme.

Learners Having Difficulty

Help students create an outline of Chapter 18 on the board. Use section titles for items with Roman numerals, subsection titles for items with letters, and topic sentences for items with Arabic numerals. Have students copy the outline and use it to study the chapter.

Auditory Learners

Organize the class into groups. Have each group debate or evaluate one of the following statements: "The *ejido* system of agriculture in Mexico needs to be changed." "New industry in Mexico City should be discouraged." "Construction in Mexico City should be halted." "Mexico is a developed country." Assign some members of each group to challenge the statement and others to defend it. Have a volunteer from each group report on the discussion to the class.

Visual Learners

Provide students with Outline Map 11. Tell them to create a map showing the six culture regions of Mexico, using the information in Section 3 and the maps on pages 189–192 for help. Students' maps should include the following: a title, culture region labels, and symbols and labels for the major cities and industries in each region. Have students write two generalizations based on the information in their maps. Display the completed maps in the classroom.

Tactile/Kinesthetic Learners

Help students list the food crops that are grown in Mexico. Select volunteers to research cookbooks for authentic recipes that use these foods, and have them choose a dish to prepare for a Mexican dinner in your classroom. Have other volunteers decorate the room with festive Mexican decorations.

Gifted Learners

Have interested students investigate the Aztec or the Mayan civilization and report their findings to the class. Ask students to focus on how Mexico's geography influenced the culture.

MULTICULTURAL FOCUS

Point out to students that about 90 percent of the people in Mexico are mestizos who mostly speak Spanish. The other 10 percent of the population, or approximately 8 million people, belong to one of the 56 ethnic groups in Mexico. These people speak their native tongue in addition to speaking, reading, and writing Spanish. Few indigenous people in Mexico, however, can read and write their native languages because these languages were never written down. Instead, the languages were passed down through oral tradition.

Some indigenous people in Mexico are interested in learning to read and write their language to perpetuate their culture. In most cases, this task involves beginning with the basics— developing letters to stand for all the sounds in the language. Once an alphabet is completed, a dictionary of words and meanings can be written.

To assist with this process, an indigenous-language school called Celiac was established in Mexico in 1987. Indigenous people from all over Middle and South America come to the school to study the fundamentals of their languages and to create a literature in their native tongues. Students are writing ethnographies, novels, books of poems, and other literature.

Discuss the importance of writing to the preservation of culture, and why it is important to preserve the cultures of indigenous peoples in Mexico and other parts of the world.

CHAPTER 18

MEXICO

Section 1: Physical Geography
Section 2: Economic Geography
Section 3: Mexico's Culture Regions
Section 4: Mexico's Future

OVERVIEW

In Chapter 18, students learn about Mexico, a land of physical, economic, and cultural contrasts.

Section 1 outlines Mexico's physical geography with its variety of landforms, climates, and types of vegetation.

Section 2 discusses the country's economy, highlighting changes in agriculture and the growth of industry.

Section 3 describes the cultural diversity of Mexico and its six culture regions.

Section 4 focuses on current developments in Mexico that hold promise for a better future.

About the Author

Octavio Paz (1914–   ) is an internationally known literary artist. Born in Mexico City, Paz is well known for his poetry and essays that have been influenced by the indigenous cultures of Mexico.

About the Artifact

Chac Mool sculptures were created around the year 1000 A.D. by the Mayan people. Some experts believe the sculptures represented the Mayan rain god.

Primary Source

Description of Change: excerpted and bracketed
Rationale: excerpted to focus on the main idea; bracketed to clarify meaning

CHAPTER 18

Mexico

"The geography of Mexico is abrupt. . . . To ascend to Mexico City from the burning sandy beaches of Veracruz is to travel through every landscape, from the suffocating vegetation of the tropics to the temperate lands of the meseta [plateau]. In the lowlands the air is warm and humid; on the high tableland it is dry and, by night, quite cool.**"**

Octavio Paz

Mexico is officially called "The United Mexican States." Its natural environments vary from deserts in the north to tropical forests in the south. Economic contrasts between wealth and poverty and between modern cities and ancient villages exist throughout the country.

Mineral resources mined beneath Mexico's large land area support a growing industrial society. With more than 90 million people, Mexico has more than twice the population of Spain and is the largest Spanish-speaking country in the world. Political and social changes within Mexico are creating a more open society. These and other developments in Mexico have an impact on the political, economic, and human geography of all of North America.

The relationship between Mexico and the United States continues to strengthen. Mexico's oil resources are important to the U.S. economy, and Mexico is the United States' third largest trading partner. In turn, the United States is Mexico's most important foreign customer. These economic connections affect thousands of workers in both countries.

GEOGRAPHY DICTIONARY

cash crop

ejido

maquiladora

inflation

monopoly

privatization

Chac Mool sculpture at **Chichén Itzá**

Ruins of a Mayan observatory at Chichén Itzá in **Yucatán, Mexico**

202

Using the Quotation

Read aloud the quotation by Octavio Paz. Ask students to explain what Paz means by: "The geography of Mexico is abrupt." *(Landforms and climates change suddenly and dramatically.)* Note the contrast between Veracruz and Mexico City, even though they are at about the same latitude. Point out that Mexico is not only a land of physical contrasts but also of economic and cultural contrasts.

Using the Geography Dictionary

Pronounce the words in the Geography Dictionary for students. Encourage them to use word analysis to determine the meaning of the English words. *(for example, **cash crop**—crop raised for money)* Have students look in the Glossary for the meanings of **ejido** and **maquiladora.**

SECTION 1
OBJECTIVES

The objectives reflect the Focus questions on this page and are assessed in the review on page 204.
1. Identify the landforms of Mexico.
2. Describe Mexico's climates and vegetation.

Using the Physical–Political Map: Mexico

Ask students to study the map on this page. Call their attention to the major landforms discussed in Section 1: Plateau of Mexico, Sierra Madres, Valley of Mexico, Isthmus of Tehuantepec, Yucatán Peninsula, and Baja California.

MEXICO: PHYSICAL–POLITICAL

PHYSICAL GEOGRAPHY

FOCUS

● *What are Mexico's major landforms?*

● *What climates and vegetation does Mexico have?*

REGION **Landforms** The landforms of Mexico are more diverse than those of some entire continents. (See the map above.) Most of the country is made up of a rugged central plateau, called the Plateau of Mexico. Its wide plains average more than 6,000 feet (1,829 m) above sea level. Isolated mountain ridges rise up to 3,000 feet (914 m) higher. Two mountain ranges—the Sierra Madre Oriental to the east and the Sierra Madre Occidental to the west—border the Plateau of Mexico.

At the south end of the Plateau of Mexico is the Valley of Mexico, where Mexico City, the capital, is located. The mountains south of Mexico City include towering snowcapped volcanoes. The highest, Mount Orizaba, rises to 18,700 feet (5,700 m). Smaller mountain ranges, volcanoes, and fertile valleys continue southward into Central America.

The outline of Mexico narrows like a funnel in the south. This is the Isthmus of Tehuantepec (tuh-WAHNT-uh-pehk), where the Pacific Ocean and Gulf of Mexico are separated by only about 137 miles (220 km). Before the Panama Canal was built, this served as one of the major routes between the Atlantic and Pacific oceans. The Yucatán Peninsula of southeastern Mexico is the country's flattest region.

The theory of plate tectonics is important in understanding Mexico's landforms. Geographers believe that millions of years ago the peninsula of Baja California in the northwest was

203

TEACHING SECTION 1

Motivator

Write the words *oriental* and *occidental* on the board. Have students suggest meanings for the terms. Then explain that they come from Latin words meaning (respectively) "east" and "west." Point out that the two terms are used in the names of Mexico's two major mountain regions.

Closure

Have students label Outline Map 11 to show the climates, vegetation, and major landforms of Mexico.

ASSESSMENT

Assign the Section 1 Review questions on this page.

Reteaching

Assign Section 1 of Reteaching Worksheet 18.

SECTION 2
OBJECTIVES

The objectives reflect the Focus questions on this page and are assessed in the review on page 208.

1. List agricultural products, industries, and resources important to Mexico.

2. Discuss patterns of land use and industrial growth in Mexico.

Chart Caption Answer

69, male; 77, female

FACTS IN BRIEF — Mexico

COUNTRY POPULATION (1994)	LIFE EXPECTANCY (1994)	LITERACY RATE	PER CAPITA GDP (1993)
MEXICO 92,202,199	69, male 77, female	90% (1993)	$8,200
UNITED STATES 260,713,585	73, male 79, female	97% (1991)	$24,700

Sources: *The World Factbook 1994*, Central Intelligence Agency; *The World Almanac and Book of Facts 1995*

What is the life expectancy of men and women in Mexico?

Section 1 Review
ANSWERS

1. Plateau of Mexico; Sierra Madre Oriental and Sierra Madre Occidental

2. High elevation, cold polar air

3. Northern Mexico: arid with desert scrub vegetation and dry grasslands; southern Mexico: hot, humid, and forested along the southeastern coast, subtropical in southern coast valleys, and inland climates varying with elevation

204

---

attached to Mexico's mainland. As Baja California detached and moved northwest and the Gulf of California opened, an ocean trench developed off Mexico's Pacific coast. The heat and pressure along this trench have created volcanoes and caused earthquakes.

Climate and Vegetation Mexico extends through the subtropical and tropical latitudes, where the savanna, steppe, and desert climates prevail. Most of northern Mexico is arid with desert scrub vegetation and dry grasslands. The forested plains along Mexico's southeastern coast are hot and humid much of the year. Summer is the rainy season.

Elevation is a key climate variable throughout much of Mexico. Away from the dry north, climates vary quickly with elevation. Many people have settled in the mild environments of the mountain valleys. The valleys along Mexico's southern coast also have pleasant, subtropical climates. The high elevations on the plains of the Plateau of Mexico, however, cause cooler temperatures than would be expected at these latitudes. Sometimes during winter, cold polar air flows southward across the Plateau of Mexico, bringing freezing temperatures as far south as Mexico City.

SECTION REVIEW

1. What is Mexico's central plateau called? What mountain ranges border it?
2. Why does Mexico City sometimes experience freezing temperatures?
3. **Critical Thinking** **Compare and contrast** the climates and vegetation in northern Mexico and southern Mexico.

ECONOMIC GEOGRAPHY

FOCUS

- *What agricultural products, industries, and natural resources are important to Mexico's economy?*

- *In what different ways has land been used in Mexico during its history? Where are the country's major industrial centers located?*

REGION **Agriculture** Agriculture is a traditional focus of life in Mexico, even though the country has limited amounts of good farmland. Nearly two-thirds of the country is mountainous, and farming here is difficult. In fact, only about 12 percent of Mexico's lands produce a crop every year.

Most of Mexico's agriculture that relies on rainfall is found in the southern part of the Plateau of Mexico and in the valleys of the south. The sediment eroded from the mountains and the ash and lava from volcanoes have created fertile valley soils. Traditional Mexican food crops such as maize (corn), beans, and squash remain important for subsistence farmers. Although slightly more than 20 percent of all Mexicans work in agriculture, farming only provides about 9 percent of the country's gross national product.

Commercial agriculture in Mexico is expanding. Yields are high on these modern farms. Nearby markets in the United States have encouraged major investments in **cash crops**. A cash crop is a crop produced primarily for direct sale in a market, rather than for other purposes. The most extensively cultivated lands produce fruits and vegetables, including tomatoes and melons. Cotton, wheat, and alfalfa for cattle feed occupy much of the land on large estates in central Mexico. The drylands of the north are important for large-scale livestock ranching.

Plantation agriculture is found in the coastal lowlands and on mountain slopes along the Gulf of Mexico. Cash crops include coffee, sugarcane, cacao, bananas, and pineapples. Despite improvements in food supply, Mexico must import basic food items in order to feed its rapidly growing population.

Building a Vocabulary

Ask students what a **maquiladora** is. Discuss the effects that an increasing number of *maquiladoras* might have on Mexico's economy and cultures. *(More* maquiladoras *might provide jobs but increase dependence on foreign markets and investment; they might bring more elements of U.S. cultures to Mexico.)*

Writing About Products

Have students work in pairs to write a script for a TV or radio commercial promoting an agricultural or industrial product from Mexico. The commercial should be aimed at an audience in the United States. Have volunteers present their scripts to the class.

Independent Practice

Have students make a chart with two columns: Hacienda System and *Ejido* System. Have students fill in the chart with descriptions of the two systems, when they have been used in Mexico, their advantages, and their disadvantages.

Closure

Have students take turns naming an agricultural product, industry, or resource of Mexico.

This Mexican teenager and his sisters cultivate a beanfield in **Guanajuato**, a state in central **Mexico**. Mexicans are debating reforms aimed at increasing farm production in their country.

Land Reform The system of land use established by the Spanish divided most of Mexico into haciendas. Workers and their families lived on the haciendas. The workers owned no land and received low wages. Less than a century ago, 96 percent of Mexico's farm families owned almost no land.

Land ownership changed after the Mexican Revolution of 1910. A major goal of this revolution was to increase the number of farmworkers who owned land. Many of the large haciendas were broken up into communal farms, called **ejidos** (eh-HEE-thos). Under this system, farmers either worked the land together or gained rights to farm individual plots of land.

Agricultural development proceeded very slowly, however. The *ejido* farmers did not own the land, and there was little money for new equipment, seeds, and fertilizers. Often, workers grew discouraged and moved to the cities seeking industrial jobs. Today, the *ejido* system is changing. Some farmers are acquiring full property rights and larger farms. Agricultural corporations also are buying up the land to build large, modern farms. The government hopes these changes will lessen rural poverty. Yet, people will continue to move to the cities in search of a better way of life and greater opportunities.

Industrial Growth Despite the traditional importance of agriculture in Mexico, more than three-quarters of Mexico's workers are employed in industrial and service jobs. Petroleum has become Mexico's major mineral resource. The country's petroleum reserves, among the largest in the world, are located primarily under the plains and offshore along the Gulf of Mexico coast.

Mining remains important in Mexico as well. Many of the gold and silver mines begun by the Aztecs have developed into major mining districts. In Mexico's northern and southern mountains, new mines have been developed that employ many workers in those regions.

Mexico City, Puebla, Guadalajara (gwahd-uh-luh-HAHR-uh), and Monterrey (mahnt-uh-RAY) are Mexico's major industrial centers. Mexico's factories produce food products, consumer goods, building materials, steel, and chemicals. In rural areas, handicrafts provide cash income.

Mexico's most rapidly growing industrial centers are the cities along the United States border, such as Tijuana (tee-uh-WAHN-uh) and Ciudad Juárez (see-u-thad HWAHR-ehz). Taking advantage of Mexico's lower wages, many U.S. and other foreign companies have established

Text continues on page 208.

205

Themes in Geography

Researching and Synthesizing

Organize students into pairs. Then have each pair of students select a pair of cities along the U.S.-Mexico border (such as Ciudad Juárez and El Paso, Tijuana and San Diego, Brownsville and Matamoros) to research. One student in each pair will research the U.S. city, and the other student will research the Mexican city. Research should focus on the cultural and economic geography of each city. Students in each pair should then work together to organize their findings into an oral report supplemented with bar graphs, charts, or other visuals that compare and contrast the two border cities.

THEMES IN GEOGRAPHY

Mexico's Northern Border Region

*B*ordering the United States, northern Mexico lies hundreds of miles north of Mexico's heartland.

Northern Mexico is a vast territory that stretches for almost 2,000 miles (3,218 km) from the Pacific Ocean to the Gulf of Mexico. It is up to 300 miles (483 km) deep and includes more than one-third of Mexico's territory and about 20 percent of its population.

Most of northern Mexico is a semiarid plateau whose eastern and western edges rise into mountain ranges. Its vegetation is mainly desert scrub or short steppe grasslands. A strip of coastline along the Gulf of Mexico in the east and part of the western mountains receive more than 20 inches (51 cm) of rain per year.

MOVEMENT Only the shallow waters of the Rio Grande separate Ciudad Juárez, Mexico, from El Paso, Texas. Each day, many people and products cross the bridge linking the two cities. This constant two-way traffic flow creates the impression that although they are in different countries, these two cities are one metropolitan area. **Name other cities along the U.S.–Mexico border whose relationship is similar to that of Ciudad Juárez and El Paso. What advantages result when this relationship exists between two countries?**

Farther west, the Gulf of California separates the main part of northern Mexico from Baja California, a long, slender peninsula.

To tie the border region more closely to the rest of the country, the Mexican government has tried to increase the region's population and improve its economic conditions. Mexico's most ambitious and successful program to develop the northern border region began in the mid-1960s. Under the Border Industrialization Program, the government encouraged foreign companies to build factories in Mexico to manufacture goods for export.

Today, there are more than 2,000 of these foreign-owned factories, or *maquiladoras*. Most are located within 20 miles (32 km) of the United States. At these plants, workers assemble foreign-made parts into automobiles, computers, sunglasses, and other products for sale outside Mexico. About two-thirds of the *maquiladoras* are owned by U.S. companies, and by the early 1990s, about 70 percent of Mexico's exports went to the United States.

Several of northern Mexico's most important cities are located close to large U.S. cities. Tijuana, one of Mexico's largest border cities, lies less than 20 miles (32 km) from San Diego, California. These two cities have nearly grown into one large metropolitan area. Other Mexican cities and their U.S. neighbors in California, Arizona, New Mexico, and Texas are separated only by the international border.

Although northern Mexico's economic growth and its growing ties with the United States have benefited both countries, these developments have caused problems as well. Because Mexico's environmental regulations often

Linking History and Geography

Display a wall map that shows northern Mexico and the southern United States. Tell students that the Mexican War of 1846–1848 concerned geographic issues. One of the causes of the war was a border dispute about lands north of the Rio Grande. In the 1848 Treaty of Guadalupe Hidalgo that ended the war, the Rio Grande became the southern boundary of Texas. Invite interested students to investigate the following: If the United States had not won in the border dispute, what would be the likely boundary between Texas and Mexico today? *(the Nueces River)*

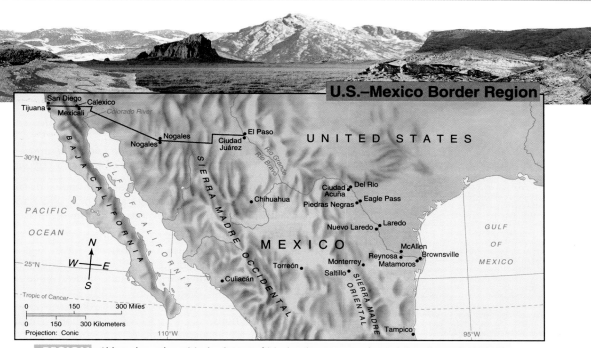

U.S.–Mexico Border Region

REGION Although northern Mexico is part of Mexico, its economy depends heavily on the United States, and its people are influenced by U.S. culture. Northern Mexicans in turn influence Americans across the border. **How does a border area such as northern Mexico differ from regions in the central part of a country?**

are not enforced, the *maquiladoras* along the U.S.–Mexico border pour industrial wastes into the air and water of both countries. In addition, population and industrial growth along the border have led to a demand for water that exceeds the available natural supply. As a result, groundwater reserves are being exhausted or destroyed from overuse.

The rapid industrialization of northern Mexico has led to additional concerns. Many Mexicans fear that the United States is "Americanizing" northern Mexico. North of the border meanwhile, some U.S. workers worry that northern Mexico's economic growth is taking place at their expense. They have seen U.S. companies close plants in the United States and relocate in Mexico to take advantage of lower wages and costs there. Both countries hope that the North American Free-Trade Agreement (NAFTA), approved by the U.S. Congress in 1993, will spur economic growth on both sides of the border.

Northern Mexico is a region where two countries and cultures meet and blend. Today and in the future, Mexico and the United States must continue to work together on trade agreements and environmental concerns to resolve issues between the two countries.

INTERACTION Many companies, such as the one that runs this plant on the Texas–Mexico border, have sought ways to limit pollution. Businesses and citizens realize that continued population growth and industrialization will cause more pollution on both sides of the border. **How can the United States and Mexico work together to solve problems caused by industrial pollution?**

Map Caption Answer

Answers will vary, but students might say that cultural and economic influences of neighboring countries often are stronger in the border regions than they are in the interior regions.

Photo Caption Answer

The countries can share information and strategies for keeping the environment clean. The governments can also coordinate efforts to establish policies that encourage environmental protection.

ASSESSMENT

Assign the Section 2 Review questions on this page.

Reteaching

Assign Section 2 of Reteaching Worksheet 18.

Extension

Students might want to research Mexico's 1910 Revolution—its causes as well as its effects.

SECTION 3
OBJECTIVES

The objectives reflect the Focus questions on this page and are assessed in the review on page 209.

1. Name Mexico's six culture regions.

2. Examine economic, political, and social characteristics of Mexico's regions.

TEACHING SECTION 3
Motivator

Ask students to recall the regions of the United States discussed in Unit 3. Have students note the factors that distinguish one region from another. *(economic, political, and social)* Tell them that in this section they will learn about the six culture regions of Mexico, distinguished by some of the same kinds of factors.

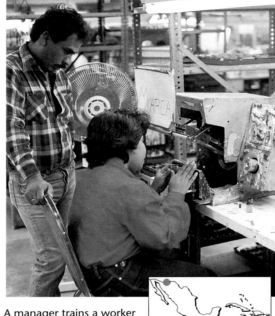

A manager trains a worker at a *maquiladora* in **Agua Prieta**, across the border from Arizona in **Mexico**. The *maquiladoras* have provided jobs and training for hundreds of thousands of Mexicans.

Section 2 Review
ANSWERS

1. The petroleum and other industries, mining, tourism, cash crops

2. In colonial times, land was divided into large estates worked by people who earned low wages and owned no land. After the 1910 revolution, haciendas were broken up into *ejidos,* on which farmers worked land together. Recently, farmers have been acquiring full property rights, and corporations are building large, modern farms.

3. U.S. markets have encouraged investments in commercial farming and cash crops. Also, U.S. companies have established factories in border cities.

208

MEXICO'S CULTURE REGIONS

F O C U S

- *Into what six culture regions can Mexico be divided?*
- *What are the economic, political, and social characteristics of Mexico's regions?*

Mexico consists of 31 states and the capital district. The country's geography can be best understood by dividing the country into six culture regions.

Greater Mexico City Greater Mexico City is the most important of Mexico's culture regions. It extends beyond the border of the capital district to include nearly 50 smaller cities. While this area is less than 100 miles (161 km) across, it holds more than one-fourth of Mexico's population. Thousands of factories and millions of motor vehicles generate some of the world's worst air pollution.

Mexico City is one of the most populous cities in the world. (See "Cities of the World" on pages 210–211.) Rural migrants are drawn to Mexico City from throughout Mexico, looking for new opportunities. Many do not find jobs, and the slums grow larger every day. At the same time, there are many modern highways, office buildings, high-rise apartments, fine museums, and major universities. Wealth and poverty often exist side by side. Economic growth, however, will likely continue as wealth from Mexico's industries, natural resources, and tax revenues flows into the Greater Mexico City region.

As Mexico's capital, Mexico City is the country's center of political power. Since the 1910 revolution, Mexico's government has been dominated by one political party. Officials of the Institutional Revolutionary Party, or PRI, control all of the government ministries and most of the labor unions.

Central Interior Northward from Mexico City and extending toward both coasts is Mexico's historic core of Spanish settlement. Many cities began here as mining and ranching centers during the colonial period. This region has many fertile valleys and small family farms. Many old-style haciendas still exist in this region, and small towns with a central square and colonial-style church are common.

factories called ***maquiladoras*** (mah-KEEL-ah-dor-ahs) in the border cities. In these plants, products are assembled for export to the United States and other nations. (See "Themes in Geography" on pages 206–207.)

Mexico also has a large tourist industry. Tourists, particularly from the United States, travel to the old colonial cities and ancient monuments of the Plateau of Mexico. Coastal cities such as Acapulco, Mazatlán, and Cancún draw tourists from around the world.

SECTION REVIEW

1. What are Mexico's major sources of income?
2. How has land been used in Mexico over time?
3. **Critical Thinking** **Discuss** the role played by the United States in Mexico's agricultural and industrial development.

Illustrating Ideas

Culture Regions Have each student choose one of Mexico's culture regions and list images he or she would use to represent the region in a geography textbook about Mexico and its regions. *(for example, Yucatán: Mayan ruins, agave plant, Caribbean resorts)* Call on volunteers to read their lists, and ask the class to guess which region is being described.

Closure

Have students list the six culture regions of Mexico and name an important fact about each.

ASSESSMENT

Assign the Section 3 Review questions on this page.

Reteaching

Assign Section 3 of Reteaching Worksheet 18.

SECTION 4
OBJECTIVES

The objectives reflect the Focus questions on page 212 and are assessed in the review on page 212.

1. Summarize economic and political challenges faced by Mexico in recent years.

2. Analyze reasons that Mexico's economic and political conditions may be improving.

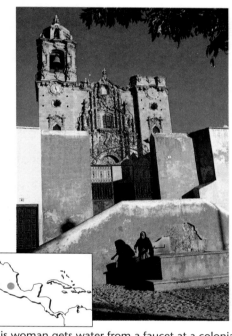

This woman gets water from a faucet at a colonial-era church in the state of **Guanajuato**, northwest of Mexico City.

The Central Interior region is losing many people to Mexico City and the northern region because the funds for improving agriculture and modernizing industry here are limited. As Mexico City becomes increasingly crowded, however, more industries are moving new factories and businesses to this region.

Oil Coast Oil, Mexico's greatest natural resource, is found along the shore of the Gulf of Mexico between Tampico and Campeche. Historically, these coastal plains were lightly settled, and they still have extensive forests. Oil production first began here in the early twentieth century. During the 1970s, new discoveries dramatically increased production and brought oil refineries and petrochemical complexes to the region.

Now-booming cities along this coast were founded during the colonial period and for centuries acted as service centers for plantations. Large tracts of forest have been cleared for commercial farming and ranching.

Southern Mexico Southern Mexico, the mountainous land south of Mexico City, is Mexico's poorest region. It has few large cities

and little industry. Basic infrastructure, including roads, telephone services, schools, and electric power, is poorly developed. It is a region of traditional villages, many of which are connected to the outside world only by dirt roads. Mayan and other indigenous languages are still spoken by about half the region's population. Subsistence agriculture is typical. Increasingly, migrants head north for new opportunities.

Northern Mexico The dry region of northern Mexico has become one of the country's most prosperous and modern. The merchants, miners, and ranchers of the region generally have cared little about events in Mexico City. Instead, they look to the United States for investment funds. Many firms here have business connections north of the border.

Monterrey is the north's industrial city, and it is developing closer economic relations with large Texas cities such as San Antonio and Houston. In the far northwest, Tijuana, Mexico, and San Diego, California, are beginning to merge into one metropolitan area. Northern Mexico's industries, such as the *maquiladoras*, and commercial farms are drawing migrants from throughout Mexico.

The Yucatán Few people live in the Yucatán, a peninsula in the southeast of Mexico. In ancient times, the Yucatán's population was probably greater than it is today. The Mayas here developed an effective agricultural system. They placed their fields on raised platforms built between drainage canals. Modern farmers are using some of the Mayas' techniques, and crop yields have improved.

Mérida is the Yucatán's major city. Its wealth came from rope made from agave fiber, which was a key product when sailing ships rigged with rope connected the world's ports. Other important centers include Cancún and Cozumel, resort cities on the Caribbean coast.

SECTION REVIEW

1. Where is each of Mexico's six culture regions located?
2. Which regions of Mexico are industrial?
3. Critical Thinking **Decide** what region of Mexico you might visit if you wanted to learn about the country's history. Support your answer.

Text continues on page 212.

THE WORLD TODAY VIDEODISC PROGRAM

Mexico: Making a Choice

1A   37681   3:39

Play   Pause

Section 3 Review
ANSWERS

1. Greater Mexico City: in and around Mexico City; Central Interior: north of Mexico City and extending to the west and east; oil coast: along the shore of the Gulf of Mexico; southern Mexico: south of Mexico City; northern Mexico: south of the U.S. border; the Yucatán: peninsula in the southeast

2. Greater Mexico City, Central Interior, oil coast, northern Mexico

3. All regions have historical value, but students may say Greater Mexico City (because of museums and universities), Central Interior (for colonial history), or southern Mexico or the Yucatán (for indigenous cultures and ruins of Mayan civilization).

209

Thinking Critically

Point out to students that this feature mentions several problems facing Mexico City. Tell students to imagine that they have just been elected mayor of Mexico City. Ask them to select one of the city's problems that, as mayor, they believe is most pressing. Have them write a speech addressing how they will solve the problem. Invite volunteers to present their speeches aloud. Encourage the rest of the class to ask questions of "the mayor."

CITIES OF THE WORLD

Mexico City

Mexico City is a rich fabric of peoples, cultures, history, and technology woven together.

Mexico City is one of the oldest cities in the Western Hemisphere. It lies high in the mountains of Mexico on a plain that was once a lake. Originally, it was the site of the Aztec city of Tenochtitlán, founded in about 1325. At that time, the city was built on a series of islands in the lake. Spaniards led by Hernán Cortés invaded the city about 200 years later. They destroyed the original city, then built a new one, Mexico City, on the site of the old. They drained the water from the lake and filled in the land so the new capital city could expand.

Building on the Past

Today, the main square of the city stands on the site of the old Aztec palace and temple. The square is named the Plaza of the Constitution, commonly called the Zócalo. A cathedral, the national palace, government buildings, museums, and commercial buildings surround the Zócalo.

To the west of the Zócalo is the wide, tree-lined boulevard Paseo de la Reforma, often called simply Reforma. At first, grand mansions lined Reforma. Now, the boulevard has become an extension of the central business district, and tall office buildings tower over both sides. Reforma passes through Bosque de Chapultepec, the largest park in Mexico City. Aztec emperors used the park hundreds of years ago.

Until the 1900s, the Zócalo contained almost the whole city. Since that time, however, migration and industrialization have led to a dramatic increase in Mexico City's population. In the 1970s, the number of people in the city rose more than 70 percent. The 1990 Census of Mexico set Mexico City's population at 15 million. By 2010, the city may contain as many as 19 million people.

The Challenges Ahead

Due to its rapid growth, Mexico City faces a number of problems, including a severe shortage of housing. Land is scarce, and construction has not been able to keep up with the need for homes. The government has provided some housing, but these efforts have helped only a small percentage of the people. As a result, large squatter settlements have grown on the outskirts of the city. Migrants from rural areas of the country pour into these settlements daily, searching for work and a better life in the city.

Mexico City's rapid population increase also has placed a great strain on the transportation system. Traffic congestion clogs the city's freeways and streets, and spews exhaust fumes into the air. The mountains surrounding the

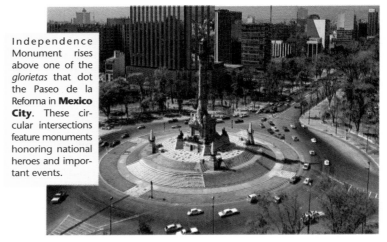

Independence Monument rises above one of the *glorietas* that dot the Paseo de la Reforma in **Mexico City**. These circular intersections feature monuments honoring national heroes and important events.

210

As the national capital, Mexico City is a transportation center and a home to numerous government offices, universities, and cultural sites. Name two buildings that are located near the Zócalo.

The Metropolitan Cathedral anchors one side of the Zócalo in **Mexico City**. Construction of the massive cathedral began in 1570 and was completed around 1810. Many striking colonial-era buildings are popular attractions in the Mexican capital, especially in the city center.

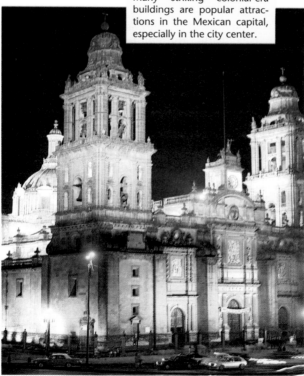

city trap the fumes as well as pollution from factories. These mountains act as a wall, blocking winds from blowing into the valley. As a result, thick smog often blankets the city.

Earthquakes present another threat to Mexico City. Because the city is built on the soft, spongy soil that was a lake bed in Aztec times, shock waves from earthquakes are magnified. When an earthquake strikes, the soil quivers almost like jelly. Tall buildings are particularly at risk. In 1985, an earthquake struck the city, resulting in some 7,200 deaths and destroying hundreds of buildings.

Another problem is that the buildings have been sinking as much as one foot (30 cm) per year into the soft ground. To combat this problem, the government has regulated the height of buildings in the city, and new buildings have special foundations that prevent them from sinking. The foundations also reduce damage from earthquakes.

Although faced with these and other challenges, Mexico City remains one of the world's major cultural centers. The city's natural beauty, rich history, and mild climate continue to attract people from around the globe.

Motivator

Recall the definition of developed and developing countries (pages 80–81). Ask for students' opinions about Mexico's status as a developed or a developing country.

Guided Practice

On the board, draw a chart with two columns. In the first column, have students list challenges Mexico has faced recently. *(debt, inflation, monopolies, political corruption)* In the second column, have students list reforms that have been made to address the challenges in the first column. *(privatization, more foreign trade)*

Closure

Ask students to summarize the changes occurring in Mexico today.

ASSESSMENT

Assign the Section 4 Review questions on this page.

Reteaching

Assign Section 4 of Reteaching Worksheet 18.

COOPERATIVE LEARNING

Organize the class into groups. Tell students that Mexico's National Solidarity Program is helping poor people improve their own neighborhoods by providing advice, building materials, and tools for refurbishing their communities. Ask groups to brainstorm ways that local action could help to solve other problems in Mexico.

Section 4 Review
ANSWERS

1. Because of high tariffs and because many were monopolies

2. Government-owned firms were sold, the *ejido* system was reformed, and foreign trade was encouraged.

3. The statement is mostly accurate. When one company controls all of a particular type of good, the company can relax standards of quality and can set prices high.

212

---

4 MEXICO'S FUTURE

FOCUS

- *What economic and political challenges has Mexico faced recently?*
- *How are conditions in Mexico improving?*

Social and Economic Issues Mexico is undergoing rapid change. During the 1980s, Mexico's economy grew little. The country had invested unwisely in previous years. Mexico's foreign debt had risen to nearly $100 billion, and the country was nearly out of money. **Inflation** skyrocketed. Inflation is the rise in prices caused by a decrease in value of a nation's currency. In 1987, prices in Mexico more than doubled in one year.

Mexico's industries, many of which were government owned, were inefficient. High tariffs protected most from foreign competition, and many companies were monopolies. A **monopoly** is a business that has no competitors. Such companies can control an industry's supply of goods and prices. For example, until recently, Mexico had four television networks owned by the same company.

Adding to the country's economic troubles was decay within the PRI. Instead of working to improve Mexico's economy, some PRI leaders and associates gained wealth through corruption. Several elections were won by fraud.

In the late 1980s, Mexico's economy began to show signs of improvement. The government sold many of its large firms to private individuals or companies. This selling of government-owned businesses or lands to private owners is called **privatization**. The *ejido* system also was reformed to increase agricultural productivity. These changes led to Mexico's developing a better relationship with the United States and other countries. Foreign competition forced many of Mexico's industries to become more efficient and productive. Mexico's government encouraged foreign trade, and many U.S. and Japanese firms began to compete in Mexico. Many Mexican firms became active in the United States.

Mexico now hopes to be considered a developed country. Vast differences, however, exist between the wealthy and the poor, and between the modern and the traditional. Ernesto Zedillo Ponce de Leon, elected president in 1994, has had to deal

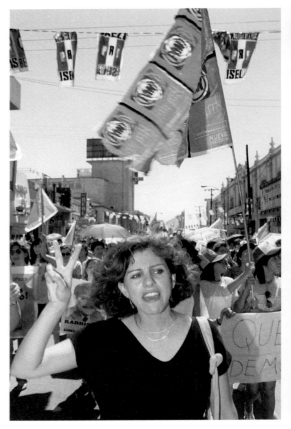

Mexican politics are more competitive after decades of one-party rule by the PRI. Opposition parties, such as the National Action Party (PAN), appeal to more Mexican voters than before.

with new economic and political crises, including a rebellion in the poor, southern state of Chiapas. Yet, the country's infrastructure is steadily improving, industries are modernizing, and agricultural production continues to increase. Politically, the nation is moving toward true democracy.

4 SECTION REVIEW

1. Why were many of Mexico's industries inefficient in the late 1980s?

2. What reforms were begun in the late 1980s?

3. Critical Thinking   Evaluate the following statement: "Monopolies do not benefit consumers."

CHAPTER 18 CHECK
ANSWERS

Building a Vocabulary
1. A crop produced for sale in a market
2. A hacienda is a large estate on which workers are paid low wages to farm the land but do not own any of it. In contrast, an *ejido* is a communal farm on which people work the land together or work individual plots.
3. Foreign-owned factories in which products are assembled for export; just south of the U.S. border
4. Inflation

Recalling and Reviewing
1. Many factories generate air pollution, which remains above the city because of the surrounding mountains.
2. To seek jobs and opportunities; increased poverty and slums near cities when migrants cannot find jobs
3. The term might apply to a political party, such as the PRI, that has little competition and can exert complete control over a country's political developments.
4. Southern Mexico has few cities and little industry and development; traditional ways of life dominate, and indigenous languages are still spoken. Northern Mexico is relatively prosperous and industrial and contains elements of U.S. culture.

Thinking Critically
1. Answers will vary. Some students might say that, with better management and a greater spirit of cooperation, an *ejido* system might work. Others might say that if farmers do not have full property rights, they will not be as productive as they could be.
2. Although brochure pictures do show certain areas of Mexico, there are many areas that are not shown.

Learning About Your Local Geography
Graphs should indicate the area's changing air quality. Students should consider geographic factors accounting for air quality patterns.

Using the Five Themes of Geography
Graphic organizers should show that place characteristics might encourage migration to cities. Mexico's location relative to the United States also enables Mexicans to emigrate to the United States.

CHAPTER 18 CHECK

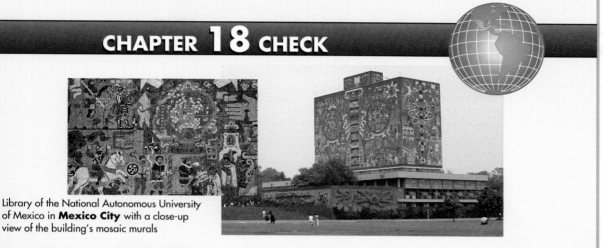

Library of the National Autonomous University of Mexico in **Mexico City** with a close-up view of the building's mosaic murals

Reviewing the Main Ideas
1. Most of Mexico consists of a large plateau bordered by two mountain ranges. Other landforms include volcanoes, plains, and valleys. The climates of Mexico range from humid tropical and subtropical to desert and highland.
2. Agriculture is widely practiced in Mexico, though good farmland is scarce. Commercial agriculture is becoming more widespread. Even so, food must be imported because of Mexico's rapidly expanding population. More than 75 percent of Mexico's workers are employed in industrial and service jobs.
3. Mexico can be divided into six culture regions. Much of Mexico's population lives in and around Mexico City, the nation's capital. Wealth and poverty often exist side by side in the urban regions of the country.
4. In the 1980s and early 1990s, Mexico experienced economic hardships. With recent political and economic reforms, officials have worked to aid development and to improve conditions.

Building a Vocabulary
1. What is a cash crop?
2. How does an *ejido* differ from a hacienda?
3. What are *maquiladoras*, and where are they often located?
4. If a country suddenly introduced a large amount of currency into its economy without changing the amount of goods available for purchase, what might occur?

Recalling and Reviewing
1. Why does Mexico City have severe air pollution? Consider the city's physical location as well as its economic activities.
2. For what reasons do many Mexicans move from rural areas to urban areas? What are some consequences of this migration?
3. Although the term *monopoly* usually refers to a business, how might it also apply to a political party? Explain, using information from the chapter.
4. Compare and contrast daily life in southern and northern Mexico.

Thinking Critically
1. Do you think an *ejido* system could ever be effective? If so, how? If not, why not?
2. Do tourist brochure pictures of Mexico's beaches, palm trees, and hotels represent an accurate portrayal of the country? Explain.

 ## Learning About Your Local Geography

Individual Project
Contact your state or local air quality board, a local environmental organization, or the weather department of a local television station. Request information on the air quality of your area. If possible, collect data going back several years so that you can see how the air quality of your area has been changing. Graph the data. What geographic factors account for the patterns you see?

Using the Five Themes of Geography
Create a graphic organizer showing how the themes of **location**, **place**, and **movement** interact in Mexico. How do relative location and place characteristics influence movement within Mexico and between Mexico and other countries?

CHAPTER 19 Central America and the Caribbean Islands

Planning Guide

TEXTBOOK	TEACHING RESOURCES	OTHER AVAILABLE MATERIALS
Section		
1. Central America Pages 215–219	Reteaching Worksheet 19.1 Skills Worksheet 19	*Study Guide for Spanish Speakers* *Lecture Notes,* Chapter 19
2. The Islands Pages 219–224	Reteaching Worksheet 19.2 Graphic Organizer 19 Critical Thinking Worksheet 19 Challenge/Enrichment Worksheet 19	*Creative Strategies for Teaching World Geography* LESSON 8 Our Disappearing Rain Forests *Study Guide for Spanish Speakers* *Lecture Notes,* Chapter 19
Chapter Review Page 225	Chapter 19 Test Chapter 19 Portfolio Assessment See also *Alternative Assessment Forms*	*Test Generator,* Chapter 19

CHAPTER RESOURCES

Lecture Notes, Chapter 19
Teaching Resources
   Chapter 19 Graphic Organizer
   Chapter 19 Skills Worksheet
   Chapter 19 Critical Thinking Worksheet
   Chapter 19 Challenge/Enrichment Worksheet

REVIEW AND ASSESSMENT

Chapter 19 Check, page 225
Chapter 19 Test, *Teaching Resources*
Chapter 19 Portfolio Assessment, *Teaching Resources*
Test Generator, Chapter 19
Alternative Assessment Forms

ADDITIONAL RESOURCES

Books for Teachers

Canby, Peter. *The Heart of the Sky: Travels Among the Maya.* New York: Kodansha America, 1994.

Covington, Paula, et al. *Latin America and the Caribbean.* Westport, CT: Greenwood Press, 1992.

Esteves, Carmen C., and Lizabeth Paravisini-Gerbert, eds. *Green Cane and Juicy Flotsam: Short Stories by Caribbean Women.* New Brunswick, NJ: Rutgers University Press, 1991.

Kulansky, Mark. *A Continent of Islands: Searching for the Caribbean Destiny.* Reading, MA: Addison-Wesley, 1992.

Wekesser, Carol, ed. *Central America: Opposing Viewpoints.* Rev. ed. San Diego, CA: Greenhaven Press, 1990.

Books for Students

Delano, Jack. *Puerto Rico Mio: Four Decades of Change.* Washington, DC: Smithsonian Institution Press, 1990.

Gelman, Rita Golden. *Inside Nicaragua: Young People's Dreams and Fears.* New York: Franklin Watts, 1988.

Hoobler, Dorothy and Thomas. *Latin America: Tradition and Change.* New York: Longman Publishing, 1991.

McCullough, David. *The Path Between the Seas: The Creation of the Panama Canal.* New York: Simon & Schuster, 1978.

Suchlicki, Jaime. *Cuba: From Columbus to Castro.* Tarrytown, NY: Pergamon Press, 1990.

Multimedia Resources

AMIGOS: A Simulation of a Race Through Latin America Developing Awareness of Its Lands, People, and History. (simulation, with teacher's and students' guides) Social Studies School Service, P.O. Box 802, Culver City, CA 90232–0802.

Castro's Cuba. (VHS, 60 minutes each, two-part program) Aims Media, 971 DeSoto Avenue, Chatsworth, CA 91311–4409.

Central America. (VHS, 20 minutes) Encyclopedia Britannica Educational Corporation, 310 South Michigan Avenue, Chicago, IL 60604–9839.

Picture Atlas of the World. (interactive maps, photo bank, animation, audio on CD-ROM) Washington, DC: National Geographic Society.

Women in Jamaica. (VHS, 58 minutes) Films for the Humanities & Sciences, P.O. Box 2053, Princeton, NJ 08543–2053.

MEETING INDIVIDUAL NEEDS

Limited English Proficiency (LEP)

Provide students with Outline Map 12. Have them create a physical-political map of Central America and the Caribbean islands, labeling the countries, the capitals, and the major landforms. Help students glue their maps to a piece of cardboard and cut them into puzzle-shaped pieces. Have students exchange their puzzles and reconstruct the maps.

Learners Having Difficulty

Use the map on page 215 and the charts on pages 216 and 224 to help students review the main concepts of the chapter. Have students identify countries in the region from clues you provide, such as the following: "I am thinking of a country that is part of the isthmus between North and South America, and whose capital city is San José." (Costa Rica)

Auditory Learners

Arrange for people who have visited or lived in a Central American or Caribbean island country to speak to the class about their experiences there. Ask the speakers to bring in pictures, postcards, or other souvenirs to help them explain the physical, human, and economic geography of the country. Allow time for students to ask questions.

Visual Learners

Have students work in pairs to create charts about the countries of Central America and the Caribbean islands. Students should use the following headings: Country, Location, Capital City, Major Landforms, Climate Types, Major Agricultural Products, and Major Industries. Display the completed charts in the classroom.

Tactile/Kinesthetic Learners

Have students work in small groups to clip newspaper and magazine articles about events and issues in Central America and the Caribbean islands. Ask students to label the places mentioned in the articles on a large outline map of the region. Have students create a bulletin board display of the map and the articles.

Gifted Learners

Have interested students investigate the issues involved in Puerto Rico's debate over becoming a U.S. state, attaining independence, or remaining a commonwealth. Have the students present a panel discussion for the class about the political future of Puerto Rico.

MULTICULTURAL FOCUS

Point out to students that two forms of popular music—calypso and reggae—originated in the Caribbean islands. Calypso is a type of folk music originally sung by African slaves who worked on the plantations in Trinidad. The slaves were not allowed to talk while they worked, so they sang songs in a French-Creole dialect to convey messages to one another.

The words of a calypso song are more important than the music. Calypso singing competitions became popular after the abolition of slavery, and champions are still crowned on the basis of their ability to choose words and make up rhymes on the spot. Two calypso songs that became American popular songs are "Day-O" and "Mary Ann."

Reggae developed in Jamaica during the 1960s as a means of communicating the need for solutions to the social problems of poor Jamaicans. Like calypso, reggae has its roots in African music. The songs are characterized by simple harmonies and short rhythmic patterns played off the beat by guitars and drums. Bob Marley of Kingston, Jamaica, was one of the most famous reggae singers. His songs have been recorded by popular performers such as Eric Clapton and Johnny Nash.

If possible, bring to class examples of calypso and reggae music for students to hear. Discuss the themes that the words express, and relate these themes to the geography of the islands.

CHAPTER 19

CENTRAL AMERICA AND THE CARIBBEAN ISLANDS

Section 1: Central America
Section 2: The Islands

OVERVIEW

In Chapter 19, students learn about the many small countries and territories that make up Central America and the Caribbean islands.

Section 1 describes the seven countries of Central America. Political strife has characterized this region for decades. Recent democratic reforms and attempts at economic cooperation promise to bring more stability to these countries.

Section 2 focuses on the Caribbean islands. The islands have a variety of cultures. Although tourists view the islands as ideal vacation spots, many of the islanders suffer from poverty and unemployment.

About the Author

Jamaica Kincaid (1949–) grew up in Antigua in the Lesser Antilles when the island was still a British colony. One of her best-known novels is *Annie John,* the story of a young girl growing up under the colonial system in the Caribbean. Kincaid currently lives in the United States.

Primary Source

Description of Change: excerpted and bracketed
Rationale: excerpted to focus on the main idea; bracketed to clarify meaning

---

CHAPTER 19

Central America and the Caribbean Islands

“**I**t is true that the early morning is the most beautiful time of day on the island [Antigua]. The sun has just come up and is immediately big and bright, the way the sun always is on an island, but the air is still cool from the night; the sky is a deep, cool blue; . . . the red in the hibiscus and the flamboyant flowers seems redder; the green of the trees and the grass seems greener.”

Jamaica Kincaid

A güiro musical instrument from the **Lesser Antilles.**

Central America and the Caribbean include 20 small countries and more than a dozen territories. The countries of Central America lie on a natural land bridge between North and South America. The Caribbean islands lie in an arc from Cuba to Trinidad and Tobago.

Cultural history helps explain the development of so many countries and territories. The Spanish, English, French, and Dutch introduced different languages, technologies, and laws to their territories in the region. These European traditions mixed with the cultures of indigenous peoples and of African slaves.

In recent decades, the region has been strongly influenced by the United States, which has offered democratic ideas, modern technology, tourist dollars, and a market for agricultural products. Most Central American and Caribbean countries now have better transport links with the United States than with each other.

Ocho Rios, **Jamaica**

214

SECTION 1

OBJECTIVES

The objectives reflect the Focus questions on this page and are assessed in the review on page 219.

1. Identify landforms and climate types of Central America.
2. Analyze the economic geography and related political issues of Central American countries.

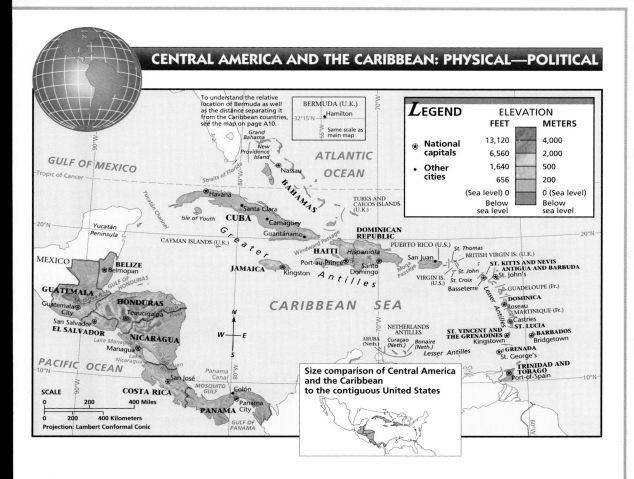

CENTRAL AMERICA AND THE CARIBBEAN: PHYSICAL—POLITICAL

CENTRAL AMERICA

FOCUS

- *What landforms and climate types are found in the countries of Central America?*
- *What activities support the economies of the Central American countries? How are economics and politics linked in the region?*

LOCATION **Isthmus Region** Central America, the isthmus between North and South America, also separates the Caribbean Sea and the Pacific Ocean. (See the map above.) Earthquakes and volcanic eruptions are frequent threats.

Along the Caribbean coast are humid-tropical plains with dense forests. Inland from the coasts, the mountains rise into mild, highland climates, where most of the people live. Coffee is the main cash crop on the humid mountain slopes. The Pacific coast of Central America has warm, sunny weather. In many areas, the savanna vegetation has been cleared for plantations and ranches.

Because the seven countries of Central America have similar environments, they grow similar export crops. Competition rather than cooperation between the countries has been typical. Free regional trade is a goal of most of the Central American countries.

Guatemala Guatemala is the most populous country in Central America. The heartland of Guatemala is found in the fertile valleys below the volcanic mountains in the southern part of the country. The capital, Guatemala City, is located here. Coffee is Guatemala's primary export. Another major export is **cardamom**, a spice grown in the central highlands. Cardamom is popular in many of the foods of Southwest Asia.

Motivator

Challenge students to list the seven countries of Central America without consulting a map. Then have them try labeling the countries on an outline map. Have them check their work with the map on page 215. Ask the following questions: Which country is the smallest? *(El Salvador)* The largest? *(Nicaragua)* Which country does not border the Caribbean? *(El Salvador)*

Building a Vocabulary

Write the word **cardamom** on the board. If possible, show students processed cardamom seeds, and explain their use as a spice. The cardamom plant, which grows 5 to 9 feet (1.5 to 2.7 m) high, is native to India.

Analyzing

Economic Geography Point out that for most of the countries in Central America, coffee is a major source of export income. Have students analyze the risks involved in heavy dependence on one commodity. Ask what kinds of things could threaten the export income from coffee. *(reduced consumer demand, competition from new growers, price wars)*

MULTICULTURAL PERSPECTIVES

Guatemala's Rigoberta Menchú received the Nobel Peace Prize in 1992 for her work in advocating the rights of Native Americans. Menchú, a Mayan of the Quiché group, led a successful campaign to have the United Nations designate 1993 as the International Year for Indigenous Populations. Review with students the meaning of *indigenous.* Then have them suggest reasons why indigenous populations seek protection and recognition of their rights.

Chart Caption Answer

Bahamas; Antigua and Barbuda, Barbados, Costa Rica, Cuba, Dominica, Jamaica, Panama, St. Vincent and the Grenadines

FACTS IN BRIEF

Central America and the Caribbean Islands

COUNTRY POPULATION (1994)	LIFE EXPECTANCY (1994)	LITERACY RATE	PER CAPITA GDP (1993)
ANTIGUA AND BARBUDA 64,762	71, male 75, female	90% (1990)	$5,800
BAHAMAS 273,055	68, male 75, female	95% (1992)	$16,500
BARBADOS 255,827	71, male 77, female	99% (1992)	$8,700 (1990)
BELIZE 208,949	66, male 70, female	93% (1991)	$2,700
COSTA RICA 3,342,154	76, male 80, female	93% (1992)	$5,900
CUBA 11,064,344	75, male 79, female	99% (1992)	$1,250 (GNP)
DOMINICA 87,696	74, male 80, female	94% (1992)	$2,100 (1992)
DOMINICAN REPUBLIC 7,826,075	66, male 71, female	83% (1991)	$3,000
EL SALVADOR 5,752,511	64, male 70, female	75% (1991)	$2,400
GRENADA 94,109	68, male 73, female	95% (1991)	$3,000 (1992)
GUATEMALA 10,721,387	62, male 67, female	55% (1991)	$3,000
HAITI 6,491,450	43, male 47, female	53% (1991)	$800
HONDURAS 5,314,794	65, male 70, female	73% (1991)	$1,950
JAMAICA 2,555,064	72, male 77, female	98% (1990)	$3,200 (1992)
NICARAGUA 4,096,689	61, male 67, female	57% (1991)	$1,600
PANAMA 2,630,000	71, male 78, female	87% (1991)	$4,500
ST. KITTS AND NEVIS 40,671	63, male 69, female	98% (1991)	$4,000 (1992)
ST. LUCIA 145,090	67, male 72, female	78% (1989)	$3,000
ST. VINCENT AND THE GRENADINES 115,437	71, male 74, female	85% (1989)	$2,000 (1992)
TRINIDAD AND TOBAGO 1,328,282	68, male 73, female	97% (1988)	$8,000
UNITED STATES 260,713,585	73, male 79, female	97% (1991)	$24,700

Sources: *The World Factbook 1994,* Central Intelligence Agency; *1995 Britannica Book of the Year; The World Almanac and Book of Facts 1995*

Which Central American or Caribbean country has the highest per capita GDP? Which countries of the region have life expectancy rates of more than 70 years for both men and women?

Land reform is a continuing issue in Guatemala. Much of the land is owned by a small percentage of farmers. Some traditional haciendas and coffee estates remain. With modern technology, sugar and cotton cultivation have increased, particularly on the Pacific slopes of the mountains.

About half of Guatemala's nearly 11 million inhabitants are indigenous peoples, most of whom live in small, isolated villages throughout the highlands. Subsistence farming is typical. Like their Mayan ancestors, farmers grow maize (corn), beans, squash, and chili peppers.

Economic inequality and discrimination against the indigenous population have led to violent politics in recent decades. Guatemalan leaders now are realizing the importance of political and economic freedom. Land hunger, or the desire of farmers to own their own land, is being reduced through settlement of the once nearly empty plains of Petén in the north.

Belize Formerly British Honduras, Belize (buh-LEEZ) is located along the Caribbean coast of the Yucatán Peninsula. Its humid-tropical, dense forests limited settlement for many years.

The country's population is made up of many kinds of people. Blacks and mulattos who speak English live along the coast. Indigenous peoples and peoples of African ancestry live in the inland forests, and settlers from Mexico and Guatemala live along the inland borders. Few crops can grow in the poor soils. Business leaders are concentrating on the fishing and tourist industries. Visitors find beautiful coral reefs and coastal resorts.

Honduras South of Belize is Honduras, a land of rugged mountains and valleys. Coffee grown in the highlands supplies most of the country's export income. Yet, small subsistence farms and large colonial-style estates dominate locally. Less than 20 percent of the land can be farmed.

About 90 percent of Honduras's people are mestizos. Most of the population lives in small mountain valleys in the highlands and in river valleys along the north coast. The rugged terrain makes transportation difficult.

The key development of modern Honduras is in the humid-tropical lowlands along the north coast. Many people have migrated out of the highlands to work on the large banana and sugarcane plantations there.

Interpreting a Chart

Have students refer to the chart of countries on page 216. Note the types of statistics shown for each country. Ask students to identify the seven countries of Central America. Then have them identify the Central American country with both the highest literacy rate and the highest per capita GDP. Discuss reasons for Costa Rica's high ranking. (*political stability, strong middle class*)

Illustrating Ideas

Have each student choose one of the countries of Central America. Ask students to find out more about the country's native customs (clothing, food, celebrations). Have them select one custom to explain to the class. Encourage students to draw or find a picture of an artifact that represents the custom—for example, a musical instrument, a handicraft, or a piece of clothing.

Guided Practice

Ask students to imagine that the Central American countries have decided to form the "United States of Central America." Discuss the advantages and disadvantages of forming one country from many. Ask students what kinds of obstacles the countries would need to overcome in order to unite. (*currency, laws, taxes*)

El Salvador El Salvador is a small country of volcanoes and plains on the Pacific side of Central America. Although volcanic ash has made El Salvador's soils the most fertile in the region, earthquakes and volcanic eruptions are continuing threats. El Salvador's climate is wet in the summer and dry in the winter.

As elsewhere in Central America, subsistence agriculture on small plots and commercial farming on large estates dominate rural areas. Coffee and cotton from large estates provide most of the export income. The majority of people, however, live a life of poverty.

El Salvador faces an increasing population and the need for effective land reform. Much of the best land is held by a few powerful families. The country's population is increasing faster than the government can provide schools and jobs.

During much of the 1980s, civil war raged in El Salvador, hindering economic development. The United States backed the Salvadoran government in the conflict. Finally, in 1992, a peace treaty ended the civil war.

Nicaragua The physical geography of Nicaragua is much like that of neighboring Honduras. There is a hot, humid plain along the Caribbean coast, low rugged mountains inland, and a volcanic plain along the Pacific coast. Nicaragua, however, also has a broad valley with large lakes inland from the Pacific coast. The valley is tectonically active with earthquake faults and volcanoes. Its fertile volcanic soils and those of the nearby mountain slopes make this area the most densely settled part of the country. The capital city, Managua, is located here.

Throughout Nicaragua's history, economic progress has been slow and political unrest typical. Most of the people continue to farm. Primary exports are coffee, beef, cotton, and bananas. The highlands contain deposits of gold and silver.

Severe problems of social inequality and land reform have always plagued Nicaragua. In 1979, the country began a period of dramatic political change. Its longtime dictator was overthrown and a revolutionary government took power. Many rich Nicaraguans fled the country. The government called itself Sandinista, named after César Sandino, a Nicaraguan patriot of the 1930s.

The United States put economic pressure on the Sandinistas and supported Nicaraguan rebels, called Contras. The economy of Nicaragua nearly collapsed. In 1990, free elections were held and the Sandinistas were defeated, although they remained the largest political party. Today, Nicaragua is a democracy with many political voices.

Costa Rica Costa Rica's central highlands with fertile soils and mild climates are the nation's heartland. Located here is San José, the capital city. Many coffee farms, which provide the country's main exports, also are found in this area.

When the European colonists first arrived in Costa Rica, most of the local peoples quickly died of diseases or migrated elsewhere. As a result, the colonists did not have local labor to develop large estates. Coffee cultivation and favorable government policies encouraged small farms. This development eventually created a large middle class

These Maya women are taking part in a religious celebration in **Guatemala**. Native American communities in the region are distinguished from one another by their distinct, bright styles of clothing.

THEME
Region

As a result of increased stability in the region, Guatemala, El Salvador, and Honduras are leading efforts to open the borders of Central American countries to free trade. Their leaders hope that by uniting their countries, they will be able to produce goods at lower costs. They also hope to gain leverage in negotiating trade agreements with Mexico, the United States, and countries in South America.

Geographical Sidelight

During the 1980s, the Sandinistas confiscated or took over much of the land, business, and other property in Nicaragua. One of the biggest problems for the current government is deciding who owns what. The country must resolve the issue of property rights in order to become politically stable.

Independent Practice

Have students label Outline Map 12 to show the countries, capital cities, vegetation and landforms of Central America. Captions should explain the information included on the map.

Closure

Divide the class into groups of seven. For each group, prepare seven sheets of paper, with the name of a different Central American country on each. Distribute the papers to group members. Have each group member write a fact about the assigned country and then pass the paper to the person on his or her right. Have groups continue in this way until the information in the text has been covered.

COOPERATIVE LEARNING

Organize the class into groups of four. Within each group, have students work in pairs to figure the population density of each Central American country. (Total population divided by total square miles equals the number of people per square mile.) Students may refer to the country charts in the Appendix of their textbooks to obtain the area of each country. One partner figures the population density of a country, while the other partner coaches. Then partners switch roles. Partners continue in this way until calculations are complete for all seven countries. Have partners compare their answers with those of the other pair in the group.

NEWS FROM AROUND THE WORLD — *Costa Rica*

Teenagers in Costa Rica Think Life Is *PURA VIDA!*

Costa Ricans live in a democratic republic where education is valued highly. Over 90 percent of Costa Ricans are literate, one of the highest rates in Middle and South America. Education is providing increased opportunities to young Costa Ricans, like those pictured above.

For most teens, weekdays are filled with a rigorous study schedule. After school, there is always work to be done around the house. Whether they are taking care of younger siblings, helping to prepare the evening

meal, or doing other chores, most teens have family obligations to fulfill. Family ties here are very important, and children learn respect, discipline, and pride at an early age.

In general, Costa Ricans have a strong interest in the United States, and many students study English. Many Costa Ricans also watch U.S. television programs that have been dubbed in Spanish. The latest U.S. fashions, music, food, and movies are all hot topics of interest among Costa Rican teenagers.

Weekends are dedicated to spending time with family and friends. There is always something to do—parties, dances, weddings, church, and more. Most of these events are family oriented. On weekends, many teens watch soccer games and soap operas on television, listen to music, talk on the phone, spend time with their friends, and dream about their futures.

If you ask Costa Rican teens what they think about life in their country, you will most likely hear *"PURA VIDA!"* Life is great!

that has supported stable democratic governments. Costa Rica has the highest standard of living and literacy rate in Central America.

Like the other countries of the region, Costa Rica has a high population growth rate. In the past, the opening of new lands closer to the coasts often relieved population pressure. Only recently have landless people and suburban slums become a serious concern. The government has encouraged industrial growth and expanded tourism, and has set aside land for national parks.

Panama Made up mostly of low hills and mountains, Panama has only narrow coastal plains. High volcanic peaks tower over the western part of the country. Most Panamanians live along the southern (Pacific) side of the nation.

To some, modern Panama seems like three separate countries. In the east is a densely forested region that is nearly uninhabited. In the middle of the country is a prosperous area surrounding the Panama Canal, with large cities. In the west are more rural areas, where subsistence farming and haciendas are common.

ASSESSMENT

Assign the Section 1 Review questions on this page.

Reteaching

Assign Section 1 of Reteaching Worksheet 19.

Extension

Have interested students do library research to learn more about the Panama Canal. Specifically, have them investigate the building of the canal, its operation, and any changes in the administration of the Panama Canal Zone. Students should prepare brief reports, including a map to show the canal's location.

SECTION 2
OBJECTIVES

The objectives reflect the Focus questions on this page and are assessed in the review on page 224.

1. Summarize the physical geography of the Caribbean islands.
2. Discuss cultural, economic, and political patterns of the Caribbean islands.

Ships move through one of the locks of the Panama Canal near the Caribbean Sea. To learn how a ship moves through a lock, see the illustration on page 142.

Panama gained its independence from Colombia in 1903. Eager to construct a canal across Panama linking the Atlantic and Pacific oceans, the United States aided Panama in its separation from Colombia. Completed in 1914, the Panama Canal allows ships and trade to travel more quickly and safely between the two oceans.

Canal fees and canal-side industries provide nearly half of Panama's gross national product. The cities at both ends of the canal—Colón on the Caribbean and Panama City on the Pacific—are major industrial centers. The United States administers the Canal Zone, a ten-mile-wide strip of land straddling the canal. In 1979, a treaty between the United States and Panama began a transition that will return control of the canal to Panama in the year 2000.

 SECTION REVIEW

1. What landforms are shared by most of the countries of Central America?
2. What crops are grown in Central America?
3. **Critical Thinking** **Suggest** a way that Central American countries might reduce political instability.

THE ISLANDS

FOCUS

● *What are the physical characteristics of the Caribbean islands?*

● *What cultural, economic, and political patterns are found on the islands?*

LOCATION **Physical Geography** To the east of Central America lie the islands Columbus called the West Indies, now called the Caribbean islands. They extend in a wide arc from south of Florida to the coast of Venezuela. Separate island groups in the Caribbean are the Greater Antilles, the Lesser Antilles, and the Bahamas. (See the map on page 215.)

The islands of the Greater Antilles are Cuba, Hispaniola, Puerto Rico, and Jamaica. The Lesser Antilles, which are smaller than the Greater Antilles, include the Virgin Islands, Barbados, Trinidad and Tobago, and many other small islands. The Bahamas, north of the Greater Antilles, include more than 700 islands.

The islands of the Caribbean have pleasant humid-tropical climates, and winters usually are drier than summers. The islands receive a yearly average of 40 to 60 inches (102 to 152 cm) of rainfall. Crops, particularly fruits and vegetables, are grown throughout the year because of the region's favorable climate. Fertile soils, however, are found only on some of the islands. On other islands, including Jamaica and Puerto Rico, there are many limestone rocks through which water rapidly drains. As a result, drought conditions are common, even though it may rain often.

Human Geography Nearly every island in the Caribbean still shows the effects of colonialism and slavery. Languages reflect the region's cultural diversity. English, Spanish, French, Dutch, Hindi, and Chinese are spoken. Blendings of these languages with African languages also can be heard. The islands' musical styles are popular inside and outside the Caribbean. Trinidad and Tobago is the home of steel-drum and **calypso** music. Jamaica is famous as the birthplace of **reggae**. The **merengue** is the national music and dance of the Dominican Republic.

Text continues on page 222.

THEME
Region

Guatemala, Honduras, Belize, El Salvador, and five Mexican states have joined together to promote travel to Mundo Maya, or Maya World. Visitors to Mundo Maya not only visit ancient ruins but also see Mayan customs still practiced today.

Section 1 Review
ANSWERS

1. Coastal plains, mountains, valleys
2. Coffee, cardamom, sugar, cotton, maize (corn), beans, squash, chili peppers, bananas
3. Possible answers: redistribution of land, increased democracy, regional cooperation

219

Planet Watch

Debating Issues

Divide the class into two groups. One group will represent the Kuna of Panama, and the other group will represent Panamanian government officials who wish to seize the Kuna's land for development. Have the groups stage a debate in which logical arguments representing each side are presented. The debate should be based on information in the feature as well as outside research. You might wish to assign a student to act as an impartial moderator.

Geographical Sidelight

The Kuna (also spelled Cuna) of northeastern Panama are one of three Native American groups in the region that have managed to retain their traditional ways of life despite the influences of the encroaching modern world. The Chocó live in the dense forests of southwestern Panama, and the Guaymi, in the country's isolated northwest.

PLANET WATCH

Case Study: Saving the Rain Forests

Along Panama's Caribbean coast, a great rain forest reaches down to the sea. Giant anteaters, nocturnal tapirs, and armadillos move along the forest floor, while eagles and migrating birds perch high in the forest's canopy. Thousands of species of plants thrive in this humid-tropical environment.

The Kuna have practiced conservation of their rain forest environment for generations.

This rain forest also is home to some 30,000 Kuna, whose villages occupy the small offshore islands. The Kuna reservation, which encompasses more than 1,100 square miles (2,849 sq. km), stretches some 100 miles (161 km) along the coast of Panama. The coastal waters provide the Kuna with fish, while the fertile mainland plains allow them to grow bananas, sweet manioc, yams, and maize (corn).

Caretakers of the Land

In 1980, a Panama official met with the Kuna. "Why do you Kuna need so much land?" he demanded. "You don't do anything with it. You don't use it. And if anyone else cuts down so much as a single tree, you shout and scream." Rafael Harris, a Kuna leader, responded. "We Kuna need the forest, and we use it and take much from it. But we take only what we need, without having to destroy everything as your people do."

In this area secluded by the rugged San Blas Mountains, the Kuna have managed to head off development and clearing of their precious rain forest. For the Kuna, the plants of the rain forest are a source of sacred medicine. The rain forest also protects their farmlands and fishing areas. Without the trees and plants that absorb rain and hold the soil in place, the land would erode and eventually destroy the fishing grounds on which the Kuna depend.

To protect the rain forest, the Kuna have set up a 230-square-mile (596-sq.-km) pro-

tected reserve on the edge of the ridge that separates the Kuna reservation from the rest of Panama. To help finance their plans, the Kuna called on international environmental groups. In addition, after building a dormitory near the rain forest, they welcomed paying visitors, including scientists, bird watchers, and the Panamanian Boy Scouts. The Kuna collect fees from tourists and sell research rights to scientists visiting the reserve. The Kuna have worked to have their reservation declared a Biosphere Reserve and a World Heritage Site. These United

A Kuna woman stitches a mola, a colorful, needlework garment. The Kuna, who live on coral islands off the northern coast of **Panama**, have maintained control over their own cultural and political ways of life.

Creating a Map

Have students work with partners to research other rain forests throughout the world that are in danger of being destroyed. Instruct them to show the location of these forests on Outline Map 1. Tell them to label the map to indicate names of the countries where the rain forests are found.

Nations classifications can help the Kuna in their struggle for international recognition as an independent people.

The Kuna have not relied on outside sources alone, using $100,000 of their own funds. "The main thing about the project is the incredible initiative that has come from the Kuna," said R. Michael Wright of the World Wildlife Fund. "They led us, we didn't lead them. Conservationists come and go, but the Kuna are going to stay." Geographers hope that the Kuna's experiment will provide a model for other peoples living in the world's rain forests.

The Rain Forests

The wettest and hottest places on Earth, tropical rain forests teem with plant and animal life. Of the 30 million or so species of plants and animals living on Earth, more than half live in rain forests. Rain forest trees and plants use carbon dioxide, helping to keep the air clean. Rain forest vegetation also holds rain-drenched soil in place, preventing it from flowing into valleys or clogging rivers. In addition, plants in rain forests provide one-fourth of all medicines found in drugstores.

Today, the world's tropical rain forests are endangered. Thirty to fifty acres (12–20 ha) of rain forest disappear each minute, which means the world is losing some 10–12 million acres (4–5 million ha) of rain forest a year. The destruction of trees and other vegetation is accompanied by the loss of animals that depend on them for habitat and food.

Several environmental groups and nations around the world are working to stop the destruction of the rain forests. They are training people in developing countries to use farming methods that do not deplete the soil. They also are educating rain forest peoples about conservation. Environmental groups are also pressuring wealthy nations to decrease their demand for the valuable hardwoods of the rain forests. As the Kuna have shown, the goal of saving the rain forests is in reach but must involve all the world's peoples.

Panama and the Kuna

Panama City and **Colón** offer job opportunities for many Kuna, a well-educated people. What is the distance from the Kuna reservation to Colón?

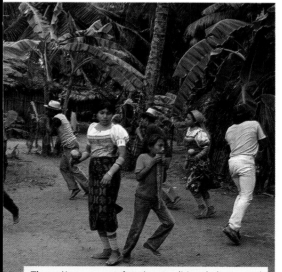

These Kuna are performing traditional dance and music in their village on an island in the **Gulf of San Blas**. Most Kuna villages depend on the fish in the coastal waters around the islands.

YOUR TURN

Many geographers wish to stop the destruction of the world's rain forests. How might you enlist the help of the people who live in, or earn a living from, the rain forests in this effort? **Plan** a conservation training workshop for a rain forest community. **Outline** the topics you will cover in your workshop, and provide a brief description of the information you will discuss for each topic.

TEACHING SECTION 2

Motivator

Play music associated with the Caribbean islands, such as calypso, reggae, or merengue. Ask students to describe images that come to mind when they think of the Caribbean islands. *(sun, blue sky, sparkling water, white-sand beaches, palm trees)* Point out that in this chapter students will learn about some of the realities behind these images.

Building a Vocabulary

Focus on the word **refugee**. Note that its base *refuge* means "a safe place." Thus, a refugee is someone who seeks a safe place. Ask students to give examples of the kinds of situations that cause people to become refugees.

Depicting Island Life

Have students work in groups to produce collages that summarize life in the Caribbean islands. Their collages can include pictures from newspapers, drawings, and pieces of fabric. They should include pictures of the island scenery, products produced in the islands, people, daily life, and anything else students find relevant. Display the collages.

Geography and Economics

In Cuba today, shortages of consumer and industrial goods make everyday life bleak for many families. People must stand in long lines for meager amounts of government-rationed sugar, rice, and bread. Meat is also very scarce. Since the government cannot afford to import oil, there are few cars on the streets. People pedal bicycles almost everywhere. Oil shortages also mean electrical-power blackouts several hours a day and a cut in television programming. Chalk, paper, and new books are scarce in Cuba's schools.

Havana, Cuba, was established by Spanish colonizers in the early sixteenth century. Havana's location just across the Straits of Florida from the United States helped make the Cuban capital a popular tourist destination before the 1959 revolution.

Much of the best land is part of large plantations or estates, many of which are controlled by foreign firms. When people abandon farming, most travel to the islands' capital cities seeking jobs in manufacturing, government, or tourism.

Beautiful beaches and sunny weather have made tourism a rapidly growing industry. The social impacts of tourism are mixed, however. Many of the jobs in tourism pay low wages. Many resorts are hidden behind large gates and high fences, isolating them from most island residents.

Even though most of the islands are politically independent, they remain economically dependent on foreign countries. The islands export raw materials and import most of their manufactured goods. In this environment of poverty, inequality of wealth, and foreign dependence, political unrest is common. The countries of the Caribbean are discovering that they have many common issues. Regional cooperation is a hope for the future.

Cuba Cuba, the largest island of the Greater Antilles, lies just 90 miles (145 km) off the southern coast of Florida. Mountains and hills cover about one-fourth of the island. The remainder of the land consists of gently rolling hills and fertile valleys.

The United States was once Cuba's most important market for exports. The Cuban economy, however, was controlled by a few wealthy people and suffered from corruption. A revolution in 1959 brought Fidel Castro to power. By 1961, Castro's government had taken over most privately owned property. Cuban society remains organized along Communist lines and has a command econ-

omy. Most farmlands have been put into **cooperatives** and large state sugarcane plantations. A cooperative is an organization owned by and operated for the mutual benefit of its members.

Much of the farmland is used to grow sugarcane, Cuba's most important crop. Cuba is one of the world's largest sugar exporters. Its success as a sugar producer is the result of several factors. First, most of Cuba's soil is fertile and holds moisture well. Second, the climate is wet for the six-month growing season but dry during the winter harvest. Third, because the agricultural land has been leveled, machinery operates easily on the sugarcane plantations. Since the 1959 revolution, the economy has depended increasingly on sugarcane. As a result, lands that once produced food now grow sugarcane. Food production cannot meet the island's needs.

Most of Cuba's industry is based on the processing of agricultural products. The country is famous for its cigars. Cuba also produces rice, coffee, vegetables, and tropical fruits.

Havana, the capital of Cuba, is the largest city in the Caribbean. Once the major port of the Spanish colonies, Havana is the country's most important trade, transportation, and manufacturing center. The city's more than 450-year history is evident in the old forts and churches and in the narrow streets. Havana also is a city of wide avenues, large parks, and modern buildings.

After Castro's revolution, Cuba became dependent on trade and aid from the Soviet Union and Eastern Europe. When the Soviet Union collapsed in the early 1990s, aid to Cuba decreased

222

dramatically. Since then, the Cuban economy has deteriorated. Large numbers of people opposed to Castro's government have fled Cuba. Many have made their new homes in the United States.

Haiti Haiti occupies the mountainous western third of the island of Hispaniola. It is the poorest and most densely populated country in the Americas. Haiti was the major French sugar-supplying colony until a slave uprising in 1791 brought independence. Since then, the country has experienced a series of corrupt governments and financial disorder. The United States sent troops to Haiti in 1994 to restore to office a democratically elected president.

Most Haitians farm small plots, which produce yams and **plantains**, a type of banana used in cooking. Coffee, cotton, and sugarcane provide some export income.

The official languages of Haiti are French and Creole. Creole is mostly French, with some Spanish, English, and African words. While Roman Catholicism is the recognized religion, **voodoo** is important throughout Haitian culture. Followers of voodoo believe that spirits of good and evil play an important part in daily life.

The capital city of Haiti, Port-au-Prince (pohrt-oh-PRINS), is the country's manufacturing center. Industries often depend on hand labor and simple machines. Because of few economic opportunities and many political problems, a large number of Haitians have tried to flee their country. Many have become **refugees** in the United States. A refugee is someone who flees his or her own country to another, often for economic or political reasons.

Dominican Republic The Dominican Republic occupies the eastern part of Hispaniola. The country was Spain's earliest colony, and the traditions of Spanish culture remain strong. Santo Domingo, the capital, was the first permanent European settlement in the Western Hemisphere.

The Dominican Republic is a land of steep mountains and fertile valleys, and its forests contain valuable tropical woods. Sugarcane and coffee are the key commercial crops, while tourism has become an important industry. The country's economy, as well as its education, health care, and housing, is more developed than that of nearby Haiti.

Puerto Rico Puerto Rico is the easternmost island of the Greater Antilles. Most of the island is

Spanish-style architecture lines the narrow streets of the older parts of **San Juan, Puerto Rico**.

mountainous, with narrow coastal plains. Once a Spanish colony, Puerto Rico is a **commonwealth**, a self-governing political unit, associated with the United States. Puerto Ricans are U.S. citizens, although they cannot vote in presidential elections.

Sugarcane is the commonwealth's most valuable crop, though coffee, tropical fruits, and tobacco also are grown. However, the most important feature of Puerto Rico's economy is its industrialization. Puerto Rico's low wages, inexpensive manufacturing, and easy access to U.S. markets have attracted U.S. firms.

Unemployment is high and incomes are low in Puerto Rico when compared to the United States. Yet, Puerto Rico's economy is much more developed than that of any other island in the Antilles. Puerto Ricans continue to debate whether the island should become a U.S. state or an independent country, or remain a commonwealth.

Closure

Have each student write a "quick" fact about the island he or she selected in the Classifying activity. Ask students to read their facts to the class, and write them on the board. Then have the class compare and contrast these facts with the images they described in the Motivator activity.

ASSESSMENT

Assign the Section 2 Review questions on this page.

Reteaching

Assign Section 2 of Reteaching Worksheet 19.

Extension

Encourage students to read *The Cay* by Theodore Taylor. This award-winning novel tells the story of Phillip—a young boy—and Timothy—an older man from a Caribbean island—who are marooned together on a barren Caribbean island during World War II. Phillip and Timothy's struggle to survive will intrigue students.

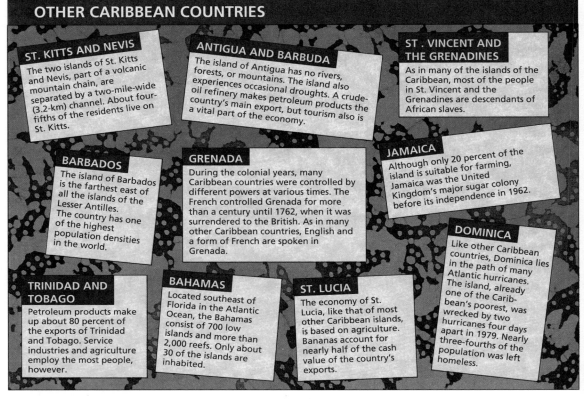

OTHER CARIBBEAN COUNTRIES

ST. KITTS AND NEVIS
The two islands of St. Kitts and Nevis, part of a volcanic mountain chain, are separated by a two-mile-wide (3.2-km) channel. About four-fifths of the residents live on St. Kitts.

ANTIGUA AND BARBUDA
The island of Antigua has no rivers, forests, or mountains. The island also experiences occasional droughts. A crude-oil refinery makes petroleum products the country's main export, but tourism also is a vital part of the economy.

ST. VINCENT AND THE GRENADINES
As in many of the islands of the Caribbean, most of the people in St. Vincent and the Grenadines are descendants of African slaves.

BARBADOS
The island of Barbados is the farthest east of all the islands of the Lesser Antilles. The country has one of the highest population densities in the world.

GRENADA
During the colonial years, many Caribbean countries were controlled by different powers at various times. The French controlled Grenada for more than a century until 1762, when it was surrendered to the British. As in many other Caribbean countries, English and a form of French are spoken in Grenada.

JAMAICA
Although only 20 percent of the island is suitable for farming, Jamaica was the United Kingdom's major sugar colony before its independence in 1962.

DOMINICA
Like other Caribbean countries, Dominica lies in the path of many Atlantic hurricanes. The island, already one of the Caribbean's poorest, was wrecked by two hurricanes four days apart in 1979. Nearly three-fourths of the population was left homeless.

TRINIDAD AND TOBAGO
Petroleum products make up about 80 percent of the exports of Trinidad and Tobago. Service industries and agriculture employ the most people, however.

BAHAMAS
Located southeast of Florida in the Atlantic Ocean, the Bahamas consist of 700 low islands and more than 2,000 reefs. Only about 30 of the islands are inhabited.

ST. LUCIA
The economy of St. Lucia, like that of most other Caribbean islands, is based on agriculture. Bananas account for nearly half of the cash value of the country's exports.

Virgin Islands The Virgin Islands, which lie just east of Puerto Rico, total only 195 square miles (505 sq. km)—about the same amount of area covered by the U.S. city of New Orleans. The Virgin Islands are politically divided between the United States and the United Kingdom. The United States territory includes the three largest islands: St. Thomas, St. John, and St. Croix (CROI). Once Denmark's sugar colony, the islands were purchased by the United States in 1917 as a base to protect the Panama Canal. Today, the United States Virgin Islands include some of the Caribbean's most popular tourist centers. Farmlands are disappearing as vacation homes cover more land. Remaining agricultural lands produce fruits, vegetables, and meat for local markets.

The British Virgin Islands include Tortola, Anegada, Virgin Gorda, Jost Van Dyke, and 32 smaller islands. The area of these islands is just 59 square miles (153 sq. km). As with the U.S. Virgin Islands, tourism is a major part of the economy.

Other Islands Between the Virgin Islands and Trinidad are more than a dozen important, though small, island territories. French possessions in the Lesser Antilles include Martinique (mahrt-uhn-EEK) and Guadeloupe (GWAHD-uhl-oop). The Dutch territories include three small tourist islands in the northeast Caribbean and the "ABC" islands—Aruba, Bonaire, and Curaçao (kyur-uh-SOE)—off the coast of Venezuela. Bermuda, a favorite tourist spot, is a prosperous possession of the United Kingdom.

Many other islands lie in the Caribbean. To learn some "quick facts" about the other independent islands, study the chart on this page.

2 SECTION REVIEW

1. How do the Greater Antilles differ from the Lesser Antilles? What economic activities are most important to the islands?
2. What is unique about Antigua's physical geography?
3. **Critical Thinking** **Compare and contrast** political and economic characteristics of Cuba and Puerto Rico.

Building a Vocabulary

1. Cardamom is a seasoning popular in the foods of Southwest Asia. Plantains are a type of banana.
2. Refugee
3. A self-governing political unit; Puerto Rico
4. Musical styles of the Caribbean islands; calypso—Trinidad and Tobago; reggae—Jamaica; merengue—the Dominican Republic

Recalling and Reviewing

1. The raising of sugarcane is important to both regions. Central American countries raise more coffee because they have the necessary mountain climates. Tourism is more important to the Caribbean islands because they have beautiful beaches and sunny weather.
2. Many of the countries were colonies at one time, and many have had unstable political histories.

CHAPTER 19 CHECK

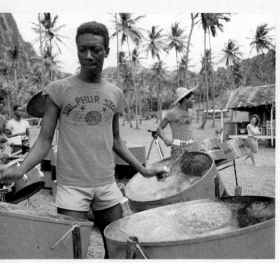

A steel-drum band in **St. Lucia**

Reviewing the Main Ideas

1. The Central American and Caribbean region includes 20 small countries and more than 12 territories. Physical and cultural characteristics have divided the region.
2. The countries of Central America have coastal plains with humid-tropical climates, inland valleys, and mountains with highland climates. Coffee, sugarcane, cotton, and bananas are major exports. Subsistence farming also is common. Population growth and the need for land reform have created economic and political challenges.
3. The Caribbean islands include the Greater Antilles, the Lesser Antilles, and the Bahamas. The climate and beauty of these islands attract many tourists. Nearly every island in the Caribbean shows the effects of colonialism and slavery. Dependence on foreign countries is a major issue, and political unrest is common.

Building a Vocabulary

1. What is cardamom? What are plantains?
2. What term refers to a person who flees a country for economic or political reasons?
3. Define *commonwealth*. Which Caribbean island is a commonwealth?
4. What are calypso, reggae, and merengue? Where did each originate?

Recalling and Reviewing

1. Compare and contrast the major economic activities of the Central American countries with those of the Caribbean islands. What factors account for the differences?
2. What political characteristics are shared by most Central American and Caribbean nations?

Thinking Critically

1. The United States purchased the Virgin Islands in 1917 as a base to protect the Panama Canal. Do you think the United States' interests in protecting the canal were primarily economic, or do you think they were primarily military?
2. Many of the countries of the Caribbean are heavily dependent upon foreign aid and imports. How might these countries become more self-sufficient?

Learning About Your Local Geography

Individual Project
You have read about calypso, reggae, and merengue. What types of music and dance are popular in your community? Interview community members about their favorite forms of music and dance. Is your community "known" for a particular form of entertainment, as Trinidad, Jamaica, and the Dominican Republic are? Design a poster advertising music and dance in your community.

Using the Five Themes of Geography

Write a paragraph about different forms of **human-environment interaction** that occur when tourists visit **regions** composed of tropical islands, such as the Caribbean. What attracts tourists to tropical island regions? How do tourists both adapt to and change the islands' environments?

Thinking Critically

1. Answers will vary. Some students will say economic, because the United States relied on the canal to carry trade between the Atlantic and Pacific oceans. Others may say military; the United States relied on the canal to move its naval ships between the Atlantic and Pacific.
2. Possible answers: focusing less on commercial agriculture might increase self-sufficiency but might detract from development. Reducing population growth might also help.

Learning About Your Local Geography

Posters should show different forms of music and dance in students' communities.

Using the Five Themes of Geography

Paragraphs will vary but might mention that it is often the islands' physical environments that attract tourists. The use of beach umbrellas would show how tourists adapt to the environments; the removal of shells from beaches would show that tourists change environments.

225

CHAPTER 20 Caribbean South America
Planning Guide

TEXTBOOK	TEACHING RESOURCES	OTHER AVAILABLE MATERIALS
Section		
1. Colombia Pages 227–229	Reteaching Worksheet 20.1 Challenge/Enrichment Worksheet 20	*Study Guide for Spanish Speakers* *Lecture Notes,* Chapter 20
2. Venezuela Pages 230–231	Reteaching Worksheet 20.2 Skills Worksheet 20	*Study Guide for Spanish Speakers* *Lecture Notes,* Chapter 20
3. The Guianas Pages 231–232	Reteaching Worksheet 20.3 Graphic Organizer 20 Critical Thinking 20	*Study Guide for Spanish Speakers* *Lecture Notes,* Chapter 20
Chapter Review Page 233	Chapter 20 Test Chapter 20 Portfolio Assessment See also *Alternative Assessment Forms*	*Test Generator,* Chapter 20

CHAPTER RESOURCES
Lecture Notes, Chapter 20
Teaching Resources
   Chapter 20 Graphic Organizer
   Chapter 20 Skills Worksheet
   Chapter 20 Critical Thinking Worksheet
   Chapter 20 Challenge/Enrichment Worksheet

REVIEW AND ASSESSMENT
Chapter 20 Check, page 233
Chapter 20 Test, *Teaching Resources*
Chapter 20 Portfolio Assessment, *Teaching Resources*
Test Generator, Chapter 20
Alternative Assessment Forms

ADDITIONAL RESOURCES

Books for Teachers
Gereffi, Gary, and Donald L. Wyman, eds. *Manufacturing Miracles: Paths of Industrialization in Latin America and East Asia.* Princeton, NJ: Princeton University Press, 1990.

Lehmann, David. *Democracy and Development in Latin America: Economics, Politics, and Religion in the Post-War Period.* Philadelphia: Temple University Press, 1990.

Lombardi, Cathryn, and John V. Lombardi. *Latin American History: A Teaching Atlas.* Madison: University of Wisconsin Press, 1983.

Winn, Peter. *Americas: The Changing Face of Latin America and the Caribbean.* Berkeley: University of California Press, 1992.

Books for Students
Ewell, Judith. *Venezuela: A Century of Change.* Stanford, CA: Stanford University Press, 1984.

Latin American History Biographies: Bolívar. New York: Chelsea House, 1985. Available through Social Studies School Service, P.O. Box 802, Culver City, CA 90232–0802.

McIntyre, Loren. *Exploring South America.* Glendale, CA: Crown, 1990.

Multimedia Resources
Folktales from Colombia. (VHS, 26 minutes) Films for the Humanities & Sciences, P.O. Box 2053, Princeton, NJ 08543–2053.

Latin America Map Game. (data sheets for individual countries, teacher's guide) Educational Materials Associates, 1985. Distributed by Social Studies School Service, P.O. Box 802, Culver City, CA 90232–0802.

Picture Atlas of the World. (interactive maps, photo bank, animation, audio on CD-ROM) National Geographic Society, Washington, DC.

The South American Series: Brazil and the Northern Countries, The Andean Region, The Southern Plains. (VHS, 14–21 minutes each) Encyclopedia Britannica Corporation, 310 Michigan Avenue, Chicago, IL 60604–9839.

Venezuela: Petroleum-Powered Economy. (VHS, 30 minutes) Films for the Humanities & Sciences, P.O. Box 2053, Princeton, NJ 08543–2053.

MEETING INDIVIDUAL NEEDS

Limited English Proficiency (LEP)

Lead students through a review of Chapter 20 to find the agricultural products of each country. Then have students create an outline map of Caribbean South America on a large piece of posterboard. Students can make an agricultural product map of the region by attaching small samples of coffee, cotton, corn, rice, and sugarcane in the appropriate places on the map. Display the maps in the classroom.

Learners Having Difficulty

Help students review the main ideas of the chapter by making a graphic organizer showing the physical and economic features of the Caribbean South American countries. Write the information on the board as students provide it. Have students copy the graphic organizer and use it as a study guide.

Auditory Learners

Assign individuals to prepare an oral report about a country of Caribbean South America. Reports should touch on the physical, human, and economic geography of each country and include information beyond what is available in the textbook. Have classmates take notes as each report is presented.

Visual Learners

Have students make crossword puzzles using the terms from the Chapter 20 Geography Dictionary and other key terms from the chapter. Give a prize for the puzzle that incorporates the most terms. Students can exchange the puzzles and solve them.

Tactile/Kinesthetic Learners

Have students work individually or in groups to create three-dimensional models of the five Andean elevation zones. Students might use plaster or a flour-water-salt mixture to form the mountains and paint each elevation zone a different color; or they might use colored clay to create the model. The models should include an elevation key showing the name and the height of each elevation. Display the completed models in the classroom.

Gifted Learners

Have interested students find newspaper and magazine articles about the illegal drug trade in Colombia and its relationship to the United States. Ask students to make a presentation of their findings to the class.

MULTICULTURAL FOCUS

Point out to students that names are often an interesting and revealing aspect of culture. Traditions for naming children vary greatly around the world: In some cultures, every person's name is unique and descriptive; in others, it is customary for parents to choose names from a set of common given names.

Customs that determine a person's last name also differ throughout the world. In Spanish-speaking countries such as Colombia and Venezuela, a person's *apellido,* or surname, is a combination of both parents' family names. A person has two last names—usually the paternal last name of the father followed by that of the mother. For example, if Jose *Salas* García marries Ana *Álvarez* Guadalupe, their son's name would be Pedro *Salas Álvarez.*

People generally use only their first (paternal) last name for everyday purposes, although some people hyphenate the names or go by their maternal last name. To indicate their lineage, people of high social rank may use both last names of both parents, such as Pedro Salas-García Álvarez-Guadalupe.

Have students consider the assumptions that are usually made in the United States about the form of a person's name. Such assumptions come into play each time they look up a name in a phone book, fill out an application form, or write a formal letter. Point out that knowing the customary style of Spanish names would thus be important to a person visiting or doing business in Colombia or Venezuela.

CHAPTER 20

CARIBBEAN SOUTH AMERICA

Section 1: Colombia
Section 2: Venezuela
Section 3: The Guianas

OVERVIEW

In Chapter 20, students will learn about the South American countries that border the Caribbean Sea.

Section 1 describes the main landform regions, economy, and peoples of Colombia and characterizes the five elevation zones of the Andes.

Section 2 focuses on Venezuela's geography. Students will learn how the discovery of oil has been an important factor in the development of modern Venezuela.

Section 3 describes the Guianas, which consist of Guyana, Suriname, and the French territory of Guiana, and explores their cultural diversity.

About the Author

Paul Theroux (1941–) is a well-known travel writer. He has written many books about his travels to South America, India, and Africa. The quotation on this page is from *The Old Patagonian Express.*

Primary Source

Description of Change: excerpted
Rationale: excerpted to focus on the main idea

CHAPTER 20

Caribbean South America

❝You think we're going to make it?" I asked the conductor.

"You will be in Bogotá tonight," he said. . . .

Soon after, the mountains appeared, the cordillera of the Andean chain; and with them the brown Magdalena River, on which men paddled dugout canoes or fished from the shore with contraptions that looked like butterfly nets. The mountains were at first scattered buttes and solitary peaks, and some were like citadels, squarish with fortresslike buildings planted around the summits.❞

Paul Theroux

Gold artifact from an early culture in present-day **Colombia**

The northern rim of South America contains Colombia, Venezuela, Guyana (gie-AN-uh), Suriname, and French Guiana. In the west, the Andes thrust skyward over 17,000 feet (5,182 m). In the east, the Guiana Highlands still reach above 7,000 feet (2,134 m) even though scientists believe the highlands have been eroding for millions of years. Between these uplands are the vast plains of the Orinoco River basin. Most of the region's cities and agricultural areas are found near the Caribbean shore and in mountain valleys.

Fertile valleys in the Andes and coastal ranches made Colombia and Venezuela key parts of Spain's colonial empire. The other former territories were colonized by European powers who arrived during the sugar boom of the 1700s. Throughout the region's history, rugged land and dense forests often have separated peoples and cultures here.

GEOGRAPHY DICTIONARY

tierra caliente

tierra templada

tierra fría

páramo

tierra helada

guerrilla

slash-and-burn farming

Andes Mountains of western **Colombia**

226

Using the Quotation

Call on a volunteer to read aloud the quotation by Paul Theroux on page 226. Ask students how the speaker is likely to be traveling. (*by train, since there is a conductor*) Using clues in the quotation and referring to a map, have students speculate on the route the train is traveling. (*perhaps south from Cartagena to Bogotá*) Ask students if they think the speaker is a native of the region or a visitor, and to explain their answers.

Using the Geography Dictionary

Write the terms *tierra helada, páramo, tierra fría, tierra templada,* and *tierra caliente* in a line down the board, with *tierra helada* at the top. Tell students that *tierra* means "land," and that the terms refer to climates. Then draw a mountain over the list of terms, and ask students to give or to suggest translations for the terms based on the diagram.

SECTION 1
OBJECTIVES

The objectives reflect the Focus questions on this page and are assessed in the review on page 229.
1. List Colombia's landform regions and the five elevation zones of the Andes.
2. Analyze Colombia's economic, human, and political geography.

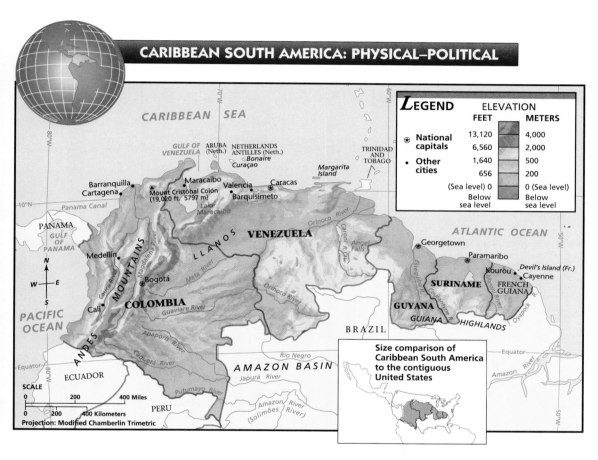

CARIBBEAN SOUTH AMERICA: PHYSICAL–POLITICAL

LEGEND   ELEVATION

	FEET	METERS
⊛ National capitals	13,120	4,000
	6,560	2,000
• Other cities	1,640	500
	656	200
(Sea level) 0		0 (Sea level)
Below sea level		Below sea level

Size comparison of Caribbean South America to the contiguous United States

SCALE
0   200   400 Miles
0 200 400 Kilometers
Projection: Modified Chamberlin Trimetric

Using the Physical–Political Map: Caribbean South America

Focus students' attention on the map on this page. Have students name the landforms and rivers of Caribbean South America. Point out that the Andes in Colombia separate into three parallel ranges, or cordilleras, like the frayed end of a rope. Have students predict which parts of Caribbean South America are the most populated and the most productive. Refer them to the maps on pages 191–192 to check their predictions.

COLOMBIA

FOCUS

- *What are Colombia's landform regions? What are the five elevation zones of the Andes?*

- *What economic activities are important to Colombia? What are the country's human and political characteristics?*

Physical Geography Colombia is the only South American country with coastlines on both the Caribbean Sea and the Pacific Ocean. (See the map above.) The Andes, however, prevent easy travel between the two coasts. With the opening of the Panama Canal, road construction, and the development of air transportation, communications have improved.

Colombia has two major landform regions: the Andes Mountains and valleys in the west, and the plains of the llanos and the Amazon Basin to the east. In the north, the plains are covered with savannas, but southward, the vegetation thickens to become dense tropical forests. Colombia's hot and humid coastal lowlands extend only a short distance inland.

LOCATION **Andes Environments** The highland climates of the Andes slopes are divided by elevation into five zones. (See the illustration on page 228.) The *tierra caliente,* or "hot country," refers to the hot and humid lower elevations between sea level and 3,000 feet (914 m). Daytime temperatures here range from 85°F to 90°F (29°–32°C). During the nights, temperatures usually range from 70°F to 75°F (21°–24°C). There is little difference between summer and winter temperatures. The *tierra caliente* produces tropical products, including bananas and cacao.

227

Motivator

Tell students that before the Revolutionary War (in North America), many people believed that America should have been named Columbia, after Christopher Columbus. Point out that although the continent (or the entire American landmass) was not named after Columbus, many U.S. towns and cities were (such as Columbus, Ohio, and Columbia, South Carolina). Colombia, a country in South America, also was named after Columbus. Tell students that they will learn more about Colombia in Section 1.

Building a Vocabulary

Point out that **guerrilla** is Spanish and is the diminutive of *guerra,* the word for "war." Have students write a sentence using the word *guerrilla*.

Analyzing Colombia's Economy

Have students work in pairs. Using the textbook and classroom resources, have one partner list Colombia's mineral resources and the other list Colombia's agricultural products. Then have partners exchange lists and check each other's work. After discussing the lists, the partners should circle the mineral and the agricultural product they think is most important to Colombia's economy and why.

Geograpical Sidelight

Eighty percent of carnations and 40 percent of roses sold in the United States come from overseas, mainly from South America. Colombia is the largest Andean exporter of cut flowers, accounting for 3 out of 5 roses flown into Miami.

Caption Answer

Tierra helada

GLOBAL PERSPECTIVES

Point out that Colombian drug smugglers often are internationally powerful. While they continue to pour cocaine into the United States, they also are finding new markets in Europe and Asia, particularly in countries such as Russia, where economic chaos makes interference less likely. Generate a class discussion about what international efforts might reduce the drug trade.

Elevation Zones in the Andes

Tierra helada:
Above 16,000 feet (4,877 m)
Daytime temperatures: 30°–35°F (-1°–2°C)
Nighttime temperatures: 15°–20°F (-9°– -7°C)
Permanently covered with snow

Páramo:
10,000 to 16,000 feet (3,048 to 4,877 m)
Daytime temperatures: 50°–55°F (10°–13°C)
Nighttime temperatures: 30°–35°F (-1°–2°C)
Potatoes, grasslands and hardy shrubs, grazing

Tree line

Tierra fría:
6,000 to 10,000 feet (1,829 to 3,048 m)
Daytime temperatures: 65°–70°F (18°–21°C)
Nighttime temperatures: 50°–55°F (10°–13°C)
Potatoes, wheat, oats, barley, beans, corn, rye

Tierra templada:
3,000 to 6,000 feet (914 to 1,829 m)
Daytime temperatures: 75°–80°F (24°–27°C)
Nighttime temperatures: 60°–70°F (16°–21°C)
Coffee, corn, wheat, cotton, potatoes, sugarcane, tobacco

Tierra caliente:
Sea level to 3,000 feet (914 m)
Daytime temperatures: 85°–90°F (29°–32°C)
Nighttime temperatures: 70°–75°F (21°–24°C)
Bananas, cacao, rice, sugarcane

Sea level

The names for Andean elevation zones vary somewhat from country to country. The names in this illustration are commonly used in Colombia. In which elevation zone is growing crops not possible?

Farther up the mountains, the air becomes cooler, and moist climates with mountain forests are typical. This zone of pleasant climates is called *tierra templada*, or "temperate country." Here, temperatures during the day range from 75°F to 80°F (24°–27°C). Nighttime temperatures range from 60°F to 70°F (16°–21°C). Many people live in this zone, where both tropical and middle-latitude types of crops are grown.

The next higher zone is the *tierra fría*, or "cold country." The *tierra fría* has a natural vegetation of cool forests and grasslands. It extends from about 6,000 feet (1,829 m) to 10,000 feet (3,048 m). Daytime temperatures here range from 65°F to 70°F (18°–21°C). During the night, the temperature range is from 50°F to 55°F (10°–13°C). The *tierra fría* is important for growing potatoes and wheat.

Above the tree line is a zone called the *páramo*, which extends from about 10,000 feet (3,048 m) to as high as 16,000 feet (4,877 m). Temperatures range from 50°F to 55°F (10°–13°C) during the day and from 30°F to 35°F (-1°–2°C) during the night. Grasslands and hardy shrubs are the usual vegetation. Llamas, alpacas, and sheep graze on these lands, while potatoes grow in the lower elevations of the *páramo*. Frost may occur on any night of the year.

The *tierra helada*, or "frozen country," is the zone of highest elevation. It is permanently covered with snow. Temperatures during the day range from 30°F to 35°F (-1°–2°C) and from 15°F to 20°F (-9° to -7°C) during the night.

Economic Geography Colombian coffee, produced mostly on family farms, is world famous.

Guided Practice

Refer students to the chart of elevation zones on page 228. Point out that the climate of Colombia is affected more by elevation than by season. Have students refer to the chart to answer these questions: What crop grows in the most altitude zones? (*potatoes*) In which altitude zone would you expect to see llamas grazing? (*páramo*) Why do bananas grow well only in *tierra caliente*? (*They require a tropical climate.*)

Closure

Have students list the advantages of Colombia's geography. (*coastlines on two oceans, proximity to Panama Canal, productive agricultural areas, mineral riches*)

ASSESSMENT

Assign the Section 1 Review questions on this page.

Reteaching

Assign Section 1 of Reteaching Worksheet 20.

Extension

Have students imagine that they live in Bogotá or Cartagena. Have them write a letter to a young person in the United States describing life in Colombia. Have students focus on both positive and negative aspects of life in modern Colombia.

The main coffee-producing area is found along the Andes ridges between the Magdalena and Cauca rivers. The country is second only to Brazil in coffee production. One of Colombia's newest and most profitable industries is flower farming. Colombia exports large amounts of cut flowers. Other commercial crops grown in Colombia include rice, maize (corn), sugarcane, tobacco, bananas, and cotton. Cattle grazing is an important occupation in the llanos.

The mountains of Colombia are rich in minerals. Ninety percent or more of the world's emeralds come from Colombia. Colombia's iron-ore deposits, however, are of more value to the country's economy. Oil is found in the eastern part of Colombia along the Venezuelan border. Recent discoveries have made oil a larger export product than coffee.

As in other Middle and South American countries, urban poverty, rapid population growth, and land reform are pressing problems. A good trade location, plentiful natural resources, and improved interior transportation, however, give Colombia the potential to prosper.

Human Geography Colombia's capital, Bogotá, is a modern city located on a plateau in the eastern Andes. Cartagena on the Caribbean

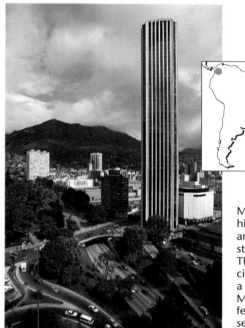

Modern high rises and highways of the banking and international district stand out in **Bogotá**. The capital and largest city of **Colombia** sits on a plateau of the Andes Mountains that is 8,563 feet (2,610 m) above sea level.

coast has preserved colonial architecture among its modern buildings.

Colombia has a history of civil war and corrupt governments. Much of Colombia is ruled more by violence and intimidation than by laws. **Guerrilla** groups often have frightened people away from elections and have controlled large regions of the country. A guerrilla is one who fights as a member of an armed band and takes part in irregular warfare, such as harassment and raids.

Much of the country's recent violence is related to a large illegal drug trade. Drugs are smuggled out of the country in great quantities. Smugglers have used their illegal profits to support private armies and corrupt judges and politicians. Income from the illegal drug trade accounts for a significant portion of Colombia's true GNP.

FACTS IN BRIEF	Caribbean South America		
COUNTRY POPULATION (1994)	LIFE EXPECTANCY (1994)	LITERACY RATE (1990)	PER CAPITA GDP (1993)
COLOMBIA 35,577,556	69, male 75, female	87%	$5,500
GUYANA 729,425	62, male 68, female	95%	$1,900
SURINAME 422,840	67, male 72, female	95%	$2,800
VENEZUELA 20,562,405	70, male 76, female	88%	$8,000
UNITED STATES 260,713,585	73, male 79, female	97% (1991)	$24,700

Source: *The World Factbook 1994*, Central Intelligence Agency

What is the most populous country of Caribbean South America?

SECTION REVIEW

1. Describe each elevation zone of the Andes.
2. What are Colombia's most important exports?
3. **Critical Thinking** **Discuss** how Colombia's physical geography relates to the country's political unrest.

Section 1 Review
ANSWERS

1. *Tierra caliente:* hot and humid zone closest to sea level; *tierra templada:* pleasant climate, supporting many crops, and with many people living there; *tierra fría:* cool forests and grasslands; *páramo:* above treeline, subject to frequent frosts; *tierra helada:* highest zone, covered with snow

2. Coffee, cut flowers, other crops, emeralds, other minerals, oil

3. The mountainous terrain blocks communication and transportation and may provide hideouts for guerrillas.

229

SECTION 2
OBJECTIVES

The objectives reflect the Focus questions on this page and are assessed in the review on page 231.

1. Describe Venezuela's landforms, climates, and vegetation.
2. Identify economic, cultural, and social characteristics of Venezuela.

TEACHING SECTION 2
Motivator

Have students look at the picture of Angel Falls on page 184. Ask what the presence of such high waterfalls in Venezuela might suggest about its landforms. *(It has regions of both high and low elevations.)* Then ask students how Venezuela's landforms might affect its climates. *(highland climates in higher elevations; warmer climates in lower areas)*

Analyzing Relationships

Organize the class into two groups. Have one group make a list of Venezuela's economic characteristics, and have the other group make a list of human characteristics. Have the two groups share their lists. Ask how items on the economic list might have influenced items on the human list *(for example, oil profits improved living conditions)*, and vice versa *(for example, cultural pride has enabled effective use of resources)*.

C☺OPERATIVE LEARNING

Organize the class into groups of four or five. The goal of the class is to create illustrated time lines of the histories of Colombia and Venezuela from colonial settlements to the present. The finished time lines for the two countries should run parallel on a bulletin board. Assign each group a segment of the time lines to prepare. Group roles include planning the time lines with other groups, researching, illustrating events, and drawing the time line segment.

2 VENEZUELA

FOCUS

● *What landforms, climates, and vegetation are found in Venezuela?*

● *What are Venezuela's economic, cultural, and social characteristics?*

LOCATION **Landforms** Venezuela is located on South America's northern Caribbean coast. The country is divided into four landform regions: (1) the northern mountains and valleys of the Andes; (2) a narrow tropical-lowland coast that is surprisingly arid in parts; (3) the llanos, the savanna plains between the Andes and the Orinoco River; and (4) the Guiana Highlands in the southeast. (See the map on page 227.)

The Orinoco River and its tributaries drain the llanos as well as the Guiana Highlands. When streams plunge over the steep cliffs in the highlands, they create many high waterfalls. Angel Falls, with a drop of 3,212 feet (979 m), is the world's highest. In contrast, parts of the llanos are so flat that the water from summer floods stands on the plains for months.

Climate and Vegetation Venezuelans experience climates that are typical of the lower latitudes. The temperatures are warm to hot year-round. A dry winter season gives way to a rainy summer season. The highlands of Venezuela, however, have many climate variations. Mountains along the north coast may receive rain throughout the year as the trade winds blow moisture inland from the Caribbean.

The coastal region has humid air, but in some locations there is very little rain. Scrub savannas with cacti are common. Thick forests were once widespread in the Andean valley, but 400 years of clearing the land has opened many areas to grasses. Farther south are the llanos, one of the world's largest short-tree and grass savannas. Huge expanses of grassland are flooded during rainy seasons. Like in nearby Brazil, the plains in the far south of Venezuela are covered with humid-tropical forests. The vegetation of the Guiana Highlands includes tall-tree savannas as well as dense forests.

Economic Geography Venezuela's mineral wealth provides the country's chief source of income. Until 1920, Venezuela was a poor country with few opportunities. Then, the vast oil deposits surrounding Lake Maracaibo (mar-uh-KIE-boh) were developed. Today, the country is a leading oil exporter. Although Venezuela has several oil refineries, much of the production is refined outside the country on offshore islands such as Curaçao and Trinidad. A large amount is also shipped to refineries in the United States.

The Venezuelan government has used oil profits to build schools, improve living conditions, and expand medical care. Venezuela's economy has become dependent on oil revenues, however. The country suffers when oil prices drop. For this reason, the government of Venezuela is encouraging new industries that process food and minerals.

The Guiana Highlands are rich in minerals, including large deposits of iron ore. Large dams have been built on the Orinoco River. Steel mills using oil and hydroelectric power are producing a steel surplus.

As for agriculture, small family farms and large commercial farms are spread throughout the

Ranchers herd cattle on the rich grazing lands of **Venezuela**.

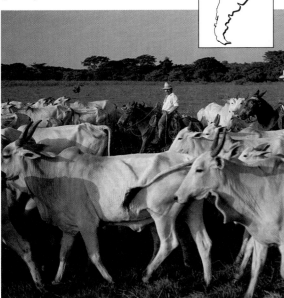

Closure

Have students list ways Venezuela can improve its economy. *(diversify industry, develop southern frontier)*

ASSESSMENT

Assign the Section 2 Review questions on this page.

Reteaching

Assign Section 2 of Reteaching Worksheet 20.

SECTION 3
OBJECTIVES

The objectives reflect the Focus questions on this page and are assessed in the review on page 232.

1. Examine physical and economic features of Guyana, Suriname, and French Guiana.

2. Summarize political and cultural characteristics of the Guianas.

TEACHING SECTION 3
Motivator

Have students locate the Guianas on the map on page 227. Ask them to speculate on the similarities and differences among these three countries. As students read Section 3, have them verify the accuracy of their assumptions.

Caracas. Rural poverty is widespread. Oil wealth and new lands to develop in the south suggest that Venezuela can continue as an economic and political leader in the region.

 SECTION REVIEW

1. What are Venezuela's four landform regions? Describe the climate of each region.
2. What resources are found in Venezuela?
3. **Critical Thinking Explain** how slash-and-burn farming might affect the environment.

 THE GUIANAS

FOCUS

● *What are the physical and economic features of the Guianas?*

● *What are the political and cultural characteristics of the Guianas?*

Guyana Guyana and its two neighbors on South America's northeast coast have similar humid-tropical environments. Nearly all agricultural lands are found on a narrow coastal plain. Guyana's main crops are rice, usually grown for food, and sugarcane, usually exported as sugar. Guyana's chief mineral resource is bauxite. Hydroelectric power enables the country to process its bauxite into aluminum for export.

Guyana is the former British Guiana. Since independence in 1966, the country's political situation has been unstable, and the government has had close relations with Cuba and other Communist countries. Guyana's economy has declined as government-run industries and businesses have collapsed.

More than half of Guyana's population is of South Asian descent. Most of these people farm small plots and run small businesses. Most of the remaining population is made up of Africans and mulattos who are descendants of African slaves. This sector of the population controls most of the large businesses and holds most of the government

The Teresa Carrena Theater is one of the many modern and beautiful structures in **Caracas**, the capital of **Venezuela**. Money from oil sales has helped finance many public works in the country.

northern part of the country. The main crops are sugarcane and coffee. Maize (corn) and wheat are grown on the cooler mountain slopes. The llanos is a land of large cattle ranches. Until recently, few people have lived in the region. With the development of irrigation projects, more farmers are moving there. Except for large settlements near mines, the Guiana Highlands are nearly empty of people. Small communities of indigenous peoples practice traditional **slash-and-burn farming**, a type of agriculture in which forests are cut and burned to clear land for planting.

Human Geography Most of the people of Venezuela live along the Caribbean coast and in the valleys of the nearby mountains. These areas were the focus of early settlement. Most Venezuelans are mestizos. Their national pride is expressed in a popular small statue, which displays a European man, an African man, and a Native American woman standing shoulder to shoulder.

Caracas (kuh-RAK-uhs) is the capital of Venezuela and the center of Venezuelan culture. It has a new subway system and a ring of modern expressways. There are modern office buildings and many job opportunities.

As in other Middle and South American cities, however, shantytowns and slums encircle

231

Creating a Chart
Physical and Human Geography
Give each student a copy of a chart with Guyana, Suriname, and French Guiana listed along the top, and Climate, Landforms, Agriculture, Resources, Government, and Culture listed along the side. Have each student (or pair of students) find out the appropriate information for one of the cells on the chart. Have the class fill in the charts as students take turns giving their assigned information.

Guided Practice
Write this sentence on the board: "The Guianas are culturally and ethnically very diverse." Call on students to provide examples from Section 3 that support this statement, and write their responses beneath it.

Closure
Organize the class into two teams for a game show about the Guianas. Have each team write questions about the Guianas' physical, economic, politi-

cal, and cultural characteristics. Then have each team answer the other team's questions.

ASSESSMENT
Assign the Section 3 Review questions on this page.

Reteaching
Assign Section 3 of Reteaching Worksheet 20.

Section 3 Review
ANSWERS

1. Physical features: humid-tropical environments, coastal plains; resources: rice, sugarcane, bauxite, hydroelectricity, fruits, lumber, cacao, bananas, gold, shrimp

2. French Guiana is a territory of France, whereas Guyana and Suriname are independent countries.

3. The conflicts within Guyana have interrelated economic and ethnic aspects. South Asians farm and run small businesses, whereas Africans and mulattos control the larger businesses.

232

---

positions. Economic differences between the two ethnic groups have created tension.

Because Venezuela claims more than half of Guyana's territory, Guyana faces an uncertain future. In Venezuela, this area is shown on maps as "occupied territory." If Guyana, already one of the poorest countries in the world, were to lose this territory, its future prosperity would be threatened.

Suriname Suriname is a former Dutch colony that gained its independence in 1975. As in Guyana, Suriname's coastal area produces sugarcane, rice, and tropical fruits. Lumber from the interior forests is exported in the form of wood and wood products. Suriname's economy depends on the country's bauxite and aluminum exports.

Suriname's population, which totals less than 500,000, is an ethnic mixture. There are Africans, mulattos, Chinese, Indonesians, and South Asians. Muslim, Hindu, Roman Catholic, and Protestant houses of worship line the streets of Paramaribo (par-uh-MAR-uh-boe), the country's capital.

Violent ethnic political disputes are common in Suriname. As various political parties repre-

A Javanese immigrant buys produce from a merchant who has immigrated to **Suriname** from India. Just over half of Suriname's population is made up of ethnic groups from South Asia and island Southeast Asia.

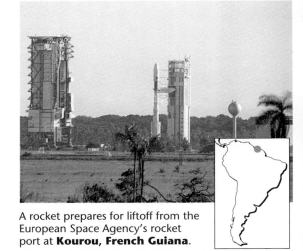

A rocket prepares for liftoff from the European Space Agency's rocket port at **Kourou, French Guiana**.

senting different ethnic groups have come to power, other ethnic groups have worked behind the scenes to support guerrilla groups.

French Guiana French Guiana is the largest of France's overseas territories. Its status as a French territory is much the same as that of a state in the United States. French Guiana sends three representatives to the National Assembly in Paris, France.

The population of French Guiana is only about 139,000, and large parts of the territory are uninhabited. Cacao, rice, sugarcane, and bananas are grown on the coastal plain. Chief exports are gold, timber, and shrimp. France used the famous offshore Devil's Island as a prison. Today, Kourou on the coast is the site of the European Space Agency's rocket port.

SECTION REVIEW

1. What physical features are shared by Guyana, Suriname, and French Guiana? What resources do these countries have?
2. How does French Guiana's political situation differ from that of Guyana and Suriname?
3. **Critical Thinking** **Agree or disagree** with the following statement: "The political conflicts within Guyana are as much related to economic factors as to ethnic factors." Support your answer.

Building a Vocabulary
1. The elevation zone of the Andes between the tree line and the *tierra helada*
2. One who fights as a member of an armed band and takes part in irregular warfare; Colombia, Suriname
3. Slash-and-burn farming

Recalling and Reviewing
1. Bogotá, Cartagena, and Caracas
2. Venezuela claims more than half of Guyana's territory.
3. Mestizos, indigenous peoples, South Asians, Africans, mulattos, Chinese, Indonesians; Venezuela

Thinking Critically
1. Answers to both parts will vary. For the first part, some students might say yes, based on the experiences of Guyana and Suriname; other students might say no, because French Guiana has such a small population. For the second part, some students might say that the freedom and dignity brought by independence are worth the troubles.
2. If solar energy became readily available, oil prices might drop because oil would be less in demand. Venezuela's economy would then suffer because it is dependent on income from oil.

Learning About Your Local Geography
Maps should show the local area divided into zones based on various characteristics. Students should understand that areas can be divided into zones in many different ways and that the zones will not always be the same.

Using the Five Themes of Geography
One illustration should show how Colombia's location bordering both Middle and South America makes the country well situated to control the movement of goods in trade. The other illustration should show that the location of the Andes prevents easy movement between Colombia's Caribbean and Pacific coasts.

CHAPTER 20 CHECK

Unloading coffee berries in **Colombia**

Reviewing the Main Ideas
1. The countries of Colombia, Venezuela, Guyana, Suriname, and French Guiana make up the northern rim of South America.
2. Unlike other South American countries, Colombia has coastlines on both the Caribbean Sea and the Pacific Ocean. Coffee is Colombia's leading crop and a major export.
3. Venezuela's chief source of income is its mineral wealth. It is one of the world's leading oil-exporting countries and is an economic and political leader in South America.
4. Guyana, Suriname, and French Guiana are situated on South America's northeast coast. Since gaining independence from Britain in 1966, Guyana has endured political instability. Suriname, a former Dutch colony, depends upon bauxite and aluminum exports. French Guiana is the largest of France's overseas territories.

Building a Vocabulary
1. What is the *páramo*?
2. Define *guerrilla*. In which countries discussed in this chapter do guerrillas play a major role?
3. What is the term used to describe the type of agriculture in which forests are cut and burned to clear land for planting?

Recalling and Reviewing
1. Name three major cities in Colombia and Venezuela.
2. What political relationship exists between Venezuela and Guyana?
3. List five ethnic groups that live in the countries discussed in this chapter. Which country's people are mostly mestizos?

Thinking Critically
1. If French Guiana became an independent country, do you think it would experience increased political violence? Why or why not? Do you think independence is worth the troubles it sometimes brings?
2. What might happen to Venezuela's economy if scientists discover an inexpensive and convenient way to produce large quantities of solar energy?

Learning About Your Local Geography
Cooperative Project
Your local area is probably too small to have different climate zones like the Andes do, but how else might you divide your area into zones? Each member of your group should select one characteristic upon which to base zoning and then should sketch a map of the area showing the different zones. As a group, compare all the maps. Do the maps look similar in terms of the zones each one shows? Or do the locations of zones on each map vary widely?

Using the Five Themes of Geography
Make two illustrations linking **location** and **movement** in Colombia. One of your illustrations should show how the location of the country helps the movement of goods in and out of the country. In the other illustration, show how the location of the Andes hinders movement within Colombia. Write captions for your illustrations.

233

CHAPTER 21 Atlantic South America
Planning Guide

TEXTBOOK	TEACHING RESOURCES	OTHER AVAILABLE MATERIALS
Section		
1. Brazil Pages 234–239	Reteaching Worksheet 21.1 Skills Worksheet 21 Challenge/Enrichment Worksheet 21	*The World Today* Videodisc Program *Lecture Notes,* Chapter 21 *Study Guide for Spanish Speakers*
2. Argentina Pages 240–241	Reteaching Worksheet 21.2 Critical Thinking Worksheet 21	*The World Today* Videodisc Program *Lecture Notes,* Chapter 21 *Study Guide for Spanish Speakers*
3. Uruguay and Paraguay Pages 241–242	Reteaching Worksheet 21.3 Graphic Organizer 21	*Study Guide for Spanish Speakers* *Lecture Notes,* Chapter 21
Chapter Review Page 243	Chapter 21 Test Chapter 21 Portfolio Assessment See also *Alternative Assessment*	*Test Generator,* Chapter 21

CHAPTER RESOURCES

Lecture Notes, Chapter 21
Teaching Resources
    Chapter 21 Graphic Organizer
    Chapter 21 Skills Worksheet
    Chapter 21 Critical Thinking Worksheet
    Chapter 21 Challenge/Enrichment Worksheet

REVIEW AND ASSESSMENT

Chapter 21 Check, page 243
Chapter 21 Test, *Teaching Resources*
Chapter 21 Portfolio Assessment, *Teaching Resources*
Test Generator, Chapter 21
Alternative Assessment Forms

ADDITIONAL RESOURCES

Books for Teachers

Correas de Zapata, ed. *Short Stories by Latin American Women.* Houston, TX: Arte Publico Press, 1990.

Dwyer, Augusta. *Into the Amazon: The Struggle for the Rain Forest.* San Francisco, CA: Sierra Club Books, 1991.

Rock, David. *Argentina.* Berkeley: University of California Press, 1985.

Slatta, Richard W. *Cowboys of the Americas.* New Haven, CT: Yale University Press, 1990.

Books for Students

Brazil. New York: Time-Life Books, 1987.

Cousteau, Jacques Yves, and Mose Richards. *Jacques Cousteau's Amazon Journey.* New York: Abrams, 1984.

Gheerbrant, Alain. *The Amazon: Past, Present, and Future.* New York: Harry N. Abrams, 1992.

Jennings, Jesse D., ed. *Ancient South Americans.* New York: W. H. Freeman, 1983.

Latin American History Biographies: Perón. New York: Chelsea House, 1987. Available through Social Studies School Service, P.O. Box 802, Culver City, CA 90232–0802.

The World Today *Videodisc Program*

A variety of still images, short videos, and activities are available for you to use as you teach this chapter. *Lecture Notes* include bar codes for specific images and videos.

Multimedia Resources

Around South America. (VHS, 57 minutes) Social Studies School Service, P.O. Box 802, Culver City, CA 90232–0802.

Brazil: New World in the Tropics. (VHS, 50 minutes) Films for the Humanities & Sciences, P.O. Box 2053, Princeton, NJ 08543–2053.

Folktales from Argentina. (VHS, 26 minutes each) Films for the Humanities & Sciences, P.O. Box 2053, Princeton, NJ 08543–2053.

Picture Atlas of the World. (interactive maps, photo bank, animation, audio on CD-ROM) National Geographic Society, Washington, DC.

STV: Rain Forest. (laserdisc, with Macintosh software) National Geographic Society, Washington, DC.

MEETING INDIVIDUAL NEEDS

Limited English Proficiency (LEP)

Organize students into groups consisting of both LEP students and students proficient in English. Give each group one of the terms in the Chapter 21 Geography Dictionary. Allow the groups time to develop a brief skit that demonstrates the meaning of their term. Have students perform their skits for the class.

Learners Having Difficulty

Have students read the Focus questions for each section and then skim the section for the answers. Ask volunteers to write the answers on the board. Then have students write a brief summary of the chapter based on the answers to the questions.

Auditory Learners

Have students discuss the meanings of the terms in the Chapter 21 Geography Dictionary. Ask questions such as the following to guide the discussion: What are the benefits and drawbacks of being a buffer state? A landlocked country? How has soil exhaustion affected the settlement patterns of Brazil? What are conditions like in Brazil's *favelas*? What is the purpose of designating a city a growth pole? What is tannin used for and where is it produced?

Visual Learners

Have students look at the cartograms on page 239, and call on volunteers to explain the information shown. Have students draw conclusions from the illustrations about the relative size and growth of the Atlantic South American countries. To extend students' skill with cartograms, you may wish to have them make their own population cartograms of several counties or states in your region and draw conclusions from the results.

Tactile/Kinesthetic Learners

Provide students with an outline map of Atlantic South America. Have students shade and label Brazil's three major landform and climate regions, Argentina's four environmental regions, and Paraguay's two climate regions. Interested students might compare Brazil's major landform and climate regions with its four economic and cultural regions. Display the maps on a bulletin board.

Gifted Learners

Have interested students investigate how international organizations are working to save the rain forests of Brazil. Ask students to present their findings to the class. Then have the class brainstorm ways in which they could help.

MULTICULTURAL FOCUS

Brazil is often called a melting pot society—one in which people from many ethnic groups are blended together into one culture. Brazilian society reflects the influence of its diverse heritage: indigenous peoples, descendants of European colonials and of African slaves, various immigrant groups from Europe and Asia, and people of mixed ancestry. In fact, it is this last group that casts the distinctive nature of Brazilian multicultural society: nearly half of the population is of combined European, African, and Native American ancestries.

The country's multicultural heritage is a point of pride for Brazilians, and although people of European background tend to have more economic advantages than other groups, racial discrimination is much less prevalent there than in many other countries.

Many Brazilians find it hard to understand the racial attitudes and divisions that are more common in North America.

Brazilian society is unified by a shared national culture, much of which is the legacy of the Portuguese, including the language, the Roman Catholic religion, and many traditional customs. (You may wish to refer students to the quotation at the beginning of Chapter 21.) However, as in any country the size of Brazil, there is also great regional variety, reflecting differences in environment, people, and economic activities.

To help students appreciate the diversity of Brazilian culture, use this information to help students draw comparisons and contrasts between Brazil and the United States in terms of human geography.

CHAPTER 21

ATLANTIC SOUTH AMERICA

Section 1: Brazil
Section 2: Argentina
Section 3: Uruguay and Paraguay

OVERVIEW

Chapter 21 examines the eastern, or Atlantic, countries of South America: Brazil, Argentina, Uruguay, and Paraguay.

Section 1 describes Brazil, the largest country in South America and the largest Portuguese-speaking country in the world. With its wealth of natural resources, Brazil has great potential, but widespread poverty and a huge foreign debt are impeding its progress.

Section 2 focuses on how Argentina's economy has been threatened by political instability over the years.

Section 3 introduces Uruguay and Paraguay, small agricultural countries that tend to be dominated by Brazil and Argentina.

About the Author

Robert Fried is an anthropologist, photographer, and travel writer from San Francisco, California.

Primary Source

Description of Change: excerpted
Rationale: excerpted to focus on the main idea

CHAPTER 21

Atlantic South America

Carnival mask from **Brazil**

GEOGRAPHY DICTIONARY

buffer state
soil exhaustion
favela
growth pole
tannin
landlocked

❝**D**espite the immense size of Brazil and the diverse ethnic backgrounds of her peoples, there is one festive occasion that unites the entire country. Each year, in the final days before Lent, Carnival erupts in an explosion of dance, music, and color. From Pôrto Alegre in the south to Macapa . . . to the north, all Brazilians join together to share in the magic and mystery of Carnival.❞

Robert Fried

If the world's continents were pieces of a puzzle, the east coast of South America and the west coast of Africa would fit together almost perfectly. This realization was the first major clue in the development of the theory of plate tectonics. As South America drifted westward from Africa, erosion gradually wore down the eastern side of the South American continent. As a result, the landforms of this region are mostly wide plains.

Brazil and Argentina, the two largest countries of eastern, or Atlantic, South America, are among the most prosperous of the world's developing countries. Sandwiched between Brazil and Argentina are Uruguay and Paraguay, two **buffer states**. A buffer state is a small country that separates two larger competing countries. Brazil and Argentina regularly assert influence over these small countries, and Uruguay and Paraguay generally are eager to accept favors from their giant neighbors.

Carnival celebration in **Rio de Janeiro, Brazil**

Using the Quotation

Ask students to describe community celebrations where they live. Compare the excitement of such events to Carnival, as described in the quotation on page 234. Explain the reference to Lent by mentioning that Brazil is a predominantly Roman Catholic nation. Note the value of community celebrations in bringing people of diverse backgrounds together.

Using the Geography Dictionary

Write the words in the Geography Dictionary on the board. Pair the terms that relate to agriculture and forestry (**tannin, soil exhaustion**), cities (***favelas,*** **growth pole**), and characteristics of countries (**buffer state, landlocked**). Then have students identify the Portuguese term (*favela*).

SECTION 1
OBJECTIVES

The objectives reflect the Focus questions on this page and are assessed in the review on page 239.

1. Analyze Brazil's physical and historical geography.
2. Describe Brazil's four economic and cultural regions, and discuss challenges facing the country.

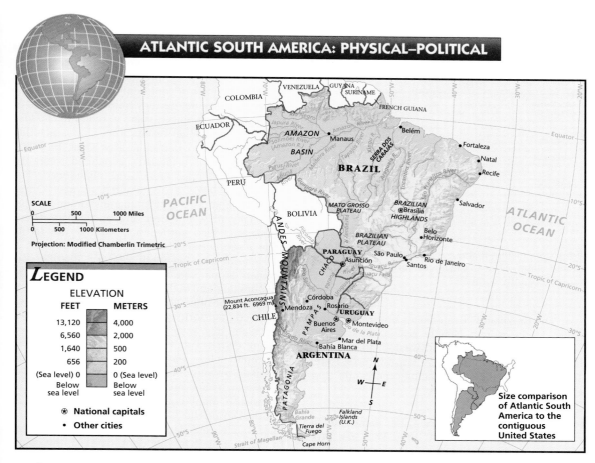

ATLANTIC SOUTH AMERICA: PHYSICAL–POLITICAL

LEGEND

ELEVATION

FEET	METERS
13,120	4,000
6,560	2,000
1,640	500
656	200
(Sea level) 0	0 (Sea level)
Below sea level	Below sea level

⊛ National capitals
• Other cities

SCALE
0   500   1000 Miles
0   500   1000 Kilometers
Projection: Modified Chamberlin Trimetric

Size comparison of Atlantic South America to the contiguous United States

Using the Physical–Political Map: Atlantic South America

Ask students to look at the map on this page. Call on a volunteer to name the countries in this region. Then have students name the capital cities of each country. Explain that Brasília is a relatively new capital and that Rio de Janeiro was the capital of Brazil until 1960.

BRAZIL

FOCUS

● *What are Brazil's physical characteristics? How has the country changed through history?*

● *Into what four economic and cultural regions can Brazil be divided? What challenges face the country?*

REGION **Physical Geography** Brazil is the largest country in South America and covers more than half of the continent's land area. It has borders with every other South American country except Ecuador and Chile. (See the map above.) Brazil is about 2,700 miles (4,344 km)

from north to south and about the same distance from east to west. This huge country is made up mostly of lowland plains and level plateaus. Forests and savanna woodlands once covered the entire country.

Brazil can be divided into three major landform and climate regions. In the north is the Amazon region of rolling plains. This is the broad lowland of the Amazon River basin where a humid-tropical climate supports the world's largest tropical rain forests.

Along the east coast are the Brazilian Highlands. Most of these old, eroded mountains are less than 9,000 feet (2,743 m) high. In the north, the coastal region is covered mostly with rain forests and savannas. A little inland from the coast, the highlands are semiarid with dry grasslands. The southeastern highlands have humid-subtropical and marine-west-coast climates. These moist southeastern environments have become Brazil's major agricultural regions.

TEACHING SECTION 1

Motivator

Play for the class some popular Brazilian music, such as that of Roberto Carlos, Milton Nascimento, or Sandra Sá. You might also play some of the music from Paul Simon's *Rhythm of the Saints* album, which was influenced by Brazilian music. Tell students that they will learn about Brazil in Section 1.

Building a Vocabulary

Write the terms **soil exhaustion** and **growth pole** on the board. Ask students to suggest meanings for both terms. As a hint, tell them that *growth pole* applies to cities. After students have suggested meanings, have them look up the two terms in the Glossary.

Identifying Physical Features

Call on volunteers to name Brazil's three physical regions. (*Amazon lowlands, Brazilian Highlands, Brazilian Plateau*) List the regions on the board. Then have students come to the board and describe each region's landforms, climates, and vegetation. Students should write their information on the board next to the name of the appropriate region.

Chart Caption Answer

Brazil's per capita GDP is a little more than one-fifth that of the United States.

236

FACTS IN BRIEF	Atlantic South America		
COUNTRY POPULATION (1994)	LIFE EXPECTANCY (1994)	LITERACY RATE (1990)	PER CAPITA GDP (1993)
ARGENTINA 33,912,994	68, male 75, female	95%	$5,500
BRAZIL 158,739,257	57, male 67, female	81%	$5,000
PARAGUAY 5,213,772	72, male 75, female	90%	$3,000
URUGUAY 3,198,910	71, male 77, female	96%	$6,000
UNITED STATES 260,713,585	73, male 79, female	97% (1991)	$24,700

Source: *The World Factbook 1994*, Central Intelligence Agency

Brazil's population is about three-fifths that of the United States. How do the figures for per capita GDP for each country compare?

The Brazilian Plateau makes up Brazil's third major landform and climate region. It is a vast area of upland plains found west of the coastal highlands and south of the Amazon forests. Here, summer rains support savanna woodlands that extend all the way to Bolivia.

Historical Geography The name *Brazil* comes from the "pau-brasil" tree. This tree provides a purple-red dye that attracted the first Portuguese colonists in the early 1500s. Brazil became the Americas' first large sugar colony. Slaves, imported from Portugal's African territories, joined Brazil's indigenous peoples in the hard work of cutting cane. Eventually, sugar plantations spread along much of the Atlantic coast, steadily replacing the tropical forests. In the northeast, Salvador and Natal grew to become Brazil's first cities. Farther inland, cattle ranches of Portuguese settlers provided hides and dried beef for world markets.

Beginning in the 1690s, gold and precious gems were discovered inland from Rio de Janeiro. A mining boom brought adventurers from around the world. São Paulo, founded in the 1500s as a mission outpost, became the economic center of southeastern Brazil. In 1763, Rio de Janeiro became the colonial capital. By the time of independence in 1822, Brazil's population of about 4 million people was twice that of Portugal. Brazil remains the largest Portuguese-speaking country in the world.

During the late nineteenth century, Brazil was well known for the coffee produced in the country's southeastern region, and it became the world's largest coffee producer. But the farms of southeastern Brazil could only support coffee for about 30 years before they showed the effects of **soil exhaustion**. Soil exhaustion refers to the loss in soil nutrients that results from always planting the same crop in a particular area. Over the years, the coffee-growing industry expanded westward as the older coffee areas were abandoned. The abandoned lands were adapted to new crops, often by new waves of immigrants from Europe, especially Italy and Germany, and Japan. These immigrants also settled and farmed the fertile soils of southernmost Brazil.

During the twentieth century, Brazil's government took the lead in South America in encouraging and redirecting economic growth. A collection of huge government-owned companies assumed authority over expanding railways, ports, and highways. These companies developed mining districts and built factories. Though the economic plan was designed to benefit the Brazilian people, the already prosperous families of politicians and military leaders gained most of the wealth. Many companies have been managed poorly. During the 1970s and 1980s, when Brazil's imports far exceeded exports, the country began to amass a huge foreign debt. By 1993, the country's foreign debt was more than $100 billion. Today, much of Brazil's export earnings go to pay just the interest on these loans. This debt burden continues to weaken the economy.

The Northeast Modern Brazil can be divided into four cultural and economic regions: the Northeast, the Southeast, the Campo Cerrado, and the Amazon.

The Northeast refers to the coastal region of Brazil that includes the colonial cities of Fortaleza, Natal, Recife, and Salvador. Plantation agriculture made this region the center of the Portuguese Empire during the 1600s. Decline set in, however, and the capital was moved from Salvador to Rio de Janeiro in 1763. After centuries of intensive cultivation, this region's land is worn out. Many farmers move every few years, seeking more fertile farmland.

Today, the Northeast is Brazil's poorest region. Literacy and health standards are low. Attempts to move industry to this region have

Understanding Chronology

Prepare a list of important events in Brazil's history. Arrange the events in scrambled order. Give all students a copy of the list, and have them arrange the events in the correct chronological order.

Analyzing Brazil's Economy

Help students recognize Brazil's wealth of natural resources and variety of economic activities. Write the words Natural Resources and Economic Activities on the board. Have students read the section for examples of both. List their responses in the appropriate column on the board.

Describing Regions

Organize the class into four groups. Assign each group one of Brazil's cultural and economic regions. Have each group outline information about their assigned region's major economic activities, cities, population characteristics, and other significant features.

been unsuccessful. Droughts in the northeast interior have caused many farms to fail, and people have moved to the cities seeking jobs. The northeastern cities are filled with huge slum areas called **favelas** (fah-VEH-lahs). With so much poverty, social unrest has become common in the region.

The Southeast The Southeast region provides a sharp contrast to the Northeast. It includes the modern cities of São Paulo and Rio de Janeiro and extends southward to the border with Uruguay. More than 120 million people, nearly 80 percent of all Brazilians, live here. It is the nation's most prosperous region, with rich natural resources, productive agriculture, and most of the country's industries.

Though the Southeast is highly urban, it is Brazil's agricultural heartland. Hot, rainy summers and dry, mild winters make the Southeast the greatest coffee-growing region in the world. Brazil's economy is no longer dependent on this single crop, however. The Southeast region also is a leader in two of Brazil's newest agricultural

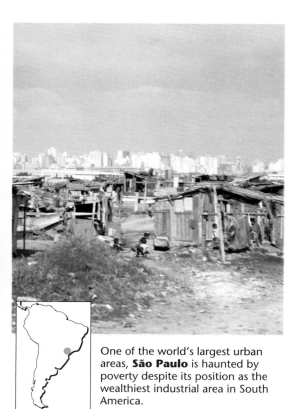

One of the world's largest urban areas, **São Paulo** is haunted by poverty despite its position as the wealthiest industrial area in South America.

exports—soybeans and orange juice. The Southeast also produces a variety of middle-latitude crops, such as cotton and maize (corn).

The modern mines of the Southeast cluster around Belo Horizonte, Brazil's iron and steel center. Huge iron-ore deposits have been developed here, and limited coal resources are found nearby. Oil deposits have been found off the coast. Belo Horizonte's steel industry has supported the development of heavy-machinery, automobile, and shipbuilding industries. Other industries in the region produce chemicals, cement, plastics, processed food, textiles, and computers. There are growing aircraft and defense industries here as well.

Rio de Janeiro is a major tourist area and an important shipping, manufacturing, and trade center. The Rio area has a population of more than 11 million. São Paulo is one of the world's largest cities. It is an awesome industrial area with a population above 20 million. São Paulo and its port, Santos, form the largest and wealthiest industrial area in South America. Although the Southeast is prosperous, many *favelas* are found here, too.

Campo Cerrado The Campo Cerrado region is a frontier land of savannas and dry woodlands that extends east to west across central and southern Brazil. The region begins in the upper São Francisco River basin and continues westward past Brasília to the Mato Grosso Plateau. Until Brazil's capital was moved to Brasília in 1960, this region was nearly empty.

Brasília, located about 600 miles (965 km) from the coast, was built in a deliberate attempt to develop the interior region of the country. Brasília features modern architecture and highways. The city was designed to accommodate 500,000 people. Today, its population exceeds 3 million. As with Brazil's other cities, Brasília has growing *favelas*.

The soils of the Campo Cerrado were once thought infertile because of high levels of acid and aluminum. After agricultural scientists discovered how to make the soils fertile, however, commercial agriculture became possible. Wealthy investors from the Southeast have established large farms, and new towns are appearing. If farming spreads throughout the region, the Campo Cerrado could become the world's greatest agricultural region within the next century.

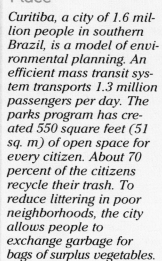

THEME
Place

Curitiba, a city of 1.6 million people in southern Brazil, is a model of environmental planning. An efficient mass transit system transports 1.3 million passengers per day. The parks program has created 550 square feet (51 sq. m) of open space for every citizen. About 70 percent of the citizens recycle their trash. To reduce littering in poor neighborhoods, the city allows people to exchange garbage for bags of surplus vegetables.

Using the Illustrations

Explain the purpose of a cartogram and help students read the cartogram on page 239. Ask: What can one learn about Brazil, Argentina, Uruguay, and Paraguay by studying the cartogram? Ask students if they think the cartogram is an effective means of providing information. Ask: What does it show that a population density map does not show?

Guided Practice

Bring in newspaper and magazine articles about the destruction of the Amazon rain forests. Distribute the articles, and after students have read them, hold a class debate about the main issues involved. In particular, have students discuss reasons for the international concern about the destruction of the forests.

Independent Practice

Have students make two columns on a sheet of paper, labeled Advantages and Disadvantages. In the Advantages column, have them list factors that will help Brazil in its future. In the Disadvantages column, have them list obstacles that Brazil must overcome. Ask students to analyze their lists and to write a brief prediction about Brazil's future.

This Jivaro man in **Brazil** is playing a traditional flute. Economic development threatens the ways of life for many Native Americans in the Amazon Basin.

The Amazon The Amazon region covers the northern and western parts of Brazil. Its humid-tropical climate and dense rain forests have made settlement and transportation difficult. Native American villages are scattered through the forests. Some of the indigenous peoples living here have had little contact with the outside world. They grow sugarcane, yams, bananas, and other crops for their own use. A few still hunt with bows and arrows or blowguns and darts.

The Amazon and its tributaries form a network of waterways thousands of miles long. The Amazon River carries more water than any other river in the world. Before the government built highways through the rain forests, river-boats were the only way to travel across the Amazon region.

Brazil's newest and largest mining district is in the Amazon. This is the Serra dos Carajás (sehr-uh dus car-uh-ZHAHS), south of the port of Belém (buh-LEHM). Here, major deposits of iron ore, copper, and tin are found, as well as major reserves of gold, bauxite, and manganese. Brazil could become the world's largest producer of these minerals.

The new roads and mining projects in the Amazon have encouraged new settlers and treasure seekers to spread into the region. Conflicts among the settlers, miners, and Native Americans have increased. Many Native Americans have lost their forest homes and have become farm workers in the developing areas.

Natural rubber, nuts, medicinal plants, and fine woods used in furniture are traditional products of the Amazon. The destruction of the Amazon forests, which is leading to the extinction of many plants and animals, has become a global issue. Some people predict that the rain forest will disappear within the next 50 years. This would be a tragic loss. The biological resources of the Amazon region are just beginning to be investigated.

Manaus, the Amazon region's largest city, is the historic center of rubber collection and is now a growing industrial area. The government of Brazil has designated Manaus as a **growth pole**. A growth pole is a metropolitan area given official help to strengthen its economic development. In Manaus, special laws encourage industrial development by foreign companies. There are few business taxes, and foreign goods enter the city with few import fees.

Brazil's Future Today, Brazil faces several social, economic, and environmental challenges. Its population exceeds that of the rest of South America. The Southeast is prosperous, but the remainder of the country is underdeveloped. While Brazil has rich natural resources and

238

Closure

Tell students to pretend that they are moving to Brazil. Have each student pick the physical or cultural and economic region to which he or she would most like to move. Ask students to write a paragraph explaining why they would like to move to the regions they have selected, and what they expect to do when they arrive there (such as, for their occupations).

ASSESSMENT

Assign the Section 1 Review questions on this page.

Reteaching

Assign Section 1 of Reteaching Worksheet 21.

Extension

Have students make an illustration to show how coffee is grown, processed, and transported around the world. For a model illustration, students might look at the charts on pages 144–145 (corn), pages 352–353 (wheat), pages 448–449 (cacao), or pages 512–513 (tea). Have students write two or three facts on their chart about coffee production in Brazil.

Population Growth in the Americas

1950

1990

 SKILL STUDY You can use cartograms to compare characteristics of countries. The size of each country in these cartograms is determined by its population and the scale in the legend. The more populous a country is, the larger it will appear in the cartogram. Cartograms also can be based on other data, such as economic or health information.

The cartogram on the left shows that the combined populations of the United States and Canada were about the same as the combined populations of Middle and South America in 1950. By 1990, however, the combined populations of Middle and South America were far larger. Which Middle or South American country had the largest population in 1950?

---

agricultural potential, unemployment is high. And as elsewhere on the continent, enormous differences still exist between the wealthy and the poor. Many of the country's people do not have enough to eat, even though Brazil is a major world food exporter.

Based on its natural and human resources, Brazil is a country of the future. The blending of cultures in Brazil has produced beautiful and vibrant art, architecture, and music. Products from Brazil are becoming common in the developed countries. In addition, Brazil has enormous mineral resources to be developed, and it has increasing supplies of fossil fuels. Hydroelectric

plants on Brazil's rivers also provide energy for the future.

SECTION REVIEW

1. What different products attracted people to Brazil from the 1500s through the 1800s?
2. Name a city found in each of Brazil's economic and cultural regions. What challenges do these cities face?
3. **Critical Thinking** **Discuss** the social and environmental consequences of the development of Brazil's interior regions.

 Skill Study
Brazil

 MAP TRANSPARENCIES

An overhead transparency of the cartograms on this page is available in *REGIONS OF THE WORLD Map Transparencies*.

Section 1 Review
ANSWERS

1. Products included the dye from trees, sugarcane, cattle hides and dried beef, gold and gems, coffee, and other crops.
2. The Northeast: Fortaleza, Natal, Recife, Salvador; the Southeast: São Paulo, Rio de Janeiro, Belo Horizonte; the Campo Cerrado: Brasília; the Amazon: Belém, Manaus; overcrowding and the growth of *favelas*
3. Social: many Native Americans have lost their homes; environmental: many plant and animal species are threatened with extinction.

239

SECTION 2
OBJECTIVES

The objectives reflect the Focus questions on this page and are assessed in the review on page 241.

1. Identify environmental regions of Argentina.
2. Summarize economic activities important to Argentina, and explain the relationship between economics and politics.

TEACHING SECTION 2

Analyzing Cause and Effect
Politics and Economics Discuss with students why Argentina's economy suffered when the government was run by dictatorships and the military. *(Leaders lacked skill and interest in developing the economy. People were afraid to invest in businesses.)*

Guided Practice

Write the following occupations on the board: tanner, oil explorer, skier, gaucho, shepherd, vintner. Call on students to choose one of Argentina's four environmental regions where each occupation might be practiced.

Independent Practice

Have students label and shade Argentina's four environmental regions on a copy of Outline Map 13.

MULTICULTURAL PERSPECTIVES

The ranchers and gauchos of the vast Argentine pampas live in relative isolation. The arrival of a visitor—even a stranded motorist— to one of their ranch houses is cause for celebration. Guests are often treated to *asado,* the traditional Argentine specialty of barbecued meat, and may be entertained by song, drink, and storytelling. The gauchos will often give their guests parting gifts of ponchos, wooden drinking vessels, or powder horns. Ask students to make a list of favorite foods, songs, stories, and gifts that they might share if a foreign visitor came to their home.

THE WORLD TODAY VIDEODISC PROGRAM

Argentina: Toward a Fertile Future

1A 9180 3:21

Play   Pause

2 ARGENTINA

FOCUS

- *What environmental regions does Argentina have?*
- *What economic activities are important to Argentina? How are economics and politics linked in the country?*

REGION **Physical Geography** Argentina is the second largest country in Middle and South America. It extends from just north of the Tropic of Capricorn to cold Tierra del Fuego at the southern tip of the continent. Argentina can be divided into four environmental regions. (See the map on page 235.)

The Andes Mountains form a long narrow region of grassy and forested highlands along Argentina's western border. These are the country's only mountains. Here, there are ski resorts and beautiful lakes. From north to south, the elevations in the Andes steadily decrease. In the north, the peaks are more than 22,000 feet (6,706 m) above sea level. In the far south, they rise less than 10,000 feet (3,048 m).

East of the Andes in northern Argentina is the Chaco (CHAHK-oh), a region of low plains covered with dry forests and savannas. These plains are so flat that water sometimes stands for months after the summer rainy season ends.

Argentina's third natural region is made up of the pampas. The pampas are wide plains with grasslands located southwest of the estuary of the Río de la Plata. The pampas stretch for almost 400 miles (644 km) and are nearly flat. With their rich soils and humid middle-latitude climates, these plains have become one of the world's major agricultural regions. They are the heartland of Argentine life and are by far the country's most densely settled region.

Argentina's fourth region, Patagonia, is found south of the pampas. Patagonia's landforms are dry plains and windswept plateaus. Because the Andes Mountains block the region from the Pacific's rain-bearing storms, the climates are arid and semiarid. Few people live in Patagonia. Oil deposits off the coast of Patagonia promise new energy resources.

INTERACTION **Economic Geography** In earlier times, the pampas were divided into ranches, some of which covered thousands of acres. Livestock was allowed to wander, much as it did in the Great Plains of the United States. Gauchos (GAU-chohz), the Argentine cowboys, managed the cattle just as the cowboys of the western United States did. When ranchers began using modern ranching methods, they fenced the livestock in. The gaucho, like the American cowboy, is vanishing.

Today, the pampas are one of the world's largest sources of beef, wheat, and vegetable oils. Maize (corn) is raised north of Buenos Aires and, like wheat, is mainly an export crop. Farmers have developed vegetable and dairy farms to feed the populations of Argentina's growing cities.

Compared to the pampas, Argentina's other regions are less developed. North of Buenos Aires between the Paraná and Uruguay rivers is a region of grassy plains that is used mainly for grazing cattle and sheep. The Chaco in northern Argentina is important for its valuable quebracho (kay-BRA-choh) forests. **Tannin,** a substance used in preparing leather, comes from quebracho trees. In the northern foothills of the Andes, farmers grow grapes and sugarcane on irrigated lands.

In Patagonia, sheep grazing on huge ranches is the chief industry. Though many crops could be grown here with irrigation, food is plentiful in Argentina, and there has been little need for developing this potential.

Cattle ranching is an important economic activity in **Argentina**. The rich soils of the pampas help make Argentina one of the world's top exporters of beef and grain.

Closure

Ask students to complete the following sentence: "Argentina's biggest obstacle to a prosperous and stable economy is...."

ASSESSMENT

Assign the Section 2 Review questions on this page.

Reteaching

Assign Section 2 of Reteaching Worksheet 21.

Extension

Have interested students research and report on similarities between the Argentine gaucho and the American cowboy. Students might focus on practices and traditions that originated with the Spanish, who brought horses, cattle, and ranch culture to the Americas.

SECTION 3
OBJECTIVES

The objectives reflect the Focus questions on this page and are assessed in the review on page 242.

1. Examine physical, cultural, and economic characteristics of Uruguay.
2. Examine physical, cultural, and economic characteristics of Paraguay.

Buenos Aires, the capital of **Argentina,** is South America's third largest metropolitan area. High-rise apartment buildings in the Retiro District, a middle-class part of the city shown here, stretch for many blocks.

ernments run by dictatorships and the military have been typical. During a particularly violent period of unrest in the 1970s, many Argentines were killed by the military government. In 1982, a brief war between Argentina and the United Kingdom broke out when Argentina invaded the Falkland Islands, a British territory in the South Atlantic. When Argentina was defeated, the military returned power to a civilian democratic government. Even so, economic progress has been slow, and the country has accumulated a large foreign debt.

2 SECTION REVIEW

1. Name Argentina's four environmental regions.
2. What are Argentina's major resources?
3. **Critical Thinking**   **Suggest** an explanation for why a country with rich natural resources might need to borrow money.

Section 2 Review
ANSWERS

1. The Andes Mountains, the Chaco, the pampas, and Patagonia
2. Agricultural resources—beef, wheat, maize (corn), grapes, sugarcane, wool—and tannin
3. Possible answer: the country might not have the ability to develop its resources, due to political or historical factors, or might not have all the resources necessary for development (such as energy resources).

Argentina competes with the United States, Canada, and Australia in selling beef and wheat on the world market. Argentina often can sell these products at lower cost because of lower production costs. The country has lower land values and pays lower wages than its competitors. Argentina also exports large amounts of wool and other products to the United States.

Even though agriculture is the country's main source of wealth, only about 10 percent of Argentina's people farm. Many more are employed in manufacturing. Most of Argentina's industry provides consumer goods, such as textiles and beverages, for local markets. Food processing is the primary export industry. Argentina has few mineral resources.

Argentina has several large cities, including Rosario (roh-ZAHR-ee-oh) and Córdoba. Buenos Aires, however, dominates the country. Its urban region has more than one-third of the nation's people and an even larger share of the industries.

Political Geography Because Argentina has abundant natural resources and great human potential, many believe that Argentina should be the richest country in South America. Unfortunately, the politics and economy of Argentina are nearly inseparable. Ineffective gov-

3 URUGUAY AND PARAGUAY

FOCUS

- *What are Uruguay's physical, cultural, and economic characteristics?*
- *What are Paraguay's physical, cultural, and economic characteristics?*

Uruguay As a buffer state, Uruguay often has been dominated by its giant neighbors. (See the map on page 235.) As in Argentina, Spanish is the national language, but many of Uruguay's people also speak Brazilian Portuguese. The beach resorts of Uruguay are popular with both Argentines and Brazilians. Tourism is a rapidly growing industry.

The Río de la Plata estuary, along Uruguay's southern edge, is the major waterway of southern South America. On the shore of the estuary is the port of Montevideo (MAHNT-uh-vuh-DAY-oh), the business and government center of Uruguay. In the southern part of the country are Uruguay's rolling plains, a continuation of the pampas. Farther north, the country becomes more hilly like neighboring parts of Brazil.

TEACHING SECTION 3

Comparing and Contrasting

Have students prepare a chart to enable them to compare and contrast Uruguay and Paraguay. Their charts should include the following headings: Climate, Vegetation, Resources, Agriculture, Industry.

Guided Practice

Have students add information about their own region to the chart that they created in the Comparing and Contrasting activity. Have students describe the similarities and differences between their local geography and that of Paraguay or Uruguay.

Independent Practice

Have students sketch the locations of Paraguay and Uruguay on a copy of Outline Map 13. Ask them to shade the major landform regions and to label the capital city of each country.

Closure

Elicit students' opinions about which country has a more promising future—Uruguay or Paraguay.

ASSESSMENT

Assign the Section 3 Review questions on this page.

Reteaching

Assign Section 3 of Reteaching Worksheet 21.

Section 3 Review
ANSWERS

1. Brazil: Portuguese language and northern hilly country; Argentina: Spanish language and the pampas
2. East: agriculture; west: grazing, tannin production
3. Cultural: most of Uruguay's people speak Spanish and/or Portuguese, whereas in Paraguay most people speak Guaraní. Political: Uruguay has a large middle class and strong democratic traditions; in Paraguay, a few wealthy families and companies influence the government. Factors accounting for the differences include Uruguay's coastal location and agricultural resources, and Paraguay's landlocked location and sparse population.

242

These Native Americans are making various crafts in **Asunción, Paraguay.** Traditional handicrafts are still practiced widely.

Uruguay is similar to the Argentine pampas in several ways. It has a humid-subtropical climate, and most of the country is a natural grassland. The soil is rich and well watered. Ranchers and farmers on the inland plains graze livestock and grow grains. In the coastal region, farmers grow vegetables and fruits. Rice has replaced wheat as the most important export crop. Uruguay has few mineral resources. Hydroelectric projects, however, provide an important energy source.

Uruguay has a high literacy rate and a large middle class. Despite occasional military dictatorships, Uruguay has strong democratic traditions and offers many economic opportunities. Uruguay's main challenge is to bring the prosperity of Montevideo into the less-developed rural areas. Many parts of the interior remain in large, unproductive haciendas. As in other countries in South America, people are abandoning rural areas. They hope the Montevideo region, where more than one-third of the country's people already live, will offer a better life.

Paraguay **Landlocked** Paraguay shares borders with Bolivia, Brazil, and Argentina. (See the map on page 235.) A landlocked country is one that is completely surrounded by land and does not have access to the ocean. The Paraguay River divides the country into two separate regions. East of the river, humid-subtropical lowlands rise to join the rolling highlands of southern Brazil. This is Paraguay's most productive agricultural region. The chief commercial crop is soybeans, though farmers grow fruits, corn, and other vegetables for local use. Rice, tobacco, sugarcane, and a tea called yerba mate (YEHR-buh mah-TAY) also are grown here.

West of the Paraguay River is the less humid region of forests. It is a continuation of the Chaco area, which begins in northern Argentina, crosses Paraguay, and extends into southeastern Bolivia. Paraguay's Chaco is used mostly for grazing. The dry forests of this region also contain tannin-producing quebracho trees. Because quebracho trees grow slowly, and because they are cut down for their tannin, the production of tannin is depleting the forests.

Paraguay has productive soil, adequate rainfall, and subtropical temperatures. Yet, the country has long been poorer than nearby Argentina and Brazil. Most of Paraguay has few people, and much of the land is underused. Because there are few industries, there are few jobs. As a result, many Paraguayans work as migrant laborers in neighboring countries.

Paraguay's people primarily are mestizos, but the indigenous influence remains strong. About 95 percent of the people speak Guaraní (gwah-ruh-NEE), an indigenous language. As in most developing countries, much of Paraguay's wealth is in the hands of a few wealthy families and large companies, which have a great deal of influence over the government.

Paraguay's future development is promising. Hydroelectric projects such as the Itaipu Dam on the Paraná River are providing Paraguay with an enormous surplus of power—far more than it can use. As a result, Paraguay is exporting power to Brazil and Argentina at a profit.

SECTION REVIEW

1. Name one thing Uruguay has in common with Brazil, and one thing Uruguay has in common with Argentina.
2. What economic activities occur east of the Paraguay River? west of the river?
3. **Critical Thinking** **Compare and contrast** cultural and political aspects of Uruguay and Paraguay. What factors account for the differences?

ANSWERS

Building a Vocabulary

1. A small country that separates two larger competing countries; Uruguay and Paraguay
2. By always planting the same crop in one area; the soil can no longer support that crop.
3. Slum areas; in Brazilian cities
4. It does not have as much access to trade routes.

Recalling and Reviewing

1. Wealth is distributed unevenly, with a few families having most of the money, and with large numbers of people having very little money.
2. Argentina and Paraguay; production of tannin
3. Brazil; development and destruction of the forests; loss of homes for Native Americans and extinction of species
4. Brazil: Portuguese; Argentina: Spanish; Uruguay: Spanish and Portuguese; Paraguay: Guaraní

CHAPTER 21 CHECK

Rio de Janeiro, Brazil

Reviewing the Main Ideas

1. Brazil, the largest country in South America, is made up mostly of plains and plateaus. Originally a Portuguese colony, the country today has four major economic and cultural regions: the Northeast, the Southeast, Campo Cerrado, and the Amazon. Brazil faces several social, economic, and environmental challenges.
2. Argentina is the second largest country in South America. It is divided into four natural regions: the western highlands, the northern Chaco, the pampas, and Patagonia. Argentina has rich agricultural resources, but political unrest has limited the country's economic progress.
3. Uruguay is a small country with strong democratic traditions. Agriculture and tourism are important industries.
4. Paraguay, a landlocked country, is sparsely populated and has much underused land and little industry. Hydroelectricity is exported for profit.

Building a Vocabulary

1. What is a buffer state? Give an example of a buffer state from this chapter.
2. How does soil exhaustion occur? What are its consequences?
3. What are *favelas*? Where are they found?
4. How might a landlocked country be at a disadvantage compared to other countries?

Recalling and Reviewing

1. How is wealth distributed among people within Brazil and Paraguay?
2. To which countries discussed in this chapter are quebracho trees important? Why?
3. In which country is the Amazon region located? How is the region changing, and what consequences might these changes have?
4. What languages are spoken in each country of Atlantic South America?

Thinking Critically

1. Do you think it is more to a country's advantage or disadvantage to be a buffer state? Use examples from the chapter to support your answer.
2. What conditions in a country might cause the military to take over the government? Do you think military takeovers should be and/or can be prevented? Explain.

Learning About Your Local Geography

Individual Project
Both Brazil and Argentina have difficulty paying the interest on their debts. Interview an economics student or check an economics textbook for an explanation of interest. Then, contact a local bank or credit union and ask for the interest rates on various types of loans, such as car loans, home improvement loans, and personal loans. Make a bar graph showing these different rates. You may want to include your graph in your individual portfolio.

Using the Five Themes of Geography

Show how the themes of **region** and **human-environment interaction** are related in the countries of Atlantic South America. Select a geographic region discussed in the chapter. Construct a chart with three columns. In the first column, list natural physical characteristics of the region you have selected. In the second column, list economic activities for which the characteristics in column 1 are suited. In the third column, list ways in which humans have changed the region's environment as a result of the economic activities in column 2.

Thinking Critically

1. Answers will vary. Some students will say a disadvantage because the buffer state may be dependent on the larger, more powerful countries. Others will say an advantage because the buffer state may be able to gain from both surrounding countries.
2. Unstable political conditions or weak economic conditions might be responsible. Answers to the second part will vary: military governments might threaten democracy, or the military can restore order to a country.

Learning About Your Local Geography

Students' graphs should show interest rates on different types of loans.

Using the Five Themes of Geography

Possible regions include southeastern and northeastern Brazil, where agriculture has led to soil exhaustion; the Amazon, where development has led to the cutting of trees and species extinction; and Paraguay's Chaco, where the production of tannin is depleting the forests.

243

CHAPTER 22 Pacific South America
Planning Guide

TEXTBOOK	TEACHING RESOURCES	OTHER AVAILABLE MATERIALS
Section		
1. Ecuador Pages 244–247	Reteaching Worksheet 22.1 Skills Worksheet 22	*The World Today* Videodisc Program *Study Guide for Spanish Speakers* *Lecture Notes,* Chapter 22
2. Peru Pages 247–248	Reteaching Worksheet 22.2 Challenge/Enrichment Worksheet 22	*The World Today* Videodisc Program *Study Guide for Spanish Speakers* *Lecture Notes,* Chapter 22
3. Bolivia and Chile Pages 248–250	Reteaching Worksheet 22.3 Graphic Organizer 22 Critical Thinking Worksheet 22	*Study Guide for Spanish Speakers* *Lecture Notes,* Chapter 22
Chapter Review Page 251	Chapter 22 Test Chapter 22 Portfolio Assessment See also *Alternative Assessment Forms*	*Test Generator,* Chapter 22
Unit Review Pages 252–253	Unit 4 Test Unit 4 Portfolio Assessment See also *Alternative Assessment Forms*	*Test Generator,* Unit 4

CHAPTER RESOURCES

Lecture Notes, Chapter 22
Teaching Resources
    Chapter 22 Graphic Organizer
    Chapter 22 Skills Worksheet
    Chapter 22 Critical Thinking Worksheet
    Chapter 22 Challenge/Enrichment Worksheet

REVIEW AND ASSESSMENT

Chapter 22 Check, page 251
Chapter 22 Test, *Teaching Resources*
Chapter 22 Portfolio Assessment, *Teaching Resources*
Test Generator, Chapter 22
Alternative Assessment Forms

ADDITIONAL RESOURCES

Books for Teachers

Bingham, Hiram. *Lost City of the Incas: The Story of Machu Picchu and Its Builders.* New York: Macmillan, 1963.

Galeano, Eduardo. *Memory of Fire: Century of the Wind. Vol. III.* New York: Pantheon Books, 1988.

Klein, Herbert S. *Bolivia: The Evolution of a Multi-Ethnic Society.* New York: Oxford University Press, 1992.

Prescott, William H. *The History of the Conquest of Mexico and the History of the Conquest of Peru.* New York: Modern Library (Random House), 1936 and 1979.

Theroux, Paul. *The Old Patagonian Express: By Train Through the Americas.* Boston: Houghton Mifflin, 1989.

Books for Students

Garcia Marquez, Gabriel. *Clandestine in Chile: The Adventures of Miguel Littin.* New York: Henry Holt & Co., 1987.

McIntyre, Loren. *The Incredible Incas and Their Timeless Land.* Washington, DC: National Geographic Society, 1975.

The World Today *Videodisc Program*

A variety of still images, short videos, and activities are available for you to use as you teach this chapter. *Lecture Notes* include bar codes for specific images and videos.

Multimedia Resources

Great Wonders of the World: Manmade—Inca Ruins. (Multimedia CD-ROM) InterOptica Publishing Ltd., 300 Montgomery Street, Suite 201, San Francisco, CA 94104.

Latin America Today: An Atlas of Reproducible Maps. Wellesley, MA: World Eagle, 1992. Available through Social Studies School Service, P.O. Box 802, Culver City, CA 90232–0802.

Picture Atlas of the World. (interactive maps, photo bank, animation, audio on CD-ROM) National Geographic Society, Washington, DC.

Valdivia: America's Oldest Civilization. (VHS, 43 minutes) Films for the Humanities & Sciences, P.O. Box 2053, Princeton, NJ 08543–2053.

MEETING INDIVIDUAL NEEDS

Limited English Proficiency (LEP)

Distinguish for students which vocabulary terms apply uniquely to this region and which are general terms. Use maps and diagrams to help explain the regional terms, and give examples from other parts of the world to elucidate the general terms.

Learners Having Difficulty

Provide students with four large index cards, and have them copy the following blank form onto each. Then tell students to scan Chapter 22 to complete a "study card" for each Pacific South American country (a sample entry is shown in parentheses).

Country Name: *(Ecuador)* Capital: *(Quito)*
Major Geographical Fact: *(on the equator)*
Physical Regions: *(forest, mountain, coastal)*
Major Challenges: *(population, transportation)*
Other Important Facts: *(oil, Galápagos)*

Auditory Learners

Organize a mock travel fair for the countries of Pacific South America. Assign students to role-play representatives of each nation's tourist bureau. These students should be prepared to promote their country as a tourist destination, with interesting facts about its sights and attractions. Other class members, acting as potential tourists, should come to the fair prepared to ask appropriate questions.

Visual Learners

Provide students with outline maps of Pacific South America, or have them make their own. Ask them to shade and label the major geographical regions of the countries and to include symbols for important characteristics of each region, such as types of industry and agriculture. Students should include a map key that explains the symbols.

Tactile/Kinesthetic Learners

Have some students pretend to be ambassadors to the United States from countries in Pacific South America. Have other students pretend to be American journalists. Ask the journalists to interview the diplomats about the major political and economic challenges facing their countries. The ambassadors should be prepared to answer the journalists' questions. The journalists should then give a brief news report based on their interview.

Gifted Learners

Have interested students investigate a past Andean civilization such as the Chavin, Nazca, Moche, Huari, Tiahuanaco, or Incan cultures. Ask students to focus on how mountain geography influenced the culture.

MULTICULTURAL FOCUS

Dwellers of mountain cities often must trade off the natural beauty and protection their location affords for accommodations in their way of life. The potential for severe weather, avalanches, rockfalls, and mud slides can put mountain cities at the mercy of their surrounding peaks. Daily life is also influenced by the physical effects of altitude. Write the following on the board:

> If it is your first day here, take it easy and rest as much as possible. It is not wise to walk even short distances.
> Avoid climbing stairs.
> Avoid drinking alcohol.

Explain to students that this advice might be given to a visitor staying in a hotel in La Paz, Bolivia. Ask students to guess why this might be so. The reason is that the city's high altitude can cause sorojche—mountain sickness—for people who are not accustomed to the altitude. In high altitudes, the thin air provides less oxygen per breath. This can cause headaches, fatigue, nausea, and other flu-like symptoms.

Even people who are physically accustomed to living in high altitudes experience the effects of the elevation. Cars and trucks cannot operate at full power because the thin air lowers their effectiveness. Since water boils at a lower temperature in high altitudes than at sea level, car and truck radiators often boil over. Cooking is also difficult because lower temperatures must be used to prevent liquids from boiling away.

CHAPTER 22

PACIFIC SOUTH AMERICA

Section 1: Ecuador
Section 2: Peru
Section 3: Bolivia and Chile

OVERVIEW

In Chapter 22, students will learn about the Andean countries of Ecuador, Peru, Bolivia, and Chile.

Section 1 describes the three landform and climate regions of Ecuador and explains how the country's economy reflects these regions.

Section 2 focuses on Peru. Students will learn about El Niño and the economic and political challenges that Peru faces.

Section 3 examines Bolivia and Chile. It highlights the major physical regions and economic resources of both countries.

About the Author

Pablo Neruda (1904–1973) is considered one of Chile's greatest and most innovative writers. His poetry won the Nobel Prize for Literature in 1971.

Primary Source

Description of Change: excerpted and bracketed
Rationale: excerpted to focus on the main idea, bracketed to clarify meaning

---

CHAPTER 22

Pacific South America

> **66** **A**nyone approaching the narrow, elongated country of Chile will find in the uppermost portion the Great North, the desert-like regions of salt and copper mines: inclement [harsh] weather, silence, and struggle. In the extreme South, the great frozen expanses, ranging from the silence of Patagonia to Cape Horn, crossed so many thousands of times by the flight of the albatross, and then, the glittering Antarctic. **99**
>
> *Pablo Neruda*

The Andes Mountains dominate the Pacific, or western, side of South America. Some mountain ridges and volcanic peaks here rise more than 20,000 feet (6,096 m) above sea level. The Andes continue to rise steadily upward due to the subduction of a crustal Pacific plate under the continent. The region's frequent earthquakes and volcanic eruptions have triggered landslides and ashfalls, changed the courses of rivers, and engulfed villages under flowing mud.

Incan vase from near **Cuzco, Peru**

The countries of Pacific South America are Ecuador, Peru, Bolivia, and Chile. Traditionally, these countries have relied on North American and European markets for trade. Today, the prosperity of the Pacific Rim has captured economic interests. Goods from East Asia are found throughout South America. South American factories are turning out products bearing Japanese labels. Meanwhile, the peoples of the region continue their struggles to reduce poverty and to reform their political systems.

GEOGRAPHY DICTIONARY

selvas

altiplano

El Niño

terrorism

coup

cartel

latifundia

minifundio

Andes Mountains in **Chile**

244

SECTION 1
OBJECTIVES

The objectives reflect the Focus questions on this page and are assessed in the review on page 247.

1. Describe physical and economic characteristics of the regions of Ecuador.
2. Analyze Ecuador's population and urban geography, and list challenges facing the country.

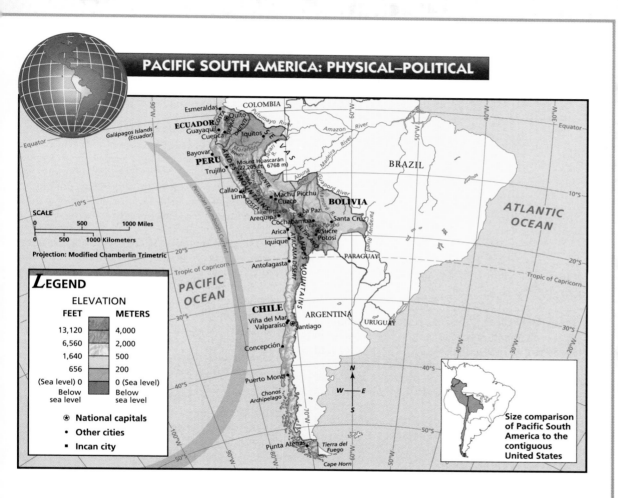

PACIFIC SOUTH AMERICA: PHYSICAL–POLITICAL

LEGEND

ELEVATION

FEET	METERS
13,120	4,000
6,560	2,000
1,640	500
656	200
(Sea level) 0	0 (Sea level)
Below sea level	Below sea level

⊛ National capitals
• Other cities
▪ Incan city

SCALE
0 500 1000 Miles
0 500 1000 Kilometers
Projection: Modified Chamberlin Trimetric

Size comparison of Pacific South America to the contiguous United States

ECUADOR

FOCUS

- *What are the physical and economic characteristics of Ecuador's regions?*
- *Where do Ecuador's peoples live? What challenges face the country?*

LOCATION **Physical Geography** The equator passes directly through Ecuador—in fact, the name *Ecuador* is Spanish for "equator." Ecuador can be divided into three landform and climate regions. (See the map above.) In the east are humid-tropical lowlands of the Amazon Basin. In Ecuador, this forested region is known as the Oriente, which means "east." South Americans call the thick rain forest vegetation found in the eastern region the *selvas*, or "forest."

The Andes form two high mountain ranges through the middle of Ecuador. The valleys in the ranges lie at elevations up to 10,000 feet (3,048 m), with peaks rising up to 10,000 feet higher. More than two dozen high volcanic peaks have permanent snowcaps, even at the equator. The high valleys have mild climates all year. About 45 percent of Ecuador's population lives in the mountain region, known as the Sierra.

Ecuador's third region is the lowlands along the Pacific coast, known as the Costa. The climates in the Costa region vary from humid tropical in the north to a narrow strip of desert in the south. A savanna region is found midway along the Costa. This region has rich agricultural potential and the fastest growing population and economy in Ecuador. About half of Ecuador's people live in the Costa.

Motivator

Have a volunteer locate Quito, the capital city of Ecuador, on a wall map and give its latitude. *(almost on the equator)* Have students guess what the climate is like in this city. Point out that although the temperature averages in the high 70s along the coastal lowlands, Quito has a moderate, spring-like climate. Ask students what factor besides latitude affects the climate here. *(elevation)*

Describing

Ecuador's Characteristics

Write the following words on the board: Andes, Costa, Galápagos, Guayaquil, Incas, Oriente, Quechua, Quito, *selvas,* Sierra. Have each student choose one of the words and write a "What Am I?" or "Where Am I?" riddle for it. Call on students to ask the class their riddles. *(for example, "I am the language of the ancient Incas. What am I?")*

Guided Practice

Have students draw a map of Ecuador on a sheet of paper. Direct them to show the country's three landform and climate regions on their maps. Within each region, have them list important agricultural products and natural resources. When the maps are completed, have students state conclusions about each region's importance to Ecuador's economy.

Chart Caption Answer

Chile

COOPERATIVE LEARNING

Organize the class into groups of four or five. Tell the groups to imagine that they are headed for the Galápagos Islands. Assign topics for each team to research, such as the islands' history, landforms, animals, and plants.

Section 1 Review
ANSWERS

1. Sierra: farming; Oriente: farming, oil drilling; Costa: farming, fishing, mining, manufacturing; Galápagos Islands: tourism

2. About 45 percent in the Sierra; half in the Costa; the remainder in the Oriente; because the Andes hinder construction and maintenance of roads

3. Guayaquil is a modern, manufacturing city on the coast, whereas Quito, the capital, is a Spanish colonial city in the Sierra.

246

FACTS IN BRIEF **Pacific South America**

COUNTRY POPULATION (1994)	LIFE EXPECTANCY (1994)	LITERACY RATE (1990)	PER CAPITA GDP (1993)
BOLIVIA 7,719,445	61, male 66, female	78%	$2,100
CHILE 13,950,557	72, male 78, female	93%	$7,000
ECUADOR 10,677,067	67, male 73, female	88%	$4,000
PERU 23,650,671	63, male 68, female	85%	$3,000
UNITED STATES 260,713,585	73, male 79, female	97% (1991)	$24,700

Source: *The World Factbook 1994,* Central Intelligence Agency

Which Pacific South American country has the highest literacy rate?

Economic Geography Ecuador's economy closely reflects its physical geography. Farmers in the Sierra grow maize (corn), potatoes, and barley. The fields are often on steep slopes where soil erosion is severe, and many of the farms are too small to support a family. Landless farmers from the Sierra are moving down the Andes slopes to clear small farms from the dense forests in the Oriente. The Oriente's greatest resource is oil, which accounts for more than 50 percent of Ecuador's export income.

The Costa has emerged during the twentieth century as Ecuador's most important agricultural region. The key export crops are bananas, coffee, and cacao. Many of the Costa's farms are large enough for prosperous family farming. Offshore from the Costa are rich fishing grounds. Natural-gas deposits have been found beneath the Costa's lowlands.

Human Geography The Sierra is a historic region of dense Native American settlement. Many people still speak Quechua (KECH-oo-ah), the language of the ancient Incas. Descendants of African slaves and of European colonists have settled primarily in the Costa region.

Guayaquil (gwie-uh-KEEL), on the Pacific coast, is Ecuador's largest city. It also is Ecuador's most important port and manufacturing center. Guayaquil's modern buildings, banks, and factories are symbols of a modern Ecuador.

Ecuador's capital city is Quito (KEE-toh), in the Sierra. A northern kingdom of the Incan Empire, it became the center for Spanish colonial

government. Despite repeated earthquakes over the centuries, Quito has survived as a classic example of a Spanish colonial city. Old churches, monuments, and open-air markets are tourist favorites.

The oil wealth of the Oriente and the agricultural progress of the Costa have moved Ecuador to world prominence. The country faces several challenges, however. Providing jobs, schools, and health care for a still rapidly growing population is an ongoing concern. Large numbers of people are moving to the Oriente, threatening the survival of the rain forest.

Another major challenge is transportation. Despite many roads and a railroad stretching from the coast to Quito, the country's regions are not well linked. The Andes make construction and maintenance of transportation systems both difficult and expensive.

After decades of corrupt governments, Ecuadoreans are having more political success. Democratically elected governments have ruled the country in recent years.

Galápagos Islands The arid Galápagos (guh-LAHP-uh-guhs) Islands, located in the Pacific Ocean about 650 miles (1,046 km) off the coast, are part of Ecuador. The animals and plants that are native to the islands are specially adapted to their habitats. Some scientists believe that the Galápagos' famous wildlife demonstrates the theory of biological evolution. The islands are home

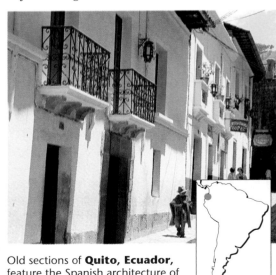

Old sections of **Quito, Ecuador,** feature the Spanish architecture of the colonial era.

Closure

Have students name Ecuador's assets and the challenges it faces. Then have them evaluate Ecuador's future.

ASSESSMENT

Assign the Section 1 Review questions on this page.

Reteaching

Assign Section 1 of Reteaching Worksheet 22.

SECTION 2
OBJECTIVES

Then objectives reflect the Focus questions on this page and are assessed in the review on page 248.
1. Identify Peruvian regions and their characteristics.
2. Discuss challenges facing Peru.

TEACHING SECTION 2
Motivator

Have students locate Peru on a map. Tell them that Peru has landform and climate regions similar to Ecuador's: the Oriente, the Andes highlands, and the coastal region. Review the characteristics of these regions in Ecuador. As students read through Section 2, have them note similarities and variations in the physical geography of the two countries.

---

to about 4,000 people, most of whom are employed in the growing tourist trade. Though profitable, the tourist trade disrupts the natural environments on the islands and is threatening the existence of the unique plant and animal species.

SECTION REVIEW

1. What economic activities are important in Ecuador's three regions and on the Galápagos Islands?
2. How is Ecuador's population distributed? Why is transportation a challenge in Ecuador?
3. **Critical Thinking** **Compare and contrast** Guayaquil and Quito.

PERU

FOCUS

- *What are the characteristics of Peru's three major regions?*
- *What challenges face Peru?*

Like Ecuador, Peru can be divided into three regions: the Oriente, the Andes highlands, and the coastal lands. (See the map on page 245.)

The Oriente The Oriente of Peru begins on the eastern side of the Andes and continues onto the Amazon lowlands. The Oriente has dense rain forests that provide lumber and tree crops such as rubber. New settlers are clearing away the forests for subsistence farming. Preserving the rain forests on this modern frontier is an international concern.

The coca plant is native to the eastern slopes of the Andes in Peru, Bolivia, Ecuador, and Colombia. The extract of coca leaves is the source of the powerful drug cocaine. Illegal drug smugglers encourage farmers here to grow coca, which has become their most profitable crop. The United States is working with the governments of the Andean countries to find substitute crops for the farmers.

The major port city of the Peruvian Amazon is Iquitos. Ships travel more than 2,000 miles

(3,218 km) on the Amazon from the Atlantic Ocean to Iquitos.

The Andes Highlands The Andes highlands make up Peru's second region. Traditional agriculture is practiced by farmers in the Andes highlands. Crops include potatoes, maize (corn), and barley. Local livestock includes the llama, alpaca, and vicuña. The easily tamed llama can be used for transport, and its coarse wool is used for making blankets. Alpacas are herded like sheep, and their wool makes fine cloth. The vicuña is a wild animal whose wooly fur is highly prized.

In southern Peru, the eastern and western ridges of the Andes separate, leaving a broad highland plain in between. This plain is called the ***altiplano*** and extends well into Bolivia. At an elevation of about 12,000 feet (about 3,658 m), the vegetation of the *altiplano* is a high-elevation grassland with few trees.

The *altiplano* and nearby valleys were the heartland of the Incan Empire. The buildings at Machu Picchu, an ancient Incan city, are impressive remains of the empire. This region continues to be one of the densest areas of Native American settlement in the Americas. Quechua is spoken by perhaps as many as 3 to 4 million Peruvians.

This gold handle from a ceremonial knife honors a legendary leader of the Chimú civilization. The Chimú were conquered by the Incan Empire in the fifteenth century.

GLOBAL PERSPECTIVES

Archaeologists believe that maize (corn) made the Incan civilization and the other great civilizations of Middle and South America possible. It provided the nourishment required for the population to thrive and build complex societies. Archaeologists also say that maize was the real "gold" that the Spanish brought back to the Old World. It spread from southern Europe to become a staple in Africa and Asia.

Building a Vocabulary

Have students use structural analysis and their knowledge of Spanish to figure out the meanings of **altiplano** (*alti* = "high" + *plano* = "plain") and **El Niño**. (*El Niño means "the male child" in Spanish and refers to "the Christ child." El Niño often occurs near Christmas.*)

Guided Practice

Have students write an editorial about current events in Peru from the point of view of one of the following people: an Andean Indian, the president of Peru, an American tourist at Machu Picchu, a resident of Lima's shantytown, a member of Peru's upper economic class. Students should look for current news articles to research their point of view.

Closure

Call on students to summarize the political and economic challenges facing Peru.

ASSESSMENT

Assign the Section 2 Review questions on this page.

Reteaching

Assign Section 2 of Reteaching Worksheet 22.

Geographical Sidelight

In 1990, Peruvians elected Alberto Fujimori, the son of Japanese immigrants, as president. Peruvians consistently tell poll takers that the country they admire most is Japan. Some are even undergoing plastic surgery to acquire Japanese physical characteristics. They hope to emigrate to Japan with papers claiming Japanese ancestry so that they can find jobs.

Section 2 Review
ANSWERS

1. The Oriente: dense forests; the Andes highlands: traditional agriculture; the coastal region: settlement in river valleys
2. Importation of food, foreign debt, and printing of more money to pay for government services
3. No, in Oregon and Washington the dry areas are on the eastern side of the mountains, not on the western side as in Peru

---

The Coastal Region West of the Andes is Peru's dry desert coast. Here, the rugged lower sides of the Andes often extend down to the sea. Settlement is concentrated in small river valleys that have a water supply from melted Andean snow.

A cool ocean current flowing northward off the coast of Peru contributes to the desert conditions. In South America, this is the Peruvian (Humboldt) Current. Air passing over the current becomes cool and very moist. Because the air masses are under high pressure, the moist air cannot travel upward. The result is little, if any, rain, although coastal fogs are common. The coastal desert is also in the rain shadow of the Andes. The Andes' slopes far to the east prevent rain by blocking the moist winds from the Amazon Basin.

Every five to ten years, the dry coasts of Peru and Ecuador are visited by unusual weather patterns called **El Niño**. The ocean near the coast warms, and most fish flee from what is normally a rich fishing area. People along the coast may suffer severe flooding from heavy rains. Scientists believe El Niño conditions are caused by the buildup of unusually warm waters in the mid-Pacific Ocean. These warm waters then slosh eastward, turning aside the cool Peruvian (Humboldt) Current. El Niño affects ocean and weather events around the world.

The coastal region of Peru is the most modern and developed. The coastal rivers have many sites suitable for development of hydroelectric power, and the region holds Peru's major mineral deposits. Fishing has become a major industry in the coastal ports. Anchovies are the primary catch.

The people who live in the coastal region are predominantly mestizo. Lima, the capital of Peru, and its port city of Callao (kuh-YAH-oh) are located in the coastal region. Both are increasingly important to the country's economy. The population of the Lima and Callao region now numbers nearly one-third of the country's people. Lima's many government jobs and manufacturing industries draw people from throughout the country, particularly from the highlands.

Political Geography Peru's problems are related to underdevelopment, poor governing, and the contrasts between the mountain and coastal regions. Large amounts of food must be imported. To manage a large foreign debt, the government

printed more money to pay for government services. The result was inflation.

Peru has made progress in recent years. The government has eased inflation problems, and a violent revolutionary movement has weakened. The government also has worked to end **terrorism,** or the use of violence as a means of political force. Poverty remains a major concern, however. Some critics also have accused the government of using undemocratic methods to solve the country's problems.

2 SECTION REVIEW

1. List Peru's three major regions, and give one characteristic of each.
2. Why has Peru experienced inflation?
3. **Critical Thinking** **Assess** whether the rain-shadow effect experienced by Peru's coastal region is the same effect experienced by Oregon and Washington in the United States.

3 BOLIVIA AND CHILE

FOCUS

- *Into what regions can Bolivia and Chile be divided? What are the features of each region?*
- *How are political and economic conditions in Bolivia similar to those in Chile?*

LOCATION **Bolivia** Bolivia is a landlocked country south of Peru. (See the map on page 245.) The country once extended west to the Pacific but lost its coastal province to Chile in 1884, following the War of the Pacific. Bolivia has the highest percentage of indigenous peoples of any South American country.

Bolivia's government has changed through revolution or military **coup** (koo) many times since the country became independent in 1825. From the French phrase *coup d'etat*, a coup is an overthrow of an existing government by another political or military group. Bolivia has had a democratically elected government, however, for more than 10 years.

Most of Bolivia can be divided into two major physical regions: the western highlands

The objectives reflect the Focus questions on page 248 and are assessed in the review on page 250.

1. Examine the various regions of Bolivia and Chile.

2. Compare political and economic conditions in Bolivia and Chile.

TEACHING SECTION 3
Motivator

Have students locate Bolivia and Chile on a map. Point out that Chile's shape would not be so elongated if Bolivia had not lost its coastal province to Chile in the War of the Pacific from 1879 to 1884. After years of strained relations, Chile has allowed Bolivia to use Arica as its main port. Bolivia also lost land in a war with Paraguay between 1932 and 1935.

Building a Vocabulary

Review the definitions of the terms **coup**, **cartel**, *latifundia*, and ***minifundio***. Remind students that *latifundia* refers both to a land tenure system and to the large estates themselves (each of which is called a *latifundium*). Have students write their own sentences using the four vocabulary terms in context.

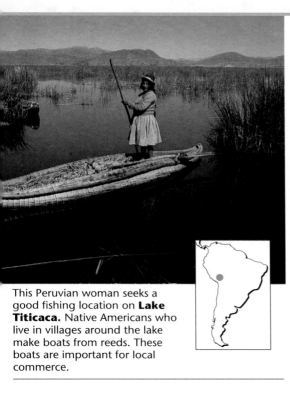

This Peruvian woman seeks a good fishing location on **Lake Titicaca.** Native Americans who live in villages around the lake make boats from reeds. These boats are important for local commerce.

good grazing lands. But there are few roads, few people, and little money for investment. Development of this region would offer the nation economic improvement.

Bolivia's greatest growth region surrounds the city of Santa Cruz in the lowlands. Santa Cruz was connected by road with La Paz in the 1950s, making it possible to provide many agricultural products to the highlands. Recent oil discoveries in eastern Bolivia may make this region more important in the future.

Chile Chile's name comes from a Native American word meaning "land's end." Chile's shape makes it one of the most unusual countries in the world. The country is 2,650 miles (4,264 km) long but only about 110 miles (177 km) wide. It lies between the Andes Mountains and the Pacific Ocean. (See the map on page 245.) From north to south, the country can be divided into three landform regions: northern, central, and southern Chile.

The northern region of Chile is the Atacama Desert, one of the driest regions in the world. South of the desert, the region becomes semiarid with just enough rain for grazing. Croplands along the valleys are irrigated from streams draining the Andes. For more than a century, the minerals of northern Chile provided most of the country's export income.

The first products to be mined in the north were guano (the droppings of seabirds) and nitrate, which were processed into fertilizer and exported to Europe and North America. While nitrate remains important, synthetic fertilizers are replacing it in most of the market. Nitrate also is an important raw material for explosives.

Northern Chile is one of the world's largest producers of copper, its key export mineral since the 1920s. Most often, the copper is exported as nearly pure bars that can be manufactured into wire and sheets. Chile's copper reserves remain very large, and its copper production dominates the world market. Even so, the country's economy rises and falls with world copper prices. Chile often has considered starting a **cartel** to control prices. A cartel is a group of business organizations that agree to limit supplies of a product and thereby keep prices high.

The main cities of the Atacama region are Arica, Iquique, and Antofagasta. Each of these cities is a port with small fishing fleets and some

Geography and Technology

Scientists are catching ground-hugging clouds and "milking" them of water in order to aid Chungungo, a fishing village on Chile's dry coast. To gather moisture, researchers set up huge nets along the Pacific. When fog blows in from the ocean, the nets capture water droplets that add 2,500 gallons (11,365 l) a day to the community's reservoir.

and the eastern lowlands. Most Bolivians live in the highlands, despite the cold temperatures and barren lands. Western Bolivia is a continuation of the Andes highlands of Peru. More than 500 miles (805 km) long, the *altiplano* contains two great lakes. These are freshwater Lake Titicaca, which Bolivia shares with Peru, and salty Lake Poopó (poh-uh-POH). Lake Titicaca covers about 3,200 square miles (8,288 sq. km). At an elevation of 12,500 feet (3,810 m), it is the highest lake in the world on which ships travel.

The Bolivian Andes are famous for their natural resources. Spanish colonial governments mined great amounts of silver near the city of Potosí. Later, tin became the key mineral export. Today, valuable mineral resources include copper, lead, and zinc. Oil and natural gas also provide income for the country.

East of Lake Titicaca, in a valley below the *altiplano*, is La Paz (luh PAHZ), the country's administrative capital. At 12,001 feet (3,658 m), it is the highest capital in the world. Despite the semiarid environment and the frequently freezing weather, La Paz is the principal industrial city of Bolivia. It is also the country's center for imports and exports.

Much of eastern Bolivia is a series of low plateaus covered with dry forests and grasslands. The plains have fertile soils, adequate rain, and

Economic Conditions
Have students tell what they think Chile or Bolivia will be like economically and politically in the year 2000. Have them base their predictions on the information in Section 3.

Independent Practice
Have each student label Outline Map 13 to show the major cities, landforms, and types of economic activity in Bolivia and Chile.

Closure
Call on volunteers to describe landform and climate regions in Chile and Bolivia.

ASSESSMENT

Assign the Section 3 Review questions on this page.

Reteaching
Assign Section 3 of Reteaching Worksheet 22.

Extension

Point out that a great food exchange occurred between Spanish and Portuguese explorers and Native Americans. This exchange profoundly influenced the diets and culture of both groups. Have students be "food detectives." Ask them to trace the origin of such common foods as wheat, peanuts, pineapples, bananas, potatoes, tomatoes, lettuce, chocolate, watermelons, peppers.

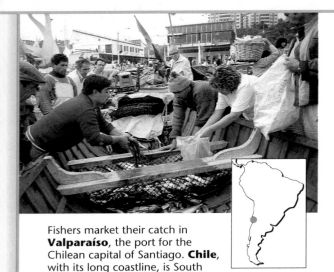

Fishers market their catch in **Valparaíso**, the port for the Chilean capital of Santiago. **Chile**, with its long coastline, is South America's chief fishing nation. Europe and the United States use fish meal and fish oil from Chile for the production of animal feed and industrial oil.

Section 3 Review
ANSWERS

1. Bolivia: western highlands—silver, tin, copper, lead, zinc; Chile: northern region—guano, nitrate, copper, other minerals
2. Both countries have experienced much political instability in the past but now have stable democratic governments.
3. Cartels are an advantage to the producing countries, which control prices, and are a disadvantage to the consuming countries, which have to pay more for a product. In both cartels and monopolies, competition is reduced so that producers have more control over the selling of products.

250

around the capital city, Santiago, and its port, Valparaíso, now totals more than one-third of all of Chile's people. As the center of the country's trade and government, the Santiago region dominates Chile's economy.

Chile's third environmental region is southern Chile, with its thick forests and towering mountains. The marine-west-coast climate brings heavy autumn and winter rains. South of the town of Puerto Montt, the region is wild, with cold, stormy weather and rugged, snowcapped mountains.

Southern Chile was settled by immigrants from Europe. Here, there are prosperous medium-size family farms. These farms provide a variety of products, including wheat, dairy products, and beef. Far southern Chile has few people. The region's forests represent a major resource for the future and could provide lumber and paper products for all of South America. There are also large deposits of low-grade coal and oil in this region. With further development, oil deposits should provide enough oil for the country as well as some for export.

Like most Middle and South American countries, Chile has been in turmoil throughout its history. After the failures of an elected Communist government in the early 1970s and a dictatorship throughout the 1980s, Chile's economic and political life has been renewed. The country's large foreign debt is under control, and prosperity has increased. Chile now has a stable democratic government, and optimism is widespread.

Although urban slums remain, Chile's healthier economy seems to be improving the status of the urban poor. After decades of little economic growth, Chile's small businesses and factories are growing rapidly. Unemployment is decreasing and wages are rising. Chile's resources are once again being developed, and the country is working to join the United States, Mexico, and Canada in NAFTA.

industries. To encourage industrial development, earlier Chilean governments gave special import rules to these port cities. These rules no longer apply, however. The Atacama is far from the markets in central Chile, and manufacturing may decline without these special rules. Landlocked Bolivia uses Arica and Antofagasta as its main ports.

Farther south is central Chile, the second environmental region and the country's heartland. Here, between the coastal mountains and the Andes, is the broad central valley. It has a Mediterranean climate and fertile soils. This region extends for almost 700 miles (1,126 km).

Central Chile was the center of Spanish colonial settlement. The land remains organized by the *latifundia* system, which was brought to the continent by European colonists. In this system, tenant farmers work part of the land for their own food in return for their labor. The term *latifundia* refers to the land tenure system as well as to the large estates on which tenants work. Today, productivity on the *latifundia* is low. Now, the trend is toward commercial farming of fruits such as grapes, peaches, and plums for export.

In the past, the government divided some large estates into small farms for families. Called *minifundios*, many of these small farms are not large enough to be profitable. In addition, social unrest often has interfered with food production, forcing Chile to import food in some years.

About 90 percent of Chile's people live in the central region. In fact, the urban population

SECTION REVIEW

1. In which regions of Bolivia and Chile is mining important? What products have been mined in these regions?
2. How are political conditions in Bolivia similar to political conditions in Chile?
3. **Critical Thinking** **Evaluate** advantages and disadvantages of cartels. How does a cartel resemble a monopoly?

Building a Vocabulary
1. Forest
2. *Altiplano*
3. A warming of the ocean; Peru, Ecuador, and other countries around the world
4. The use of violence as a means of political force
5. No, because a coup is an overthrow of an existing government by another political or military group, whereas a revolution involves more people and more change

Recalling and Reviewing
1. Threats to the survival of the rain forest, floods caused by El Niño
2. Bolivia is landlocked, so it does not have a coastal region. Chile is long and narrow and is situated between the Andes Mountains and the Pacific Ocean. Therefore, Chile is more easily divided into regions from north to south than from east to west.
3. Along the coast

Thinking Critically
1. The city was well hidden and protected by the Andes and may have been abandoned by its inhabitants; it was not destroyed by war.
2. Bolivia would be less dependent on Chile's ports, and its economy might improve.

Learning About Your Local Geography
Reports should name the last major road or highway built in the area and should address the reasons it was built and the effects it has had.

Using the Five Themes of Geography
Maps should show the Pacific Rim and should show arrows going in all directions among the South and North American Pacific coasts and the East Asian coasts.

CHAPTER 22 CHECK

Southern **Chile**

Reviewing the Main Ideas
1. Ecuador has forest, mountain, and coastal regions. The country's major challenges are population growth and the lack of a good transportation system.
2. Like Ecuador, Peru has forest, mountain, and coastal regions. The coastal region is the most developed. Political unrest continues to be an issue in Peru.
3. Bolivia is a landlocked country. La Paz is the administrative capital. Bolivia's western highlands hold most of the country's people. The eastern lowlands have development potential.
4. Chile is a long, narrow country situated between the Andes Mountains and the Pacific Ocean. The minerals of northern Chile provide much of the country's export income. Chile has a stable government, and the country's economy is improving.

Building a Vocabulary
1. What does *selvas* mean?
2. What term is used to describe the plain located between the eastern and western ridges of the Andes?
3. What is El Niño? What countries does it affect?
4. Define *terrorism*.
5. Is a coup the same as a revolution? Explain.

Recalling and Reviewing
1. What environmental concerns do Ecuador and Peru have?

2. Both Ecuador and Peru have an inland forest region, a highland region, and a coastal region. Why are Bolivia and Chile divided differently?
3. Where are most of the major industrial cities of Pacific South America located?

Thinking Critically
1. The ancient Incan city of Machu Picchu was discovered not by Spanish explorers but by Hiram Bingham in 1911. The stone structures of the city, although empty of people, were still in very good condition. What do these facts suggest about the city and its inhabitants?
2. Chile has agreed to grant Bolivia a strip of land that would give Bolivia access to the coast. The two countries, however, have not been able to agree on exact details. How do you think conditions in Bolivia would change if the country were no longer landlocked?

Learning About Your Local Geography

Individual Project
One of the major challenges facing the countries of Pacific South America is a lack of good transportation. Contact your state or local department of transportation to find out about the last major road or highway built in your area. Request information about why this road was built, when it was built, and how it has affected the area as a whole. Synthesize this information into a brief written report. You may want to include your report in your individual portfolio.

Using the Five Themes of Geography
How is the Pacific coast **location** of the countries discussed in this chapter likely to affect future **movement** of people and goods into and out of these countries? What global **region** are these countries a part of because of their location? Sketch a map of this region and draw arrows to indicate patterns of movement.

251

Unit 4 Review

SKILLS

Guide students through the steps of the geographic method. Point out that the geographic method can be used to answer questions about local geography as well as the geography of faraway places. As an example, have students use the geographic method to learn about why their school is located where it is. Have them form a hypothesis, such as that the school's location is in the center of the area from which it draws students. Then have them identify, collect, and analyze the information they need to test their hypothesis, such as by taking a survey of where fellow students live and mapping out the findings. The class should come to a conclusion about whether the hypothesis was correct.

Practicing the Skill
ANSWERS

Reports should indicate that students have followed the geographic method to come to a conclusion regarding a geographic topic or issue in Middle and South America.

UNIT 4 REVIEW SKILLS

Using the Geographic Method

You may have learned in a science class about the "scientific method." Similarly, geographers use the "geographic method" to collect and analyze geographic information and to make sense of that information.

1. *Ask a question.* What would you like to know more about? Most questions start with one of the following words: who, what, where, when, why, or how. For example, you might ask the question "Why does Peru import food?"

2. *Form a hypothesis.* An educated guess is called a hypothesis. A hypothesis gives you something to focus on in answering your question. There may be several reasons that a country imports food, and your hypothesis should identify one or more of these reasons. Your hypothesis might be "Peru imports food because its agricultural methods are not efficient enough to feed the country's population."

3. *Identify what information you need in order to determine whether your hypothesis is correct.* For example, to test the hypothesis that Peru must import food because of its agricultural methods, you would want to find out what those methods are. You might also want to find out the size of Peru's population.

4. *Collect the information necessary to test your hypothesis.* Once you have identified what information you need, you must decide how to collect this information. Collecting information can take many forms. Ideally, you would visit a place and observe it firsthand. This is called field work. Unfortunately, field work is not always possible. Instead, you can gather information at the library. To find out about Peru's agricultural methods and its population, you might look in an atlas, an almanac, or in a book about Peru. Your information might be in the form of maps, graphs, charts, or photographs.

5. *Analyze the information you have collected.* Look for patterns or relationships in your information, particularly if your source is a map, graph, chart, or photograph. Is the information you have collected sufficient to test your hypothesis? If not, you may need to return to Step 3.

6. *Draw a conclusion.* Based on the information you have analyzed, was your hypothesis correct? If not, you might want to go back to Step 2 and form a new hypothesis. If your hypothesis was correct, you may want to present your information to others.

PRACTICING THE SKILL

Identify a topic in this unit about which you would like to learn more, such as political instability, deforestation, or the growth of cities in Middle and South America. Then, follow the steps of the geographic method listed above to come to a conclusion. Write a report summarizing your findings and the process you followed.

BUILDING YOUR PORTFOLIO

Individually or in a group, complete one of the following projects to show your understanding of the geography concepts involved.

A Middle and South America: A Regional Approach

As you know, **region** *is one of the Five Themes of Geography. In what ways can the concept of region be applied to the countries of Middle and South America? For this project, you will sketch a series of maps, each map showing a different aspect of region with regard to Middle and South America.*

1. Some geographers consider Mexico to be part of the North American region, rather than part of the Middle and South American region. Decide whether you agree or disagree, and sketch a map to defend your opinion. Use symbols and shading to show regional features.
2. Your next map should show the regional divisions within any country of Middle or South America. With symbols and a legend, show how the country's regions differ in physical, economic, or human characteristics.
3. Because people move among different regions, the geographic theme of movement is closely related to the theme of region. For your third map, select any country in Middle or South America, and with arrows, show methods and routes of transportation within that country.
4. One type of movement that occurs in many Middle and South American countries is migration from rural regions to urban regions. Select a country in which this movement is happening and, on your fourth map, indicate migration to the country's cities. Next to the map, write several sentences explaining the causes and effects of this type of movement.
5. Another type of migration that occurs in some Middle and South American countries is from crowded developed regions into regions that have few people. Select a country where this type of migration has taken place. For your fifth map, sketch the country, showing the migration. Include several sentences explaining the causes and effects of this type of migration.

Present and explain your maps to the rest of the class.

B Land Use and Reform

A theme common to many of the countries of Middle and South America has to do with the ways in which land is used, distributed, and reformed. Construct a series of models showing these different patterns, including one model showing what you think is the best use for land. Your models may be physical representations, or they may be diagrams on paper as explained in the Unit 3 Skill.

1. Your first model should show subsistence agriculture in Middle and South America. Include with the model a list of the crops grown in subsistence agriculture, where subsistence agriculture occurs, and who practices subsistence agriculture.
2. Another model should show agribusiness and plantation agriculture in Middle and South America. Again, include with the model a list of the crops grown in agribusiness and plantation agriculture, where agribusiness and plantation agriculture occurs, and who practices agribusiness and plantation agriculture. Who owns the land, and who provides the labor?
3. A third model should show the hacienda system in Middle and South America. Include with the model a description of the hacienda system. Also include a statement comparing the hacienda system to the *latifundia* system. Conduct research to find out what similar systems are called in Brazil and in Argentina.
4. Your next set of models should show how any of the systems you have modeled above have been reformed as *ejidos* (Mexico), cooperatives (Cuba), or *minifundios* (Chile). Include with these models statements of how, when, where, and why the land has been reformed, and how successful the reforms were or have been.
5. Finally, decide what form of land use you think is best. You may wish to consider several factors, including how productive the land is, how the land is divided among various people, and environmental effects. Your model may be one you have already made from the steps above, or it may be a new one altogether.

Present and explain your models to the rest of the class.

253

UNIT 5 EUROPE *CHAPTERS 23–27*

OVERVIEW

In Unit 5, students will learn about the landforms and cultures of one of the most densely settled continents.

Chapter 23 • INTRODUCING EUROPE provides an overview of Europe, featuring its landforms, resources, history, and cultural diversity.

Chapter 24 • NORTHERN EUROPE describes the United Kingdom, the Republic of Ireland, and the industrialized Nordic countries.

Chapter 25 • WEST CENTRAL EUROPE focuses on France, Germany, Belgium, the Netherlands, Luxembourg, Switzerland, and Austria.

Chapter 26 • SOUTHERN EUROPE explores the countries located on the Mediterranean Sea.

Chapter 27 • EASTERN EUROPE describes the political and economic changes and ethnic unrest that have characterized Eastern Europe in recent years.

OBJECTIVES

1. Describe the major landforms and climates of Europe.
2. Explain the historical development of European nations.
3. Analyze how land, climate, and resources have played an important role in the economic development of Europe.
4. Identify different cultural groups in Europe, and understand how they interact with one another.
5. Discuss Europe's influence in international affairs.
6. Explain how European nations have attempted to develop policies of mutual political and economic cooperation.
7. Interpret special-purpose maps to analyze relationships among climates, population patterns, and economic resources in Europe.
8. Learn to observe and identify elements of physical geography and human geography in fine art.

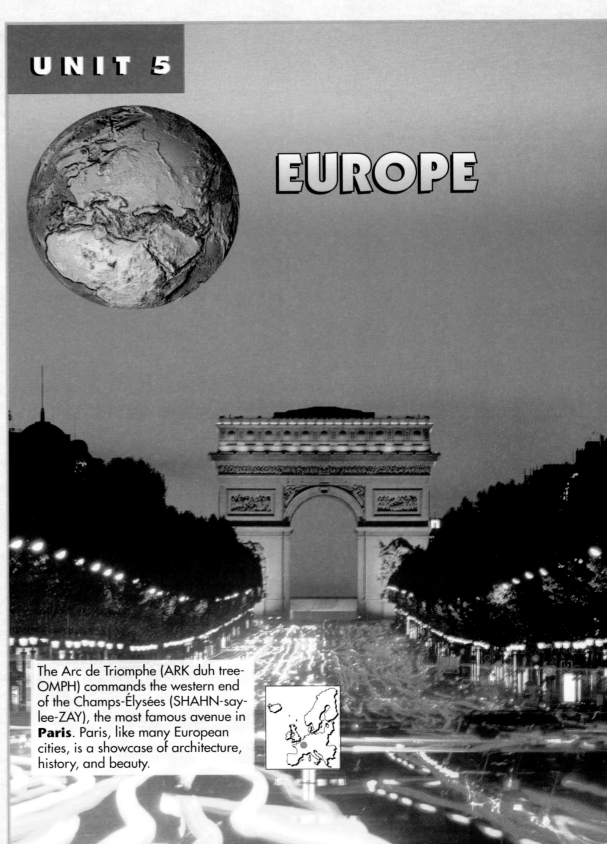

UNIT 5

EUROPE

The Arc de Triomphe (ARK duh tree-OMPH) commands the western end of the Champs-Élysées (SHAHN-say-lee-ZAY), the most famous avenue in **Paris**. Paris, like many European cities, is a showcase of architecture, history, and beauty.

Using the Illustrations

Direct students' attention to the photograph of the Arc de Triomphe ("Arch of Triumph") on page 254. Point out that this structure is an important historical monument in France, built in the early 1800s as a tribute to the army of emperor Napoléon Bonaparte. Beneath the arch is the Tomb of the Unknown Soldier, honoring unidentified French soldiers who died during World War I.

Refer students to the photographs on this page. Have them locate the main physical features mentioned in the captions (Danube River, the Alps) on the physical-political map of Europe on page 259. They should trace the course of the Danube River through Europe and observe the relative location of the Alps in France, Italy, Switzerland, and Austria.

Teaching with Sketch Maps

Have students draw a map of Europe from memory. Direct them to form a mental picture of the region and to transfer what they remember to paper. Tell them to label any bodies of water, physical features, political names, and boundaries they know. Students might add this map to their portfolios and compare it with another they will draw after they have completed Unit 5.

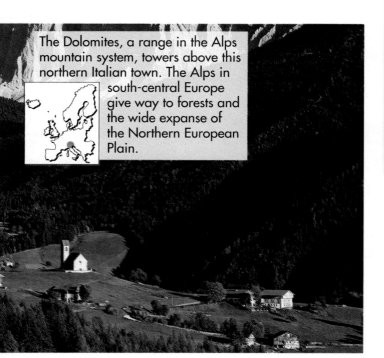

The Dolomites, a range in the Alps mountain system, towers above this northern Italian town. The Alps in south-central Europe give way to forests and the wide expanse of the Northern European Plain.

UNIT 5 CHAPTERS

23 Introducing Europe

24 Northern Europe

25 West Central Europe

26 Southern Europe

27 Eastern Europe

Gold Spanish coin

THE WORLD TODAY VIDEODISC PROGRAM

The videodisc profiles the following countries in Unit 5: Great Britain, France, Germany, Spain, and Hungary.

Early Viking mask

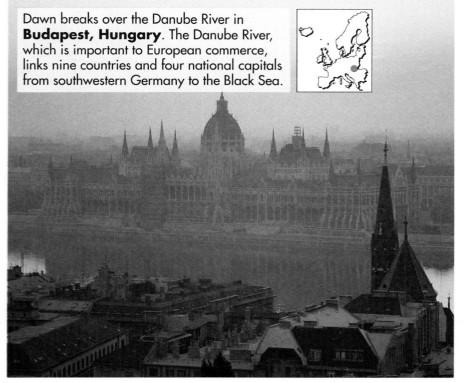

Dawn breaks over the Danube River in **Budapest, Hungary**. The Danube River, which is important to European commerce, links nine countries and four national capitals from southwestern Germany to the Black Sea.

255

One World, Many Voices

LITERATURE

Introduce the excerpt below by informing students that *Don Quixote* is considered a classic of European literature. You might provide a historical context for this selection by asking students what they know about the Middle Ages, chivalry, and knighthood, and then filling in the gaps for them. Ask them to read first the biographical paragraph about Cervantes. Tell students that although the author clearly intended to portray the hero Don Quixote in a comical way, the character over time has come to represent an idealist who nobly fights for an impossible cause.

More About Miguel de Cervantes

After serving in the Spanish army, Cervantes was held captive by Barbary pirates in Algeria for several years. Following his release, he returned to Spain and pursued his writing career. Despite the popularity of *Don Quixote* soon after it was published, Cervantes received very little money for his novel and died penniless.

ONE WORLD, MANY VOICES

LITERATURE

Geography and Literature

The complete title of Cervantes's famous novel is The Ingenious Gentleman Don Quixote of La Mancha. *La Mancha is a dry, treeless plateau in south-central Spain. Cervantes chose this region to depict a rather poor and isolated area.*

Primary Source

Description of Change: excerpted and bracketed
Rationale: excerpted to focus on the main idea; bracketed to clarify meaning

---

Interpreting Literature

Answers

1. Windmills on the Spanish plain
2. He tries to persuade Don Quixote that the giant's "arms" are really the wings on the windmills, and he explains their function.

"At this point they caught sight of thirty or forty windmills which were standing on the plain there, and no sooner had Don Quixote laid eyes upon them than he turned to his squire [attendant] and said, 'Fortune is guiding our affairs better than we could have wished; for you see there before you, friend Sancho Panza, some thirty or forty lawless giants with whom I mean to do battle. I shall deprive them of their lives, and with the spoils from this encounter we shall begin to enrich ourselves; for this is righteous warfare, and it is a great service to God to remove so accursed a breed from the face of the earth.'

'What giants?' said Sancho Panza.

'Those that you see there,' replied his master, 'those with the long arms some of which are as much as two leagues in length.'

'But look, your Grace, those are not giants but windmills, and what appear to be arms are their wings which, when whirled in the breeze, cause the millstone to go.'

'It is plain to be seen,' said Don Quixote, 'that you have had little experience in this matter of adventures. If you are afraid, go off to one side and say your prayers while I am engaging them in fierce, unequal combat.'

Saying this, he gave spurs to his steed [horse] Rocinante, without paying any heed to Sancho's warning that these were truly windmills and not giants that he was riding forth to attack. Nor even when he was close upon them did he perceive what they really were, but shouted at the top of his lungs, 'Do not seek to flee, cowards and vile creatures that you are, for it is but a single knight with whom you have to deal!'"

Miguel de Cervantes (1547–1616) was a Spanish writer from Madrid. Many people of his day enjoyed reading romantic adventure stories about medieval knights. In these tales, the heroes traveled to distant lands, slayed dragons, and battled evil foes. Cervantes poked fun at this literary style by writing *Don Quixote*. The excerpt at left from this comic masterpiece follows the exploits of a mentally unstable, but likable, aging landowner who believes he is actually a knight from medieval times. Don Quixote, accompanied by his wise servant Sancho Panza, rides across Spain trying to perform daring deeds.

Interpreting Literature

1. What does Don Quixote see on the Spanish plain that he mistakes for "giants"?
2. How does Sancho Panza attempt to reason with Don Quixote?

256

FOR THE RECORD

Muslim geographers, such as Abu Abdallah Mohammed Idrisi, Al-Muqaddasi, and Ibn Battuta, traveled extensively during the ninth through fourteenth centuries. They sought geographic knowledge about the world partly because of their religious beliefs. Identifying cardinal directions for north, south, east, and west from precise locations was important because Islam encourages Muslims to make a pilgrimage to the holy city of Mecca at some point during their lifetimes, and to turn towards the city during daily prayers. Students might be interested in knowing that an astrolabe (shown below) is a scientific instrument that dates back to the ancient Greeks. During the Middle Ages it was refined as a navigational tool to determine latitude.

More About Abu Abdallah Mohammed Idrisi

Idrisi served for more than 20 years as the official geographer in the court of Roger II, Norman king of Sicily. His accomplishments included a map of the world that was engraved in silver, a world map divided by climate zones and longitude lines, and several geography texts.

Primary Source
Description of change: excerpted
Rationale: to focus on the main idea

FOR THE RECORD

Abu Abdallah Mohammed Idrisi (1100–1165) was a famous Arab geographer from Morocco. His ancestors were among the Muslim princes and caliphs that ruled Spain after the Muslim conquest of the Iberian Peninsula in 7ll. Idrisi's interest in geography took him to many corners of western Europe, including Spain, Portugal, England, Italy, and the Atlantic coast of France. Below is a passage from Idrisi's greatest geographical work, which described what Arabs and Europeans knew about the world by the twelfth century. The text combined the findings of Greek and Arab geographers, eyewitness accounts, and Idrisi's own firsthand observations about foreign lands.

Eighteenth-century brass astrolabe

"**N**orway is a large island, but most of it is a desert. This island has two capes, of which one in the west is opposite the island of Denmark, a half-day's sail away, while the other is opposite the coast of Finmark.

It is said that there are three flourishing cities in Norway, two are opposite Finmark and the third opposite Denmark. These cities look all the same, they are not very populous, and rather poor, for it rains a great deal in this country and fog prevails a good deal of the time. The Norwegians, having sown their grain, harvest it while it is still green and take it to their dwellings to dry it by the fire, for there is but little sunshine. There are a great many trees in this country, with immense trunks, that are very heavy. They also state that there lives there a race of savages, whose heads are next to their shoulders so that they do not have any neck at all. They live in the midst of dark forests where they build their homes and live entirely on nuts and chestnuts."

Coastal **Norway**

Analyzing Primary Sources

1. Read about the physical geography and climate of Norway on pages 277–278 in your text. In what ways is Idrisi's description of Norway incorrect?
2. Study the map on page 259. Why might early geographers have assumed Norway was an island?

CHAPTER 23 Introducing Europe
Planning Guide

TEXTBOOK	TEACHING RESOURCES	OTHER AVAILABLE MATERIALS
Section		
1. Physical Geography Pages 262–263	Reteaching Worksheet 23.1	*REGIONS OF THE WORLD Map Transparencies:* Europe *Creative Strategies for Teaching World Geography* LESSON 10 The Drama of the European Union *Study Guide for Spanish Speakers* *Lecture Notes,* Chapter 23
2. Natural Resources Pages 263–264	Reteaching Worksheet 23.2 Graphic Organizer 23 Challenge/Enrichment Worksheet 23	*Study Guide for Spanish Speakers* *Lecture Notes,* Chapter 23
3. Historical Geography Pages 265–266	Reteaching Worksheet 23.3 Critical Thinking Worksheet 23	*Study Guide for Spanish Speakers* *Lecture Notes,* Chapter 23
4. Human Geography Pages 266–270	Reteaching Worksheet 23.4 Skills Worksheet 23	*Creative Strategies for Teaching World Geography* LESSON 11 Dream Vacation: Europe *Study Guide for Spanish Speakers* *Lecture Notes,* Chapter 23
Chapter Review Page 271	Chapter 23 Test Chapter 23 Portfolio Assessment See also *Alternative Assessment Forms*	*Test Generator,* Chapter 23

CHAPTER RESOURCES

Lecture Notes, Chapter 23
Teaching Resources
    Chapter 23 Graphic Organizer
    Chapter 23 Skills Worksheet
    Chapter 23 Critical Thinking Worksheet
    Chapter 23 Challenge/Enrichment Worksheet

REVIEW AND ASSESSMENT

Chapter 23 Check, page 271
Chapter 23 Test, *Teaching Resources*
Chapter 23 Portfolio Assessment, *Teaching Resources*
Test Generator, Chapter 23
Alternative Assessment Forms

ADDITIONAL RESOURCES

Books for Teachers

Bryson, Bill. *Neither Here Nor There: Travels in Europe.* New York: William Morrow & Co., 1992.

Hoffman, George W., et al. *Europe in the 1990s: A Geographic Analysis.* New York: John Wiley & Sons, 1989.

Jordan, Terry G. *The European Culture Area: A Systematic Geography.* New York: Harper & Row, 1973.

Laver, Michael, and Norman Schofield. *Multiparty Government: The Politics of Coalition in Europe.* New York: Oxford University Press, 1990.

Books for Students

Attenborough, David. *The First Eden: The Mediterranean World and Man.* New York: Little, Brown & Co., 1987.

Cardini, Franco. *Europe 1492: Portrait of a Continent Five Hundred Years Ago.* New York: Facts on File, 1989.

The World Today *Videodisc Program*

A variety of still images, short videos, and activities are available for you to use as you teach this chapter. *Lecture Notes* include bar codes for specific images and videos.

Multimedia Resources

Europe. (16mm film/VHS, 26 minutes) Physical Geography of the Continents series. National Geographic Society, Educational Services, Washington, DC 20036.

Europe Alive GUIDisc. (Multimedia CD-ROM, interactive tour) MediAlive, 766 San Aleso Avenue, Sunnyvale, CA 94086.

Mapping Europe. (Lesson plans with reproducible pages) Available through Social Studies School Service, 10200 Jefferson Blvd., P.O. Box 802, Culver City, CA 90232–0802.

MEETING INDIVIDUAL NEEDS

Limited English Proficiency (LEP)

Help students create a graphic organizer for each section of the chapter. Use the Focus questions at the opening of the section as the basis of the organizer. On the board, have student volunteers help complete the organizers. Students might copy the graphic organizers and use them to study the chapter.

Learners Having Difficulty

Review with students the meanings of the terms in the Geography Dictionary. Ask volunteers to explain each term's significance in Europe, as you write a context sentence that explains the meaning of the term. Next have students use the sentences to write brief summaries of the chapter.

Auditory Learners

If possible, obtain audiotapes of languages spoken in Europe. Possible language selections include Spanish, Portuguese, Italian, French, German, Czech, Dutch, Polish, Swedish, Norwegian, and Finnish. Play as many different tapes as possible so that students can appreciate the variety of languages spoken in Europe. Discuss the need for people living in the region to be multilingual.

Visual Learners

Have students work in small groups. Provide each group with pictures of scenes from Europe. If possible, also provide information about each picture. Then have members of each group write captions identifying the content of each picture and describing the picture in terms of the Five Themes of Geography. Students might refer to the pictures and captions shown in Chapter 1 for examples.

Tactile/Kinesthetic Learners

Have students examine a world map and focus on the fact that Europe and North America are in similar latitudes. Have students use a climate map of the world to compare the climate where they live with the climate in the parts of Europe with a similar latitude.

Gifted Learners

Have interested students research energy sources used in Europe, such as ocean tidal power, solar energy, and nuclear power. Then ask students to hold a panel discussion about whether or not these energy sources could be developed further in the United States to help decrease our dependence on foreign oil.

MULTICULTURAL FOCUS

Write the following names on the board:

Johann Christian Bach
Béla Bartók
Ludwig van Beethoven
Claude Debussy
George Frideric Handel
Joseph Haydn
Wolfgang Amadeus Mozart
Giacomo Puccini
Gioacchino Rossini
Franz Schubert
Richard Wagner

Ask students if they know what these individuals have in common. (*They were Europeans who composed classical music for operas.*) Point out that these names represent just a small number of the composers of classical music who are from Europe. Classical music is written mainly for operas, ballets, religious ceremonies, and concerts. Explain to students that opera—a play in which the players sing, instead of saying, their lines—originated in Italy in the late 1500s. By the late 1600s, other European countries had begun to produce operas. Although operas are currently performed throughout the world, most of the best-known opera companies perform in European opera houses.

If possible, obtain a videocassette of an opera such as *Carmen, The Marriage of Figaro,* or *The Magic Flute* and play portions for students. Ask students to describe the themes of the opera and the mood and style of the music. Have students compare and contrast opera themes, mood, and style with the themes, mood, and style of American musicals or other types of popular music in the United States.

CHAPTER 23
INTRODUCING EUROPE

Section 1: Physical Geography
Section 2: Natural Resources
Section 3: Historical Geography
Section 4: Human Geography

OVERVIEW

Chapter 23 provides students with an overview of Europe.

Section 1 describes Europe's major landform regions and climate types.

Section 2 discusses the region's forest, soil, mineral, and water resources. It focuses on the importance of Europe's rivers to its economic development.

Section 3 highlights European history, noting the effects of two world wars as well as recent events.

Section 4 analyzes European population patterns and focuses on the cultural diversity of the region. Students will also learn of the economic, social, and environmental challenges Europe faces today.

CREATIVE STRATEGIES FOR TEACHING WORLD GEOGRAPHY

LESSON 10 The Drama of the European Union (Geography Students Are the Worst Hams)

CHAPTER 23

Introducing Europe

Fifteenth-century illuminated French manuscript

Geographers define Europe as the region stretching from the Atlantic Ocean to the Ural Mountains and from the Arctic Ocean to the Mediterranean Sea. Europe is actually the western end of the world's largest landmass, known as **Eurasia**. Eurasia is made up of the continents of Europe and Asia. According to the theory of plate tectonics, about 300 million years ago, the two continents collided and formed the Ural Mountains at their joining.

The first geography textbooks referred to Europe as the "peninsula of peninsulas." This observation is still true. While Europe contains many peninsulas of its own, it also can be viewed as a giant peninsula of Eurasia. Europe consists of more than 35 countries, some of which are among the most economically developed in the world. Several are tiny **microstates**, or very small countries.

In this textbook, the geography of Europe is divided into two parts. The western part is discussed in the five chapters that follow. The eastern part is discussed in Unit 6, which covers Russia and the other countries that once were part of the Soviet Union.

GEOGRAPHY DICTIONARY

Eurasia
microstate
natural boundary
navigable
icebreaker
imperialism
alliance
balance of power
reunification
famine
multilingual
economic association
European Union (EU)

Clifden Castle, **County Galway, Ireland**

Using the Geography Dictionary

Refer students to the list of terms in the Geography Dictionary on page 258. Have them look up each term in the Glossary. Then ask students to create a classification system by which the terms might be grouped (for example, physical, political, economic, cultural).

MAPPING THE REGION

Using the Physical-Political Map

Remind students of the reference to Europe in the Chapter 23 introduction as the "peninsula of peninsulas." On a wall map, trace the outline of Europe so that students can observe how the continent is a peninsula. Then ask them to study the map on this page and identify five countries or pairs of countries that are peninsulas. (*Spain and Portugal, Italy,* *Denmark, Norway and Sweden, Greece*) Have students locate the four major landform regions of Europe: Northwest Highlands, Northern European Plain, the Central Uplands, and the Alps.

MAP TRANSPARENCIES

Overhead transparencies from the maps on pages 259–261 are available in the *REGIONS OF THE WORLD Map Transparencies.*

MAPPING THE REGION

EUROPE

PHYSICAL–POLITICAL

LEGEND

ELEVATION

FEET		METERS
13,120		4,000
6,560		2,000
1,640		500
656		200
(Sea level) 0		0 (Sea level)
Below sea level		Below sea level

Ice caps

⊛ National capitals

• Other cities

SCALE
0   250   500 Miles
0   250   500 Kilometers
Projection: Azimuthal Equal Area

MAPPING THE REGION

Using the Climate Map

Direct students' attention to the climate map on this page. Ask one student to call out the names of the individual countries that make up Europe, and have other students name the main climate types in each country. Then have students compare this map to the population map below it. Ask students which climates predominate in the areas of lowest population density. *(ice cap, tundra, subarctic)*

MAPPING THE REGION

Using the Population Map

As students study the population map on this page, point out that many European cities are located on or near rivers. Ask students to identify the countries with the greatest population density. *(United Kingdom, Netherlands, Belgium, Germany)* Point out that the sparsely populated island of Greenland is part of Denmark.

Continues on page 261

CLIMATE

Answers

1. Ireland, United Kingdom, Denmark, Czech Republic, Netherlands, Belgium, Luxembourg
2. Alps
3. Herding, mining, timber (logging), hydroelectric power generation, manufacturing and trade

POPULATION

Answers

1. Manchester, Birmingham, London, Madrid
2. Germany
3. Tundra, subarctic, humid continental

MAPPING THE REGION

EUROPE

CLIMATE

LEGEND

- Steppe
- Mediterranean
- Humid subtropical
- Marine west coast
- Humid continental
- Subarctic
- Tundra
- Ice cap
- Highland

POPULATION

LEGEND

POPULATION DENSITY

Persons per sq. mile	Persons per sq. km
520	200
260	100
130	50
25	10
3	1
0	0

- Metropolitan areas with more than 2 million inhabitants

Have them identify the countries with at least three metropolitan areas with more than 2 million people. *(United Kingdom, Germany, Italy)*

MAPPING THE REGION

Using the Economy Map

Help students compare the energy resources in Europe. Have them look closely at the economy map on this page. Ask: Where are oil and natural-gas deposits found in Europe? *(North Sea, Romania)* Where is hydroelectric power used? *(United Kingdom, Ireland, Norway, Sweden, France, Switzerland, Austria, Greece)* Nuclear power? *(United Kingdom, France, Spain, Switzerland, Germany, Hungary)* Geothermal power? *(Iceland)*

MAPPING THE REGION

EUROPE

ECONOMY

LEGEND

- Herding
- Livestock raising
- Commercial farming
- Dairying
- Commercial fishing
- Forests
- Manufacturing and trade
- Limited economic activity

- Coal
- Natural gas
- Oil
- Nuclear power
- Hydroelectric power
- Geothermal power
- Uranium
- Other minerals
- Timber

CLIMATE

◆ A marine-west-coast climate dominates most of Northern and West Central Europe.

◆ Southern Europe, including most of Italy and all of Greece, has a Mediterranean climate.

◆ Away from the oceanic influence of the mild North Atlantic Drift, much of Eastern Europe has a humid-continental climate.

Compare this map to the physical–political map on page 259.

❓ Which countries have an entirely marine-west-coast climate?

❓ Name the major mountain ranges where highland climates are shown.

Compare this map to the economy map above.

❓ What economic activities are common in the subarctic and tundra climates of northern Scandinavia?

POPULATION

◆ One of Europe's most crowded areas is in a belt stretching through Belgium, the Netherlands, and central Germany.

◆ Except along coastal areas, the northern Scandinavian countries of Norway, Sweden, Finland, and Iceland are sparsely populated.

◆ Italy has many of the most crowded areas in Southern Europe.

❓ Which metropolitan areas west of the prime meridian (0°) have populations of more than 2 million?

Compare this map to the physical–political map on page 259.

❓ Which country has the most metropolitan areas with populations of more than 2 million?

Compare this map to the climate map on page 260.

❓ Which climate regions of Scandinavia are most sparsely populated?

ECONOMY

◆ Europe's irregular coastline has created many natural harbors that have become important manufacturing and trade centers.

◆ The rivers of Europe are of great importance for trade, water resources, and hydroelectricity.

◆ Petroleum and natural gas have been found beneath the waters of the North Sea claimed by Norway and the United Kingdom. Most other European countries must import all of their oil and gas.

Compare this map to the physical–political map on page 259.

❓ Name the manufacturing and trade centers shown in Spain and Portugal.

❓ What area of western Germany is an important manufacturing region?

❓ Which country of Eastern Europe is a producer of petroleum?

SECTION 1
OBJECTIVES

The objectives reflect the Focus
questions on this page and are
assessed in the review on page 263.

1. Identify Europe's landform
regions.

2. Describe Europe's climate types.

TEACHING SECTION 1
Motivator

Organize the class into groups of
four. Tell students that Europe has
four main landform regions. Have
each group study the physical-
political map on page 259 and determine what these four regions are. Then
have the groups check their answers
with information in Section 1.

Building a Vocabulary

Write the term **natural boundary** on
the board. Indicate on a wall map
how the Pyrenees mountains form a
natural boundary between Spain and
France. Then ask students to recall
the natural boundary between the
United States and Mexico. *(the Rio
Grande River)* Have students identify
any natural boundaries that may be
found between their state and neighboring states, between nearby counties, or between local communities.

THEME
Interaction

*The Alps have been
described as "Europe's
water towers." Snow and
ice from Alpine slopes
feed Europe's four major
rivers. Excessive eco-
nomic development in
the form of ski runs and
resorts, mountain roads
and tunnels, and hydro-
electric installations
threatens to interfere with
the natural water flow.*

PHYSICAL GEOGRAPHY

FOCUS

● *What are Europe's four major landform regions?*

● *What climate types are found in Europe?*

Landforms Europe west of the former Soviet
Union can be divided into four major landform
regions: the Northwest Highlands, the Northern
European Plain, the Central Uplands, and the
Alpine mountain system. (See the map on page
259.) Landforms in Europe vary over very small
distances.

The Northwest Highlands is an ancient,
eroded region of rugged hills and low mountains.
The region includes the hills of Ireland and
England, the Scottish Highlands, Brittany in northwestern France, most of the Iberian Peninsula,
and the Kjølen (CHUHL-uhn) Mountains in
Norway and Sweden. Much of the region was
affected by the continental ice sheets during the
last ice age. When the ice melted, it left thousands
of lakes, poor river drainage, and thin soils.

The largest landform region in area is the
Northern European Plain. This broad coastal plain
extends from the Atlantic coast of France into

Russia and borders the North Atlantic Ocean, the
North Sea, and the Baltic Sea. Most of the region
is less than 500 feet (152 m) above sea level.

The third landform region is the Central
Uplands. This ancient, hilly area of Europe con-
sists of the Massif Central (ma-SEEF sehn-TRAHL)
of France and the Jura Mountains on the French-
Swiss border. It also includes the Black Forest and
the Bavarian Plateau of Germany, the Bohemian
Highlands of Germany and the Czech (CHEHK)
Republic, and the Ardennes (ahr-DEHN), in the
Benelux countries.

The Alps are western and central Europe's
highest mountain range. They stretch from
the Mediterranean coast of France through
Switzerland, Austria, and northern Italy to the
Balkan Peninsula. Many peaks in the Alps reach
more than 14,000 feet (4,268 m). The highest peak,
France's Mont Blanc (mohn BLAHN), reaches to
15,771 feet (4,807 m). Because of their high elevations, the Alps have large glaciers and frequent
avalanches due to heavy winter snowfall.

Other ranges in the Alpine system include
the Carpathian Mountains in Eastern Europe and
the Apennines, which run the length of the Italian
peninsula. The Pyrenees (PIR-uh-neez) form a
natural boundary between Spain and France.
When physical features such as mountain ranges
and rivers determine political borders, they are
called natural boundaries.

REGION Climate Compared to world regions
of similar latitude, much of Europe has mild conditions throughout the year. The moderating influence of the North Atlantic Ocean helps prevent
extreme temperatures. (See the map on page
260.) The high latitudes mean long summer days
and long winter nights.

Europe has three major climate types: marine
west coast, humid continental, and Mediterranean.
The marine-west-coast climate is found throughout
most of Northern and West Central Europe. The
prevailing westerly winds blowing over the warm
ocean current called the North Atlantic Drift bring
mild temperatures and rain. Frequent Atlantic
storms bring clouds and even more rain to the
region. Rainfall averages between 20 and 80 inches
(51 and 203 cm) a year. The windward western
slopes of the coastal mountains of Scotland and
Norway receive the region's highest rainfall totals
due to the orographic effect. Occasional snow and
frosts occur in winter.

The Rhine River cuts a path through
western Europe toward the North
Sea. Ships can sail the Rhine all the
way from Switzerland to the sea.

262

Guided Practice

Review the three major climate types of Europe with the class. Have students create illustrations that depict economic activity within each climate region. For example, they might choose to depict farms in the marine-west-coast climate region, or resorts in the Mediterranean climate area. Discuss the effects of both climate and landforms on ways of life and economic activity.

Closure

Call on students to identify the main landform region and climate of each European country.

ASSESSMENT

Assign the Section 1 Review questions on this page.

Reteaching

Assign Section 1 of Reteaching Worksheet 23.

SECTION 2
OBJECTIVES

The objectives reflect the Focus questions on this page and are assessed in the review on page 264.

1. Explain the importance of Europe's location near various bodies of water.

2. List Europe's major resources.

To the east and inland of the marine-west-coast climate is a humid-continental climate region. This region is located away from the direct influence of the mild Atlantic currents. It stretches across central Sweden and Finland and southward across most of Eastern Europe as far south as northern Bulgaria. The humid-continental climate has four distinct seasons, including a cold, snowy winter and a humid summer. Winters are severe from Finland to Poland. Periodic summer droughts affect Hungary and Romania.

High mountains, particularly the Alps, separate the marine-west-coast and humid-continental climates from Europe's third major climate region, the Mediterranean climate. This climate region includes most of Portugal, Spain, Italy, Albania, Macedonia, and Greece, and parts of France, Croatia, and Bulgaria. (See the map on page 260.) The Mediterranean region receives between 10 and 30 inches (25 and 76 cm) of rainfall a year. Most of the rainfall comes during the mild winter when occasional North Atlantic storms pushed by the westerly winds bring rain into the Mediterranean basin. A long sunny, dry summer is typical.

Pockets of minor climate types cover parts of Southern and Northern Europe. A small area of humid-subtropical climate is located south and southeast of the Alps, mainly in the Po Valley of northern Italy and extending into the Balkan region between the Adriatic and Black seas. In some areas of Spain, where high mountains block moist ocean air from reaching farther inland, a dry steppe climate prevails. A subarctic climate, meanwhile, stretches across northern Norway, Sweden, and Finland. The very northern portions of these countries and the island nation of Iceland have a tundra climate.

NATURAL RESOURCES

FOCUS

● *How are various bodies of water important to Europe?*

● *What are Europe's major resources?*

MOVEMENT **Water** Europe's long, irregular coastline is indented with hundreds of excellent ice-free natural harbors. Because these harbors generally are located near the mouths of **navigable** rivers, Europe is ideally situated for trade by sea. A navigable river is one that is deep and wide enough for ships.

Of the seas bordering Europe, the Mediterranean is the largest. Although it suffers from pollution, the Mediterranean supports some fishing. The Mediterranean is connected to the Black Sea by the strait of Bosporus (BAHS-puh-ruhs) in Turkey. This strait gives Bulgaria and Romania access to world sea routes. Geographers consider the Bosporus to be a boundary between Europe and Asia.

Other major European seas are the North Sea and the Baltic Sea. The North Sea is vital for trade and fishing, and beneath its waters lie Europe's major oil reserves. The Baltic Sea ices over during the winter months, however. Ships

The docks of **Hamburg, Germany,** are crowded with products ready for export to world markets.

SECTION REVIEW

1. To which landform regions does France belong?
2. What are Europe's three major climate types? Name one characteristic of each climate type.
3. **Critical Thinking**   **Determine** which areas of Europe are likely to be agriculturally productive, based on landforms and climates. Explain your reasoning.

Section 1 Review
ANSWERS

1. All of Europe's four major landform regions

2. Marine west coast, humid continental, and Mediterranean; characteristics might include what various seasons are like.

3. Possible answer: the Northwest Highlands area has thin soils and poor drainage, and the Alpine mountain system is mountainous, so neither is suited to agriculture. Any of the three major climates can support agriculture, so areas where they overlap with the Northern European Plain and the Central Uplands are likely to be agriculturally productive.

263

Motivator

Refer students to the physical–political map of Europe on page 259. Then ask students questions such as the following: What bodies of water create the coastlines of Europe? *(Baltic Sea, North Sea, Norwegian Sea, Atlantic Ocean, Mediterranean Sea, Black Sea)* What countries or microstates are landlocked? *(Andorra, San Marino, Liechtenstein, Vatican City, Austria, Switzerland, Czech Republic,* *Slovakia, Hungary, and others)* On what river is London located? *(Thames)* Which countries are located near bodies of water that probably freeze in winter months? *(Sweden, Finland, and the Baltic countries)*

Building a Vocabulary

Write the term **navigable** on the board. Note that the suffix *-able* is used to form adjectives meaning "capable of being." Thus, *navigate + -able* = "capable of being navigated."

Guided Practice

List Europe's natural resources on the board. Then have students brainstorm occupations related to these resources.

Independent Practice

Provide students with Outline Map 14. Have them label bodies of water surrounding Europe and the region's major rivers and lakes.

COOPERATIVE LEARNING

Have students work in pairs on a project exploring Europe's important water routes. Give each pair a list of European cities divided into departure points and destinations (for example, Paris to Strasbourg). Have them plot a commercial water route between the cities.

Section 2 Review
ANSWERS

1. Seas: Mediterranean, Black, North, Baltic; rivers: Rhine, Danube; they provide access to trade routes, hydroelectricity, and other water resources.

2. Iron, coal, oil, natural gas, nuclear power, hydroelectricity, ocean tidal power, solar energy; because Europe's own oil and gas deposits do not satisfy its industrial demands

3. Deforestation in both regions has occurred because of cutting of trees for timber or clearing of land for farming. Most deforestation in Europe occurred centuries ago; deforestation in Middle and South America is recent. Many trees in Europe also have died from pollution.

called **icebreakers** are needed to keep sea lanes open to Sweden, Finland, and the Baltic countries of Estonia, Latvia, and Lithuania. The warm North Atlantic Drift keeps Europe's other important seaways free of ice during winter.

The rivers of Europe are of great importance for trade, water resources, and hydroelectricity. Most are connected by a series of canals. Even cities far from the coast have access to the sea through this extensive river and canal system. The Rhine and Danube are Europe's most developed rivers. Unfortunately, massive amounts of pollution enter the sea via these and other rivers. Cleaning up and controlling pollution in Europe's rivers is a serious environmental challenge.

INTERACTION **Forests and Soil**
Most of Europe's original forests were cut for timber or cleared for farming centuries ago. In recent years, air pollution and acid rain have killed many of the remaining trees. Only in Sweden and Finland do large areas of timber-producing forest still exist. Clearing and overgrazing have destroyed nearly all of the Mediterranean's original oak forests, leaving only a scrub-plant community covering the hillsides. Europe must now import wood to satisfy its timber needs. Most European nations, however, have begun reforestation programs.

Europeans have made good agricultural use of their soil, and farmers grow a wide variety of crops. Intensive farming techniques and modern technology have made Europe's crop yields among the highest in the world. Much of Eastern Europe, however, lags in agricultural production, because its farming technology is not as modern as that in other European countries.

Minerals and Energy To meet its current industrial and energy needs, Europe relies heavily on imports. The United Kingdom, Germany, the Benelux countries, and the Czech Republic, however, all have deposits of iron and coal, which were important in the growth of these industrial countries. Sweden and France also have large iron deposits.

The oil reserves and natural-gas deposits beneath the North Sea and the natural-gas

Some scientists believe acid rain killed much of the dense tree growth that once surrounded this home in the Black Forest of **Germany**.

deposits in the Netherlands do not satisfy Europe's industrial demands. Imported oil and gas from Southwest Asia, Africa, and Russia supply more than 80 percent of the energy needs of most European countries.

Europe has other energy sources, however. Nuclear power fills some energy needs, particularly in France, Belgium, Great Britain, and Hungary. Hydroelectricity, meanwhile, is plentiful in the mountainous nations such as Norway and Switzerland. France has been successful in producing ocean tidal power and in developing and using solar energy.

2 SECTION REVIEW

1. Name three seas and two rivers found in Europe. Why are seas and rivers important to Europe?
2. What mineral and energy resources does Europe have? Why must Europe import oil and gas?
3. **Critical Thinking** **Compare and contrast** deforestation in Europe with the deforestation that is taking place in Middle and South America.

Closure

Have students name resources exported and imported by European countries and by what possible routes.

ASSESSMENT

Assign the Section 2 Review questions on page 264.

Reteaching

Assign Section 2 of Reteaching Worksheet 23.

SECTION 3
OBJECTIVES

The objectives reflect the Focus questions on this page and are assessed in the review on page 266.

1. Summarize Europe's history through World War II.

2. Discuss changes in Europe from World War II to the present.

TEACHING SECTION 3
Motivator

Show students a map of Europe from before 1990. Have them compare this map with the map on page 259 and note differences in the names of countries and political boundaries. Ask students if they know reasons for these differences. Emphasize that political boundaries in Europe have changed many times over the centuries.

HISTORICAL GEOGRAPHY

FOCUS

- *What were the major historical developments in Europe prior to World War II?*
- *How has Europe changed since World War II?*

Rise of Nations Throughout history, different people have fought for the control of Europe. From about 900 to 300 B.C., while the Greek civilization was flourishing in the Mediterranean area, warring tribes ruled the rest of Europe. Wars and conquests continued throughout the expansion of the Roman Empire during the second century B.C. The Roman Empire at its peak included northern Africa, southwestern Asia, and most of western and central Europe.

The fall of the Roman Empire marked the beginning of a period in Western European history, from about A.D. 500 to 1500, called the Middle Ages. During the first part of this period, various Germanic groups established a number of new kingdoms. The most important of these was the Frankish kingdom. It reached its height in about 800, under the reign of the emperor Charlemagne (SHAHR-luh-mayn).

During the latter part of the Middle Ages, Europeans increasingly ventured out onto the oceans, dominating world trade for centuries. From 1000 to 1500, the Mediterranean ports of Venice and Genoa were the main centers of trade.

By 1500, after centuries of bitter fighting among kingdoms, France, Spain, Portugal, and England had become nations. France rose as a strong power in Europe in the mid-1600s, and by the early 1800s, it controlled most of western Europe. Then in 1815, the British defeated Napoléon's French army at Waterloo, a town in Belgium. For nearly a century after Waterloo, Britain was Europe's leading political and economic power.

Imperialism During the sixteenth and seventeenth centuries, Spain and Portugal became the first European nations to establish large foreign colonies. By the second half of the nineteenth century, the British, French, Dutch, and Belgians also had established colonies in the Americas, Africa, Asia, and the Pacific. The policy of gaining control over territory outside a nation is called **imperialism**.

Britain's colonial empire became the largest in the world. Its trade connections, skilled population, and control of the seas made Britain the world's leading power during the nineteenth century. At its height, the British Empire included one-fourth of the world's population and one-fifth of the world's land area. It was true that "the sun never set on the British Empire" because, with the earth's rotation, some part of the empire was always in daylight.

The World Wars Germany, which did not become a unified nation until 1871, was late to enter the colonial race. But by the end of the nineteenth century, Germany, too, had obtained colonies in Africa and the Pacific. By then, Germany was one of the strongest military and industrial powers in Europe. Its military buildup frightened many other European nations.

In this atmosphere of fear and mistrust, the European nations formed military **alliances**. An alliance is an agreement between countries to support one another against enemies. Countries joined in an alliance are called allies.

These European alliances helped keep a **balance of power** in the region for several years. A balance of power exists when members of a group have equal levels of strength. This balance in Europe, however, was lost in the early 1900s. World War I broke out between the two main alliances of Germany and Austria-Hungary, and Great Britain, France, and Russia.

Germany lost World War I and faced harsh peace terms from Great Britain and its allies. Bitterness over the war combined with severe economic hardship led to World War II only 21 years later. Germany and its ally Italy conquered much of Europe before being defeated by the allied forces of Great Britain, the United States, the Soviet Union, and several other nations.

Post–World War II After the war, Germany was divided into two nations, East Germany and West Germany. East Germany and the rest of Eastern Europe came under Communist rule and the control of the Soviet Union. The end of World War II marked the beginning of the period known as the Cold War.

The military alliance of the North Atlantic Treaty Organization (NATO) formed in 1949 to

GLOBAL PERSPECTIVES

Have students create a map showing empires of Europe during the age of colonization and imperialism. Provide students with Outline Map 1. Have them show areas of the world controlled by European nations during the second half of the nineteenth century. (Students might use encyclopedias to find this information.) Have them color code colonial powers and their possessions. When the maps are completed, lead a discussion about whether imperialism exists today.

Building a Vocabulary

Discuss the concept of a "word family," a group of words built on the same base. Explain that knowing the meaning of the base word in the family helps in understanding the meaning of related words. Demonstrate with the word family *ally,* **alliance,** *allied.* Then have students form a word family based on *unify.* Explain the meaning of each word. *(unify, unified, unification,* **reunification***)*

Sequencing European History

Select 10 events from European history, and write each one on a slip of paper (for example, British defeat Napoléon's French army at Waterloo; East and West Germany become reunified). Make additional sequence packs so that several pairs of students can work on the task simultaneously. Shuffle the slips of paper in each pack, and give one pack to each pair of students. Have the pairs arrange the slips of paper in chronological order.

Closure

Ask students to bring in and discuss newspaper articles that focus on recent political developments in Europe.

ASSESSMENT

Assign the Section 3 Review questions on this page.

Reteaching

Assign Section 3 of Reteaching Worksheet 23.

Section 3 Review
ANSWERS

1. France, from the mid-1600s to 1815, and Britain for the rest of the nineteenth century

2. To protect the countries of western Europe from possible invasion by the Soviet Union and its Eastern European allies; because the Cold War has ended

3. Many students will agree with the statement. From 900 to 300 B.C., warring tribes ruled most of Europe. Wars and conquests continued during the Roman Empire and its decline. Various Germanic groups then formed new kingdoms, which later fought bitterly. Then France dominated but was defeated by the British in 1815. Two world wars occurred in the first half of the twentieth century, and the Cold War lasted for much of the second half of that century.

War has helped shape the history and landscape of Europe. The cross-covered beaches of **Normandy, France,** are evidence of the sacrifices and the costs of these wars.

provide for the common defense of western Europe. Its goal was to protect its members from possible invasion by the Soviet Union and its Eastern European allies, who in 1955 formed their own alliance called the Warsaw Pact. Members of NATO include 13 European nations plus Turkey, Canada, and the United States.

During the early 1990s, Europe again experienced historic changes. As the Soviet Union began to break up, the Cold War thawed and East and West Germany underwent **reunification**. Reunification is the process of reuniting, or rejoining, into one unit. Czechoslovakia (chehk-uh-sloh-VAHK-ee-uh) peacefully split into two nations, while the three major ethnic groups in the new nations of former Yugoslavia engaged in a terrible civil war. With the end of the Cold War, the Warsaw Pact dissolved. The role of NATO likely will change as well.

SECTION REVIEW

1. What countries dominated Europe from the 1600s through the 1800s?
2. Why was NATO formed? Why will its role likely change?
3. **Critical Thinking** **Agree or disagree** with the following statement: "The history of Europe is a history of wars and conflict."

HUMAN GEOGRAPHY

FOCUS

- *What are the demographic and cultural characteristics of Europe?*
- *How are the economies of Europe's country's linked?*
- *What challenges does Europe face?*

Population Changes The population of Europe grew rapidly in the years between 1000 and 1300. Between 1300 and 1450, however, population growth in Europe experienced major declines due to diseases, **famines,** and wars. A famine is an extreme shortage of food.

Then during the seventeenth and eighteenth centuries, agricultural advances led to another increase in population. As the Industrial Revolution took hold, trade increased and living standards rose. Workers took factory jobs and moved from farms and villages to the cities. Today, Europe's main centers of population reflect the urbanization that took place during the Industrial Revolution. London, Paris, Milan, Amsterdam, and Berlin all grew dramatically during the 1800s.

Over the past several centuries, Europe's surge in population led in part to increasing emigration. Millions of Europeans have immigrated to the United States, Canada, and Australia. Others have moved to overseas colonies or settled in South American nations such as Brazil, Argentina, and Chile. Many of these emigrants were seeking new opportunities or were escaping religious persecution, wars, famine, and poverty.

Today, mainly due to economic progress, much of western Europe is attracting immigrants to its shores. Foreign laborers, called guest workers, have moved to western European countries from poorer Eastern and Southern European countries and from Turkey, North Africa, and Asian nations such as India and Pakistan. In some urban areas, overcrowding has led to slum conditions and increasing ethnic tension.

Even with immigration, however, European population growth rates are the lowest in the world. In fact, some European countries, such as Hungary and Germany, are experiencing a population decline. In addition, life expectancy rates

OBJECTIVES

The objectives reflect the Focus questions on page 266 and are assessed in the review on page 270.

1. Examine Europe's demographic and cultural characteristics.
2. Analyze Europe's economic geography.
3. Discuss challenges facing Europe.

TEACHING SECTION 4
Motivator

Point out that one reason the United States and Canada are concerned about events in Europe is the historical, cultural, political, and economic ties between the two regions. Ask students for examples in each category.

Building a Vocabulary

Review how adding a number prefix changes the meaning of a word. Demonstrate with *lingual*. Attach the prefixes *mono-*, *bi-*, and *multi-*, and note how the meaning changes (monolingual—speaking one language, bilingual—speaking two languages, **multilingual**—speaking many languages). Ask students for other examples using number prefixes. (*unicycle, bicycle, tricycle; unilateral, bilateral, multilateral*)

---

that are among the highest in the world coupled with low birthrates means that the region is becoming a population of older people.

Language Most of the languages spoken in Europe are of the Indo-European family. This language family includes the Germanic, Celtic, Romance (Latin or Italic), Slavic, and Greek language groups among others. (See the chart below.)

Although English is the most widely spoken language in Europe, German and French are also widespread. Ninety percent of all Europeans between the ages of 15 and 24 speak a second language. Some countries, including Luxembourg, Switzerland, the Netherlands, and Belgium, are considered **multilingual**. That is, many people speak three or more languages.

Major Languages of Europe

Indo-European ▶	Germanic ▶	English German Dutch-Flemish Swedish Norwegian Danish Icelandic
	Romance ▶ (Latin or Italic)	Portuguese Spanish French Italian Romanian
	Celtic ▶	Irish Gaelic Scottish Gaelic
	Baltic ▶	Lithuanian Latvian
	Slavic ▶	Russian Ukrainian Belorussian Polish Czech Slovak Serbo-Croatian Slovene Macedonian Bulgarian
	Greek ▶	
	Albanian ▶	
Uralic ▶	Finno-Ugric ▶	Estonian Finnish Hungarian

Languages from two major language families are spoken across much of Europe. Languages that are related are grouped together within these language families. Find the language group to which English belongs. What other languages are part of that language group?

Religion Many different religions are practiced in Europe. Roman Catholicism is the dominant religion in the Southern European countries such as Spain, Italy, and Portugal. Catholics also are in the majority in Poland, France, Austria, Ireland, and Belgium. Most of Northern and Central Europe is mainly Protestant, including the United Kingdom, the Scandinavian countries, and Germany. Small numbers of Jews live in many parts of western Europe. Significant numbers of Muslims live in southeastern Europe and among the new immigrants and guest workers from North Africa, Turkey, and Southwest Asia.

REGION **Economic Cooperation** World War II put severe stress on the economies of the European nations. Moreover, after the war, several European colonies in Asia and Africa gained their independence. The colonies had been major sources of inexpensive raw materials that supported industrial growth in some European countries. To strengthen their economies, the nations of western Europe began forming **economic associations**. An economic association is an organization formed to break down trade barriers among member nations.

The most important of these cooperative economic associations has been the **European Union**, commonly known as the **EU**. The EU, once called the European Community, was formed in 1957 by Belgium, the Netherlands, Luxembourg, France, West Germany, and Italy. Since then, nine other countries have joined: the United Kingdom, Ireland, Denmark, Greece, Spain, Portugal, Austria, Finland, and Sweden. Most other European nations, as well as Turkey, are seeking membership.

Combining the natural resources and industries of its members, the EU has become the world's largest exporter. The present 15-nation organization has a population far exceeding that of the United States and Japan, its two main trade rivals. The EU seeks a common European currency, and its ultimate goal is a unified region with a free flow of people and goods. In 1993, several European nations passed the Maastricht Treaty, which outlines steps toward further economic and political integration in the EU.

Switzerland, Norway, Liechtenstein, and Iceland belong to the European Free Trade Association (EFTA). The EFTA holds trade agreements with the EU.

Text continues on page 270.

Geography and Language

Close to 50 million Europeans speak a minority language such as Sorbish, Frisian, or Basque. The European Union supports several organizations that work to protect these languages.

Geographical Sidelight

Although an agreement to establish one European currency was proposed in 1991, a single currency may not emerge before the end of the century. A single currency would allow residents of an EU country to travel to the other 14 nations of the EU without paying complicated and expensive exchange fees. At present, European travelers must, for example, still exchange French francs into Belgian francs or British pounds into German marks and pay a fee for every transaction.

Chart Caption Answer

German, Dutch-Flemish, Swedish, Norwegian, Danish, Icelandic

Global Connections

Linking Tourism and Economics

Organize the class into two groups. Have the members of one group research the tourism industry in the United States. Have the members of the other group research the tourism industry in their state. Instruct students to investigate the following: (a) approximate number of visitors per year (U.S. group will use numbers for foreign tourists; state groups will include both foreign and out-of-state tourists); (b) dollars generated by tourism; and (c) dollars spent promoting tourism. Each group should present its findings in the form of a bar or line graph or other visual.

CREATIVE STRATEGIES FOR TEACHING WORLD GEOGRAPHY

LESSON 11 Dream Vacation: Europe (This lesson is appropriate for any European country.)

Geographical Sidelight

The United States is a popular destination for tourists from around the world. Nearly 46 million foreign tourists visited the United States in 1993, and each tourist spent an average of $1,624 in this country. Altogether, foreign tourists spent more than $74 billion in the United States in 1993. Canadian and Mexican tourists were the most numerous in 1993, followed by tourists from Japan, the United Kingdom, and Germany.

GLOBAL CONNECTIONS

Tourism

Tourism is growing so rapidly that it is now the world's largest industry. In 1993, 500 million people traveled the world. By 2000, the number will have exploded to 650 million.

Responsible tourism requires a balance between economic concerns and cultural and environmental concerns.

With an estimated 8 million jobs depending on tourism, the European Union (EU) has pushed specific plans to attract visitors to member countries. Nearly 300 million tourists visited Europe in 1993, spending more than $160 billion. Despite the importance of tourism to the region's economy, some people question whether European nations are wise to encourage more of it. For years, France's Lascaux caves have been closed to tourists after the discovery that such traffic endangered the prehistoric cave paintings. Restrictions may also be placed on entrance to Paris's Cathedral of Notre Dame. More than 11 million people shuffled through the cathedral in 1990.

Throngs of tourists are flocking to the rest of Europe and the globe as well. More than 19 million visitors poured into Britain in 1993. Venice, Italy, has played host to as many as 100,000 tourists on peak summer days. In East Asia, meanwhile, tourism in Thailand, Malaysia, Singapore, and Japan has made the region the world leader in growth in international tourism.

The Tourist Industry

Despite tourism-related problems such as pollution and overcrowding, most nations continue to encourage tourism. In some countries, tourism brings in billions of dollars. Italy, for example, took in more than $21 billion in 1992.

Money spent by tourists creates jobs. Hotels and restaurants employ a variety of workers, including bellhops, waitpersons, and cooks. Tourism also supports other industries. The building of a hotel provides jobs for construction workers, plumbers, electricians, and the makers of brick, glass, and steel. The opening of a restaurant creates new customers for butchers, dairy workers, farmers, and packagers of fresh and frozen foods. For some developing nations, tourism may mean the difference between poverty and a standard of living that provides citizens with basic education and health care.

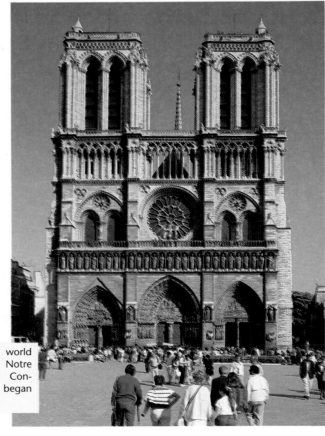

Tourists from around the world flock to the Cathedral of Notre Dame in **Paris, France.** Construction of the cathedral began in A.D. 1163.

Sometimes, however, nations reap few rewards from tourism. When foreign developers build hotels and restaurants that serve tourists, much of the money from tourism flows out of the country. Tourism also can disrupt cultures. For example, hula skirts and shell necklaces are alien to a number of Polynesian cultures. Yet because tourists expect to see these items throughout the Pacific, many Polynesians have adopted the skirts and jewelry to satisfy visitors. Although some areas do not go so far as to adopt new customs, they often mass-produce souvenirs such as machine-made baskets that cheapen the fine handicrafts their cultures have developed over centuries.

Unrestrained tourism also has left its mark on much of the world's natural environment. Desperate for photographs, some tourists disobey the rules of African wildlife preserves, persuading guides to drive as close as possible to rare and endangered animals. Tourism has been blamed for spoiling the French Riviera, a region along the Mediterranean coastline. Overconstruction has led to overcrowding, and the dumping of wastes has polluted the Mediterranean Sea. Now, the Alps and the Himalayas also are threatened by tourism.

Responsible Tourism

Tourists as well as host nations gradually are learning important lessons. Both realize that a balance must be reached between economic concerns and a responsibility to a region's culture and environment. Nations know that tourists will eventually turn away from places where indigenous cultures have disappeared and natural environments have been destroyed. The French Riviera, for example, experienced a 30 percent decline in tourism in recent years.

Many tourists have become more sensitive travelers. Some are signing up for "ecotours" that promote tourism to wilderness areas and wildlife reserves while encouraging conservation of the environment. On most ecotours, tourists must be willing to do without the conveniences of fancy hotels and luxury rental cars.

Responsible tourists approach other nations and cultures with respect for languages, customs, and ways of life different from their own. Responsible host nations recognize that, in the long run, the economic rewards of tourism will multiply when cultures are preserved and natural environments are left unspoiled.

Tourists look over crafts at a "shell market" on a beach in the **Yasawa Islands**, part of the South Pacific country of **Fiji**. Natural beauty and mild climates draw many tourists to South Pacific islands.

This climber nears the top of the world on one of the mighty peaks of the Himalayas near **Mount Everest** in South Asia. The Himalayas, which have some of the highest mountains in the world, are popular destinations for mountain climbers and hikers.

▼ YOUR TURN

Plan a trip in which you will visit three different areas of the world. **Write** a brief description of each place, and **explain** why you have selected each location. Also explain how you will plan your travels so that you show respect for cultures and do not harm natural environments. With your plans, include an outline map of the world with each of your chosen locations labeled.

270

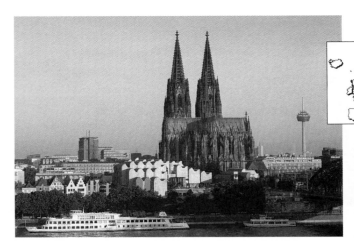

Historic Cologne Cathedral rises above modern buildings in this city along the Rhine River in **Germany**. Many European cities feature historic buildings nestled among modern structures in areas once leveled by fighting in World War II.

European Ways of Life Today, European city skylines include shopping centers, fast-food restaurants, and high-rise buildings next to their centuries-old architecture. Europe is a society of consumers, much like the United States. Teenagers on both sides of the North Atlantic Ocean share similar tastes in clothing, music, movies, and sports.

Advanced transportation and communication networks crisscross much of Europe. The region's railroads, highways, seaports, river-canal systems, and airline connections are generally modern and efficient. Europeans are also leaders in the development of high-speed rail and aerospace technology. In some European countries, however, particularly in Eastern and Southern Europe, transportation is not as well developed.

Most European nations provide many social services to their citizens. Social programs often provide for the health care, education, and welfare of citizens throughout their lives. All these programs are supported by taxes that are high by U.S. standards. Taxes collected by governments also provide public transportation and support education, sports, and the arts.

Issues Americans and Europeans face some similar challenges, such as unemployment, crime, traffic congestion, and limited energy resources. Environmental pollution is another concern among Europeans. Antinuclear movements and environmental political parties such as the "Greens" influence the politics of some European nations. Overcrowding, loss of wilderness areas, and a long history of environmental abuse have brought environmental issues to the top of Europe's list of concerns.

Some minority groups in Europe, such as the Basques in Spain and various ethnic groups in the former Yugoslavia, are seeking independence. In some instances, these movements have led to violence. In addition, many European cities have large foreign populations of guest workers. Violence toward these immigrants has broken out, especially in former East Germany, where unemployment is high. In addition, terrorist groups have carried out bombings, assassinations, hijackings, and kidnappings.

Other European concerns are at the international level. The political and economic changes and ethnic tensions in the former Soviet Union and in the Balkan nations are affecting the entire region. Europe also faces continued economic competition from Japan and the United States.

To face these challenges, Europe is seeking to strengthen economic and political unity among its many countries. While political unity remains a distant vision, greater economic cooperation promises increased prosperity for all of Europe.

4 SECTION REVIEW

1. Name five language groups found in Europe.
2. What is the European Union?
3. **Critical Thinking** **Explain** the causes of modern-day violence in Europe.

CHAPTER 23 CHECK
ANSWERS

Building a Vocabulary
1. The landmass on which Europe and Asia are located
2. A very small country
3. A natural boundary
4. An agreement between countries to support one another against enemies; allies
5. They were reunified.
6. Many people might die as a result of an extreme shortage of food.

Recalling and Reviewing
1. The Baltic Sea; Sweden, Finland, and the Baltic countries; humid continental, subarctic, and tundra
2. Forests: most original forests were cut centuries ago, and in recent years pollution and acid rain have killed more trees. Rivers: many of Europe's rivers are polluted, and the rivers carry this pollution to the sea.
3. Some European economies were hurt by the loss of the colonies because they had been sources of inexpensive raw materials.
4. Luxembourg, Switzerland, the Netherlands, and Belgium

CHAPTER 23 CHECK

Produce market in **Athens, Greece**

Reviewing the Main Ideas
1. Europe has four major landform regions and three major climate types. The largest landform region is the Northern European Plain. Most of Europe has mild climate conditions.
2. Europe's navigable rivers and long coastline make it ideally situated for trade. Europe also has agricultural, mineral, and energy resources, although some imports are necessary.
3. Throughout history, various groups have fought for control of Europe and of the world's oceans and trade. Europe's control of overseas territories, or imperialism, reached its peak during the late nineteenth century. Two world wars were fought during the first half of the twentieth century. From World War II until the 1990s, much of Eastern Europe was under Communist rule.
4. Europeans speak many different languages and practice several religions. The European Union is one of the world's most important economic associations. Most of Europe is prosperous and has a high level of social services, but Europeans still face many economic, cultural, political, and environmental challenges.

Building a Vocabulary
1. To what does the term *Eurasia* refer?
2. What is a microstate?
3. What type of boundary do the Pyrenees form?
4. Define *alliance*. What are the members of an alliance called?

5. What happened to East and West Germany in 1990?
6. Why might a famine cause a population decline?

Recalling and Reviewing
1. Which sea in Europe requires the use of icebreakers during the winter? What countries border this sea, and what climates do they have?
2. What environmental issues surround the use of Europe's forests? its rivers?
3. How did the loss of overseas colonies affect some of Europe's economies?
4. Identify four European countries that are multilingual.

Thinking Critically
1. Would you say that Europe is a region of increasing unity, or do you think it is a region of increasing disunity? Support your answer.
2. Some people believe that one must understand Europe's history in order to understand the rest of the world. Do you agree? Why or why not?

Learning About Your Local Geography

Individual Project
You read that Europe's social programs are supported by taxes that are high by U.S. standards. Make a list of the different taxes that people in your area must pay. For what purposes is each tax used? Conduct a brief survey of your community members asking their feelings about the taxes they pay. Do they think their taxes are low or high? How do they feel about the way their tax money is spent? You may want to include your survey in your individual portfolio.

Using the Five Themes of Geography
Sketch a map of the world. With arrows, symbols, and shading, show how **location** and **movement** interacted to make Europe an economic leader. Be sure to include a legend for any symbols you use. Write a caption explaining the map's contents.

Thinking Critically
1. Answers will vary. Some students may cite Germany's reunification and the formation of the European Union as examples of increasing unity. Others may disagree, pointing to the breakups of Czechoslovakia and Yugoslavia.
2. Answers will vary. Some students may say that because of Europe's history of imperialism, it is true that one must understand Europe's history to understand the rest of the world. Others may say that other countries have their own histories and cultures, despite Europe's involvement, and can be studied apart from Europe.

Learning About Your Local Geography
Students should demonstrate that they have completed the survey.

Using the Five Themes of Geography
Maps might indicate that Europe's location close to many bodies of water and its mastery of how to move along those bodies of water gave Europe access to many resources worldwide.

271

CHAPTER 24 Northern Europe
Planning Guide

TEXTBOOK	TEACHING RESOURCES	OTHER AVAILABLE MATERIALS
Section		
1. The United Kingdom Pages 273–275	Reteaching Worksheet 24.1 Challenge/Enrichment Worksheet 24 Skills Worksheet 24	*The World Today* Videodisc Program *Study Guide for Spanish Speakers* *Lecture Notes,* Chapter 24
2. The Republic of Ireland Pages 276–277	Reteaching Worksheet 24.2 Critical Thinking Worksheet 24	*Study Guide for Spanish Speakers* *Lecture Notes,* Chapter 24
3. The Nordic Countries Pages 277–280	Reteaching Worksheet 24.3 Graphic Organizer 24	*Study Guide for Spanish Speakers* *Lecture Notes,* Chapter 24
Chapter Review Page 281	Chapter 24 Test Chapter 24 Portfolio Assessment See also *Alternative Assessment Forms*	*Test Generator,* Chapter 24

CHAPTER RESOURCES

Lecture Notes, Chapter 24
Teaching Resources
    Chapter 24 Graphic Organizer
    Chapter 24 Skills Worksheet
    Chapter 24 Critical Thinking Worksheet
    Chapter 24 Challenge/Enrichment Worksheet

REVIEW AND ASSESSMENT

Chapter 24 Check, page 281
Chapter 24 Test, *Teaching Resources*
Chapter 24 Portfolio Assessment, *Teaching Resources*
Test Generator, Chapter 24
Alternative Assessment Forms

ADDITIONAL RESOURCES

Books for Teachers

Eames, Andrew. *Four Scottish Journeys.* London: Hodder & Stoughton, 1991.

Elton, Geoffrey. *The English.* Cambridge, MA: Blackwell Publishing, 1992.

Newby, Eric. *Round Ireland in Low Gear.* New York: Viking, 1988.

Olsen, Gregg Mathew. *The Struggle for Economic Democracy in Sweden.* Brookfield, VT: Aldershot, 1992.

Books for Students

Gies, Frances, and Joseph Gies. *Life in a Medieval Castle.* New York: Harper & Row, 1990.

Hintz, Martin. *Norway.* Chicago: Children's Press, 1982.

Meyer, Carolyn. *Voices of Northern Ireland: Growing Up in a Troubled Land.* San Diego, CA: Harcourt Brace Jovanovich, 1987.

Sutherland, Dorothy B. *Scotland.* Chicago: Children's Press, 1985.

Zickgraf, Ralph. *Sweden.* New York: Chelsea House, 1988.

The World Today *Videodisc Program*

A variety of still images, short videos, and activities are available for you to use as you teach this chapter. *Lecture Notes* include bar codes for specific images and videos.

Multimedia Resources

The Anglo-Saxons. (Multimedia CD-ROM, history of England) Cambrix Publishing, 6269 Variel Avenue, Suite B, Woodland Hills, CA 91367.

The British Way of Life. (16mm film/VHS, 24 minutes) AIMS Media, 9710 DeSoto Avenue, Chatsworth, CA 91311–4409.

Discovering Ireland. (VHS, 50 minutes) Available through Social Studies School Service, 10200 Jefferson Blvd., P.O. Box 802, Culver City, CA 90232–0802.

Environmental Problems in the United States and Norway. (VHS, 27 minutes) Agency for Instructional Technology, Box A, Bloomington, IN 47402–0120.

Picture Atlas of the World. (interactive maps, photo bank, animation, audio on CD-ROM) National Geographic Society, Washington, DC.

MEETING INDIVIDUAL NEEDS

Limited English Proficiency (LEP)

Have students look at the physical-political map of Northern Europe on page 273 as you describe the locations, elevations, major cities, landforms, and rivers of each country. Then ask students to use the map to help them name the country that fits a specific description, such as "the smallest Nordic country located on the Jutland Peninsula," or "the country whose capital is Helsinki."

Learners Having Difficulty

Help students create on the board an outline of Chapter 24. Call on volunteers to contribute information for the outline as you write it on the board. Have students copy the outline and use it to study the chapter. Then have students work in pairs to ask each other questions about Northern Europe based on the information in the outline.

Auditory Learners

Ask students to identify the Northern European countries that are members of the European Union. Discuss the trade advantages of membership in such an alliance. Then ask students to compare aspects of socialism with aspects of capitalism. Finally, discuss how Sweden's neutrality in foreign policy might have helped it become one of the most prosperous nations in the world.

Visual Learners

Divide the class into eight groups. Assign each group a country in Northern Europe. Have each group create travel posters enticing tourists to travel to the assigned country. Tell students that the posters should include the following: name of country, drawings or pictures of famous places in the country, and a catchy phrase or slogan about the country. Display the posters in the classroom.

Tactile/Kinesthetic Learners

If possible, arrange to have students tour a local industry. Ask students to find out why the industry is located in or near their community. Then use this field trip as a basis for a discussion about what industries are currently important in Northern Europe and why.

Gifted Learners

Have interested students do research to compare the welfare system, tax rate, and standard of living in Sweden, Great Britain, and the United States. Have students present their results to the class.

MULTICULTURAL FOCUS

The countries in Northern Europe have longstanding seafaring traditions. Many ballads and shanties, or songs sung by sailors as they worked, have been kept alive and passed down since the days of sailing ships.

Sea shanties were sung with a rhythm that helped alleviate the boredom and drudgery of work on sailing ships as sailors manipulated heavy ropes and sails. Themes of the tunes varied. They might be fanciful and humorous, describe the realities of life on board ship, or show admiration for a particular vessel.

Read to the students the following verse and refrain from an English song titled "The Eddystone Light," about the son of a lighthouse keeper:

*Me father was the keeper of the Eddystone
  Light,*
He married a mer-my-aid one night;

Out of the match came children three—
Two was a fish and the other was me.

Jolly stories, jolly told
When the winds is bleak and the nights is cold;
No such life can be led on the shore
As is had on the rocks by the ocean's roar.

Next read aloud the following verse from an Irish song called "The Irish Rover":

*In the year of our Lord eighteen hundred and
  six,*
We set sail from the fair Cove of Cork.
*We were bound far away with a cargo of
  bricks*
For the grand City Hall in New York.
*We'd an elegant craft, she was rigged fore and
  aft,*
Oh, how the trade winds drove her;
*She has twenty-three masts, and she stood
  fearful blasts,*
And they called her the Irish Rover.

CHAPTER 24

NORTHERN EUROPE

Section 1: The United Kingdom
Section 2: The Republic of Ireland
Section 3: The Nordic Countries

OVERVIEW

Chapter 24 describes the countries of Northern Europe, a region of many islands and peninsulas.

Section 1 explores the United Kingdom (England, Scotland, Wales, and Northern Ireland). Students will learn of England's past as the center of a global empire and the present role of the United Kingdom in the European Community and in the rest of the world.

Section 2 focuses on the independent Republic of Ireland, which is struggling to overcome a depressed economy.

Section 3 discusses the Nordic countries. It describes the region's economic and human characteristics, such as its generally high level of industrialization and high standard of living.

About the Author

Nikos Kazantzakis (1883–1957) was born on the Greek island of Crete. He was a versatile writer. His work included poems, dramas, essays, and novels, one of which was the well-known *Zorba the Greek.* The quotation on this page is from *England,* an account of Kazantzakis's travels in the British Isles.

Primary Source

Description of Change: excerpted
Rationale: excerpted to focus on main idea

CHAPTER 24

Northern Europe

66 **T**he English love . . . the pleasant tranquillity of the countryside. But deeper down, they feel the sea as their native land, their own domain. . . . Nothing is more enticing for the English than strange tales of the sea, pirate adventures, sea heroes. From the time they are small children, they dream of distant travels and sea-washed glory. The sea satisfies the romantic, anxious necessities of the English spirit. 99

Nikos Kazantzakis

Gold Celtic jewelry

Northern Europe is a land of many islands and narrow peninsulas where fingers of the sea cut into the long coastlines. The nations of Northern Europe have been heavily influenced by their locations. Their nearness to the sea and, for some, their far north location, have greatly affected their climate, cultural heritage, and economic development.

Northern Europe can be divided into two regions: the British Isles and Scandinavia. The British Isles lie on the continental shelf off the European coast and consist of Great Britain and Ireland. These islands are separated by the Irish Sea and are divided politically into two nations, the United Kingdom and the Republic of Ireland.

Scandinavia, or the Nordic countries, covers a broad area stretching from the western Russian border to the North Atlantic Ocean. This region includes the nations of Norway, Sweden, Denmark, Iceland, and Finland.

GEOGRAPHY DICTIONARY

constitutional monarchy
loch
nationalize
bog
peat
fjord
socialism
neutral
uninhabitable
geyser

Southern coast of the **United Kingdom**

272

NORTHERN EUROPE: PHYSICAL–POLITICAL

To understand the relative location of Greenland as well as the vast distance separating it from Europe, see the map on pages A4 and A5.

Size comparison of Northern Europe to the contiguous United States

LEGEND

ELEVATION		
	FEET	METERS
Ice caps		
National capitals	13,120	4,000
	6,560	2,000
Other cities	1,640	500
	656	200
	(Sea level) 0	0 (Sea level)
	Below sea level	Below sea level

THE UNITED KINGDOM

FOCUS

- *What are the United Kingdom's physical and economic characteristics?*
- *What role does the United Kingdom play in the global community?*

REGION The United Kingdom (UK) includes England, Scotland, and Wales (all on the island of Great Britain) and Northern Ireland. Because the people of the United Kingdom live mainly on the island of Great Britain, they are often referred to as the British. The United Kingdom is one of the most populous nations in Europe. It also has one of Europe's highest population densities.

The United Kingdom is a **constitutional monarchy**. That is, it has a reigning king or queen, but a parliament serves as the lawmaking branch of government. The royal family is a symbol of history and patriotism to the British.

Physical Geography Great Britain can be divided into two landform regions, lowland and highland Britain. (See the map above.) The southeastern half of the island is lowland Britain. This region is part of the Northern European Plain. Only the narrow English Channel separates the British portion from the rest of the Northern European Plain. Highland Britain is in the north and west. The major highland areas are made up of ancient eroded mountains. Highland Britain includes the beautiful English Lake District and the large Scottish lakes, or **lochs**. Lochs are long, deep lakes carved by glaciers.

For centuries, the rivers of Britain have been important for trade. They connect the major

Motivator

Remind students that the United States was once part of Great Britain's colonial empire. Have students discuss some aspects of American culture that can be traced to a British heritage. *(language, democratic principles, literature, religion)* Ask students to suggest what the United States might be like today if the original 13 colonies had belonged to a different country.

Building a Vocabulary

Write the term **nationalize** on the board. Explain that the suffix *-ize* means "to become or make into." *Nationalize* means "to make privately owned business into government-owned business." Note that *privatize* means the opposite of *nationalize*.

Illustrating

Economic Geography   Organize the class into groups, and assign one of the following time spans to each group: 1700s, 1800–1850, 1850–1900, 1900–1950, 1950–present. Direct students to draw pictures depicting industry in Great Britain during the assigned period. Have students arrange their illustrations in chronological order on the bulletin board.

GLOBAL PERSPECTIVES

Have students brainstorm ways in which the Industrial Revolution changed life in Great Britain. *(People no longer worked at the same occupation generation after generation. They left farms to work in factories in big cities. They no longer handcrafted all their own goods.)* Then point out that many developing nations in the world today are undergoing these same changes as they seek to industrialize their economies.

Chart Caption Answer

United Kingdom

Geography and History

The development of the British textile industry during the Industrial Revolution was aided by new machines that greatly sped up the fabric-making process. James Watt's steam engine ended manufacturing's dependence on waterpower. Richard Arkwright invented a carding machine to process rough wool fibers. James Hargreaves's spinning machine turned wool fibers into yarn, and Edward Cartwright's machine loom wove the yarn into fabric.

274

FACTS IN BRIEF   Northern Europe

COUNTRY POPULATION (1994)	LIFE EXPECTANCY (1994)	LITERACY RATE (1991)	PER CAPITA GDP (1993)
DENMARK 5,187,821	73, male 79, female	99%	$18,500
FINLAND 5,068,931	72, male 80, female	99%	$16,100
ICELAND 263,599	77, male 81, female	99% (1992)	$16,000
IRELAND 3,539,296	73, male 79, female	99%	$13,100
NORWAY 4,314,604	74, male 81, female	99%	$20,080
SWEDEN 8,778,461	75, male 81, female	99%	$17,600
UNITED KINGDOM 58,135,110	74, male 80, female	99%	$16,900
UNITED STATES 260,713,585	73, male 79, female	97%	$24,700

Sources: *The World Factbook 1994*, Central Intelligence Agency; *The World Almanac and Book of Facts 1995*

Which is the most populous country in Northern Europe?

ports of London, Liverpool, and Glasgow (GLAS-koh) to the sea. The most important river is the Thames (TEHMZ), which connects London to the North Sea.

Even though Great Britain is relatively far north (50° to 60° north latitude), it has a marine-west-coast climate. The prevailing westerly winds blowing off the North Atlantic Drift help keep its climate mild. Britain is known for its rainy, cloudy, and foggy weather. Britain's west coast, especially in Scotland and the Hebrides Islands, averages more than 60 inches (152 cm) of rainfall per year.

INTERACTION   **Economic Geography** In the 1700s, Great Britain had abundant coal and iron deposits, a large labor force, and an excellent transportation network of rivers and canals. By the early 1800s, the British had built the first railroads. All these resources helped Britain lead the Industrial Revolution.

Three of Great Britain's early industries were iron and steel, shipbuilding, and textiles. Deposits of iron ore and coal found in the central part of England fueled the iron and steel industry. During the middle of the nineteenth century, with Great Britain's steel industry already established, steel

replaced wood as a material for building ships. The island's coastal cities were a perfect home for this industry. The British textile industry began mainly in northern England. The invention of spinning machines and the harnessing of steam power replaced home weaving. Local wool and, later, cotton from the United States and India, as well as wool from Australia, supplied the textile industry.

Although Great Britain was first to industrialize, it lost its world leadership role to foreign competition. By 1900, Germany and the United States outproduced Great Britain in many industries. In recent decades, Great Britain's coal mines and traditional industries have suffered rapid decline. Unemployment, especially in the cities of northern England and Scotland and in the mining towns of the Midlands and Wales, is high.

To curb its post–World War II industrial decline and to increase economic equality, the British government **nationalized** many of its industries, such as steel and shipbuilding. A nationalized industry is owned and operated by the government. Because of limited success, the government has in recent years sold some of these industries back into private ownership. Some industries, such as railroads, remain nationalized, as do services such as the postal and health services.

Agriculture employs less than 2 percent of the British work force yet produces more than half of the nation's food. The southern part of Great Britain has prosperous, modernized farms. In contrast, most of highland Britain has limited farming because of the rocky, glaciated terrain. Grazing of cattle and sheep is common on the island.

The United Kingdom has more fossil fuels than any other European nation. In 1969, oil was discovered under the North Sea, and the United Kingdom became Europe's major oil producer and exporter. The oil deposits will last only a few more decades, however. The United Kingdom also depends on coal, offshore natural-gas deposits, and hydropower to generate electricity. In addition, it was one of the first countries to use nuclear power; about one-fifth of its electricity is nuclear generated.

Today, most British are employed in service industries, especially banking, insurance, communications, and government. In education, British universities are considered some of the best in the world. The tourist industry is also important to the economy.

A British police officer provides security during a celebration by immigrants from South Asia in **London**. Many immigrants to the **United Kingdom** are from former colonies of the British Empire.

PLACE **London** London is the United Kingdom's capital city and largest urban area. It is home to the Houses of Parliament and Buckingham Palace, the residence of the British monarch, and many cathedrals and museums. London is also a center of trade, industry, and services, such as banking and insurance. Its early economic importance was due to its location on the Thames River. London is a major theater, retail, and tourist center as well.

Like most other large cities, London suffers from urban decay, pollution, and traffic. The government has made effective efforts to clean up London's air and the waters of the Thames. A vast rail network helps ease some traffic congestion. The government also maintains a "greenbelt" of parks and woodlands that circle the city.

Northern Ireland The six northern counties of Ireland are called Northern Ireland or sometimes Ulster. Part of the United Kingdom, Northern Ireland sends representatives to Parliament in London. The majority of Northern Ireland's population is descended from Scottish and English Protestants who arrived during the seventeenth century. Irish Roman Catholics form almost one-third of the population.

There has been bitter fighting between the Protestant majority and the Roman Catholic minority of Northern Ireland. The Roman

Catholics are seeking more social, political, and economic opportunities. Many would like Northern Ireland to join the Republic of Ireland, which is mostly Roman Catholic. The Protestants, however, fear becoming the minority on the island and want to remain a part of the UK.

Due to the violence, British troops have been stationed in Northern Ireland since 1969. Several thousand people have been killed by fighting and terrorist activities. Open talks to end the conflict began after the Irish Republican Army (IRA), one of the most active of the terrorist groups, agreed to stop terrorist activities in 1994.

A Global Role At its height during the nineteenth and early twentieth centuries, the British Empire ruled the seas. The empire's colonies provided raw materials for British industries and served as markets for finished goods. When World War II ended in 1945, many colonies sought independence. New nations appeared on the map, including India, Pakistan, Jamaica, Kenya, and Nigeria. The United Kingdom still holds some territories scattered around the world.

The once-powerful British Empire has been replaced by a commonwealth. In this context, a commonwealth is an association of self-governing states with similar backgrounds and united by a common loyalty. The Commonwealth of Nations is a voluntary association of 50 former British colonies. Member nations meet to discuss economic, scientific, and business matters.

The United Kingdom joined the European Union in 1973. The British now look to other European nations for trade rather than to their distant former colonies. For instance, sugar may come from France instead of Jamaica, and dairy products come from Denmark rather than New Zealand. The UK remains a major international trading nation, exporting more than 30 percent of its goods.

SECTION REVIEW

1. Into what two landform regions can Great Britain be divided?
2. What is the Commonwealth of Nations?
3. **Critical Thinking** **Explain** how Britain's location, resources, and industry combined to make Britain the leading world power during the nineteenth century.

275

OBJECTIVES

The objectives reflect the Focus questions on this page and are assessed in the review on page 277.

1. Describe Ireland's physical features.

2. Identify human and economic characteristics of the Republic of Ireland.

TEACHING SECTION 2
Motivator

Call on a volunteer to locate Ireland on a wall map. Emphasize that only Northern Ireland is part of the United Kingdom. Note that one of Ireland's most famous natural features is its green countryside. Ask students to suggest how Ireland's landforms and climate produce this lush countryside.

Building a Vocabulary

Explain that **peat**, found in **bogs**, is often used for fertilizer and fuel. In Ireland, peat is the only source of energy that does not have to be imported. The country's dependence on energy imports is an economic challenge. Have students write a sentence using both **peat** and **bog**.

---

2 THE REPUBLIC OF IRELAND

FOCUS

- *What are Ireland's physical features?*
- *What human, political, and economic characteristics does the Republic of Ireland have?*

Physical Geography Across the Irish Sea from Great Britain is Ireland, sometimes called the "Emerald Isle" because of its green countryside. (See the map on page 273.) The entire island has a marine-west-coast climate. Rainfall varies from 30 to 100 inches (76 to 254 cm) per year.

Rolling green fields and pastures cover the island's central part. The Shannon River, the longest river in the British Isles, flows through the middle of the island. Low, rocky mountains form a rugged coastline. Much of Ireland is either rocky or boggy. **Bog** is soft ground that is soaked with water. For centuries, **peat** deposits from the bog areas have been used for fuel. Peat is composed of decayed vegetable matter, usually mosses. Less than 5 percent of the island is forested.

Human Geography The Irish, like the Welsh and Scots, are descendants of the Celts who settled in Ireland around 400 B.C. Irish Gaelic, a Celtic language, is still considered an official language but is spoken by only a small percentage of the Irish. Today, English is the main language.

In the thirteenth century, the Norman rulers from Britain conquered Ireland. Eventually, the Protestant British controlled most of the land, reducing the Roman Catholic Irish to landless farmers. Lack of economic opportunity coupled with a potato famine during the nineteenth century caused many Irish to emigrate, particularly to the United States.

For several hundred years, the Irish rebelled against British control. Then in 1921, Ireland became a dominion, similar to Canada, called the Irish Free State. In the 1930s, the Irish government cut most ties with Britain, and a 1937 constitution renamed Ireland as Eire (EHR-uh). The completely independent Republic of Ireland, recognized by Britain, was formed in 1949. Today, the Irish Republic is made up of 26 counties and covers about 80 percent of the island.

Economic Geography Although Ireland's farming population has declined in recent decades, farm income has risen. Since Ireland joined the European Union in 1973, Irish farmers have modernized agriculture and now grow a wider variety of crops. The country produces dairy products, beef, poultry, potatoes, and grains both for export and home consumption. More than half of Ireland's farm products are exported to the EU nations.

Most Irish workers are employed in manufacturing. Some of Ireland's industries are based on food products, including dairies and breweries. Other factories manufacture glassware and fine textiles. The greatest industrial development has

Traffic along O'Connell Street crosses the River Liffey in **Dublin, Ireland**. O'Connell Street is the Irish capital's main road and is part of the central shopping district. Statues of famous Irish people line the center of the street.

Explaining

Human Geography   Call attention to the division of the "Emerald Isle" into the Republic of Ireland and Northern Ireland. Be sure that students understand the history behind this division. Discuss the similarities and differences between the two areas. Review the religious conflict in Northern Ireland. Ask students to suggest how this conflict might affect teenagers living in Ireland and Northern Ireland.

Closure

Have students write questions and answers about Ireland on index cards and then quiz each other.

ASSESSMENT

Assign the Section 2 Review questions on this page.

Reteaching

Assign Section 2 of Reteaching Worksheet 24.

SECTION 3
OBJECTIVES

The objectives reflect the Focus questions on this page and are assessed in the review on page 280.
1. Examine the physical geography of the Nordic countries.
2. Describe economic and human characteristics of the Nordic countries.

occurred around the seaport city of Dublin. Nearly one-third of the nation's population now lives in Dublin, the capital and center for education, banking, and shipping. The country's other, smaller cities lie mainly along the coast.

Except for peat deposits used to generate some electricity, Ireland is an energy-poor island, with little oil, natural gas, or coal. Its high taxes and high unemployment rates are depressing the economy. People continue to emigrate from Ireland to seek jobs elsewhere, especially in the United Kingdom, the United States, Canada, and Australia. In addition, by borrowing money to help its unemployed, Ireland has acquired a huge debt. The country's future economic development depends on more foreign investment, new industries, and increased tourism.

SECTION REVIEW

1. Why is Ireland sometimes called the "Emerald Isle"?
2. What economic activities are important to Ireland?
3. **Critical Thinking**   **Suggest** three reasons that the Irish rebelled against British control in the nineteenth century.

THE NORDIC COUNTRIES

- *What are the environments of the Nordic countries?*
- *What are the economic and human characteristics of the region?*

Once the home of fierce, warlike Vikings, today the Nordic countries are peaceful, prosperous, and industrialized nations. In general, the people are urbanized and well educated and have some of the highest standards of living in the world. Because of their high health standards, long lifespans, and low birthrates, they also have some of the oldest populations in the world. Except for Finnish, all the Scandinavian languages are closely related, and all the Nordic nations are Protestant (Lutheran) in religion.

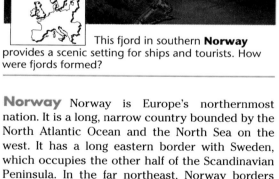

This fjord in southern **Norway** provides a scenic setting for ships and tourists. How were fjords formed?

Norway   Norway is Europe's northernmost nation. It is a long, narrow country bounded by the North Atlantic Ocean and the North Sea on the west. It has a long eastern border with Sweden, which occupies the other half of the Scandinavian Peninsula. In the far northeast, Norway borders Finland and Russia. The country also includes thousands of islands. (See the map on page 273.)

Norway's climate is influenced by the westerly winds and the warm North Atlantic Drift. Most of coastal Norway has a marine-west-coast climate, and the ocean waters remain ice-free all year. The far north is Arctic tundra and barren mountains. In winter, snow blankets northern Norway under almost continuous darkness. In contrast, there is almost continuous daylight during the summer. This is why the area north of the Arctic Circle is called the "land of the midnight sun."

Norway is one of Europe's most mountainous nations. In fact, transportation by sea or over mountains on skis was once the only way to reach many Norwegian towns. The Kjølen Mountains tower over most of the country. These steep mountains and the country's wet climate make Norway a major producer of hydroelectricity. Its inexpensive electricity has attracted many industries.

Most Norwegians live on narrow coastal plains or on the shores of the **fjords** (fee-AWRDS).

Section 2 Review
ANSWERS

1. Much of the island is covered with green fields.
2. Agriculture, manufacturing, tourism
3. Reasons may have been religious, in that many Irish were Roman Catholic whereas the British were mostly Protestant; economic, in that the British controlled most of the land; and/or political, in that the Irish may simply have desired independence for its own sake.

Photo Caption Answer

Ice-age glaciers carved them out of the mountains.

TEACHING SECTION 3

Motivator

Write the names of the Nordic countries on the board. Have students locate the countries on a map. Then play "Mystery Country" with the class. Provide a clue, such as "smallest of the Nordic countries" *(Denmark),* about each of the countries, and have students deduce to which country the clue applies.

Building a Vocabulary

Write the term **uninhabitable** on the board. Help students analyze its structure. The prefix *un-* means "not." The base *inhabit* means "to reside in." The suffix *-able* means "capable of being." *Uninhabitable* describes a place not capable of being resided in. Ask students to suggest conditions that might make a place uninhabitable.

Synthesizing

Physical and Economic Geography
Write the names of the Nordic countries across the board. Ask students to name terms that describe the physical features of each country, as you list them under the appropriate country. To get the class started, refer students to the photograph of the fjord (page 277), and ask with which country fjords are associated. *(Norway)* Then have students relate economic activities to physical features.

From Viking days to the present, Norwegians have looked to the sea as a source of income. Thus, Norway vigorously opposes an international ban on commercial whaling. Government officials argue that coastal nations should have the right to make use of available water resources, just as nations are free to make use of their land resources. Norway's position is at odds with that of the United States and other members of the International Whaling Commission.

Sweden leads Europe in innovative programs to move unemployed workers back into the labor force. Under one program, the government pays unemployed persons up to six months of their previous wages to start a new company. To encourage companies to hire workers, another government program pays as much as 60 percent of a new employee's salary for the first six months. Various retraining programs also help unemployed workers move into new jobs.

A family ventures across a frozen bay in **Stockholm, Sweden**. In exchange for high taxes, Swedes receive many welfare benefits, including incentives for raising children.

Fjords are narrow, deep inlets of the sea between high, rocky cliffs. Ice-age glaciers that stretched down to the sea carved the fjords out of the mountains. Spectacular waterfalls dropping from vertical cliffs along the coast attract tourists to this area each summer.

The long, rugged coastline, sheltered harbors, and the closeness of fishing grounds make Norway Europe's leading fishing nation. Fishing is one of the oldest industries in Norway and provides major export products. Shipping also is important to the Norwegian economy. Even Norway's most valuable resource, oil, comes from the sea. Norway's North Sea oil fields, developed during the 1970s, are expected to produce for only a few more decades, however. Today, the country exports most of its oil to other European nations.

More than half of Norway's population lives in the very southeast corner of the nation. Oslo, the capital, cultural center, and leading seaport and industrial center, is located here. This modern city lies at the end of a wide fjord. Well-planned parks, wide avenues, and houses built along the hillsides are typical. Government buildings and the royal palace also grace the city. Oslo is close to hiking and skiing trails. Business people are often seen taking winter lunch breaks on cross-country skis through the city parks.

Sweden Sweden is the largest Scandinavian country, both in area and in population. Except for a marine-west-coast climate along its southern tip, Sweden is divided into two cold climates. The southern-central portion has a humid-continental climate, while the northern portion is subarctic. Unlike ice-free Norway, Sweden's east coast on the Gulf of Bothnia usually freezes during winter. The Kjølen Mountains line Sweden's western border with Norway. The rest of the country is generally a rolling, glaciated plain with thin soils and thousands of lakes. (See the map on page 273.)

Sweden's four main sources of wealth are forestry, farming, mining, and manufacturing. Forests cover more than half of Sweden and provide large quantities of wood pulp for paper and building materials. Sweden is the world's third largest exporter of wood products. To help sustain the forests, the Swedish government limits cutting and encourages replanting.

The richest farmland in Sweden is in the southern region, called Skåne. Here, the soil is fertile, and the climate is mild. Wheat, rye, potatoes, sugar beets, cattle, and dairy products are sources of income. Less than 5 percent of the Swedes are farmers, yet they produce nearly all of the country's food.

Northern and central Sweden have rich iron-ore deposits. Kiruna, in the north, is the major iron-mining center. From here the ore is exported to other European countries from the port of Narvik in northern Norway. Hydroelectric power from Sweden's rivers provides more than half of the nation's electricity.

Sweden's industries are highly diversified, technologically advanced, and well managed. The nation produces a variety of products for export, such as its high-quality steel. The country also produces automobiles, aircraft, household appliances, chemicals, furniture, glassware, and a variety of electrical products. It is a world leader in the production and use of industrial robots.

Sweden is one of the most prosperous and democratic nations in the world. The country has a mixed economy with some elements of **socialism**, an economic system in which the government owns and controls the means of producing goods. Sweden's government controls about 10 percent of the nation's industries and services, including telephones, railroads, airlines, and some electric utilities. The remaining 90 percent of the country's economy is in the private sector. A number of Swedish businesses are run by cooperatives owned by their customers.

Sweden's socialist system includes an extensive social-welfare program. Government pro-

Making Inferences

Human Geography Refer students to the chart on page 274. Call on volunteers to read the facts about each Nordic country and to draw conclusions about its prosperity. Ask how the Nordic countries—with limited resources and limited habitable areas—have become successful. *(Students might mention aspects of Nordic culture, history, and society that have helped the people make the most of their resources.)*

Guided Practice

Have students complete a Nordic countries chart. Label the following columns across the top of the chart: Features, Denmark, Finland, Iceland, Norway, Sweden. In the Features column, write phrases such as "Smallest country," "People originated from Central Asia," "Contains many islands." Direct students to place a check under the name of each country to which the phrase applies.

Independent Practice

Provide students with a copy of Outline Map 14. Have them label the map to show the countries, capital cities, and major landforms and bodies of water of the Nordic countries. Captions should explain the climates and economic activities found in the region.

grams take care of almost all the educational and medical needs of Swedish citizens. Such programs are expensive and take up about one-third of the national budget. The Swedes pay for these services with some of the world's highest taxes.

About 85 percent of Sweden's population lives in cities and towns, mainly in the country's southern part. Stockholm, the capital, is the nation's main commercial center. It is a beautiful city of islands and forests.

Sweden has a **neutral** foreign policy and is not a member of NATO. A country is neutral when it chooses not to take sides in international conflicts. Sweden ended its membership in the EFTA and joined the EU in 1995.

Denmark Denmark is the smallest of the Nordic countries. Most of the country's land area lies on the low Jutland Peninsula, which borders Germany on the south. More than 400 islands make up the rest of the country, including the large island of Greenland. Denmark faces the North Sea on the west and the Baltic Sea on the east. Narrow straits separate Denmark from Norway and Sweden. The country has benefited economically from this central location between land and water routes. (See the map on page 273.)

Denmark is the most densely populated Nordic country. One-fourth of its population lives in the area surrounding Copenhagen, the capital city. Copenhagen lies on an island along a narrow strait leading to the Baltic Sea. The city is a center of trade, shipping, and manufacturing.

Part of the Northern European Plain, Denmark's flat lowlands are excellent for farming. In fact, about 60 percent of the country's land is used for agriculture. The nation's mild marine-west-coast climate and rich pastureland support widespread dairy farming and hog raising. The Danes are famous for their fine butter, cheese, and ham. Together, dairy and meat products account for most of the country's farm income.

Although Denmark is a largely agricultural nation, it also is modern and industrialized. The Danes have succeeded in producing quality goods for world markets, such as machinery, furniture, medicines, packaged foods, and electronics. All of Denmark's industry is privately owned.

Denmark has a large fishing fleet to take advantage of its access to the sea. Danish North Sea waters also hold natural-gas and oil deposits, which may decrease Denmark's dependence on imported fuels.

Denmark is one of the most prosperous of the EU nations, and the Danes have one of the world's highest per capita incomes. The country's social-welfare system includes medical care, child care, and care for the aged. High taxes help fund these and other welfare programs.

Greenland Greenland lies between the North Atlantic and Arctic oceans. Athough located nearer to North America, it is a self-governing province of Denmark. Greenland is not a descriptive name for this island as thousands of feet of ice cap cover nearly 85 percent of it. The entire icy interior of the island is **uninhabitable**. An uninhabitable region is one that cannot support human life and settlements. Only along the coasts are there areas of tundra climate where people can live. Even much of the coastal area is glaciated barren rock.

Most of the people of the island are Eskimo, also called Inuit, or are of European-Eskimo descent. Most Greenlanders live along the island's southwestern coast. Fishing is the chief occupation, but some Eskimos still practice traditional hunting of seals and small whales.

Iceland The island nation of Iceland lies in the middle of the North Atlantic Ocean just south of the Arctic Circle. (See the map on page 273.) Situated on the active Mid-Atlantic Ridge where the North American and Eurasian plates are separating, Iceland is Europe's westernmost nation.

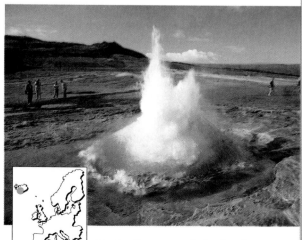

An erupting geyser in **Iceland** scatters some wary visitors. Iceland uses its underground heat for geothermal energy.

COOPERATIVE LEARNING

Organize the class into groups of four or five. Assign one of the Nordic countries to each group. Direct groups to tape-record a narration for a "Day in the Life of. . ." documentary for their country. Members should decide together what places and activities to feature. Each group member should then take responsibility for writing and recording a description of one or more places or activities. Group members can also find or draw pictures to go with their tape. When the narrations are complete, have groups play them for the class.

Closure

Reveal the identity of each "mystery country" from the Motivator activity on page 278. Have students write clues for their own Nordic mystery country. Then have them read their clues to one another and deduce the country being described.

ASSESSMENT

Assign the Section 3 Review questions on this page.

Reteaching

Assign Section 3 of Reteaching Worksheet 24.

Extension

Have interested students find out more about one or more of the major cities of the Nordic countries—Oslo, Stockholm, Copenhagen, Reykjavik, and Helsinki. Students might research the cities' important sites, buildings and monuments, bodies of water, economic activities, histories, and so on. Ask students to create maps of the cities.

THEME
Movement

Before World War II, Vyborg, Russia, was Finland's second-largest city. Viipuri, as the city was called, and the surrounding province of Karelia surrendered to the Soviets in 1944. Rather than live under Soviet rule, 400,000 Finns fled to western Finland, leaving behind an almost empty province. Communist social planners imported thousands of Soviet men, women, and children to repopulate the abandoned farms and houses. Today, a growing number of Finns are returning to Vyborg.

Section 3 Review
ANSWERS

1. Norway and Sweden
2. Its fishing waters and its hydroelectric and geothermal energy sources; Finland must find new markets because the Soviet Union, which used to buy manufactured goods from Finland in return for oil and gas, collapsed.
3. Advantage: citizens' medical, educational, and other needs are taken care of by the government. Disadvantage: taxes are high.

This is a land of lava rock and more than 100 volcanoes. Three-quarters of Iceland is mountainous and uninhabitable. More than 100 glaciers cover the high mountain interior, while much of the coast is extremely rugged and battered by waves. The name Iceland is misleading, since coastal temperatures are mild and remain above freezing. Due to its mid-ocean location, Iceland has mainly a marine-west-coast climate, which is wet and cloudy much of the year.

Iceland's most important natural resource is the rich North Atlantic fishing waters that surround the island. More than 70 percent of the nation's export income comes from fishing. Iceland's other major resource is energy. Because of its mountains and wet climate, Iceland produces large amounts of hydroelectricity.

Iceland also harnesses geothermal energy. As a result of volcanic activity, underground water rises all over the island as steam, forming **geysers**. The word *geyser* is an Icelandic term for hot springs that shoot hot water and steam into the air. This hot water heats most Icelandic homes, as well as many greenhouses, where crops such as tomatoes, grapes, and cucumbers grow year-round.

The world's northernmost capital, Reykjavik (RAYK-yuh-veek) contains more than half of Iceland's population. The country has little crime or unemployment.

Finland Most of Finland lies between two arms of the Baltic Sea: the Gulf of Bothnia to the west and the Gulf of Finland to the south. Finland borders Russia on the east and Sweden and Norway in the far north. (See the map on page 273.) Finland's landforms show the effects of the last ice age, when a thick sheet of ice covered the whole country. Finland is mainly a glaciated plain with about 50,000 lakes, extensive areas of bog and wetlands, and thousands of small offshore islands.

Because the Gulf of Bothnia freezes during most winters, Finland has more severe climates than the other Nordic countries, which are near ice-free seawater. Finland has three major climate regions. In the far south is a humid-continental climate with summer temperatures that permit agriculture. The central part of the country has a subarctic climate of evergreen forests. The northern tip of the country has a tundra climate.

The severe climates and glaciated terrain make only 10 percent of Finland suitable for farm-

ing. Hay and oats are grown to feed cattle, and potatoes and grains are important food crops. Forests, which cover about 60 percent of the country, are Finland's greatest natural resource and make the nation a world leader in paper and other forest products. Finland also produces electronics and is a major shipbuilder. Finnish exports include televisions, mobile telephones, icebreakers, and cruise ships.

Finland's major trade partners are other Nordic nations and the nations of the European Union. The Soviets used to buy manufactured goods from Finland in exchange for oil and gas. In an effort to adjust to the market loss resulting from the collapse of the Soviet Union, Finland joined the EU in 1995.

About one-fifth of the population of Finland lives in or around Helsinki, which lies along the Gulf of Finland. Helsinki is the Finnish capital, leading industrial area, and major seaport.

The Finns came from northern Asia. Their language belongs to the same language family as Hungarian and Estonian and is unrelated to other languages of Europe. Finland's second official language is Swedish, which almost all Finns learn in school.

Lapland Across northern Finland, Sweden, and Norway is a region known as Lapland. This tundra region is populated by the Lapps, or Sami as they call themselves. These reindeer herders probably originated from hunters of northern Asia.

The Lapps have managed to maintain some of their culture, even with a modern Nordic way of life in their midst. Although only about 5 percent of their population is involved in traditional reindeer herding, the herders are important to the Lapps as the keepers of their culture. Most Lapps now earn a living from tourism.

SECTION REVIEW

1. What two countries do the Kjølen Mountains border?
2. What are Iceland's major resources? How has Finland's economic situation changed in recent years?
3. **Critical Thinking** **Specify** one advantage and one disadvantage of the social-welfare systems in Sweden and Denmark.

ANSWERS

Building a Vocabulary
1. Constitutional monarchy
2. A long deep lake carved by glaciers; Scotland
3. Soft ground that is soaked with water; peat
4. A narrow, deep inlet of the sea between high cliffs; Norway
5. Unable to support human life and settlement; Denmark (in Greenland) and Iceland

Recalling and Reviewing
1. In both countries, the government owns and operates some industries and services (such as railroads), while other industries are privately owned.
2. Large percentages of the people live in and around cities; London, United Kingdom; Dublin, Republic of Ireland; Oslo, Norway; Stockholm, Sweden; Copenhagen, Denmark; Reykjavik, Iceland; and Helsinki, Finland
3. Coal, oil, hydroelectricity, natural gas, nuclear power, peat, geothermal energy

---

CHAPTER 24 CHECK

Big Ben and the Houses of Parliament along the Thames River, **London**

Reviewing the Main Ideas
1. Great Britain can be divided into two landform regions and has a marine-west-coast climate. Although Great Britain was the world's first industrialized nation, its traditional industries have declined. Northern Ireland is part of the United Kingdom. The British Empire has been replaced by the Commonwealth of Nations, which includes former British colonies.
2. Ireland is covered with rolling green fields. Once under British control, the southern 80 percent of the island of Ireland is now the independent Republic of Ireland. Most Irish workers are employed in manufacturing.
3. The Nordic countries, including Norway, Sweden, Denmark, Iceland, and Finland, have some of the highest standards of living in the world. Landforms range from mountains to plains, and climates are cold. In general, the economies of the Nordic countries are industrial.

Building a Vocabulary
1. What term describes Britain's government?
2. What is a loch? Where are lochs found?
3. Define *bog*. What fuel product comes from bogs?

4. What is a fjord? Which country has many fjords?
5. Define *uninhabitable*. Which countries discussed in the chapter have large uninhabitable regions?

Recalling and Reviewing
1. Compare the economic systems of the United Kingdom and Sweden.
2. How is the population of each Northern European country distributed, particularly with regard to cities? Name a major city in each country.
3. What energy resources are found in the Northern European countries?

Thinking Critically
1. Why do you think Northern Ireland remains part of the United Kingdom when the rest of Ireland is an independent republic? What are the causes of conflict in Northern Ireland?
2. Many sports that people enjoy today began as activities essential to everyday life. How do you think skiing has been essential to daily life in the Nordic countries? What other sports might have originated from activities that people performed to meet their basic needs?

Learning About Your Local Geography

Cooperative Project
In Sweden, government programs take care of the educational and medical needs of Swedish citizens. In your own area, who is responsible for paying the costs of people's educational and medical needs? With your group, research how each set of needs is addressed. Present your information in a chart.

Using the Five Themes of Geography
Using Iceland as an example, write a paragraph discussing the effects that **location** and **place** characteristics can have on **human-environment interaction**. How have Iceland's location and place characteristics affected the ways in which the Icelandic people interact with their environment?

Thinking Critically
1. Possible answer: whereas Roman Catholics form the majority of the Republic of Ireland, the majority of Northern Ireland's population is Protestant. The violence in Northern Ireland has been caused by social, political, and economic differences between the Roman Catholic minority and the Protestant majority.
2. Answers will vary, but students should mention the importance of skis for transportation; hunting, fishing, and horse racing

Learning About Your Local Geography
Charts should show how local educational and medical needs are met.

Using the Five Themes of Geography
Iceland's location as an island just south of the Arctic Circle, and its place characteristics of rich fishing waters and volcanic activity and geysers, allow the Icelandic people to fish for a living and to use geothermal energy for heating (human-environment interaction).

CHAPTER 25 West Central Europe

Planning Guide

TEXTBOOK	TEACHING RESOURCES	OTHER AVAILABLE MATERIALS
Section		
1. France Pages 283–285	Reteaching Worksheet 25.1 Challenge/Enrichment Worksheet 25	*The World Today* Videodisc Program *Creative Strategies for Teaching World Geography* LESSON 11 Dream Vacation: Europe *Study Guide for Spanish Speakers* *Lecture Notes,* Chapter 25
2. Germany Pages 286–287	Reteaching Worksheet 25.2 Critical Thinking Worksheet 25 Challenge/Enrichment Worksheet 25	*The World Today* Videodisc Program *Study Guide for Spanish Speakers* *Lecture Notes,* Chapter 25
3. The Benelux Countries Pages 290–291	Reteaching Worksheet 25.3	*Study Guide for Spanish Speakers* *Lecture Notes,* Chapter 25
4. The Alpine Countries Pages 291–294	Reteaching Worksheet 25.4 Skills Worksheet 25 Graphic Organizer 25	*Study Guide for Spanish Speakers* *Lecture Notes,* Chapter 25
Chapter Review Page 295	Chapter 25 Test Chapter 25 Portfolio Assessment See also *Alternative Assessment Forms*	*Test Generator,* Chapter 25

CHAPTER RESOURCES

Lecture Notes, Chapter 25
Teaching Resources
    Chapter 25 Graphic Organizer
    Chapter 25 Skills Worksheet
    Chapter 25 Critical Thinking Worksheet
    Chapter 25 Challenge/Enrichment Worksheet

REVIEW AND ASSESSMENT

Chapter 25 Check, page 295
Chapter 25 Test, *Teaching Resources*
Chapter 25 Portfolio Assessment, *Teaching Resources*
Test Generator, Chapter 25
Alternative Assessment Forms

ADDITIONAL RESOURCES

Books for Teachers

Ardagh, John. *Writers' France: A Regional Panorama.* New York: Hamish Hamilton (Viking Penguin), 1989.

Fisher, Marc. *After the Wall: Germany, the Germans, and the Burden of History.* New York: Simon & Schuster, 1995.

Marsh, David. *The Germans: A People at the Crossroads.* New York: St. Martin's Press, 1990.

Books for Students

Greene, Carol. *Austria.* Chicago: Children's Press, 1986.

Hills, C. A. R. *The Seine.* Morristown, NJ: Silver Burdett, 1981.

Moss, Peter, and Thelma Palmer. *France.* Chicago: Children's Press, 1986.

Schrepfer, Margaret. *Switzerland, the Summit of Europe.* New York: Dillon Press, 1989.

Spencer, William. *Germany Then and Now.* New York: Watts, 1994.

The World Today *Videodisc Program*

A variety of still images, short videos, and activities are available for you to use as you teach this chapter. *Lecture Notes* include bar codes for specific images and videos.

Multimedia Resources

The Berlin Wall: 1961. (VHS, 20 minutes) Films for the Humanities & Sciences, P.O. Box 2053, Princeton, NJ 08543–2053.

The German Way of Life. (16mm film/VHS, 25 minutes) AIMS Media, 9710 DeSoto Avenue, Chatsworth, CA 91311–4409.

Let's Visit France. (Multimedia CD-ROM, interactive tour) Queue, 338 Commerce Drive, Fairfield, CT 64305.

Picture Atlas of the World. (interactive maps, photo bank, animation, audio on CD-ROM) National Geographic Society, Washington, DC.

MEETING INDIVIDUAL NEEDS

Limited English Proficiency (LEP)

Pair LEP students with students who are proficient in English. Have the pairs answer the Focus questions in each section. Then, using the answers to the Focus questions, help students create graphic organizers on the board about the countries of West Central Europe. Have students copy the graphic organizers and use them to study the chapter.

Learners Having Difficulty

Divide the class into four groups. Assign each group a section of the chapter. Ask each group to write a brief summary of the assigned section. Then have a volunteer from each group read the summary to the class. Make a list on the board of the main ideas presented in each summary. Then discuss the main ideas of the chapter with the class.

Auditory Learners

Have students describe and discuss the role of each West Central European country in the global community. Then hold a panel discussion about the challenges facing countries in this region and ways in which these challenges might affect other countries in the world.

Visual Learners

Have students work in pairs to make land-use maps of West Central Europe, using information provided in the chapter. Then have students compare their land-use maps with physical maps and climate maps of the region.

Tactile/Kinesthetic Learners

During one week, have students browse through newspapers and magazine advertisements, watch television commercials, and look in grocery and department stores for products from West Central European countries sold in the United States. Have students keep a log of the goods and the countries where they are made. At the end of the week, make a chart on the board that shows students' findings.

Gifted Learners

Assign students a different West Central European country to research the country's main trading partners, imports, and exports. Then provide each student with Outline Map 1 of the world. Have each student create a trade map of his or her country, using arrows to show the major trading partners and the flow of imports and exports to and from the assigned country.

MULTICULTURAL FOCUS

Write the following information on the board in the form of a time line:

1790: first wooden, scooter-like bicycle, called a *célérifère*, invented by Comte Mede de Sivrac of France

1816: bicycle, called a *draisine*, with a steering bar connected to the front wheel, invented by Karl von Drais of Germany

1839: foot pedals added to the *draisine* by Kirkpatrick Macmillan of Scotland

1885: first safety bicycle with wheels of equal size, invented by J. K. Starley of England

Tell students that the history of the invention of the bicycle makes it appropriate that the most popular international bicycle race, the Tour de France, occurs in Europe.

The three-week long race is a much-anticipated event. Towns and villages bid for the opportunity to be one of the many finishing points on the route. The Tour de France draws riders and tourists from all over the world and people line the roads to cheer on the contestants.

The precise route changes each year, but the rugged 2,500 mile (4,000 km) course always follows a path through the Alps and Pyrennes mountain ranges and finishes in Paris.

Discuss how sporting events reflect cultural ties within a community. *(Students might mention the civic pride that results from baseball, hockey, or football teams being affiliated with a town or city.)*

CHAPTER 25

WEST CENTRAL EUROPE

Section 1: France
Section 2: Germany
Section 3: The Benelux Countries
Section 4: The Alpine Countries

OVERVIEW

In Chapter 25, students will learn about West Central Europe, the industrial and agricultural heartland of Europe.

Section 1 describes France, the region's largest nation and an economic leader.

Section 2 covers Germany, an industrial and trade giant whose reunification has had both positive and negative effects.

Section 3 discusses the Benelux countries (Belgium, the Netherlands, and Luxembourg), modern, prosperous nations active in world trade and business.

Section 4 features the Alpine countries of Switzerland and Austria, two highly developed nations with a central European location.

About the Author

Peter Mayle (1939–), a former advertising executive, left England to fulfill his dream of living in rural France. The quotation on this page is from *A Year in Provence*, Mayle's best-selling book about how he and his wife adjusted to their new environment.

CREATIVE STRATEGIES FOR TEACHING WORLD GEOGRAPHY

LESSON 11 Dream Vacation: Europe (This lesson is appropriate for any European country.)

Primary Source

Description of Change: excerpted and bracketed
Rationale: excerpted to focus on main idea; bracketed to clarify meaning

CHAPTER 25

West Central Europe

❝**M**eanwhile, a thousand miles to the north, the wind that had started in Siberia was picking up speed for the final part of its journey. We had heard stories about the Mistral. . . . Typical Gallic [French] exaggeration, we thought. . . . And so we were poorly prepared when the first Mistral of the year came howling down the Rhone valley, turned left, and smacked into the west side of the house with enough force to skim roof tiles into the swimming pool and rip a window that had carelessly been left open off its hinges. The temperature dropped twenty degrees in twenty-four hours.❞

Peter Mayle

The history of West Central Europe is rich and varied, and the region's influence has extended over distant parts of the world. Today, West Central Europe is the continent's industrial heartland and center of population and agricultural production. This core region includes France; Germany; the Benelux countries of Belgium, the Netherlands, and Luxembourg; and the Alpine countries of Switzerland and Austria.

France is a leading industrial and agricultural nation, whose people share a great cultural pride in their country. Germany is one of the world's largest industrial powers. West of Germany, along the North Sea, lie the modern and industrialized Benelux countries. The landlocked, rugged Alpine countries, meanwhile, have prospered by taking advantage of their central European location.

Shield and helmet of King Charles IX of **France** (1550–1574)

GEOGRAPHY DICTIONARY

mistral
primate city
city-state
loess
cosmopolitan
dike
polder
confederation
canton
foehn
arable

Geneva, Switzerland

Using the Quotation

Write the term **mistral** on the board. After students have read the quotation on page 282, ask them to speculate about the meaning of the word. (*a type of wind*) Then have them read the quotation again, noting the characteristics of the mistral. (*strong, cold; comes from the north, blows through southern France*) Point out that *mistral* means "masterful."

Using the Geography Dictionary

Pronounce for students the terms in the Geography Dictionary. Have students work in pairs to look up the terms in the Glossary. Direct the partners to group the terms based on their meanings. Call on the partners to explain their word connections. **(mistral, foehn; primate city, city-state, cosmopolitan; loess, arable; dike, polder; confederation, canton)**

SECTION 1
OBJECTIVES

The objectives reflect the Focus questions on this page and are assessed in the review on page 285.

1. Analyze France's physical and economic geography.

2. Identify France's human and urban characteristics and its role in the global community.

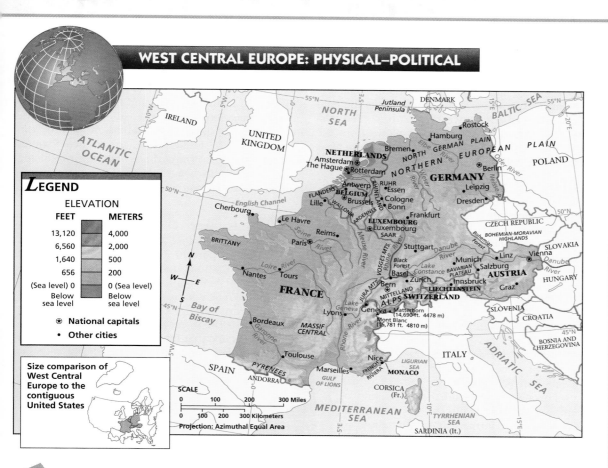

WEST CENTRAL EUROPE: PHYSICAL–POLITICAL

LEGEND

ELEVATION

FEET		METERS
13,120		4,000
6,560		2,000
1,640		500
656		200
(Sea level) 0		0 (Sea level)
Below sea level		Below sea level

⊛ **National capitals**

• **Other cities**

Size comparison of West Central Europe to the contiguous United States

SCALE
0   100   200   300 Miles
0   100   200   300 Kilometers
Projection: Azimuthal Equal Area

Using the Physical–Political Map: West Central Europe

Direct students' attention to the map on this page. Ask a volunteer to name the countries in this region. Ask questions such as, What bodies of water border these countries? (*Baltic Sea, North Sea, Atlantic Ocean, Mediterranean Sea*) Which is the largest country? (*France*) Which countries are landlocked? (*Luxembourg, Switzerland, Austria*) What physical features do various countries share? (*Rhine and Danube Rivers, Lake Geneva, the Alps*)

FRANCE

FOCUS

● What is France's physical geography? economic geography?

● What human and urban characteristics does France have? What role does France play in the global community?

REGION **Physical Geography** France's landforms vary from high mountains to hills and flat plains. Its three coastlines face the North Atlantic Ocean to the west, the English Channel to the northwest, and the Mediterranean Sea to the southeast. Numerous rivers and two major climate regions add to the country's diverse geography. (See the map above.)

Except for Brittany, which is part of Europe's Northwest Highlands landform region, most of

northern France is part of the Northern European Plain. This lowland area is the largest landform region of France.

The mostly hilly Central Uplands lie mainly in central and eastern France and include the Massif Central, a volcanic area in south central France. The Ardennes, Vosges, and folded Jura mountain ranges form France's eastern borders with Germany and Switzerland. The rocky and rugged Mediterranean island of Corsica also falls in this landform region.

The mountainous region of southern France includes the Pyrenees and the Alps. The beautiful French Riviera is located where the Alps meet the Mediterranean coast.

France has four important rivers: the Seine, the Loire, the Garonne, and the Rhone. The canals that connect these rivers are being modernized to provide a better network for transporting goods to neighboring nations.

Except for the high mountain areas, almost all of France has a mild climate. Northern France

Ask students to call out words or images that come to mind when they think of France. *(Paris, fashions, sidewalk cafes, Eiffel Tower)* Continue until students have run out of ideas. Tell them to make a list of additional images as they study the section.

Building a Vocabulary

Write the term **primate city** on the board. Note that *primate* is related to *primary,* meaning "first." Explain that Paris is a primate or primary city—a city that is first in a nation in population and economy. Ask students to tell how all these terms mean "first": *primary election, primary school, primary color, primary tooth.* Note that *prime, primer,* and *primeval* also belong to this word family.

Describing

Physical, Economic, and Urban Geography Organize the class into three sections: Northern France, Central France, and Southern France. Ask students to be ready to describe the landforms, rivers, climate, natural resources, economic activities, and major cities of their assigned area. Emphasize France's diverse landforms, and ask students to tell how these landforms affect ways of life.

THE WORLD TODAY VIDEODISC PROGRAM

France: Culture Clashes

1B 32271 4:56

Play   Pause

has a marine-west-coast climate with mild temperatures and plentiful rainfall. Southern France has a sunny and warm Mediterranean climate. The **mistral,** however, sometimes breaks the mild Mediterranean winter weather pattern. The mistral is a strong, cool wind that blows from the Alps toward the Mediterranean coast. These powerful winds follow the valley of the Rhone River.

Economic Geography Although few of the French people are farmers, agriculture remains a vital part of the economy. Second only to the United States in farm exports, France is by far the largest farm exporter in the European Union.

In the north and interior of France, the leading agricultural products are wheat, sugar beets, potatoes, dairy products, and meat. The typical Mediterranean crops of grapes, olives, and vegetables thrive in southern France. Here, however, farmers must use some irrigation. France leads the world in the variety of wines produced and in export income from wine.

Lorraine, in the northeast, is one of Europe's largest sources of iron ore. Deposits here formed the foundation of France's early steel industry and were a major export. The traditional steel, coal, and textile industries are being replaced by new industries, however.

Some French industries, particularly those using high technology, have expanded rapidly. France produces satellites, jet aircraft, and helicopters. The nation's aerospace industry, which requires aluminum, is centered in southern France near the region's bauxite mines.

The country's main economic concerns lie in the area of energy resources. The French use some hydroelectricity and solar energy, and operate a tidal power plant on the north coast. Still, the nation relies on nuclear energy for more than half of its electricity and must import oil as well.

The French are famous for the production and export

of perfumes, jewelry, glassware, and furniture. These finely crafted French goods are sold in stores around the world. France also remains a leader in fashion design, and Paris is one of the world's fashion centers.

Tourism is another important part of the economy. The French Alps, the French Riviera, the coast of Brittany, the many wine-producing regions, and the beautiful city of Paris attract millions of foreign visitors each year.

PLACE **Human Geography** France has a rich culture unified by language and religion. About 75 percent of the French people are Roman Catholic. In addition, the French government budgets money to maintain and promote French culture and language.

Paris, France's capital and largest urban area, is the **primate city** of France. A primate city is one that ranks first in a nation in terms of population and economy. Paris contains almost one-fifth of the nation's total population in its metropolitan area. The city of Paris was founded more than 2,000 years ago on an island in the middle of the Seine River. This island was at the crossroads of water and land routes that were important to France and the rest of Europe. As Paris grew, it spread from the island onto both banks of the Seine.

Paris, a nodal region, is the nation's center for education, government, communications, banking, business, and tourism. As the transportation center of France, Paris is the hub of all major rail, canal, air, and highway systems in the country. The city is served by a modern subway system called the *métro,* two major airports, and the nearby seaport cities of Le Havre (luh HAHVR) and Cherbourg (SHEHR-bur). Like other major cities, Paris suffers from air pollution, traffic, noise, and urban sprawl.

The busy seaport of Marseilles (mahr-SAY) on the Mediterranean is France's oldest and second largest city. Marseilles also has a major French

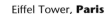

Eiffel Tower, **Paris**

naval base, and is an industrial center. Lyons (lee-OHN), on the Rhone River in eastern France, is the nation's third largest city. Lyons also is an important manufacturing center.

Issues Like most industrialized nations, France faces problems such as unemployment and international competition in industry and agriculture. Dependence on imported oil is a continuing concern for the nation.

France enjoys close ties with most former colonies through an organization called the French Community. Immigration from these former colonies and elsewhere is a heated issue in France. Politicians have debated the effects of immigration on the economy and French culture.

The country still maintains overseas territories such as Guadeloupe in the Caribbean and Réunion in the Indian Ocean, which send representatives to France's parliament. The people of the Mediterranean island of Corsica have a degree of self-government, yet some want complete independence from France. Meanwhile, independence

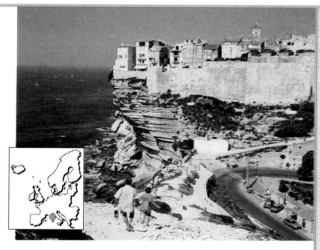

The historic town of **Bonifacio** overlooks the Mediterranean Sea at the southern tip of **Corsica**. The French island's coastline consists of high cliffs with few natural harbors.

movements on the French island of New Caledonia in the Pacific have led to violence.

France has had a long relationship with the United States. The French helped the American colonies fight for their independence from Britain. Over a century and a half later, during World War II, the United States helped drive the Germans out of France. Signs of the countries' special relationship mark both cultures. For example, the United States' Declaration of Independence and the French Declaration of the Rights of Man express similar ideas. Perhaps the most obvious symbol is the Statue of Liberty in New York harbor, a gift from France to the United States.

Militarily, France is strong, with its own nuclear-defense system. Although France withdrew its armed forces from NATO in the 1960s, the country has kept some ties with the alliance for the defense of Europe. France is one of the most forceful influences in the European Union.

SECTION REVIEW

1. What landform regions and climates does France have? What economic activities are found in different areas of France?
2. How are Paris, Marseilles, and Lyons important?
3. **Critical Thinking** **Compare** France's relationships with its former and present colonies to those of the United Kingdom with its former and present colonies.

FACTS IN BRIEF — West Central Europe

COUNTRY POPULATION (1994)	LIFE EXPECTANCY (1994)	LITERACY RATE	PER CAPITA GDP (1993)
AUSTRIA 7,954,974	73, male 80, female	99% (1992)	$17,000
BELGIUM 10,062,836	74, male 80, female	98% (1992)	$17,700
FRANCE 57,840,445	74, male 82, female	99% (1991)	$18,200
GERMANY 81,087,506	73, male 80, female	99% (1991)	$16,500
LIECHTENSTEIN 30,281	74, male 81, female	100% (1992)	$22,300 (1990)
LUXEMBOURG 401,900	73, male 81, female	100% (1989)	$20,600
MONACO 31,278	74, male 82, female	99% (1989)	$16,000 (1991)
NETHERLANDS 15,367,928	75, male 81, female	99% (1991)	$17,200
SWITZERLAND 7,040,119	75, male 82, female	99% (1991)	$21,300
UNITED STATES 260,713,585	73, male 79, female	97% (1991)	$24,700

Sources: *The World Factbook 1994,* Central Intelligence Agency; *The World Almanac and Book of Facts 1995*

The German economy is recovering from the high costs of reunification in the early 1990s.

SECTION 2
OBJECTIVES

The objectives reflect the Focus questions on this page and are assessed in the review on page 287.

1. Describe Germany's physical and economic features.
2. List characteristics of Germany's regions, and identify challenges facing the country.

TEACHING SECTION 2
Motivator

Ask students to imagine that the Civil War in the United States had created two "Americas." Ask students to suggest what problems would occur with reunification today. Briefly explain the very different reasons for the division of Germany, and tell students that the joys and problems of Germany's reunification are discussed in this section.

Building a Vocabulary

Write the term **loess** on the board. Tell students that a diagram can aid in recalling a word's meaning. Read a description of loess from an encyclopedia. Then work with the class in creating a diagram for it. The diagram might show wind acting on exposed rock, carrying away tiny mineral particles and depositing them in another place.

COOPERATIVE LEARNING

Organize the class into five groups. Assign one of Germany's five economic regions to each group. Tell the groups that they are to prepare for a trade fair in one of Germany's cities. Groups should decide what products they will bring to the fair to represent the economic activity of their region. Students should draw pictures of or find magazine ads for these products. Have groups display their products. Groups may want to make signs and maps for their displays.

Photo Caption Answer

The North German Plain, which is part of the Northern European Plain

2 GERMANY

FOCUS

- *What are Germany's physical and economic features?*
- *What are Germany's five major regions? What challenges does Germany face?*

The German federal system consists of 16 *Lander* (states), of which three are **city-states**: Berlin, Bremen, and Hamburg. A city-state is a self-governing city and its surrounding area. Although the seat of government presently remains in Bonn, Berlin is the capital of Germany.

REGION **Physical Geography** Germany is divided into three main landform regions: the North German Plain, the Central Uplands, and the Alps. (See the map on page 283.) The elevation of Germany increases from north to south. The lowest region is the North German Plain, which has coastlines on the North and Baltic seas and is part of the Northern European Plain. The German portion of the Central Uplands contains the hilly Black Forest and Bohemian Forest. The Alps rise along the southwestern border of Germany.

Germany's major rivers include the Rhine, the Danube, the Elbe, and the Weser. An excellent canal system links the rivers to form a network of waterways. The Rhine is particularly vital to Germany's transport system and industry.

Most of Germany has a marine-west-coast climate. Landforms and distance from the sea cause the temperature and rainfall to vary. The northern part of the country, which is affected by moist and mild ocean winds, has mild winters and cool summers. The region to the south and east, toward the interior of Germany, experiences a more continental influence. Here, summers are warmer, and winters are colder. Most of Germany receives enough rainfall for productive farming.

INTERACTION **Economic Geography** Although Germany is one of the world's giant industrial nations, it does not have a wealth of natural resources. Coal deposits boosted its industrialization in the last century and continue to supply about one-third of the nation's energy. Germany is Europe's leading steel producer, yet it must import iron ore. Germany also imports most of its oil and natural gas.

Germany's forests are mostly in the Central Uplands. Conservation and careful harvesting of timber is strictly practiced. The nation's recent forest losses have resulted mainly from air pollution rather than from logging or land development. Tree loss from pollution and acid rain is of national concern to Germans.

Germany's industrious population is still its most important economic resource. Its workers, among the world's best paid and best educated, are considered world leaders in scientific, technological, and organizational skills.

German industry is extremely diverse. The country is a world leader in the production and

Historic castles are nestled throughout the Bavarian Plateau of southern **Germany**. Look at the physical–political map on page 283. What landform region dominates northern Germany?

Writing

Germany's Challenges Have students write help-wanted ads that might appear in newspapers in eastern Germany. Students should think of workers needed to bring the region up to the economic level of western Germany, to repair its infrastructure, and to deal with its environmental problems. Have students explain their ads. Ask: Why is unemployment a problem in eastern Germany?

Guided Practice

On the board, draw a rough sketch of Germany's elevation from north to south. Review the three main landform regions with the class, showing how these physical features correspond to the rise in elevation.

Independent Practice

Have students make a graphic organizer of Germany's five major geographical regions.

Closure

Have students list the challenges facing Germany and suggest ways these challenges might be addressed.

ASSESSMENT

Assign the Section 2 Review questions on this page.

Reteaching

Assign Section 2 of Reteaching Worksheet 25.

export of automobiles, electronics, chemicals, and optical and surgical instruments. Its transportation network, with modern railroads, airports, and *autobahns* (freeways), is admired around the world.

Reunification has allowed the nation to combine many valuable resources. Reunified Germany is a large country with varied geographical characteristics. The country can be divided into five major regions.

Northwestern Germany Northwestern Gemany lies along the windy and damp coast of the North and Baltic seas. Soils are poor here, and sand dunes cover most of the offshore islands. Farther south and inland are some of Europe's richest soils. This productive agricultural belt is created by thick deposits of **loess** (LEHS), or dust-sized soil particles deposited by the wind. Grains, potatoes, and sugar beets are important crops. Hamburg, on the Elbe River, is Germany's major seaport and second largest city.

The Ruhr The Ruhr in western Germany is the industrial heart of Europe. The region developed around Europe's largest coal deposits and became Germany's most densely populated and most polluted area. The Ruhr has become almost one continuous belt of cities and industry, called the Ruhrstadt (Ruhr City). This "smokestack region" manufactures chemicals, pharmaceuticals, plastics, and automobiles. To the south, along the Rhine River, are the cities of Cologne and Bonn, the seat of government.

The Rhineland South of the Ruhr is the fertile Middle Rhine Valley, also called the Rhineland. This is both an agricultural and an industrial area. The Rhine Valley is famous for its castles, forests, vineyards, and farms lining the steep river gorge. This region's main city, Frankfurt, is Germany's transportation and financial center. West of the Rhine River, along the French border, is an old industrial area called the Saar. The region's coal deposits and steel industry helped it prosper. Also an agricultural area, the Saar is famous for its wine and cheese.

Southern Germany Southern Germany includes Bavaria, the largest German state, and the major city and manufacturing center of Munich (MYOO-nik). Munich produces electrical and scientific equipment, automobiles, textiles, and toys. Munich and nearby Stuttgart have been growing

rapidly due to new high-tech industries. Southern Germany also has rich farmland and mountains, forests, and lakes.

East Germany The former East Germany lies mainly on the Northern European Plain. Its soils are fertile, with sugar beets, grains, and potatoes the most important crops. Inefficient government-run farms in the region are being returned to private landowners. In the southeast, coal deposits lie beneath the hilly region of the Central Uplands. The city of Rostock in the north is the region's major Baltic port, while Berlin is the region's largest city and the country's capital. (See "Cities of the World" on pages 288–289.)

One of the most serious challenges facing eastern Germany is cleaning up pollution. The former East German Communist government used low-grade coal for its energy needs and dumped chemical wastes into the region's rivers.

Because East Germany lagged behind West Germany in transportation, housing, and industry, a "wealth gap" exists between eastern Germany and the rest of the country. Eastern Germany's services and its inefficient and outdated industries are being modernized at great cost.

Issues Although unification was a goal of Germany for more than 40 years, the financial and emotional realities of reunification have been challenging. Many Germans did not realize the costs of elevating eastern Germany to western standards. The nation must deal with eastern Germany's serious environmental problems, crumbling infrastructure, and high unemployment rate. Many eastern Germans have moved to western Germany to seek employment.

Germany is a key member of the EU and NATO. As an economic powerhouse of Europe, Germany will continue to play an important role in Europe's future.

② SECTION REVIEW

1. What are Germany's major rivers, and how are they economically important? What products does Germany export?
2. Which of Germany's regions are agricultural? Which are industrial? Which have coal resources?
3. **Critical Thinking** **Explain** both the positive and negative aspects of Germany's reunification.

Text continues on page 290.

THEME
Movement

Germany's newest canal links the Main, a tributary of the Rhine, with the Danube. The canal crosses Europe's continental divide and connects hundreds of inland North Sea ports with the Black Sea. The emperor Charlemagne tried to have such a canal dug in A.D. 793, but the project failed.

Section 2 Review
ANSWERS

1. Rhine, Danube, Elbe, Weser; they are vital to transportation and industry; steel, automobiles, electronics, chemicals, optical and surgical instruments

2. Agricultural: northwestern, southern, and eastern Germany, the Rhineland; industrial: the Ruhr, the Rhineland, southern Germany; coal: the Ruhr, the Rhineland, eastern Germany

3. Positive aspects: reunification was a cultural goal, and it has allowed the country to combine resources. Negative aspects: the country must deal with eastern Germany's relative lack of modernization, environmental problems, poor infrastructure, and high unemployment.

Cities of the World

Using Geography Themes

Have students work independently or in small groups to draw three maps illustrating the following: Berlin as the capital of Germany in 1871, Berlin as a divided city after World War II, and a reunified Berlin today. For each map, instruct students to provide a caption relating the geographic theme of region to Berlin during that period of time in the city's history. Encourage students to consult outside references (such as historical atlases) and to make their maps as detailed as possible.

Geography and History

Allied aerial bombardment during World War II practically leveled Berlin. Women known as the Trummerfrauen ("rubble women") removed the huge amount of debris in the city left by bomb damage. At the war's end, approximately 70 percent of Berlin's population was female.

Berlin

Berlin, once again the capital of a united Germany, is home to many symbols of historic German power and influence.

Historically, Berlin has symbolized Germany's greatest triumphs and achievements as well as its worst tragedies and crimes. Today, Berlin is again Germany's capital. For many, the city represents the end of the Cold War that divided both Germany and Europe for nearly 45 years.

The Heart of Germany

Present-day Berlin officially became a city in 1307 when two towns on opposite banks of the Spree River merged. At this time, Germany was divided into hundreds of kingdoms, or states. Munich, Cologne, and Dresden, all well-established cultural centers, were among the major cities of the German states. Smaller Berlin, a part of Prussia, grew slowly during the next 300 years.

Rapid development characterized Berlin in the second half of the seventeenth century. By the 1680s, the city had several new cultural institutions, including its first public library. In 1701, the King of Prussia officially made the city his royal capital.

In the decades that followed, the Prussian kings continued to build, rebuild, and expand their capital. Berlin soon became an important political, cultural, and economic center. The Brandenburg Gate and a wide, tree-lined avenue called Unter den Linden are among the famous Berlin landmarks built by the Prussians.

By 1815, Berlin had a population of about 200,000. Growth intensified as the Industrial Revolution lured hundreds of thousands more

The *Quadriga of Victory*, a statue of a horse-drawn chariot, commands the top of the Brandenburg Gate at the western end of Unter den Linden. The Brandenburg Gate was completed in 1791.

people to the city to work in the new factories. In 1871, Prussia completed its task of creating a unified Germany. Berlin became the capital of the German domain and Europe's strongest economic and military power. By the early 1900s, Berlin was a large metropolis of more than 2 million people.

Berlin remained the capital of Germany through two world wars. From 1933 to 1945, when Adolph Hitler and the Nazi party ruled the country, Berlin represented in the eyes of the free world the horrors that Nazi Germany had unleashed upon the world.

After Germany's defeat in World War II, Berlin, like the country itself, was divided. East Berlin became the capital of East Germany, a Communist state controlled by the Soviet Union. West Berlin became part of democratic West Germany, an ally of the United States. Bonn, a small city on the Rhine River, became West Germany's capital. West Berlin was entirely surrounded by East German territory.

Divided Berlin immediately arose as a crisis spot in the Cold War between the United States

288

Writing About History

Ask students to write an eyewitness account of the tearing down of the Berlin Wall. Instruct them to write it from the perspective of a participant in the event, a journalist, or a government official in either East or West Germany. Tell them to use information in the feature as well as their historical imagination to write the accounts.

and the Soviet Union. Between 1945 and 1961, almost 3 million East Germans, fleeing Soviet control, escaped to the West. Many crossed from East Berlin into West Berlin.

To stop this mass flight, Soviet-backed East Germany built the Berlin Wall in 1961. This barrier of concrete, electric fences, barbed wire, floodlights, trenches, armed soldiers, and attack dogs eventually surrounded all of West Berlin. In time, the massive wall came to symbolize the division between East and West.

For 28 years, democratic West Berlin stood out as a prosperous industrial center highlighted by many cultural attractions. In contrast, Communist East Berlin, though a showplace city of Eastern Europe, experienced slower economic growth and a deteriorating infrastructure. This division existed until 1989

The historic Brandenburg Gate is in central **Berlin** just east of where the Berlin Wall once stood. The wall cut across many of Berlin's busiest thoroughfares.

when the collapsing East German government opened the Berlin Wall. News reports throughout the world showed Berliners tearing down the barrier. What was once a symbol of division now symbolized freedom and unity.

The Reunited City

Today, reunified Berlin is home to more than 3 million people. Except for a few fragments, the Berlin Wall is gone, and landmarks in what was once East Berlin are being restored. Berlin holds many museums, universities, and theaters. Parks, woodlands, and waterways occupy a large part of its land area.

Berlin, however, faces many challenges. Rebuilding the eastern part of the city is very expensive. In addition, many former East Berliners have been slow to adjust to a capitalist economy. Still, Berlin's location and well-educated population promise to attract many new businesses to the city. With its difficulties related to reunification, combined with high hopes for the future, Berlin continues as a symbol for all of Germany.

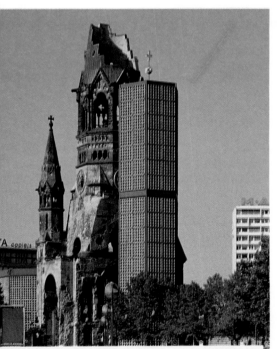

The remains of the old Kaiser Wilhelm Memorial Church and its towering, modern replacement stand as a reminder of the devastation of World War II and the postwar rebirth of **Berlin**.

Geographical Sidelight

For months after the Berlin Wall was officially opened, souvenir hunters and others wishing to sell bits of history continued to chip away at the wall's remaining fragments. These scavengers became known as "wallpeckers."

THE WORLD TODAY VIDEODISC PROGRAM

Germany: Moving Home

1B   17236   4:19

Play   Pause

SECTION 3
OBJECTIVES

The objectives reflect the Focus questions on this page and are assessed in the review on page 291.

1. Examine physical and economic characteristics of the Benelux countries.

2. Summarize urban and cultural aspects of the Benelux countries.

TEACHING SECTION 3
Motivator

Have students look for Benelux on the map of West Central Europe on page 283. They will discover that Benelux is not on the map. Explain that the name *Benelux* is an acronym for *Be*lgium, the *Ne*therlands, and *Lux*embourg. The Benelux Economic Union became operative in 1948. Ask students to suggest common characteristics of these countries.

Building a Vocabulary

Help students remember the meaning of **cosmopolitan** by breaking it down. Point out that *cosmos* means "world" or "universe," and *-polis* means "city." *Cosmopolitan* literally refers to a "city of the world." Note that New York City can be described as cosmopolitan. Have students name features of a cosmopolitan city. (*many foreign visitors, ethnic neighborhoods and restaurants, attitude of openness to outside influences*)

GLOBAL PERSPECTIVES

The International Court of Justice, established by the UN Charter in 1945 and located in the Netherlands, has rendered verdicts in areas such as boundaries, treaties, and human rights. Compliance with the court's rulings is not mandatory, and only about four dozen countries have chosen to abide by its verdicts. Ask students why some nations might be more interested than others in accepting the court's rulings. (*Smaller, less powerful countries can obtain a more equal legal foothold with larger nations.*)

3 THE BENELUX COUNTRIES

FOCUS

- *What are the physical and economic characteristics of the Benelux countries?*
- *What are the urban and cultural characteristics of the Benelux countries?*

Belgium Belgium can be divided into three landform regions: a coastal region, a central plain, and the Ardennes. Much of the coastal plains, which are used mainly for farming, is reclaimed from the sea and from wetlands. The central plain between the Schelde and Meuse rivers contains Belgium's most productive soils. In the southeast, the forested hills of the Ardennes reach to 2,283 feet (696 m). (See the map on page 283.)

Belgium has two language regions. In the northern coastal part of the country, people speak Flemish, a language related to Dutch. This region is called Flanders and contains about 70 percent of the country's population. In the southern part of Belgium, the French-speaking people, called Walloons, make up most of the region's population. This language region is called Wallonia. Most of the nation's citizens view themselves as either Flemish or Walloon rather than as Belgians. The government now has assemblies for each language region.

Belgium's main resource is its skilled and productive labor force, which produces high-quality machinery and textiles. Belgium is known for quality carpets, cut diamonds, and fine chocolate as well. With its small population, Belgium depends on international trade to maintain its high standard of living. Its small but efficient farms provide most of the nation's food and produce some exports.

Brussels, Belgium's capital and largest city, is one of Europe's most **cosmopolitan** cities. A cosmopolitan city is one that is characterized by its many foreign influences. Brussels' central location in Europe and excellent transportation connections have made the city a center for finance. Numerous organizations have their headquarters there, including the EU and NATO.

INTERACTION **The Netherlands** *Netherlands* means "low lands," and indeed, nearly one-half of the Netherlands lies below sea level. (See the map on page 283.) These flat, low-lying lands were once covered by wetlands or the North Sea. First, the country built **dikes**, or walls to keep out water. Then it constructed a system of canals to remove the water. Finally, the land was drained, or pumped dry, by windmills, which became a symbol of the country. Lowland areas that have been drained are called **polders**. Today, polders are the sites of farms, cities, industries, and airports. (See "Themes in Geography" on pages 292–293.) The name Holland is sometimes used to refer to the Netherlands. Its people are known as the Dutch.

Although the country is well known for its agriculture, the Netherlands is a highly industrialized nation. The diversified Dutch industry is geared toward the export market, and the EU nations are its major customers. One of the world's largest oil refineries and the headquarters for the world's largest oil company are located in the Netherlands. Also important are food processing, dairy products, and breweries. Important crops include potatoes, sugar beets, and flowers, especially tulips.

The Netherlands must import almost all of its industrial raw materials. Though there are coal deposits in the south, it is cheaper for the Netherlands to import coal than to mine it. The

Brussels is the capital of **Belgium** and headquarters for many international organizations, including NATO.

Classifying

Economic Characteristics Make two columns on the board: Exports and Imports. Have students name exports and imports of the Benelux countries as you list them in each column. By each item, write B, N, and/or L to indicate whether it is an export or import of Belgium, the Netherlands, or Luxembourg. Then ask students to draw conclusions about the importance of trade to the prosperity of these countries.

Closure

Have students make a list of the cultural features of the Benelux countries.

ASSESSMENT

Assign the Section 3 Review questions on this page.

Reteaching

Assign Section 3 of Reteaching Worksheet 25.

SECTION 4
OBJECTIVES

The objectives reflect the Focus questions on this page and are assessed in the review on page 294.

1. Discuss relationships between Switzerland's political, cultural, physical, economic, and urban characteristics.

2. Identify important characteristics of Austria.

The Dutch have little breathing room in this crowded city in the southwestern part of the country. The **Netherlands** is one of Europe's most densely populated countries.

country has large natural-gas deposits. About half of this gas is exported by pipeline to other European nations. The country imports almost all of its oil.

The old port city of Amsterdam is the Netherlands' official capital. Famous for its canals, museums, and art, Amsterdam also is home to international banking and modern industries. The North Sea Canal connects Amsterdam to the North Sea.

The city of Rotterdam to the southwest serves the trade traffic that enters and exits Europe on the Rhine River. Rotterdam is the world's busiest seaport and the site of oil refineries, chemical plants, and shipbuilding. The Hague is the seat of government, where parliament, the royal family, and the International Court of Justice are all located.

The Netherlands is a modern democratic society. Dutch is the national language, but most people speak English, German, or French as well. The country has Europe's most expensive social security system. Government benefits for education, unemployment, and medical care are among the highest in the world. Dutch workers receive high wages and are considered to be some of the world's most productive and skilled.

Luxembourg Tiny Luxembourg is a prosperous country bordered by France, Germany, and Belgium. (See the map on page 283.) Luxembourg's population is almost all Roman Catholic, and the people speak a Germanic language with many French influences. Most citizens are multilingual and well educated.

Luxembourg's iron and steel industry is located in the south near the iron-ore mines. This was once the nation's major industry. In recent decades, however, the nation has attracted international banking and rubber and chemical industries.

SECTION REVIEW

1. What is Belgium's major resource? What is significant about the Netherlands' elevation?

2. What are Belgium's two language regions?

3. Critical Thinking **Suggest** three reasons for Luxembourg's prosperity.

4 THE ALPINE COUNTRIES

FOCUS

● *How are Switzerland's political, cultural, physical, economic, and urban characteristics linked?*

● *What are Austria's main characteristics?*

Switzerland Switzerland is a **confederation** of 26 **cantons**, or states. A confederation is a group of states joined together for a common purpose. Each canton governs itself, except for military defense and international relations, which are under federal control.

Although Switzerland appears to be a culturally divided land, it is one of the world's most stable countries. Most Swiss speak German, but many also know French, Italian, and Romansh. (See the map on page 294.) The nation is peacefully and equally divided into two major religious groups, Protestant and Roman Catholic.

Switzerland contains three landform regions: the forested Jura Mountains, the Alps, and the Swiss Plateau. (See the map on page 283.) Most of Switzerland is mountainous. Glaciers, long narrow lakes, and deep winter snows dominate the spectacular Alpine scenery. Many of Europe's major rivers, including the Rhine and the Rhone, begin in the Swiss Alps.

The Swiss Plateau, or Mittelland, separates the Jura Mountains and the Alps. The Mittelland

Text continues on page 294.

Themes in Geography

Critical Thinking

After students have read the feature, elicit comments by asking several questions: Are the Dutch working with or against their environment? Are the Dutch fighting a losing battle with a traditional and hostile enemy (the sea)? Do the economic benefits from agricultural development outweigh the expense and difficulties of repairing dikes and maintaining polders? Encourage students with opposing opinions to elaborate on their views.

THEME
Interaction

Land reclamation in the Netherlands has been costly and sometimes even futile. It is estimated that nearly 2,720 square miles (7,045 sq. km) of land were reclaimed between 1200 and 1970. During these same years, however, approximately 2,000 square miles (5,180 sq. km) were lost to the sea. Today, about 12 square miles (30 sq. km) of land are reclaimed each year in the Netherlands.

Photo Caption Answer

Any cities that lie below sea level, which include Rotterdam and The Hague

Polders of the Netherlands

The building of polders is an example of people adapting the environment to suit their needs.

Much of the Netherlands was once covered by shallow sea, inland lakes, and swamps. As the demand for agricultural and industrial land increased, the Dutch found that they could reclaim, or take back, large areas of valuable land from the sea if they drained the water from low-lying areas. They believed that reclamation was necessary because most of the Netherlands consists of low-lying and flooded land, with few areas of higher elevation.

At first, the Dutch built walls, called dikes, around the area to be drained. Then they pumped the water from this area into canals that drained into the North Sea. In the early days, the Dutch used windmills to generate the power to operate the pumps. The windmill became a symbol of the Netherlands. Later, they used steam power and, eventually, electric power. The areas of land that were once submerged beneath the sea or inland waters are called polders. Because the polders lie below sea level, they have no natural drainage. All excess water must be pumped from the polders into the canals.

As a result of the creation of the polders, and because of the poor natural drainage of the land, the Dutch landscape is covered by a network of canals. These are important for drainage, for irrigating farmlands, and as waterways for transporting goods. The polders have provided very fertile farmland.

The Netherlands is one of Europe's most densely populated countries. Because it is also a very small country, there is much competition among farming, industry, and cities for use of the reclaimed land. Many of the cities, including Amsterdam, the capital and largest city, have been built on polders.

LOCATION The city of **Amsterdam**, the capital of the Netherlands, is built on land that was once below water. Using the map on page 293, identify other Dutch cities that are built on or near reclaimed land.

Since the first construction of the dikes and creation of polders, the Dutch have struggled constantly to keep the sea from flooding their land. Storm tides from the North Sea have broken through the dikes many times in the country's history. If it were not for the dikes and sand dunes that line the shores, half the country would be flooded at high tide. Major projects are under way to improve the dikes along the country's coastline.

The cost of repairs is high, however. The Netherlands spends several hundred million dollars each year to repair dikes and drain water.

292

This high cost combined with sinking lowlands and other environmental damage has prompted the Netherlands to reconsider its use of the land. The Dutch have made plans to restore a portion of the land to wetlands, lakes, and forests.

In addition, the Dutch have recently been reminded that flooding can come from inland areas as well as from the North Sea. Heavy rainfall in early 1995 caused flooding that forced more than 300,000 people, mostly Dutch, from their homes in parts of West Central Europe. Northeastern France, western Germany, and the Benelux countries received some of the heaviest rainfall. Development along rivers, such as the Rhine River in western Germany and the Netherlands, kept much of the rainwater from being absorbed in marshes and flood-plains instead of spilling into populated areas. Europeans, and especially the Dutch and their neighbors, are searching for ways to prevent similar disasters in the future.

REGION Large areas of land in the Netherlands lie below sea level. This means that excess water must be drained artificially by the use of an extensive system of canals, dikes, and pumping stations. The photograph above shows land after the polders were created. **To where is water from polders drained?**

Polders of the Netherlands

FRISIAN ISLANDS
Wadden Zee
6°E
4°E
53°N
Groningen
NORTH SEA
N
W E
S
Lake Ijssel
52°N
Amsterdam
The Hague
Utrecht
Rotterdam
Waal R.
Rhine R.
Ijssel R.
Maas R.
GERMANY
BELGIUM
51°N
⊛ Capitals
• Other cities
--- Canals
▨ Polders
0   25   50 Miles
0   25   50 Kilometers
Projection: Conic

INTERACTION Since the first polders were built, the Dutch people have fought a constant battle with the waters of the North Sea. A barrier dike protects Lake Ijssel and surrounding polders from the Wadden Zee and the North Sea. Look at the map of the Netherlands to the left. **What areas would be flooded if the coastal dunes and dikes were removed?**

Building a Vocabulary
Review the meaning of **confederation,** and note that Switzerland is a confederation. Ask students to name another country they have studied with this type of government. (*Canada*) Draw a diagram on the board to illustrate the difference between a federal system of government (as in the United States) and a confederation system.

Comparing and Contrasting
Have students prepare a chart that compares and contrasts landform regions, natural resources, and economic activities in the Alpine countries.

Independent Practice
Have students use Outline Map 14 to label mountains, major rivers and lakes, and major cities in the Alpine countries.

Closure
Call out facts such as "Contains the Mittelland," and ask if the fact applies to Switzerland, to Austria, or to both.

ASSESSMENT

Assign the Section 4 Review questions on this page.

Reteaching
Assign Section 4 of Reteaching Worksheet 25.

Map Study
German

THEME
Interaction
Although the Alps already contain dozens of road and rail tunnels, Switzerland is planning two more railway tunnels. They will carry freight and high-speed passenger trains through the Central Alps. Switzerland expects that a unified Europe will create a greater need for trans-Alpine travel.

Section 4 Review
ANSWERS

1. Because Switzerland is mountainous, tourism is important to the economy, and agriculture is not. Because of the country's location, its people are culturally diverse.

2. Agricultural and forest resources, hydroelectricity, scenery

3. Its mountainous terrain, its political and economic stability, its diverse population, and its nonmembership in international organizations; increased economic stability and the presence of international organizations and conferences

294

Languages of Switzerland

LEGEND

- German
- French
- Italian
- Romansh language
- ⊛ National capital
- • Other cities

SCALE
0   50   100 Miles
0   50   100 Kilometers
Projection: Lambert Conformal Conic

MAP STUDY
Maps can be used to show areas where specific languages are spoken. Most people in Switzerland are multilingual. The colors on this map show where various languages are dominant. In which language region is the Swiss capital located?

---

stretches from Lake Geneva on the western border with France to Lake Constance on the northern border with Germany.

Switzerland's mainly highland climate has variable temperatures and precipitation within short distances. The mountains surrounding the country protect much of it from Northern Europe's cold winter air. In the Alpine valleys and the Mittelland, the warm southern **foehn** (FUHRN) keeps winter temperatures mild. A foehn is a warm, dry wind that blows down the slopes of the Alps.

Except for hydroelectric power, Switzerland is resource poor. The country's mountainous terrain and poor soils mean there is little **arable** land, or land that is suitable for growing crops. Dairy farming and grape growing in the Mittelland are the country's main agricultural activities.

Switzerland's economy depends largely on foreign trade. Most Swiss goods are exported, mainly to EU nations. Swiss industries include machinery, chemicals, and watches. The Swiss also are famous for their chocolate and cheese. The snow-covered Alps support a year-round tourist industry.

International banking, however, brings the most foreign funds into Switzerland. The country's peaceful history has made it a stable world finance center. To remain neutral, Switzerland has not joined the United Nations, the EU, or NATO.

All of Switzerland's major cities are in the Mittelland. Zurich, the largest, is a leading world-banking center. Bern, the nation's capital, is centrally located between the country's German-speaking and French-speaking populations. The

city of Geneva is known for diplomacy. It is home to many international organizations and conferences.

Austria Austria is about twice as large as neighboring Switzerland. (See the map on page 283.) Almost all of the population is German-speaking and Roman Catholic. The Alps loom over the western two-thirds of the nation. Vienna, Austria's capital city and heart of commerce and culture, lies along the banks of the Danube River.

Austria is mainly an industrial country, but agriculture also is important. Austrian farmers supply almost all of the nation's food. Austria has forest resources and hydroelectricity but is limited in other energy resources. The nation's most valuable natural resource is its scenic beauty, which supports a vital tourist industry.

Austria's diversified industry produces machinery, chemicals, and steel for trade mainly within the EU. The country became a member of the EU in 1995 but also is a major trading partner of Hungary, the Czech Republic, and Slovakia.

4 SECTION REVIEW

1. How has Switzerland's physical geography influenced its culture and economy?
2. What are Austria's major resources?
3. **Critical Thinking Determine** what features have allowed Switzerland to maintain neutrality. What effects has neutrality had on the country?

Building a Vocabulary
1. A city that ranks first in a nation in terms of population and economy; Paris
2. Dust-sized soil particles deposited by the wind; northwestern Germany
3. Dikes were built to keep water out of the lowland areas; the drained areas are polders.
4. It means that Switzerland is a group of states joined together for a common purpose.

Recalling and Reviewing
1. France, Germany, Switzerland, Austria
2. Both countries have agricultural resources. France has more mineral resources than does Germany. Both countries have many industries. Both countries must import oil.
3. Both Belgium and Switzerland are culturally diverse, have specialty industries (such as chocolate), depend on foreign trade, have centers for finance, and are home to international organizations.
4. Possible answers: Paris, on the Seine; Lyons, on the Rhone; Hamburg, on the Elbe; Cologne and Bonn, on the Rhine; Vienna, on the Danube

CHAPTER 25 CHECK

Sidewalk cafe, **Paris**

Recalling and Reviewing

1. In which countries discussed in this chapter are the Alps located?
2. Compare and contrast the economic geography of France and Germany.
3. How are Belgium and Switzerland similar?
4. Name four West Central European cities that are located along rivers. Name the rivers.

Thinking Critically

1. How do you think the French people would react to the idea of France joining with several other European countries to become one large, unified country?
2. Can a nation with few resources or with challenging physical conditions become prosperous? Explain, using two examples from the chapter.

Learning About Your Local Geography

Individual Project
France's mistral and Switzerland's foehn influence weather and climate patterns in West Central Europe. How is the climate of your local community influenced by various winds? Contact the weather bureau or the weather department of a local television station, and request information on your area's winds over the past week—where the winds came from, how strong they were, how they affected the area. Graph your information, and include captions explaining the information shown. Present your graph to the class. You may want to include your graph in your individual portfolio.

Using the Five Themes of Geography
Sketch a map showing the different language **regions** of West Central Europe. With shading or symbols, indicate how a region's **location** determines which languages are spoken in that region. In particular, consider the countries that border any given region or country. How are the languages spoken in that region or country influenced by the bordering countries?

Reviewing the Main Ideas
1. France's landforms vary from high mountains to hills and flat plains. Paris is France's capital and largest urban area. The country is one of the most forceful influences in Europe.
2. Germany is a major industrial power, although it lacks many natural resources. The country can be divided into five major regions: northwestern Germany, the Ruhr, the Rhineland, southern Germany, and East Germany. Reunification has posed some serious challenges for the country.
3. The Benelux countries are Belgium, the Netherlands, and Luxembourg. These three industrialized countries share many geographic similarities.
4. The Alpine countries include Switzerland and Austria. Tourism, banking, manufacturing, and agriculture are major economic activities.

Building a Vocabulary
1. What is a primate city? Give an example of a primate city.
2. Define *loess*. Where in Germany is it found?
3. How were dikes used in the development of polders?
4. What does it mean to say that Switzerland is a confederation of cantons?

Thinking Critically
1. Possible answer: many French people might be reluctant to give up France's independence. The French take great cultural pride in their country, which is unified by language and religion.
2. Possible answer: yes. Neither Germany nor Switzerland is rich in natural resources, and Switzerland's mountainous terrain is not suited for agriculture. Much of the Netherlands lies below sea level and has been drained. Yet all of these countries are prosperous.

Learning About Your Local Geography
Students' graphs should show their areas' recent wind patterns.

Using the Five Themes of Geography
Maps should show the various language regions of the countries of West Central Europe. With shading or symbols, maps should show that a region often uses the languages of bordering or surrounding regions. For example, French is spoken in southern Belgium.

295

CHAPTER 26 Southern Europe
Planning Guide

TEXTBOOK	TEACHING RESOURCES	OTHER AVAILABLE MATERIALS
Section		
1. Spain and Portugal Pages 297–300	Reteaching Worksheet 26.1 Critical Thinking 26	*The World Today* Videodisc Program *Study Guide for Spanish Speakers* *Lecture Notes,* Chapter 26
2. Italy Pages 300–304	Reteaching Worksheet 26.2	*Study Guide for Spanish Speakers* *Lecture Notes,* Chapter 26
3. Greece Pages 304–306	Reteaching Worksheet 26.3 Graphic Organizer 26 Challenge/Enrichment Worksheet 26 Skills Worksheet 26	*Study Guide for Spanish Speakers* *Lecture Notes,* Chapter 26
Chapter Review Page 307	Chapter 26 Test Chapter 26 Portfolio Assessment See also *Alternative Assessment Forms*	*Test Generator,* Chapter 26

CHAPTER RESOURCES

Lecture Notes, Chapter 26
Teaching Resources
    Chapter 26 Graphic Organizer
    Chapter 26 Skills Worksheet
    Chapter 26 Critical Thinking Worksheet
    Chapter 26 Challenge/Enrichment Worksheet

REVIEW AND ASSESSMENT

Chapter 26 Check, page 307
Chapter 26 Test, *Teaching Resources*
Chapter 26 Portfolio Assessment, *Teaching Resources*
Test Generator, Chapter 26
Alternative Assessment Forms

ADDITIONAL RESOURCES

Books for Teachers

Di Scala, Spencer M. *Italy: From Revolution to Republic:*
    1700 to the Present. Boulder, CO: Westview Press,
    1995.
Fuentes, Carlos. *The Buried Mirror: Reflections on Spain*
    and the New World. Boston: Houghton Mifflin, 1992.
Harrison, Barbara Grizzuti. *Italian Days.* New York:
    Weidenfeld & Nicholson, 1989.
Hillaby, John D. *Journey to the Gods.* London: Constable,
    1991.
Proper, Datus C. *The Last Old Place: A Search Through*
    Portugal. New York: Simon & Schuster, 1992.

Books for Students

Carrick, Noel. *San Marino.* New York: Chelsea House, 1988.
Italy. Alexandria, VA: Time-Life Books, 1986.
Lyle, Keith. *Passport to Spain.* New York: Franklin Watts,
    1987.
Stein, R. Conrad. *Greece.* Chicago: Children's Press, 1987.

The World Today *Videodisc Program*

A variety of still images, short videos, and activities are
available for you to use as you teach this chapter. *Lecture
Notes* include bar codes for specific images and videos.

Multimedia Resources

Great Wonders of the World: Manmade—The City of
    Venice. (Multimedia CD-ROM) InterOptica Publishing
    Ltd., 300 Montgomery Street, Suite 201, San Francisco,
    CA 94104.
Greece: Children of the Peloponnese. (VHS, 26 minutes)
    Films for the Humanities & Sciences, P.O. Box 2053,
    Princeton, NJ 08543–2053.
Lisbon. (VHS, 45 minutes) Films for the Humanities &
    Sciences, P.O. Box 2053, Princeton, NJ 08543–2053.
Madrid. (VHS, 40 minutes) Films for the Humanities &
    Sciences, P.O. Box 2053, Princeton, NJ 08543–2053.
Picture Atlas of the World. (interactive maps, photo bank,
    animation, audio on CD-ROM) National Geographic
    Society, Washington, DC.

MEETING INDIVIDUAL NEEDS

Limited English Proficiency (LEP)

On a physical-political wall map of the world, point out to students the location of Southern Europe in relation to other regions of the world. Help students see that most of Southern Europe is located on peninsulas and that it borders the Mediterranean Sea along with parts of Asia and Africa. Explain to students that this location made seaports in Southern Europe important trading centers.

Learners Having Difficulty

Help students review the main ideas of the chapter by creating a chart on the board using the following headings: Physical Characteristics, Historical Characteristics, Political Characteristics, and Cultural Characteristics.

Auditory Learners

Explain to students that many famous philosophers were from Greece, and many of their ideas on government helped form the basis of democracy and have endured for more than 2,000 years. If possible, obtain a copy of *The Republic* by Plato or *Politics* by Aristotle, and read excerpts describing what those men believed constituted an ideal government. Then have students discuss how closely the United States government conforms to those ideals.

Visual Learners

Divide the class into six groups. Assign each group a country or the microstates discussed in Chapter 26 in order to create agricultural product maps. Have students skim the chapter to make lists of the agricultural products grown in each country or microstate. Then provide students with poster board to make outline maps of their assigned countries or microstates. Next, using pictures of food products from grocery store advertisements, have students cut and paste the products grown in each country or microstate onto the outline map. Arrange the maps on a bulletin board in the class.

Tactile/Kinesthetic Learners

Have students work with partners to create an illustrated time line about significant events discussed in Chapter 26 that are related to the history of Spain, Portugal, Italy, and Greece. Display the completed time lines in the classroom.

Gifted Learners

Have interested students research the Renaissance in Europe. Ask them to write a report that explains the changes in Europe brought about by this period of rebirth.

MULTICULTURAL FOCUS

Point out to students that the most popular professional sport in Southern Europe is football—better known in the United States as soccer. Although the sport's exact origins are unknown, an early form of soccer was played in ancient Rome. During the Middle Ages the game traveled to many parts of Europe, but participation was highest in Italy. Neighboring towns competed with one another in a very rough version of the sport; teams included hundreds of players. In Italy today, the daily newspaper with the greatest distribution is the *Gazetta dello Sport*, or the *Sports Gazette*. This newspaper publishes the results of the latest professional soccer games, the statistics and standings of the teams in the three Italian leagues, and interesting information about the players. Italy occasionally hosts the internationally famous World Cup soccer championship.

Besides Italy's professional teams, hundreds of amateur teams compete with each other on a weekly basis. The beginning of the rivalries between some of these local teams can be traced back several hundred years.

Have interested students research how soccer has become an international sport as a result of cultural diffusion.

CHAPTER 26

SOUTHERN EUROPE

Section 1: Spain and Portugal
Section 2: Italy
Section 3: Greece

OVERVIEW

Chapter 26 explores the countries of Southern Europe, a region defined by the Mediterranean Sea.

Section 1 describes economic, political, and social conditions in the nations that share the Iberian Peninsula—Spain and Portugal.

Section 2 features Italy, whose culture has enriched Western civilization for centuries. Students will learn that although Italy today is a leading industrial nation, it faces political problems and economic inequities.

Section 3 focuses on Greece's ancient culture and describes how modern Greece is struggling for political and economic stability.

About the Author

Jan Morris (1926–) is an English writer. She has written travel books and essays about many regions of the world. The quotation on this page is from *Spain*, published in 1979.

Primary Source

Description of Change: excerpted and bracketed
Rationale: excerpted to focus on main idea; bracketed to clarify meaning

---

CHAPTER 26

Southern Europe

❝This is still a country of local specialties: wines vary widely from town to town, and so do foods. In San Sebastian they make small pastries designed to look like ham and eggs. In Toledo they make marzipan. Oviedo is famous for its stews, richly compounded of vegetable-broth and black pudding, and Vigo for its eel pies. . . . Uniformity has not yet fallen upon Spain . . . this one [town] makes paper, this one lace, this one swords, this one cars, this one wooden figures of Don Quizote, and the little town of Jijona, near Alicante, makes nothing else but nougat. **❞**

Jan Morris

Floor mosaic from a former Roman villa in **Spain**

The countries of Southern Europe, also known as Mediterranean Europe, are Spain, Portugal, Italy, Greece, and several microstates. The four larger nations occupy peninsulas on Europe's southern edge.

The Mediterranean Sea stretches some 2,320 miles (3,733 km) from east to west. *Mediterranean* comes from a Latin word meaning "middle of the land." Indeed, in ancient times, the Mediterranean Sea was considered the center of the Western world, as it is surrounded by Europe, Africa, and Asia. The Strait of Gibraltar (juh-BRAWL-tuhr) links the Mediterranean to the North Atlantic Ocean.

Small waves, weak currents, and low tides characterize the Mediterranean. As a result, its poorly circulated waters are easily polluted. A growing population, coastal industrialization and a flood of tourists continue to stress the Mediterranean environment.

Mijas, southern **Spain**

GEOGRAPHY DICTIONARY

Meseta
plaza
cork
dialect
autonomy
Renaissance
coalition government
balance of trade

296

Using the Quotation

After reading the quotation, ask students what characteristics make Spain so attractive to the writer. *(diverse, quaint towns; local specialties)* Then ask whether reading this passage makes students want to visit Spain themselves. Point out that the writer implies a contrast between the local diversity of Spain and the "uniformity" of other countries.

Using the Geography Dictionary

Write these headings on the board: Physical Geography, Economic Geography, Political Geography, Cultural Geography. Ask volunteers to find the meanings of the vocabulary terms in the Glossary. Then have students classify the terms under the appropriate heading. *(physical—***Meseta;** *economic—***cork, balance of trade;** *political—***autonomy, coalition government;** *cultural—***plaza, dialect, Renaissance)**

SECTION 1
OBJECTIVES

The objectives reflect the Focus questions on this page and are assessed in the review on page 300.

1. Describe landforms and climates of Spain and Portugal.

2. Analyze the economic and urban geography of Spain and Portugal.

3. Summarize the history and culture of Spain and Portugal.

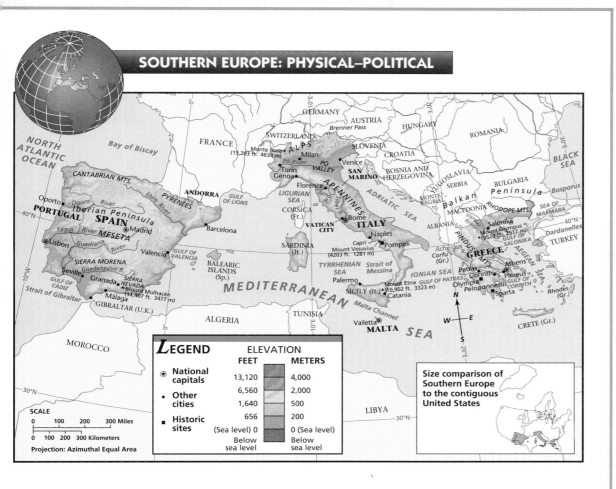

SOUTHERN EUROPE: PHYSICAL–POLITICAL

Using the Physical–Political Map: Southern Europe

Have students locate the region's large nations, and microstates and Gibraltar. (See the chart on page 306.) Then call on volunteers to name the major landforms, including peninsulas, islands, seas, mountain ranges, plateaus, and rivers. Ask students to make generalizations about how the physical geography of this region has influenced its history and economic activity.

**THE WORLD TODAY
VIDEODISC PROGRAM**

Spain: Modern Life Transforms a Traditional Village

1B   601   4:36

Play   Pause

SPAIN AND PORTUGAL

FOCUS

● *What landforms and climates are found in Spain? in Portugal?*

● *What economic activities are important in Spain and Portugal? What are each country's major cities?*

● *What are the historical, political, and cultural characteristics of Spain and Portugal?*

REGION **Physical Geography** Spain and Portugal lie on the Iberian (ie-BIR-ee-uhn) Peninsula. Spain covers over 80 percent of the

peninsula, while Portugal occupies the remaining western portion. The peninsula is bounded on the north and west by the North Atlantic Ocean and on the south and east by the Mediterranean Sea. (See the map above.) Spain also includes the hilly Balearic (bal-ee-AR-ik) Islands in the Mediterranean and the volcanic Canary Islands in the North Atlantic Ocean.

Plateaus dominate the interiors of both countries. A high plateau called the **Meseta** (muh-SAYT-uh) covers much of Spain. This rocky and treeless plateau averages 2,100 feet (640 m) in elevation and is almost surrounded by mountains. It is the driest and least populated part of Spain. Portugal's interior plateau is cut by rivers, which have formed steep valleys that slope westward toward narrow coastal lowlands.

Spain is one of Europe's most mountainous nations. The Cantabrian (kan-TAY-bree-uhn) Mountains rise up in the northwest while the

297

TEACHING SECTION 1

Motivator

Have students locate areas of the world that were once Spanish and Portuguese colonial possessions. Then help students find and list the places in the world where Spanish and Portuguese are spoken today. Point out that Spain and Portugal were once major powers in the world. Their influence is still felt extensively around the globe.

Building a Vocabulary

Write the term **autonomy** on the board. Focus on the prefix *auto-*, meaning "self." Ask students how the meaning of this prefix relates to the meaning of *autonomy* (*nomos* means "law"). Tell students that Spain has 17 autonomous regions. Ask them to explain what this means.

Illustrating Ideas

Provide encyclopedias and books on Spain and Portugal for students to use as references. Have students work in pairs to design various scenic postcards that a visitor might send from those countries. Students should write an appropriate caption to accompany their drawings. Encourage students to select scenes that depict each country's physical, economic, historical, and cultural geography.

FACTS IN BRIEF — Southern Europe

COUNTRY POPULATION (1994)	LIFE EXPECTANCY (1994)	LITERACY RATE (1990)	PER CAPITA GDP (1993)
ANDORRA 63,930	75, male 81, female	99% (1992)	$14,000 (1992)
GREECE 10,564,630	75, male 80, female	93%	$8,900
ITALY 58,138,394	74, male 81, female	97%	$16,700
MALTA 366,767	75, male 79, female	90% (1988)	$6,600 (1992)
PORTUGAL 10,524,210	71, male 78, female	85%	$8,700
SAN MARINO 24,091	77, male 85, female	97% (1991)	$16,000 (1992)
SPAIN 39,302,665	74, male 81, female	95%	$12,400
VATICAN CITY 821	Not available	Not available	Not available
UNITED STATES 260,713,585	73, male 79, female	97% (1991)	$24,700

Sources: *The World Factbook 1994*, Central Intelligence Agency; *The World Almanac and Book of Facts 1995*

Which country has the lowest literacy rate?

Sierra Morena and Sierra Nevada ranges are located in the south. The Pyrenees stretch from the Bay of Biscay to the Mediterranean with peaks over 10,000 feet (3,048 m) high. Historically, these mountain ranges have isolated the peoples of Spain from each other. Even today, regional differences exist, partly because of this geographical separation.

The most populated lands are the coastal plains and river valleys. The Ebro, the Douro, the Tagus, the Guadiana (gwahd-ee-AHN-uh), and the Guadalquivir (gwahdl-KWIV-uhr) are the most important rivers on the Iberian Peninsula. The Guadalquivir in southern Spain is one of the country's few navigable rivers.

Spain's climate types include marine west coast, Mediterranean, and steppe. The marine-west-coast climate is found in the north between the Cantabrian Mountains and the Atlantic coast. This area of lush forest and pasture is the country's coolest and wettest region.

A Mediterranean climate bathes southern and central Spain and all of Portugal. Here, the summers are sunny, hot, and dry, while the winters are mild and wet. Portugal's northern region is at higher elevations and is somewhat cooler, wetter, and forested.

The Meseta of central Spain has a semiarid steppe climate with very hot summers and very cold winters. Because the surrounding mountains block moist ocean air from reaching inland, the Meseta receives little rainfall. Some of the plateau suffers from desertification.

Agriculture Because much of Spain is rugged and dry, there is little arable land. Even so, Spanish farmers make up about 10 percent of the work force and produce a wide variety of crops. Spain has three agricultural regions, which are based on its climates. In the north, ranchers raise dairy and beef cattle, and farmers grow vegetables and fruit. In the Mediterranean region, olives, almonds, citrus fruits, grapes, and vegetables are important products. The high and dry Meseta is not as productive, although farmers can grow grains, olives, and grapes where irrigation and good soils allow. Much of the rest of the Meseta is used for grazing sheep and goats.

Many Spanish farmers live in small villages and travel each day to the fields. Activity in small agricultural villages centers around a **plaza**, a public square in the center of town. Often the site of a marketplace, the plaza usually serves as a gathering place for people.

Portugal is a leading exporter of wine. The dry, warm climate south of the Tagus River also supports citrus fruits and olives. North of the Tagus River, farmers grow cereal grains, especially wheat and corn.

Economic Geography Spanish agriculture supports a large food-processing and export industry and is a major part of the national economy. The country also has a good natural resource base for industrialization. The iron-ore mines in northern Spain support the country's steel industry.

Some industries, such as automobiles, machine tools, and processed food, are supported by foreign investment and have been very successful. Spain is a world leader in automobile manufacturing, and automobiles account for a large share of the nation's exports. Other industries, however, especially steel and coal mining, have suffered labor cutbacks. Spain also must import almost all of its oil.

Spain has one of the world's largest tourist industries. Its sunny beaches and rich cultures

Guided Practice

Have students work in pairs. While one partner calls out the names of rivers, mountain ranges, plateaus, and cities in Spain and Portugal, the other partner should locate them on a map. Have partners reverse roles and repeat the procedure.

Independent Practice

Provide student pairs with an outline map of Spain and Portugal, or have them make their own. Using the maps on pages 260–261 and other school resources, have each pair create a map that shows the climate regions and areas of agricultural activity in the two countries. Then have students write a generalization about the relationship between the climate regions and agricultural activity.

Closure

Have each student name a place in Spain or Portugal he or she would like to visit and tell why. Have students show the location of the place on a map and describe its physical setting, climate, and economy.

have made the country Europe's leading tourist destination. As a result, construction is booming along the country's Mediterranean coast and in the Canary Islands. Such growth has caused pollution and overcrowding, however. The many high-rise buildings block out some of Spain's most scenic coasts. The tourist industry as a whole also competes with agriculture for the region's limited water supplies.

Portugal imports more goods than it exports. The country's income from its tourist industry, however, helps offset its trade deficit. A mineral-poor nation, Portugal must import almost all of its energy resources. Much of the nation's industry and exports are based on food processing, textiles, shoes, and electronics. The Portuguese also have a long tradition of fishing the North Atlantic, and seafood remains an important economic resource to the nation.

The Iberian Peninsula is a leading producer of **cork**, which is a bark stripped from the trunk of the Mediterranean oak. Cork is used to make a variety of products, including insulation and flooring. Its main use is for bottle stoppers.

PLACE Urban Geography Madrid, Spain's capital, is located on the Meseta. Although it is separated from other populated areas, Madrid serves as Spain's transportation, financial, and administrative center. Spain is investing large sums of money in high-speed rails and highways to better connect Madrid to other Spanish cities and to the rest of Europe.

Barcelona (bahr-suh-LOH-nuh) is Spain's second largest city. Located on the Mediterranean coast with a good harbor, it is a leading seaport and industrialized urban area. This ancient Roman port is a popular tourist destination as well.

Southwest of Barcelona on the Mediterranean coast lies Valencia, Spain's third largest city. Valencia is the center of a rich agricultural area and is known for its citrus fruits, especially oranges. The city is eastern Spain's railroad center and has a booming beach resort and tourist industry.

Portugal's two largest cities are Lisbon and Oporto (oh-POHRT-oh). Lisbon, the nation's capital and largest city, is located at the mouth of the Tagus River. It serves as a major transportation, industrial, and cultural center. Oporto lies in northern Portugal on the Douro River. It is a major seaport and industrial area and is the center for bottling and exporting wine.

Historical Geography From 200 B.C. to A.D. 400, Iberia was a province of Rome. In 711, the Moors, a Muslim people from North Africa, invaded the peninsula. They ruled much of the Iberian Peninsula for more than 700 years. During those years, their art, architecture, agriculture, and science influenced much of the region. Grenada, the last city held by the Moors, was reconquered by Spain in 1492.

From about the fifteenth to the seventeenth century, Spain and Portugal were the most powerful European nations and controlled much of the ocean trade. Their explorations for new trade routes brought them to the Americas, Asia, and Africa. Both nations built large colonial empires, bringing back gold, spices, and other valuable materials from these lands. To their colonies,

After stripping the Mediterranean oak of its bark, workers in western **Spain** load trucks to take their harvest to the factory. Workers at the factory steam the bark to produce cork stoppers for bottles.

Extension

Select students to research the work of a Spanish artist, such as El Greco, Velázquez, Murillo, Miró, Zurbarán, Goya, Picasso, or Dali. Ask students to give a presentation to the class on the ways in which the particular artist's work reflects or has influenced Spanish life and history. The Unit 5 skill on page 316 may provide students with a starting point.

SECTION 2
OBJECTIVES

The objectives reflect the Focus questions on this page and are assessed in the review on page 304.

1. Examine Italy's physical and historical geography.
2. Compare and contrast Italy's northern, central, and southern regions.

Section 1 Review
ANSWERS

1. Plateaus; Mediterranean
2. Both countries must import energy resources. Madrid is the capital of Spain, and Lisbon is Portugal's capital.
3. Possible answer: the Pyrenees may have isolated the Basques from speakers of other languages. Also, the Basques lived in the area before the French and Spanish settled nearby, so it is not surprising that Euskara is not related to French or Spanish.

300

Spain and Portugal brought their languages, religion, and culture traits.

Political Geography   Today, Spain is a prosperous democracy with a king. About 75 percent of the population speaks Castilian, the official national language. Castilian is a Spanish **dialect** from the Castile region around Madrid. A dialect is a regional variety of a major language. Many dialects are spoken throughout Europe.

Although the Roman Catholic religion and the Spanish language are unifying forces, many divisions exist in Spanish society. In fact, the Spanish government has granted **autonomy** to 17 regions within Spain. Autonomy is the right to self-government. The central government, however, controls national matters such as defense.

The status of the Basques, an ethnic minority, is a major concern for the future of Spain. The Basques live primarily along the Bay of Biscay and the Pyrenees in northeastern Spain and an adjoining area of France. They lived in the area long before the French or the Spanish settled near them. Their language, which they call Euskara (ehws-kuh-RAH), deepens the mystery surrounding them. Euskara is not related to any other known language. Although the government has granted the Basques autonomy, some Basques want complete independence.

In the past few decades, social, economic, and political conditions have improved in Spain and Portugal. Yet the nations still suffer from an unequal distribution of wealth. Spain and Portugal joined the European Union in 1986. While membership in the EU is improving economic conditions, it also is changing the way of life of many rural peoples by pressuring them to abandon traditional farming methods.

SECTION REVIEW

1. What landform features do Spain and Portugal share? Which climate type is found in both countries?
2. What resources must Spain and Portugal import? What is each country's capital?
3. **Critical Thinking**   **Suggest** two reasons that the Basques speak a language unrelated to other languages. Consider both physical and historical factors.

2   ITALY

FOCUS

- *What is Italy's physical geography? What role has Italy played in Europe's history?*
- *How are Italy's regions similar and different?*

Physical Geography   Italy is a boot-shaped peninsula stretching southward from Europe some 760 miles (1,223 km) into the Mediterranean Sea. It is bounded on the west by the Tyrrhenian Sea and on the east by the Adriatic Sea. Italy also includes two large islands, Sicily and Sardinia. (See the map on page 297.)

The Alps form Italy's northern borders with France, Switzerland, Austria, and Slovenia. The Apennines extend from the Alps on the northwest coast to the southern tip of Italy, creating a rugged spine down the center of the peninsula. Italy's largest plain is the Po Valley, situated in northern Italy between the Alps and the Apennines. The Po River flows eastward through the center of the valley to the Adriatic Sea. The Arno and the Tiber in central Italy flow to the west coast.

La Scala, in **Milan, Italy**, is one of Europe's leading opera houses. The origins of modern opera, a drama in which characters sing most of their lines, lie in Italy in the late 1500s.

TEACHING SECTION 2
Motivator

Ask students to consider the role Italy has played in the history of the United States. Have students name some past and present contributions of Italy to the creation and evolution of American society. (*law, language, art, architecture, literature, immigration to the United States, food, fashion*)

Building a Vocabulary

Write *renaissance* and **Renaissance** on the board. Point out the distinction between *renaissance,* the generic word meaning "rebirth," and *Renaissance,* the historical period. Provide students with sentences using each term: "During the *Renaissance,* classical art, literature, and learning were revived in Europe." "Our community is experiencing a cultural *renaissance.*"

Researching Italy's Regions

Assign groups of students one of the following: Natural Regions, Northern Italy, Central Italy, Southern Italy, Rome. Allow the groups 20 minutes to research and prepare a 3–5-minute presentation responding to these questions: What are three facts about your topic? What are the significant dates, people, places, and products? Provide several resource books in the classroom, and encourage students to use wall maps.

The great power and wealth of the ancient Roman Empire can be seen in the ruins of the Colosseum of **Rome**. The Colosseum, which seated about 50,000 spectators, is one of many Roman ruins scattered throughout Europe.

Most of Italy has a Mediterranean climate. The Alps and the Apennines prevent the cold winds of Northern and Eastern Europe from reaching the country. Italy's surrounding seas also help prevent temperature extremes in winter and summer. The Po Valley has a humid-subtropical climate with warm, humid summers and cool, rainy winters. Rainfall and winter snows are heavy in the Alps and northern Apennines. In contrast, overgrazing and deforestation have eroded the drier southern mountains. When it does rain, floods and mud slides may occur.

Historical Geography Italy's rich culture dates back as far as 500 B.C., when Rome was a thriving city. At its peak in about A.D. 117, the Roman Empire stretched from Britain to Mesopotamia. Unable to protect such a huge area from invaders, Rome collapsed in the fifth century. Italy was divided into many small kingdoms, city-states, and papal (church) states. San Marino and Vatican City still exist as separate independent countries.

Almost 1,000 years after the fall of the Roman Empire, Italy became the center of the **Renaissance** (REHN-uh-sahnts), which means "rebirth." During this period, a renewed interest in learning spread throughout Europe. It was during this period that powerful Italian city-states, such as Florence, Genoa, and Venice, developed. Rivalry among these city-states, and invasions by Spain and Austria, prevented Italy from becoming a unified nation until the late nineteenth century. Although Italy has been unified for over 100 years, differences between its north and south regions remain strong.

Northern Italy The Po Valley in northern Italy is the country's economic heartland. Warm

summers, year-round water resources, and rich soils make the valley the "breadbasket" of Italy.

Industry is also important in the Po Valley. The large cities of Milan, Turin, and Genoa form an industrial triangle that benefits from its nearness to the rest of Europe. This industrialization, however, has caused severe air pollution, which blankets much of the area.

Italy's Alps provide water, hydroelectric power, and some timber to the industrial areas. The Alps are also a major tourist region, with their glacial lakes, ski resorts, and vineyards. The eastern Po Valley is less industrialized. Venice, near the mouth of the Po River, is the main city here. (See "Planet Watch" on pages 302–303.)

Central Italy Central Italy is the country's political and cultural center. Rome, Italy's capital and largest city, lies in this region. The city has spread out over both banks of the Tiber River. Ancient Rome was the center of the once-vast Roman Empire. The Forum, the Colosseum, and other Roman ruins still stand in the city today. Rome handles Italy's political, educational, scientific, and international affairs.

The city of Florence, on the Arno River, is an important art and cultural center. A beautiful and historic city, Florence is one of Europe's leading tourist destinations. The Apennines' summer and winter resorts and the region's volcanic lakes also attract thousands of tourists.

Southern Italy Southern Italy, the country's poorest region, is characterized by rural poverty and small farms. Because the area has very dry summers, water is scarce. Throughout history, southern Italy has suffered from natural disasters, including earthquakes, volcanic eruptions, floods, and droughts. Human destruction of valuable forest and soil cover has contributed to desertification in the region.

Unlike northern Italy, southern Italy has few industrial centers. With mixed results, the

Text continues on page 304.

THEME
Location

As a center of government, culture, and commerce, Italy has both assimilated and dispersed many influences—including food. Arabs from western Asia and Africa introduced cane sugar, ice cream, and candy bases. It is believed that the Venetian traveler Marco Polo brought spaghetti to Italy from China. Corn, tomatoes, and red peppers came from the Americas.

COOPERATIVE LEARNING

Organize the class into groups of four or five. Assign each group a major Italian city. The group will guide the rest of the class on a tour of the city's sights. The presentation should include photographs or drawings of the city's sights, a map of their location, and a narrative explaining their significance. Individual roles might include researcher, writer, artist, and guide.

Planet Watch

Using Geography Themes

Point out that most of the problems described in "Saving Venice" are the result of human-environment interaction. List some of these problems on the board, and identify their causes and effects for students. *(Air pollution: cause—industrial wastes from factories; effects—damage to buildings and to art. Water pollution: cause—chemical wastes from factories; effects—algae growth, harm to tourist industry. Sinking infrastructure: cause—wells dug beneath lagoon; effect—buildings damaged further by exposure to water and tide)*

PLANET WATCH

Saving Venice

Venice is famous for its historic architecture— and the efforts now being waged to save these treasures.

The city of Venice is built on a group of about 120 islands in the middle of a lagoon. For hundreds of years, the water separating the islands from the Italian mainland protected Venetians from invasion and wars that rocked the rest of Europe. By the early fifteenth century, the city was Europe's center of trade with Asia.

Although Venetians benefited from the sea, they knew it could destroy all they had built. Venice is located along the Adriatic Sea, only a few feet above sea level. Floods caused by high tides are a constant danger. As a result, for much of Venice's early history, a watchful government kept the canals free of silt. They also prevented people from building anything in the lagoon that might interfere with the natural flow of the tides.

After a terrible flood in the late 1800s, the sea caused few problems. Then, in 1966, Venice faced disaster. On November 4, strong winds prevented the morning tide from flowing out of the lagoon. As the afternoon tide flowed in and heavy rains pounded the city, the high waters burst the city's fuel oil storage tanks, releasing thick black oil into the lagoon. Oily black water threatened historic churches and palaces in and around St. Mark's Square.

The devastating flood of 1966 almost drowned the city, along with some of the world's most famous art and architecture. Suddenly, people began to look closely at the city,

and many did not like what they saw. Venice was decaying. Paintings, sculptures, and ancient buildings had been neglected, and the city was sinking. Industrial air pollution made these problems worse.

A Victim of Growth

The rapid growth of industry in post–World War II Italy brought wealth to the nation but was harmful to Venice. In the 1940s, industry began to expand facilities at the Marghera site, an industrial complex built on the mainland near Venice. Soon, Marghera became one of Italy's largest industrial centers.

Developers bought up mainland farms so they could build cheap apartment complexes for industrial workers. To make room for large ships and oil tankers, new channels were dug in the lagoon. Two artificial islands also were created in the lagoon to house more industry.

By the 1960s, Venice was showing the effects of this development. Smoke belched from factories, polluting the air and damaging

Geographical Sidelight

Millions of wooden piles (poles) support the foundations of Venice's buildings. One historic church rests on two million piles. To construct this and other buildings and to support a thriving shipbuilding industry, Venice nearly depleted the forests of northern Italy.

Deep-water channels provide access to the open sea for shipping traffic from **Venice** and other cities around the Lagoon of Venice. The inset map offers a close-up view of Venice and the canals that cut through the city.

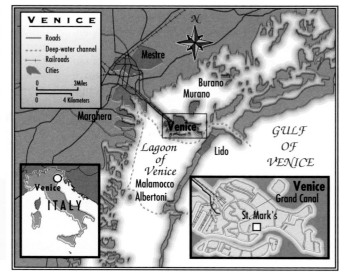

302

Interpreting Maps

Refer students to the map on page 302. Point out the short distance between Marghera and Venice. Ask how this contributed to Venice's environmental problems. (*Venice could not evade the chemical wastes coming from Marghera's factories by air and water.*) Remind students that the floods that occur periodically from high tides also have contributed to Venice's environmental problems. Ask: How do city planners propose to control flooding in Venice in the future? (*by installing gates to hold back water from the Gulf of Venice*) Have students locate on the map the places where such barriers could be installed. (*at the openings between the barrier islands where the ship channels are*)

historic palaces and churches. Wells that had been dug beneath the lagoon to provide industry with water had caused the city to sink nearly a foot (31 cm). Chemicals polluted the waters, spurring the growth of algae. As the algae rotted, the lagoon's murky waters gave off a horrible odor. In addition, pollutants had severely damaged the foundations of buildings along the waterways.

Plans for Saving Venice

Critics claim that the city and national governments have been slow to respond to Venice's problems. In an effort to repair the damage caused by years of neglect, the city has dug new channels to replace those destroyed, and the mayor has promised to clean the silt from the canals. During this process, water will be pumped from waterways, and owners of buildings along them will be forced to restore foundations. A ban on automobiles in the city's center remains.

The Italian government, which is short of funds, has acted even more slowly than the city. In 1984, the government organized a group of companies to develop plans and carry out the work of saving Venice. One of the group's suggestions is that sea barriers be built at openings to the lagoon. Acting as gates, the barriers will lie flat on the bottom of the lagoon, allowing ships in and out and tides to flow freely. When high tides threaten flooding,

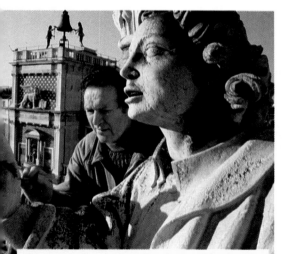

This man is cleaning a statue damaged by decades of air pollution in **Venice**. Time, pollution, and the heavy flow of tourists all take their toll on the city's famous architecture and artwork.

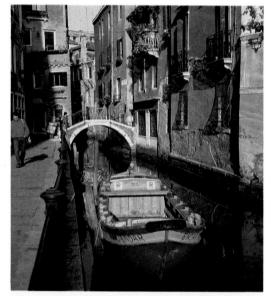

Boats and canals replace cars and streets as the primary means of transportation in **Venice**. About 400 permanent bridges span the city's canals.

the barriers will be raised to hold back the water. The group also hopes to focus attention on the need to protect the region's fragile ecosystem.

Plans like these depend on money. A $1.8 billion loan approved by the national government will pay for part of the work. The fund-raising efforts of Save Venice, Inc., an international organization formed shortly after the 1966 flood, continue to focus on restoring the city's historic structures. Venetians are hopeful that Italy is serious about saving and protecting their city. They would like to use their experience to inform other people about environmental issues. Some hope that a restored Venice can one day become a center of environmental research.

▼ YOUR TURN

As a visiting environmentalist from Venice, **prepare** a speech that you will present to high school students around the United States. Include the problems that Venice faces, the attempts to solve them, and reasons that students should look to Venice as an environmental case study.

Your Turn

Speeches should include references to the problems of air and water pollution. They should describe efforts to solve the problems, such as by banning automobiles in the city, cleaning silt from canals, and strengthening the foundations of many buildings. Although aspects of Venice's environmental problems are somewhat unique, the economic choices made at the expense of environmental concerns provide a useful example to other cities.

303

Guided Practice

Create a gallery of Renaissance art in your classroom. Hang prints or photos of paintings, sculpture, and architecture, featuring the work of Leonardo da Vinci, Raphael, and Michelangelo. Conduct a tour of the "gallery," explaining the pieces to the class. Ask students to suggest why Italy was the center of the Renaissance. (*flourishing cities, wide contact with other cultures, economic growth*)

Independent Practice

Have students look in magazines for pictures of Italian products or Italian-influenced products. (*leather handbags and shoes, sports cars, pizza, spaghetti sauce, wine, olives*) Then have students write one or two sentences about why these images do or do not present a complete picture of Italy's economy and culture.

Closure

Have students review with a partner information they learned about Italy's geography by making an outline.

ASSESSMENT

Assign the Section 2 Review questions on this page.

Reteaching

Assign Section 2 of Reteaching Worksheet 26.

Geographical Sidelight

Christopher Columbus, the Western world's most famous sailor, was born in Genoa in 1451. Growing up in this busy harbor city doubtless influenced his decision to become a navigator.

Section 2 Review
ANSWERS

1. Mountain ranges are the Alps and the Apennines. The Po Valley is located between them in northern Italy.
2. Northern Italy; all three regions attract tourists.
3. After the Roman Empire fell, Italy was divided into many small kingdoms and states. During the Renaissance many city-states became powerful. Rivalry among the city-states and invasions by Spain and Austria kept Italy from becoming a unified nation until the late nineteenth century.

304

Italian government has attempted to develop southern Italy's economy. It has introduced nationalized industries, such as steel mills and refineries. Naples, the largest city in southern Italy, is the region's leading manufacturing center and major port. The nearby island of Capri, Mount Vesuvius, and the ancient city of Pompeii attract many tourists.

The narrow Strait of Messina separates the island of Sicily from the southern tip of the Italian peninsula. Mount Etna, an active volcano, rises 10,902 feet (3,323 m) above the eastern part of the island. Because Sicily's interior is mountainous, most of the island's population lives along the coast. The rocky island of Sardinia is located across the Tyrrhenian Sea southwest of the Italian peninsula. Tourism, farming, and sheep grazing are the main economic activities.

Political Geography Historical, social, and economic differences exist between northern and southern Italy. At the same time, the Italian language, strong family loyalties, and the Roman Catholic Church bind Italians together.

Italy's politics are fragmented, and many political parties exist. No single political party has been strong enough to gain control or form a stable **coalition government**. A coalition government is one in which several political parties join together to run the country. Italy has had more than 50 governments since World War II.

Today, the government must deal with poverty and unemployment in the nation's south, serious pollution problems, and government corruption. Also, nearly all of Italy's energy resources must be imported, which has a major effect on the country's **balance of trade**. Balance of trade is the relationship between the value of a country's exports and its imports. A favorable balance of trade exists when the value of a country's exports is higher than the value of its imports.

② SECTION REVIEW

1. What mountain ranges are found in Italy? Where is the Po Valley?
2. Which of Italy's regions is most agriculturally productive? Which regions attract tourists?
3. **Critical Thinking** **Explain** the historical background of Italy's regional differences.

GREECE

FOCUS

- *How have Greece's physical features influenced the country's economy and history?*
- *What are Europe's microstates?*

LOCATION **Physical Geography** Greece is located at the southern end of the Balkan Peninsula, which extends into the Mediterranean Sea. The Ionian (ie-OH-nee-uhn) Sea lies to the west and the Aegean (i-JEE-uhn) Sea to the east. This strategic Mediterranean location is close to major trade routes and the entrances to the Black and Red seas. (See the map on page 297.)

Greece consists of many peninsulas, islands, and rugged mountains. Hundreds of narrow bays and inlets form the country's jagged coastline. About one-fifth of the area of Greece is made up of more than 2,000 islands. The largest of these is Crete. Mountains divide mainland Greece into many isolated valleys. The highest peak in Greece is Mount Olympus, with an elevation of 9,570 feet (2,917 m).

Economic Geography The dry climate and mountainous terrain limit agriculture to less than 30 percent of the land. Even so, agriculture supports about 20 percent of the Greek work force. The food-processing and food-export industries are based on agricultural products such as olive oil, tobacco, grapes, and citrus fruits. Oil refining is a major industry; however, Greece is heavily dependent on imported oil.

Shipping is a vital part of the Greek economy, and its fleet includes oil tankers, cargo ships, cruise ships, and fishing boats. Ancient ruins, beautiful beaches, and the sunny climate make the tourist industry an important part of the Greek economy as well. While financial aid from the European Union has helped improve its economic situation, Greece remains one of the poorest nations in the EU.

Historical Geography Ancient Greece was vital to the development of Western culture. More than 2,500 years ago, the Greeks developed ideas that today are the roots of our democratic system

The objectives reflect the Focus questions on page 304 and are assessed in the review on page 306.

1. Analyze the effects of physical features on Greece's economy and history.
2. Name Europe's microstates.

TEACHING SECTION 3
Motivator

Tell students that ancient Greece—and particularly the city-state of Athens—is referred to as "the cradle of democracy." Call on volunteers to explain what this means. Then ask if they can name other contributions made by the Greeks to Western civilization (*Olympics, architecture, sculpture, drama, literature, mythology*)

Classifying

Write facts about northern, central, and southern Greece on cards—for example, "location of Salonika" or "most populated region." Write one fact per card. Make several packs of "fact cards." Have students separate the cards into piles representing the regions of Greece. When a student's classifying has been checked, have him or her mix up the cards and pass them along to another student.

Development plans call for rerouting a major part of the Achelous River—Greece's largest—from one side of the country to the other. The object is to bring irrigation and hydropower to Thessaly. Greek environmentalists condemn the project, arguing that it will be an ecological disaster for the plant and animal life that depend on the river's natural course.

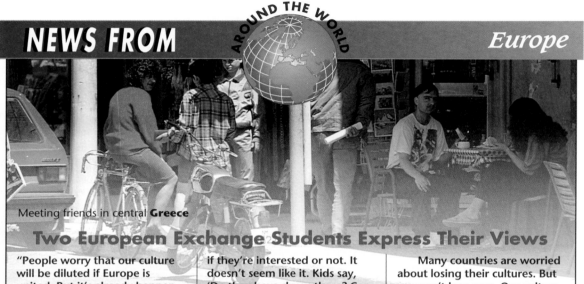

NEWS FROM
AROUND THE WORLD
Europe

Meeting friends in central **Greece**

Two European Exchange Students Express Their Views

"People worry that our culture will be diluted if Europe is united. But it's already happening. Holland is trying to be like America.

There are some things I like about America. Everyone is so polite. . . . But I think America is too big. You can travel 1,000 miles this way and that way and you're still in America speaking English. In Europe you'd go through five countries.

It's more important for Americans to go to Europe than for Europeans to go to America. In school here [the United States], nobody knows anything about my country. I don't know

if they're interested or not. It doesn't seem like it. Kids say, 'Do they have drugs there? Can you drink there? Can you say something in your language?' And that's the end of their questions. That's all they want to know."

Alice van Exel,
the Netherlands

"European political alliances have been disastrous in the past. But maybe we can learn from the past. Every country is very willing. We want to see a united Europe.

Many countries are worried about losing their cultures. But we won't lose ours. Our culture is centuries old, it would be very difficult to lose that. Even in America, Greek people still have their customs, even if they've lived here for generations.

After unification, it will be more difficult to find a job, much harder because you'll be competing with people from all over Europe. You will have to know more languages, have a better education, have skills that will separate you from everyone else."

Yolanda Grammaticaki,
Greece

GLOBAL PERSPECTIVES

The original Greek Olympic Games declined in the fourth century A.D. after Rome gained control of Greece. In 1896, however, the games were revived through the efforts of a French educator and have continued through the present day. Provide all students with Outline Map 1. Have them mark and label the sites of Winter and Summer Olympics since 1896. This information is available in almanacs. Emphasize that the Olympic Games present many opportunities for building international goodwill.

of government. Greek art, philosophy, and science formed much of the basis of Western civilization.

Although it was the birthplace of democracy, Greece has had little democratic government in its long history. Its location at the crossroads of the Mediterranean exposed the country to invasions. Through the centuries, Persians, Romans, and Turks have all ruled Greece. Finally, in 1829, Greece became a free country after 400 years of Turkish rule. During World War II, the country was occupied by Germany. Then, immediately after the occupation ended, a bloody civil war broke out. After several governments, including a military dictatorship, the Greek government became a parliamentary republic in 1974.

Southern Greece Joined to the mainland by the Isthmus of Corinth, the Peloponnesus (pehl-uh-puh-NEE-suhs) Peninsula forms southern Greece. The peninsula's Mediterranean climate makes the region ideal for growing a variety of crops. Along with agriculture, fishing and tourism are important to the economy.

The ancient cities of Olympia and Sparta are in southern Greece. Olympia was the site of the first Olympic Games, which were held in 776 B.C. Sparta was famous for its athletes and warriors.

Crete lies directly south of the Peloponnesus Peninsula. Greece's largest and most populated island, Crete is primarily agricultural. To the west, in the Ionian Sea, are a number of scenic islands,

Independent Practice

Have each student choose one of the microstates listed in the chart on this page and write a "mini-report." The report should include information on the physical, economic, historical, and cultural geography of the selected country.

Closure

Hold a "Class Olympics" based on what students have learned about Greece and Europe's microstates. Events might be organized around physical geography, economic geography, historical geography, and regions. There might also be events such as map drawing, essay writing, and so on. Some events might be team events while others are individual events.

ASSESSMENT

Assign the Section 3 Review questions on this page.

Reteaching

Assign Section 3 of Reteaching Worksheet 26.

306

OTHER COUNTRIES IN EUROPE

ANDORRA
Andorra has no railway system, and its location in the Pyrenees does not permit air transportation. Roads link Andorra to neighboring Spain and France.

LIECHTENSTEIN
A huge wooden cross marks the spot where visitors to Naafkopf Peak in southern Liechtenstein can sit in three different countries at one time: Liechtenstein, Austria, and Switzerland.

MALTA
Malta's location in the Mediterranean Sea between Sicily and North Africa has made the islands important in struggles between Mediterranean powers throughout history.

VATICAN CITY
Tiny Vatican City is an independent state surrounded by the city of Rome. The pope, the head of the Roman Catholic Church, has absolute political authority within the city.

GIBRALTAR
Gibraltar, a British possession since 1713, sits at the entrance to the Mediterranean Sea from the Atlantic Ocean. The narrow peninsula at the southern tip of Spain is dominated by a limestone and shale ridge known as the Rock.

MONACO
Monaco draws people from around the world to its Mediterranean beaches, port, casino, and cultural attractions. Two well-known auto races, the Monte Carlo Rally and the Grand Prix de Monaco, take place in this tiny country.

SAN MARINO
Tradition has it that San Marino was founded early in the fourth century by Christians escaping persecution. It is one of the world's oldest and smallest republics.

Andorra, Liechtenstein, Monaco, San Marino, and Vatican City are microstates. Gibraltar is a dependent territory of the United Kingdom.

such as Corfu. The islands to the east, in the Aegean Sea, extend Greece's borders nearly to the mainland of Turkey. Most of these Greek islands are rocky and dry and depend on tourism for income. Greece and Turkey, rivals for centuries, both claim the offshore waters of the Aegean Sea, where there may be gas and oil deposits. Tension over the island of Cyprus remains a major issue. Although both nations are members of NATO, their relations with each other remain poor.

Central Greece Central Greece is the country's most populated region. Athens, the nation's capital and largest city, is located here. Athens was chosen as the capital in 1834 because of its historical importance. Most of the ancient city is located in the Acropolis, on a high hilltop overlooking the modern part of the city. The ancient marble temple, called the Parthenon, is one of the world's most photographed and famous buildings. After surviving almost 2,500 years of warfare and weather, it is now suffering from the effects of air pollution. Athens and its seaport of Piraeus (pie-REE-uhs) have become the country's largest center of urban and industrial growth.

Northern Greece The Pindus Mountains form a spine dividing northern Greece. To the west of the mountains, there are forests and pasture to graze sheep and cattle but almost no flat land to grow crops. This area is the wettest part of the nation.

In contrast, the northeast has the country's flattest land. Here, the plain of Thessaly and the Vardar River valley are the most productive agricultural areas. Salonika, located in this agricultural area, is the nation's second largest city and the major seaport for northern Greece.

SECTION REVIEW

1. How does Greece's physical geography affect economic activities? What effect has location had on Greece's history?
2. Which microstates border two other countries?
3. **Critical Thinking** **Propose** a way to protect the Parthenon from further damage caused by pollution.

Building a Vocabulary
1. The Meseta
2. It is often the site of a marketplace and serves as a social gathering place.
3. A bark stripped from Mediterranean oak trees
4. No, a dialect is a regional variety of a major language. Euskara is unrelated to other languages.
5. The right to self-government; all of them
6. A government run by several political parties

Recalling and Reviewing
1. Answers may vary, but "mountainous" would be appropriate.
2. Olives, nuts, citrus fruits, grapes, vegetables, and grains; oil
3. The colonial empires of Spain and Portugal brought their languages, religion, and many culture traits to their colonies. Italy's Renaissance spread a renewed interest in learning throughout Europe. Greece's ancient cultures, science, and ideas about democracy formed much of the basis of Western civilization.
4. Barcelona and Oporto

CHAPTER 26 CHECK

The ancient Acropolis above **Athens, Greece**

Reviewing the Main Ideas
1. Spain covers over 80 percent of the Iberian Peninsula. It is a mountainous nation with three main climate types. Spain has a good natural resource base for industrialization and is one of Europe's leading tourist destinations.
2. Portugal occupies the western part of the Iberian Peninsula. Its climate is primarily Mediterranean. Portugal must import many goods and almost all of its energy resources.
3. Italy is a long, boot-shaped peninsula in the Mediterranean region of Europe. It was the center of the Roman Empire and the Renaissance. Northern Italy is industrialized and prosperous, while southern Italy is dry and poor.
4. Greece is a mountainous country with many islands and peninsulas. Agriculture, shipping, and tourism are important to the economy.
5. Europe's microstates are Andorra, Liechtenstein, Monaco, San Marino, and Vatican City. Gibraltar is a British possession at the southern tip of Spain. Malta is an independent island country.

Building a Vocabulary
1. What is Spain's high interior plateau called?
2. How is the plaza significant to agricultural villages in Spain?
3. What is cork?
4. Is Euskara a dialect? Explain.
5. Define *autonomy*. Which countries discussed in this chapter have autonomy?
6. What is a coalition government?

Recalling and Reviewing
1. What one word would you use to describe the landforms of Spain, Italy, and Greece?
2. For what crops is a Mediterranean climate suited? What natural resource must be imported by most of the countries in Southern Europe?
3. What cultural and political influences have the Southern European countries had on the rest of Europe and on the world?
4. What are two major port cities of the Iberian Peninsula?

Thinking Critically
1. Using what you learned in this chapter, write two sentences about the colonization of Middle and South America from the point of view of a Spanish or Portuguese colonist. Then, use what you learned in Unit 4 to write two sentences about the same experience from the point of view of one of the indigenous peoples.
2. The first Olympic Games were held in Greece in 776 B.C. What purposes do you think these games served? What purposes do today's Olympic Games serve?

Learning About Your Local Geography

Cooperative Project
Many dialects are spoken throughout Europe. In the United States, different regions often have their own accents. With your group, brainstorm and make a list of words that are pronounced differently in your area than they are in other areas. Using the pronunciation guide in this book, write out the local pronunciations for these words. What factors contribute to the growth of dialects and accents?

Using the Five Themes of Geography
Create a graphic organizer linking **location** and **human-environment interaction** in the countries of Southern Europe. How does the Mediterranean location influence human activities in these countries? How do these activities affect the physical environment?

Thinking Critically
1. Answers will vary. Sentences from the point of view of the colonists may emphasize the economic advantages of colonization to Spain or Portugal. Sentences from the point of view of the indigenous people may emphasize some of the more negative aspects of colonization, such as the introduction of certain diseases.
2. The ancient games may have served to showcase athletic talents and to provide a healthy outlet for rivalry between regions. The games today may have similar purposes.

Learning About Your Local Geography
Groups' lists should show local or regional pronunciations of various words. Factors may include cultural differences in the regions' origins and settlement.

Using the Five Themes of Geography
Graphic organizers will vary. The Mediterranean location is suited for agriculture, tourism, and shipping. These activities, in turn, cause desertification, pollution, and water shortages.

CHAPTER 27 Eastern Europe
Planning Guide

TEXTBOOK	TEACHING RESOURCES	OTHER AVAILABLE MATERIALS
Section		
1. An Overview of Eastern Europe Pages 309–311	Reteaching Worksheet 27.1 Critical Thinking 27	*The World Today* Videodisc Program *Study Guide for Spanish Speakers* *Lecture Notes,* Chapter 27
2. Countries of Eastern Europe Pages 311–313	Reteaching Worksheet 27.2 Challenge/Enrichment Worksheet 27	*Study Guide for Spanish Speakers* *Lecture Notes,* Chapter 27
3. A Region in Conflict Pages 313–314	Reteaching Worksheet 27.3 Graphic Organizer 27 Skills Worksheet 27	*Study Guide for Spanish Speakers* *Lecture Notes,* Chapter 27
Chapter Review Page 315	Chapter 27 Test Chapter 27 Portfolio Assessment See also *Alternative Assessment Forms*	*Test Generator,* Chapter 27
Unit Review Page 316–317	Unit 5 Test Unit 5 Portfolio Assessment See also *Alternative Assessment Forms*	*Test Generator,* Unit 5

CHAPTER RESOURCES

Lecture Notes, Chapter 27
Teaching Resources
    Chapter 27 Graphic Organizer
    Chapter 27 Skills Worksheet
    Chapter 27 Critical Thinking Worksheet
    Chapter 27 Challenge/Enrichment Worksheet

REVIEW AND ASSESSMENT

Chapter 27 Check, page 315
Chapter 27 Test, *Teaching Resources*
Chapter 27 Portfolio Assessment, *Teaching Resources*
Test Generator, Chapter 27
Alternative Assessment Forms

ADDITIONAL RESOURCES

Books for Teachers

Drakulic, Slavenka. *How We Survived Communism and Even Laughed.* New York: W. W. Norton & Co., 1992.

Havel, Václav. *Summer Meditations.* New York: Knopf, 1992.

Laufer, Peter. *Iron Curtain Rising: A Personal Journey Through the Changing Landscape of Eastern Europe.* San Francisco: Mercury House, 1991.

Books for Students

Carran, Betty. *Romania.* Chicago: Children's Press, 1988.

Leone, Bruno, series ed. *Eastern Europe: Opposing Viewpoints.* San Diego: Greenhaven Press, 1990.

Sandak, Cass. *Poland.* New York: Franklin Watts, 1986.

The World Today *Videodisc Program*

A variety of still images, short videos, and activities are available for you to use as you teach this chapter. *Lecture Notes* include bar codes for specific images and videos.

Multimedia Resources

Budapest. (VHS, 40 minutes) Films for the Humanities & Sciences, P.O. Box 2053, Princeton, NJ 08543–2053.

Picture Atlas of the World. (interactive maps, photo bank, animation, audio on CD-ROM) National Geographic Society, Washington, DC.

Poland. (16mm film/VHS, 25 minutes) The Changing Faces of Communism series. National Geographic Society, Educational Services, Washington, DC 20036.

Warsaw. (VHS, 40 minutes) Films for the Humanities & Sciences, P.O. Box 2053, Princeton, NJ 08543–2053.

MEETING INDIVIDUAL NEEDS

Limited English Proficiency (LEP)

Help students create charts on the board that compare the characteristics of each country of Eastern Europe before and after Communist control. Use the headings Economic Geography, Political Geography, and Cultural Geography. Students may come to the board to add characteristics or call them out as you write them on the board.

Learners Having Difficulty

Have students work in small groups. Assign individuals or pairs the Focus questions for each section. Ask students to find and write down answers to the questions assigned them. Then have them share their work with the group. Have students work together to prepare group answers to each Focus question.

Auditory Learners

Ask students to name the challenges facing the countries of Eastern Europe, as you write them on the board. Then have the class evaluate the following statement: "Nationalism and religious differences have become hallmarks of the countries of Eastern Europe."

Visual Learners

Refer students to the economy map of Europe on page 261 in their textbook. Have them compare the economic geography of Eastern Europe with that of the other regions of Europe. Ask students to draw conclusions from their comparisons. Then ask volunteers to read their conclusions to the class. Finally, hold a discussion about the conclusions students have drawn.

Tactile/Kinesthetic Learners

Have students work in small groups to write skits about life in Eastern Europe before and after Communist rule. Each member of the group should have a part in their skit. Assign each group one of the following areas on which to focus their skit: economic geography, political geography, cultural geography. After students have presented their skits to the class, discuss the issues presented.

Gifted Learners

Have interested students do research to find out how the boundaries of Poland have shifted since 1700. Ask students to make maps showing the boundary changes and to write captions that explain the reasons for the changes.

MULTICULTURAL FOCUS

Point out to students that in some areas of Eastern Europe cultural traditions that have been passed down from generation to generation are still followed, especially in rural villages. One such tradition in Romania surrounds wedding preparations and the wedding ceremony, although traditions may vary slightly from village to village.

First, a horseman rides through the village proclaiming the date of the wedding celebration and inviting everyone to attend. Weddings in Romanian villages are always held on Sundays.

On the day of the wedding, the bride's hair is braided. She wears an ornamental headband decorated with flowers, colored ribbons, leaves, and gems. The bridegroom wears a hat covered with flowers, feathers, and small branches from a fir tree. The best man symbolically shaves the bridegroom as a sign that he has left bachelorhood and has joined the brotherhood of married men. The wedding party is usually transported in a horse-drawn carriage to the church for the ceremony.

During the wedding ceremony, a ritual is performed called the forgiveness ceremony, in which the couple asks family members to pardon them for leaving their family units.

A wedding feast follows the ceremony. The bride and groom traditionally share a large loaf of bread, while water is sprinkled on the bread and wedding guests throw grains of corn on the couple. This ritual symbolizes a long and prosperous marriage.

Ask students to compare wedding traditions in rural Romania with traditions in the United States, with the traditions of their religious group, or with their own family wedding traditions.

OVERVIEW

In Chapter 27, students will learn about Eastern Europe. In recent years, this region has experienced revolutionary political and economic changes as well as catastrophic ethnic conflict.

Section 1 describes the physical geography of Eastern Europe and provides an overview of its political, ethnic, and economic history, focusing on the rise and fall of Soviet control.

Section 2 examines Poland, the Czech Republic, Slovakia, Hungary, Romania, Bulgaria, and Albania and their adjustment to democratic and free-market reforms.

Section 3 explains the breakup of Yugoslavia and the terrible civil war that ensued.

About the Author

Vaclav Havel (1936–) is the former president of Czechoslovakia and a well-known playwright. His election in 1989 ended decades of Communist control of the presidency. The quotation on this page is from *Summer Meditations,* Havel's biographical memoir about political events in Czechoslovakia and his work as an activist for human rights.

Primary Source

Description of Change: excerpted and bracketed
Rationale: excerpted to focus on main idea; bracketed to clarify meaning

CHAPTER 27

Eastern Europe

❝ **W**e [in the former Czechoslovakia] are what history has made us. We live in the very centre of Central Europe, in a place that from the beginning of time has been the main European crossroads of every possible interest, invasion, and influence of a political, military, ethnic, religious, or cultural nature.❞

Vaclav Havel

GEOGRAPHY DICTIONARY

regional specialization
legacy
complementary region

Eastern Europe stretches from the often cold, stormy shores of the Baltic Sea south to the warmer and sunnier beaches along the Adriatic and Black seas. Eastern Europe is actually located in the middle of the European continent. Parts of the former Soviet Union occupy the true eastern part of the continent.

Historically, the countries of Eastern Europe have found themselves caught between larger and more powerful neighbors, including the Turks, Austrians, Germans, and Russians. The region's many ethnic groups often survived by becoming allies of nearby empires.

Conflicts among the peoples of Eastern Europe have also surfaced throughout the region's history. Ethnic and religious differences and extreme nationalism have become hallmarks of the region. Such conflicts continue in the region today.

Peasant gravestone from **Serbia**

Charles Bridge, **Prague, Czech Republic**

Using the Quotation

Ask a volunteer to read aloud the quotation by Vaclav Havel. Refer students to the map of Eastern Europe on this page. Then call on students to describe the relative locations of the Czech Republic and Slovakia and of the rest of Eastern Europe. Ask students to explain how its physical geography has helped make the former Czechoslovakia a "crossroads." (*in the center of the continent, few natural barriers to invasion or migration*)

Using the Geography Dictionary

As students read the chapter and learn the terms' meanings, have them write two sentences for each term. In the first sentence, students should use their own words to define the term. In the second sentence, students should use the term in context, giving examples that show the term's meaning.

SECTION 1
OBJECTIVES

The objectives reflect the Focus questions on this page and are assessed in the review on page 311.

1. Describe the physical setting and cultural background of the Eastern European countries.

2. Explain the effects of Soviet control and of the end of such control on the region.

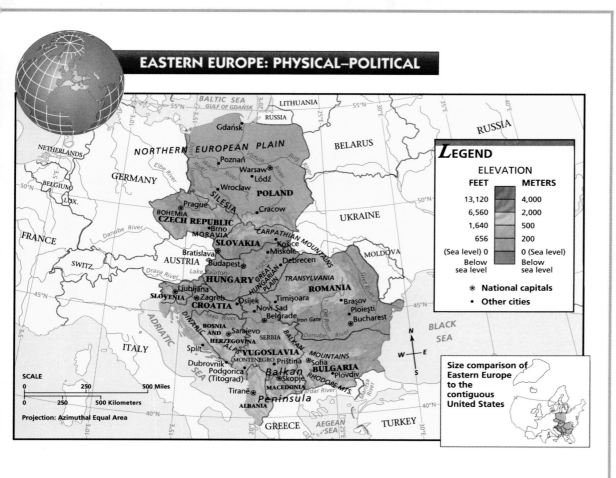

EASTERN EUROPE: PHYSICAL–POLITICAL

Using the Physical–Political Map: Eastern Europe

Call on a volunteer to name the countries of this region. Point out that the Danube River is vital to several countries in Eastern Europe. Have students trace its course on the map. Ask: For which countries in this region does the Danube form a natural boundary? (*Slovakia, Hungary, Croatia, Serbia, Romania, Bulgaria*)

AN OVERVIEW OF EASTERN EUROPE

FOCUS

- *What is Eastern Europe's physical geography? cultural background?*

- *How did Soviet control affect Eastern Europe politically and economically? What effects has the end of Soviet control had?*

REGION **Physical Geography** Eastern Europe is a region of varied landforms, including mountain ranges, plains, and plateaus. The plains of Poland are part of the vast Northern European Plain. To the south, the curving Carpathian Mountains stretch from the Czech Republic eastward across southern Poland to Ukraine, before

heading south and west into Romania. Large plains lie between the mountain ranges. One of the largest of these is the Great Hungarian Plain. Another large plain area lies along the lower Danube River in southern Romania.

Low, rugged mountains and valleys characterize the Balkan Peninsula. The Dinaric Alps extend inland from the eastern shore of the Adriatic Sea, while the Rhodope and Balkan mountains rise up in Bulgaria. Communication among the peoples of the mountains has often been difficult. This isolation has contributed to the survival of many different ethnic groups in the region.

Two major climate types are found in Eastern Europe. In the north is the humid-continental climate. Winters are cold and snowy, and summers are warm and rainy. Farther south, winters are mild and summers become drier. The Mediterranean climate found along the Adriatic coast draws vacationers for much of the year.

TEACHING SECTION 1

Motivator

Ask students to report what they know about the events that have occurred in Eastern Europe since the late 1980s. Have students suggest what was the major change in these countries. (*the end of Soviet control*) Ask: Do you think that the outlook for the countries of this region is generally favorable or unfavorable? Why?

Building a Vocabulary

Write the term **legacy** on the board. Ask a volunteer to read the definition of the term from a dictionary. Point out that the root *leg-* comes from the Latin word for "law," and thus it refers in its primary sense to a legal inheritance or gift. Then refer students to the passage on page 311 where the term is used. Have students name both positive and negative types of legacies.

Guided Practice

Organize the class into several groups. Assign one of the countries of Eastern Europe to each group. Have members look in other resources to find examples of migrations and invasions in the country's history. Have groups list their findings on paper. Ask students how these migrations and invasions contributed to the region's ethnic diversity. Emphasize that political boundaries do not always correspond to cultural boundaries.

FACTS IN BRIEF — Eastern Europe

COUNTRY POPULATION (1994)	LIFE EXPECTANCY (1994)	LITERACY RATE	PER CAPITA GDP (1993)
ALBANIA 3,374,085	70, male 77, female	92% (1989)	$1,100
BOSNIA AND HERZEGOVINA 4,651,485	72, male 78, female	90% (1991)	Not available
BULGARIA 8,799,986	70, male 77, female	98% (1990)	$3,800
CROATIA 4,697,614	70, male 77, female	96% (1991)	$4,500
CZECH REPUBLIC 10,408,280	69, male 77, female	99% (1992)	$7,200
HUNGARY 10,319,113	67, male 76, female	98% (1992)	$5,500
MACEDONIA 2,213,785	72, male 76, female	90% (1990)	$1,000
POLAND 38,654,561	69, male 77, female	98% (1991)	$4,680
ROMANIA 23,181,415	69, male 75, female	96% (1991)	$2,700
SERBIA AND MONTENEGRO 10,759,897	71, M; 76, F 77, M; 82, F	90% (1991)	$1,000
SLOVAKIA 5,403,505	69, male 77, female	99% (1993)	$5,800
SLOVENIA 1,972,227	70, male 78, female	90% (1991)	$7,600
UNITED STATES 260,713,585	73, male 79, female	97% (1991)	$24,700

Sources: *The World Factbook 1994*, Central Intelligence Agency; *The World Almanac and Book of Facts 1995*

Considerable fighting makes statistics for Bosnia and Herzegovina unreliable.

MOVEMENT **A Cultural Crossroads** For centuries, waves of migrating peoples have swept into Eastern Europe, making the region a cultural crossroads. The new residents brought their own languages, religions, and other culture traits. By the start of the twentieth century, the human geography of Eastern Europe was extremely diverse.

The languages of Eastern Europe reveal the history of migrations and the many different people who settled in the region. The most widespread languages in Eastern Europe belong to the Slavic language group. Other languages spoken in the region include Romanian, Albanian, and Hungarian.

Religion also has played a part in separating the many ethnic groups. People in the eastern parts of the region primarily follow the traditions of the Eastern Orthodox church. Northern peoples are mostly Roman Catholic. Islam spread into southern sections of the region with the Turkish (Ottoman) invasions of the 1300s.

Throughout much of its history, Eastern Europe was controlled by neighboring empires. By the end of World War I, however, the countries of the region had gained independence. Many people migrated to the country in which their ethnic group was the majority. This was not possible for everyone, however. As a result, large numbers of ethnic minorities remained in each country.

Borders between Eastern European countries changed again after World War II. For example, the eastern portions of Czechoslovakia and Romania were joined to Ukraine. The eastern and western borders of Poland were moved as much as 125 miles (201 km) to the west. Many people again moved or were forced to move to where their ethnic group was dominant. These border changes and the resulting problems for groups left behind are continuing sources of political turmoil in the region.

Soviet Control After World War II, the Soviet Union came to dominate Eastern Europe. The Soviets controlled the nations of the region through military treaties and the placement of Soviet military bases in most of the countries. Communist governments took power and controlled almost all aspects of people's lives. The western boundary of Eastern Europe became known as the "Iron Curtain."

Beginning with Poland in 1989, the countries of Eastern Europe were successful in establishing independent governments. As Communist rule collapsed, personal freedoms increased and market economies emerged. The countries of Eastern Europe are now seeking closer ties to democratic countries around the world. The new political order in the region, however, has allowed old ethnic conflicts to resurface.

Economic Geography During their years under Communist rule, the countries of Eastern Europe had command economies. In most countries, farmers were forced to merge their lands to create large farms controlled by the government. The Council for Mutual Economic Assistance, or COMECON, oversaw economic ties between the Soviet Union and Eastern Europe.

Closure

Have students work in pairs to answer the Focus questions on page 309.

ASSESSMENT

Assign the Section 1 Review questions on this page.

Reteaching

Assign Section 1 of Reteaching Worksheet 27.

Extension

Have students write a short script of a "You Are There" TV show. The subject of the show should be the collapse of communism in one of the Eastern European countries. Students can find information by reading magazine and newspaper articles from the early 1990s. Have students present their shows.

SECTION 2
OBJECTIVES

The objectives reflect the Focus questions on this page and are assessed in the review on page 313.

1. Identify economic characteristics of individual countries in Eastern Europe.

2. Examine historical and cultural features of individual countries in Eastern Europe.

Different religious cultures meet in **Hungary**, where this former Muslim mosque now is a Christian church. On top of the dome, note the cross, a symbol of Christianity, that has been added to the crescent, a symbol of Islam. Most Hungarians are Christians.

Under COMECON, **regional specialization** was the main plan. That is, each country specialized in products that it could make or grow best. COMECON gave the Eastern European countries access to the minerals and raw materials of the Soviet Union at discount prices. Many consumer goods then were shipped to the Soviet Union. Roads, railroads, and pipelines still connect the countries of Eastern Europe to Russia and Ukraine.

With the breakup of the Soviet Union, economic reorganization among Eastern European countries has varied. Poland, Hungary, and the Czech Republic have made rapid progress toward building free-market economies. Other countries, such as Bulgaria and Romania, have been slower to change. For all former Soviet countries, the end of Communist-controlled economies has meant some degree of financial distress. Many factories have closed, and unemployment has increased. Today, privately owned businesses are beginning to appear, offering workers new opportunities.

Another problem facing Eastern Europe is environmental decay, a **legacy** of the Communist era. A legacy is something received from a previous era. During the Communist era, production was more important than environmental concerns. As a result, air, soil, and water pollution, deforestation, and the destruction of natural resources were widespread. Many countries have begun the long and expensive task of cleaning up their environments.

Many Eastern European countries are middle-income nations, somewhere between the world's developed and developing countries. The region has good education and health-care facilities, highways, and railroads. The peoples of Eastern Europe look forward to greater economic progress and political freedom.

 SECTION REVIEW

1. How has Eastern Europe's physical geography influenced the region's cultural background? What factors have caused ethnic conflicts in Eastern Europe?
2. How has the breakup of the Soviet Union affected daily life in Eastern Europe?
3. **Critical Thinking   Suggest** why the words *iron* and *curtain* were used to describe the western boundary of Eastern Europe.

 COUNTRIES OF EASTERN EUROPE

FOCUS

● *What are the economic characteristics of each country in Eastern Europe?*

● *What is the historical and cultural background of each country?*

Poland Poland is Eastern Europe's largest and most populous country. Despite some industrial progress, Poland is well known as an agricultural region.

Warsaw, the country's capital, is located on the plains. (See the map on page 309.) With about

Section 1 Review
ANSWERS

1. With many rugged mountains, communication among peoples who live in the mountains has been difficult, and this isolation has contributed to the survival of many ethnic groups. Factors causing ethnic conflicts include isolation, the large number of languages spoken, the many religions practiced, and frequent boundary changes that have left large numbers of ethnic minorities in certain regions.

2. Personal freedoms have increased, and economic opportunities are increasing.

3. Because western Europe was not Communist but Eastern Europe was, there was little interaction and movement between the two regions, as if there were some kind of physical barrier, such as a curtain made of iron, between them.

TEACHING SECTION 2

Motivator

Direct students to review the climate, population, and economic maps of Europe on pages 260–261. Ask them where in Eastern Europe they would expect to find farming, where they would expect to find mining, and where they would expect to find industry. Have them check their predictions with information in Section 2.

Building a Vocabulary

Write the homophones *complementary* and *complimentary* on the board, and distinguish between them. *(complementary—completing something or supplying what is lacking; complimentary—giving praise)* Review the meaning of **complementary region,** and ask students to give examples. *(an industrial region and an agricultural region; an oil-exporting region and an oil-refining region)*

Describing

Have students choose one of the countries of Eastern Europe and write a paragraph about the country from the point of view of a young person living there. The paragraph should give historical and cultural facts about the country and should include the young person's views on the changes in his or her country. Have students gather in small groups and read their paragraphs to each other.

MULTICULTURAL PERSPECTIVES

Eastern Europe is home to many ethnic groups and subgroups. These include Poles, Kashubians, Silesians, Czechs, Slovaks, Slovenes, Hungarians, Gypsies, Romanians, Pomaks, Serbs, Albanians, Bosnian Muslims, Croats, Dalmatians, and Jews. Each of these groups has unique cultural traditions and practices, and many have a unique language. Ask students to research one of these ethnic groups and to prepare a short report on some of its customs.

THE WORLD TODAY VIDEODISC PROGRAM

Hungary: A Clean Slate

1B 8900 4:41

Play   Pause

312

Visitors look over the art and crafts for sale in the Old Town Market Square in **Warsaw**. The homes on the square have been rebuilt to recapture their original fifteenth-century style.

1.6 million people, it is the cultural, political, and historical center of Polish life. Its location on the Vistula River makes Warsaw the center of the national transportation and communications networks as well.

In 1980, an independent organization of workers, called Solidarity, formed in Poland. Solidarity threatened the authority of the Communist party and became a popular focus for Polish nationalism and anti-Soviet activities. In 1989, the organization succeeded in bringing democratic elections to the country. Today, Poland is trying to model its economic and political system after the developed countries of Europe.

Czech Republic and Slovakia
In 1993, Czechoslovakia peacefully divided into the Czech Republic and Slovakia. Czechoslovakia had formed in 1918 from the combination of Czech and Slovak territories. Czech lands held mineral resources important for industrialization, while Slovakia was primarily agricultural. This is an example of what geographers call **complementary regions**. The merging of the two regions benefited both.

Czechoslovakia gained its independence from Soviet control in 1989. This split took place so peacefully that it has been called the "velvet revolution." Cultural differences between the Czechs and Slovaks remained, however. The

Slovaks, who had never had their own independent country, soon voted for separation.

Prague is the capital and industrial heartland of the Czech Republic. This prosperous city features historic buildings and one of Europe's oldest universities. Prague is the center of Czech life and is undergoing a cultural rebirth.

Slovakia is more rural than the Czech lands, and incomes are lower than in the Czech Republic. Bratislava, the capital of Slovakia, is on the Danube River. As the cultural, economic, and industrial center of Slovakia, Bratislava draws rural Slovaks seeking better-paying jobs. The Slovaks have been less enthusiastic than the Czechs in embracing a free-market economy.

Hungary Hungary is a major agricultural producer. Hungary's past Communist government put most of the nation's farmland into state-controlled, cooperative farms. Local farm managers sometimes were allowed to make key business decisions, however. They kept technology current, chose their crops, and marketed their products. This experience is helping Hungary improve its economy today.

Budapest is Hungary's capital city as well as the country's political, cultural, and economic center. About 20 percent of Hungary's people live in the Budapest region, where manufacturing is most developed.

Hungary's future looks promising. The country is making cultural and economic links with the developed countries of Europe and with the United States. Russia and Ukraine, meanwhile, remain a ready market for Hungarian products and an important source of raw materials.

Romania Although the Communist government expanded industry after World War II, Romania remains a developing country. With the exception of oil, most of Romania's potential mineral resources are undeveloped. The famous Ploiesti (plaw-YEHSHT) oil fields are some of the richest in Eastern Europe. Even so, Romania's oil resources do not meet the country's needs.

Romania's cultural, industrial, and commercial center is Bucharest (BOO-kuh-rehst), the capital city. About 70 percent of all Romanians belong to the Romanian branch of Eastern Orthodox Christianity. A large number of ethnic Hungarians live in northern Romania.

Closure

Have students make a chart showing each country's economic, historical, and cultural features.

ASSESSMENT

Assign the Section 2 Review questions on this page.

Reteaching

Assign Section 2 of Reteaching Worksheet 27.

SECTION 3
OBJECTIVES

The objectives reflect the Focus questions on this page and are assessed in the review on page 314.

1. Summarize the history of Yugoslavia, and list the territory's new republics.

2. Discuss the role that ethnic conflicts have played in the countries of the former Yugoslavia.

TEACHING SECTION 3
Motivator

Ask students what they've seen on television or read in the newspapers about the conflicts in the former Yugoslavia. Remind students of the difference between a civil war and a war fought between separate countries. Tell students that in Section 3 they will learn how Yugoslavia was formed and why it broke apart.

For over 20 years, Romania had a particularly repressive Communist dictatorship. A violent uprising in 1989 overthrew the government. The new Romanian government promises a parliamentary democracy and a market economy.

Bulgaria Bulgaria occupies the eastern side of the Balkan Peninsula. During the Communist era, the Bulgarians were the most loyal followers of the Soviet Union. With extensive Soviet economic aid, Bulgaria became more prosperous than some other countries of Eastern Europe.

Important industries are concentrated near Sofia, the capital and largest city. Sofia is also the historic center of Bulgaria's cultural and political life. Today, it is a bustling, modern city.

The new government contains many leaders from the Communist era, and life in Bulgaria is changing only gradually. The people do enjoy more freedoms, and a market economy is growing slowly.

Albania Physically and culturally isolated from the rest of Eastern Europe, Albania is Europe's least developed and poorest country. Albania's capital city, Tirane (ti-RAHN-uh), the site of the country's few factories, has fewer than 300,000 people.

After World War II, a Communist government took power, but the Albanian Communists feuded with the Soviet Union and received no economic aid. A command economy coupled with isolation meant little economic progress for the country.

In 1990, the Communist government collapsed, and Albanians began to welcome foreign contact. However, economic change and movement toward democracy are coming only slowly. Economic and political progress could slow even more if Albania becomes involved in ethnic conflicts with neighboring countries.

 SECTION REVIEW

1. What economic activities support the countries discussed in this section?
2. What roles are played by Eastern Europe's capital cities?
3. **Critical Thinking** **Compare and contrast** Poland's and Bulgaria's relationships with the Soviet Union during the Communist era.

3 A REGION IN CONFLICT

FOCUS

● *How did Yugoslavia form, and why did it break up? How is the territory divided now?*

● *What role have ethnic conflicts played in the region?*

The Breakup of Yugoslavia Along the western half of the Balkan Peninsula lie the new nations of the former country of Yugoslavia. Yugoslavia was created after World War I mainly by U.S., British, and French policy makers. *Yugoslavia* means "land of the southern Slavs." Lands populated by several Slavic groups were combined from pieces of the Turkish (Ottoman) Empire and the Austro-Hungarian Empire. The plan was that the region's many Slavic groups would share common goals that would benefit the new country. The main Slavic groups in the former Yugoslavia were the Serbs, Croats, Bosnians, Slovenians, and Macedonians.

Although Yugoslavia became a Communist country after World War II, its leader, Josef Broz Tito steered a course independent from the Soviet Union. Tito kept a tight grip to keep the country unified, and indeed, Yugoslavia survived as a single nation until 1991. The strong central government had been able to force the ethnic groups to live together. The Serbs and Croats, however, have thought of themselves as separate peoples since the seventh century. With Tito's death and, later, the end of communism in the Soviet Union, the country soon began to unravel violently.

Serbia and Montenegro together have claimed the name Yugoslavia. Bosnia and Herzegovina, Croatia, Slovenia, and Macedonia have all declared their independence. Due to ethnic and religious differences, civil war broke out in Bosnia between Serbs, Croats, and Bosnians. Most Croats and Slovenians are Roman Catholic, the Serbs and Macedonians are Eastern Orthodox, and many Bosnians are Muslims. Because of the mixture of ethnic groups in many regions, additional border changes are possible. As these ethnic struggles continue, the economy of the region is in steady decline, along with the quality of people's lives.

COOPERATIVE LEARNING

Organize the class into small groups. Direct each group to research and make a poster about a major city of Eastern Europe, such as Warsaw, Budapest, Sofia, or Prague. Group members should compile a list of the city's historical and cultural attractions and share their findings with each other. Individual roles might include researchers, writers, and artists.

Section 2 Review
ANSWERS

1. Agriculture, industry
2. The cities are their countries' cultural, political, historical, and economic centers.
3. Poland rebelled against the Communist government and the Soviet Union, whereas Bulgaria was the most loyal follower of the Soviet Union.

313

Guided Practice

Refer students to the map of Eastern Europe on page 309, and indicate to them the areas that were formerly Yugoslavia. Encourage them to study this map as they read Section 3. Then have students work in pairs to identify the major groups in conflict with one another (based on religion or ethnicity) within each area.

Independent Practice

Provide students with Outline Map 15. Have them label the new republics and indicate which regions are in conflict.

Closure

Have students make maps of the former Yugoslavia region showing how it might be reorganized so as to reduce conflict.

ASSESSMENT

Assign the Section 3 Review questions on this page.

Reteaching

Assign Section 3 of Reteaching Worksheet 27.

Section 3 Review
ANSWERS

1. After World War I by U.S., British, and French policy makers who thought the region's many Slavic groups would share common goals; a strong central Communist government

2. Serbs, Croats, Bosnians, Slovenians, Macedonians; much tension exists between many of the groups, and there has been violence among them; in Serbia and Montenegro, Bosnia and Herzegovina, and Croatia

3. Yugoslavia's breakup became quite violent, whereas Czechoslovakia's was peaceful.

314

Serbia and Montenegro Serbia is in the eastern plains of former Yugoslavia. In the southwest, Serbia borders Montenegro, a mountainous coastal territory north of Albania. The capital of Serbia is Belgrade, also the capital of former Yugoslavia. Located on the Danube River, Belgrade is one of the major cities of Eastern Europe.

The Serbs have been demonstrating extreme nationalism since the late 1980s. Radical Serbs maintain that no Serb should be governed by someone from another ethnic group. Of particular concern is the future status of the Vojvodina and Kosovo regions. Vojvodina is part of the Great Hungarian Plain north of Belgrade. Its people are mostly Hungarians who fear Serbian nationalism. Many have fled Vojvodina into Hungary. Kosovo is a southern Serbian region along the border with Albania. Most of the people here are Muslim Albanians. Friction between Kosovo's Albanians and Serbs has escalated.

Bosnia and Herzegovina Bosnia and Herzegovina generally are referred to as Bosnia. This is a mountainous territory located between the plains of Serbia and the coastal mountains of Croatia. Within the mountains of Bosnia lie isolated valleys where rural life has changed little in the past century. The once-beautiful capital of Sarajevo (SAIR-uh-yay-voe) has been destroyed in the civil war. No single ethnic group is a majority.

Croatia Croatia occupies the northern part of the Dinaric Alps and has a long, narrow coastal region along the Adriatic Sea. Zagreb is the capital and cultural center. About 12 percent of the people in Croatia are ethnic Serbs who would like to see Yugoslavia reestablished. Tensions between Croatia and Serbia will likely continue.

Slovenia Slovenia is a former Austrian territory situated in the mountains northwest of the Balkan Peninsula. Slovenians have long thought of themselves as part of Western rather than Eastern Europe. Few minorities live in the region, and independence came with little bloodshed. The territory remains largely agricultural, and the farms have always been privately owned. Ljubljana (lee-yoo-blee-AHN-uh), the capital city and principal industrial center, lies on the country's major plain.

Macedonia Macedonia refers to both a province of northern Greece and the southern

Ethnic Albanians worship at a Muslim mosque In **Pristina**, the capital of the Kosovo province in **Serbia**. Although a minority in Serbia, ethnic Albanians make up most of the population of Kosovo.

region of former Yugoslavia. The capital of the latter is the city of Skopje (SKAWP-yeh). About 65 percent of the people in Macedonia are Macedonians. The largest minority, about 20 percent, are Muslim Albanians. Many of the farms in this agricultural land provide little more than subsistence incomes. With few jobs available, many young Macedonians seek work in other European countries. There is a history of cooperation between Macedonians and Serbs in Yugoslavian affairs. As a result, this region has been spared much of the impact of the civil war. The region's economic future remains uncertain, however.

Future Changes The history of the Balkan region is being rewritten every day. It is difficult to understand the turmoil and bitterness that has plagued the Balkans for centuries. Many attempts have been made by the United States, the EU, and the United Nations to negotiate peace, but there is no clear or easy solution to the conflict.

3 SECTION REVIEW

1. When and why was the country of Yugoslavia formed? What held the country together until 1991?
2. What ethnic groups live in the region? How do these groups relate to each other? Where have ethnic conflicts been most extreme?
3. **Critical Thinking** **Compare and contrast** the breakups of Yugoslavia and Czechoslovakia.

Building a Vocabulary

1. Regional specialization, in which each country specialized in the products it could make or grow best
2. Something received from a previous era; environmental decay in Eastern Europe from the Communist era
3. Regions that produce different products that can be combined to benefit both or all of the regions; the Czech lands had industrial potential whereas Slovakia was mostly agricultural.

Recalling and Reviewing

1. Humid continental and Mediterranean
2. Eastern European languages include Slavic languages, Romanian, Albanian, and Hungarian. Religions include Eastern Orthodox, Roman Catholic, and Islam.
3. Both are developing countries.

CHAPTER **27** CHECK

Black Sea vacationers in **Bulgaria**

Reviewing the Main Ideas

1. Eastern Europe is a region of varied landforms and cultures. The borders of Eastern European countries have changed many times. From World War II to the late 1980s, the Soviet Union controlled Eastern Europe and introduced communism to the region. Today, the countries are becoming democratic and are building free-market economies.
2. Some of the countries of Eastern Europe are Poland, the Czech Republic, Slovakia, Hungary, Romania, Bulgaria, and Albania. The economic activities in these countries vary, as do the ways in which the countries have responded to independence from Soviet control.
3. The country formerly called Yugoslavia began to unravel violently when communism in the Soviet Union ended and the central government could no longer control ethnic tensions. Rivalries between ethnic groups wrack Serbia and Montenegro, Bosnia and Herzegovina, and Croatia.

Building a Vocabulary

1. What term describes the main plan of COMECON for the countries of Eastern Europe? Describe the plan.
2. Define *legacy*, and give an example of a legacy.
3. What are complementary regions? How were the Czech and Slovak territories complementary regions?

Recalling and Reviewing

1. What climates are found in Eastern Europe?
2. What languages do Eastern Europeans speak? What religions do they practice?
3. What do Romania and Albania have in common economically?

Thinking Critically

1. How might regional specialization be a disadvantage for a country?
2. The ethnic tensions in the countries that were formerly called Yugoslavia are centuries-old. Why do you think that even after 40 years of repression by a Communist government, ethnic identities were still alive?

Learning About Your Local Geography

Cooperative Project
As the countries of Eastern Europe become more open to the other countries of the world, Eastern Europeans are eager to learn about the ways of life of people in these other countries. As a group, write a letter to a teenager in Eastern Europe about your daily lives. You might want to include what you do for fun, what your favorite foods are, and what you study in school.

Using the Five Themes of Geography

Make a flowchart showing how physical and human **place** characteristics have influenced peoples' **movement** in the countries of Eastern Europe. You might want to make your flowchart from the point of view of an Eastern European trying to decide whether to move to a different area within the region. What factors might keep you from moving? What factors might encourage you to move?

Thinking Critically

1. It might increase a country's dependence on other countries for certain products or materials. If anything were to disrupt the country's relations with its providers, it might be at a loss for vital materials.
2. Possible answer: ethnicity is very basic to a people's identity and cannot be forgotten or removed even over a long period of time.

Learning About Your Local Geography

Letters should give information about students' daily lives.

Using the Five Themes of Geography

Flowcharts should show that rugged mountains have hindered movement in Eastern Europe. On the other hand, border shifting has encouraged or forced people to move to certain places.

315

SKILLS

Point out that paintings are not the only form of art that can reflect geography. Sculpture, music, and dance can also show geographical influences. Play for students some selections of Claude Debussy's *La Mer*. Tell students that Debussy was an important French composer, and that *La Mer* means "the sea" in French. Have students describe the images of ocean and water that the music conjures up. Also point out that works of art do not come solely from Europe but come from all world regions. Mention that geography can influence how people study art, in addition to how people create art.

Practicing the Skill
ANSWERS

1. The painting shows agriculture, probably the harvesting of wheat. *Glean* means to pick up the materials that have been cut by reapers, so the characters are probably picking up wheat cuttings.
2. The painting shows a plains region and wheat farming, so it probably depicts northern France.
3. It is significant in that although women were agricultural workers at the time of the painting, they probably did not usually work alongside men in the fields doing the same work as men did. While men probably cut the grain, women gathered it.
4. Almost certainly not; although agriculture is a major part of France's economy and wheat is still raised in the north, few French people are farmers today, and modern agriculture probably uses much more technology and less human labor.

316

UNIT 5 REVIEW **SKILLS**

Linking Geography and Fine Art

Throughout the centuries, artists have been interested in the physical world around them. By studying works of art, we can learn much about the physical and human geography of a particular region.

Peter Graham, a Scottish artist who lived from 1836 to 1921, often painted scenes from the Scottish highlands. As you know from Chapter 24, these highland areas contain ancient, eroded rocky hills, as shown in the painting to the right. Also shown clearly is the cloudy weather for which Great Britain is known. The painting even portrays the economic activities for which the region is suited: grazing and fishing. If you were to visit the area shown in the painting today, you would probably find a similar scene.

Wandering Shadows, Peter Graham

The Gleaners, Jean-Francois Millet

Jean-Francois Millet was a famous French painter who lived from 1814 to 1875. Most of his paintings, such as this one at left, show scenes from French rural life.

PRACTICING THE SKILL

Study the painting *The Gleaners* above, and then answer the following questions.

1. What economic activity is shown in the painting? Use a dictionary to identify the meaning of the word *glean*.
2. Based on the landform and on the activity shown in the painting, what part of France does this painting probably show?
3. The main characters of the painting are women. How is this significant?
4. Millet completed this painting in 1857. If you were to visit the place portrayed in the painting today, would you find a similar scene? Why or why not?

BUILDING YOUR PORTFOLIO

Decide whether students will work individually or in groups on the unit projects described below. Students should then choose which project they will work on. Provide students with copies of the Portfolio Assessment Worksheets in *Teaching Resources*. Chapters 23–27 worksheets and the Unit Portfolio Assessment Worksheet will guide students step-by-step to complete either project.

Teaching with Sketch Maps

Have students sketch a map of Europe from memory. Tell them to include as many details as they can about place names, landforms, and boundaries of regions they studied in Unit 5. Students should compare this map with the sketch map they completed at the beginning of Unit 5 to note their improvement in identifying geographic features.

◤ BUILDING YOUR PORTFOLIO

Individually or in a group, complete one of the following projects to show your understanding of the geography concepts involved.

Ⓐ *A Unified Europe?*

In this unit you have read much about disunity in Europe. Some countries, such as Czechoslovakia and Yugoslavia, have broken up into smaller countries. In addition, minority groups in some countries, such as the Roman Catholics in Northern Ireland and the Basques in Spain, want independence. Yet, as you have also read, many Europeans hold an ideal of a united Europe. Write a draft of what a constitution of the new united Europe might say. Your constitution will consist of four written articles, each outlining a different aspect of the new united Europe.

1. Your first article will address political aspects of the new united Europe. What form of government will it have? How will leaders be chosen? Will individual regions or countries have any political powers?
2. The second article of your constitution will be about the new united Europe's military. How will the region defend itself? Will it have a national military force, and if so, who will serve in it?
3. Your third article will relate to the new united Europe's economic system. Will industries be publicly or privately owned? What currency will be used? How will taxes be collected, and what will they be used for? How will trade among the different regions be regulated?
4. The constitution's fourth article will address cultural matters. What will be the new united Europe's official language or languages? Will all cultural groups have the same rights and privileges? How would any ethnic conflicts be handled?

Present your constitution to the rest of the class. In doing so, address the following questions. Do you think it is desirable for the countries of Europe to unite? Do you think such unification is possible?

Ⓑ *Water Pollution in Europe*

As you know, water is one of Europe's most important resources. Rivers and harbors are used for trade and transportation, for fishing, and for hydroelectricity. Yet many of Europe's waters suffer from pollution. What are the causes of this pollution? Which bodies of water are particularly polluted, and why? How might the problem of pollution be solved? Investigate the answers to these questions, and write a script for a brief documentary film presenting your findings. Include with your script illustrations, maps, and other visual materials that will help your audience understand your points. Your documentary will be organized according to Europe's regions.

1. First, consider water pollution in Northern Europe. You might want to focus on Great Britain's rivers, or the Irish, North, or Baltic sea.
2. Second, discuss the pollution of rivers in West Central Europe, such as the Rhine or the Seine.
3. Next, address the pollution of the Mediterranean Sea. Include a discussion of the environmental effects of tourism.
4. Finally, your documentary should discuss water pollution in Eastern Europe, such as along the Danube.

Present your script and supporting visual materials to the rest of your class.

Project A:

"Constitutions" should indicate that students have carefully thought about the political, military, economic, and cultural aspects of increased European unification. In presentations, students should assess the desirability of European unification and whether such unification would be possible.

Project B:

Scripts and supporting visual materials should address water pollution in Europe's regions. Focuses should be on the causes of the pollution (for example, industrial wastes, oil spills, ship and barge transportation, tourism, or lack of regulation), bodies of water that are particularly polluted, and how pollution might be curbed.

Regions in Transition

Understanding Regional Connections

In a line down the left side of the board, write Sweden, Romania, Poland, and Finland. Then write Estonia, Latvia, Lithuania, and Moldova down the right side. Call on four students to come up to the board and "match" one of the countries on the right to one of the countries on the left by drawing a line between them. After each student draws his or her line, ask other students in the class to say whether the match is a good one, and why or why not. *(Students may cite geographical proximity or similarity of cultural characteristics as reasons why certain countries should be matched.)* Have students correct inappropriate matches. *(Estonia should be matched to Finland, Latvia to Sweden, Lithuania to Poland, and Moldova to Romania.)*

Chart Caption Answer

Lithuania

Geographical Sidelight

Since they became independent of the Soviet Union, crime in the Baltic nations has increased dramatically. Crime gangs have made these countries their headquarters, largely due to the nations' locations. Because they lie directly between Western Europe, Eastern Europe, and Russia, the countries are well situated for illegal trading and smuggling activities. Furthermore, the newly independent Baltic states do not have the resources to stop or even systematically track the activities of the crime gangs. Evidence of the increased criminal activity can be seen in the city of Tallinn, Estonia's capital. In the early 1990s, that city had one of the highest murder rates in the world.

REGIONS IN TRANSITION

Redrawing the European Map

The collapse of the Soviet Union marks one of the most significant events of the twentieth century. Geographers are defining new regions in Europe that include countries that were once part of the Soviet Union. Among these are Moldova, sandwiched between Romania and Ukraine, and the three small Baltic countries of Lithuania, Latvia, and Estonia.

The Baltic Countries

Lithuania, Latvia, and Estonia were once part of the Russian Empire. They enjoyed brief independence after the Communist Revolution in 1917, but in 1940, they were forced into the Soviet Union. The Baltic states regained their political independence in 1991. Achieving economic and cultural independence, however, could take years for several reasons. First, Estonia and Latvia include large numbers of ethnic Russians. Second, Russian army and navy bases in the region have long been important to the Baltic states' local economies. And third, transportation and business links tie the Baltics to the former Soviet Union.

Each of the Baltic countries is particularly dependent on minerals and fossil fuels from Russia and Ukraine. Similarly, the traditional markets for Baltic goods are in Russia and

FACTS IN BRIEF — The Baltic Countries and Moldova

COUNTRY POPULATION (1994)	LIFE EXPECTANCY (1994)	LITERACY RATE	PER CAPITA GDP (1993)
ESTONIA 1,616,882	65, male 75, female	100% (1989)	$5,480
LATVIA 2,749,211	64, male 75, female	99% (1992)	$4,810
LITHUANIA 3,848,389	67, male 76, female	98% (1989)	$3,240
MOLDOVA 4,473,033	65, male 72, female	96% (1992)	$3,650
UNITED STATES 260,713,585	73, male 79, female	97% (1991)	$24,700

Sources: *The World Factbook 1994,* Central Intelligence Agency; *1995 Britannica Book of the Year*

Which country has life expectancy rates closest to that of the United States?

Ukraine. For these reasons, continued cooperation with the other regions of the former Soviet Union is necessary. Establishing strong economic links to the rest of the world will come gradually.

As the Baltic countries reexamine their cultural makeup and economies, geographers ponder how best to include these countries in the subregions of Europe. Perhaps Estonia and Latvia will be identified as Scandinavian countries of Northern Europe. The Estonian language is related to Finnish, and ties between Estonia and Finland are growing stronger. Many Estonians watch Finnish television broadcasts, and ferries link the capitals of Tallinn, Estonia, and Helsinki, Finland. Latvia, like Estonia, was heavily influenced by Sweden in the past. Lutheranism is the traditional religion of Latvia, Estonia, and Scandinavia.

In contrast to Estonia and Latvia, Lithuania shares a border with Poland and has ancient ties to that country. About 7 percent of the people in Lithuania are ethnic Poles. In both countries, Roman Catholicism is the dominant religion. Geographers speculate that many of these cultural ties will grow, and that Lithuania, like Poland, will be considered part of Eastern Europe.

These Latvian children in local dress are playing a traditional stringed musical instrument from their country. People in the Baltic countries proudly maintained their cultural traditions during the Soviet era.

318

Using Geography Themes

Have students study Kaliningrad to reinforce the geographic theme of location. Organize students into small groups, and have them consider the following questions: What countries does Kaliningrad border? Through what countries must one pass to travel between Kaliningrad and Russia? In what ways would it be advantageous for Russia to keep Kaliningrad? What logistical and other problems might arise for either

Russia or Kaliningrad as a result of their physical separation? Compare all groups' answers in a classwide discussion.

Geographical
Sidelight

THE BALTIC COUNTRIES AND MOLDOVA: PHYSICAL–POLITICAL

LEGEND

ELEVATION

	FEET	METERS
⊛ National capitals	13,120	4,000
	6,560	2,000
• Other cities	1,640	500
	656	200
	(Sea level) 0	0 (Sea level)
	Below sea level	Below sea level

Size comparison of the Baltic Countries and Moldova to the contiguous United States

Kaliningrad

The future status of the Kaliningrad region along the Baltic Sea also is of interest to geographers. This territory remains part of Russia, even though it is now separated from Russia by independent Lithuania. Before World War II, this region was part of Germany. After the war, many of the region's ethnic Germans were forced to move to Siberia and Central Asia. Ethnic Russians replaced the German population. Kaliningrad could become valuable to Russia as a window on Europe. Some geographers believe, however, that it will be next to demand its independence.

Moldova

Throughout history, control of Moldova has shifted many times. It has been dominated by the Turks, Polish princes, Austria, Hungary, Russia, and Romania. The Soviet Union acquired Moldova by force from Romania during World War II. Like the Baltic states, Moldova became independent of the Soviet Union in 1991.

Moldova's ethnic composition is diverse. Ethnic Romanians make up about two-thirds of the population, but the country also has many Russians and Ukrainians. Ethnic conflicts among these groups have emerged and

remain to be settled. Gradually, geographers may include Moldova in the same region as Romania. It is even possible that the two countries will eventually merge.

Mapping the New Europe

As countries realign themselves culturally and politically, borders may shift again and again. The new Europe may stretch across the geographical boundaries of the European continent, though ties to the former Soviet Union will remain. Geographers are hopeful that these changes can be made peacefully and with respect for the rights of all ethnic groups.

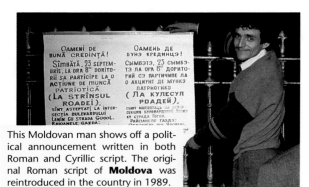

This Moldovan man shows off a political announcement written in both Roman and Cyrillic script. The original Roman script of **Moldova** was reintroduced in the country in 1989.

Several factors are hindering the possible reunification of Moldova and Romania. First, ethnic conflicts in Moldova's Trans-Dniester region threaten the country's stability. Many people were killed in or forced to flee from this region.

Second, it is unclear whether Moldovan citizens want reunification. Although 64 percent of Moldova's people are ethnic Romanians, 1993 polls showed that only 7 percent of the population favored reunification.

Third, Romania is concerned about integrating Moldova's economy with its own, as free-market reforms are progressing even more slowly in Moldova than in Romania.

Due to these obstacles, Moldova and Romania are unlikely to merge in the near future. Nevertheless, Moldova continues to drift further away from the other former Soviet republics, avoiding closer political and economic integration with them.

UNIT 6 RUSSIA AND NORTHERN EURASIA

CHAPTERS 28-31   OVERVIEW

In Unit 6, students will learn about the landforms and people of Russia and Northern Eurasia, a vast region that is undergoing many political and economic changes.

Chapter 28 • INTRODUCING RUSSIA AND NORTHERN EURASIA describes the historical and political development of northern Eurasia, from the earliest immigrants through czarist rule and the rise and fall of Soviet communism.

Chapter 29 • RUSSIA introduces the landforms, climate regions, natural resources, and economic and industrial centers of the largest country of the former Soviet Union. It also discusses the political issues confronting the Russian people today.

Chapter 30 • UKRAINE, BELARUS, AND THE CAUCASUS describes the physical geography and human geography of the newly independent countries that lie along the western edge of Russia.

Chapter 31 • CENTRAL ASIA discusses the Central Asian countries of Kazakhstan, Kyrgyzstan, Uzbekistan, Turkmenistan, and Tajikistan. These countries are facing political and economic restructuring.

OBJECTIVES

1. Examine the relationship between events in Russian history and the creation of the Soviet Union and Communist state.

2. Analyze the strengths and weaknesses of the Communist system and the reasons for the breakup of the Soviet Union.

3. Identify nations in Northern Eurasia that were once part of the Soviet Union, and describe their ethnic makeup.

4. Link the natural resources of Russia and Northern Eurasia with the economic development of the region.

5. Describe the physical features and climate patterns of Russia and Northern Eurasia.

6. Interpret special-purpose maps to analyze relationships among climate, population patterns, and economic resources in Russia and Northern Eurasia.

7. Analyze how political events can affect the economic stability of a region.

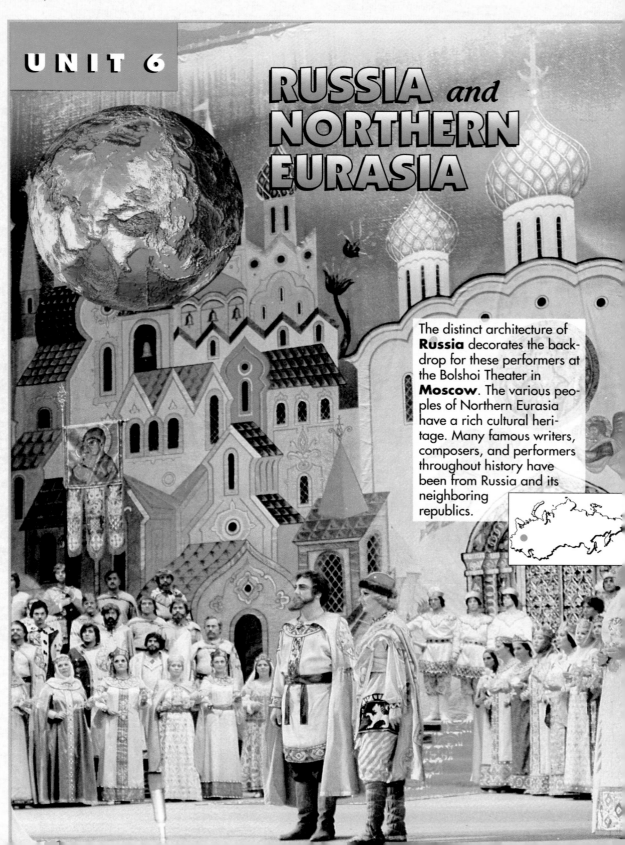

UNIT 6

RUSSIA and NORTHERN EURASIA

The distinct architecture of **Russia** decorates the backdrop for these performers at the Bolshoi Theater in **Moscow**. The various peoples of Northern Eurasia have a rich cultural heritage. Many famous writers, composers, and performers throughout history have been from Russia and its neighboring republics.

Using the Illustrations

Ask students to look at the photographs on these pages. Ask: What photograph shows physical geography? *(the steppe)* Point out that the steppe is described often in Russian literature. Refer students to the description of the steppe in the literature selection on page 322 and in the Chapter 30 opening quotation on page 350. Point out that one reason the steppe is a familiar theme is that it stretches a great distance from Ukraine to the Ural Mountains, and it has played an important role in migration across Russia and Northern Eurasia. Ask: What photograph shows human-environment interaction? *(the house)* What images show cultural traits? *(the house, the play at the Bolshoi Theater)*

Teaching with Sketch Maps

Have students draw a map of Russia and Northern Eurasia from memory. Tell them to try to include as many political names or boundaries as they can that refer to areas that were formerly part of the Soviet Union. Students might add this map to their portfolios and compare it with another they will draw after they have completed Unit 6.

UNIT 6 CHAPTERS

28 *Introducing Russia and Northern Eurasia*

29 *Russia*

30 *Ukraine, Belarus, and the Caucasus*

31 *Central Asia*

A gold comb from the early Scythian kingdom

The rolling plains of the steppe sweep toward the distant Ural Mountains. Stretching from **Ukraine** to **Central Asia**, the vast steppe is the most productive agricultural region of Northern Eurasia.

Russian *matryoshka* nesting doll

Gingerbread trim decorates this *dacha* near **Kursk, Russia**. Country homes such as this are popular vacation locations for many Russians. Some *dachas* are small, one-room houses or even rented rooms in a larger house.

THE WORLD TODAY VIDEODISC PROGRAM

You might wish to use the profile of Russia as you teach Unit 6.

Geographical Sidelight

The breakup of the Soviet Union has created changes in place-name references. National Geographic's January 1993 map of Russia is the main source for place-names in Unit 6. The first spelling of a place-name in this unit is the newest transliteration; it is followed by the former Soviet spelling in parentheses.

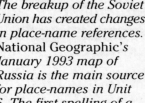

One World, Many Voices

LITERATURE

After students have read the literary excerpt below, refer them to the photograph on page 321 and to the climate map of Russia and Northern Eurasia on page 326. Use these visual elements to indicate physical characteristics of the Russian steppe and the location and extent of the steppe climate in the region.

More about Anton Chekhov

Anton Chekhov combined a busy medical practice and literary career with an interest in travel. Between 1887 and 1892 he journeyed to the Ukraine and to the Crimean peninsula, traveled across Siberia, and returned home via China, India, and Egypt.

ONE WORLD, MANY VOICES

LITERATURE

Russian Railroad

Primary Source

Description of Change: excerpted
Rationale: excerpted to focus on main idea

"The Donetz Railroad. A cheerless station, quiet and lonely, gleaming white on the steppe, with walls hot from the sun, with not a speck of shade and, it appears, with not a single human being. The train which brought you here has left; the sound of it is scarcely audible and at last dies away. The neighborhood of the station is deserted, and there are no carriages but your own. You get into it— this is so pleasant after the train—and you roll along the road through the steppe, and by degrees, a landscape unfolds such as one does not see near Moscow—immense, endless, fascinating in its monotony. The steppe, the steppe, and nothing else; in the distance an ancient grave-mound or a windmill; oxcarts laden with coal file by. Birds fly singly low over the plain and the monotonous beat of their wings induces a drowsiness. It is hot. An hour or two passes, and still the steppe, the steppe, and still in the distance the grave-mound. The driver rambles on telling you some long-drawn-out, irrelevant tale, frequently pointing at something with his whip, and tranquillity takes possession of your soul, you are loath to think of the past."

Interpreting Literature

Answers

1. Semiarid steppe; treeless plain
2. The narrator describes the scenery as "endless" and refers to the landscape's "monotony." Repetition of "the steppe, the steppe" also conveys an unchanging landscape.

Interpreting Literature

1. What type of climate and vegetation region does Chekhov describe?
2. What words does the narrator use to indicate that the scenery changes very little throughout the carriage ride?

Anton Chekhov (1860–1904) is one of Russia's most important literary artists. Ill health from tuberculosis, along with a love of writing, caused him to give up a career in medicine. He wrote many plays and hundreds of short stories. Most of his work concerns the everyday lives of people from many classes of pre-revolutionary Russian society in the late 1800s. Chekhov's tales reveal his understanding of human nature and also express his affection for the Russian physical landscape. The selection on this page is from a story entitled "At Home."

322

FOR THE RECORD

Laurens van der Post (1906–) is a noted travel writer and novelist. He has said that "one never really knows another country" unless one observes it firsthand through the lives of its citizens. Many of his books describe Africa, where he was born, and his travels to England, France, and Japan. Van der Post was curious about Russia because he was not satisfied with the images of the country he received through newspapers, magazines, and television. In 1961, he began the first of four lengthy trips across the Soviet Union. This excerpt is from *Journey into Russia*, published three years later.

"This Russian now told me that it was right to begin with a feeling of the immensity of the land. I must build on that as perhaps the single greatest fact for understanding his people and his native country. But he feared that it would take time before a stranger could realize how immense the land was. . . .

So wide apart are the two extremities of land that it is day in one when it is still night in another. He himself had once flown in a fast jet aircraft from Baikal [Baykal] to Moscow, made four landings on the way and each time as they landed the sun had set and each time they took to the air the sun had risen again. . . . Already in Turkmenia [Turkmenistan], ploughing and sowing would be done and the spring nearly over, but in Uzbekistan where we would land in a few hours we would find the earth aglow with purple heliotrope, scarlet poppies and wild red tulips. Yet the apple trees around Moscow would not blossom until late in May and in the far north of Siberia the tundra rivers would not be free of ice until the end of June. There by August the ice and frost would start their march in the south. In the extreme south the bamboo shot up at a rate of a yard a day. On the northern Siberian shore it took the straggling larches a century to grow the thickness of a man's middle finger. In the south they would already be harvesting the grain while in Kamchatka the snow was still falling. That, measured in the seasons, was the physical frame of the Soviet Union."

Analyzing Primary Sources

1. What does this selection suggest is most significant about Russia's physical geography?
2. How does Van der Post illustrate Russia's vast size?

Analyzing Primary Sources

Answers

1. Its huge dimensions
2. By focusing on seasonal extremes and time differences that occur throughout Russia

CHAPTER 28 Introducing Russia and Northern Eurasia

Planning Guide

TEXTBOOK	TEACHING RESOURCES	OTHER AVAILABLE MATERIALS
Section		
1. A Diverse Heritage Pages 328–329	Reteaching Worksheet 28.1	*REGIONS OF THE WORLD Map Transparencies:* Russia and Northern Eurasia *Study Guide for Spanish Speakers* *Lecture Notes,* Chapter 28
2. The Soviet Union Pages 329–331	Reteaching Worksheet 28.2 Graphic Organizer 28 Critical Thinking Worksheet 28	*Study Guide for Spanish Speakers* *Lecture Notes,* Chapter 28
3. A New Geography of Northern Eurasia Pages 331–332	Reteaching Worksheet 28.3 Challenge/Enrichment Worksheet 28 Skills Worksheet 28	*Study Guide for Spanish Speakers* *Lecture Notes,* Chapter 28
Chapter Review Page 333	Chapter 28 Test Chapter 28 Portfolio Assessment See also *Alternative Assessment Forms*	*Test Generator,* Chapter 28

CHAPTER RESOURCES

Lecture Notes, Chapter 28
Teaching Resources
   Chapter 28 Graphic Organizer
   Chapter 28 Skills Worksheet
   Chapter 28 Critical Thinking Worksheet
   Chapter 28 Challenge/Enrichment Worksheet

REVIEW AND ASSESSMENT

Chapter 28 Check, page 333
Chapter 28 Test, *Teaching Resources*
Chapter 28 Portfolio Assessment, *Teaching Resources*
Test Generator, Chapter 28
Alternative Assessment Forms

ADDITIONAL RESOURCES

Books for Teachers

Kaiser, Robert J. *The Geography of Nationalism in Russia and the USSR.* Princeton: Princeton University Press, 1994.

Komarov, Boris. *The Geography of Survival: Ecology in the Post-Soviet Era.* Armonk, NY: M.E. Sharpe, 1994.

MacKenzie, David, and Michael W. Curran. *A History of Russia, the Soviet Union, and Beyond.* Belmont, CA: Wadsworth Publishing, 1993.

Books for Students

Anderson, Madelyn Klein. *Siberia.* New York: Dodd, Mead, 1988.

Barbour, William, and Carol Wekesser, eds. *The Breakup of the Soviet Union: Opposing Viewpoints.* San Diego: Greenhaven Press, 1994.

Hautzig, Esther. *The Endless Steppe.* New York: Harper & Row, 1968.

The World Today *Videodisc Program*

A variety of still images, short videos, and activities are available for you to use as you teach this chapter. *Lecture Notes* include bar codes for specific images and videos.

Multimedia Resources

Communism and the Cold War. (laserdisc, HyperCard software) Coronet/MTI Film & Video, 108 Wilmot Road, Deerfield, IL 60015.

Picture Atlas of the World. (interactive maps, photo bank, animation, audio on CD-ROM) National Geographic Society, Washington, DC.

The Volga. (VHS, 45 minutes) Films for the Humanities & Sciences, P.O. Box 2053, Princeton, NJ 08543–2053.

MEETING INDIVIDUAL NEEDS

Limited English Proficiency (LEP)

Use a physical–political wall map of the world to help students appreciate the enormous size of the former Soviet Union. Point out the northernmost and southernmost latitudes and the easternmost and westernmost longitudes of the region. Direct students to compare the former Soviet Union's size with that of other countries.

Learners Having Difficulty

Help students review the main concepts of the chapter by making a causes-events-effects diagram. Draw a three-column chart on the board, and in the center column write a list of historical events relating to northern Eurasia. Then call on volunteers to supply the causes and effects of these events.

Auditory Learners

Have groups of students write scripts for a radio documentary titled "A Free-Market Economy: Yes or No?" Tell groups that their scripts should explore the thoughts and opinions of the following people regarding changes in the former Soviet Union: Russian workers, former Communist leaders, and potential American investors. Have groups present their documentaries to the class.

Visual Learners

Have students create a poster or a bulletin board display in the style of a newsmagazine's "People" or "Newsmakers" section, featuring prominent figures in the history of Russia and the Soviet Union. Assign pairs of students a personality from Chapter 28 about whom to write a short, descriptive blurb accompanied by a photograph or a drawing.

Tactile/Kinesthetic Learners

Divide the class into five groups. Ask each group to research, prepare, and present a short skit about citizens of Russia or the Commonwealth of Independent States in one of the following situations: living at home, working, attending school, receiving health care, or enjoying personal freedoms.

Gifted Learners

Have interested students investigate the major challenges facing the countries of northern Eurasia today, such as pollution and environmental damage, economic development, and political reforms. Direct students to form a panel to discuss these challenges and brainstorm possible solutions. You may wish to have volunteers write letters to state leaders in the region describing their ideas.

MULTICULTURAL FOCUS

The Russian people have long had a reputation as devoted readers. During the Soviet era, people had access to a large variety of printed materials—books, newspapers, and magazines.

However, the communications media were carefully censored by editorial boards appointed by the Communist party. Criticism of the Communist government and economic system was not tolerated, and materials containing such criticism were condemned and prevented from being published or were eliminated after publication.

Have students participate in the following activity to help them understand Soviet censorship. Divide the class into several groups, and provide each group with a copy of a U.S. daily newspaper or weekly newsmagazine. Have students act as editorial censors by crossing out all passages in the publication that criticize the U.S. government, question its policies, or suggest that life in the United States is not perfect. When all the groups have finished, have the class discuss the kinds of ideas they eliminated and the kinds they allowed to remain.

You may wish to lead a discussion about the importance of a free press in a democracy. If possible, obtain a copy of a translated newspaper or newsmagazine from the Commonwealth of Independent States, and discuss with students the kinds of materials published there today.

CHAPTER 28

INTRODUCING RUSSIA AND NORTHERN EURASIA

Section 1: A Diverse Heritage
Section 2: The Soviet Union
Section 3: A New Geography of Northern Eurasia

OVERVIEW

In Chapter 28, students will learn about historical and political development in northern Eurasia, from early migrations to events following the breakup of the Soviet Union.

Section 1 explains how migrations over the course of several centuries brought together diverse cultures. It also describes the economic and social conditions that brought an end to czarist rule and ignited the Communist Revolution.

Section 2 focuses on the social, political, and economic impact of Communist rule in the Soviet Union.

Section 3 discusses the forces that broke the Soviet Union apart and the challenges that confront the region today.

About the Artifact

Peter Carl Fabergé (1846–1920), a Russian goldsmith and jeweler, created many famous decorative items made from gold, silver, and gems. Decorative Easter eggs, such as the one shown here, were among his most famous creations. This egg also functioned as a clock.

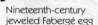

CHAPTER 28

Introducing Russia and Northern Eurasia

Nineteenth-century jeweled Fabergé egg

For most of the twentieth century, much of the vast region of northern Eurasia was dominated by the Union of Soviet Socialist Republics (USSR), commonly known as the Soviet Union. The Soviet Union's beginnings reach back to 1917 when the Russian Empire ended in violent revolution.

The main goal of the Soviet Communist government was to achieve economic equality for all. In trying to reach this goal, however, the government used oppressive methods of control on its citizens. It limited economic and social freedoms and suspended basic political rights. This command economy system could not feed or house all of the country's people and fell behind the Western free-market economies.

After the collapse of the Soviet Union in 1991, the political geography of northern Eurasia began to change dramatically. The land that was once the Soviet Union is now organized into 15 independent countries, including Russia, the world's largest country. There are new borders, new political systems, new place-names, and new goals for the future. And perhaps most important, the many peoples of northern Eurasia are rediscovering their ethnic heritages.

GEOGRAPHY DICTIONARY

czar

abdicate

soviet

autarky

superpower

client state

glasnost

perestroika

St. Basil's Church in **Moscow**

Using the Geography Dictionary

Pronounce the terms in the Geography Dictionary. Ask students to identify familiar words. (Students might choose **czar**, **soviet**, *glasnost*, and *perestroika*.) Explain that the Russian ruler prior to the Communist Revolution was called the czar (tsar). Point out the phonetic connections between *czar*, *caesar* (ruler in ancient Rome), and *kaiser* ("emperor" in German). Tell students that these terms can be traced to the Latin word for "emperor." Suggest ways that *czar* is used today to describe someone who is in charge of something. (Students may be familiar with references to the "energy czar" or the "drug czar"— heads of U.S. governmental agencies.)

MAPPING THE REGION

Using the Physical-Political Map

Refer students to the definition of Eurasia from Chapter 23 (page 258). Then tell them to study the map on this page and to locate the landform that forms the dividing line between Europe and Asia. (*Ural Mountains*) Use the maps on pages 326–327 to point out contrasts between the eastern and western (European/ Asian) parts of this region.

MAP TRANSPARENCIES

Overhead transparencies from the maps on pages 325–327 are available in the *REGIONS OF THE WORLD Map Transparencies*.

MAPPING THE REGION

RUSSIA AND NORTHERN EURASIA

PHYSICAL–POLITICAL

LEGEND

⊛ National capitals

• Other cities

ELEVATION	
FEET	**METERS**
13,120	4,000
6,560	2,000
1,640	500
656	200
(Sea level) 0	0 (Sea level)
Below sea level	Below sea level

SCALE

0 — 500 — 1000 Miles

0 — 500 — 1000 Kilometers

Projection: Two-Point Equidistant

MAPPING THE REGION

Using the Climate Map

Have students examine the climate map of Russia and northern Eurasia on this page. Ask them to name the climates that dominate Russia. (*subarctic, tundra*) Then have them compare the climate map to the population map below it. Ask: In what climate is the population density highest? (*humid continental*) In what three types of climate is the population density lowest? (*tundra, subarctic, desert*)

MAPPING THE REGION

Using the Population Map

Direct students' attention to the population map on this page. Ask them to write a brief summarizing statement about population density patterns of this region. (*Most people live in the western part of the region; the eastern part is sparsely populated.*) Volunteers might read their statements aloud. Call on students to suggest reasons for the patterns of population density.

CLIMATE

Answers

1. 60° N
2. Caucasus region and eastern parts of Central Asian republics
3. Desert

POPULATION

Answers

1. Yekaterinburg, Chelyabinsk, Omsk, Almaty, Toshkent, Novosibirsk
2. Russia, Ukraine, Uzbekistan
3. Tundra

MAPPING THE REGION

RUSSIA AND NORTHERN EURASIA

CLIMATE

LEGEND

- Desert
- Steppe
- Mediterranean
- Humid subtropical
- Humid continental
- Subarctic
- Tundra
- Highland

SCALE
0   500   1000 Miles
0   500   1000 Kilometers
Projection: Two-Point Equidistant

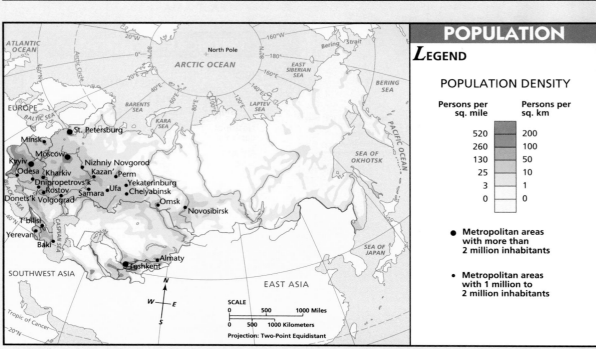

POPULATION

LEGEND

POPULATION DENSITY

Persons per sq. mile	Persons per sq. km
520	200
260	100
130	50
25	10
3	1
0	0

● Metropolitan areas with more than 2 million inhabitants

● Metropolitan areas with 1 million to 2 million inhabitants

SCALE
0   500   1000 Miles
0   500   1000 Kilometers
Projection: Two-Point Equidistant

MAPPING THE REGION

Using the Economy Map

Direct students' attention to the economy map on this page. Focus on the region's natural resources that are located in the west near population centers. Point out that many natural resources are also found in Siberia. Refer students to the physical–political map on page 325 to locate Siberia. Ask students to consider the obstacles that hinder developing these resources to benefit the region's economy. (*vast distance from population centers, difficulties in transporting industrial products across the region*)

MAPPING THE REGION

RUSSIA AND NORTHERN EURASIA

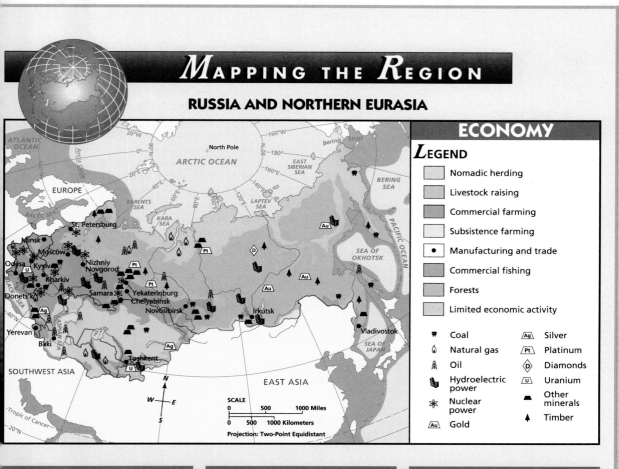

ECONOMY

LEGEND

- Nomadic herding
- Livestock raising
- Commercial farming
- Subsistence farming
- • Manufacturing and trade
- Commercial fishing
- Forests
- Limited economic activity

♠ Coal	Ag Silver
◊ Natural gas	Pt Platinum
♠ Oil	◊ Diamonds
Hydroelectric power	U Uranium
✳ Nuclear power	━ Other minerals
Au Gold	♠ Timber

SCALE
0   500   1000 Miles
0   500   1000 Kilometers
Projection: Two-Point Equidistant

CLIMATE

◆ Most of Russia, the largest of the republics of the former Soviet Union, lies in the cold subarctic and tundra climate regions.

◆ The dry steppe and desert climates, with hot summers and cold winters, dominate much of the Central Asian republics.

◆ Ukraine, a large agricultural producer, has a humid-continental climate in the northwest and a warmer and drier steppe climate in the southeast.

? Regions with tundra climates lie no farther south than which latitude?

Compare this map to the physical–political map on page 325.

? Where are highland climates located?

? Which climate type dominates the region surrounding the Aral Sea?

POPULATION

◆ Moscow and St. Petersburg are the two largest of many metropolitan areas in the region.

◆ The largest cities and many of the more densely populated regions of eastern Russia lie along the route of the Trans-Siberian Railroad.

◆ Large areas of the Central Asian republics and Siberia are sparsely populated.

? Which metropolitan areas east of 60°E longitude have populations of more than 1 million?

Compare this map to the physical–political map on page 325.

? Which republics have metropolitan areas with populations of more than 2 million?

Compare this map to the climate map on page 326.

? Which climate region in Russia is the most sparsely populated?

ECONOMY

◆ Many of the manufacturing centers are located near large deposits of natural resources.

◆ The most productive farming lands are in the steppe regions of Ukraine, Russia, and Kazakhstan.

◆ Although the climate and terrain discourage large-scale farming in much of Siberia, the region has a rich supply of other resources.

Compare this map to the physical–political map on page 325.

? What energy resource is found in the Donets and Kuznetsk basins?

? Name the manufacturing and trade centers in the three republics of the Caucasus region.

Compare this map to the climate map on page 326.

? What do you think is necessary to support farming in the desert region that surrounds the Aral Sea?

327

OBJECTIVES

The objectives reflect the Focus questions on this page and are assessed in the review on page 329.

1. Summarize Russia's early history, and discuss the growth of the Russian Empire.

2. Examine the factors that led to the Communist Revolution.

TEACHING SECTION 1

Building a Vocabulary

Remind students that **czar** is from the Latin word that means "emperor or ruler." Ask them to read Section 1 to identify the date when the first czar began to rule *(1547)* and the year when the last czar abdicated *(1917)*. Tell students that **abdicate** comes from two Latin word parts: *ab-* ("away from") and *dicare* ("to proclaim").

Role Playing

Russia's Early History Have students play the role of a foreign correspondent assigned to Russia in 1917. Tell them to write a newspaper article describing the conditions and factors leading up to the revolution, doing additional research as needed. Have students write an appropriate headline for their article and share it with the class.

Map Study
1689

....

Geography and Language

In addition to differences in religion between Russia and western Europe, the two regions also used different alphabets. While most of western Europe wrote with the Roman alphabet, the Russians used the Cyrillic alphabet. The use of these different alphabets continues today.

A DIVERSE HERITAGE

FOCUS

● *What peoples inhabited and ruled the Russian steppe before the fifteenth century? What was the Russian Empire?*

● *What led to the Communist Revolution?*

MOVEMENT **Invaders of the Steppe** The roots of the Russian Empire lie deep in the grassy plains of the steppe. The steppe stretches from what is now Ukraine west to the Ural Mountains. (See the map on page 325.)

For thousands of years, people moved west across the steppe. These migrants drove their herds away from droughts and wars in regions to the east. In time, the immigrants mixed with the settled population and became farmers and traders. Each wave of immigrants brought new ways of life, languages, and religions to the region.

In the ninth century, the city of Kiev (now called Kyyiv) emerged in the region as an important trading center between the Mediterranean and Baltic seas. Kiev's first leaders were Viking traders from Scandinavia who called themselves Rus (ROOS). The word *Russia* comes from the name of these early traders.

During the following centuries, missionaries from the Balkans brought Orthodox Christianity to the region. Merchants traveling into the forests farther north founded new towns there as well. Cities usually developed on high riverbanks near important stream junctions.

In 1223, the fiercest invaders of the steppe appeared on the eastern horizon. They were the Mongols led by Batu, the grandson of Genghis Khan, the Mongol chieftain. These warriors on horseback won victory after victory against the peoples of the steppe. They destroyed Kiev in 1240 and made the region the western outpost of their growing empire.

The Russian Empire The Mongols demanded taxes but ruled through local leaders who

---

MAP STUDY Historical maps can be used to show changes in political geography. This map uses colors to show land acquired by the Russian Empire, and later by the Soviet Union, by particular years. The breakup of the Soviet Union at the end of 1991 reversed this expansion. By what date had Russia acquired much of Siberia in the East?

History of Russian Expansion

Finland: Russian territory 1809–1918

Baltic republics: independent 1918–1940 and 1991

Poland: Russian territory 1815–1918

North Pole

ARCTIC OCEAN

St. Petersburg

Moscow

Russian territory 1871–1881

LEGEND

Russia 1462–1533

Territory acquired by
1689
1725
1801
1945

Maximum extension of Russian territory

Russian boundary in 1993

Alaska, settled by Russians beginning in 1784, was sold to the United States in 1867.

SCALE
0   500   1000 Miles
0   500   1000 Kilometers
Projection: Two-Point Equidistant

Locate the Russian steppe on a wall map, and remind students why it attracted migrants and invaders over the centuries (for grazing and herding, trade opportunities, empire building). Have students work in pairs to create a time line identifying the peoples that settled in or invaded the steppe and including events that led to the creation of the Russian Empire.

Closure

Ask students to name similarities and differences between the American and Russian revolutions.

ASSESSMENT

Assign the Section 1 Review questions on this page.

Reteaching

Assign Section 1 of Reteaching Worksheet 28.

SECTION 2
OBJECTIVES

The objectives reflect the Focus questions on this page and are assessed in the review on page 331.

1. Identify the key Soviet political and economic principles.

2. Describe daily life in the Soviet Union.

3. Explain the international significance of the Soviet Union.

---

controlled different areas. As a result, several states emerged, the strongest of which was Muscovy, north of Kiev. Its chief city was Moscow. In the late 1400s, Ivan III, the prince of Moscow, won control over parts of Russia from the Mongols.

Then in 1547 at age 16, Ivan IV, who became known as "Ivan the Terrible," crowned himself **czar** (ZAR) of all Russia. The word *czar*, or *tsar*, comes from the Latin *Caesar* and means "emperor." The Russian Empire under Ivan IV stretched from north of Kiev to the Arctic Ocean and east to the Urals.

Ivan's victories over neighboring rivals allowed Russian fur trappers, hunters, and pioneers to migrate eastward into Siberia. By 1637, these explorers reached the Russian Pacific coast. The Russian czars then built a land-based empire larger than any in the world. In the 1860s, the Russians spread into Central Asia, where much of the population was Muslim. Russian territory even spread eastward to what is now Alaska and California.

Several factors, however, hindered Russia's expansion into Europe. The Russian tradition of political control conflicted with western European ideals of government limited by laws and customs. Religions of the two regions also clashed as Russia was mainly Eastern Orthodox and most of Europe was either Protestant or Roman Catholic. In addition, Russia faced the powerful kingdoms of Sweden and Poland to the west and the Ottoman Empire to the south.

Despite these factors, Russia was able to gain some European territory. Czar Peter the Great, who ruled from 1682 to 1725, acquired lands along the Baltic Sea. Catherine the Great, who ruled from 1762 to 1796, took the northern side of the Black Sea. By this time, many non-Russian peoples found themselves within Russian borders.

The Empire Falls Although industrialization had begun in Russian cities by the late 1800s, Russia remained largely a country of poor farmers. Only a small percentage of the people were royalty, landowners, or wealthy merchants. A slightly larger number were factory workers and craftspeople. Most, about 85 percent, were peasants. The poor economic condition of these people was Russia's greatest problem.

In the 1890s, life for many Russians worsened. Droughts led to food shortages, and a global economic depression slowed trade and reduced exports. In 1905 and 1906, riots broke out in several large cities. Workers destroyed factories, and peasants burned crops and seized many estates.

Many people were forced to move to Siberia; thousands more went there willingly to escape the violence. The government began new social and economic programs, but they were too little and too late. By the start of World War I in 1914, the foundations of Russian society were on shaky ground.

Russia was unprepared to fight in this war that spread across Europe. The loss of life and the high costs of World War I, combined with Russia's social and economic problems, caused the czar to **abdicate**, or resign, in early 1917. A temporary republic was set up but met with little success. In the fall of 1917, the small Bolshevik party overthrew the government in what became known as the Communist Revolution.

SECTION REVIEW

1. What factors hindered the westward expansion of the Russian Empire?

2. What was the Communist Revolution? What were its causes?

3. Critical Thinking Assess the extent to which physical geography played a role in the region's early history.

THE SOVIET UNION

FOCUS

● *On what principles was the Soviet Union based? What were the country's political and economic realities?*

● *What was Soviet daily life like?*

● *How was the Soviet Union internationally significant?*

The Soviet Union The Bolsheviks, led by V.I. Lenin, wanted to remake Russia following the ideas of Karl Marx, a nineteenth-century German philosopher. Marx believed that the people of the working classes were victims of capitalism. He

Section 1 Review
ANSWERS

1. Differences between the Russian Empire and the countries of Europe in political ideas and religions, and the presence of powerful European kingdoms

2. It was the name given to the overthrow of Russia's government by the Bolshevik party in 1917. Its causes included an uneven distribution of wealth, droughts and food shortages in the 1890s, and the loss of life in and the costs of World War I.

3. Possible answer: the plains of the steppe enabled the movement of many groups of people across the region and allowed the growth of Kiev, the Mongol Empire, and the Russian Empire.

TEACHING SECTION 2

Building a Vocabulary

Remind students that the prefix *auto-* or *aut-* means "self." Then have a volunteer read aloud the paragraph on page 330 that defines **autarky.** Point out that the main idea behind this economic policy is self-sufficiency or independence. Call on students to suggest how the term **client state** suggests the opposite idea.

Guided Practice

Write the following words or phrases on the board: Private Property, Religion, The Cold War, Free Speech, Industrial Production, Agricultural Production. Have students take turns citing specific examples of Soviet economic or political policies that relate to each.

Independent Practice

Have students write a short essay describing the effects of Soviet economic and political policies on the daily lives of Soviet citizens.

---

argued that capitalist business leaders kept their workers poor to maintain their own power and wealth. Lenin believed the solution was for Russia's government to own and control all farms and factories. Many disagreed with Lenin's ideas, however, and a civil war raged until 1921, when the Bolsheviks finally gained control.

The country was renamed the Union of Soviet Socialist Republics (USSR), commonly known as the Soviet Union. Each republic had its own local governing council, or **soviet**. The republics were founded on the idea that, where possible, each of the large ethnic groups should have its own territory. In reality, however, the republics had little power. Communist party leaders in Moscow made all economic and political decisions. The Soviet Union became a one-party, totalitarian state in which there were few freedoms.

After Lenin's death in 1924, Joseph Stalin took power. His brutal dictatorship lasted until 1953. Both Lenin and Stalin tried to promote a single culture. Names of cities and streets were changed to honor communism's heroes, and Russian became the official language. Many other languages and cultures survived, however.

Soviet Economic Geography

The enormous land area of the Soviet Union stretched from the Arctic Sea south to the mountains of Iran and east from the Baltic Sea to the Pacific Ocean. To develop this huge country, Soviet leaders set goals based on detailed five-year economic plans. Government officials decided what crops would be grown, where industries would be built, and how much workers would be paid.

After bitter struggles with independent farmers, particularly in Ukraine, almost all the farmland in the Soviet Union came under government control. The government, however, invested little in agriculture. As a result, food production was poor on the government-run farms. By the late 1950s, the nation had to import large amounts of grain. Private plots, which farm families were allowed to work in their spare time, produced about one-fourth of the country's food.

Dramatic industrial growth, however, did take place under Stalin's plans. The government developed mines and built dams. Huge factories specializing in heavy industries, such as steel and military goods, sprang up in many cities.

Soviet planners followed a policy of **autarky** (aw-tahr-kee), under which a country strives for self-sufficiency in economic production. That is, the Soviet Union would try to produce all the goods that the country needed. Planners decided that only a few huge factories were needed to accomplish this goal. Indeed, the Soviet Union became a world leader in the production of many minerals and industrial products. Without competition, however, efficiency and product quality were often low. The production of consumer goods and service industries fell well behind that of the United States and western Europe.

Life in the Soviet Union

By U.S. standards, life in the Soviet Union was dismal. Working conditions were often poor, and food supplies were limited in quality and quantity. The Soviet people often waited in line for hours to buy everyday products. For many families, limited housing meant crowding into cramped two-room apartments. Yet many believed that such sacrifices were necessary to build the Communist ideal. And for most people, life in the Soviet Union by the mid-1950s was as good as it had ever been.

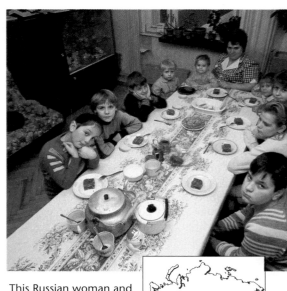

This Russian woman and 9 of her 11 children sit down to eat in their crowded six-room apartment. As **Russia** lurches toward a free-market economy, consumer prices, especially for food, have risen dramatically. Those pushing for a free-market economy believe that reforms will eventually improve living conditions over those under communism.

330

The government did achieve some successes in education and health care. Under the czars, only about 25 percent of the population could read and write. By the 1980s, well over 90 percent could. Much of Soviet education stressed science and technology, because training in these fields could help increase Soviet production. Many people also became doctors, and basic health care was widely available.

Personal freedoms, however, were strictly limited in the Soviet Union. Citizens who disagreed with government leaders could be jailed. Under Stalin, millions were sent to dreaded prison camps in the wilds of Siberia and the Arctic. Soviet leaders also set out to eliminate religious worship so that it could not compete with people's devotion to communism. Many Christian, Jewish, and Muslim houses of worship were closed or destroyed.

The Cold War Although they were World War II allies, the United States and the Soviet Union became bitter rivals after that war. These two huge, powerful nations, or **superpowers**, competed in the production of military arms and in the development of space programs. This intense political struggle became known as the Cold War and for more than 40 years created the dangerous potential for nuclear war.

The United States and the Soviet Union also competed for the support of other countries around the world. Both gained allies by supplying military goods and economic aid to these countries. Many of these allies became **client states** of the superpowers. A client state is a country that is politically, economically, or militarily dependent on a more powerful country. Violent struggles among these client states nearly drew the United States and the Soviet Union into direct conflict several times.

SECTION REVIEW

1. What economic policies did the Soviet Union follow? What economic problems did the country have?
2. List five characteristics of daily life in the Soviet Union.
3. **Critical Thinking** **Explain** why the rivalry between the Soviet Union and the United States was called the Cold War.

A NEW GEOGRAPHY OF NORTHERN EURASIA

FOCUS

● *Why did the Soviet Union break apart?*
● *What challenges does the region face?*

Collapse of the Soviet Union During the 1980s, the Soviet economy continued to decline. In 1985, Mikhail Gorbachev became what turned out to be the last leader of the Soviet Union. Gorbachev saw the need for major reforms in his country. He promoted a policy of **glasnost** (openness), which allowed more open discussion of the country's problems. He also initiated major economic reforms called **perestroika** (restructuring).

Under Gorbachev, the Communist party began to lose much of its authority. The Soviet people grew tired of sacrifice without gain, and as their demand for consumer goods rose, their frustrations increased. With the central Soviet government losing its authority, its client states in Eastern Europe collapsed in 1989.

Then in 1991, the Soviet Union itself broke apart. Officially, the Soviet Union was replaced by the Commonwealth of Independent States (CIS). This organization does not have a strong central government. Instead, it provides a way for the various regions of the former Soviet Union to address shared problems. The first democratically elected president of Russia, Boris Yeltsin, has called for more democracy and a market economy. Politics in the former Soviet Union, however, has been unstable. Tensions between supporters and opponents of reform and between ethnic groups have made the political future of the former Soviet republics uncertain.

Human Geography Life has changed rapidly for the peoples of the former USSR. News from around the world now flows more freely. Religious and democratic rights are expanding.

Within the 15 independent countries of the former Soviet Union, however, ethnic disputes continue. The Soviet Union contained more than 100 ethnic groups, and the region's division into separate countries still leaves many ethnic minorities. Civil wars are now raging in some areas.

TEACHING SECTION 3

Building a Vocabulary

Have students review the definitions of the terms *glasnost* and *perestroika*. Ask how the two concepts worked hand in hand to bring about reforms in the Soviet Union. Have students speculate about what areas of Soviet life, besides politics and economics, might have benefited from *glasnost*. (*Students might mention the arts, communications, or religion.*)

Guided Practice

Have students work in small groups to create a leaflet concerning an environmental challenge for the former Soviet Union. On one side of the leaflet, students should provide information about specific environmental dangers and suggest actions that could correct the problems. On the other side, they should draw a map and locate and label problem areas. Provide time for groups to share their leaflets with the rest of the class.

Closure

Ask students to name and discuss challenges that face the former Soviet Union. List their responses on the board.

ASSESSMENT

Assign the Section 3 Review questions on this page.

Reteaching

Assign Section 3 of Reteaching Worksheet 28.

GLOBAL PERSPECTIVES

Have students consider the role played by information technology in the collapse of the Soviet Union. Discuss the ways they think that exposure to outside ideas helped *glasnost* and *perestroika* to occur. Ask students if they think it is possible for any leader today to cut a country off from the rest of the world.

Section 3 Review
ANSWERS

1. His policies of open discussion and economic restructuring led to a decline in the Communist party's authority.
2. Economic: little money available for investment, increased unemployment and crime, uneven distribution of resources; environmental: deforestation, pollution, toxic waste
3. Answers will vary. Some students may say unity, pointing to the economic interdependence of different regions. Other students may say disunity, pointing to ethnic tensions and regionalism.

332

Economic Geography Despite these dividing forces, economic geography may create unity. Soviet planners made the various regions dependent on each other. For example, the Baltic republics relied on the oil and minerals of Russia. Cotton grown in Central Asia was made into cloth in Belarus and Ukraine. And the grains from Central Asia were made into bread in Moscow and St. Petersburg (formerly Leningrad). Breaking such well-established links may make little sense.

The economic geography of the region is changing, however. Although little money is available for investment, individually owned farms and small businesses are appearing. Some factories are now corporations in which the workers own some stock. Some former Soviet republics, especially Russia, have made important progress in privatizing formerly government-owned businesses.

The country's economic upheaval has caused some problems. Unemployment and crime have increased, troubles that some blame on capitalism. For much of the population, the Soviet system represented order, stability, and security. Many remain unsure about the future.

The new-found freedom of local communities and regions to make their own decisions has created some conflict as well. Many resource-rich regions believe that the income from selling their products should remain in their own territories.

These **Moscow** demonstrators reflect the political confusion that has followed the collapse of the Soviet Union. Images of Jesus Christ, Lenin, and Czar Nicholas II all compete for attention in this 1993 anti-government demonstration.

Resource-poor regions, on the other hand, fear that their economies will suffer greatly under such a system. In all regions, finding people who can become effective local leaders is a continuing problem after so many years of the central government in Moscow making all decisions.

In the past, officials in Moscow also filtered the republics' contact with foreign countries. Today, however, communities all over the former Soviet Union are beginning to develop international ties of their own. For example, the city of Novosibirsk (noh-voh-suh-BIRSK) has built an international airport, hoping to profit as the midpoint on the many flights between western Europe and East Asia.

INTERACTION **Environmental Issues** A disturbing legacy of the Soviet Union is its record of environmental pollution and destruction. Soviet industrial leaders often felt pressured to produce immediate results at the lowest cost. As a consequence, factories and industries seldom bothered with controlling pollution or preserving the environment.

Throughout the region, there is evidence of deforestation, river and lake pollution, and the improper disposal of toxic wastes. For example, the Soviet Union dumped several nuclear reactors and large amounts of radioactive waste in the Arctic Ocean. Many sewers still dump directly into rivers and lakes as well.

Even though geographers and environmentalists are drawing worldwide attention to these problems in the former Soviet Union, solutions still lie far in the future. The various republics have little money to invest in the expensive and overwhelming job of cleaning up their environment.

SECTION REVIEW

1. What role did the reforms of Mikhail Gorbachev play in the collapse of the Soviet Union?
2. What economic and environmental challenges face the region?
3. **Critical Thinking** **Predict** which is more likely to characterize the region over the next decade: unity or disunity. Support your answer.

Building a Vocabulary
1. Emperor; no
2. Resign; Russia's social and economic problems, combined with the losses during World War I
3. Soviets
4. So as not to be economically dependent on other countries
5. A country that is politically, economically, or militarily dependent on a more powerful country

Recalling and Reviewing
1. In the steppe
2. An organization that consists of the regions of the former Soviet Union and serves as a forum for leaders to address shared problems
3. Soviet industrial leaders felt pressured to produce immediate results at the lowest cost, which often meant disregarding the environment. Deforestation, river and lake pollution, and toxic waste problems face the region today.

CHAPTER 28 CHECK

Old country church in **Ukraine**

Reviewing the Main Ideas

1. Northern Eurasia has been settled by many groups of people over time. The Mongol Empire covered the area for most of the thirteenth, fourteenth, and fifteenth centuries. The czars of the Russian Empire then ruled until 1917, when the Communist Revolution took place.
2. Following the Communist Revolution, the country was renamed the Union of Soviet Socialist Republics (USSR), commonly known as the Soviet Union. The government had total economic and political control. After World War II, the United States and the Soviet Union became bitter rivals.
3. The Soviet Union collapsed in 1991. Freedom, democracy, and capitalism are spreading. As it adjusts to change, however, the region faces numerous challenges.

Building a Vocabulary

1. What does *czar* mean? Does Russia still have a czar?
2. Define *abdicate*. Why did the Russian ruler abdicate in 1917?
3. What were the governing councils of the Soviet Union's republics called?
4. Why might a country want to follow a policy of autarky?
5. What is a client state?

Recalling and Reviewing

1. Where did the roots of the Russian Empire lie?
2. What is the Commonwealth of Independent States?
3. Why did the Soviet Union have little concern for the environment? What environmental problems face the region today as a result of policies followed by the Soviet Union?

Thinking Critically

1. Agree or disagree with the following statement: "Modern-day ethnic conflicts in the former Soviet Union have their roots in the region's early history."
2. Despite the many hardships of daily life in the Soviet Union, the country survived for nearly 70 years. In your opinion, what factors were responsible for keeping the Soviet Union together?

Learning About Your Local Geography

Cooperative Project
Largely because of their physical geography, the plains of the Russian steppe have been subject to many invasions over the centuries. Organize your group into two teams. One team should design a "plan of invasion" of your local area. The other team should design a "plan of defense." Both teams should consider the area's physical geography. Teams should share their completed plans.

Using the Five Themes of Geography

Over time, northern Eurasia has been divided into several different **regions**. What roles did **location** and **movement** play in the changes to regional boundaries before 1917? Sketch one or more maps of the area to illustrate your answer to the question.

Thinking Critically
1. Students should agree with the statement. Many different groups of people settled in the region in its early history. Then when the Russian Empire expanded, it came to include many non-Russian peoples.
2. Possible answers include the fact that even though daily life was difficult, it was still as good as it had ever been; people's belief in Communist ideals and willingness to make sacrifices; the Soviet Union's repression of dissent; and the order, stability, and security of life in the Soviet Union.

Learning About Your Local Geography
All teams' plans should take into account physical factors that protect against invasion (such as surrounding mountains) or make invasion easier (for example, wide plains).

Using the Five Themes of Geography
Maps should show that the area's location in both Asia and Europe caused shifting boundaries as different peoples moved in and gained control, and as area empires expanded outward.

333

CHAPTER 29 Russia
Planning Guide

TEXTBOOK	TEACHING RESOURCES	OTHER AVAILABLE MATERIALS
Section		
1. Physical Geography Pages 335–337	Reteaching Worksheet 29.1 Skills Worksheet 29	*Creative Strategies for Teaching World Geography* LESSON 12 Physical Geography of Russia *Study Guide for Spanish Speakers* *Lecture Notes,* Chapter 29
2. Resources and Agriculture Pages 337–341	Reteaching Worksheet 29.2	*Study Guide for Spanish Speakers* *Lecture Notes,* Chapter 29
3. Economic and Industrial Regions Pages 341–346	Reteaching Worksheet 29.3 Graphic Organizer 29 Critical Thinking Worksheet 29	*Study Guide for Spanish Speakers* *Lecture Notes,* Chapter 29
4. Political Geography Pages 346–348	Reteaching Worksheet 29.4 Challenge/Enrichment Worksheet 29	*Study Guide for Spanish Speakers* *Lecture Notes,* Chapter 29 *The World Today* Videodisc Program
Chapter Review Page 349	Chapter 29 Test Chapter 29 Portfolio Assessment See also *Alternative Assessment Forms*	*Test Generator,* Chapter 29

CHAPTER RESOURCES

Lecture Notes, Chapter 29
Teaching Resources
    Chapter 29 Graphic Organizer
    Chapter 29 Skills Worksheet
    Chapter 29 Critical Thinking Worksheet
    Chapter 29 Challenge/Enrichment Worksheet

REVIEW AND ASSESSMENT

Chapter 29 Check, page 349
Chapter 29 Test, *Teaching Resources*
Chapter 29 Portfolio Assessment, *Teaching Resources*
Test Generator, Chapter 29
Alternative Assessment Forms

ADDITIONAL RESOURCES

Books for Teachers

Bobrick, Benson. *East of the Sun: The Epic Conquest and Tragic History of Siberia.* New York: Poseidon Press, 1992.

Khanga, Yelina, and Susan Jacoby. *Soul to Soul: The Story of a Black Russian American Family, 1865–1992.* New York: W. W. Norton, 1992.

Matthiessen, Peter. *Lake Baikal: Sacred Sea of Siberia.* San Francisco: Sierra Club Books, 1992.

Raeff, Marc. *Russia Abroad: A Cultural History of the Russian Emigration, 1919–1939.* New York: Oxford University Press, 1990.

Thomas, Bill. *Red Tape: Adventure Capitalism in the New Russia.* New York: Dutton, 1992.

Books for Students

Brown, Archie, et al, eds. *The Cambridge Encyclopedia of Russia and the Former Soviet Union.* New York: Cambridge University Press, 1994.

Troyat, Henri. *Ivan the Terrible.* New York: Dutton, 1984.
Troyat, Henri. *Peter the Great.* New York: Dutton, 1987.

The World Today *Videodisc Program*

A variety of still images, short videos, and activities are available for you to use as you teach this chapter. *Lecture Notes* include bar codes for specific images and videos.

Multimedia Resources

Baikal: Blue Eye of Siberia. (VHS, 107 minutes) Films for the Humanities & Sciences, P.O. Box 2053, Princeton, NJ 08543–2053.

Chernobyl: The Taste of Wormwood. (VHS, 52 minutes) Films for the Humanities & Sciences, P.O. Box 2053, Princeton, NJ 08543–2053.

Great Wonders of the World: Manmade—Trans-Siberian Railway. (Multimedia CD-ROM) InterOptica Publishing Ltd., 300 Montgomery Street, Suite 201, San Francisco, CA 94104.

MEETING INDIVIDUAL NEEDS

Limited English Proficiency (LEP)

Help students create graphic organizers on the board that outline the characteristics of Russia's six economic regions. Encourage students to refer to the organizers as they study Chapter 29.

Learners Having Difficulty

Organize the class into groups of four. Within each group, assign individuals a section of Chapter 29. Direct students to write a brief summary of their assigned section and then share it with their group. Call on volunteers to read their summaries to the class. You may wish to duplicate the individual summaries so that each group member receives a complete chapter summary.

Auditory Learners

Students might participate in a panel discussion to explore the problems Russia faces in changing to a free market economy and moving toward a more open society. Have each panel member assume a different role: factory worker, government official, soldier, farmer, teacher. Encourage the class to ask questions of each member.

Visual Learners

Display a map of the Russian Federation (such as the National Geographic Society's 1993 "Russia and the Newly Independent Nations of the Former Soviet Union") that shows the country's major political subdivisions. Use the map to point out the location and size of the autonomous republics listed on text page 346.

Tactile/Kinesthetic Learners

To help students understand the pace of change occurring in Russia, have students collect newspaper and magazine articles about Russian current events and summarize them on index cards. On a bulletin board divided into three sections—for economic, political, and cultural geography—ask students to post their articles and summaries in the appropriate category.

Gifted Learners

Have interested students research the history of agricultural production in Russia since the 1700s. Students should focus on one of the following: agriculture produced by the serfs under the czars, collective farms during the Soviet era, agriculture today under private ownership.

MULTICULTURAL FOCUS

Point out to students that among the many Russian institutions that are reviving in the post-Soviet era is the Russian Orthodox church. From the time Rus rulers adopted Christianity in 988 until the Communist Revolution of 1917, the Orthodox faith was a central force in Russian culture. Russian Orthodox was the state church of the czars and empresses, and it spread with the expanding empire through Siberia and into Alaska.

Students may be aware that Russian Orthodox is a denomination of the Eastern Orthodox church, a major branch of Christianity whose beliefs and traditions differ somewhat from Roman Catholicism and Protestantism. You may wish to assign several students to do further research on Russian Orthodox beliefs, practices, and church buildings and to present their findings to the class.

During the Communist era, Soviet leaders sought to eliminate religious worship. Most Orthodox churches and monasteries were closed or destroyed. Some were converted into factories or public buildings; the 600-year-old Tolgstoy Convent, for example, served as a Soviet prison.

Despite the obstacles to religious worship, a 1991 poll found that some 50 to 60 million Russians considered themselves to be Russian Orthodox. The end of Soviet restrictions has brought a resurgence of religious expression in Russia. January 7, the Russian Orthodox Christmas, has been declared a national holiday.

RUSSIA

Section 1: Physical Geography
Section 2: Resources and Agriculture
Section 3: Economic and Industrial Regions
Section 4: Political Geography

OVERVIEW

In Chapter 29, students will learn about Russia, the largest country of the former Soviet Union.

Section 1 describes Russia's landforms, climate, and vegetation.

Section 2 discusses Russia's vast natural resources and their potential for providing a prosperous future for the nation.

Section 3 focuses on Russia's economic and industrial regions, including the Moscow and St. Petersburg regions, the Volga and Urals regions, Siberia, and the Russian Far East.

Section 4 discusses the Russian people and the political issues they are confronting.

About the Author

Aleksandr Solzhenitsyn (1918–) received the Nobel Prize for Literature in 1970. His famous novels are highly critical of the repressive aspects of the former Soviet system.

Primary Source

Description of Change: excerpted
Rationale: excerpted to focus on main idea

CHAPTER 29

Russia

GEOGRAPHY DICTIONARY

habitation fog
taiga
serf
light industry
heavy industry
smelter

66 Traveling along country roads in central Russia, you begin to understand why the Russian countryside has such a soothing effect. It is because of its churches. They rise over ridge and hillside, descending towards wide rivers like red and white princesses, towering above the thatch and wooden huts of everyday life with their slender, carved, and fretted belfries. From far away they greet each other; from distant, unseen villages they rise towards the same sky. 99

Aleksandr Solzhenitsyn

Russia, officially "the Russian Federation," is a giant among the nations of the world. It stretches 6,000 miles (9,654 km) from the center of Europe and the shores of the Baltic Sea to the eastern tip of Asia and the coast of the Bering Sea. Russia's northern border is on the icy shores of the Arctic seas, while its southernmost border, 2,000 miles (3,218 km) away, climbs the rocky peaks of the Caucasus Mountains. The country's 150 million people live in a territory more than 6.5 million square miles (16.8 million sq. km) in area, one-ninth of the entire land surface of the planet. Russia is nearly twice the size of Canada, the world's second largest country. Because Russia covers so much of Europe and Asia, for the past several centuries it has played a major role in the affairs of both continents.

The painting on this late-eighteenth-century lacquer box, titled "Princess of the North Wind," is based on a Russian fairy tale.

Snowy taiga forest of **Russia**

334

Using the Quotation

Call on a volunteer to read aloud the quotation that introduces the chapter. Then refer students to the photograph of a Russian church on page 333. Remind students that religion has remained an important aspect of Russian life and culture, despite attempts by the former Communist government to close down houses of worship.

Using the Geography Dictionary

Refer students to the background photograph on page 334 of a **taiga** forest. Although taiga forests are often associated with Russia, they are also found in other places in Asia, Europe, and North America that have cold winters and short growing seasons. *Taiga* is a Russian word. Ask students for examples of other words they may know from the Russian language. (*tundra, czar, borscht, cosmonaut*)

SECTION 1

OBJECTIVES

The objectives reflect the Focus questions on this page and are assessed in the review on page 337.

1. Identify Russia's landform regions.

2. Describe Russia's climate types and vegetation.

Using the Physical–Political Map: Russia

Have students examine the map on this page. Then call on individual students to describe the general physical characteristics of Russia. (*mountains, flat lowlands, rivers, lakes*) Based on what they know about the latitude, climate, and physical geography of the region, ask students to suggest which areas in Russia would be most densely populated and which would be most sparsely populated. Have them check their responses by referring to the population map on page 326.

📖 **CREATIVE STRATEGIES FOR TEACHING WORLD GEOGRAPHY**

LESSON 12 Physical Geography of Russia: Scrambled Notes

RUSSIA: PHYSICAL–POLITICAL

LEGEND

ELEVATION

	FEET	METERS
⊛ National capital	13,120	4,000
• Other cities	6,560	2,000
	1,640	500
	656	200
	(Sea level) 0	0 (Sea level)
	Below sea level	Below sea level

Size comparison of Russia to the contiguous United States

PHYSICAL GEOGRAPHY

FOCUS

- What landform regions are found in Russia?

- What types of climates and vegetation does Russia have?

REGION **Landforms** Plains are the dominant landform in Russia. Extending west of the Urals is the Russian Plain, a part of the Northern European Plain that connects with plains in Poland and northern Germany. (See the map above.) This region is where most of the Russian people live. The Ural Mountains, which divide

Europe from Asia, are high rolling hills. Because the Urals run north to south, they provide no protection from the cold winds that sweep out of the Arctic each winter. Nor did the Urals ever stop the invaders who came from inner Asia over the centuries.

The lands east of the Urals are called Siberia. Directly east of the Urals is the Western Siberian Lowland, a large, flat area with widespread marshes. The huge Ob River system drains the area. In the spring, the Ob and other rivers of Siberia that drain toward the Arctic Ocean thaw first in the south. Downstream to the north, however, the rivers remain frozen much longer. Ice jams here block the passage of the melting snow and ice upstream. As a result, annual floods can be expected.

Situated between the Yenisei (yehn-uh-SAY) and Lena rivers is the Central Siberian Plateau.

Motivator

Remind students that Russia is the largest country in the world. Then ask them to consider the advantages and disadvantages of such enormous size. *(wide range of resources and climates and wide access to trade; difficulties in communication, defense, distributing goods and services)*

Building a Vocabulary

Write the term **habitation fog** on the board. Have a volunteer read aloud the definitions of *habitat* and *habitation* from a dictionary. Then ask students to consider how these words can help them deduce the meaning of the geography term on the board. Ask them for examples of human activities that could cause habitation fog.

Guided Practice

Review with students the vegetation and climate regions of Russia. Refer to the climate map on page 326. Have students work in pairs to create a chart that compares Russia's landform regions with its climate regions and vegetation. Indicate to students that they should expect overlaps between landform regions and climates.

Geography and History

The harsh Russian winter has played a role in devastating the armies of invaders. One famous example is the disaster that befell French emperor Napoléon's troops on their retreat from Moscow in 1812. Winter set in, and Napoléon was unable to resupply his soldiers with clothing or food so far from home. Napoléon lost more than 500,000 of 600,000 soldiers—many suffering from starvation and exposure—on their long march out of Russia.

COOPERATIVE LEARNING

Organize the class into small groups. Tell students to imagine they are taking a trip on the Trans-Siberian Railroad. Assign each group a different landform region of Russia, and have students work together to create a poster that illustrates elements of physical geography they might observe. Group members could focus on the climate, vegetation, or landforms of their region. One member might write a short travel narrative to present to the class.

336

FACTS IN BRIEF	The Russian Federation		
COUNTRY POPULATION (1994)	LIFE EXPECTANCY (1994)	LITERACY RATE	PER CAPITA GDP (1993)
RUSSIA 149,608,953	64, male 74, female	99% (1993)	$5,190
UNITED STATES 260,713,585	73, male 79, female	97% (1991)	$24,700

Sources: *The World Factbook 1994*, Central Intelligence Agency; *The World Almanac and Book of Facts 1995*

Russia's population is only about three-fifths that of the United States, although Russia's land area is nearly twice as large. Russia is the largest of the republics of the former Soviet Union.

This land of upland plains and valleys holds many undeveloped mineral deposits of lead, iron ore, and coal, which should provide income for centuries.

While flat lowlands make up Russia's heartland, soaring mountains stand guard along much of its eastern rim. Mountains stretch all the way along the border with Mongolia and then to the Pacific coast. These mountain peaks are comparable in size to those of the Rocky Mountains in the United States. The mountains north of Mongolia also contain Lake Baykal (bie-KAWL), the world's deepest lake. Known as the "Jewel of Siberia," Lake Baykal is home to many endangered species, including the world's only freshwater seal. (See "Planet Watch" on pages 338–339.)

Along Russia's Pacific coast is the peninsula of Kamchatka. This is a land of mountains much like the volcanic Cascade Mountains of the U.S. Pacific Northwest. The mineral riches that might exist in the east could be important to future economic development in Russia and East Asia.

Russia's many rivers are important resources, forming a network of valuable waterways for transportation and the production of hydroelectric power. The Volga, Europe's longest river,

covers 2,293 miles (3,689 km) on its gently sloping journey from the plains north of Moscow to its mouth on the Caspian Sea. Canals between rivers connect the Volga with the Baltic Sea in the northwest, while the Volga-Don Canal allows barges on the southern Volga to reach the Black Sea. In Siberia, the dam at Bratsk on the Angara River downstream from Lake Baykal is one of the world's largest.

REGION **Climate** Russians sometimes joke that winter lasts for 12 months and then summer begins. Nearly all of Russia is located at high latitudes that correspond to those of Canada and Alaska. Most of the country lies in the humid-continental and subarctic climate regions. (See the map on page 326.) Winters are cold and long, and winter days are very short. During the five coldest months of the year, it is possible to ice-skate on frozen rivers and canals from St. Petersburg (formerly Leningrad) on the Baltic Sea south to Astrakhan on the Caspian Sea.

Russian winters are surprisingly dry. The snow remains on the ground for many months, continuously blowing into new drifts. Snow and rain in Russia come mostly from moisture that has evaporated from the Atlantic Ocean far to the west. Only the coastal areas of eastern Russia receive rain-bearing winds from the Pacific Ocean.

Winters are particularly severe in Siberia, where temperatures often drop below -40°F (-40°C). During winter, **habitation fogs** hang over cities. A habitation fog is a fog that is caused by fumes and smoke from cities. The Siberian summer often lasts for less than three months. Surprisingly, temperatures can rise as high as 100°F (38°C) during this short season. Russians take advantage of the warmer weather to tend gardens, attend sporting events, and stroll through parks.

The long winter isolates Russia from the rest of the world. Travel between cities is difficult, and

The Shilka River meanders between mountain ranges in southeastern **Siberia** near the Mongolian border.

Locating Landforms

Call on individual students to go to a classroom wall map and locate the following sites discussed in Section 1: Bering Sea, Ural Mountains, Russian Plain, Western Siberian Lowland, Ob River, Yenisei River, Lena River, Central Siberian Plateau, Lake Baykal, Volga River, Caspian Sea, Baltic Sea, Black Sea, Angara River, Sea of Japan, Caucasus Mountains.

Closure

Have students label Outline Map 16 using the charts they created in the Guided Practice activity.

ASSESSMENT

Assign the Section 1 Review questions on this page.

Reteaching

Assign Section 1 of Reteaching Worksheet 29.

SECTION 2
OBJECTIVES

The objectives reflect the Focus questions on this page and are assessed in the review on page 341.
1. List resources found in Russia.
2. Summarize physical, historical, and political forces that have shaped Russian agriculture.

This nuclear-powered icebreaker is clearing a path through the frozen **Kara Sea**. Icebreakers allow ships to travel along the frozen northern Russian coast throughout the winter.

even friends inside the same city rarely visit each other. The country has few ports that are free of ice year-round. Except in the summer, the Arctic Ocean freezes all the way to Russia's northern shores. Barge traffic here requires the use of icebreakers. Similarly, the Pacific shore and most of the Baltic coast are iced over for more than four months each year. Russia's only large, ice-free port is Murmansk, located in the far northwest where the warm waters of the North Atlantic Drift reach around northern Norway.

The cold climates of Siberia and the Arctic create permanently frozen ground beneath the surface of the earth. This permafrost causes many problems. Tree growth is stunted because the roots cannot penetrate the frozen soil. When the surface layer melts in summer, buildings tilt, highways buckle, and railroad tracks slip sideways.

Vegetation Vegetation in Russia varies from north to south as the climates vary. Along Russia's northern coast, temperatures rise above freezing for only a few weeks each year. Tundra vegetation of mosses, wildflowers, and stunted trees grows here.

To the south is the **taiga**, a huge forest of tall, evergreen needle-leaf trees such as spruce, fir, larch, and pine. The dark taiga provides the setting in much of Russian literature. Historically, the taiga has served as protection against foreign invaders.

The taiga provides a large supply of softwood for building products and paper pulp. Steady logging of the taiga west of the Ural Mountains has left many cleared areas, but in Siberia, most of the taiga remains intact. With good management, it will provide forest resources for a long time. Although farming is possible in the taiga, yields are low because of the short growing season.

Farther south in European Russia is an area of forests with deciduous, hardwood trees. This type of forest also is found along the Sea of Japan in eastern Russia. Still farther south are the drier grasslands known as the steppe. The steppe stretches west into Ukraine, east into Kazakhstan, and south to the Caucasus Mountains. Many of the forests here have been cut for their lumber or cleared for farms. The grasslands now are plowed under for crops or used for grazing.

 SECTION REVIEW

1. Where are Russia's plains located? its mountains? What are some of the country's major rivers?
2. What is the dominant characteristic of Russia's climates? How do the climates influence Russia's vegetation?
3. **Critical Thinking** **Explain** how landforms, climates, and vegetation have influenced Russia's relationships with people outside the country.

2 RESOURCES AND AGRICULTURE

FOCUS

● *What resources are found in Russia?*

● *What forces have shaped Russian agriculture?*

Resources Russia has more forest, energy, and mineral resources than any other country in the world. Unfortunately, the former Soviet government wasted many of these resources. For example, the taiga west of the Urals is badly depleted. Because the trees of the taiga grow slowly, it will be many years before these forests grow back. As a result, forest products must now be transported

Text continues on page 340.

Section 1 Review
ANSWERS

1. Plains: west and east of the Urals and between the Yenisei and Lena rivers; mountains: on the border between Europe and Asia, along the Mongolian border to the Pacific coast, on the Kamchatka Peninsula; rivers: Ob, Yenisei, Lena, Volga, Angara
2. Cold; vegetation varies from north to south according to climate.
3. The Urals are too low to have prevented invasion from inner Asia, although the taiga has offered protection. Also, because of the cold climate and frozen harbors, parts of Russia are cut off from much of the world.

Planet Watch

Problem Solving

Organize students into three or four groups. Ask each group to develop a community-based plan for balancing economic development concerns and the protection of the Lake Baykal ecosystem. Encourage students to create maps, charts, and other visuals to illustrate their ideas. Plans should address the concerns of different groups (government, local residents, various industries such as petroleum and fishing) and should include a time line for implementing the plans. Have groups present their plans to the class and discuss them.

Geographical Sidelight

To help students grasp the great depth and enormous volume of water in Lake Baykal, direct them to a wall map that shows the five Great Lakes of North America. Point out that although Lake Baykal's surface is about the size of Lake Superior, Baykal holds a volume of water equivalent to that of all the Great Lakes combined—some 5,439 cubic miles (22,659 cubic kilometers) of water.

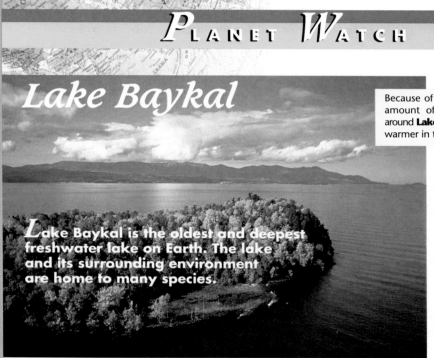

PLANET WATCH

Lake Baykal

Lake Baykal is the oldest and deepest freshwater lake on Earth. The lake and its surrounding environment are home to many species.

Because of the moderating effect of the large amount of water in the lake, temperatures around **Lake Baykal** are cooler in the summer and warmer in the winter than areas farther inland.

North of the Gobi lies crescent-shaped Lake Baykal. Grayling, pike, perch, and many other species of fish swim through its cold waters as eagles, falcons, and osprey fly overhead. Moose, wolves, and brown bears roam the surrounding forest. The lake's islands serve as a resting place for silver-furred nerpas, the world's only known freshwater seals.

Set in a basin seven times as deep as the Grand Canyon, Lake Baykal holds as much water as all of North America's Great Lakes. Formed 25 million years ago by the movement of plates along the earth's crust, Lake Baykal sits on active faults along which small earthquakes occur daily.

Baykal is like no other lake. In parts of the lake floor, warm-water springs sustain a variety of plant and animal life. Oxygen traveling down to the lake's deepest parts allows flatworms and freshwater shrimp, snails, and sponges to survive on the lake bottom. One species of freshwater shrimp helps keep the water pure by filtering algae and bacteria.

The Human Presence

The Buryat, a Mongol people, have lived along Baykal's eastern and northern shores for more than 1,700 years. Conquered by Russia in the late 1600s, the Buryat were part of that nation until the 1900s. In 1923, their lands became a separate Soviet republic. With Soviet rule came new attitudes toward Baykal. Soviet planners believed that Baykal's water would be ideal for producing cellulose, a material used in jet airplane tires. In 1957, the government announced plans for construction of the Baykalsk Cellulose-Paper Plant at one end of Lake Baykal.

Despite the risks of criticizing the government, scientists, environmentalists, and local residents spoke out. Industrialization, they claimed, would spoil the clean waters of Baykal. Even so, the government forged ahead, completing Baykalsk in 1966. A Buryat woman contrasted Buryat with Soviet attitudes. "In ancient times all life was considered sacred. Now those times are gone, nobody thinks about it anymore."

Critics' fears have been realized. Chemicals dumped into the lake by the Baykalsk plant have damaged Baykal's southern part. Pollutants from its smokestacks have killed trees in hundreds of square miles of forest nearby.

The Baykalsk Cellulose-Paper Plant spews pollution into the environment around **Lake Baykal.** In Russia, as elsewhere around the world, people must balance environmental protection with economic development.

Using Geography Themes

To reinforce the theme of human-environment interaction, have students write a paragraph discussing the importance of protecting natural resources such as Lake Baykal. Ask them to explain how the preservation or destruction of such resources can affect other parts of the world. Ask volunteers to read their paragraphs aloud. Students may want to include their paragraphs in their individual portfolios.

Geographical Sidelight

Although temperatures moderate somewhat around Lake Baykal, the surface of the lake still freezes over during the winter. Truck convoys travel over ice in the winter to supply remote towns. Before completion of the Trans-Siberian Railroad, railway track was laid over the frozen lake during the Russo-Japanese War in 1904 so that supplies could reach the battlefront.

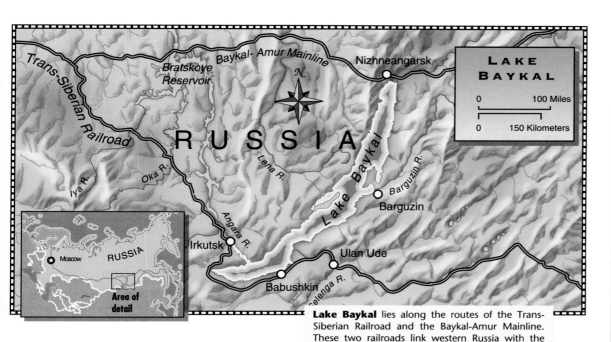

Lake Baykal lies along the routes of the Trans-Siberian Railroad and the Baykal-Amur Mainline. These two railroads link western Russia with the Russian Far East.

After building Baykalsk, the Soviet government was eager to develop industry throughout Siberia. New coal-burning factories gave off pollutants that produced acid rain in the region. Factories dumped chemicals and waste into the Selenga River, which flows into Lake Baykal.

With industrialization came the need for more transportation and fuel sources. Construction of the Baykal-Amur Mainline, a railroad linking Siberia with the Pacific Ocean, caused soil erosion along Lake Baykal's northern shore. Oil tankers began to appear on the lake despite the threat of oil spills.

These events have endangered the plants and animals that live in and around Baykal, many of which are found nowhere else on Earth. The number of shrimp is decreasing, as is the population of *omul,* a whitefish important to Baykal's fishing industry. To the dismay of scientists around the world, the lake's more than 60,000 nerpa also face an uncertain future due to the buildup of toxic chemicals in the food chain.

Protecting Lake Baykal

As industry's effects on the lake became apparent and the Soviet Union changed policies under Mikhail Gorbachev, environmentalists began to win small but important victories. They stopped a proposed Soviet-Italian project to mine limestone near Baykal. They also defeated a government plan to transport Baykalsk's toxic waste through pipes and dump it into one of Siberia's

large rivers. In 1987, their actions finally encouraged the government to issue a decree protecting Lake Baykal.

Dramatic events have shaken the world since 1987, however. The Soviet Union has broken up, and Russia's economy is struggling. It is feared that Baykal's fate is no longer a priority among those who have the power to save Russia's "Sacred Sea."

Used to overcoming obstacles, environmentalists continue to work to protect Baykal, and concern about the lake's future is spreading among the Russian population. An American writer visiting Lake Baykal saw Russian citizens demonstrating against the destruction of the lake, a sight that would have been rare just a decade or so ago. "We're here for our children," said one Russian woman. "We must save Baykal for our children."

▼ YOUR TURN

Disputes often arise when environmental protection conflicts with economic development. Think of examples of such conflict that have taken place in the United States. Work with a partner to **write** a debate that might take place between an environmentalist and an executive of a company that is about to construct an industrial plant in an area of environmental concern. In your debate, **explain** how the conflict will be resolved.

Your Turn

Students might mention conflicts between environmentalists and logging companies. Debates should reflect an awareness of both sides' concerns and interests.

339

Remind students that an abundance of natural resources is only one factor in building a country's prosperity. Ask students what it takes to turn those resources into wealth for a country. *(Answers might include investment in mining, plants, and equipment; education; and economic policies that allow people to start their own businesses.)*

Building a Vocabulary

Write the word **serf** on the board. Explain that the word *serf* is related to the word *serve*. Ask for other words formed from the same root. *(servant, service, servitude)* Point out that Russian serfs labored to benefit the czar and not themselves.

Mapping Resources

Provide students with Outline Map 16. Have them indicate on the map the locations of Russia's natural resources as well as major cities such as St. Petersburg and Moscow. Then ask volunteers to suggest some transportation methods and routes that might bring raw materials to these cities. Ask what transportation problems might exist in Russia, given its physical geography.

Geographical Sidelight

The agricultural reforms that Russia is undertaking to increase output include privatizing land, reducing the size of farms, and increasing crop prices. However, the reforms face many hurdles, such as a lack of capital for seed and equipment, inadequate transportation, and outdated food-processing plants. Farmers who have worked all their lives on collective farms are finding it hard to adapt to the ideas of private land ownership and free enterprise.

Workers lay a pipeline that will transport natural gas westward from **Siberia**. Although it has many economic problems, Russia is rich in natural resources.

these metals are iron ore, copper, manganese, gold, platinum, and nickel. Russia produces specialized, high-technology metals such as tungsten and vanadium as well. Russia also is a major producer of diamonds. Large areas of Siberia await exploration for even more minerals. Efficient development of natural resources could provide future prosperity.

long distances from Siberia. Many of Russia's other remaining natural resources also are located in remote areas. Large, new investments will be necessary to continue developing these resources.

Russia has been a major oil producer since the early twentieth century. At first, most of the oil came from the shore of the Caspian Sea that today belongs to Azerbaijan. After World War II, however, oil fields were developed between the Volga River and the Ural Mountains. As these reserves were depleted, huge gas and oil discoveries were made in the lower Ob River region in northwest Siberia. About 70 percent of Russia's oil and natural gas now comes from this region. Pipelines carry the oil and gas to Moscow and St. Petersburg. Natural-gas pipelines run even farther west to Germany and France. Smaller oil and gas fields are located on Sakhalin Island and on the Kamchatka Peninsula. Soon, new fields will be developed in Siberia, in the Arctic, and north of the Caucasus.

Russia also has large coal reserves. Low-grade coal lies beneath the region south of Moscow, but the most developed coal fields are in the Kuznetsk (kuz-NEHTSK) Basin in Siberia. Even more coal lies undeveloped in the Tunguska and Lena river basins. Reserves here could last for centuries. Unfortunately, these deposits are far from Russian cities and markets. They also are distant from potential export customers around the Pacific Rim. Even so, companies from Japan and South Korea have plans to help develop mines in these regions.

Russia also has the potential to become a major supplier of more than a dozen metals. Among

INTERACTION **Agriculture** Russia's climate limits the country's agricultural potential. The growing season is short, and the soils under much of the taiga are poor. Russia's finest soil, the black earth of the steppe, often does not receive enough rainfall. In addition, a long history of political problems related to agriculture slowed Russia's agricultural success. Before the middle of the nineteenth century, much of the farming in Russia was done by **serfs**, peasants who were bound to the land. Even after the serfs were freed in 1861 and agricultural improvements were made, food production continued to fall behind needs. In 1891, a crop failure led to one of the worst famines in Russia's history.

The Communist government, which took power in 1917, wanted to modernize the country's agriculture. During the late 1920s and early 1930s, the government seized farmers' lands to form collective and state farms. On collective farms, workers were to share any profits the farm earned. In reality, however, the Soviet government controlled the farms and took most of what was produced. Farmers were paid such low prices for agricultural goods that they could barely survive. State farms were larger and were run like factories. The farmers worked the land for a salary. Because farmers were not given any reward for increased production, agriculture under the Soviets was unproductive and inefficient.

Returning land ownership to the farmers is a primary goal of the Russian government today. Officials now believe that the farmers will produce the highest yields when they own the land themselves.

Russia's most productive agricultural region is found south of a line drawn from St. Petersburg

340

Have students use school library or classroom resources to label Russia's primary agricultural areas on their copy of Outline Map 16. Ask them to write answers to these questions: Between what parallels of latitude does most of the farming land lie? (*45°N and 60°N*) What parts of North America lie between these parallels? (*a strip of the northern United States and much of Canada*) Why is the Russian growing season so short? (*high latitudes*)

Closure

Review with students the obstacles to developing Russia's natural resources and agricultural potential.

ASSESSMENT

Assign the Section 2 Review questions on this page.

Reteaching

Assign Section 2 of Reteaching Worksheet 29.

SECTION 3
OBJECTIVES

The objectives reflect the Focus questions on this page and are assessed in the review on page 346.

1. Analyze industrial and economic changes occurring in Russia.

2. Compare and contrast Russia's economic regions.

Village children watch this Russian farmer work his small plot of land south of **St. Petersburg**. Where were the most productive farmlands of the former Soviet Union?

3 ECONOMIC AND INDUSTRIAL REGIONS

FOCUS

● *What economic changes is Russia undergoing?*

● *How are each of Russia's economic regions similar and different?*

The Soviet Union was an industrialized country able to produce all of the products a modern country needs. Generally, however, Soviet industrialization concentrated on military goods. As a result, consumer goods usually were expensive and in short supply. Because Soviet industrial firms were monopolies and thus had little competition, quality was often poor.

Today, Russian industry is rebuilding and restructuring so that the country can compete in the world market. Former defense-related industries are beginning to produce consumer goods, and new businesses are developing. The companies that have focused on one product now are diversifying. While the country's move to a free-market economy is a very long and difficult process, the Russian people are eager for a better selection of consumer goods.

The Moscow Region Moscow, with its massive Kremlin and onion-shaped church domes, has symbolized Russia for centuries. The city became the home of the Russian Orthodox Church in the fourteenth century and Russia's capital when the country was unified in the fifteenth century. Even during the 200 years that St. Petersburg was the country's capital, most Russians still looked to Moscow as their country's heart and soul.

Today, with more than 9 million people, Moscow is the country's largest city and the heart of Russia's most important industrial region. As the national capital and a cultural center, Moscow's economic advantages are many. Moscow also is the center of communications and transportation for the nation. Roads, railroads, and air routes radiate from the capital to all points in Russia.

The Moscow industrial region stretches 250 miles (402 km) to the east and includes the city of

Photo Caption Answer

The steppe, especially in Ukraine but stretching also to northwestern Kazakhstan

Section 2 Review
ANSWERS

1. Forest products, oil, natural gas, coal, metals, and diamonds; mineral and energy resources could be important in the future if further developed.

2. Possible answers: grains, south of a line from St. Petersburg to Kazan' to Novosibirsk; fruits and vegetables, grown in gardens near cities; meats and eggs, grown on farmers' private plots

3. Students' time lines should include farming done by serfs prior to 1861; crop failure in 1891; formation of collective and state farms in the 1920s and 1930s; and return of land ownership to farmers in the 1990s.

to Kazan' to Novosibirsk. It is an area of mixed farming, focusing mainly on growing grains and raising livestock. Near cities, small gardens provide fresh fruits and vegetables for summer markets. Cattle ranching is important in the dry lands north of the Caucasus.

Much of the food that Russia produces still comes from the small plots that were established during the Soviet era. The farmers use these one- or two-acre plots as their private land and sell the products at competitive prices on the open market. This intensive cultivation produces a variety of summer fruits and vegetables as well as high-grade pork, beef, and eggs. The Russian government hopes to increase food production with the privatization of the larger farms. When the Soviet Union broke up, however, Russia lost the productive farmlands of Ukraine and northern Kazakhstan. For now, Russia must continue to rely on imports to feed its people.

2 SECTION REVIEW

1. What are Russia's major resources? What resources are likely to be important in the future?

2. Name three agricultural products raised in Russia. Where is each one produced?

3. Critical Thinking **Construct** a time line showing how agriculture has changed in Russia over the past two centuries.

341

TEACHING SECTION 3
Motivator

Ask students to think of regions of the United States that are known for certain products. *(examples: Detroit—automobiles; Silicon Valley—computers; Florida—citrus fruits)* Tell students that in Section 3 they will learn about the economic and industrial regions of Russia and the products for which they are known.

Building a Vocabulary

Write the terms **light industry** and **heavy industry** on the board. Briefly explain the difference. Have students suggest examples of each. List the examples on the board under the appropriate term. Call on a volunteer to read aloud the Glossary definition of **smelter** and link it to light or heavy industry. *(heavy)*

Brainstorming

Ask students to imagine that they have the opportunity to start a new business in Russia. Have them offer ideas for products or services that Russians may need or desire. Select a few ideas, and focus on the conditions that have to exist for these products or services to be created, marketed, and sold.

Photo Caption Answer

St. Petersburg

GLOBAL PERSPECTIVES

Point out to students that one of the keys to economic growth is communication, which enables companies to deal with suppliers and markets worldwide. In Russia, the communication infrastructure is underdeveloped. In 1993, there was a 32-year waiting list for telephone lines, and some 360,000 villages had no phone service at all. The international switchboard could handle only 124 calls to the United States at a time. U.S. and European telephone companies are helping to upgrade the system, but it will take years.

Ask students what other global communication links Russian businesses might need. *(computer networks, satellite television, radio)* Elicit a discussion about what implementing these systems might entail.

The colorful, onion-shaped domes of St. Basil's Church and the walled Kremlin draw tourists to Red Square. The Kremlin is the political heart of **Moscow**, the national capital. Where was the Russian capital before it was moved to Moscow?

Nizhniy Novgorod (Gorky) as well as smaller cities surrounding Moscow. The region has more than 50 million people, about one-third of the country's total population. Many people here work as government officials.

To try to ease industrial pollution problems in Moscow, planners decided that **light industry**, rather than **heavy industry**, would dominate. Light industry focuses on producing lightweight goods, such as clothing. Heavy industry usually involves manufacturing based on metals such as steel. Advanced-technology and electronics industries also have been developed here, and mineral resources are found nearby.

The St. Petersburg Region

While Moscow symbolizes Russia's values and traditions, St. Petersburg represents the country's desire to modernize and adapt ideas and practices from the West. Located on the Gulf of Finland about 420 miles (676 km) northwest of Moscow, St. Petersburg has been called the "Venice of the North." The city is built on the delta of the Neva River and has many canals instead of streets. With more than 5 million people, St. Petersburg is the second largest city in Russia. (See "Cities of the World" on pages 344–345.)

St. Petersburg is situated in a region of low, poorly drained land. As a result, the city is compact and does not sprawl outward like Moscow. The surrounding area has few natural resources, so fuel and raw materials must be brought in from other regions.

The city's harbor, canals, and rail connections provide good facilities for trade. The harbor and Gulf of Finland are kept open during winter by icebreakers. The city's chief industrial products are textiles, ships, chemicals, and machinery. St. Petersburg benefits from its location close to the innovative industrial centers of Europe. The city hopes the number of high-technology industries will grow.

The Volga Region

Russia's third major industrial region stretches along the middle part of the Volga River. Easy transportation on the Volga, energy from hydroelectric plants, and fossil fuel resources benefit the region. During World War II, many factories were moved to the Volga region to keep them safe from German invaders. Through much of its course, beginning near the city of Kazan', flowing past Samara (Kuibyshev) and Saratov (suh-RAHT-tuhf) and on to Volgograd (once known as Stalingrad), the Volga River is more like a chain of lakes.

The Volga region is famous for its heavy-machine industries and giant factories that produce such goods as motor vehicles, chemicals, and food products. Famous Russian caviar, a delicacy, comes from the fishery based at the ancient city of Astrakhan on the Caspian Sea.

The Urals Region

Mining in the southern Urals began more than two centuries ago, and the rich mineral deposits laid the base for the region's industrial development. Nearly every important mineral except coal and oil has been discovered here. Copper and iron **smelters**, factories that process the ores, are still important in the region.

The Soviet government built many industries in the Urals region during its drive to industrialize the country in the 1930s. As with the Volga region during World War II, whole factories were transported to the Urals ahead of the advancing Germans. The great cities of the Urals, such as Yekaterinburg (formerly Sverdlovsk), Chelyabinsk (chehl-YAH-buhnsk), and Magnitogorsk (mag-NEET-uh-gawrsk), started as commercial centers for mining districts. Today, these cities manufacture machinery and metal goods. As coal from Siberia and Kazakhstan was brought in to process iron ore into steel, heavy industry grew.

342

Guided Practice

Discuss how the resources of the various regions of Russia are important to the country as a whole. Ask students to consider what would happen to Russian industry if the country did not have oil and coal resources in the Urals. Remind students that Ukraine was the Soviet Union's breadbasket. Ask what effect they think Ukraine's independence has had on Russia.

Independent Practice

Have students make a chart with the headings Resources and Industries across the top, and with the six economic and industrial regions in Section 3 listed down the left-hand side. Have students complete the chart based on the information in Section 3. Ask for volunteers to share their lists with the class.

Closure

Remind students of why the collapse of the Soviet Union has had an enormous economic impact on Russian industry.

ASSESSMENT

Assign the Section 3 Review questions on page 346.

Siberia Siberia stretches from the Urals to the Pacific Ocean. The popular image of Siberia is that of a vast, frozen wasteland. In many ways, this image is accurate. Large areas of Siberia are almost uninhabited and have little industry. This enormous region has been a rich frontier luring Russian adventurers for more than 400 years. It continues to do so today.

Settlement in Siberia mostly follows a line along the route of the Trans-Siberian Railroad. Construction of this railway started in 1891, and by 1916, it connected Moscow to Vladivostok on the Sea of Japan. More than 5,700 miles (9,171 km) long, the Trans-Siberian Railroad is the longest single rail line in the world. For many towns in the region, the train provides the only transportation link to the outside world.

Many natural resources have been developed in Siberia during the twentieth century. So that these resources could be more easily transported, a more direct railway, called the Baykal-Amur Mainline (BAM), was completed across eastern Siberia in 1989. The building of BAM was an impressive construction effort. Built by youth organizations, BAM crosses five mountain ranges and 17 major rivers. It is connected by four tunnels and more than 3,000 bridges. Most of the construction of BAM took place in winter so that trucks could be driven over frozen soils.

Less than 10 percent of Russia's industry is in Siberia. Government plans for settling and developing Siberia have not gone smoothly, partly because of the harsh climate and difficult terrain. Many people would rather live in the cities of European Russia even though wages are higher in Siberia.

Lumbering and mining are the region's most important industries. Most of the lumbering and mining towns are connected by rail or river routes to the Trans-Siberian Railroad or to BAM. Siberia's mineral ores and forest products are needed by consumers and industries in western Russia.

Rich coal deposits have made the Kuznetsk Basin, or the Kuzbas, Siberia's most important industrial region. *Kuznetsk* means "blacksmith" in Russian. The Kuzbas is located in southwestern Siberia between the Ob and Yenisei rivers. Coal from the Kuzbas is sent to the Urals region and even farther west. Novosibirsk is the manufacturing and transportation center of Siberia. Its population of almost 1.5 million makes it the largest city in Siberia. As the resources of Siberia are further developed, Novosibirsk and nearby cities will gain in importance.

The Russian Far East Russia has more coastline on the Pacific Ocean than any other country. This area bordering the Sea of Okhotsk and the Sea of Japan is known as the Russian Far East. The climate of the Russian Far East is less severe than that of the rest of Siberia. The summer weather is mild enough for some successful farming, particularly in the Amur River valley. Yet, the region cannot produce enough food for itself. Much of the land remains heavily forested, and the region's minerals are only beginning to be developed.

Khabarovsk (kuh-BAHR-uhfsk), on the Amur River, is the principal city of the Russian Far East. Its central location enables it to process forest and mineral resources from throughout the region. Vladivostok, which means "lord of the east" in Russian, is the chief seaport of the Russian Far East. It is a major naval base and the home port for a large fishing fleet. The city of Komsomol'sk-na-Amure produces iron and steel. Russia hopes that the Russian Far East will become part of the Pacific Rim trading network and find new markets for its natural resources and products.

Sakhalin Island, with its petroleum and mineral resources, lies off the eastern coast of Siberia

A plant worker at a fishery in **Astrakhan** on the Caspian Sea prepares caviar for market. Russia and Iran produce nearly all of the world's true caviar from fish in the Black and Caspian seas.

Text continues on page 346.

> **Geographical Sidelight**
>
> *The Trans-Siberian Railroad was completed in several sections. The last section allowed the railroad line to bypass a link that ran through Manchuria in China. With this last section completed, the railroad was located entirely within Russia's borders. Today, an "express" train takes seven days to travel from Moscow to Vladivostok over the railroad.*

Cities of the World

Interpreting Views

Tell students that the Russian poet Joseph Brodsky described St. Petersburg as a "stalled train bound for eternity." Brodsky was labeled a dissident in the USSR for his criticisms of the former Soviet government and was targeted by the secret police there. In 1964, he was sentenced to five years of compulsory work at a remote collective farm in the former Soviet Union. In 1972, he was forced to emigrate to the United States. Given his background, ask students what Brodsky might have meant by his description of St. Petersburg. (*St. Petersburg is a great and historic city ["bound for eternity"], but its forward progress was "stalled" under the Soviet government.*) Have students suggest their own metaphors for St. Petersburg based on the information in the feature and in the text.

Geographical Sidelight

Peter the Great pushed to modernize Russia and to make it one of Europe's great powers. Alexis, Peter's son and heir, opposed his father's reforms. Peter had his son arrested, tortured, and sentenced to death. Alexis died in prison before he could be formally executed.

CITIES OF THE WORLD

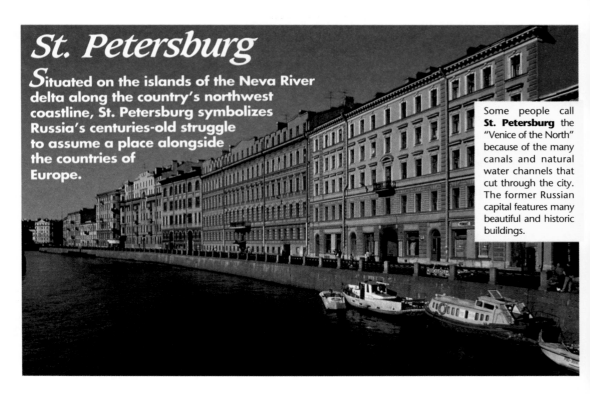

St. Petersburg

Situated on the islands of the Neva River delta along the country's northwest coastline, St. Petersburg symbolizes Russia's centuries-old struggle to assume a place alongside the countries of Europe.

Some people call **St. Petersburg** the "Venice of the North" because of the many canals and natural water channels that cut through the city. The former Russian capital features many beautiful and historic buildings.

St. Petersburg is Russia's second largest city, a major transportation center, and one of the most beautiful urban centers in Europe. The city was founded in 1703 by Peter the Great, one of Russia's most brilliant but brutal rulers. The site that Peter chose for the city was poor. The land was swampy and the climate unhealthy. Though thousands of forced laborers died from malaria and other diseases while building the new city, Peter the Great was determined to build a city and port on the Baltic coast. This city would serve as Russia's "window on Europe." Through that window, Russia could adopt the scientific and technical skills of the countries of Europe and begin to modernize.

To emphasize this plan, Peter gave his city the German-sounding name St. Petersburg. In 1712, St. Petersburg replaced Moscow as the capital of the Russian Empire. Remembering the thousands who died building the city, Russian people called St. Petersburg "the city built on bones." But as Peter imported some of Europe's finest craftsworkers and architects to St. Petersburg, a well-planned city soon took shape. Peter's ambitious plan was continued by his successors. By the late eighteenth century, St. Petersburg was a large and impressive city in which Russians took great pride.

During the nineteenth and early twentieth centuries, St. Petersburg rose to become the center of modern Russian culture. Great nineteenth-century writers such as Alexander Pushkin and Fyodor Dostoyevsky set some of their most important works here. Peter Tchaikovsky, a famous Russian composer, wrote his music in St. Petersburg. The city has also been home to some of Russia's most famous ballet dancers.

Many of Russia's most significant historical events have taken place in the heart of St. Petersburg. In March 1917, the uprising that ended the Russian monarchy began here. Eight

344

Using Geography Themes

Have students work in pairs to create a list of St. Petersburg's human and physical place characteristics, using information in the feature and on page 342 of the text. Remind students that the map and photographs in the feature and on page 349 also are good sources of information about the city's place characteristics. Have volunteers write their lists on the board and compare them.

Once the Russian capital, **St. Petersburg** boasts numerous historical sites. In which section of the city are the Winter Palace and Summer Palace located?

Russians stroll the streets in the whitish twilight of 11 P.M. in summertime **St. Petersburg**. Because of the city's northern location, St. Petersburg's citizens emerge from long, dark winter nights into the "white nights" of early summer, when the sun barely sets.

Map Caption Answer

Central City

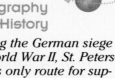

Geography and History

During the German siege in World War II, St. Petersburg's only route for supplies was across frozen Lake Ladoga during the winter. An oil pipeline and electrical cables along the lake bed provided some relief as the war dragged on. During the war, fighting, starvation, and displacement dramatically lowered St. Petersburg's population. The city's population did not regain its prewar size until the 1960s.

Geography and Economics

Trade and shipbuilding were the main economic activities in St. Petersburg's early years. The city later developed into one of the most important industrial centers in Russia. Point out to students that after the collapse of the Soviet Union in the early 1990s, the people of St. Petersburg faced huge layoffs in the city's defense-related industries and an annual inflation rate of 1000 percent.

months later, the Bolshevik party led by V.I. Lenin seized control of the city and went on to establish the Soviet Union. Moscow became the new capital.

During World War II, St. Petersburg heroically resisted Nazi Germany. For 900 days, beginning in September 1941, the Germans held the city under siege. More than 800,000 civilians died during the struggle, most from starvation. Yet the city held out until the siege finally ended in January 1944.

St. Petersburg has had three names in its history. In 1914, when Russia was at war with Germany, St. Petersburg was changed to the more Russian-sounding Petrograd. After Lenin's death in 1924, the Soviet Union's rulers renamed the city Leningrad. In 1991, immediately after the collapse of the Soviet Union, the city's people voted to change the name back to St. Petersburg.

Today, St. Petersburg is divided into four sections. The Central City, St. Petersburg's central hub, is situated along the south bank of the Neva River. The Vyborg and the Petrograd quarters lie along the river's north bank. Vasilievsky Island is situated between the Neva's north and south forks.

St. Petersburg is home to 17 professional theaters and more than 60 museums, including the renowned Hermitage. The Peter and Paul Fortress, originally built to defend the city, stands at the mouth of the Neva River. For many years, the fortress was Russia's most dreaded prison. The city's main street and shopping area is a wide avenue called the Nevsky Prospekt. St. Petersburg's most famous statue, the giant *Bronze Horseman,* honors Peter the Great.

St. Petersburg is one of the world's northernmost cities. This location gives the city a unique attraction: its famous "white nights." Each year from mid-June to early July, the sun barely sets. Instead, for 30 to 40 minutes each day the city is bathed in a whitish twilight. The people of St. Petersburg mark this special time with a music and art festival.

Reteaching
Assign Section 3 of Reteaching
Worksheet 29.

Extension

Have interested students research
and write a report on the process of
privatization of Russian industry.
They should focus on why foreign
companies or individuals might or
might not want to invest capital in
Russian enterprises.

SECTION 4
OBJECTIVES

The objectives reflect the Focus
questions on this page and are
assessed in the review on page 348.

1. Examine political and cultural
characteristics of the Russian
Federation.

2. Discuss current and future issues
facing Russia.

TEACHING SECTION 4
Motivator

Call on students to tell what they
know from newspaper articles or
television about Russia or the
Russian people and their way of life.
Write their observations on the
board. Ask students for reasons why
they think that the Russian people
are better or worse off than they
were 10 years ago. Ask them to speculate about the Russian people 10
years from now.

A tram takes people
up one of the steep
hills surrounding
Golden Horn Bay
in **Vladivostok**.
Icebreakers keep
the harbor open
when it freezes dur-
ing the winter.

Section 3 Review
ANSWERS

1. Industries are diversifying and are produc-
ing more consumer
goods. Russians
will have a greater
selection of con-
sumer goods.

2. Industrial regions:
Moscow, St. Peters-
burg, Volga, and
Urals regions; re-
gions with mineral
resources: Urals and
Far East regions and
Siberia; trade: St.
Petersburg and Far
East regions

3. Possible answer: all
regions will be
affected, but the
Moscow and St.
Petersburg regions
are likely to be most
affected because they
are the most popu-
lous regions.

in the Sea of Okhotsk. The Kuril (KYUR-eel)
Islands, important for commercial fishing, are
located east of Sakhalin. Russia took the islands
from Japan at the end of World War II, and Japan
claims it still owns some of the southern Kurils.

 SECTION REVIEW

1. How are Russia's industries changing? How
will these changes affect the daily lives of
the Russian people?

2. Which of Russia's regions are industrial?
Which have mineral resources? Which are
well situated for trade?

3. **Critical Thinking** **Predict** which
regions are likely to be most affected by
the country's economic changes.

4 POLITICAL GEOGRAPHY

FOCUS

● *What are the political and cultural characteristics of the Russian Federation?*

● *What current and future issues does
Russia face?*

The ancient heart of Russia is the upper Volga
River basin. The Russian Orthodox Church fled
here after Kyyiv (formerly Kiev) fell to the
Mongols in 1240. By collecting the Mongol
ruler's taxes, the princes of Moscow became the
leading power in the region. When the Mongols

finally were defeated in the late 1400s, Russian
authority spread outward. Military expeditions
and land-hungry adventurers increased Moscow's
territory in all directions and brought Russian language, laws, and customs to neighboring peoples.
The church often followed close behind. The
expansion of the Russian state was much like the
colonial developments of England and France.
Russia, however, invested little to expand its
empire overseas.

PLACE **The Russian Federation** The
Russian Federation's system of government
includes a strong president and a parliament,
called the Federal Assembly. The president and
members of parliament are elected by Russian citizens. The existence of many political groups in
the parliament often has slowed efforts to move
Russia toward a market economy. Some legislators think reform has occurred too fast. In addition, many oppose granting more freedoms to
some of Russia's numerous ethnic groups.

During the Soviet era, many minority ethnic
groups in Russia lived within their own territorial
subdivisions. Although these territories had some
local power, they were in fact tightly controlled
from Moscow. In the aftermath of the fall of the
Soviet Union, some of these territories became
part of partially self-governing republics within
Russia. Twenty-one such republics now make up
about 15 percent of Russia's land area. Each
republic has its own legislature and leaders, but
its degree of authority is much debated.

Republics of the Russian Federation	
Adygea	Karelia
Bashkortostan	Khakassia
Buryatia	Komi
Chechnya	Mari El
Chuvashia	Mordvinia
Dagestan	North Ossetia
Gorno-Altay	Tatarstan
Ingushetia	Tuva
Kabardino-Balkaria	Udmurtia
Kalmykia	Yakutia
Karachay-Cherkessia	

These republics of the Russian Federation have large
non-Russian populations. The central government in
Moscow has given the republics some control over
local affairs. Even so, some people in the republics are
calling for more freedom and even independence.

346

Creating a Culture Map

Have students create a large map of Russia and mount it on a bulletin board. Assign students particular ethnic regions to research. Ask them to illustrate the culture of the region using photos or their own drawings. Post the illustrations on the appropriate parts of the map.

Guided Practice

Russia and the United States are both multicultural and multiethnic societies. Point out that American cultural diversity stems largely from immigration, whereas in Russia it results mainly from expansion and migration. Lead a discussion about ways that ethnic diversity affects a society.

Independent Practice

Refer students to the map on this page. Have them analyze the pie graphs to determine which countries appear to have the most ethnically diverse populations. Then have students rank the countries according to the percentage of other ethnic groups within their borders, from the country with the largest percentage (*Tajikistan*) to those with the smallest (*Ukraine and Armenia*).

Ethnic Groups of the Former Soviet Republics

Source: *The World Factbook 1994*, Central Intelligence Agency

LEGEND

Percentage of population that is of the official nationality

Percentage of population that is ethnic Russian

Percentage of population that is of other ethnic groups

SKILL STUDY The pie graphs above show the ethnic population breakdown of the former Soviet republics. Centuries of Russian domination and decades of Soviet policies resulted in large ethnic Russian populations in many of the republics. Conflicts between ethnic Russian populations and large minority populations have created problems within the now-independent republics and within the Russian Federation itself. Efforts to ease ethnic tensions must respect minority rights. After Russia, which republic's population has the largest percentage of ethnic Russians?

Although about 82 percent of its current population is ethnically Russian, Russia remains a diverse country with dozens of ethnic groups. All of the Russian republics supposedly were formed to have a dominant minority ethnic group. In many of these republics, however, ethnic Russians form a majority of the population, just as they did in the former Soviet republics. (See the map and graphs above.)

The most populous republic in the federation is Bashkortostan, southwest of the Urals. Yakutia in Siberia has the largest land area. Some republics, such as Chechnya in the south, have used the instability in Russia to push for independence. Fighting between the Russian army and Chechen rebels increased sharply in 1994, claiming thousands of lives. Other ethnic groups, such

as the Tartars along the Volga River, also have sought independence.

Many Russians fear that such desires for independence threaten the future of the federation. Some leaders, however, believe they can bring better government and prosperity to their regions.

Issues Despite the high hopes that followed the breakup of the Soviet Union, bringing freedom and prosperity to the peoples of Russia has been difficult. The reliable certainties of the past have been replaced with an exciting but uncertain present. The concepts of free-market economics, individual liberty, and personal profit are still viewed with suspicion by many people. Nevertheless, progress has been made as increasing numbers of Russians learn about capitalism and

---

Skill Study
Kazakhstan

THE WORLD TODAY VIDEODISC PROGRAM

Russia: A New Russian Business

1B 41154 4:29

Play   Pause

Closure

Ask students to speculate about what political changes are still to come in Russia, based on what they have read or heard. Go around the room and have each student state and explain his or her prediction. Ask students whether they think that the United States should take an active role in helping Russia stabilize its political relationships, and if so, what role.

ASSESSMENT

Assign the Section 4 Review questions on this page.

Reteaching

Assign Section 4 of Reteaching Worksheet 29.

Extension

Have interested students research important political events that have occurred in the Russian Federation since 1991. Have them make a time line showing these events, or have them focus on one event and write a paragraph explaining its significance.

Section 4 Review
ANSWERS

1. Thousands of Chechens and Russians have died in fighting since Chechnya declared its independence from Russia.

2. Possible answers: the existence of many political groups in the parliment slowing reform efforts; holding the different regions together

3. Possible answer: the people of the Russian Federation will have a better selection of consumer goods and a greater voice in government.

NEWS FROM *Russia*

AROUND THE WORLD

Teenagers in Russia Have More Free Time

When Russia was part of the Soviet Union, teenagers spent much of their free time in government-sponsored groups. Almost all children joined the Young Pioneers when they reached nine years of age. At age 14, those who did not join Komsomol (the Young Communist League) were considered disloyal to the government and could not expect to attend a university or receive good jobs as adults.

Now that Russia is no longer under communism, few teenagers join political groups. Some Young Pioneer groups still exist, although the organization suffers from lack of funds. Those attending the summer camps run by the Young Pioneers have noticed big changes. No longer intent on promoting communism, Young Pioneer groups

have become far less rigid. Boys and girls used to resemble young soldiers as they saluted and marched with red scarves secured around their necks. "There are no songs, no marching, no red ties anymore," said Artyom V. Chibisov, a 13-year-old boy from Moscow. One young woman had a practical reason for not missing the scarves. "It means less ironing," she said.

As groups such as the Young Pioneers and Komsomol weaken or disappear, young Russians are left with more free time and are seeking new ways to spend it. Although Russian teenagers rarely worked at after-school jobs, Russia's economic crisis offers them even less chance to do so now. Many spend their free time socializing with friends.

Moscow teenagers who enjoy playing rollerball, like those pictured above, arrive each day in Gorky Park to meet friends and practice their skills. Other teens have become swept up in Western music, spending so much time in music-related activities that they are called *rokry* (rockers) or *metallisty* (heavy metal fans).

Many Russian teens, however, have other interests besides sports and music. Some join large groups that work to clean up or conserve the environment. Others remain in the International Clubs that schools organized as the old Soviet Union became more open. In these clubs, teenagers plan parties and meetings and correspond with pen pals in other nations.

democracy. Stable political relationships are slowly being created.

Russia's greatest resource is its people, who are well educated. They have survived the challenges of living in a country characterized by isolation and a harsh environment. During the Soviet era, government officials controlled the lives of the people. Now the people can make their own decisions and determine their own futures. The Russian people have much to offer themselves and the world.

 SECTION REVIEW

1. What is significant about the republic of Chechnya?
2. Name two political challenges facing the Russian Federation.
3. **Critical Thinking   List** ways in which a free-market economy and a democratic government might benefit the Russian Federation.

348

Building a Vocabulary

1. Over cities; because they are caused by the cities' fumes and smoke
2. A huge evergreen forest
3. Peasants bound to the land; no
4. Light industry produces lightweight products whereas heavy industry produces heavy products.
5. A factory that processes ores

Recalling and Reviewing

1. Lake Baykal; because it is the world's deepest lake and is home to many endangered plants and animals
2. Development has been difficult because of the distances resources must be transported.
3. Most populated areas: cities in the Moscow and St. Petersburg regions; least populated area: Siberia
4. Private ownership of farms, diversification of industries, and production of consumer goods are increasing. The transition is difficult, but people will have access to more products.

Thinking Critically

1. Some students may say Europe because the largest populations are found on the European side of the Urals. Others may say Asia because most of the country's territory is east of the Urals.
2. Possible advantage: there may be less conflict if most people share a similar ethnic background. Possible disadvantage: ethnic diversity in a nation can be positive.

Learning About Your Local Geography

Students' lists should include local youth organizations and their projects. Students who wish to volunteer to work for organizations should be encouraged to do so provided that parents/guardians and youth organizations have been informed and have granted permission.

Using the Five Themes of Geography

Journals might mention cold climate as a place characteristic affecting places visited; icebreakers might be necessary for movement. Another entry might discuss how movement in much of Siberia is possible only by railroad.

CHAPTER **29** CHECK

Summer Palace in **St. Petersburg**

Reviewing the Main Ideas

1. Russia is a huge country dominated by plains, although it also has hills and mountains. Winters are cold and long, making travel difficult. Vegetation varies from forest to steppe.
2. Russia has rich forest, energy, and mineral resources. Its climate, however, limits the country's agriculture, which has undergone much change over the past two centuries. Today, much of the food that Russia produces comes from farmers' private plots. Fruits, vegetables, and meats are among the products raised.
3. Russian industry is in the process of rebuilding and restructuring as the country makes the transition to a free-market economy. Economically, Russia can be divided into Siberia, the Far East, and the regions around Moscow, St. Petersburg, the Volga River, and the Urals.
4. The Russian Federation's government is divided over reform. About 15 percent of its territory is divided into 21 largely self-governing republics. Major national issues include building stable political relationships and bringing freedom and prosperity to the country's peoples.

Building a Vocabulary

1. Where are habitation fogs found, and why?
2. What is the taiga?
3. What are serfs? Do they still exist in Russia?
4. What is the difference between light industry and heavy industry?
5. Define *smelter*.

Recalling and Reviewing

1. What large lake is found in Siberia? Why is this lake significant?
2. How has Russia's size influenced development of the country's resources?
3. Where do most of Russia's people live? What is the least populated area?
4. How have agricultural and industrial activities in Russia changed as a result of the breakup of the Soviet Union? What effect do these changes have on the country's people?

Thinking Critically

1. Do you think Russia is best considered a country of Europe or a country of Asia? Keep in mind demographic, cultural, and physical factors.
2. As you know from the chapter, the republics of the Russian Federation are based to some extent on the location of ethnic groups. What are some advantages and disadvantages of forming political units, such as republics or countries, in this way?

Learning About Your Local Geography

Individual Project

Siberia's Baykal-Amur Mainline (BAM) was built largely by youth organizations. Using school or community resources, investigate youth organizations in your own area and the projects with which they have been involved. Make a list of each organization, and next to each item, write down some of its projects. If possible, volunteer to spend an afternoon working with one of the local youth organizations. You may want to include your list in your individual portfolio.

Using the Five Themes of Geography

Imagine you are taking a week-long tour of Russia in the winter. Keep a journal of your **movement**, of the physical **place** characteristics you encounter, and of evidence of **human-environment interaction**. How do place characteristics affect what places you visit and how you move between them? What forms of human-environment interaction have made your movement possible?

CHAPTER 30 Ukraine, Belarus, and the Caucasus

Planning Guide

TEXTBOOK	TEACHING RESOURCES	OTHER AVAILABLE MATERIALS
Section		
1. Ukraine Pages 351–354	Reteaching Worksheet 30.1 Skills Worksheet 30	*Study Guide for Spanish Speakers* *Lecture Notes,* Chapter 30
2. Belarus Pages 354–355	Reteaching Worksheet 30.2	*Study Guide for Spanish Speakers* *Lecture Notes,* Chapter 30
3. The Caucasus Pages 355–356	Reteaching Worksheet 30.3 Graphic Organizer 30 Critical Thinking Worksheet 30 Challenge/Enrichment Worksheet 30	*Study Guide for Spanish Speakers* *Lecture Notes,* Chapter 30
Chapter Review Page 357	Chapter 30 Test Chapter 30 Portfolio Assessment See also *Alternative Assessment Forms*	*Test Generator,* Chapter 30

CHAPTER RESOURCES

Lecture Notes, Chapter 30
Teaching Resources
    Chapter 30 Graphic Organizer
    Chapter 30 Skills Worksheet
    Chapter 30 Critical Thinking Worksheet
    Chapter 30 Challenge/Enrichment Worksheet

REVIEW AND ASSESSMENT

Chapter 30 Check, page 357
Chapter 30 Test, *Teaching Resources*
Chapter 30 Portfolio Assessment, *Teaching Resources*
Test Generator, Chapter 30
Alternative Assessment Forms

ADDITIONAL RESOURCES

Books for Teachers

Denbar, Rachel. *Bloodshed in the Caucasus: Escalation of the Armed Conflict in Nagorno Karabakh.* New York: Human Rights Watch, 1992.

Gaddis, J. L., and John Lewis. *The United States and the End of the Cold War.* New York: Oxford University Press, 1992.

Kuropas, Myron P. *The Ukrainian Americans: Roots and Aspirations.* Toronto: University of Toronto Press, 1991.

Walker, Christopher, ed. *Armenia and Karabagh: The Struggle for Unity.* London: Minority Rights Publications, 1991.

Books for Students

The Armenians. New York: Rizzoli, 1986.

Boyette, Michael. *Soviet Georgia.* New York: Chelsea House, 1987.

Shaw, Warren. *The World Almanac of the Soviet Union: From 1905 to the Present.* New York: World Almanac, 1990.

The World Today *Videodisc Program*

A variety of still images, short videos, and activities are available for you to use as you teach this chapter. *Lecture Notes* include bar codes for specific images and videos.

Multimedia Resources

Picture Atlas of the World. (interactive maps, photo bank, animation, audio on CD-ROM) National Geographic Society, Washington, DC.

The Soviet World in Transition. Part I: Geography; Part II: History. (16mm film/VHS, 20 minutes each) National Geographic Society, Educational Services, Washington, DC 20036.

Steel Town. (VHS, 20 minutes) Coronet/MTI Film & Video, 108 Wilmot Road, Deerfield, IL 60015.

MEETING INDIVIDUAL NEEDS

Limited English Proficiency (LEP)

Have students work in small groups to create an illustrated poster of one of the countries discussed in Chapter 30. Posters should depict the physical features, resources, economic activities, agricultural products, and human geography of the country. Use the posters as a basis for a discussion comparing these countries.

Learners Having Difficulty

On a sheet of paper, have students make a three-column chart with the following headings: What I Already Know, What I Expect to Learn, What I Learned. Students should complete the first two columns after reading the Focus questions for each section. Then have students complete the third column after reading each section. Students may use their completed charts as a study aid.

Auditory Learners

Play for the class some music from the region discussed in Chapter 30. Possible works might include those of classical composer Aram Khachaturian, who was from the Caucasus, or traditional folk music from Ukraine, Belarus, or the Caucasus. Ask for students' responses to the music. Discuss their impressions of specific qualities in the music that might suggest the physical or human geography of the region.

Visual Learners

Provide students with newspaper and magazine articles about current events in Ukraine, Belarus, and the Caucasus. Direct students to make graphic organizers of the main ideas in the articles. Display the diagrams in the classroom.

Tactile/Kinesthetic Learners

Assign each student one of the countries in Chapter 30—Ukraine, Belarus, Georgia, Armenia, Azerbaijan. Ask students to write the name of their country on one side of an index card and a description of the country that relates to one of the Five Themes of Geography on the other side. Collect the cards, shuffle them, and deal them out with the names of the countries facedown. Call on students to read aloud the description and guess the name of the country.

Gifted Learners

Have interested students investigate one of the severe earthquakes that have rocked the Caucasus and report their findings to the class.

MULTICULTURAL FOCUS

There is an old Eastern European folktale that explains an ancient rivalry between the Cossacks of Ukraine and the Tartars (Tatars), a Turkish people of southwestern Russia and Ukraine.

According to the tale, there were two men named Cossack and Tartar who once were such good friends that people used to say "as loyal as Tartar to Cossack" and "as gentle as Cossack to Tartar." Cossack, however, had finer belongings than Tartar, and jealousy between them grew as the two friends traveled together. Finally, they began to squabble over food. The story ends with this thought: Cossack and Tartar are bitter enemies to this day, and the pity of it is, they were once the best of friends.

Explain to students that in folk legends such as this, passed on orally from generation to generation, the characters often represent groups of people, and the simple story and emotions presented may reflect larger themes. For example, this folk tale reflects not only the rivalry between the Cossacks and the Tartars but also the themes of economic inequality and uneven distribution of resources.

Have students relate folk tales or legends they may be familiar with. Or, if possible, you may wish to locate other folktales from the region to read aloud to the class. Ask: What groups of people do the characters represent? What themes are presented? Have students speculate on why and when the stories might have come into existence.

CHAPTER 30

UKRAINE, BELARUS, AND THE CAUCASUS

Section 1: Ukraine
Section 2: Belarus
Section 3: The Caucasus

OVERVIEW

In Chapter 30, students will learn about the geographical features and the cultures represented in the newly independent countries that lie along the western rim of the Russian Federation.

Section 1 examines Ukraine's rich agricultural and mineral resources.

Section 2 describes the cultural and economic history of Belarus and its present political structure.

Section 3 focuses on the countries of the Caucasus, an area with several landform regions and diverse cultures located between the Black and Caspian seas.

About the Author

Nikolay Gogol (1809–1852) was a Russian novelist, dramatist, and short story writer. He was born in the Ukraine region. The quotation on this page is from *Taras Bulba,* his famous epic novel about the Cossacks.

Primary Source

Description of Change: excerpted and bracketed
Rationale: excerpted to focus on main idea; bracketed to clarify meaning

---

CHAPTER 30

GEOGRAPHY DICTIONARY

Cossack

homogeneous

Ukraine, Belarus, and the Caucasus

❝ **T**he farther they [the Cossacks] penetrated the steppe, the more beautiful it became. . . . Nothing in nature could be finer. The whole surface resembled a golden-green ocean, upon which were sprinkled millions of different flowers. . . . Oh, steppes, how beautiful you are! ❞

Nikolay Gogol

Door woodcarving from **Georgia**

The political geography of Eurasia has changed dramatically since the collapse of the Soviet Union. The former Soviet republics have become independent countries, making previous internal Soviet boundaries international frontiers. Border posts and customs stations dot once-deserted roads. Gradually, a new system of political and economic relationships is developing as local political leaders take on the authority that was once centered in Moscow. The new geography is being tested and adjusted, again and again.

The countries of Ukraine, Belarus, Georgia, Armenia, and Azerbaijan form the western rim of the Russian Federation. The latter three countries are part of a region called the Caucasus. All of these countries are working to build new societies. In many cases, they have combined ideas and values from their own ancient histories with those from the former Soviet system and from the democratic countries of western Europe.

Ukraine field

Using the Quotation

Have a volunteer read aloud the quotation from Nikolay Gogol. Refer students to the photograph of the steppe on page 321. Ask students how they think the steppe region may have changed since Gogol wrote this passage in the 1800s. *(Farms and cities have replaced much of the grasslands.)*

Using the Geography Dictionary

Have a volunteer read a dictionary definition of **Cossack** aloud. Point out that these frontier people were known for their skill with horses and that *Cossack* means "free person" in Turkic. Next, have a volunteer read the definition of **homogeneous** on page 355. Note that few countries have a truly homogeneous population.

SECTION 1
OBJECTIVES

The objectives reflect the Focus questions on this page and are assessed in the review on page 354.
1. Identify physical features, resources, and economic activities of Ukraine.
2. Discuss Ukraine's human geography.

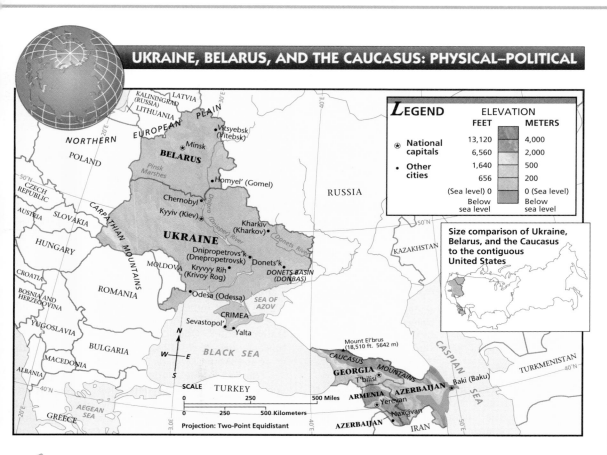

UKRAINE, BELARUS, AND THE CAUCASUS: PHYSICAL–POLITICAL

Using the Physical–Political Map: Ukraine, Belarus, and the Caucasus

Have students examine the map on this page. Call on individual students to locate the Dnipro River, the Donets River, and the Black Sea. Point out why navigable waterways are important in agricultural and industrial regions such as Ukraine. *(for shipping agricultural products, manufactured goods, and mineral resources to markets or processing plants)*

UKRAINE

FOCUS

- *What physical features, resources, and economic activities are found in Ukraine?*
- *What is the human geography of Ukraine?*

REGION **Physical Geography** A portion of the Carpathian Mountains runs through western Ukraine, but most of the country lies in a region of plains and low hills that stretch north from the Black Sea. (See the map above.) In plains regions such as this, climate and vegetation are linked in broad, east-to-west bands. The climates here become warmer and drier from north to south, though winters throughout the country are cold. Ukraine's latitudes are the same as those of southern Canada.

Deciduous forests were once widespread in northern Ukraine. In the south, the forests opened onto the grasslands of the steppe. The capital city of Kyyiv grew near the former forest-steppe boundary, where the products of the steppe and forests could be traded. Today, little of Ukraine's original vegetation remains. Farm fields now cover much of the land.

Resources Since ancient times, rich farmlands have been Ukraine's greatest natural resource. The region's fertile black soils and summer rains produce high yields of grains and vegetables. Indeed, Ukraine was the breadbasket for the Russian Empire and, later, for the Soviet Union. Even though farming under the Soviet system was inefficient, Ukraine's farmlands helped make the Soviet Union the world's largest producer of wheat, oats, barley, and sugar beets.

Ukraine's level plains also permit commercial farming using modern machinery. Irrigation water for large-scale agriculture comes from

Compare Ukraine with the Midwest as a vital agricultural region. Have students imagine the impact on the United States if the Midwest suddenly became an independent country.

Building a Vocabulary

Refer students to the discussion about **Cossacks** on page 353. Ask a volunteer to give four facts about Cossacks. *(independent, warlike, controlled the steppe region beginning in the sixteenth century, fought Russia's southward expansion)* Have another student consult an encyclopedia to find additional information about Cossacks and share it with the class.

Guided Practice

Write the following places in Ukraine on the board: Dnipro River, Donets Basin, Kryvyy Rih, Crimea, Kyyiv. Have students work with partners to identify the natural resources and economic activities of Ukraine and to match them with the appropriate place.

Chart Caption Answer

Ukraine

GLOBAL PERSPECTIVES

The legacy of the Cold War includes intercontinental ballistic missiles and long-range bombers that remain in Ukraine. Tell students that the United States and Russia have agreed on a general disarmament pact. In 1994, the United States agreed to help Ukraine dismantle its nuclear weapons. Ukraine had cited high costs as a reason for not dismantling the weapons. Begin a class discussion by asking: Do the former Cold War superpowers have a responsibility to financially help Ukraine dismantle the weapons? *(Some students might say yes since weapons were placed there as a result of U.S./Soviet rivalry and before Ukraine became independent.)*

FACTS IN BRIEF — Ukraine, Belarus, and the Caucasus

COUNTRY POPULATION (1994)	LIFE EXPECTANCY (1994)	LITERACY RATE	PER CAPITA GDP (1993)
ARMENIA 3,521,517	69, male 76, female	99% (1992)	$2,040
AZERBAIJAN 7,684,456	67, male 75, female	97% (1992)	$2,040
BELARUS 10,404,862	66, male 76, female	98% (1992)	$5,890
GEORGIA 5,681,025	69, male 77, female	99% (1990)	$1,390
UKRAINE 51,846,958	65, male 75, female	98% (1993)	$3,960
UNITED STATES 260,713,585	73, male 79, female	97% (1991)	$24,700

Sources: *The World Factbook 1994*, Central Intelligence Agency; *1995 Britannica Book of the Year*

Because of major economic changes since 1990 in the republics of the former Soviet Union, per capita GDP figures might be unreliable. Which country in the region has the largest population?

hydroelectric dams on the region's major river, the Dnipro (formerly Dnieper).

Beneath the plains of Ukraine lie large deposits of mineral resources. The Russian Empire's first important industrial region developed near the still-productive coal fields of the Donets Basin (also called the Donbas) in eastern Ukraine. Iron and steel industries were established here in the late 1800s.

During the Soviet period, the government developed many additional mining areas in Ukraine. The most important is at Kryvyy Rih (formerly Krivoy Rog) between Kyyiv and the Black Sea. Here, a huge open-pit iron-ore mine supports a region of heavy industry. The area also has rich deposits of manganese and other metals. Ukraine has only small oil and gas deposits, however.

INTERACTION **Economic Geography** Ukraine is well known for its production of agricultural machinery, transport vehicles, and food processing equipment. South of Kyyiv, a group of industrial cities has developed around the giant dam on the Dnipro River. This dam, one of the largest in Europe, supplies electricity to the metal and chemical industries throughout the Donets region. The region's most important city is Donets'k, while the city of Kharkiv (formerly Kharkov) is the administrative center. Odesa (formerly Odessa) is the busiest port on the Black Sea and is expected to grow as Ukraine expands its world trade.

Since Ukraine's independence, the country's economic progress has been slow. As elsewhere in the former Soviet Union, the country has had difficulties establishing new trade links. In addition, holdover leaders from the Communist era delayed reforms for moving the economy away from central planning. As a result, the transition to a free-market economy has been slow.

A major concern for Ukraine is making its farmland more productive. Agriculture is still

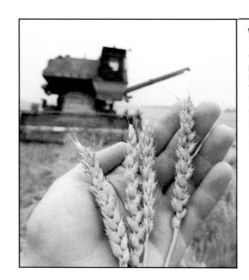

Wheat: From Field to Consumer

• The head of the wheat plant contains the wheat kernels, wrapped in husks. The kernel includes the bran or seed coat, the endosperm, and the germ from which new wheat plants grow.

Bran (seed coat)

Endosperm

Germ

Harvesting

Independent Practice

Provide students with Outline Map 16. Have them highlight the border of Ukraine and label the place names and physical features discussed in Section 1.

Closure

Have students write a descriptive paragraph that summarizes the physical and human geography of Ukraine.

ASSESSMENT

Assign the Section 1 Review questions on page 354.

Reteaching

Assign Section 1 of Reteaching Worksheet 30.

Extension

Have interested students do research on the city of Kyyiv or Odesa. Some may wish to make a time line focusing on historical events; others may wish to concentrate on the modern aspects, cultural attractions, or economic activities. Ask students to share their reports or time lines with the class.

organized in the inefficient state farms and collectives of the Soviet period. Ukraine's farm population also is in decline as younger Ukrainians move to the cities.

Human Geography In area and population, Ukraine is about the same size as France. Kyyiv serves as the country's political capital as well as its cultural and historical center. The city's churches, museums, and opera houses are stately symbols of Ukrainian life.

A large number of ethnic Russians live in Ukraine, mostly in the Donets region and along the Russian border. Ukrainians and Russians share similar cultures. As a result, many Russians call Ukrainians—to their displeasure—"little Russians" and talk of Ukraine as if it were part of Russia.

Ukrainians, however, believe there are important differences between themselves and Russians. Nationalism is on the rise among many Ukrainians, who have long cherished a separate identity. Some Ukrainians claim that they are descended from the independent, often warlike **Cossacks**. The Cossacks controlled the steppe region beginning in the sixteenth century and fiercely fought the southward expansion of the Russian Empire.

Major religions practiced in Ukraine include Eastern Orthodox Christianity and Roman Catholicism. Many Ukrainians also belong to the Uniate Church, whose members share the beliefs

Through centuries of Russian domination, **Ukraine** has held onto its traditions. Ukrainian folk musicians, such as these, sing ballads and historical songs.

of Eastern Orthodox Christianity but recognize the authority of the Roman Catholic Pope. This church arose in part because the lands of western Ukraine were once part of Poland.

Crimea Debate in Ukraine continues over the status of Crimea, a small peninsula that juts southward into the Black Sea. Although Crimea was part of the Ukrainian republic in the Soviet Union, it has never thought of itself as part of Ukraine. Instead, Crimea's ties have been to Russia. In fact, about

Geography and History

Some areas of western Ukraine were part of Poland when Adolf Hitler and Joseph Stalin divided Poland in 1939. Independence-minded partisans joined the Ukrainian Insurgent Army and fought against both Germany and the Soviet Union. Western Ukraine soon came under Soviet control, however. Today, Ukrainian survivors of World War II, many of whom were sent to labor camps in Siberia and were then exiled for fighting the Soviets, have lived to see their dreams of an independent Ukraine realized.

Stored in grain elevators

- Whole wheat flour contains all the parts of the kernel. White flour is produced by grinding only the endosperm. Vitamins are added to some white flour, then called enriched flour, to replace vitamins found in the bran and germ.

Cleaning

Water

Conditioning

Whole wheat grinding

Course grinding

Whole wheat flour

Grain parts are separated

Endosperm

Endosperm is ground

Whole wheat flour & bread

Bran (seed coat)

Germ

Cereal

Wheat germ

• People use wheat to make breads, pastas, and breakfast foods. Wheat by-products also are used in many other foods. Wheat often is used for livestock feed.

Add vitamins

Enriched white flour

OBJECTIVES

The objectives reflect the Focus questions on this page and are assessed in the review on page 355.

1. Identify the main physical and economic characteristics of Belarus.

2. Examine cultural characteristics of Belarus.

TEACHING SECTION 2
Motivator

Have a volunteer find Belarus (Byelorussia) on a wall map. Ask students to suggest how the country's location might have influenced its climate *(cool, humid continental)* and its cultures *(tied to both Russia and Poland).*

Guided Practice

Discuss notable points about the land in Belarus that lies between Moscow and Warsaw. *(best farmland; invasion route of Napoléon and Hitler)* Ask what challenges the Belorussians face in developing their economy. *(few mineral resources, poor soils, lost trade links)*

Section 1 Review
ANSWERS

1. Fertile soils, summer rains, and level plains allowing commercial farming using modern machinery; coal, iron ore, other minerals, and hydroelectricity

2. Many Ukrainians cherish an identity separate from Russia. Crimea is a peninsula that juts out from Ukraine but has many cultural ties to Russia.

3. Ukraine was the breadbasket for the Soviet Union, and Sevastopol, in Crimea, was the site of an important Soviet naval base. Legacies from the Soviet era include Communist leaders, central planning, and inefficient agriculture.

70 percent of the population is ethnic Russian. The Russian royal family built summer palaces in Crimea, and during the Soviet era, the area was a favorite vacation spot of Russian officials.

Vineyards and fruit orchards thrive in Crimea's Mediterranean climate, and the region's beaches are still popular resorts. Valuable iron-ore deposits in the area have led to the development of an iron-and-steel industry at the city of Kerch, and the port of Sevastopol' (suh-VAS-tuh-pohl) was the site of an important Soviet naval base. Many Russians would like to see Crimea become part of Russia.

SECTION REVIEW

1. Why is Ukraine agriculturally productive? What resources have helped industrialization in the region?
2. Why do many Ukrainians dislike being called "little Russians"? What is significant about Crimea?
3. **Critical Thinking** **Explain** how Ukraine was significant to the Soviet Union. What legacies are left in Ukraine from the Soviet era?

2 BELARUS

FOCUS

- *What are the physical and economic characteristics of Belarus?*
- *What are the cultural characteristics of Belarus?*

INTERACTION **Physical Geography** Belarus, formerly the Soviet republic of Byelorussia (bee-ehl-oh-RUSH-uh), is a region of plains located north of Ukraine. (See the map on page 351.) The country has a cool, humid-continental climate similar to that of neighboring Poland and Russia.

Although farming is important to Belarus, swamps and marshes cover much of the country. The best farmland and highest ground is found in an east-to-west band across the middle of the country. This hilly countryside, made up of sandy, glacial moraine, formed during the last ice age.

Shoppers in this market in **Minsk, Belarus,** look over melons imported from the Central Asian republic of Uzbekistan.

The railroad from Moscow to Warsaw, Poland, runs along this landform. It is part of the route used to invade Russia by Napoléon in 1812 and by Hitler in 1940.

Economic Geography With few mineral resources and generally poor soil, Belarus has relied on its traditions of education and technical training to build its economy. Enough of its forests remain to support some wood products industries. Peat, however, is still used as a fuel, even though its burning creates severe air pollution. The country's industries, many of which produce consumer goods, are concentrated in Minsk, the capital. Minsk is also an important transportation center. During the Soviet era, most of Belarus's products went to other Soviet republics. Replacing these lost trade links is proving difficult.

Human Geography Throughout history, the lands of Belarus have been culturally connected to Poland as well as to Russia. The people of Belarus, often called White Russians (*byelo* means "white" in Russian), speak a Slavic language that is similar to both Polish and Russian. And the Belarus people practice both Eastern Orthodox and Roman Catholic religions. As in other regions of the former Soviet Union, the once-large Jewish population of Belarus has dwindled to very few, in part due to recent emigration.

Before 1991, Belarus had never been an independent country. In 1995, Belarussians voted to strengthen economic ties with Russia. Belarus has

354

Closure

Contrast the physical, economic, and human geography of Belarus with that of Ukraine.

ASSESSMENT

Assign the Section 2 Review questions on this page.

Reteaching

Assign Section 2 of Reteaching Worksheet 30.

The objectives reflect the Focus questions on this page and are assessed in the review on page 356.

1. Describe the physical features of the Caucasus.
2. Analyze the conflicts occurring within and among Georgia, Armenia, and Azerbaijan.

TEACHING SECTION 3
Motivator

Tell students that in 1988, Armenia experienced an earthquake that killed some 25,000 people. Then mention that a 1989 earthquake of similar magnitude in San Francisco killed only 63 people. Have students suggest reasons for the difference. Tell them that in Section 3 they will learn about both physical and cultural turbulence in the Caucasus.

---

not moved forward with important economic reforms, and many former communists remain in power in the country.

SECTION REVIEW

1. Why is agriculture limited in much of Belarus?
2. How has Belarus's location influenced its cultural geography?
3. **Critical Thinking** **Suggest** a reason that Belarus has not broken with many traditions from the Soviet era.

THE CAUCASUS

FOCUS

- *What are the physical features of the Caucasus?*

- *What conflicts are taking place within and among Georgia, Armenia, and Azerbaijan? What are the sources of these conflicts?*

REGION **Physical and Human Geography**
The Caucasus region is the broad isthmus of mountains and isolated valleys that separate the Black Sea and the Caspian Sea. (See the map on page 351.) The Caucasus can be divided into three landform regions. In the north along the border with Russia is a wide mountain range. The region's highest peak is Mount El'brus at 18,510 feet (5,642 m)—the highest elevation in Europe. Lowland regions, meanwhile, lie along the Black and Caspian seas. In the south of the Caucasus is a rugged, volcanic plateau with mountains. Both earthquakes and volcanic eruptions threaten this tectonically active region.

Ethnically, the Caucasus is one of the most complex regions in the world. Dozens of different languages are spoken. The Caucasus is also rich in mineral resources, including manganese, lead, zinc, and copper.

Since ancient times, the armies of invading powers swirled around the Caucasus. The region's people often retreated into the mountain valleys to defend themselves against the invaders. As a

result, the region has traditions of fierce independence and ethnic pride.

Today, the Caucasus is organized into three independent countries: Georgia, Armenia, and Azerbaijan. Future divisions based on ethnicity are possible, however. Georgia recognizes some additional ethnic subregions. Nonetheless, ethnic violence has spread throughout the Caucasus. Permanent peace in the region may take years to achieve.

Georgia Georgia stretches inland from the lowlands along the eastern side of the Black Sea. The mild climate of the Georgia coast, which is much like that of the Carolinas in the United States, has made it home to several busy resorts. Citrus fruits and tea are grown in the region.

The Georgians have long been famous for their enterprise and business skills. Vegetables from Georgia still appear for sale in Moscow's street markets. An independent Georgia must extend its trading links, however. The country has little oil or other fossil fuels, and its hydroelectric power does not meet its needs.

Georgia's capital of T'bilisi, located at the crucial pass between the Black and Caspian seas, has been the center of Georgian culture for nearly 2,000 years. During the Soviet era, industrialization around T'bilisi and other cities in Georgia expanded, resulting in serious air pollution. Georgia's most immediate concern since independence, however, is violent unrest among its ethnic groups. In some areas, Islamic groups have longstanding ethnic quarrels with the Georgians, who are traditionally Eastern Orthodox. Russian troops have tried to intervene in the fighting, which has strained Russia's relations with Georgia.

Armenia Landlocked Armenia lies in the volcanic plateaus and mountains of the southern Caucasus, where the average elevation is more than a mile (1.6 km) above sea level. Severe earthquakes have rocked the country.

Compared to the rest of the Caucasus, the people of Armenia are unusually **homogeneous** (hoe-muh-JEE-nee-uhs), or of the same kind. More than 90 percent of the population is ethnically Armenian, and most people are Christian as well. In fact, Armenians claim to be the first nationality that adopted Christianity. The Islamic countries bordering Armenia—Turkey, Azerbaijan, and Iran—historically are its enemies.

Section 2 Review
ANSWERS

1. Much of the country is covered with swamps and marshes.
2. Because Belarus is situated between Russia and Poland, the people speak a language that is similar to both Polish and Russian, and they practice both Eastern Orthodox and Roman Catholic religions.
3. Possible answer: the country had never been independent before 1991, so it may have lacked historical patterns to create its own political and economic system.

355

Making a Comparison Chart

Countries of the Caucasus Write Georgia, Armenia, and Azerbaijan across the top of the board. Down the side, write Crops, Minerals, and Industries. Have students copy these headings on a sheet of paper and fill in information under each country. Then have them work in pairs to add new categories to their charts—for example, Religion or Reasons for Conflicts.

Closure

List on the board the obstacles the Caucasus countries must overcome. *(ethnic tensions, pollution, regional rivalries)* Have students suggest possible solutions to the obstacles.

ASSESSMENT

Assign the Section 3 Review questions on this page.

Reteaching

Assign Section 3 of Reteaching Worksheet 30.

Extension

Have interested students research one of the conflicts mentioned in this section. Ask them to trace the origins of the conflict and to draw conclusions about the effects it has on life in the region today. Encourage students to illustrate their written reports with maps.

COOPERATIVE LEARNING

Organize the class into groups of four. Tell groups that they are to present a 5- to 10-minute panel discussion on Caucasus regional issues. Within each group, one student should represent each country, and the fourth student should act as moderator. After the groups have had time to prepare, ask them to present their panel discussions to the rest of the class.

Section 3 Review
ANSWERS

1. Mountains, valleys, lowlands, and a volcanic plateau
2. Georgia: unrest among and quarrels between ethnic groups; Armenia and Azerbaijan have fought each other over the Nagorno-Karabakh region, which is located in Azerbaijan but whose population is Armenian.
3. Armenia is landlocked between countries that are its historic enemies, and it is situated in a region subject to severe earthquakes.

356

This Christian monastery is nestled in the Caucasus Mountains of **Armenia**. Nearly all the residents of the country are Armenian Orthodox Christians.

Armenia's roots go back to ancient times. Written Armenian literature can be traced to the fifth century, and Armenian states once existed in what is now eastern Turkey. Armenian communities could be found throughout Southwest Asia as well. As the Ottoman Empire began to collapse, however, a civil war raged in Turkey, and about 1.5 million Armenians were killed. Some fled to the United States and other Western countries, but most retreated to their ancient homeland in the south Caucasus. Here they found protection in the windswept, grassy countryside and the ruins of ancient Armenian churches.

Armenia industrialized during the Soviet era. The heavy industries of the capital of Yerevan produced some of the Soviet Union's worst air pollution. Today, Armenia is developing its economy and democratic government. Traditionally, Armenians have valued education and community life.

The most pressing issue in Armenia is the status of Nagorno-Karabakh, a region that was part of ancient Armenia. This rural territory, which is populated by Armenians, now lies entirely within the neighboring country of Azerbaijan. Armenia and Azerbaijan have fought over this region. Until the conflict is resolved, Armenia's economy is unlikely to move forward much.

Azerbaijan Azerbaijan stretches west from the southern end of the Caspian Sea. Because much of the country is in the rain shadow of the mountains of Georgia and Armenia, steppe and semiarid environments are typical. The land is mostly low mountains, foothills, and plains. About three-fourths of the Azerbaijan people are Azeri. They speak a Turkic language and first arrived in the region during the eleventh century. Most are farmers and herders.

Baku, Azerbaijan's capital city, is the center of a large oil-refining industry. Development here began during the late 1800s, and the oil reserves should last well into the future. Indeed, new deposits under the shallow Caspian Sea will provide income for many years. In the past, Baku's oil products were shipped north into Russia. Today, they are also exported to world markets.

Azerbaijan has strong ties to parts of Iran to the south. In fact, more than twice as many Azeri people live in Iran as in Azerbaijan. In addition, the Azeri are rediscovering Islam. Many Azeri students are studying Islam in Iran and other Islamic countries.

SECTION REVIEW

1. What landforms make up the Caucasus?
2. What is the main source of the violence in Georgia? Why have Armenia and Azerbaijan fought each other?
3. **Critical Thinking Explain** how Armenia's location has presented problems for the country.

CHAPTER 30 CHECK

ANSWERS

Building a Vocabulary
1. The Cossacks
2. Of the same kind; Armenia

Recalling and Reviewing
1. The agricultural and mineral resources of Belarus are not as rich as those of Ukraine.
2. There are many different ethnic groups in the region. Redividing or subdividing the countries might help reduce ethnic conflict.
3. Christian: Ukraine, Belarus, Georgia, Armenia; Islamic: Georgia, Azerbaijan

---

CHAPTER 30 CHECK

Leaving flowers and Easter eggs at the grave of a loved one, a tradition in **Georgia**

Reviewing the Main Ideas

1. Ukraine has rich agricultural and mineral resources. Although industry is well developed in its cities, economic progress has been slow since the country became independent. Nationalism is increasing among Ukrainians, but some parts of Ukraine, such as Crimea, still identify with Russia.
2. Belarus, north of Ukraine, has poor soils and few mineral resources. The people of Belarus have been culturally connected to Poland and to Russia. Now independent for the first time, Belarus has made little progress with economic and political reform.
3. The Caucasus is the broad isthmus of mountains and valleys that separate the Black and Caspian seas. Ethnically, the Caucasus is one of the most complex regions in the world. The Caucasus countries are Georgia, Armenia, and Azerbaijan.

Building a Vocabulary

1. From what group of people do some Ukrainians claim to be descended?
2. Define *homogeneous*. Which Caucasus country has the most ethnically homogeneous population?

Recalling and Reviewing

1. How do Belarus's resources differ from Ukraine's?
2. Why might the countries of the Caucasus be redivided or subdivided in the future?
3. Which of the countries discussed in this chapter have large Christian populations? Which have large Islamic populations?

Thinking Critically

1. The word *ukraine* comes from a Slavic word meaning "borderland." How might the word also be applied to other countries discussed in this chapter?
2. The United States has many immigrants from the former western Soviet Union republics, including Belarus and Armenia. Why do you think these people immigrated to the United States?

Learning About Your Local Geography

Individual Project
Investigate the dam nearest to your community, and write a paragraph about its origins and uses. When and why was it built? How big is the dam? What purposes and areas does it now serve? You may want to include your paragraph in your individual portfolio.

Using the Five Themes of Geography

Sketch a map of the countries discussed in this chapter. With symbols or shading, show how the **location** of each country has influenced its human **place** characteristics. Consider each country's location relative to other countries and regions. What effects does location have on the religions practiced in these countries? on the languages spoken by their peoples? What other cultural characteristics of these countries have been determined or influenced by location? Do any political issues in these countries concern their locations?

Thinking Critically

1. Belarus, like Ukraine, borders both Europe and Russia, and Armenia and Azerbaijan border countries of Southwest Asia as well as each other and other former republics of the Soviet Union.
2. Possible answers include a desire to practice religion freely, to escape ethnic persecution, or to seek greater economic opportunity.

Learning About Your Local Geography
Paragraphs should discuss the origins and uses of a local dam.

Using the Five Themes of Geography
With symbols or shading, maps might indicate that Ukraine's human geography has been influenced by the country's location next to Russia; that of Belarus by its location between Russia and Poland; and that of the Caucasus countries by their location between Russia and Southwest Asia.

357

CHAPTER 31 Central Asia
Planning Guide

TEXTBOOK	TEACHING RESOURCES	OTHER AVAILABLE MATERIALS
Section		
1. Physical Geography Pages 359–360	Reteaching Worksheet 31.1	*Creative Strategies for Teaching World Geography* LESSON 13 Understanding Climates Through Climographs *Study Guide for Spanish Speakers* *Lecture Notes,* Chapter 31
2. Human Geography Pages 360–361	Reteaching Worksheet 31.2 Challenge/Enrichment Worksheet 31	*Study Guide for Spanish Speakers* *Lecture Notes,* Chapter 31
3. Countries of Central Asia Pages 361–364	Reteaching Worksheet 31.3 Graphic Organizer 31 Critical Thinking Worksheet 31 Skills Worksheet 31	*Study Guide for Spanish Speakers* *Lecture Notes,* Chapter 31
Chapter Review Page 365	Chapter 31 Test Chapter 31 Portfolio Assessment See also *Alternative Assessment Forms*	*Test Generator,* Chapter 31
Unit Review Page 366–367	Unit 6 Test Unit 6 Portfolio Assessment See also *Alternative Assessment Forms*	*Test Generator,* Unit 6

CHAPTER RESOURCES

Lecture Notes, Chapter 31
Teaching Resources
    Chapter 31 Graphic Organizer
    Chapter 31 Skills Worksheet
    Chapter 31 Critical Thinking Worksheet
    Chapter 31 Challenge/Enrichment Worksheet

REVIEW AND ASSESSMENT

Chapter 31 Check, page 365
Chapter 31 Test, *Teaching Resources*
Chapter 31 Portfolio Assessment, *Teaching Resources*
Test Generator, Chapter 31
Alternative Assessment Forms

ADDITIONAL RESOURCES

Books for Teachers

Banuazizi, Ali, and Myron Weiner, eds. *The New Geopolitics of Central Asia and Its Borderlands.* Bloomington: Indiana University Press, 1994.

Diuk, Nadia. *The Hidden Nations: The People Challenge the Soviet Union.* New York: Morrow, 1990.

Edwards-Jones, Imogen. *The Taming of Eagles: Exploring the New Russia.* London: Weidenfeld & Nicolson, 1992.

Richards, Susan. *Epics of Everyday Life: Encounters in a Changing Russia.* New York: Viking, 1991.

Books for Students

Franck, Irene M., and David M. Brownstone. *The Silk Road: A History.* New York: Facts on File, 1986.

Milner-Gulland, Robin. *Cultural Atlas of Russia and the Soviet Union.* New York: Facts on File, 1990.

The World Today *Videodisc Program*

A variety of still images, short videos, and activities are available for you to use as you teach this chapter. *Lecture Notes* include bar codes for specific images and videos.

Multimedia Resources

A Day in the Life of the Soviet Union. (16mm film/VHS, 16 minutes) AIMS Media, 9710 DeSoto Avenue, Chatsworth, CA 91311–4409.

Picture Atlas of the World. (interactive maps, photo bank, animation, audio on CD-ROM) National Geographic Society, Washington, DC.

MEETING INDIVIDUAL NEEDS

Limited English Proficiency (LEP)

Pair LEP students with students proficient in English. Have each pair create a graphic organizer focusing on one of these main concepts relating to Central Asia: Water, Agriculture, People. Students should use information in the chapter for terms and ideas to complete their organizer.

Learners Having Difficulty

Organize the class into three groups, and assign each group a section of Chapter 31. Direct the groups to scan the text and to write questions about the main ideas in the section, providing the answers as well. The groups should then pass their questions and answers on to another group. Have one member of each group ask the other members the questions they received, checking their responses. The groups may then pass the questions and answers again, so that each group will have received the questions of the other two groups.

Auditory Learners

Obtain a copy of *The Travels of Marco Polo,* edited by Manuel Komroff. Read students some of Polo's descriptions of Central Asia.

Visual Learners

To help students understand the influence the former Soviet Union had on Central Asia, have them create a chart of the five countries with the headings Before, During, and After. Organize the class into five groups, assigning a country to each group. Students should provide information about the economy of their assigned country before, during, and after the Soviet era. Then assemble the portions together into one chart titled "Central Asia: Before, During, and After the Soviet Union." Have students draw conclusions about the region's economies over time from the information in the chart.

Tactile/Kinesthetic Learners

Have students use the information on pages 359–360 in their text to create an illustration about irrigation systems in Central Asia.

Gifted Learners

Have students investigate and report on the status of the space programs in the Baikonur Cosmodrome and at the John F. Kennedy Space Center in Cape Canaveral, Florida. Students might obtain information from articles in newspapers, magazines or scientific journals or by writing to the centers.

MULTICULTURAL FOCUS

Hundreds of years before the Russian Empire spread into the region, the heart of Central Asia was part of an area known as Turkistan, named for the groups of Turkic peoples who inhabited it. During the Middle Ages, Turkistan was a cultural crossroads, traversed by traders who brought with them a variety of political, social, and religious ideas and customs. Several cities along the Silk Road became important centers of Islamic culture and medieval thought, learning, and art.

Following the Communist Revolution, the Bolsheviks tried to suppress emerging pan-Turkic and pan-Islamic movements by dividing Central Asia along ethnic lines into autonomous regions (the present-day republics). As in other parts of the Soviet Union, religious practice was severely restricted. Many mosques and Islamic colleges were destroyed or forcibly closed.

Newfound self-determination and the strong resurgence of Islam following the collapse of the Soviet Union raise many questions about the future. Elicit a class discussion about the conflicts that Central Asians face as they rebuild their societies. Will the countries follow the model of Turkey, a secular state with connections to both Europe and Asia, or that of Iran, a more inward-looking Islamic state?

One vision that has begun to re-emerge is that of uniting the Turkic republics into a new Turkistan. Ask students to hypothesize about what cultural characteristics this nation would have and what challenges it might face.

OVERVIEW

In Chapter 31, students will learn about the arid region of Central Asia, which includes the newly independent countries of Kazakhstan, Kyrgyzstan, Uzbekistan, Turkmenistan, and Tajikistan.

Section 1 describes the landforms and climate of the region and focuses on the issue of water resources.

Section 2 identifies the people of Central Asia and discusses their cultural background, economic history, and economic future.

Section 3 explores the political and economic challenges each country is facing since independence.

About the Author

Marco Polo (1254–1324) of Venice was only 17 when he embarked on an extensive trading journey throughout Asia with his father and uncle. His widely read account of their travels was the alleged inspiration for many explorers, particularly Christopher Columbus. The quotation is from *The Travels of Marco Polo.*

Primary Source

Description of Change: excerpted and bracketed
Rationale: excerpted to focus on main idea; bracketed to clarify meaning

CHAPTER 31

Central Asia

❝ **F**or twelve days the course is along this elevated plain, which is named Pamer [Pamir Plateau] . . . . So great is the height of the mountains, that no birds are to be seen near their summits; and however extraordinary it may be thought, it was affirmed, that from the keenness of the air, fires when lighted do not give the same heat as in lower situations, nor produce the same effect in cooking food. ❞

Marco Polo

Handwoven textile from **Turkmenistan**

The dry lands of Central Asia lie east of the Caspian Sea and north of the world's highest mountains. The nearest ocean to Central Asia is the Indian Ocean, but the towering mountains block the humid ocean air blowing inland that might bring needed rains.

Throughout its history, Central Asia often has been a buffer zone for neighboring empires. In ancient times, the region marked the eastern edge of Greek civilization under Alexander the Great. A few centuries later, it became the western outpost of China's empire. In the nineteenth century, the British colonial empire in India and the expanding Russian Empire fought for control of the area. The Soviets took control of the region after the Communist Revolution.

With the collapse of the Soviet Union, Central Asia's five former Soviet republics became independent countries. These are Kazakhstan (kuh-ZAK-stan), Kyrgyzstan (kir-JEEZ-stan), Uzbekistan (uz-BEHK-is-tan), Turkmenistan (turk-MUHN-is-tan), and Tajikistan (tah-JIK-is-tan).

GEOGRAPHY DICTIONARY

oasis

mosque

dryland agriculture

clan

Spring tulips in the Pamirs of **Tajikistan**

CENTRAL ASIA: PHYSICAL–POLITICAL

Size comparison of Central Asia to the contiguous United States

LEGEND

	ELEVATION	
	FEET	METERS
⊛ National capitals	13,120	4,000
	6,560	2,000
• Other cities	1,640	500
	656	200
	(Sea level) 0	0 (Sea level)
	Below sea level	Below sea level

SCALE
0   250   500 Miles
0   250   500 Kilometers

Projection: Two-Point Equidistant

PHYSICAL GEOGRAPHY

FOCUS

- *What are the landforms and climates of Central Asia?*
- *What effects have various irrigation methods had on the region?*

REGION **Landforms and Climate** Plains and low plateaus cover much of Central Asia. To the south and east, the spectacular mountain peaks of the Kopet Mountains, the Pamirs, and the Tian Shan rim the region. Summer rains and winter snows in the mountains provide water for irrigation far out on the plains.

Central Asia is a region of mainly steppe, desert, and highland climates. Its latitudes compare to those of Nevada and Idaho, and hot summers and cold winters are typical. Climates become drier from north to south, as reflected in the region's vegetation. North of the Aral Sea, dry steppe vegetation is usual, whereas south of the sea, desert shrubs and shifting sands are common. Grasslands reappear on the lower mountain slopes and give way to forests at higher elevations.

INTERACTION **Water Resources** In dry regions such as Central Asia, water resources must be managed carefully. The main sources of water in the region are the narrow streams draining north from the mountains. These streams have been used for irrigation for thousands of years. The Syr Darya (sir DAHR-yuh) and the Amu Darya (ahm-oo DAHR-yuh) drain the snowcapped Pamirs and carry their water all the way to the deserts near the Aral Sea. Irrigation projects and natural oases are nearly continuous along both channels. An

Physical Geography   Have students make a flowchart showing links between Central Asia's climates, landforms, and water resources. (Example: the desert lands near the Aral and Caspian seas are irrigated by the Kara Kum Canal which carries water from the Amu Darya [which originated in the Pamirs]. The Aral Sea has shrunk as a result of so much irrigation.)

Closure

Have students sketch a map showing Central Asia's mountain ranges, climates, and bodies of water.

ASSESSMENT

Assign the Section 1 Review questions on this page.

Reteaching

Assign Section 1 of Reteaching Worksheet 31.

SECTION 2
OBJECTIVES

The objectives reflect the Focus questions on this page and are assessed in the review on page 361.

1. Examine Central Asia's cultural and economic background.
2. Identify challenges facing the region.

Section 1 Review
ANSWERS

1. Plains, plateaus, mountains; steppe, desert, highland
2. The sea has lost much water over the past few decades due to wasteful irrigation methods. Its level has dropped, it has shrunk in area, and its deltas no longer host wildlife.
3. To restore the sea, all irrigation would have to stop for several decades, which might hurt the region's economy. Stabilizing the level of the sea would allow some irrigation.

MULTICULTURAL PERSPECTIVES

A Turkmenistan legend claims that when God created the world, the Turkmens were first in line for sunny land, but last in line for water. Have students make up similar legends that describe their own region.

oasis is a site in the desert where water can be found in springs or wells or under dry riverbeds.

The Soviet government, seeking increased farm production, invested heavily in water developments in the region. The more than 500-mile-long (805-km) Kara Kum Canal, one of the world's largest and longest canals, carries water from the Amu Darya to the deserts near the Caspian Sea. Unfortunately, poor irrigation methods and over-watering wasted large amounts of water. In addition, the Soviets directed farmers to plant cotton, which uses a great deal of water. As the irrigation water evaporated, salts built up in the fields, decreasing crop growth.

The Aral Sea in particular has suffered from these policies. Once the world's fourth largest inland body of water, the Aral Sea has lost more than 60 percent of its water since 1960. Its level has dropped by 50 feet (15 m) and continues to fall. Aral'sk, once the sea's major fishing port, was by 1993 more than 60 miles (97 km) from the shore. The deltas of the Amu Darya and Syr Darya, once rich in wildlife, are now dry and desolate. Windstorms sweep the dry seafloor, blowing salty dust hundreds of miles downwind.

To restore the waters of the Aral Sea, all irrigation in the region would have to cease for several decades. Such a regional solution is difficult to achieve. Indeed, the countries of the region have not been able to agree on reducing their use of water headed toward the sea in order to stabilize it, let alone restore it.

A cemetery of rusted ships lies on the dried bottom of what once was part of the **Aral Sea** in **Central Asia**.

SECTION REVIEW

1. What landforms and climates are found in Central Asia?
2. What has happened to the Aral Sea? Why did it happen?
3. **Critical Thinking** **Suggest** why the Central Asian countries might decide to try to stabilize the Aral Sea's level rather than restore it to earlier levels.

HUMAN GEOGRAPHY

FOCUS

● *What is the cultural background of Central Asia? the economic background?*

● *What challenges does the region face?*

Most of Central Asia's major groups speak Turkic languages. One exception is the Tajiks, whose language is related to Iranian. As for religion, Islam has been the dominant faith for nearly 1,000 years and is once again central to people's lives. In 1989, the region had only 160 functioning **mosques**; by 1993, it had more than 10,000. A mosque is an Islamic place of worship. Each of the region's largest ethnic groups has its own country now; however, there are many minority ethnic groups.

INTERACTION **Economic Geography** For thousands of years, life in Central Asia was sustained by fields of grains and vegetables, orchards of apricots and apples, and herds of sheep, goats, and horses. Some communities practiced subsistence farming using simple irrigation methods; others were involved in herding.

Many of the cities of southern Central Asia began as trading centers along the Silk Road, an overland trade route from Southwest Asia to China. (See "Global Connections" on pages 362–363.) Caravans along the route stopped to rest and trade in towns in the fertile, irrigated valleys at the base of the mountains. These towns developed into great cultural and economic centers.

During the Soviet era, government planners attempted to modernize the region's economy.

Cultures, Economy, and Issues
Have students make a chart with three columns. In the first column, have them fill in Central Asia's cultural characteristics. The second column should be filled in with details about the region's economic background. In the third column, have students list challenges facing the region.

Closure

Ask what the economy of Central Asia might be like today if it had not been part of the Soviet Union.

ASSESSMENT

Assign the Section 2 Review questions on this page.

Reteaching

Assign Section 2 of Reteaching Worksheet 31.

SECTION 3
OBJECTIVES

The objectives reflect the Focus questions on this page and are assessed in the review on page 364.

1. Explain how each country of Central Asia has reacted to independence and the Soviet Union's collapse.

2. Identify each country's economic and cultural characteristics.

This knife grinder practices his trade in **Bukhoro, Uzbekistan**. Traditional as well as modern ways of life are found throughout the republics of Central Asia, the least developed region of the former Soviet Union.

Herders were forced onto collectives and state farms, while massive irrigation projects attempted to expand the production of cotton. Cities grew dramatically, as wide avenues and rows of high-rise apartments surrounded the old mud-brick buildings and narrow streets. The government built factories to process nearby minerals.

Issues Central Asia is the least-developed region of the former Soviet Union. Incomes are low, and birthrates and illiteracy rates are high. Although Central Asians welcome their political independence from Moscow, the break has caused an economic decline in the region. Central Asia receives foreign aid from the world's developed countries as well as from oil-rich Islamic countries such as Libya and those in Southwest Asia.

Economic growth will depend on increased exports of cash crops and minerals. Development will take place only slowly, however. The combination of decreasing water resources, increasing populations, economic decline, and traditional ethnic conflicts will heighten and prolong conflict in the region.

SECTION REVIEW

1. How did people's economic activities change during the Soviet era?
2. What social and economic challenges does the region face?
3. **Critical Thinking** **Decide** whether Central Asia is a culturally homogeneous region. Explain your answer.

COUNTRIES OF CENTRAL ASIA

FOCUS

● *How has each Central Asian country reacted to independence and the collapse of the Soviet Union?*

● *What are each country's economic characteristics? cultural characteristics?*

Kazakhstan Kazakhstan is the second largest territory of the former Soviet Union, second only to the Russian Federation. The country is gradually dismantling the Soviet system by privatizing collectives and state farms and by selling the government's assets.

The country's major industrial region is found around Qaraghandy (formerly Karaganda) in the north. Here lies a huge coal field, with copper, lead, and zinc mines nearby as well. Kazakhstan also has major undeveloped oil deposits along the Caspian Sea.

During the Soviet era, huge areas in northwest Kazakhstan were plowed for **dryland agriculture**, which depends on rainfall rather than irrigation to water crops. The Soviets' goal was to use modern methods and machinery to increase grain production on huge state farms. But dust storms and frequent periods of drought hindered this goal.

Ethnic Kazakhs make up only about 40 percent of the population. Russians make up the country's second largest ethnic group and are particularly numerous in the districts along the Russian border, in the Qaraghandy region, and in the capital city of Almaty (formerly Alma-Ata). The former Soviet Union's great space port, the Baikonur Cosmodrome, is near the lower Syr

Text continues on page 364.

1. Before the Soviet era, most people were subsistence farmers or herders. The Soviets forced people onto collectives and state farms and developed some industry.

2. Low incomes, high birth and illiteracy rates, economic decline, and traditional ethnic conflicts

3. Possible answer: although Islam and Turkic languages dominate, the region is not homogeneous—there are many minority ethnic groups.

GLOBAL PERSPECTIVES

Foreign companies are investing in Kazakhstan, a country with some of the largest unexplored oil, gas, and mineral deposits in the world. Many are interested in oil drilling, communications, pharmaceuticals, retail, and fast food. Ask students how Kazakhstan might manage foreign investment.

Global Connections

Relating Geography and History

Organize students into groups of four or five. Have each group research Polo's journey to China. Students should imagine that they are part of Polo's expedition and have just returned to Europe. In a short account of their travels, have students name and describe the trading cities they visited along the route. They should also describe the physical features and landforms of the route. Each group should create a map that traces the route of their journey. Maps should also include the cities and landforms along the route. Have group members read their accounts to the class and point out each part of the route on their maps as they describe it.

GLOBAL CONNECTIONS

The Silk Road

For centuries, goods, ideas, technology, and religions traveled back and forth from China to the Mediterranean Sea over an ancient route that came to be called the Silk Road. The 5,000-mile (8,045-km) trade route linked China and the Western world.

The Silk Road's beginnings reach far into the past, to a time when groups of nomads roamed Central Asia. These nomads eventually began trading goods produced in Iran, Mesopotamia, China, and India. While carrying gold, precious stones, and jade between countries, the nomads also spread ideas and technology. They may have introduced metalworking and the wheel to eastern Asia.

The nomads were warriors as well as traders. Around 200 B.C., one fierce group, the Huns, moved from the Asian steppes to threaten China. In 138 B.C., China sent an expedition westward to form alliances with other

Many of Central Asia's cities began as trading centers where caravans traveling the Silk Road stopped to trade and rest in the region's fertile valleys.

nomads against the Huns. Soon, other groups traveled westward to gather allies and obtain horses for the Chinese army. The expeditions included diplomats and brought livestock, gold, and silk to trade. The Chinese finally defeated the Huns and established important trade links with their neighbors.

The Golden Age of the Silk Road

By about 100 B.C., Chinese silk had traveled as far west as the Mediterranean, and the Silk Road had entered its golden age. Trade flourished as great caravans traveled across Asia. Gold, ivory, amber, and linen flowed eastward to China on the backs of camels and in carts pulled by oxen or mules. Westward, the caravans carried pottery, iron weapons, and, of course, silk. The various rulers of the region controlled the caravans that traveled through their territories.

Along with traded goods came ideas and religions. In the first century, Buddhist missionaries from India introduced their religion to China. Buddhism later spread to Korea and Japan. In the seventh century, Christian missionaries carried their form of Christianity to eastern Asia.

Perils of the Silk Road

Travelers and traders on the Silk Road encountered natural geographic obstacles. From China, caravans wound westward around the Great Wall and then from oasis to oasis across the desert. From there, they continued through icy mountain passes and over a great plateau. Caravans then descended onto the plains of Mesopotamia, to arrive at trading cities near the Tigris and Euphrates rivers. The Silk Road usually ended in Antakya (Antioch), a Mediterranean seaport. From there ships carried goods to Europe. Other land routes branched off from the Silk Road, reaching ports along the Arabian Sea and the Persian Gulf.

These merchants show off beautiful silk fabrics manufactured in **Samarqand, Uzbekistan**. Silk is a popular natural fiber often used in the production of clothing, upholstery, and curtain materials.

362

Using Geography Themes

Have students list the Five Themes of Geography and choose the theme that best relates to the feature. *(movement)* Call on volunteers to explain which theme they chose and why. Have students write one or two sentences applying the geographic theme of movement to the Silk Road. Point out that not only material goods but also ideas, technology, and culture traits move along trading routes such as the Silk Road.

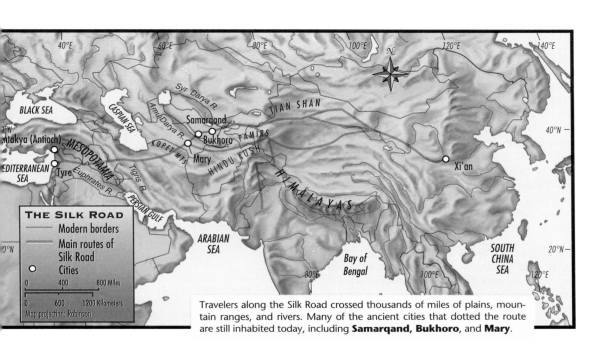

THE SILK ROAD
— Modern borders
— Main routes of Silk Road
○ Cities

0   400   800 Miles
0   600   1200 Kilometers
Map projection: Robinson

Travelers along the Silk Road crossed thousands of miles of plains, mountain ranges, and rivers. Many of the ancient cities that dotted the route are still inhabited today, including **Samarqand, Bukhoro,** and **Mary**.

Although few people traveled over the entire route of the Silk Road, each part was hazardous. Bandits often raided the caravans. Some travelers lost their way in the desert and died of thirst in the blistering heat. Others perished in sandstorms, and some were buried by avalanches, never to emerge from the mountains.

The Decline of the Silk Road

From the seventh to the mid-thirteenth century, religious and political turmoil often blocked the Silk Road. Muslims, followers of the growing religion Islam, fought with the peoples of Persia and China. From Chinese prisoners, Muslims learned paper-making techniques that had long been kept secret. They also successfully smuggled silkworms out of China. Once these techniques spread to Europe, the need for trade with Asia decreased. Besides, the collapse of the Roman Empire left western Europe too weak and too poor to trade much with eastern Asia.

Thus, by the time Marco Polo journeyed to China over the Silk Road in the late thirteenth century, the route had experienced a decline. Upon his return to Europe, Polo described great trading cities and resting areas that had deteriorated into rubble.

Although the road remained open through the middle of the fourteenth century, in 1368 it again fell victim to political forces. Determined to isolate China from the rest of the world, China's Ming Dynasty closed the eastern parts of the route. A centuries-old tradition of trade between China and the Western world came to an end.

The deep, rich appearance of dyed silk makes beautiful clothing. This woman sells fabrics made from silk and other materials in **Bukhoro, Uzbekistan.**

▼ YOUR TURN

Imagine you are a trader during the early days of the Silk Road. **Write** an entry for a daily journal. **Explain** what goods you are trading, and tell from where they came and to where you are taking them. Then **describe** the natural environment you encounter on your journey.

GLOBAL PERSPECTIVES

China's isolationism and the decline of the Silk Road hindered trade between China and the outside world for well over 200 years. But when tea drinking as a social custom emerged in Europe in the early seventeenth century, the demand for tea launched a renewed European interest in trade with China. Tea drinking had been part of social life in China since the fifth century. Before that time, tea had been used there as a medicine.

Your Turn

Journal entries should include appropriate details from the map on this page and the feature.

363

TEACHING SECTION 3

Motivator

Point out that, like the rest of the former Soviet Union, Central Asia is a region in transition. The nations each face inner turmoil but also have many common problems. Ask students to recall what they learned about the issues facing Eastern European and northern Eurasian countries since the Soviet breakup. Have students suggest what problems might confront the Central Asian countries.

Analyzing

Economy and Culture Have students work in pairs to create a chart of the five Central Asian countries. The charts should list area, population, resources, and economic activities. You may wish to provide resources for students to use to obtain additional information. Then have the pairs answer this question for each country: Which resources or economic activities have potential for the country's future development?

Closure

Have students write sentences about the extent to which each republic has changed since independence.

ASSESSMENT

Assign the Section 3 Review questions on this page.

Reteaching

Assign Section 3 of Reteaching Worksheet 31.

COOPERATIVE LEARNING

Organize the class into five groups, and assign each group a Central Asian country. Tell each group that it is in charge of publicity for a touring exhibition highlighting the culture of its country. Have each group create a poster that reflects its country's location, history, and culture. Specify different group roles, such as researcher, illustrator, writer, and presenter.

Chart Caption Answer

Uzbekistan

Section 3 Review

ANSWERS

1. Uzbekistan and Turkmenistan
2. Kazakhstan: Kazakhs, Russians; Kyrgyzstan: Kyrgyz, Russians, Uzbeks; Uzbekistan: Uzbeks, Tajiks, Kazakhs; Turkmenistan: Turkmens; Tajikistan: Tajiks, Uzbeks; mineral and oil deposits
3. The major space port used by the Soviet Union is in Kazakhstan. With the breakup, many astronauts may have feared losing access to the space port.

364

Darya. Russia's continuing space program requires cooperation with the Kazakh government.

Kyrgyzstan Kyrgyzstan is a landlocked country of the western Tian Shan. Traditionally, the Kyrgyz have lived as herders and farmers, but many Russians came to Kyrgyzstan to develop mines and hydroelectric projects. Many Uzbeks also live in this country.

Today, the Kyrgyz people barely form a majority in their own country. Newspapers and radio broadcasts are in both Kyrgyz and Russian. Kyrgyzstan's new government is working to restore traditional land management and further develop the country's mineral resources.

Uzbekistan Uzbekistan stretches east from the Aral Sea. Its landforms include the foothills of the Tian Shan as well as the lower Amu Darya, with many wide desert plains in between. Toshkent (formerly Tashkent), the capital, is one of the most industrialized cities in Central Asia. Many of its industries, however, were run by Russians, most of whom have left the region. Large numbers of Uzbeks, Tajiks, and Kazakhs remain.

Few changes have taken place in the country since independence, and the economy has declined steadily under the strict central government. The government hopes that expansion of the mining and oil industries can bring new income.

Turkmenistan During the Soviet era, Turkmenistan became a country of two desert waterways: the channel of the Amu Darya and the Kara Kum Canal, which passes the capital city of Ashgabat (formerly Ashkabad). Much of the rest of the country is desert with large areas of shifting sands. Turkmens make up a little more than 70 percent of the population.

Turkmenistan's Soviet system has changed little since independence. The government still owns nearly everything, and the collectives and state farms are still operating. Turkmenistan has little industry. Its economic future lies in the large oil deposits that have been discovered along the eastern shore of the Caspian Sea.

Tajikistan Within Tajikistan's borders lie the mineral-rich Pamirs and the great valleys of the upper Amu Darya system. More than half of the country is about 10,000 feet (3,048 m) above sea level. The mountain passes across Tajikistan form the major route connecting Central Asia to India.

FACTS IN BRIEF — Central Asia

COUNTRY POPULATION (1994)	LIFE EXPECTANCY (1994)	LITERACY RATE	PER CAPITA GDP (1993)
KAZAKHSTAN 17,267,554	63, male 73, female	97% (1992)	$3,510
KYRGYZSTAN 4,698,108	64, male 72, female	97% (1993)	$2,440
TAJIKISTAN 5,995,469	66, male 72, female	98% (1992)	$1,180
TURKMENISTAN 3,995,122	62, male 69, female	98% (1992)	$3,330
UZBEKISTAN 22,608,866	65, male 72, female	97% (1992)	$2,430
UNITED STATES 260,713,585	73, male 79, female	97% (1991)	$24,700

Sources: *The World Factbook 1994*, Central Intelligence Agency; *1995 Britannica Book of the Year*

As in other republics of the former Soviet Union, major economic changes since 1990 might make per capita GDP figures unreliable. Which is the most populous country in the region?

Traditionally, Tajiks are a mountain people who are divided into many **clans**. A clan is a group of people descended from a common ancestral family. Tajiks and Uzbeks make up most of the population. Tajiks' historic ties are to the south in Afghanistan.

The 10-year Soviet occupation of Afghanistan disrupted Tajik life. As the Soviet armies retreated, they left behind unsettled conditions and many weapons. Since the collapse of the Soviet Union, Tajikistan has known only turmoil. Traditional and modern Tajiks have been fighting each other, and the country's limited industry has nearly stopped.

3 SECTION REVIEW

1. In which Central Asian countries has there been little change since the Soviet Union's collapse?
2. What ethnic groups live in each Central Asian country? What resources hold the most economic potential for these countries?
3. **Critical Thinking** **Imagine** that you are a Russian astronaut. Why might you have had some reservations about the breakup of the Soviet Union?

CHAPTER 31 CHECK
ANSWERS
Building a Vocabulary

1. An oasis is a site in the desert where water can be found in springs or wells or under riverbeds. Oases are sources of water in dry regions.

2. An Islamic place of worship

3. Rainfall

4. Clan

Recalling and Reviewing

1. Prior to the Soviet era, irrigation methods were simple and controlled by communities. The Soviets designed massive irrigation projects that wasted much water.

2. It was an overland trade route from Southwest Asia to China. Because caravans stopped to rest and trade in Central Asia, some of its towns became major cultural and economic centers.

3. It has experienced much turmoil, with fighting among Tajiks and the stopping of industry.

Thinking Critically

1. Plans will vary. Although the assistance of other countries and the investments of multinational companies may be necessary to get development of oil deposits started, the Central Asian countries might want to limit such outside involvement so that most of the profits from the development remain in the region.

2. Possible answer: the Soviet Union had sufficient grain from other areas, such as Ukraine. No other area in the country, however, was more suited to growing cotton, and the Soviet Union wanted to be as self-sufficient as possible.

Learning About Your Local Geography

Groups' lists and sentences should indicate an understanding that local scientific activities can have national or international significance.

Using the Five Themes of Geography

Diagrams should show the shrinking of the Aral Sea due to poor irrigation methods in the region.

CHAPTER 31 CHECK

Mosque in **Kazakhstan**

Reviewing the Main Ideas

1. Plains and low plateaus cover much of Central Asia, and climates become drier from north to south. Poor irrigation methods, combined with a growing population, have threatened water resources and shrunk the Aral Sea.

2. Many different ethnic groups live in Central Asia. Most of the major groups speak Turkic languages, while Islam is the dominant religion. Traditional economic activities include herding and subsistence farming. Soviet planners tried to modernize the economy by introducing state-run farms and industry. Even so, Central Asia is the least-developed region of the former Soviet Union and faces many challenges.

3. The countries of Central Asia are Kazakhstan, Kyrgyzstan, Uzbekistan, Turkmenistan, and Tajikistan. With development, mineral resources and oil could provide income to these countries.

Building a Vocabulary

1. Define *oasis.* Why are oases important?

2. What is a mosque?

3. What is dryland agriculture's major source of water?

4. What term describes a group of people descended from a common ancestral family?

Recalling and Reviewing

1. How did irrigation methods in Central Asia change when the Soviets took control of the region?

2. What was the Silk Road? How did it affect Central Asia's towns?

3. What has happened in Tajikistan since it became independent?

Thinking Critically

1. Imagine you are an official in one of the Central Asian countries. Design a plan for the development of oil deposits in your country. Consider the extent to which you would invite the involvement of other countries and multinational oil companies.

2. Although cotton has been the major cash crop of much of Central Asia, the region's landforms and climates might be better suited to growing grains such as wheat. Why do you think the Soviets focused on cotton instead?

Learning About Your Local Geography

Cooperative Project

With your group, investigate a major scientific facility, one comparable to Kazakhstan's Baikonur Cosmodrome, in your area or state. Make a list of some of the projects being completed at the facility. Below the list, write a few sentences explaining the national or international significance of the projects.

Using the Five Themes of Geography

Sketch a series of diagrams of the Aral Sea shrinking over time to show the effects that **human-environment interaction** can have on a **region**. You may wish to draw cross-sectional diagrams that indicate the drop in the Aral Sea's level in addition to its decrease in area. Below your diagrams, write several sentences explaining what factors have been responsible for the Aral Sea's loss of water.

365

Unit 6 Review
SKILLS

Point out that a major economic challenge Russia faced after the collapse of the Soviet Union was inflation, evidenced by rapidly rising prices on most products. Ask students if they have ever witnessed a rise in prices on goods they bought or considered buying over a span of time. Have interested students monitor and graph the prices on several specific commodities (such as food, gasoline, clothing items, entertainment products) over one month. Afterwards, ask whether their graphs indicate a period of inflation or economic change in general.

Geography and Economics

At the beginning of 1992, one U.S. dollar could purchase 180 Russian rubles. By mid-1993, however, one U.S. dollar was worth 800 rubles.

Practicing the Skill
ANSWERS

Reports should discuss the forms of international economic support to the former Soviet Union. Possible forms of support include direct monetary aid, reduced prices on goods exported to the region, lowered trade barriers on goods imported from the region, and investment in the region's industry and infrastructure. Debates among experts usually deal with which forms of support are best. Countries giving support often must weigh the needs of their own people against the needs of the former Soviets.

366

UNIT 6 REVIEW ⊕ SKILLS
Linking Geography and Economics

As you know, Russia and several other former Soviet republics are now making the transition from a command economy to a free-market one. No handbooks exist to smoothly guide these new countries through the process. In addition, the region's unique geography and history make the transition a step into unknown territory.

Gaining a clear perspective on the current changes may take years, even decades. Already experts disagree on exactly what effects the move to a system of free enterprise is having on the region. Some geographers and economists point to the region's high inflation, rising crime rates, and monthly wages equivalent to less than 50 U.S. dollars. These experts claim that economic conditions in the region have in fact worsened since the Soviet Union's collapse.

For example, a textile factory in the Russian city of Ivanovo once produced one-quarter of the entire Soviet Union's fabric needs. Since the Soviet Union's collapse, however, the factory has had a shortage of raw cotton, as its links to the cotton-growing Central Asian republics have broken. Also, the Soviet government bought much of the factory's cloth. When the breakup took place, the factory lost its major customer. Of the 6,500 workers employed by the factory at one time, only 2,000 workers were needed by the beginning of 1993. At that time, each earned about $12 a month.

Yet other geographers and economists point out that while wages are low by Western standards, basic goods and services, such as heat and public transportation, are much cheaper in Russia than they are elsewhere. These experts also claim that while wages now may be lower than they were several years ago, there was little to buy before the breakup, so higher wages were meaningless. In addition, many Russians have private gardens in which they can raise some of their own foods. This reduces the amount of money they must spend on groceries. While admitting that the former Soviet people are experiencing economic hardships during the transition, many experts believe that in the end the free market will be of much greater benefit than harm.

There are already some people for whom the economic changes have been undeniably positive. One "success story" has to do with housing. During the Soviet era, most people lived in cramped apartments, and the government declared that private homes could be no larger than 20 feet by 20 feet (6 m by 6 m).

Today, in places such as Desna, a village not far from Moscow, more land is becoming available for private ownership. People are starting to build their own—larger—homes. Construction is often expensive and usually must start from scratch, including the building of access roads and the installation of plumbing and utilities. Yet many people believe that home ownership is well worth the costs and the troubles. And as more and more building takes place, it in turn sparks economic growth for many other people and businesses—suppliers of lumber and other building materials, equipment manufacturers, construction workers, real estate agencies, and so on.

Sorting out all of the different issues regarding the economic transition to a free-market economy is not easy. Yet one thing is clear. There are few world regions in which geography and economics are changing as much as, or are more closely linked than, they are in northern Eurasia.

▊ PRACTICING THE SKILL

The economic success of the former Soviet Union is linked not only to the region's own geography but also to the world's geography. Several countries, including the United States, want to show their support for the former Soviet Union's economic and political shifts.

Research the issue of international economic support to the former Soviet Union. What forms has the support taken? What are the major debates among experts regarding economic assistance to the region? Summarize your findings in a brief written report.

BUILDING YOUR PORTFOLIO

Decide whether students will work individually or in groups on the unit projects described below. Students should then choose which project they will work on. Provide students with copies of the Portfolio Assessment Worksheets in *Teaching Resources*. Chapters 28–31 worksheets and the Unit Portfolio Assessment Worksheet will guide students step-by-step to complete either project.

Teaching with Sketch Maps

Have students sketch a map of Russia and Northern Eurasia from memory. Tell them to include as many details as they can about place names, landforms, and boundaries of regions they studied in Unit 6. Students should compare this map with the sketch map they completed at the beginning of Unit 6 to note their improvement in identifying geographic features.

BUILDING YOUR PORTFOLIO

Individually or in a group, complete one of the following projects to show your understanding of the geography concepts involved.

A A World of Change

While the geographic changes we read about in magazines, newspapers, and textbooks may seem far away from our own lives, in reality these changes affect people all around the world. Using the breakup of the Soviet Union as an example, prepare a report on the consequences, both personal and global, and both far and near, of geographic change. To complete your report, consider the following questions.

1. How has the collapse of the Soviet Union affected the daily lives of people who live in the region? Imagine that you live in one of the countries of the former Soviet Union. Write, in the form of a journal entry, about a typical day in your life before the Soviet Union collapsed. Then write another journal entry about a typical day in your life now.
2. How did the collapse of the Soviet Union and the end of the Cold War affect people in other parts of the world? Interview some members of your community about how they personally have been affected by the changes in northern Eurasia. How have their views of the former Soviet Union changed? Did their economic livelihoods change as a result of military base closings or the decline of military industries? Write a summary of your findings from your interviews.
3. How did the collapse of the Soviet Union and the end of the Cold War have an impact on global political and economic relationships? Did political relationships between countries change? How was international trade affected? Present this information using charts or graphs.
4. What areas of the world are likely to experience geographic change within the near future? Select one of the areas, and write one page explaining why the area is likely to change, what forms the change will take, and what consequences the change will have.

Organize your materials and present your report to the rest of your class.

B Economic Systems: A Closer Look

The collapse of the Soviet Union and the transition of some of its former republics from communism to a free-market economy has raised many questions about the world's different economic systems—not only communism but also capitalism and socialism. You already know a little bit about these systems but now research them in greater depth. Make a poster-sized chart showing the information you collect on the systems. Your chart will be divided into three rows: one for capitalism, one for communism, and one for socialism. The chart will have four columns, as explained below.

1. The first column should be called "Origins." For each economic system, investigate how it arose. Where did each system originate? What groups or individuals played a major role in the origins and development of each system, or in the ideas behind it? What social and economic conditions set the stage for the formation of each system and its ideas?
2. The chart's second column should discuss the "Key Ideas" of the theory behind each economic system. On what assumptions is each system based? What does each system seek to do, and what are its methods? What other ideas are important in each system?
3. The third column, "Practice and Reality," should discuss examples of how the system has worked in practice. In which countries has each system been tried? How closely do the systems in practice resemble the systems in theory (as described in the chart's second column)? What challenges has each system faced? In places where any of the systems have failed, what were the reasons for the failure?
4. The fourth column should be called "Interaction with Politics." Although each system is technically an economic system, economics and politics are often closely related. What roles do governments play in each system? How does each system involve political processes?

Present your chart to the rest of your class.

Project A:

Projects should indicate that students have considered the effects of the collapse of the Soviet Union on people who live in the region (in the form of "before" and "after" journal entries), on students' local community members (in a written summary of interviews), and on global political and economic relationships (shown with charts or graphs). Students should also present written discussions of areas likely to experience geographic change in the near future.

Project B:

Charts should give information on the origins and key ideas of capitalism, communism, and socialism. Charts should also show how and where each system has been tried, the extent to which practice has resembled theory for each system, and how each system has involved governments and political processes.

*R*EGIONS IN *T*RANSITION

Facing the Past and Present

Patterns of trade and culture can change quickly in our modern world. For example, until just a few decades ago, the United States looked primarily to Europe for trade. Many innovations in culture and technology and most of America's immigrants came from Europe. Today, the American focus on Europe has faded. U.S. trade with Japan and other Pacific Rim nations now far exceeds U.S. trade with Europe. New ideas, technology, and immigrants to the United States now come from all around the globe.

Central Asia

With the collapse of the Soviet Union, similar changes are taking place in Central Asia. In the past, many ties linked Central Asia to the Soviet Union. For instance, the economies of the two regions were integrated. Central Asia exported cotton and oil to Russia and the countries of Eastern Europe. In exchange, Central Asia received a variety of manufactured goods from these areas. The Soviet Union also heavily influenced the culture of Central Asia, and many Central Asians learned to speak Russian.

Looking South

Today, Central Asia's links to the former Soviet Union have weakened. At the same time, its ancient ties to Southwest Asia have grown stronger. The Silk Road, and the travelers that came before it, formed valuable trade links between the oasis cities of Central Asia and the busy Southwest Asian ports on the Mediterranean. Now the peoples of Central Asia are looking southward once again. For example, links are forming between the region's new republics and Turkey. Many people in Central Asia are traditionally Turkic in culture and language. As a result, Turkey's business leaders are working to expand their industries in Central Asia. Regular air travel from Turkey to the capitals of Central Asia now is possible as well.

Religion provides another firm link between Central Asia and Southwest Asia. Islam first arrived in Central Asia in the eighth century, but its role diminished during the Soviet era. Missionaries from the Arab countries and Iran are now committed to reestablishing this connection. In fact, Iran is spending millions of dollars to build roads and rail lines to Central Asia.

Central Asia and Southwest Asia also share a similar climate, environment, and way of life. In both of these dry regions, water conservation and irrigation are basic to daily life. Many of both regions' people grow cotton and herd animals, though this way of life may soon change. Both Central Asia and Southwest Asia are experiencing an invasion of modern Western culture and are struggling to deal with its impact. Many people believe that the ideas and trends that arrive daily through compact discs, videotapes, and satellite television are a threat to their traditional beliefs and ways of life. Such shared fears of cultural loss are bringing Central Asia and Southwest Asia even closer together.

These children are learning about Islam in **Dushanbe, Tajikistan**. Although the former Communist government discouraged the practice of religion, today Islam flourishes in the independent Central Asian republics.

Language Groups of Southwest and Central Asia

LEGEND

Dominant languages:
- Turkic
- Iranic
- Semitic
- Greek
- Other
- Sparsely populated

MAP STUDY This map uses colors to show the major language groups that link ethnic peoples throughout Southwest and Central Asia. Turkic languages, such as Azerbaijani and Uzbek, and Iranic languages, such as Kurdish and Persian, are spoken by ethnic peoples throughout Central Asia and northern Southwest Asia. Jews and Arabs also are linked by language: both Hebrew and Arabic are Semitic languages. Very often, however, the links between ethnic groups are overshadowed by the differences in culture and history.

Defining the Region

For all these reasons, geographers are constantly reexamining this and other regions of the world. Will they decide to include the new countries of Central Asia in the region of Southwest Asia? Or will Russia regain control of Central Asia and once again push aside the region's ancient economic and cultural links to Southwest Asia? The geographers are watching and waiting.

In the tradition of many Turkic communities, these community elders are meeting to discuss local schools in **Tajikistan**. The Central Asian republics share many traditions with both northern Eurasia and Southwest Asia.

UNIT 7 SOUTHWEST ASIA

CHAPTERS 32–34

OVERVIEW

In Unit 7, students will learn about the geography, ancient history, and rapid modernization of a region often referred to as the Middle East.

Chapter 32 • INTRODUCING SOUTHWEST ASIA surveys the physical geography, economic geography, historical geography, and human geography of the region.

Chapter 33 • THE EASTERN MEDITERRANEAN describes the landforms, climates, economies, resources, and histories of Turkey, Cyprus, Israel, Syria, Lebanon, and Jordan.

Chapter 34 • INTERIOR SOUTHWEST ASIA features Iraq, Iran, Afghanistan, Saudi Arabia, Kuwait, the United Arab Emirates, Qatar, Bahrain, Oman, and Yemen. It focuses on the role that oil has played in changing this region's human geography.

OBJECTIVES

1. Describe the physical geography and economic geography of Southwest Asia.
2. Understand how climate and landforms affect the way people live in Southwest Asia.
3. Trace the history of and the development of religion in Southwest Asia.
4. Identify the political, social, and environmental challenges confronting the countries of Southwest Asia.
5. Understand the significance of the region's strategic location.
6. Interpret special-purpose maps to analyze relationships among climate, population patterns, and economic resources in Southwest Asia.
7. Analyze relationships between geography and history.

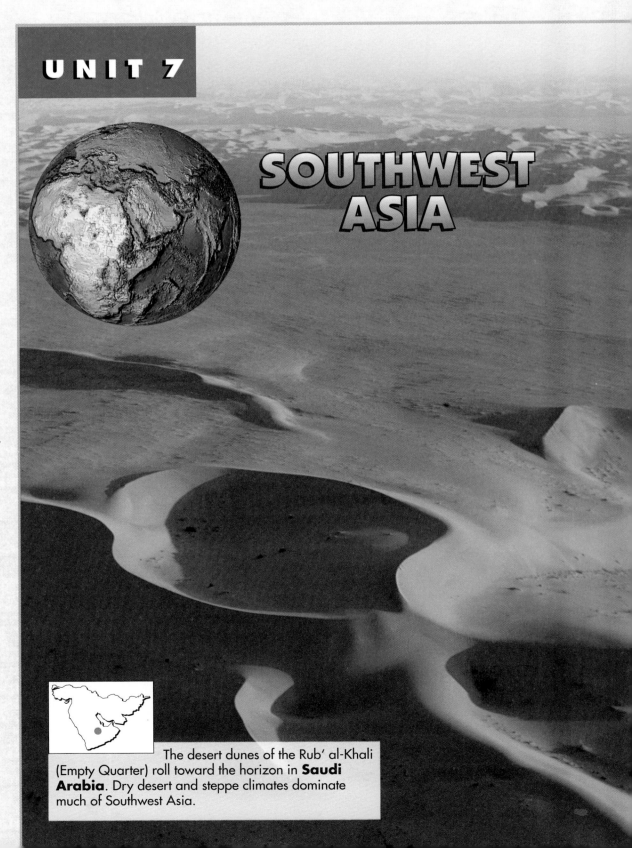

UNIT 7

SOUTHWEST ASIA

The desert dunes of the Rub' al-Khali (Empty Quarter) roll toward the horizon in **Saudi Arabia**. Dry desert and steppe climates dominate much of Southwest Asia.

Using the Illustrations

Tell students that they will learn in Unit 7 that two major influences on the history and culture of Southwest Asia have been climate and religion. Ask students to study the photograph of the desert in Saudi Arabia on the opposite page. Then refer them to the climate map on page 376 that shows the extent of the desert and steppe climates in Southwest Asia. Also refer students to the pho-tographs on this page, the photographs of the Jewish menorah on page 374, the mosque on page 381, the Dead Sea Scrolls on page 386, and the minaret on page 406. Point out how each photograph relates to Judaism or Islam.

Teaching with Sketch Maps

Have students draw a map of Southwest Asia from memory. Tell them to try to visualize this region in relationship to its location near Africa and Europe. Students might wish to add this map to their portfolios and to compare it with another they will draw after they have completed Unit 7.

UNIT 7 CHAPTERS

32 Introducing Southwest Asia

33 The Eastern Mediterranean

34 Interior Southwest Asia

The sun sets over **Izmir**, along the Aegean coast of **Turkey**. The towers of a mosque rise at left-center. Although Turkey has historical ties to Christianity, nearly all Turks today are Muslim.

Pages from the Qur'an (Koran)

Persian carpet

THE WORLD TODAY VIDEODISC PROGRAM

The videodisc profiles the following countries in Unit 7: Turkey, Israel, and Jordan.

More than a million Muslims make a pilgrimage *(hajj)* each year to **Mecca, Saudi Arabia**, Islam's holiest city. Here, thousands wait to enter the packed Great Mosque.

One World, Many Voices

LITERATURE

Tell students that the literature excerpt below depicts both the physical geography and the human geography of the Arabian Peninsula and the Arabian perimeter. Discuss how the story portrays the cultural life of the village as intertwined with the economic activity that supports the village. *(The villagers pray that the pearl hunt will be successful and that the divers will return safely. The whole village celebrates the return in a ceremony that includes dancing and singing.)*

More About Khalil al-Fuzay

Khalil al-Fuzay's short stories about life in Saudi Arabia have been published in several anthologies. The story from which the excerpt on this page was chosen appeared in *The Literature of Modern Arabia,* a collection of short fiction, drama, and poetry by contemporary writers from the Arabian Peninsula.

ONE WORLD, MANY VOICES

LITERATURE

Khalil al-Fuzay (1944–) is a journalist and short story writer from Saudi Arabia. The selection below is from a story entitled "Scattered Voices." It reflects aspects of Saudi Arabian culture and traditional ways of life in the coastal villages on the Persian Gulf. Decades before this country's economy was transformed by the discovery of vast oil reserves, the waters around southern Saudi Arabia, Kuwait, and the island of Bahrain were noted for their many pearl-producing oyster beds.

Pearls from offshore **Bahrain**

Primary Sources

Description of Change: excerpted and bracketed *Rationale:* excerpted to focus on main idea; bracketed to clarify meaning

Interpreting Literature

Answers

1. Pearl diving—harvesting pearls from oyster beds
2. References to the sounds of the Gulf waves and the village's location near a seashore

"The peace-loving village keeps a hold on hope at all times. When the men set off for the pearl-hunt, the women send them off with prayers; and they welcome them back with sky-piercing ululations [wailing]. The traditional diving-songs place no value on composure: at the return of his son, the old man dances; the mother sings at the return of hers, while the wife's happiness is such that it brings tears to her eyes. This happens when they come back with plenty of pearls, but also when they come back with nothing. Reunion was always a joyous occasion.

After sunset the men gather in a semi-circle. Their dark faces flicker in the light of the fire at which they warm their drums and tambourines. The communal singing begins. It is exquisite [beautiful] in both its joy and its sorrow. The dancers dance themselves into a frenzy. In the pauses between songs, you can hear the sounds of the Gulf waves as they enfold the golden sands of the shore."

Translated by Ahdaf Soueif and Thomas G. Ezzy

Interpreting Literature

1. How do the people in this story make their living?
2. What details in this selection indicate the village's location near the Persian Gulf?

372

FOR THE RECORD

You might begin by drawing students' attention to the background photograph on this page. Point out that during ancient and medieval times most of Lebanon was covered with dense forests. Although the cedars of Lebanon were an important economic resource in Southwest Asia for centuries, they were nevertheless a nonrenewable resource. Emphasize that the author describes the trees' value in terms of their relationship to the environment.

More About William O. Douglas

William O. Douglas, a well-known champion of civil liberties during his long career on the United States Supreme Court, was also an ardent defender of the environment. His love of wilderness areas led him to devote leisure time to numerous backpacking trips all over the world. Douglas made his home in Goose Prairie, Washington.

FOR THE RECORD

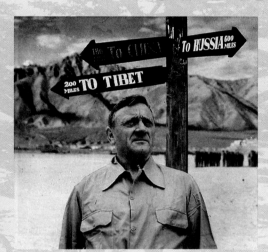

William O. Douglas (1898–1980) was a justice on the United States Supreme Court for more than 35 years. He was also a noted conservationist, outdoorsman, and world traveler. The passage below is from *Strange Lands and Friendly People*, a book about his travels throughout Southwest Asia. The beautiful cedars of Lebanon he writes about have been described by many writers since biblical times. Although most of Lebanon's forests were cut centuries ago, some of the cedars can still be found in the northern Lebanon Mountains.

Geography and History

Wood from the cedars of Lebanon has been prized over the centuries because of its durability, beauty, and pleasant fragrance. Early civilizations in Southwest Asia imported timber from the heavily forested areas of Lebanon for shipbuilding and for construction of temples and palaces.

"We had walked four days under a sweltering sun. There was no cover of any kind—for man, birds, or beasts. The Mediterranean sky was cruel. When at last I reached the cedars at Hadeth, . . . I lay on my back under the first cedar and thanked God for trees. I felt as if I had reached home, as if protective arms were around me. I had escaped an enemy that pursued me with a hot breath. I, a refugee from the sun who was burned to a frazzle, was at last safe. . . .

These cedars rise a hundred feet or more. While the trunks are usually four or five feet thick, like our tamarack [a North American tree], some are over thirty feet in diameter. These are the monarchs of the forest, perhaps a thousand years old.

The branches are long and swooping and cover the ground for fifty feet or more around. A forest of them is so closely woven with branches that it is difficult to get a picture of a single tree. At Bsharreh there are only about four hundred cedars left, packed together in a small ten-acre tract, the remnants of a mighty race."

Analyzing Primary Sources

1. What connection is shown in this selection between the region's climate and admiration for the cedars of Lebanon?
2. Why does the author believe these trees are important?

Analyzing Primary Sources

Answers

1. The author admires the beauty and size of the cedars and welcomes the relief they provide from the sun's rays in the hot climate.
2. He describes their significance as very old trees—"monarchs of the forest"—and points out that very few trees from the original cedar forests are still standing.

CHAPTER 32 Introducing Southwest Asia

Planning Guide

TEXTBOOK	TEACHING RESOURCES	OTHER AVAILABLE MATERIALS
Section		
1. Physical and Economic Geography Pages 378–379	Reteaching Worksheet 32.1 Skills Worksheet 32	*REGIONS OF THE WORLD Map Transparencies:* Southwest Asia *Creative Strategies for Teaching World Geography* LESSON 14 Contributions from the Countries of Southwest Asia and North Africa *Study Guide for Spanish Speakers* *Lecture Notes,* Chapter 32
2. Historical and Human Geography Pages 379–384	Reteaching Worksheet 32.2 Graphic Organizer 32 Challenge/Enrichment Worksheet 32 Critical Thinking Worksheet 32	*Study Guide for Spanish Speakers* *Lecture Notes,* Chapter 32
Chapter Review Page 385	Chapter 32 Test Chapter 32 Portfolio Assessment See also *Alternative Assessment Forms*	*Test Generator,* Chapter 32

CHAPTER RESOURCES

Lecture Notes, Chapter 32
Teaching Resources
    Chapter 32 Graphic Organizer
    Chapter 32 Skills Worksheet
    Chapter 32 Critical Thinking Worksheet
    Chapter 32 Challenge/Enrichment Worksheet

REVIEW AND ASSESSMENT

Chapter 32 Check, page 385
Chapter 32 Test, *Teaching Resources*
Chapter 32 Portfolio Assessment, *Teaching Resources*
Test Generator, Chapter 32
Alternative Assessment Forms

ADDITIONAL RESOURCES

Books for Teachers

Held, Colbert C. *Middle East Patterns: Places, People, and Politics.* Boulder, CO: Westview Press, 1994.

Issar, Arie S. *Water Shall Flow from the Rock: Hydrogeology and Climate in the Lands of the Bible.* New York: Routledge, 1990.

Melamid, Alexander. *Oil and the Economic Geography of the Middle East and North Africa.* Princeton, NJ: Darwin Press, 1991.

Schmandt-Besserat, Denise. *Before Writing: From Counting to Cuneiform.* Austin: University of Texas Press, 1992.

Books for Students

Leone, Bruno, series ed. *Islam: Opposing Viewpoints.* San Diego: Greenhaven Press, 1995.

Leone, Bruno, series ed. *The Middle East: Opposing Viewpoints.* San Diego: Greenhaven Press, 1992.

Marshall, Bruce, ed. *The Real World: Understanding Our World Through the New Geography.* Boston: Houghton Mifflin, 1991.

Messenger, Charles. *The Middle East.* New York: Franklin Watts, 1988.

The World Today *Videodisc Program*

A variety of still images, short videos, and activities are available for you to use as you teach this chapter. *Lecture Notes* include bar codes for specific images and videos.

Multimedia Resources

The Five Pillars of Islam. (VHS, 30 minutes) Films for the Humanities & Sciences, P.O. Box 2053, Princeton, NJ 08543–2053.

Holy Land: Judaism, Christianity, and Islam in the Middle East. (VHS, 25 minutes) Encyclopedia Britannica Educational Corporation, 310 South Michigan Avenue, Chicago, IL 60604–9839.

Islamic Art. (VHS, 30 minutes) Films for the Humanities & Sciences, P.O. Box 2053, Princeton, NJ 08543–2053.

Picture Atlas of the World. (interactive maps, photo bank, animation, audio on CD-ROM) National Geographic Society, Washington, DC.

MEETING INDIVIDUAL NEEDS

Limited English Proficiency (LEP)

Pair LEP students with students who are proficient in English. Have pairs of students make a set of flash cards with Geography Dictionary terms on one side of the cards and their definitions on the other side. Each pair should use the cards until both students have mastered the vocabulary terms.

Learners Having Difficulty

Provide students with a partially completed outline of Chapter 32 that includes the section titles and subheads within each section. Pair students to work together to complete the outlines.

Auditory Learners

Have interested students present an oral report about a major city of Southwest Asia, such as Istanbul, Beirut, Aleppo, Damascus, Tel Aviv, Tehran, Isfahan, Baghdad, Mecca, or Medina. Have students discuss the city's historical or modern significance.

Visual Learners

Have students copy the Focus questions at the beginning of each section of Chapter 32. Then tell them to leaf through Unit 7 to find a photograph or photographs that relate to each question. Help them get started by providing an example. (The Focus question for Section 1 about the physical geography of Southwest Asia could be paired with the photographs on pages 370, 401, 403, and 407.)

Tactile/Kinesthetic Learners

Have students work in small groups to create models, diagrams, or dioramas depicting the irrigation systems used in Southwest Asia. Groups should also write an extended caption that explains how the systems work. Display the models in the classroom.

Gifted Learners

Students might do research on the *Code of Hammurabi*, the set of laws written during the reign (1792–1750 B.C.) of the Babylonian king Hammurabi and one of the first legal codes in history. Then discuss how U.S. laws are similar to or different from these ancient laws. Students might present their findings to the rest of the class, explaining how the laws influenced the early civilizations of Southwest Asia.

MULTICULTURAL FOCUS

The unifying impact of Islam over many peoples and places in Southwest Asia is exemplified by Islamic art, a distinctive style that flourished across the Muslim realm. Shaped by Islamic beliefs and customs, the style blends the artistic traditions of many cultures, including Egyptian, Syrian, Persian, Roman, and Byzantine.

A primary characteristic of Islamic art is its decorative and abstract quality, often involving intricate geometric and floral designs. Although not expressly forbidden in the Qur'an, Islamic tradition generally rejects representations of religious figures and living beings in art. Strict Islamic theology taught that Allah was the sole creator of life; artists who depicted life realistically were considered to be infringing on Allah's work.

Islamic artists sought to beautify all aspects of daily life. Intricate designs in carvings, tile work, rugs, and textiles adorned mosques, shops, and homes. Remarkable effort and skill also went into decorating utilitarian objects such as those for cooking and eating. Ornamental calligraphy, also a hallmark of Islamic art, beautified manuscripts as well as art objects and walls of mosques. Features of Islamic architecture include elaborate archways and the distinctive minarets of mosques.

If possible, arrange a field trip to a museum that displays Islamic art; as an alternative, you might bring in illustrated books about Islamic art. Discuss how art and architecture reflect characteristics of a culture. Call on students to provide examples from their own experiences.

OVERVIEW

In Chapter 32, students will learn about Southwest Asia, often called the Middle East. The chapter surveys the physical, economic, historical, and human geography of the region.

Section 1 explores the physical and economic geography of Southwest Asia, focusing on the region's landforms, rivers, resources, and industries.

Section 2 examines the history of several of the region's civilizations. Students will learn how ethnic and religious diversity have contributed to political conflicts and territorial wars and threatened economic stability.

CREATIVE STRATEGIES FOR TEACHING WORLD GEOGRAPHY

LESSON 14 Contributions from the Countries of Southwest Asia and North Africa: Did You Know You Were Speaking Arabic?

CHAPTER 32

Introducing Southwest Asia

Jewish menorah

Southwest Asia is the crossroads where Asia, Europe, and Africa meet. The region stretches from Turkey and the eastern shore of the Mediterranean to the high plateaus of Iran (i-RAN), and south to the Indian Ocean coast of the Arabian Peninsula. Southwest Asia is often called the Middle East, a term that originated with the European colonial powers. The Europeans viewed the region as midway to the Pacific shores of eastern Asia, which was called the Far East.

The ancient foundations of both farming and city life were laid in Southwest Asia. This is where people first learned to read and write, and where societies first built great monuments. Three of the world's major religions—Judaism, Christianity, and Islam—first appeared here as well. Today, the discovery of oil, modernization, and the growing influence of Western cultures are changing traditional ways of life in Southwest Asia.

GEOGRAPHY DICTIONARY

exotic river

furrow

fossil water

Judaism

Christianity

Islam

Muslim

Qur'an

monotheism

secularism

drip irrigation

Bedouin

374

Irrigation canal in **Jordan**

Using the Geography Dictionary

Ask a volunteer to name the terms in the Geography Dictionary that are not related to religion. (***exotic river, furrow, fossil water, drip irrigation, Bedouin***) Call on another to name the three terms that identify major religions. (***Judaism, Christianity, Islam***) Ask students to suggest why these three terms are defined in this chapter. (*All three religions originated in Southwest Asia.*)

Then call on a volunteer to give the meaning of the word parts *mono-* and *the-* or *theo-* (*one; god*). Use this information to define **monotheism**. Discuss the relationship between this term and the three religions listed. Ask students if they can name the sacred scripture of each religion. (*Torah, Bible, **Qur'an***)

MAPPING THE REGION
Using the Physical–Political Map

Ask students to examine the physical–political map of Southwest Asia on this page. Have volunteers list the mountain ranges. Ask: Where are most of the ranges located relative to countries' political borders? (*near or along them*) What other factors might have determined borders in the region? (*other physical features, political agreements*)

MAP TRANSPARENCIES

Overhead transparencies from the maps on pages 375–377 are available in the *REGIONS OF THE WORLD Map Transparencies.*

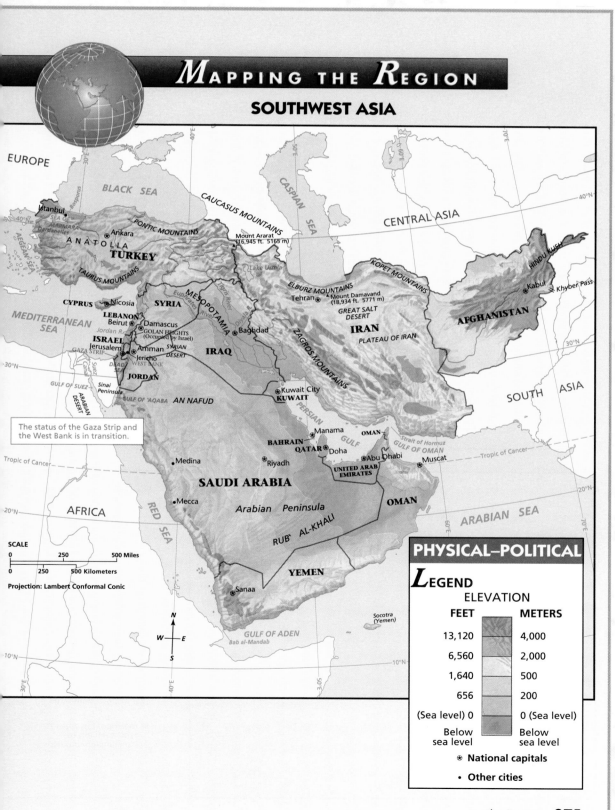

MAPPING THE REGION

SOUTHWEST ASIA

EUROPE

BLACK SEA

CAUCASUS MOUNTAINS

CASPIAN SEA

CENTRAL ASIA

Istanbul
Bosporus
MARMARA SEA
Dardanelles
AEGEAN SEA
PONTIC MOUNTAINS
Ankara
ANATOLIA
TURKEY
TAURUS MOUNTAINS

Mount Ararat
(16,945 ft. 5165 m)

Lake Urmia

KOPET MOUNTAINS

HINDU KUSH

Kabul
Khyber Pass

ELBURZ MOUNTAINS
Tehran
Mount Damavand
(18,934 ft. 5771 m)
GREAT SALT DESERT

AFGHANISTAN

CYPRUS Nicosia
SYRIA
MESOPOTAMIA
Euphrates River
Tigris River
LEBANON
Beirut Damascus
GOLAN HEIGHTS
(Occupied by Israel)
Jordan R.
ISRAEL
Jerusalem Amman
GAZA STRIP
DEAD SEA
Jericho
WEST BANK
JORDAN

Baghdad
IRAQ
SYRIAN DESERT

ZAGROS MOUNTAINS

IRAN
PLATEAU OF IRAN

MEDITERRANEAN SEA

GULF OF SUEZ
Suez Canal
Sinai Peninsula
GULF OF 'AQABA
ARABIAN DESERT
AN NAFUD

Kuwait City
KUWAIT

PERSIAN GULF

SOUTH ASIA

The status of the Gaza Strip and the West Bank is in transition.

Manama
BAHRAIN
QATAR Doha
GULF
OMAN
Strait of Hormuz
GULF OF OMAN
Abu Dhabi
Muscat
UNITED ARAB EMIRATES

Tropic of Cancer

Medina
Riyadh

SAUDI ARABIA

OMAN

Mecca

AFRICA

RED SEA

Arabian Peninsula

RUB' AL-KHALI

ARABIAN SEA

SCALE
0   250   500 Miles
0   250   500 Kilometers

Projection: Lambert Conformal Conic

YEMEN

Sanaa

Socotra (Yemen)

N
W — E
S

GULF OF ADEN
Bab al-Mandab

PHYSICAL–POLITICAL

*L*EGEND
ELEVATION

FEET		METERS
13,120		4,000
6,560		2,000
1,640		500
656		200
(Sea level) 0		0 (Sea level)
Below sea level		Below sea level

⊛ **National capitals**

• **Other cities**

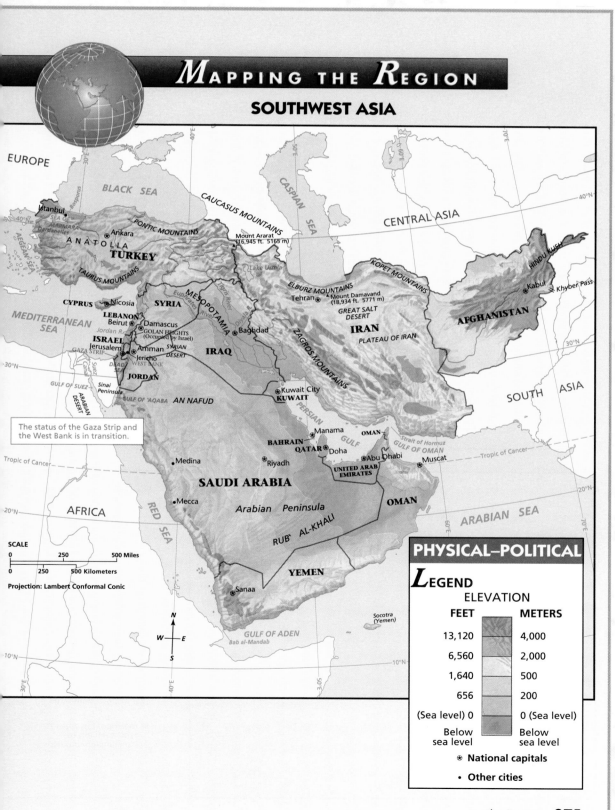

MAPPING THE REGION

Using the Climate Map

Have students look at the climate map of Southwest Asia on this page. Ask them to rank the region's climate types in terms of the amount of area each one covers, with the climate type covering the most area listed first. (*desert, steppe, Mediterranean, highland, humid subtropical*) Refer students to the physical–political map on page 375, and ask which countries they would choose to live in or visit based solely on climate.

MAPPING THE REGION

Using the Population Map

Have students examine the population map of Southwest Asia on this page. Ask a volunteer to compare it to the physical–political map on page 375 to name the most densely populated country in the region. (*Israel*) Point out countries in the region with sizable areas that are sparsely populated. (*Saudi Arabia, Jordan, Syria, Iraq, Iran, Afghanistan, Turkey*)

Continues on page 377

CLIMATE

Answers

1. Desert, steppe, Mediterranean, humid subtropical, highland
2. Desert
3. Desert, Mediterranean

POPULATION

Answers

1. Riyadh, Jidda
2. Rub`al-Khali
3. Mediterranean, steppe

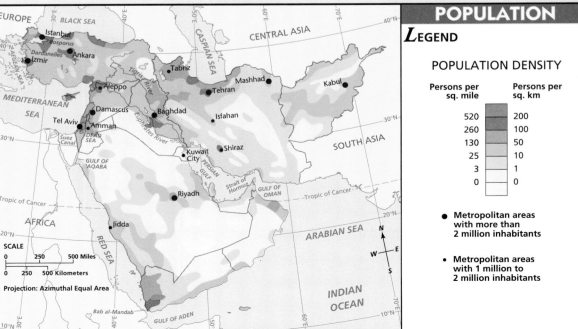

Tell students to use the climate map and the physical–political map to suggest reasons why these areas are so sparsely populated. (*Most are desert areas, the others are mountainous terrain.*)

MAPPING THE REGION

Using the Economy Map

Ask students to look at the economy map of Southwest Asia on this page. Have them identify the economic activities and resources that seem to dominate the region. (*nomadic herding, oil*) Note the areas marked as having limited economic activity, and ask students to suggest why economic activity is limited in these areas. Point out that students may wish to refer to the physical–political, climate, and population maps to get ideas or to check hypotheses. Have students refer to the population map to identify what type of farming takes place in the areas of greatest population density. (*commercial farming*)

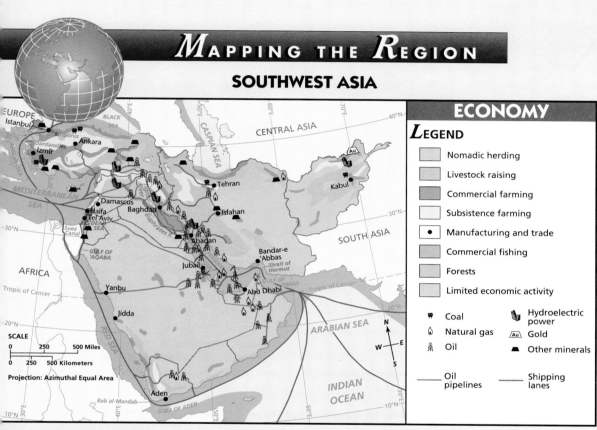

Mapping the Region

SOUTHWEST ASIA

ECONOMY

LEGEND

- Nomadic herding
- Livestock raising
- Commercial farming
- Subsistence farming
- • Manufacturing and trade
- Commercial fishing
- Forests
- Limited economic activity

Coal	Hydroelectric power
Natural gas	Au Gold
Oil	Other minerals
—— Oil pipelines	—— Shipping lanes

SCALE
0 250 500 Miles
0 250 500 Kilometers
Projection: Azimuthal Equal Area

ECONOMY

Answers

1. The Persian Gulf provides access to world markets for oil, which is heavily produced in the surrounding region.
2. Haifa, Tel Aviv
3. Extensive irrigation

CLIMATE

◆ Desert and steppe climates are found throughout most of Southwest Asia.

◆ A Mediterranean climate dominates the coastal areas along the Mediterranean Sea and parts of northern Iran.

◆ Afghanistan lies at the edge of a large highland climate region that spreads east across Asia.

(?) Which climate types are found in Southwest Asia?

Compare this map to the physical–political map on page 375.

(?) Which climate type dominates nearly all of the Arabian Peninsula?

(?) Which climate types are found in Israel?

POPULATION

◆ Some of the most densely populated areas of Southwest Asia are along the Mediterranean coast and along the Tigris and Euphrates rivers.

◆ The Arabian Peninsula, despite large sparsely populated areas, has several cities with more than 1 million inhabitants.

◆ Most of the largest cities of Southwest Asia are in the northern and western parts of the region.

(?) Which Saudi Arabian cities have populations of more than 1 million?

Compare this map to the physical–political map on page 375.

(?) Name the physical region where the population of the Arabian Peninsula is the least densely populated.

Compare this map to the climate map on page 376.

(?) Which climate types dominate the more densely populated northern and western regions of Southwest Asia?

ECONOMY

◆ Oil production is the major economic activity for many Southwest Asian countries, particularly those along the Persian Gulf and in Mesopotamia.

◆ Pipelines carry oil across large areas of Southwest Asia to ports on the Persian Gulf and the Mediterranean Sea.

◆ The irrigated lands along the Tigris and Euphrates rivers form an important agricultural region.

(?) Why do you think many manufacturing and trade centers are located along the Persian Gulf?

(?) Name the major manufacturing and trade centers shown along the Mediterranean coast.

Compare this map to the climate map on page 376.

(?) How do you suppose most agriculture can be successful in some of the dry regions of Southwest Asia?

377

SECTION 1
OBJECTIVES

The objectives reflect the Focus questions on this page and are assessed in the review on page 379.
1. Describe the physical geography of Southwest Asia.
2. Describe the economic geography of the region.

TEACHING SECTION 1
Motivator

Have students name a climate type and a natural resource identified with Southwest Asia. (*desert climate and oil*) Ask them to predict the main economic activity in the region in terms of the number of people involved in it. (*agriculture*) Students may be surprised by the answer. Tell students that they will learn how agriculture is possible there despite the dry climate and limited water resources.

Building a Vocabulary

Write the term **exotic river** on the board. Point out that *exotic* is often used to mean "strange" or "out of the ordinary." Geographically, it means "not native to the place where it is found." On a wall map, trace the course of the exotic Tigris and Euphrates rivers from the mountains of Turkey across the desert to the Persian Gulf.

Geographical Sidelight

Satellite images have detected traces of an ancient river that apparently flowed from the Hejaz Mountains of western Saudi Arabia 500 miles (805 km) eastward across the desert, forming a large delta that once covered what is now Kuwait. The river, which may have been three miles (5 km) wide in places, probably dates to a relatively wet historical period that occurred between five and eleven thousand years ago. Similar discoveries of hidden and dried-up rivers in North Africa have yielded new sources of groundwater. That possibility exists for the ancient Saudi Arabian river also.

PHYSICAL AND ECONOMIC GEOGRAPHY

FOCUS

- *What is the physical geography of Southwest Asia?*
- *What is the region's economic geography?*

REGION **Physical Geography** According to the theory of plate tectonics, Southwest Asia lies at the intersection of the African, Eurasian, and Arabian plates. The grinding collision of these plates has created a mixture of rugged mountains, valleys, upland plateaus, and narrow gulfs and seas bordered by coastal plains. Frequent earthquakes in the region are reminders of the earth's continuing tectonic activity.

As a result of prevailing subtropical high pressure, a desert climate covers most of Southwest Asia. In fact, Southwest Asia is at the center of the world's largest arid, or dry, region. Dry lands stretch from the Atlantic coast of the Sahara in North Africa to western China. Steppe climates are found along desert edges and on high interior plateaus. Winter rains are common in these areas. A Mediterranean climate dominates the coastal areas along the Mediterranean Sea.

In many of the desert areas, rain falls only a few times each year, and surface water is rare. People can live in the desert only where there is water. Springs, where water rises naturally from cracks in the surrounding rock, are an important water resource, often forming oases. Over the years, many springs in the region have been made into productive wells. Water can also be found beneath dry riverbeds. The arid lands between these sources of water are usually uninhabited, visited only by migrating herders.

The desert traveler can suffer from both extreme heat and cold. The sun is able to heat the ground quickly because the skies are clear and there is little vegetation for shade. Afternoon summer temperatures can range from 100°F to 120°F (38°–49°C). During the night, however, the temperature can drop more than 30°F (17°C). Frosts occur often in winter.

Mountains receive most of the region's precipitation, both as summer rain and winter snow.

A herding village is nestled within the Kopet Mountains of northeastern **Iran**. The cooler, wetter climates of many of the mountainous regions stand in contrast to the dry, hot conditions of the large desert areas of Southwest Asia.

Mountain environments are wetter because of the orographic effect and because the cooler temperatures reduce evaporation. In contrast to desert areas, some of which receive an average of less than four inches (10 cm) of rain per year, some mountain peaks receive more than 50 inches (127 cm) of rain each year. The streams that drain these highlands provide water to the dry valleys below. Streams that flow from humid environments across deserts into seas are called **exotic rivers**.

INTERACTION **Agriculture** Despite arid climates and rugged landforms that make farming difficult, agriculture is a major occupation in much of Southwest Asia. Fertile soils are found only in the river valleys, on high mountain plateaus, and in a few oases. In many of the wetter parts of the region, overgrazing has led to increased soil erosion. The few mountain forests have been cut to near destruction as well. The countries of the region have begun attempts at conserving their soil and plant resources.

Most farmers raise only enough crops and animals to support their own families. Any extra food and animal products are sold for cash at city markets. Barley and wheat are the important grains in the area, while citrus fruits, olives, figs, nuts, and grapes are grown in the Mediterranean coastal areas. Livestock, mostly sheep, goats, and some cattle, is raised on some farms.

Analyzing Maps

Physical Geography   Review with students the theory of plate tectonics, and explain its influence on the physical geography of Southwest Asia. Have students compare the map on page 47 with the map on page 375 to identify physical features of the region that lie along plate boundaries. (*Red Sea, Persian Gulf, Zagros Mountains*)

Guided Practice

Emphasize to students that in recent years oil has shaped Southwest Asia's economic geography. Have students suggest as many uses as they can for oil, and help them make a list of petroleum-based products. Ask students to consider, given the large number of uses for oil and petroleum-based products, how the economies of Southwest Asian countries might rise and fall with the price of oil.

Closure

Have students work with partners to answer the Focus questions on page 378.

ASSESSMENT

Assign the Section 1 Review questions on this page.

Commercial agriculture is practiced throughout Southwest Asia but is widespread only in Israel (IZ-ree-uhl), where citrus fruits and other crops are grown for sale to European markets. The other countries of the region are developing commercial agriculture, particularly to provide fruits and vegetables for local markets.

Irrigation first developed in Southwest Asia, and it has steadily expanded over the centuries. In the mountains, most irrigation developments are small in scale. Often, water is simply diverted into a small canal or **furrow** (narrow trench) that leads directly to a field. Large-scale irrigation usually depends on an exotic river such as the Tigris (TIE-gruhs) or Euphrates (yoo-FRAY-teez), both of which begin in the mountains of Turkey. Wide, deep canals lead away from the exotic rivers. Smaller and smaller canals then lead into individual fields. These large irrigation systems are much larger than one community can maintain.

In many places, deep wells can provide irrigation. Such wells require modern drilling technology and powerful pumps. The groundwater in these wells is usually **fossil water**, or water that is not being replaced by rainfall. Eventually, overuse will deplete this water source.

Resources and Industry   Many people think of oil when they think of Southwest Asia. Indeed, oil is the region's richest mineral resource. The main oil deposits lie along the shores of the Persian Gulf (also called the Arabian Gulf) and in Iraq (i-RAHK).

Since many of the oil-rich countries have small populations and little industry, most of their oil is available for export. As a result, the economies of the oil-exporting countries rise and fall with the price of oil. When oil prices fall, unemployment and business failures increase in these countries. To have more control over world oil prices, the oil-rich countries of the region joined with other oil-rich countries around the world to form the Organization of Petroleum Exporting Countries (OPEC). OPEC was quite successful in the 1970s, but its effectiveness has since declined. Nevertheless, world demand for oil has continued to rise. Large profits should flow into the oil-rich countries as long as their enormous oil reserves last.

Other than oil, the countries of Southwest Asia have few resources that are important for developing industry. Turkey does have large deposits of coal, iron ore, and chromium, a metal

used to coat and protect other metals. Iran has a copper industry, and the Dead Sea on the border between Israel and Jordan contains salts that are used in the chemical industry.

Of the countries of Southwest Asia, only Israel is considered a developed country. The oil-rich countries have high incomes, but their industry is limited mainly to oil refining and related chemicals. Most other industries in the region center on food processing and consumer goods.

SECTION REVIEW

1. What climate type dominates most of Southwest Asia?
2. What crops are grown in Southwest Asia? What is the region's most valuable resource?
3. **Critical Thinking   Imagine** that you will be traveling through a desert in Southwest Asia. Given the desert environment, what items might you want to bring with you?

2 HISTORICAL AND HUMAN GEOGRAPHY

FOCUS

- *What were some of Southwest Asia's ancient civilizations, and how were they significant? What powers have controlled the region in its history?*

- *What religious and ethnic groups live in Southwest Asia?*

- *What challenges face the region?*

Historical Geography   The world's first civilizations developed in the area known as the Fertile Crescent. The Fertile Crescent is a wide arc of productive land that runs along the eastern shore of the Mediterranean Sea, through the plains along the Tigris and Euphrates rivers, to the Persian Gulf. The ancient name for these plains is Mesopotamia. Many of the plants and animals found on farms throughout the world today were first domesticated in the Fertile Crescent.

Section 1 Review
ANSWERS

1. Desert
2. Barley, wheat, citrus fruits, olives, figs, nuts, grapes; oil
3. Possible answers include warm clothing and blankets for cold nights, a hat and sunscreen for the daytime, and much water.

379

Extension

Have interested students find out more
about OPEC. Ask them to put together
an oral presentation about when OPEC
was formed, who its members are,
what purposes it serves, and why its
effectiveness has declined.

SECTION 2
OBJECTIVES

The objectives reflect the Focus
questions on page 379 and are
assessed in the review on page 384.

1. Summarize the history of
Southwest Asia.

2. Identify the major religious and
ethnic groups living in the region.

3. Discuss challenges facing the
region.

TEACHING SECTION 2
Motivator

Use a wall map to introduce Southwest Asia as an important cultural
crossroads. Point out land and water
routes connecting the region to
Europe, Asia, and Africa. Ask volunteers to share what they know about
early civilizations or places in this
region. Challenge students to think of
current events that show the area's
global importance.

The period from about 3500 B.C. to 600 B.C.
saw the rise and fall of many important civilizations in Southwest Asia. Each culture left a part of
its knowledge and way of life for later civilizations.
In Mesopotamia, the Sumerians developed writing. Later, the Hittites developed iron weapons.
Along the Mediterranean shore, Phoenician merchants formed a large trading network.

After 600 B.C., most of the region was controlled by one empire after another. These included
the Persian, Macedonian (Greek), and Roman
empires. By A.D. 800, Muslim armies had conquered
most of Southwest Asia and had ventured far to
the east and west. Following these conquests,
Southwest Asia divided into many small empires. A
political and religious prince known as either a
caliph, emir, or sultan ruled each territory.

In Turkey, an influential leader named Osman
founded the Ottoman Empire in about 1300. In
1453, the Ottomans captured the city of
Constantinople, now called Istanbul, and made it
the empire's capital. At its peak in the 1600s, the
Ottoman Empire stretched in the east from the
Persian Gulf to the Caucasus, and in the south
along the eastern Mediterranean, around the Red
Sea, and across much of North
Africa. In the north, the em-
pire included Europe's Balkan
Peninsula and Crimea.

As Ottoman power de-
clined in the 1700s and 1800s,
the powers of Europe ex-
panded their control over the
region. The Ottoman Empire
finally collapsed at the end of
World War I in 1918. Some
parts of the empire then
became colonies of France,
Britain, and Italy. The Europeans were interested in the
natural resources and transportation routes of Southwest
Asia. They guarded these
important shipping routes
through their military bases in
the region.

Most of the present-day
countries of Southwest Asia
achieved their independence
during the past 60 years. Today, the leaders of these countries vary from traditional

kings, sheiks, and emirs to dictators. There also
are a few democracies. Since independence,
nationalism and modernization have been the
main forces shaping the cultures of the region.

Religion Three of the world's major religions
originated in Southwest Asia. **Judaism**, the religion of the Jews, appeared during the period
between 3500 B.C. and 600 B.C. in a region called
Palestine. Judaism was the first major religion to
center around the belief in a single god.

Christianity developed out of Judaism and
spread during the Roman era. It is a religion
founded on the teachings of Jesus Christ, whom
Christians believe was the son of God. Although
the Romans persecuted Christians for many
years, Christianity eventually became the main
religion of the Roman Empire. After the Roman
Empire divided into the Eastern Roman Empire
and the Western Roman Empire in the fourth century, Christianity split into the Eastern Orthodox
Church, based in Constantinople, and the Roman
Catholic Church, based in Rome. In the 1500s,
another group of European Christians, called
Protestants, broke away from the Roman Catholic
Church. Today, Christianity in
its many variations is the dominant religion of the Western
Hemisphere.

Islam was founded by the
prophet Muhammad (moe-
HAM-uhd), who lived from
about A.D. 570 to 632. Muhammad worked as a mer-
chant in Mecca, a trading city in
western Saudi Arabia. When he
was about 40 years of age, he
often went to nearby mountains
to think about the social injustices in Mecca. Muhammad said
that during one of his trips, he
was told by the angel Gabriel, a
messenger of God, to preach
religion to the Arabs.

After its founding, Islam
spread rapidly. In 10 years,
Muhammad was able to unite
the peoples of western Saudi
Arabia, and within a century of
his death, Islam had spread
much farther. A person who follows Islam is called a **Muslim**.

This tablet honors the founding of the
Babylonian Sun Temple in the ninth
century B.C. near the present-day city
of **Baghdad, Iraq**.

Building a Vocabulary

Some of the vocabulary terms in the section have an Arabic origin. Point out that **Islam** is from *al-islam,* which means "the act of submitting oneself unreservedly to God." A **Muslim** is "one who surrenders." **Qur'an** (better known by the transliteration *Koran*) means "recital" in Arabic.

Diagramming Historical Geography

Students working in groups can create a time line of events in Southwest Asia. Assign a different section of the time line to each group. Have students use library resources to chart wars, internal upheavals, migrations, and the discovery of oil. Results should be combined and transferred to a sheet of butcher paper posted in the classroom.

Guided Practice

Have students bring to class newspaper and magazine articles about current events in Southwest Asia. Call on volunteers to summarize their stories for the class and to identify the major issues involved. Write the issues on the board; after several have been noted, ask volunteers to categorize them. (*for example, disputes over resources and territory, religious and ethnic differences*)

The words that Muhammad received from God (whom Muslims call Allah) were recorded by his followers in the **Qur'an** (Koran), Islam's sacred book of writings. Islam remains a unifying culture trait of Southwest Asia.

Islam shares many of the traditions of Judaism and Christianity. One of these shared traditions is **monotheism,** or the belief in one god. Muslims believe in one god, Allah. The Qur'an also has several passages similar to the Old Testament of the Bible. Like Christianity, however, Islam has split over the years into different groups. This division has resulted from varying interpretations of the Qur'an and from questions of leadership throughout the Islamic world. There are two major branches of Islam, the Sunni and the Shi'a (SHEE-ah). The Sunni emphasize the basic Islam of the Qur'an and personal prayer. The Shi'a follow imams (i-MAHMS), who are religious as well as political leaders.

Today, about 90 percent of Muslims are Sunni, and 10 percent are Shi'a. The Shi'a are concentrated in Iran, southern Iraq, Yemen, and Lebanon (LEHB-uh-nuhn). The Shi'a are particularly powerful in Iran and hope to spread their political ideas throughout Southwest Asia.

In recent years, Islamic fundamentalism has grown. Muslim fundamentalists stress family unity and the importance of the oldest male family member. They also wish to reduce foreign influences on the region. Equal opportunity for women and the independence of children are seen as a reflection of Western culture and thus are considered wrong. Similarly, **secularism,** an indifference to or rejection of religion, is considered evil.

Some Muslim leaders teach that religion is more important than laws and customs created by governments. This belief, however, conflicts with the growing nationalism throughout Southwest Asia. Sorting out these cultural issues is difficult because Islam has no single powerful leader. The Muslim leaders within each country often act independently.

Culture and Conflict As a result of the region's long history of many empires and religions, there are several ethnic groups in Southwest Asia. Arabs are the dominant group in all countries of the region except Turkey, Cyprus, Israel, Iran, and Afghanistan. Nearly all Arabs are Muslims. Therefore, the many Arab countries share several common cultural and religious

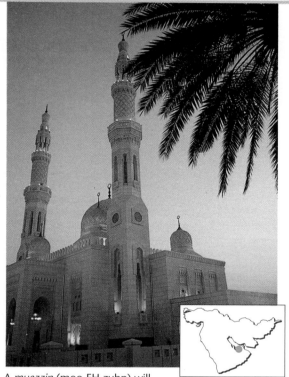

A *muezzin* (moo-EH-zuhn) will call fellow Muslims to prayer from the tower of this mosque in **Dubai, United Arab Emirates**.

views. Although both the Turks and the Iranians are Muslim, their histories and cultures are distinctive and non-Arab.

The Jews in Israel are another important group in the region. After the spread of Islam in the seventh century, only small communities of Jews remained in Southwest Asia. Then in the late 1800s, Jews from Europe began moving to the ancient homeland in Palestine. In 1948, the Jewish state of Israel formed in Palestine, and Jews from Europe and elsewhere began moving to Israel in large numbers. The people of Israel view themselves as descendants of the ancient Hebrews, whose kingdom thrived here some 3,000 years ago.

Before 1948, the lands of Israel had been the home of the Arab Palestinian people. This Arab-Israeli conflict over homelands has led to several wars and has dominated political issues in Southwest Asia. In 1993 and 1995, Israel and the Palestinians reached major agreements providing for gradual and limited Palestinian self-rule in certain areas. Israel and Jordan signed a peace treaty in 1994, and efforts toward peace in the region have continued.

Text continues on page 384.

MULTICULTURAL PERSPECTIVES

Although it is usually identified with Arabic cultures, Islam is practiced by people of many different ethnic groups. In addition to Southwest Asia, Islam is the predominant religion of northern and eastern Africa, the Central Asian republics, Indonesia, and Malaysia. Muslims everywhere, regardless of ethnicity or nationality, are united by the Five Pillars of Faith that govern actions in their daily lives: (1) the profession of Muslim faith; (2) ritual prayer, performed five times a day; (3) almsgiving to the poor and needy; (4) fasting during the holy month of Ramadan; and (5) the pilgrimage to the city of Mecca, which is to be made at least once by every Muslim. Have interested students learn more about a Pillar of Faith and report their findings to the class. Use their presentations as a basis for a class discussion on how these principles compare to those of other major religions.

Planet Watch

Analyzing Issues

Have students work in pairs to make a chart in which they will weigh the costs of "water security" for each country in Southwest Asia against political, economic, and environmental concerns. Across the top of the chart will be the following headings: Reasons for Shortages, Possible Solutions, and Obstacles/Challenges. Down the side of the chart, students should list each country in the region. Have volunteers present their charts to the class.

382

PLANET WATCH

The Politics of Water

Did the ancient civilization of Mesopotamia in Southwest Asia disappear because it did not have enough water to survive? Will competition for water rock this already tense region of the world today? Control of the region's water resources is a key political issue. One expert predicts, "In the year 2000, water will be more expensive than oil."

Water Resources

The countries of Southwest Asia obtain water from different sources. In Saudi Arabia, the wealth from the nation's oil resources helps fund the desalinization plants along the Persian Gulf. Without desalinization, there would be little fresh water available for cities or irrigation farming. Many other countries in the region, however, do not have the money to build and run such plants.

Israel, with a vast expanse of desert, depends on several methods to obtain fresh water. Since 1964, much of Israel's water has been supplied by a pipeline called the National Water Carrier, which carries water from the Sea of Galilee to Tel Aviv and down the coastal plain toward the Negev Desert. Israel also relies on desalinization plants and conserves water in agriculture through use of **drip irrigation**. Drip irrigation is a process by which pipes placed near plant roots drip only as much water as the plants need to grow. Even so, droughts and a population explosion, resulting from a large number of refugees immigrating to Israel from the former Soviet Union and elsewhere, have strained the country's water resources.

Droughts and a growing population have strained water supplies to the west in Jordan as well. Even though Jordan and Israel share some water resources, including the Jordan River, political conflicts in the past kept the two countries from reaching agreements to share water effectively. Leaders in Israel and Jordan hope their 1994 peace treaty will help them find joint solutions. In addition, Jordan has begun using drip irrigation methods and is searching for deeper groundwater sources.

Even though they border many seas, the countries of Southwest Asia face a severe water shortage.

Qanat Irrigation System

Mountains

Water from mountains seeps down to water table

Head well

Ventilation & access shafts

Water tunnel

Irrigated fields

Surface canal

Distribution canals

Water table

In the dry climates of **Iran** and **Afghanistan**, some farmers continue to use an ancient method of irrigating their crops. Runoff from mountain precipitation is channeled through underground tunnels, called *qanats* (kuh-NAHTZ), to fields below the mountains. To learn more about *qanats*, see page 401.

Using Geography Themes

Use the following activity to reinforce the geographic theme of region. Tell students that the government of Turkey has promoted a "peace pipeline" that would transport water from two of the country's major rivers southward through Syria, Jordan, Saudi Arabia, and other countries of the region. Ask students to suggest why Turkey might promote such a project. *(Turkey has strong historical, cultural, and economic ties to the rest of Southwest Asia.)* Ask students why the project is called a peace pipeline. *(Students might suggest that the pipeline's success depends on peace among these countries. In turn, it might promote peace in the region through cooperation on the project.)*

Syria, with a large amount of desert land, gets most of its water from the Euphrates River. Droughts, rapid population growth, and wasteful water use, however, have led to shortages. In addition, harmful pesticides and chemicals have been dumped into the Euphrates, making it a dangerous source of drinking water. Syria also has neglected its irrigation and sewage systems, allowing pipes to crumble and leak.

Conservation Efforts

Faced with growing populations, periodic droughts, and limited resources, the nations of Southwest Asia are looking closer at cutting their water use. For some nations, however, water conservation involves difficult economic decisions. For example, some crops grown in the region, such as bananas and cotton, need large amounts of water. Nations must weigh the cost of using the water to grow these crops against the cost of saving the water and importing the crops from other countries.

Effective conservation methods must be accompanied by policies that protect the environment. Countries are realizing that pesticides and chemical fertilizers can make their way into rivers, aquifers, and wells. Also, unrestricted digging of wells can lower the water table and allow seawater to enter aquifers.

Cooperation Among Nations

Because most rivers and aquifers do not begin or end at a nation's borders, Southwest Asia's water problems cannot be solved without regional cooperation. If countries upstream take too much

Leaders of the United States, Israel, and Jordan gathered in 1994 for the signing of a peace treaty between the two Southwest Asian countries. Israelis and Jordanians hope the peace treaty, signed along the border between their two countries, will aid them in finding common ground on important water issues.

water from or pollute a river, countries downstream suffer. For example, because the headwaters of the Tigris and Euphrates rivers are in Turkey, Turkey's use of the rivers affects Syria and Iraq downstream.

Before the countries of Southwest Asia can solve their water problems, however, they must solve or set aside their political differences. As long as conflicts remain, the threat of war hovers over the region and makes nations unwilling to invest in pipelines and desalinization plants that could be attacked. Future peace negotiations in the region will likely be linked to water agreements. In turn, peace could set the stage for easing the growing water crisis.

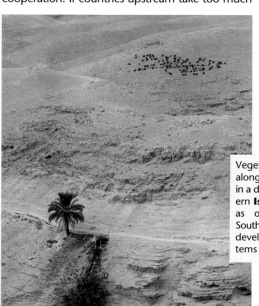

Vegetation clings to life along this irrigation canal in a desert region of southern **Israel**. Israel, as well as other countries of Southwest Asia, uses well-developed irrigation systems for agriculture.

Geographical Sidelight

Israel has held negotiations with private companies in Turkey about importing water from Turkey to Israel via flexible barges or "floating bags." Have interested students research this method and discuss its potential role in helping to resolve Israel's water problem.

▼ **YOUR TURN**

Draw a map of Southwest Asia and label all the countries, bodies of water, and major rivers of the region. Then **assess** each country's natural water resources, noting if any of these resources are shared among countries. **Write** a paragraph outlining what might be each country's concerns about use of the region's water resources. **Describe** how a regional water plan might address these concerns.

Your Turn

Students' paragraphs should take each country's natural water resources into account. Students' regional plans should address each country's concerns.

383

384

This modern home in **Saudi Arabia** is furnished with many items of Arab culture. Many Arab men, such as the one pictured here, wear a *kaffiyeh* (kuh-FEE-yuh), a cloth headdress that offers protection against the sun and wind.

Many other ethnic and religious minorities live in Southwest Asia. In Lebanon, there is a large community of Christians. In Iran, there are Zoroastrians (zohr-uh-WAS-tree-uhns), who follow the religion of ancient Persia. The Kurds, who are also Muslim but non-Arab, are a large ethnic group living in the bordering regions of Iraq, Iran, Syria, Armenia, and Turkey. The Kurds do not have their own country.

Migrating herders, known as nomads or **Bedouins**, still roam the outer parts of the deserts. Many years ago, caravans of camels carried traders and their goods from oasis to oasis. Camels, long a symbol of life in Southwest Asia, are now used for transport in isolated areas only. Some people still use donkeys as well, but for the most part, trucks have taken the place of camels and donkeys. Some wealthy families in the region keep camels in much the same way as some Americans keep horses.

Issues Although the Arab-Israeli conflict has dominated the political geography of Southwest Asia, many other challenges face the region. Modern health care practices together with a tradition of large families have led to rising popula-

tions throughout the region. These growing populations need food and want more basic consumer goods and opportunities.

Also, as populations in the region continue to grow, so will conflicts over water resources. In most of Southwest Asia, farms and cities depend on careful management of rivers, springs, and wells for water. Many communities are miles away from the rivers that provide their water. To make sure that everyone receives water, governments must maintain cooperation among communities. In ancient times, wars often led to the collapse of irrigation systems, and starvation resulted. Today, all of the important exotic rivers of Southwest Asia cross international borders, further complicating the issue. (See "Planet Watch" on pages 382–383.)

Another issue of international significance concerns the region's oil resources. The importance of oil in world trade has given the oil-rich countries of Southwest Asia more political power. For example, Arab countries used their oil supplies to influence international attitudes toward the Arab-Israeli conflict during a war in 1973. Arab countries would not sell oil directly to countries that supported Israel in the war.

In the long run, the oil wealth of the region, as well as foreign aid, will change the region more than any issue or event since the spread of Islam. Whole industrial complexes are being built, including chemical and oil-related manufacturing plants. New schools, housing, and Western influences are appearing throughout the region. The enormous changes taking place in the countries of Southwest Asia make predicting the region's future difficult.

2 SECTION REVIEW

1. What were three major innovations developed by the ancient civilizations of Southwest Asia? What empires have existed in the region?

2. What three major religions originated in Southwest Asia? What ethnic and religious groups live in the region today?

3. **Critical Thinking** **Discuss** the roles that natural resources can play in a region's conflicts. Use two examples from Southwest Asia.

CHAPTER 32 CHECK
ANSWERS
Building a Vocabulary

1. Exotic river; because of the orographic effect and because cooler temperatures reduce evaporation
2. The belief in one god; Judaism, Christianity, Islam
3. The Qur'an
4. Because secularism is an indifference to or rejection of religion, and religion is very important to Muslim fundamentalists

Recalling and Reviewing

1. Small-scale irrigation, in which water runs through a small canal directly from its source to a field; large-scale irrigation, in which a series of several canals leads from a river into fields; irrigation provided by pumping from wells; drip irrigation, in which pipes drip water to plant roots
2. It runs from the Mediterranean Sea, along the Tigris and Euphrates rivers, to the Persian Gulf. It supported the world's first civilizations.
3. Because of its resources and transportation routes
4. Sunni and Shi'a
5. The conflict between Arabs and Israelis

Thinking Critically

1. Possible answers: people's different needs based on location, time period, occupation; differing interpretations of religious texts
2. Possible answer: the region's countries might now seek regional solutions to problems and might be less inclined to enter into conflicts.

Learning About Your Local Geography

Reports should discuss the extent to which students' community members are dependent on gasoline (oil). Students should understand that because much oil comes from Southwest Asia, events there can have worldwide effects. Community members might try to reduce oil use by driving less (forming car pools, using mass transportation).

Using the Five Themes of Geography

The intersection at and movement of the African, Eurasian, and Arabian plates make the region tectonically active. The region's history and cultures reflect the movement of peoples through Asia, Europe, and Africa.

CHAPTER 32 CHECK

Mounds of potash, taken from the **Dead Sea**, will be used to make fertilizer.

Reviewing the Main Ideas

1. Southwest Asia is at the center of the world's largest arid region. Agriculture is a major occupation in the region, but oil provides the most income.
2. The world's first civilizations developed in Southwest Asia. After 600 B.C., a series of empires controlled most of the region. When the last empire fell in 1918, parts of the region came under European control. Most countries here achieved independence during the past 60 years.
3. Several religious and ethnic groups live in Southwest Asia. Tensions between these groups, rising populations, scarce water resources, and oil wealth all are factors that will shape the region's future.

Building a Vocabulary

1. What term describes a stream that flows from a humid environment, such as mountains, across a desert into a sea? Why are mountain environments wetter?
2. Define *monotheism*. In what three major religions is monotheism a central belief?
3. What is Islam's sacred book of writings?
4. Why is secularism considered evil by Muslim fundamentalists?

Recalling and Reviewing

1. What types of irrigation are used in Southwest Asia?
2. Where is the Fertile Crescent? How is it historically significant?
3. Why were Europeans interested in Southwest Asia following the collapse of the Ottoman Empire?
4. What are Islam's two major branches?
5. What conflict has dominated political issues in Southwest Asia?

Thinking Critically

1. Both Christianity and Islam are divided into different branches; certain divisions exist among the followers of Judaism as well. What factors might cause divisions within a major religion?
2. During the Cold War, some countries in Southwest Asia sought support from the United States, while others turned to the Soviet Union. How do you think the end of the Cold War affects political events in the region as a whole?

Learning About Your Local Geography

Individual Project
Gasoline is a major oil product. Interview a few community members to find out how much they drive each week and how often they buy gasoline for their cars. Summarize your findings in a written report. Conclude with a few statements discussing how people in your community might be affected by a conflict in Southwest Asia that reduced or cut off oil supplies to the United States. What actions might community members take to reduce the local impact of such a conflict?

Using the Five Themes of Geography

Write a paragraph discussing how Southwest Asia's **location** near Europe, Africa, and the rest of Asia has influenced the region's physical and human **place** characteristics. Also consider the role that **movement** has played in determining these characteristics.

385

CHAPTER 33 The Eastern Mediterranean
Planning Guide

TEXTBOOK	TEACHING RESOURCES	OTHER AVAILABLE MATERIALS
Section		
1. Turkey Pages 387–389	Reteaching Worksheet 33.1 Skills Worksheet 33	*The World Today* Videodisc Program *Study Guide for Spanish Speakers* *Lecture Notes,* Chapter 33
2. Israel Pages 389–394	Reteaching Worksheet 33.2	*The World Today* Videodisc Program *Study Guide for Spanish Speakers* *Lecture Notes,* Chapter 33
3. Syria, Lebanon and Jordan Pages 394–396	Reteaching Worksheet 33.3 Critical Thinking Worksheet 33 Graphic Organizer 33 Challenge/Enrichment Worksheet 33	*The World Today* Videodisc Program *Study Guide for Spanish Speakers* *Lecture Notes,* Chapter 33
Chapter Review Page 397	Chapter 33 Test Chapter 33 Portfolio Assessment See also *Alternative Assessment Forms*	*Test Generator,* Chapter 33

CHAPTER RESOURCES

Lecture Notes, Chapter 33
Teaching Resources
    Chapter 33 Graphic Organizer
    Chapter 33 Skills Worksheet
    Chapter 33 Critical Thinking Worksheet
    Chapter 33 Challenge/Enrichment Worksheet

REVIEW AND ASSESSMENT

Chapter 33 Check, page 397
Chapter 33 Test, *Teaching Resources*
Chapter 33 Portfolio Assessment, *Teaching Resources*
Test Generator, Chapter 33
Alternative Assessment Forms

ADDITIONAL RESOURCES

Books for Teachers

Friedman, Thomas L. *From Beirut to Jerusalem.* New York: Anchor, 1989.

Jayyusi, Salma Khadra, ed. *Anthology of Modern Palestinian Literature.* New York: Columbia University Press, 1992.

Korn, David A. *Stalemate: The War of Attrition and Great Power Diplomacy in the Middle East.* Boulder, CO: Westview Press, 1992.

Books for Students

Beaton, Margaret. *Syria.* Chicago: Children's Press, 1988.

Israel. Alexandria, VA: Time-Life Books, 1986.

Leone, Bruno, series ed. *Israel: Opposing Viewpoints.* San Diego: Greenhaven Press, 1994.

Whitehead, Susan. *Jordan.* New York: Chelsea House, 1988.

The World Today *Videodisc Program*

A variety of still images, short videos, and activities are available for you to use as you teach this chapter. *Lecture Notes* include bar codes for specific images and videos.

Multimedia Resources

Ancient Egypt, Middle East, and Greece. (Multimedia CD-ROM, history to 1945) Queue, 338 Commerce Drive, Fairfield, CT 64305.

Homeland: Israel and Palestine. (VHS, 25 minutes) Encyclopedia Britannica Educational Corporation, 310 South Michigan Avenue, Chicago, IL 60604–9839.

In the Holy Land. (laserdisc, HyperCard software) Coronet/MTI Film & Video, 108 Wilmot Road, Deerfield, IL 60015.

Israel. (16mm film/VHS, 25 minutes) Nations of the World series. National Geographic Society, Educational Services, Washington, DC 20036.

Jerusalem: Within These Walls. (VHS, 59 minutes) National Geographic Society, Educational Services, Washington, DC 20036.

Picture Atlas of the World. (interactive maps, photo bank, animation, audio on CD-ROM) National Geographic Society, Washington, DC.

MEETING INDIVIDUAL NEEDS

Limited English Proficiency (LEP)

Pair LEP students with students who are proficient in English. Have the pairs complete the Roman numeral headings for an outline of the chapter, leaving space for details. Then have pairs exchange their outlines and complete them. Call on a volunteer to write one of the outlines on the board. Discuss it with the students, making additions or corrections as necessary.

Learners Having Difficulty

Help students create a time line of important events discussed in Chapter 33. Begin a time line on the board, and ask students to copy it in their notebooks and to fill in important dates and events as they read the chapter. Use the time line to help students review the main ideas of the chapter.

Auditory Learners

Assign each student a city mentioned in the chapter. Tell each student to pretend that he or she is the city's official host and translator for visiting dignitaries. Students should review the chapter and present a brief oral report to the class about their city. The class should role-play the part of dignitaries and ask the host questions.

Visual Learners

Provide students with Outline Map 17 of Southwest Asia in *Teaching Resources*. Have students use the information in Chapter 33 to create land-use maps of the Eastern Mediterranean. Students should label countries, landforms, and major rivers and include a title and a legend. Ask students to write three generalizations about land use in the region, based on their maps. Call on volunteers to read their statements aloud. Display the maps on a bulletin board.

Tactile/Kinesthetic Learners

Have students draw illustrations of items mentioned in the chapter. They might choose to draw sites, foods, crops, or economic activities. Ask them to include with their illustrations captions explaining the illustrations' content and significance in the Eastern Mediterranean region.

Gifted Learners

Have interested students investigate the way of life on an Israeli kibbutz. Then ask students to imagine that they have gone to Israel to work on a kibbutz, and have them write a letter home to their family describing their experiences.

MULTICULTURAL FOCUS

Judaism was the first major religion based on a belief in one God who wants people to act in a just and compassionate manner. This tenet is the most important teaching of Judaism. The Torah—the first five books of the Hebrew Bible—contains this and all other fundamental laws and teachings of Judaism.

The Talmud, another sacred writing of Judaism, describes Jewish history, folklore, laws, and ethics. It gives Jews guidance in matters of civil and religious laws.

Unlike many religions, Judaism does not have one person appointed as its religious leader. Rather, each local assembly chooses its own rabbi and administers its own affairs. Rabbis perform worship services, teach and interpret Jewish laws, and provide their congregations with spiritual guidance.

The synagogue, or temple, is the Jewish house of worship; many synagogues are designed so that their congregation faces the holy city of Jerusalem during worship services. Jews do not limit their worship to the synagogue, however. They also observe religious holidays and worship daily at home through prayer and special rituals performed on the Sabbath. As in other religions, however, there is considerable variation in particular practices among Jews.

You might read aloud passages from books or articles about Jewish communities or ways of life, as in the work of Isaac Bashevis Singer. Lead a discussion about the ideas expressed in these passages.

CHAPTER 33

THE EASTERN MEDITERRANEAN

Section 1: Turkey
Section 2: Israel
Section 3: Syria, Lebanon, and Jordan

OVERVIEW

In Chapter 33, students will learn about the physical, economic, and human geography of the countries of the Eastern Mediterranean.

Section 1 explores the landforms, economy, and major cities of Turkey and describes the social and political changes that have shaped modern Turkish society. It also discusses the divided island of Cyprus.

Section 2 focuses on Israel's landforms, agriculture, resources, and industry and its unique history.

Section 3 describes the political and economic challenges facing Syria, Lebanon, and Jordan.

About the Author

Golda Meir (1898–1978) was the prime minister of Israel from 1969 to 1974. As a young woman, she emigrated to Palestine from the United States. The quotation on this page is from her autobiography, *My Life.*

About the Artifact

The Dead Sea Scrolls have provided scholars with historical and biblical information dating back as far as the fourth century B.C. The discovery of these manuscripts and manuscript fragments in desert caves of present-day Israel began in 1947.

Primary Source

Description of Change: excerpted
Rationale: excerpted to focus on main idea

CHAPTER 33

The Eastern Mediterranean

66 The kibbutz to which we came in the autumn of 1921 consisted of a few houses and a cluster of trees. . . . There were no orchards, no meadows, no flowers, nothing, in fact, except wind, rocks and some sun-scorched fields. . . . Today it is quite an impressive forest, and whenever I pass it, I remember how we dug endless holes in the soil between the rocks and then carefully planted each sapling, wondering whether it would ever grow to maturity there and thinking how lovely the roadside, the whole country in fact, would be if only those trees of ours managed to survive. 99

Golda Meir

Pieces of the Dead Sea Scrolls

Each day members of three different religions go to the Cave of Machpelah (mak-PEE-luh) in the Israeli-occupied West Bank. To Muslims, the cave is a mosque. To Christians, it is a cathedral. To Jews, it is a holy temple. Many believe that Abraham and Sarah, biblical ancestors honored by Jews, Muslims, and Christians, are buried there.

Such complex interactions of history, religion, and geography are typical in the Eastern Mediterranean. The peoples of this region have built their lives among the ruins of past civilizations. Like their ancestors, they remain committed to strong beliefs that shape their lives and the future of the region.

A grove of pine saplings in **Israel**

SECTION 1
OBJECTIVES

The objectives reflect the Focus questions on this page and are assessed in the review on page 389.

1. Summarize Turkey's physical, economic, and urban geography.
2. Discuss the formation of modern Turkey and issues facing the country today, including the status of Cyprus.

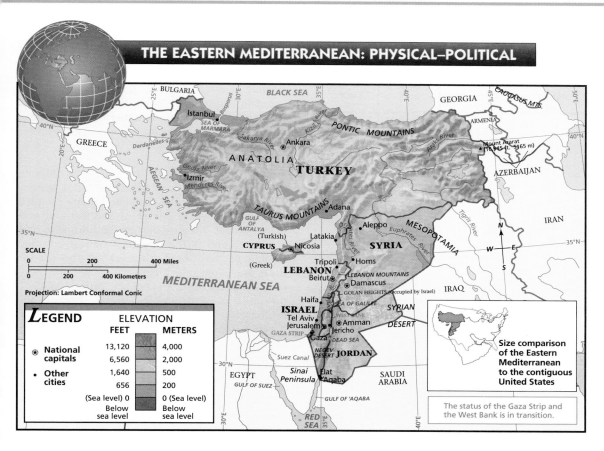

THE EASTERN MEDITERRANEAN: PHYSICAL–POLITICAL

LEGEND

ELEVATION

	FEET	METERS
⊛ National capitals	13,120	4,000
• Other cities	6,560	2,000
	1,640	500
	656	200
	(Sea level) 0	0 (Sea level)
	Below sea level	Below sea level

SCALE
0   200   400 Miles
0   200   400 Kilometers
Projection: Lambert Conformal Conic

Size comparison of the Eastern Mediterranean to the contiguous United States

The status of the Gaza Strip and the West Bank is in transition.

TURKEY

FOCUS

- *What are the physical and economic characteristics of Turkey? What are the country's major cities?*

- *How did Turkey change during the 1920s? What issues face Turkey today? What is the status of Cyprus?*

REGION **Physical Geography** Turkey is often described as a country of two peninsulas. A small part of Turkey lies on a corner of Europe's Balkan Peninsula. The larger part of Turkey occupies the Asian peninsula of Anatolia. Two narrow straits, the Bosporus (BAHS-puh-ruhs) and the Dardanelles (dahrd-uhn-EHLZ), divide the two parts of Turkey, thereby dividing Europe from Asia. These straits, along with the Sea of Marmara,

also connect the Black Sea with the Mediterranean Sea. (See the map above.)

The landforms of European Turkey are mainly rolling plains and hills. Anatolia, meanwhile, has rugged coastlines, broad interior plateaus, and mountains that rise steadily in the east. Central Turkey has two main mountain ranges: the Taurus (TAWR-uhs) Mountains in the south and the Pontic Mountains in the north along the Black Sea coast. Farther east, rows of rugged mountains and narrow valleys cross the country. These mountains merge with the Caucasus Mountains. Along Turkey's border with Iran and Armenia, the higher of the two peaks of Mount Ararat rises up to 16,945 feet (5,165 m).

Turkey has two climate types. The coasts of the Mediterranean and Black seas have winter rains and summer droughts typical of the Mediterranean climate. Western Turkey has a similar climate, but its good rains make it the country's most productive agricultural region and one of the wettest areas of Southwest Asia. The

TEACHING SECTION 1

Motivator

On a wall map, indicate the locations of Turkey and Cyprus. Ask students to consider why this region has been of strategic interest throughout history. *(It is a bridge between Europe and Asia.)* Point out that while most of Turkey is in Asia, the country has had many political, economic, and cultural ties to Europe.

Using Geography Themes

Defining Regions   Have students refer to the map on page 387, and suggest ways of dividing Turkey into regions. *(Black Sea, straits, Mediterranean, Central, Eastern)* Organize the class into groups according to the regions you agree on. Have students use their textbooks and classroom or library resources to prepare a handbook of facts about their region, focusing on landforms, climate, economy, cities, and population.

Guided Practice

Begin an outline on the board titled "Modern Turkey." Have students take turns going to the board to write details under the following main categories: Religion, Language, Daily Life, Government, Foreign Relations. *(Responses should emphasize Turkey's Western characteristics and include details about its conflict with Cyprus.)*

MULTICULTURAL PERSPECTIVES

Review with students Turkey's historically strategic position as a bridge between Asia and Europe. Point out that although Islam is the predominant religion in Turkey, Turkish Jews and Muslims have coexisted there for centuries. Tell students that this cooperation dates back to 1492. At that time, more than 100,000 Jews—who had been driven out of Spain by the Inquisition—moved to Istanbul, capital of the Ottoman Empire. Their culture was allowed to flourish there under the favor of the sultans. Tell students that Turkey has been influenced by the many peoples, religions, and empires (Arabic, Persian, Roman, Byzantine, Ottoman, and Greek) that have existed within its borders.

FACTS IN BRIEF
The Eastern Mediterranean

COUNTRY POPULATION (1994)	LIFE EXPECTANCY (1994)	LITERACY RATE	PER CAPITA GDP (1993)
CYPRUS 730,084	74, male 79, female	95% (1991)	$11,390 (Greek area) $3,130 (Turkish area) (1992)
ISRAEL 5,050,850	76, male 80, female	92% (1991)	$13,350
JORDAN 3,961,194	70, male 74, female	80% (1990)	$3,000
LEBANON 3,620,395	67, male 72, female	80% (1990)	$1,720
SYRIA 14,886,672	65, male 68, female	64% (1990)	$5,700
TURKEY 62,153,898	69, male 73, female	81% (1990)	$5,100
UNITED STATES 260,713,585	73, male 79, female	97% (1991)	$24,700

Sources: *The World Factbook 1994,* Central Intelligence Agency; *The World Almanac and Book of Facts 1995*

Which country in the region is the most populous?

---

interior of Anatolia has a drier steppe climate. To the east, the climate becomes even drier and the land more rugged. Rain and snow in the eastern mountains form the sources of the Tigris and Euphrates rivers.

Economic Geography   With its rugged landforms, less than half of Turkey is suitable for farming. The most densely settled parts of the country are around the Sea of Marmara and along the Aegean coast. Here, farmers grow typical Mediterranean crops. Wheat and barley are the important grains, while tobacco, hazelnuts, cotton, and citrus fruits are grown for export. The drier and cooler climate of the high interior plateau limits farming to grains and raising livestock.

The eastern mountains are the least developed part of Turkey. Most people here live in small, isolated villages. The region also has many ethnic Kurds, whose relations with the Turks have always been cautious. The central government hopes that the construction of large dams will bring progress to eastern Turkey. Deposits of oil and mineral ores are being developed here as well. In the meantime, mohair, the wool of Angora goats, brings in some cash income.

Because Turkey has a better network of roads and railroads than most other countries in Southwest Asia, industrial development has steadily increased since the collapse of the Ottoman Empire after World War I. Turkey hopes eventually to become a member of the European Union, and many Turks are already guest workers in EU nations.

Urban Geography   Combining the old with the new, Istanbul and Ankara are the leading cities of Turkey. As the Ottoman Empire's capital city, Istanbul was a center of political and economic power. Today, it lies in the European part of Turkey and is the country's largest city and leading seaport. Its harbor, called the Golden Horn, opens onto the Bosporus.

When the Turkish republic formed in 1923, the more centrally located Ankara became the capital city. Today, Ankara is home to some of Turkey's most important industries. It is also a center for railroad transportation and for the development of the eastern part of the country.

Turkish cities were among the first in the Muslim world to modernize. Industrial development, education, and better health care and housing are improving the standard of living in Turkey. Contact with people from the cities is slowly changing the way of life among people in rural areas as well.

Modern Turkey   The roots of modern Turkey reach back to the early 1920s, when a revolution reorganized Turkish society and opened it to Western ideas and modernization. Although most people remained Muslims, Islam lost its status as the state religion. The government took over the many schools previously controlled by religious groups. The Roman alphabet also replaced the Arabic alphabet, and laws changed to reflect European codes. Even the wearing of European-style clothing was required. In addition, Turkey became the first Muslim country to begin recognizing women's rights. Since the revolution, Turkey has distanced itself from Arab political affairs and from other Muslim countries. After several periods of military dictatorship, democratic traditions are on the rise.

Other than conflict with ethnic Kurds, which included large-scale military operations in 1995, Turkish foreign affairs focus on two main issues. The first is the bitter, centuries-old conflict between

Closure

Have students write a paragraph about ways in which Turkey might be considered a "region in transition." (*It has many ties to both Europe and Asia; industry is increasing; its standard of living is improving.*)

ASSESSMENT

Assign the Section 1 Review questions on this page.

Reteaching

Assign Section 1 of Reteaching Worksheet 33.

Extension

Have interested students research and report on Kemal Atatürk, the army officer who was elected as the country's first president in 1923. Atatürk was the principal force behind Turkey's modernization.

SECTION 2
OBJECTIVES

The objectives reflect the Focus questions on this page and are assessed in the review on page 394.

1. Describe Israel's physical and economic geography.

2. Explain how the state of Israel was created, and identify challenges the country has faced since its formation.

Istanbul, Turkey sits along the Bosporus, an important route for shipping between the Black Sea and the Aegean and Mediterranean seas beyond. Turkey's largest city has been known by several names and ruled by many empires over a period of more than 2,500 years.

the Turks and the Greeks over the island of Cyprus and the potential oil resources in the Aegean Sea. The sea's resources are a matter of pride as well as economic gain to both countries. The second issue is cooperation with the independent countries of Central Asia. During the periods of the Russian Empire and the Soviet Union, Turkey had little contact with Central Asia. Today, Turkey is leading development efforts there. In many ways, the country is still a physical and cultural bridge between Asia and Europe.

Cyprus Cyprus is a rocky, mountainous island located in the eastern Mediterranean Sea. (See the map on page 387.) A wide, fertile plain separates two major mountain ranges on the island.

Cyprus gained its independence from Britain in 1960. The island, however, has experienced cultural conflict for decades. About 78 percent of the people of Cyprus consider themselves to be Greeks. About 18 percent think of themselves as Turks. Each group is suspicious and resentful of the other. As a result of the struggle, the Greeks now live in the south part of the island, and the Turks live in the north. Their dividing line passes through the capital city of Nicosia (NIK-uh-SEE-uh). Each community has declared its own republic and has its own government. The intense dispute alternates between periods of violent bit-

terness and peaceful discussion. United Nations forces help keep peace on the island.

The Greek part of Cyprus, which has more industry and tourist facilities, has three times the per capita income of the more rural Turkish part. Yet all of the people of Cyprus know that without peace, the island will attract few tourists and its economy cannot grow. Many hope stable political relationships soon will emerge in Cyprus.

> ### SECTION REVIEW
>
> **1.** Into what two parts is Turkey physically divided, and what are the landforms of each part? Which region of Turkey is the least economically developed? What are the country's two major cities?
> **2.** List four social or political changes that have taken place in Turkey since the early 1920s.
> **3. Critical Thinking Explain** why Cyprus might be considered part of Europe rather than part of Asia.

 ISRAEL

FOCUS

● *What are Israel's major physical and economic features?*

● *When and how did Israel become a country? What challenges have faced Israel since its formation?*

LOCATION Physical Geography Israel lies along the eastern shore of the Mediterranean Sea on what once was the ancient land of Palestine. (See the map on page 387.) This small country has many environments. A strip of fertile land with a Mediterranean climate lines the country's coast. Inland is the hilly, semiarid region of Galilee. Beyond this region is the hot, arid valley of the Jordan River and the Dead Sea. The Dead Sea is nearly 1,302 feet (397 m) below sea level, the lowest point on any continent. The Dead Sea, which has no outlet to another body of water, is so salty that swimmers cannot sink in it. Southwest of the Dead Sea is the rocky Negev Desert, which stretches to the Gulf of 'Aqaba.

Section 1 Review
ANSWERS

1. The part on Europe's Balkan Peninsula (rolling plains and hills) and the part on the peninsula of Anatolia (rugged coastlines, broad interior plateaus, mountains); the eastern mountains; Istanbul and Ankara

2. Islam lost its status as the state religion; the government took over religious schools; the alphabet changed; laws changed; women's rights were recognized; democracy has increased.

3. About three-quarters of its people consider themselves to be Greeks (Greece is part of Europe).

TEACHING SECTION 2

Motivator

Help students locate the Sinai Peninsula, the Gaza Strip, the West Bank, East Jerusalem, and the Golan Heights on the map on page 387. Call on volunteers to share what they know about these places. Have students suggest why these areas have been important to Israel. *(Students may suggest strategic, political, historical, or cultural reasons.)*

Building a Vocabulary

Help students further understand **kibbutz, Zionism,** and **Holocaust.** *Kibbutz* means "gathering" in Hebrew. *Zion* refers to the Jewish people or the Jewish homeland, and the suffix *-ism* means "a practice or process." The general meaning of *holocaust* is "a thorough destruction by fire." Link this information to the Glossary definitions of the terms.

Summarizing

Challenges and Change Bring in recent news articles about Israel's relationship with its Arab neighbors, and distribute them among students. Call on volunteers to identify the main issues involved, and help students relate these to the challenges Israel has faced since its formation. Then ask students to summarize the current situation in the country.

THE WORLD TODAY VIDEODISC PROGRAM

Israel: Water, a Scarce Resource

2A 23341 4:55

Play   Pause

390

These Palestinian women near **Jerusalem** are sorting olives for packing, while other workers (background) pick the olives from trees. Olive cultivation is common in countries along the Mediterranean Sea.

Agriculture Israel is an economically developed country. Unlike in other countries of Southwest Asia, only about 5 percent of Israel's people work in agriculture. Modern commercial agriculture is typical, and Israel's farms are some of the most productive in the world. The best farmland lies along the coastal plain and on irrigated farms farther inland. Israel exports fruits, particularly oranges, as well as vegetables and cut flowers to European markets.

The **kibbutz**, or Israeli collective farm, remains an important element of Israeli identity. These farms were important in the development of the country. Today, many people from foreign countries, including the United States, travel to Israel to work with Israelis on a kibbutz.

Israel has one of the world's most sophisticated water management systems and uses nearly every drop of its water resources. The main system is the National Water Carrier, whose pipelines carry water from the Sea of Galilee throughout the country, even to the Negev Desert's northern borders. Here, irrigation agriculture has replaced dry farming and herding. In the coastal region, modern wells provide additional water.

Resources and Industry Israel has almost no oil or coal, little hydroelectricity, and relatively few minerals. The Dead Sea does contain potash, magnesium, bromide, and other salts, which are used for fertilizers and chemical products. One Israeli company has found another use for the minerals. Sales of its Dead Sea mud cosmetics are booming. The firm was not the first to use the mineral-rich mud, however. The Egyptian queen Cleopatra came to the Dead Sea for her facials centuries ago. For the most part, though, Israel's economic growth relies on the skills of its well-educated population.

Israel has developed high-technology industries, such as the manufacturing of computers, complex military weapons, and precision engineering equipment. Rough diamonds are imported, cut, polished, and then exported around the world.

Tourism is another major industry. Many people come to visit the holy places and the sunny beaches along the Mediterranean Sea and the Gulf of 'Aqaba. In addition, Israel's economy is boosted by aid from the United States and from Jews worldwide.

MOVEMENT **Historical Geography** The Jews established their first kingdom, with its capital at Jerusalem, in the eastern Mediterranean more than 3,000 years ago. Over time, however, a series of conflicts forced most of the Jews to leave the region, and it became populated largely by Arabs.

In the nineteenth century, a movement known as **Zionism** began to grow among Jews outside Palestine who longed for a homeland. The roots of Zionism lay in Europe, where Jews suffered persecution for many centuries. The Zionists believed that a Jewish state should be established in Palestine. They also believed that Jews from around the world should move there. As a result of this movement, more Jews began settling in Palestine in the late 1800s and early 1900s.

After World War II, many nations throughout the world supported the idea of creating the modern nation of Israel as a Jewish homeland. These nations were sympathetic to the persecution, known as the **Holocaust**, that Jews had suffered

Guided Practice

Write the heading "Israel's Economy" on the board. Under the heading, make a chart with three columns: Agriculture, Industry, Tourism. Have volunteers come to the board to fill in items in each column. Students might include locations, descriptions, or reasons for success.

Independent Practice

Have students work in groups to create a map of Israel. Maps should note important physical features, climate regions, and cities. Students should also label the occupied territories and the areas with a degree of Palestinian autonomy. Have each group write an extended caption about Israel's physical and political characteristics.

Closure

Organize students into pairs. Give each student several blank cards. Have students write on each of their cards a word, name, or phrase associated with or related to Israel. Then have the students in each pair exchange cards and write a sentence explaining the significance of each of the phrases written by the partner. Have pairs compare their sentences.

NEWS FROM AROUND THE WORLD — Israel

Israeli Teens Live with the Ancient and the Modern

Israel covers less than 8,000 square miles (20,720 sq. km), roughly the size of New Jersey. Yet nearly 5 million people crowd into the country. Jewish peoples have immigrated to Israel from almost every part of the world, bringing their languages and traditions with them. As a result, Israeli cities are a mixture of many cultures. American, Chinese, French, Greek, Hungarian, Indian, Iraqi, Italian, Moroccan, Turkish, and Vietnamese restaurants can be found even in small cities.

Most Israelis grow up in the cities, where the population density is highest. Because space is scarce, many city dwellers live in crowded high-rise apartment buildings. Families can find these apartments particularly cramped. Bunk beds are popular with teenagers, who often must share bedrooms with younger brothers or sisters.

For Jewish teens in Israel, schools are a unifying force, bringing together students of many different backgrounds. Arab teens attend their own schools, however, where subjects are taught in Arabic rather than Hebrew. This division is a daily reminder of the religious, cultural, and political beliefs that separate the peoples in this troubled part of the world.

In Jerusalem, Israeli teenagers can look down from the rooftops of their apartment buildings on the holy places of three religions—Christianity, Judaism, and Islam. The rooftops and the walls surrounding and dividing the city's quarters are a kind of transportation network for young Israelis. They race across the buildings and explore the rooms and stairways hidden within the walls.

Although daily life in Israel is often overshadowed by political issues and turmoil, Israeli teens, like those pictured above, are much like most young people in the United States. Television, rock music, and other symbols of American popular culture are widespread, and many young Israelis hope someday to travel or study in the United States.

during the war. The word *holocaust* means "widespread destruction." Nazi Germany persecuted and killed millions of Jews during the Holocaust. As a step toward creating a Jewish homeland, Britain, which then controlled Palestine, gave the region its independence after the war. How the area was to be divided was unclear, however.

Finally, in 1948, a United Nations plan divided the lands west of the Jordan River between Jews and Arabs. The Arab lands west of the Jordan River, called the West Bank, were joined to the country of Jordan. At this time, Jews occupied little land in the region and totaled only about one-third of the population.

Neighboring Arab countries and the Arab people of Palestine, called Palestinians, rejected the establishment of Israel. The Palestinians and the Arab countries believed that Israel's existence denied Arabs their essential rights to their homeland. As a result, war broke out between Arabs and Israelis in 1948. The Israeli forces defeated the Palestinians and their Arab allies.

Many Palestinians fled to refugee camps in nearby Arab states. From these camps, some Palestinians continued to attack Israeli settlements in hopes of regaining the land they felt was theirs. Wars again broke out between the Arab countries and Israel in 1956, 1967, and 1973.

Text continues on page 394.

Cities of the World

Making a Time Line

Have students work in small groups to create a wall-size, illustrated time line of Jerusalem, from the city's beginnings to present day. Each group should be assigned to complete a specific section of the time line. Students should use the information in the feature as well as other classroom and library resources to complete the time line. Display the finished time line, and have each group explain its section to the rest of the class.

CITIES OF THE WORLD

Jerusalem

The capital of ancient Israel 3,000 years ago, Jerusalem is the capital of modern Israel today. The city's past and present merge in its skyline of ancient ruins, medieval towers, and modern high-rise buildings.

Jerusalem is of central importance to people of the Jewish faith. For centuries, Jews living outside the city have turned toward Jerusalem in prayer three times each day. Hundreds of millions of non-Jews also care deeply about the city. Christians cherish it as the place where Jesus lived, preached, and was crucified. Muslims honor the site in Jerusalem where they believe the prophet Muhammad rose to heaven.

Perhaps more than in any other city, the threads of human history wind through Jerusalem's streets.

An Ancient History

Because it is sacred to so many, several groups have fought for control of Jerusalem in its 5,000-year history. It became the center of Jewish life in 1003 B.C., when King David made it his capi-tal. Thirty-three years later, David's son Solomon built his magnificent Temple there. Although the Babylonians conquered Jerusalem and destroyed the Temple in 586 B.C., the Israelites recovered the city and built a second Temple.

Jerusalem was under Roman control when Jesus lived and preached there. A generation later, in A.D. 66, the Jews revolted against Roman rule. The Romans crushed the revolt, however, and in A.D. 70 destroyed the second Temple. The Temple's only remnant, known today as the Western Wall, remains Judaism's most sacred religious shrine.

After the Romans were finally driven out, Jerusalem came under the control of a series of groups, including the Byzantines, Arabs, Christian Crusaders, Egyptians, and Turks. In the late seventh century, the Arabs built the Dome of the Rock and the al-Aqsa Mosque where the Jewish Temple once stood. Later, during the sixteenth century, the Turks constructed the high stone walls that still surround what today is called the Old City.

Modern Jerusalem

The modern history of Jerusalem begins in the 1860s, when the city's growing Jewish population built the first neighborhoods outside the city's walls. Jerusalem continued to spread outward, especially after the British wrestled the city from the Turks in 1917.

When modern Israel formed in 1948, another round in the struggle for Jerusalem began. Neighboring Arab states attacked Israel, and Jordan occupied the Old City. Israel recovered the western part of the city in 1949 and made it the country's capital. The eastern part, including the Old City, remained under Jordanian control.

Jerusalem was reunified when Israel defeated Jordan, Egypt, and Syria in the Six-Day War in 1967. Today, the Old City is part of a metropolis of more than 500,000 people. Within

Visitors view the Children's Memorial, part of the Holocaust Memorial, or Yad Vashem, in **Jerusalem**. Israel was founded in 1948, three years after the end of World War II and the defeat of Nazi Germany.

392

Using Geography Themes

Organize students into small groups. To reinforce the geography themes of place and location, have each group research and prepare two side-by-side maps: a detailed map of Jerusalem as a divided city between 1949 and 1967 and another map showing Jerusalem reunified in 1968. Students should use different colors to show the areas controlled by Jordan and by Israel. Both maps should include the city's major landmarks, a legend, and a caption comparing the two maps.

Time has left **Jerusalem** with many historic as well as modern structures scattered throughout the city. In which part of Jerusalem would a visitor find holy sites for three major world religions?

Map Caption Answer

The Old City

its walls are the major Jewish, Christian, and Muslim religious sites, access to which is guaranteed to all by Israeli law. Jewish, Christian, and Muslim authorities control their respective sacred sites.

The focal point of the Old City remains the Temple Mount, a huge 30-acre (12 ha) platform on which the Temple once stood. Muslims pray in the mosques located on the Temple Mount, while Jews gather at the Western Wall that forms one of the Temple Mount's edges. Nearby, Christians retrace the "Way of the Cross," which leads to the Church of the Holy Sepulcher.

Jerusalem's western part is home to more modern attractions such as the Israel Museum, the Hebrew University, Israel's parliament building, and popular shops and cafes. Near the city's western edge is Yad Vashem, Israel's memorial to the 6 million Jews killed during the Holocaust. Even in its modern sections, Jerusalem remains a city cloaked in history.

Worshippers welcome the Jewish Sabbath on Friday evening at the Western Wall. The golden Dome of the Rock, sacred to Muslims, rises to the left.

Extension

Encourage interested students to investigate the history of the Palestinian people since 1948 and to write a report detailing their arguments for a homeland. Students may wish to use Outline Map 18 to show where Palestinian settlements are located.

SECTION 3

OBJECTIVES

The objectives reflect the Focus questions on this page and are assessed in the review on page 396.

1. Identify physical and economic characteristics of Syria, Lebanon, and Jordan.

2. Examine the relationships between culture and politics in Syria, Lebanon, and Jordan.

Section 2 Review
ANSWERS

1. It is the lowest point on any continent and is extremely salty; potash, magnesium, bromide, other salts

2. Many Jews were persecuted and killed during the war, spurring support for a Jewish homeland (Israel). In the 1967 war, Israel occupied several territories, some of which are still a source of conflict.

3. Possible answers: Israel receives aid from other countries and Jews worldwide; many Israelis emigrated from developed countries; Israel's population is well educated.

Geography
and Archaeology

Tell Leilan, an archaeological site in Syria, was the center of the Akkadian empire, one of the earliest known civilizations. A city of over 10,000 flourished here more than 4,000 years ago. Researchers have found evidence in soil samples that a 300-year drought may have led to the society's downfall.

394

As a result of the war in 1967, called the Six-Day War, Israel occupied land from Egypt, Syria, and Jordan. This land included all of the Sinai (SIE-nie) Peninsula west to the Suez Canal, the Gaza (GAHZ-uh) Strip, the West Bank of the Jordan River, East Jerusalem, and the Golan (goh-lahn) Heights. (See the map on page 387.) The Sinai was returned to Egypt in 1982. In 1993 and 1995, Israel and the Palestinians agreed to limited Palestinian self-rule in Gaza and parts of the West Bank. Negotiations on expanding this agreement and on the status of the Golan Heights have continued.

Population and Urban Geography

Today, Israel is a modern, urban nation with a democratic government. Nearly 85 percent of its people are Jewish, and large numbers of Jews from both developed and developing countries continue to immigrate to Israel. Integrating Jews from many different cultures has not always been easy. In addition, many Palestinians live in Israel. Even though the Palestinian population in Israel is growing rapidly, the country will remain a primarily Jewish state for years to come.

As Israel's population has grown, so have the country's cities. Tel Aviv (tehl uh-VEEV) is the nation's largest city. Haifa is the country's major port and an important center of oil-refining and chemical industries. Jerusalem, the capital of Israel, is the nation's cultural and political center. (See "Cities of the World" on pages 392–393.)

Issues Israel's most serious problem is the ongoing conflict with its Arab neighbors. A crucial question is the fate of the Palestinian people and the territories Israel has occupied since 1967. Many Israelis have settled in these lands, further complicating the situation. Although Israel and the Palestinians have reached some agreements, the road to a stable and lasting peace is a long one.

Palestinians living in Israel continue to stage demonstrations seeking their independence. Palestinians living outside Israel also long to return to the land they view as their home. The Palestinian Liberation Organization (PLO) is the leader of the Palestinian cause and is pursuing further negotiations for Palestinian self-rule with Israel. The PLO has many internal divisions, however, which may threaten its stability and effectiveness.

Because of the threat of war and frequent border clashes, Israel maintains a large army. In fact, all young Israeli men and women must serve

in the military. The cost of supporting this military force places a burden on the country's economy, even though Israel receives a large amount of foreign aid from the United States and other countries. Despite agricultural and industrial successes, Israel's future remains troubled until permanent peace comes to countries of the Eastern Mediterranean.

SECTION REVIEW

1. How is the Dead Sea unusual? What resources does it contain?
2. What role did World War II play in the creation of the state of Israel? How are the effects of the 1967 Six-Day War felt today?
3. **Critical Thinking** **Suggest** an explanation for why Israel is more economically developed than other countries in the Eastern Mediterranean.

SYRIA, LEBANON, AND JORDAN

FOCUS

- *What are the physical and economic characteristics of Syria, Lebanon, and Jordan?*

- *How are culture and politics related in these three countries?*

Syria Syria stretches east from the mountainous Mediterranean coast to the plains of Mesopotamia. (See the map on page 387.) The Syrian Desert lies in the center and southeast of the country. Accordingly, most Syrians live in the western one-fourth of the country, which has a Mediterranean climate. Settlement farther east is dependent on the water resources of oases and of the Euphrates River.

Syria is trying to develop modern agriculture, but despite programs to encourage high-technology farming, most of the agricultural land is made up of small subsistence farms and large inefficient estates. In the mountain valleys of western Syria, farmers grow citrus fruits and other Mediterranean crops. In the drier environments east of the coastal region, cotton, wheat,

Ask students to bring in news stories about Syria, Lebanon, and Jordan. Use these stories as a basis for discussing the current issues confronting these countries. Tell students that they will learn in this section about the physical, economic, and cultural geography that forms the background to current events in this region.

Building a Vocabulary

Point out to students the photo on page 381. Note that the tall towers are called **minarets,** from an Arab word for "lighthouse." Ask students to suggest how this definition reflects the minarets' role as a symbol of a community to Muslims. (*A lighthouse is a symbol of guidance and safety to ships at sea just as a mosque and its minarets offer guidance and symbolize a community to Muslims.*)

Guided Practice

Write Syria, Lebanon, and Jordan on the board. Have students name characteristics of each country that hinder agricultural and industrial production. Write their responses on the board beneath the appropriate heading. Then have volunteers label each item as being related to physical geography or human geography.

---

and barley are key crops, and livestock is important. In the Syrian Desert, a few Bedouin herders still move their flocks with the seasons. Even though about one-third of Syria's people work in agriculture, the country must import food.

Agricultural progress depends on developing water resources. Syria's irrigation water comes mostly from the Euphrates River in the northeast and the Orontes (aw-RAHNT-eez) River in the northwest. A huge dam on the Euphrates River has brought thousands of acres of land into production. The dam also provides electric power for industry in the northern part of the country. The government hopes this area will become a center for commercial agriculture.

Although traditional livestock raising, farming, textiles, and crafts are still the main occupations in Syria, the country has some industry and is producing more basic consumer goods. In addition, Syria has a small amount of oil available for export, and new oil deposits are being developed in eastern Syria. Public education has spread widely in recent years, and all large cities have universities.

Damascus, Syria's capital, was one of the great oasis cities of the ancient world as well as a center of Islamic civilization. The city lies on the plains east of the mountains of Lebanon. The water from the rains and snows in the mountains have supplied Damascus and the surrounding farmlands for centuries.

Modern Damascus is a city of contrasts. Wide boulevards remain from French colonial times. These are mixed with the narrow streets of the old city and marketplaces, called **suqs** (sooks). Traditional textiles and metalwork from the Damascus *suqs* are famous throughout the world for their beauty and design. Ancient Roman ruins also still stand in various parts of the city. And rising over the city's mosques are dozens of **minarets**. A minaret is the tower of a mosque from which a *muezzin* (crier) calls Muslims to pray five times a day.

Syria has been a leader in the Arab opposition to Israel, and its relations with Turkey and Iraq are strained as well. Syria claims some Turkish territory along the Mediterranean, while Iraq fears that Syria will divert too much water from the Euphrates River.

Once aided by the Soviet Union, Syria has changed its foreign policy since the end of the Cold War and has proposed peace negotiations with Israel. It now seeks aid from Europe and the

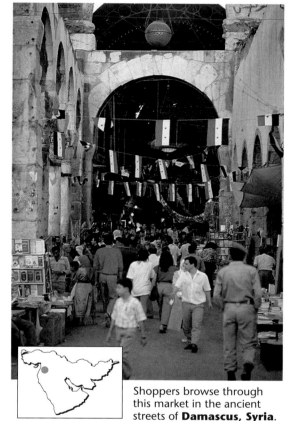

Shoppers browse through this market in the ancient streets of **Damascus, Syria**. The Syrian capital is one of the world's oldest cities.

United States. Syria's leading role in Southwest Asian affairs is likely to last for many years.

Lebanon Lebanon lies between Israel and Syria, wedged between the Mediterranean coast and the Lebanon Mountains. (See the map on page 387.) Its shore was the ancient home of the Phoenician traders. Later, it was an important province of the Roman and Ottoman empires. Unlike in any other country in the region, about one-fourth of the people of Lebanon are Christian. The beautiful mountains and coast of Lebanon have attracted many other groups as well. These include people representing more than four different varieties of Islam. Traditionally, the Lebanese have one of the most modern and Westernized cultures in Southwest Asia.

Beirut (bay-ROOT), the capital, used to be called the "Paris of the Middle East." Its schools and universities once attracted students from

GLOBAL PERSPECTIVES

Some experts argue that the seeds of reconciliation in Southwest Asia lie in economic cooperation among Israel, the Palestinians, the Arab nations, and Europe, with the goal of boosting the standard of living throughout the region. They believe that prosperity provides less fertile ground for the discordant forces of ethnic intolerance and fundamentalism. One Israeli commentator writes: "Here, as elsewhere in the world, democracy emerges faster, and works better, where steady incomes, reasonable housing, accessible education, sufficient food, and adequate health care are widespread."

Discuss with students whether or not they agree with this statement, not only for Southwest Asia, but for other countries where democracy has or has not taken hold.

395

Independent Practice

Have students write a statement describing what they think is the most important industry for each country discussed in the section to develop for the future. Call on volunteers to read and explain their statements to the class.

Closure

Discuss the future prospects of Syria, Lebanon, and Jordan, in light of their physical and economic characteristics and the political tensions in the region.

ASSESSMENT

Assign the Section 3 Review questions on this page.

Reteaching

Assign Section 3 of Reteaching Worksheet 33.

Extension

Have interested students investigate the history of the civil war in Lebanon and the effects it has had on the city of Beirut and on the country as a whole. Encourage students to create maps and charts as needed to illustrate their reports.

THE WORLD TODAY VIDEODISC PROGRAM

Jordan: Women Balancing Tradition and Modernity

2A 32201 3:50

Play   Pause

Section 3 Review
ANSWERS

1. Jordan; Syria
2. The wars have been caused by fighting among and within the country's many religious groups, including Christians and Muslims. The presence of Palestinian refugees has further complicated the situation.
3. Syria has been a leader in the Arab opposition to Israel but now has proposed peace negotiations with Israel. Lebanon and Jordan both host hundreds of thousands of Palestinian refugees as a result of the conflict.

396

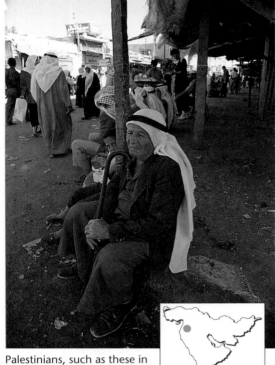

Palestinians, such as these in **Jordan**, are a people without a country of their own. In recent years, however, Palestinians have won the right to limited self-rule in some territories occupied by Israel since 1967.

throughout Southwest Asia, and tourists flocked to its beaches and historic sites. Unfortunately, periodic civil wars that began in the late 1950s and intensified between 1975 and 1990 have wracked Lebanon. Christians and the various Muslim groups living in Lebanon are the main opponents. Fighting has also broken out among factions within these religious groups.

The Arab-Israeli conflict has further complicated the situation in Lebanon, as several hundred thousand Palestinians have settled in refugee camps there. Foreign countries and the United Nations have intervened many times in Lebanon to resolve the conflict. The most heated intervention has come from Syria and Israel. Both countries have invaded and occupied areas of Lebanon, mainly to keep the civil war from spilling over into their countries and because of the Palestinian refugee issue.

Lebanon's economy lies largely in ruins, and most of the international business community has left the country. Today, after years of fighting, peace may be on the horizon. A stable government, however, has yet to come to power.

Jordan Jordan stretches east from the Dead Sea and the Jordan River into the rocky Arabian Desert. Only the northwest corner of the country is not arid. (See the map on page 387.)

When Jordan became an independent kingdom in 1946, fewer than 400,000 people lived east of the Jordan River. They were mostly migrating herders and subsistence farmers. When Israel formed in 1948, nearly a million Palestinians took refuge in Jordan. Even more refugees came after the 1967 Arab-Israeli war. As a result, Palestinians in Jordan outnumber the country's original inhabitants. Relations between these groups often have been tense. The large refugee population has strained the country's resources and placed Jordan on the front line of the Arab-Israeli conflict.

Jordan has few minerals or other natural resources, and the dry environment makes farming difficult. In fact, most of Jordan's desert lands can never be cultivated. Large-scale irrigation is only possible in the northwestern area. As a result, much of the country's food must be imported. Aid from many countries, including the United States and the oil-rich Arab states, has helped the country develop. Negotiations between Jordan and Israel over water rights and other issues are progressing.

Modern Jordan centers around the capital city of Amman (a-MAHN). Jordan's port city of 'Aqaba on the Gulf of 'Aqaba also is an important trade center and the nation's only seaport. Goods headed for Damascus and northern Iraq travel through the port daily. The country is trying to build more industries. Food products, simple consumer goods, and building materials are produced for local markets. Jordanians hope that peace in the region will help the country's economic development.

SECTION REVIEW

1. Which country discussed in this section does not border the Mediterranean Sea? To which country does the Euphrates River provide irrigation water and hydroelectric power?
2. How have cultural factors played a role in Lebanon's civil wars?
3. **Critical Thinking Discuss** the ways in which Syria, Lebanon, and Jordan have been involved in the Arab-Israeli conflict.

ANSWERS

Building a Vocabulary

1. An Israeli collective farm
2. The movement in favor of the establishment of a Jewish state in Palestine to which Jews from around the world should move; in Europe
3. The Holocaust

Recalling and Reviewing

1. Around the Sea of Marmara and along the Aegean coast; wheat, barley, tobacco, hazelnuts, cotton, citrus fruits
2. The organization that leads the Palestinian cause (independence and the return to Palestine); internal divisions
3. Much of the rest of the country is desert; the development of water resources
4. About one-fourth of its people are Christian.
5. Relations between the refugees and Jordan's original inhabitants, whom the refugees outnumber, have been tense. Also, the refugee population has strained the country's resources and has placed Jordan at the forefront of the Arab-Israeli conflict.

CHAPTER 33 CHECK

Hagia Sophia in **Istanbul, Turkey**

Reviewing the Main Ideas

1. Turkey lies on two peninsulas and has two climate types. Less than half of the country is suitable for farming. Ankara became the capital of Turkey in 1923 when a revolution reorganized Turkish society.
2. Cyprus, a Mediterranean island, is divided between Greeks and Turks. Each community has declared its own republic.
3. Israel lies along the eastern shore of the Mediterranean Sea. Israel has few resources but is economically developed, with commercial agriculture and high-tech industries. The country was formed in 1948 as a Jewish homeland. Arab countries and people rejected Israel's establishment. Israelis and Arabs have reached some agreements on Palestinian self-rule, but conflicts continue.
4. Syria, which stretches from the Mediterranean coast to Mesopotamia, is trying to modernize its agriculture. Lebanon lies between Israel and Syria and has been wracked by civil wars since the late 1950s. Jordan, stretching east of Israel, hosts many Palestinian refugees.

Building a Vocabulary

1. What is a kibbutz?
2. Define *Zionism*. Where did the movement originate?
3. What name has been given to the persecution and killing of Jews during World War II?

Recalling and Reviewing

1. What are the most densely settled areas of Turkey? What crops are grown in these areas?
2. What is the Palestinian Liberation Organization? What challenges does it face?
3. Why do most Syrians live in Syria's western region? On what does Syria's agricultural progress depend?
4. How is Lebanon culturally distinct in the region?
5. How have Jordan's Palestinian refugees presented a challenge for the country?

Thinking Critically

1. Would you describe the Arab-Israeli conflict as a religious conflict? Why or why not?
2. Using three examples from the chapter, discuss the effects that political conflicts can have on a country's economy.

Learning About Your Local Geography

Cooperative Project

On a kibbutz, members share equally all the responsibilities for its operation. To what extent does your classroom or school function in the same way? With your group, make a list of all of the maintenance chores required to keep your classroom or school a pleasant environment. Then, make a chart showing who is responsible for completing each chore. If certain chores are not already assigned, complete these chores with your group (if possible).

Using the Five Themes of Geography

Sketch a map of Israel, Lebanon, and Jordan. With arrows, show how the relative **locations** and borders of these countries have affected peoples' **movement** within the **region**.

Thinking Critically

1. Possible answer: the conflict has religious elements. Many Israelis are Jews and many Arabs are Muslims, and the conflict began when Israel was created as a Jewish homeland. However, the conflict is also a debate about homelands, as both the Israelis and Arabs see the region as being their homeland.
2. In Cyprus, conflicts between Greeks and Turks discourage tourism, a major source of income. In Israel, the cost of supporting a large army places a burden on the country's economy. In Lebanon, civil wars have driven out the international business community, and the economy is in ruins.

Learning About Your Local Geography

Groups should have considered the extent to which responsibilities are or might be shared in the classroom or school.

Using the Five Themes of Geography

Arrows should show the movement of Palestinian refugees from Israel to Lebanon and Jordan.

397

CHAPTER 34 Interior Southwest Asia
Planning Guide

TEXTBOOK	TEACHING RESOURCES	OTHER AVAILABLE MATERIALS
Section		
1. Iraq Pages 399–401	Reteaching Worksheet 34.1 Challenge/Enrichment Worksheet 34	*Study Guide for Spanish Speakers* *Lecture Notes,* Chapter 34
2. Iran Pages 401–403	Reteaching Worksheet 34.2 Critical Thinking Worksheet 34	*Study Guide for Spanish Speakers* *Lecture Notes,* Chapter 34
3. Saudi Arabia Pages 404–405	Reteaching Worksheet 34.3 Skills Worksheet 34	*Study Guide for Spanish Speakers* *Lecture Notes,* Chapter 34
4. The Arabian Perimeter Pages 405–406	Reteaching Worksheet 34.4 Graphic Organizer 34	*Study Guide for Spanish Speakers* *Lecture Notes,* Chapter 34
Chapter Review Page 407	Chapter 34 Test Chapter 34 Portfolio Assessment See also *Alternative Assessment Forms*	*Test Generator,* Chapter 34
Unit Review Pages 408–409	Unit 7 Test Unit 7 Portfolio Assessment See also *Alternative Assessment Forms*	*Test Generator,* Chapter 34

CHAPTER RESOURCES

Lecture Notes, Chapter 34
Teaching Resources
    Chapter 34 Graphic Organizer
    Chapter 34 Skills Worksheet
    Chapter 34 Critical Thinking Worksheet
    Chapter 34 Challenge/Enrichment Worksheet

REVIEW AND ASSESSMENT

Chapter 34 Check, page 397
Chapter 34 Test, *Teaching Resources*
Chapter 34 Portfolio Assessment, *Teaching Resources*
Test Generator, Chapter 34
Alternative Assessment Forms

ADDITIONAL RESOURCES

Books for Teachers

Blair, Sheila, and Jonathan Bloom, eds. *Images of Paradise in Islamic Art.* Austin: University of Texas Press, 1991.

Bowen, Donna Lee, and Evelyn A. Early, eds. *Everyday Life in the Muslim Middle East.* Bloomington: Indiana University Press, 1993.

Hawley. T. M. *Against the Fires of Hell: The Environmental Disaster of the Gulf War.* San Diego, CA: Harcourt Brace Jovanovich, 1992.

Rubin, Barnett R. *The Fragmentation of Afghanistan.* New Haven: Yale University Press, 1995.

Books for Students

Batchelor, John, and Julie Batchelor. *The Euphrates.* Minneapolis, MN: Silver Burdett Press, 1981.

Clark, James I. *The Middle East.* Evanston, IL: McDougal, Littel, and Co., 1989.

Cook, Michael. *Muhammed.* New York: Oxford University Press, 1983.

The World Today *Videodisc Program*

A variety of still images, short videos, and activities are available for you to use as you teach this chapter. *Lecture Notes* include bar codes for specific images and videos.

Multimedia Resources

Arabia: Sand, Sea & Sky. (VHS, 59 minutes) National Geographic Society, Educational Services, Washington, DC 20036.

Family Matters: The Role of the Family in the Middle East. (VHS, 25 minutes) Encyclopedia Britannica Educational Corporation, 310 South Michigan Avenue, Chicago, IL 60604–9839.

Picture Atlas of the World. (Interactive maps, photo bank; animation, audio on CD-ROM) National Geographic Society, Washington, DC.

Revolution: Iraq and Iran. (VHS, 25 minutes) Encyclopedia Britannica Educational Corporation, 310 South Michigan Avenue, Chicago, IL 60604–9839.

MEETING INDIVIDUAL NEEDS

Limited English Proficiency (LEP)

Help students create a cause-event-effect graphic organizer of the Interior Southwest Asian conflicts discussed in Chapter 34. Write a column of events on the board, and call on volunteers to complete the organizer by filling in the causes and effects. Students may wish to copy the graphic organizers to help them understand the region's challenges.

Learners Having Difficulty

Help each students create a chart with the following headings across the top: Economic Geography, Physical Geography, Political Geography. Have students write the names of the Interior Southwest Asian countries along the left side of their charts. Call on volunteers to read the text aloud while students complete their charts.

Auditory Learners

Have students write radio broadcasts about news events that have occurred or might occur in Interior Southwest Asia. Broadcasts might be about the Persian Gulf War, conflicts among various ethnic groups, or events of economic importance (such as those concerning the region's oil resourses). Have students present their broadcasts.

Visual Learners

Assign a group of students to create a map that shows the world's top 10 oil-producing countries, and assign another group to create a map that shows the world's top 10 oil-consuming countries. Use the maps as a basis for discussing the global importance of Interior Southwest Asia.

Tactile/Kinesthetic Learners

Before class, prepare five clipboards with five sheets of paper each, and label each clipboard with one of the geography themes. Organize students into five groups, assign each group a section of Chapter 34. Start each group with a clipboard; have students find ideas in their section that relate to the geography theme, as a member of each group records them on the top sheet. After a few minutes, have the groups place their sheets at the bottom, and pass the clipboard on to the next group. Continue the activity until the groups have completed all five clipboards, and then compare the lists.

Gifted Learners

Have interested students investigate the roles and rights of women in the Interior Southwest Asian countries and report their findings to the class.

MULTICULTURAL FOCUS

Muslims believe that the Holy Qur'an was given to the prophet Muhammed in a series of revelations by the angel Gabriel. Islamic history does not reveal whether the book was first transcribed during Muhammed's life or after his death in A.D. 632, but Muslims consider Allah, and not Muhammed, to be the author of the book. They feel that the incomparability of the text's poetic language to any human writings provides conclusive evidence for the book's divine origins.

The Qur'an teaches that there is only one God, Allah. It also mentions Allah's prophets, who included Abraham, Moses, Jesus, Muhammed, and others. The book foretells a judgment day for all humans and defines a code of conduct for daily living.

Most Muslims memorize all parts of the Qur'an, making it one of the most widely read books in the world. Even those who cannot read learn the text by ear, and the art of properly reciting the book has been handed down through generations. The Qur'an is also the final authority in Islamic law.

For centuries, Muslims rejected the idea of translating the Qur'an into languages other than Arabic because they believed that Allah's words should remain in their original form. As a result, people who converted to Islam learned Arabic. After the early 1900s, however, translations of the Qur'an began to appear. Have students obtain one of these translations and choose a passage to interpret and present to the class.

CHAPTER 34
INTERIOR SOUTH-WEST ASIA

Section 1: Iraq
Section 2: Iran
Section 3: Saudi Arabia
Section 4: The Arabian Perimeter

OVERVIEW

Chapter 34 describes the countries of interior Southwest Asia, focusing on the role that oil has played in changing the region's human geography.

Section 1 examines the influence of Iraq's physical geography on settlement and highlights Iraq's economic and political characteristics.

Section 2 describes the physical, economic, and human geography of Iran. It also examines Afghanistan, one of the world's least developed nations.

Section 3 focuses on Saudi Arabia's physical geography and the impact of oil on this country.

Section 4 examines the smaller states on the Arabian perimeter—Kuwait, the Gulf states, Oman, and Yemen.

About the Author

Ameen Rihani (1876–1940) was a poet and translator of Arabic literature. Rihani, of Syrian heritage, wrote extensively about Southwest Asia. The quotation on this page is from *Around the Coasts of Arabia*.

Primary Source

Description of Change: excerpted
Rationale: excerpted to focus on main idea

CHAPTER 34

Interior Southwest Asia

> ❝ We meet with many of them on the road to Mecca. From Java and Sumatra, from India and Afghanistan, from Morocco . . . they seek the fountain of eternal life and bliss. Colour and race and rank, like the distance of land and sea, yield to the supreme purpose, make the world of the pilgrim move in a common faith toward a goal unique, a shrine in the desert. ❞

Ameen Rihani

Clay cuneiform tablet from ancient **Mesopotamia**

According to the theory of plate tectonics, a new ocean is being born under the Red Sea as Arabia and Africa slowly separate. While the mountains of western Saudi Arabia push upward, the earth's crust is buckling beneath the plains of Mesopotamia and the Persian Gulf.

The human geography of Interior Southwest Asia, however, is changing much more quickly than the landforms. Oil is the key reason. Beneath Mesopotamia and the Persian Gulf lie the world's largest deposits of petroleum. The discovery of this "black gold" in the region has brought wealth and increased Western contact. The inheritors of this vast treasure are now struggling to deal with the resulting changes in their societies.

Muslims at prayer in **Mecca, Saudi Arabia**

GEOGRAPHY DICTIONARY
fallow
theocracy
pilgrimage
perimeter
sultanate

Using the Quotation

Point out that the quotation refers to the Islamic pilgrims who travel to Mecca. Have a volunteer locate Mecca on a wall map. Ask students to summarize the meaning of the quotation. *(The pilgrims are of different ethnic groups and economic levels and are separated by distance, but they are united in their faith.)* Discuss the unifying influence that a shrine or other holy place can have on followers of a particular religion.

Using the Geography Dictionary

Have students classify the words in the Geography Dictionary according to the following categories: Religion, Government, Agriculture. Tell students that one term can be assigned to two categories (**theocracy**— Government and Religion) and that one term will not be classified (**perimeter**). Have students check their answers with a partner.

SECTION 1
OBJECTIVES

The objectives reflect the Focus questions on this page and are assessed in the review on page 401.

1. Discuss the effects of Iraq's physical geography on its settlement patterns.

2. Analyze Iraq's economic and political geography.

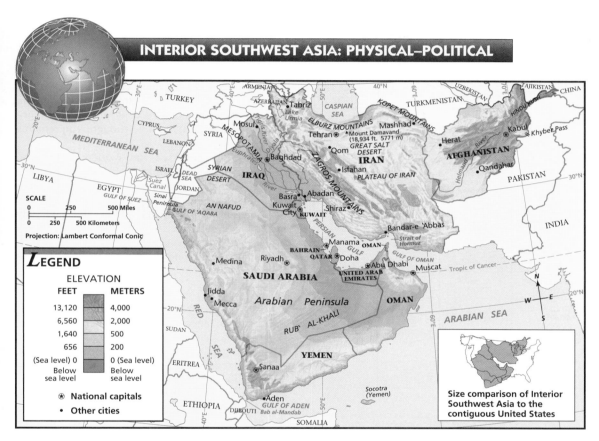

INTERIOR SOUTHWEST ASIA: PHYSICAL–POLITICAL

LEGEND

ELEVATION

FEET	METERS
13,120	4,000
6,560	2,000
1,640	500
656	200
(Sea level) 0	0 (Sea level)
Below sea level	Below sea level

⊛ National capitals
• Other cities

SCALE
0   250   500 Miles
0   250   500 Kilometers
Projection: Lambert Conformal Conic

Size comparison of Interior Southwest Asia to the contiguous United States

Using the Physical–Political Map: Interior Southwest Asia

Refer students to the map on this page, and have them suggest which country or countries might be politically dominant in the region. *(Students might say Saudi Arabia because of its size or Iraq because of its central location.)* Ask students which countries might be disadvantaged by their locations or physical features.

Have a volunteer name the major bodies of water in the region. Point out the Strait of Hormuz and the Bab al-Mandab. Ask why some or all of the region's bodies of water might be of strategic or economic importance. Then ask how countries that border these bodies of water might be able to use their locations to their advantage in the region or in the world.

IRAQ

FOCUS

● How has Iraq's physical geography affected the country's settlement patterns?

● What is the economic geography of Iraq? the political geography?

LOCATION **Physical Geography** Iraq, the ancient land of Mesopotamia, lies at the southeastern end of the Fertile Crescent. The country's rich plains along the Tigris and Euphrates rivers are bounded by the Syrian and the An Nafud deserts in the west and by the Zagros Mountains of Iran in the east. In the north is a mountainous area, similar to nearby parts of Iran and Turkey. Most of Iraq's rain falls in these northern mountains. The plains, deserts, and mountains form Iraq's three landform regions.

Population In the northern part of Iraq, most of the people live in the mountain valleys and grow grains such as wheat and barley. Sheep, goats, and cattle graze on the mountain slopes. While some valleys have small irrigation projects, rain is the sole source of water for most crops. Many people in the north are Kurds, who continue to seek independence.

In the western deserts and dry grasslands, a few Bedouins still graze their flocks over wide areas. In the dry seasons, the nomadic Bedouins stop at oases and river settlements to trade their animal products for grain and manufactured goods. The traditional life of the nomad is ending, however. More people are settling in villages and on permanent pastures, as well as in cities.

The most populated areas of Iraq are the regions of farming villages located on the plains along rivers and irrigation canals. The capital city of Baghdad on the Tigris River is the country's largest city. It is a center of manufacturing,

Motivator

Write Iraq on the board. Have students call out people, phrases, or events that come to mind. (*Student responses are likely to concern the 1991 Persian Gulf War: invasion of Kuwait, oil fires, Saddam Hussein*) Write students' responses on the board. Discuss how they might be divided into categories.

Building a Vocabulary

Have a student volunteer paraphrase the meaning of **fallow**. Ask students why leaving a field fallow might present problems for some farmers. (*loss of income*)

Analyzing Settlement Patterns

Ask students to draw their own outline maps of Iraq, indicating major landforms and rivers. Then ask them to show major settlement areas and, using arrows and short captions, explain the relationship between physical geography and these settlement patterns.

MULTICULTURAL PERSPECTIVES

Point out that the region in southern Iraq that was once called Mesopotamia was home to the world's earliest civilizations. People living between the Tigris and Euphrates rivers (*Mesopotamia* means "between the rivers") developed an advanced civilization based on agriculture. Have students do research on the ancient Sumerians to discover some of their important achievements. (*first writing system, advances in mathematics and astronomy*) Then hold a class discussion on the legacy of this ancient civilization in the modern world.

Chart Caption Answer

High revenues and relatively small populations

FACTS IN BRIEF — Interior Southwest Asia

COUNTRY POPULATION (1994)	LIFE EXPECTANCY (1994)	LITERACY RATE (1990)	PER CAPITA GDP (1993)
AFGHANISTAN 16,903,400	46, male 44, female	29%	$200 (1989)
BAHRAIN 585,683	71, male 76, female	77%	$12,000
IRAN 65,615,474	65, male 67, female	54%	$4,780
IRAQ 19,889,666	65, male 67, female	60%	$2,000 (GNP)
KUWAIT 1,819,322	73, male 77, female	73%	$15,100
OMAN 1,701,470	66, male 70, female	41% (1992)	$10,000
QATAR 512,779	70, male 75, female	76% (1992)	$17,500
SAUDI ARABIA 18,196,783	66, male 70, female	62%	$11,000
UNITED ARAB EMIRATES 2,791,141	70, male 74, female	73% (1992)	$24,000
YEMEN 11,105,202	50, male 53, female	38%	$800
UNITED STATES 260,713,585	73, male 79, female	97% (1991)	$24,700

NOTE: Recent GDP figures for Afghanistan are unavailable because of continuing conflict and instability.

Sources: *The World Factbook 1994*, Central Intelligence Agency; *1995 Britannica Book of the Year*

Why do you suppose the per capita GDP figures for some of these countries are as high as they are?

---

government services, and commerce. Baghdad is also an ancient city of Islamic culture and learning. The government is attempting to make Baghdad the grandest city in the Arab world. It has built impressive public buildings and freeways that contrast against the narrow, crowded streets of the old parts of the city.

Economic Geography Even though less than 15 percent of Iraq's land is arable, about one-third of the country's people are involved in agriculture. To conserve moisture and nutrients, croplands must be left **fallow**, or unused, every two or three years.

Irrigation has always been vital for agriculture in the region, and irrigation maintenance—clearing mud from canals and repairing flood damage—is a constant task for Iraq. Soil salinization presents another challenge to the region's

agriculture. As irrigation water evaporates, salts build up in the soils. Over the centuries, a large part of Iraq's irrigated fields has become too salty for most crop growth. Some crops, however, such as barley and date palms, can grow in slightly salty soil. With the introduction of modern farming practices, the country could increase its agricultural production. For now, though, Iraq remains a major food importer.

Iraq has some of the world's largest oil deposits. Oil fields lie both in the south near the Persian Gulf and in the northern part of the country. Pipelines carry the oil to Mediterranean and Red Sea ports. Unfortunately, the country's highly mechanized oil industry provides few jobs. Iraq's other industries concentrate on food products, consumer goods, and building materials such as bricks and cement.

Political Geography In recent years, Iraq has been an aggressive state. After agriculture, the military is the country's largest employer. Hoping to gain new territory, Iraq invaded Iran in 1980. A decade-long war between the two countries resulted. Neither side gained significant territory in the devastating conflict.

In 1990, Iraq invaded the tiny, oil-rich country of Kuwait. The United States and other nations sent troops to keep Iraq from invading neighboring Saudi Arabia. Although the United Nations demanded that Iraq withdraw from

Although most of the population of **Iraq** lives in urban areas, these Arabs live in the marshlands of the southeastern part of the country.

Guided Practice

Write the headings Economic Geography and Political Geography on the board. Have volunteers come to the board to write, under each heading, descriptive phrases that relate to Iraq. When they have finished, ask students to suggest possible connections between the two columns. (*Both the war with Iran and the invasion of Kuwait may have had economic components.*)

Closure

Ask students to write summary statements about Iraq's physical, economic, or political geography.

ASSESSMENT

Assign the Section 1 Review questions on this page.

Reteaching

Assign Section 1 of Reteaching Worksheet 34.

SECTION 2
OBJECTIVES

The objectives reflect the Focus questions on this page and are assessed in the review on page 403.

1. Describe Iran's physical features, economic activities, and cities.
2. Identify Iran's cultural characteristics and summarize the country's recent political history.
3. Identify key features of Afghanistan.

Kuwait, Iraq refused. Finally, after a brief air and ground conflict, known as the Persian Gulf War, Iraq suffered a crushing defeat, and Kuwait was freed. Iraq's totalitarian government still clings to its ambition to become the leader of all of the Arab nations.

SECTION REVIEW

1. Which of Iraq's three landform regions is the most populated? Why?
2. What agricultural challenges does Iraq face? Where are the country's oil fields?
3. **Critical Thinking** **Suggest** why the United States became involved in the prevention of Iraq's expansion in 1990.

This Iranian man is drawing water from a well that is part of a *qanat* in northeastern **Iran**. Runoff from the mountains in the background provides water for the surrounding farmlands and communities.

IRAN

FOCUS

- *What are the physical features of Iran? economic activities? towns and cities?*
- *What cultural characteristics does Iran have? How has the country changed politically over the past several decades?*
- *What are Afghanistan's key features?*

LOCATION **Physical Geography** Iran, formerly called Persia, is a country of arid highland plains and rugged mountains. (See the map on page 399.) The heart of the country is the high Plateau of Iran. The Zagros Mountains rise to the southwest, while to the north are the high Elburz Mountains and the Kopet Mountains. Iran's many mountains are steadily rising as a result of the region's tectonic activity, which also causes frequent earthquakes.

High pressure dominates the atmosphere over Iran most of the year, resulting in mainly dry, hot weather. Steppe and desert climates are typical. Central and southeastern Iran are particularly desolate. Winds blowing over the many dry lake basins create huge dust storms here. Rains come mostly during the winter and spring when middle-latitude storms sweep across the country. The wettest part of Iran lies along the Caspian Sea.

INTERACTION **Economic Geography** About one-third of Iran's people battle rugged land and difficult climates to make their living from farming. Barley and wheat are the major grains produced. Tree crops, particularly pistachio (puh-STASH-ee-oe) nuts and almonds, are grown in the southeast. In fact, Iran is the world's largest producer of pistachios. The lands along the shore of the Caspian Sea also are agriculturally important, producing fruits, tea, cotton, rice, and mulberry trees. The mulberry trees feed silkworms, which support Iran's carpet and clothing industries.

Rain-fed agriculture is possible only in the high plateaus and mountains of northern Iran. Elsewhere in the country, the waters of the mountains must be diverted to the fertile valley soils. The Iranian government has built many large-scale irrigation projects over the years. Individual communities manage thousands of additional small-scale irrigation projects. Iran also is famous for its *qanats* (kuh-NAHTZ). These long tunnels carry water from springs at the foot of the mountains to the plains. *Qanats*, which need a great deal of maintenance, are very deep near the mountains but rise to the surface near the farms. Water shortages and dry *qanats* have caused many people to migrate to towns from villages. (See "Planet Watch" on pages 382–383.)

Geographical Sidelight

Iran was once known to the Western world as Persia. In 1935, the country's government changed the name to Iran, which means "land of the Aryans." About two-thirds of Iran's people are descended from Aryan tribes that migrated from central Asia nearly 4,000 years ago.

TEACHING SECTION 2

Motivator
Tell students that until 1935, Iran was called Persia. Ask students to think of phrases with the term *Persian* in them. (*Persian Gulf, Persian Empire, Persian rugs*) Ask students what they can surmise about Iran's physical, historical, economic, and human geography based on these phrases.

Building a Vocabulary
Point out that Iran is a **theocracy**. Remind students that *theo-* means "god" and *-cracy* means "rule" or "government." Ask students to contrast a theocracy with a democracy.

Guided Practice
After reviewing Iran's human geography (pages 402–403), help students create a graphic organizer that classifies elements of Iran's culture and political history.

Independent Practice
Provide students with Outline Map 17. Have them locate the following on the map: Iran; Afghanistan; Plateau of Iran; Zagros, Elburz, and Kopet Mountains; Hindu Kush; Khyber Pass; Great Salt Desert; Persian Gulf; Caspian Sea; Gulf of Oman; Tehran; Abadan; Tabriz; Kabul. Ask students to write a short paragraph about both Iran and Afghanistan that explains how each country's physical geography affects its economic activity.

THEME
Place

Tehran is a city of enormous contrasts. The city's population has grown rapidly over the last decade to make it one of the most densely populated cities in the world. Pollution, inflation, unemployment, and crime are major problems. Yet, although poverty is increasing, there is still widespread support among the urban poor for the Islamic regime. At the same time, a new middle class has emerged—merchants who prospered from the Islamic revolution. In contrast to the fundamentalist laws of the regime, however, this rising class is increasingly embracing Western culture and behavior. Forbidden Western clothing styles, music, films, and alcohol are found in Tehran's middle-class homes and are sometimes seen in public. Although "morality police" can still arrest people for listening to the wrong music, watching videotapes, or drinking alcohol, laws are enforced irregularly, and most citizens act freely inside their own homes.

This petrochemical plant near **Abadan, Iran**, is one of many similar industrial facilities found along the shores of the Persian Gulf. The oil business has provided some jobs for people at plants like these in the region.

Land reform programs have led to widespread individual farm ownership. Yet many farming communities still use the methods and tools of past centuries. The government has provided few funds for farmers and has kept food prices low. As a result, farmers cannot afford to invest in technology, and yields remain low by world standards. Like Iraq, Iran is a major food importer.

Other farming traditions remain as well. For example, a few peasants take part in a seasonal migration of people and livestock. Hoping for good winter rains, they plant grain on the plateaus during the fall. After the spring harvest, they move to mountain pastures, where their herds graze during the summer.

Iran was the first Persian Gulf country to develop its oil deposits, and in 1951, it nationalized its oil industry. Today, refineries, petrochemical plants, and busy ports provide some jobs. Iran's mountain regions also hold deposits of metallic ores, many of which were discovered recently through the analysis of satellite images. Iran once had an important fishery in the Caspian Sea, but pollution has caused its decline.

In many of Iran's smaller towns, local wool, silk, and metals are used in handmade products sold around the world. Persian rugs in particular have been prized for centuries. The wool for making rugs is cleaned, dyed, combed, and sent to rural homes, where it is spun into yarn. The yarn is then sent to skilled artisans, who knot designs that are hundreds of years old. Many Iranians buy rugs as investments.

PLACE Urban Geography As in other developing countries, rural-to-urban migration is typical in Iran. In the older sections of the towns and cities, life goes on today much as it did centuries ago. As in the Arab countries, a central marketplace, shielded from the sun and rain, displays goods such as rugs, cloth, household utensils, and spices. A mosque is always nearby. In contrast, the newer sections of Iranian cities have modern high-rise buildings and shopping centers clustered on wide avenues.

Tehran (tay-uh-RAN), Iran's capital city, lies on the northern edge of the Plateau of Iran at the foot of the Elburz Mountains. The city is more than 5,000 feet (1,524 m) above sea level. Tehran is the site of many government agencies and is Iran's most industrialized city. The city's population is over 7 million.

Abadan was founded along the Persian Gulf to market and transport Iran's oil exports. The once-empty desert land around the city became a busy industrial center with modern housing, hospitals, and schools. Abadan, however, was nearly destroyed in the Iran-Iraq war and will take years to rebuild. Iran's other important cities include Tabriz (tuh-BREEZ), Isfahan (is-fuh-HAHN), and Qom (KOHM). These cities are most famous as historic sites and religious centers, with beautiful mosques and religious schools and shrines.

Human Geography The Shi'a branch of Islam is the dominant religion in Iran and provides a key source of national unity. The population of Iran is ethnically diverse, however. Ethnic Persians make up about half the country's people. They speak Farsi, an Indo-European language. Many smaller ethnic groups live in the country's mountain valleys, where the surrounding mountains have protected them from repeated waves of invaders during past centuries.

Stormy politics have long characterized Iran. In ancient times, city-states and empires battled for control of the countryside. Feuds between the many ethnic groups raged almost continuously. More recently, in the decades after World War II, Iran's ruler, titled the Shah, used the country's oil income to launch a program of modernization and industrialization. The Shah was popular among developed countries, who benefited from Iran's industrial development.

Opposition to the Shah, however, grew steadily inside Iran. The Shah's economic changes

Closure

Write the headings Iran and Afghanistan on the board. Ask students to provide key facts about the physical, economic, and human geography of each country. When possible, make comparisons or draw contrasts between the two countries.

ASSESSMENT

Assign the Section 2 Review questions on this page.

Reteaching

Assign Section 2 of Reteaching Worksheet 34.

Extension

Encourage interested students to investigate what happened to the Shah of Iran after he was ousted in 1979. Have them report their findings to the class and discuss the international repercussions of this event.

SECTION 3
OBJECTIVES

The objectives reflect the Focus questions on page 404 and are assessed in the review on page 405.

1. Relate Saudi Arabia's physical geography to its settlement patterns.

2. Explain how oil has affected Saudi Arabia and its global relationships.

widened the gap between the country's rich and poor. A small number of Iranians became wealthy and began enjoying Western ways of life. The growing number of poor Iranians had little to show for the country's new wealth. Also, the Shah's Westernization of Iran angered Iran's powerful Shi'a leaders, who believed the Shah's policies were not in line with traditional Islamic views.

After increasing unrest and violence, the Shah fled Iran in 1979. A fundamentalist Islamic government then came to power under the Ayatollah Khomeini. Among the Shi'a, an ayatollah is a religious leader of the highest rank and greatest respect.

Peace eventually returned to Iran through the use of religious courts. People who disobeyed religious law were imprisoned; some were executed. Many skilled and educated Iranians left the country. Khomeini died in 1989, but Iran remains a **theocracy**, a country governed by religious law.

The Islamic government of Iran is struggling to redirect the country's future. Its population is rising much faster than its economic growth, and poverty is increasing. Many industries also run irregularly. In addition, Iran's relations with many foreign countries, including some of its Arab neighbors and the United States, are strained.

LOCATION **Afghanistan** Afghanistan is a dry, mountainous country landlocked between Iran, Pakistan, and the Central Asian republics. The towering Hindu Kush range, which reaches up to 25,000 feet (7,620 m), cuts across northern

Afghanistan. The country's location on the route between India and Central Asia has attracted nearby empires throughout its history. Merchants, warriors, and missionaries have all used the Khyber Pass on Afghanistan's border with Pakistan for thousands of years. (See the map on page 399.)

Afghanistan is one of the world's least developed countries. Farming and grazing livestock are the main occupations, though the country's steppe and desert climates limit agriculture to isolated, irrigated regions. Some Afghan farmers also raise opium poppies as a cash crop. This flower is the source of the powerful drugs heroin and morphine.

Kabul is the country's capital and largest city. Afghan cities employ people mainly in food-processing industries and government. Traditional crafts, though, remain an important source of cash income. Although Afghanistan's mountains likely hold important minerals, only the country's natural-gas deposits are developed.

The Pathans make up the largest ethnic group in Afghanistan. They are closely related to the Tajiks to the north. There are dozens of other ethnic groups as well. As for religion, almost all Afghans are Muslim. Disputes between and among Shi'a and Sunni are common.

Building a modern state in this culturally diverse region has been difficult. A civil war, which included intervention by the Soviet Union, raged during much of the 1980s. Although the Soviets withdrew from Afghanistan in 1989, fighting between rival political and religious groups has continued. An end to the fighting and an easing of political tensions are required for rebuilding the war-torn country.

Residential neighborhoods in **Kabul**, the capital and largest city of **Afghanistan**, climb the slopes of the surrounding Hindu Kush ranges. Much of northern and central Afghanistan is mountainous.

SECTION REVIEW

1. What landforms dominate in Iran? In what part of the country is rain-fed agriculture possible? What is Iran's capital?

2. What branch of Islam is widespread in Iran? Why were the leaders of this branch offended by the Shah's policies?

3. **Critical Thinking** **Agree or disagree** with the following statement: "Afghanistan has more in common with its neighbors to the north than with its neighbors to the west." Explain your answer.

Geographical Sidelight

Relations between Afghanistan and Iran have often been strained over the years. Periodic disputes have arisen over water rights concerning the Helmand River (Afghanistan's largest), which flows from the mountainous region near Kabul across Afghanistan to swamps on the Iranian border.

Section 2 Review
ANSWERS

1. Highland plains and rugged mountains; northern Iran; Tehran

2. Shi'a branch; because the Shah's policies were Westernizing Iran and traditional Islamic views wish to limit Western influences

3. Possible answer: Afghanistan may have more ties to the north economically (no oil), culturally (Pathans related to Tajiks), and politically (Soviet intervention in the 1980s). But unlike the countries to the north, Afghanistan was not a Soviet republic.

403

TEACHING SECTION 3

Motivator

Ask students what images come to mind when they think of Saudi Arabia. Many students will probably say "desert" and "oil." Tell students that they will learn in Section 3 about how people have adapted to Saudi Arabia's desert environment and how they have been affected by its oil resources.

Using a Chart

Have students review the "Facts in Brief" chart on page 400 and compare Saudi Arabia with other Southwest Asian countries. Ask students to combine facts from the text with information in the chart to determine how oil has affected Saudi Arabia both inside the country and globally.

Guided Practice

On the board, make a chart titled Saudi Arabia. Have columns labeled West, Interior, and East; have rows labeled Physical Description, Extent of Settlement, Major Cities, and Reasons for Settlement. Have volunteers fill in the chart.

Independent Practice

Have students write a paragraph discussing Saudi Arabia's seawater desalinization—how it is significant and how it is affordable.

GLOBAL PERSPECTIVES

While Mecca's primary significance is religious, the city also sometimes plays an international political role. For example, a 1993 peace agreement among factions in Afghanistan, monitored by the Organization of the Islamic Conference, was reaffirmed in Mecca. Also in 1993, Saudi Arabia banned Libyans from flying to Mecca for their annual pilgrimage due to international restrictions on Libya. Lead a class discussion about cities in the United States that have global political significance.

Skill Study

As happened in the United States in the 1970s, higher oil prices and the limited availability of oil can have very negative effects on economies of countries dependent on imported oil.

SAUDI ARABIA

FOCUS

● How is Saudi Arabia's physical geography related to settlement in the country?

● How has Saudi Arabia's oil affected the country, internally and globally?

REGION **Physical Geography** Saudi Arabia occupies much of the Arabian Peninsula. (See the map on page 399.) Most of the country is desert plains, which stretch from the Persian Gulf to the mountains along the Red Sea.

In the southeastern corner of Saudi Arabia lies one of the world's largest areas of windswept sand. It is the Rub' al-Khali, also called the Empty Quarter. This area is uninhabited, visited only occasionally by Bedouin herders. Beyond the sands rise the western mountains, the most prominent landforms in Saudi Arabia. These mountains line the Red Sea coast for more than 1,000 miles (1,609 km). Peaks in the southern end of the range reach nearly 10,000 feet (3,048 m).

Population The people of Saudi Arabia live where there is either water or oil. Mecca, built on an ancient oasis spring site in the western mountains, is the most important city of Islam. It was in Mecca that Muhammad lived and began teaching Islam. Muslims are required to make a **pilgrimage** (journey) to Mecca at least once in their lifetimes. Medina, also in the western mountains, is another holy city. Today, more than 2 million pilgrims from all parts of the world visit the holy cities each year. A more recent area of settlement in Saudi Arabia lies near the oil fields along the Persian Gulf. These underground fields may contain as much as one-fourth of the world's remaining oil.

Between the western mountains and the shores of the Persian Gulf, only isolated oases can support human settlement. The largest oasis is in the area surrounding Riyadh (ree-YAHD), the capital. The oases of the Saudi Arabian interior have been settled for thousands of years by nomadic Bedouins.

Economic Geography The Saudi economy is based on petroleum. Oil sales provide nearly all the government's income, and the total is almost

Who has the most oil?

The Top 12 (in billions of barrels)

257.8 Saudi Arabia
100 Iraq
98.1 UAE*
94 Kuwait
92.9 Iran
62.7 Venezuela
57 CIS**
51.3 Mexico
24.7 USA
24 China
22.8 Libya
17.9 Nigeria

The World (by percentage)

World's Top 5 64.4% (all in Southwest Asia)
Rest of Top 12 26.1%
Rest of world 9.5%

*United Arab Emirates
**Commonwealth of Independent States

Estimated proven oil reserves (as of January 1, 1993)
Source: *Oil and Gas Journal*

SKILL STUDY The bar graph shows the 12 countries in the world with the largest estimated proven reserves of oil. The country with the most oil, Saudi Arabia, has more than two-and-a-half times the oil of Iraq, a distant second. The bars for the Top 5 countries, all of which are located along the Persian Gulf in Southwest Asia, are shown in orange. The bars for the remaining Top 12 countries are shown in green.

The pie graph shows that the Top 5 countries, shown in orange, have nearly two-thirds of the world's estimated proven reserves of oil. In all, the Top 12 countries control more than 90 percent of the world's oil. Some of the world's oil-producing countries cooperate with each other in setting prices and production goals for their oil. What effect would such an organization have if it decided to increase oil prices or lower the amount of oil produced, or both?

404

more than the country can spend. Government-paid social services, including education, are available to all.

Saudi Arabia's modern oil fields and related chemical industries, however, provide few jobs. To increase employment, the country has encouraged industry and the modernization of agriculture. Despite the country's desert environment, Saudi Arabia now boasts modern dairies and ranches. Dates, wheat, and barley are the main crops. Construction of seawater desalinization plants has been particularly important to the country's economy. Saudi Arabia produces more desalinized seawater than any other country in the world.

Young Saudis can be found behind desks as office workers, businesspeople, and bankers. Meanwhile, foreign workers from Egypt, Yemen, and South Asia perform much of the manual labor in the country. Before the development of the Saudi petroleum industry, the government received most of its money by taxing visitors to the holy cities. Pilgrims still provide a large amount of money for the government.

A Global Role As the world's largest oil exporter, Saudi Arabia has a powerful role both in Southwest Asia and in the world. By increasing or decreasing their oil exports, the Saudis have the power to influence oil prices worldwide.

The Saud family, which has ruled the country since its founding in 1932, is devoutly Muslim. In fact, Saudi society is organized under Islamic law. As a rich Muslim state, Saudi Arabia gives economic and military aid to the poorer states of the region. Although Saudi Arabia strongly defends the rights of the Palestinians, it has been patient in working for a peaceful settlement of the Arab-Israeli issue.

SECTION REVIEW

1. What parts of Saudi Arabia support human settlement?
2. How has the government of Saudi Arabia used its oil wealth? What has the oil industry not provided to most Saudis?
3. **Critical Thinking Predict** what might happen to world oil prices if Saudi Arabia decided to drastically reduce its oil exports. Explain your answer.

4 THE ARABIAN PERIMETER

FOCUS

- *What are the countries of the Arabian perimeter?*
- *What are the physical, economic, and political characteristics of these countries?*

Along the southern and eastern **perimeter** (outer boundary) of the Arabian Peninsula is a string of small, traditional Arab states. (See the map on page 399.) Most consist of a strip of coastline, one or two port cities, and an isolated area of dry, undeveloped countryside. All of these small countries have significant locations in relation to world shipping. They also hold important oil deposits.

Kuwait Kuwait is an oil-rich slice of desert land on the northern Persian Gulf between Iraq and Saudi Arabia. The country's population has grown rapidly with the development of its oil deposits. The government also receives substantial income from foreign investments.

Wealthy Kuwaiti families have ruled the country for many years, and foreign laborers perform much of the country's work. Before the 1990–1991 Persian Gulf War, many of these workers were Palestinians. Now, workers come mostly from poorer Arab countries and from South and Southeast Asia.

Soon after the war, Kuwait began rebuilding from the devastation caused by the Iraqi army. The country again has one of the highest per capita incomes in the world, as well as an extensive social welfare program. Seawater desalinization plants, oil fields, and highways are functioning again.

Gulf States Along the Persian Gulf between Saudi Arabia and Oman are the United Arab Emirates, Qatar, and Bahrain. All of these Gulf states are oil rich. Except for a few oases, most of the land here is desert.

Until oil was discovered in the region, life in the thinly populated states of this part of the Arabian Peninsula had changed little over the centuries. Traditional occupations include fishing, pearl diving, and trading. Oil wealth has brought rapid economic development, however, changing people's lives. Seaports, roads, modern housing,

A minaret stands out amid the distinct architecture of the old section of **Sanaa, Yemen**. Sanaa was founded before the first century A.D.

airports, and irrigation systems are springing up. The United Arab Emirates and Bahrain in particular have become Westernized. They are popular destinations for many tourists from Kuwait and Saudi Arabia.

Oman Oman, formerly known as Muscat and Oman, is a **sultanate**. A sultanate is a country that is ruled by a Muslim monarch (sultan). Oman has long been isolated from the modern world, and most of the country's people live by fishing and traditional agriculture. The government, however, has begun to use the income from the country's large oil deposits to improve its citizens' standard of living.

Yemen Yemen lies in the mountainous southwestern corner of the Arabian Peninsula. North Yemen and South Yemen merged in 1990 to form one country. South Yemenis sought independence again in 1994 but were defeated in a short war. Yemen has a strategic location next to the Bab al-Mandab, the narrow strait that connects the Red Sea with the Indian Ocean. Many ships pass through the strait on their way to the Suez Canal and the Mediterranean Sea.

Yemen is not an industrialized country. Instead, traditional agriculture, raising sheep and

goats, and small-village life are typical. Crops grown on the coastal drylands include dates and grains. Humid mountain slopes support cultivation of fruit trees and vegetables. Coffee, though, is Yemen's most famous crop. In fact, the word *coffee* has its origin in Yemeni Arabic, and some believe that coffee may have been first cultivated here. Today, however, Yemen's coffee is not particularly successful on the world market. It is considered of low quality in comparison to the coffee from countries with better growing conditions.

Yemen has two principal cities. Its capital, Sanaa, lies on the high plateau east of the mountains. This ancient, walled city has a few factories that provide basic goods for the population. On the south coast of Yemen is the port city of Aden. Aden has an oil refinery and ancient trading links to the Persian Gulf and East Africa.

Many of Yemen's people exhibit more loyalty to their families and clans than to the national government. Conflicts among these groups take place frequently, hindering Yemen's development. The country's per capita income ranks as one of the world's lowest, and about 60 percent of the people are illiterate. Many young people seek unskilled jobs in Saudi Arabia and the small Gulf states. The money they send home is crucial to the country's economy.

In the early 1980s, oil was discovered in Yemen. Money gained from oil exports is providing Yemen with increasing financial independence. Yemen's foreign policy now stresses an independent course and seeks a more active role among the Arab nations.

SECTION REVIEW

1. Which Arabian perimeter countries border the Persian Gulf? Which country is next to the Bab al-Mandab?
2. What is the major source of income in the region? What kinds of political systems exist here?
3. **Critical Thinking** **Determine** what effects newly discovered oil deposits may have on the daily lives of people in Oman and Yemen. Explain your answer.

CHAPTER 34 CHECK
ANSWERS

Building a Vocabulary
1. Unused; to conserve moisture and nutrients
2. A journey; Mecca
3. A country ruled by a Muslim monarch; Oman

Recalling and Reviewing
1. The Kurds
2. They carry springwater from mountains to plains; Iran
3. Iraq and Iran
4. Afghanistan
5. Riyadh; because a large oasis is there
6. It has given Yemen greater financial and thus political independence.

CHAPTER **34** CHECK

Khyber Pass between **Afghanistan** and **Pakistan**

Reviewing the Main Ideas

1. Iraq lies at the southeastern end of the Fertile Crescent. Baghdad is the capital and largest city. Iraq has large oil reserves, but many of the country's people are involved in agriculture. In recent years, Iraq has been politically aggressive.
2. East of Iraq lies Iran. Iran was the first Persian Gulf country to develop its oil deposits. Rural-to-urban migration is common. Iran's population is ethnically diverse, and turbulent politics have long characterized the country.
3. Afghanistan is a dry, mountainous country located east of Iran and south of the Central Asian republics.
4. Saudi Arabia occupies much of the Arabian Peninsula and holds one of the world's largest areas of windswept sand as well as the most important city of Islam, Mecca. The Saudi economy is based on petroleum.
5. The countries of the eastern and southern Arabian perimeter include Kuwait, the Gulf states, Oman, and Yemen. All of these countries have valuable oil deposits.

Building a Vocabulary

1. Define *fallow*. Why must Iraq's croplands be left fallow occasionally?
2. What is a pilgrimage? To what city must Muslims make a pilgrimage at least once in their lifetimes?

3. Define *sultanate*. Which Arabian perimeter country is a sultanate?

Recalling and Reviewing

1. What people of northern Iraq seek their independence?
2. What purpose do *qanats* serve? Which country is famous for its *qanats*?
3. Which two countries discussed in the chapter engaged in a decade-long war during the 1980s?
4. Which country discussed in this chapter is not oil rich?
5. What is Saudi Arabia's capital? Why is the city located where it is?
6. How has the recent discovery of oil in Yemen affected the country as a whole?

Thinking Critically

1. What factors do you think have prevented Iraq and Iran from using the profits from their oil to modernize agriculture?
2. Is Saudi Arabia a theocracy? Support your answer.

Learning About Your Local Geography

Individual Project
Persian rugs are world famous. Conduct a survey of community members to find out what craft industries are important in your community. Is your community "famous" for any crafts? What crafts do community members produce? Do they sell them, and if so, where? Does the community have many crafts shops? Draw some illustrations of different crafts produced by your community members. You may want to include your illustrations in your individual portfolio.

Using the Five Themes of Geography

Write a paragraph about the theme of **human-environment interaction** in the countries discussed in this chapter. How have people changed the natural environment? How have they adapted to it? What forms of interaction do you think will take place in the future?

Thinking Critically

1. Possible answer: the long war between the two countries may have drained their economies. Also, Iran's fundamentalist leaders may wish to keep technology out of the country.
2. Yes. The ruling family is devoutly Muslim, and Saudi society is organized under Islamic law.

Learning About Your Local Geography

Illustrations should show crafts produced in students' communities.

Using the Five Themes of Geography

Changes to the environment include the construction of irrigation systems and the removal of oil from deposits. Adaptations include the nomadic way of life and settlement only in certain areas.

SKILLS

Use the history of the Arab-Israeli conflict to demonstrate the links between geography and history. Have students sketch a series of chronological maps showing how boundaries, divisions, and populations within the Eastern Mediterranean region have changed over time. Maps might show the region in ancient times (populated by Jews), in the eighth century (populated by Arabs), in the late 1800s (with an increasing Jewish population), in 1948 (with the creation of the state of Israel), in 1967 (with the Israeli occupation of new territories), in 1982 (with the return of Sinai to Egypt), and in 1993 (with increasing Palestinian self-rule in some territories).

UNIT 7 REVIEW SKILLS

Linking Geography and History

No understanding of the present can be complete without an understanding of the past. Only by examining the history of a region can we make sense out of its modern geography. Similarly, only by studying the history of a group of people—where they originated and how and where they spread—can we understand the group's human geography today.

For example, to explain the current human geography of the Muslim world, we must study how Islam and Arab culture spread in earlier centuries. At the start of the seventh century, Arabs were a small, divided group of peoples scattered across the arid Arabian Peninsula. By 750, however, Arabs controlled an area stretching from Spain, across North Africa and western Asia, to India, and from Central Asia to the Arabian Peninsula. In just over a century, the Arabs had built the largest empire the world had ever seen.

What brought the Arabs together and allowed them to expand over such a vast area during those 150 years? The unifying force was the new religion of Islam. Muhummad, Islam's founder, began preaching in Mecca in 610. Muhummad left Mecca in 622 and settled in the town of Medina 200 miles (322 km) away. He returned to Mecca several years later, now at the head of an army of devoted followers. The simple message of Islam—that there was only one God, Allah, whose prophet was Muhammad—had great appeal in the troubled world of the seventh century. Islam was not just a system of religious beliefs, but a way of life.

Inspired by their new religion, Arabs swept out of the Arabian Peninsula to spread Islam. Partly as a result of internal divisions among the peoples they encountered, the Arabs met with great success. In some places, peoples who had been the victims of religious persecution welcomed the often more tolerant Muslims. During the seventh and eighth centuries, Islam became a major world religion and Christianity's main rival.

The spread of Islam and the Arabic language gave unity to a vast area and created a vibrant Islamic culture. The center of the Islamic world was the newly built city of Baghdad in Mesopotamia. Baghdad also became a center of education, art, and trade. Tigers, rubies, and ebony came to the city from India. Porcelain, paper, and ink came from China. Ivory, gold, and slaves came from Africa.

Islamic culture drew on a number of ancient civilizations. The philosophy and science of the Greeks, the artistic and administrative skills of Persia, and the mathematical discoveries of India mingled in the Islamic world. Scholars built on the knowledge of these older civilizations to make important progress in astronomy, geography, physics, chemistry, and mathematics.

Although Islamic power began to decline in the thirteenth century, the spread of Islam continued. Muslim traders and missionaries traveled east to Malaysia, Indonesia, and the Philippines and south farther into Africa. The Mongols, who built a Eurasian empire in the thirteenth century, converted to Islam and carried it deeper into Asia. The Turkish Ottoman Empire's capture of Constantinople in the fifteenth century and other Ottoman conquests spread Islam into Eastern Europe's Balkan Peninsula. In the sixteenth century, Muslim rulers created the powerful Mogul Empire in India that lasted for 300 years.

Islam continues to thrive. Today, it is the religion of the majority of the population throughout North Africa, in most of western and Central Asia, and in much of southern and southeastern Asia. Islam also has a large following in parts of Africa south of the Sahara and keeps a foothold in Europe's Balkan Peninsula. The United States has a large and growing Muslim population as well.

Practicing the Skill
ANSWERS

Maps, illustrations, and captions should indicate that students have researched how geography was involved in a particular local, regional, or national historical event.

408

PRACTICING THE SKILL

Select a local, regional, or national historical event in which you are interested. Research the event. How was geography involved? Did land or water play a role? What types of movement took place to cause the event or as an effect of it? Did the event change the physical or human geography of any areas? Use maps and illustrations, with captions, to show how the event demonstrates the link between geography and history.

BUILDING YOUR PORTFOLIO

Individually or in a group, complete one of the following projects to show your understanding of the geography concepts involved.

A Three Religions

You have read that Southwest Asia was the birthplace of three major world religions—Judaism, Christianity, and Islam. Find out more about each of these three religions, beyond what you read in the unit. Make a poster to display the information you collect. To make your poster, you will need to do the following.

1. Investigate the origins and histories of the three religions. Exactly where in Southwest Asia did each one form? How did each of the three religions spread to other parts of the world? What people were involved in the formation of each one, and what roles did they play? For your poster, sketch a map showing the places of origin and the spreading of each religion. Under the map, include a caption discussing other aspects of the religions' origins and histories.
2. Find out more about particular beliefs and practices of each religion. What are the central ideas of each religion? What customs do the followers of each religion have? What forms of worship do they participate in? How do their beliefs and customs affect their daily lives? Sketch drawings showing some of the different beliefs or practices of each religion. Next to each drawing, write several statements explaining the belief or practice. Put your drawings and statements on your poster.
3. Explore the different groups and branches within each of the three religions. How is each religion divided internally? How did these divisions arise? How do each religion's branches differ in beliefs or practices from the other branches of the same religion? Make a chart showing this information and include it on your poster.

Present your poster to the rest of your class.

B Oil Makes the World Go Round

Southwest Asia contains some of the world's largest oil deposits. The region's oil resources are of great economic value and political significance, not only in the region but around the world. Write the script for a brief documentary film about the region's oil. Include with your script illustrations, maps, and other visual materials that will help your audience understand your points. To prepare your script, think about the following questions.

1. Where are Southwest Asia's major oil resources located? How do people remove the oil from the earth, and how do they process it afterwards? To where is the oil exported, and how is it transported from Southwest Asia to its destinations? Your documentary film might use a map to show this information.
2. How is oil used, in the region and in the world? How do the countries of Southwest Asia use the money they make from oil? To what extent has oil wealth changed the daily lives of most people in these countries? Make rough sketches of the images your documentary film might use to show oil-based products and the effects of oil wealth in Southwest Asia.
3. How does Southwest Asia's oil affect political events, both in the region and around the world? What role has oil played in recent conflicts within the region? Your documentary might include a made-up interview with or dialogue between experts discussing these questions.

Present your script and supporting materials to the rest of the class.

Project A:
Posters should display information on the origins, histories, beliefs, practices, and branches of Judaism, Christianity, and Islam.

Project B:
Documentary film scripts and supporting materials should include information on the location of oil resources, the removal of oil from the earth, the transportation of oil, the uses of oil and oil wealth, and the effects of oil on political events.

Regions in Transition

Mapping a Transition Zone

Have students work in pairs to draw maps of Africa. Maps should highlight the transition zone between North Africa and the rest of the continent. Students should label the countries in the transition zone and indicate the dominant climate type and economic activities with shading and symbols as needed. Maps should also show major cities. Have pairs write an extended caption explaining what characteristics define this region of Africa as a transition zone. Call on volunteers to present and explain their maps to the class.

REGIONS IN TRANSITION

Differences and Connections

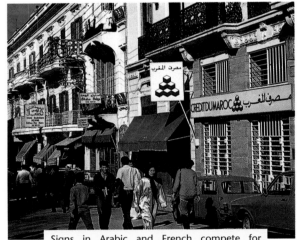

Signs in Arabic and French compete for attention in the North African country of **Morocco**. Arabic is spoken throughout North Africa and many parts of Southwest Asia.

In this textbook the countries of Africa are grouped together to emphasize the region's connections. Yet in any region covering an area as large as Africa, there are important differences within the region. Thus, Southern Africa is different from West Africa, and West Africa is different from North Africa. This is no less true of other large regions of the world. Venezuela, for example, is different from Argentina in many important ways.

North Africa and Southwest Asia

One subregion of Africa that geographers often include in another region is North Africa. These geographers see greater connections between North Africa and Southwest Asia than they see between North Africa and the rest of the continent. For example, in both areas Arabic is the main language, and Islam is the major religion. Political issues also tie North Africa to its eastern neighbors, as does the physical geography of the region. The countries of North Africa and Southwest Asia are all a part of a vast desert region. They face common issues such as water conservation and management.

The African Continent

There are just as many reasons for placing North Africa in a region with the rest of the African continent. People in this part of the world share important historical, cultural, and economic ties. For example, the ancient Egyptians maintained close contact with peoples farther up the Nile River, in the interior of Africa. In turn, cultures south of the Sahara contributed to the development of the great Nile Valley civilizations of Egypt. Farther west, Mediterranean peoples have a long history of trading with West African kingdoms south of the Sahara. The Islamic religion, meanwhile, spread southward through contact between the peoples of North Africa and the rest of the continent. Today, mosques can be found as far south as Nigeria and Tanzania.

North Africa also has political connections to the rest of Africa. Many African countries face similar political and economic issues that

The Blue Mosque, so called because of its interior blue tiles, is one of many beautiful places of worship for Muslims in **Istanbul, Turkey**. Although various religions are practiced in North Africa and Southwest Asia, most people in the two regions are Muslims.

410

Using Geography Themes

To reinforce the theme of movement, have students create a class time line about North Africa and Southwest Asia. Organize the class into groups of four or five. Assign a section of the time line to each group. Groups should use their textbook and other classroom and library resources to research events in the region's history that involve movement, such as the spread of Islamic merchants' trading networks and the expansion of early African kingdoms. Once groups have completed their sections, assemble the time line on a classroom wall. Have each group choose a spokesperson to present its section of the time line and to explain how the events affected the present-day characteristics of North Africa and Southwest Asia.

Geographical Sidelight

For centuries, Cairo, Egypt, has been a center of Islamic learning and culture for the entire North Africa and Southwest Asia region. Established around 970, al-Azhar University in Cairo is one of the oldest in the world. It continues to be a major center for religious, and especially Islamic, studies.

CREATIVE STRATEGIES FOR TEACHING WORLD GEOGRAPHY

LESSON 15 Contributions from the Countries of Southwest Asia and North Africa: Conversations—Why Study Southwest Asia and North Africa?

Regional Links

Linguistic, religious, and climatic characteristics are only some of the links between the regions of North Africa and Southwest Asia. This map shows Arabic speakers, Muslims, and desert climates are dominant throughout much of the two regions and in parts of the surrounding transition zones.

stem in part from their shared colonial history. To resolve some of these issues, the countries of North Africa are working with other African nations through associations such as the Organization of African Unity.

Transition Zones

The differences and connections between North Africa and the rest of the continent are perhaps most evident in a band of countries that lie just south of the Sahara. These countries, from Mauritania in the west to Ethiopia and Sudan in the east, form a transition zone because they are as similar to North Africa as to Africa south of the Sahara. In Chad and Sudan, for example, there are strong political and cultural differences between north and south. The northern areas of these countries are tied closely to North Africa, while the southern parts of these countries are tied closely to African nations to the south.

In effect, North Africa and the countries in the transition zone have important connections to two major world regions. How they are organized depends on the geographer's point of view. Understanding the differences among countries and their connections to other areas of the world is more important than determining where on a map a region begins or ends.

This Arab woman and her child trek through one of the desert regions that cover much of North Africa and Southwest Asia.

UNIT 8 AFRICA *CHAPTERS 35–40*

OVERVIEW

In Unit 8, students will learn that Africa is a region of enormous diversity.

Chapter 35 • INTRODUCING AFRICA identifies the continent's landforms and climates and traces the environment's impact on African history.

Chapter 36 • NORTH AFRICA explains the significance of the Nile River, the Sahara, and oil in shaping economic and political patterns in North Africa.

Chapter 37 • WEST AFRICA focuses on the physical and cultural geography of the region.

Chapter 38 • EAST AFRICA emphasizes the region's physical and human geography and East Africa's significance as the site of the earliest human settlements.

Chapter 39 • CENTRAL AFRICA analyzes the political and economic development of Central African nations.

Chapter 40 • SOUTHERN AFRICA explains how European colonialism influenced Southern Africa.

OBJECTIVES

1. Describe the landforms and climates found in Africa.
2. Identify significant events in Africa's history.
3. Analyze how natural resources affect the political and economic development of Africa.
4. Explain the economic, political, and social challenges facing African nations.
5. Compare the cultures and ways of life of peoples in different regions of Africa.
6. Analyze special-purpose maps of Africa to interpret data about climate, population, and economy.
7. Learn how to interpret photographs and to use them as geographic tools.

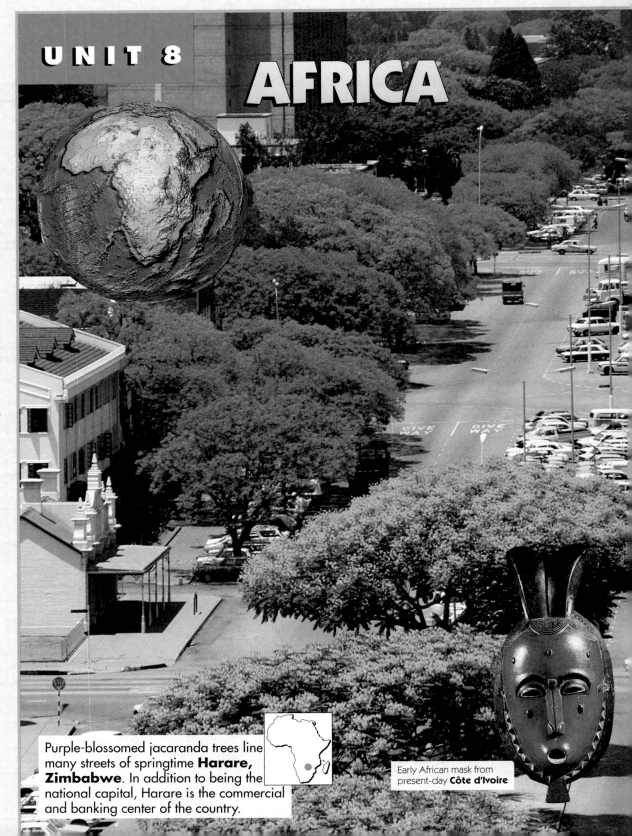

UNIT 8 AFRICA

Purple-blossomed jacaranda trees line many streets of springtime **Harare, Zimbabwe**. In addition to being the national capital, Harare is the commercial and banking center of the country.

Early African mask from present-day **Côte d'Ivoire**

TEACHING UNIT 8

Using the Illustrations

Ask students to compare the photographs of the rural village and the modern city on these two pages. Have students identify an important natural resource in the rural scene, and ask how it is being used. *(water, for washing clothes)* Then ask students to choose the photograph that shows the location where water is more likely to be a plentiful resource and to explain why. *(A city could not develop without a plentiful water supply; the rural scene shows water scarcity, dry climate.)*

Teaching with Sketch Maps

Have students draw a map of Africa without consulting any resources. Tell them to form a mental picture of the African continent and to transfer what they remember to paper. Students should add this map to their portfolios. Tell them that they will compare this map with another they will draw as they conclude their study of Unit 8.

UNIT 8 CHAPTERS

35 Introducing Africa

36 North Africa

37 West Africa

38 East Africa

39 Central Africa

40 Southern Africa

Water, a valuable resource everywhere, is scarce in some parts of Africa. This nearly dry riverbed is near **Bobo-Dioulasso** in southwestern **Burkina Faso** in the Sahel. Burkina Faso is a landlocked country that was plagued by severe drought throughout the 1970s and 1980s.

PORTFOLIO ASSESSMENT

"Building Your Portfolio" on page 489 provides a choice of two unit projects for students to work on as they study Unit 8. The unit projects are supported by Chapter Portfolio Assessment worksheets in *Teaching Resources.*

THE WORLD TODAY VIDEODISC PROGRAM

The videodisc profiles the following countries in Unit 8: Egypt, Nigeria, Kenya, and the Republic of South Africa.

Kilimanjaro towers above the wild animals of Amboseli National Park in **Kenya**. Located just across the border in northeastern **Tanzania**, the extinct volcano is Africa's highest point. At 19,340 feet (5,895 m), Kilimanjaro's highest peak is covered in ice and snow year-round.

One World, Many Voices

LITERATURE

Ask a volunteer to read "Souffles" aloud. Point out that the poem has a cultural message because it illustrates a traditional African religious belief—that the spirits of deceased loved ones can communicate with the living, and that life is found everywhere in the physical world. Explain to students that the poem also has a political message. The poem was published in 1960, a time when many Africans were striving for independence from European nations. By honoring ancestors' traditions, some Africans believed that they could cast off the effects of decades of European colonialism.

More About Birago Diop

Birago Diop was of Wolof ancestry, an ethnic group of West Africa. While practicing veterinary medicine in rural Senegal, he collected authentic Wolof folktales and translated them into French. Their publication in 1947 and 1958 contributed to the preservation of Wolof culture. Diop was appointed ambassador to Tunisia from Senegal in 1960.

ONE WORLD, MANY VOICES

LITERATURE

SOUFFLES

(IT IS THE BREATH OF THE ANCESTORS)

LISTEN MORE OFTEN TO THINGS THAN TO BEINGS
    HEAR THE FIRE'S VOICE,
    HEAR THE VOICE OF WATER.
    HEAR, IN THE WIND, THE SOBBING OF THE TREES.
IT IS THE BREATH OF THE ANCESTORS.

THE DEAD ARE NOT GONE FOREVER
THEY ARE IN THE PALING SHADOWS,
THEY ARE IN THE DARKENING SHADOWS.
THE DEAD ARE NOT BENEATH THE GROUND,
THEY ARE IN THE RUSTLING TREE,

IN THE MURMURING WOOD,
THE FLOWING WATER,
THE STILL WATER,
IN THE LONELY PLACE, IN THE CROWD;
THE DEAD ARE NEVER DEAD.

LISTEN MORE OFTEN TO THINGS THAN TO BEINGS.
    HEAR THE FIRE'S VOICE.
    HEAR THE VOICE OF WATER.
    IN THE WIND HEAR THE SOBBING OF THE TREES.
    IT IS THE BREATH OF THE ANCESTORS.
    THEY ARE NOT GONE
    THEY ARE NOT BENEATH THE GROUND
    THEY ARE NOT DEAD.

THE DEAD ARE NOT GONE FOREVER.
    THEY ARE IN A WOMAN'S BREAST,
    A CHILD'S CRY, A GLOWING EMBER.
    THE DEAD ARE NOT BENEATH THE EARTH,
    THEY ARE IN THE FLICKERING FIRE,
    IN THE WEEPING PLACE, THE HOME.
    THE DEAD ARE NEVER DEAD.

LISTEN MORE OFTEN TO THINGS THAN TO BEINGS
    HEAR THE FIRE'S VOICE,
    HEAR THE VOICE OF WATER.
    HEAR, IN THE WIND, THE SOBBING OF THE TREES.
    IT IS THE BREATH OF THE ANCESTORS.

Translated by Samuel Allen
From Poems from Africa

Birago Diop (1906–1989) is one of Africa's best-known writers. He was born in Senegal, a former French colony. While completing his university studies in Paris, Diop joined other French-speaking African writers in a literary movement that emphasized African traditions. Diop feared that European colonialism threatened the existence of many unique African cultures. When he returned to his homeland, he developed his literary talents while pursuing a career as a veterinary surgeon. The poem at left reminds modern Africans to remember the teachings and culture of their ancestors. It also reflects important themes in contemporary African literature—respect for nature and the connection between the physical world and the spiritual realm.

Interpreting Literature

Answers

1. Through what the living can observe and experience in nature; by what is passed down from one generation to the next
2. Answers will vary. Students might mention ways that many societies worldwide preserve the traditions of previous generations through language, dress, cooking, religion, crafts (art), and music.

Interpreting Literature

1. How do the ancestors mentioned in the poem pass on their teachings and culture to the living?
2. How might the message in this poem apply to cultures outside of Africa?

414

FOR THE RECORD

Captain Harry Dean (1864–1935) was an African American sea captain. Before he was 15 years old, Dean had sailed around the world with his uncle. On his first voyage to Africa, the young man formed an emotional bond with Africa that lasted his lifetime. Dean had studied the accomplishments and glories of powerful ancient African trading kingdoms. He longed to restore "Africa for the Africans" and undo the effects of European colonialism. As a grown man, Dean sailed his own ship, *Pedro Gorino*, to Africa. He believed the continent's development depended on establishing a great African merchant fleet. The excerpt below is from his autobiography, *Umbala*. Dean recorded his thoughts during an overland journey across southern Africa in the early 1900s.

As we were travelling less than thirty miles a day we had plenty of time for thought. In three days of dreaming I rebuilt the Ethiopian Empire. In three days of dreaming I recaptured Africa for the Africans. . . . The ruins of Zimbabwe were no longer ruins, but stately masonry. The sons of the ancient race who raised those piles of stone to forgotten gods once more were proud possessors of all they surveyed. And those dark descendants of the Phoenicians, still worshipping the crane and the ram, reattained the genius of their ancestors, sailing their ships to every country, bearing the wealth of Africa. As in the ancient days, precious stones and metals poured from Sheba northward through all Arabia, and westward down the wide rivers of the jungle. Nowhere were there slaves, or poverty, or ignorance. In three days of dreaming I dammed the rivers to water the karroo [a dry region of southern Africa] until the desert bloomed like the rose. I built cities in trackless thickets, and from the forests of Africa constructed such a fleet of graceful ships as the world has never seen.

*isher casts a net on the **Zaire River**.*

Analyzing Primary Sources

1. What natural resources and physical geography features of Africa does Captain Harry Dean mention?
2. What dreams does Dean have for Africa?

Geography and History

The Phoenicians were a seafaring and trading society that flourished north of Egypt around 1200 B.C. They established trading centers and colonies throughout the Mediterranean region.

Geographical Sidelight

The Great Zimbabwe was a mining city in southern Africa. The ruins Dean mentions refer to what is left of the city—a massive granite tower and a fortress wall. Zimbabwe means "house of stone."

Primary Source
Description of Change: excerpted and bracketed
Rationale: excerpted to focus on main idea, bracketed to clarify meaning

Analyzing Primary Sources

Answers

1. Natural resources: precious stones and metals; physical features: rivers, jungle, desert, forest
2. To eliminate non-African influence in African affairs; for Africa to become a major trading power by building a great merchant fleet

415

CHAPTER 35 Introducing Africa
Planning Guide

TEXTBOOK	TEACHING RESOURCES	OTHER AVAILABLE MATERIALS
Section		
1. Physical Geography Pages 421–423	Graphic Organizer 35 Reteaching Worksheet 35.1	*REGIONS OF THE WORLD Map Transparencies:* Africa *Creative Strategies for Teaching World Geography* LESSON 16 How Much Do You Know About Africa? *Study Guide for Spanish Speakers* *Lecture Notes,* Chapter 35
2. Historical Geography Pages 423–426	Reteaching Worksheet 35.2 Critical Thinking Worksheet 35 Challenge/Enrichment Worksheet 35	*Study Guide for Spanish Speakers* *Lecture Notes,* Chapter 35
3. Human and Economic Geography Pages 426–428	Reteaching Worksheet 35.3 Skills Worksheet 35	*Study Guide for Spanish Speakers* *Lecture Notes,* Chapter 35
Chapter Review Page 429	Chapter 35 Test Chapter 35 Portfolio Assessment See also *Alternative Assessment Forms*	*Test Generator,* Chapter 35

CHAPTER RESOURCES

Lecture Notes, Chapter 35
Teaching Resources
    Chapter 35 Graphic Organizer
    Chapter 35 Skills Worksheet
    Chapter 35 Critical Thinking Worksheet
    Chapter 35 Challenge/Enrichment Worksheet

REVIEW AND ASSESSMENT

Chapter 35 Check, page 429
Chapter 35 Test, *Teaching Resources*
Chapter 35 Portfolio Assessment, *Teaching Resources*
Test Generator, Chapter 35
Alternative Assessment Forms

ADDITIONAL RESOURCES

Books for Teachers

Davidson, Basil. *Modern Africa: A Social and Political History.* New York: Longman, 1994.

Glickman, Harvey, ed. *Political Leaders of Contemporary Africa South of the Sahara: A Biographical Dictionary.* Westport, CT: Greenwood Press, 1992.

Rau, Bill. *From Feast to Famine.* UK: Zed Books, 1991; distributed by Humanities Press, Atlantic Highlands, NJ.

Books for Students

Chiasson, John C. *African Journey.* New York: Macmillan (Bradbury Press), 1987.

Chu, Daniel, and Elliott P. Skinner. *A Glorious Age in Africa: The Story of Three Great African Empires.* African World Press, 1990. Distributed by Social Studies School Service, P.O. Box 802, Culver City, CA 90232–0802.

Murray, Jocelyn. *Africa.* (Cultural Atlas for Young People Series) New York: Facts on File, 1990.

Nabwire, Constance R. and Bertha Vining Montgomery. *Cooking the African Way.* Minneapolis: Lerner Publications, 1988.

Multimedia Resources

Africa. (interactive software) Educational Activities, 1983. Social Studies School Service, P.O. Box 802, Culver City, CA 90232–0802. Background, tutorial, and tests on African history and geography.

Africa: A New Look. (16mm film, 27 minutes) International Film Foundation, 155 W. 72nd Street, New York, NY 10023.

Welcome to Africa. (Multimedia CD-ROM) Walnut Creek CDROM, 4041 Pike Lane, Suite D, Concord, CA 94520.

What Is a Resource? Africa Project. (activity book: hand-outs, maps, discussion questions, simulations) Social Studies School Service, P.O. Box 802, Culver City, CA 90232–0802.

MEETING INDIVIDUAL NEEDS

Limited English Proficiency (LEP)

Pair LEP students with students who are proficient in English. Have the pairs skim the chapter to find the answers to the Focus questions in each section. Then help students create a chart about Africa on the board. Rephrase the Focus questions as the headings in the chart, and ask students to fill in the chart using the answers to the Focus questions.

Learners Having Difficulty

At the conclusion of each section in Chapter 35, ask students to list in note form a summary of the information presented. At the end of the chapter, help students create on the board an outline of the chapter from the summaries of each section.

Auditory Learners

If possible, arrange the classroom chairs in a circle. From an anthology of African literature, select an appropriate poem that reveals something about African culture and daily life, politics, or geography. Call on a volunteer who reads well to read the poem aloud. Then begin a class discussion about what the poem expresses about Africa.

Visual Learners

Review with students the Five Themes of Geography as you write them on the board. Ask students to study the photographs throughout Chapter 35. Have them work in pairs to identify the geography themes that are expressed in the images.

Tactile/Kinesthetic Learners

Have students create a classroom bulletin board about Africa. Have them work in groups to collect recent photographs with captions or newspaper and magazine articles about Africa. Ask students to plot and label on a large outline map of Africa the location of place names shown in the photographs and mentioned in the articles.

Gifted Learners

Have interested students research the lives of African independence and political leaders such as Kwame Nkrumah, Patrice Lumumba, Jomo Kenyatta, and Nelson Mandela. Students might elaborate on the leaders' goals, obtain excerpts from speeches, and collect photographs. Suggest that students focus on events or people that influenced the leaders' lives, motivating their work for independence. Have students make an oral presentation to the class.

MULTICULTURAL FOCUS

Write the following sayings from the Akan ethnic group of West Africa on the board:

When a man descends from heaven, he descends into a human society.

When the left hand washes the right and the right the left, then both hands will be clean.

Call on students to volunteer the meanings of these proverbs. Then explain that the sayings illustrate a cultural tradition of many African peoples—that of strong ties to one's community or family and awareness that cooperative effort is vital to society.

Many aspects of African daily life and religion focus on communal bonds. A person is responsible to a community or kinship group that includes not only deceased ancestors and living relatives but also those who are yet to be born. In villages, people often settle conflicts with one another by seeking the advice of elders who are recognized as authority figures for the community. Traditionally, folktales that are repeated orally deliver moral messages to set ethical standards for a community. In many parts of Africa, particularly in rural settings, work and entertainment are structured to benefit the group.

Have students think of circumstances in which their school acts as one unified community. Ask students to recall occasions in their daily life when they have put aside their own individual interests to advance group goals.

CHAPTER 35

INTRODUCING AFRICA

Section 1: Physical Geography
Section 2: Historical Geography
Section 3: Human and Economic Geography

OVERVIEW

In Chapter 35, students will examine Africa's physical geography, its history, and its rich cultural traditions.

Section 1 describes Africa's landforms, its climates and vegetation, and the challenges posed by its environment.

Section 2 describes early African civilizations and the effects of European colonialism. It also discusses the drive for independence by African nations beginning in the 1950s.

Section 3 identifies the language groups and religions of Africa. Students will also learn of important issues that affect the political and economic development of African nations.

CREATIVE STRATEGIES FOR TEACHING WORLD GEOGRAPHY

LESSON 16 How Much Do You Know About Africa?

CHAPTER 35

Introducing Africa

GEOGRAPHY DICTIONARY

escarpment

rift

Sahel

animism

Cave paintings, **South Africa**

Africa contains the world's oldest archaeological sites. They tell us about the places where humans first gathered around campfires to share food and tell stories. The African cultures descended from these early societies have given the world a wealth of art, architecture, music, and folktales. Africa also is rich in natural resources that are a key part of the world economy.

Africa stretches some 5,000 miles (8,045 km) from north to south. The shifting sands and gravel plains of the Sahara separate Africa's Mediterranean coast from the greater part of Africa lying farther south. Current political boundaries divide the region into more than 50 countries. Modern Africa also includes several hundred ethnic groups with different languages and religions. This diversity adds a richness to African cultural life but has also led to conflicts among and within countries.

Africa has modern cities with skyscrapers and superhighways. These symbols of economic development, however, are often overshadowed by more common signs of poverty. Many Africans have too little to eat, poor health services, and few opportunities to go to school. Populations are rising rapidly. Creating jobs for the increasing number of people is a struggle all African governments share.

A young girl near **Timbuktu, Mali**

Using the Geography Dictionary

Write the words in the Geography Dictionary on the board, and pronounce them for the class. Have students look at the photograph of the dry riverbed on page 413 and find the reference to **Sahel** in its caption. Point out that the Sahel is not a country but a region bordering the Sahara that is defined by physical characteristics. Ask students to speculate about what these characteristics are.

(Students should guess that the Sahel is generally dry and lacks lush vegetation.) Then refer students to photographs that illustrate **rift** (page 454) and **escarpment** (page 478). Have them compare the Glossary definitions of both terms with details in the photographs.

MAPPING THE REGION

Using the Physical-Political Map

Have students **study the map on this page**. As you **name several countries**, call on students to **identify their location and type of terrain**, and to tell something they may know about the countries. For example, after finding Somalia, a student might answer, "In recent years, Somalia experienced a civil war in which many people died."

MAP TRANSPARENCIES

Overhead transparencies from the maps on pages 417–420 are available in the *REGIONS OF THE WORLD Map Transparencies*.

MAPPING THE REGION

AFRICA

PHYSICAL–POLITICAL

LEGEND

ELEVATION

FEET		METERS
13,120		4,000
6,560		2,000
1,640		500
656		200
(Sea level) 0		0 (Sea level)
Below sea level		Below sea level

⊛ National capitals

• Other cities

MAPPING THE REGION

Using the Climate Map

Have students review the climate map of Africa on this page and identify the climate region that most resembles their own local climate. Have students compare their location in terms of latitude with that of the corresponding African climate. Then ask students to write a statement relating climate and latitude. (*Latitude affects climate. Places in low latitudes generally have a hot climate, those in middle latitudes have a temperate climate, and those in high latitudes generally have a cold climate.*)

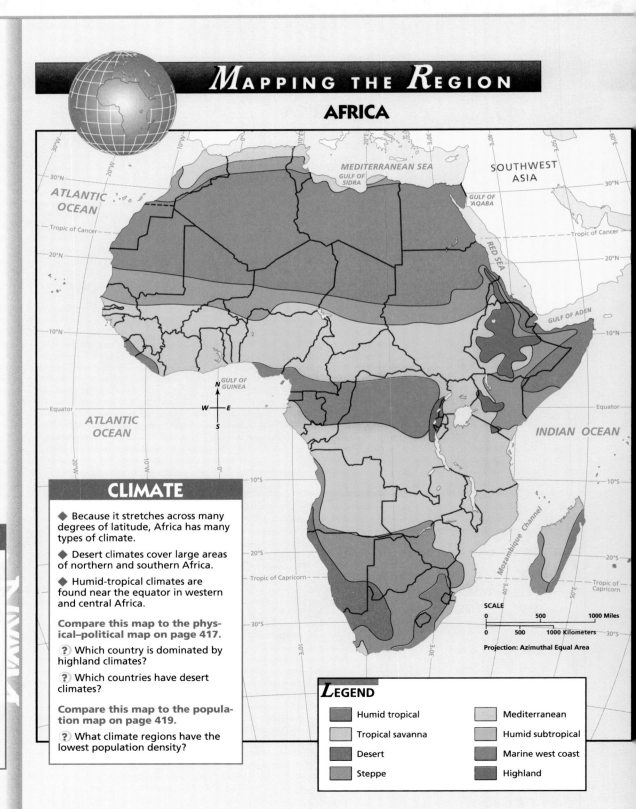

MAPPING THE REGION

AFRICA

CLIMATE

◆ Because it stretches across many degrees of latitude, Africa has many types of climate.

◆ Desert climates cover large areas of northern and southern Africa.

◆ Humid-tropical climates are found near the equator in western and central Africa.

Compare this map to the physical–political map on page 417.

? Which country is dominated by highland climates?

? Which countries have desert climates?

Compare this map to the population map on page 419.

? What climate regions have the lowest population density?

LEGEND

- Humid tropical
- Tropical savanna
- Desert
- Steppe
- Mediterranean
- Humid subtropical
- Marine west coast
- Highland

SCALE
0   500   1000 Miles
0   500   1000 Kilometers

Projection: Azimuthal Equal Area

CLIMATE

Answers

1. Ethiopia
2. South Africa, Botswana, Namibia, Angola, all countries of North Africa, Sudan, Mali, Mauritania, Niger, Chad, Eritrea, Somalia, Ethiopia, Kenya, Djibouti
3. Desert, some steppe regions

MAPPING THE REGION

Using the Population Map

Have students examine the population map of Africa on this page and identify the areas with the highest population density. (*Answers should reflect relative location, such as "near the Gulf of Guinea."*) List their responses on the board. Then direct students to find these areas on the physical–political map of Africa on page 417. Ask them to name the country or countries that correspond to these areas of high population density, and write their responses next to the first list. Then ask students to name the large physical feature in northern Africa with a low population density. (*Sahara*)

MAPPING THE REGION

AFRICA

POPULATION

◆ The population of Africa is distributed unevenly throughout the continent.

◆ Large areas of Africa are sparsely populated or uninhabited.

◆ Africa has many metropolitan areas with more than 2 million inhabitants.

? Which metropolitan areas have more than 2 million inhabitants?

Compare this map to the economic map on page 420.

? Why do you think manufacturing and trade centers have large populations?

? Why do you think some areas are sparsely populated?

LEGEND

POPULATION DENSITY

Persons per sq. mile	Persons per sq. km
520	200
260	100
130	50
25	10
3	1
0	0

● Metropolitan areas with more than 2 million inhabitants

● Metropolitan areas with 1 million to 2 million inhabitants

SCALE
0   500   1000 Miles
0   500   1000 Kilometers

Projection: Azimuthal Equal Area

POPULATION

Answers

1. Algiers, Casablanca, Tripoli, Tūnis, Alexandria, Cairo, Dakar, Abidjan, Lagos, Kinshasa, Khartoum, Addis Ababa, Nairobi, Luanda, Maputo, Cape Town

2. Lure of potential jobs

3. Undeveloped or little economic activity, harsh climates

Have students look closely at the economy map on this page. Call on volunteers to name specific mineral resources that are most plentiful in Africa. *(gold, diamonds, oil)* Organize the class into small groups. Assign several mineral resources to each group, and ask students to create a chart listing the countries in which these resources are plentiful. Have them discuss how the availability of these resources might affect the economies of these countries. *(Abundant mineral resources generally benefit a nation's economy.)*

SECTION 1
OBJECTIVES

The objectives reflect the Focus questions on page 421 and are assessed in the review on page 423.

1. Identify major landforms of Africa.

2. Describe the climate and vegetation of Africa.

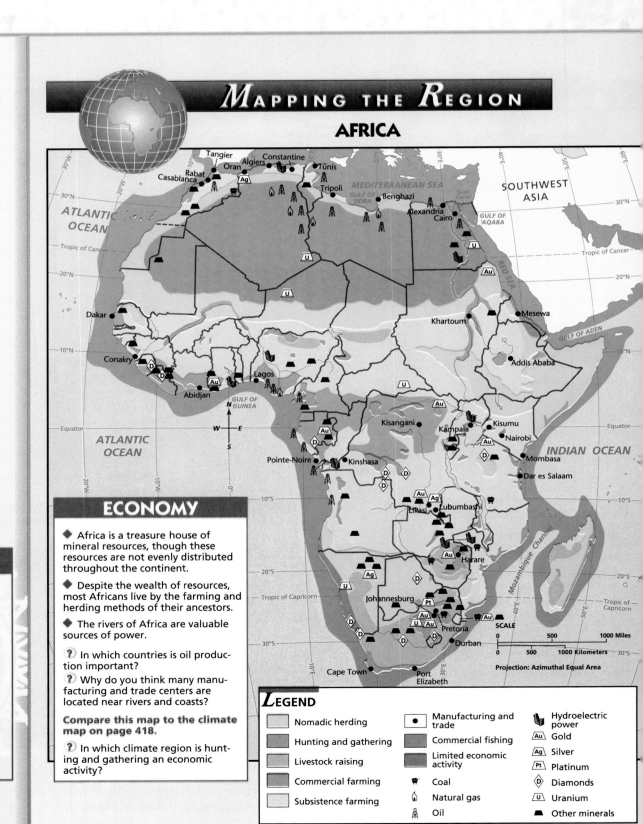

MAPPING THE REGION
AFRICA

ECONOMY

◆ Africa is a treasure house of mineral resources, though these resources are not evenly distributed throughout the continent.

◆ Despite the wealth of resources, most Africans live by the farming and herding methods of their ancestors.

◆ The rivers of Africa are valuable sources of power.

⁇ In which countries is oil production important?

⁇ Why do you think many manufacturing and trade centers are located near rivers and coasts?

Compare this map to the climate map on page 418.

⁇ In which climate region is hunting and gathering an economic activity?

LEGEND

▢ Nomadic herding	• Manufacturing and trade	⚒ Hydroelectric power
▢ Hunting and gathering	▢ Commercial fishing	Ⓐ Gold
▢ Livestock raising	▢ Limited economic activity	Ⓐ Silver
▢ Commercial farming	⚒ Coal	⟨Pt⟩ Platinum
▢ Subsistence farming	🜊 Natural gas	◇ Diamonds
	⚒ Oil	Ⓤ Uranium
		▲ Other minerals

ECONOMY

Answers

1. Libya, Morocco, Algeria, Nigeria, Egypt, Cameroon, Gabon, Angola

2. Some rivers are important transportation routes. Goods also are shipped to and from ports along the coasts.

3. Humid tropical

420

Motivator

Ask students what images, in terms of physical features and climate, come to mind when they think of Africa. *(Students are likely to mention stereotypical images such as jungle terrain and hot weather.)* Tell students that in this section they will learn about the geographical diversity of Africa.

Building a Vocabulary

Write the terms **escarpment** and **rift** on the board. Have students look up the meanings of these terms in the Glossary. Ask how these two terms are related. *(They are both landforms.)* Locate the Drakensberg Escarpment and the Eastern and Western Rift valleys on a wall map.

Analyzing Maps

Africa's River System Have students use the physical-political map of Africa on page 417 to trace the courses of the following rivers: Nile River, Niger River, Zambezi River, and Zaire River. Have them list the countries through which the rivers run.

PHYSICAL GEOGRAPHY

FOCUS

- *What are the major landforms of Africa?*
- *What types of climate and vegetation are found in Africa?*

LOCATION **Landforms** Africa is a continent mostly of high plateaus and wide plains. Erosion has shaped Africa's landforms for millions of years. There are few major mountain ranges. Only the Atlas Mountains north of the Sahara and the mountains of Ethiopia (ee-thee-OE-pee-uh) compare to the mighty ranges on other continents.

The most famous river in Africa is the Nile, an exotic river that has its sources in the highlands of East Africa, where there is abundant rainfall. More than 95 percent of Egypt's (EE-juhpt) people depend directly on the Nile.

Africa has several other great rivers with curious courses. (See the map on page 417.) The Niger (NIE-jer) begins in the Fouta Djallon (FOOT-uh juh-LONE) highlands along Africa's west coast. Instead of flowing toward the coast, it flows northeast to its inland delta in the southern Sahara before turning south toward the Atlantic. The Zambezi (zam-BEE-zee) River, meanwhile, heads south toward the Kalahari (KAL-uh-HAHR-ee) but then turns east over the Victoria Falls to the Indian Ocean. In Southern Africa, the Orange River begins in a plateau edge called the Drakensberg Escarpment near the east coast. An **escarpment** is a steep slope capped by a nearly flat plateau. The river then flows west all the way to the Atlantic. Africa's rivers typically have waterfalls or rapids near the coast. As a result, traveling inland by river can be

Victoria Falls are on the Zambezi River in **Southern Africa**. Over a mile (1.6 km) wide, the falls produce a deafening roar and mist that can be seen for nearly 40 miles (64 km). Local inhabitants called the mighty falls "Smoke That Thunders."

difficult. The Zaire (zah-IHR) (Congo) River is navigable only inland above the falls. Geographers think these stream patterns can be traced to when Africa was the center of Gondwanaland millions of years ago.

Africa has five huge depressions on its surface. These depressions, or basins, are each more than 625 miles (1,006 km) across and drop up to 5,000 feet (1,524 m) below the surrounding highlands. Three of these basins lie along the southern edge of the Sahara. In the west is El Djouf (ehl JOOF), which contains the inland delta of the Niger River. In the center is the Chad Basin, which holds shallow Lake Chad. In the east is the Sudan Basin, with the Sudd swamps at its center. The two remaining basins are in Central and Southern Africa. They are the forested Zaire Basin in Zaire and the arid Kalahari Basin in Botswana (baht-SWAHN-uh).

Over long periods of time, these basins have filled with sediment eroded from the surrounding highlands. Before Gondwanaland broke up, these basins were the ending points for major rivers. Since then, however, Africa's four greatest rivers—the Nile, the Niger, the Zambezi, and the Zaire—have cut channels to the sea.

The eastern third of Africa is the continent's highland region. The highlands are cut by two deep troughs called **rift** valleys. The Eastern Rift Valley and the Western Rift Valley begin near Lake Malawi (Lake Nyasa) and continue north into the Red Sea

The Atlas Mountains, which stretch for about 1,200 miles (1,931 km) across northwestern Africa, are divided into several ranges. Shown here is part of one of the southern chains, the High Atlas Mountains, located in **Morocco**.

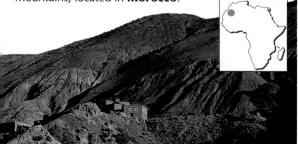

Geography and Geology

The rift valleys of East Africa provide stark evidence for the theory of plate tectonics. Geologists are closely studying the Afar Triangle in Ethiopia, which lies at the northern end of the Eastern Rift Valley. The triangle marks an active area of tectonic motion where continental plates are apparently separating. Some geologists believe that the Horn of Africa, located to the east of the rift system, may eventually separate from the rest of the continent, a process that will take some 30 million years.

Refer students to Chapter 5, page 48, to review the continental drift theory. Ask students to explain how the theory applies to the development of the African continent as well as the landforms on the continent. Encourage students to draw diagrams that illustrate their explanations.

Provide each student with Outline Map 19. Have students work in pairs and use the information in Section 1 to create their own vegetation map of Africa. Students' maps should illustrate the location of rain forests, savannas, and deserts. Students should also label countries.

Have students write a paragraph describing how aspects of Africa's physical geography can make living there challenging.

C⊙⊙PERATIVE LEARNING

Organize the class into groups of four or five. Groups will produce illustrated reports about different aspects of the savanna environment. Begin by using the climate map on page 418 to emphasize the extent of the savanna in Africa. Assign each group a separate topic to research, such as (1) the different types of savannas and their location, (2) wildlife, (3) trees and vegetation, and (4) economic activities. Group members may contribute by researching, writing, illustrating, or presenting their report in class.

and then Syria. In many areas, the rifts have narrow valley floors with mountainous slopes on each side. In other areas, such as the Serengeti Plain in Tanzania (tan-zuh-NEE-uh), the rifts are wide and not as deep. Volcanoes have been active along both rifts. In fact, the highlands of Ethiopia are made up of layers of hardened lava.

REGION **Climate** Africa is the second largest continent in area. Because it stretches across many degrees of latitude, it has many climate types. (See the map on page 418.) Humid-tropical climates are found near the equator in the Zaire River basin and along the coast of the Gulf of Guinea (GIN-ee). These regions receive rain more than 200 days each year, and temperatures are the same day after day. Rain forests here have tall hardwood trees with leaves that stay green all year. The soils of the rain forests are poor, however, and farming is difficult. Few people live in the central parts of the rain forests.

Much larger regions with tropical-savanna climates surround the rain forests. Instead of rain year-round, these areas have dry weather during the winter. In regions with good rains, the savanna trees and grasses are tall. These regions are called tall-tree savannas. Farther away from the equator, seasonal drought is common. The trees become shorter and are better adapted to water shortages. These drier bush savannas have low, thorny trees and short, drought-resistant grasses. Clearing the savannas has created some of Africa's prime agricultural land.

Along the southern edge of the Sahara is a region of bush savannas and grasslands called the **Sahel** (suh-HAYL). The Sahel sometimes experiences years of drought. In the past, the people of the Sahel coped with droughts by moving south with their cattle, goats, and sheep. Today, however, moving is difficult because governments stop people from crossing their borders. As a result, a growing population is crowding the available grazing lands in the Sahel. The desert is expanding due to overgrazing, and many people and farm animals have died. Many more people barely survive in refugee camps with little hope for the future.

Africa's deserts are found in both the north and the south. The Sahara is the world's greatest desert, both in area and in climatic extremes. The Namib (NAHM-ib) and Kalahari deserts are in Southern Africa. Like the driest parts of the

These vineyards thrive in the Mediterranean climate region of **South Africa**. Africa has a great variety of climates. Some provide ideal conditions for agriculture, but others discourage farming. Produce from mild African climates is found in European markets when crops there are out of season.

Sahara, the Namib is a severe coastal desert with sparse plant and animal life. In contrast, the Kalahari has low grasses and shrubs. Parts of it receive up to 10 times as much rain as the Namib.

Much of Southern Africa lies about the same distance from the equator as the southern United States. For example, Durban (DUHR-buhn), South Africa, is about as close to the equator as San Antonio, Texas. Southern Africa has many different climate regions. The Kalahari's eastern edge is a bush savanna that receives about 10 inches (25 cm) of rain each year. Rains increase toward the east coast, and tall-tree savannas gradually become dominant. In summer, rainstorms move into Southern Africa from the Indian Ocean, producing regions of humid-subtropical and marine-west-coast climate in the southeast. The southern tip of Africa has a Mediterranean climate, making this region a center for fruit growing.

Africa's highlands stretch from the Cape of Good Hope to Ethiopia. These high plateaus and mountains have cooler climates than are usually found at these latitudes. The plateau region of South Africa, for example, is not tropical like the lowlands along the east coast. Instead, subhumid to semiarid grasslands are present. Even though the

422

ASSESSMENT

Assign the Section 1 Review questions on this page.

Reteaching
Assign Section 1 of Reteaching Worksheet 35.

Extension
Point out that geographers disagree about whether the Kalahari is really a desert. Have interested students research the arguments on both sides and then debate the issue.

SECTION 2
OBJECTIVES

The objectives reflect the Focus questions on this page and are assessed in the review on page 426.
1. Identify and describe early African civilizations.
2. Explain the effects of European colonization on Africa.

equator is near, the highlands of Kenya (KEN-yuh) have mild weather all year.

The northern part of Africa lying along the Mediterranean Sea has a Mediterranean climate with good winter rains. The long, dry summer means that productive agriculture requires irrigation and wise conservation of winter rainwater. Vegetation ranges from desert plants in the northeast to grasslands and Mediterranean scrub forest in the northwest. The Sahara begins only a short distance south from the Mediterranean shore.

Natural Challenges Life in Africa can be difficult. Almost every year, some regions suffer from drought while others are ruined by floods. Earthquakes and volcanic eruptions occur along the rift valleys. These news-making events, however, cannot compare with the natural disadvantages and hazards Africans must face each day.

Africa has little fertile land compared with areas of the world of similar size. Much of the soil in dry regions has too much salt or lime to be agriculturally productive. In humid regions, soils often are not fertile enough for good crop yields. Good soils are common only in the highlands of eastern and southern Africa and along river valleys. Swarms of insects sometimes wipe out crops.

Diseases are a constant threat in Africa. Malaria and tuberculosis are widespread. In areas with intense irrigation, such as Egypt, certain parasites cause disease. In the humid tropics, sleeping sickness, carried by the tsetse (TSEHT-see) fly, attacks people and cattle. And in many areas, poor diets encourage a host of related diseases. AIDS (Acquired Immune Deficiency Syndrome) also is destroying a large part of the African population. By the beginning of 1995, the World Health Organization (WHO) estimated that as much as 60 percent of the global adult population infected with the AIDS virus lived in Africa.

SECTION REVIEW

1. What are the major landforms in Africa?
2. What type of vegetation would you find in a tropical-savanna climate with good rains?
3. **Critical Thinking   Compare and contrast** the climates of North Africa and Southern Africa.

Early African Kingdoms

Numerous African civilizations flourished long before Europeans set out to build colonial empires. This map shows many early African kingdoms. Which West African kingdom existed from about 1350 to 1600?

HISTORICAL GEOGRAPHY

FOCUS

- *What economic activity supported most early African civilizations?*
- *What effect did European colonization have on the land and peoples of Africa?*

MOVEMENT Early African Civilizations
Archaeologists have discovered more evidence of early humanity in Africa than on any other continent. More than 8,000 years ago, residents of Africa began planting barley and herding animals along the Nile River. Early in their history, Egyptians learned to use water from the Nile for irrigation. The people of the Nile Valley soon created one of the world's first great civilizations.

Later, cities along the African Mediterranean traded with the ancient Greeks and Phoenicians.

GLOBAL SKILL BUILDER

You might wish to use **Using Different Types of Maps** from the interactive *Global Skill Builder* CD-ROM to reinforce students' understanding of historical maps, such as the maps on this page and on page 425.

Map Caption Answer
Songhai

Section 1 Review
ANSWERS

1. High plateaus and wide plains
2. Tall trees and grasses
3. Both northern and southern areas have deserts. Southern Africa has many climate regions, including savanna, humid-subtropical, marine-west-coast, and Mediterranean climates. In contrast, most nondesert areas of North Africa have a Mediterranean climate.

423

TEACHING SECTION 2

Motivator

Ask students to recall family stories passed down from their parents or grandparents. Explain that much of Africa's early history was preserved in the same way, memorized by elderly oral historians called *griots.* Tell students that they will learn about significant events in Africa's ancient history that were kept alive by *griots.*

Creating a Chart

Early Civilizations   Have students work in small groups to complete a chart describing the early kingdoms in Africa. The column heads should be Kush, Axum, Ghana, Mali, Songhai, and Kanem-Bornu. The row headings should be Time Period, Location, Economic Activity, and Significance of Kingdom. Students might illustrate their charts and display them in the classroom.

Guided Practice

Discuss with students how Europeans set political boundaries in Africa. *(without regard for Africa's landforms, climate regions, or cultures)* Then have students hold a panel discussion in which they discuss the effects of these boundaries on independent African nations.

Carthage, in modern Tunisia (too-NEE-zhuh), was a commercial power in the western Mediterranean. Alexandria, at the western mouth of the Nile, was a city of learning and culture before the Roman Empire. The cities of North Africa also flourished during the expansion of Islam after A.D. 700. Cairo (KIE-roe) remains an important city of Islamic learning and culture.

Instead of keeping written records, most African peoples kept oral histories. Stories of governments and families were memorized and passed verbally from one generation to the next. We know of several great civilizations and ancient kingdoms that developed in Africa. (See the map on page 423.)

The kingdom of Kush in present-day Sudan had close connections to ancient Egypt and controlled much of the middle Nile River valley. The capital of Kush was Meroë (MEHR-uh-wee). After a period of decline, Kush was conquered by rulers of powerful Axum. Axum, located south of Kush in the highlands of present-day Ethiopia, was founded as a city of traders and merchants. Today, its ruins reveal the innovation of dry-stone construction, by which walls were built without cement or mortar.

Other empires arose at different times in other parts of Africa. Perhaps the most famous ones were founded in the dry savannas of West Africa. These early empires grew powerful by controlling trade across the Sahara between tropical Africa and the Mediterranean coast. Caravans brought salt from the Sahara and goods from Mediterranean cities. They also brought gold, ivory, and copper from the coastal forests to the south. Probably the most famous city of this region was Timbuktu, a major trading center.

While empires were developing in the east and west, civilizations were growing in Central and Southern Africa as well. These kingdoms also were involved in establishing trade routes across Africa. More than 1,000 years ago, Central African king-

This mask was crafted from ivory in the first half of the sixteenth century in the early African forest kingdom of **Benin**.

doms were trading gold, spices, precious woods, and slaves with Arabia, Persia, India, and China.

In Southern Africa, there are ruins of ancient cities. One of these cities was Great Zimbabwe (zim-BAHB-way), which began as a village of iron-workers during the third century A.D. Impressive stone structures built on a hilltop around the thirteenth century still stand. Civil wars within the region and new trading patterns brought about by contact with Europe led to the empire's decline.

Contact with Europe   A dramatic change in the history of Africa began with the arrival of the European explorers in the fifteenth century. Most Europeans first dealt with African ethnic groups from trading posts along the African coast.

After Europeans began settling in the Americas, workers were needed for the plantations the colonists were building there. Africans, sold as slaves, became a major source of this labor. The main slave-trading ports stretched from the southern edge of Mauritania (mahr-uh-TAY-nee-uh) to the northern edge of the Namib Desert. The trading posts were located at sites where there was a harbor along with a river or trail leading inland.

Often, European merchants bought the slaves from coastal African leaders. These leaders were paid with European manufactured goods, particularly guns. Many of the slaves had been captured inland. As the demand for slaves grew, slave raids to the interior increased. The lives of millions of Africans were destroyed by the horrific practice of enslavement. The slave trade left a lasting imprint on Africa.

European exploration of the African interior did not begin until a few hundred years after the slave trade began. Much of this exploration centered on the search for the sources of Africa's rivers. Among the early explorers was Mungo Park. In 1795 and 1805, he traveled to West Africa,

Independent Practice

Provide students with Outline Map 19. Ask them to mark the sites of ancient African civilizations and to label each according to the European powers that controlled the same areas in 1900. Have them indicate the present-day nations that now occupy these areas.

Closure

Remind students of the legacies of Africa's early kingdoms. *(innovations in construction and agricultural techniques; trading empires sparking contact with Europe)* Contrast these with the later effects of European colonization on Africa. *(Political boundaries set by Europeans led to ethnic conflicts and environmental problems today. Africans also lost economic and personal autonomy.)*

ASSESSMENT

Assign the Section 2 Review questions on page 426.

Reteaching

Assign Section 2 of Reteaching Worksheet 35.

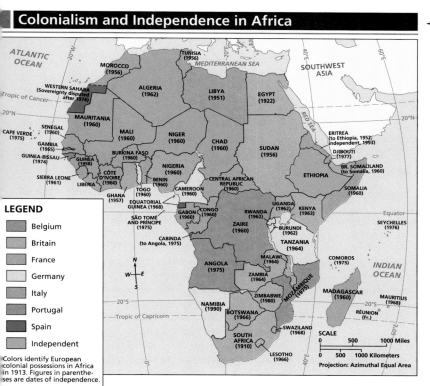

Colonialism and Independence in Africa

LEGEND
- Belgium
- Britain
- France
- Germany
- Italy
- Portugal
- Spain
- Independent

Colors identify European colonial possessions in Africa in 1913. Figures in parentheses are dates of independence.

 MAP STUDY

Often, historical maps use colors to identify important features and dates to highlight events. In this map, colors identify European colonial possessions in Africa in 1913. The dates tell when colonies became independent. Note that South Africa became a self-governing country within the British Empire in 1910. It achieved complete independence in 1931. Which African countries had not been colonized by 1913?

the rule of one European country or another by 1900. The Europeans imposed their culture, technology, and economic policies on Africa. The consequences of this process still exist today.

When the Europeans settled in Africa, they set up plantations to grow cash crops such as coffee, cotton, cacao, palm products, and peanuts. The Europeans also mined gold, iron ore, and copper. In contrast, most Africans still practiced traditional subsistence agriculture.

The Europeans also built roads, schools, ports, and hospitals in the areas where there were exportable resources. All of Africa's modern capital cities developed during this colonial era. The colonizers also paid for the European education of a small number of Africans. Many of these Africans later became leaders of independence movements.

Independence Africans began struggling for their independence as soon as the colonial powers appeared. After World War II, pressure for independence grew. Beginning with Libya (LIB-ee-uh) in 1951, country after country became independent. Today, all African countries are independent. Ties remain, however, to Europe and other developed countries.

For the most part, Africa's political boundaries today are those set by the Europeans. These boundaries have created many problems. Some African people, such as the Somalis of Kenya, Ethiopia, and Somalia (soe-MAHL-ee-uh), have

exploring the upper parts of the Niger River. In 1857, Richard Burton and John Speke went to Africa in search of the source of the Nile River in the African interior. Four years later, Speke returned to the interior with J.A. Grant and discovered the White Nile's source, which he named Lake Victoria in honor of Queen Victoria.

European Colonization Tales of Africa's beauty and treasures quickly reached Europe, and mapmakers began charting the continent. With little consideration for the Africans living there, European powers scrambled to claim territories throughout Africa. To settle conflicts among themselves, the Europeans divided Africa at a conference in Berlin in 1884–1885. They took a map of Africa and drew boundaries without regard for Africa's landforms, climate regions, or cultures. European boundaries divided ethnic groups and cut through paths of migration and agricultural lands.

The Africans had little say about either these boundaries or the colonization. Except for Liberia (lie-BIR-ee-uh) and Ethiopia, all of Africa was under

Map Study
Ethiopia and Liberia

GLOBAL PERSPECTIVES

Ask students to name other nations that have struggled for independence. *(Students might mention Mexico, South American nations, the original 13 British colonies in North America, Vietnam, and India.)* Have students research and report on the reasons that these nations sought their freedom from colonial powers.

Geography and Politics

From the 1950s through the 1980s, Africa was a major geopolitical battleground of the Cold War. The United States and Soviet Union poured economic aid into the continent as a means of gaining influence and undermining each other's political power. Since the end of the Cold War, foreign investment of all kinds in Africa has dropped considerably.

Extension

Have interested students research the government of an African country of their choosing. They might present reports that describe the type of government, identify specific leaders, and evaluate how effective the government has been in addressing specific problems.

SECTION 3
OBJECTIVES

The objectives reflect the Focus questions on this page and are assessed in the review on page 428.

1. Analyze problems related to a growing African population.

2. Identify economic resources important to the development of Africa.

TEACHING SECTION 3
Motivator

Have students name languages that are spoken in the United States and suggest why so many are present. *(because many ethnic groups have settled here)* Point out that in Section 3 they will learn why many different languages are spoken in African countries.

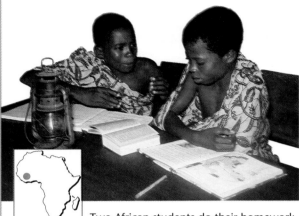

Two African students do their homework in a village in **Côte d'Ivoire**, a country on the Atlantic coast of West Africa.

HUMAN AND ECONOMIC GEOGRAPHY

FOCUS

● *What effect does an increasing population have on daily life in Africa?*

● *What resources in Africa are important to Africa's economic development?*

Human Geography Africans are divided into hundreds of ethnic groups by different languages, religions, and ways of life. One way to sort Africa's ethnic groups is to study the different language families. Most people in North Africa speak Arabic. Along the southern Sahara are peoples who speak Semitic languages related to those of the Berbers and Arabs living farther north. These are part of the Afro-Asiatic family. South of the Sahara, many black Africans speak one of the Niger (Bantu) languages. These languages are spoken from West Africa through the Zaire Basin to the southeastern coast of South Africa. In some inland areas of East Africa and West Africa, people speak Nilo-Saharan languages.

Another language family is the ancient Khoisan, which is still spoken by isolated peoples in Southern Africa, such as the San (Bushmen). These people, now few in number, are not black Africans. Their yellow-brown skin, short stature, and unique blood type make them unlike any other people. Madagascar (mad-uh-GAS-kuhr), off the east coast of Africa, meanwhile, was settled by migrants from southeast Asia more than a thousand years ago. The language here is related to those of the southwestern Pacific region.

The colonial period brought European languages to Africa. In South Africa, the Cape Dutch language developed into Afrikaans. Many African countries have adopted French or English as an official language. In the countries of North Africa, however, Arabic is the official language.

Religion and family traditions are particularly important for modern Africans. Christian churches and Islamic mosques can be found in most African towns. Africans often combine these organized religions with more ancient forms of traditional worship. Many people believe in the spirits of the waters, animals, trees, and mountains. In most cultures, the memories of ancestors are honored, and

---

found themselves divided among several countries. Other groups were thrown together under one rule. Nigeria (nie-JIR-ee-uh), for example, has more than 200 ethnic groups. This merging of different cultures has caused conflict.

European boundaries also created landlocked states. Because these countries do not have direct access to any oceans, they often have economic and political problems importing and exporting goods across the boundaries of other countries. Finally, some countries were carved out of only one climate region, limiting the kinds of crops that can be grown.

African nations are trying to squeeze into a short period of time progress that has taken developed countries many years to achieve. Most Africans appreciate the value of Western democracy, and democracy has spread in recent years. Yet some African leaders feel that a one-party government can move faster to achieve national goals. Other governments are dictatorships. Often, unrest—and sometimes revolution—threaten the stability of many African countries.

SECTION REVIEW

1. How did early African empires gain power in a region?

2. What were the effects of European colonization?

3. **Critical Thinking** **Compare and contrast** the permanence of oral histories and written records.

Section 2 Review
ANSWERS

1. Through control of trade routes

2. When the Europeans divided Africa, the boundaries cut across ethnic groups and did not consider climate or landform regions. Europeans also transferred European culture, technology, and policies to Africa, set up mines and plantations, built roads, schools, ports, and hospitals in some areas, and promoted some education.

3. Neither is permanent. Oral histories may be forgotten or changed. Written records can be lost, stolen, soiled, or destroyed, as by fire.

426

Building a Vocabulary

Write the word **animism** on the board. Ask students to find the definition in the Glossary. Tell them that the word comes from the Latin word for "soul" and is closely related to the words *animal* and *animate*. Then have students read the poem on page 414. Discuss how the poem illustrates animism.

Illustrating Ideas

Creating a Language Map Have students create a language map of Africa. Call on students to name the major language groups of Africa. (*Afro-Asiatic, Niger-Congo, Nilo-Saharan, Khoisan, and Malayo-Polynesian*) Then provide students with Outline Map 19, and have them indicate the areas in which these languages are spoken.

Guided Practice

Have students work in small groups to prepare the script for a "documentary film" on the economy of Africa. The script should describe Africa's natural resources, manufacturing, and agriculture. It should also analyze the problems involved in Africa's economic development. Encourage students to use visuals to illustrate their scripts. Have them read their scripts to the class.

LANGUAGE FAMILIES OF AFRICA

LANGUAGE FAMILY:	AFRO-ASIATIC	NIGER-CONGO	KHOISAN	NILO-SAHARAN	MALAYO-POLYNESIAN
Language: English	Arabic	Swahili	Nama	Kanuri	Malagasy
mother	umm	mama mzazi	//gûs	yâ	reny
child	walad	mtoto	/gôaï	táda	zaza
head	rā`s	kichwa	tanas	kəlâ	loha
water	mā`	maji	/gami	njî	rano
tree	šajarah	mti	heis	kəská	hazo
house	bayt	nyumba	omi	fáto	trano
red	aḥmar	-ekundu	/awa	cimê	mena
eat	akala	la	‡û	búkin	homana
go/walk	ḍahaba/mašã	enda	!gû	lengîn	mamìndra

Africans speak a wide variety of languages. These languages originated from different language families. The chart above shows examples from five African language families. Nine English words are translated into five languages. For the Nama language, /, //, !, and ‡ represent different clicks, or sounds made by the tongue.

their spirits are believed to be powerful forces in daily life. This traditional belief in nature and spiritual beings is called **animism**. (See "One World, Many Voices" on pages 414–415.)

During the past 25 years, populations have increased in Africa. The use of modern health practices is largely the reason for this growth. Families have become larger and people are living longer. Many farming communities are now overcrowded, however. Overgrazing by cattle and goats has caused severe soil erosion. The traditional rural cultures of Africa are undergoing dramatic change as people move to cities to seek jobs.

Unfortunately, the number of people moving to the cities is usually greater than the number of jobs. The unemployed and those with temporary work live in the slums that surround African cities. Providing jobs and better housing is the most pressing issue for many African governments. Governing under these conditions is difficult.

REGION **Resources** Africa is a treasure house of mineral resources. Gold, copper, chromium, manganese, uranium, and cobalt are some of the important minerals. Africa's mineral resources are not evenly distributed, however, and some countries are much richer than others. Most of Africa's mineral wealth is exported to industrial countries of the Northern Hemisphere. As a result, the resource-rich countries of Africa are often targets of economic competition among more powerful countries.

Africa's oil deposits also are distributed unevenly. The greatest oil and natural gas deposits lie in Libya, Nigeria, and Algeria (al-JIR-ee-uh). Oil also is found offshore between Cameroon (kam-uh-ROON) and Angola (an-GOE-luh). Because oil is so important to modern industrial nations, all African countries have active oil-exploration programs. More oil deposits will be developed along the west coast of Africa and in the Sudan.

The rivers of Africa are a valuable source of power. Hydroelectric power stations have been built or are planned at almost every rapids or waterfall. Perhaps 30 percent of the world's future hydroelectric potential exists in Africa.

Despite its resources, Africa remains the least developed continent. Many Africans live by the farming and herding methods of their ancestors. South of the Sahara, cattle and goats are the most important animals. In fact, in many African cultures, cattle are a measure of wealth. Teenagers discuss cows in much the same way that teens in wealthy countries talk about cars.

Manufacturing often is limited to producing simple consumer goods and processed food to meet the needs of local markets. Few products manufactured in Africa travel to world markets. Economic progress is limited mainly by the region's poor educational systems, a lack of trained business managers, and little money for investment.

Some African countries are small in population and land area, which can create an economic

MULTICULTURAL PERSPECTIVES

The peoples of Africa have rich, varied, and ancient musical heritages. During the 1980s, African popular music began gaining a wider audience in the United States and elsewhere abroad. Modern "Afropop" combines African rhythms with American, European, and Middle Eastern styles. Youssou N'Dour (Senegal), King Sunny Adé (Nigeria), Malathini and the Mahotella Queens and Ladysmith Black Mambazo (South Africa), and Les Têtes Brulées (Cameroon) are among the performers who have a worldwide audience.

Point out that American music forms have been influenced by African music. (*jazz, gospel, rock and roll*)

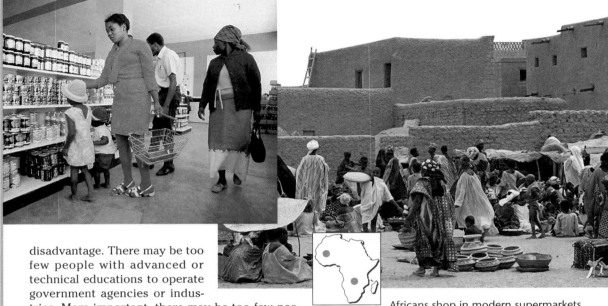

Africans shop in modern supermarkets, such as in **Zambia** (left), and in open-air markets, such as this one in **West Africa** (right). Are there any open-air markets in your hometown?

disadvantage. There may be too few people with advanced or technical educations to operate government agencies or industries. More important, there may be too few people to buy the goods produced by factories. A solution is for neighboring countries to cooperate economically. Goods can then be sold in several countries. Workers can seek jobs in other nations as well. Recently, organizations such as the Economic Community of West African States have developed in each of Africa's regions. These kinds of economic ties represent a break from the trading patterns of the colonial past.

REGION **Issues** Most countries of Africa face several challenges. First, struggling against often illogical colonial boundaries, African leaders must build stable countries from territories that contain peoples of different languages, religions, and cultures.

A second challenge is slow economic progress. African economists explain that African agricultural exports are not always treated fairly on the global market. African countries do not receive the same trade protection as other countries. Another cause of slow economic development is poor government economic policies. For example, governments often want their people to grow valuable export crops such as cotton and coffee. This discourages food production in rural areas. As a result, earnings from exports are spent on food imports instead of on development. Money for investment is in short supply. Furthermore, foreign aid from the developed countries often helps the city dwellers rather than the rural people.

A third challenge is Africa's rapidly increasing population. As a result of its current rate of population growth, Africa is becoming a continent of children who need to be cared for and educated. Also, the quality of Africa's natural environments is declining as more areas are cleared for farming, and overgrazing increases.

Improving people's lives in Africa will not be easy. Foreign aid is needed, mainly to improve agriculture and to develop appropriate manufacturing industries. Income from mining and export crops is necessary to improve infrastructure and transportation and to provide better schools. Cooperation among African countries will help in finding answers to these shared problems.

SECTION REVIEW

1. Why are many Africans moving from rural to urban areas?
2. How are Africa's rivers important economic resources?
3. **Critical Thinking** **Explain** how a country might be disadvantaged by either a large or a small population.

CHAPTER 35 CHECK
ANSWERS
Building a Vocabulary
1. A steep slope topped by an almost flat plateau; the Drakensberg Escarpment in Southern Africa
2. Eastern Rift Valley and Western Rift Valley
3. The Sahel is a region of dry savannas and grasslands along the southern edge of the Sahara. In recent years, it has become overgrazed and the desert has expanded.
4. The belief in nature and spiritual beings

Recalling and Reviewing
1. El Djouf, Chad Basin, and Sudan Basin on the southern edge of the Sahara; Zaire Basin in Zaire; Kalahari Basin in Botswana
2. Borders were drawn with little consideration for the African people. Borders cut through climate zones, paths of migration, agricultural lands, and ethnic groups.
3. Gold, copper, chromium, manganese, uranium, cobalt, oil, natural gas, hydroelectric power

CHAPTER 35 CHECK

Senegal village

Reviewing the Main Ideas
1. Africa is a continent primarily of high plateaus and wide plains. Soils are generally poor for agriculture. The climate varies from humid tropical to desert.
2. Before European colonization in the late nineteenth century, Africa was home to great empires. Europeans divided Africa with little regard to ethnic, economic, or climate regions. Most African countries did not receive their independence from colonial rule until the 1960s.
3. The countries of Africa have many characteristics in common, but they have many differences as well. Several hundred ethnic groups inhabit the region, each with its own language, religion, and customs.

Building a Vocabulary
1. Define *escarpment*. Give an example from this chapter.
2. What are the two deep troughs called that are found in the highlands of East Africa?
3. Describe the Sahel. Where in Africa is it found? How has the region changed since the late 1960s?
4. What is animism?

Recalling and Reviewing
1. Name the five huge basins in Africa. Where are they located?
2. What were the major effects of European colonization on the lands and peoples of Africa?
3. What economic resources are important to development in Africa?

Thinking Critically
1. Describe what the development of Africa might have been like if European colonization had not taken place. Consider the formation of the countries and their economies.
2. How would you explain to an African village leader the economic advantages for his village if the farmers produced more food crops instead of cash crops for export?

Learning About Your Local Geography
Cooperative Project
With your group, research and report on the original boundary of your town or city. Who drew the boundary? Does the original boundary make sense geographically? Why or why not? Draw a map of the original boundary to include in the report.

Using the Five Themes of Geography
Carefully examine the photographs on this page. For each photograph, write two or three sentences that explain how one or more of the Five Themes of Geography are evident in that photograph.

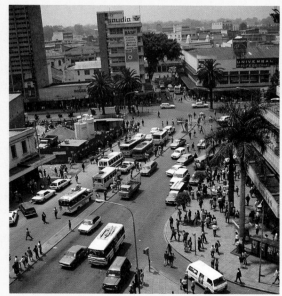

Nairobi, Kenya

Thinking Critically
1. Answers will vary. It is possible that some countries would be more stable, if they had not been politically and culturally fractured by the colonial powers and thus might have stronger economies.
2. Answers will vary, but students should point out that if all crops are exported, then food for the village has to be imported. Money spent on imported food might be better spent on developing other resources such as hydroelectric power.

Learning About Your Local Geography
Groups' reports will vary but should provide answers to the suggested questions and include a map of the original boundary of the town or city.

Using the Five Themes of Geography
Top photo: answers will vary but might include human-environment interaction (fishing) and movement (boat in foreground). Bottom photo: answers will vary but might include place (crowded, busy city) and movement (people walking, driving, and taking buses).

CHAPTER 36 North Africa
Planning Guide

	TEXTBOOK	TEACHING RESOURCES	OTHER AVAILABLE MATERIALS
Section			
1.	Physical Geography Pages 431–432	Reteaching Worksheet 36.1	*Creative Strategies for Teaching World Geography* LESSON 14 Speaking Arabic *Study Guide for Spanish Speakers* *Lecture Notes,* Chapter 36
2.	Egypt Pages 433–435	Reteaching Worksheet 36.2 Graphic Organizer 36 Challenge/Enrichment Worksheet 36	*The World Today* Videodisc Program *Study Guide for Spanish Speakers* *Lecture Notes,* Chapter 36
3.	Libya and Tunisia Pages 435–437	Reteaching Worksheet 36.3	*Study Guide for Spanish Speakers* *Lecture Notes,* Chapter 36
4.	Algeria and Morocco Pages 437–438	Reteaching Worksheet 36.4 Skills Worksheet 36 Critical Thinking Worksheet 36	*Study Guide for Spanish Speakers* *Lecture Notes,* Chapter 36
Chapter Review Page 439		Chapter 36 Test Chapter 36 Portfolio Assessment See also *Alternative Assessment Forms*	*Test Generator,* Chapter 36

CHAPTER RESOURCES

Lecture Notes, Chapter 36
Teaching Resources
    Chapter 36 Graphic Organizer
    Chapter 36 Skills Worksheet
    Chapter 36 Critical Thinking Worksheet
    Chapter 36 Challenge/Enrichment Worksheet

REVIEW AND ASSESSMENT

Chapter 36 Check, page 439
Chapter 36 Test, *Teaching Resources*
Chapter 36 Portfolio Assessment, *Teaching Resources*
Test Generator, Chapter 36
Alternative Assessment Forms

ADDITIONAL RESOURCES

Books for Teachers

Asante, Molefi K., and Kariami W. Asante, eds. *African Culture: The Rhythms of Unity.* Trenton, NJ: Africa World, 1989.

Bowen, Donna Lee, and Evelyn A. Early, eds. *Everyday Life in the Muslim Middle East.* Bloomington: Indiana University Press, 1993.

Oliphant, Margaret. *The Atlas of the Ancient World.* New York: Simon & Schuster, 1992.

Phillipson, David W. *African Archaeology.* New York: Cambridge University Press, 1985.

Books for Students

Hintz, Martin. *Morocco.* Chicago: Children's Press, 1985.

Jenkins, Ernestine. *A Glorious Past: Ancient Egypt, Ethiopia, and Nubia.* New York: Chelsea House, 1995.

Moorehead, Alan. *The Blue Nile.* New York: Harper & Row, 1972.

Sanders, Renfield. *Libya.* New York: Chelsea House, 1987.

The World Today *Videodisc Program*

A variety of still images, short videos, and activities are available for you to use as you teach this chapter. *Lecture Notes* include bar codes for specific images and videos.

Multimedia Resources

Africa: Northern Region. (VHS, 15 minutes) Encyclopedia Britannica Educational Corporation, 310 South Michigan Avenue, Chicago, IL 60604–9839.

The Ancient Egyptian. (16mm film/VHS, 27 minutes) International Film Foundation, Inc., 155 West 72nd Street, New York, NY 10023.

Morocco: Berber Profile. (VHS, 30 minutes) Films for the Humanities & Sciences, P.O. Box 2053, Princeton, NJ 08543–2053.

Oasis: Living in a Desert Community. (VHS, 20 minutes) Encyclopedia Britannica Educational Corporation, 310 South Michigan Avenue, Chicago, IL 60604–9839.

Sahara: Before the Desert. Films for the Humanities & Sciences, P.O. Box 2053, Princeton, NJ 08543–2053.

MEETING INDIVIDUAL NEEDS

Limited English Proficiency (LEP)

Pair LEP students with students who are proficient in English. Have the pairs skim the sections to find the answers to the Section Review questions. Then have them use the answers to write a brief summary of the chapter. Have students compare and contrast their summaries with those in Reviewing the Main Ideas in the Chapter 36 Check on page 439.

Learners Having Difficulty

Help students create a chart about North Africa. Write the following headings across the top of the board: Physical Characteristics, Cultural Characteristics, Economic Characteristics. Then write the following country names down the left side of the chart: Egypt, Libya, Tunisia, Algeria, Morocco. Have volunteers give information from the chapter to complete the chart, as you write it on the board.

Auditory Learners

Review with students the old and new methods of irrigation used along the Nile. As students describe basin irrigation and perennial, or continual, irrigation, draw a sketch of each on the board. Then discuss the pros and cons of each method.

Visual Learners

Have students study the physical-political map of North Africa on page 431. Ask them to discuss the locations of major cities in this region relative to landforms and waterways. Point out reasons why North Africa became a crossroads of trade between Europe and Asia.

Tactile/Kinesthetic Learners

Provide students with an outline map of North Africa. Have them make an agricultural product map of the region. Ask students to compare and contrast their agricultural product map with a climate map of the region. Ask them to write generalizations about the kinds of agricultural products grown in the North African climate.

Gifted Learners

Have interested students work in pairs to research the Egyptian pyramids. They might investigate the cultural significance of the pyramids or research the materials and the methods used to build them. Have pairs present an oral report about the pyramids or build a scale model of the pyramids.

MULTICULTURAL FOCUS

The Sahara in North Africa is one of the few areas of the world where nomadism still plays a significant cultural role. Movement is a permanent way of life for approximately one-third of the people who live in the Sahara. Desert nomads travel with their families, belongings, and livestock in a seasonal cycle—over routes they have traveled all their lives—seeking grazing lands and oases. Thus have they adapted to the limited water and vegetation available in the desert.

Historically, the nomadic peoples of the Sahara controlled vast areas of the desert and the trade routes through it. Changes beginning in the twentieth century, however, have profoundly affected their way of life. For example, roads built across the Sahara have bypassed or destroyed nomads' traditional trading routes. Closed borders, government policies discouraging nomadism, and desertification in the Sahel have also limited the areas through which nomadic peoples can roam.

These changes have contributed to a steady decline in the number of people who follow the nomadic life. Many are obliged to move to cities to seek work. Yet others have adapted their traditional ways to the modern world, using trucks to lead their herds and to transport goods across Saharan highways.

NORTH AFRICA

Section 1: Physical Geography
Section 2: Egypt
Section 3: Libya and Tunisia
Section 4: Algeria and Morocco

OVERVIEW

In Chapter 36, students will learn about the countries of North Africa.

Section 1 describes the major landforms of the region, including the Sahara and the Nile River.

Section 2 focuses on Egypt, detailing the importance of the Nile to the country. It also discusses economic, urban, and political issues.

Section 3 discusses agriculture and industry in Libya and Tunisia, analyzing the importance of the discovery of oil in Libya.

Section 4 profiles Algeria and Morocco. It examines Algeria's colonial history and economic strengths and describes the variety and contrasts of Morocco's geography and people.

About the Author

Farouk El-Baz is a geologist. He led a scientific research team across the Sahara in 1978 to study the desert's geology, botany, and archaeology.

CREATIVE STRATEGIES FOR TEACHING WORLD GEOGRAPHY

You might wish to use the Lesson 14 activity about speaking Arabic (Unit 7) as you teach this chapter.

Primary Source

Description of Change:
excerpted and bracketed
Rationale: excerpted to focus on main idea; bracketed to clarify meaning

---

CHAPTER 36

North Africa

66 A dune is a curiously dynamic creature. It can form behind any impediment—a bush or rock—or simply as the child of wind currents. Once formed, a dune can grow. It can change shape and move with the wind. It can even breed new dunes. Some of these offspring may be carried on the back of the mother dune. Others are born and race down wind, outpacing their parents. Some dunes in the Western Desert [of the Sahara in Egypt] move as much as 100 meters a year. 99

Farouk El-Baz

The Suez Canal cuts through the arid isthmus separating the Red Sea and the Mediterranean Sea. Some geographers mark this human construction as the boundary between Africa and Asia. Completed in 1869, the canal was the great engineering work of its time. It is a 100-mile-long (161-km), sea-level ditch that saves ships thousands of miles on their route from the Mediterranean Sea to the Indian Ocean. The Suez Canal passes through the North African country of Egypt and is one of the world's busiest waterways.

Along with Egypt, the countries of Africa's northern rim include Libya, Tunisia, Algeria, and Morocco. The countries west of Libya often are called the *Maghreb*, which in Arabic means "west." Culturally, North Africa is predominantly Arabic and Islamic. The region is a key part of the Arab world.

GEOGRAPHY DICTIONARY

erg
reg
basin irrigation
perennial irrigation
silt
capital
free port

Alabaster image of King Tutankhamen of **Egypt** from the fourteenth century B.C.

The Suez Canal, **Egypt**

Using the Quotation

Ask students what images the writer elicits when talking about sand dunes. *(living creatures, a family, movement)* What sense of the desert does the writer convey through these images? *(Students may say that, like living things, dunes are always changing.)* Point out that the desert terrain of North Africa has influenced the region's culture and history.

Using the Geography Dictionary

Write the terms **erg** and **reg** on the board. Point out to students that these are Arabic words for which English has no direct equivalents. Ask them why they think this might be. *(English-speaking countries do not commonly have these landforms.)*

SECTION 1
OBJECTIVES

The objectives reflect the **Focus** questions on this page and are assessed in the review on **page 432.**

1. Describe the physical features of the Sahara, and explain how they influence human settlement.
2. Explain the historical and agricultural significance of the Nile River.

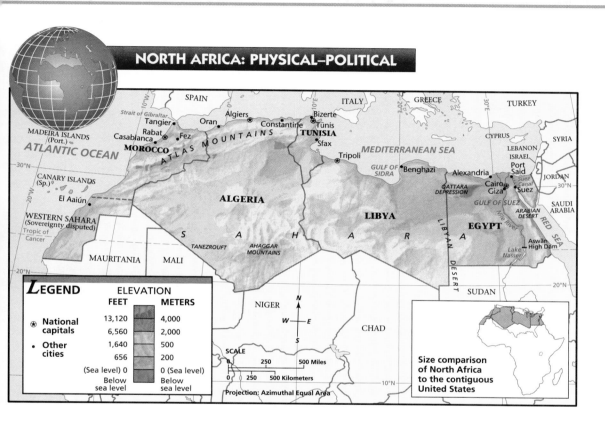

NORTH AFRICA: PHYSICAL–POLITICAL

Using the Physical–Political Map: North Africa

Ask students to identify the terrain that most typifies North Africa. *(desert)* Tell them that the Sahara is classified as a desert region by climate. Ask them to suggest other ways it could be classified. *(Answers will vary but could include ethnic groupings, religions, political boundaries, or economic activities.)*

PHYSICAL GEOGRAPHY

FOCUS

- *How do the Sahara's physical characteristics influence settlement?*
- *Why is the Nile River important, historically and agriculturally?*

The countries of North Africa share similar environments. Along their Mediterranean coasts, summers are dry, and winter rains are welcomed. Oak trees and grasses cover much of the plains and plateaus here, and irrigation is crucial for agriculture. From Tunisia westward, the Atlas Mountains parallel the coast. Within the mountains are pleasant valleys as well as snow-covered peaks. North Africa holds two of the world's geographic wonders—the immense Sahara and the Nile River.

LOCATION **The Sahara** The word *sahara* comes from the Arabic word for "desert." The

Sahara, combined with the deserts of the Arabian Peninsula, Iran, and Central Asia, forms the world's largest arid region. Deserts may be created by two separate factors. The first is a weather pattern dominated by regional high pressure. In high-pressure cells, the air spirals downward toward the earth's surface. Even if the descending air is damp, rain does not form. The second factor is the absence of moist sea air. Most of the world's rain starts as evaporation from an ocean. A desert results if an ocean is far away, or if the ocean air is blocked by mountains. In the Sahara, both of these factors are at work. In parts of central Libya and Algeria, summer temperatures are regularly above 120°F (49°C).

Because it has little water, the Sahara has few plants. Therefore, wind and a rare rain can be very effective in eroding the land. Bare rock surfaces are common in the Sahara, particularly in highland regions such as the Ahaggar (uh-HAHG-uhr) Mountains in southern Algeria. The basins below the rocky ridges fill with the eroded sediment. Some basins are covered with high, shifting sand dunes. Such a sea of sand is called an **erg**. In other

TEACHING SECTION 1

TEACHING SECTION 1

Motivator

Ask students if the United States has a desert region. *(the Southwest)* Ask volunteers who have visited or read about desert areas to describe them. Tell students that in this section they will learn about the Sahara and other landforms in North Africa.

Building a Vocabulary

Point out that all rivers and streams carry **silt.** Have students write a new caption for the photograph of the Nile Valley on this page and to include the word *silt.*

Guided Practice

Help students analyze characteristics of the Sahara and the Nile River. Write these general statements on the board: "The Sahara's climate affects its population density and economic activities" and "Population is dense and economic activities are varied along the Nile River." Have students use information from the maps on pages 418–420 to suggest facts that extend each statement. Write

Continues on page 433

Geography and Art

Prehistoric engravings and paintings on rock surfaces are found throughout the Sahara. Over 30,000 examples of rock art have been discovered that were created between 4,000 and 7,000 years ago. They consist of a variety of styles and subject matter, evidence of successive ages of innovation in the Sahara. Today, some Saharan nomadic people continue to paint and engrave on desert rocks.

Section 1 Review
ANSWERS

1. The desert is very dry, making human settlement difficult.

2. The fertile land near the Nile River was a good place for ancient Egyptians to settle and for their culture to thrive. People in modern-day Egypt still depend on the Nile River for irrigation.

3. Possible answer: the Nile is in the desert, so it loses more water to evaporation than does the Amazon in the rain forest.

432

This town and vegetation mark the location of an oasis and the vital water it provides in the Sahara of central **Algeria**.

areas, the wind blows the sand and dust away, leaving a gravel-covered plain. This landform is called a **reg**. The Tanezrouft Reg in southern Algeria is a featureless gravel plain several hundred miles across. With no landmarks as a guide, the occasional traveler through the plain can easily become lost.

The Sahara also has broad depressions. In western Egypt, the Qattara Depression, 436 feet (133 m) below sea level, forms a vast wilderness of quicksand and salt marsh. Other basins have large, dry lake beds where rainwaters briefly accumulate. In Arabic, a dry lake bed is called a *sebka*. *Sebka* surfaces are severely eroded by winds. Dust storms seem like fog and can even blot out the sun. Sometimes the dust travels very far. Saharan dust occasionally falls with rain in the Caribbean.

The Sahara is almost uninhabited. Human settlement is limited to sites where there is water. Ancient oasis towns are found at springs and where wells can reach water below dry riverbeds. The few roads crossing the Sahara follow old caravan routes from oasis to oasis.

The Nile River In striking contrast to the Sahara, the valley and delta of the Nile contain thousands of acres of rich, fertile land. On the banks of this famous river, ancient Egyptians developed their science and art and built pyramids, temples, and towns.

At 4,187 miles (6,737 km) long, the Nile is the longest river in the world. It is formed by the union of two rivers: the White Nile and the Blue

---

Much of the fertile **Nile River valley** south of Cairo is bordered by steep, dry cliffs. The valley's width narrows to a few hundred yards in some places, but is as much as 14 miles (23 km) wide in other areas.

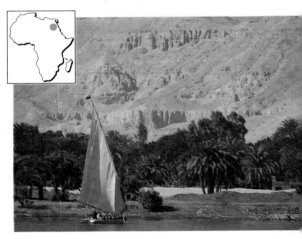

Nile. The source of the Nile was unknown to the ancient Egyptians. They explained it as a gift from the gods. For many centuries, explorers searched for the Nile's source. Not until the latter half of the eighteenth century, however, was the source of the Blue Nile discovered in the highlands of Ethiopia. Nearly a century later, the source of the White Nile was found in the tributaries of Lake Victoria in East Africa. The two rivers meet at Khartoum, Sudan, to form the Nile River.

The Nile River valley is like a long oasis in the middle of a desert. The fertile lands along the river are a product of the Nile itself. Before dams were built upstream, the Nile regularly overflowed its banks in late summer, restoring richness to the soil century after century. In addition to water, these floods brought rich, fertile **silt** to the farm fields. Silt is sediment that is coarser than clay yet finer than sand. It feels similar to flour.

 SECTION REVIEW

1. Why is settlement sparse in the Sahara?

2. How has the Nile River been important to people from ancient times to the present?

3. Critical Thinking   Suggest a reason why the Nile carries less water than the Amazon, even though the Nile is longer.

their responses under the appropriate sentence. (Examples might include, "Most areas in the Sahara have fewer than 3 people per square mile" or "The Nile is a source for hydroelectric power.")

Independent Practice

Have students draw pictures that depict a scene along the Nile River in ancient times. (Their art might focus on flooding and irrigation.)

Closure

Contrast for students how the Sahara and the Nile River affect agriculture in North Africa.

ASSESSMENT

Assign the Section 1 Review questions on page 432.

Reteaching

Assign Section 1 of Reteaching Worksheet 36.

The objectives reflect the Focus questions on this page and are assessed in the review on page 435.

1. Explain how different forms of irrigation and agriculture have presented challenges for Egypt's growing population.

2. Analyze the two major political forces at work in modern Egypt.

---

2 EGYPT

FOCUS

● *What forms of irrigation and agriculture have been used in Egypt, and what challenges have they presented for the growing population?*

● *What two major political forces are influencing Egypt today?*

Almost 2,500 years ago, a Greek geographer wrote that "Egypt is an acquired country, the gift of the river." This statement is still true today. The Nile River remains Egypt's main resource. In Egypt, the Nile flows north through 1,000 miles (1,609 km) of desert and then empties into the Mediterranean Sea. (See the map on page 431.)

Little rain falls in Egypt. During some years, there is no rainfall at all in the southern part of the country. Cairo receives about 1.5 inches (3.8 cm) a year. Coastal areas around the city of Alexandria receive from 4 to 8 inches (10 to 20 cm) a year. For this reason, irrigation from the Nile River is of utmost importance.

INTERACTION **Agriculture** Nearly 97 percent of Egypt's cultivated land is irrigated by the Nile. For many centuries, the Egyptians used **basin irrigation**. Farmers divided their land into shallow basins by building walls of earth up to about six feet (2 m) high. When the Nile overflowed each year, water and silt were deposited on the fields. The earth walls helped trap the water on the land. Seeds were then sown in the rich mud. Under the strong sun, the moist, fertile soil yielded good crops. By using basin irrigation, fields up to about nine miles (14 km) from the Nile were irrigated. Only crops that would ripen during the short time that the land was moist, however, could be grown. Beans, wheat, barley, and onions were the traditional food crops grown with basin irrigation, but only one crop could be grown each year.

In most cases, basin irrigation has given way to **perennial** (continual) **irrigation**. This type of irrigation was first practiced by pumping water directly from the Nile. Today, huge dams stop the Nile's annual flood by storing the floodwater. Then the water is released steadily throughout the year and carried to the fields by networks of canals.

FACTS IN BRIEF	North Africa		
COUNTRY POPULATION (1994)	**LIFE EXPECTANCY** (1994)	**LITERACY RATE** (1990)	**PER CAPITA GDP** (1993)
ALGERIA 27,895,068	67, male 69, female	57%	$3,300 (1992)
EGYPT 60,765,028	59, male 63, female	48%	$2,400
LIBYA 5,057,392	62, male 66, female	64%	$6,600
MOROCCO 28,558,635	66, male 70, female	50%	$2,500
TUNISIA 8,726,562	71, male 75, female	65%	$4,000
UNITED STATES 260,713,585	73, male 79, female	97% (1991)	$24,700

Source: *The World Factbook 1994*, Central Intelligence Agency

Which North African country has the highest GDP per person? Why do you think this is so?

Thus, two or three crops can be grown instead of only one. As a result of perennial irrigation, rice and cotton have become important crops in Egypt.

The Nile's largest dam is located about eight miles (13 km) south of Aswān (a-SWAHN). The Aswān High Dam is 364 feet (111 m) high and about two miles (3.2 km) long. It can irrigate more than 2 million acres (809,400 ha) of farmland. The dam provides about one-third of Egypt's electricity.

Although the Aswān High Dam provides electric power and water throughout the year, the dam has created problems. The Nile no longer deposits the fertile silt needed to enrich the land, because the sediment is trapped behind the dam. Expensive artificial fertilizers are now used. The fishing industry in the Mediterranean also has suffered because the trapped sediment no longer provides nutrients for the fish. The fishing industry has been helped, however, by fishing in the new lake created by the dam. Evaporation from this lake, though, reduces the amount of water available to farmers.

A serious problem associated with perennial irrigation is an increase in a disease produced by small organisms that invade the liver and blood of people. These organisms are carried by snails that live in the irrigation canals. Because people are in close contact with irrigation water, it is difficult to keep the disease from spreading.

INTERACTION **Population Growth** Population density in Egypt's rural areas is among the

Chart Caption Answer

Libya; substantial revenue from large oil deposits and a population smaller than that of other North African countries

THE WORLD TODAY VIDEODISC PROGRAM

Egypt: Crossroads of Cultures

2A   601   3:49

Play   Pause

Motivator

Ask students whether they have ever seen museum exhibits on ancient Egypt. Discuss what they have seen or know about ancient Egypt. (*pyramids, the Sphinx, mummies, the Nile*) Tell students that in Section 2 they will learn about present-day Egypt and the challenges it faces.

Building a Vocabulary

Contrast **basin irrigation** and **perennial irrigation**. Point out that *basin* can mean "bowl." Scooped-out farmlands held Nile floodwaters much like water in a bowl. *Perennial* means something that continues without stopping, which relates to the systematic release of stored floodwaters to agricultural fields. Then create a word web on the board to list the advantages and drawbacks of each method.

Guided Practice

Ask students to work in small groups to hold a panel discussion, with volunteers to assume the roles of an Arab regionalist, an Egyptian nationalist, and an Islamic fundamentalist. Tell students that the panel discussion should concern the conflicts among the three regarding the direction that the Egyptian government should take.

GLOBAL PERSPECTIVES

Bring resources to class that show Egypt's ancient architectural wonders (pyramids, the Great Sphinx). Point out that the Great Sphinx at Giza, Egypt, carved in about 2500 B.C., has been severely eroded by thousands of years of exposure to wind-blown desert sand. Efforts to restore the Sphinx have involved scientists from many nations. An architectural team used American- and German-made maps and British-owned fragments of the giant sculpture to create a three-dimensional computer model of the original Sphinx. Ask students why people are interested in preserving cultural treasures in foreign countries. (*Students might say that everyone benefits from the preservation of such cultural treasures because they chronicle civilization and human endeavor.*)

highest in the world. Egypt's cities are growing rapidly, too. As a result, the amount of productive land per person is steadily declining. Nearly all of Egypt's agricultural land is found along the Nile, particularly in the Nile Delta. New farmland is increasing slowly as irrigation canals are built farther from the river and as salt marshes in the delta are reclaimed.

Still, it is difficult to put new land and technology into use fast enough to keep pace with Egypt's rapid population growth. With such intensive land use, there is danger that the soil will become unproductive. More and more fertilizer is needed. In addition, overwatering has brought to the surface salts that are harmful to crops. Once self-sufficient in food, Egypt now must import more than half of its food.

Cotton is Egypt's main cash crop and agricultural export. Egyptian cotton, prized around the world, once was grown mainly on large estates. Today, it is produced primarily on small farms, some no bigger than gardens. Many farmers now proudly work the land they own. By cooperating in seed purchases, marketing, and fertilizer applications, small farmers can use modern technology with their hand labor.

PLACE **Urban Geography** Cairo, Egypt's capital, has more people than any other city in Africa. In fact, it is one of the world's fastest growing cities. About 10 million people crowd into the city and its surrounding region, which sprawl along the banks of the Nile River. Cairo was founded more than 1,000 years ago on high ground east of the river at the head of the Nile Delta. From this location, Cairo's political leaders commanded both the delta and the river valley. In the late 1800s, railways connected the city to ports along the Mediterranean and Gulf of Suez. The city spread south along the river.

Modern Cairo is a mixture of building styles, reflecting the cultural diversity of the city. High-rise offices, department stores, hotels, and government buildings tower over historic Islamic mosques and universities. Much of the city, however, consists of tiny, mud-brick houses situated on maze-like streets. Overcrowding, traffic jams, and inefficient services plague the city.

Despite limited housing available in Cairo and other crowded Egyptian cities, migration from rural to urban areas continues. Such population pressure has resulted in the growth of satellite

cities, which are connected to Cairo by commuter rail lines. The once-sleepy village of Giza, across the Nile near the pyramids, is now an industrial suburb with more than 1 million people.

Egypt's second largest city is Alexandria, located just west of the delta along the Mediterranean. Founded by Alexander the Great in 332 B.C., Alexandria has about 3 million people and handles much of the country's trade.

Industry and Trade Although Egypt is one of Africa's most industrialized countries, it must expand and modernize its industries to meet the economic needs of its growing population. Textile and yarn factories have been developed, and Egypt now exports these products. Other industrial growth has been only partly successful.

Egypt has limited natural resources. It has no forests and only small coal deposits. Oil, Egypt's main export, is found on both sides of the Gulf of Suez. As Egypt develops, more of this oil will be needed inside the country, and less will be available for export. Egypt has manganese and iron ore, which form the basis for a growing steel industry. Gypsum and other minerals are used for making cement, and phosphate and nitrates are used in fertilizers.

A lack of skilled workers also has contributed to Egypt's slow industrial development. Skilled,

---

The Nile Delta in **Egypt** stands out sharply from the surrounding desert in this satellite photograph. Cultivated areas, shown in red, follow a narrow path south along the Nile River.

Closure

Review with students past and present challenges facing Egypt.

ASSESSMENT

Assign the Section 2 Review questions on this page.

Reteaching

Assign Section 2 of Reteaching Worksheet 36.

SECTION 3
OBJECTIVES

The objectives reflect the Focus questions on this page and are assessed in the review on page 437.

1. Explain how the discovery of oil has affected Libya.

2. Describe factors that have made Tunisia a relatively successful country in North Africa.

TEACHING SECTION 3
Motivator

Write the word *oil* on the board. Ask students to describe the ways that oil is important to the United States and to students in particular. List student responses on the board. Tell them that in this section they will learn how the discovery of oil resulted in many national and international changes for Libya.

The building styles of **Cairo** (right) are a mix of old and new. West of the city, across the Nile, is the suburb **Giza**, where ancient Egyptians built many great pyramids, including the one shown below right.

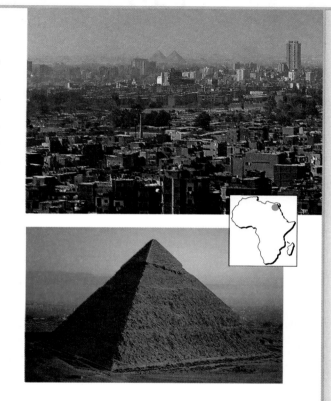

educated Egyptians often choose to work on development projects in the oil-rich Arab states. In addition, like other developing countries, Egypt lacks a good home market. Most people can afford to buy only the most necessary items.

Another problem is Egypt's balance of trade. The total value of Egypt's imports is greater than the value of its exports. As a result, little money is available for investment, and foreign aid from the developed countries and from the rich Arab nations makes up an important part of the government budget. Workers living out of the country often send money to their families in Egypt.

Issues Two political forces are pulling at modern Egypt. The first is the struggle between Arab regionalism and Egyptian nationalism. Egypt has significant influence in the Arab countries. Regional as well as international issues of concern to Arabs often are settled in Egypt. At odds with these international and regional concerns are the local issues emphasized by Egyptian nationalists. Most Egyptians prefer that their government focus on economic issues that will assist Egyptians. For example, they would like to see improvements made in the many inefficient government-owned industries. Egyptian farmers want more aid from their government.

The second political force tugging at Egypt is Islamic fundamentalism. Egyptians are divided on whether their government should change to closely follow Islamic rules. If the movement toward Islamic fundamentalism strengthens, traditional religious laws would make social life more conservative and, for many, less tolerable.

SECTION REVIEW

1. How have different forms of irrigation and agriculture presented problems for Egypt's rapidly growing population?
2. Over what two political issues are Egyptians divided, and why?
3. Critical Thinking **Imagine** that you are an Egyptian farmer. Why might you migrate to Cairo?

LIBYA AND TUNISIA

FOCUS

● *How did the discovery of oil in Libya affect the country?*

● *Why has Tunisia been a relatively successful country in North Africa?*

Libya Situated along the most desolate shore of the Mediterranean, Libya was ignored for much of its history. (See the map on page 431.) Then, sudden oil wealth and a radical government dramatically changed the country and its image. Major oil deposits were discovered in the southern desert in 1959, and Libya is now a major oil exporter. Libyan crude oil is particularly valuable because it contains little sulphur and, therefore, causes less pollution. Libya also exports natural gas to the nearby markets of Europe.

Unlike Egypt, Libya has no great river and is sparsely populated. More than 75 percent of the nation's people live within 10 miles (16 km) of

Section 2 Review
ANSWERS

1. Irrigation: basin irrigation supports one crop each year; perennial irrigation supports more crops, but the needed dams and canals prevent natural soil enrichment, hurt fishing in the Mediterranean, and spread disease. Agriculture: intensive land use may make the soil unproductive, and overwatering has harmed crops.

2. Conflict between Arab regionalism and Egyptian nationalism: some want Egypt to focus on regional concerns, and others want to focus on the needs of the Egyptian people. Islamic fundamentalism: Egyptians disagree over how strictly religious the government should be.

3. Possible answer: if land use has made the soil unproductive, Cairo might offer greater economic opportunities.

435

Guided Practice

Discuss the factors that prevented Libya from developing its economy prior to 1959. Then have students write "newspaper headlines" that evaluate the impact of the discovery of oil in Libya. *(for example, "Oil Revenue Used for Public Health Programs")* Students might make a collage of their headlines on a poster titled "Oil and Libya."

Independent Practice

Have students create a travel brochure in which they describe and illustrate Tunisia's location, climate, and other factors that make it a desirable country to visit or to live in.

Closure

Compare Libya and Tunisia's locations, natural resources, and agriculture.

ASSESSMENT

Assign the Section 3 Review questions on page 437.

Reteaching

Assign Section 3 of Reteaching Worksheet 36.

Extension

Have interested students report on an aspect of Tunisia's history, such as ancient Carthage.

COOPERATIVE LEARNING

Organize the class into small groups. Have the groups create a chart comparing the literacy rates, infant survival rates, life expectancies, and per capita GDP of Libya and Tunisia. Ask each group to use the data to write a statement comparing the quality of life in the two countries. Specify roles such as researchers, writer, and chart maker.

Section 3 Review
ANSWERS

1. Oil profits gave Libya the means to establish social services and gain political power.
2. A strategic location, a high percentage of land suitable for agriculture, a good transportation system, a variety of minerals to export, a good school system
3. Possible answer: Tunisia is close to Europe, and thus to European democratic traditions, while Libya is located along the most desolate portion of the Mediterranean shore.

436

Humans and machines carve a path through the desert to create the Great Manmade River Project in **Libya**.

the Mediterranean Sea because the country is 93 percent desert. Agriculture is possible, but the summer heat and desert winds make farming difficult. Along the northeastern coast are the plateaus of Cyrenaica (sir-uh-NAY-uh-kuh), where herding sheep and goats has long been the traditional occupation. The major city in this region is Benghazi (ben-GAHZ-ee). To the west is the region of Tripolitania, which receives slightly more rain. Tripoli, Libya's capital and major city, is located in this region. Typical Mediterranean crops, including wheat, citrus fruits, and olives, are grown here.

Libya's oil wealth is large in relation to its small population. The government has used oil revenues to establish generous programs for public health, housing, and transportation. Educational opportunities and health conditions continue to improve. Oil profits also have been spent on irrigation projects, the most spectacular of which is called the "Great Manmade River Project." During oil exploration in the country's southwestern desert, a large aquifer in the Sahara was discovered. Huge pipes now transport this water northward to the coast. Because this precious water from beneath the Sahara is not replaced by rains, however, it will be depleted some day.

Oil wealth has given Libya political power. Libya's leader, Muammar al-Qaddafi, is an avid Arab nationalist who has supported radical causes around the world. He has tried to overthrow governments in neighboring countries and has engaged in border disputes with Egypt. People opposed to his policies have been killed. For these reasons, many Western governments view Libya as a potential threat and pay close attention to events there.

LOCATION **Tunisia** Tunisia has a strategic location between the eastern and western basins of the Mediterranean Sea. (See the map on page 431.) For thousands of years, this location made Tunisia a center of Mediterranean culture. In ancient times, the city of Carthage, now in ruins, controlled trade throughout the Mediterranean. Today, European tourists flock to these sunny shores.

Most Tunisians live on coastal lands, which have a mild Mediterranean climate. In these lowlands, farmers grow wheat and olives, and olive oil is an important export. More than 25 percent of Tunisia's land is suitable for agriculture. This is a large amount compared with that of other North African countries whose territories extend much farther into the Sahara. Plateaus in western Tunisia mark the beginning of the Atlas Mountains.

While agriculture in Tunisia is important, most of its people live in cities. Tunisia's transportation system is good, and the government has encouraged industrial expansion. Tunisia has a variety of minerals to export, including some oil. Many industries rely on hand labor, which provides more jobs.

Tūnis, the capital, and the other major cities of Sfax and Bizerte, are very European. In fact, as a former French colony, Tunisia maintains ties to France. Many Tunisians work in France and send part of their income home. Influences of French culture are common, and the ideals of European democracy are widely appreciated. For example, Tunisian women have the right to vote, and the national school system is envied by neighboring countries. Literacy is relatively high. These ideals earn Tunisia large foreign aid contributions.

Tunisians visit a marketplace in **Tūnis**. About 1 million people live in Tūnis.

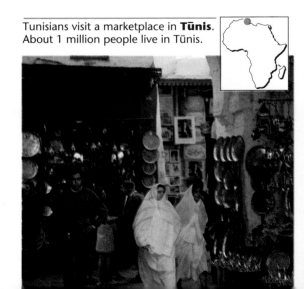

SECTION 4
OBJECTIVES

The objectives reflect the Focus questions on this page and are assessed in the review on page 438.

1. Describe physical, cultural, and economic characteristics of Algeria.
2. Analyze aspects of Morocco that make it a nation of contrasts.

TEACHING SECTION 4
Motivator

Ask students to describe cities they live in or have visited that have both old and new elements. Tell students that in this section they will learn about Morocco, a country that reflects the old and the new.

Building a Vocabulary

Write the terms **capital** and **free port** on the board. Ask students how they think a free port might encourage the flow of capital into a country. *(Capital is attracted to markets where imported products are not taxed.)* Then refer students to the definition of *capitalism* in Chapter 8 (page 81). Have them restate in their own words the connection between *capitalism* and *capital.*

As in other areas of North Africa, the population in Tunisia is increasing faster than the number of jobs. Migration from rural areas is bringing people into the cities. Islamic fundamentalism is gaining in popularity, although it is repressed by the government.

SECTION REVIEW

1. Why was the discovery of oil in Libya important?
2. What advantages does Tunisia have over some of its neighbors?
3. **Critical Thinking** **Explain** why Libya and Tunisia are different politically even though they are both well off economically.

4 ALGERIA AND MOROCCO

FOCUS

- *What are the physical, cultural, and economic characteristics of Algeria? What problems does Algeria face?*

- *Why can Morocco be considered a nation of contrasts?*

Algeria Like Egypt and Libya, Algeria is a country both of the Mediterranean rim and of the Sahara. (See the map on page 431.) More than 80 percent of all Algerians live in the north along the Mediterranean coast, where lands are fertile. To the south is the desolate Ahaggar, the highlands of the Sahara.

Modern Algerians are a blend of Arab, Berber, and European cultures. The Berbers were Algeria's original residents, and their language still is spoken along the Atlas Mountains. A few Berbers can be seen leading their flocks of sheep and goats on annual migrations in and out of the Atlas valleys.

During 132 years of French colonization, large investments and more than a million French settlers made Algeria one of the most prosperous regions in Africa. After a bitter and bloody war, the Algerians ousted the French, and Algeria gained its independence in 1962. France left behind the infrastructure that is the basis of modern Algeria. The United States and the European Community are now Algeria's most important trading partners.

Most of Algeria's income comes from oil and natural gas fields scattered across the Sahara. Pipelines carry these products to the coast where they are shipped to markets in Europe and eastern North America. A new pipeline under the Mediterranean Sea will increase Italy's dependence on Algeria's oil and gas resources.

Algerians can proudly claim that much of the profits from petroleum are invested in education and industrial developments. Today, Algeria has a higher proportion of factory workers than any other Arab country. A wide range of consumer goods are produced for the local market. Industries are concentrated in three main cities: Algiers, the capital and major port; Oran, another major port; and Constantine, located farther inland.

Algeria shares many problems with its neighboring countries, however. Many of these problems stem from its rapid population growth. More than half of the country's food is imported, and the schools are producing twice as many graduates as there are jobs. Some people believe that the government should encourage more free enterprise, which could lead to rapid job growth. Also, Islamic fundamentalism is gaining popularity. Since 1992, violence in the conflict between fundamentalists and their opponents, especially the government, has claimed thousands of lives.

LOCATION **Morocco** Morocco is an Islamic kingdom on both the Mediterranean and the Atlantic coasts. (See the map on page 431.) In the

Women at this plant in **Algeria** work on parts for televisions and computers. Although the Algerian economy is dependent on oil and natural-gas production, manufacturing continues to grow in importance.

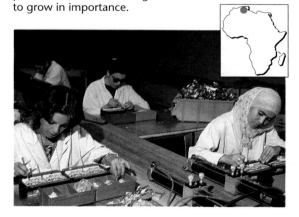

MULTICULTURAL PERSPECTIVES

Islamic fundamentalism in Algeria has been growing. In late 1991, the military stepped in and annulled democratic elections that would likely have given fundamentalists control of parliament. Many military and civilian leaders feared that the fundamentalist victory would lead to a government based on Islamic law.

Ask students why they think that the Islamic fundamentalist movement has a large following in North Africa. *(Answers will vary, but students might indicate that the majority of people in North Africa are Muslims and that Islam is a unifying element there.)*

Classifying

Algeria's Characteristics   Point out why Algeria's proximity to Europe and the Mediterranean and its location among the Arab nations of North Africa have influenced its history and culture. Have students classify Algeria's characteristics by studying the maps on pages 417–420. Tell them to complete an outline about Algeria with four main headings: Location, Climate, Economy, and Population.

Illustrating Ideas

Analyzing Morocco's Geography Have students in small groups create posters titled "Morocco—A Study in Contrasts." The posters might include illustrations and captions that show the contrasts in landforms, natural resources, architecture, and population. Display completed posters and discuss them.

Closure

Have students name physical, cultural, and economic facts about Algeria and Morocco.

ASSESSMENT

Assign the Section 4 Review questions on this page.

Reteaching

Assign Section 4 of Reteaching Worksheet 36.

Irrigated farms, such as these outside the city of **Safi** on the Atlantic coast of **Morocco**, produce citrus fruits, grapes, and vegetables.

THEME
Location

The Strait of Gibraltar, which is bordered on the south by Morocco, measures only eight miles (13 km) across at its narrowest point. Its location between the Atlantic Ocean and the Mediterranean Sea has made it a vital shipping route for Southern Europe, North Africa, and Southwest Asia.

is the *Kasba*, the palace of the local sultan or lord. Nearby, the modern "new town" features wide streets and large European-style houses. The new and old towns were originally separated by open land or by cemeteries. Gradually, the two towns grew together.

north is the Strait of Gibraltar, which separates Europe and Africa by only eight miles (13 km). Inland, a series of plateaus rises toward the Atlas Mountains. The soils and rocks of the mountains act as a reservoir, storing water for the streams that flow south. In sharp contrast to Morocco's mountain region, which has the wettest climate of the Arab nations, the Saharan coast in southern Morocco is extremely dry.

While Morocco has only limited deposits of oil and metal ores, the country contains three-fourths of the world's total reserves of phosphate, the raw material in many fertilizers. This valuable mineral is recovered from limestone found along the lower slopes of the Atlas Mountains.

Despite its rich natural resources, the country has a shortage of skilled labor and **capital** for investment. Capital is any source of wealth used to produce more wealth. For example, the money people invest in a savings account is capital because it produces more money through the interest it earns. The machinery used in an industry also is considered capital because it is used to produce goods that will, in turn, create new wealth. A shortage of capital is common in developing countries.

Morocco's location has encouraged communication with Europe and the Mediterranean region throughout the country's history. Rabat, the capital, and Casablanca are busy centers for Atlantic trade. The port city of Tangier is located on the Strait of Gibraltar. This beautiful Muslim city is a brief ferry ride for tourists traveling from Spain. Tangier is a **free port**, which means that almost no taxes are placed on goods unloaded there.

Sharp contrasts of Moroccan life are evident in the cities. In the ancient sections, sometimes called "old town," narrow streets are lined with rectangular, mud-brick houses that seem stacked against each other. In the center of town is a mosque and a covered market. Also in the old town

The population of Morocco includes Arabs and Berbers. Jews, people of European descent, and some black Africans also live here. Despite this diversity, the Arabic language and Islamic faith dominate the culture. The Moroccan way of life can be described as traditional versus modern. Traditional Morocco is evident in rural areas where about half of the country's people are subsistence farmers. They till small plots of barley, wheat, and vegetables on the plateaus below the Atlas Mountains. The best example of modern Morocco is found in Casablanca, the country's largest city. Its population is about 3 million. The wide city streets are lined with office towers, shopping districts, and apartments. Slums on the city's edge are filled with recent rural migrants.

Morocco's king has personal influence throughout the Islamic world because he is a direct descendant of the prophet Muhammad. Morocco hopes to become a powerful Arab state. This goal was evident in Morocco's annexation of the former Spanish colony of Western Sahara, which holds large phosphate reserves. The king of Morocco also has allied the country with the United States and western Europe. As a result of this alliance, Morocco has received large amounts of foreign aid. The government also encourages trade and cultural exchange with developed countries.

Section 4 Review
ANSWERS

1. Oil and natural gas; to invest in education and in industry
2. Morocco has a wet mountain region and a dry desert region; limited deposits of oil and metal but much phosphate; rich natural resources but limited skilled labor and capital for investment; and a way of life that is a mixture of modern and traditional.
3. Answers will vary. Students are likely to express approval or disapproval of government involvement in religion.

438

CHAPTER 36 CHECK
ANSWERS

Building a Vocabulary

1. Erg: a large area covered by sand; reg: a gravel plain from which the sand and dust have been blown away; the Sahara

2. Silt; from the floods of the Nile River; rich and fertile for agriculture

3. Money or goods used to produce more money or goods; needed to build an economy

4. Goods unloaded at a free port are not taxed.

Recalling and Reviewing

1. Without the Nile River, Egypt would have very little land suitable for agriculture.

2. Because Tunisia does not extend very far into the Sahara, a large proportion of its land can be farmed.

3. Both countries have profited from large supplies of oil in the Sahara. Both countries border the Mediterranean.

4. Possible answers include Arabs, Muslims, Berbers, Europeans, Jews, and black Africans.

Thinking Critically

1. Answers will vary. A trade agreement might provide jobs to both regions and a reliable supply of oil and natural gas to the European countries. Obstacles: there would be many countries involved, and North Africa and Europe have many cultural differences.

2. Possible answers include stemming population growth, improving education and infrastructure for industrialization, and resolving conflicts.

Learning About Your Local Geography

Tables should indicate careful and thorough research. Lists should show that students have thought about how their location relative to a big city affects their lives.

Using the Five Themes of Geography

Maps might show landform or climate regions or regions based on resources, historical background, or culture.

CHAPTER 36 CHECK

Cropland near **Aswān, Egypt**

Reviewing the Main Ideas

1. The Sahara, a desert, and the Nile River are two of North Africa's most important physical features. They have had a major impact on the settlement and economic patterns of people in the region.

2. Most of Egypt's agriculture depends on water from the Nile River. Egypt is currently undergoing rapid population growth and urbanization. Regional, national, and religious issues influence Egypt's politics.

3. The discovery of oil in Libya brought the country wealth and power. Tunisia's location and appreciation of democracy have helped the country. Like its neighbors, Tunisia faces the challenges of rapid population growth and religious extremism.

4. Algeria has substantial natural resources, on which some European countries are becoming increasingly dependent. Algeria is more industrialized than other countries in North Africa. Morocco is a nation of physical, economic, and social contrasts.

Building a Vocabulary

1. Define *erg* and *reg*. Where would one find them?

2. What is the term for sediment that is coarser than clay yet finer than sand? From where does this sediment come, and why is it significant?

3. What is capital, and why can a shortage of capital be a problem?

4. How is a free port different from other ports?

Recalling and Reviewing

1. How has the Nile River been important to Egypt?

2. Why is Tunisia's relative location with regard to the Sahara significant?

3. What do Libya and Algeria have in common?

4. Name three different cultures represented by inhabitants of the countries of North Africa.

Thinking Critically

1. Do you think that there should be a free trade agreement between the countries of Europe and the countries of North Africa? What advantages would such an agreement have, and what obstacles might prevent it from occurring?

2. Make a list of strategies the governments of North African countries might use to boost their economies.

Learning About Your Local Geography

Individual Project

Use an almanac to find your state's five most populous cities. Make a table listing these cities and their populations. Do you live in one of these cities? If so, how often do you leave it, and for what purposes? If not, how often do you visit one of them, and for what purposes? List some ways in which living in or out of a city affects your life. You may want to include your table in your individual portfolio.

Using the Five Themes of Geography

Using **place** characteristics as a guide, divide North Africa into nonpolitical **regions**. You might base your regions on landforms, climates, natural resources, historical background, cultures, or anything else you think relevant. Sketch a map showing your regional divisions.

CHAPTER 37 West Africa

Planning Guide

TEXTBOOK	TEACHING RESOURCES	OTHER AVAILABLE MATERIALS
Section		
1. Physical Geography Pages 441–442	Graphic Organizer 37 Reteaching Worksheet 37.1	*Study Guide for Spanish Speakers* *Lecture Notes,* Chapter 37
2. Human Geography Pages 443–445	Reteaching Worksheet 37.2 Skills Worksheet 37	*Study Guide for Spanish Speakers* *Lecture Notes,* Chapter 37
3. Sahel Countries Pages 445–446	Reteaching Worksheet 37.3	*Study Guide for Spanish Speakers* *Lecture Notes,* Chapter 37
4. Atlantic Coast Countries Pages 446–448	Reteaching Worksheet 37.4	*Study Guide for Spanish Speakers* *Lecture Notes,* Chapter 37
5. Guinea Coast Countries Pages 448–450	Reteaching Worksheet 37.5 Critical Thinking Worksheet 37 Challenge/Enrichment Worksheet 37	*The World Today* Videodisc Program *Study Guide for Spanish Speakers* *Lecture Notes,* Chapter 37
Chapter Review Page 451	Chapter 37 Test Chapter 37 Portfolio Assessment See also *Alternative Assessment Forms*	*Test Generator,* Chapter 37

CHAPTER RESOURCES

Lecture Notes, Chapter 37
Teaching Resources
    Chapter 37 Graphic Organizer
    Chapter 37 Skills Worksheet
    Chapter 37 Critical Thinking Worksheet
    Chapter 37 Challenge/Enrichment Worksheet

REVIEW AND ASSESSMENT

Chapter 37 Check, page 451
Chapter 37 Test, *Teaching Resources*
Chapter 37 Portfolio Assessment, *Teaching Resources*
Test Generator, Chapter 37
Alternative Assessment Forms

ADDITIONAL RESOURCES

Books for Teachers

Graham, Ronnie. *The Da Capo Guide to Contemporary African Music.* New York: Da Capo Press, 1988.

Harden, Blaine. *Africa: Dispatches from a Fragile Continent.* New York: Norton, 1990.

Stock, Robert. *Africa South of the Sahara: A Geographical Interpretation.* New York: Guilford Press, 1995.

Books for Students

Barrett, Jeannie M. *Ghana.* New York: Chelsea House, 1988.

Koslow, Philip. *Centuries of Greatness, 750-1900: The West African Kingdoms.* New York: Chelsea House, 1995.

Lutz, William. *Senegal.* New York: Chelsea House, 1987.

The World Today *Videodisc Program*

A variety of still images, short videos, and activities are available for you to use as you teach this chapter. *Lecture Notes* include bar codes for specific images and videos.

Multimedia Resources

Black Sugar: Slavery from an African Perspective. (16mm film/VHS, 26 minutes) Films for the Humanities & Sciences, P.O. Box 2053, Princeton, NJ 08543–2053.

Discovering the Music of Africa. (VHS, 22 minutes) Encyclopedia Britannica Educational Corporation, 310 South Michigan Avenue, Chicago, IL 60604–9839.

Welcome to Africa. (Multimedia CD-ROM) Walnut Creek CDROM, 4041 Pike Lane, Suite D, Concord, CA 94520.

MEETING INDIVIDUAL NEEDS

Limited English Proficiency (LEP)

Pair LEP students with students who are proficient in English. Have the pairs outline a section of Chapter 37, making sure that all sections are covered. When the pairs have finished, have them share their outlines and put the outlines together to form a chapter outline. Have students copy the chapter outline, and encourage them to use it to study the chapter.

Learners Having Difficulty

Help students review the main ideas of Chapter 37 by asking them to make generalizations about the following topics: What are the climate and vegetation patterns in West Africa? What are the main ethnic groups and religions of West Africa? What are the predominant economic activities in West Africa? What are serious issues facing West Africa today? Allow students to skim the chapter to help them develop generalizations.

Auditory Learners

Select appropriate music from West Africa to play for students. Discuss with students the features of the music, such as kinds of instruments and rhythms. Ask students what the music expresses about West Africa, as you write their comments on the board.

Visual Learners

Arrange a field trip to an art museum that has examples of West African artwork, or provide students with several photographs of West African artwork. Ask students to describe the artwork as you write the adjectives they use on the board. Then begin a class discussion about what West African artwork suggests about the culture of the region.

Tactile/Kinesthetic Learners

Ask students to create graphic organizers showing ethnic groups in each country of West Africa. Then have students draw an outline map of West Africa and use their graphic organizers to help them draw new boundaries to divide West Africa into countries based on ethnic groupings. Have students make new names for their newly formed countries. Have students share their graphic organizers and maps of West Africa with the class.

Gifted Learners

Have interested students do research to find French influences that remain in West Africa, such as language, architecture, art, music, government, and literature. Have students present evidence of the influences in the form of posters, collages, dramatic readings, or graphic organizers.

MULTICULTURAL FOCUS

You might introduce the work of writer Wole Soyinka (b. 1934) to students. Soyinka, from Nigeria, received the Nobel Prize in Literature in 1986. Although he is primarily known as a dramatist, his work also includes short stories, poetry, essays, novels, and works of an autobiographical nature.

Soyinka writes about many aspects of West African life, but he focuses on the dual influences of European culture and traditional African cultures on African peoples. Although Soyinka's parents were converts to Christianity and he grew up in a missionary compound, he was also profoundly influenced by the culture of the Yoruba peoples who lived in the province of Abeokuta, where he grew up. The Yoruba are the largest ethnic group in Nigeria. In *Ake: The Years of Childhood,* Soyinka describes how his boyhood experiences near a Yoruba village expanded his awareness of the ties he had to both European and African customs and beliefs.

Examples of Wole Soyinka's work might be found in your school library or community library. *A Dance in the Forest,* a play published in 1960, celebrates the independence of Nigeria from England.

CHAPTER 37

WEST AFRICA

Section 1: Physical Geography
Section 2: Human Geography
Section 3: Sahel Countries
Section 4: Atlantic Coast Countries
Section 5: Guinea Coast Countries

OVERVIEW

In Chapter 37, students will learn about the issues facing West Africa.

Section 1 describes the physical geography of the region. It identifies three environmental zones and describes the human activities that occur in each.

Section 2 discusses cultural and settlement patterns and the economic challenges facing the region.

Section 3 focuses on the Sahel countries and the effects of drought.

Section 4 describes the Atlantic coast countries and their economies.

Section 5 examines the Guinea coast countries and their economies.

About the Author

Sanche de Gramont (1932–) of France also writes as Ted Morgan, a name he assumed on obtaining American citizenship. Gramont lived in Morocco for several years and traveled throughout Africa. The quotation on this page about the Niger River is from *The Strong Brown God.*

Primary Source

Description of Change: excerpted
Rationale: excerpted to focus on main idea

CHAPTER 37

West Africa

❝The Niger is the cradle of West Africa. It is a moving path into the heart of the continent, a long, liquid magic wand, that makes fertile the soil it touches. Cattle drink its water and graze on its green banks. Its fish and game birds provide food. Its trees give wood for dugouts. The river means an end to hunger, thirst, and isolation. ❞

Sanche de Gramont

The rich history and remarkable culture of West Africa are two of the major characteristics that define this area as a region. The great kingdoms of Mali, Ghana, Songhai, and Benin (buh-NIN), among others, were sophisticated and highly developed. Today, West Africa is famous for the art that came out of these early civilizations.

The interior kingdoms of Mali, Ghana, and Songhai were well known for their earthenware. The Sapi, ancestors of ethnic groups in present-day Sierra Leone (see-ER-uh lee-ONE), were expert ivory and stone sculptors. The quality of Sapi artwork led to trade with the Portuguese in the early sixteenth century. The city of Ife (EE-fay), in present-day western Nigeria, was famous for its metal and clay sculptures. In the forest kingdom of Benin, craftsworkers and artists lived in special neighborhoods and created art only for the Oba, or king. Benin was especially well known for its beautiful brass sculptures.

The vitality of West Africa was interrupted first by the slave trade and then by European colonization. Today, that vitality is being renewed by talented artists and writers.

GEOGRAPHY DICTIONARY

zonal
pastoralism
millet
sorghum
staple
cassava
lingua franca

Early antelope headpiece worn by the Bambara people of present-day **Mali**

The Niger River, **Niger**

SECTION 1
OBJECTIVES

The objectives reflect the Focus questions on this page and are assessed in the review on page 442.
1. Identify the environmental zones of West Africa.
2. Describe the physical geography of each zone.

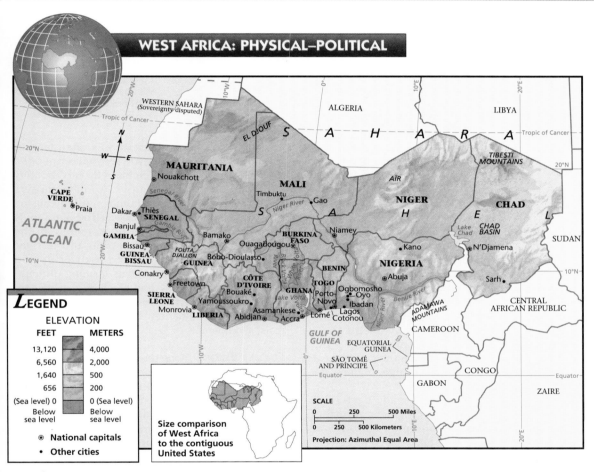

WEST AFRICA: PHYSICAL–POLITICAL

LEGEND

ELEVATION

FEET		METERS
13,120		4,000
6,560		2,000
1,640		500
656		200
(Sea level) 0		0 (Sea level)
Below sea level		Below sea level

⊛ National capitals

• Other cities

Size comparison of West Africa to the contiguous United States

SCALE

0   250   500 Miles

0   250   500 Kilometers

Projection: Azimuthal Equal Area

PHYSICAL GEOGRAPHY

FOCUS

- *What are the three environmental zones of West Africa?*

- *What are the physical features of West Africa's environmental zones?*

REGION **Climate** West Africa is lower in elevation than eastern and southern Africa. It is mostly wide, rolling plains, with a few highland areas. There are no major mountain ranges here to break up the region's climate pattern. Thus, geographers refer to the climate regions of West Africa as **zonal**. This means that, when mapped, the climate pattern runs in bands from east to west. (See the map on page 418.) West Africa con-tains three main zones, each with distinct climates and vegetation. In the north are the drylands of the southern Sahara, the areas dominated by desert and steppe climates. In the middle is the tropical-savanna zone, with winter drought and summer rains. In the south is the forest zone, with humid-tropical climates and dense forests of the West African coastal belt.

Dryland Zone The dryland zone lies along the southern edge of the Sahara. Here, rainfall varies greatly from year to year. Climatic variations are typical in desert margins. In some years, there is no rain. In others, rain fills the dry lake beds, and rivers flow across the desert. During the 1950s and 1960s, the Sahel had many years of good rains. Grazing was good and crops grew well. Health standards improved and populations increased dramatically.

A long period of drought began in the Sahel during the 1970s, however. Herds of cattle, sheep,

441

Use some of the illustrations in Unit 8 to introduce the physical geography of West Africa. Tell students to examine the photographs of Burkina Faso (page 413), Senegal (page 429), and Mali (on this page). Ask them to look for details in the photographs that indicate the type of climate in each of these locations.

Building a Vocabulary

Write the word **zonal** on the board. Have a volunteer read aloud the definition in the Glossary. Point out how it relates to *zone,* an area with unique characteristics that set it apart from areas that are near it. Then refer students to the climate map on page 418 to locate the three climate zones that run horizontally across West Africa.

Generalizing

Environmental Zones   List on the board the three environmental zones of West Africa. Then have students write three generalizations about each zone. You may wish to have students compare the West Africa map on page 441 to the World Biome map on page 57 and review the characteristics of forest, savanna, and desert biomes described in Chapter 6.

Geography and Economics

The economies of countries of the Sahel, such as Niger, fluctuate with the weather. During periods of sufficient rainfall, the land can support as many as 10 million cattle, sheep, and goats. However, the droughts of the 1970s and 1980s destroyed all but one million animals, bringing famine to the country.

Section 1 Review
ANSWERS

1. Zonal; the climate pattern runs in bands, or zones, from east to west.

2. The dryland zone is subject to drought and desertification. The wet climate of the forest zone is better suited to human settlement.

3. People might move from the upland plains and from the dryland zone to the river valleys of the savanna zone, because the valleys have good rains and fertile soils.

People make their way through a dust storm in **Mali,** in West Africa's desert region. Some of the dust will be carried hundreds of miles out over the Atlantic.

and goats quickly overgrazed the savanna and desert grasslands, and many animals died. The Sahel's inhabitants cut trees for firewood. As the vegetation was stripped, the Sahara spread southward. During this period of desertification, crops failed for several years. People often ate their seed grain instead of saving it to grow crops the next year. With little or no food, many people died. Many more became refugees, migrating to the large cities in search of food from the government and relief agencies. Today, refugee camps, or shantytowns, surround the cities.

Some years in the past decade have been rainy. It is doubtful, however, that the droughts have ended for good. Geographers know that droughts in the Sahel have taken place repeatedly in past centuries. Fewer people and fewer animals existed then, however. Herds could be moved to the better lands in the south. This movement is not possible now. The lands to the south are blocked by settled farmers and the political boundaries of new countries. Still, the people of the Sahel who depend on farming and grazing for their livelihood have adapted to this unpredictable climate. They have discovered new methods of conserving water and growing crops. They have also improved trading relationships with their southern neighbors.

Savanna Zone   In general, the rains in Africa increase toward the equator. The thorny, bush savannas of the southern Sahel gradually give way to tall trees and lush grasses. When the summer rains are good, the farmers of the savanna zone prosper.

Despite good rains and fertile soils, however, this region has few people. In the past, raids for the slave trade devastated the population. Today, the region is home to the dreaded tsetse fly and black fly. The tsetse fly carries sleeping sickness, a dis-

ease that is often fatal to cattle as well as people. The black fly carries a tiny parasitic worm from rivers to humans, often causing blindness. The black fly is common along rivers; therefore, people in the savanna zone often avoid settling in the river valleys. Instead, they live in the drier upland plains, where poor soil makes farming difficult. Although the tsetse fly and black fly can be controlled using powerful insecticides, the chemicals are expensive and harmful to the environment, animals, and humans.

Forest Zone   Along the coast of West Africa are forested plains. Here, it rains almost daily for up to 10 months a year. This is West Africa's most densely populated region. As populations have increased during the last decades, much of the forest has been cut. The effects of commercial lumbering, especially in Côte d'Ivoire (kote deev-WAHR) (Ivory Coast), Nigeria, and Ghana, have been particularly damaging.

West Africa's coastal plains have few natural harbors. The coastline in many places is muddy and lined with mangrove trees. These trees form dense thickets that make transportation difficult. Elsewhere along the coast, sandy beaches with coconut palms and fishing villages are typical. The sunny beaches attract vacationers to newly built coastal resorts.

SECTION REVIEW

1. What do geographers call the pattern of West Africa's climate and vegetation regions? Why?

2. Why do more people live in the forest zone than in the dryland zone?

3. *Critical Thinking*   **Predict** how West African settlement patterns might change if the diseases carried by the tsetse fly and black fly could be easily prevented.

Closure
Have students take turns at a wall map naming elements of the physical geography of West Africa.

ASSESSMENT

Assign the Section 1 Review questions on page 442.

Reteaching
Assign Section 1 of Reteaching Worksheet 37.

SECTION 2
OBJECTIVES

The objectives reflect the Focus questions on this page and are assessed in the review on page 445.
1. Describe cultural and settlement patterns in West Africa.
2. Analyze the economic challenges facing West African countries.

TEACHING SECTION 2
Motivator

Tell students that most West African countries were once European colonies and that the countries that exist now contain many different ethnic groups. Have students suggest how these factors might affect life and politics in West African countries.

2 HUMAN GEOGRAPHY

FOCUS

- *How do the ways of life of people in West Africa relate to the area in which they live?*
- *What economic challenges face West Africans?*

REGION **Cultural Patterns** West Africa is one of the most densely settled parts of Africa. People are concentrated in the coastal region and in the dry savannas of the southern Sahel. (See the map on page 419.) There are more than 500 African ethnic groups living in the region. Most countries contain at least four different ethnic groups. Some countries have as many as 50. Nigeria is home to more than 200 ethnic groups. Most of these ethnic groups were forced to live together in the new countries formed by European political boundaries. Today, many of the ethnic groups get along. Sometimes, however, there are religious, cultural, and land use conflicts.

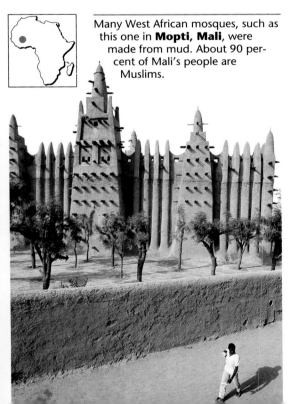

Many West African mosques, such as this one in **Mopti, Mali**, were made from mud. About 90 percent of Mali's people are Muslims.

West Africa is a region of religious diversity as well as ethnic diversity. Generally, the peoples of the dry northern lands are Muslims. Islam came with trade across the Sahara. In the south, people are primarily Christians. A smaller number of animists are found throughout West Africa. In many cases, animism has blended with Christianity and Islam.

To a large extent, cultural and religious diversity in West Africa is related to location and place. For example, the northern drylands are wide open, and travel is easy. Resources necessary for daily life are often far apart. Thus, in the past, a few ethnic groups could control large territories. In contrast, in the forested south travel was more difficult and less necessary. Many ethnic groups controlled much smaller areas. The coastal peoples, meanwhile, had more contact with European traders and explorers, and European missionaries spread Christianity inland. Africans often adapted European Christianity into Christian organizations of their own.

Economic and Settlement Patterns

Economic patterns that developed in West Africa during the colonial period continue. The region exports its raw materials and agricultural products mainly to developed countries. Most manufactured goods, in turn, are imported from these nations. All West African countries receive some form of foreign aid from developed countries and from international agencies.

West Africa is primarily an agricultural region, although only a small percentage of the land is arable. In the northern areas, pastoral nomads move their herds of cattle, sheep, and goats, following the rains. **Pastoralism** is an agricultural way of life that involves raising livestock and usually includes some farming. Often, pastoralists will grow crops to tide them over between the seasonal rains. **Millet** and **sorghum** are grain crops most often grown in the north because they are resistant to drought. The terrible drought in the 1970s forced many pastoralists to give up their way of life and move to cities in search of jobs. This migration of people, especially of young men, to the cities has lowered agricultural productivity. With fewer workers, traditional agriculture has declined.

In the savanna and forest zones of West Africa, small, compact farming villages are common. Most farmers practice traditional agriculture. The **staple**, or main, foods include root crops such as yams and **cassava**. Cassava has fleshy roots that provide a nutritious carbohydrate.

CƠOPERATIVE LEARNING

Organize the class into groups of four or five. Have students research manufacturing in three countries of West Africa, such as Nigeria, Niger, and Ghana. Have them focus on what is produced and who buys the products. Specify different tasks, such as making a chart of the countries researched, drawing pictures of the products, and making oral presentations of their findings in class.

Building a Vocabulary

Point out how **pastoralism** relates to other words students may know. *Pastor* comes from a Latin word meaning a shepherd or one who herds animals. A *pasture* is land where animals graze. Have students write sentences that pair these terms: **millet** and **sorghum, staple** and **cassava.**

Guided Practice

Contrast West African cultural and settlement patterns. Start two lists on the board, labeled North and South. Ask students to suggest characteristics of daily life or of the peoples in each region. *(North: Muslim, nomadic culture, dry climate; South: Christian, farming villages, forests, rain)*

THEME
Movement

Some African crops—such as cassava, peanuts, and rubber—are not native to Africa but originated in South America. They were introduced to Africa during the colonial period. Bananas, on the other hand, were brought from Africa to the Americas, where they have thrived.

Geography and Economics

To address economic issues, West African nations formed the Economic Community of West African States in 1975. The organization was formed to negotiate customs and trade agreements between member nations and to coordinate economic policies.

Geographical Sidelight

The Tuareg are nomads who constitute about 8 percent of Niger's population. They are one of the few West African peoples that are matrilineal—families are traced through the mother's side.

NEWS FROM

AROUND THE WORLD

Niger

Tuareg tending their livestock in **Niger**.

Tuareg Teenagers on the Move

Imagine growing up in a traditional nomadic society, moving constantly. The Tuareg (TWAH-reg), a Berber group who live mainly in Niger, Mali, and Algeria, spend their lives moving from place to place in search of water and grazing lands. At a young age, Tuareg boys and girls learn to ride dromedaries, Arabian one-humped camels. Children also learn to care for the goats, sheep, and cattle of the Tuareg caravans.

The history of the Tuareg is passed down by the women, who share stories of the Arab invasion of West Africa centuries ago and of French colonizers who later left their imprint on the region. Reli-gious education is generally left to the men, who recite from the Qur'an, the holy book of Islam.

Although most Tuareg are Muslims, their daily lives reflect a freedom and independence from traditional Muslim societies. Tuareg women may own and inherit property, and many women are wealthier than their husbands. Also, the faces of Tuareg women are not veiled, while those of Tuareg men are. Forbidden to show their faces in public, Tuareg males wear a long scarf wound around their heads, covering the mouth, nose, and chin.

Tuareg society is divided into social classes. A girl is encouraged to wed a Tuareg of her family's social position or of a higher rank. Men often are judged by their dromedaries—the animals' appearance and the fineness of the saddles they bear.

The Tuareg come into contact with other peoples when they travel to villages to trade cattle for rice, sugar, tea, and other supplies. In these settled communities, Tuareg teenagers might participate in a soccer game and mingle with French-speaking teens. Such visits, though, are only a pause in the lives of most Tuareg. To support their nomadic way of life, they must move with the changing seasons.

West Africa has large-scale commercial farms, but crops are grown for export, not for feeding the people. To increase income from exports, some African leaders have sought to increase the export of cash crops and minerals. This economic strategy, however, forces African countries to continue to buy more goods from the developed countries. It also forces the economies of the African countries to be dependent on the global market. Even though a West African country might have a good crop yield, the price that other countries are willing to pay for that crop might be extremely low that year.

One result of this practice is that progress has been slow for the people in rural areas who produce cash crops for export rather than grow food for the people. Many West African countries must now import basic food supplies. Thus, there is even less money to invest in the education of the growing population and in infrastructure.

Economic growth remains the most challenging issue facing the countries of West Africa. Some of the countries have rich resources. Given more education and training programs, their chance for economic improvement increases. For the countries without adequate resources, more foreign aid and regional cooperation are the keys to growth. Indeed, at economic summits, African leaders meet to discuss the economic challenges facing their countries. The leaders have concluded that export-

Independent Practice
Have students assume the role of a West African journalist. Have them write a newspaper article about the ways West African countries are trying to improve their economies. Ask volunteers to share their work.

Closure
Call on students to list facts that show how West African culture, economy, and settlement reflect the environmental zones of the region.

ASSESSMENT
Assign the Section 2 Review questions on this page.

Reteaching
Assign Section 2 of Reteaching Worksheet 37.

Extension
Have interested students find recipes or pictures that show how cassava, millet, and sorghum can be prepared.

SECTION 3
OBJECTIVES

The objectives reflect the Focus questions on this page and are assessed in the review on page 446.
1. Identify the effects of drought on the Sahel countries.
2. Explain why the Niger is important to the Sahel countries.

led development is hurting Africa's chances to provide a better life for its people. They hope to change this economic pattern.

SECTION REVIEW

1. What challenges do the pastoral nomads face in the Sahel?
2. What happens when a poor country focuses on growing cash crops for export?
3. **Critical Thinking   Compare and contrast** northern West Africa and southern West Africa.

SAHEL COUNTRIES

FOCUS

- *What effects did the drought of the 1970s and 1980s have on the Sahel countries?*

- *Why is the Niger River important to the countries of the Sahel?*

REGION   The Sahel countries of West Africa are Mauritania, Mali, Niger, Burkina Faso (boor-KEE-nuh FAH-soe), and Chad. (See the map on page 441.) These countries share many common features. Most of the people live in the savanna region to the south, where herding and farming are the traditional occupations. In contrast, the stony plains and shifting sands of the Sahara are nearly empty of people. Struggles between the desert people and the settled farmers are frequent.

In the developing countries of the Sahel, there are few railroads or good roads. French is the ***lingua franca*** of these former French colonies. A *lingua franca* is a common language spoken and understood by the majority of the people, although other languages may be spoken at home. Arabic is often heard, a sign of the importance of Islam in the region. Many people speak one of the many languages of the diverse ethnic groups in the region.

The drought of the 1970s and early 1980s was devastating in the Sahel. In Niger, for example, nearly 90 percent of the herd animals died. Desertification forced one-fourth of the population to move to nearby countries. Recovery has been slow, and refugee settlements surround all of the capital cities in the Sahel countries.

Mauritania, Mali, and Niger Mauritania stretches east from the Atlantic coast into the Sahara. About two-thirds of the population are Moors, an Arab and Berber mixture. Almost all of the people are Muslims. Arabic and French both are official languages. The capital of Mauritania is the coastal city Nouakchott (nu-AHK-shaht). Nouakchott had 5,000 people when it was founded in 1960. Migration from rural and drought-stricken areas has increased the population to more than 500,000. Most of the people live in refugee camps. International aid,

FACTS IN BRIEF   **West Africa**

COUNTRY POPULATION (1994)	LIFE EXPECTANCY (1994)	LITERACY RATE (1990)	PER CAPITA GDP (1993)
BENIN 5,341,710	50, male 54, female	23%	$1,200
BURKINA FASO 10,134,661	46, male 48, female	18%	$700
CAPE VERDE 423,120	61, male 65, female	66% (1989)	$1,070 (1991)
CHAD 5,466,771	40, male 42, female	30%	$500
CÔTE D'IVOIRE 14,295,501	47, male 51, female	54%	$1,500
GAMBIA 959,300	48, male 52, female	27%	$800
GHANA 17,225,185	54, male 58, female	60%	$1,500
GUINEA 6,391,536	42, male 46, female	24%	$500
GUINEA-BISSAU 1,098,231	46, male 49, female	36%	$800
LIBERIA 2,972,766	55, male 60, female	40%	$800
MALI 9,112,950	44, male 48, female	17%	$650
MAURITANIA 2,192,777	45, male 51, female	34%	$1,050 (1992)
NIGER 8,971,605	43, male 46, female	28%	$650
NIGERIA 98,091,097	54, male 57, female	51%	$1,000
SENEGAL 8,730,508	55, male 58, female	38%	$1,400
SIERRA LEONE 4,630,037	44, male 49, female	21%	$1,000
TOGO 4,255,090	55, male 59, female	43%	$800
UNITED STATES 260,713,585	73, male 79, female	97% (1991)	$24,700

Source: *The World Factbook 1994*, Central Intelligence Agency

In which West African countries is the literacy rate at least 50 percent?

Section 2 Review
ANSWERS

1. The threat of drought and the loss of labor due to migration to the cities
2. The country has to import food, and there is less money to invest in education and in infrastructure.
3. In the north, most people are Muslims, lands are wide open, travel is easy, resources are far apart, and, historically, a few groups controlled large territories. In the south, most people are Christians, lands are forested, travel is more difficult, and groups controlled smaller areas.

Chart Caption Answer
Cape Verde, Côte d'Ivoire, Ghana, Nigeria

445

Motivator

Ask students to turn to the map on page 441 and trace the Niger River's route through the Sahel countries. Have students brainstorm about ways that a drought affects a farming and herding region such as the Sahel. Tell them that they will learn how a drought in recent years changed population patterns in the Sahel.

Comparing and Contrasting

Write the headings Niger River and Nile River on the board. Help students compare and contrast the importance of the Niger River with that of the Nile River (Chapter 36), focusing on irrigation and the river as a lifeline to the interior. Write the responses under the appropriate heading.

Closure

Have students write a paragraph with this topic sentence: "Drought in the 1970s and 1980s devastated the Sahel."

ASSESSMENT

Assign the Section 3 Review questions on this page.

Reteaching

Assign Section 3 of Reteaching Worksheet 37.

Section 3 Review

ANSWERS

1. Herd animals died; people were forced to move to other countries and to cities; Lake Chad shrank.
2. Mali and Niger
3. Possible answer: a *lingua franca* is a common language that the majority of people speak and understand. It provides a means of communication among various groups that have their own languages.

Motorbikes (mopeds) are assembled in **Ouagadougou, Burkina Faso**. Two-wheel commuters choke the city's streets during rush hour.

particularly from the oil-rich Islamic countries, provides much of the country's food.

Because its northern half is in the Sahara, Mali sometimes seems to be two countries. Bamako, the capital, is in the savanna lands along the Niger River. Nearly all of the people of Mali live in this southern region. The government of Mali hopes to use the Niger's waters to expand irrigation. Potential also exists to increase tourism to Mali's ancient trading cities of Timbuktu and Gao.

Landlocked Niger consists of a small savanna region along the Niger River valley and a large expanse of the Sahara and Sahel to the northeast. The capital is Niamey (nee-AHM-ay), situated on the Niger River. Only 2 percent of the country's land is arable, all of it located along the river. Ethnic groups include the Hausa, Djerma, Fulani, and Tuareg. As in the other northern countries of West Africa, Islam is the predominant religion. Democratic reforms are beginning to take place.

Burkina Faso and Chad Burkina Faso, formerly Upper Volta, is a landlocked country of dry savannas situated between Mali and Niger. Ethnic groups and subgroups include (among many others) the Mossi, Gurunsi, Senufo, Lobi, Mande, and Fulani. Most of the people are animists, and cattle raising is traditional here. Before the severe drought of the 1970s and 1980s, great cattle drives moved herds to countries to the south. Today, many workers migrate to the coastal countries in search of jobs. Near the center of the country is the capital city of Ouagadougou (wahg-uh-DOO-goo). Great slums have grown around the city since the drought, and the search for firewood has removed nearly every tree within 25 miles (40 km). The government is supported mainly by aid from developed countries.

Chad stretches from the central Sahara to the tall-tree savanna region in the south. Chad has more than 200 ethnic groups, and more than 100 local languages are spoken. The few people in the north are mostly Islamic herders with Arabic customs. The numerous people in the south are Christian and animist farmers. As a result, the politics of Chad are complex. Many groups have fought for control of the government in N'Djamena (ehn-juh-MAY-nuh), the capital. Meanwhile, the drought of the 1970s and 1980s reduced Lake Chad to one-third the size it was in 1950. An important fishery there has nearly collapsed. With few resources and internal conflict, Chad has little chance for economic prosperity in the near future.

 SECTION REVIEW

1. How did the drought of the 1970s and 1980s affect the Sahel countries?
2. To which countries is the Niger River an important resource?
3. **Critical Thinking** **Determine** how a *lingua franca* might unite the people of a country.

ATLANTIC COAST COUNTRIES

FOCUS

- *How do people in the Atlantic coast countries earn a living?*
- *How are the countries of the Atlantic coast alike? How are they different?*

REGION The Atlantic coast countries of West Africa are Senegal, Gambia (GAM-bee-uh), Guinea, Guinea-Bissau (GIN-ee bis-OW), Liberia, and Sierra Leone. (See the map on page 441.) The dry savannas of Senegal give way to forests in Sierra Leone and Liberia. Much of the original forest has been cleared by farmers, however. To the south, the wetter climates permit the growing of bananas, coffee, and cassava. Farmers grow rice in river valleys

446

SECTION 4
OBJECTIVES

The objectives reflect the Focus questions on page 446 and are assessed in the review on page 448.

1. Describe the agricultural and manufacturing activities in the Atlantic coast countries.
2. Compare and contrast the Atlantic coast countries.

TEACHING SECTION 4
Motivator

List the Atlantic coast countries on the board. Pronounce their names for students. Use a wall map to point out general characteristics that these countries share, such as location, relatively small size in comparison with other West African countries, and population density.

Creating a Map

Have students work in small groups. Assign one Atlantic coast country to each group. Ask students to create a map of their assigned country showing major agricultural and manufacturing activities. Each map should have a caption based on information from the chart on page 445. Display the maps to show the similarities and differences of these countries.

throughout the region. In the wettest areas, farmers tend oil palms, rubber trees, and vegetable gardens.

Rains are heavy on the mountain slopes. Supposedly, Sierra Leone (which means "mountain lion" in Portuguese) was named by Portuguese navigators who thought that the rainstorms coming out of the mountains sounded like lions. Mountain areas in Sierra Leone have received more than 240 inches (600 cm) of rain in one year.

Gambia and Senegal

The boundary between Gambia and Senegal is one of the most unusual in the world. Gambia, a former British colony, is composed of the land 15–30 miles (24–48 km) on both sides of the Gambia River. The country is 295 miles (475 km) long, and apart from a tiny coastline, entirely surrounded by Senegal. Senegal and Gambia have proposed to merge, but they have many issues to settle. In Gambia, the largest ethnic group is the Mandinka, and Islam is the predominant religion. Gambia is primarily an agricultural country, and peanuts are the major crop.

Senegal, a former French colony, is more developed with better commercial traditions. For centuries, peanuts have been the dominant cash crop in Senegal. Cotton and corn also are important. Senegal's capital, Dakar (duh-KAHR), was a major administrative center for the nearby French colonies. There, manufacturing income is growing. Dakar is a modern city, with an international airport, sports stadiums, cinemas, and museums. Factories process peanuts, cotton, rice, fish, and phosphate ores. French is the official language, and Wolof (also the name of the largest ethnic group) is widely spoken. Most of the population is Muslim. Senegal's political stability encourages tourism.

Guinea

After Guinea gained independence from France, the new country's first leader was a cruel dictator. Initially promising prosperity, he came to control both the lives of the people and the economy. More than 2 million people fled the country; many others died in prisons. Poverty increased in the rural areas, and hardships continued for 25 years until the dictator died.

Today, the government of Guinea is making progress. Conakry (KAHN-uh-kree), the capital of Guinea, is the center of education for the country, while most industrial activity takes place northeast of the capital. Particularly important to the economy are Guinea's large reserves of bauxite, the ore of aluminum. These reserves total about one-third of the world's supply. The products of the oil palm

African and French cultures have influenced the development of **Dakar**. Once the capital of French West Africa, Dakar is one of Africa's major cities and the capital of independent **Senegal**.

also are important to Guinea's economy. Palm oil is used for making cosmetics and soap and in cooking, palm kernels are processed for starch and food, and palm fiber is used for making floor mats and rope.

The major ethnic groups in Guinea include the Fulani, Malinke, and the Susa. Islam is practiced by more than 75 percent of the population; about 5 percent are animists.

Guinea-Bissau and Cape Verde

Guinea-Bissau and Cape Verde were once Portuguese colonies. In fact, the city of Bissau was the major port for the Portuguese West African slave trade. The colonies were often administered together, and they shared common paths to establish Communist governments after independence. Having abandoned communism, both countries are now attempting to establish democratic governments. The ethnic makeup in Guinea-Bissau is diverse. More than 50 percent of the population is animist. The people survive on subsistence agriculture, growing rice and cassava. Poverty in Guinea-Bissau is widespread, however,

Cape Verde is a group of islands located 400 miles (640 km) off the coast of Senegal. When the Portuguese arrived in 1456, Cape Verde was uninhabited. The 10 islands are volcanic in origin and appear stark and barren. There is little farmland.

The Portuguese brought African slaves to the islands in the 1460s. Intermarriage between the Africans' descendants and Europeans

THEME
Movement

Since all-weather roads have been built along both sides of it, the Gambia River is no longer the most important form of passenger transportation in Gambia. However, the deep river is still vital to Gambia's international trade. Oceangoing ships can travel 150 miles (241 km) upstream, as far as the town of Kuntaur.

Closure

Review with students the similarities and differences among the Atlantic Coast countries.

ASSESSMENT

Assign the Section 4 Review questions on this page.

Reteaching

Assign Section 4 of Reteaching Worksheet 37.

Extension

In late 1989, a bloody civil war broke out in Liberia. Have students do research to report on the causes of the war and the response of other West African countries to it.

SECTION 5
OBJECTIVES

The objectives reflect the Focus questions on this page and are assessed in the review on page 450.

1. Describe the historical, cultural, and economic geography of Nigeria, and explain its leadership role in the region.
2. Describe the economic geography of the Guinea coast countries.

Section 4 Review
ANSWERS

1. Cash crops: peanuts, cotton, maize, oil-palm products, rubber; manufacturing centers: Dakar, Senegal; northeast of Conakry, Guinea
2. Both countries began as settlements for freed slaves.
3. Senegal has commercial crops, manufacturing, a modern city, political stability, and a largely Muslim population. In Guinea-Bissau, however, poverty is widespread, subsistence agriculture is common, and much of the population is animist.

THE WORLD TODAY VIDEODISC PROGRAM

Nigeria: Growth with Style

2A 39102 4:26

Play   Pause

448

---

produced most of the present population. Many islanders have emigrated to the United States.

Liberia and Sierra Leone Liberia and Sierra Leone are unusual in that both countries were founded as settlements for freed slaves. The name *Liberia* comes from the Latin word *liber*, which means "free." Liberia's settlers came from the United States beginning in 1822. The country became a republic in 1847 and has continued to maintain close relations with the United States. Sierra Leone was founded in 1787 for freed slaves from the British Empire. It remained a British colony until 1961. In Liberia, the descendants of African American slaves are the most prosperous people in the country, though they total only about 2 percent of the country's population. In Sierra Leone, the descendants of freed slaves, who call themselves Creoles, make up about 2 percent of the population. Recent politics in both countries have been violent. The descendants of the freed slaves are often resented by the majority of the people.

Liberia and Sierra Leone are poor countries. In its long history, Liberia has received little economic support from foreign countries. Sierra Leone, meanwhile, was probably the least developed of Britain's African colonies. In Liberia, rubber plantations, most of which were owned by a single U.S. company, once dominated the country's economy. Today, manufacturing in Sierra Leone and Liberia is centered around processed food, textiles, and building supplies. With large reserves of iron ore, diamonds, and bauxite, both countries hope expanded mining will improve their economies.

SECTION REVIEW

1. What are the major cash crops in the Atlantic coast countries? Where are the major manufacturing centers?
2. What is unusual about Liberia and Sierra Leone?
3. **Critical Thinking** **Compare and contrast** Senegal and Guinea-Bissau.

GUINEA COAST COUNTRIES

FOCUS

● *Why do the Guinea coast countries look to Nigeria for leadership?*

● *What is the economic situation in the various Guinea coast countries?*

REGION The coastline of West Africa that faces south is called the Guinea coast. The countries here are Nigeria, Ghana, Côte d'Ivoire, Togo (TOE-goe), and Benin. These humid countries are densely settled near the coasts. Climates become drier inland.

Nigeria Compared to the other countries of West Africa, Nigeria is enormous, and it is by far Africa's most populous country. Nigeria stretches

---

Cacao: From Field to Consumer

• Cacao beans are enclosed in pods that grow 4 to 15 inches long. Pods ripen fully in three to six months. Each pod contains 20 to 60 beans.

• Harvesters ferment the beans in trays, on banana leaves, or in sweating boxes for a week or more.

Cacao seeds

Cacao pod

Fermenting

TEACHING SECTION 5

Motivator

Locate on a wall map the Guinea coast countries of West Africa, and then refer students to the population map of Africa on page 419. Point out the area of dense population that fans out inland from the Gulf of Guinea. Locate for students the major ports of Abidjan and Lagos.

Synthesizing

Divide the class into three groups. Assign each group a different subject to research about Nigeria: history, economy, culture. Ask each student to list three facts about Nigeria, relevant to their assigned subject. Have them take turns stating aloud a different fact. Then use the statements to help students understand why Nigeria is a political and economic leader of the region.

Guided Practice

Discuss the economic geography of the Guinea coast countries. Use the illustrations on page 448–449 to identify the export crop common to each country. *(cacao)* Ask how the photograph on page 449 illustrates one reason Nigeria is the economic leader of the region. *(oil reserves and oil production)* Have students suggest additional economic similarities and differences among the countries.

Oil production is an important economic activity in the Niger River delta of **Nigeria**. Nigeria has some of the largest oil reserves in the world.

from the humid climate of the Gulf of Guinea to the dry savannas along Lake Chad. In the tropical south, oil-palm products, cacao, and rubber are important. In the north, peanuts and cotton are grown.

Nigeria's greatest mineral resources are the large oil deposits of the Niger River delta. Nigeria's economy depends so heavily on oil, however, that it can be severely hurt when world oil prices are low. Coal deposits also are located in the delta. Inland, there are large iron-ore and manganese mines. Nigeria holds about half of the world's supply of columbite, a metal used to harden stainless steel. The Niger River, meanwhile, is a major hydroelectricity and fishing resource.

When Nigeria became independent from Britain in 1960, a network of roads and railroads

was already in place. Lagos (LAY-gahs), Nigeria's former capital city, is an industrial and commercial center. Its bustle and congestion are known worldwide. The government has moved the capital to the new city of Abuja (ah-BOO-juh), in the center of Nigeria. New factories also are being built throughout the country. Nigeria has truck and automobile assembly plants in addition to other factories that export goods throughout the region.

Cities have been important in Nigeria for centuries. In the dry north, walled cities such as Kano (KAH-noe) were once centers for prosperous Muslim kingdoms. Kano now has an international airport and is the largest city in the northern region. Ibadan (ee-BAHD-uhn) in the south, whose origins remain a mystery, was an important town before colonization. Today, this hilly city has a prominent university and is second in size only to Lagos.

There are four main ethnic groups in Nigeria. The two large groups in the south are the Yoruba of the Lagos-Ibadan region and the Ibo of the lower Niger River region. These and other southern groups were those most influenced by Christian missionaries and European ideas. The two large groups in the north are the Fulani and the Hausa. These and most other people in the northern regions are Muslims who follow traditional ways of life.

Nigeria has been the stage for ethnic and religious conflict. The Ibo in the southeastern province of Biafra (bee-AF-ruh) tried to secede from the rest of the country in the late 1960s. A terrible civil war resulted. Each of the four large ethnic groups, with its own customs, religion, and language, is suspicious of the others. English has become the *lingua*

GLOBAL PERSPECTIVES

Introduce the term *reparations* to the class. Note that the governments of some African countries, most notably Nigeria, believe that they should be compensated for the loss of millions of people during the slave trade of the 1400s–1800s. The slave trade, they say, denied Africa the labor it needed for modern economic growth. Reparations might be in the form of aid packages or debt forgiveness.

Ask students if they think that the United States and other countries that participated in the slave trade (Portugal, Spain, England, and others) should pay reparations. *(pro: governments have paid reparations to other exploited groups; con: the slave trade occurred long ago and was carried out primarily by individuals, not governments)*

• Cacao beans are then cleaned, roasted, and hulled.

Shells can be used for fertilizer, organic mulch, and the

production of cocoa butter. Roasted beans are crushed,

producing a thick, dark liquid.

• The thick liquid can be combined with sugar and vanilla to make sweet chocolate. Cocoa butter can be used in the preparation of medicines and other consumer products, such as soaps and cosmetics.

• Pressing the liquid to remove most of the cocoa butter, or cacao fat, produces powdered cocoa. Chocolate is mostly cacao fat, while cocoa can contain as little as 10 percent fat.

Roasting   Winnowing   Coarse grinding   Milling   Kneading   Pressing

Cocoa butter

Chocolate

Soap

Lotion

Bar chocolate

Grinding

Cocoa powder

Press-cake

Hot cocoa

Independent Practice

Ask students to work with a partner to complete a chart comparing the natural resources, agricultural products, and manufactured products of each of the Guinea coast countries.

Closure

Call on students to evaluate the economies of the countries discussed in Section 5.

ASSESSMENT

Assign the Section 5 Review questions on this page.

Reteaching

Assign Section 5 of Reteaching Worksheet 37.

Extension

Students can work with partners to make a time line showing political leadership (dictators, military rulers, elected officials) and events (coups, civil wars, free elections) in a Guinea coast country since it gained independence. Discuss the relationship between the country's political history and economic progress.

Section 5 Review
ANSWERS

1. It is the most populous country in West Africa and is rich in natural resources.
2. Agriculture
3. Possible answers: to unite the country's peoples, as in Nigeria, or to encourage development in certain regions of the country, as in Côte d'Ivoire

franca because it is not the language of any single group. Despite these problems, the idea of a united Nigeria is becoming stronger. Moving the capital from Lagos to Abuja, where no ethnic groups have land claims, may further unite the Nigerian people.

Nigeria influences West Africa militarily, politically, and economically. Because of its large population and its oil wealth, Nigeria has often assumed a leadership role in Africa. However, conflict between supporters and opponents of democratic reforms has added to Nigeria's political instability.

Ghana and Côte d'Ivoire Côte d'Ivoire and Ghana have similar climates, soils, and natural resources. Ghana took its name from the former African empire. (The word *ghana* means "war chief.") When Ghana gained independence from Britain in 1957, the country had prosperous farms, good roads, industrial opportunities, and a promising economy. Since then, its economy has had ups and downs. Côte d'Ivoire became independent from France in 1960 with fewer economic opportunities. Nonetheless, it is one of Africa's most prosperous countries today.

During the colonial period, Ghana was called the "Gold Coast." European traders built forts at the country's natural harbors and began trading inland for gold, ivory, and slaves. In 1919, the British began building roads, harbors, and agricultural processing centers. Ghana soon became the world's largest cacao producer. After independence, the poor economic policies of several dictatorships and elected governments caused the economy to weaken. The governments of Ghana failed to invest in agriculture. Farmers were not given fair prices for their products and, in turn, could not reinvest in their farms. As a result, yields declined. Cacao production dropped by 80 percent.

The current government in the capital of Accra (uh-KRAH) has realized these past errors, and recent economic reforms have improved the economic outlook. Ghana has rich gold mines and the ability to produce high-quality cacao. Another hope for Ghana's future is continued development of the Volta River Project. This great dam on the Volta River has created one of the world's largest human-made lakes. The hydroelectric power generated here, easy shipping access, and irrigation opportunities could help restore Ghana's economic position.

The healthy economy of Côte d'Ivoire is related to its history of political stability. The French developed prosperous farms and large plantations here. After independence, Félix Houphouet-Boigny, a rich African planter and Boule chief, was elected president and stayed in power for more than 30 years. The stable government invested in agriculture, and Côte d'Ivoire became the largest producer of cacao. The country maintained good relations with France and managed its finances well. Côte d'Ivoire has relied on free enterprise and foreign investment to encourage economic growth.

The economic heartland of Côte d'Ivoire is the southern forested zone, although much deforestation has occurred. Lumber, coffee, and cacao are grown here for export. The coffee crop is the largest in Africa and an important part of the world market. Abidjan (ab-i-JAHN), the capital, is a modern city and major port. With many small and large industrial firms, it has the largest manufacturing output of all the former French African cities. Abidjan is crowded, however. In order to encourage development in the central and northern parts of the country, the government has considered moving the capital to the inland town of Yamoussoukro (yahm-i-SOO-kro).

Togo and Benin Togo and Benin are small, poor countries situated between Ghana and Nigeria. Both have savanna environments with tropical forests along the coast and on the uplands. As in the other countries along the Guinea coast, coffee, cacao, and palm oil are important products.

Togo was a German colony before World War I. After that war, it came under French control until gaining independence in 1960. Benin, the former French colony of Dahomey (duh-HOE-mee), is named after the kingdom of Benin. Togo and Benin both have numerous ethnic groups, and most of the people are animists or practice African religions. Both are countries of small farmers who provide little tax revenue for their governments. Manufacturing industries are few and small, and there are not enough jobs.

5 SECTION REVIEW

1. Why is Nigeria such an important country in the region?
2. What economic activity is important to every Guinea coast country?
3. **Critical Thinking** **Suggest** two reasons that countries move their capitals.

CHAPTER **37** CHECK

Construction of Notre Dame de la Paix Cathedral (Our Lady of Peace Cathedral) in **Yamoussoukro, Côte d'Ivoire**, was completed in 1990. The cathedral is one of the largest Christian churches in the world.

Reviewing the Main Ideas

1. The climate and vegetation pattern in West Africa is divided into three zones. Most people live in the forest zone and in the dry savannas of the southern Sahel.
2. West Africa is a region of religious and ethnic diversity. More than 500 ethnic groups live here. West Africa has been influenced by the spread of Islam and by Christianity, though many people still retain their traditional beliefs.
3. Agriculture is the predominant economic activity in West Africa. In the north, pastoral nomads raise livestock. Villages of small farmers are common in the savanna and forest zones.
4. The most serious issue facing West Africa today is economic development. Development of resources, education and training programs, foreign aid, and regional cooperation are the keys to economic improvement.

Building a Vocabulary

1. Why do geographers use the word *zonal* to describe West Africa's climate and vegetation pattern?
2. What activities does pastoralism involve?
3. What is a staple food? Give two examples of staple foods in West Africa?
4. What are millet and sorghum? Why are they suitable for West Africa?

Recalling and Reviewing

1. Describe West Africa's three environmental zones.
2. What effects has the drought of the 1970s and 1980s in the Sahel had on West Africa?
3. What two Atlantic coast countries were founded as settlements for freed slaves?
4. Why are the Guinea coast countries of Ghana, Côte d'Ivoire, and Nigeria more prosperous than Togo and Benin?

Thinking Critically

1. One problem in West Africa is that countries often compete rather than cooperate with one another. What specific steps do you think the leaders of West Africa could take to bring about regional cooperation?
2. Using three examples from the chapter, discuss the various ways by which countries get their names. Also, suggest a reason that the people or government of a country might want to change that country's name.

Learning About Your Local Geography

Cooperative Project
With your group, conduct research to determine how the population density in your area has changed over the last 30 years. What factors have contributed to this change? Present your information using charts, graphs, or illustrations.

Using the Five Themes of Geography

Create a graphic organizer that illustrates the relationships among the geographic themes of **location**, **place**, **movement**, and **human-environment interaction** in the Sahel.

451

CHAPTER 38 East Africa
Planning Guide

TEXTBOOK	TEACHING RESOURCES	OTHER AVAILABLE MATERIALS
Section		
1. Physical Geography Pages 453–454	Reteaching Worksheet 38.1 Challenge/Enrichment Worksheet 38	*Study Guide for Spanish Speakers* *Lecture Notes,* Chapter 38
2. Human Geography Pages 454–455	Reteaching Worksheet 38.2	*Creative Strategies for Teaching World Geography* LESSON 17 Exploring Culture Through Language: African Charades *Study Guide for Spanish Speakers* *Lecture Notes,* Chapter 38
3. Countries of East Africa Pages 455–461	Reteaching Worksheet 38.3	*The World Today* Videodisc Program *Study Guide for Spanish Speakers* *Lecture Notes,* Chapter 38
4. The Horn of Africa Pages 461–462	Graphic Organizer 38 Reteaching Worksheet 38.4 Skills Worksheet 38 Critical Thinking Worksheet 38	*Study Guide for Spanish Speakers* *Lecture Notes,* Chapter 38
Chapter Review Page 463	Chapter 38 Test Chapter 38 Portfolio Assessment See also *Alternative Assessment Forms*	*Test Generator,* Chapter 38

CHAPTER RESOURCES

Lecture Notes, Chapter 38
Teaching Resources
    Chapter 38 Graphic Organizer
    Chapter 38 Skills Worksheet
    Chapter 38 Critical Thinking Worksheet
    Chapter 38 Challenge/Enrichment Worksheet

REVIEW AND ASSESSMENT

Chapter 38 Check, page 463
Chapter 38 Test, *Teaching Resources*
Chapter 38 Portfolio Assessment, *Teaching Resources*
Test Generator, Chapter 38
Alternative Assessment Forms

ADDITIONAL RESOURCES

Books for Teachers

Mostert, Noel. *Frontiers: the Epic of South Africa's Creation and the Tragedy of the Xhosa People.* New York: Knopf, 1992.

Pouwels, Randall Lee. *Horn and Crescent: Cultural Change and Traditional Islam on the East African Coast 800–1900.* New York: Cambridge University Press, 1987.

Smith, A. *The Great Rift: Africa's Changing Valley.* New York: Sterling, 1989.

Vansina, Jan. *Paths in the Rainforests.* Madison: University of Wisconsin Press, 1990.

Books for Students

Amin, Mohamed, and Duncan Willett. *Last of the Maasai.* UK: Bodley Head, 1990.

East Africa. Alexandria, VA: Time-Life Books, 1987.

Jenkins, Ernestine. *A Glorious Past: Ancient Egypt, Ethiopia, and Nubia.* New York: Chelsea House, 1995.

The World Today *Videodisc Program*

A variety of still images, short videos, and activities are available for you to use as you teach this chapter. *Lecture Notes* include bar codes for specific images and videos.

Multimedia Resources

Africa: Central and Eastern Regions. (VHS, 15 minutes) Encyclopedia Britannica Educational Corporation, 310 South Michigan Avenue, Chicago, IL 60604–9839.

ECO East Africa. (Multimedia CD-ROM, environmental simulation of wildlife preserve) IVI Publishing, 1380 Corporate Center Curve, Suite 305, Eagan, NM 55121.

Independence. (VHS, 25 minutes) Encyclopedia Britannica Educational Corporation, 310 South Michigan Avenue, Chicago, IL 60604–9839.

The Rift Valley. (VHS, 44 minutes) Films for the Humanities & Sciences, P.O. Box 2053, Princeton, NJ 08543–2053.

MEETING INDIVIDUAL NEEDS

Limited English Proficiency (LEP)

Provide students with outline maps of East Africa, or have them make their own. Then have students divide East Africa into its three language categories—Nilo-Saharan, Cushite, and the Niger-Congo (Bantu). Students should use the information in the textbook to help them make their maps.

Learners Having Difficulty

Help students study this chapter by reviewing information presented in the physical-political map of East Africa on page 453. Ask students to write two questions about information presented in the map. Then allow students time to ask their questions and to call on other students to answer them.

Auditory Learners

You might obtain some foreign language instruction audio tapes or recordings in Swahili and play them for the class. Have students take turns learning to pronounce words and phrases.

Visual Learners

Divide the class into three groups. Assign each group one of the following maps to create: Natural Wonders of East Africa, Climate of East Africa, and Agricultural Products and Natural Resources of East Africa. Students should use the information in Chapter 38 and the maps of Africa in Chapter 35 to help them complete their own maps. Their maps should include map titles, country boundaries and names, and map keys.

Tactile/Kinesthetic Learners

Have students make illustrated time lines showing major events in the history of East Africa. Then have students write critical thinking questions based on the information presented in the time lines.

Gifted Learners

Have interested students research information about the archaeological site at Olduvai Gorge in Tanzania. Have them make dioramas showing the site and the artifacts found there by archaeologists. Students might pretend to be the archaeologists who discovered the site, and then write a journal entry describing their discoveries.

MULTICULTURAL FOCUS

Ethiopia in East Africa was among the first regions in the African continent to convert to Christianity. Syrian missionaries introduced the religion there between the 4th and 6th centuries. Eventually the Ethiopian Orthodox church split with the Roman Catholic church in Rome over religious doctrine and became aligned with the Egyptian Coptic church. Today, approximately 40 percent of the population in Ethiopia is Christian.

Historically, Christianity in Ethiopia has absorbed influences from other religions, including Judaism, Islam, and the animist beliefs of traditional ethnic groups in East Africa. Many Ethiopian Christians believe that they are descendants of King Solomon of Israel and the Queen of Sheba, whose lives are chronicled in the Bible. *Kebra Nagast* (The Glory of Kings) was an epic written by Ethiopian monks in the 13th century. It describes a traditional legend that Ethiopians, and not Jews, are God's chosen people.

Political change in Ethiopia in recent years affected the influence of the Christian church. A revolution in 1974 that overthrew the Ethiopian emperor also resulted in the persecution of many Christian priests and deacons in rural areas. Interested students might wish to research the status of Christianity in present-day Ethiopia.

CHAPTER 38

EAST AFRICA

Section 1: Physical Geography
Section 2: Human Geography
Section 3: Countries of East Africa
Section 4: The Horn of Africa

OVERVIEW

In Chapter 38, students will learn about the countries of East Africa.

Section 1 provides an overview of the region's physical geography.

Section 2 discusses the ethnic makeup of the region as well as the three major language groups. The economy of the region is also covered.

Section 3 discusses the history and economic development of Kenya, Tanzania, Rwanda, Burundi, Uganda, and Sudan.

Section 4 focuses on the countries of the Horn of Africa, including Ethiopia, Eritrea, Somalia, and Djibouti.

About the Author

Elspeth Huxley (1906–) spent her childhood in Kenya, where her father was an officer in the British colonial army. In 1913, her family moved to Thika, north of Nairobi. The quotation on this page is from her autobiography, *The Flame Trees of Thika.*

Primary Source

Description of Change: excerpted
Rationale: excerpted to focus on main idea

CHAPTER 38

East Africa

“**O**n our left rose a long, dark-crested mountain range from which sprang rivers that watered a great part of the Kikuyuland. These rivers, no larger than streams, had dug down through soil red as a fox and rich as chocolate to form steep valleys whose sides were now green with young millet and maize. ”

Elspeth Huxley

Ceramic mural in **Burundi**

Deep beneath the scenic landscapes of East Africa, the earth's mantle is churning, its intense heat struggling to escape to the surface. Active volcanoes and hot springs here tell us that the processes of plate tectonics are trying to break Africa apart. So far, the continent has arched up and split along the rift valleys.

As the rifts opened, the geography of the region was dramatically altered. People settled near newly formed lakes in the rift valley. They brought cattle and goats to the region and began farming there. These settlers found greatest agricultural success in the fertile highlands. Later, the highland peoples discovered that the ivory tusks of elephants were prized by merchants of the Mediterranean and India. Ivory was traded for goods from as far away as China.

Today, tourists travel from throughout the world to admire the beauty of East Africa's national parks. Important to Africa's economy, tourism creates jobs in the region's hotels, restaurants, and transportation facilities. Best of all, the tourists take pictures, not ivory.

Agricultural plots in **Kenya**

GEOGRAPHY DICTIONARY

Swahili
gum arabic
sisal

Using the Quotation

Help students obtain a sense of place about the land described in the quotation on page 452. Refer students to the background photograph on that page and to the illustration of the rift valleys on page 454. Ask students how they imagine the East African soil, as described by the author, looks and feels. Tell them that the East African highlands in Kenya have some of the richest soil in Africa.

Using the Geography Dictionary

Ask students to find the definitions for the terms in the Geography Dictionary. Ask them to identify a general category for each word. (*Swahili*—language, *gum arabic*—food additive, *sisal*—nonfood agricultural product)

SECTION 1
OBJECTIVES

The objectives reflect the Focus questions on this page and are assessed in the review on page 454.

1. Describe the landform patterns in East Africa.
2. Identify climate patterns in East Africa.

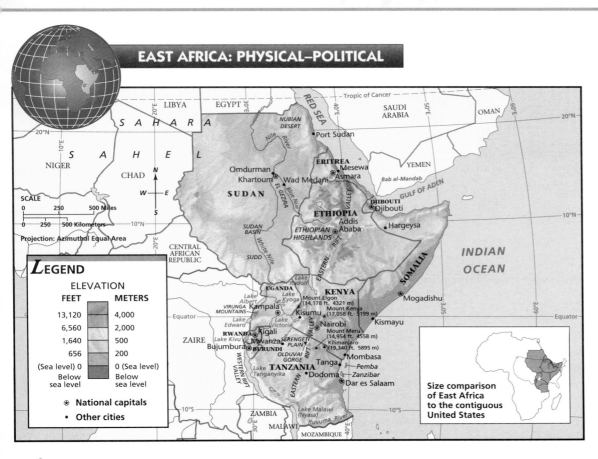

EAST AFRICA: PHYSICAL–POLITICAL

Using the Physical–Political Map: East Africa

Have students examine the map on this page and identify the countries of the region and their capitals. Then have students describe major landforms—such as rifts, mountains, and lakes—and identify the countries in which they are located.

PHYSICAL GEOGRAPHY

FOCUS

- *What major landforms are found in East Africa?*
- *What is the climate like in East Africa?*

REGION **Landforms and Climate** East Africa is a region mostly of high plains and plateaus. The plateaus are topped by snow-capped volcanoes and cut by the Eastern and Western Rift valleys. Much of East Africa's scenery is spectacular. Mount Kenya in central Kenya is more than 17,058 feet (5,199 m) high. Mount Elgon (EHL-gahn) on the Kenya-Uganda (yoo-GAN-duh) border is 14,178 feet (4,321 m) high. The highest point, Kilimanjaro, rises to 19,340 feet (5,895 m). Even though Kilimanjaro is near the equator, it is so high that its twin peaks are always covered with snow.

From above, the rift valleys appear as giant scars on Africa's surface. The rift walls are usually steep cliffs, with the flat valley floors sometimes more than 10,000 feet (3,048 m) below. The Eastern Rift begins in southern Tanzania and continues northward to Ethiopia and onto the floor of the Red Sea. (See the map on page 417.) The dry plains along the Eastern Rift in Tanzania and Kenya are sites of famous game parks. The more humid Western Rift is filled with a chain of lakes, which are often located between high volcanoes. Lake Victoria, the largest lake in Africa, occupies a shallow, saucer-shaped basin on the plateau between the Eastern and Western rifts. To the east, the land slopes downward to the Indian Ocean. The sandy beaches and beautiful lagoons of the East African coastlines are beginning to attract tourists.

East Africa is not only a region of magnificent landforms, but also one of great climate contrasts. The region contains some of Africa's driest deserts and wettest forests. The deserts of northern Sudan are like those of neighboring Egypt. The Nile River forms a long oasis within the surrounding desert of

TEACHING SECTION 1

Guided Practice

Have students look at the map of East Africa on page 453, noting the chain of lakes from Lake Albert in Uganda to Lake Malawi. Ask students to say why these lakes form a chain. Tell them that they are looking at surface evidence of the East African rift system, where the Somali plate is separating from the African plate.

Closure

Have students relate the orographic effect (Chapter 3) to the climate and vegetation of East Africa.

ASSESSMENT

Assign the Section 1 Review questions on this page.

Reteaching

Assign Section 1 of Reteaching Worksheet 38.

SECTION 2
OBJECTIVES

The objectives reflect the Focus questions on this page and are assessed in the review on page 455.

1. Describe the human geography of East Africa, identifying cultural characteristics.
2. Describe the economic geography of East Africa, identifying land use patterns.

The rift valleys are a major physical feature of **Kenya.** The Western and Eastern Rift valleys cut across East Africa.

THEME
Location

Lake Tanganyika is the second deepest lake in the world, after Lake Baykal in Russia. It is a trough in the rift. Neighboring Lake Malawi is much shallower, leading geologists to conclude that Lake Tanganyika is older, has been rifting faster, or both.

CREATIVE STRATEGIES FOR TEACHING WORLD GEOGRAPHY

LESSON 17 Exploring Culture Through Language: African Charades

Section 1 Review
ANSWERS

1. Spectacular landforms, including the rift valleys, snow-capped Kilimanjaro, Lake Victoria, sandy beaches
2. Desert, steppe, tropical savanna, highland
3. Movement is easy and visibility is good.

454

---

bare rocks and shifting sands. Farther south, the lands along the Nile become increasingly wet. The desert gives way to the more common thorny-shrub desert. Eventually, a grassy-shrub savanna appears. Still farther south, near Uganda, vegetation changes to tall-tree savanna. The drylands of Sudan and Ethiopia suffered desertification during the Sahel droughts.

The moist climates of the Ethiopian plateau stand high above the deserts of Somalia and northern Kenya. Along the coast of the Indian Ocean is a tropical-savanna climate with low grasslands and woodlands. Inland, dry savannas with a steppe climate are typical up to the forested highlands. The greatest climate changes occur along the sides of the rift valleys. The floors of the rifts are usually semi-arid, with thorn scrub and dry grasslands. In contrast, the surrounding mountains have a humid highland climate with dense forests. The mountains enjoy rains caused by the orographic effect, while the rift valleys are in rain shadows.

SECTION REVIEW

1. Why can it be said that East Africa has some of the most spectacular scenery in all of Africa?
2. What are the various climate types found in East Africa?
3. **Critical Thinking** **Explain** why plains make good game parks.

---

2 HUMAN GEOGRAPHY

FOCUS

● *Into what three language categories can peoples of East Africa be divided?*
● *What are the main economic activities in East Africa?*

REGION **Ethnic Heritage** The many hundreds of modern ethnic groups in East Africa can be divided into three language categories. The Nilo-Saharan speakers of East Africa live primarily on the plains of southern Sudan. Among these peoples are the cattle-keeping Dinka and the Nauer. Several Nilo-Saharan speaking groups, such as the Masai, migrated south into the highlands only a few centuries ago.

In the second language category are Cushite speakers. Cushite is a subdivison of the Afro-Asiatic language family. Cushite speakers live in areas stretching from the coast of the Red Sea across the Horn of Africa. The Horn of Africa is made up of Ethiopia, Eritrea, Somalia, and Djibouti (juh-BAHT-ee). The name refers to the rhinoceros-horn shape of this part of Africa. (See the map on page 453.) Within this language group are the Amhara of the Ethiopian highlands and the Somali of the coastal region.

The third language group is the Niger-Congo (Bantu) language family of East Africa. These Bantu languages are similar to those of West and Southern Africa. Included in this group are the Kikuyu of Kenya and the Sukuma of Tanzania.

In part, the wide ethnic diversity in East Africa is due to its location. East Africa has always attracted visitors and settlers. The ancient Egyptians regularly sent trading and military expeditions up the Nile. Later, Arab traders established trading cities along the east coast. Trading ships carried gold, ivory, and slaves from the interior to the ports of the Persian Gulf and India.

One of the legacies of this period is **Swahili** (swah-HEE-lee). Swahili developed as the key trade language along the coast. The language contains many Arabic words within a distinctly African grammar. Through the centuries, Swahili has developed a rich literature and has become the common language of the region. Swahili is the official language of Kenya and Tanzania.

TEACHING SECTION 2

Building a Vocabulary

Have students read the paragraph about **Swahili** on page 454. Ask them what Swahili and the term **gum arabic** suggest about the region's past. (*Arabic influence*)

Categorizing

Have students complete a chart of the agricultural products in the various climate regions of East Africa.

Closure

Have students work in small groups to answer the Focus questions on page 454.

ASSESSMENT

Assign the Section 2 Review questions on this page.

Reteaching

Assign Section 2 of Reteaching Worksheet 38.

SECTION 3
OBJECTIVES

The objectives reflect the Focus questions on this page and are assessed in the review on page 461.

1. Identify physical and human characteristics of Kenya, Tanzania, Rwanda, Burundi, Uganda, and Sudan.
2. Describe the political and economic geography of these countries.

During the colonial period, immigrants from India and Europe settled in the cities as merchants and skilled workers. East Africa's many ethnic groups make the cities of East Africa particularly interesting. Every sort of clothing can be seen, and in the evenings, the spicy aromas of many different foods float through the streets.

Economic Geography Although East Africa has several modern cities, most of the people live in rural areas as traditional farmers. There, economic opportunities vary with the natural environment. In the drylands, herding is important. Cattle are highly valued and often are a measure of wealth. Major food crops include maize, sorghum, and millet. **Gum arabic**, which comes from the sap of acacia trees, is an important cash crop. It is used in some medicines and also gives many candies their chewy texture. Cotton grown on irrigated lands is the most valuable crop of the drylands.

In the humid highlands, peas, beans, and cassava are chief crops. Bananas and tomatoes also grow year-round. The most valued cash crop in the highlands is coffee. Though coffee is grown on large plantations, a dozen coffee trees can provide a small cash income for a family.

East Africa's cities have factories that produce consumer goods, processed foods, and building materials. As in West Africa, however, many manufactured goods are imported from developed countries.

Graduating students at the University of Nairobi in **Kenya** reflect the ethnic diversity of East Africa.

SECTION REVIEW

1. What are the three major language groups of East Africa?
2. How do land use patterns in the drylands of East Africa differ from those of the humid highlands?
3. **Critical Thinking** **Assess** the extent to which peoples' movements shape their languages. Give an example from the section.

3 COUNTRIES OF EAST AFRICA

FOCUS

- *What are the physical and human environments of the countries of East Africa?*
- *What economic and political challenges face the countries of East Africa?*

Kenya Kenya's first cities were founded along the coast of the Indian Ocean by Arab merchants from the Persian Gulf. Beginning in the 1500s, Portugal controlled this coast for about 200 years until Arab forces recaptured it. British merchants began trading on the coast during the 1800s. In the 1890s, Britain extended its sphere of influence inland as far as the Western Rift valley. This territory included modern Kenya and modern Uganda. The British soon built a railway from Mombasa to Lake Victoria.

In 1900, the railway reached the highlands. There, near some railroad repair shops, the city of Nairobi (nie-ROE-bee) emerged. The local people, the Kikuyu, had recently moved from these rich farmlands because of a terrible smallpox epidemic in 1898 and 1899. British administrators encouraged settlers from England to move to the "empty" area. Soon, the fertile highlands were covered by large, productive farms whose products were shipped back to Europe. Along with the British settlers came workers and businesspeople from India and Pakistan. The remaining Kikuyu and similar groups had little choice but to move out of their traditional areas or to become farm workers. Others found menial jobs in the cities.

After World War II, black Africans protested British colonial rule. There were peaceful demonstra-

Text continues on page 458.

Geographical
Sidelight

The word Swahili *is from an Arabic word meaning "of the coast." Swahili has become one of the most commonly used languages in Africa, a* lingua franca *rivaling English and French.*

Section 2 Review
ANSWERS

1. Nilo-Saharan, Cushite, and Niger-Congo (Bantu) family
2. Herding is important in the drylands, and grain crops are grown. In the humid highlands, people grow a wider variety of crops.
3. Movement has great power to shape language, as with Swahili, which developed as a trade language and contains both African and Arabic elements.

455

After students have read the feature, ask students if they were inspired by Wangari Maathai's story. Begin a class discussion about ways that single individuals can be effective agents for change in their communities. Encourage students to talk about individuals they may know who address local problems such as Maathai has. Ask students to consider why Maathai's program was successful. *(The people who were directly affected by the problem of soil erosion were encouraged to participate in solving the problem.)*

Geographical Sidelight

Students who are interested in learning more about the Green Belt Movement may write:

The Green Belt Movement P.O. Box 67545 Nairobi, Kenya

For information on a film about the Green Belt Movement, write:

African Development Foundation 1400 I Street, NW Washington, DC 20005

PLANET WATCH

Women Tree Planters of Kenya

Many of the trees in the highlands of **Kenya** have been cleared for crops.

Avocado, mango, and acacia trees thrive in the Kenyan highlands, where not long ago farm workers could not even find shade.

Ten million young trees scattered throughout the country offer new hope to Kenyan farmers and families. The trees were planted not by foresters or government workers but by the many women who farm the lands. "We are planting trees to ensure our own survival," says Wangari Maathai, founder of Kenya's Green Belt Movement. The organization encourages tree planting in Kenya's rural communities as a way of fighting deforestation and desertification.

Deforestation

Like many other developing nations, Kenya is a country mainly of farmers and herders. The most productive farmlands are in the highlands, an area once green with trees and lush vegetation. Over the past century, however, much of the land has been stripped. Today, only about 10 percent of the original forests remain. Many trees were cut for firewood, which farm families use in open-hearth cooking.

One by one, the trees began to disappear from the highlands. Without tree roots to hold the soil in place, the land eroded. Erosion of banks along streams and rivers then threatened the water supplies.

The land was losing its ability to sustain the region's growing population as well as its fertility. Farmers moved down to the savannas in search of better land. These plains are home to nomadic peoples and much of Kenya's wildlife.

Wangari Maathai recognized what was happening to her country. The daughter of a farm worker, she is a biologist and the first woman in Kenya to receive a PhD. "When I would visit the village where I was born," she says, "I saw whole forests had been cleared for cultivation and timber. People were moving onto hilly slopes and riverbeds and marginal areas that were only bush when I was a child. Springs were drying up." Maathai was shocked to find children suffering from malnutrition. "[M]y community was supposed to be a rich, coffee-growing area." Instead of eating nutritious, traditional foods, such as beans and corn, she explains, people were relying on refined foods such as rice because they need less cooking—and thus less firewood.

The Green Belt Movement

On June 5, 1977, in honor of World Environment Day, Maathai and a few supporters planted seven trees in Nairobi, Kenya's capital. This small tree planting began the Green Belt

456

Movement, which soon swept through the Kenyan highlands and captured the attention of people around the world.

From the beginning, Maathai knew that the success of her efforts depended mainly on women. In Kenya, as in much of Africa, men tend cash crops such as coffee and cotton, while women collect firewood, tend livestock, and grow corn, beans, and other food crops. It is the women, then, who provide food for their families and who were the first to see the connection between poor soil and famine.

The movement's organizers encouraged women to plant trees by pointing out that they would no longer have to walk miles to collect firewood. They would have their own wood available for fires, fences, and buildings. If the tree seedlings survived, the women also would be rewarded with a small sum of money.

First, a few small nurseries were established to distribute free seedlings. The nurseries are staffed by local women, who are paid for their work. Nurseries also train and pay local people called Green Belt Rangers to visit farms, check on seedlings, and offer advice.

Soon, nurseries were appearing in communities throughout the highlands. Kenyan women talked to friends and neighbors, sharing the benefits of planting seedlings. Five years after neighbors encouraged Esther Wairimu to plant seedlings, her fields were surrounded by mango, blue gum, and other trees. "I have learned that a tree, in another way altogether, is life," she says. Today, the advantages of tree planting are apparent. Once-barren farms are ringed by greenery. Farmers now have fuel and shade trees. Even more important, the soil is being protected from erosion.

The number of Green Belt nurseries has grown to 1,500, and nearly all of them are run by women. "My greatest satisfaction is to look back and see how far we have come," Wangari

Workers from the Green Belt Movement teach schoolchildren in **Kenya** how to care for tree seedlings.

Maathai says. "Something so simple but meaning so much, something nobody can take away from the people, something that is changing the landscape."

Expanding the Movement

Maathai believes that local people must work together to protect the environment. She stresses that the Green Belt Movement did not rely on outside experts but on Kenyan farmers. Furthermore, the movement receives little support from the government. Most of its funds come from small donations of money from people around the world and from gifts from groups like the United Nations Development Fund for Women.

Now Maathai dreams of spreading her movement to other African nations. Recognizing the obstacles she and other environmentalists face, she says, "[W]e must never lose hope. When any of us feels she has an idea or an opportunity, she should go ahead and do it. . . . One person *can* make a difference."

Wangari Maathai (right) has encouraged Kenyan women to plant millions of trees. Women are paid, although very little, for seedlings they plant that survive.

▼ YOUR TURN

Working in groups of three or four, think of an environmental issue facing your community or state. **Develop** a plan that demonstrates how "one person can make a difference." **Present** your group's plan to the rest of the class.

Motivator

Ask students to name animals they may have seen in television nature programs, in photography books, or on calendars about African wildlife. Point out that many of the programs are filmed in East Africa because of the abundance of animals there and the protection the animals are afforded in nature preserves.

Building a Vocabulary

Write the word **sisal** on the board. Tell students that sisal is grown in many parts of the tropics, including Mexico, and that the word comes from the town of Sisal in southeast Mexico.

Evaluating

Economic Geography Have students work in small groups. Assign one of the countries discussed in Section 3 to each group, and have students evaluate the economy of that country. Have representatives from the groups join in a panel discussion to present their findings. Students might present ideas for improving or expanding the economies.

Geography and Economics

Tourism is an important industry in Kenya. In 1990, about 889,000 visitors came to Kenya, bringing in about $467 million. The major attractions for these visitors are Kenya's game reserves and national parks, such as Amboseli and Tsavo national parks, where tourists can observe elephants and other big game.

THE WORLD TODAY VIDEODISC PROGRAM

Kenya: Coming of Age

2A 15751 4:13

Play   Pause

tions as well as an outbreak of violence by the Mau Mau. The Mau Mau was a secret guerrilla organization made up of Kikuyu. At the heart of this conflict were different attitudes toward land. The British viewed land as personal wealth, power, and property. In contrast, the Kikuyu valued land for the amount of food it could produce. For them, land could not be bought or sold.

Since independence from Britain in 1963, Kenya's leaders have provided a stable, though not a fully democratic, government. Encouraging tourism, controlling population growth, and improving Kenya's small-farm agriculture have been the government's chief goals.

Despite the productive natural environments in the highlands, most of Kenya is too dry for farming. Only about 13 percent of the country is suitable for agriculture. Agriculture in the highlands is increasingly concentrated on small farms. Many large farms

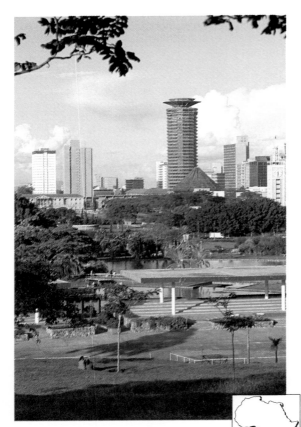

A modern city, **Nairobi** plays an important political role in East Africa. The city also hosts some United Nations offices.

that were developed during the colonial period have been divided into small holdings. The key cash crops are tea, coffee, cotton, and **sisal**. Sisal is a strong fiber used to make rope and twine.

Kenya's industry is concentrated in small factories located primarily in Nairobi and Mombasa, the country's two largest cities. Kenyan canned foods and some manufactured goods are sold throughout the region. Kenya is Africa's favorite tourist attraction. It features cool, green highlands with rich farmlands, forests, and popular game parks on its dry savanna plains. There are also wide vistas of the Eastern Rift and snowcapped mountains. Tourism is a major source of income.

Kenya's greatest challenge is its rapidly increasing population. Due to this population growth, there are no empty farmlands in most of the highlands. Many people migrate to the towns to look for work. Kenya's capital city, Nairobi, is filled with young people, many of whom have only farming skills, little education, and high hopes. Very few will find a good job. Some may eventually return to the family land. Most will remain in the city, however, living on the income of occasional work. This increasing number of people has also put pressure on the game parks. Many people would like to farm these lands. If the game parks were converted to farmland, however, the tourist industry would suffer, along with the animals that the game parks protect.

Tanzania Along Kenya's southern border is the United Republic of Tanzania. (See the map on page 453.) It was created in 1964 when Tanganyika (tan-guhn-YEE-kuh), which had been independent since 1961, merged with the small island country of Zanzibar (ZAN-suh-bahr). Zanzibar had been independent since 1963. Tanzania is mostly upland plateaus covered by bush and tall-tree savannas. Humid forests are found where elevations rise above 6,000 feet (1,829 m). The greatest landform in Tanzania is Kilimanjaro on the border with Kenya. The mountain's southern slopes are a rich agricultural region much like the Kenya highlands.

The rift valleys also pass through Tanzania. In the Western Rift is deep Lake Tanganyika. Between the Eastern Rift and Lake Victoria is the large Serengeti Plain, where herds of antelope and zebras still roam free. Nearby is the famous archaeological site of Olduvai (old-duh-way) Gorge, where evidence of some of the earliest humanlike settlements has been discovered.

458

Guided Practice

Have the class create a comparison chart of the physical, cultural, political, and economic geography of Kenya, Tanzania, Rwanda, Burundi, Uganda, and Sudan. Draw a blank chart on the board, using the names of the countries discussed in this section as the column heads and these row heads: Land, People, Economy, and Government. Ask students to provide information from Section 3 to complete the chart.

Independent Practice

Have students create a time line about the history and politics of the countries discussed in Section 3.

Closure

Review how the physical geography and human geography of the countries studied in Section 3 affect the economy and politics of these countries.

Most of the people of the Tanzanian mainland belong to one of many Bantu-speaking groups. Along the coast and on the islands of Zanzibar and Pemba are many Arabs. These traditional Islamic societies maintain a separation from Tanzanian life. Zanzibar has the world's major source of cloves.

Tanzania is a country of subsistence farmers. Infertile soils and limited technical support hinder farm productivity. Tanzania has many minerals, but the deposits are neither large nor rich. Most of Tanzania's factories make goods such as foods and cement for the local market. Dar es Salaam (dahr ehs suh-LAHM) is the country's major city and chief port. A modern railroad connects the city with the copper fields in the neighboring country of Zambia (ZAM-bee-uh). Tanzania is in the process of moving its capital from Dar es Salaam to Dodoma (DOHD-uh-mah), a new town inland.

After independence, Tanzania launched an ambitious program of village modernization to make the country more self-sufficient. The government pressured farmers to work in cooperative villages in which farmers would share the farmlands,

Olduvai Gorge, in northern **Tanzania**, is a canyon about 30 miles (48 km) long and 295 feet (90 m) deep. Many human fossils, some of which are more than 1 million years old, have been found there.

field work, processing of products, and profits. For numerous reasons, this plan failed. Some farmers complained of unequal work loads or resented losing their private land. Some people began to rely on government assistance. As a result, production dropped and food imports steadily increased. Most of the cooperative villages have closed, and the lands have returned to private control.

Rwanda and Burundi In the pleasant highlands east of the Western Rift are the small land-locked countries of Rwanda (ru-AHN-duh) and Burundi (bu-ROON-dee). (See the map on page 453.) Once a German colony, these territories were administered by the Belgians after World War I. At independence in 1962 they were divided into two countries. Rwanda and Burundi have the densest rural settlement in Africa. Foreign aid has led to improvements in farming technology and better public health.

Coffee and tea, grown on small farms on the fertile mountain slopes, are the main exports. Rwanda also has some tin mines, and Burundi has nickel deposits. Rwanda is popular with tourists, who come to see the mountain gorillas.

Politics in these two countries are complex. Most of the people are Hutu, a Bantu-speaking group. When the Germans arrived, the Hutu were controlled by the Tutsi. The Tutsi are related to the Masai and the peoples along the Nile. The Hutu are farmers; the Tutsi are cattle keepers. The Hutu staged an uprising

FACTS IN BRIEF East Africa

COUNTRY POPULATION (1994)	LIFE EXPECTANCY (1994)	LITERACY RATE (1990)	PER CAPITA GDP (1993)
BURUNDI 6,124,747	38, male 42, female	50%	$700
DJIBOUTI 412,599	47, male 51, female	48%	$1,200
ERITREA 3,782,543	46 (all) (1993)	20% (1993)	$500
ETHIOPIA 54,927,108	51, male 54, female	24% (1991)	$400
KENYA 28,240,658	51, male 55, female	69%	$1,200
RWANDA 8,373,963	39, male 41, female	50%	$800
SOMALIA 6,666,873	54, male 55, female	24%	$500
SUDAN 29,419,798	53, male 55, female	27%	$750
TANZANIA 27,985,660	42, male 45, female	94% (1991)	$600
UGANDA 19,121,934	37, male 38, female	48%	$1,200
UNITED STATES 260,713,585	73, male 79, female	97% (1991)	$24,700

Sources: *The World Factbook 1994*, Central Intelligence Agency; *The World Almanac and Book of Facts 1995*

Large-scale ethnic conflict makes statistics for Rwanda unreliable.

Extension

Kilimanjaro is an extinct volcano, but many volcanoes in East Africa are active. Have interested students research volcanic activity in the region and write a script about volcanoes for a television program on the Great Rift Valley. Have students present their programs to the class.

SECTION 4
OBJECTIVES

The objectives reflect the Focus questions on page 461 and are assessed in the review on page 462.

1. Identify characteristics of the physical and human environments of the Horn of Africa.

2. Describe the political and economic geography of the Horn of Africa.

Hydroelectric power provided by Owen Falls Dam has enabled **Jinja** to become **Uganda's** main industrial center. The town is located where the Nile River flows out of Lake Victoria.

in 1959, and after independence in 1962, violence between the Tutsi and Hutu in both countries claimed more than a half million lives. In 1994, hundreds of thousands of Rwandans, mostly Tutsi, were killed in renewed ethnic violence. Millions of Rwandans fled to refugee camps outside the country. Ethnic tensions and periodic violence have continued in both Rwanda and Burundi.

Uganda The railroad from Mombasa on the Indian Ocean goes beyond Nairobi to Kampala (kahm-PAHL-uh), the capital of landlocked Uganda. (See the map on page 453.) Most of Uganda is a high plateau, and the country's highlands are some of Africa's most beautiful farmlands. Inland swamps and lakes cover about 20 percent of the country. The lands along the northern shore of Lake Victoria were part of the ancient kingdom of Buganda. Dams across the upper Nile provide hydroelectricity.

Unfortunately, the promise of prosperity for Uganda at independence in 1962 was destroyed. The country's decline began in 1971 when Idi Amin seized power after a military takeover. People who disagreed with Amin often were murdered. Successful farmers as well as British and Asian businesspeople fled. Foreign investment stopped, and the country's economy collapsed. Though Amin was driven out in 1979 by an invasion of Ugandan and Tanzanian forces, civil unrest continued until 1988 when peace was restored.

Rebuilding Uganda has not been easy, but the return of foreign aid, investment, and peace offers Ugandans new hope. The country relies heavily on

its agricultural potential. Adding to Uganda's woes is the country's battle against AIDS, which has hit Uganda particularly hard. International agencies continue to offer support.

Sudan Sudan is Africa's largest country in area. (See the map on page 453.) It also is one of Africa's least developed countries. The word *sudan* refers to the dry savannas that extend across the middle of the country. Northern Sudan is in the dryland zone. Here, summer dust storms make life miserable. In the south, Sudan has moist savannas. The Nile runs the length of the country, providing a crucial water resource.

For the most part, modern agricultural development in Sudan has been the result of irrigation projects. The largest of these projects is the Gezira (juh-ZIR-uh) Scheme near the capital city of Khartoum. El Gezira is the area between the White Nile and the Blue Nile near their junction. Former grazing land, this area has been transformed into a major producer of cotton and food crops. Although Sudan is mainly an agricultural country, it has developed some of its mineral resources. Oil has been discovered in the south but is undeveloped.

Like other countries along the southern edge of the Sahara, Sudan includes strong influences from Arab as well as black African cultures. The Nubians and Arabs of the north total about 50 percent of the people. They are Islamic herders who have a long history of contact with nearby Southwest Asia. The people in southern Sudan are primarily Christians

---

A worker surveys part of an irrigation system in a sugarcane field in **Sudan**. Most of Sudan's commercial crops are grown in the irrigated northern and eastern areas of the country.

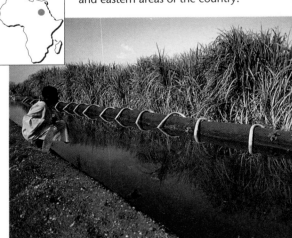

Motivator

Ask students what they know about the famines in Ethiopia and Somalia. Discuss whether they think the famines are caused by environmental conditions, by humans, or by a combination of both. Point out that Section 4 discusses the political and economic conditions in the Horn of Africa.

Analyzing
Physical and Economic Geography

Have students work in small groups to create a poster showing the ways of making a living in each country discussed in this section. Have them analyze how the environment of the countries helps determine these occupations. Ask groups to share their posters and analyses with the class.

Guided Practice

Work with students to develop two web diagrams comparing the political unrest in Ethiopia and Somalia. The hub of one diagram should read "Civil War in Ethiopia"; the other should read "Civil War in Somalia." Have students complete the diagrams by providing the causes of the civil wars and their effects on the countries.

---

who were influenced by European missionaries. Farming is their agricultural tradition.

Uniting these peoples is the most difficult political problem for Sudan. Since independence in 1956, Sudan has experienced many periods of unrest. The government in Khartoum has tried to force Islam, as well as Arabic ways of life, on the southern region. The southern people have resisted and have been in rebellion against the central government since 1980. Development has stopped and will not continue until this conflict is settled. Meanwhile, life for everyone in Sudan deteriorates as desertification increases and civil war continues.

3 SECTION REVIEW

1. What physical features in Kenya and Tanzania might attract tourists?
2. Compare and contrast economic and political issues in Uganda and Sudan.
3. **Critical Thinking   Categorize** the causes of conflicts experienced over the past 50 years by the countries discussed in this section.

4 THE HORN OF AFRICA

FOCUS

- *What is the physical environment of the Horn of Africa?*

- *What is the economic and political situation in the Horn of Africa?*

Ethiopia   Dry lowlands of savannas and deserts surround Ethiopia's humid mountain region. (See the map on page 453.) The densely populated highlands are the core of Ethiopian life. Yet while the rich soils of the mountain slopes and upland plateaus offer agricultural promise, there has been little progress. Life for the herders in the drylands below has been harsh. In fact, with more than 50 million people, Ethiopia is one of the world's poorest countries. The capital city of Addis Ababa (ad-uh-SAB-uh-buh) holds more than 1 million people.

Typical of mountain regions, Ethiopia's climates and agricultural products vary with elevation. The hot, semiarid lowlands reach up to about 5,000 feet (1,524 m). With irrigation, bananas, dates,

and other fruits can be grown. Cattle and goats are grazed on the dry savannas. From about 5,000 feet (1,524 m) up to about 9,000 feet (2,744 m), the land was once forested. Wild coffee, grains, oranges, and other Mediterranean-type crops now grow here. Farther up, wheat and barley are grown.

In the last 30 years, droughts have hit Ethiopia hard. Despite aid from around the world, several million people starved during the 1980s. The country's mountainous terrain makes communication difficult, and the few roads are often impassable. In most of the lowlands, there are no roads.

Except for the period between 1935 and 1942, the Ethiopian highlands have never been under foreign rule. The mountainous slopes have protected the interior from foreign invaders. Missionaries brought a form of Christianity to the highlands between the fourth and sixth centuries, but the Ethiopian Orthodox Church survived several attempted invasions by Islamic armies. Today, the highland people practice Christianity, while most of the lowland people are Muslims.

Cooperation among Ethiopia's many ethnic groups is limited. While most of the people are Gallas and Amharas, there are other groups, including

---

Christians visit a religious shrine in **Ethiopia**. More than one-third of Ethiopians are Christian.

Independent Practice

Provide students with Outline Map 19. Have them create a map showing the ethnic makeup of the countries in the Horn of Africa. Students should include a map key of symbols representing the different ethnic groups. Display completed maps in the classroom.

Closure

Compare the physical, political, and economic characteristics of the Horn of Africa countries with the countries discussed in Section 3.

ASSESSMENT

Assign the Section 4 Review questions on this page.

Reteaching

Assign Section 4 of Reteaching Worksheet 38.

MULTICULTURAL PERSPECTIVES

Ethiopia is one of the few places in the world where the lyre, an ancient stringed instrument, is still played. One type is used by Ethiopian Christians to accompany sacred songs; other types provide music for everyday life. Ask students to name musical instruments associated with their own or another culture.

Section 4 Review
ANSWERS

1. Both have desert and savanna regions, but Ethiopia also has a humid mountain region.
2. Drought, civil wars, and mountainous terrain have hindered communication and transportation.
3. Eritrea recently gained independence from Ethiopia, whereas Djibouti has maintained independence but is wanted by Ethiopia and Somalia. Eritrea has suffered from drought, famine, and war and is relatively poor; Djibouti's location has made the country relatively prosperous. Eritrea is more culturally uniform than is Djibouti.

462

---

some Somalis. Nearly half of the people speak Amharic, once the official language. English is taught as a second language. Tigre (TEE-gray) is spoken in the northern, Islamic region of Ethiopia.

Since the days of the powerful ancient kingdom of Axum, Ethiopia's history often was one of quarreling small states. The majority of people were landless peasants. A ruling class, along with the royal family and the church, held most of the economic and political power. They controlled most of the good farmland, industry, and the government.

This system collapsed in 1974 when the combination of drought and ethnic struggles led to a major change of government. A Communist government took over, seizing most lands, industries, and businesses. Many rural people were forced to move to new villages where the government could control them. Eventually, a multisided civil war spread across the country. In 1991, the Communist government was overthrown. Ethiopians hope a general election held in 1995 will aid in the rebuilding of their country.

Despite its political tensions, Ethiopia is the home of the Organization of African Unity (OAU). The members of the OAU believe that African states can solve their problems through cooperative efforts. Some of these problems involve disputes among member nations, threats to independence, lack of foreign investment, and economic and social issues associated with modernization.

Eritrea Eritrea (ehr-uh-TREE-uh), on the Red Sea coast north of Ethiopia, was once an Italian colony. In the 1960s, the Eritreans began a long fight to separate from Ethiopia. During the struggle, Eritrea's infrastructure was destroyed, and drought and famine took their toll on the province. Eritreans gained control of their country in 1991, but full independence was not granted until 1993. Agricultural output has increased, and roads, bridges, hotels, and factories are being repaired. There may be oil and natural gas deposits off the coast. Although the country is still poor, Eritreans are enjoying peace for the first time in many years.

Somalia Somalia is a land of deserts and dry savannas. Only along the coastal edge and in the far southern part of the country, near the capital of Mogadishu (mahg-uh-DISH-ooh), is there enough rainfall for farming. Most Somalis are nomadic herders. Livestock and hides are the main exports. The trees of the limited coastal woodlands pro-

duce frankincense and myrrh, which are marketed to Christian churches around the world.

Somalia differs from most other African countries in that it contains mostly one group of people, the Somali. Most Somali have the same culture, religion (Islam), language (Somali), and way of life (herding). Many geographers falsely believed that this cultural uniformity would save Somalia from ethnic conflict. But during the early 1990s, Somalia was devastated by civil war.

Longstanding disputes between various clans over water rights and political authority led to fighting. The violence came during a period of severe drought. Livestock died and crops failed. Nearly a whole generation of children died, with their suffering documented on television and in newspapers around the world. The United Nations sent troops to Somalia to help stabilize the country and to provide aid. Somalis have worked to stop fighting that resumed after UN troops left the country in 1995.

Djibouti Although Djibouti is a small desert country, it is strategically located on the Bab al-Mandab (bab uhl-MAN-duhb), the strait that connects the Red Sea and the Indian Ocean. All the ships that travel between the Mediterranean Sea and the Indian Ocean must pass nearby. The city of Djibouti was built primarily as a French colonial port. Though independence came in 1977, the French government still contributes economic support to the country. The city of Djibouti also is the main port for the railway that runs into Addis Ababa. This railway carries about half of Ethiopia's trade. Foreign aid and trade have made Djibouti more prosperous than its neighboring countries.

The people of Djibouti are almost equally divided between the Issa, who are closely tied to the people of Somalia, and the Afars, who are linked to leading families in Ethiopia. Somalia and Ethiopia have vied for control of Djibouti, but so far the country has maintained self-government.

4 SECTION REVIEW

1. How does the physical environment of Somalia differ from that of Ethiopia?
2. Why has economic development in Ethiopia been slow despite its rich agricultural potential?
3. **Critical Thinking** **Compare and contrast** Eritrea and Djibouti.

CHAPTER 38 CHECK
ANSWERS
Building a Vocabulary
1. Swahili
2. A product of the acacia tree; in the production of medicine and candies
3. No; Kenya

Recalling and Reviewing
1. Possible answers: valleys, snowcapped mountains, lakes, plains, scenic coastlines
2. Nilo-Saharan speakers, Cushite speakers, and Niger-Congo (Bantu) language family speakers
3. Kenya promotes tourism, population control, and small-farm agriculture. The government in Tanzania pressured farmers to work together in cooperative villages. The plan failed, and much of the land was returned to private control.
4. It has made outside invasion difficult, but it has hindered communication and transportation.

 CHAPTER 38 CHECK

Tourists take in the sights of a **Kenya** wildlife preserve.

Reviewing the Main Ideas
1. Magnificent landscapes and climate contrasts characterize East Africa.
2. East Africa is a region of great ethnic diversity. Most of the ethnic groups can be divided into three main language categories.
3. Most East Africans are traditional farmers, though industrialization has begun in the region.
4. In order for the pace of economic development to increase in East Africa, ethnic groups and nations must cooperate.

Building a Vocabulary
1. What is the official language of Kenya and Tanzania?
2. What is gum arabic? For what is it used?
3. Is sisal a food crop? Where is it grown?

Recalling and Reviewing
1. What physical features make East Africa the most scenic region in Africa?
2. Into what three language groups can East Africans be divided?
3. What techniques for economic development have been used in Kenya and Tanzania?
4. How has Ethiopia's mountainous terrain both helped and hindered the country?

Thinking Critically
1. What evidence is there of cultural diffusion in East Africa? Has cultural diffusion been helpful or harmful? Explain.
2. What effects might civil war have on a country's infrastructure and economy?

Learning About Your Local Geography

Individual Project
Create a travel brochure that highlights the physical features of your local environment that would attract tourists. With your classmates, create a bulletin board display using the brochures. You may wish to include your brochure in your individual portfolio.

Using the Five Themes of Geography
Identify three different patterns or types of **human-environment interaction** mentioned in the chapter. Make an illustration of each. Write a caption for each illustration, explaining how the illustration shows the relationship between human-environment interaction and the **movement** of people and goods.

Farmers harvest wheat in **Ethiopia**.

Thinking Critically
1. Possible answers include Swahili, the spread of Islam, and the effects of colonialism. Answers to the second part will vary but might mention that East Africa often has been a region of unrest.
2. Possible answer: violence causes damage to roads, buildings, livestock, and fields of crops. Much money is needed to help a country rebuild after a civil war.

Learning About Your Local Geography
Brochures should show the physical features in the area that might attract tourists.

Using the Five Themes of Geography
Possible subjects for illustrations include cattle herding, highland farming, farming in drylands, and mining. Students may also try to illustrate different philosophies of land use, such as land for personal gain, land to provide food, the sharing of land, and land as a refuge for wildlife. Captions should explain relationships between the two themes.

Chapter 39 Central Africa
Planning Guide

TEXTBOOK	TEACHING RESOURCES	OTHER AVAILABLE MATERIALS
Section		
1. Physical and Human Geography Pages 465–466	Reteaching Worksheet 39.1 Skills Worksheet 39	*Study Guide for Spanish Speakers* *Lecture Notes,* Chapter 39
2. Zaire Pages 467–471	Reteaching Worksheet 39.2 Challenge/Enrichment Worksheet 39	*The World Today* Videodisc Program *Study Guide for Spanish Speakers* *Lecture Notes,* Chapter 39
3. Northern Central Africa Pages 471–472	Reteaching Worksheet 39.3 Critical Thinking Worksheet 39	*Study Guide for Spanish Speakers* *Lecture Notes,* Chapter 39
4. Southern Central Africa Pages 473–474	Reteaching Worksheet 39.4 Graphic Organizer 39	*Study Guide for Spanish Speakers* *Lecture Notes,* Chapter 39
Chapter Review Page 475	Chapter 39 Test Chapter 39 Portfolio Assessment See also *Alternative Assessment Forms*	*Test Generator,* Chapter 39

CHAPTER RESOURCES

Lecture Notes Chapter 39
Teaching Resources
    Chapter 39 Graphic Organizer
    Chapter 39 Skills Worksheet
    Chapter 39 Critical Thinking Worksheet
    Chapter 39 Challenge/Enrichment Worksheet

REVIEW AND ASSESSMENT

Chapter 39 Check, page 475
Chapter 39 Test, *Teaching Resources*
Chapter 39 Portfolio Assessment, *Teaching Resources*
Test Generator, Chapter 39
Alternative Assessment Forms

ADDITIONAL RESOURCES

Books for Teachers

Matthiessen, Peter. *African Silences.* New York: Random House, 1991.

Nugent, Roy. *Drums Along the Congo.* Boston: Houghton Mifflin, 1993.

Vansina, Jan. *Paths in the Rainforests.* Madison: University of Wisconsin Press, 1990.

White, Landeg. *Magomero: Portrait of an African Village.* New York: Cambridge University Press, 1990.

Books for Students

Gellhorn, Martha. *Weather in Africa.* New York: Hippocrene Books, 1989.

Jefferson, Margo and Elliott P. Skinner. *Roots of Time: A Portrait of African Life and Culture.* Trenton, NJ: Africa World Press, 1990.

Lamb, David. *The Africans.* New York: Random House, 1987.

The World Today *Videodisc Program*

A variety of still images, short videos, and activities are available for you to use as you teach this chapter. *Lecture Notes* include bar codes for specific images and videos.

Multimedia Resources

Africa: Central and Eastern Regions. (VHS, 15 minutes) Encyclopedia Britannica Educational Corporation, 310 South Michigan Avenue, Chicago, IL 60604–9839.

Afropop Worldwide Listener's Guide. (companion to National Public Radio's series showcasing contemporary music from Africa, the Caribbean, and the Americas) World Music Productions, 131 Park Place, Brooklyn, NY 11217.

Mysterious Elephants of the Congo. (VHS, 60 minutes) PBS Video, Public Broadcasting Service, 1320 Braddock Place, Alexandria, VA 22314–1698.

MEETING INDIVIDUAL NEEDS

Limited English Proficiency (LEP)

Pair LEP students with students who are proficient in English. Have students create graphic organizers that compare and contrast the problems that hinder economic development in Central African countries. Then have students write one or two sentences outlining possible solutions to the economic problems of each country. Have pairs share their work with the class.

Learners Having Difficulty

Divide students into two groups—one group to answer the Focus questions and the other group to answer the Section Review questions. Have students skim the chapter for the answers to the questions, as volunteers in each group act as recorders and write down the answers.

Auditory Learners

Discuss the legacy of European colonialism on countries of Central Africa. Discussion should focus on the political, cultural, and economic changes that occurred as a result of European colonialism in the region.

Visual Learners

Divide students into small groups. Have students create land-use maps of Central Africa based on the information presented in the chapter. Maps should include manufacturing centers, agriculture, mining, and herding.

Tactile/Kinesthetic Learners

Have students do research to make a transportation map of Central Africa that includes major highways, railways, inland waterways, sea routes, seaports, and airports. Then have them generalize about the kinds of transportation networks that are used in Central Africa and in the landlocked countries in particular.

Gifted Learners

Have interested students choose a country in Central Africa and identify a challenge currently facing the country. Then have them conduct a panel discussion in which each member of the panel presents his or her solution to the challenge. At the end of the discussion, other students might vote on the solution they feel best solves the challenge.

MULTICULTURAL FOCUS

Discuss the difference between a nuclear family—parents and children—and an extended family. Point out to students that although Central Africa has hundreds of ethnic groups with various customs and traditions, there is one cultural trait that many of the groups share—family structure based on extended families. African families have a strong bond with their extended families, who often live together in the same household. Extended families provide each other with financial support, employment, medical care, social interaction, and religious education.

Ask students to identify family members who make up an extended family (grandparents, aunts, uncles, cousins). Write the names on the board as volunteers call them out. Then explain to students that many African groups classify extended family members in ways that differ from their own experience. Different terms may be used to designate a sister of one's father and a sister of one's mother. Terms for family members such as cousins are often determined by specific family ties.

Discuss ways that living in the same household or neighborhood with extended family members might strengthen family ties.

CHAPTER 39

CENTRAL AFRICA

Section 1: Physical and Human Geography
Section 2: Zaire
Section 3: Northern Central Africa
Section 4: Southern Central Africa

OVERVIEW

In Chapter 39, students will learn about the countries of Central Africa, and about the various challenges they face.

Section 1 describes the physical geography of the region as a whole, including the ways of life of its peoples.

Section 2 focuses on Zaire, the largest country in Central Africa. It covers Zaire's history, natural resources, and main cities.

Section 3 explores economic activities in the northern part of Central Africa: the Congo, the Central African Republic, Cameroon, São Tomé and Príncipe, Equatorial Guinea, and Gabon.

Section 4 describes the challenges and prospects of the countries of southern Central Africa: Angola, Zambia, and Malawi.

About the Author

Mbonu Ojike wrote *My Africa* in 1945, from which the quotation on this page is taken. Ojike's autobiographical work describes Africa's rich geographic and cultural diversity. He was concerned that Europeans and Americans were not well informed about Africa, and he hoped to provide a realistic portrait of the continent.

Primary Source

Description of Change: excerpted
Rationale: excerpted to focus on main idea

CHAPTER 39

Central Africa

GEOGRAPHY DICTIONARY

periodic market
trust territory
exclave

66**W**esterners always ask me about the jungles, as if there were no other kind of home for Africans. On the contrary, more than half of Africa is covered with steppe, savanna, or grassland; one-third is desert and lakes; while but one-sixth is forest or bush. These forest and bush lands sometimes grow into dense jungles along the equator where heaviest rainfall makes the soil most fertile. 99

Mbonu Ojike

The languages of the Niger-Congo family, often called the Bantu languages, are found throughout Africa south of the Sahara. Over a period of 2,000 years, Bantu speakers spread into Central and Southern Africa from the border of present-day Nigeria and Cameroon. This expansion is often referred to as the Great Bantu Migration.

Evidence indicates that the Bantu speakers were ironworkers as well as farmers and cattle keepers. As they moved into the lands of hunters and gatherers, they either absorbed these peoples, teaching them a Bantu language and culture, or displaced them. Great Bantu speaking states emerged in Central Africa, including the Kongo kingdom, and the Kuba, Luba, Lunda, and Lozi states.

From study of this great expansion, we learn that although the ethnic groups of Central Africa are extremely diverse, common threads tie their histories together. Linguists have found that the Bantu languages, which number around 300, are no more different from each other than are the Germanic languages of northern Europe. Today, Bantu languages are spoken throughout much of the African continent.

Wood-carving and beadwork image honoring an early Bamileke leader in present-day **Cameroon**

Rice fields near the coast of **Angola**

Using the Quotation

Have students look at the physical–political and climate maps of Africa on pages 417–418 and decide whether the writer's description is generally correct. Use the photographs on pages 466, 467, and 474 to show the physical diversity of Central Africa.

Using the Geography Dictionary

Have students locate the definitions of **periodic market, trust territory,** and **exclave** in the Glossary. Have a volunteer read each definition aloud. Write sentences on the board using the terms.

The objectives reflect the Focus questions on this page and are assessed in the review on page 466.

1. Describe the physical geography of Central Africa.

2. Analyze the economic geography of Central Africa.

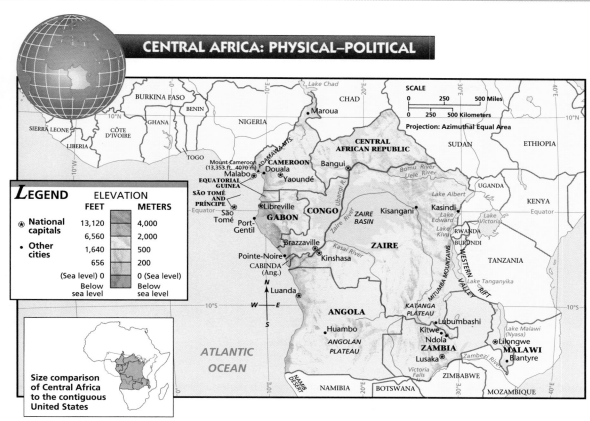

CENTRAL AFRICA: PHYSICAL–POLITICAL

LEGEND

ELEVATION

	FEET	METERS
	13,120	4,000
	6,560	2,000
	1,640	500
	656	200
	(Sea level) 0	0 (Sea level)
	Below sea level	Below sea level

⊛ National capitals
• Other cities

Size comparison of Central Africa to the contiguous United States

Using the Physical–Political Map: Central Africa

Have students examine the map on this page. Ask them to describe the main physical features of the region. Their descriptions should include references to rivers, to elevation, and to the location of cities. Draw students' attention to the Zaire River, which is a vital transportation route of Central Africa.

PHYSICAL AND HUMAN GEOGRAPHY

FOCUS

● *What is the physical geography of Central Africa?*

● *What is the economic geography of Central Africa?*

REGION **Landforms and Climate** The heart of Africa has dense tropical forests, wide plains covered with savanna woodlands, and pleasant forested highlands. Humid-tropical climates stretch from the Atlantic coast inland to the mountains along the Western Rift. Much of Central Africa has a tropical-savanna climate with wet summers and dry winters that favor savanna vegetation. Small desert areas lie along the northern edge of Cameroon near Lake Chad and in southwestern Angola along the edge of the Namib Desert.

Central Africa is mostly plateaus and rolling plains. Along the coast are narrow lowlands that connect to several escarpments inland. The greatest highlands in Central Africa are the Adamawa Mountains in Cameroon and Mitumba Mountains along the western side of the rift-valley lakes. Similar mountains cross Malawi (muh-LAH-wee) and continue south of the Zambezi River. (See the map above.)

Central Africa has two important river systems—the Zaire and the Zambezi. The Zaire River, also called the Congo, leads out of the huge interior Zaire Basin into the Atlantic. From Kinshasa (kin-SHAHS-uh) upstream, riverboats can navigate the Zaire for about 1,085 miles (1,736 km) before reaching the falls near Kisangani (kee-suhn-GAY-nee). The Zambezi River drains Zambia and northern Zimbabwe before heading across Mozambique to its mouth on the Indian Ocean.

Ways of Life Population density is low across most of Central Africa. Although most Central Africans are subsistence farmers, the economy is shifting to commercial agriculture. Cash crops are

Motivator

Ask students what they might expect from a region that has fertile farmland, rich mineral deposits, and a mighty river system. *(a prosperous region and agricultural production sufficient for the population's needs)* Point out that although Central Africa has these attributes, most of its people are subsistence farmers.

Building a Vocabulary

Refer students to the photographs on pages 468–469. Ask if they know of **periodic markets** where they live. *(farmers' or flea markets)*

Describing

Have students write descriptive paragraphs about the physical geography of Central Africa. *(Paragraphs should include references to climate, river systems, and landforms.)*

Guided Practice

Work with students to create a list of the economic problems facing Central African countries. Discuss the list with the class.

Independent Practice

Have students assume the role of a teenager in Central Africa. Have students write a diary entry that explains the reasons someone they know left a rural area for the city and the impact on the community.

MULTICULTURAL PERSPECTIVES

Traditional family networks in Central Africa include not just the nuclear family but an entire clan. The oldest living male member of the family is the head, or chief, of the extended family. Children are often cared for by aunts, uncles, and other relatives. Ask students what benefits such a family system might have.

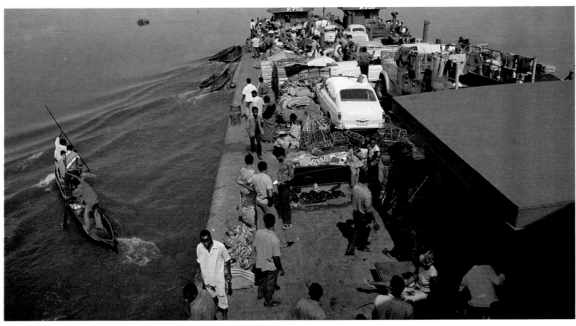

A boat transports passengers and freight along the **Zaire River**.

Section 1 Review
ANSWERS

1. Vegetation: tropical forests, savanna woodlands, forested highlands; climate: humid tropical, tropical savanna, desert; landforms: plains, plateaus, escarpments, highlands, rivers, basins

2. Jobs in farming, mining, manufacturing, tourism, small businesses, government

3. Possible answer: jobs with cash wages may be less available to women, so men migrate to look for such jobs; women are left behind to tend farms and households.

produced on large commercial farms and on small plots. Depending on local customs, farmers either live in small villages or on their farms. Many farmers have fields for cash crops as well as food crops. As in many African countries, women are the primary farmers, and they perform most of the household duties as well.

In rural areas, people trade their products at **periodic markets**. A periodic market is an open-air trading market held regularly at a crossroads or in a town. It also is a social event where people visit with their friends and share news. (See "Themes in Geography" on pages 468–469.)

As in other countries in Africa, many people in Central Africa are migrating away from rural areas to seek jobs with cash wages. These migrants are usually men, and their absence strains African family life. Some find jobs in the many mines, while others look for work in the tourist industry and at game parks. Most of the new jobs are found in factories and businesses in the cities and towns, where the government may offer jobs. Central Africa's capital cities are overflowing with new residents. City governments struggle to provide basic services as slums sprawl ever outward.

The countries of Central Africa import most manufactured goods from developed countries. International aid provides some of the funds with

which to purchase these foreign goods. The export of cash crops and minerals remains the main source of income, however. In some cases, governments have encouraged cash crops to the extent that production of food crops has decreased. At times, world and government-controlled prices are so low that farmers earn little for their food crops. When this happens, the farmers usually raise only enough food for their own communities and have nothing extra to sell to the towns.

Food production often has not kept up with rapidly increasing populations. All Central African countries must now import food despite the region's many fertile farmlands. Investment in agriculture must become a priority in order to improve food production throughout the region.

 SECTION REVIEW

1. Describe the vegetation, climate, and landforms of Central Africa.
2. What various types of jobs are available in Central Africa?
3. **Critical Thinking** **Suggest** possible reasons that farmers in Central Africa often are women while migrants to cities are usually men.

466

Closure

Have students locate on a wall map the Central African geographic features discussed in this section.

ASSESSMENT

Assign the Section 1 Review questions on page 466.

Reteaching

Assign Section 1 of Reteaching Worksheet 39.

SECTION 2
OBJECTIVES

The objectives reflect the Focus questions on this page and are assessed in the review on page 471.

1. Identify economic challenges faced by Zaire.

2. Explain the importance of water, energy sources, and other natural resources in Zaire.

TEACHING SECTION 2
Motivator

Have students locate Zaire on a map of Africa and point out the physical features of the country. Then ask them to name some of the economic activities they might expect the country to have, based on its physical features. (*farming; manufacturing, with hydroelectric power provided by the river*) Point out that Section 2 examines Zaire's economic development.

2 ZAIRE

FOCUS

- *What problems hinder economic development in Zaire?*

- *What resources does Zaire have that can be used to its advantage?*

Zaire is the largest and most populous country in Central Africa. Its physical geography is dominated by the plains along the Zaire River. The river's many navigable stream channels allow transportation by riverboats and canoes. The surrounding uplands are the Angolan and Katanga (kuh-TAHNG-uh) plateaus to the south and the Mitumba Mountains to the east. The Zaire River tumbles down steep falls near the coast before it reaches its mouth at the Atlantic Ocean. Zaire's Atlantic coastline is a short 25 miles (40 km) long. (See the map on page 465.)

Zaire is home to some 250 different ethnic groups, whose staple crops and economic activities vary with the local climate and vegetation. People in the open lands of the savannas primarily are grain farmers and cattle keepers. In the humid forests, coffee and cassava are harvested. Many large and small communities in Zaire are often isolated because of inadequate transportation and communication facilities. The country's poor roads mean that a 60-mile (96-km) journey by truck can take more than a day. Even on a curvy paved road in the United States, a trip of the same length would take less than two hours. Telephone systems are not linked nationwide, and the postal system is inefficient. Throughout Zaire, electricity service is unreliable. In many rural areas, kerosene lamps provide the only light after sunset.

A woman carries her child through fields in the Virunga Mountains of eastern **Zaire**. Although about 75 percent of Zaire's people work in agriculture, most of them are subsistence farmers who use traditional methods.

Historical Geography Zaire once was called the Belgian Congo. During the past century, its history has been one of Africa's most turbulent. The people of Zaire have suffered alternating periods of poor government or almost no government.

The impact of European control was severe. The Berlin Conference of 1884–1885 awarded the Belgian Congo to King Leopold II of Belgium. The king considered the area his personal empire. The peoples of the Congo were forced to labor under threat of injury and often were treated as slaves.

Public outrage forced King Leopold to turn over the territory to the Belgian government in 1908. Under the government, workers received better treatment and had better living conditions. Belgian businesses invested heavily in the territory, and prosperous copper mines opened. Many Belgians came to work in the Congo and soon held nearly all of the jobs in government and business. Many Africans, however, had little contact with the European customs and economic system. Few efforts were made to train Africans for the new jobs.

After World War II, running the Congo began to cost the Belgian government more and more money. Unemployment increased dramatically when copper prices fell in the late 1950s. Adding to the discontent, the people in neighboring French and British colonies were gaining their independence. Riots broke out in the capital of Leopoldville (now called Kinshasa), and unrest increased. The colony finally won its independence from Belgium, and in June 1960, with only six months' notice, the Democratic Republic of the Congo formed.

Text continues on page 470.

THEME
Movement

In parts of Zaire, where there are few roads, the river is the main highway. Riverboats have carried goods and passengers between the capital, Kinshasa, and towns and villages deep in the interior. In 1993, however, Zaire shut down its riverboats. It could no longer afford fuel and spare parts. As a result, thousands of villages lost their main supply line and means of transportation.

Themes in Geography

Cultural Connections

Outdoor flea markets and "trade days" held weekly or monthly are popular in many communities in the United States. Ask students if they have ever visited or participated in events such as these. Invite a volunteer to describe a flea market or trade day, including the types of goods that are usually sold. Ask students to point out similarities between these events and an African periodic market. *(Students might suggest that, like periodic markets, flea markets or trade days allow access to a variety of goods, encourage bargaining for lower prices, and promote social interaction.)*

Photo Caption Answer

Differences: periodic markets are held outdoors and usually begin in early morning; people may bargain over the price of goods. Similarities: both are social centers, and they provide a place to sell needed goods.

468

THEMES IN GEOGRAPHY

Periodic Markets

Periodic markets are a regular part of rural life in much of Africa. They are held one or more days each week in a village, at a crossroad, or in a convenient open field. These markets give people the opportunity to shop, sell their products, and share news about families and friends. They are events that break up the steady pace of village life.

Because markets usually start early in the morning, people often begin their walk to market before sunrise. Most of their marketing is completed before the heat of the afternoon. The markets are well organized by a village leader or a local government official.

Sellers can rent stalls and display tables for a small fee. Some spread their wares on the ground or on makeshift tables. People selling the same products usually set up next to each other so shoppers can compare the quality and prices of their offerings. Buyers bargain over prices, always hoping to make a better deal. Children may have

Periodic markets shape the daily economic, social, and cultural life of many African communities.

jobs wrapping parcels and delivering them to customers.

Traveling merchants also come to the periodic markets. They usually have trucks so they can transport their goods to a different market each day. Merchants have an assortment of manufactured goods to sell. One may offer drugstore items such as cosmetics, soap, detergents, brushes, and music cassettes. Another merchant may sell household goods such as cloth, pails, lanterns, batteries, and blankets.

As populations in Africa have increased in the last decade, some periodic markets have become daily markets with permanent residents. Some have become towns. Kasindi (kuh-SIN-dee), in eastern Zaire, is an example of a small village that has developed as the result of periodic markets. It began as a border outpost on the main road between Zaire and Uganda. This location made it accessible to sellers and to customers.

Kasindi lies in the Western Rift Valley, just south of the Ruwenzori (roo-wuhn-ZOHR-ee)

MOVEMENT Before sunrise, people can be seen walking along the roadside on their way to the **Kasindi** market. Merchants from near and far bring their goods to sell. Buyers walk from table to table, inspecting goods and bargaining over prices. How is a periodic market different from a U.S. supermarket or shopping mall? How is it similar?

Using Geography Themes

Ask students to name the geography theme that best applies to the development of African periodic markets. *(location)* Have students sketch their own neighborhood or community and designate a convenient place for such a market. Volunteers might show their sketches to the rest of the class. Have them explain why they chose a particular location.

mountain range and north of Lake Edward. The area can supply varied products for its periodic markets. The immediate environment is a tall-tree savanna; however, environments vary widely over short distances along the rift valley, permitting many products to be grown.

Kasindi's location on an important transportation route makes it accessible to merchants who bring manufactured products such as gasoline. Traffic across the border also affects Kasindi. In recent decades, neighboring Uganda often has experienced serious unrest. At these times, products from Zaire have helped the Ugandans survive. In turn, many of the manufactured goods needed by Zaire come through Uganda from Nairobi, Kenya, an international trade center.

PLACE **Kasindi** is located near the Semliki River, shown here, and an international border. Officials from **Zaire** and **Uganda** often stop in Kasindi on government business. It is a place where goods from both countries can be traded. What might be the eventual result of this interchange?

LOCATION **Kasindi's** absolute location is 0°5' north latitude, 29°45' east longitude. The market's relative location is on the road between **Zaire** and **Uganda**, north of Lake Edward and south of Mount Margherita. Can you describe the absolute and relative locations of your favorite places to shop?

Map Caption Answer

Students' answers will vary.

Photo Caption Answer

It could become a cultural crossroads where cultural traits might be passed between the countries over time.

Guided Practice

Discuss with students the natural resources found in Zaire. Then ask volunteers to conduct a panel discussion in which they suggest ways that the resources can be used to improve Zaire's economy. Have the class evaluate the suggestions presented in the panel discussion.

Independent Practice

Ask students to work in pairs and use the information in the section to list the causes and effects of Zaire's underdeveloped economy. Ask pairs to share their lists, and discuss them with the rest of the class.

Closure

Remind students of influences on Zaire's economy: colonialism, political strife, and underdevelopment of resources.

ASSESSMENT

Assign the Section 2 Review questions on page 471.

Reteaching

Assign Section 2 of Reteaching Worksheet 39.

Geography and Economics

By most measures of potential, Zaire should have one of the highest standards of living in Africa. Instead, starvation is prevalent throughout the country. Many analysts blame the declining economy on the policies of Zaire's corruption-plagued government. Even within Kinshasa, the capital, where food is abundant in the central market and many people are civil service employees, people have a difficult time feeding their families. Food costs for basic items have risen so dramatically that the price of a sack of corn flour can be twice the monthly salary of the average government worker.

Geography and Language

Shaba, *as in Shaba province, is the Swahili word for copper.*

Chart Caption Answer

Gabon has a rich supply of natural resources.

470

Although the Belgians left behind school buildings and many hospitals, they did not leave behind education or medical systems. The Belgians had trained few African educators or medical personnel. When the Belgians fled the country after independence, Zaire was left with few teachers or doctors. Nor did the Belgians train Africans in government jobs, and most Africans lacked the experience to manage the country effectively.

After independence, several political parties battled to control the central government. The idea of a united Congo was not widely supported. People often thought first of their own ethnic groups and region. Lawlessness and rebellion were widespread. To make matters worse, the mineral-rich province of Shaba (then called Katanga) tried to withdraw from the new nation. Civil war resulted as several armies stormed across

FACTS IN BRIEF Central Africa

COUNTRY POPULATION (1994)	LIFE EXPECTANCY (1994)	LITERACY RATE (1990)	PER CAPITA GDP (1993)
ANGOLA 9,803,576	44, male 48, female	42%	$600
CAMEROON 13,132,191	55, male 59, female	55%	$1,500
CENTRAL AFRICAN REPUBLIC 3,142,182	41, male 44, female	27%	$800
CONGO 2,446,902	46, male 49, female	57%	$2,900
EQUATORIAL GUINEA 409,550	50, male 54, female	50%	$700
GABON 1,139,006	52, male 58, female	61%	$4,800
MALAWI 9,732,409	39, male 41, female	25% (1989)	$600
SÃO TOMÉ AND PRÍNCIPE 136,780	61, male 65, female	50% (1988)	$450 (1990)
ZAIRE 42,684,091	46, male 49, female	72%	$500
ZAMBIA 9,188,190	44, male 45, female	73%	$800
UNITED STATES 260,713,585	73, male 79, female	97% (1991)	$24,700

Sources: *The World Factbook 1994*, Central Intelligence Agency; *The World Almanac and Book of Facts 1995*

From what you have read in the chapter, why does Gabon have the highest per capita GDP?

the country. Economic conditions worsened, and famine threatened.

After several changes of government, the commander of the central army, Mobutu Sese Seko, took control. Many industries owned by Europeans were given to Mobutu's friends and relatives. Such corruption hurt the economy. The Mobutu government also changed the country's name to Zaire in 1971, and in 1975, it changed the names of cities and provinces in an effort to remove signs of the colonial past. The capital Leopoldville, for example, was changed to Kinshasa. Mobutu succeeded in keeping the country together into the 1990s, but the country's economic troubles have continued.

Resources If successfully developed, Zaire's rich resources could bring about a better economic life for the people. There are vast areas that could be developed as game parks for tourists. Commercial crops also can be grown successfully.

Zaire's mineral resources are even more substantial than its agricultural ones. The copper deposits in the Shaba province in the southeast are among the largest in the world. Deposits of cobalt, iron ore, manganese, gold, and tin are nearby. Combined with adjacent areas in Zambia, the copper belt of Shaba is the second largest mining region in Africa. Zaire also has enormous reserves of industrial diamonds. The rivers of Zaire have the greatest hydroelectric potential in Africa, and oil has been found off Zaire's narrow Atlantic coast.

Life in Zaire Zaire has three major cities. The first is the capital, Kinshasa, located on the Zaire River near the Atlantic coast. The river docks in Kinshasa receive goods from throughout the Zaire Basin. Kinshasa is a congested city where traditional markets and modern stores exist side by side. Colonial buildings, high-rise apartments, and shantytowns often share the same neighborhood. Kinshasa is the center of political power in Zaire.

Another important city is Lubumbashi (loo-boom-BAHSH-ee), the industrial center of the Shaba copper belt. The railroad from Lubumbashi carries copper to ports in South Africa. Many manufactured goods arrive on return trips. A third important city is Kisangani, located on the great falls of the Zaire River. Industries use the electricity from the power plant there to process products from surrounding forests and mines.

SECTION 3
OBJECTIVES

The objectives reflect the Focus
questions on this page and are
assessed in the review on page 472.
1. Describe economic activities in
the countries in the northern part
of Central Africa.
2. Compare and contrast the
economies of these countries.

TEACHING SECTION 3
Motivator

Call on students to provide examples
of ways a government can affect a
nation's economy. *(controlling debt,
providing loans and incentives for agricultural and industrial production, promoting education)* Point out that weak
governments contribute to the economic problems of some countries in
northern Central Africa.

Classifying Economies

Have students assume the role of
investors. Ask them to write a report
classifying the countries in the northern part of Central Africa as successful or unsuccessful in their attempts
to develop. The report should justify
their classifications and should recommend the country that would provide the best return for their
investment. Ask volunteers to read
their reports, and discuss them.

Zaire's only experience with public service
was that imposed by colonial powers. Today, people take jobs in the government to gain personal
income. Indeed, most government ministers
become wealthy. Meanwhile, teachers and government workers in distant areas sometimes wait
months for their paychecks. Skilled workers often
keep a small farm to ensure that food is available
for their families.

There were some 80,000 miles (128,720 km)
of good roads at the time of independence. Today,
paved roads total less than 12,000 miles (19,308
km) and are in poor repair. The ruins of colonial
factories and farms stand silent as much land
returns to subsistence agriculture.

 SECTION REVIEW

1. How has Zaire's history affected its economic development?
2. What economic potential does Zaire have
in its natural resources?
3. **Critical Thinking** **Construct** a time
line showing political and economic
changes in Zaire.

 NORTHERN CENTRAL AFRICA

FOCUS

- *What are the major economic activities in
the northern part of Central Africa?*
- *What are the similarities and differences
among the economies of the countries in
northern Central Africa?*

Congo Situated in the forested lands west of
northern Zaire is the People's Republic of the
Congo, usually called "the Congo." (See the map
on page 465.) Brazzaville (BRAZ-uh-vill), its capital, is located across the Zaire River from
Kinshasa. In the Congo, the Zaire River is referred
to as the Congo River.

As the capital of French Equatorial Africa
during the colonial period, Brazzaville grew into a
cosmopolitan city. After independence in 1960,
the city became less important, and the Congo
has lost its leadership role in the region.
Economic development has been uneven. Foreign

aid has supported agricultural projects, though
with little success.

Two-thirds of the people live in the southern
part of the country between Brazzaville and the
coast. Here, most villages are within a day's walk
of the Congo-Ocean Railway, which connects the
port city of Pointe-Noire (pwant-nuh-WAHR) with
Brazzaville. This transportation route is essential
to the economic development of the Congo.
Farmers can get their crops to market with less
spoilage, and imports and exports move easily.
The transportation route also allows the government to provide better schools, farmer education,
and health services.

Away from the Congo-Ocean Railway, the
Congo seems desolate. Beautiful hardwoods were
exported from the country's forests for many
years. Once-forested areas are now bare. On the
coastal plain, several small oil fields have been
developed, and an oil refinery at Pointe-Noire
serves the region. Improvements in transportation
services along the Congo-Ubangi (oo-BANG-ee)
river system would benefit the interior of the
Congo and the Central African Republic to the
north.

Central African Republic North of the
Congo and Zaire is the landlocked Central African
Republic. (See the map on page 465.) The northern half of the country is savanna, while dense
forest covers the southern half. More than 80 percent of the people are subsistence farmers with
little government contact, although only about 2
percent of the land is farmed. During the colonial
period, the country was exploited by foreign companies. These companies had complete authority
over the region. Programs of forced labor did not
end until after 1950, and the local Africans came
to view European business ventures with distrust.

The country's traditional exports are coffee
and cotton. Industrial diamonds are the only commercial minerals. Most overseas exports travel by
river to Brazzaville and then by rail to the coast.
The economy, however, is barely functioning
because of poor governments, and the country is
dependent on foreign aid.

Cameroon Much like neighboring Nigeria,
Cameroon stretches all the way from the humid-
tropical forests along the Atlantic coast into the
dry savanna regions. (See the map on page 465.)
The country's farmers grow a wide range of forest
and savanna products. Mount Cameroon, a

471

472

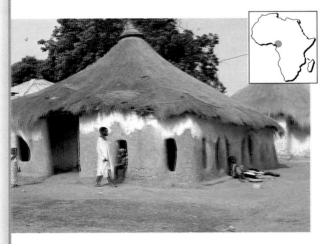

Traditional forms of architecture are common outside the modern cities of Africa. In rural areas and villages of northern **Cameroon,** many homes, such as these, are made from clay.

volcanic peak, rises 13,353 feet (4,070 m) and can be seen from the Atlantic coast.

Cameroon was a German colony that was divided between France and Great Britain after World War I. It later became part of a trust territory. A **trust territory** is a region that is placed under the control of another nation until it can govern itself. In 1960, the larger, French part of the trust territory received its independence and became Cameroon. The British part was divided between Nigeria and Cameroon. Both French and English are used by the government, and schoolchildren study both languages.

Cameroon has dozens of ethnic groups, yet the country enjoys relatively stable political leadership. Although dictatorships often have governed the country, the economy has prospered. There are many skilled business managers and prosperous towns. Coastal oil deposits have been developed and now provide almost half of the country's foreign income. Even so, the government has always maintained that agricultural development is most important to the country's welfare. Most of the people have small farms and receive government loans and up-to-date information on farming technology. Food crops are given such importance that Cameroon is nearly self-sufficient in food. Manufacturing and mining also are important to economic development.

The Island Countries The Central African region includes two small island countries near the Atlantic coast. One of these, São Tomé (sow tuh-MAY) and Príncipe (PREEN-sip-uh), is a tiny country made up of two volcanic islands. Although the country has dense forests and

fertile land, the economy is struggling. The country completed its move toward democracy in the early 1990s.

Equatorial Guinea is a former Spanish colony made up of a mainland province and a group of offshore islands. Equatorial Guinea is underdeveloped, with low health standards. A dictator led the country to financial disaster during the 1970s. He had many people killed as others fled the country or into the forests. Today, the people live as subsistence farmers and hope for a better future. Spain and France provide some aid and technical support.

INTERACTION **Gabon** Rich natural resources have made Gabon (guh-BONE) the most prosperous country in Central Africa, although the country's wealth is not evenly distributed. Three-fourths of the people of Gabon are subsistence farmers.

Gabon is nearly covered with dense rain forests, and few people have settled in these areas. Only 1 percent of the country's land is farmed. Major cash crops include cacao and coffee. In the past, Gabon's main resource was hardwoods that were exported to Europe and used to manufacture furniture. Deforestation has taken its toll, however, and today, oil provides most of the government income. Gabon also has active gold, diamond, manganese, and iron mines. New roads and railroads are being built to carry these resources to the coast. The wealth in the capital of Libreville (LEE-bruh-vill) and the city of Port-Gentil (pawr-zhahn-TEE) continues to grow. The government provides educational opportunities and public-health facilities.

3 SECTION REVIEW

1. How is the Congo-Ocean Railway important to the Congo?
2. What are the reasons for Cameroon's relative economic prosperity?
3. **Critical Thinking** **Compare and contrast** the economic geography of the Central African Republic with that of Gabon.

SECTION 4
OBJECTIVES

The objectives reflect the Focus questions on this page and are assessed in the review on page 474.

1. Identify the challenges confronting Angola, Zambia, and Malawi.

2. Evaluate economic prospects for Angola, Zambia, and Malawi.

TEACHING SECTION 4
Motivator

Tell students that the photographs in Section 4 provide clues about the economies of Angola, Zambia, and Malawi. Then use a wall map to contrast Angola's coastal location with Zambia and Malawi's landlocked positions. Point out that students will learn how location affects these countries' ability to engage in international trade.

Building a Vocabulary

Write the word **exclave** on the board. Explain that the prefix *ex-* means "out." Ask a volunteer to read aloud the definition of *exclave* from the Glossary. Have students draw a picture that illustrates the meaning. Then have students locate the Angolan exclave of Cabinda on the physical-political map of Central Africa on page 465.

SOUTHERN CENTRAL AFRICA

FOCUS

- *What are some of the challenges facing Angola, Zambia, and Malawi?*
- *What are the economic prospects for Angola, Zambia, and Malawi?*

LOCATION **Angola** Angola lies along Africa's west coast between Zaire and Namibia. (See the map on page 465.) Luanda (looh-ANN-duh), Angola's capital, is a harbor city founded by Portuguese traders in 1575. From the sea, Luanda's high-rise buildings and industrial facilities are impressive. But a closer look reveals decades of strife and neglect. From the 1960s through the early 1990s, Angola was a war zone.

During the colonial period, Portugal had little money to invest in the colony. Many poor Portuguese people came to Angola as settlers, and they competed directly with local Africans for jobs and land. Relations between the European settlers and rural Angolans remained tense and sometimes violent. When Angola became independent of Portugal in 1975, the new leaders chose a Communist form of government. Groups from the southern part of the country rebelled against the Communist government, and Angola

Copper mines near **Kitwe, Zambia**, are some of the country's richest. Factories near the mines process the copper.

became a center of the rivalry between the world powers. South Africa and the United States supported the rebels, and Cuba and the Soviet Union backed Angola's Communist government. Although a peace settlement was reached in 1994, chances for peace were unclear.

Angola has substantial natural resources that could aid its development. One of Africa's largest petroleum reserves lies along the coast north of Luanda and in Cabinda (kuh-BIN-duh), an **exclave** north of the Zaire River mouth. An exclave is a part of a country that is separated from the rest of the country and surrounded by foreign territory. Angola also has rich deposits of minerals and fertile agricultural lands. The problems of poverty, unemployment, and poor health services need to be addressed.

Zambia Zambia lies on a savanna-covered high plateau. (See the map on page 465.) The country's copper and iron have been mined for more than 2,000 years. In fact, Zambia's mineral deposits in the copper belt are even richer than those in Zaire, and the country is a leading copper producer. Zinc, lead, manganese, cobalt, gold, and vanadium are mined as well. Vanadium is a gray mineral that can be combined with iron to make very strong steel. Mines provide needed jobs for Zambians and supply much of the government's income.

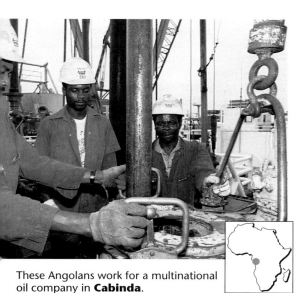

These Angolans work for a multinational oil company in **Cabinda**.

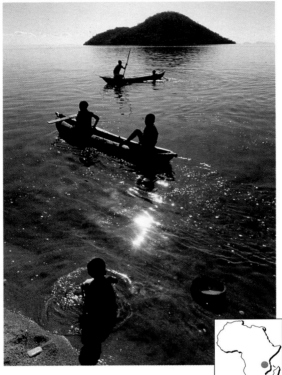

People fish and swim in **Lake Malawi (Lake Nyasa)**. The lake, which drains into the Zambezi River, is about 360 miles (579 km) long but averages only 25 miles (40 km) in width.

Much of Zambia's land is suitable for agriculture and grazing. Farming on large commercial fields as well as on small plots remains important. About half of Zambians live in cities, however, and the country must import food.

Zambia, formerly the British colony of Northern Rhodesia (roe-DEE-zhuh), made the change to independence with ease. Large revenues from copper mines helped the country quickly achieve one of the highest per capita GNPs in Central Africa. In recent years, however, Zambia has experienced economic decline.

As a landlocked country, Zambia's economic stability depends on its ability to move imports and exports. Zambia is most dependent on South Africa but also relies on other neighboring countries. During the civil war in Angola, the railway leading to the Atlantic Ocean often was blocked. Minerals could not be easily exported, and imports became difficult to get. Zambia also depended on Zimbabwe for many manufactured

goods whose production was disrupted by civil war in that country.

Several periods of severe drought and fluctuating copper prices in the world market also have affected life in Zambia. That the country's copper reserves are steadily declining adds to the discontent. Today, however, Zambians are hopeful because democracy has returned to the government. Neighboring countries are at peace and are rebuilding Zambia's vital transportation links to the ocean. A railway and oil pipeline now go to Tanzania as well.

Malawi In contrast to Zambia, Malawi is densely populated. With fewer natural resources, Malawi must base its economic development on its agricultural potential. Most of the country's foreign income comes from tobacco, its most important export. Some new factories have been built to produce items such as textiles and farm tools, but industrialization is taking place slowly.

Malawi's major challenge is its landlocked location. (See the map on page 465.) If important minerals are discovered in the future, it would be difficult and costly to move ores to the coastal ports. For this reason, Malawi's cautious government has worked hard at getting along with its neighbors.

Meanwhile, the government continues to invest in farm improvements. Both commercial farms and small farms are supported with road and railroad construction. The government sells fertilizers and seeds to farmers at low prices.

Malawi's major undeveloped resources include its beautiful forested highlands and mountains, as well as Lake Malawi (Lake Nyasa). The rift valley, lined with grass-covered plateaus, crosses part of the country. The government is expanding the national parks and forest areas. With new roads and hotels, Malawi could profit from tourism.

4 SECTION REVIEW

1. How does being landlocked affect the economic situation of Zambia and of Malawi?
2. What natural resources can be developed in Angola, Zambia, and Malawi?
3. **Critical Thinking** **Discuss** the conflicts experienced by Angola since 1975. Why did many other countries become involved?

CHAPTER 39 CHECK
ANSWERS
Building a Vocabulary
1. An open-air trading market; they allow people to visit friends and share news.
2. Trust territory
3. A part of a country that is separated from the rest of the country and surrounded by foreign territory; Cabinda

Recalling and Reviewing
1. The Zaire and the Zambezi
2. A colonial economic system that benefited few Africans, civil war, corruption in government
3. It has relatively stable political leadership; oil deposits have been developed; the government concentrates on agricultural development. It is almost self-sufficient in food instead of growing only cash crops and having to import most food.
4. Recovering from effects of civil war; addressing poverty, poor health care, and unemployment

---

CHAPTER 39 CHECK

Cameroon rock band

Reviewing the Main Ideas
1. Central Africa is a plains region covered with tropical forests and savannas. Most of the region has a tropical-savanna climate.
2. Central Africa imports many manufactured goods from developed countries. International aid and exports of minerals and cash crops provide income to pay for these goods. Most of the countries in the region must import food.
3. Many people in Central Africa are migrating to the cities in search of work. Cities are overcrowded, and many people live in slums surrounding the cities.
4. Most Central African countries have rich natural resources. Tourist industries can be developed around the beautiful lands and lakes. The development of these resources and the move toward democracy provide Central Africans with hope for the future.

Building a Vocabulary
1. Define *periodic market*. How do periodic markets serve a social as well as an economic function?
2. What term is used to describe a region that is under control of another nation until it can govern itself?
3. What is an exclave? Give an example of an exclave from the chapter.

Recalling and Reviewing
1. What are the two important river systems of Central Africa?
2. What political problems have caused Zaire's economy to suffer?
3. Why is Cameroon's economic situation better than that of most of the other countries in Central Africa?
4. What challenges face Angola?

Thinking Critically
1. Using examples from the chapter, explain how the quality and availability of transportation is related to the economic development of an area.
2. Predict which country in Central Africa will enjoy the most economic success by the year 2000. Give reasons for your prediction.

Learning About Your Local Geography
Cooperative Project
Plan a periodic market. In your group, make a list of items grown or produced locally that could be sold or bartered at the market. Your group should decide the best time and location in your community for the market and provide reasons for your choices. Share your group's list of items for sale or barter and the choice for time and location for the market with the whole class. Work with other groups to make a final decision for the time and place to hold the market.

Using the Five Themes of Geography
Make a list of the landlocked countries discussed in this chapter. Describe the relative **location** of each in terms of neighboring countries. Make a list of common problems faced by these countries due to their relative location. Explain how **movement** is affected by each country's relative location.

Thinking Critically
1. Answers will vary. Students should note that transportation systems are necessary for movement of imports and exports. Students might compare economies of countries with river systems (Congo, Zaire) and coastlines (Angola, Congo) to landlocked countries (Malawi, Zambia).
2. Answers will vary. Likely choices are Gabon, Cameroon, and Zaire. Students should support their answers with evidence from the textbook.

Learning About Your Local Geography
Make sure that groups select items that are locally available.

Using the Five Themes of Geography
Landlocked countries in Central Africa are the Central African Republic, Zambia, and Malawi; each is surrounded by neighboring countries. Problems are the high cost of moving exports and imports and dependence on reliable transportation routes in neighboring countries. All movement into or out of the landlocked countries involves neighboring countries.

475

Chapter 40 Southern Africa
Planning Guide

TEXTBOOK	TEACHING RESOURCES	OTHER AVAILABLE MATERIALS
Section		
1. Physical and Historical Geography Pages 477–479	Reteaching Worksheet 40.1	*Study Guide for Spanish Speakers* *Lecture Notes,* Chapter 40
2. Republic of South Pages 479–481	Reteaching Worksheet 40.2 Critical Thinking Worksheet 40	*The World Today* Videodisc Program *Study Guide for Spanish Speakers* *Lecture Notes,* Chapter 40
3. Other Countries of Southern Africa Pages 481–486	Reteaching Worksheet 40.3 Graphic Organizer 40 Challenge/Enrichment Worksheet 40 Skills Worksheet 40	*Study Guide for Spanish Speakers* *Lecture Notes,* Chapter 40
Chapter Review Page 487	Chapter 40 Test Chapter 40 Portfolio Assessment See also *Alternative Assessment Forms*	*Test Generator,* Chapter 40
Unit Review Pages 488–489	Unit 8 Test Unit 8 Portfolio Assessment See also *Alternative Assessment Forms*	*Test Generator,* Unit 8

CHAPTER RESOURCES

Lecture Notes, Chapter 40
Teaching Resources
    Chapter 40 Graphic Organizer
    Chapter 40 Skills Worksheet
    Chapter 40 Critical Thinking Worksheet
    Chapter 40 Challenge/Enrichment Worksheet

REVIEW AND ASSESSMENT

Chapter 40 Check, page 487
Chapter 40 Test, *Teaching Resources*
Chapter 40 Portfolio Assessment, *Teaching Resources*
Test Generator, Chapter 40
Unit 8 Portfolio Assessment, *Teaching Resources*
Alternative Assessment Forms

ADDITIONAL RESOURCES

Books for Teachers

Akenson, Donald Harmon. *God's Peoples.* Ithaca, N.Y.: Cornell University Press, 1992.

Illustrated History of South Africa. New York: Reader's Digest Association, 1989.

Lanting, Francis. *Okavango: Africa's Last Eden.* New York: Chronicle Books, 1993.

Modisane, Bloke. *Blame Me on History.* New York: Simon & Schuster, 1990.

Morris, Donald. *The Washing of the Spears: The Rise and Fall of the Zulu Nation.* New York: Touchstone Books, 1986.

Books for Students

Laure, Jason. *South Africa: Coming of Age Under Apartheid.* New York: Farrar, Straus & Giroux, 1980.

Mallaby, Sebastian. *After Apartheid: The Future of South Africa.* New York: Times Books, 1992.

Otfinoski, Steven. *Nelson Mandela: The Fight Against Apartheid.* Brookfield, CT: Millbrook Press, 1992.

Tessendorf, K. C. *Along the Road to Soweto: A Racial History of South Africa.* New York: Atheneum, 1989.

The World Today *Videodisc Program*

A variety of still images, short videos, and activities are available for you to use as you teach this chapter. *Lecture Notes* include bar codes for specific images and videos.

Multimedia Resources

Africa: Southern Region. (VHS, 15 minutes) Encyclopedia Britannica Educational Corporation, 310 South Michigan Avenue, Chicago, IL 60604–9839.

Eyewitness: South Africa. (16mm film/VHS, 35 minutes) Aims Media, 9710 DeSoto Avenue, Chatsworth, CA 91311–4409.

The Limpopo Line. (VHS, 40 minutes) Films for the Humanities & Sciences, P.O. Box 2053, Princeton, NJ 08543–2053.

Mandela. (VHS, 52 minutes) Films for the Humanities & Sciences, P.O. Box 2053, Princeton, NJ 08543–2053.

MEETING INDIVIDUAL NEEDS

Limited English Proficiency (LEP)

Pair LEP students with students who are proficient in English. Have the pairs use index cards to write down and define the vocabulary words in this chapter, with sketches or short phrases. Then have them use the index cards as flash cards until they have mastered the meaning of the vocabulary words. Finally, ask students to explain why these words are used in the context of Southern Africa.

Learners Having Difficulty

Help students review the information presented about the countries of Southern Africa by making charts of each country based on the Five Themes of Geography. Create the charts on the board as students volunteer the information. Have students copy the charts and use them to study the chapter.

Auditory Learners

Assign each student a different country in Southern Africa to study. Ask students to give oral reports about each country that includes the following information: location in Southern Africa, physical environment, settlement patterns, political geography, economic situations, challenges, and hopes for the future.

Visual Learners

Have students collect newspaper and magazine articles about race relations in Southern Africa. Then, on an outline map of Southern Africa, have students locate the places discussed in the newspaper and magazine articles. Students might also create a bulletin board display using the articles.

Tactile/Kinesthetic Learners

Divide students into small groups. Have each group create a matching game in which players match details in Chapter 40 with the correct country in Southern Africa. Students should make the components needed to play the game as well as write a set of rules for their game. When all groups have finished, have them exchange and play the games.

Gifted Learners

Have interested students research the African National Congress and make a time line of significant events in the organization's history. Have students analyze the time lines and write critical thinking questions based on the information presented. Display the time lines and questions in the classroom.

MULTICULTURAL FOCUS

Call on students to relate what they know about the racial and ethnic composition of the South African population. Most students will probably mention two broad categories—black Africans and white Africans of European descent. Point out that the cultural influences in South Africa are varied and complex. Write on the board the names of some of the groups students will learn of in Chapter 40 that have shaped the history of the country—San, Khoi, Zulu, Xhosa, Tswana, Venda, Sotho, English, German, Dutch (Boers) and Coloureds (of mixed racial ancestry).

Nelson Mandela has spoken many times of the challenges facing South Africa and his vision of uniting South Africa's diverse peoples who have historically been divided. The principles of the 1955 Freedom Charter of the African National Congress emphasized the goals necessary for transforming South Africa into a democratic nation that is racially tolerant and democratic:

1. The nation of South Africa belongs to *all* of its citizens.
2. The authority of the government derives from the will of its people.
3. The destiny of South Africa as a free nation depends on the ability of people to attain their birthright of land, peace, and liberty, without respect to color, ethnicity, gender, or beliefs.

Elicit a class discussion about how much the country has fulfilled the vision expressed by Mandela and the Freedom Charter.

CHAPTER 40

SOUTHERN AFRICA

Section 1: Physical and Historical Geography
Section 2: Republic of South Africa
Section 3: Other Countries of Southern Africa

OVERVIEW

In Chapter 40, students will learn about Southern Africa.

Section 1 describes Southern Africa's physical environment and its early settlement and colonial history.

Section 2 focuses on the Republic of South Africa. It discusses the history and effects of apartheid as well as the country's agriculture, industry, and population centers.

Section 3 examines the economies and political structures of the countries of Namibia, Botswana, Zimbabwe, Mozambique, and Madagascar.

About the Author

Alan Paton (1903–1988) wrote many books about racial strife in South Africa, including the well known novel *Cry the Beloved Country*. Paton, a white South African, grew up in the eastern region of South Africa formerly known as the province of Natal.

Primary Source

Description of Change: excerpted and bracketed *Rationale:* excerpted to focus on main idea, bracketed to clarify meaning

---

CHAPTER 40

Southern Africa

GEOGRAPHY DICTIONARY

veld
apartheid
embargo
sanction
exile
copra

66 I was born in that country known as the Union of South Africa. The heart of it is a great interior plateau that falls on all sides to the sea. But when one thinks of it and remembers it, one is aware not only of mountains and valleys, not only of the wide rolling stretches of the veld, but of solemn and deep undertones that have nothing to do with any mountain or any valley, but have to do with men [people]. 99

Alan Paton

Southern Africa has been the scene of clashes between Africans and Europeans since the 1600s. The struggle between cultures has dominated the history of South Africa, the most powerful and influential country in this part of the world. Because of South Africa's importance, events there have affected all other nations in Southern Africa.

Beneath the region's surface are some of the world's greatest mineral deposits. The wealth that comes from these rich natural resources has long been controlled by people of European ancestry. Blacks have struggled to share in control of the land here and the wealth it has generated. The struggle is most clear in the cities of South Africa, where wealth and poverty stand side by side.

Domination of the governments and economies of Southern Africa by whites has been breaking down in recent years. The great changes taking place throughout the region are being watched around the world.

Wood sculpture from **Madagascar**

Southern coast of **South Africa**

Using the Quotation

Point out that the quotation on page 476 refers to both physical *and* human geography. After students have read the quotation, ask them what the "solemn and deep undertones" might refer to. (*Students might mention the racial tensions that exist in the Republic of South Africa.*) Use students' prior knowledge about individuals or events in South Africa to introduce the chapter.

Using the Geography Dictionary

Write the terms in the Geography Dictionary on the board, and pronounce them for students. Call on volunteers to identify terms with which they may be familiar, and ask them to explain the context. (*Students are likely to mention* **apartheid, embargo,** *and* **sanction** *because of newsworthy events in recent years or from a history class.*)

SECTION 1
OBJECTIVES

The objectives reflect the Focus questions on this page and are assessed in the review on page 479.

1. Explain the relationship between landforms and climate in Southern Africa.
2. Explain the historical development of Southern Africa.

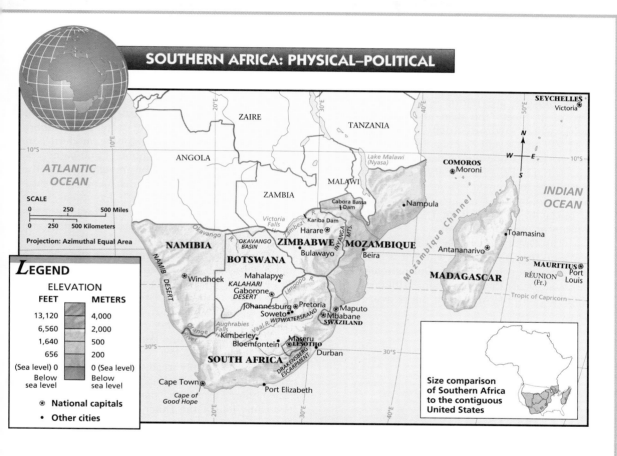

SOUTHERN AFRICA: PHYSICAL–POLITICAL

Using the Physical–Political Map: Southern Africa

Have students locate significant physical features, landforms, or landmarks in Southern Africa discussed in Section 1: Drakensberg Escarpment, Orange River, Zambezi River, Victoria Falls, and Kariba Dam. Point out how the Zambezi forms a natural boundary of northwestern Zimbabwe.

PHYSICAL AND HISTORICAL GEOGRAPHY

FOCUS

● *How does the Drakensberg Escarpment affect Southern Africa's climate?*

● *What various groups of peoples settled in Southern Africa?*

LOCATION **Landforms and Climate** Southern Africa stretches from the subtropical low latitudes south to the edge of the middle latitudes. Climates range from tropical savanna in the north to Mediterranean in the south. Except for a rim of lowlands around the coast, Southern Africa is mostly a region of high plains, with plateaus rising above 4,000 feet (1,219 m).

The Drakensberg Escarpment rises near the coast of the Indian Ocean and reaches up to 11,425

feet (3,482 m). (See the map above.) From the east, the Drakensberg looks like a mountain, but its western side actually forms the high edge of a wide region of plains. Moisture flows into Southern Africa primarily from the Indian Ocean, where it is forced upward by the Drakensberg Escarpment and Inyanga Mountains. Beyond these rainy eastern slopes, climates become drier farther inland.

The term used for South Africa's various natural regions is *veld* (vehlt). The grassy, high plains in central South Africa north and west of Lesotho make up the highveld. The bushveld is a savanna region of short trees and bushes that is found at lower elevations. Still another of South Africa's regions is the lowveld, which includes the dry, tall-tree savannas as well as the moist forests at the base of the Drakensberg.

The major river systems of Southern Africa are the Orange and the Zambezi. Water resource development along these rivers is important to the region. Dams, irrigation canals, and tunnels deliver water and hydroelectric power to nearby farms and distant cities.

TEACHING SECTION 1

Motivator

Write the terms *Southern Africa* and *South Africa* on the board. Point out the difference between the two terms.

Building a Vocabulary

Write the word **veld** on the board. Tell students that it is an Afrikaans word, related to the English word *field*. Have them suggest meanings for *highveld*, *bushveld*, and *lowveld*.

Analyzing

Climate and Landforms   Provide students with Outline Map 19. Have them create a climate map of Southern Africa by using the climate map of Africa on page 418 as a reference. Instruct them to draw and label landforms as well, and to write captions that explain how Southern Africa's landforms affect climate.

Guided Practice

Organize the class into two groups. Have a student in one group name a physical feature of Southern Africa described in Section 1, and have a student in the second group locate the feature on a wall map of Africa. Students will continue this sequence until all of the significant features have been identified and located.

THEME
Movement

The Great Trek began with a handful of Boer families moving northward from the Cape Colony in 1834. Their numbers swelled to between 12,000 and 14,000 by the early 1840s. The migrants walked alongside ox-driven carts, heavily loaded with all their belongings, and settled in the highveld lands in present-day northern and central South Africa.

Geography and History

Shaka (1787–1828), the powerful founder of the Zulu Empire, was a legendary figure even in his own time. An imposing 6'3" warrior, Shaka ruled mainly through the fear he inspired. He engaged the small Zulu clan—one of hundreds in southern Africa—in a series of tribal wars that resulted in the Zulus becoming the dominant group in the region. The innovations he developed in weaponry and in organizing his army contributed to Shaka's battlefield success.

The Zambezi winds southward from Central Africa before turning toward the coast to form the northern border of Zimbabwe. The Zambezi's first great drop is at Victoria Falls. Farther downstream, it meets the dam at Kariba (kuh-REE-buh), where hydroelectricity is generated for both Zimbabwe and Zambia. From there, it flows toward rapids at Cabora Bassa in Mozambique (moe-zihm-BEEK). Here, another dam produces electricity for consumers as far away as South Africa. Finally, the river crosses the coastal plain of Mozambique to end its 1,700-mile (2,735-km) journey at the Indian Ocean.

The Orange River begins its westward journey atop the Drakensberg where it is fed by summer rains and winter snows. Downstream, it joins the Vaal (vahl) River, its major tributary. Before emptying into the Atlantic, it pours over the high

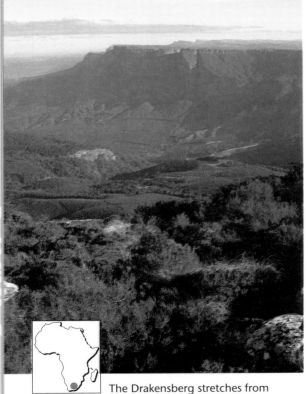

The Drakensberg stretches from northeastern **South Africa** to the southern part of the country.

Aughrabies (aw-GRAHB-eez) Falls near the border with Namibia (nuh-MIB-ee-uh).

MOVEMENT   **Historical Geography**   Archaeologists believe that the early peoples of Southern Africa were closely related to the modern San. Early San peoples lived as hunters and gatherers. Beginning more than 4,000 years ago, most of the San peoples were steadily displaced by Bantu-speaking settlers from the north. Eventually, these Bantu speakers spread to all but the southern rim of Africa. Their farming societies raised cattle and sorghum, and they were expert ironworkers.

The region's first permanent European settlers were from the Dutch East India Company. Drawn to a natural harbor near the Cape of Good Hope, they founded Cape Town in 1652. Here, they encountered the San and the Khoi (Hottentots). Unlike the foraging San, the Khoi were cattle herders and knew about farming. The Dutch business leaders encouraged farming in the region to supply fresh food for ships passing to and from Southeast Asia.

Other Europeans, particularly French and Germans, joined the Dutch settlers at the Cape. Slaves were brought from Southeast Asia as well. Far removed from Europe, the European settlers soon called Africa home and called themselves Afrikaners. Over the years, many of the Europeans developed into frontier farmers, or "Boers." They spoke a mixture of Dutch, words learned from Malayan slaves, and Bantu languages. This language eventually became modern Afrikaans. Today, about 70 percent of white South Africans are descendants of these first European settlers.

As a result of political shifts in Europe, the region of the Cape of Good Hope at the tip of Southern Africa became a British colony in 1814. The British administrators made the English language, English laws, and English education the rule. English traders and missionaries spread across the region. Many Boers were unhappy with the British takeover. In the 1830s, Boers began moving farther into the interior, a migration sometimes called the Great Trek.

Before the Boers began their northward trek, the Zulus and other groups of Bantu speakers were migrating southward. By the early 1800s, the Zulu Empire extended into Southern Africa. The Zulus were led by Shaka, a great warrior who forged his armies into powerful forces. As the

478

Independent Practice

Have students create a time line of the historical geography of Southern Africa. Time lines should include the names of settler groups and dates of significant events related to settlement.

Closure

Call on students to describe the waves of settlement in Southern Africa through the nineteenth century.

ASSESSMENT

Assign the Section 1 Review questions on this page.

Reteaching

Assign Section 1 of Reteaching Worksheet 40.

Extension

Have students research the history of the Zulu Empire in the 1800s and prepare a short report.

SECTION 2
OBJECTIVES

The objectives reflect the Focus questions on this page and are assessed in the review on page 481.

1. Explain the system of apartheid.
2. Discuss how people protested the system of apartheid.
3. Give reasons for South Africa's economic development.

Boers moved northward, they clashed with the Zulus over control of the land. The Zulus resented the loss of their land to people they saw as invaders. Eventually, the Zulu Empire was defeated by the British and the Boers.

The Boers established two independent countries in the region—the Orange Free State and the Transvaal. Meanwhile, the British were establishing the colony of Natal (nuh-TAHL) on the Indian Ocean coast. Many people from India came there to work as laborers in the growing sugar industry.

At first, few Europeans penetrated what they thought was the barren interior of Southern Africa. Then, in 1868, diamonds were found near Kimberley. About 20 years later, the world's largest gold deposits were discovered near present-day Johannesburg (joe-HAHN-uhs-buhrg). The vast mineral wealth brought many British settlers to the Southern African interior. Black Africans also began moving to the cities and mines to look for work. Soon Maputo (muh-POO-toe), the tiny capital of Portuguese Mozambique, became a booming port for shipments from Johannesburg.

Competition for mineral wealth and the Boers' distrust of British expansion led to the Anglo-Boer War (1899–1902). The Boers lost this bitter struggle. In 1910, the Transvaal and the Orange Free State were joined with the Cape and Natal colonies to form the Union of South Africa. The government, however, was nearly all white. Pretoria (prih-TAWR-ee-uh), in the Transvaal, became the administrative capital. Bloemfontein (BLOOM-fahn-tayn), in the Orange Free State, became the site of the judicial capital, where the Supreme Court is located. Cape Town, in the Cape Province, became the legislative capital, where the parliament meets. The four original provinces were divided into nine in 1994.

SECTION REVIEW

1. What is Southern Africa's single greatest landform, and how does it affect the region's climate?
2. Beginning with the San, list the various groups of people who have settled in Southern Africa.
3. Critical Thinking **Explain** why a country might want to have three separate capitals.

REPUBLIC OF SOUTH AFRICA

FOCUS

● *What was the system of apartheid, and what were its effects?*

● *How did people protest apartheid?*

● *Why is South Africa the continent's most economically developed country?*

Political Geography Despite the country's beauty and rich natural resources, political issues dominate the geography of South Africa. After the 1948 elections, South Africa's white-run government formally established the separation of the country's many ethnic groups. This policy of "separate development" became known by its Afrikaans name, **apartheid** (uh-PAHR-tayt). The idea behind apartheid was that each ethnic group would progress best under its own culture and traditions. Ideally, each group would be equal. In practice, however, separate could not be equal.

Apartheid was a series of laws that separated the races where they lived, worked, and went to school. It affected the daily lives of all people. Status, power, and wealth belonged to white South Africans, who controlled the national government and the economy. Coloureds—South Africans of mixed race—occupied a middle status under apartheid. Coloureds included descendants of the San and Khoi as well as people of Asian ancestry. At the bottom of the racial divisions were black South Africans, who make up about 75 percent of the population. South Africa became divided into black and white lands.

The black population was further divided into territories based upon African ethnic groups such as Zulu, Xhosa, Tswana, Venda, Sotho, and others. These territories, called homelands, extended in a broad horseshoe of lands along the east coast and in northern South Africa. Under the homelands policy, blacks were given citizenship in the homeland of their ethnic group, whether they lived there or not. The homelands policy created less-developed regions with limited medical care, educational opportunities, and progress. Many families were separated. Often, men found work in the cities, leaving women and children behind in the impoverished rural areas.

THE WORLD TODAY VIDEODISC PROGRAM

South Africa: Long Troubles, Slow Healing

2B	7652 3:49

Play   Pause

Section 1 Review
ANSWERS

1. The Drakensberg Escarpment; it blocks moist air from moving inland from the Indian Ocean, so the interior of Southern Africa is dry.

2. San, Bantu speakers, Dutch, Khoi, French, Germans, Southeast Asians, British, Indians

3. Possible answer: the country's government might avoid the appearance of favoritism among different provinces or areas of the country; each of South Africa's capital cities is located in a different province.

TEACHING SECTION 2

Motivator

Tell students that, in 1993, Nelson Mandela and South Africa's then-president, F.W. de Klerk, were jointly presented with the Nobel Peace Prize. Engage students in a discussion about what was particularly remarkable about this event, given the political situation in South Africa just 10 years before.

Building a Vocabulary

Help students understand how the Dutch term **apartheid** relates to racial segregation. *Apart* means separate, and *-heid* is similar to the English suffix *-hood,* meaning "a condition or quality." Therefore, apartheid is the condition of being separate. Point out that the policy of apartheid meant not only segregation of the races but also political and economic discrimination.

Evaluating

Opposition to Apartheid On the board, list different forms of opposition to apartheid. Identify the ones that were external to South Africa *(excluding South African athletes from international competition, boycotts against South African products, sanctions)* and internal *(civil disobedience).* Call on students to select the ones that they think were most effective and to explain why.

GLOBAL PERSPECTIVES

Opposition to apartheid took many forms. Some multinational companies doing business in South Africa came under heavy criticism and decided to cease trading with the nation. Some universities, too, came under pressure to divest their portfolios of stocks in companies doing business in South Africa. Have students investigate a particular organization that ceased dealing with South Africa and determine what factors influenced the decision.

Geography and History

Nonviolent civil disobedience was one form of opposition to apartheid. Interestingly, the idea of mass resistance through nonviolence was perfected in South Africa by Mohandas Gandhi, who used it in 1906 to fight discrimination against Asian Indians there.

The national government gave some of the homelands status as independent countries. The government then claimed that many black South Africans living outside their designated homelands were not citizens of South Africa. They could be treated like foreigners without civil rights in cities where they may have been born. The independence of the homelands was never recognized by other countries.

Apartheid made South Africa an outcast among other nations. In 1961, foreign opposition to apartheid forced the Union of South Africa to leave the British Commonwealth. The country became the Republic of South Africa. South African sports teams and products were refused in other countries. Some countries placed **embargoes** on South African products. An embargo is a legal restriction against the movement of freight. Such economic measures are called **sanctions**, which are put in place by other nations in the hopes of forcing a nation to cease an illegal or immoral act. Countries that placed sanctions argued that economic pressure would cause South Africa to find better and quicker solutions to apartheid. While sanctions affected the white community, the economic pressure also increased unemployment among black workers and created hardships for the country's poorest people.

Protests Even before apartheid was established, organizations in South Africa were fighting against racial discrimination. The most well known of these organizations is the African National Congress (ANC), founded in 1912. In the 1950s, the ANC openly protested apartheid by disobeying apartheid laws. The South African government responded quickly and brutally. The ANC leaders, including the well-known black lawyer Nelson Mandela, believed that, in self-defense, blacks would have to confront violence with violence. In response, the government banned the ANC, charged its leaders with treason, and gave Mandela and other ANC leaders sentences of life in jail. Other black South Africans fled to nearby countries, becoming **exiles**, people who have been forced to leave their country.

Ending Apartheid Apartheid did not work for many reasons. Not only did it discriminate against the majority of the people, it failed economically. No country can prosper by holding back most of its people. Apartheid began to dis-

Nelson Mandela, an important leader in the long struggle against apartheid, was elected president of **South Africa** in 1994.

appear as official government policy in 1989. At that time, the government opened relations with the ANC. This generally peaceful dialogue resulted in the repeal of South Africa's apartheid laws. In 1990, Nelson Mandela and others who had been jailed for disagreeing with apartheid were released. Finally, in 1994, Mandela and the ANC won the country's first all-race elections.

Today, public facilities are being integrated, and people of different races meet each other socially more often. Although there has been resistance to school integration among some white South Africans, public schools must now accept students of all races. Memories of the injustices of apartheid live on, however, and race relations remain tense. Also, cultural and political differences between various black ethnic groups, such as the Zulus and the Xhosas, are more apparent now that they are no longer overshadowed by the black-white divisions of the nation. The tensions between these ethnic groups have influenced the negotiations involved in drafting a new constitution for South Africa by 1999.

Economic Geography South Africa faces numerous challenges, many of which have roots in the old system of apartheid. For example, as apartheid ended, about three-fourths of black South Africans were farmers. Many of them were subsistence farmers in the former homelands. In contrast, one-fifth of white South Africans were farmers and ranchers. Agriculture among white South Africans has been as modern and productive as any other in the world. The new government has had to battle problems caused by severe

Guided Practice

Discuss why minerals are the basis for South Africa's economic development, and point them out on the economy map on page 420.

Independent Practice

Have students assume the role of a black South African and write a letter to an American friend explaining the ways that apartheid affected him or her and how things are different now.

Closure

Discuss with students how South Africa's economy was negatively affected by its political policies in recent decades.

ASSESSMENT

Assign the Section 2 Review questions on this page.

Reteaching

Assign Section 2 of Reteaching Worksheet 40.

Extension

You might play some recordings of South African singer Miriam Makeba for the class. Discuss the political and personal themes in her work.

soil erosion in the former homelands and high unemployment among black South Africans. To improve the economic status of black South Africans, government plans have included jobs programs and other employment reforms.

South Africa is the continent's most economically developed country. Although the country has only 6 percent of Africa's population, it has more than half of Africa's industrial firms. South Africa's economic activity also accounts for about half of Africa's total output. The basis for South Africa's wealth is its minerals, which are exported throughout the world. South African mines use the most advanced technology. Mineral resources include gold, uranium, copper, platinum, iron ore, coal, chromium, and diamonds. South Africa has nearly every mineral necessary for a modern, industrial society except bauxite and petroleum. South African factories manufacture almost all the industrial products the country needs.

PLACE **Urban Geography** The largest industrial region on the African continent is the Witwatersrand (WIT-wawt-uhrz-rahnd). Johannesburg, the capital, is located almost at its center. (See "Cities of the World" on pages 482–483.) Although most industries are South African owned, many large U.S., European, and Japanese companies are represented as well. Many of South Africa's automobiles are built near Port Elizabeth, on the southeast coast. Cape Town and Durban in Natal are other active ports.

Now that apartheid has ended, cities are open to all races, and the integration of neighborhoods has begun. There is, however, a continuing common pattern of settlement in South Africa. Many white South Africans live close to city centers and in suburbs away from industrial districts. White neighborhoods are similar to those in American suburbs, with modern-style houses and shopping centers.

Black areas are usually on the outer edges of cities, often across industrial districts and away from the white neighborhoods. Often, blacks must commute long distances to work. Coloured and Indian neighborhoods are usually situated between the black and white districts. Black neighborhoods vary in style from small, government-built modern houses to slum settlements built from scrap lumber and sheet metal. The new government has worked to improve utility service to existing

homes and to help black South Africans build hundreds of thousands of new homes.

South Africa's Future The effects of apartheid will remain for years. Black South Africans have had little access to their country's riches and economic opportunities, while white South Africans have the best educations and job skills. They enjoy higher incomes, hold the best jobs, own most of the good land, and live in the best neighborhoods. Blacks only recently have moved into positions of authority.

South Africa's black population is growing at a very fast rate, yet its white population is barely increasing. There are not enough schools in black neighborhoods, while schools in white neighborhoods are sometimes nearly empty. South Africa's economy will prosper only when equal educational, economic, and political opportunities exist for all the country's people.

SECTION REVIEW

1. What was apartheid, and how did it affect the daily lives of South Africans?
2. Why is South Africa the continent's most economically developed nation?
3. **Critical Thinking** **Evaluate** the different ways that apartheid was protested, both in and out of South Africa.

OTHER COUNTRIES OF SOUTHERN AFRICA

FOCUS

- *How are the histories of the countries of Southern Africa alike and different?*

- *What are the major economic activities in the countries of Southern Africa?*

The conflicts in South Africa often have overshadowed other countries of the region. Although these countries are strongly affected by South Africa, they are culturally diverse, independent nations. Mineral resources give these countries potential for industrial development.

Text continues on page 484.

Section 2 Review
ANSWERS

1. It was a series of laws that separated the races. It gave power, status, and wealth to whites and forced blacks to live in impoverished homelands. It discriminated against black South Africans.
2. It has valuable mineral resources.
3. Students should list and evaluate various ways that apartheid was protested, including international economic sanctions against South Africa and the ANC's refusal to obey apartheid laws.

Cities of the World

Geography and Economics

Have students work independently or in pairs to research other cities in the world that grew up around a gold-mining industry. Have them report on the importance of gold to the cities' current economies. Encourage students to use graphs, charts, or other visuals to illustrate their findings. Allow time for several students to present their reports to the class.

Geographical Sidelight

South Africa produces more than 30 million ounces (933 million g) of gold each year. In comparison, the United States—the world's fourth largest producer—produces less than 2 million ounces (56 million g) of gold annually.

Geographical Sidelight

In August 1993, Jacqui Mofokeng of Soweto became the first black woman ever crowned Miss South Africa. Her reign brought controversy in a still-divided South Africa. Even so, Mofokeng was determined to "help the people in [my] country . . . learn to live with each other in a peaceful way."

CITIES OF THE WORLD

Johannesburg

"Gold! Gold! Gold!" The cry of "gold!" marked the beginning of Johannesburg.

Unlike many other major cities around the world, Johannesburg, South Africa, owes its existence and growth largely to gold. In 1886, a wandering prospector came across gold-bearing rocks on a farm near a range of low hills known as the Witwatersrand, or "ridge of white waters." The news of the gold strike spread far and wide. Thousands of fortune seekers from around the world made their way to the area. People came mainly from Europe, America, Australia, and other parts of southern Africa. A dusty mining camp soon sprang up, its streets filled with a jumble of languages. As more gold was discovered and more people arrived, the small town of Johannesburg began to expand and take shape.

JOHANNESBURG

- ▬▬ Major highways
- ═══ Minor highways
- ─── Other roads
- ▬▬ Railroads
- ⋅⋅⋅ City limits
- ✈ Airports/ Airfields

0   4 miles
0   4 kilometers

Kempton Park
Alexandria
Edenvale
Johannesburg
Germiston
Boksburg
Soweto
Alberton

AFRICA
Johannesburg

Johannesburg is at the heart of a bustling urban area that includes other modern cities and black townships. Many residents from these cities and townships, such as Soweto, commute to Johannesburg for work.

The Mining Town

In Johannesburg's earliest years, gold mining was by far the most important part of its economy. But as the town grew, so did its need for other goods and services. Factories supplied machinery for the mines. New businesses provided food and clothing for the region's growing population. Banks, schools, and hospitals were built. Today, a century after gold was first discovered, Johannesburg forms the heart of a bustling, modern metropolitan area.

These South African workers are mining for gold more than one mile (1.6 km) underground. Many gold-mining companies have headquarters in **Johannesburg**, the country's leading business center.

Using Geography Themes

Challenge students to summarize the feature by writing a paragraph using each of the Five Themes of Geography.

Role Playing

Provide students with the following scenario: "You are a black South African living in Soweto in 1997. Write a letter to the Johannesburg daily newspaper telling what it is like to live in a city 'free of apartheid.'" Call on volunteers to read their letters aloud.

Gold and businesses related to gold are still important to the city's economy, and the Witwatersrand region is a major source of gold. In fact, nearly half of all the world's gold mined during the past 100 years has come from this region. Although most of the mines have closed, miners continue to work as much as two miles underground to haul out gold-bearing ore east and west of the city in the 62-mile-long (100-km) Witwatersrand.

A Changing City

Today, Johannesburg is the financial and commercial capital of South Africa. It is a city of modern skyscrapers, busy industrial districts, and sprawling suburbs. Most of Johannesburg's residents now work in jobs that have little to do with mining.

Although most blacks still live in vast townships, housing projects, and slums on the outskirts of the city, many have moved to the city center. In contrast, most whites live in comfortable homes in pleasant suburbs of the city. But some blacks, South Africans of mixed race (called coloureds), and those of South Asian descent also can now afford to live in suburbs that once were all white. In addition, schools have become more integrated.

With the end of apartheid, schools in **South Africa**, such as the one attended by these students outside **Johannesburg**, are open to South Africans of all races.

As more blacks move into positions of authority in government and employment, the townships also have begun to change. The more than 2 million residents of Soweto, a large township southwest of downtown Johannesburg, once had to travel to the main city to do even basic shopping. Today, Soweto's homes have better utility services, such as electricity, and many stores and other businesses have opened in the township.

Downtown Johannesburg, once exclusively white, was declared open to all races when apartheid ended in the early 1990s. South Africans of all races can now live in the same apartment buildings, eat in the same restaurants, and work and go to school together. Divisions remain, but Johannesburg is leading South Africa into a future free of apartheid.

Vans from this minibus station carry many **Soweto** residents to their jobs in Johannesburg. Road and rail lines link Soweto to the capital city.

Geographical Sidelight

Economic sanctions were imposed on the Republic of South Africa as a means of pressuring the country to give up its apartheid policies. South Africa lost an estimated $27 billion in trade and investment because of such sanctions. In September 1993, African National Congress leader Nelson Mandela made a historical plea before the anti-apartheid committee of the United Nations General Assembly, asking that remaining economic sanctions against South Africa be lifted. Mandela's speech came the day after South Africa's parliament voted to allow blacks to participate in the country's politics for the first time.

The objectives reflect the Focus questions on page 481 and are assessed in the review on page 486.

1. Compare and contrast the histories of the nations of Southern Africa.
2. Identify economic activities of the nations of Southern Africa.

TEACHING SECTION 3
Motivator

Call on volunteers to cite details about Southern Africa. If most responses refer to the Republic of South Africa, point out that people often make the mistake of associating the entire region with that one nation. Have students try to name Southern African countries without consulting the textbook.

Building a Vocabulary

Write the word **copra** on the board, and have students read the definition on page 485. Ask them whether they have ever eaten anything that contains oil derived from copra. Point out that many commercially produced cakes, cookies, and candies contain coconut oil. Have them look for this ingredient on product wrappers.

MULTICULTURAL PERSPECTIVES

The !Kung San, who live in the Kalahari Desert in Botswana, speak a language that uses different click sounds made by the tongue and the roof of the mouth. The *!* in the name *!Kung San* stands for a type of click. Have students use library resources to investigate other languages spoken in Southern Africa. Provide a space on the bulletin board for them to list their findings.

Chart Caption Answer

Zimbabwe

Namibia Namibia lies on the Atlantic coast north and west of South Africa. (See the map on page 477.) Its road to independence was slow and violent. Once a German colony, it was called German Southwest Africa. During World War I, South African forces drove out the Germans. It then became South West Africa. South Africa's administration of the region was challenged repeatedly by the United Nations. In 1990, South Africa finally granted the territory its independence as Namibia.

The new government is democratic and is represented by several political parties in the parliament. At one time, apartheid was introduced into Namibia. However, lands once reserved for whites and blacks, as well as for Khoisan peoples, have been opened to settlement by all groups. With independence, the laws of discrimination have disappeared.

Most Namibians live in the far north, where there is dry savanna. Along the coast is the Namib Desert; inland is the Kalahari Desert. People also live in the central highlands, where the climates are cooler and not as arid. Ranching is the only commercial agriculture. The fishing fleet of this territory also provides important income. The economy is most dependent, however, on mining. Namibia has enormous mineral reserves, including literally mountains of copper, lead, zinc, and uranium. Along the southern coast are large diamond deposits.

Botswana Like Gabon in Central Africa, Botswana is becoming one of Africa's success stories. Formerly the British territory of Bechuanaland (bech-uh-WAHN-uh-land), Botswana is an arid, landlocked country occupying much of the Kalahari Desert. (See the map on page 477.) Britain took control of the area in 1886 to connect the Cape Colony with Rhodesia. Britain invested little in Botswana, however. At the time of independence in 1966, there were no paved roads, no electricity, and only one important factory, a meat-canning plant.

Today, the country earns much income from its enormous diamond and copper deposits and has one of the fastest-growing economies in Africa. Cattle herding and mining are the principal economic activities. Botswana also is one of the more democratic countries in the region and has opened its doors to refugees from neighboring countries. A new capital stands at Gaborone.

FACTS IN BRIEF	Southern Africa		
COUNTRY POPULATION (1994)	LIFE EXPECTANCY (1994)	LITERACY RATE (1990)	PER CAPITA GDP (1993)
BOTSWANA 1,359,352	60, male 66, female	23%	$4,500
COMOROS 530,136	56, male 60, female	15% (1989)	$700
LESOTHO 1,944,493	60, male 64, female	59%	$1,500
MADAGASCAR 13,427,758	52, male 56, female	80%	$800
MAURITIUS 1,116,923	67, male 75, female	80%	$7,800
MOZAMBIQUE 17,346,280	47, male 50, female	33%	$600
NAMIBIA 1,595,567	59, male 64, female	72%	$2,500
SEYCHELLES 72,113	66, male 73, female	85% (1991)	$5,900 (1992)
SOUTH AFRICA 43,930,631	62, male 68, female	76% (1992)	$4,000
SWAZILAND 936,369	52, male 60, female	65%	$2,500
ZIMBABWE 10,975,078	40, male 44, female	67%	$1,400 (1991)
UNITED STATES 260,713,585	73, male 79, female	97% (1991)	$24,700

Sources: *The World Factbook 1994*, Central Intelligence Agency; *The World Almanac and Book of Facts 1995*

Which country in Southern Africa has the lowest life expectancy rates?

The Tswana (from which Botswana gets its name) is the largest ethnic group, and many of these people are cattle herders and corn farmers. Most people live along the wetter eastern edge of the country, where dry savannas are the typical vegetation. In the Kalahari Desert are the remaining groups of hunting-and-gathering San.

Botswana's major river is the Okavango (oe-kuh-VANG-oe), which starts in the wet highlands of Angola and flows into the Okavango Basin. There, it forms the rich wetland forests of the inland Okavango Delta. Large game reserves are located in the region. These wetlands in the desert could become a major tourist attraction.

Zimbabwe Zimbabwe takes its name from the ancient walled ruins in the central part of the country. There, a city of stone-walled buildings and ceremonial structures was built between the twelfth and sixteenth centuries. These ruins are a symbol of the achievements of African cultures before the arrival of Europeans.

Comparing and Contrasting

Southern African Countries
Have students work in small groups to compare and contrast the countries discussed in Section 3. Have them create a chart that lists similarities and differences between these countries in terms of landforms, economic resources, history, and development. Ask a volunteer to draw one group's chart on the board. Have students check their chart against the one on the board.

Guided Practice
Organize the class into small groups, and assign one of the countries discussed in this section to each group. Have students create a poster illustrating the resources and the ways of making a living found in their assigned country. Display completed posters in the classroom.

Independent Practice
Have students write newspaper editorials explaining the effects of colonialism on the countries discussed in this section. Ask volunteers to read their editorials to the class.

Southern Africa: Transportation Network

LEGEND
- —— Highways
- —— Railroads
- ┄┄ Navigable rivers
- ✈ Airports
- ⚓ Seaports
- ⊛ National capitals
- • Other cities

MAP STUDY Some maps, such as the one above, provide information that shows how people and goods are transported between locations. The legend shows symbols for major roads, railroads, airports, and seaports that are vital to trade and travel. In some places, rivers also are important parts of transportation networks.

During the colonial period, Zimbabwe, Zambia, and Malawi were British territories. The British had combined them into the Federation of Rhodesia and Nyasaland. The members were to profit economically, because each region would offer something important to the federation. Malawi, formerly Nyasaland, was to provide a labor force. Zambia, formerly Northern Rhodesia, was to provide mineral wealth. Zimbabwe, formerly Southern Rhodesia, would provide industrial strength. The few Europeans, mainly in Southern Rhodesia, controlled most of the economic and political affairs of the federation, however.

Northern Rhodesia and Nyasaland objected to the influence of white-dominated Southern Rhodesia. In 1964, Nyasaland and Northern Rhodesia became independent as the separate countries of Malawi and Zambia. Southern Rhodesia, calling itself Rhodesia, remained a self-governing British colony. In 1965, the white-dominated parliament of Rhodesia declared itself independent, an illegal action under British law. Furthermore, most of the people living in Rhodesia were black Africans who were not being equally represented by the white government. In response, the United Nations ordered a trade embargo against Rhodesia. Only white-dominated South Africa openly traded with the country.

In 1979, this economic pressure, combined with successes of guerrilla fighters, brought black majority rule to Rhodesia. Soon after, Rhodesia became Zimbabwe, but the long period of conflict weakened the once prosperous economy.

In the years since the civil war, Zimbabwe's leaders have focused on rebuilding the country. The many factories started during the embargo are exporting goods to neighboring countries. The country has food for export when the region is not suffering from drought, and the commercial agriculture on farms owned by whites is prosperous again. Although the majority of commercial farms in Zimbabwe are owned by whites, the government recently passed legislation that would equalize the distribution of land among black and white farmers. Other economic and political reforms are taking place as well.

Mozambique Mozambique, located on the east coast of Africa between the Republic of South Africa and Tanzania, has a good base of economic resources. The country produces tea, sugarcane, cotton, cashew nuts, sisal, and **copra**. Copra is the dried meat of the coconut, from which coconut oil is obtained. Because coal in Africa is uncommon, the country's coal reserves are valuable. Other sources of income are tax collections at the ports of Maputo, the capital, and Beira (BAY-ruh). Beira's port handles many products from interior countries, particularly Malawi, Zimbabwe, and Zambia.

Until this century, there was little economic development in this former Portuguese colony. Today, the government has several development plans under way. One is an irrigation project in the Limpopo River valley. Plans call for growth of an extensive network of irrigation canals to support small farms. Also important is the dam in the Zambezi River valley at Cabora Bassa.

Geography and History
Nationalist movements spread throughout Africa after World War II. The first nationalist movement in Zimbabwe began in the 1950s. ZAPU, led by Joshua Nkomo, worked peacefully for change. In 1963, some black nationalists, disappointed with Nkomo's moderate position, formed the more militant ZANU, led by Robert Mugabe, and prepared for guerrilla warfare to achieve black rule. Mugabe became Zimbabwe's prime minister in 1980.

GLOBAL SKILL BUILDER

You might wish to use **Using Different Types of Maps** from the interactive *Global Skill Builder* CD-ROM to reinforce student's understanding of transportation maps.

Closure

Ask students to list ways the countries discussed in this section might best continue to develop their economies, taking their varied histories into account. Have them share their lists, and discuss the suggestions.

ASSESSMENT

Assign the Section 3 Review questions on this page.

Reteaching

Assign Section 3 of Reteaching Worksheet 40.

Extension

Have students research the Namib Desert and the Kalahari Desert. They should find out the size of the deserts, the economic activities taking place, and the presence of mineral resources. Students should compile their information into an illustrated written report.

COOPERATIVE LEARNING

Organize the class into groups of four or five. Have each group complete an urban profile of a different city in Southern Africa (excluding Johannesburg). Assign individuals in each group the following tasks: researching, writing, illustrating, presenting.

Section 3 Review
ANSWERS

1. Namibia: Germany; Botswana and Zimbabwe: Britain; Mozambique: Portugal; Madagascar: France
2. Mining: Namibia; agriculture: Zimbabwe, Comoros, Seychelles, Mauritius; both: Botswana, Swaziland, Lesotho
3. Possible answer: the conflicts in both countries began around the time of independence but thereafter took different turns. In Zimbabwe, the conflict concerned inequalities between blacks and whites; in Mozambique, the conflict was between the Communist government and rebels.

486

Other Countries of Southern Africa

Swaziland
Economic Resources
Timber is an important resource in Swaziland, which also mines gold, coal, asbestos, and iron ore. The country also has valuable grazing lands, and it exports sugarcane and citrus fruits.

At a Glance
Although the vast majority of Swaziland's population is Swazi, Europeans own nearly half the land. Income from European-owned mining operations is vital to the landlocked country.

Seychelles
Economic Resources
Seychelles' economy is based on agriculture. Its main products are coconuts, bananas, and cinnamon. Tourists also provide important income for the islands.

At a Glance
Seychelles, which is made up of 92 islands in the Indian Ocean, is a popular destination for tourists.

Lesotho
Economic Resources
Lesotho's limited resources and poor land for farming produce small exports of wheat, diamonds, and wool. The building of large dams will provide hydroelectricity and water for export in the future.

At a Glance
Lesotho, which is completely surrounded by South Africa, is located on the high plateau of the Drakensberg Escarpment.

Comoros
Economic Resources
The economy of Comoros revolves around tourism, fishing and farming, and the export of perfumes, spices, and copra.

At a Glance
The poor, rocky soil of Comoros forces the islands, located in the Mozambique Channel, to import most of their food.

Mauritius
Economic Resources
The economy of Mauritius is heavily dependent on sugar production. The cultivation of sugarcane was begun during the period of European control, first by the Dutch, then the French, and finally the British.

At a Glance
The volcanic island of Mauritius, once a home of the now-extinct dodo bird, is almost completely surrounded in the Indian Ocean by reefs.

Mozambique has suffered continuing violence and rebellion. It began with the war for independence in the 1960s and has continued since independence in 1975. Rebels trying to oust the Communist government cut roads and bombed rail lines, nearly collapsing the economy. In the early 1990s, pressure from neighboring countries forced an uneasy peace treaty between the government and rebel forces.

Madagascar The island countries off the east coast of Africa share connections with Africa as well as with Asia. The largest of these islands is Madagascar. (See the map on page 477.)

Madagascar is rich in natural wonders. Along Madagascar's east coast is a narrow, humid, coastal plain and a steep escarpment. Atop the escarpment are a mild highland climate and rich farmlands. In the west, the island becomes more arid as it slopes down to the coast on the Mozambique Channel. The Southeast Asians who settled Madagascar more than 1,000 years ago brought Asian farming systems and crops. Later, Arabs traded between Madagascar and Africa's east coast. The Malagasy language has a rich literature and is spoken throughout Madagascar.

When the French took control of Madagascar in the 1890s, they often neglected this isolated island. Schools and roads were built, but little else. Since independence in 1960, industrial development has been slow. The rapidly increasing population has meant that more and more available farmland is being used. Soil erosion has been severe, and deforestation is increasing in the highlands. The government has nationalized many businesses, and the economy has steadily declined. The country is dependent on foreign aid.

3 SECTION REVIEW

1. What European countries colonized each country discussed in this section?
2. Which South African countries have economies based on mining? on agriculture? on both mining and agriculture?
3. **Critical Thinking** **Assess** whether the conflicts of the 1960s–1980s in Zimbabwe and Mozambique had similar causes.

CHAPTER 40 CHECK
ANSWERS

Building a Vocabulary

1. Highveld: grassy, high plains; bushveld: short trees and bushes; lowveld: savannas and moist forests
2. Laws that separated races in where they lived, worked, and went to school
3. Legal restrictions on moving freight into a country; to force them to cease an illegal or immoral act
4. Exile
5. Dried coconut meat; Mozambique

Recalling and Reviewing

1. The Drakensberg Escarpment blocks moist air from the ocean, so climates inland are drier.
2. Apartheid laws separated the races. Whites held most status, power, and wealth. Blacks were moved to homelands with poor medical or educational facilities and few job opportunities.
3. It has gold, diamonds, and other mineral resources needed for industry.
4. Britain invested little in Botswana and left it poor and underdeveloped. Today, the country has a rapidly growing economy.
5. Mining

CHAPTER **40** CHECK

Early African ruins, **Zimbabwe**

Reviewing the Main Ideas

1. The region of Southern Africa has a long history of settlement by various groups of people. San (Bushmen), Bantu speakers, Dutch Boers, Indians, the British, and Southeast Asians were some of the peoples who settled the area.
2. Under apartheid, South Africa was divided into black and white regions, and political and economic power rested with the white minority. With the end of apartheid, blacks increasingly hold positions of authority in politics and employment.
3. Most of the countries of Southern Africa are rich in mineral resources, although some countries have not been able to develop these resources to their advantage.

Building a Vocabulary

1. Name and describe the three natural regions of Southern Africa.
2. What was the system of apartheid?
3. What are embargoes? For what reasons do countries impose sanctions on other countries?
4. What is the term for a person who has been forced to leave his or her country?
5. Define *copra*. Where in Southern Africa is it produced?

Recalling and Reviewing

1. Describe the relationship between landforms and climate in Southern Africa.
2. How did apartheid affect daily life in South Africa?
3. Why is South Africa the wealthiest country on the African continent?
4. Compare the economic situation in Botswana immediately following independence to its economic situation today.
5. What is the major economic activity in Southern Africa?

Thinking Critically

1. In your opinion, what are the positive and negative aspects of industrialization? If you were an economic adviser to Namibia, for example, would you advise the country to develop its mineral resources? Why or why not?
2. Imagine you have been living on Madagascar and are moving to Lesotho. Write a paragraph comparing the physical and economic geography of the two countries. In which country would you prefer to live? Why?

Learning About Your Local Geography

Cooperative Project

As a class, make a list of the mineral resources found in or near your area. Then, organize into groups of four. Each group should select a different mineral resource and research where it is found, how it is mined, to where it is transported, and for what it is used. Groups should present their information to the rest of the class.

Using the Five Themes of Geography

Write a paragraph about how the geographic themes of **region** and **place** apply to the countries of Southern Africa. Into what kinds of political and social regions have these countries been divided? How do the various regions differ in place? Discuss how certain ways of defining regions can put some groups of people at a disadvantage.

Thinking Critically

1. Answers will vary. Possible answers: industrialization generates investment capital within the country. Or, industrialization can bring environmental and other problems. Students may suggest agricultural development over industrial development.
2. Paragraphs should cite differences in physical geography and economic geography. Madagascar has plains, highlands, and rich farmlands. Lesotho is located on the high plateau of the Drakensberg Escarpment and has poor soil. Answers to the second part of the question will vary.

Learning About Your Local Geography

Reports will vary, depending on the mineral resource assigned to each group.

Using the Five Themes of Geography

Students should discuss apartheid and systems that divide people into regions based on their ethnic background.

Unit 8 Review

SKILLS

Have students refer to the following pairs of photographs: Harare, Zimbabwe, on page 412 and Nairobi, Kenya, on page 458; Nile River valley on page 432 (top right) and Nile Delta on page 434; and Zaire mountain fields on page 467 and Zambia copper mines on page 473 (upper right). Have students identify the main idea of each photograph. Also have them identify the photographer's point of view and location relative to the subject of each photograph and (where applicable) the foreground, middle ground, and background. Then have students compare and contrast the two photographs in each pair.

Interpreting Photographs

Imagine this textbook without any photographs. Not only would it be less interesting, but you also would not have any visual images of the places you are studying. Photographs are important tools for studying geography. They provide information about a location's physical features, economy, people, and climate.

There are several different kinds of photographs, all serving different purposes. Aerial photographs, taken from above, are useful for showing land use and settlement patterns, and they aid cartographers in making maps. Landsat photographs, taken from satellites, provide much the same information, but they can show more land area because they are taken from farther away. In contrast, photographs taken closer to Earth reveal a location's vegetation, landforms, and weather, as well as culture traits. Cultural geographers study photographs to discover how people live—their

housing, their dress, and ways in which they have adapted to or changed their environment.

Photographs record things as they happen—effects of floods or severe drought, crops at harvest time, a highway being built. Photographs also record changes over time, such as the effects of deforestation or urbanization. Historical photographs can be compared to modern-day photographs to find out how a place has changed.

How can we analyze photographs to obtain useful geographic information? One way to examine a photograph is to seek its main idea or theme. Ask, "What does this photograph tell me about this location?" How a photographer arranges a subject in the camera's view is a good clue. If you were asked to take only one photograph of the city or town in which you live, you might choose a wide view from far away to show its size, or perhaps you might show people at work to reveal something about the local economy.

When you study a photograph, note what is in the foreground, the middle ground, and the background. What part of the photograph provides the clearest details? Try to estimate the sizes of various features by comparing them to other objects in the same part of the photograph. Remember, however, that photographs have limitations. The distance between the camera and its subject often distorts what you see—the actual distance between the foreground and the background can be difficult to judge. Because of this, geographers do not use photographs to determine exact distance or size.

Identifying geographic characteristics from photographs takes practice, because the camera captures insignificant details as well as important ones. Learning to interpret photographs effectively will enhance your study of geography.

Practicing the Skill
ANSWERS

1. A tall-tree savanna
2. Foreground: grasses; middle ground: trees; background: mountains
3. Grasses would appear to be much farther away and would have less detail.

PRACTICING THE SKILL

The photograph above was taken on an animal reserve in Southern Africa. Using what you have learned, examine the photograph for geographic information. Then answer the following questions.

1. What biome is depicted in the photograph?
2. What is in the foreground? the middle ground? the background?
3. If the photograph had been taken from the top of the mountains, how would the view change?

488

BUILDING YOUR PORTFOLIO

Decide whether students will work individually or in groups on the unit projects described below. Students should then choose which project they will work on. Provide students with copies of the Portfolio Assessment Worksheets in *Teaching Resources.* Chapters 35–40 worksheets and the Unit Portfolio Assessment Worksheet will guide students step-by-step to complete either project.

Teaching with Sketch Maps

Have students sketch a map of Africa from memory. Tell them to include as many details as they can about place names, landforms, and boundaries of regions they studied in Unit 8. Students should compare this map with the sketch map they completed at the beginning of Unit 8 to note their improvement in identifying geographic features of Africa.

BUILDING YOUR PORTFOLIO

Individually or in a group, complete one of the following projects to show your understanding of the geography concepts involved.

A Redrawing the Map of Africa

The year is 1900, 16 years after the Berlin Conference. You are an African who has been asked to attend the Africa Conference. (Reminder: No such conference took place, and Africans were not invited to participate in the Berlin Conference.) The goal of this conference is to create a new map of Africa, one that will better serve the needs of Africans.

Your report to the Africa Conference will consist of a map of Africa with borders and country names that you think best, along with written and visual materials that support your decisions. Be prepared to explain any changes you have made to the European map of Africa. To prepare your report to the Africa Conference, you will need to do the following.

1. Identify the countries that were landlocked after European boundaries were set. What resources do these countries have? How do the countries transport these resources to market? You might want to present this information in a chart.
2. Identify the minerals and other natural resources under European control. How are these resources used? How would you use these resources differently? Present your information using flowcharts.
3. Locate maps of Africa that show language and cultural groups before European colonization. Compare to a post-colonization map. Make your own map that illustrates how cultures and/or language groups were divided or thrown together.
4. Create a new map of Africa, including boundaries and names of the African nations. Write two or three paragraphs explaining any changes you have made to the European map of Africa.

Organize the information you have gathered, and present your information to the other groups at the Africa Conference (the rest of the class).

B Investigating Desertification

The United Nations has estimated that 40 percent of Africa's nondesert land is in danger of desertification. Imagine that you are on the Planet Watch committee. Your subcommittee has been assigned to document the causes and effects of desertification and to suggest a solution to the problem. Although your report will focus on Africa, you will need to conduct research to find out more about desertification and its effects worldwide. Complete the following steps to prepare for your presentation to Planet Watch.

1. Prepare a poster that illustrates the definition of *desertification.* Include on the poster a written definition of *desertification.*
2. Draw a cross-sectional diagram or flowchart illustrating how desertification occurs. Use labels to explain what is happening. Be sure to include characteristics of the physical environment that contribute to desertification.
3. Create two maps. One should show countries in Africa most affected by desertification. The other map should be a world map illustrating the worldwide extent of desertification.
4. Consider the following questions: What will happen if desertification continues at its present rate? What will have to change in Africa to control desertification? Work the answers to these questions into a proposed solution to desertification.

Present your research to the Planet Watch committee (the rest of the class). Your proposed solution to desertification should be the conclusion of your presentation.

Project A:

Students should present a chart showing landlocked countries and their resources, a flowchart identifying minerals and other resources under European control, a map illustrating how colonization divided or threw together cultures and/or language groups, and a new map of Africa to better serve the needs of Africans. The new map should be accompanied by two or three paragraphs explaining changes. Students' new maps may try to preserve a more equitable distribution of resources among countries. They may also create boundaries along cultural lines.

Project B:

Presentations should include a poster illustrating the definition of desertification, a diagram or flowchart showing how desertification occurs, maps showing desertification in Africa and worldwide, and a proposed solution to desertification.

UNIT 9 EAST AND SOUTHEAST ASIA   *CHAPTERS 41–45*

OVERVIEW

In Unit 9, students will learn that the East and Southeast Asia region has large, diverse populations, unique land-forms, and ancient histories.

Chapter 41 • INTRODUCING EAST AND SOUTHEAST ASIA identifies the physical geography, economic geography, and human geography of a region that is home to about one-third of the world's population.

Chapter 42 • CHINA, TAIWAN, AND MONGOLIA explains the history, culture, and geography of those three nations.

Chapter 43 • JAPAN AND KOREA describes the land-forms, climates, resources, agriculture, industries, and interrelated histories of Japan and Korea.

Chapter 44 • MAINLAND SOUTHEAST ASIA examines the physical, economic, and cultural characteristics of Vietnam, Laos, Cambodia, Thailand, and Burma. It also introduces the physical geography and human geography of the region's island nations.

Chapter 45 • ISLAND SOUTHEAST ASIA discusses the nations of Malaysia, Singapore, Brunei, Indonesia, and the Philippines.

OBJECTIVES

1. Describe how climate and landforms affect the way people live in East and Southeast Asia.

2. Identify the political, economic, social, and cultural issues that exist in East and Southeast Asia.

3. Explain the importance of East and Southeast Asia as a major region of the world.

4. Understand links between historical geography, cultural geography, and physical geography in East and Southeast Asia.

5. Interpret special-purpose maps to analyze relationships among climate, population patterns, and economic resources in East and Southeast Asia.

6. Learn to detect cultural bias and generalizations.

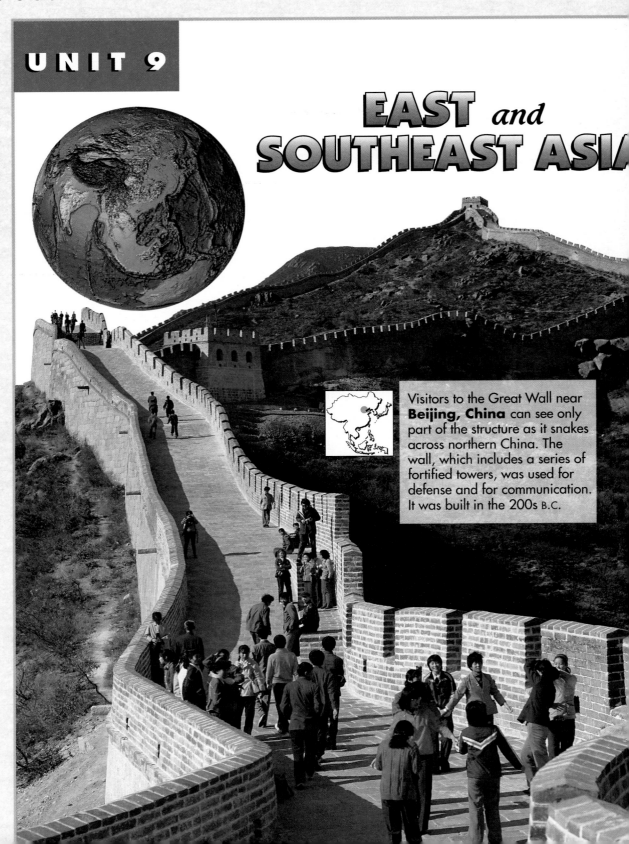

UNIT 9

EAST and SOUTHEAST ASIA

Visitors to the Great Wall near **Beijing, China** can see only part of the structure as it snakes across northern China. The wall, which includes a series of fortified towers, was used for defense and for communication. It was built in the 200s B.C.

490

Using the Illustrations

Direct students' attention to the photographs on these pages. Have them select the photographs that illustrate the geographic theme of human-environment interaction. (the Great Wall; Hong Kong) You might also refer students to the map on page 507 to locate the Great Wall in northern China. Ask students to suggest why the construction of the Great Wall was a remarkable engineering achievement. (the length of the wall and its path through mountainous terrain) Point out that the photographs of the Balinese dancers and the Japanese and Chinese artifacts illustrate cultural characteristics of East and Southeast Asia.

Teaching with Sketch Maps

Have students draw a map of East and Southeast Asia from memory. Tell students to include as many political boundaries, country and city names, and bodies of water as they can. Students might add this map to their portfolios and compare it to another they will draw after they have completed Unit 9.

Dancers perform the *legong*, a popular folk dance in **Bali, Indonesia**. *Legong* dancers wear elaborate costumes and use their dance to tell stories about such things as adventures, battles, and love.

Chinese paper-cut fan

UNIT 9 CHAPTERS

41 Introducing East and Southeast Asia

42 China, Taiwan, and Mongolia

43 Japan and Korea

44 Mainland Southeast Asia

45 Island Southeast Asia

Japanese lacquer bowl and tray

THE WORLD TODAY VIDEODISC PROGRAM

The videodisc profiles the following countries in Unit 9: China, Japan, Thailand, Singapore, and the Philippines.

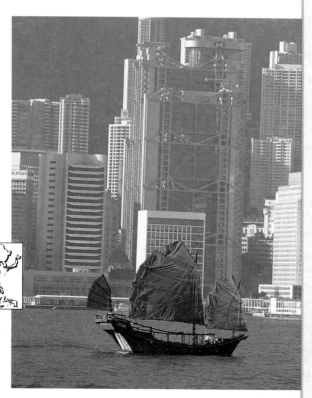

A Chinese junk sails past the skyscrapers of **Hong Kong**, one of East Asia's major commercial centers. China is to take possession of the prosperous British territory in 1997.

491

One World, Many Voices

LITERATURE

You might provide students with the historical context of the government land reform program for which Yuan-tsung Chen worked. Tell students that for centuries Chinese peasants had labored on their own small subsistence farms, or had paid high rents to landlords who owned large amounts of property. As a result, most Chinese farmers lived in poverty and were unable to produce enough food for their families. The intent of land reform was to improve agricultural production in China and to raise the peasants' standard of living.

More About Yuan-tsung Chen

Yuan-tsung Chen was a translator and film editor in China. She came to the United States in 1972 and taught Chinese language at Cornell University for several years. Currently, Yuan-tsung Chen is writing a sequel to *The Dragon's Village*, the novel excerpted on this page. She resides in Berkeley, California.

ONE WORLD, MANY VOICES
LITERATURE

Yuan-tsung Chen (1932–) was born in Shanghai, China. When the People's Republic of China formed in 1950, Chen and many other young people were enlisted by the Communist government to work toward land reform. She spent 20 years helping peasants in agricultural cooperatives. The passage below is from a novel she wrote about her experiences in Gansu, a province in northwestern China. It describes the physical geography of the remote region.

Primary Source

Description of Change: excerpted and bracketed
Rationale: excerpted to focus on the main idea, bracketed to clarify meaning

"We had traveled for a whole day over this ancient land. . . . At first, we had marveled at the strangeness of the landscape. It was a plain riven [cut] by deep gullies so that the dirt road either meandered wildly to avoid the slits in the earth or plunged zigzag down and up the sides of the ravines that couldn't be avoided. The earth had been ravaged and made desolate. I knew from my history books that these eroded lands were once pastures and forested plateaus. Then the pastures had been ploughed up to grow crops and the forests cleared for farmland. The natural rhythm of nature had been disturbed. The animals had disappeared and so had their dung. War had devastated the farms. Marauding [plundering] warlord armies had chopped down the remaining trees for firewood. Without vegetation to hinder them, the rain and run-off rivulets [small streams] of centuries had eaten into the fields and carried them away. The green clothing of the earth had been filched [stolen] and the naked earth was dying of cold.

I sighed disconsolately [unhappily]."

Interpreting Literature

Answers

1. Most of the natural vegetation had been either plowed up or cut down.
2. Severe erosion caused by the loss of pastureland and forests had created a barren landscape, and the ecology of the region had been disturbed.

Interpreting Literature

1. What human activities have changed or damaged the land in Gansu?
2. Why is the narrator saddened by what she observes there?

Farming in Gansu province

492

FOR THE RECORD

Introduce the primary source below by discussing the significance of the fifteenth century as an age of exploration and discovery. Tell students that a new global awareness developed as a result of voyages undertaken by Chinese, Portuguese, Spanish, and Muslim geographers and explorers during the 1400s. Explain that what motivated these expeditions was more a desire to seek trade opportunities than an interest in acquiring geographic knowledge.

More About Ma Huan

Ma Huan grew up in the vicinity of Hangzhou Bay, an important center of navigation in China during the 1400s. Little is known about his early life or education. His writings indicate that he adopted the Muslim faith. Ma Huan acquired knowledge of Arabic script and fluency in the Persian language, which no doubt influenced his selection as translator on the Cheng Ho voyages.

FOR THE RECORD

Ma Huan (c. 1380–c. 1460) of China was the official interpreter for Cheng Ho, a diplomat and explorer who led several naval expeditions to many places in Asia and Africa during the fifteenth century. The expeditions were undertaken to expand China's influence as a trading power throughout the lands bordering the "Western Ocean"—all the bodies of water between China and the east coast of Africa. Ma Huan recorded his impressions of the people and places they encountered. The passage below describes an empire in central Vietnam called Champa.

Cheng Ho's ship

Primary Source

Description of Change: excerpted and bracketed
Rationale: excerpted to focus on the main idea, bracketed to clarify meaning

"**Y**ou come to the city where the king resides; its foreign name is Chan city. The city has a city-wall of stone, with openings at four gates, which men are ordered to guard. The king of the country is a So-li man, [and] a firm believer in the Buddhist religion. . . . The house in which the king resides is tall and large. It has a roof of small oblong tiles on it. The four surrounding walls are ornately constructed of bricks and mortar, [and look] very neat. The doors are made of hard wood, decorated with engraved figures of wild beasts and domestic animals.

The houses in which the people live have a covering made of thatch; the height of the eaves cannot exceed three *ch'ih* (36.7 inches), [people] go in and out with bent bodies and lowered heads; a greater height is an offense.

As to the colour of their clothing; white clothes are forbidden, and only the king can wear them; for the populace, black, yellow, and purple coloured [clothes] are all allowed to be worn; to wear white clothing is a capital offense."

Translated by Feng Ch'eng-Chün

Analyzing Primary Sources

1. What is the first thing Ma Huan observes about the city of Chan?
2. How does Ma Huan note the differences between royalty and common people in Champa?

Analyzing Primary Sources

Answers

1. It was a fortified city surrounded by a large stone wall.
2. By differences in the quality, materials, and dimensions of their homes; by differences in the clothing worn by royalty and commoners

CHAPTER 41 Introducing East and Southeast Asia

Planning Guide

TEXTBOOK	TEACHING RESOURCES	OTHER AVAILABLE MATERIALS
Section		
1. Physical Geography Pages 498–499	Reteaching Worksheet 41.1 Skills Worksheet 41	*REGIONS OF THE WORLD Map Transparencies:* East and Southeast Asia *Study Guide for Spanish Speakers* *Lecture Notes,* Chapter 41
2. Economic Geography Pages 500–502	Reteaching Worksheet 41.2 Graphic Organizer 41 Critical Thinking Worksheet 41	*Study Guide for Spanish Speakers* *Lecture Notes,* Chapter 41
3. Human Geography Pages 502–504	Reteaching Worksheet 41.3 Challenge/Enrichment Worksheet 41	*Creative Strategies for Teaching World Geography* LESSON 18 Asian Holiday Calendar *Study Guide for Spanish Speakers* *Lecture Notes,* Chapter 41
Chapter Review Page 505	Chapter 41 Test Chapter 41 Portfolio Assessment See also *Alternative Assessment Forms*	*Test Generator,* Chapter 41

CHAPTER RESOURCES

Lecture Notes, Chapter 41
Teaching Resources
    Chapter 41 Graphic Organizer
    Chapter 41 Skills Worksheet
    Chapter 41 Critical Thinking Worksheet
    Chapter 41 Challenge/Enrichment Worksheet

REVIEW AND ASSESSMENT

Chapter 41 Check, page 505
Chapter 41 Test, *Teaching Resources*
Chapter 41 Portfolio Assessment, *Teaching Resources*
Test Generator, Chapter 41
Alternative Assessment Forms

ADDITIONAL RESOURCES

Books for Teachers

Gibney, Frank. *The Pacific Century: America and Asia in a Changing World.* New York: Charles Scribner's Sons, 1992.

Mathews, Bruce, and Judith Nagata, eds. *Religion, Values, and Development in Southeast Asia.* Singapore: Institute of Southeast Asian Studies, 1986.

McNeely, Jeffrey A. *Soul of the Tiger: Searching for Natural Answers in Southeast Asia.* Garden City, NJ: Doubleday, 1988.

SarDesai, D.R. *Southeast Asia; Past and Present.* Boulder, CO: Westview Press, 1994.

Books for Students

Clark, James I. *Southeast Asia.* Evanston, IL: McDougal, Littel, 1989.

Hinton, Harold C. *East Asia and the Western Pacific.* Washington, DC: Stryker-Post, 1988.

Marshall, Bruce, ed. *The Real World: Understanding Our World Through the New Geography.* Boston: Houghton Mifflin, 1991.

Pan, Lynn. *Into China's Heart: An Emigre's Journey Along the Yellow River.* New York: Weatherhill, 1985.

The World Today *Videodisc Program*

A variety of still images, short videos, and activities are available for you to use as you teach this chapter. *Lecture Notes* include bar codes for specific images and videos.

Multimedia Resources

Asia Alive GUIDisc. (Multimedia CD-ROM, interactive tour) MediAlive, 766 San Aleso Avenue, Sunnyvale, CA 94086.

The Holy Mountain of Tibet. (VHS, 43 minutes) Films for the Humanities & Sciences, P.O. Box 2053, Princeton, NJ 08543–2053.

Picture Atlas of the World. (interactive maps, photo bank, animation, audio on CD-ROM) National Geographic Society, Washington, DC.

MEETING INDIVIDUAL NEEDS

Limited English Proficiency (LEP)

Organize students into three groups, and assign each group one of the sections of Chapter 41. Have group members draw pictures to illustrate the main concepts of the section. Students should write captions for their illustrations. Choose volunteers from each group to compile the illustrations into a bulletin board display titled "The Physical, Economic, and Human Geography of East and Southeast Asia."

Learners Having Difficulty

Have pairs of students list the main topics of Chapter 41 on a sheet of paper, leaving space between each. Then have pairs exchange papers, and have students add subtopics beneath each main topic. Call on volunteers to read parts of their outlines as you copy their suggestions on the board. Fill in any topics or subtopics that have been left out. Ask students to correct their outlines and to use them as study aids.

Auditory Learners

Assign each student a country in East and Southeast Asia. Have students create an oral history of a particular period in the assigned country's history. Students should use information presented in the chapter. Have volunteers present their oral histories to the class.

Visual Learners

To help students appreciate the extremes of rainfall that characterize much of East and Southeast Asia, have them graph the average yearly rainfall totals given in Section 1 of Chapter 41. Then provide students with the data for the area in which they live, and have them add it to their graphs. Ask students to describe how the amount of rainfall a region receives affects the ways of life in the region.

Tactile/Kinesthetic Learners

Have students construct two- or three-dimensional models showing various economic activities in East and Southeast Asia. Subjects for models might include paddy agriculture, aquaculture, logging, industry, and trade.

Gifted Learners

Have interested students investigate East and Southeast Asian endangered species and give an oral report describing the actions taken by countries in the region to save these animals.

MULTICULTURAL FOCUS

Point out that a society's institutions—such as laws, religion, family, and education—are based on philosophical ideas. The ideas of the Chinese philosopher Confucius (551–479 B.C.) have shaped ways of life in East and Southeast Asia.

In Confucius's time, during the decline of the Zhou dynasty, Chinese states were perpetually at war with one another. Confucius feared that the constant upheavals in Chinese society were destroying standards of morality, behavior, and social organization and threatening the very existence of civilization. In order to heal the society, Confucius believed, people should aspire to a high moral character, both in private and in public.

Confucius felt that political leaders, in particular, should act with virtue, so that their conduct would be emulated by their followers. He said, "Only if the right sort of men had charge of a country for a hundred years would it become really possible to stop cruelty and do away with slaughter." The Confucian ideal of political rule by a learned elite and emphasis on the individual's respect for tradition, order, and duty to society still have great influence in East and Southeast Asia.

Ask students if they can identify some of the philosophical underpinnings on which U.S. society and government are based. Ask what personality traits students consider most important in U.S. citizens and leaders.

OVERVIEW

Chapter 41 introduces students to the region of East and Southeast Asia. Approximately one-third of the world's population lives in this region.

Section 1 describes the physical geography of East and Southeast Asia, where rugged landforms and harsh climates limit areas of settlement.

Section 2 highlights the region's economic geography. It explains the role of agriculture as the region's primary economic activity and describes the effects of rapid industrialization.

Section 3 examines the region's human geography, focusing on diverse languages, religions, and cultures and the effects of colonialism on current political systems.

CREATIVE STRATEGIES FOR TEACHING WORLD GEOGRAPHY

LESSON 18 Asian Holiday Calendar

CHAPTER 41

Introducing East and Southeast Asia

GEOGRAPHY DICTIONARY

paddy

aquaculture

newly industrialized country (NIC)

pictogram

Buddhism

Taoism

Confucianism

Shinto

East and Southeast Asia is home to some 2 billion people, roughly one-third of Earth's human population. With its forbidding mountains, blazing deserts, fertile river valleys and deltas, and volcanic islands, this vast region is one of great geographic as well as cultural diversity. Here, traditional farming villages, huge modern cities, and some of the planet's last large wilderness areas all compete for space.

Korean temple wall painting

The diverse resources and peoples of East and Southeast Asia are connected to the world community. Each day, automobiles, televisions, cameras, computers, bicycles, clothing, and canned foods from Japan, South Korea, Indonesia, China, and Taiwan (TIE-WAHN) pour into the United States, Europe, and other countries around the globe. At the same time, however, some of the world's poorest nations struggle to survive and develop their economies in this region.

494

Celebrating the Chinese New Year in **Singapore**

Using the Geography Dictionary

Pronounce for students the terms in the Geography Dictionary on page 494. Have students look up terms they do not know in the Glossary. Then ask them to suggest a classification system for the terms. *(For example: two terms relate to raising* food—**paddy, aquaculture;** *four terms name religions*—**Buddhism, Taoism, Confucianism, Shinto;** *one term concerns economics*—**newly industrialized country (NIC);** *one term is linked to language*—**pictogram***)*

As students study the map on this page, point out significant features: (1) volcanic islands of Japan, the Philippines, and Indonesia on the Pacific Ring of Fire; (2) natural borders (Himalayan and Tian Shan ranges); (3) mountain ranges in western China that are the source of major rivers.

MAP TRANSPARENCIES

Overhead transparencies from the maps on pages 495–497 are available in the *REGIONS OF THE WORLD Map Transparencies.*

MAPPING THE REGION

EAST AND SOUTHEAST ASIA

PHYSICAL–POLITICAL

LEGEND

ELEVATION

FEET		METERS
13,120		4,000
6,560		2,000
1,640		500
656		200
(Sea level) 0		0 (Sea level)
Below sea level		Below sea level

⊛ National capitals

• Other cities

SCALE

0 — 500 — 1000 Miles

0 — 500 — 1000 Kilometers

Projection: Two-Point Equidistant

MAPPING THE REGION

Using the Climate Map

As students examine the climate map on this page, point out that the region stretches from 50° north of the equator to 10° south. Ask students what effect this wide latitude range would be likely to have on the climate of the region. (*a wide range of climates*) Have them discuss the extremes of climates found in the region, which range from subarctic to humid tropical. Have students identify other countries that lie as far north as Mongolia (*Canada, for example*) and as far south as Indonesia (*Brazil, for example*).

MAPPING THE REGION

Using the Population Map

Have students examine the population map on this page. Point out that almost one-third of the world's population lives in this region. Have students discuss the uneven distribution of population. Ask them to use what they have learned from the physical-political map of this region, and from the climate map, to suggest reasons for this uneven distribution.

Continues on page 497

CLIMATE

Answers

1. Himalayas, Kunlun Mountains, Plateau of Tibet
2. Borneo, New Guinea
3. Singapore, Malaysia, Philippines

POPULATION

Answers

1. Osaka, Tokyo
2. Desert and highland climate regions discourage economic development and widespread settlement.
3. Java

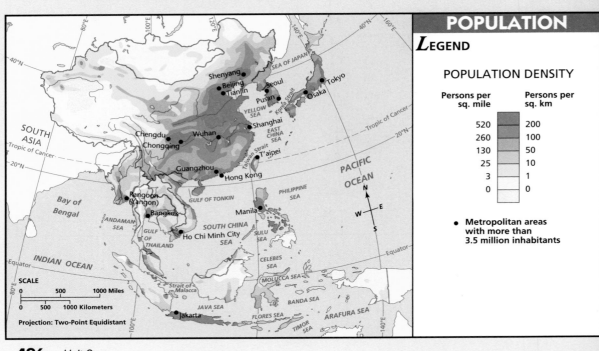

MAPPING THE REGION

EAST AND SOUTHEAST ASIA

CLIMATE
LEGEND

- Humid tropical
- Tropical savanna
- Desert
- Steppe
- Humid subtropical
- Humid continental
- Subarctic
- Highland
- ← Wet monsoon air flow
- ← Dry monsoon air flow

POPULATION
LEGEND

POPULATION DENSITY

Persons per sq. mile	Persons per sq. km
520	200
260	100
130	50
25	10
3	1
0	0

● Metropolitan areas with more than 3.5 million inhabitants

MAPPING THE REGION

Using the Economy Map

As students study the economy map on this page, point out that most of the people are involved in agriculture. Ask students which type of agriculture is practiced in the most densely populated areas. (*commercial farming*) What kind of economic activity is practiced in Mongolia? (*nomadic herding*) in the Malay Archipelago? (*forestry*)

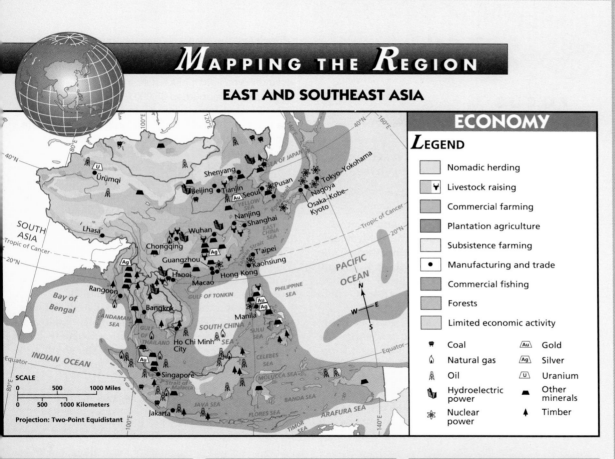

MAPPING THE REGION

EAST AND SOUTHEAST ASIA

ECONOMY

LEGEND

- Nomadic herding
- Livestock raising
- Commercial farming
- Plantation agriculture
- Subsistence farming
- Manufacturing and trade
- Commercial fishing
- Forests
- Limited economic activity

Coal		Au	Gold
Natural gas		Ag	Silver
Oil		U	Uranium
Hydroelectric power			Other minerals
Nuclear power			Timber

SCALE
0   500   1000 Miles
0   500   1000 Kilometers
Projection: Two-Point Equidistant

ECONOMY

Answers

1. Singapore, Jakarta, and Manila are all in the Malay Archipelago.
2. Humid subtropical, steppe, humid continental
3. Warm, wet climate is ideal for rice farming.

CLIMATE

◆ Dry desert and variable highland climates dominate large regions of northern and western China.

◆ The coasts of Southeast Asia and much of Indonesia have a humid-tropical climate, with warm, rainy weather all year.

◆ A humid-continental climate, with cold winters and warm, moist summers, dominates northern Japan. Southern Japan, with warm, moist summers and mild winters, has a humid-subtropical climate.

Compare this map to the physical–political map on page 495.

❓ What blocks the southern flow of moisture to the desert areas of western and northern China?

❓ Which two large islands in the Malay Archipelago have highland climates?

❓ Which countries have only a humid-tropical climate?

POPULATION

◆ The huge population of East and Southeast Asia is unevenly distributed, with large sparsely populated areas as well as densely populated regions and islands.

◆ The most crowded part of China, the world's most populous country, is its eastern half.

◆ Many metropolitan areas with populations of more than 3.5 million each are located throughout East and Southeast Asia.

❓ Which Japanese metropolitan areas have populations of more than 3.5 million?

Compare this map to the climate map on page 496.

❓ Why do you think western areas of China are less populated than eastern regions of the country?

Compare this map to the physical–political map on page 495.

❓ Which Indonesian island is most densely populated?

ECONOMY

◆ Large deposits of coal and minerals such as tungsten, tin, and iron ore can be found in East and Southeast Asia.

◆ Tropical-forest lumber, especially valuable hardwoods, are major exports of Southeast Asia.

◆ In the desert and high-plateau areas, herders raise sheep and other livestock.

❓ Name two major manufacturing and trade centers of the Malay Archipelago.

Compare this map to the climate map on page 496.

❓ Which climate types dominate the regions used for commercial farming in China?

❓ Why do you think rice farming is an important economic activity in many parts of East and Southeast Asia?

The objectives reflect the Focus questions on this page and are assessed in the review on page 499.

1. Identify landforms and climates of East and Southeast Asia.
2. Explain how landforms and climates relate to patterns of human settlement in East and Southeast Asia.

Remind students that some 2 billion people live in East and Southeast Asia. Focus their attention on the population density map on page 496. Note that the population is not evenly distributed. Ask students to speculate on reasons for this. Tell them that they will learn how the region's physical features influence population distribution.

Locating Landforms

Organize the class into three groups: Mountains, Coastal Plains, and Plateaus. Tell students to be ready to locate examples of their landform on a map of East and Southeast Asia. Ask for positive and negative factors associated with living on or close to a particular landform. *(For example, an advantage of living in the Huang He valley would be rich soil for farming. A disadvantage would be the threat of flooding.)*

Geographical Sidelight

The volcanic eruptions in East and Southeast Asia often have global effects. For example, the 1883 eruption of the Indonesian volcano Krakatau generated explosions that were heard 2,200 miles (3,540 km) away in Australia. Krakatau also spewed enough ash and dust to drift around the earth several times and cause beautiful red sunsets all through the following year. Furthermore, the collapse of the volcano triggered a series of huge sea waves that affected places as far away as Hawaii and South America.

PHYSICAL GEOGRAPHY

FOCUS

● *What landforms and climates are found in East and Southeast Asia?*

● *How have these landforms and climates influenced human settlement in the region?*

LOCATION **Landforms** East and Southeast Asia stretches from Burma (also called Myanmar) in the southwest to Japan in the northeast. The region borders Russia on the north and India on the southwest. The Indian Ocean lies to the south and the Pacific Ocean to the east. (See the map on page 495.)

Mountainous volcanic islands, such as Japan, the Philippines (FIL-uh-peenz), and Indonesia, line East and Southeast Asia to the south and east, forming part of the Pacific Ring of Fire. Tectonic activity caused by the subduction of the oceanic plates under the continental plates in this area has formed the world's largest concentration of active volcanoes. Eruptions from these volcanic mountains can be deadly, but they are also the source of rich soils.

Paralleling the island chains are some of the world's deepest marine trenches. The differences in elevation between these trenches and the summits of the islands' volcanic peaks just a few hundred miles away can be more than 35,000 feet (10,668 m).

Rugged mountains, plateaus, and hills dominate much of mainland East and Southeast Asia. The great mountain ranges and plateaus of the region are natural borders and often serve as political and cultural boundaries as well. The forbidding Himalayas, the world's highest mountain system, form a distinct boundary between East Asia and South Asia. Similarly, the high Tian Shan (tee-ahn SHAN) range separates northwestern China from Russia and some of the Central Asian republics. The Kunlun (KOON-LOON), Qin Ling, and Greater Khingan (SHING-AHN) ranges all lie within China. Numerous smaller ranges fan out from north to south across the Southeast Asian mainland.

Between the mountains lie isolated plateaus and basins. Few people inhabit the plateaus because they are very dry with extremely cold winters. Between the Himalayas to the south and the Kunlun Mountains to the north lies the

The Japanese mold the natural landscape into beautiful gardens in public places and private residences throughout the country. This garden is located on the island of **Shikoku**.

Guided Practice

Have students work in pairs to complete a climate chart for East and Southeast Asia. Along the left side of the chart, list questions such as these: What landforms or bodies of water affect this climate area? What is the population density? What are the major economic activities? Across the top of the chart, have students list the different climate types. Have them answer each question for each climate type.

Independent Practice

Provide all students with Outline Maps 20 and 21. Have them complete the maps to show the major landforms and rivers of East and Southeast Asia. Then have students write a paragraph summarizing the relationship of population density to climate and landforms in East and Southeast Asia.

Closure

Call out the names of mountain ranges, rivers, plateaus, plains, and basins in East and Southeast Asia, and have students see how quickly they can locate them on the map on page 495.

ASSESSMENT

Assign the Section 1 Review questions on this page.

Plateau of Tibet, the world's highest plateau. Its elevation averages 16,000 feet (4,877 m) above sea level. In contrast, part of the arid Tarim Basin located between the Kunlun and the Tian Shan ranges drops below sea level. To the east of the Tarim Basin is the Gobi, the desert that makes up part of the plateau of Mongolia.

The region's high mountains are the source of all of East and Southeast Asia's major rivers, including two of the world's longest, the Huang He (Yellow River) in northern China and the Chang (Yangtze) in central China. The region's river valleys and deltas are the most densely populated places on Earth. When the rivers overflow their banks during floods, rich alluvial soil is spread over the surrounding land. Some of this fertile land has been under cultivation for more than 5,000 years and continues to feed the region's huge population. The river waters also are important for transportation and hydroelectricity.

The region's limited lowlands hold the other major centers of population. People crowd onto the narrow coastal plains and the interior areas of the North China Plain, the Manchurian Plain, and the Sichuan Basin.

INTERACTION Climate Climate factors play a part in the high population density of the river valleys and plains of East and Southeast Asia. The climates outside these lowlands, especially those of the desert areas of the interior, cannot support large human populations. For example, the Gobi of Mongolia and northern China and the Taklimakan Desert of northwestern China have hot summers and bitterly cold winters. The rain-shadow effect of the surrounding mountains prevents moist ocean air from bringing rain to these desolate areas. Likewise, areas in the highland climates of interior East Asia and in the severe subarctic climate along the borders of Mongolia, China, and Russia are too cold for much human settlement.

Other climates of the region include a steppe climate in areas of China and a humid-continental climate in parts of northeastern China, North Korea, and northern Japan. The cold, snowy winters and hot summers of the humid-continental regions limit agriculture. Southern Japan, South Korea, Taiwan, and southern China have a humid-subtropical climate, with warm, moist summers and mild winters. This climate region lies in the path of severe tropical storms, or typhoons.

These storms form in the tropical North Pacific Ocean and flow to the northwest, following the warm waters of the Japan Current (Kuroshio). They usually strike the Philippines, Japan, South Korea, southern China, or Taiwan.

The western islands of Indonesia and the coasts of Southeast Asia have a humid-tropical climate. Prior to recent deforestation, these areas were completely covered by dense tropical rain forests. Temperatures are warm, and rainfall is heavy all year. Some of the eastern islands of Indonesia and the interior mainland of Southeast Asia have a tropical-savanna climate with a distinct dry season.

Rainfall Extremes of rainfall characterize much of East and Southeast Asia. In some coastal and island locations of Southeast Asia, yearly rainfall totals average more than 80 inches (203 cm). Around 250 inches (635 cm) of rain fall in some mountain areas of Irian Jaya in Indonesia. In contrast, in the far continental interior, such as in the Taklimakan and Gobi deserts, annual rainfall totals less than 10 inches (25 cm). Interior cities, such as Beijing (BAY-JING), China, receive from 10 to 20 inches (25 to 51 cm) of rain per year.

Much of the rainfall in East and Southeast Asia is seasonal, due to the monsoon system. The winter months are dry, as winds blow from the north and west off the Asian continent. The summer months are humid and rainy, with winds bringing in moisture from the tropical waters of the Indian and Pacific oceans.

Unlike East Asia's mainland areas, island nations of the region, such as Japan and the Philippines, receive rain even during the winter. Dry winds flowing out of Asia must cross over the Sea of Japan or the South China Sea. Here they pick up moisture before reaching the islands.

SECTION REVIEW

1. Name five mountain ranges found in East and Southeast Asia. Between which ranges is the Tarim Basin located?
2. List five climate types found in East and Southeast Asia.
3. **Critical Thinking** **Explain** how the mountains of East and Southeast Asia both limit and support human settlement in the region.

Section 1 Review
ANSWERS

1. Himalayas, Tian Shan, Kunlun, Qin Ling, Greater Khingan; between the Kunlun and the Tian Shan
2. Desert, highland, subarctic, steppe, humid continental, humid subtropical, humid tropical, tropical savanna
3. Volcanic mountains can produce deadly eruptions; interior mountain ranges are too high and too cold to live in and block rain from reaching some areas. But the eruptions from volcanic mountains and the rivers that originate in the interior mountains are the sources of rich soils that support large populations.

Reteaching

Assign Section 1 of Reteaching Worksheet 41.

Extension

Have students use almanacs or other resources to identify the highest mountain peak in each of these ranges: the Himalayas, Tian Shan, Kunlun, Qin Ling, Greater Khingan. Then have them make a bar graph comparing the heights of these peaks. (Instead of bars, students might draw "peaks.")

SECTION 2
OBJECTIVES

The objectives reflect the Focus questions on this page and are assessed in the review on page 502.

1. List major resources of East and Southeast Asia, and analyze the use of these resources.

2. Discuss the region's rapid industrial growth.

TEACHING SECTION 2
Motivator

Have students examine product labels in their clothes and on their school supplies. On the board, list items that were manufactured in East and Southeast Asia. Use this quick inventory to point to the growing industrialization of the region's countries.

---

2 ECONOMIC GEOGRAPHY

FOCUS

- *What are the major resources of East and Southeast Asia? How have these resources been used?*

- *What effects has rapid industrialization had on the region?*

INTERACTION **Agriculture** East and Southeast Asia's land area is huge. Because of terrain and climate factors, however, only a small percentage of the land is suitable for farming. Yet most of the people of East and Southeast Asia are involved in agriculture. This combination of many farmers and relatively little arable land means that most of the region's farms are tiny. For example, in Japan the average farm size is about two and one-half acres (1 ha). In contrast, the average size of a farm in the U.S. state of Iowa is 328 acres (133 ha).

The type of farming practiced varies with the climate. In humid-tropical and humid-subtropical climates, rice is the major food crop. In general, rice is grown in water-covered fields known as **paddies** (also spelled *padis*). These flooded fields produce most of the food consumed by the region's large population.

In tropical climates, commercial planters produce export crops such as rubber, tea, coffee, coconuts, sugarcane, and various spices. In the cooler and drier humid-continental and steppe climates of China, wheat is the main food crop. In the desert and high-plateau areas, herders keep flocks of sheep or other animals.

Fishing Farming in seas and ponds, called **aquaculture**, is an important local activity throughout much of the region. Seafood such as shrimp and oysters are commercially farmed in protected bays and river mouths. For a local food supply, many nations in East and Southeast Asia depend upon freshwater fish from lakes, rivers, and flooded rice paddies.

Surrounded partially or entirely by water, several nations in the region, especially Japan, Taiwan, and South Korea, have large fishing fleets. In supplying the great demand for seafood, these fleets have exhausted local waters and now fish across the entire Pacific Ocean. Although vast, even the Pacific Ocean has limited resources. Parts of its fish population are now threatened.

Forests The forests of the region are generally divided into two types, middle-latitude and tropical. The middle-latitude forests lie mainly in Japan, China, and North and South Korea. The tropical rain forests of Thailand (TIE-land), Indonesia, Burma, and the Philippines once stretched from border to border in these countries.

Logging of these forests has increased at an amazingly rapid rate because of the demand in Japan and other countries for tropical-forest lumber, especially valuable hardwoods. In Thailand, for example, only one-fifth of the country's original forests remain. If this rate and the destructive methods of removal continue, most of the region's tropical-hardwood forests will not survive beyond the next few decades. In addition, deforestation leads to more flooding and soil erosion. It also threatens the habitats of the region's endangered animals, such as the panda, Asian rhino, and orangutan.

Mineral and Energy Resources Some of East and Southeast Asia's most abundant mineral resources are tin, tungsten, coal, and iron ore. More

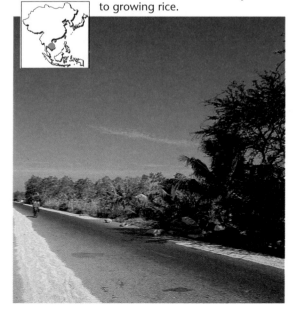

Harvesters have left rice to dry along this road in southern **Vietnam**. The wet, warm climates of East and Southeast Asia are ideally suited to growing rice.

500

Building a Vocabulary
Use word analysis to help students learn the meaning of **aquaculture.** Explain that the prefix *aqua-* means "water." *Culture* as a suffix refers to the growing of plants or animals with the intent to improve their development. Have students think of other words containing *aqua-* or *-culture.* (*aquamarine, aquarium, agriculture, horticulture, bee culture*)

Guided Practice
Direct students' attention to the bar graph on this page. Point out factors that have fueled economic growth in the countries listed. (*development of natural resources, low costs for labor and land encourage investors*) Then discuss ways that rapid industrialization has affected the region. (*increased migration to cities creates slums; increased air and water pollution*)

Independent Practice
Have students create a graphic organizer to depict information in Section 2. The organizer might provide an overview of the region's resources, focus on one resource, or link agricultural products to climate types. Encourage students to share their organizers with the rest of the class.

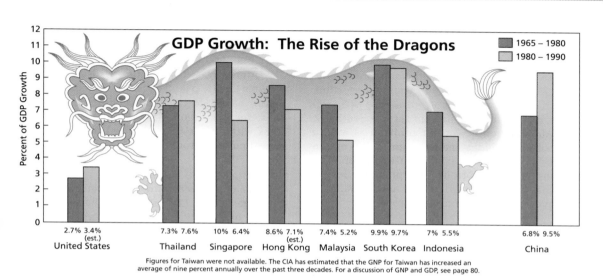

GDP Growth: The Rise of the Dragons

Legend: 1965 – 1980 / 1980 – 1990

Percent of GDP Growth

United States	Thailand	Singapore	Hong Kong	Malaysia	South Korea	Indonesia	China
2.7% 3.4% (est.)	7.3% 7.6%	10% 6.4%	8.6% 7.1% (est.)	7.4% 5.2%	9.9% 9.7%	7% 5.5%	6.8% 9.5%

Figures for Taiwan were not available. The CIA has estimated that the GNP for Taiwan has increased an average of nine percent annually over the past three decades. For a discussion of GNP and GDP, see page 80.

Source: *World Development Report 1992: Development and the Environment,* The World Bank

 SKILL STUDY These bar graphs show how the economic growth of the "Four Dragons" and "Little Dragons" of Asia outpaced that of the United States in the 25 years after 1965. Although the United States still had the largest economy in the world, the economies of these East and Southeast Asian countries and China grew much faster. Which two countries experienced the fastest rate of growth during the 1980s?

than half of the world's supply of tin is found in the region, mainly in Malaysia, Thailand, and Indonesia. The region also holds more than one-third of the world's supply of tungsten, which is used in the production of electronics and lighting materials.

Parts of the region hold large oil deposits. Indonesia, Malaysia, and Brunei (broo-NIE) are the major producers and exporters of this resource. Although there are natural-gas and oil deposits in China, Burma, and Thailand, none of these countries appears to have enough to support major exporting. China has developed several medium-size petroleum fields and may become a more important oil exporter as more reserves are discovered and developed.

Melting snow from the region's high mountains and large amounts of rainfall provide dependable sources of hydroelectric power. Japan has effectively developed its hydroelectric resources. China and other countries in the region have the potential to do the same.

Industrialization
Japan is the region's major economic power and the world's second leading industrial nation after the United States. It is also a leader in foreign investment and international banking, and is a major producer and exporter of automobiles and high-tech products such as electronics. Although Japan's exports are far greater than its imports, the nation is a major importer of raw materials, such as coal and oil.

Rapid industrialization now characterizes several other countries of East and Southeast Asia. These nations are known as **newly industrialized countries (NICs).** Collectively called the "Four Dragons" or "Four Tigers," the NICs of Taiwan, Singapore, Hong Kong, and South Korea have experienced explosive economic growth and are now very competitive, even against Japan. Ironically, much of their growth has been fueled by Japanese investors and industries seeking lower labor and land costs.

As industrialization and trade expand, other NICs have appeared. In recent years, Malaysia, Thailand, and Indonesia, called the "Little Dragons" or "Little Tigers," have shown impressive economic growth. Some nations of East and Southeast Asia, however, remain isolated and poor. Vietnam, Laos, Cambodia, Burma, and Mongolia all have struggling economies.

COOPERATIVE LEARNING
Organize the class into groups of four or five. Assign each group to research the major trading partners of one of the newly industrialized countries mentioned on this page. Each group should draw a map illustrating its country's trading links and attach a caption explaining the map. Roles for individual group members might be researchers, note takers, mapmakers, and caption writers. Afterward, create one large outline map of the world on butcher paper or on the board. Have each group come up one by one and draw its country's trade links on the outline map. Each group should use different-colored markers or chalk.

 Skill Study
South Korea, China

GLOBAL SKILL BUILDER

You might wish to use **Reading Graphs** from the interactive *Global Skill Builder* CD-ROM to reinforce students' understanding of bar graphs.

501

Closure

Have students work in pairs to list resources of East and Southeast Asia. Then have partners quiz each other on their lists.

ASSESSMENT

Assign the Section 2 Review questions on this page.

Reteaching

Assign Section 2 of Reteaching Worksheet 41.

Extension

Have interested students write a report on aquaculture in East and Southeast Asia. Limited land resources and a huge population make aquaculture a vital food source for the region. Aquaculture also holds promise as a food source for other areas of the world.

SECTION 3

OBJECTIVES

The objectives reflect the Focus questions on this page and are assessed in the review on page 504.

1. Summarize the region's population characteristics, including growth, cultures, languages, and religions.
2. Examine the region's historical and political geography.

Section 2 Review
ANSWERS

1. Rice, rubber, tea, coffee, coconuts, sugarcane, spices, wheat; tin, tungsten, coal, iron ore, oil, natural gas, hydroelectricity
2. The growth of shantytowns, sewage problems, traffic congestion, water and air pollution
3. Both have been used at a rapid rate in response to economic demands, and both are now threatened.

Geography and Language

More than 900 million people worldwide speak Mandarin Chinese, more people than speak any other language. It is followed by English, Hindi, Spanish, and Russian. On a world languages map, compare the areas of the world where these five languages are spoken. Ask students if there are places outside of China where Chinese is regularly spoken. (Students may mention cities outside of China with a large Chinese population.)

The region's rapid industrial growth has caused a large-scale movement of people from rural areas to the cities, resulting in more shantytowns, sewage problems, and traffic congestion. Industrialization also has led to severe pollution of the rivers and coastlines of East and Southeast Asia. Air pollution is a major problem as well; a cloak of smog hides many major cities in the region. Singapore, however, is an exception. The city has greatly controlled its pollution levels. Other nations, particularly Japan and Taiwan, are now investing in environmental protection.

SECTION REVIEW

1. What major crops are raised in East and Southeast Asia? What mineral and energy resources are important to the region?
2. What challenges face the region as a result of its industrial growth?
3. **Critical Thinking** **Compare** ways in which people of East and Southeast Asia have used fish resources to the ways they have used forest resources.

HUMAN GEOGRAPHY

FOCUS

- *In what ways is the population of East and Southeast Asia diverse?*
- *How have foreign countries played a role in the region's history? What different political systems are found in the region?*

Population and Culture Roughly one-third of the world's people live in East and Southeast Asia. The population is not distributed evenly among the region's countries, however. China has more than 1 billion people, while tiny Brunei has fewer than a million.

Population growth in much of the region continues at a rapid rate. Birthrates have increased and death rates have decreased because of improved medical care. Some nations now have programs to control their population growth, and Japan and Singapore have extremely low birthrates. Even China's population growth rate has slowed somewhat, though some officials

estimate the country's population may reach more than 1.5 billion in the next 30 years.

A great variety of cultures exists among the region's many peoples. Some nations, such as Japan, have one dominant culture. Other countries, such as Indonesia, have a diversity of cultures within their borders. The Chinese form the largest cultural group.

Language The people of East and Southeast Asia speak hundreds of different languages. Some languages, like Chinese, are character-based. Instead of a small alphabet of letters, the written Chinese language has nearly 50,000 characters. Many scholars believe some of the characters are symbols or **pictograms**, simple pictures of the objects and ideas they represent. For example, many of the Chinese characters, such as the characters for mountain and tree, reflect the physi-

Mountain

Tree

Scholars have studied how the Chinese written language reflects the country's geography. Here, you can see that mountain peaks are suggested in the character for *mountain*. Similarly, the character for *tree* resembles the trunk and bare limbs of a tree.

Have students look at the photographs on pages 490, 491, and 494. Ask volunteers to read the captions aloud. Ask: What do these photographs tell us about the cultures and histories of East and Southeast Asia? Tell students that the images in the photographs show only a small part of the region's human geography, which will be studied in Section 3.

Building a Vocabulary

Focus first on the suffix *-ism*. Explain that this suffix can form words meaning "a doctrine, theory, system, or principle." **Buddhism, Taoism,** and **Confucianism** are examples. Then have a volunteer read aloud the Glossary definition of **pictogram**. Have students try to create their own pictograms for physical features such as a river, a forest, and a mountain.

Guided Practice

Have pairs of students create a three-column chart summarizing the history and political status of each of the countries of this region. In the three columns, students should write the name of the past colonizing nation (if any), independence date, and current type of government. Provide students with additional resources if necessary. Discuss why the political systems in the region are so varied.

---

cal geography of China. (See the illustration on page 502.) Characters can also be combined. Two tree characters placed side by side become the character for *forest*. Three tree characters grouped together may mean "thick forest," "many trees," or "dark." The characters and their meanings have gradually changed over time. Tracing their origins and development reveal part of the cultural history of China and its people.

Within a single country in East and Southeast Asia, dozens of languages may be spoken. As a result, some countries have adopted one official language to unify the people, as in China and Indonesia. Even in these countries, however, many regional dialects exist. In contrast, Japan and Korea each are relatively uniform in terms of languages. The Japanese language dominates in Japan, and Korean is the major language used in North and South Korea. South Korea even has a public holiday to celebrate its written alphabet, called *han'gul*, which was developed in the fifteenth century.

English and French are still used in many East and Southeast Asian countries with colonial histories. English also has become the business language of several nations in the region.

Religion Most of the world's major religious systems are represented in East and Southeast Asia. Islam is the main religion in Indonesia, Malaysia, and Brunei. In fact, Indonesia has more Muslims than any other country in the world. Christianity is practiced primarily in the areas of the region, such as the Philippines, that were once controlled by European countries. South Korea also has a fairly large Christian population.

Buddhism (BOO-diz-uhm) is the major religion in Thailand, Burma, Tibet (in China), Laos, Cambodia, Vietnam, and Mongolia. Buddhism was founded in India around 500 B.C. by a teacher called the Buddha. *Buddha* means "Enlightened One," and Buddhists seek enlightenment through meditation (deep thought). Buddhism is divided into a few major branches, which are practiced in different regions. The two main branches are Theravada (ther-uh-VAHD-uh) Buddhism and Mahayana (mah-huh-YAHN-uh) Buddhism. Theravada Buddhism, which is practiced mainly in Southeast Asia, is the earlier branch. It believes that humans must look only to themselves for salvation. In contrast, Mahayana Buddhism, practiced in China, Korea, and Japan, holds that

humans may also seek the aid of spiritual beings. There are also many local varieties of Buddhism.

China's religions are a mixture of Buddhism, Taoism, and Confucianism. **Taoism** (DOW-iz-uhm) was founded in China in the sixth century B.C. by a man named Lao-Tzu. He taught that there is a natural order to the universe, called the *Tao*, or "way." The basic idea of Taoism is to live a simple life close to nature in order to be in perfect harmony with the *Tao*. **Confucianism**, which is based on the teachings of Confucius, a Chinese philosopher who lived from 551 to 479 B.C., spread throughout China and later to Korea and Japan. It is more a code of ethics than an organized religion. The Confucian code centers around family, social relationships, and duty.

Unlike Judaism, Christianity, and Islam, Buddhism, Taoism, and Confucianism do not claim to have been revealed by a god. Instead, they stem from the teachings of individual

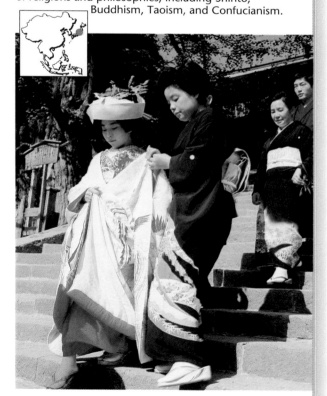

This Japanese bride prepares for her Shinto wedding. The people in East and Southeast Asia follow a variety of religions and philosophies, including Shinto, Buddhism, Taoism, and Confucianism.

COOPERATIVE LEARNING

Organize the class into small groups. Assign one of the main religions of East and Southeast Asia to each group member. Each student should research the origins and beliefs of the assigned religion. Have students share their findings with their groups. Then have each group make a chart comparing the religions.

MULTICULTURAL PERSPECTIVES

Not all cultures use the Gregorian calendar. In the Chinese calendar, for example, years come in cycles of 60. Each 60-year cycle is divided into five mini-cycles of 12, in which each year is named for an animal. The animals, in order, are the rat, ox, tiger, hare, dragon, snake, horse, sheep, monkey, rooster, dog, and pig. The year 2000 will be the 17th year in the 78th cycle. As the fifth year in the second mini-cycle, it will be the year of the dragon. Have students compute the animals associated with the years in which they were born.

Closure

Students might make and write the clues for a crossword puzzle relating to the human geography of East and Southeast Asia. Have students complete each others' puzzles.

ASSESSMENT

Assign the Section 3 Review questions on this page.

Reteaching

Assign Section 3 of Reteaching Worksheet 41.

Extension

Have students make a mobile depicting East and Southeast Asia. For example, they might draw or cut out pictures of foods, manufactured products, animals, clothing styles, and religious symbols that characterize the region and then hang them in a mobile.

Section 3 Review
ANSWERS

1. Islam (Indonesia, Malaysia, Brunei), Christianity (South Korea, countries once controlled by European countries), Buddhism (Thailand, Burma, China, Laos, Cambodia, Japan, the Koreas, Vietnam, Mongolia), Taoism (China), Confucianism (China, Korea, Japan), and Shinto (Japan)

2. Great Britain, the Netherlands, France, Portugal, Spain; the United States controlled the Philippines for over 45 years and became involved in the Korean and Vietnam wars.

3. English and French are spoken and Christianity is practiced in nations with colonial histories.

504

human beings. Meanwhile **Shinto** (SHIN-toe), a religion practiced in Japan, does not claim to have any founder at all. Shintoists believe that gods, called *kami*, inhabit natural objects such as rivers, trees, rocks, and mountains. The worship of these gods involves a loosely structured set of rituals focused around community activities at local shrines. An important symbol of Shinto is the *torii*, a gateway marking the entrance to these shrines. The daily lives of most Japanese people mix Shinto with Buddhism. For example, many Japanese weddings are Shintoist, whereas funerals might be Buddhist.

MOVEMENT **Colonial History** In the eighteenth and nineteenth centuries, many parts of East and Southeast Asia came under the control of foreign nations. Great Britain controlled the area that now contains the countries of Burma, Malaysia, Singapore, and Brunei. It also gained the small colony of Hong Kong on the southeast coast of China. The Netherlands oversaw the thousands of islands of the Dutch East Indies, known today as Indonesia. France controlled an area known as French Indochina, which later became Cambodia, Laos, and Vietnam. Portugal acquired small possessions on the island of Timor, in Southeast Asia, and Macao (muh-KAU), on the coast of China. And for over 45 years, the United States controlled the Philippines, which it had won from Spain in 1898 as a result of the Spanish-American War.

China also was a target of foreign control. Europeans dominated its coastal ports in the nineteenth century, and the Japanese invaded China during the 1930s and 1940s. Japan had also colonized Taiwan in 1895 and Korea in 1910. During World War II, Japan expanded its empire to control much of East and Southeast Asia.

Post–World War II After World War II, the countries of East and Southeast Asia sought independence and an end to colonialism. Maps of the region were redrawn to include the new nations of the Philippines, Indonesia, Burma, Vietnam, Cambodia, and Laos. In China in 1949, after a bitter civil war, the Communists established the People's Republic of China. The Nationalists, who opposed the Communists, escaped complete defeat by forming a government-in-exile on the island of Taiwan. The country's official name is the Republic of China.

The region continued to be a site of international interest and conflict into the second half of the twentieth century. In the early 1950s, the Korean War pitted South Korean, U.S., and United Nations forces against North Korea and China. After more than three years of fighting, the two sides agreed to a truce and Korea remained divided into two countries.

In the late 1960s and early 1970s, the Vietnam War again focused world attention on the region. This war, in which South Vietnamese and U.S. troops battled North Vietnamese forces, ended in the unification of North and South Vietnam under a Communist government.

Political Geography As a result of the countries' varied histories and experiences, political systems vary widely among the nations of East and Southeast Asia. While Japan has had a democracy for decades, other nations, such as the Philippines and South Korea, have only recently gained democracy. China and North Korea are Communist countries, while others, such as Burma, remain under the control of strict military dictatorships. Hong Kong and Macao are colonies of the United Kingdom and Portugal, respectively, but both will become part of China in the late 1990s.

Today, tensions among and within the countries of the region still dominate the political landscape. Indeed, thousands of refugees have had to leave their homelands as a result of conflicts in these countries. Border disputes and claims to various islands continue to affect relations between China and Vietnam, and between the Philippines and Malaysia. Vietnam has occupied parts of Cambodia and Laos.

SECTION REVIEW

1. What religions are practiced in East and Southeast Asia? In what countries is each one practiced by a large part of the population?

2. What European countries have been involved in East and Southeast Asia? How has the United States played a role in the region?

3. **Critical Thinking** **Assess** the extent to which East and Southeast Asia was influenced culturally by colonialism.

Motivator

Place a map of the United States alongside a map of China. Briefly summarize the climates, landforms, and population density of U.S. regions. Ask students to speculate on where similar features might be found in China. For example, where in China might a humid-subtropical climate be found, as in the southeastern United States? (*Southern China*)

Building a Vocabulary

Tell the class that grouping words by theme helps in recalling their meanings. The terms for this section relate to benefits and improvements. Ask students what benefits are gained by **double cropping** and **Special Economic Zones (SEZs)**. (*double cropping—provides two or more crops per season; SEZs—facilitate economic investment*)

Identifying

Physical Geography and Cities

Write the names of China's major landforms, rivers, and cities on slips of paper. Fold the slips of paper, and put them in an open container. Have each student choose one of the slips and then locate the landform or city on a wall map. If necessary, supply students with variant spellings of place names (for example, Chang [Yangtze] River).

---

demonstrations in Beijing, China's capital. Chinese army troops and tanks crushed the protests, killing and injuring hundreds of demonstrators in the city's Tiananmen Square.

SECTION REVIEW

1. Name, in order, China's major dynasties. How and when did the last dynasty end? What enabled Japan to take control of parts of China in the 1930s?
2. What goals did Mao Zedong have for China? What were the Great Leap Forward and the Cultural Revolution? How has China changed since Mao's death, and how has it remained the same?
3. **Critical Thinking** **Suggest** an explanation for why the Chinese were not prepared for attacks by foreign powers in the 1800s, even though China had earlier been the most advanced nation in the world.

CHINA'S REGIONS

FOCUS

- *What is the physical geography of each of China's regions? the economic geography?*
- *What are the major cities of southern and northern China?*

REGION To study a country of China's huge size and vast regional differences, geographers divide the country into four geographic regions. These four regions are southern China, northern China, northeastern China, and western China. (See the map on page 507.)

Southern China Southern China is bordered on the north by the Qin Ling range, on the west by the Plateau of Tibet, and on the east by the East China and South China seas. Vietnam, Laos, and Burma lie to the south. All of southern China has a humid-subtropical climate except for the island of Hainan, which has a humid-tropical climate.

Southern China is the nation's most productive region and contains more than one-fourth of its population. The most densely populated areas are the river basins along the southeastern coast, especially near the mouths of the large Chang and Xi rivers.

Despite continuous cultivation for more than 4,000 years, the soil of southern China remains fertile. The alluvial deposits left by flooding rivers and careful management renew the soils. Southern China is often called "China's Rice Bowl." Indeed, the production of rice in this region has helped make China the world's leading rice producer and a rice exporter.

The Chang Delta is one of China's major rice-growing regions. The humid-subtropical climate here provides a long growing season that allows **double cropping** on the same land. This practice raises two crops in one year. In the far south, three crops may be grown, usually two crops of rice and one of vegetables.

Rice is usually grown on small terraced paddies. In hilly terrain, each paddy is built at a different level, or terrace, from the next, increasing the amount of area available for cultivation. The paddy is plowed, fertilized, and flooded to make it ready for the seedlings. Traditionally, water buffalo have supplied the power for plowing, but small mechanical tillers are now replacing buffalo power.

Southern China's other important crops include tea and cotton. The mountainous region between the Xi and Chang rivers is the center of tea growing. Production here has helped make China the world's second largest producer of tea. Because most Chinese tea is used in China, however, the country ranks only third as a tea exporter, after India and Sri Lanka. Cotton is the main raw material for China's textile industry. The major cotton-growing area is the Chang River valley, where farmers often practice double cropping, alternating cotton with a food crop.

In addition to agriculture, much of China's industry and trade lies along the Chang River. Shanghai, the country's largest city and major seaport, is also located on the river's delta. With more than 50 universities and colleges, Shanghai is an important educational as well as industrial center.

Upstream lie the industrial cities of Nanjing and Wuhan. Their early industrialization was based on local iron-ore and coal deposits. The city of Chongqing, farther inland along the upper Chang River, lies in the Sichuan (Red) Basin, which has rich soils as well as coal and minerals.

Describing China's Regions

Organize students into several small groups. Assign one of China's four regions to each group. Tell the groups to pretend to be the assigned region's "Chamber of Commerce." Have them list factors that would attract industry or tourism to their region. For example, the Himalayas might attract mountain climbers to western China. Forests might attract lumber mills to northeastern China. Have groups share their lists and "promote" their regions.

Guided Practice

Draw a chart on the board that lists Southern China, Northern China, Northeastern China, and Western China across the top and Landforms, Climate, Population, Resources, and Economic Activity along the side. Have the class work together to complete the chart comparing China's four regions. Students may wish to copy the chart in their notebooks.

Independent Practice

Provide all students with a copy of Outline Map 20. Have them label the map to show the major landforms, rivers, and cities of China. Then have students write a paragraph comparing one of China's regions to the region where they live.

Geographical Sidelight

Silk is the filament used by silkworms to make their cocoons. The Chinese have been raising silkworms for thousands of years. For a long time, the wearing of silk clothing was allowed only to the Chinese nobility, and knowledge of how to raise silkworms was carefully guarded. Today, nearly all the world's silk comes from Asia, with China the major producer.

COOPERATIVE LEARNING

Organize the class into groups of four or five. Tell students that many people in China are starting their own businesses. Have students imagine that they are part of this wave of new entrepreneurs. Have each group write a plan for a new business. Each group member could be responsible for a different part of the plan, such as the type of business, location, advertising, personnel, or funding. Have groups share their plans and discuss why establishing a business might be difficult for many people in China.

Rice, sugarcane, and soybeans are grown here. The region is also a major silk producer.

Farther south, at the mouth of the Xi River, is the famous trading center of Guangzhou (Canton). Guangzhou is now the largest city south of Shanghai and has attracted much industrial and commercial development. In addition, the region's warm, wet climate allows a variety of crops to be grown in the fertile delta area surrounding the city. More resources may be found near Guangzhou's front door—large, untapped oil fields may lie just offshore in the shallow waters of the South China Sea.

Located on the southeast coast of China just south of Guangzhou is the British crown colony of Hong Kong. It is one of the world's busiest ports and largest financial centers. Much of China's trade and foreign investment passes through this port. Major industries include textiles, electronics, plastics, food processing, and tourism. All of this trade and industry has helped make Hong Kong one of the world's most densely populated places. The colony faces an ongoing struggle to house its more than 5 million people. Great Britain and China have agreed that Hong Kong will return to Chinese control in 1997. A capitalist system will continue, and the former colony will have some autonomy.

Near Hong Kong lies the Portuguese province of Macao. Its half million people are crowded onto only six square miles (15.5 sq. km) near the mouth of the Xi River. Textiles, fireworks, fish exports, and tourism provide the province's main income. After more than 400 years of Portuguese rule, Macao will be returned to China in 1999. Like Hong Kong, it will be allowed to keep a capitalist system.

Along the coast of southern China are **Special Economic Zones (SEZs)** designed to attract foreign companies and investment to China. The SEZs nearest to Hong Kong have experienced rapid economic growth. Areas that were rice fields a few years ago now bustle with modern factories, high-rise buildings, traffic jams, and new freeways. In these "open-door" ports, foreign goods and money circulate freely. Most foreign investment pours in from Taiwan, South Korea, and Japan, although investments from the United States are increasing.

Northern China Northern China lies north of the city of Nanjing and the Qin Ling range. It includes the Huang He valley and the fertile North China Plain. Chinese culture first developed here, and the region remains the nation's center of culture and political power. About one-half of China's population lives in this region. In fact, the North China Plain, which is about the size of Texas, has a population larger than that of the entire United States.

Unlike humid southern China, the drier and cooler north sometimes experiences droughts and dust storms. The climate is mainly steppe, with

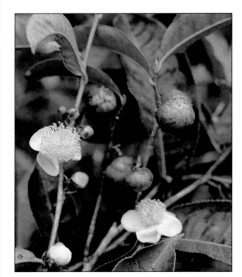

Tea: From Plantation to Consumer

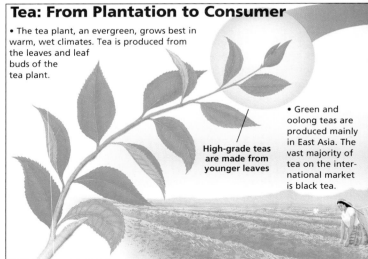

• The tea plant, an evergreen, grows best in warm, wet climates. Tea is produced from the leaves and leaf buds of the tea plant.

High-grade teas are made from younger leaves

• Green and oolong teas are produced mainly in East Asia. The vast majority of tea on the international market is black tea.

512

Closure

Write the names of China's four regions on the board. Through class discussion, summarize each region's physical features, climate, population, resources, and economic activity.

ASSESSMENT

Assign the Section 2 Review questions on page 515.

Reteaching

Assign Section 2 of Reteaching Worksheet 42.

Extension

Have students make a chart illustrating how rice is cultivated and processed. They can use the tea chart on pages 512–513 as a model. Display completed charts in the classroom. Students might also research customs, superstitions, and festivals related to rice.

very cold winters, especially in the interior. Along the coastal areas and on the Shandong Peninsula, the climate is humid continental. Here, winters are severe, but summer rainfall is reliable.

The Huang He begins in Tibet, flows across northern China, and empties into the Yellow Sea. It has changed its course many times and overflowed its banks, causing serious flood damage. Thousands have drowned over the centuries in the river's flood waters. For this reason, the Huang He has been called "China's sorrow." The Chinese have worked to tame the river's waters for irrigation and hydroelectric power.

Wind-blown deposits of fine yellowish-brown soil, called loess, cover much of northern China, making its lands yellowish. In fact, the dust turns houses, clothing, and water in the region all the same shade. The Yellow Sea and the Huang He (Yellow River) take their names from this yellow-colored loess.

The steppe climate of northern China has made this region the nation's wheat belt and made the country one of the world's largest wheat growers. Yet, because of its huge population, China imports wheat, especially from the United States, Canada, and Australia. Other grains grown in this region include millet and corn.

Beijing is northern China's largest city as well as China's cultural center and national capi-

Tiananmen Square lies near the heart of **Beijing** within the Inner City. Chinese troops violently ended weeks of massive demonstrations for political reform here in 1989.

tal. *Beijing* means "north capital." The city, which dates back more than 3,000 years, has served as China's capital almost continuously since the thirteenth century. The ancient, walled part of Beijing is divided into two sections, the Outer City and the Inner City. Within the Inner City lies the Imperial City, from which the government of the emperors ruled. Nested in the Imperial City itself was the Forbidden City, where the emperors lived, secluded from the world.

Geographical Sidelight

China had hoped that Beijing would host the Olympic Summer Games in the year 2000. The International Olympic Committee, which selects the Olympic sites, however, voted in 1993 to hold the 2000 games in Sydney, Australia, instead. A major factor in the committee's decision was concern about China's human rights abuses and repression of political dissent. Lead a discussion about why a country or city might want to host the Olympics. (The games bring prestige and economic benefits to the host.)

THE WORLD TODAY VIDEODISC PROGRAM

China: Community Ties

2B 22531 4:23

Play   Pause

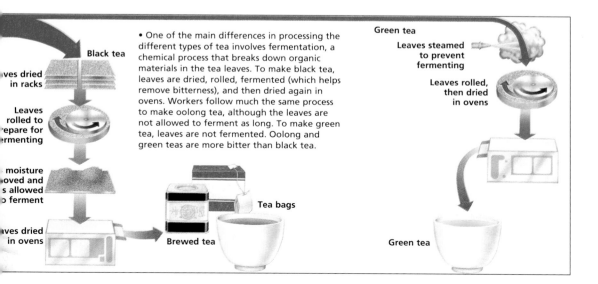

Black tea

ves dried in racks

Leaves rolled to prepare for fermenting

moisture oved and s allowed o ferment

ves dried in ovens

Brewed tea

Tea bags

• One of the main differences in processing the different types of tea involves fermentation, a chemical process that breaks down organic materials in the tea leaves. To make black tea, leaves are dried, rolled, fermented (which helps remove bitterness), and then dried again in ovens. Workers follow much the same process to make oolong tea, although the leaves are not allowed to ferment as long. To make green tea, leaves are not fermented. Oolong and green teas are more bitter than black tea.

Green tea

Leaves steamed to prevent fermenting

Leaves rolled, then dried in ovens

Green tea

SECTION 3
OBJECTIVES

The objectives reflect the Focus questions on page 515 and are assessed in the review on page 516.

1. Analyze Taiwan's physical geography and human geography.
2. Discuss challenges facing Mongolia.

TEACHING SECTION 3
Motivator

Direct students' attention to the map of China, Taiwan, and Mongolia on page 507. Ask students to describe the outstanding difference in physical geography between Taiwan and Mongolia. *(Taiwan is an island. Mongolia is a landlocked nation.)* Tell students that in this section they will discover many other contrasts between the two countries.

Building a Vocabulary

Write the term **martial law** on the board. Tell students that *martial* is derived from *Mars,* the name for the Roman god of war. Ask students how *martial* relates to its root *Mars.* *(Martial law is military rule of a civilian population, often during an emergency such as a war.)* Then ask students to explain the meaning of *martial arts.* *(skills related to self-defense)*

Section 2 Review
ANSWERS

1. Southern China has a humid-subtropical climate and grows rice, vegetables, tea, cotton, sugarcane, and soybeans. Northern China has steppe and humid-continental climates, and wheat, millet, and corn are grown. Southern China, northeastern China, and western China all have mineral resources.

2. Shanghai, Nanjing, Wuhan, Guangzhou, Hong Kong (U.K.), Macao (Portugal); Beijing, in northern China

3. Possible answer: yes, because the two provinces have different backgrounds from China and adjustment will not be easy; also, China will have to deal with the overcrowding in these cities.

514

The Potala Palace commands the heights around **Lhasa, Tibet**. The palace, with more than 1,000 rooms, is the former home of the Dalai Lama.

Beijing grew beyond its original walls and spread out across the plains. Today, it is a modern city of industry, subways, department stores, hotels, and wide boulevards. Parliament, housed in the Great Hall of the People, and the Museum of the Chinese Revolution surround Beijing's Tiananmen Square.

Northeastern China Northeastern China consists of the three provinces that were once known as Manchuria, and the autonomous region of Inner Mongolia. The main landforms of the northeastern provinces are the Manchurian Plain and a surrounding rim of mountains. The region has a humid-continental climate with severe winters. Forested areas of subarctic climate cover the very far northern sections. The climate here is similar to that of neighboring Siberia (in Russia). Northeastern China's short summer limits agriculture to mostly wheat, corn, and soybeans.

Oil, coal, iron ore, zinc, lead, and manganese are plentiful in the region. In particular, the oil fields here have helped make China self-sufficient in energy. They produce more than 40 percent of China's oil, some of which is exported to Japan. The region also contains China's remaining forest resources. Other important industries include iron and steel, chemicals, paper, textiles, food processing, and shipbuilding. The largest city and the center of the iron-and-steel industry is Shenyang.

Just to the west, behind the Greater Khingan Mountains, is Inner Mongolia, an autonomous region of China. The area was originally populated by Mongolians, but today the Han Chinese outnumber the Mongolians many times over. This dry region is at the southern fringes of the Gobi. Irrigation farming is the major activity.

Western China The two large, autonomous regions of Tibet (officially called Xizang) and Xinjiang (SHIN-jee-AHNG) make up most of Western China. These regions have some local government and retain their own culture and languages. Both Tibet and Xinjiang were originally populated by people who were not Han Chinese. Han Chinese colonists, though, may soon outnumber the Tibetans and other minority groups.

Western China covers more than one-third of the nation's territory. The land, however, is too high, too dry, or too cold to support a large population. Most of the people are either nomadic herders or irrigation farmers.

Tibet has been occupied by China since 1950. Taking up most of the Plateau of Tibet, it is one of the highest and most barren regions in the world. It averages over 14,500 feet (4,420 m) in elevation. In the south, the towering Himalayas separate Tibet from India and Nepal. Mt. Everest, the world's highest peak, looms over the border between China and Nepal.

The huge Buddhist palace of the Dalai Lama towers above Lhasa, the capital of Tibet. The Dalai Lama is the spiritual leader of the Tibetans, who practice Lamaism, a form of Buddhism. The Dalai Lama fled to India in 1959 after an unsuccessful revolt against Chinese control.

The Chinese are allowing Tibetan Buddhism to be practiced again after decades of banning it. Even so, Tibetans continue to call for independence. The Tibetans fear that they will become an oppressed minority in their own land because economic development has brought thousands of Han Chinese into the region. Although Tibet remains one of the poorest and most isolated parts of China, it has unexplored mineral wealth.

Xinjiang, north of Tibet, is populated mostly by Muslim Turkic people, mainly the Uygurs and Kazakhs. Mountains ring the dry desert basins of Xinjiang on every side except the east, where they meet the Gobi. The largest and driest basin of Xinjiang is the Tarim. Its lowest point, the Turpan Depression, drops to 505 feet (154 m)

Human Geography Ask two small groups of students to play the role of the People's Republic of China and Taiwan for the class. The groups should describe their country's government, economic system, political leaders, and place in world affairs. They should also explain how there came to be two Chinas. Help them prepare for their roles. After the presentations, have the class discuss whether or not the two Chinas should reunite.

Organize the class into two groups: Taiwan and Mongolia. Tell students to be ready to report on the landforms, climate, population, cities, resources, and economic activities of their assigned country. Make a chart on the board. As students respond, fill in the chart.

Have students work in pairs to learn more about the histories of Taiwan and Mongolia and the challenges that face these countries today. Have one student in each pair make a time line showing Taiwan's history while the other makes a list of challenges facing Taiwan. Have them discuss how Taiwan's history and current challenges are related. Then have them switch roles to study Mongolia's history and challenges.

---

below sea level. The Tarim Basin also holds the Taklimakan Desert and Lop Nur, a huge, dry lake bed and salt flat.

Xinjiang has minerals, including oil, coal, and iron ore. Ürümqi, its capital city, produces textiles, steel, and other products. Most Uygurs and Kazakhs, however, live as they have for centuries, herding animals.

2 SECTION REVIEW

1. What climates dominate southern and northern China, and what crops are grown in these two regions? Which of China's regions have major mineral resources?
2. What major cities in southern China are industrial? What is the capital of China, and where is it located?
3. **Critical Thinking** **Predict** whether the return of Hong Kong and Macao will present any challenges for China. Explain your answer.

3 TAIWAN AND MONGOLIA

FOCUS

- *What is the physical geography of Taiwan? the human geography?*
- *What challenges does Mongolia face?*

LOCATION **Taiwan** Early Portuguese explorers called the island of Taiwan *Formosa*, meaning "beautiful island." Taiwan lies 100 miles (161 km) across the Taiwan Strait from the southeast coast of China. The island and 86 smaller islands nearby make up the Republic of China. (See the map on page 507.)

Taiwan has a humid-subtropical climate with year-round rainfall and occasional typhoons. The eastern half of the island is mountainous, with peaks soaring 13,113 feet (3,997 m). These mountains form a rugged coastline. The western half of the island consists of fertile plains.

Because of its warm, wet climate and fertile soil, Taiwan is self-sufficient in rice and has a large food-processing industry. Major crops are rice, fruits, and vegetables. Taiwan also has a large fishing fleet.

Today, the population of Taiwan is about 85 percent Taiwanese. The ancestors of these people migrated to the island from China over hundreds of years. A small percentage of the population is descended from the island's original Malay settlers. The remainder is made up of "mainlanders," or Nationalist Chinese, who escaped from mainland China after the Communists took over in 1949. The Nationalist Chinese party controlled Taiwan under **martial law** (military rule) for 38 years. Only in the past few years have democratic rights expanded.

Taiwanese near **T'aipei** celebrate at a festival honoring the dead. Elements of Taoism and Buddhism are part of the celebration.

GLOBAL PERSPECTIVES

Ask students to discuss whether they think that Taiwan should be admitted to full membership in the United Nations. Point out that Taiwan's loss of UN membership in 1971 strengthened China's claim that Taiwan is a renegade Chinese province. Ask how recognizing Taiwan as an independent nation might have serious repercussions for U.S.-Chinese relations.

Closure

Distribute to the class a list of statements such as the following: "experiences extreme temperature changes"; "export-oriented country"; "Ulaan-baatar is the capital." Have students put a mark next to each statement to identify to which country, Mongolia or Taiwan, it applies. Go over the statements as a class.

ASSESSMENT

Assign the Section 3 Review questions on this page.

Reteaching

Assign Section 3 of Reteaching Worksheet 42.

Extension

Have students do research to make a shoe box diorama illustrating nomadic life in Mongolia. The scene should include the herders' circular, felt tents (called yurts), as well as yak and other livestock. Students might use stones and sticks to depict rocky terrain and sparse vegetation.

Geographical Sidelight

In the spring of 1993, delegates from the People's Republic of China and Taiwan signed agreements pledging increased cooperation between the two countries. Substantial barriers remain, however. China and Taiwan each maintained their own texts of the agreements, written in different types of Chinese characters. The year was left out of both documents, since for China the year was 1993 whereas for Taiwan it was 82. In Taiwan's calendar, the year 1 is marked by the establishment of the Nationalist government under Sun Yat-sen in 1912.

Section 3 Review
ANSWERS

1. Its warm, wet climate and fertile soil enable Taiwan to be self-sufficient in rice and to support a food-processing industry.

2. Mongolia faces blizzards in the winter, brushfires during the summer, and food shortages due to limited agriculture.

3. Possible answer: because Taiwan is more economically successful than mainland China, reunification would probably benefit China more than it would benefit Taiwan.

516

Much of the population of Taiwan is urbanized. T'aipei, on the north end of the island, is the largest city, a financial center, and the nation's capital. Millions of motorbikes clog the streets of this modern city each day. On the south end of the island is Kaohsiung, the country's second largest city, the major seaport, and a center of heavy industry.

Taiwan is one of Asia's most prosperous and industrialized nations. Export oriented, the country is a world leader in the production of computers, calculators, scientific instruments, and sports equipment. Taiwan's major trading partners are the United States, Japan, and China (through Hong Kong).

The average per capita income of Taiwan is more than 10 times that of China. Taiwan's rapid industrial success has led to environmental problems, however. The government has initiated an expensive clean-up program. In addition, the building of new highways, rapid transit, sewage plants, and recreational facilities promises to improve the quality of life for Taiwan's people.

Both the People's Republic of China and the Republic of China, or Taiwan, claim to be the true government of China. Until 1971, Taiwan represented China in the United Nations. Its membership was then replaced by the People's Republic of China. This loss, along with the loss of other international memberships, has not affected Taiwan's trade and successful industrialization.

The People's Republic of China seeks reunification and, should it occur, is willing to grant some political autonomy to Taiwan. Presently, Taiwan is not considering immediate reunification. The Taiwanese, however, are major investors in the rapidly industrializing regions of coastal China. This increasing economic interdependence between the two countries is drawing them closer together.

LOCATION **Mongolia** Mongolia is a large landlocked nation surrounded by China and Russia. It includes the Gobi, vast rocky plateaus covered with steppe grasses, and several mountain ranges. (See the map on page 507.) Due to its continental location, Mongolia experiences extreme temperatures. In the winter, temperatures may drop as low as -54°F (-48°C), and severe blizzards occasionally hit the country. Blistering temperatures during the dry summers can lead to raging brushfires.

During the thirteenth century, Genghis Khan led his fierce horse-soldiers out of Mongolia to conquer much of Asia, including China. Their Mongol Empire also reached into Europe and India. A few centuries later, Mongolia fell under the rule of China. In 1911, Outer Mongolia, with Russian support, broke away from China and declared its independence. Inner Mongolia remains part of China.

In 1924, Outer Mongolia installed a Communist government and became the Mongolian People's Republic. At this time, the country became economically dependent on the Soviet Union. Soviet aid to Mongolia, however, ended with the collapse of the Soviet Union in the early 1990s. Mongolia now seeks foreign aid and investment from other countries, especially Japan and South Korea.

Mongolia's natural resources include coal, copper, and perhaps oil. About half the nation's exports come from copper mining. Industrialization, meanwhile, is limited to textiles and carpets. There is little farming, and the country faces food shortages. The capital and only city is isolated Ulaanbaatar (oo-lahn-BAH-tawr).

Mongolia is three times larger than France but has a population of only 2 million. Today, about half of the population still depends on herds of sheep, goats, cattle, camels, horses, and yak (oxen) to earn a living. In fact, livestock outnumbers people many times over. The country's major religion is Buddhism. Much of the Mongolian culture, however, vanished under Soviet influence.

In 1990, free elections were held in Mongolia, and the government began allowing some free enterprise. Although Mongolia is one of the world's poorest and most isolated countries, there is hope for development of its natural resources and increased trade.

SECTION REVIEW

1. How does Taiwan's physical geography influence its economic geography?
2. What challenges does Mongolia face as a result of its physical geography?
3. **Critical Thinking** **Suggest** a reason why the People's Republic of China seeks reunification with Taiwan but Taiwan is not presently considering reunification.

CHAPTER 42 CHECK
ANSWERS
Building a Vocabulary

1. A government controlled by outside forces; Manchuria
2. Growing two crops in one year on the same land; because of its long growing season
3. Special Economic Zones (SEZs)
4. Military rule; 38 years

Recalling and Reviewing

1. Northern China
2. The People's Republic of China is the large Communist country located on the Asian mainland. The Republic of China, located mainly on the island of Taiwan, was formed by Nationalist Chinese.
3. Southern China: the Chang and Xi rivers; northern China: the Huang He (Yellow River); because the region's yellow soil makes the river yellow
4. The Himalayas
5. The Soviet Union provided aid to Mongolia, so now Mongolia is seeking aid from other countries.

CHAPTER **42** CHECK

Inside the Forbidden City in **Beijing**

Reviewing the Main Ideas

1. China has the world's oldest continuous civilization. For more than 2,000 years, a series of dynasties ruled China. In 1912, the last of these dynasties was overthrown. In 1949, China became a Communist country. Today's leaders are promoting more economic freedoms but are resisting political reforms.
2. China's four geographic regions include the fertile valleys of southern China; northern China, the most populated part of the country; northeastern China, rich in mineral resources; and dry, rugged western China.
3. Taiwan, an island off the southeast coast of China, is one of Asia's most prosperous and industrialized countries. Many of Taiwan's people are Chinese who left China since the Communists took over in 1949.
4. Mongolia is a large landlocked nation surrounded by China and Russia. It is one of the world's poorest and most isolated countries.

Building a Vocabulary

1. What is a puppet government? In what region of China did Japan establish a puppet government in the early 1930s?
2. Define *double cropping*. Why is southern China able to practice double cropping?

3. Along the coast of southern China are five areas designed to attract foreign investment. What are these areas called?
4. What is martial law? For how long was Taiwan under martial law?

Recalling and Reviewing

1. In what region did Chinese culture first develop?
2. Explain the difference between the People's Republic of China and the Republic of China.
3. What two major rivers run through southern China? What major river runs through northern China, and how did it get its name?
4. What mountains separate Tibet from India and Nepal?
5. How was Mongolia affected by the collapse of the Soviet Union?

Thinking Critically

1. Do you think that Maoism had any positive aspects? Explain your answer.
2. In what ways is the political situation in Tibet similar to the situation in Northern Ireland?

Learning About Your Local Geography

Cooperative Project

Under China's communism, the role of women in the work force changed. What economic opportunities are open to women in your area? First, identify the meaning of the term "equal opportunity employment." Then, identify several different companies or businesses in your area. Each member of your group should select one of these companies and research its hiring policies. To do this, contact the personnel department of the company. You may also ask to interview a female employee of the company about her career. As a group, construct a chart that shows the information you have gathered.

Using the Five Themes of Geography

Construct a chart with four columns, one for each major **region** of China. In each column, explain how the region's **location** and **place** characteristics have influenced forms of **human-environment interaction** in the region.

Thinking Critically

1. Answers will vary. Possible positive aspects of Maoism are that it aimed to feed the population and that it gave women equal status to men.
2. In both regions there is a call for independence. Also, as Tibetans fear that they will become an oppressed minority as more Han Chinese move in, so too do Irish Roman Catholics feel that they are an oppressed minority in Northern Ireland.

Learning About Your Local Geography

Groups' charts should indicate that groups have researched the employment policies of local companies with regard to gender.

Using the Five Themes of Geography

Charts might show such items as how climates support or limit agriculture in each region.

CHAPTER 43 Japan and Korea

Planning Guide

TEXTBOOK	TEACHING RESOURCES	OTHER AVAILABLE MATERIALS
Section		
1. Japan Pages 519–526	Reteaching Worksheet 43.1 Critical Thinking Worksheet 43 Skills Worksheet 43	*The World Today* Videodisc Program *Creative Strategies for Teaching World Geography* LESSON 19 Japanese Arts/Origami *Study Guide for Spanish Speakers* *Lecture Notes,* Chapter 43
2. North and South Korea Pages 526–528	Reteaching Worksheet 43.2 Challenge/Enrichment Worksheet 43 Graphic Organizer 43	*Study Guide for Spanish Speakers* *Lecture Notes,* Chapter 43
Chapter Review Page 529	Chapter 43 Test Chapter 43 Portfolio Assessment See also *Alternative Assessment Forms*	*Test Generator,* Chapter 43

CHAPTER RESOURCES

Lecture Notes, Chapter 43
Teaching Resources
   Chapter 43 Graphic Organizer
   Chapter 43 Skills Worksheet
   Chapter 43 Critical Thinking Worksheet
   Chapter 43 Challenge/Enrichment Worksheet

REVIEW AND ASSESSMENT

Chapter 43 Check, page 529
Chapter 43 Test, *Teaching Resources*
Chapter 43 Portfolio Assessment, *Teaching Resources*
Test Generator, Chapter 43
Alternative Assessment Forms

ADDITIONAL RESOURCES

Books for Teachers

Ekiguchi, Kunio, and Ruth S. McCreery. *A Japanese Touch for the Seasons.* New York: Kodansha, 1987.

Morton, W. Scott. *Japan: Its History and Culture.* New York: McGraw Hill, 1994.

Pezeu-Massabau, Jacques. *The Japanese Islands: A Physical and Social Geography.* Rutland, VT: Tuttle, 1978.

Books for Students

Besher, Alexander. *The Pacific Rim Almanac.* New York: Harper & Row, 1991.

Bowring, Richard, and Peter Kornicki. *The Cambridge Encyclopedia of Japan.* New York: Cambridge University Press, 1993.

Clark, James I. *Japan.* Evanston, IL: McDougal, Littel, 1989.

Japan. Alexandria, VA: Time-Life Books, 1985.

The World Today *Videodisc Program*

A variety of still images, short videos, and activities are available for you to use as you teach this chapter. *Lecture Notes* include bar codes for specific images and videos.

Multimedia Resources

Beyond the Rising Sun: Discovering Japan. (software, simulation) Educational Activities. Available through Social Studies School Service, 10200 Jefferson Blvd., P.O. Box 802, Culver City, CA 90232–0802.

Exotic Japan: An Introduction to Japanese Culture and Language. (Multimedia CD-ROM) The Voyager Co., 578 Broadway, Suite 406, New York, NY 10012.

Japan: Of Tradition and Change. (16mm film/VHS, 23 minutes) National Geographic Society, Educational Services, Washington, DC 20036.

Korea: Ancient Treasure, Modern Wonder. (VHS, 25 minutes) Films for the Humanities & Sciences, P.O. Box 2053, Princeton, NJ 08543–2053.

A Look at Japanese Culture Through the Family. (33 case studies, teacher's guide) Social Science Educational Consortium. Available through Social Studies School Service, 10200 Jefferson Blvd., P.O. Box 802, Culver City, CA 90232–0802.

Picture Atlas of the World. (interactive maps, photo bank, animation, audio on CD-ROM) National Geographic Society, Washington, DC.

MEETING INDIVIDUAL NEEDS

Limited English Proficiency (LEP)

The following demonstration may help students appreciate Japan's proximity to Korea. On a wall map of the United States, use chalk to draw a circle with a radius of 150 miles (to scale) with your community as the center. Point out locations on the circumference of the circle, and explain to students that they are as close to your community as Korea roughly is to Japan.

Learners Having Difficulty

Help students visualize the physical, political, economic, and cultural characteristics of Japan, North Korea, and South Korea. On the board, create a graphic organizer showing the similarities and differences among the countries. One useful type of organizer might be a Venn diagram, in which two or more circles overlap each other in certain places but are distinct in others.

Auditory Learners

Read aloud excerpts from a travel book about Japan or Korea that describe some of the country's points of interest. Then ask students what they would include on their itinerary if they were to travel to that country and why.

Visual Learners

Have students create three graphs comparing Japan, South Korea, and North Korea. The three graphs should have these titles: Physical Geography, Population, Economy. Allow students to decide what information (for example: rainfall, population growth, GDP) they will graph and what types of graphs they will produce.

Tactile/Kinesthetic Learners

Organize students into three groups, and assign each group one of the countries described in this chapter. Have students write and perform skits about the daily lives of the people of their country.

Gifted Learners

A brief simulation may give students insight into Japanese business management techniques. Students should first research these techniques. Then have students imagine the following: They are employed in an automobile factory that has been given an order for 300 new cars to be delivered in two weeks. The factory is currently capable of producing only 250 cars in that time. Have students role-play a meeting between factory managers and employees in which they brainstorm possible solutions to the problem.

MULTICULTURAL FOCUS

Japanese art has traditionally reflected an awareness of nature and its simple beauty. In traditional Japanese architecture, buildings are designed to be in harmony with their natural surroundings. Many of the finest examples of this style are Buddhist temples and Shinto shrines, noted for their graceful lines, simple style, and sense of proportion.

Shinto shrines, often surrounded by woods and mountains, are constructed using wooden frames set on pillars. Ceiling beams extend well beyond the walls, forming distinctive roofs with long overhangs and dark, quiet spaces under the eaves.

Traditional Japanese home design is influenced by these shrines. Thin interior walls are not designed for support, but rather to provide privacy, and they can be removed to enlarge a room's size. Exterior walls often contain sliding doors. Traditional walled gardens that surround many houses are one of the oldest and most highly developed of Japanese art forms.

Lead students in a discussion about how the following factors might influence the style of architecture in a region: climate, natural resources, religion, space, traditions, and aesthetics. Ask students to provide examples of the ways that architecture in their community and region reflects some of these factors.

CHAPTER 43

JAPAN AND KOREA

Section 1: Japan
Section 2: North and South Korea

OVERVIEW

Chapter 43 describes Japan and Korea. Because of their physical proximity, the history of these countries frequently has been interwoven.

Section 1 explores Japan's landforms, climate, history, agriculture, resources, industry, and culture. Students will learn that Japan rebuilt its industry after World War II to become the second leading economic power in the world.

Section 2 describes the Korean peninsula, contrasting the political and economic conditions in North and South Korea. Students will learn that reunification of the two Koreas is a future possibility.

About the Author

Makoto Ōoka is a Japanese poet. The quotation on this page is from a personal essay he wrote for *Mt. Fuji*, a photographic book that showcases the physical beauty and cultural significance of the famous mountain.

Primary Source

Description of Change: excerpted
Rationale: excerpted to focus on main idea

CHAPTER 43

Japan and Korea

❝ **B**ecause I was born in Mishima in Izu, which is connected to the southern base of Mt. Fuji, Fuji was for me part of the home that I was used to seeing every day. . . . The Fuji that is the most familiar and nostalgic to me is the water from the melted snow on the mountain that passes through the rock beds deep in the earth and gushes forth in the western and northern parts of the town of Mishima. The water is plentiful, and it is in this clear river water that I have my Fuji. Once when asked, 'What does Mount Fuji mean to you?' I answered, 'Mt. Fuji is water.' ❞

Makoto Ōoka

Less than 150 miles (241 km) of water separate the Korean peninsula, on the eastern edge of the Asian mainland, from the islands of Japan. As a result, the two nations of the Korean Peninsula share with Japan several geographic and cultural characteristics. In other ways, however, the three countries are very different.

Japan's small land area is very mountainous. The country is subject to floods, typhoons, and earthquakes, and has few mineral and energy resources. Yet despite these disadvantages, Japan has become one of the world's leading industrial and economic powers.

In contrast to Japan, much of the Korean peninsula contains valuable mineral deposits. Political geography, however, has split the peninsula into two very different countries, Communist North Korea and industrialized South Korea.

GEOGRAPHY DICTIONARY

tsunami

annex

subsidy

work ethic

trade surplus

urban agglomeration

export economy

demilitarized zone (DMZ)

Japanese sandals

Mount Fuji, **Japan**

Using the Quotation

Have students look closely at the photograph of Mount Fuji on page 518. Tell them that Mount Fuji is Japan's highest peak and a national and religious symbol. Then read aloud the quotation by Makoto Ōoka. Point out the down-to-earth significance of the mountain to the author, who grew up near it.

Using the Geography Dictionary

Read aloud the terms in the Geography Dictionary. Note that five of the terms are open compound words **(work ethic, trade surplus, urban agglomeration, export economy, demilitarized zone)**. Have students speculate on the meanings of these terms by putting together their knowledge of one or both words in the compound. Then have them check their responses by consulting the Glossary.

SECTION 1
OBJECTIVES

The objectives reflect the Focus questions on this page and are assessed in the review on page 526.

1. Analyze Japan's physical geography and historical geography.

2. Discuss issues surrounding Japan's economic success.

3. Describe Japanese cultural features and daily lives.

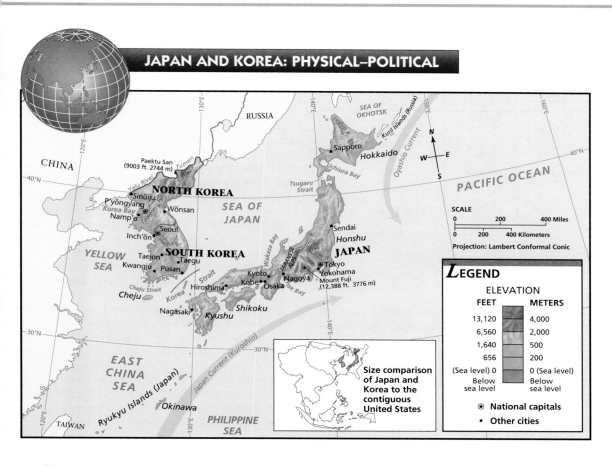

JAPAN AND KOREA: PHYSICAL–POLITICAL

Using the Physical–Political Map: Japan and Korea

Have students examine the map on this page. Ask volunteers to name the principal landforms that Japan and North and South Korea occupy and the bodies of water that surround them. Then, comparing the latitude of these countries to the U.S. East Coast, have students predict their climates. Point out the Oyashio and Japan currents. Ask students to speculate on how these affect Japan's climate. *(The icy Oyashio Current cools the north; the tropical Japan Current warms and brings moisture to the south.)*

JAPAN

FOCUS

● *What are the physical features of Japan? What relationships has Japan had with other countries throughout its history?*

● *In what ways has Japan been economically successful?*

● *What are the cultural characteristics of Japan's people? What are their daily lives like?*

REGION **Landforms** Japan is a country about the size of California. The Sea of Japan separates the island nation from mainland Asia. (See the map above.) The country is made up of four large "home" islands and more than 3,000 smaller ones. The home islands from north to south are

Hokkaido (hah-KIDE-oh), Honshu (HAHN-shoo), Shikoku (shi-KOH-koo), and Kyushu (kee-YOO-shoo). Honshu is the largest and most populated island. The Ryukyu (ree-OO-kyoo) Islands to the south are also part of Japan. Okinawa is the largest of these. Japan claims the southern Kuril Islands located northeast of Hokkaido as well. Russia has held the Kurils since the end of World War II.

More than 70 percent of Japan is mountainous. The longest mountain range is the Japanese Alps, which form a volcanic spine through the island of Honshu. About one-fourth of Japan's 200 volcanoes are active. Mount Fuji (FOO-jee), Japan's highest peak and national symbol, is an inactive volcano. It rises 12,388 feet (3,776 m).

Japan lies along a subduction zone, where the Pacific plate dives beneath the Eurasian and Philippine plates. Earthquakes and volcanic eruptions rock the region and have killed thousands of people. The 1995 earthquake in Kobe killed more than 5,000 people, for example. Large

Motivator

Locate Japan on a wall map for the class. Point out some of the country's physical characteristics: it is a relatively small, mountainous island nation located on the Ring of Fire; it has few mineral and energy resources; and it is subject to natural disasters. Tell students that all of these factors have influenced Japanese history and culture.

Building a Vocabulary

Write the term **tsunami** on the board, and review its meaning (a large-scale sea wave produced by an undersea volcanic eruption or an earthquake). Point out that it is a Japanese word. On the board, write other Japanese words that have been absorbed into English: *geisha, kimono, Kabuki, sushi, tofu, sumo, haiku, origami.* Have volunteers define these terms.

Responding to Challenges

On the board, list issues facing the United States, such as urban crime, aging of the population, treatment of minorities, day care, illiteracy, inflation, and unemployment. Ask students which of these issues Japan also faces and how the country is responding to these issues. Have them find examples and information in the text, in newspaper and magazine articles, or on television and radio news broadcasts.

THE WORLD TODAY VIDEODISC PROGRAM

Japan: Teenage Rites of Passage

2B   14520   4:27

Play   Pause

Chart Caption Answer

Japan

THEME
Interaction

Climbing Mount Fuji is a popular activity in Japan. The ascent has long been a pilgrimage for Japanese Buddhists, who regard the peak as sacred. Nowadays, pleasure climbers can drive halfway up the mountain and walk up the rest of the way. The rising numbers of automobile tourists and climbers, however, are creating significant problems with erosion, garbage, pollution, and traffic.

520

FACTS IN BRIEF	Japan and Korea		
COUNTRY POPULATION (1994)	LIFE EXPECTANCY (1994)	LITERACY RATE	PER CAPITA GDP (1993)
JAPAN 125,106,937	76, male 82, female	99% (1991)	$20,400
NORTH KOREA 23,066,573	67, male 73, female	99% (1990)	$1,000 (1992)
SOUTH KOREA 45,082,880	67, male 74, female	96% (1990)	$9,500
UNITED STATES 260,713,585	73, male 79, female	97% (1991)	$24,700

Sources: *The World Factbook 1994,* Central Intelligence Agency; *The World Almanac and Book of Facts 1995*

Which country has the highest life expectancy rates?

sea waves caused by tectonic activity also kill many Japanese. Known as **tsunami,** these waves can travel across an ocean at hundreds of miles per hour and rise up to a height of 100 feet (30 m) or more by the time they reach shore. A 1993 earthquake in northern Japan created waves and caused fires that destroyed hundreds of homes.

Plains, which make up only about 20 percent of Japan's land area, lie mainly along the Pacific coast. Cities, farmland, and Japan's population compete for space on these small coastal plains. Here, a population four times that of California crowds into a land area about one-fifth the size of that state.

Climate   Japan lies at about the same latitudes as the east coast of the United States, and its climates are similar to that region. Hokkaido and northern Honshu have a humid-continental climate with summers cooled by the Oyashio Current from the north. Icy winds blowing off the Asian continent make winters in northern Japan cold, and the windward Sea of Japan coast experiences heavy snowfall.

Southern Japan, which includes Kyushu, Shikoku, and the southern part of Honshu, has a humid-subtropical climate. Here, winters are mild, and summers are warm and humid. The Japan Current, which flows from the tropical Pacific, warms southern Japan. The area receives rain year-round and has a long growing season. Some areas receive more than 80 inches (203 cm) each year. In late summer and early fall, typhoons bring destructive winds but necessary rainfall.

MOVEMENT   **Historical Geography**   Japan's first inhabitants were the Ainu, who probably arrived from Central Asia several thousand years ago. The Ainu were driven farther and farther north by Mongoloid invaders from Asia who arrived in Japan around 300 B.C. Within a few hundred years, these invaders had settled all the major islands.

Many elements of Japanese culture are tied to the cultures of the Asian mainland. For example, the Japanese language strongly resembles Korean and Chinese. Buddhism, one of Japan's major religions, was introduced from China in the sixth century, and Japan still has many Chinese-style Buddhist temples. Also, Chinese ideas and practices laid the basis for Japan's early political systems.

Despite these influences, the Japanese developed a distinct culture over the centuries. The Shinto religion existed in Japan long before the introduction of Buddhism and still survives today. And in the eighth century, Japan began to move further away from Chinese ideas in the development of a new political system. For hundreds of years, military leaders, or *shoguns,* ruled Japan. They were supported by warriors, called *samurai.* Japan's island location and the *shoguns'* military power kept out Asian invaders.

In the mid-1500s, the Portuguese arrived in Japan seeking trade. The Spanish and Dutch soon followed, resulting in the introduction of Christianity. Soon after, however, European traders and missionaries were either driven out of Japan or restricted to the port of Nagasaki. Japanese leaders feared that foreign ideas might cause instability in Japanese society.

Japan remained cut off from the world from the 1600s to the mid-1860s, when merchants and other groups in Japanese society pressed for greater contact with the outside world. As a result, the United States and many European nations soon entered into treaties with Japan.

By this time, Japan had begun its modernization period, called the Meiji (MAY-jee) Restoration *Meiji* means "enlightened rule." The *shoguns* were overthrown, and an emperor took their place. Japan's modernization took many forms, including rapid industrialization and changes in education, law, and government. By 1890, Japan had a constitution and parliamentary system of government.

Japan soon became recognized as a world power. In 1895, it defeated China in a war over

Problem Solving

Have students calculate the average population density of Japan (people per square mile) using the country charts in the Appendix in the back of their textbook. Ask students to provide examples of ways in which the Japanese people have adapted their industry and way of life to limited space. Have them relate Japan's tradition of cooperation and group loyalty to population density.

Identifying Main Ideas
Physical and Economic Geography
Have students sketch a map of Japan. Then have them write statements that tell a main idea about Japan's physical features, agriculture, natural resources, and industry. *(For example, "Japan's terrain consists mostly of mountains"; "Japanese farmers are extremely efficient"; "Japan has few natural resources"; "Japan is the second largest industrial power in the world.")* Have students read their statements to a partner.

Guided Practice

As a class, make a chart to summarize Japan's economic activity. Across the top, write Activity, Products, Advantages, Disadvantages, Issues. Down the side, write Agriculture, Fishing, Aquaculture, Forestry, Manufacturing. Project the chart on a transparency. Call on students to provide information for the appropriate spaces. Note that some spaces will be left blank.

control of Korea. In 1905, it defeated Russia in the Russo-Japanese War. Japan **annexed**, or formally joined, Korea to its empire in 1910. By 1920, Japan had further expanded its borders to include many North Pacific islands.

Then in 1940, Japan signed an alliance with Germany and Italy and, in 1941, entered World War II by attacking the U.S. naval base at Pearl Harbor, Hawaii. Japan conquered much of Southeast Asia and numerous Pacific islands before being pushed back by U.S. and allied forces.

The end of World War II marked the end of Japan's military empire. Japanese cities had suffered massive bombing during the war. Two major cities, Hiroshima and Nagasaki, were almost completely destroyed by atomic bombs dropped by the United States.

After World War II, Japan established a democratic government. With U.S. aid, Japan also began to rebuild into a major world industrial power. Today, Japan is a constitutional monarchy with several political parties. The government is made up of an elected law-making body, called the Diet (DEE-it), and a prime minister. Japan's emperor remains a symbol of the nation but has no real political power.

INTERACTION **Agriculture** Less than one-fifth of Japan's land is arable. Yet Japanese crop yields are some of the highest in the world, and the nation produces about 40 percent of its food needs. With a population half that of the United States, Japan has twice as many farmers.

Japanese agricultural success is due to terraced cultivation and modern farming methods. The farmers own their land and live in small villages. The size of the average Japanese farm is only two and one-half acres (1 ha). Many Japanese farms are under constant pressure from expanding urban areas.

Sixteenth-century *samurai* armor

Almost half of Japan's farmland, mostly in southern Japan, is used to grow rice. Tea, mulberry trees (for silkworms), soybeans, and a variety of fruits and vegetables also grow well in this warmer climate. Northern Japan's shorter growing season produces wheat, barley, potatoes, and vegetables. The cool island of Hokkaido is home to a successful dairy industry.

Japan imports about half of its food, especially grains and meat. Food imports have increased because many Japanese have added more Western foods, such as meat, bread, and potatoes, to their usual diet of rice, fish, and vegetables. Packaged foods and fast-food restaurants also are on the rise. French fries are now so popular that the nation imports $85 million in frozen fries each year, about half of which come from the U.S. state of Washington. Other foreign products are imported under a quota or completely banned.

The Japanese farmer is protected by high government **subsidies**, especially for rice. A subsidy is financial support given by a government. Subsidies give Japanese farmers an economic advantage over foreign competitors in the Japanese market.

Resources Japan's location near seas rich in marine life, as well as its long coastline, has helped create the largest fishing industry in the world. In fact, Japan makes over 10 percent of the world's fish catch. The country has a local fishing fleet of thousands of small boats as well as a large oceangoing fleet. Aquaculture programs also supply Japanese markets with fish, shellfish, seaweed, and pearls (from oysters). The Japanese have continued to practice commercial whaling despite international protests.

COOPERATIVE LEARNING

Organize the class into groups of five. Give each group a large poster board or a sheet of newsprint. On each poster, start a graphic organizer with one of these topics written at the top: Physical Geography, History, Agriculture, Resources, Industry, Trade, Government, Cities, Society. Below the title, write the Five Themes of Geography as headings across the sheet. Each member of the group should be responsible for finding facts about the topic related to one theme, and then completing that part of the graphic organizer. You may wish to provide time and resources in class or assign the research as homework. Display the completed organizers in class.

521

Independent Practice

Have students make a thematic time line of Japan's history. Provide them with names for various periods in Japan's history: Discovery and Settlement, Invasion, Shogun Rule, Meiji Restoration, Military Empire, Democracy. Have students arrange the themes chronologically along the time line. Underneath each theme, have them list major events of the period.

Then have students write a paragraph comparing Japan and Great Britain. *(Both are influential island nations with powerful mainland neighbors; both are constitutional monarchies; both established empires; both are highly industrialized.)*

Closure

Tell students to imagine that they have recently moved to Japan. Have them write a letter to a friend in the United States about Japan. They should describe where in Japan they live, the local physical features, and the local economy.

THEME
Interaction

The world's best-known pearl manufacturer, Mikimoto, is located at Ago Bay on southern Japan's Kii Peninsula. The company was founded by Kokichi Mikimoto, who invented the "cultured pearl" 100 years ago. Mikimoto discovered that carefully tended oysters could be coaxed to produce nearly perfect pearls around an implanted nucleus, a process that takes two years to accomplish. Westerners at first believed that Mikimoto's "cultured pearls" were fakes, but the technique eventually gained widespread acceptance. Japanese pearl farms now produce about 70 tons (63 metric tons) of pearls each year, despite pollution problems and stiff competition from rapidly growing Chinese manufacturers.

Forests cover more than two-thirds of Japan, making it one of the world's most forested nations. The cutting of Japan's timber is carefully controlled to prevent the loss of forests, limit soil erosion, and provide national parks. As a result, Japan imports timber from the United States, Canada, and Southeast Asia.

Japan has few energy resources and industrial raw materials such as iron ore. Oil is the country's major import. Japan also imports coal from Australia and the United States. Hydroelectric power plants provide about 12 percent of Japan's electricity needs. Nuclear power plants provide another 25 percent. Conservation efforts and new nuclear power plants have helped lower the country's dependence on imported oil.

Industry Despite its dependence on imported natural resources, Japan produces goods that are considered some of the best in the world. The country is the second largest industrial power in the world, after the United States. It is a world leader in the production of cameras, televisions, videocassette recorders, radios, watches, and motorcycles. The country ranks high in the pro-

Workers feed mulberry leaves to silk-worms on the island of **Honshu, Japan**. Silk comes from the cocoons these worms will spin when they begin the process of changing into moths.

duction of steel, computers, telephones, and many household products. The Japanese are also leaders in such high-tech fields as electronic miniaturization and industrial robotics.

Many nations are looking at Japan to find the reasons for its economic success. Some reasons lie in Japan's cultural geography. Nearly everyone in Japan is of Japanese origin. This makes communication and the setting of common goals much easier. Also, Japan's people have a strong **work ethic**, a belief that work is good in and of itself. Many Japanese work on weekends and until late at night.

Other reasons for Japan's success lie in the relationship between employers and employees. Workers often participate in decision making in their companies. In general, Japanese workers are loyal to their employers, and some keep the same jobs their whole lives. Strikes are rare by Western standards. In addition, the Japanese are familiar with the markets and nations that buy their exported products and are generally well educated. In fact, Japan has one of the world's highest rates of literacy and of high school graduation.

The Japanese government, through the Ministry of International Trade and Industry (MITI), works closely with businesses and offers financial aid and import protection. Also, Japan's domestic market is one of the most competitive in the world. A Japanese product that survives in the home market will probably compete successfully in international markets as well.

International Japan The Japanese have succeeded in moving whole industries overseas. For example, there are several Japanese-owned factories in the United States producing automobiles and televisions. The Japanese have also invested in foreign industries, such as entertainment and real estate. Many of the buildings and hotels in cities such as Los Angeles and Honolulu are Japanese owned. Most of the world's largest banks are Japanese-owned as well.

At the same time, however, Japan is feeling competition from other newly industrialized Asian countries, especially China, the "Four Dragons," and the "Little Dragons." Many of Japan's older industries, such as shipbuilding, steel, and textiles, are suffering from this competition. Also, the country has built up a huge **trade surplus**. A trade surplus exists when a nation exports more than it imports. For instance, Japanese exports to the

Extension

Have interested students write a diary entry for a day in the life of a Japanese student. Students will need to research Japanese schools and curriculum. The diary entry should reflect the rigorous schedule of a typical school day. For example, the school day in Japan is longer than in the United States, and students may attend after-school academies for further drill and instruction.

United States are two times greater than its imports from the United States. This trade imbalance could cause some nations to set up trade barriers to Japanese goods.

Urban Heartland The core of Japan's cities and industry lies on Honshu's eastern coastal plains. The three nodal areas are the Tokyo Bay area, the Kansai region, and the Nagoya area. A dense network of roads, high-speed rails, seaports, and air routes connects the entire region.

With more than 25 million people, the Tokyo Bay area is the largest **urban agglomeration** in the world. An urban agglomeration is a densely inhabited and contiguous region surrounding a central city. Tokyo is Japan's capital and center of communications, government, banking, education, and trade. It is the site of the Imperial Palace and the *Ginza*, the world's largest and busiest shopping district. Because little land is available, Tokyo's real estate prices are among the highest in the world. (See "Cities of the World" on pages 524–525.) Yokohama (yoh-kuh-HAHM-uh), located just south of Tokyo, is one of the world's busiest seaports.

The Kansai region has three major industrial centers: Osaka (oh-SAHK-uh), Japan's second largest city; the nearby port city of Kobe (KOH-bee); and Kyoto (kee-OHT-oh), the ancient capital. Kyoto's many temples and traditional handicrafts attract large numbers of visitors. The Nagoya region, meanwhile, lies on Ise (EE-say) Bay. Nagoya is its major city and industrial center.

All of Japan's major urban areas are faced with unaffordable housing, air pollution, and traffic congestion. Unlike in large U.S. cities, however, crime rates in Japan are low.

Japanese Way of Life Japan has many millionaires as well as urban poor, but most Japanese belong to the middle class. Most families live in suburbs and spend several hours each day commuting to and from jobs in the cities.

A typical middle-class home in Japan is small—usually about the size of a small apartment in the United States—and expensive. Sliding wooden screens separate sparsely furnished rooms. People generally sit on cushions or mats placed on the floor. Families are used to living in crowded spaces, and most children must share their bedrooms. Western-style homes, recently introduced to Japan, have proved popular, though they are very expensive.

Japanese society has undergone many changes over the past several decades. In urban and suburban areas, Western clothing, such as blue jeans, has largely replaced the traditional kimono. Rock music and Western soft drinks are popular, and baseball, golf, and skiing are favorite forms of recreation.

The birthrate in Japan is as low as that of Europe, and the Japanese now boast the world's highest life expectancy—82 years for women and 76 for men. As the average age of the population becomes higher, many worry about the increased demands on social services. The roles of Japanese women also are changing as women become more educated and urbanized. More than half of Japanese women are in the work force and are seeking a larger role in Japanese society.

Japan's traditional culture, however, remains an important part of Japanese society. Japanese artists practice traditional forms of poetry, music,
Text continues on page 526.

These traveling businessmen in **Osaka, Japan**, are spending the night in one of the country's "capsule hotels." Many businesspeople find these hotels more convenient than commuting long distances to and from crowded Japanese cities.

Cities of the World

A View of the Future

According to the project's design engineers, Tokyo's "Alice City" project will offer subterranean homes, offices, and stores as an answer to the city's severe overcrowding. Many large cities such as Paris and Houston already have large underground shopping complexes, but the concept of an entire subterranean community of residential, business, and retail elements has yet to be tried. Have students write a paragraph explaining why they would like to live in such a community or why they would not. Call on volunteers to read their paragraphs to the class. Students may want to include their paragraphs in their individual portfolios.

......................●
Geographical
Sidelight

Only major streets in Tokyo are named. Many city addresses are determined by dividing the city into wards and other districts. A ward is divided into sections, which in turn are further divided into subdivisions. Each block of each subdivision is numbered, as is each house in each block. Japanese identify an address by using a system that gives the number for the block, the number for the house, the number for the subdivision, the name of the section, and the name of the ward. A writer might send a letter to the following fictional address: 1–2, Yamabuki 3-chome, Chiyoda-ku. That address means "block 1, house 2, subdivision 3 in the Yamabuki section of Chiyoda ward."

CITIES OF THE WORLD

Tokyo

*T*okyo is Japan's largest industrial and urban area. The city is the center of the country's government and finance.

In the fifteenth century, a Japanese feudal lord built his castle overlooking a natural harbor on the island of Honshu. In time, the city of Edo grew up around the castle. In 1868, a new emperor changed the city's name to Tokyo, meaning "eastern capital," and made it the capital of Japan.

The City Rebuilt
The growth of Tokyo followed the original city plan, with the castle in the center. The town

Illuminated signs, in English and Japanese, scream at traffic and shoppers crowding into the Ginza district. The district is famous for its stores and entertainment.

Metropolitan **Tokyo** sprawls outward from the Imperial Palace. Stone walls and a moat insulate the palace from the bustling city.

expanded outward in a circle, similar to the growth patterns of Moscow and Paris. Most of present-day Tokyo, however, has been rebuilt since the end of World War II, when bombing destroyed about half of the city.

Earthquakes also have disrupted the Tokyo skyline, as in 1923 when a huge earthquake devastated large areas of the city. On average, three tremors shake the foundations of Tokyo each day! The city's modern skyscrapers have been specially designed to endure these frequent quakes.

Tokyo and its neighboring cities have become so large that their boundaries overlap. Tokyo, Nagoya, and Osaka form one huge, continuous urban area that stretches along the narrow plain of Japan's east coast and includes

524

the port cities of Yokohama and Kawasaki.

Looking for Space

The chief concern in Tokyo, as in the rest of Japan, is space. The city has expanded to fill all the available land on the narrow coast, in addition to landfill areas. The severe shortage of housing leads some to look upward rather than outward to find space.

Indeed, Tokyo is already crowded with skyscrapers, and more are springing up all the time. Tall buildings often tower over smaller neighbors, blocking out the natural light. As a result, a law called *nisshoken,* or "the sunshine law," requires developers to pay money to those who live in the shadows of their towering buildings.

Tokyo sits at the heart of a sprawling metropolitan area. Commuters from outlying cities use railway transportation to travel to and from their jobs in Tokyo.

A weekend getaway to the country might be on the minds of these Japanese commuters being crammed onto trains in crowded **Tokyo**. Workers are employed to shove as many passengers onto commuter trains as possible.

Only some 10 percent of the land area in Tokyo is open space, much of which is devoted to beautiful public gardens and parks featuring Shinto shrines and Buddhist temples. The Japanese take great pride in their parks and strive to keep them clean and safe. The crime rate in Tokyo is much lower than that of other large cities of the world.

As in most other major cities, however, traffic congestion is a problem in Tokyo. Efficient transportation networks and high-speed commuter trains help ease traffic problems and cut down pollution. The city used to be badly polluted from automobiles and factories. Since 1965, though, the government has worked to cut the level of air pollution and keep the city clean. As a large industrial city, Tokyo still faces many challenges, but its officials and citizens have taken important steps toward managing several pressing urban issues.

Geographical Sidelight

Recreation facilities in Tokyo are as crowded as housing facilities and public transportation systems. On a holiday afternoon, a public swimming pool may have up to 20,000 city dwellers swarm to its waters.

Geographical Sidelight

Although Tokyo has worked to reduce the city's air pollution levels, the city's rivers show signs of severe industrial pollution. In fact, the amount of chemicals in the city's rivers makes it possible to develop photographic negatives simply by putting the negatives in the water.

The objectives reflect the Focus questions on this page and are assessed in the review on page 528.

1. Identify physical characteristics of the Korean peninsula.
2. Explain the division of the Korean peninsula, and compare and contrast North and South Korea.

TEACHING SECTION 2
Motivator

Use a world wall map to show students the Korean peninsula. Point out that Florida is about the same size. Mention that unlike Florida, however, Korea is divided into two separate countries. Ask students to suggest how this division might have arisen. Tell them that they will learn in Section 2 about the Korean peninsula and its two countries.

Building a Vocabulary

Write **demilitarized zone** on the board, and circle the prefix *de-*. Explain that *de-* means "removal" or "reversal." When *de-* is added to *militarized*, *demilitarized* means "the removal of the military." Point out Korea's demilitarized zone on a map, and explain that no troops are allowed in this area. Ask students if they know of other demilitarized zones. (*between North and South Vietnam, created in 1954*)

Section 1 Review
ANSWERS

1. Most of the country is mountainous, so there is little area suitable for farming; the few plains areas are quite crowded. Also, earthquakes, volcanic eruptions, and tsunami have killed many people in Japan. Nineteenth-century changes included Japan's reopening to foreign countries; rapid industrialization; changes in education, law, and government; and emergence as a world economic power.

2. Uniform cultural geography, strong work ethic, industry in which workers have a voice and remain loyal to employers, an educated population, government policies

3. Ways of life are becoming increasingly Western. A possible reason is increased communication and trade between Japan and Western countries.

A heater under the table and a blanket provide heat during dinner for three generations of this family in **Kyoto, Japan**. This dinner includes rice and fish, important parts of Japanese diets.

painting, and theater, and traditional ways of life are still followed in rural areas. A land of contrasts, modern Japan blends old and new, rural and urban, East and West.

SECTION REVIEW

1. What challenges has Japan faced as a result of its physical geography? What changes did Japan experience in the nineteenth century?
2. What are the major reasons for Japan's economic success?
3. **Critical Thinking** **Explain** how Japanese ways of life are changing today. What do you think accounts for the changes?

NORTH AND SOUTH KOREA

FOCUS

- *What landforms and climate types are found on the Korean peninsula?*

- *How is the peninsula politically divided? How do North and South Korea differ?*

REGION **Physical Geography** Korea is a 600-mile-long (965-km) peninsula, about the size of Florida. The narrow Korea Strait and the Sea of Japan separate the peninsula from Japan. The peninsula is divided into two nations, North Korea and South Korea. The Yalu River on the north and the Yellow Sea on the west separate North Korea from China. North Korea shares its northeastern border with Russia. (See the map on page 519.)

The landforms of Korea are mostly hills and low mountains. The most mountainous region is in the northeast, where several peaks rise more than 8,000 feet (2,439 m). On the east coast, steep mountains plunge into the sea, while a coastal plain hugs the western coast. This crowded plain, which covers about one-fifth of the peninsula, holds the richest farmland and most of the population.

Korea has two climate regions. In the south, the climate is humid subtropical. Summers are hot and humid, and heavy rain usually falls during the summer monsoon season. The area is subject to typhoons during the summer and fall. Winters are mild along the coast but cold in the mountains. In the north, the climate is humid continental. Here, the growing season is shorter, and winters are colder. The continental influence of Asia brings cold air and snow during winter.

MOVEMENT **Historical Geography** The Koreans are descendants of a number of peoples who migrated to the peninsula from Central Asia and northern China. Separated from other lands by sea and mountains, the Korean culture evolved in isolation for centuries, even though China and Japan invaded Korea many times.

The last such major invasion occurred in 1895, when Japanese forces defeated the Chinese and annexed Korea. The Japanese removed the Korean monarchy in 1910 and took complete control, making Korea an economic colony of Japan. The Japanese controlled Korea until the end of World War II.

After World War II, the victorious powers agreed to temporarily divide the Korean peninsula along the 38th parallel. The Soviet Union occupied the northern part, and the United States occupied the southern part. Political differences between the two regions, however, prevented the expected reunification. In the north, the Soviets set up a Communist government, and North Korea became the Democratic People's Republic of Korea. In the south, the United Nations supervised a government election, and South Korea became the Republic of Korea. Soviet and U.S. occupation forces withdrew in 1949.

Writing

Have students imagine that they live in North Korea. Ask them to write a letter to a pen pal in South Korea describing life in the Communist country. Suggest that students read more about Communist rule in North Korea before writing their letters.

Guided Practice

Have students work with partners and use classroom or library resources to make an outline that compares the two Koreas. One student should work on North Korea and the other on South Korea. The outlines should include the following headings under each country's name: Natural Resources, Agricultural Products, Industrial Products, Gross Domestic Product, Per Capita Income, Major Trading Nations. Then lead a class discussion based on the outlines. Ask: Which nation is more productive? What factors have hampered North and South Korean development since the 1950s?

Independent Practice

Provide students with Outline Map 20. Have them shade and label the major elements of the Korean peninsula's physical geography, including its landform and climate regions.

In 1950, North Korea invaded South Korea, sparking the Korean War. The United Nations sent troops to defend South Korea, while Communist China sent forces to North Korea. Eventually, a truce line was established near the original 38th-parallel boundary. The war devastated much of Korea's land and left more than 1 million people dead. The Koreans remained a divided people.

North Korea North Korea is a strict Communist state with few political and religious freedoms. Kim Il Sung, who ruled for more than 40 years, died in 1994. It is unclear whether his son can hold onto power. North Korea has a large military, and the United States has worked to prevent the country from obtaining atomic weapons.

The country has several hydroelectric power sources and large deposits of minerals, including coal, iron, and copper. Major industries include steel and textiles, and all industry is government owned. As for agriculture, about 40 percent of North Koreans work on state farms; private citizens can own only small garden plots. Rice, potatoes, and wheat are the main crops. The capital and largest city is P'yŏngyang (pee-AWNG-yahng).

South Korea South Korea is slightly smaller than North Korea but has more than twice as many people. The flatter land and warmer climate of South Korea allow for greater agricultural development and a wider variety of crops. Only about 15 percent of South Koreans farm, yet the nation is self-sufficient in most foods. Crops include rice, cabbage, sweet potatoes, and many fruits and vegetables. Fishing is an important economic activity as well.

With the help of the United States, South Korea rebuilt its industry after the Korean War. The nation progressed rapidly from an agricultural nation to an industrial one. South Korea has an **export economy**, which means most of its manufactured products are made for export to other nations. Indeed, trade accounts for a large part of the South Korean economy. The country's major industries are electronics, steel, automobiles, shipbuilding, and textiles. It is also a major producer of televisions, microwave ovens, fax machines, and computers.

South Korea's industrial base is heavily diversified and competitive, and the work force is literate and well trained. Lower wages in South Korea have helped the country to produce goods of high quality but with a lower price than those from Japan. South Korea is trying to expand its nuclear power and mining to help make up for large imports of oil and raw materials.

The country's recent economic progress has led to much social change. Industrialization and prosperity have created a large urban middle class interested in consumer goods. Seoul (SOLE), the capital and largest city, is a huge, rapidly growing industrial, commercial, and cultural center. Small agricultural villages are disappearing as young people move to the cities and as industrial suburbs and factories expand outward.

South Koreans have many political and religious freedoms. About 20 percent of the population are Buddhist, and another 20 percent are Christian. Although government leaders have enjoyed nearly unlimited power, demands by the people have recently led to the expansion of democratic rights.

Korean Reunification Koreans have long wished for peace and unity. In 1972 and again in 1991, North and South Korea agreed to seek

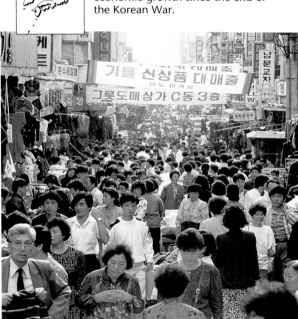

Shoppers throng this shopping district in **Seoul, South Korea**. South Koreans have experienced dramatic economic growth since the end of the Korean War.

> ## Geographical Sidelight
>
> *South Korea is chipping away at Japan's dominance in the consumer electronics market. Korean companies' lower-priced products are especially popular in Eastern Europe, the former Soviet Union, Pakistan, Thailand, and Vietnam.*
>
> ## Geographical Sidelight
>
> *In 1977, North Korean President Kim Il Sung announced that his son, Kim Jong Il, would succeed him upon Il Sung's death or retirement. This succession occurred in 1994, in effect creating the world's first Communist dynasty, although it is unclear how long it will last. Little is known about Kim Jong Il outside of this closed country.*

Closure

Have students brainstorm about things North Korea and South Korea might do to pave the way for reunification. *(open trade between the two Koreas; exchange visits between politicians, businesspeople, and teachers; allow correspondence and visits between family members separated by the division of Korea)* Note that although the two countries have vastly different present-day cultures, they share a 1,000-year history.

ASSESSMENT

Assign the Section 2 Review questions on this page.

Reteaching

Assign Section 2 of Reteaching Worksheet 43.

Extension

Have interested students research aspects of the Korean War. Students might profile important personalities or significant battles or locations, or they might describe how and why different foreign countries participated in this conflict.

GLOBAL PERSPECTIVES

Have students debate whether the United States should maintain a military presence in Korea. *(Pro: North Korea's army and nuclear program threaten South Korea. Con: South Korea can afford to defend itself. Relations in the region have changed in recent years.)*

Section 2 Review
ANSWERS

1. Mountainous regions: in the northeast and along the east coast. Plains region: along the western coast. Climate regions: humid subtropical in the south and humid continental in the north

2. Since World War II. North Korea has been a Communist nation whose people have little freedom. South Korea has become an industrial power; its people have many freedoms and democratic rights.

3. After World War II, both Korea and Germany were divided into two countries, one Communist and one controlled by Western democratic powers. Germany was reunified in 1990, but Korea remains divided.

528

NEWS FROM AROUND THE WORLD — South Korea

Teenagers in South Korea Prepare for the Future

Today, as in the past, teenagers in South Korea are taught to honor tradition, respect their elders, and place family loyalty above all other ties. Now, however, they must also adjust to great changes taking place in their society.

Instead of parents, grandparents, children, uncles, and aunts living under one roof, most youngsters live with only their parents. The size of families has decreased, and many young South Korean adults live in cities, where family bonds are weakening.

Education remains highly valued in South Korean society, and the opportunities for education are expanding. Schools reflect a mixture of traditional and modern elements. Students must study English as well as the Korean language. They participate in centuries-old sports, such as *ssirum*, a form of wrestling, and the martial art of *tae kwon do.* (See the photograph above.) Students also play sports introduced into the country only in the last century or so, such as soccer, volleyball, and basketball.

Korean teenagers are serious students, and about ninety percent of South Koreans attend high school. Those planning to attend a university must take an entrance examination after graduation from high school. Study for this examination begins as early as seventh grade. By the time students are high school seniors, they usually attend school six days a week. Most of their waking hours are spent in class or studying. About 40 percent of all students go on to college.

Sending a child to college is a priority for South Korean parents. "Our ancestors would sell their cows, their oxen and even their paddies, if necessary, to pay for an education for their children. And it was the best investment they ever made," claims one father whose 18-year-old son is studying day and night for the college examination.

peaceful reunification. However, many obstacles lie in the way of this goal, and the process has not moved forward significantly.

There is almost no trade or communication between North and South Korea. Their border is a **demilitarized zone (DMZ)**, a buffer zone just under two and one-half miles (4 km) wide into which military forces may not enter. Yet just a few miles to the north stands North Korea's large army, and to the south are South Korean troops and thousands of U.S. soldiers. In many respects, the two countries remain worlds apart.

2 SECTION REVIEW

1. Where are Korea's mountainous regions? Where is its plains region? What are the peninsula's two climate regions?
2. When was the peninsula divided into North and South Korea? What differences do these two countries have?
3. **Critical Thinking** **Compare and contrast** post–World War II political developments in Korea to those in Germany.

CHAPTER 43 CHECK
ANSWERS
Building a Vocabulary
1. Large sea waves caused by earthquakes and volcanic eruptions; they can travel fast and rise high.
2. It is formally joining the territory to itself.
3. Financial support given by a government
4. A densely inhabited and contiguous region surrounding a central city; the Tokyo Bay area
5. Demilitarized zone (DMZ); a zone into which military forces are not allowed

Recalling and Reviewing
1. Marine and forest resources; industrial raw materials and energy resources
2. After World War II, the Soviet Union occupied the northern part, and the United States occupied the southern part. During the Korean War, China aided North Korea, and the United Nations aided South Korea. The truce left the peninsula divided.
3. Its mountainous terrain more closely resembles that of North Korea; its economic and political characteristics are more like those of South Korea.

CHAPTER 43 CHECK

Children playing in traditional dress in **South Korea**

Reviewing the Main Ideas
1. Mountainous Japan is made up of four large islands and more than 3,000 smaller ones. Most of Japan's people live in the few coastal plains areas. The country is a major world economic power. Despite rapid modernization during this century, Japan has held on to much of its traditional culture.
2. The Korean peninsula is divided into two countries, Communist North Korea and democratic and industrial South Korea. Although the two countries have discussed reunification, they have not reached any agreements.

Building a Vocabulary
1. What are tsunami? Why are they dangerous?
2. When one country annexes a territory, what is it doing?
3. Define *subsidy*.

4. What is an urban agglomeration? Give an example of an urban agglomeration discussed in the chapter.
5. What term describes the border between North and South Korea? What does the term mean?

Recalling and Reviewing
1. What are Japan's major resources? What resources must the country import?
2. What role did foreign countries play in the division of the Korean peninsula?
3. Which country does Japan resemble more—North Korea or South Korea? Why?

Thinking Critically
1. In your opinion, should the United States adopt some of Japan's ideas about management and economic development? Why or why not?
2. What obstacles do you think lie in the way of Korean reunification?

Learning About Your Local Geography
Cooperative Project
Western foods, such as french fries, are more and more common in Japan. What "foreign" foods are popular in your area? With your group, make a list of the foods that you like to eat whose ingredients or recipes probably originated in another country. If your group has access to and permission to use kitchen facilities and supplies, prepare one of the items on your list. Share it with the rest of the class and explain its origins.

Using the Five Themes of Geography
Using symbols or shading, sketch a map showing how **location** has influenced human **place** characteristics in Japan, South Korea, and North Korea. Consider economic, cultural, and political characteristics of the three nations. Include a legend of the symbols and shading you use, and write captions explaining the relationship between location and place in these countries.

Thinking Critically
1. Answers will vary. Some students may think that Japan's ideas are the major reason for that country's success and should be adopted. Other students may think that Japan's ideas will work only in the context of Japanese culture and society and will not apply to other countries.
2. Possible answer: the two countries' different historical backgrounds and political systems

Learning About Your Local Geography
Groups' lists should mention popular foreign foods.

Using the Five Themes of Geography
Maps might indicate that Japan's location is suitable for world trade, that Japan and Korea share several cultural characteristics because Japan is located near the Korean peninsula, or that North Korea's location bordering Russia and China may relate to North Korea's communism.

CHAPTER 44 Mainland Southeast Asia

Planning Guide

TEXTBOOK	TEACHING RESOURCES	OTHER AVAILABLE MATERIALS
Section		
1. Physical and Human Geography Pages 531–533	Reteaching Worksheet 44.1 Critical Thinking 44 Skills Worksheet 44	*Study Guide for Spanish Speakers* *Lecture Notes,* Chapter 44
2. The Countries of Mainland Southeast Asia Pages 534–538	Reteaching Worksheet 44.2 Challenge/Enrichment Worksheet 44 Graphic Organizer 44	*Study Guide for Spanish Speakers* *Lecture Notes,* Chapter 44
Chapter Review Page 539	Chapter 44 Test Chapter 44 Portfolio Assessment See also *Alternative Assessment Forms*	*Test Generator,* Chapter 44

CHAPTER RESOURCES

Lecture Notes, Chapter 44
Teaching Resources
    Chapter 44 Graphic Organizer
    Chapter 44 Skills Worksheet
    Chapter 44 Critical Thinking Worksheet
    Chapter 44 Challenge/Enrichment Worksheet

REVIEW AND ASSESSMENT

Chapter 44 Check, page 539
Chapter 44 Test, *Teaching Resources*
Chapter 44 Portfolio Assessment, *Teaching Resources*
Test Generator, Chapter 44
Alternative Assessment Forms

ADDITIONAL RESOURCES

Books for Teachers

Chandler, David P. *A History of Cambodia.* Boulder, CO: Westview Press, 1992.

Clark, Cal, ed. *The Evolving Pacific Basin in the Global Political Economy: Domestic and International Linkages.* Boulder, CO: Lynn Rienner Publishers, 1992.

Hayslip, Le Ly. *When Heaven and Earth Changed Places: A Vietnamese Woman's Journey from War to Peace.* New York: Doubleday, 1989.

Books for Students

Chandler, David P. *The Land and Peoples of Cambodia.* New York: HarperCollins, 1991.

Ho, Minfong. *Rice without Rain.* New York: Lothrop, Lee & Shepard Books, 1990.

The Land and People of Cambodia. Philadelphia: Lippincott, 1991.

Mabie, Margot C. J. *Vietnam There and Here.* New York: Henry Holt & Co., 1985.

Ngor, Haing, and Roger Warner. *A Cambodian Odyssey.* New York: Macmillan, 1988.

The World Today *Videodisc Program*

A variety of still images, short videos, and activities are available for you to use as you teach this chapter. *Lecture Notes* include bar codes for specific images and videos.

Multimedia Resources

Cambodia: Year Ten. (VHS, 58 minutes) Films for the Humanities & Sciences, P.O. Box 2053, Princeton, NJ 08543–2053.

Great Wonders of the World: Manmade—Ankgor Wat. (Multimedia CD-ROM) InterOptica Publishing Ltd., 300 Montgomery Street, Suite 201, San Francisco, CA 94104.

The Mekong. (VHS, 57 minutes) Churchill Films, 12210 Nebraska Avenue, Los Angeles, CA 90025.

Picture Atlas of the World. (interactive maps, photo bank, animation, audio on CD-ROM) National Geographic Society, Washington, DC.

Vietnam. (16mm film/VHS, 25 minutes) The Changing Faces of Communism series. National Geographic Society, Educational Services, Washington, DC 20036.

MEETING INDIVIDUAL NEEDS

Limited English Proficiency (LEP)

Locate Mainland Southeast Asia on a wall map. Point out that the region is located on the Indochina Peninsula, which lies between India and China. Explain to students how bordering these two areas—home to some of world's earliest civilizations—has influenced Southeast Asian culture throughout history. Then call attention to the region's location between the Pacific and Indian oceans. Use this information as a springboard for helping students focus on the physical, economic, and political characteristics of Southeast Asia.

Learners Having Difficulty

Organize students into small groups. Divide the groups in half, assigning each half one of the sections in Chapter 44. Direct students to write five questions about the main ideas of their assigned section and to pose the questions to the other half of their group.

Auditory Learners

Divide students into five groups. Assign each group one of the countries of Mainland Southeast Asia. Instruct each group to organize and present a radio news report discussing political events in or the economies of its assigned country.

Visual Learners

Have students design postage stamps for a country of Mainland Southeast Asia. The stamps might depict wildlife, plants, famous people, historical events, customs, and natural resources of these countries. Allow students to use encyclopedias and other resources as necessary for additional information. Display their designs on the bulletin board.

Tactile/Kinesthetic Learners

Have groups of students make models, dioramas, or diagrams of the three major types of farming methods used in Southeast Asia: slash-and-burn agriculture, intensive paddy agriculture, and plantation agriculture. Display students' completed work in the classroom.

Gifted Learners

Direct interested students to investigate recent political, economic, and environmental developments in Mainland Southeast Asia. Students might use current newspaper and magazine articles or social studies journals. Have students compile their information into a television news show, and allow students time to present their news stories to the class.

MULTICULTURAL FOCUS

Point out to students that art forms often provide a window into a culture's past. Shadow-puppet theater—a form of drama performed in Thailand, Cambodia, Indonesia, and other parts of Southeast Asia—reflects the historical range of Hindu influence over the region. Shadow-puppet theater is still deeply embedded in the cultures of Southeast Asia.

First performed for royalty hundreds of years ago, the plays are sometimes based on the *Ramayana*—a Hindu epic from India that tells of a prince named Rama who rescues his wife from evil kidnappers. Since it was considered inappropriate for human characters to portray mythic figures, the tradition arose of using leather puppets to represent the Hindu deities.

The bodies of the puppets are cut with intricate designs to allow light to pass through them. Very large pieces of leather with whole settings cut out of them are used as scenery. The puppeteer presses the puppets, which are attached to bamboo poles, against the back of a thin cotton or silk screen. A single light behind the stage casts shadows of the puppets on the screen, which move to accompanying narration and music. The flickering shadows emphasize the mystic and symbolic elements of the story. Shadow-puppet theater performances often last from early evening until dawn.

Ask students for suggestions about how they would dramatize events from their own history or culture.

OVERVIEW

In Chapter 44, students will learn about Southeast Asia, a gateway region between the Indian and Pacific oceans.

Section 1 describes the physical and human geography of both the mainland and island nations of Southeast Asia. It explains how the region's location and natural resources have attracted many waves of settlement.

Section 2 examines the physical, economic, and cultural characteristics of Vietnam, Laos, Cambodia, Thailand, and Burma. Students will learn that after years of political turbulence, these nations are struggling to develop their economies.

About the Author

Le Ly Hayslip grew up in a small village in Vietnam in the midst of the Vietnam War. She and her children moved to the United States in the mid-1970s. The quotation on this page is from *When Heaven and Earth Changed Places,* Hayslip's memoir about Vietnam, the war, and a 1986 journey she made to Vietnam.

Primary Source

Description of Change: excerpted
Rationale: excerpted to focus on main idea

CHAPTER 44

Mainland Southeast Asia

GEOGRAPHY DICTIONARY

shifting agriculture

intensive agriculture

defoliant

      "I glance out of the window. . . . As we bank toward the south, climbing over the sandy coastline, I see the upturned thimbles of Marble Mountain and the lazy snakes of the Song Han delta. . . . I spot a loamy patch of ground covered with thatchroofed houses rising among the paddies like a mother's knee. There is a scraggly jungle on two sides—fighting . . . to come back after a decade of fire and poison to become the lush paradise I once knew. **"**

Le Ly Hayslip

Southeast Asia lies between the Indian and Pacific oceans. Key sea routes joining these oceans cross the region, including the narrow Strait of Malacca and the South China Sea. These important waters, as well as the region's valuable resources, have attracted Chinese, Indians, Arabs, and Europeans for centuries. In recent decades, much of Southeast Asia has undergone rapid industrialization. Some of the world's fastest growing economies are found here.

    Southeast Asia has a diverse cultural geography. Because it borders both China and India, the region shows influences of these two large neighbors. Nonetheless, mountains, peninsulas, river valleys, and islands have isolated the region to an extent and nurtured several distinct cultures. This chapter examines the region's physical, cultural, and economic geography and then explores the nations of mainland Southeast Asia. The region's island countries are explored further in Chapter 45.

Thai mask

Farming village in northern **Vietnam**

Using the Quotation

Note that the author describes an aerial view of Vietnam and that she has not seen her homeland for several years. Have students identify the images that describe mountains and rivers. *(upturned thimbles, lazy snakes)* Then ask them to point out words or phrases that indicate wartime damage to Vietnam's environment. *(scraggly jungle, fire and poison)*

Using the Geography Dictionary

Read aloud the terms in the Geography Dictionary. Have students look up each term in the Glossary. Then have students suggest conditions or events that might lead to the use of **shifting agriculture** *(soil exhaustion),* **intensive agriculture** *(high population density, little technology),* and **defoliants** *(war).*

SECTION 1
OBJECTIVES

The objectives reflect the Focus questions on this page and are assessed in the review on page 533.

1. Describe the landforms and climates of Southeast Asia.
2. Summarize the region's historical geography and economic geography.

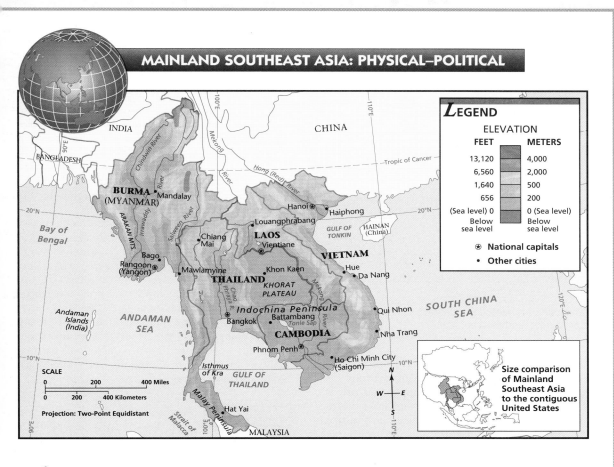

MAINLAND SOUTHEAST ASIA: PHYSICAL–POLITICAL

LEGEND
ELEVATION

FEET	METERS
13,120	4,000
6,560	2,000
1,640	500
656	200
(Sea level) 0	0 (Sea level)
Below sea level	Below sea level

⊛ National capitals
• Other cities

SCALE
0   200   400 Miles
0   200   400 Kilometers
Projection: Two-Point Equidistant

Size comparison of Mainland Southeast Asia to the contiguous United States

Using the Physical–Political Map: Mainland Southeast Asia

Referring to the map on this page, point out the five countries of mainland Southeast Asia, their capital cities, the two main peninsulas, and the bodies of water that surround the region. Call on volunteers to name a mountain chain *(Arakan),* a plateau *(Khorat),* an isthmus *(Kra),* a lake *(Tonle Sap),* and four deltas *(Irrawaddy, Chao Phraya, Mekong, Hong).*

1 PHYSICAL AND HUMAN GEOGRAPHY

FOCUS

● *What landform regions, rivers, and climates are found in Southeast Asia?*

● *What groups of people have settled in Southeast Asia over time? What is the region's economic geography?*

REGION **Landforms** Southeast Asia contains three landform regions. (See the map above.) In the north, mountain ranges fan out from the high Himalayas and the Plateau of Tibet into Burma, Thailand, Laos, and Vietnam. A central region of flat plains and low plateaus reaches between the mountain ranges. It then stretches across southern Thailand, Cambodia, and Borneo. The third landform region consists of thousands of islands. Some islands formed from single volcanoes that grew from the sea bottom. Larger islands, such as Java, have hundreds of volcanoes, many of them active. Most of Indonesia and all of the Philippines are volcanic and part of the Pacific Ring of Fire. Earthquakes and volcanic eruptions shake this region frequently.

Mainland Southeast Asia is also known as the "land of great rivers." Four important rivers flow from Asia's mountainous interior through the region. They are (1) the Irrawaddy in Burma; (2) the Chao Phraya (chow PRIE-uh) in Thailand; (3) the Mekong (MAY-KAWNG), which borders Thailand and flows through Laos, Cambodia, and Vietnam; and (4) the Hong (Red) River in northern Vietnam. The valleys and deltas of these rivers support rich agriculture.

Climate All of Southeast Asia has tropical climates. The Philippines, the western islands of Indonesia, the Malay Peninsula, and most of the

Motivator

Ask each student to list reasons why the region of Southeast Asia—both mainland and islands—is important to the world. Then call on students to share their ideas. Create a composite list on the board. Add, delete, or modify items on the list as the class learns about Southeast Asia.

Building a Vocabulary

Write the terms **shifting agriculture** and **intensive agriculture** on the board. Read phrases such as "uses much human labor," "abandon the fields," "paddy agriculture," and "slash and burn." Have students tell whether each phrase refers to shifting agriculture or to intensive agriculture. Ask whether the two terms are opposites. *(no)* Point out that, for example, shifting agriculture can be intensive.

Identifying Economic Resources

Organize the class into several large circles. Tell one person in each group to name an export product of Southeast Asia, such as spices. The person on the right should then repeat "spices" and name a different export product. The game should continue in this way around the circle, each person stating in order the export products named previously and adding a new one to the list.

Chart Caption Answer

Cambodia

Geography and History

The fact that Thailand never became a European colony was largely due to the efforts of King Mongkut in the mid-1800s. Mongkut, who was the model for the king in the Broadway musical The King and I, *successfully concluded trade treaties with Western powers, including Great Britain and the United States. Although Thailand did have to make some concessions in these treaties, the agreements helped the country remain independent. Mongkut also shrewdly balanced the interests of Great Britain and France against each other, another strategy that helped keep Thailand free of colonial rule.*

532

FACTS IN BRIEF	Mainland Southeast Asia		
COUNTRY POPULATION (1994)	LIFE EXPECTANCY (1994)	LITERACY RATE (1990)	PER CAPITA GDP (1993)
BURMA 44,277,014	58, male 62, female	81%	$950
CAMBODIA 10,264,628	48, male 51, female	35%	$600
LAOS 4,701,654	50, male 53, female	64% (1993)	$900
THAILAND 59,510,471	65, male 72, female	93%	$5,500
VIETNAM 73,103,898	63, male 68, female	88%	$1,000 (GNP)
UNITED STATES 260,713,585	73, male 79, female	97% (1991)	$24,700

Sources: *The World Factbook 1994*, Central Intelligence Agency; *The World Almanac and Book of Facts 1995*

Which country ranks last in figures for life expectancy, literacy rate, and per capita GDP?

coasts of the mainland are humid tropical. These areas have hot, year-long growing seasons, and heavy rainfall supports dense tropical rain forests.

Most of the interior of the mainland countries and the eastern islands of Indonesia have a tropical-savanna climate and are affected by the seasonal monsoon rains and dry seasons. Deadly typhoons, which bring high seas, strong winds, and torrential rainfall, are common in the entire region, especially in the Philippines and Vietnam.

MOVEMENT **Settlement** Fossils found on the island of Java indicate early human settlement in Southeast Asia. Some tribal groups still live on the rugged mountainous islands, including Mindanao, Borneo, and New Guinea. These dark-skinned peoples were driven out of most of the region or mixed with the Caucasians and Mongoloids who migrated to Southeast Asia thousands of years ago.

The Burmese migrated to Southeast Asia from China around 200 B.C. Soon, the Annamese people arrived from northern China and settled in Vietnam. The largest and most advanced culture was that of the Khmer (kuh-MEHR), who dominated present-day Cambodia beginning in the ninth century. By the twelfth century, they had spread their empire to include most of mainland Southeast Asia. In the 1200s, the Thais migrated from southern China to areas settled by the Khmers.

Peoples from beyond Southeast Asia's borders continued to pour into the region for the next several centuries. Indians set up trading cities and brought a Buddhist-Hindu influence to the region. Arab traders, although they did not settle in the region, brought a Muslim influence. European colonization began in the early 1500s with the arrival of the Portuguese, who were followed by the British, Dutch, French, and Spanish. Only Siam (sie-AM), now called Thailand, was not colonized by the Europeans. Many Chinese also migrated to the region during the colonial period.

During World War II, the Japanese conquered all of Southeast Asia. After the war, several Asian nationalist groups sought independence. Soon, new countries appeared on world maps. The Philippines became independent of the United States. French Indochina split into three nations: Vietnam, Cambodia, and Laos. The Dutch East Indies became Indonesia, and the many British colonies and protectorates became independent countries. These included Burma, Malaya and North Borneo (now Malaysia), Singapore, and Brunei.

INTERACTION **Agriculture** Agriculture provides many nations of Southeast Asia with their most valuable exports. Three major types of farming are common in the region: slash-and-burn agriculture, intensive paddy agriculture, and plantation agriculture.

Farming is difficult in the poor soils of the region's rugged, forested mountains. In addition, the slash-and-burn techniques used in many areas force farmers to abandon the fields after only two or three years due to soil depletion. They must move to a new area to start over. Because of the movement from area to area, this form of farming is called **shifting agriculture**. Most farmers in these forested areas are subsistence farmers who grow a variety of crops, including cassava, yams, rice, bananas, pineapples, beans, and sugarcane.

In the lowlands, farmers grow rice on fertile slopes and along river banks. Rice accounts for more than half of the region's agricultural land use, and the tropical climates allow two or even three crops of rice per year. People move the seedlings by hand from seedbeds to the flooded paddy fields. Harvesting also is done by hand. Because paddy agriculture uses much human labor, it is called **intensive agriculture**. This farming method is especially successful in areas such

Guided Practice

Write the combination *2-3-4* on the board. Explain that Southeast Asia has two climates, three landform regions, and four important rivers. Have students try listing the climates, landform regions, and rivers, independently. Then have them check their work by referring to the textbook. Call on volunteers to locate the landform regions and rivers on a wall map.

Independent Practice

Provide all students with Outline Map 21. Have them label the map to show the countries and capital cities of mainland Southeast Asia. Ask them to list near each country the names of groups that settled there over time.

Closure

Write these topics on the board: Landforms, Climate, Settlement, Agriculture, Resources. Have each student choose one of these topics and write a summary statement. Call on volunteers to read their statements.

as India, China, and Japan, where there is little arable land. Rice farmers also raise animals in the paddy fields, particularly ducks, water buffalo, pigs, and fish.

During the colonial period, a plentiful supply of low-cost labor, tropical climates, and nearness to sea routes made Southeast Asia an ideal site for large European-owned plantations. Today, most plantations are controlled by the local population. Crops grown for export include rice, coconuts, oil palms, spices, sugarcane, and rubber, the most valuable commercial crop in Southeast Asia. In fact, Malaysia, Indonesia, and Thailand produce most of the world's supply of natural rubber. Coffee also is a valuable export crop, especially that produced in the rich soils of the mountainous islands of Indonesia.

Resources Southeast Asia has large areas of tropical forest, which hold valuable hardwoods such as mahogany, ebony, and teak (an evergreen tree valued for furniture making). As a result, large areas of these forests are being cut for their woods. In most cases, there is little concern about replanting the forests.

This continued clearing of the region's tropical rain forests has already led to serious environmental problems, including loss of habitat for wildlife, soil erosion, and water shortages. It could also affect climates. Some countries in the region now recognize the effects of this deforestation. These countries have set aside national park areas and have instituted logging export bans. For some countries' forests, however, it is almost too late.

Southeast Asia has valuable mineral deposits, particularly tin and oil. The world's major tin deposits are found in Malaysia, Thailand, and Indonesia. Other important minerals include tungsten, iron ore, and manganese. In addition, Thailand has natural-gas fields, Vietnam and Indonesia have coal deposits, and Indonesia, Malaysia, and Brunei export petroleum.

Regional Cooperation Although the Southeast Asian countries are far from a unified group, several are members of the Association of Southeast Asian Nations (ASEAN). These countries—Thailand, Malaysia, Singapore, Brunei, Indonesia, the Philippines, and Vietnam—have treaties of cooperation with each other. Except for the Philippines, these nations are experiencing rapid economic growth. The region's poorer nonmember nations now also are considering joining. The ASEAN countries, meanwhile, are exploring the formation of a free-trade area which will remove tariffs on some products moving among member nations.

 SECTION REVIEW

1. What are Southeast Asia's three landform regions? What are its two major climates?
2. What are the region's most important exports?
3. **Critical Thinking   Explain** how Southeast Asia's physical geography and history may have hindered unity in the region today.

Elephants haul timber in the tropical rain forest of northern **Thailand**. Elephants are used in the timber industry in areas where the terrain is too difficult to use machines.

Section 1 Review
ANSWERS

1. Landform regions: northern mountain ranges, central plains and plateaus, thousands of islands; climates: humid tropical and tropical savanna
2. Agricultural products (rice, coconuts, oil palms, spices, sugarcane, rubber, coffee), lumber (hardwoods), minerals (tin, oil, tungsten, iron ore, manganese, natural gas, coal)
3. Physical geography: mountainous terrain and many islands may have limited communication and transportation. History: the region's large number of ethnic groups and outside influences may have reduced the common ground among different areas and groups.

533

Extension

Have students think of a name and design a logo for a Southeast Asian plantation or company that produces a commercial crop. The logo would appear on the product's packaging and advertising. Show students the logo of a large U.S. agribusiness, forestry, or mining company to get them started.

SECTION 2
OBJECTIVES

The objectives reflect the Focus questions on this page and are assessed in the review on page 538.

1. Identify key characteristics of each country of mainland Southeast Asia.
2. Discuss the political backgrounds and current political situations in the region's countries.

This street vendor carts his products through the streets of **Ho Chi Minh City, Vietnam**. Vietnamese officials hope that some free-enterprise reforms will help improve the country's economy.

2 THE COUNTRIES OF MAINLAND SOUTHEAST ASIA

FOCUS

- *What are the physical, economic, and cultural characteristics of each country in mainland Southeast Asia?*

- *What are the political backgrounds of these countries? What are their current political situations?*

From east to west, the countries of mainland Southeast Asia are Vietnam, Laos and Cambodia (with Laos to the north and Cambodia to the south), Thailand, and Burma. (See the map on page 531.)

Vietnam Vietnam is a long, narrow country that occupies the eastern portion of the Indochina Peninsula. The country's east coast faces the South China Sea. Laos and Cambodia form its rugged western border, and China is to the north. Mountains dominate Vietnam's interior.

Vietnam's population is heavily concentrated in the deltas formed by the Hong River in the north and the Mekong River in the south, and on the coastal plain between the two deltas. The Vietnamese, who arrived from China more than 2,000 years ago, make up more than 85 percent of the country's population. The largest minority groups are the Chinese, who live mainly in the cities, and the Montagnards, who live in the mountains.

Vietnam, along with Laos and Cambodia, was part of French Indochina. When the French returned to Southeast Asia after World War II, they found that Communist forces (the Vietminh) had gained power in the north. The Vietminh, under the leadership of Ho Chi Minh, demanded independence.

After eight years of fighting, the Vietnamese defeated the French in the north. As a result, in 1954 Vietnam was divided along the 17th parallel. North Vietnam established a Communist government, with its capital at Hanoi (ha-NOI). South Vietnam established a government friendly to the United States, with its capital at Saigon (sie-GAHN). Both governments wanted reunification but could not agree on the terms. In 1963, North Vietnamese troops invaded South Vietnam. The U.S. military supported the South Vietnamese forces.

During the Vietnam War, more than 55,000 Americans and more than 2 million Vietnamese were killed. Due largely to the high human costs of the war, the United States withdrew its troops

534

Motivator

Before class, prepare one large cutout shape each of Vietnam, Laos, Cambodia, Thailand, and Burma. Do not identify the countries. Distribute individual cutouts in envelopes to various students around the classroom. Explain that each student with a cutout is expected to come to the board, identify the country, and place the cutout in its correct orientation and relative location to all other countries in Southeast Asia. Students should not refer to a map during the activity. The countries should fit together at the conclusion of the exercise.

Building a Vocabulary

Write **defoliant** on the board. Recall that *de-* means "removal." Tell students that *-foliant* relates to foliage, or leaves. Thus, a defoliant is a substance that removes leaves from plants.

Understanding Chronologies

As a class, construct a time line of political events in Vietnam from 1954 through 1976. Discuss how the events shown by the time line had an influence on Vietnam's present conditions.

---

in 1973. Soon after, the South Vietnamese government collapsed. In 1975, North Vietnamese forces occupied Saigon and renamed it Ho Chi Minh City. The country was officially united in 1976 as the Socialist Republic of Vietnam. More than 1 million South Vietnamese tried to escape the Communist takeover. Seeking a new homeland, many of these refugees found their way to the United States, Australia, and other Western nations.

The long war devastated Vietnam's economy. Forests and other vegetation over about one-fifth of South Vietnam's area still show the effects of **defoliants** sprayed by U.S. forces. A defoliant is a chemical that makes plants' leaves drop off, usually killing the plants. In addition, much of Vietnam's infrastructure, especially roads and airports, still must be rebuilt. To speed economic development, the Communist government recently opened the nation to a free-market economy and has welcomed outside investment. The United States government ended a 19-year ban on trade with Vietnam in 1994 and restored diplomatic relations with the country the following year.

Vietnam has the resources necessary to build a strong economy. In the north, there are coal, oil, and mineral resources, while the delta of the Hong River is a fertile agricultural region. The south has the productive rice fields of the Mekong Delta as well as rubber plantations. Vietnam is now one of the world's largest rice exporters after Thailand and the United States. Other major exports include oil, rubber, and seafood. Although Vietnam remains one of Asia's poorest nations, it has the potential to become strong and self-sufficient.

Laos Landlocked and mountainous, Laos is one of the poorest countries in the world. It was first unified in the fourteenth century under a Lao prince. From the 1890s until 1954, Laos was under French rule.

Although the country was officially neutral during the Vietnam War, major supply routes for the North Vietnamese crossed Laos. The United States bombed parts of the country in an attempt to block these routes. In 1975, Communist forces took over the monarchy and formed the Lao People's Democratic Republic.

Laos's most important crop is rice, but the country also grows tobacco, corn, cotton, and coffee. Hydroelectric power exported to neighboring

Thailand is the nation's major source of income. Lumber is the other leading export. In 1990, in an effort to conserve its tropical forests, Laos banned the export of logs. Illegal logging by corrupt military leaders for export to Japan and other nations continues, however. The country has almost no industry and survives largely on foreign aid.

More than half of the population is made up of Lao, who are closely related to the Thais. The Lao are Buddhists and live mainly in rural valleys, especially along the Mekong River. The remainder of the population consists of highlanders and isolated mountain groups, such as the Hmong people. In total, Laos has more than 50 ethnic groups. The capital and only large city is Vientiane in the Mekong River valley.

The main problem facing Laos is poverty. More than 80 percent of the people are subsistence farmers who live in small, isolated villages. Most of these villages are not served by roads. There is little government spending on health care, and life expectancy is nearly 20 years less than that in neighboring Thailand.

The government, though Communist, allows privately owned businesses and religious freedom. Since the Vietnam War, more than 10 percent of the Laotian population has fled the country. Many now live in the United States and Australia. Thousands more remain in refugee camps in Thailand.

Cambodia Cambodia, the successor of the Khmer Empire, lies between Vietnam and Thailand. Most of its population are Khmer and Buddhist. The people are concentrated in Cambodia's fertile rice-growing region along the Mekong River. Rice and fish are the main food sources. The capital is Phnom Penh, located in the rice-growing region of the south.

Located in the northwest is Tonle Sap, Southeast Asia's largest lake. Tonle Sap supports a fishery and acts as a reservoir during the monsoon season. It helps reduce flooding downstream by filling with overflow water from the Mekong River. After the monsoon, water drains back into the Mekong.

Since independence from France in 1953, Cambodia has experienced much political unrest. During the Vietnam War, the United States bombed North Vietnamese supply lines in Cambodia. In 1975, Communist forces called the Khmer Rouge seized power. The Khmer Rouge

GLOBAL PERSPECTIVES

As of 1990, Vietnam's exports went mostly to Japan, Hong Kong, the Philippines, and Thailand, while imports came from Japan, Germany, Hong Kong, and France. Vietnam's new market economy presents rich opportunities for foreign investment. U.S. government sanctions, however, prevented U.S. businesses from participating in this growth before 1994. The United States has sought resolution of issues regarding U.S. soldiers that were taken as prisoners of war or were listed as missing in action during the Vietnam War. Lead a discussion about how Vietnam and the United States might cooperate to each other's benefit.

Identifying Political Conditions

Have students make a chart describing the current political situations of the five mainland Southeast Asian countries.

Guided Practice

Point out that the countries of mainland Southeast Asia face many challenges. Ask students to name two major challenges facing each country today (as discussed in Section 2). Write their responses on the board. *(examples: Vietnam—wartime damage to environment and damage to infrastructure; Laos—lack of industry, dependence on foreign aid; Cambodia—legacy of genocide, political unrest; Thailand—deforestation,* refugees; Burma—illegal drug trade, international isolation) Then ask students to name other countries they have studied that are facing the same issues. *(for example, Colombia—illegal drug trade)* Have students discuss possible reasons for the similarities. *(similar physical environments, similar levels of economic development)* As a class, propose ways that some or all of the countries might meet these challenges.

MULTICULTURAL PERSPECTIVES

Theravada Buddhism, an ancient school of Buddhism that emphasizes the individual pursuit of spiritual progress, is a central force in Thai and Burmese society. Theravadans believe that following the disciplined life of poverty, meditation, and study in a monastery is a means of renewing the soul and earning spiritual merit. Most Buddhist men in Thailand and Burma become monks or priests, at least for a short period of time. In Burma particularly, every boy is expected to join the order temporarily as a novice, usually during one rainy season from July to October. The initiation signals the youth's entry into adulthood. Elicit a class discussion about the purposes of rites of passage such as this.

These fishers check the haul in this fish trap at Tonle Sap, a lake in western **Cambodia**. The freshwater lake supports an important fishing industry in the country.

government isolated Cambodia and launched a program to establish a "Khmer peasant nation."

The goals of the Khmer Rouge were similar to those of the Cultural Revolution in China. The methods of achieving a simple, pure society were much more brutal in Cambodia, however. Cities were emptied, the educated were executed, families were separated, and citizens were forced into field labor. More than 1 million Cambodians and others, such as Vietnamese, were killed by their own government. At this time, the nation was called Kampuchea.

Vietnamese forces invaded Cambodia in 1978. They overthrew the Khmer Rouge and set up a government friendly to Vietnam. Today, despite periodic violence, the nation is trying to rebuild after years of warfare. Thousands of Cambodians remain in refugee camps in neighboring Thailand. United Nations troops are stationed in Cambodia to oversee the country's peaceful transition to democracy. Cambodia has sought economic aid from Japan and other countries.

Thailand *Thailand* means "land of the free," and it is the only Southeast Asian country that was not a European colony. The Kingdom of Thailand occupies the central part of the Southeast Asian mainland and extends south into the Malay Peninsula. Rugged, forested mountains form Thailand's border with Burma in the north and west. The Mekong River forms its eastern boundary with Laos. To the southeast is Cambodia.

The plains along the Chao Phraya River in central Thailand are the most populated and productive part of the nation. The climate is mostly tropical savanna and is driest on the Khorat Plateau in the northeast portion of the country. In the south along the narrow Malay Peninsula, the climate is humid tropical. All of Thailand is subject to seasonal monsoonal rainfall.

Thailand leads the world in rice exports and is a major fish exporter. Sugarcane, corn, and rubber are other export crops. Many of the country's food exports go to Japan, Thailand's major trade partner. Forests provide valuable tropical hardwoods, but extensive forest loss has affected the water supply and has caused other serious environmental problems. Thailand's minerals include tin, tungsten, and lead. The country also has oil resources and large offshore natural-gas deposits in the Gulf of Thailand.

The economy of Thailand has grown rapidly with the help of foreign investment, especially from Japan. The textiles, electronics, and automobile assembly industries have experienced rapid growth, while tourism provides one of the nation's largest sources of foreign income. Major imports include automobiles and machinery from Japan.

Bangkok, Thailand's largest city and capital, is built on the Chao Phraya. Much of Bangkok is connected by canals, called *klongs*, for which the city is famous. Bangkok also is well known for its palaces and Buddhist monasteries. In recent years, it has become known for its traffic jams and air pollution. As more and more water is drawn out of wells, the city is slowly sinking.

Thailand is a constitutional monarchy with a king, and popular support for the royal family is strong. Through much of Thailand's history, however, political power has been in the hands of military leaders.

More than three-quarters of the population is made up of Thais, who are Buddhist. The remaining population includes Chinese, Lao, and many mountain peoples. Along Thailand's eastern border areas are thousands of refugees from Cambodia, Laos, and Vietnam. This border region has also been an area of tension between Thailand and neighboring Cambodia and Laos.

Burma Burma occupies the mountainous northern and western part of the Southeast Asian mainland. It borders China on the north, India and Bangladesh on the west, and Thailand and Laos on the east. The country's southernmost tip extends to the narrow Isthmus of Kra along the Andaman Sea.

All of Burma is tropical and strongly influenced by the monsoon. The southern part of the country has a humid-tropical climate and receives up to 110 inches (279 cm) of rainfall annually, mostly during monsoonal downpours between June and September. Burma's northern part, which falls in a mountain rain shadow, is drier and has a tropical-savanna climate. Because mountains block most of the summer monsoon rains from the Indian Ocean, annual rainfall here can be as little as 32 inches (81 cm).

The broad north-south Irrawaddy River valley, much of which is cultivated, dominates the central part of Burma. A series of mountain ranges runs north to south over most of the rest of the country. Tropical forests cover much of the nation and contain the world's largest reserves of teak.

Burmans form more than half of the population. They are Buddhists and have a culture rich in art, architecture, and literature. Many ethnic minorities live in the mountains, but most of the population is crowded into the Irrawaddy River valley. The country's capital and largest city is Rangoon (also called Yangon).

THE WORLD TODAY VIDEODISC PROGRAM

Thailand: Ancient Traditions and Modern Prosperity

2B　　601　　3:55

Play　　　　Pause

Geographical Sidelight

Aung San Suu Kyi was awarded the Nobel Peace Prize in 1991 for her efforts in opposing Burma's military junta. At the time of the award, she had been under house arrest without a telephone for more than two years.

This *klong* (canal) in **Bangkok, Thailand**, features a floating market. Bangkok's system of canals is important to the local transportation of people and goods.

The Shwe Dagon pagoda, a Buddhist place of worship, is one of the most famous buildings of **Rangoon, Burma**. Nearly nine of every ten Burmese are Buddhists.

Rice, Burma's chief crop, is grown on the river floodplains. The country was once Asia's largest rice exporter, but today rice accounts for only a small part of the country's exports. Instead, lumber and seafood are the major exports. Burma also supports a large illegal opium trade. Smugglers distribute the drug as heroin in the United States, Europe, and many other parts of East and Southeast Asia.

Oil production in the lower Irrawaddy River valley is limited and goes to meet local needs. Mines in the northern mountains contain jade, and the central and southern parts of the country are rich in other minerals, including silver, lead, zinc, copper, tin, and tungsten. Offshore gas deposits have been discovered but not yet developed.

Burma gained its independence from Britain in 1948. In 1962, a military government took over all business and trade and created a socialist economy cut off from foreign influence. Indians and Chinese who lived in the country were driven out to neighboring nations.

In 1989, the military government changed the country's name from Burma to Myanmar, but the U.S. government still recognizes the country as Burma. Although pro-democracy protests have taken place in Burma, the military government remains in power. Government troops continue to battle minority and rebel groups in the northern mountains, and thousands have fled into Bangladesh to the west. To strengthen its position and the economy, the government now seeks foreign investment.

SECTION REVIEW

1. What ethnic groups live in each country of mainland Southeast Asia? To which countries is the production of rice important, and how?

2. How would you characterize political events in the region over the past few decades? What type of government does each country have today?

3. **Critical Thinking**   **Compare and contrast** post–World War II political developments in Vietnam to those in Korea.

538

CHAPTER 44 CHECK

ANSWERS

Building a Vocabulary

1. Shifting agriculture; slash-and-burn techniques deplete soil of its nutrients and eventually make it unproductive.
2. Agriculture that requires much human labor; rice is an example

Recalling and Reviewing

1. The Irrawaddy, Chao Phraya, Mekong, and Hong (Red); the river valleys and deltas are agriculturally productive and densely populated.
2. The forests contain woods that are valuable for industries such as furniture making. Continued deforestation may hurt wildlife, erode the soil, cause water shortages, and affect climates.
3. Supply routes for the North Vietnamese crossed the two countries. The routes were bombed by the United States.
4. Vietnam: Ho Chi Minh City (formerly Saigon); Laos: Vientiane (capital); Cambodia: Phnom Penh (in south); Thailand: Bangkok (canals); Burma: Rangoon (largest city)

Angkor Wat temple ruins in northwestern **Cambodia**

Reviewing the Main Ideas

1. Southeast Asia contains three landform regions and has tropical climates. Many different groups of people have settled in Southeast Asia over time. The region has valuable forest and mineral resources. Economic cooperation among the countries of the region is growing.
2. The countries of mainland Southeast Asia are Vietnam, Laos, Cambodia, Thailand, and Burma. These countries have experienced much political unrest over the past few decades.

Building a Vocabulary

1. What term describes the form of farming in which farmers periodically move to new areas? Why is this movement necessary, given the techniques the farmers use?
2. What is intensive agriculture? Give an example of a crop whose cultivation often involves intensive agriculture.

Recalling and Reviewing

1. What are the four major rivers of Southeast Asia? How are these rivers important to the region?

2. Why are large forest areas of Southeast Asia being logged? What effects might continued deforestation have?
3. How were Laos and Cambodia involved in and affected by the Vietnam War?
4. Name one city from each country of mainland Southeast Asia. List a characteristic of each city.

Thinking Critically

1. Thailand has numerous refugees from other countries in Southeast Asia. Why have these refugees left their own countries? Why do you think have they gone to Thailand? How might Thailand's history be related to the reasons refugees go there?
2. For what reasons might a country use a defoliant during war?

Learning About Your Local Geography

Individual Project
Interview community members and read old newspapers (available at the public library) to find out how your local community was affected by the Vietnam War. How did community residents feel about the war at the time it was happening? Did many people from the community fight in the war? If so, how did the experience affect them? Put your findings together into a brief written report. You may want to include your report in your individual portfolio.

Using the Five Themes of Geography

Draw two illustrations of different forms of **human-environment interaction** in Southeast Asia. How has the **region's** absolute **location** influenced human-environment interaction? In particular, you might want to focus on different forms of agriculture that are made possible by the region's location, or on the use of certain natural resources.

Thinking Critically

1. Refugees may have left their own countries because of instability and violence. Thailand attracts refugees because it is one of the region's most politically stable countries and has a rapidly growing economy. It also is the only country in the region that was never a European colony; this may be one reason why the country has been more stable.
2. To clear dense vegetation and increase visibility

Learning About Your Local Geography

Reports should discuss how students' communities were affected by the Vietnam War.

Using the Five Themes of Geography

Illustrations might show slash-and-burn, intensive, or plantation agriculture or the cutting of tropical forests. All of these are allowed by the region's tropical location.

539

CHAPTER 45 Island Southeast Asia
Planning Guide

TEXTBOOK	TEACHING RESOURCES	OTHER AVAILABLE MATERIALS
Section		
1. Malaysia, Singapore, and Brunei Pages 541–543	Reteaching Worksheet 45.1 Skills Worksheet 45	*The World Today* Videodisc Program *Study Guide for Spanish Speakers* *Lecture Notes,* Chapter 45
2. Indonesia Pages 543–547	Reteaching Worksheet 45.2 Challenge/Enrichment Worksheet 45	*Study Guide for Spanish Speakers* *Lecture Notes,* Chapter 45
3. The Philippines Pages 547–550	Reteaching Worksheet 45.3 Graphic Organizer 45 Critical Thinking 45	*Study Guide for Spanish Speakers* *Lecture Notes,* Chapter 45 *The World Today* Videodisc Program
Chapter Review Page 551	Chapter 45 Test Chapter 45 Portfolio Assessment See also *Alternative Assessment Forms*	*Test Generator,* Chapter 45
Unit Review Pages 552–553	Unit 9 Test Unit 9 Portfolio Assessment See also *Alternative Assessment Forms*	*Test Generator,* Unit 9

CHAPTER RESOURCES

Lecture Notes, Chapter 45
Teaching Resources
    Chapter 45 Graphic Organizer
    Chapter 45 Skills Worksheet
    Chapter 45 Critical Thinking Worksheet
    Chapter 45 Challenge/Enrichment Worksheet

REVIEW AND ASSESSMENT

Chapter 45 Check, page 551
Chapter 45 Test, *Teaching Resources*
Chapter 45 Portfolio Assessment, *Teaching Resources*
Test Generator, Chapter 45
Alternative Assessment Forms

ADDITIONAL RESOURCES

Books for Teachers

Ariff, Mohamed. *The Malaysian Economy: Pacific Connections.* New York: Oxford University Press, 1991.

Malaysian Studies—Archaeology, Historiography, Geography, and Bibliography. DeKalb, IL: Northern Illinois University Center for Southeast Asian Studies, 1986.

Nance, John. *The Gentle Tasaday: A Stone Age People in the Philippine Rain Forest.* New York: Harcourt Brace Jovanovich, 1975.

SarDesai, D.R. *Southeast Asia; Past and Present.* Boulder, CO: Westview Press, 1994.

Books for Students

Clark, James I. *Southeast Asia.* Evanston, IL: McDougal, Littel, 1989.

The Land and People of Malaysia and Brunei. Philadelphia: Lippincott, 1991.

Lawson, Don. *The New Philippines.* New York: Franklin Watts, 1986.

Wee, Jessie. *Philippines.* New York: Chelsea House, 1988.

The World Today *Videodisc Program*

A variety of still images, short videos, and activities are available for you to use as you teach this chapter. *Lecture Notes* include bar codes for specific images and videos.

Multimedia Resources

How to Study Cultures. (16mm film/VHS) United Learning. Available through Social Studies School Service, 10200 Jefferson Blvd., P.O. Box 802, Culver City, CA 90232–0802.

MECC DATAQUEST: Asia and Oceania. (Apple software, teacher's guide, reproducible handouts) Available through Social Studies School Service, 10200 Jefferson Blvd., P.O. Box 802, Culver City, CA 90232–0802.

Picture Atlas of the World. (interactive maps, photo bank, animation, audio on CD-ROM) National Geographic Society, Washington, DC.

Southeast Asia. (16mm film/VHS, 60 minutes) Educational Design. Available through Social Studies School Service, 10200 Jefferson Blvd., P.O. Box 802, Culver City, CA 90232–0802.

MEETING INDIVIDUAL NEEDS

Limited English Proficiency (LEP)

On the board, sketch and label a peninsula and an archipelago to help students understand the difference between the two landforms. Have students use the map on page 541 to locate the Malay Archipelago. Direct students to identify islands as you make statements such as "I am thinking of an island that is located at 16° north latitude, 121° east longitude." (Luzon).

Learners Having Difficulty

Have students create a question for each of the ideas in the Reviewing the Main Ideas section of the Chapter 45 Check (page 551). Then ask pairs of students to quiz one another about the chapter using the questions.

Auditory Learners

Students will have read in Chapter 45 that Indonesia's national motto is "Unity in Diversity." Assign each student one of the other countries in Southeast Asia, and ask each to imagine that he or she is a public relations officer charged with making up a motto for the country. Call on volunteers to name their country, to say its motto, and to explain why they chose it.

Visual Learners

Refer students to the article "Indonesia: Two Worlds, Time Apart" in the January 1989 issue of *National Geographic*. Point out that the opening page of the article summarizes Indonesia's geographical challenges and advantages. Have students find examples of these challenges and advantages in the article's photographs and text.

Tactile/Kinesthetic Learners

Organize the class into small groups. Have students imagine that they have been chosen to persuade all Indonesian citizens to use the national language, Bahasa Indonesia. Have groups develop and present their promotional ideas—such as slogans, billboards, advertisements—to the class.

Gifted Learners

Tell students that in the early 1970s, news emerged from the Philippines about a group called the Tasaday, who reportedly followed a Stone Age way of life with no knowledge of the outside world. Since 1986, these first reports have been called into serious question. Have interested students investigate the controversy surrounding the Tasaday discovery.

MULTICULTURAL FOCUS

The music of Southeast Asia reflects the region's tropical climate and vegetation, as well as the settlement patterns of the people. Music concerts in most parts of Southeast Asia are held in open settings such as courtyards and public lawns.

Musical instruments are often made of naturally occurring materials. They include bamboo mouth harps, flutes, and pipes; drums made from tropical woods; and buffalo horns. Local instruments from some of the island nations include horns made of bones, shells, skulls, fruits, seeds, crab claws, coconut, and shark bones.

The music produced by these instruments often bears a resemblance to sounds heard in the tropical forests of Southeast Asia. Balinese music, for example, includes special vocal techniques and mouth harp ensembles that sound like the croaking of frogs and other animal sounds.

Instruments and styles of music differ widely between rural and urban areas. In metropolitan areas of Southeast Asia, large groups of metal gongs are played for court and state ceremonies. In towns and outlying areas, smaller groups of musicians use fewer instruments in music for village ceremonies and everyday activities.

Lead a discussion with students about how the music they listen to reflects their culture and geographic region. You may wish to have interested students investigate the history of familiar musical instruments and the materials from which they are constructed.

CHAPTER 45

ISLAND SOUTHEAST ASIA

Section 1: Malaysia, Singapore, and Brunei

Section 2: Indonesia

Section 3: The Philippines

OVERVIEW

In Chapter 45, students will learn about the island nations of Southeast Asia.

Section 1 describes Malaysia, Singapore, and Brunei. It discusses Malaysia's rapid industrialization as well as the prosperity of Singapore and Brunei.

Section 2 examines Indonesia, a country of many islands and the largest country in Southeast Asia, and focuses on its industrialization.

Section 3 discusses the Philippines, an economically troubled country that faces a number of social and political challenges.

About the Author

John Nance is an American journalist. The quotation on this page is from *The Gentle Tasaday,* a book he wrote about the ways of the Tasaday, an indigenous group living in the Philippine rain forest.

Primary Source

Description of Change: excerpted

Rationale: excerpted to focus on the main idea

CHAPTER 45

Island Southeast Asia

66 The frill of shrubby ferns and grass at the edge quickly became swallowed in tangles of vines, bamboo, rattan, palms, acacias, and tree ferns over which towering mahoganies and oaks cast a ragged, sun-shielding canopy. The moist jungle sprang up sawtooth slopes and down into ravines, across limestone cliffs, and broad plateaus, climbing 5,000-foot peaks before it slanted onto the sandy shores of the Celebes Sea. 99

John Nance

Part of a dance costume from **Bali, Indonesia**

In 1991, Mount Pinatubo, on the Philippine island of Luzon, began to erupt for the first time since 1380. The volcano shot ash over 9 miles (14 km) into the air and sent 1,500°F (816°C) lava flowing down the mountain at more than 60 miles per hour (97 kph).

Luzon is just one of the some 20,000 tropical islands of Southeast Asia. Physical geographers call the region the Malay Archipelago (ahr-kuh-PEHL-uh-goh). An **archipelago** is a large group of islands. These islands first attracted the seafaring Malay people centuries ago. The area's riches also have attracted Chinese, Indians, and various European colonial powers, who dubbed the region the East Indies. They also called part of the region the Spice Islands because of its valuable spices, such as cinnamon and pepper. Today, island Southeast Asia has a wealth of cultures, languages, foods, music, and art.

Bamboo forest on **Sumatra, Indonesia**

GEOGRAPHY DICTIONARY

archipelago

entrepot

kampong

sovereignty

540

Using the Quotation

Ask students to imagine what it might be like to enter a tropical rain forest. Then read aloud the quotation on page 540. Ask how the author conveys the density of the forest. *(Every space seems to be filled with vegetation.)* Note that the author is describing a rain forest in the Philippines but that rain forests are characteristic of all of island Southeast Asia.

Using the Geography Dictionary

Read aloud to the class the terms in the Geography Dictionary. Point out various **archipelagoes** (Greater and Lesser Antilles, Northwest Territories, island Southeast Asia, Aleutian Islands) on a wall map, and let students form their own definition of the word.

SECTION 1
OBJECTIVES

The objectives reflect the Focus questions on this page and are assessed in the review on page 543.

1. Identify physical, cultural, and economic characteristics of Malaysia.

2. Discuss the reasons for and the effects of the prosperity of Singapore and Brunei.

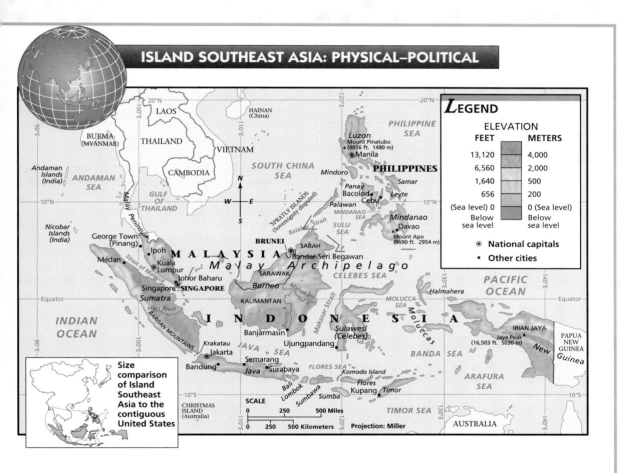

ISLAND SOUTHEAST ASIA: PHYSICAL–POLITICAL

Using the Physical-Political Map: Island Southeast Asia

Direct students' attention to the map on this page. Have them locate Malaysia, Singapore, Brunei, Indonesia, and the Philippines. Ask students to identify physical features that are common to several of the countries. *(mountains and islands)* Have students note unusual characteristics of the physical geography and the size of these nations. *(Malaysia is split between a peninsula and part of an island. Singapore and Brunei are very small nations.)*

Size comparison of Island Southeast Asia to the contiguous United States

MALAYSIA, SINGAPORE, AND BRUNEI

FOCUS

● *What are the physical, cultural, and economic characteristics of Malaysia?*

● *What factors account for the prosperity of Singapore and Brunei? How are the countries' citizens affected by the prosperity?*

Malaysia The South China Sea divides the country of Malaysia into two parts. Western Malaysia occupies part of the narrow Malay Peninsula and is connected to mainland Southeast Asia. Eastern Malaysia lies on the northern part of the island of Borneo. (See the map above.)

The British governed Malaya, on the Malay Peninsula, until 1957. Then in 1963, Malaya joined with three other former British colonies—Sarawak and Sabah, both on north Borneo, and Singapore—to form a new country called the Federation of Malaysia. Singapore left the Federation in 1965 and became an independent country.

All of Malaysia lies near the equator and has a humid-tropical climate. The interior of the country is forested and mountainous. Its population is concentrated along the Malay Peninsula's western coastal plain, where the capital city of Kuala Lumpur (kwahl-uh LUM-pur) lies.

Slightly more than half of the country's people are Muslim Malays. The largest minority groups are the Chinese, who make up one-third of the population, and Indians, who make up one-tenth. Other minority groups include the Dayaks, a tribal people of Eastern Malaysia. The Malays control the nation's political power, while the Chinese, who live mainly in urban areas,

541

Motivator

Have students locate Singapore and Brunei on the physical–political map of East and Southeast Asia on page 495. Tell them that they will learn why these nations, the smallest in the region, are also the wealthiest. Refer students to the photographs on this page and pages 544–545 for evidence of prosperity in Singapore and Brunei.

Identifying

Ask each student to write one physical, one cultural, and one economic characteristic for Malaysia. Ask volunteers to read their statements.

Guided Practice

On the board, have students list factors that contribute to the success of Singapore and Brunei. Then ask: How have economic conditions affected the citizens of each country?

Closure

Have students make a chart showing facts about Malaysia, Singapore, and Brunei.

ASSESSMENT

Assign the Section 1 Review questions on page 543.

Reteaching

Assign Section 1 of Reteaching Worksheet 45.

Geographical Sidelight

Since 1969, the Malaysian government has enacted race relations policies that favor native Malays. For example, Malays are given preference in government jobs and in college admissions. The intent of these policies is to raise the economic status of the Malays to meet that of the more economically successful Chinese minority.

Chart Caption Answer

Singapore

COOPERATIVE LEARNING

Organize the class into groups of four. Supply the groups with additional resource material on Malaysia. Have one pair in each group create a graphic organizer to describe the physical geography, economic geography, and cultural geography of Western Malaysia. Have the other pair do the same for Eastern Malaysia. Have the pairs compare the information in each graphic organizer.

FACTS IN BRIEF — Island Southeast Asia

COUNTRY POPULATION (1994)	LIFE EXPECTANCY (1994)	LITERACY RATE (1990)	PER CAPITA GDP (1993)
BRUNEI 284,653	69, male 73, female	95% (1987)	$9,000 (1991)
INDONESIA 200,409,741	59, male 63, female	77%	$2,900
MALAYSIA 19,283,157	66, male 72, female	78%	$7,500
PHILIPPINES 69,808,930	63, male 68, female	90%	$2,500
SINGAPORE 2,859,142	73, male 79, female	88%	$15,000
UNITED STATES 260,713,585	73, male 79, female	97% (1991)	$24,700

Sources: *The World Factbook 1994*, Central Intelligence Agency; *The World Almanac and Book of Facts 1995*

Which country in island Southeast Asia has the highest per capita GDP?

---

hold most of the financial power. Calming the ethnic tension between these two groups is an important concern in Malay society. The national language is Malay, though English is commonly spoken as a result of the country's British colonial history.

Malaysia is rich in natural resources and is a major exporter of rubber, palm oil, timber, tin, and oil. Rubber plantations and tin mines are located mainly on the Malay Peninsula, while offshore oil and natural-gas deposits line the Strait of Malacca. The economy of Eastern Malaysia depends heavily on forestry and petroleum.

Industrialization in Malaysia is taking place at a rapid pace. The country's stable government, educated work

force, and good transportation and telecommunications networks have attracted much foreign investment. Major industrial exports include electronics and textiles. Malaysia also continues to attract high-tech industries and has expanded its tourism industry.

Singapore and Brunei The two smallest nations in Southeast Asia are also the wealthiest. Industrialized Singapore lies just off the southern tip of the Malay Peninsula. Oil-rich Brunei lies on the north coast of the island of Borneo. (See the map on page 541.)

Singapore is the name of the nation, its major city, and the island on which the city is located. The country owes much of its prosperity to its location and productive population. (See "Global Connections" on pages 544–545.)

Singapore's harbor faces the bustling Strait of Malacca and is one of the busiest ports in the world. Singapore is a modern city of high-rise hotels, banks, and shopping malls. The city is the center of trade and banking for Southeast Asia, and its industries produce refined-oil products, ships, textiles, electronics, and many high-tech products. Its people are generally prosperous and well educated. More than 75 percent of the population are Chinese. The remainder is Malay and Indian.

Brunei is an independent, oil-rich sultanate. The government provides its citizens with basic education, health care, and some housing. About two-thirds of the population is Malay, while about

This mosque in **Bandar Seri Begawan, Brunei**, is evidence of the wealth provided by the small country's oil industry. Revenues from the oil industry account for more than half of the country's GDP.

SECTION 2
OBJECTIVES

The objectives reflect the Focus questions on this page and are assessed in the review on page 547.

1. Describe Indonesia's physical geography and major islands.
2. Examine Indonesia's economic geography, cultural geography, and political geography.

TEACHING SECTION 2
Motivator

Tell students to imagine that the 48 contiguous U.S. states were actually 48 separate islands. Ask what effects this might have on internal U.S. economic, political, and cultural relationships. Then point out that Indonesia is made up of over 10,000 islands that stretch almost the same distance as from Los Angeles to New York.

Building a Vocabulary

Read aloud the definition of **kampong** from a dictionary. *(a traditional village in a Malay-speaking country)* Ask students to consider why this word also refers to slum areas near the large city of Jakarta. *(Many people living in the urban* kampongs *migrated from rural villages.)* Discuss why rapid movement into an urban area often results in slum conditions.

one-fifth is Chinese. Thousands of guest workers from other Southeast Asian nations also live in the country.

Brunei's economy is based primarily on petroleum exports. Knowing that its oil supplies will run out one day, the government has invested for the future. It has plans for agricultural expansion and, to attract new industry, it has a tax policy favorable to businesses.

SECTION REVIEW

1. On what two major land features does Malaysia lie? Why is there tension between the country's two largest ethnic groups?
2. How does the overall prosperity of Singapore and Brunei benefit the countries' citizens?
3. **Critical Thinking** **Decide** which country's economy has a stronger foundation for future growth: that of Singapore or of Brunei. Explain your reasoning.

INDONESIA

FOCUS

- *What is the physical geography of Indonesia? What are the country's major islands?*

- *What economic, cultural, and political characteristics does Indonesia have?*

For more than 300 years, Indonesia was ruled by the Dutch. The colony was known as the Netherlands East Indies (also called the Dutch East Indies). Indonesia became independent in 1950. Today, the country has one of the world's most rapidly increasing populations.

REGION **Physical Geography** Indonesia is the largest country in Southeast Asia. Its 13,660 islands stretch some 3,000 miles (4,827 km) between the Indian and Pacific oceans, nearly the same distance as that between Los Angeles and New York. (See the map on page 541.)

All of Indonesia lies along the equator, giving the entire nation a humid-tropical climate. The islands to the west receive rain year-round, while the islands to the east experience more seasonal monsoon rainfall. Most of the islands, especially Kalimantan and Irian Jaya, are covered by vast tropical rain forests. A wide variety of wildlife lives on the islands, including the Komodo Island dragon, a lizard that weighs up to 300 pounds (136 kg) and grows up to 12 feet (3.6 m) in length. Most coastal areas have dense mangrove swamps and many coral reefs.

Its location at the junction of several major tectonic plates makes Indonesia the most active volcanic region on Earth. The nation has more than 150 volcanoes, about 50 of which are active. In 1883, the volcanic island of Krakatau erupted, killing thousands of people on nearby Java and Sumatra. Although the volcanoes present a danger, they also provide the lava and ash that enrich the islands' soils.

Major Islands From west to east, the largest islands are Sumatra, Java, Kalimantan (part of Borneo), Sulawesi (also called Celebes), Timor, and Irian Jaya (the western part of New Guinea). Holding most of the nation's industry and political power, Java is the core island of Indonesia. Although Java makes up only seven percent of the country's land area, it is home to two-thirds of the population. It is one of the world's most densely populated areas and is home to the huge, rapidly growing capital city of Jakarta (juh-KART-uh). Vast slums are expanding around the city.

The larger island of Sumatra lies at the far western edge of Indonesia. The island has rich natural resources, including volcanic soils, dense tropical forests, and large oil and natural-gas deposits. Many different cultures exist on the island.

Bali, Java's island neighbor to the east, has a rich cultural background. The Balinese practice a Hindu-Buddhist religion, and thousands of elaborate temples and shrines dot the island. The islanders are known in particular for their art, festivals, and dances. Bali supports one of Indonesia's largest tourist industries.

The island of Timor was once split between the Dutch, who controlled west Timor, and the Portuguese, who controlled east Timor. Indonesia took over the western half of the island when it gained its independence from the Netherlands. East Timor, however, was seized by Indonesia

Text continues on page 546.

Geography and Language

The word amok—as in "everything's gone amok"—is borrowed from the Malay language. It aptly describes Jakarta's traffic jams and overloaded telephone system—signs of the city's explosive growth.

Section 1 Review
ANSWERS

1. The Malay Peninsula and the island of Borneo; because the Malays control most of the political power whereas the Chinese hold most of the financial power
2. They are provided with services such as education and health care.
3. Possible answer: Singapore, because it has a diversified economy and a location suited to trade, whereas Brunei's economy is based on oil deposits that will eventually run out

Global Connections

Simulating World Trade

Organize the class into two groups, each on opposite sides of the room. Give each group a number of slips of paper with phrases written on them such as "Grain worth $1 million," "Textiles worth $15 million," and "Barrels of oil worth $50 million." Also give each group $100 million in "bills" of fake money.

Tell students that they will be trading with each other. They must trade through you, in the center of the room, and they must give you a fee equal to 10 percent of the value of the goods they are trading. Explain that you, like Singapore, are acting as an entrepot, and you are profiting from your location between the two groups.

After several trades, ask each group to total their assets. Then show them how much you have collected as an entrepot. Ask students why it might be to their economic advantage to relocate part of their operations to the center of the room. Point out that many foreign companies now have operations in Singapore.

Geographical Sidelight

Other punishable offenses in Singapore include driving without a seat belt and not flushing toilets in public buildings. Those caught littering must not only pay a fine but also undergo counseling.

Geographical Sidelight

Because Singapore's government operates much like an efficiently run corporation, Singapore is often referred to as a "nation-corporation." It has also been dubbed "Singapore, Inc."

GLOBAL CONNECTIONS

Singapore

In just a few decades, Singapore has risen from a poor country to become an economic force on the Pacific Rim.

Travelers visiting Singapore should take care to read the rules posted all over the city. Laws prohibit spitting, jaywalking, littering, smoking in restaurants, and eating or drinking on the subways. Fines over the equivalent of 300 U.S. dollars can be the price for breaking one of these rules. These laws help make Singapore one of the cleanest nations in the world. It is also one of the most efficient, and there is little unemployment, poverty, or crime. Life in Singapore was not always so ordered and prosperous, however.

Cargo ships gather behind the towering skyscrapers of **Singapore**, one of the world's most important shipping and trade centers.

The Importance of Location

A tiny island-nation, Singapore has little land and few natural resources. Its one economic advantage is its location. Situated between the Indian Ocean and the South China Sea, Singapore is at a crossroads of Southeast Asian trade. Its location attracted the British to Singapore in 1819, and Singapore soon became a busy Asian port and a major British naval base.

When Singapore gained its independence in 1959, many people were still poor farmers. The nation lacked adequate housing, and about half the population had never attended school. For a few years, Singapore was united with Malaysia, but the two separated in 1965. Then Britain decided to close down its naval base in Singapore. With little industry, Singapore's economy seemed too weak to support its people.

Singapore was still able to profit from its location, however. The island served as an **entrepot** (AHN-truh-poe), or intermediate port, for nations sending goods between the Pacific and Indian oceans. These nations used Singapore to temporarily store their goods before shipping them to other countries. Southwest Asian nations such as Indonesia also began shipping their oil to Singapore, where it was refined before being exported to countries around the world.

Warehousing and oil refining were not enough to support a strong economy, however. Singapore's then Prime Minister Lee Kuan Yew realized that sustained economic growth depended on the development of industry. The government set out to create new economic resources where there were none before.

The government launched a long-term industrialization program. Lacking the money to build large factories, Lee sought to lure foreign investment. To encourage multinational companies to locate in Singapore, the government improved transportation facilities, educated workers, and offered low tax rates on businesses. Construction began on roads, railroads, power stations, housing for workers, a new port, and an airport.

By the early 1990s, 3,000 foreign companies had operations in Singapore. Of these, 800 were American, making the United States the largest of Singapore's foreign investors. Today, Singapore is one of the world's busiest ports. It is southern Asia's banking center, the world's third largest oil refiner, and a leader in communications and information technology.

Use the following activity to reinforce the geographic theme of region. Have students use the information presented in their textbooks to create several maps that illustrate the various regions in which Singapore might be included. Suggest to students that maps might show a physical region, a cultural region, and an economic region. Each map should have an extended caption explaining how and why Singapore is a part of each region.

MAJOR WORLD SHIPPING ROUTES

NORTH AMERICA
EUROPE
ASIA
AFRICA
SOUTH AMERICA
AUSTRALIA
ANTARCTICA
PACIFIC OCEAN
INDIAN OCEAN
ATLANTIC OCEAN

Rotterdam • Hamburg
Marseilles
Haifa
Al Ahmadi
Ras Tanura • Abu Dhabi
Suez Canal
Durban
Pusan • Tokyo–Yokohama
Kobe–Osaka
Kaohsiung
Hong Kong
Singapore
Sydney
New York
New Orleans
Los Angeles–Long Beach
Houston
Panama Canal
Maracaibo
Santos
Valparaíso

LEGEND
• Major ports
— Major shipping routes
The width of the lines are in proportion to the amount of ocean trade along the routes.

MAP STUDY Maps can provide useful information to businesses trying to get products to markets. The thick bands on this map show major shipping routes throughout the world. The thicker the width of a band, the greater the amount of goods that are shipped along the route. A prime location helps make Singapore a hub of international shipping. Which two canals on this map are important links along shipping routes?

Another Side to Success

Singapore is the product of a focused government with significant economic and political control. In fact, the government controls nearly every aspect of life in Singapore. Elections mean little because one political party is so strong that it almost always wins. The government owns many of the nation's companies and controls the press, making sure that newspapers do not criticize government decisions. It also bans certain movies and music.

According to one human-rights organization, "Singapore is not a country in which individual rights have significant meaning." This may be the price of Singapore's economic success. But for now, most of Singapore's people are content to enjoy the rewards of a government "dedicated to the pursuit of capitalism."

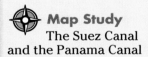

Singapore residents enjoy their beautiful city as they relax at this festival market. Strict antilittering laws help keep Singapore one of the world's cleanest cities.

▼**YOUR TURN**

Choose a country in Southeast Asia and **investigate** its economic resources, political system, and history. Imagine you are a government official of this country, and **write** a report suggesting policies for developing the economy. Discuss the policies' positive and negative consequences.

THE WORLD TODAY VIDEODISC PROGRAM

Singapore: Prosperity at a Price

2B 30437 4:10

Play   Pause

Map Study
The Suez Canal and the Panama Canal

Your Turn
Students' reports should recommend policies appropriate to their chosen countries. Reports should discuss both the positive and negative consequences of the policies.

545

Guided Practice

Write the names of Indonesia's main islands in one column on the board: Bali, Irian Jaya, Java, Kalimantan, Sulawesi, Sumatra, Timor. Refer students to the maps on pages 495–497 and 541. Have them work with partners to identify aspects of physical geography that these islands share. *(humid-tropical climate, tropical rain forests, active volcanoes, mountains)*

Independent Practice

Have students create graphic organizers for Indonesia that classify facts about the country into three categories: Economic Geography, Cultural Geography, Political Geography. When the organizers are complete, have pairs of students compare their work.

Closure

Have students sketch a map of Indonesia's main islands and write a few facts about each one.

ASSESSMENT

Assign the Section 2 Review questions on page 547.

Reteaching

Assign Section 2 of Reteaching Worksheet 45.

Geography and Natural Science

Wallace's Line is an imaginary line between Kalimantan and Sulawesi that divides Indonesia's wildlife into two groups. The plants and animals of the western islands resemble those of mainland Southeast Asia, and the plants and animals of the eastern islands resemble those of Australia. Scientists believe that the western islands were once part of the Asian mainland. The eastern islands, however, were always separated from the Asian continent and therefore did not develop the same wildlife.

MULTICULTURAL PERSPECTIVES

While many people in Western cultures tend to think of dance, theater, or music as being forms of entertainment, such "art" often takes on greater meaning in other cultures. In Indonesia's famous puppet theater, for example, performances often are presented to mark important life moments, such as birthdays and marriages. Futhermore, the stories might hold ethical messages and commentary. Have students create and present their own puppet shows.

Crowded residential areas continue to push outward from **Jakarta, Indonesia**. Like many other large cities in developing countries, Jakarta's population is rapidly increasing and is expected to approach 14 million by the year 2000.

from Portugal in 1975. Protest of the invasion by the Portuguese-speaking Catholic population continues. The Indonesian military's efforts to suppress this resistance have resulted in the loss of thousands of lives.

The three-fourths of the large island of Borneo that is part of Indonesia is called Kalimantan. The population of this island is made up of numerous groups with different languages and cultures. To the east of Kalimantan, across the Makasar Strait, is the island of Sulawesi. Two peninsulas jutting out to the east and southeast from a north-south backbone make Sulawesi look a little bit like a capital K. With 3,404 miles (5,477 km) of coastline, no place on the island is more than 70 miles (113 km) from the sea.

Irian Jaya, the former Dutch colony on the western half of the huge island of New Guinea, became part of Indonesia in 1969. Irian Jaya is extremely rugged and forested. Its snow-capped mountain spine reaches an elevation of 16,500 feet (5,029 m). Various peoples live in the highlands, forests, and swamps of this undeveloped land. New Guinea's eastern half is the independent country of Papua New Guinea. (See Unit 11: The Pacific World and Antarctica.)

Economic Geography Indonesia remains a poor agricultural country, but it has taken steps toward increased industrialization and prosperity. Agriculturally, Java and Sumatra are the most pro-

ductive islands. Just 20 years ago, Indonesia was the world's largest importer of rice. (See "Planet Watch" on pages 548–549.) Today, the country is self-sufficient in rice. Because of the mountainous terrain of the islands, different crops are grown at different elevations. Coconuts, sugarcane, rice, and rubber are grown along the coastal plains. Tea and coffee thrive on the mountain slopes.

After Brazil, Indonesia has the world's largest remaining areas of tropical rain forest. The market for tropical hardwoods, however, makes logging profitable. As a result, huge tracts of Indonesia's rain forest have been destroyed. To conserve forest resources and develop more local wood industries, Indonesia recently banned the exporting of logs.

Indonesia's other natural resources include large deposits of tin, coal, and oil. Petroleum and natural gas are major exports, with petroleum being the country's main source of income. Most of the oil deposits are on Sumatra, Java, Kalimantan, and Irian Jaya. Petroleum deposits also lie under the Java Sea.

Industrialization in Indonesia is in an early stage of development. The goal is to decrease the country's dependence on oil exports and to further develop industry. Most of Indonesia's present factories produce textiles, food and wood products, and chemicals related to petroleum. Indonesia seeks to attract more high-tech industries from the developed nations. Its supply of labor and low wages—some Indonesian factory workers earn less than the equivalent of one U.S. dollar a day—have already drawn many multinational corporations. In addition, Indonesia hopes to modernize the economy by developing a reliable communications system for all the islands.

Most industry is centered in a few cities on Sumatra and Java, especially in Jakarta. Millions of people have crowded into Jakarta, many of whom live in the **kampongs** that ring the city. A *kampong* is a traditional village, but it has also come to be the term for the slums around Jakarta.

SECTION 3
OBJECTIVES

The objectives reflect the Focus questions on this page and are assessed in the review on page 550.

1. List characteristics of the islands, population, and economy of the Philippines.

2. Discuss challenges facing the Philippines.

TEACHING SECTION 3
Building a Vocabulary

Write the term **sovereignty** on the board. Have students read the passage on page 550 that defines *sovereignty*. Ask students to recall the definition of *autonomy* from page 300. Ask: Do the two terms mean the same thing? *(no)* How are they different? *(autonomous powers have self-government but not complete or ultimate authority)*

Human Geography Almost 90 percent of Indonesians are Muslims, giving this area the largest Muslim population in the world. Hindus, Buddhists, Christians, and animists also live in Indonesia.

Indonesia's complex cultural geography is evident in the more than 300 ethnic groups and more than 350 languages and dialects spoken on the various islands. To help unify the nation's population, the government has started a program to persuade all citizens to use a national language called Bahasa Indonesia. Basic education for all Indonesians is another major goal of the government. The national motto is "Unity in Diversity."

Like in many other Southeast Asian countries, however, a serious "wealth gap" exists among the different groups in Indonesian society. Chinese Indonesians are among the wealthier groups, as they own most of the nation's businesses. The economic status and Christian religion of the Chinese in Indonesia have led to resentment and even violent attacks against them by Muslim Indonesians.

Since independence in 1950, Indonesia has had only two political leaders, both of whom have supported strong military control of the nation.

This man crafts a guitar at a small factory in the **Philippines**. Manufacturing, even in small industries, is an important source of jobs in the country.

Java dominates the country politically and economically. The Indonesian government's major goal is economic development.

 SECTION REVIEW

1. Name the major islands of Indonesia and list one main characteristic of each.

2. What is Indonesia's main source of income? Why have multinational corporations been attracted to the country?

3. Critical Thinking   **Explain** the meaning of the motto "Unity in Diversity." Do you think the motto applies to the United States? Why or why not?

 THE PHILIPPINES

FOCUS

● *What are the major characteristics of the Philippines' islands, population, and economy?*

● *What challenges do the Philippines face?*

LOCATION **Physical Geography** The Republic of the Philippines lies north of Indonesia and east of mainland Southeast Asia across the South China Sea. (See the map on page 541.) The Philippines include two main islands, Luzon in the north and Mindanao in the south. Between these two large islands are more than 7,000 smaller ones. Only about 700 are inhabited.

The Philippine islands lie on the Pacific Ring of Fire, and most of the islands are volcanic. Most of the Philippines' rugged terrain is covered by dense tropical rain forests. All the islands have a humid-tropical climate and lie in the path of dangerous Pacific typhoons.

Population The central plain of Luzon holds the greatest concentration of people. The capital, Manila, is located here and is a major seaport and the largest industrial area. This already crowded urban area of Luzon continues to grow, drawing workers from the rural areas of all the islands.

The population of the Philippines, called Filipino, is of Malay origin, with strong Chinese

Text continues on page 550.

Section 2 Review
ANSWERS

1. Java: two-thirds of the nation's population; Sumatra: rich natural resources; Bali: art, festivals, and dances; Timor: split into western and eastern parts; Borneo: many cultural groups; Sulawesi: shape of capital K; Irian Jaya: rugged, forested

2. Petroleum; low wages, a large work force

3. The motto means that even though a country's people may have different cultural characteristics, they can still come together as a single national group; answers will vary.

THE WORLD TODAY VIDEODISC PROGRAM

Philippines: Illegal Logging

2B   37948   3:54

Play   Pause

Role Playing

Have the class or groups of students stage a play that might be presented in an Indonesian village. Some students might play farmers, some might represent rice crops, some might play planthoppers, some might play other insects such as the wolf spider, and some might represent pesticides. Whatever the role assigned, players should tell how their "character" fits into the rice farm ecosystem in Indonesia. A second act of the play might demonstrate the introduction of IPM methods and their effects.

Geographical Sidelight

Point out that pesticide use harms more than just insects. In 1972, the U.S. government severely restricted the use of DDT, first used as an insecticide in 1939. DDT was used against various insects that carry deadly diseases or are harmful to agriculture.

Unfortunately, many insect species develop resistance to DDT, and many beneficial insects also are affected by the pesticide. In addition, DDT is a very stable compound, which means that it remains in the environment for a long time. Consequently, the chemical enters the food chain, ultimately contaminating fish, birds, other wildlife, and humans. The pesticide has been recorded in high levels in humans where DDT still is widely used, particularly in developing countries. Disease and mortality rates from pesticide poisoning in those areas also are high.

PLANET WATCH

Rice Growers Take a Stand

In 1986, 1,000 agriculture experts covered much of Indonesia on motorcycles to bring new techniques to farmers and save the nation's rice crops. In that one year alone, the brown planthopper, an insect that causes rice to dry out and rot, had wiped out enough food to feed 3 million people. Part of an ambitious and highly successful program called Integrated Pest Management (IPM), these field trainers helped bring the planthopper plague under control.

Fighting Pests with Chemicals

In the late 1960s, Indonesia was one of the world's largest importers of rice. To become self-sufficient in rice, the government set out to help farmers improve crop yields. To do this, the government built modern irrigation systems for rice paddies and introduced new strains of rice. To protect crops from insects and disease, the government spent as much as $150 million a year on pesticides. Often, farmers had to agree to use these chemicals before the government would loan them money.

> *The belief that increased rice production depended on increased use of pesticides outweighed the environmental and health concerns about pesticide use.*

At first, the plan seemed to work. By 1983, the nation no longer had to import rice. Success proved to be short-lived, however. By 1986, about 600,000 tons of rice, ravaged by planthoppers, lay rotting in farmers' fields. The pests seemed unstoppable, even though farmers were using more and more pesticides. The Indonesian government then learned of research conducted by the United Nations Food and Agriculture Organization (FAO). FAO research showed that pesticides were not the solution to the planthopper plague. Instead, the chemicals were a large part of the problem.

Managing Pests a New Way

Peter Kenmore, the FAO manager and researcher whose work brought new rice-growing methods to Indonesia, says, "Trying to control population outbreaks with insecticides is like pouring kerosene on a house fire." Kenmore found that uncontrolled spraying of pesticides—some farmers sprayed their fields up to 15 times a month—was in fact helping the planthoppers to thrive. The chemicals were killing insects that could control the population of planthoppers naturally, such as wolf spiders, ladybugs, and wasps. Furthermore, pesticides had brought about changes in the planthoppers, allowing them to reproduce in greater numbers.

The FAO offered Indonesia a long-term solution to its problem through Integrated Pest Management. Farm experts toured the country, identifying places with planthopper problems and training farmers to farm without pesticides. They visited the small theaters that are at the center of life in most Indonesian villages. There,

Farmers on **Java, Indonesia**, learn about efficient pest-control methods at this class sponsored by the United Nations Food and Agriculture Organization. These classes help prevent the overuse of chemical pesticides.

548

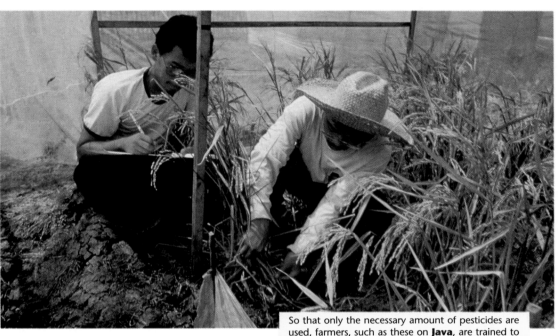

So that only the necessary amount of pesticides are used, farmers, such as these on **Java**, are trained to identify and count insects. Pesticides also kill beneficial insects that otherwise would help keep the numbers of harmful pests in check.

villagers staged plays, acting out the roles of farmers and pests. In "Farmer Field Schools," small groups met weekly to learn IPM methods. "By going out into the rice paddies," Kenmore says, "the trainers showed farmers how to diagnose problems, calculate the ratio of good bugs to bad and decide how much damage the crop could take without harming the yield." Farmers also learned to identify whether crop damage was caused by pests or by other factors. Armed with this knowledge, farmers could determine whether pesticide use would be helpful or harmful.

The Benefits of IPM

Support from farmers and the government has made the Integrated Pest Management program a huge success in Indonesia. Today, Indonesian farmers use pesticides only when absolutely necessary and in small amounts.

A lady beetle, a beneficial insect, feasts on an aphid that was attacking an alfalfa.

Limited spraying has allowed beneficial insects like the wolf spider to survive, and careful planning of crops has denied the planthoppers a continuous source of food. In the years since the government banned the use of most pesticides, the planthopper plague has disappeared and rice yields have increased. Government dollars once spent on pesticides are now used for IPM programs that teach farmers new ways of managing crops.

Now farmers want IPM techniques for crops other than rice. "They are demanding integrated pest management for their other crops and asking why we don't have a system for beans or cabbage," Kenmore points out. "I tell them it will just take more research."

Guided Practice

On the board, write key words from the first three subsections of Section 3. *(For example: Luzon, Manila, Filipino, Pilipino, Tagalog, plantations, deforestation, petroleum)* Have a student choose a word and explain its importance to the Philippines. If the student is correct, he or she can erase the word. Call on another student, and continue in this way until all the words have been erased.

Ranking

Have students name issues or challenges facing the Philippines. List their suggestions on the board. Then have students imagine that they are leaders in a democratic Philippine government. Ask each student to plan an agenda for working on these issues. Have students copy the items from the board in order of priority. Call on students to read and explain their agendas.

Closure

Have students find and discuss answers to the Focus questions on page 547 with a partner.

ASSESSMENT

Assign the Section 3 Review questions on this page.

Reteaching

Assign Section 3 of Reteaching Worksheet 45.

GLOBAL PERSPECTIVES

Point out that the U.S. military presence in the Philippines ended in 1992 when the Philippine Senate failed to renew a military-bases agreement. To some Filipinos, the closing of U.S. military bases in 1992 signaled a victory for Filipino nationalism over U.S. colonialism. To 80,000 Filipinos who earned their living in base-related jobs, however, the closings meant hard economic times. Hold a class discussion about the impact of colonialism on the Philippines and other nations.

Section 3 Review
ANSWERS

1. Luzon; sugar, coconut oil, other agricultural products, lumber, iron ore, copper, gold, silver, electrical products, textiles
2. Physical: volcanic eruptions, typhoons; political: government corruption, rebel movements, Spratly Islands dispute; economic: foreign debt, unemployment, uneven distribution of wealth
3. The islands were colonized by Roman Catholic Spain.

550

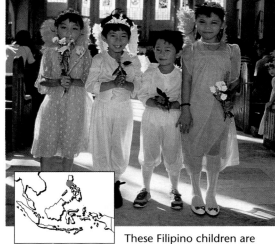

These Filipino children are dressed in their church clothes for Mass at a Roman Catholic church. The Spanish colonial past of the **Philippines** has resulted in a population that is mainly Roman Catholic.

and Spanish influences. About 85 percent of the population is Roman Catholic, with a small minority of Muslims living in the south. The national language is Pilipino, though English is used widely in education and business. Many Filipinos also speak an island language called Tagalog (tuh-GAHL-uhg).

Economic Geography The Philippines is a primarily agricultural country. Many Filipino farmers, however, are poor and own no land. A small group of rich landlords controls most of the best land. Huge sugarcane and coconut plantations cover the islands, and commercial farming is widespread. Sugar and coconut oil are the major agricultural exports, but small farms also produce tobacco, tropical fruits, and corn.

The forested mountains of the Philippines support a tropical hardwood lumber industry. Deforestation along steep slopes on some of the islands, however, has led to serious environmental problems. Minerals mined include iron ore, copper, gold, and silver. Unlike in neighboring Indonesia, little petroleum has been discovered in the Philippines. Rather, the nation's economy depends on its exports of electrical products and textiles. Its major trade partners are the United States and Japan.

Historical and Political Geography The Philippines have a long colonial history. Spain controlled the islands for 300 years, until

the Spanish-American War ended in 1898. At that time, the United States took over the islands. Japanese forces occupied the Philippines for three years during World War II, before the islands became an independent republic in 1946. The new republic's government was modeled after that of the United States.

Corruption, however, has troubled the Philippine government. Despite attempts at land reform, most of the land and industry remain in the hands of a few powerful families. In addition, violent Muslim rebel movements in the south and attacks by Communist rebel guerrilla forces in the north are ongoing problems.

In 1986, a popularly supported military revolt led to the fall of the 20-year government of Ferdinand Marcos, which had set aside basic political rights. A democratic government formed under the nation's first female president, Corazon Aquino.

Although democracy has returned to the Philippines, prosperity has not. The nation's huge foreign debt and high unemployment have weakened the economy. In fact, the Philippines is one of the poorest nations in Southeast Asia. The gap between the rich and poor of this island nation is steadily widening.

The Spratly Islands The Spratly Islands, a collection of more than 100 islands and reefs in the South China Sea, are one of the most disputed territories in the world. The Philippines, China, Vietnam, Malaysia, and Taiwan all have claimed **sovereignty**, or authority to rule, over the Spratly Islands. The intense interest in these tiny islands stems from the oil and natural-gas deposits lying off the islands' coasts.

3 SECTION REVIEW

1. On which Philippine island do most Filipinos live? What are the country's major resources and exports?
2. What physical, political, and economic challenges have the Philippines faced recently?
3. **Critical Thinking** **Suggest** an explanation for why the Philippines have a much higher percentage of Roman Catholics than do the other countries of East and Southeast Asia.

Building a Vocabulary

1. A large group of islands; the Malay Archipelago
2. Traditional villages, also the term for a slum in parts of island Southeast Asia: Jakarta
3. Authority to rule

Recalling and Reviewing

1. Singapore, just off the Malay Peninsula, is located nearer to Western Malaysia. Brunei is located on Borneo near Eastern Malaysia.
2. Both consist of many islands, are volcanically active, and have a humid-tropical climate.
3. Malaysia: Malay, English; Indonesia: Bahasa Indonesia and more than 350 other languages and dialects; the Philippines: Pilipino, English, Tagalog

CHAPTER 45 CHECK

Collecting latex from a rubber tree in
Malaysia

Reviewing the Main Ideas

1. The South China Sea divides Malaysia into two parts. More than half of Malaysia's people are Muslim Malays. The country is rich in natural resources, and industrialization is occurring at a rapid pace.
2. Singapore and Brunei are the smallest and wealthiest nations in Southeast Asia.
3. Indonesia is Southeast Asia's largest country and the world's most active volcanic region. Java is its core island. Indonesia remains mostly agricultural but is becoming more developed. The country has more than 300 ethnic groups.
4. The two main islands of the Philippines are Luzon and Mindanao. Like Indonesia, the Philippines is primarily agricultural. It has a long colonial history. The country faces several political and economic challenges.

Building a Vocabulary

1. What is an archipelago? Give an example of an archipelago.
2. What are *kampongs*? In what city are many *kampongs* found?
3. Define *sovereignty*.

Recalling and Reviewing

1. Describe the locations of Singapore and Brunei in relation to the two parts of Malaysia.
2. What physical characteristics do Indonesia and the Philippines have in common?
3. What languages are spoken in Malaysia, in Indonesia, and in the Philippines?

Thinking Critically

1. As you know, European colonists called the islands of the Malay Archipelago the East Indies. You also know, from Unit 4, that Europeans called the Caribbean islands the West Indies. Why do you think these two terms are no longer commonly used?
2. Why do you think the Spratly Islands are claimed both by Malaysia and the Philippines but not by Indonesia?

Learning About Your Local Geography

Individual Project

Despite many years of political problems, democracy has recently returned to the Philippines. Democracy, however, carries responsibilities. People must make informed choices. Contact local political organizations to find out when your area's next election will be held, and who the candidates for any election race are. Contact the campaign headquarters of each candidate to collect information on their positions. Decide which candidate best represents your own point of view on the issues. If possible, volunteer to work for the campaign of that candidate. Give a report to your class outlining your candidate's platform. You may want to include your report in your individual portfolio.

Using the Five Themes of Geography

Write a paragraph discussing how the theme of **human-environment interaction** in the islands of Southeast Asia is related to their **location** on the Pacific Ring of Fire. How have volcanoes both hurt and helped the people of the islands?

Thinking Critically

1. Because they describe the locations of the two sets of islands in relation to Europe (the Caribbean Islands were to the west of Europe and the Malay Archipelago was to the east)
2. The Spratly Islands are located in the South China Sea, so the northern part of Borneo (Eastern Malaysia) separates them from Indonesia. Also, unlike the Philippines, Indonesia has major oil deposits, and the interest in the Spratly Islands stems from their oil and natural-gas deposits.

Learning About Your Local Geography

Reports should indicate that students have researched the positions of major candidates in an upcoming election.

Using the Five Themes of Geography

Paragraphs should mention that although eruptions from volcanoes can kill people and destroy property, they also provide lava and ash that enrich the islands' soils.

551

SKILLS

Lead a discussion based on the following questions: What effects can cultural bias have on a group of people or on a society? How might the negative effects of cultural bias be avoided? How might cultural bias itself be reduced? Can cultural bias ever have positive effects? What is the difference between cultural bias and cultural pride? Review with students the various types of cultural bias. Ask for volunteers to name examples of cultural bias from United States history.

Practicing the Skill
ANSWERS

Students should present examples of each of the four types of cultural bias listed, as witnessed in books, magazines, and other sources. Each example should be accompanied by a few sentences explaining how it shows cultural bias.

552

UNIT 9 REVIEW   SKILLS

Detecting Cultural Bias

Recognizing cultural bias is an important skill in the study of human geography. Cultural bias is the influence that a person's own culture has on his or her view toward other cultures and peoples. Cultural bias can be expressed in many ways.

Sometimes, a person's actions reflect a cultural bias. At its most extreme, cultural bias can be expressed in violent actions against members of a particular culture group. Other culturally biased actions might include attempts to change another people's cultural beliefs. For example, many of the earliest Europeans to visit East and Southeast Asia believed that Asian beliefs and practices were inferior to their own. As a result, they tried to force some elements of their own cultures on the region's peoples. Similarly, many early Asians, such as the Chinese, believed that foreigners were culturally inferior. But the Chinese chose a different set of actions to express this cultural bias. Wanting to keep "barbarians" out of their society, Chinese leaders tried to close China off to foreigners by limiting trade with them.

Cultural bias, however, is often expressed not in actions, but in words. Ma Huan, the Chinese explorer about whom you read in "For the Record" on page 493, made a number of judgments about the peoples he encountered on his travels in the fifteenth century. About some of the peoples on the island of Java he said, "they have very ugly and strange faces" and "the food which these people eat is very dirty and bad." Both of these remarks, which may seem like simple observations, are actually judgments, as indicated by the words "ugly," "strange," and "bad." Even if a judgment seems positive, using words such as "beautiful," "normal," or "good," it may still express cultural bias.

A less obvious form of cultural bias is a **stereotype**, or a generalization about a group of people. Stereotypes tend to ignore differences within groups, instead treating everyone in a group as being alike. For example, after visiting the island of Sumatra, Ma Huan said "The people are very rich and prosperous." While it may have been true that some of the people Ma Huan met in Sumatra were "rich and prosperous," it is unlikely that all Sumatrans were. In fact, Ma Huan probably did not meet all of the people of Sumatra, and he probably did not stay long enough to gain a good understanding of the people he did meet. Thus his statements were based on limited, and possibly inaccurate, information.

Probably the hardest form of cultural bias to detect has to do with perspective, or point of view. When we use our own culture and experiences as a point of reference from which to make statements about others, we are showing a form of cultural bias called **ethnocentrism**. A good geographical example of ethnocentrism is in how people have named places. Europeans called East and Southeast Asia "the Far East," which is a culturally biased term because it describes East and Southeast Asia in terms of its location relative to Europe. Similarly, in Ma Huan's accounts of his travels abroad, he refers to his homeland, China, as "the Central Country." This reference reflects his belief that China was the geographical center of the world.

Detecting cultural bias, with its many forms, is a great challenge. In our study of geography and of the world's peoples, we must constantly be on the lookout for such bias. Too often it can cloud our view of the world and hide its rich diversity of peoples and cultures.

PRACTICING THE SKILL

Using books, magazines, newspapers, movies, television shows, news reports, and observations in your own school and community, identify one example of each of the following:
   (1) an action that shows cultural bias
   (2) a culturally biased judgment
   (3) a stereotype
   (4) ethnocentrism
List each example and write a few sentences explaining how it shows cultural bias.

BUILDING YOUR PORTFOLIO

Individually or in a group, complete one of the following projects to show your understanding of the geography concepts involved.

A Following the Dragons

You are a government official in one of the world's poorer developing countries. Your job is to assess your country's ability to successfully and quickly industrialize, and to decide whether such industrialization is desirable. To make these assessments, it will be helpful to look at Japan and the newly industrialized countries of East and Southeast Asia—Taiwan, Singapore, Hong Kong, South Korea, Malaysia, Thailand, and Indonesia. Your study of these countries will require you to do the following.

1. Research the population characteristics of each country. Make a graph showing each country's population size. Under your graph, write a caption identifying other population characteristics of each country, including quality and availability of education, general wage level, and degree of cultural uniformity.
2. Make a list of each country's resources. Consider location and infrastructure as well as agricultural, mineral, and energy resources.
3. Make a table comparing the historical and political characteristics of each country. What historical background does each country have? What type of government and political system are in place in each country now, and how stable have each country's political affairs been in recent decades? To what extent have foreign countries and companies been involved in each country?
4. Write a paragraph discussing the effects of industrialization in each country. How has industrialization affected the environment? Has the process benefited all groups within each country equally?
5. Summarize the information you have gathered by constructing a large chart comparing the population characteristics, resources, historical and political characteristics, and the effects of industrialization in each country you have studied. Examine the chart. Are there any characteristics that are shared by every country, without which industrialization would not have been possible? Does industrialization have any unavoidable negative effects?

Present your chart to the leaders of your country (the rest of the class). Discuss the implications of your findings for the industrial growth of your developing country.

B Speak, Write, Read, Listen

As you know, communication plays an important role in geography. It is one of the major links between peoples across and within regions. Create a "Communications Almanac" to discuss the spoken languages, writing systems, literacy rates, and telecommunications of East and Southeast Asia. To prepare your almanac, you will need to do the following.

1. Research and make a list of the region's major language families and language groups. Write a short paragraph describing the origins of each language group, and how each one relates to other languages of the world or region.
2. Make a language map of the region. Sketch where language groups are, and note on the map the official languages of each country.
3. Investigate different types of writing systems. Identify and write down definitions of the following words: ideography, logography, phonetization, syllabary, alphabetic writing. Research Chinese, Japanese, and Korean writing systems. Use the terms you have defined to compare and contrast the three systems in a paragraph.
4. Make a bar graph showing the literacy rates of each country of East and Southeast Asia.
5. Find out how widespread telecommunications systems are in each country. You may wish to focus on the numbers of televisions, radios, and/or telephones per person in each country. Present this information using charts or graphs.

Organize your materials into your almanac, and present it to the rest of the class.

553

UNIT 10 SOUTH ASIA *CHAPTERS 46–48*

OVERVIEW

In Unit 10, students will learn about the geographical and cultural contrasts of South Asia, one of the most heavily populated regions on Earth.

Chapter 46 • INTRODUCING SOUTH ASIA provides an overview of the landform regions, climates, cultural history, agriculture, and industry of this region. It also describes the serious economic and environmental challenges facing the area.

Chapter 47 • INDIA explores the geography and cultural diversity of the largest country in South Asia.

Chapter 48 • THE INDIAN PERIMETER profiles the countries that surround India: Pakistan, Nepal, Bhutan, Bangladesh, Sri Lanka, and the Maldives.

OBJECTIVES

1. Describe how climate and landforms affect the way people live in South Asia.
2. Trace the history of South Asia and identify cultural groups that have lived in the region since ancient times.
3. Describe the natural resources of South Asia.
4. Identify the political, environmental, economic, and social issues in South Asia.
5. Interpret special-purpose maps to analyze relationships among climate, population patterns, and economic resources in South Asia.
6. Contrast how people in different regions adapt to elements of physical geography.

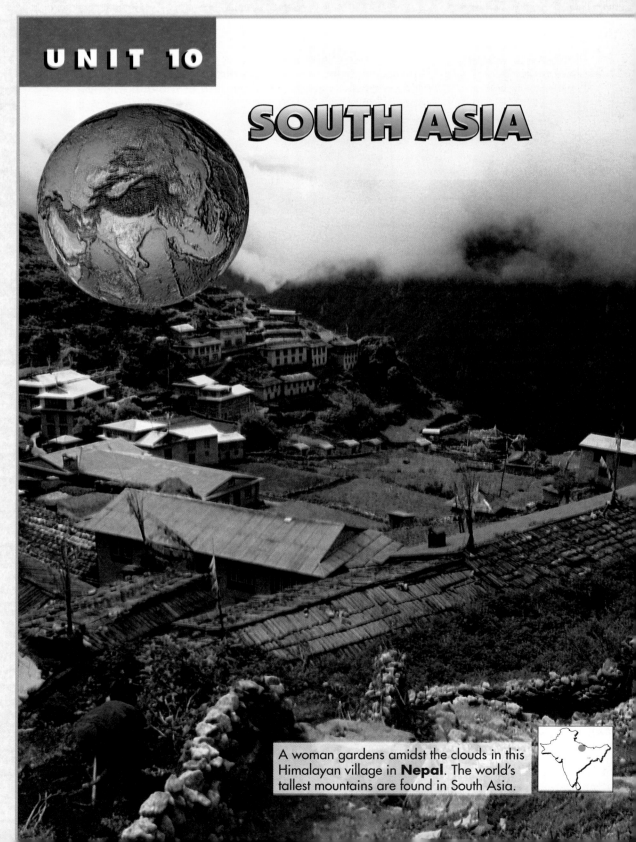

UNIT 10

SOUTH ASIA

A woman gardens amidst the clouds in this Himalayan village in **Nepal**. The world's tallest mountains are found in South Asia.

554

Using the Illustrations

Direct students' attention to the photograph on the opposite page. Tell them to cover the caption and to describe the village. *(Students should notice that the rural village is built on the side of a hill in a mountainous location.)* Then ask them to read the caption and locate Nepal on the map on page 559. Indicate Nepal's relative location near the Himalayas. Point out that Nepal's proximity to the Himalayas affects its economic geography. Rivers that originate in the mountain range benefit agriculture, and the Himalayan peaks attract tourists. Point out that the Taj Mahal (below) is a major tourist attraction in India.

Teaching with Sketch Maps

Have students draw a map of South Asia from memory. Tell students to include as many political boundaries, country and city names, and bodies of water as they can. Students might add this map to their portfolios and compare it with another they will draw after they complete Unit 10.

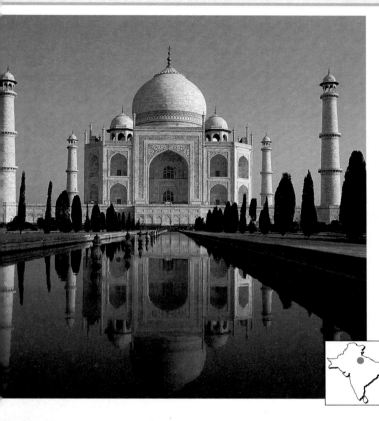

UNIT 10 CHAPTERS

46 *Introducing South Asia*

47 *India*

48 *The Indian Perimeter*

A *vina* (VEE-nuh), a stringed musical instrument from **South Asia**

Completed in 1643, the Taj Mahal in **Agra, India**, is one of India's most famous landmarks.

THE WORLD TODAY VIDEODISC PROGRAM

You might wish to use the profile of India as you teach Unit 10.

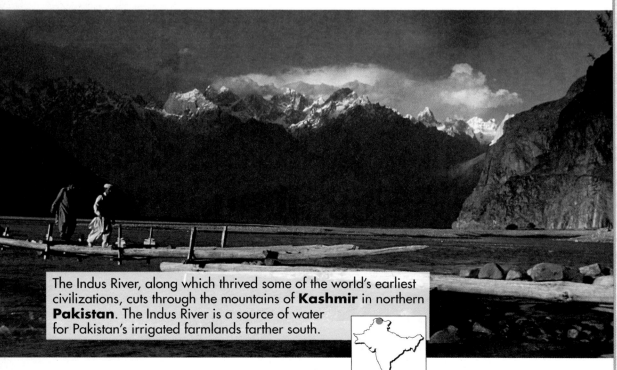

The Indus River, along which thrived some of the world's earliest civilizations, cuts through the mountains of **Kashmir** in northern **Pakistan**. The Indus River is a source of water for Pakistan's irrigated farmlands farther south.

One World, Many Voices

LITERATURE

You might introduce this feature by telling students that in Unit 10 they will learn about the many contrasts between rural and urban life in India. Ask students to suggest specific differences between rural and urban settings in general, and write their responses on the board. *(Students might mention urban areas as having more places to shop, greater availability of goods or merchandise, and more people)* Have them match the examples on the board with details in the literature selection below.

More About R. K. Narayan

In addition to writing fiction, R. K. Narayan has also translated several Hindu epics, including the *Mahabharata*. His contribution to Indian literature was recognized when he won India's most prestigious literary prize for *The Guide*, a novel published in 1958.

Primary Source

Description of Change: excerpted and bracketed
Rationale: excerpted to focus on the main idea; bracketed to clarify meaning

ONE WORLD, MANY VOICES

LITERATURE

"Kavadi was a wonderful place for one like me from the village—a street full of all sorts of shops, sewing machines rattling away, coloured ribbons streaming down from shop-fronts. My father had his favourite shop. . . . He would buy us each a toy—a ball, a monkey dangling at the end of a rubber-piece, and a doll, and invariably [always] an exercise book and a pencil for my elder brother, declaring that he was past the age of having toys, a reminder which made him smart every time he heard it. The road would be ankle deep in bleached dust and the numerous cattle and country carts passing along stirred it up so much that a cloud always hung over the road, imparting an enchanting haze to the whole place, though, by the time we started back, so much of this dust settled on our skins and hair that our mother had to give us a bath as soon as we reached home."

R. K. Narayan (1906–) has written many novels and short stories about life in India. The passage at left is from his novel *The English Teacher*, published in 1945. It describes a man's pleasant boyhood memories of visits his family made to a town near their rural village.

Interpreting Literature

Answers

1. It seemed more exciting than his village because of the different shops and activities in the town that he could not see at home.
2. The town has dirt roads. He describes the movement of people, carts, and animals from the country stirring up lots of dust on the roads.

Interpreting Literature

1. What does the boy like about this town that is different from his rural village?
2. What type of roads does the town have? What details does the author use to show that many people from rural areas visit this town?

FOR THE RECORD

Point out that although the majority of India's population lives in rural areas, its urban areas are generally very crowded. Bombay and Calcutta are among the 10 most populated cities in the world. After students have read the selection below, tell them to visualize the scene and to identify details that suggest noise, commotion, and movement. Call attention to the author's phrase, "always in India this feeling of a crowd." Ask students to name places they may have lived in or visited that convey the same feeling.

More About V. S. Naipaul

V. S. Naipaul left Trinidad in 1949 to seek an education at Oxford University. After working in England for several years, he embarked on travels to Africa, Southwest Asia, and India. His books about India were written partly from his desire to understand his ancestral home, a country he knew of only through the recollections of East Indian immigrants in the Caribbean.

FOR THE RECORD

V. S. Naipaul (1932–) is a well-known writer of fiction and travel books. Naipaul's family moved from India to the Caribbean island of Trinidad before he was born. He did not visit India until he was a grown man. The selection below from Naipaul's travel book *India* describes his impressions of Chandigarh, a city in northern India.

"The traffic was of all sorts: buses, trucks with towering loads, packed three-wheeler taxi-buses with about 20 people each (I counted), mule carts, tractors with trailers, some of the trailers carrying very wide loads of straw in sacking, or carrying logs placed crosswise, so that they occupied a good deal more of the width of the road than you thought from a distance. There seemed to be no limit to a load. . . . Many bicycles carried two or three people each: the cyclist proper, someone on the cross bar, someone sitting sideways on the carrier at the back. A motor-scooter could carry a family of five: father on the main saddle, one child between his arms, another behind him holding on to his waist, mother on the carrier at the back, sitting sideways, with the baby.

Always in India this feeling of a crowd, of vehicles and services stretched to their limit: the trains and the aeroplanes never frequent enough, the roads never wide enough, always needing two or three or four more lanes. . . . Hooters and horns, from scooters and cars and trucks, sounded all the time, seldom angrily. The effect was more that of celebration, as with a wedding procession."

Analyzing Primary Sources

1. What examples does Naipaul use to show the variety of transportation methods in Chandigarh?
2. What type of human geography does this excerpt concern?

Primary Source
Description of Change: excerpted
Rationale: excerpted to focus on the main idea

CHAPTER 46 Introducing South Asia
Planning Guide

TEXTBOOK	TEACHING RESOURCES	OTHER AVAILABLE MATERIALS
Section		
1. Physical Geography Pages 562–564	Reteaching Worksheet 46.1 Graphic Organizer 46 Skills Worksheet 46	*REGIONS OF THE WORLD Map Transparencies:* South Asia *Creative Strategies for Teaching World Geography* LESSON 20 South Asia Bingo *Study Guide for Spanish Speakers* *Lecture Notes,* Chapter 46
2. Historical and Human Geography Pages 564–566	Reteaching Worksheet 46.2 Critical Thinking Worksheet 46	*Study Guide for Spanish Speakers* *Lecture Notes,* Chapter 46
3. Economic Geography Pages 566–568	Reteaching Worksheet 46.3 Challenge/Enrichment Worksheet 46	*Study Guide for Spanish Speakers* *Lecture Notes,* Chapter 46
Chapter Review Page 569	Chapter 46 Test Chapter 46 Portfolio Assessment See also *Alternative Assessment Forms*	*Test Generator,* Chapter 46

CHAPTER RESOURCES

Lecture Notes, Chapter 46
Teaching Resources
    Chapter 46 Graphic Organizer
    Chapter 46 Skills Worksheet
    Chapter 46 Critical Thinking Worksheet
    Chapter 46 Challenge/Enrichment Worksheet

REVIEW AND ASSESSMENT

Chapter 46 Check, page 569
Chapter 46 Test, *Teaching Resources*
Chapter 46 Portfolio Assessment, *Teaching Resources*
Test Generator, Chapter 46
Alternative Assessment Forms

ADDITIONAL RESOURCES

Books for Teachers

Chatterji, A. K. *Monsoons, Floods, and Cyclones in India.* New Delhi: Radiant Publishers, 1992.

Frater, Alexander. *Chasing the Monsoon.* New York: Knopf, 1991.

Rathore, L. S. *The Poems on the Thar.* Jodhpur, India: Thar Bliss, 1986.

Schmidt, Karl J. *An Atlas and Survey of South Asian History.* Armonk, NY: M.E. Sharpe, 1995.

Sen, David N. *Environment and Plant Life in the Indian Desert.* Jodhpur, India: Geobios International, 1982.

Books for Students

Great Rivers of the World. Washington, DC: National Geographic Society, 1987.

Marshall, Bruce, ed. *The Real World: Understanding Our World Through the New Geography.* Boston: Houghton Mifflin, 1991.

Masami, Minocheher Rustom. *We Indians.* New York: Oxford University Press, 1989.

The World Today *Videodisc Program*

A variety of still images, short videos, and activities are available for you to use as you teach this chapter. *Lecture Notes* include bar codes for specific images and videos.

Multimedia Resources

Asia Alive GUIDisc. (Multimedia CD-ROM, interactive tour) MediAlive, 766 San Aleso Avenue, Sunnyvale, CA 94086.

Astonishing Asia. (Multimedia CD-ROM) InterOptica Publishing Ltd., 300 Montgomery Street, Suite 201, San Francisco, CA 94104.

The Ganges. (VHS, 49 minutes) Films for the Humanities & Sciences, P.O. Box 2053, Princeton, NJ 08543–2053.

Picture Atlas of the World. (interactive maps, photo bank, animation, audio on CD-ROM) National Geographic Society, Washington, DC.

South Asia: Why Are Forests Disappearing? (VHS, 15 minutes) Agency for Instructional Technology, Box A, Bloomington, IN 47402–0120.

MEETING INDIVIDUAL NEEDS

Limited English Proficiency (LEP)

Use the photographs in Chapter 46 as a means of conveying the main ideas of the chapter. As students examine each photograph, call on volunteers to describe what they see. Help students relate their descriptions to physical or human geography.

Learners Having Difficulty

Organize the class into groups of three. Within each group, assign individuals a section of Chapter 46. Direct students to write a brief summary of their assigned section and then to share it with their group. Call on volunteers to read their summaries to the class. You may wish to duplicate the individual summaries so that each group member receives a complete chapter summary.

Auditory Learners

Allow students a few minutes to examine the physical–political map of South Asia on page 559. Tell them to write several statements about South Asia's physical geography based on the map (for example, "The Himalayas are a mountain chain that stretches across the northern part of the region."). Call on students in random order to read one of their statements. Students should not duplicate any statements previously given.

Visual Learners

Organize students into three groups. Assign each group a section of Chapter 46. Instruct each group to create a cause-event-effect chart for the section, but to complete only the "event" column of the chart. Then have groups exchange charts and complete them. Call on volunteers to copy their group's completed chart on the board. Have the class critique the charts.

Tactile/Kinesthetic Learners

Provide students with Outline Map 22. Instruct them to consult their textbooks to label and shade the map to indicate the areas related to cultural groups of South Asia from ancient times to the present. Tell them to include dates from each group identified in their maps.

Gifted Learners

Have interested students research Mohenjo-Daro or Harappa, sites of two ancient civilizations in the Indus River valley. Instruct them to investigate either city's architectural plan, construction techniques and materials, water and sewer systems, and public facilities. Students might wish to illustrate this information. Call on volunteers to present a report of their findings to the class.

MULTICULTURAL FOCUS

The people of South Asia hold several festivals, some of which are national celebrations. On August 15, India observes the anniversary of the day in 1947 when the country gained its independence from Great Britain. Indians set aside January 26 to celebrate the 1950 formation of the Republic of India. On this holiday, people dressed in regional costumes march in large parades.

Many other festivals are religious, and their variety reflects the different religions practiced in South Asia. Hindus in the region celebrate Holi, a spring festival. People mark the end of winter by singing, dancing, holding plays, and lighting huge bonfires. The fires represent the annihilation of Holika, an evil spirit. On the second day of Holi, called Color Day, people throw brightly colored powders and squirt colored liquids at one another to chase away evil spirits.

Diwali, the Indian New Year, marks the culmination of the monsoon season. During this time, many people thoroughly clean and paint their homes and pay off debts. One of the main highlights of Diwali is a show of fireworks called the Festival of Lights. Oil lamps are lit on rooftops and windowsills to greet Lakshmi, the goddess of wealth. It is believed that she favors those people who set off the longest and most boisterous firecrackers.

Call on students to describe festivals in their communities that are linked to religion, the seasons, or history. (Some students might mention Cinco de Mayo, July 4, or Kwanzaa celebrations.)

OVERVIEW

Chapter 46 introduces students to South Asia, a region of many contrasts that faces serious economic and environmental challenges.

Section 1 describes the landform regions and climates of South Asia, focusing especially on the influence of the monsoons.

Section 2 identifies the cultural groups that have inhabited or controlled South Asia from ancient times through the colonial era. It also explains how more recent events, including independence and the division of the subcontinent, have affected the region's peoples.

Section 3 discusses agriculture and industry in South Asia, noting the many challenges faced by South Asian countries in both these areas.

CREATIVE STRATEGIES FOR TEACHING WORLD GEOGRAPHY

LESSON 20   South Asia: Bingo

CHAPTER 46

Introducing South Asia

A representation of the Hindu god Vishnu

South Asia, sometimes called the Indian **subcontinent**, is a large triangular peninsula that juts into the Indian Ocean. A subcontinent is a large landmass that is smaller than a continent. The region is one of physical extremes. Barren deserts contrast with flooded river deltas and tropical islands. The world's highest mountains rise up in the north, earning this area the nickname "the roof of the world."

South Asia is a region of great cultural contrasts as well. More than one-fifth of the world's people live on the South Asian peninsula, making it one of the most populated regions on Earth. The region's long cultural history has resulted in a mixture of religions, languages, and other culture traits.

In recent decades, South Asia has experienced rapid population growth but limited economic development. Some of the world's poorest nations struggle here with serious economic and environmental challenges.

Historic Brhadisvara temple in southern **India**

Using the Geography Dictionary

Ask students to identify the two terms in the Geography Dictionary that refer to religions. *(Hinduism, Sikhism)* Explain the meaning of **subcontinent** by breaking it down. Ask students to think of other words that begin with the prefix *sub-*. *(submarine, substandard, subzero)* Point out that the prefix *sub-* can mean "below" or "under" or describe a secondary position in relationship to something larger. Use a wall map to show India's size and its location on the Asian continent. Then have volunteers read aloud the Glossary definitions of the rest of the terms.

MAPPING THE REGION
Using the Physical-Political Map

Have students examine the map on this page and identify South Asia's three major landform regions. *(Himalayas, Indo-Gangetic Plain, Deccan Plateau)* Then have them trace the courses of the three main rivers that flow from the Himalayas. *(Indus, Ganges, Brahmaputra)* Point out that the Indo-Gangetic Plain is fertile but is also prone to flooding.

MAP TRANSPARENCIES

Overhead transparencies from the maps on pages 559–561 are available in the *REGIONS OF THE WORLD Map Transparencies.*

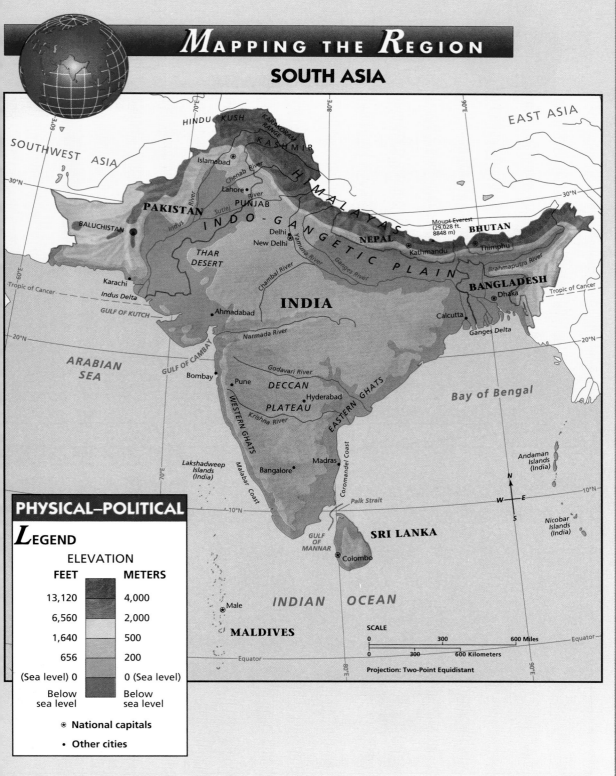

MAPPING THE REGION
SOUTH ASIA

PHYSICAL–POLITICAL

LEGEND

ELEVATION

FEET		METERS
13,120		4,000
6,560		2,000
1,640		500
656		200
(Sea level) 0		0 (Sea level)
Below sea level		Below sea level

⊛ National capitals

• Other cities

SCALE

Projection: Two-Point Equidistant

MAPPING THE REGION

Using the Climate Map

Have students look closely at the climate map on this page, and at the physical–political map on page 559, to answer the following questions: What is the dominant climate in Pakistan? *(desert)* Which country in the region is mostly humid tropical and humid subtropical? *(Bangladesh)* What two climates are found on the Deccan Plateau? *(steppe, tropical savanna)* Where else in the region is the tropical-savanna climate found? *(Sri Lanka)* In what climate is the city of Calcutta? *(tropical savanna)* Karachi? *(desert)* Dhaka? *(humid tropical)*

MAPPING THE REGION

Using the Population Map

As students study the population map on this page, ask them if they can discern a general population pattern for this region. *(South Asia is a densely populated region with many large cities.)* Point out the band of very dense population in the northern part of the region. Tell them to locate this same area on each map on pages 559–561. Point out why the Indo-

Continues on page 561

CLIMATE

Answers

1. Humid tropical, tropical savanna, desert, steppe, humid subtropical, highland
2. Pakistan's
3. Deccan Plateau

POPULATION

Answers

1. Dry desert and steppe climates discourage large-scale agriculture and settlement.
2. Indo-Gangetic Plain
3. Indus River, Ganges River

MAPPING THE REGION

SOUTH ASIA

CLIMATE

LEGEND

- Humid tropical
- Tropical savanna
- Desert
- Steppe
- Humid subtropical
- Highland
- ← Wet monsoon air flow
- ← Dry monsoon air flow

POPULATION

LEGEND

POPULATION DENSITY

Persons per sq. mile	Persons per sq. km
520	200
260	100
130	50
25	10
3	1
0	0

- ● Metropolitan areas with more than 2 million inhabitants
- ● Metropolitan areas with 1 million to 2 million inhabitants

560

Gangetic Plain might be considered a uniform region. *(The area is densely populated; its climate type is generally humid subtropical; farming is the primary economic activity.)*

MAPPING THE REGION

Using the Economy Map

Have students examine the economy map on this page. Direct their attention to the main commercial farming areas. Ask: What is significant about their locations? *(They are located along rivers.)* Point out the large land area that is devoted to subsistence farming. To help students understand the vast numbers of Indians who are involved in agriculture, have them compare this map to the population map on page 560. Direct students' attention to the legend. Ask: What are the main population densities in the areas of subsistence farming? *(25–130 people per square mile [10–50 per sq. km] and 130–260 people per square mile [50–100 per sq. km])*

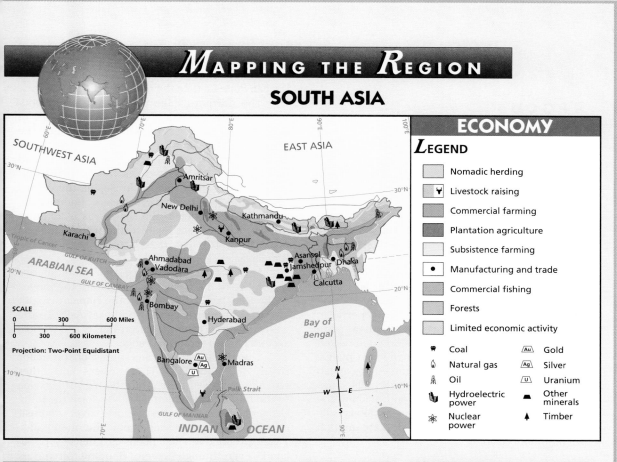

MAPPING THE REGION
SOUTH ASIA

ECONOMY

LEGEND

- Nomadic herding
- Livestock raising
- Commercial farming
- Plantation agriculture
- Subsistence farming
- Manufacturing and trade
- Commercial fishing
- Forests
- Limited economic activity

Coal	Au Gold
Natural gas	Ag Silver
Oil	U Uranium
Hydroelectric power	Other minerals
Nuclear power	Timber

SCALE
0   300   600 Miles
0   300   600 Kilometers

Projection: Two-Point Equidistant

CLIMATE

◆ Climate patterns in South Asia are greatly influenced by the monsoons that alternately bring wet and dry seasons to most of the region.

◆ Much of the northernmost areas of South Asia are dominated by variable highland climates.

◆ The driest climates of South Asia are found in the northwestern portions of the region.

(?) What climate types are found in South Asia?

Compare this map to the physical–political map on page 559.

(?) Other than those dominated by highland climates, which country's climate is least influenced by the monsoons?

(?) What physical landform region in central India is dominated by a steppe climate?

POPULATION

◆ Much of northern India is densely populated.

◆ Many densely populated areas and large metropolitan areas are located in coastal regions and along major rivers.

◆ The most sparsely populated areas of South Asia are in the western and northern parts of the region.

Compare this map to the climate map on this page 560.

(?) Why do you think the western part of Pakistan is not densely populated?

Compare this map to the physical–political map on page 559.

(?) What is the name of the landform region in densely populated northern South Asia?

(?) Which two major rivers flow through densely populated areas in the northwestern and northeastern parts of the region?

ECONOMY

◆ South Asia's most heavily cultivated regions are located along major rivers, especially the Indus and Ganges rivers.

◆ India's supply of uranium ore supports a number of nuclear power plants and has enabled the country to develop an atomic bomb.

◆ Many major manufacturing and trade centers are located along the coasts and in regions along the Indus and Ganges rivers.

(?) Name the manufacturing and trade centers shown along the coast of South Asia.

Compare this map to the climate map on page 560.

(?) One of Bangladesh's major crops is rice. Why do you suppose this is so?

Compare this map to the population map on page 560.

(?) Why do you suppose northern India is so densely populated?

561

SECTION 1
OBJECTIVES

The objectives reflect the Focus questions on this page and are assessed in the review on page 564.

1. Describe the three landform regions of South Asia, and explain how they were formed.
2. Identify South Asia's climates, and examine the monsoon's effects.

TEACHING SECTION 1
Motivator

Ask students to think of images, places, or elements of physical geography that they associate with India and its neighboring countries. *(Students might mention high mountains, the Taj Mahal, the Ganges River, crowded cities, and monsoons.)* Point out that South Asia is a region of great physical contrasts.

Building a Vocabulary

Have a volunteer read aloud the paragraph on page 564 that includes the term **tropical cyclone**. Point out the similarity between the word *cyclone* and words such as *cycle* and *circle*. Ask students to try to make a connection between these words and the vocabulary term. *(The winds in such a storm move in a circular motion.)*

Tell students that Hindus consider the Ganges the most sacred of rivers. They believe that its waters have the power to wash away earthly sins, to bring wealth and fertility, and to cure illnesses. Annual festivals draw many pilgrims, who bathe in the Ganges to cleanse and purify themselves. Mourners cast the ashes of their dead into the Ganges to ensure safe passage to paradise. Hold a class discussion on the various ways water is used and viewed in different religions, drawing on students' own backgrounds and experiences.

PHYSICAL GEOGRAPHY

FOCUS

- *What are the three landform regions of the Indian subcontinent? How was each region formed?*
- *What are South Asia's climates? How do the monsoons affect the region?*

REGION Tectonic Origins Scientists believe that India was once part of the supercontinent of Gondwanaland. Over 65 million years ago, the Indian plate broke away from other portions of the supercontinent and became an island. It drifted north toward the Eurasian plate at a rate of about three to four inches (8–10 cm) per year. About 40 million years ago, the Indian plate slammed into Eurasia. This collision compressed and lifted the sea floor between the two plates, forming mountains. India continued to push under the Eurasian plate for about 1,200 miles (1,931 km), thickening the crust and pushing the mountains even higher. Earthquakes along this mountain rim indicate continued plate motion and mountain building.

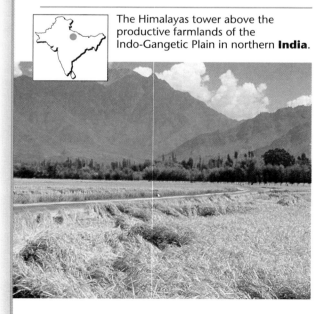

The Himalayas tower above the productive farmlands of the Indo-Gangetic Plain in northern **India**.

Landforms South Asia can be divided into three major landform regions: the Himalayan Mountain rim, the Indo-Gangetic Plain, and the Deccan Plateau. (See the map on page 559.)

The northern edge of the Indian subcontinent contains the world's highest mountains, of which the Himalayas (him-uh-LAY-uhz) are the longest and highest system. This northern mountain rim crosses most of Nepal (nuh-PAWL) and Bhutan (boo-TAN), as well as the northern portions of India and Pakistan. In Nepali, the official language of Nepal, *Himalaya* means "home of the snows." One glacier in the Himalayas is 20 miles (32 km) long. Mount Everest, the world's highest peak at 29,028 feet (8,848 m), is part of the Himalayas.

Most of the people of this mountainous region live in the valleys that separate the ranges of the Himalayas. To the east, the Himalayas split into the various ranges of Southeast Asia. To the west, they join the Karakoram range, the world's second highest mountain range.

The Himalayas are the source of the major rivers of South Asia. These rivers and other streams carry sediment from the high Himalayas to the base of the mountain rim to form the Indo-Gangetic Plain. Here, the flooding rivers deposit rich alluvial soil eroded from the mountains. These productive river plains are densely populated, even though severe flooding often has destroyed crops and killed many people here.

Three major rivers—the Indus, the Ganges (gan-JEEZ), and the Brahmaputra—cut through the region. Their vast alluvial plains stretch south from the Himalayas to the Arabian Sea in the west and to the Bay of Bengal in the east. The Indus River flows southward through Pakistan and is the lifeblood of that desert and mountain country. The Ganges flows eastward across northern India, where half of India's huge population lives. The Brahmaputra originates north of the Himalayas in Tibet and flows eastward. It then makes a sharp turn to the south, entering India in the country's far northeastern corner. Where the Ganges and Brahmaputra meet, they form Bangladesh's low delta plain.

The ancient Deccan Plateau, the third landform region, covers much of the southern two-thirds of the Indian subcontinent. The plateau tilts downward toward the east. In the west are the Western Ghats (GAWTS), the steep, rugged hills that face the Arabian Sea. In the east are the more gentle slopes of the Eastern Ghats. These hills face the Bay of Bengal.

Part of the Deccan Plateau is a shield of granite formed more than 600 million years ago. The younger portion of the Deccan is made up of volcanic layers of lava, which are more than 10,000 feet (3,048 m) thick. Weathering of the volcanic rocks has produced fertile soils in the region. With irrigation, these areas of the Deccan support productive grain fields. The plateau also holds most of India's mineral resources.

Between the Ghats and the coastlines of the Arabian Sea and the Bay of Bengal are narrow coastal plains. These fertile plains are productive agricultural areas and are heavily populated. Offshore to the southeast is the large island of Sri Lanka (sree LAHNG-kuh). To the southwest is a chain of low coral islands called the Maldives.

Climate South Asia has several climate types. A humid-tropical climate dominates in Bangladesh and along the southwest coasts of India and Sri Lanka. Most of interior India, which includes most of the Deccan Plateau, has a tropical-savanna climate. A humid-subtropical climate covers the foothills along the southern edges of the mountains of northern India and northern Pakistan. Highland climates are found in the mountains of the northern rim.

To the northwest lie the drier climates. A desert climate stretches from the Thar (TAHR) Desert, or Great Indian Desert, in northwest India and eastern Pakistan across the irrigated Indus Valley of Pakistan to the country's borders with Afghanistan and Iran. Part of the central Deccan Plateau, meanwhile, has a steppe climate. The Western Ghats block moist winds off the Arabian Sea from reaching the area.

REGION **The Monsoon** The monsoons strongly influence the climates of South Asia. These seasonal winds blow from the oceans in summer and from the interior of the Asian continent in winter.

The wet summer monsoon usually begins around mid-June. The high summer sun rapidly heats the interior of the Asian continent, causing the air above the land to warm and rise. This creates a large low-pressure area that pulls moist air, the summer monsoon, inland from the Indian and Pacific oceans. As the summer monsoon flows inland, it brings high humidity, heavy rains, and

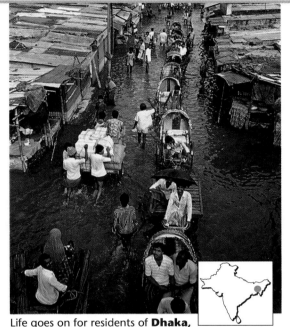

Life goes on for residents of **Dhaka, Bangladesh**, a regular flooding victim of the heavy rains brought by the annual wet monsoon.

thunderstorms. The heaviest rains fall where the warm, moist winds meet the Western Ghats and the foothills of the Himalayas. Some areas receive as much as 400 inches (1,016 cm) of rain, and flooding threatens many villages. In the fall, a period of hot, still weather brings some relief from the rains.

During the dry winter monsoon, the interior of Asia becomes bitterly cold. This heavy, cold air causes a huge high-pressure center to form over the interior of Asia. Cold, dry winds flow toward the surrounding oceans. As the winds descend down the mountains toward India, they warm and become even drier, making South Asia's winter dry and warm as well. In the spring, the air again becomes hot, humid, and still. Schools and some businesses close for vacation at this time.

The timing of the monsoon is critical to the farmers and to the economy of South Asia. Seeds are planted in the spring so that they have time to develop and take hold in the soil. If the monsoon rains come too soon, the seeds will wash away. If the monsoon rains come late or are in short supply, the seedlings will die from a lack of moisture.

The Thar Desert and the lower Indus Valley of Pakistan are not in the path of the wet monsoon winds. Thus, the amount of annual rainfall in

Chapter 46 **563**

Closure

Have students work in pairs to create an outline summarizing South Asia's landforms and climates.

ASSESSMENT

Assign the Section 1 Review questions on this page.

Reteaching

Assign Section 1 of Reteaching Worksheet 46.

SECTION 2
OBJECTIVES

The objectives reflect the Focus questions on this page and are assessed in the review on page 566.

1. Summarize the history of South Asia from ancient times through the end of the colonial era.

2. Discuss the effects of the 1947 division of the subcontinent.

TEACHING SECTION 2
Motivator

Have students study the photographs on this page and on page 565. Call on volunteers to read the captions aloud. Ask: How do these photographs show links between physical geography and human geography? What do they tell us about South Asia's cultures and background? Point out that Section 2 is about South Asia's history and human geography.

Section 1 Review
ANSWERS

1. By the collision of the Indian and Eurasian plates; they are the source of the rivers that carry sediment to form the Indo-Gangetic Plain; the Deccan Plateau

2. In the northwest and on part of the central Deccan Plateau; it provides rain for crops.

3. In the summer, warm air rises to create a low-pressure area that pulls moist air inland. In the winter, cold air causes a high-pressure area that makes cold dry winds flow toward the oceans. In the summer, the Northern Hemisphere tilts toward the sun, warming the air over Asia. The opposite is true in winter.

these areas is low, and irrigation is needed to grow crops. The Thar Desert receives as little as five inches (13 cm) of rain per year. Here, dust storms are more common than rain storms.

The coastal regions of the Bay of Bengal are occasionally struck by severe **tropical cyclones** during the spring and fall. These hurricane-like storms produce winds above 100 miles per hour (161 kph) and dangerous high seas. The storms have taken hundreds of thousands of lives, especially in the low delta area of Bangladesh.

SECTION REVIEW

1. How were the mountains of the Indian subcontinent formed, and what role have these mountains played in forming the Indo-Gangetic Plain? To which landform region do the Western and Eastern Ghats belong?

2. Where are the subcontinent's drier climates found? How is the summer monsoon important to South Asia's agriculture?

3. **Critical Thinking** **Compare and contrast** the origins of the summer and winter monsoons. How is the earth's position relative to the sun a factor?

HISTORICAL AND HUMAN GEOGRAPHY

FOCUS

● *What cultural groups lived in or controlled South Asia from ancient times through the colonial era? How did the region achieve its independence?*

● *What effects did the 1947 division of the subcontinent have on its peoples?*

MOVEMENT **Cultural Influences** The Indus civilization, one of the world's first advanced civilizations, developed in South Asia's Indus River valley around 2500 B.C. This early society was based on irrigation agriculture and had a written language with an alphabet of 250 to 500 characters. It also produced stone and bronze sculp-

tures, metal tools, and pottery. The Indus civilization declined around 1750 B.C. for reasons not completely clear. Soil salinization and overpopulation may have been part of the cause.

In about 1500 B.C., nomads from Central Asia, called Aryans, conquered the Indus area. By 1000 B.C., the Aryans controlled northern India as well. The local farming peoples, the Dravidians, were driven to the south. Some of them remained, however, and mixed with the Aryan population. This new Indo-Aryan culture was centered near the upper Ganges River.

The Indo-Aryan culture produced many distinctive culture traits of South Asia, including the Hindu religion. **Hinduism**, whose roots date to prehistoric times in South Asia, is the major religion of India and Nepal. Hindus worship many gods and believe that salvation is offered through *dharma*, or good conduct. They also believe in reincarnation, or rebirth after death. Sanskrit, an Indo-Aryan language which flourished from about 500 B.C. to A.D. 1000, is the classical literary language of Hinduism. Hindu scholars still use Sanskrit today.

Hindu pilgrims crowd into the Ganges River in **Varanasi, India**. Bathing in the sacred Ganges is an important spiritual activity for Hindus.

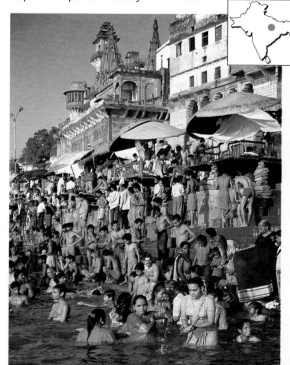

Another religion, Buddhism, had begun to form in northern India by about 500 B.C. At about the same time, a series of Indian empires and kingdoms rose to power in the region and ruled for the next 1,000 years. The Mauryan Empire, which ruled during the third century B.C., helped spread Buddhism throughout much of Asia. After A.D. 500, the Indian empires declined as peoples from Central Asia again invaded northern India.

Beginning around A.D. 1000, several Muslim invasions in the northwest brought Islam into the region. During the late fifteenth century, Muslim teachings blended with Hinduism to form yet another religion, **Sikhism** (SEEK-iz-uhm), in the region. Today, Sikhs make up a small minority. They are concentrated in the state of Punjab in northwestern India.

The last Muslim invasion took place in 1526 when the Mogul Empire established Muslim control over most of India. Today, Muslims are the majority in Pakistan and Bangladesh and form a large minority in India.

Colonialism Europeans arrived in India during the sixteenth century. The Portuguese came first, seeking to form new trade links, expand their empire, and spread Christianity. Today, millions of Indians are Christians.

The Dutch, the French, and the British soon followed the Portuguese. For over a century, large trading companies from these countries sought to earn profits in India. At first, the products available in India were not directly valuable to the Europeans back in their home countries. Instead, the products were valued as goods to be traded for yet other goods elsewhere. For example, the Europeans longed for spices from the East Indies (present-day Indonesia). In exchange for the spices, the Europeans needed goods such as cotton to trade, and India was a ready source of that raw material. The European companies developed many such trading networks.

In time, however, the Dutch came to monopolize the East Indies spice trade, and the French and British began to take more of an interest in India for its own resources. And as Europe indus-

English architectural styles are common in **Simla**, located in the Himalayan foothills of northwestern **India**. The cool climate and scenic setting made the city the summer capital of British colonial India.

trialized, Great Britain developed a textile industry that could make direct use of Indian cotton.

For most of this period, the Europeans had competed peacefully for the region's trade. A series of conflicts, however, began in the 1740s. British victories over other European powers as well as over native Indian powers steadily increased British political control of the region. As a result, in 1858 almost all of the Indian subcontinent officially became part of the British Empire, controlled by the British government rather than by trading companies. British India included present-day India, Pakistan, Bangladesh, and Sri Lanka.

For the next several decades, the Indian subcontinent was considered the "jewel in the crown" of the British Empire. The region's many products, especially cotton, traveled to Great Britain's factories. Other important products shipped to Great Britain included tea, indigo, coffee, and **jute**. Jute is a plant fiber used to produce coarse fabrics, such as burlap, and twine.

India also served as a market for goods manufactured in Great Britain. To ensure an efficient flow of goods, the British built railroads, roads, and ports in the region. In addition, they brought the English language, English law and education, and the English political system.

Independence Demands for independence increased in the early twentieth century,

Geographical Sidelight

The period of the Gupta Empire, a northern Indian dynasty that flourished from the fourth to the sixth century A.D., is considered the golden age of India. Cultural activity and learning flourished at this time, supported by a wealthy upper class. The Gupta era produced the decimal system of notation as well as epics in Sanskrit and classics of Hindu philosophy. Students from as far as China and Java came to study at Gupta universities.

Closure

Have students work in pairs to answer the Focus questions on page 564.

ASSESSMENT

Assign the Section 2 Review questions on this page.

Reteaching

Assign Section 2 of Reteaching Worksheet 46.

Extension

Encourage interested students to research tensions in South Asia between Hindus and Muslims by reading news magazines and newspapers from recent years. Ask students to try to discover the sources of tension, the ways groups have expressed their frustrations, and the methods used by political leaders to handle these conflicts. Have them present their findings to the class. Hold a class discussion on this subject.

SECTION 3
OBJECTIVES

The objectives reflect the Focus questions on this page and are assessed in the review on page 568.

1. Identify agricultural challenges facing South Asia.

2. Explain what has hindered South Asia in its industrialization.

Section 2 Review
ANSWERS

1. Indus civilization, Aryans, Dravidians, Indo-Aryans, Hindus, Buddhists, Muslims, Sikhs; the British Empire; the person who led the main independence movement against British rule

2. It was divided to prevent war between Hindus and Muslims. Minorities were surrounded by hostile majority populations, and violence increased.

3. Time lines should show various periods and key dates and events in the subcontinent's history.

particularly after the Massacre at Amritsar in 1919. British troops fired on a large crowd of people protesting British policies. Some 400 Indians were killed and 1,200 wounded.

The main independence movement in British India was led by Mohandas Gandhi. Gandhi believed that all life is sacred and thus should never be harmed. Related to this philosophy, called *ahimsa*, was Gandhi's belief that a policy of "nonviolent noncooperation" was the best way to bring about political and social change, including independence. Thus Gandhi's methods often took the forms of boycotts, marches, and fasts, but never violence.

Gandhi's efforts were effective. Britain granted independence to the colony in 1947, ending the colonial era. First, however, the British divided the subcontinent into the countries of India and Pakistan. Pakistan itself had two parts: West Pakistan and East Pakistan (present-day Bangladesh). The two parts were separated by 1,000 miles (1,609 km) of Indian land.

The British divided the region because many feared that war might break out between Hindus and Muslims if they were forced to live in the same country. Hindus were in the majority in India, while Muslims formed a majority in Pakistan. A mountainous region called Kashmir was split between the two nations, even though the local population wanted independence and was mainly Muslim. The island of Ceylon (now called Sri Lanka) gained its independence in 1948, just one year after the region broke away from Great Britain. Sri Lanka is a Buddhist country but has a large Hindu minority.

The political division of the subcontinent at independence was traumatic for millions of people in South Asia. Minorities found themselves surrounded by hostile majority populations, and violence broke out. Millions of India's Muslims fled to Pakistan, just as millions of Hindus and Sikhs fled from Pakistan to India. Hundreds of thousands of people died in riots and massacres. Even Gandhi, who had millions of followers, was not spared by the violence he so detested. He was shot in 1948 by a Hindu extremist.

In 1971, East Pakistan, with the help of India's military and with much bloodshed, broke away from West Pakistan and formed the independent nation of Bangladesh. In 1973, a new constitution made Pakistan, formerly West Pakistan, an Islamic republic.

Today, religious and ethnic tension is still one of the most pressing issues facing South Asia. Independence movements among different groups have led to continuing violence.

SECTION REVIEW

1. List three different groups of people that lived in South Asia before the colonial era. What European empire did the subcontinent become part of in the 1800s? Who was Mohandas Gandhi?

2. Why was the subcontinent divided into India and Pakistan? What problems did this division create?

3. Critical Thinking **Construct** a time line showing the cultural and political history of the region.

ECONOMIC GEOGRAPHY

FOCUS

● *What agricultural challenges does South Asia face?*

● *What has hindered industrialization in much of South Asia?*

INTERACTION **Agriculture** Most of the people of South Asia are farmers who live in small villages. The small, inefficient farms, the high cost of fertilizers, and a lack of mechanization keep most South Asian countries from achieving consistent food production. Also, crop disease and natural disasters such as monsoon floods, droughts, tropical cyclones, dust storms, and locust swarms damage crops.

To improve agricultural output, new types of rice and wheat that produce higher yields and require shorter growing seasons were developed. At first, these new types of grain produced such high yields that their introduction was known as the **Green Revolution**. Natural disasters and population increases, however, reduced the gains that had been made. Also, these new improved grains needed larger amounts of water and costly fertilizers, which many farmers could not afford.

Building a Vocabulary

Write the term **Green Revolution** on the board. Point out that it refers to scientific methods that helped improve crop yields in developing areas, including South Asia. Have students suggest how the meanings of *green* and *revolution* relate to this term. *(Green refers to plant growth, revolution to fundamental change.)*

Guided Practice

Write the following headings on the board: Old Challenges, Proposed Solutions, and New Challenges. Organize the Section 3 information on agriculture under these three headings. Discuss how South Asia might meet its new challenges through population control and other means.

Today, there is little land left to develop in South Asia, and soil erosion and salinization threaten existing farmlands. As the region's population grows, supplying the demand for more farmland and for more wood is increasingly difficult. Often, the nearby forests are destroyed in the process. This deforestation, in turn, causes a loss of forest resources, destroys wildlife habitat, and creates increased flooding and soil loss. The cycle is hard to break.

REGION **Resources and Industry** Much of South Asia has experienced a slow rate of industrialization. Part of the explanation lies in the region's colonial history. During the colonial period, the British did not promote industry in South Asia because they did not want the region's industries to compete with British industries. To the British colonists, then, the region was mainly a source of raw materials. As a result, some Indian industries declined during the colonial era.

A good example of the British policy and its devastating effects on Indian industry involved **calicoes**. Calico, a fine fabric woven from cotton and printed with colorful patterns, originated in the subcontinent's southwestern town of Calicut (now called Kozhikode) in the eleventh century. By the 1600s, the handcrafted calicoes were popular throughout Europe, and India was the major producer and exporter. By 1700, however, British cloth manufacturers had successfully fought for a ban on imports of India's calicoes. Even though the British cloths were not of as high quality, the ban kept the Indian calicoes from competing and forced their decline. The ban not only boosted the British industries but also gave them an even greater supply of raw cotton from South Asia because the region was using less cotton itself.

Then during the mid-nineteenth century, events halfway around the world affected India's cotton market. When cotton exports from the southern United States dropped dramatically during the U.S. Civil War, Great Britain's textile factories ran short of raw cotton. To supply their cotton needs, the British relied more heavily on, and paid more for, India's cotton. When the U.S. Civil War ended, though, Great Britain turned back to U.S. supplies. This sudden loss of income devastated India's economy, to which cotton exports had become key. Farmers who had stopped growing food in order to grow cotton found themselves with neither food nor money with which to buy

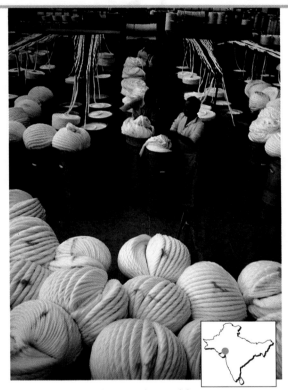

Cotton is prepared for market and for use in various industries at this mill in **Ahmadabad, India**. India is one of the world's major producers of cotton.

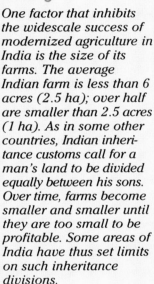

food. For the next 35 years, the people of India suffered a series of severe famines.

After independence, the new countries of South Asia focused on feeding their growing populations. As a result, few resources could be devoted to developing industry. The region's leaders sought foreign investment, but political turmoil, social unrest, and natural disasters interfered.

Even so, some industry did develop, particularly in India. Emphasis was placed on state-owned heavy industries, such as iron and steel, supported by the region's coal and iron-ore deposits. Cotton clothing and textile industries once again sprang up and became the region's most successful industries. Most of this industrialization took place in the cities.

A lack of industrial resources, especially petroleum, may hinder future development in the region. Although India has increased petroleum production and Pakistan has expanded development of its natural-gas supplies since the 1980s, petroleum imports are still a major economic burden to South Asian countries.

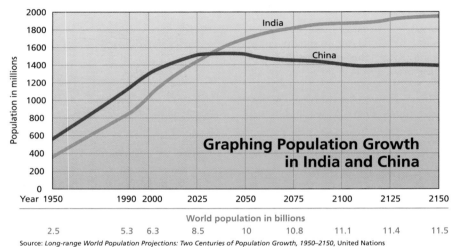

Source: *Long-range World Population Projections: Two Centuries of Population Growth, 1950–2150*, United Nations

SKILL STUDY   This line graph shows past and projected population increases of India and China over time. Despite government efforts to hold down birthrates, the combined populations of China and India are projected to climb to more than 3.3 billion by 2150. When is India projected to replace China as the world's most populous country?

Despite these challenges, the potential for growth is strong. The region has a huge low-wage labor force, which could draw labor-intensive manufacturing industries. These industries make small consumer goods, clothing, chemicals, and some high-technology items, such as computers and telecommunication systems. With industrialization, a South Asian middle class could grow. In turn, this growth would increase the demand for more locally manufactured goods. This economic growth may help cut the huge foreign debts that most of the region's countries have acquired over the years.

Population Growth All of the nations of South Asia are faced with rapid population growth. The population of India alone is more than 900 million. Its growth rate of about 2 percent means that the country gains more than 15 million people each year. Most other South Asian nations will double their populations in 20 to 40 years. Poverty in the region continues to grow along with the population.

Although the region's population is mainly rural, millions of people have migrated to the larger cities, seeking jobs. Several South Asian cities have some of the largest populations in the world. Unfortunately, these large cities are now surrounded by huge shantytowns with few city

services. Unemployment is high, and many people struggle to survive by working at temporary jobs. Urban poverty has led to increased crime.

Part of the answer to South Asia's problem of balancing people and resources might be to control population growth. This solution faces many obstacles, however. For example, poor communication systems keep government family-planning programs from reaching many of the rural areas, where more than 70 percent of the population live. In addition, most rural farmers want large families to help work the fields. The region's low literacy rates add to the problem. Education must be expanded to reach the majority of the population before meaningful change can take place.

SECTION REVIEW

1. What agricultural obstacles must South Asia overcome?
2. What factors have hindered South Asia's industrialization?
3. **Critical Thinking   Agree or disagree** with the following statement: "The Green Revolution proves that technology can provide complete answers to many of the problems of developing countries."

CHAPTER 46 CHECK

ANSWERS

Building a Vocabulary

1. A large landmass that is smaller than a continent
2. Hurricane-like storms that produce high winds and seas; the coastal regions of the Bay of Bengal
3. Many gods, salvation through *dharma*, reincarnation; Sikhism was formed as a blend of Hinduism and Muslim teachings.
4. Jute

Recalling and Reviewing

1. The Indus, the Ganges, the Brahmaputra; they originate in the northern mountain rim and flow to the Indo-Gangetic Plain; the Ganges
2. Trade links and profit, empire expansion, spread of Christianity, raw materials, markets for manufactured goods
3. If the rains come too soon, seedlings will be washed away; if the rains come too late, seedlings will die from lack of moisture.
4. Coal and iron ore; Deccan Plateau

Thinking Critically

1. Answers will vary. Students might point out that Gandhi's methods were effective in achieving the region's independence.
2. Whereas the Europeans divided Africa into colonies for their empires, the British divided South Asia to prevent war shortly before granting it its independence. The divisions of both regions, however, caused conflicts.

Learning About Your Local Geography

Paragraphs should describe the seasons and their effects on students' daily lives.

Using the Five Themes of Geography

Arrows might show the movement of Aryans from Central Asia to South Asia, the movement of Buddhism from South Asia to East and Southeast Asia, the movement of Muslims into South Asia from the northwest, the movement of raw materials from the region to Europe, or the movement of ideas and manufactured goods from Europe to the region.

CHAPTER 46 CHECK

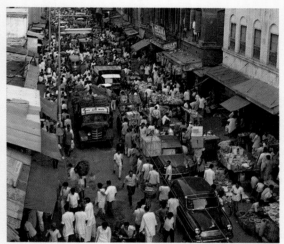

Calcutta, India

Reviewing the Main Ideas

1. South Asia, once part of the supercontinent of Gondwanaland, can be divided into three major landform regions and has several climate types. The monsoons strongly influence life in the countries of South Asia.
2. South Asia has had many cultural influences, from ancient times to the present. Europeans arrived in South Asia during the sixteenth century. Britain later came to dominate much of the region. Partly through the efforts of Mohandas Gandhi, the region gained its independence in 1947. The division of the subcontinent into India and Pakistan was traumatic for millions of people.
3. Many South Asians are farmers who live in small villages. The region's agriculture faces several challenges. South Asia also lags behind much of the world in industrialization, though the potential for growth is strong. All of the countries of South Asia are experiencing rapid population growth, which strains their resources.

Building a Vocabulary

1. Define *subcontinent*.
2. What are tropical cyclones? What regions of South Asia do tropical cyclones tend to hit?
3. State three beliefs of Hinduism. How is Sikhism related to Hinduism?
4. What is the name of the plant fiber used to produce coarse fabrics and twine?

Recalling and Reviewing

1. What are the three major rivers of South Asia? Through which landform regions do they flow? Near which river was the Indo-Aryan culture centered?
2. What interests did European colonists have in South Asia?
3. How is the timing of the monsoon critical to South Asia's farmers?
4. What resources supported the region's early heavy industries? Where are most of India's mineral resources located?

Thinking Critically

1. Mohandas Gandhi believed that nonviolent non-cooperation was the best way to bring about political and social change. Do you agree? Why or why not?
2. Compare and contrast the twentieth-century British division of the Indian subcontinent with the nineteenth-century European division of Africa. Did the divisions of both regions take place for the same reasons? Did they have similar effects? Explain your answer.

 ## Learning About Your Local Geography

Individual Project

To many South Asians, the change of the seasons is of great significance to daily life. Write a paragraph describing your area's seasons in terms of average temperatures, precipitation, and humidity. Also discuss the extent to which your daily activities change with the seasons over the course of one year. You may want to include your paragraph in your individual portfolio.

Using the Five Themes of Geography

Sketch a map of South Asia. With arrows, show how the region's **location** has influenced the **movement** of peoples, ideas, and goods in and out of the region over time. Label each arrow and include a caption explaining what exchanges and events are taking place on the map.

CHAPTER 47 India

Planning Guide

TEXTBOOK	TEACHING RESOURCES	OTHER AVAILABLE MATERIALS
Section		
1. Physical Geography Pages 571–572	Reteaching Worksheet 47.1 Critical Thinking Worksheet 47	*Study Guide for Spanish Speakers* *Lecture Notes,* Chapter 47
2. Economic Geography Pages 572–576	Reteaching Worksheet 47.2 Skills Worksheet 47	*Study Guide for Spanish Speakers* *Lecture Notes,* Chapter 47
3. Human Geography Pages 576–582	Reteaching Worksheet 47.3 Graphic Organizer 47 Challenge/Enrichment Worksheet 47	*The World Today* Videodisc Program *Creative Strategies for Teaching World Geography* LESSON 21 Culture Traits: The Caste System in India *Study Guide for Spanish Speakers* *Lecture Notes,* Chapter 47
Chapter Review Page 583	Chapter 47 Test Chapter 47 Portfolio Assessment See also *Alternative Assessment Forms*	*Test Generator,* Chapter 47

CHAPTER RESOURCES

Lecture Notes, Chapter 47
Teaching Resources
    Chapter 47 Graphic Organizer
    Chapter 47 Skills Worksheet
    Chapter 47 Critical Thinking Worksheet
    Chapter 47 Challenge/Enrichment Worksheet

REVIEW AND ASSESSMENT

Chapter 47 Check, page 583
Chapter 47 Test, *Teaching Resources*
Chapter 47 Portfolio Assessment, *Teaching Resources*
Test Generator, Chapter 47
Alternative Assessment Forms

ADDITIONAL RESOURCES

Books for Teachers

Blank, Jonah. *Arrow of the Blue-Skinned God: Retracing the Ramayana through India.* Boston: Houghton Mifflin, 1992.

Brown, Judith M. *Modern India: The Origins of an Asian Democracy.* New York: Oxford University Press, 1994.

Darian, Stephen. *A Ganges of the Mind: A Journey on the River of Dreams.* New Delhi: Ratna Sager, 1988.

Singh, Tej Vir, ed. *Geography of the Mountains.* New Delhi: Heritage Publishers, 1990.

Books for Students

Clark, James I. *India.* Evanston, IL: McDougal, Littel, 1989.

Currimbhoy, Nayana. *Indira Gandhi.* New York: Franklin Watts, 1985.

Easwaran, Eknath. *Gandhi the Man.* New York: Random House, 1978.

Johnson, Donald J., and Jean E. Johnson. *Through Indian Eyes.* Center for Training and Education, 1992. Available through Social Studies School Service, 10200 Jefferson Blvd., P.O. Box 802, Culver City, CA 90232–0802.

Sarin, Amita Vohra. *India: An Ancient Land, a New Nation.* New York: Dillon Press, 1985.

The World Today *Videodisc Program*

A variety of still images, short videos, and activities are available for you to use as you teach this chapter. *Lecture Notes* include bar codes for specific images and videos.

Multimedia Resources

Great Wonders of the World: Manmade—Taj Mahal. (Multimedia CD-ROM) InterOptica Publishing Ltd., 300 Montgomery Street, Suite 201, San Francisco, CA 94104.

India. (16mm film/VHS, 60 minutes) Available through Social Studies School Service, 10200 Jefferson Blvd., P.O. Box 802, Culver City, CA 90232–0802.

India After Independence. (VHS, 30 minutes) Films for the Humanities & Sciences, P.O. Box 2053, Princeton, NJ 08543–2053.

India Speaks. (VHS, 21 minutes) Available through Social Studies School Service, 10200 Jefferson Blvd., P.O. Box 802, Culver City, CA 90232–0802.

Picture Atlas of the World. (interactive maps, photo bank, animation, audio on CD-ROM) National Geographic Society, Washington, DC.

MEETING INDIVIDUAL NEEDS

Limited English Proficiency (LEP)

Help students create a time line on the board of events in India's history. As students read the chapter, have them note dates and events in India's history. After students have finished the chapter, call on volunteers to name dates and events, as you record them on a time line on the board. Students may want to copy the completed time line and include it in their individual portfolios.

Learners Having Difficulty

As students finish each section, have them answer the Focus questions in newspaper headline form. For example, for the Section 1 question "What environmental challenges does India face?," an answer might read, "Ganges Polluted by Industrial Wastes." Call on volunteers to read their headlines aloud, and have the class discuss details that might be included in accompanying newspaper articles.

Auditory Learners

Have students work with partners to answer the review questions at the end of each section in Chapter 47. One partner asks the question, and the other partner responds with the answer. Tell students to take turns asking the questions and providing the answers.

Visual Learners

Have students work in small groups to research and create posters that support the Indian environmental movement called "Embrace the Trees." Remind students that this movement is helping to slow the pace of deforestation in India. Have students share their completed posters with the class. Ask students how they might revise their posters to use them for a global movement with the same goals.

Tactile/Kinesthetic Learners

Organize the class into small groups. Provide each group with a copy of the "Yellow Pages" phone directory of your community. Have students leaf through the directory to identify businesses that could be cottage industries.

Gifted Learners

Direct interested students to investigate the environmental issues facing India by reading current newspaper and magazine articles. Have students pretend to be environmentally aware Indian citizens, and have them write a letter to the editor of a city newspaper in India. Call on volunteers to read their letters aloud.

MULTICULTURAL FOCUS

Although Mohandas Gandhi had studied in London, he was often critical of Western values. In particular, he questioned the attitude (often expressed to justify colonialism) that more advanced civilizations are measured by their industrial and technological progress. Instead, Gandhi idealized the way of life of the Indian civilization that had evolved for millennia before colonialism. Gandhi was thus wary of attempts to industrialize and modernize India, believing that the changes would destroy Indian culture.

You may wish to hand out or write on the board the following excerpt written by Gandhi in 1946:

Our ancestors . . . saw that happiness was largely a mental condition. A man is not necessarily happy because he is rich, or unhappy because he is poor. The rich are often seen to be unhappy, the poor to be happy. . . . Observing all this, our ancestors dissuaded us from luxuries and pleasures. We have managed the same kind of plow as existed thousands of years ago. We have retained the same kind of cottages. . . . It was not that we did not know how to invent machinery, but our forefathers knew that, if we set our hearts after such things, we would become slaves and lose our moral fibre. . . . They saw that our real happiness and health consisted in a proper use of our hands and feet.

Ask students to consider the pros and cons of industrialization. Do people ever become "slaves" to machines? How might happiness and health be connected to "a proper use of our hands and feet"?

569B

OVERVIEW

In Chapter 47, students will learn about India, the largest country in South Asia.

Section 1 describes India's landforms and climates and discusses the environmental challenges that face India today.

Section 2 focuses on the Indian economy, exploring changes in India's agricultural and industrial activity.

Section 3 examines India's diverse cultural and political characteristics and contrasts its rural and urban life.

About the Author

Priit J. Vesilind is an editor at *National Geographic* magazine. The quotation on this page is from an article he wrote about the monsoon.

CREATIVE STRATEGIES FOR TEACHING WORLD GEOGRAPHY

LESSON 21 Culture Traits: The Caste System in India

Primary Source

Description of Change: excerpted
Rationale: excerpted to focus on main idea

CHAPTER 47

India

“ In India the monsoon is predictably unpredictable. . . . Rains can skip entire regions. It can rain for days, even weeks, until the earth is squirming with life and walls are slippery with mildew. For another week the skies may be clear, but then rain will come again, relentless, finally tedious, until memories of cracked, dry earth and fierce sun fade into wishful thinking. ”

Priit J. Vesilind

One out of every six people in the world lives in India, the largest country in South Asia. In fact, India is second only to China in population. By the middle of the twenty-first century, India could become the most populous country in the world. Today, India's more than 900 million people crowd into an area about one-third the size of the United States.

India is one of the world's most complex culture regions. Its peoples practice several major world religions and speak languages from more than 50 language groups. The 25 states of this democratic republic are for the most part divided along ethnic lines. Economic contrasts between rural and urban, rich and poor, and industrial and agricultural also define India's cultural landscape.

Inlaid stonework from **Agra, India**

GEOGRAPHY DICTIONARY

sanctuary
cottage industry
caste system

Farming during the wet monsoon in southwestern **India**

Using the Quotation

Have students examine the background photograph on page 570 as you read aloud the quotation about the monsoon. Point out that although the monsoon is vital to Indian agriculture, the author uses words or phrases that paint a somewhat unfavorable picture of the rains (unpredictable, relentless, tedious). Ask students if there are weather patterns in their own locale that elicit mixed feelings.

Using the Geography Dictionary

Write the terms in the Geography Dictionary on the board. Encourage students to suggest the meanings of any of the terms that are familiar to them. Point out that **cottage industry** concerns the economy, **caste system** reflects culture, and **sanctuary** (as it is used in Chapter 47) pertains to the environment.

SECTION 1
OBJECTIVES

The objectives reflect the Focus questions on this page and are assessed in the review on page 572.

1. Describe India's relative location, major rivers, and climate types.
2. Discuss environmental challenges facing India.

INDIA: PHYSICAL–POLITICAL

Size comparison of India to the contiguous United States

LEGEND

ELEVATION

FEET	METERS
13,120	4,000
6,560	2,000
1,640	500
656	200
(Sea level) 0	0 (Sea level)
Below sea level	Below sea level

⊛ National capital
• Other cities

SCALE
0   300   600 Miles
0   300   600 Kilometers
Projection: Two-Point Equidistant

Using the Physical–Political Map: India

Direct students' attention to the map on this page. Have students name the major bodies of water that border India. Then ask them to name and describe the location of major landforms. Note the variety of landforms and elevations in India. Ask students where they think population is most heavily concentrated and why. *(on the Indo-Gangetic Plain and along rivers, because of the availability of water)*

PHYSICAL GEOGRAPHY

FOCUS

● *Where is India located relative to other countries and to seas and oceans? What are India's major rivers and climate types?*

● *What environmental challenges does India face?*

LOCATION **Landforms** The shape of India resembles an arrowhead pointing south. The towering Himalayas in the north form the base of the arrowhead and mark a natural boundary between India and China. The Himalayas also cross the small kingdoms of Nepal and Bhutan in the northeast. These countries have served as buffer states between India and China's autonomous region of Tibet. To the east, India borders Burma in the hill country and almost surrounds Bangladesh. To the west of India lies Pakistan, while in the far northwest is the disputed Kashmir region, a territory claimed by both India and Pakistan.

Water surrounds much of India. The southern tip of the country plunges into the Indian Ocean. The eastern Coromandel Coast faces the Bay of Bengal, and the western Malabar Coast borders the Arabian Sea. India also includes the Andaman and Nicobar islands in the Bay of Bengal and the Lakshadweep Islands in the Arabian Sea. (See the map above.)

Water is vital to the interior of the country as well. Here, India's major rivers act as lifelines. The Ganges, which flows from the Himalayas southeastward to the Bay of Bengal, is called the "mother river" and is sacred to Hindus. Over one-third of the country's population lives in the Ganges' fertile plains region. Other important

Motivator

Have students note India's shape on the map on page 571. Point out how physical features provide a natural boundary (the Himalayas) or separate India from other countries (the Thar Desert and the oceans and seas that surround the country).

Locating

Give each student a copy of Outline Map 22. Have students label the countries and bodies of water that border India as well as its major rivers. Students should also indicate India's major climate regions by shading areas of the map with different colors and providing a legend.

Guided Practice

Discuss how India's environment has been extensively damaged in recent years. On the board, label two rows: Water, Land. Label two columns: Environmental Problem, Cause. Tell students to copy this chart structure and to work with a partner to fill in the appropriate categories. After they have finished, discuss students' charts and ways that India might protect its environment.

Section 1 Review
ANSWERS

1. China, Nepal, Bhutan, Burma, Bangladesh, Pakistan; tropical savanna
2. Poor regulation of industry, lack of enforcement of pollution controls, deforestation, overgrazing, soil erosion
3. Possible answer: the Ganges is the lifeline of northern India; its plains region is agriculturally rich and provides much food.

FACTS IN BRIEF   India

COUNTRY POPULATION (1994)	LIFE EXPECTANCY (1994)	LITERACY RATE (1991)	PER CAPITA GDP
INDIA 919,903,056	58, male 59, female	52%	$1,300 (1994)
UNITED STATES 260,713,585	73, male 79, female	97%	$24,700 (1993)

Sources: *The World Factbook 1994*, Central Intelligence Agency

India's population is more than three times that of the United States. India's per capita GDP, however, is a fraction of the per capita GDP of the United States.

rivers include the Brahmaputra in northeastern India, the Narmada in western India, and the Godavari and Krishna rivers, which drain the Deccan Plateau and flow to the Bay of Bengal.

Climate   Most of India lies in the tropics and is strongly influenced by the monsoon. The Malabar Coast, the Western Ghats, and the upper Bay of Bengal coastal area are all humid tropical. In contrast, most of the interior of India, the Coromandel Coast, and the Eastern Ghats are tropical savanna. A region of steppe climate stretches over the central Deccan Plateau, while a humid-subtropical climate covers the northern edges of India along the base of the mountains. A desert climate extends along the western border with Pakistan and includes portions of the Thar Desert.

INTERACTION   **Environmental Issues**   A major challenge for India is to clean up and protect its environment. As much as 70 percent of India's surface water may be polluted. Even the sacred Ganges has been polluted by industrial wastes. Industry is poorly regulated in India, and pollution controls often are not enforced. This lack of regulation has had disastrous results. For example, in 1984 in the city of Bhopal (boh-PAHL), several thousand people were killed when a chemical plant accidentally released a cloud of poisonous gas.

Deforestation, overgrazing, and soil erosion have caused India's most serious environmental damage. Indeed, much of the country's natural habitat has disappeared. In some places, the government has created **sanctuaries**, areas in which forests and endangered animals are protected. (See "Planet Watch" on pages 574–575.) An environmental movement called "Embrace the Trees"

is helping to slow the pace of deforestation in India. Protecting and restoring India's forests remains a long and massive project, however.

SECTION REVIEW

1. What countries border India? What climate does most of India's interior have?
2. What are the causes of environmental damage in India?
3. **Critical Thinking   Suggest** a reason why Hindus consider the Ganges River to be sacred.

2 ECONOMIC GEOGRAPHY

FOCUS

- *Why has India's agricultural production been limited? Why is it now improving?*
- *What mineral and energy resources does India have? How has India's industry changed?*

INTERACTION   **Agriculture**   About three-fourths of India's people are involved in agriculture, mostly as subsistence farmers. The country does, however, produce many cash crops for export, including tea, cashews, tobacco, and coffee. India's cotton fields support the country's large textile industry.

Even though much of India's soil is fertile, crop yields are very low. The country's poor yields stem from a lack of widespread irrigation and modern farming methods. Most modern farm equipment and fertilizers cost too much for many farmers to buy. Cow dung, which is readily available, could be used for fertilizer. Farmers often burn it, however, for heating and cooking, because firewood and charcoal are in short supply.

Cattle are used as work animals and as a source of dung and milk. Cows are sacred in India, however, and many are allowed to wander freely. Farmers also use water buffalo to work the land. In addition, they raise sheep for wool and meat, and goats for meat and milk.

Closure

Have students answer the Focus questions on page 571.

ASSESSMENT

Assign the Section 1 Review questions on page 572.

Reteaching

Assign Section 1 of Reteaching Worksheet 47.

SECTION 2
OBJECTIVES

The objectives reflect the Focus questions on page 572 and are assessed in the review on page 576.

1. Explain why India's agricultural production has been limited but is now increasing.

2. Identify India's mineral and energy resources, and examine India's industrial development.

TEACHING SECTION 2
Motivator

Direct students' attention to the photographs on pages 570 and 580. Ask students to make a generalization about agriculture in India based on these illustrations. *(Students should note the unmechanized nature of agricultural techniques.)*

This cow threatens to stop traffic in **Jaipur, India**. Cows, considered sacred by Hindus, can be found roaming freely in many urban and rural areas.

One of India's main goals is to make better use of its agricultural resources. To this end the Indian government and foreign countries have aided several rural agricultural development projects. New irrigation systems are increasing farmland, and flood-control and soil-erosion programs are saving precious soil. Better grain-storage facilities reduce damage by pests, and the use of high-yield crops allows higher production and shorter growing seasons. Mechanization also makes the farms more efficient, though it sometimes adds to rural unemployment because fewer workers are needed. All of these improved agricultural methods have helped make India self-sufficient in food production in most years. Although poverty persists, the people of India no longer suffer major famines.

Mineral and Energy Resources India has many mineral resources, including large deposits of iron ore, bauxite, and coal. Much of the iron ore and bauxite is exported. India has some petroleum deposits but must import oil to meet its consumption needs. The country's supply of uranium ore supports a number of nuclear power plants and has enabled India to develop an atomic bomb.

India's great river systems are a source of hydroelectric power, though only a small part of this resource is developed. A large hydroelectric-irrigation project planned for the Narmada River will be among the world's largest, with 30 major dams and hundreds of smaller ones.

Industry Even though India is largely agricultural, it ranks about tenth among the world's industrial countries. Manufacturing generates around one-fourth of India's income.

In recent decades, the government wanted to make India self-sufficient in manufactured goods. Many large, state-owned industries and some private ones formed to produce chemical fertilizers, steel, railroad and farm machinery, automobiles, and many other items. The iron and steel mills in the Jamshedpur region, west of Calcutta, grew to become some of the world's largest. The cotton textile mills in Bombay and the jute mills in Calcutta also became major production centers. Although the government's approach did decrease India's dependence on imported goods, the nation still acquired a huge trade deficit. In addition, inefficiency and corruption marked the state-owned industries.

The government has since changed its policies. Today, small privately owned industries are the fastest-growing sector of the economy. India has also opened its doors to foreign capital and industry, which are attracted by India's huge pool of low-cost labor. The government has been particularly successful at promoting high-tech industry in the country. For instance, Bangalore, a city in southern India, has attracted so many foreign computer companies it is called India's "Silicon Valley."

To draw more foreign industry, India must strengthen its infrastructure. The country's roads, railroads, airlines, electricity, and telecommunications systems, little changed from colonial times, are inadequate and unreliable. In addition,

Text continues on page 576.

GLOBAL PERSPECTIVES

Many U.S. companies have increased their operations in India, sensing that India's markets may hold more potential for growth than domestic markets. The companies have invested millions of dollars in factories that manufacture products such as medical equipment, plastics, and appliances. By late 1993, one U.S. company alone had sales in India of $400 million. Not only do India's markets appeal to foreign companies, but India's labor force—particularly its supply of inexpensive yet talented engineers—also is important in the design of new products.

573

Planet Watch

Problem Solving

Tell students that they are Indian farmers living near a wildlife preserve. They are meeting to find a solution that allows their villages to survive and grow and that also protects the tigers and their habitat. Have students create a list of their specific concerns *(for example, need for farmland and wood, protection against tiger attacks)*. Then ask them to suggest ways of addressing these concerns. Write students' suggestions on the board. Then have students examine each suggestion's consequences for the tiger and reject any that do not help protect the animal. Students should then work in pairs to refine the remaining solution(s) on the board into a written proposal.

PLANET WATCH

Case Study: Saving the Tiger

The tiger is a solitary animal that usually stalks its prey at night, deep in the forests. Tigers hunt deer and pigs and sometimes young elephants and rhinoceroses. The tiger is prey, though, as well as hunter. Humans have killed tigers both for sport and profit. Rugs made from tigers' skins were once popular throughout much of the world, as were clothes trimmed with tiger fur. Even today, laws designed to protect the tiger are ignored by poachers, who kill the animals for the thousands of dollars their skins will bring.

> *Once roaming over much of Asia, the tiger has become one of Earth's most endangered species.*

The Tiger in India

According to one source, tigers in India in the early 1800s were "so numerous it seemed to be a question as to whether man or the tiger would survive." Exactly how many tigers existed in India is not clear, but by 1900, they numbered only 40,000 due to hunting. Among India's British rulers as well as the country's maharajas, or princes, a fallen tiger was a symbol of status.

To slow the decline of the Indian tiger population, the British government limited hunting to certain areas. They also set up seasons, or time periods, during which hunting tigers was illegal. Laws against poaching were enforced so harshly that few people dared to hunt illegally.

As a result of these policies, India's tiger population thrived for decades. Then, in 1947, India gained its independence from Britain. Free of foreign rule, people began to ignore many of the British hunting laws, and poaching increased.

At this same time, India's tigers faced an added threat. Between 1947 and 1990, India's population swelled from less than 500 million to more than 800 million. As the number of people soared, the forests began to disappear rapidly. The amount of India's forested lands decreased by almost 40 percent from 1900 to the early 1990s. Farmers began to plant crops and graze buffalo and other farm animals on lands once covered by trees and roamed by tigers.

With their habitat shrinking, large numbers of tigers fell prey to hunters, starved, or attacked humans and were killed in return. By 1972, fewer than 2,000 tigers remained in India. The Caspian, Javan, and Balinese tigers are already extinct.

Project Tiger

In 1973, the Indian government took steps to prevent the extinction of the tiger. With support from the World Wildlife Fund, nine tiger reserves were created as part of Project Tiger. Since then, six more reserves have been added. The core

A tiger makes its way through a river in one of India's Project Tiger reserves. The total core area of the reserves, where people are not allowed, is more than 3,000 square miles (7,770 sq. km).

Using Geography Themes

Have students choose one of the Five Themes of Geography and sketch an illustration showing how that theme relates to the material presented in the feature. For example, a student might show a tiger in a farm field, illustrating the tiger's loss of habitat due to human-environment interaction. Students' illustrations should include extended captions explaining the relationship between the chosen theme and the feature material.

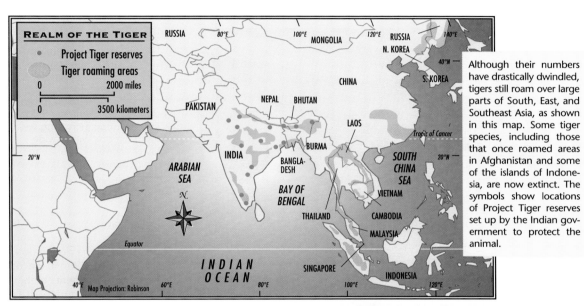

Although their numbers have drastically dwindled, tigers still roam over large parts of South, East, and Southeast Asia, as shown in this map. Some tiger species, including those that once roamed areas in Afghanistan and some of the islands of Indonesia, are now extinct. The symbols show locations of Project Tiger reserves set up by the Indian government to protect the animal.

Geographical Sidelight

Tiger attacks in India kill 50 to 60 people each year. Because tigers generally attack from behind, some fishers wear masks on the back of their heads to confuse the animals. Forest officials often wear suits of armor.

areas of the reserves, which are surrounded by less-restrictive buffer zones, are closed to people. No one is allowed to enter these areas unless accompanied by a park ranger. In addition, tens of thousands of acres have been set aside as national parks and wildlife preserves.

Many poor, rural Indians have reacted negatively to the reserves. Scarcely able to feed their families, villagers have little sympathy for the tiger. They argue the reserve lands are desperately needed for grazing livestock, collecting firewood, and gathering wild honey. Many believe that tigers are being protected at the expense of people.

A wildlife expert from India's Ranthambhore National Park understands the concerns of India's rural poor. "They're right that the park does not yet benefit them. It's our job to change that, to help them see that their survival and that of the park are linked, that if the forest is destroyed, their lives and all our lives will be destroyed."

He and other environmentalists belong to a foundation that reaches out to villagers to show them that they and the reserves can co-exist. In an effort to reverse deforestation, the foundation distributes more than 40,000 tree seedlings a year. To deal with the problem of too much livestock and too little grazing land, it has introduced new types of buffalo that need little land. It has tried to aid the local economy by forming a crafts cooperative, in which

Indian villagers are preparing a human-like dummy that will be a decoy for tigers. These dummies are outfitted with electric-shock devices whose purpose is to discourage tiger attacks on humans.

women make scarves and other handicrafts for sale in urban areas.

Since 1973, the number of India's tigers has doubled to about 4,000. This number is still dangerously low, however, and poaching remains a problem. As many as 20 tigers were killed by poachers in Ranthambhore in 1992. Environmentalists, however, are determined to succeed in their efforts to save the tiger and its habitat.

YOUR TURN

Imagine you are a worker on a tiger reserve in India. **Write** a brief talk explaining to villagers why it is important to all Indians to save the tiger and its habitat. **Address** the specific concerns of the farmers.

Your Turn

Students' talks should explain how the survival of the tiger and its habitat is linked to the villagers' own survival.

575

Analyzing Agricultural Production

Organize the class into pairs. Have one member of each pair identify the problems that have limited India's agricultural production. Have the other member find the solutions India is using to address these problems. Then, as a class, discuss ways India could increase agricultural production.

Guided Practice

On the board, create a chart with three columns: Past, Present, Future. Have volunteers fill in the chart with regard to the development of India's resources and industry.

Closure

Have each student imagine that he or she works in India as a farmer, irrigation specialist, miner, manufacturer, government worker, foreign business-person, and so on. Have students write advertisements for or announcements of their goods or services.

ASSESSMENT

Assign the Section 2 Review questions on this page.

Reteaching

Assign Section 2 of Reteaching Worksheet 47.

A worker at a factory in central **India** checks heavy machinery manufactured for export to Australia. As in other developing countries, officials in India hope increasing levels of exports will provide faster economic growth in their country.

manufacturing employs only a small percentage of the working population. Nonetheless, India's overall economy is expected to grow at a rapid pace.

 SECTION REVIEW

1. What effects have recent agricultural development projects had on India?
2. What mineral and energy resources does India have? How did India's government change its policies regarding industry, and why did it do so?
3. **Critical Thinking** **Explain** how India's agricultural and mineral resources supported the country's industrialization.

3 HUMAN GEOGRAPHY

FOCUS

● *What are India's cultural and political characteristics?*

● *How does rural India differ from urban India?*

REGION **Language and Religion** India is one of the most culturally complex countries in the world. Eighteen major languages and many others are spoken in the country. Although Hindi

is the national language, it is spoken by only about one-third of the population. English is the language of India's government, higher education system, and business community. It is spoken mainly in urban areas. Most people of the Indo-Aryan north speak Hindi, while most people in the nation's south speak Dravidian. Both languages have many dialects.

India is also complex in terms of religion. The country is about 80 percent Hindu, 14 percent Muslim, 3 percent Christian, and 2 percent Sikh. Millions of Buddhists, animists, and followers of other religions also live in India.

For hundreds of years, religious tensions have plagued India. Ancient hostility still exists between the country's Hindu and Muslim populations. Hindu nationalism has grown stronger in recent years, leading to violent riots between the two groups. In two months alone, from December 1992 to January 1993, violence among members of various sects killed more than 1,700 people, after militant Hindus destroyed a Muslim mosque in northern India. The Hindus charged that the mosque was built on the site where a Hindu god was born and where a Hindu temple had stood.

Independence movements have also led to violence. In the Punjab province, hundreds of people have died in conflicts between Hindus and Sikh separatists. Meanwhile, separatists in Kashmir have been fighting for their independence.

Political Geography Under the leadership of Jawaharlal Nehru following independence, India adopted a constitution making the country a democratic republic. The nation is made up of a union of states held together by a strong central

Section 2 Review
ANSWERS

1. They have helped make India self-sufficient in food in most years. Mechanization, however, sometimes adds to unemployment.

2. Iron ore, bauxite, coal, petroleum, uranium ore, hydroelectricity; while much industry was formerly state-owned, private and foreign industries and investment are now increasing; because the state-owned industries were inefficient and corrupt

3. Cotton supported a textile industry; iron ore and other minerals supported iron and steel industries.

576

OBJECTIVES

The objectives reflect the Focus questions on page 576 and are assessed in the review on page 582.

1. List cultural and political characteristics of India.

2. Compare and contrast rural India and urban India.

TEACHING SECTION 3
Motivator

Ask students to list features that signify cultural diversity within a society. *(differences in languages, religion, ethnic background, nationality)* Have them name countries that they believe to be culturally diverse. *(Answers will vary, but students are likely to name the United States as the primary example.)* Tell students that they will learn why India is considered to be a very culturally diverse nation.

Building a Vocabulary

Ask students to define *cottage. (small dwelling)* Note that **cottage industries** are those conducted in a home or a small workshop, not in a large factory. Have students give examples of economic activities that they think would work well as cottage industries.

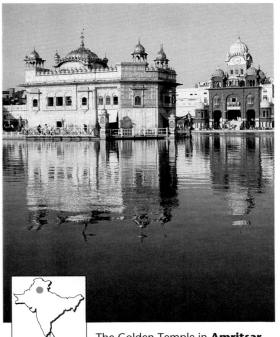

The Golden Temple in **Amritsar, India**, is the chief house of worship for Indian Sikhs. Sikh demands for an independent state in Punjab, and opposition to such demands by the Indian government, has at times brought violence to the region.

government and headed by a prime minister and a parliament.

Nehru died in 1964 and was succeeded as prime minister two years later by his daughter, Indira Gandhi. Indira Gandhi, India's first female prime minister, remained in office until 1984, when she was assassinated by Sikhs who disliked her resistance to Sikh independence. Indira Gandhi's son, Rajiv Gandhi, then became prime minister. In 1991, he too was assassinated, bringing an end to the Gandhi family leadership and throwing India's politics into even greater turmoil. Although India's government has tried to keep politics and religion separate, a movement in favor of a Hindu nation has been growing stronger.

Rural Life The differences between rural and urban life are striking. About three-quarters of India's people are poor farmers who live in rural villages. (See "Themes in Geography" on pages 580–581.) Many have seen little progress in their lifetimes, and some 60 percent cannot read or write.

Most villages are not connected to paved roads. Electricity, telephones, and running water are rare, and sanitation is poor. Often, one doctor must try to serve thousands of people. Rural houses are built of local materials. In the alluvial river valleys, straw-roofed homes are made of mud. In other areas, the houses might be made of dried cattle-dung plaster, brick, or bamboo.

Families tend to be large. In fact, one of every five children born in the world today is born in India. Most of India's rapid population growth occurs in rural areas, where children are needed to help in the fields. In addition, both the Hindu and Muslim cultures value large families and have resisted family planning.

Despite government attempts at land reform, landless field workers make up more than 30 percent of India's rural population. Those who do own small plots of land are often forced into debt to meet farm and living expenses. Village moneylenders may charge as much as 300 percent interest on loans. Other farmers work the land of large landowners and must give a large share of the crops they grow to the owners.

Rice is the staple food in the south, wheat in the north, and millet in the Deccan Plateau region. The diets of most Indians contain little meat. The Hindu religion forbids the eating of beef, and the Muslim religion does not allow the eating of pork. Some religions in India call for a completely vegetarian diet.

Village industries are limited mostly to **cottage industries**. In cottage industries, workers make small consumer items in homes and small workshops. These products include handwoven carpets, metal tools, locks, jewelry, and shoes and other leather goods.

Most Indian men in rural areas wear a *dhoti* (DOHT-ee), a simple white cloth wrapped around the hips and between the legs. Hindu women wear a *sari*, a wide piece of cloth draped to cover the body from shoulder to ankle. Indians usually can tell what part of the country a Hindu woman is from by the way she drapes her *sari*.

Rural India is a male-dominated society. When a woman marries, she is no longer part of her own family, but rather becomes a member of her husband's family. A rural wife shares field work with her husband and must gather drinking water and firewood for the family. She may also care for the cattle, as well as raise the children. Literacy rates among rural Indian women are very low.

Geographical Sidelight

Indira Gandhi was one of the most important political figures of her time. The only child of Jawaharlal Nehru, she showed political aptitude as early as age 12, establishing a youth organization to help with the National Congress movement. After her father became prime minister in 1947, she accompanied him on many of his trips to foreign countries. By the mid-1950s, she had begun to establish her own political career, gaining seats on influential government committees. Although her terms as India's prime minister were marred by power struggles and accusations of corruption, she guided the country through a war with Pakistan and promoted social welfare and international diplomacy.

577

Analyzing

Social Systems Call on volunteers to describe the caste system in their own words. On the board, have students list the various castes, describe each one, and put them in order from highest to lowest. Discuss why abolishing a traditional social system such as the caste system is so difficult.

Mapping

India's Diverse Cultures Assign half the class the topic of language and the other half the topic of religion. Have students in the language group create a language map of India, while students in the other group construct a map showing India's religions. Display the maps side by side, and note similarities and differences.

Describing

India's Cities Draw an outline map of India on the board. Have volunteers identify and mark the locations of New Delhi, Bombay, Calcutta, and Madras. Distribute among the class cards with phrases such as "Hollywood of India," "modern subway system," "capital," and "automobile assembly." Have cardholders come to the board and copy their cards' phrases next to the names of the appropriate cities.

COOPERATIVE LEARNING

Organize the class into groups of four or five. Tell each group to imagine that it is an Indian family. Have half of the groups be rural families and the other half urban families. Give each group 15 minutes to identify the ways of life of its family, based on information in the text, and to decide on what role within the family each group member will take. Then have each rural family meet with an urban family for 15 minutes. Afterwards, lead a discussion about how the groups' interactions revealed differences between rural and urban India.

THE WORLD TODAY VIDEODISC PROGRAM

India: The Sacred Ganges

2B   44966   4:01

Play   Pause

578

This Indian woman is wearing a *sari*, a light, loose-fitting garment made of a brightly colored cloth, such as silk. Many Indians wear light, cool clothing because of the warm climates that dominate much of the country.

The Caste System Most rural villages are still based on the **caste system**. The Hindu Aryans developed this social system thousands of years ago. It divided India's society into four major classes, or castes, by occupation. The Brahmins were the highest caste. They were priests and intellectuals and the only people who could read and write. Below the Brahmins in order of rank were the Kshatriyas (kuh-SHA-tree-uhz), or warriors; Vaisyas (VISHE-yuhz), the traders and merchants; and Sudras, the farmers and laborers.

Below these four classes were the untouchables, who were considered so lowly as not to be part of any caste. They were not allowed to enter a temple and did only the most unpleasant tasks. They were forbidden from having contact with Indians of any caste.

Under the caste system, a person is born into a caste and cannot move into another. Some Indians believe that taking up the occupation of another caste or marrying into another caste is a violation of moral code, punishable by rebirth in another life as an untouchable or a non-human form of life.

Since independence from British rule, the government has worked hard to abolish the caste system. It declared that the poor treatment of untouchables was illegal. Mohandas Gandhi, who helped free India from British control, saw the untouchables as human beings who deserved to be treated with respect and dignity, just as people of other classes were. Gandhi called the untouchables *harijans*, or "children of God."

Yet, regardless of the law, the caste system remains a part of Hindu life, and violence against the untouchables still takes place. As improved education and transportation reach the villages, the caste system will likely change, as it has in many cities.

City Life In general, Indian city dwellers wear European-style clothing, read English-language newspapers, and work in factories and office buildings. Smog, traffic, and noise are part of their daily lives. Education in universities and technical schools is available in the cities, and urban women work with men in professional and factory jobs. Some are members of trade unions and seek pay equal to that of men.

The cities are home to a small number of wealthy industrialists and landowners. These prosperous few live in large walled-in homes or guarded high-rise apartments. The working middle class, however, makes up a larger, and growing, sector of urban society. Its members have become part of India's growing consumer economy. Most live in small apartments. Some can afford televisions, motor scooters, and electric kitchen appliances. They can also afford entertainment in the form of movies, soccer games, shopping, and restaurants.

Despite some economic progress, however, most of the urban population remains poor. Many are homeless and live in the streets. Many others live in slums in makeshift homes built of scraps of wood, cardboard, and cloth. They do not have clean water, sewage systems, or garbage collection. They look for work each day for whatever pay they can get.

Guided Practice

Draw a time line on the board to illustrate key events that have shaped India's political geography in the twentieth century. Include Gandhi's campaign of nonviolent noncooperation, division of the subcontinent, independence from Great Britain, and subsequent Indian leadership. Discuss with students the conflicts surrounding these political events.

Independent Practice

Have students write a paragraph discussing what India's human geography might be like in 20 years. Paragraphs might focus on the status of religious conflicts, on India's government or political system, on conditions in rural and/or urban areas, or on the caste system.

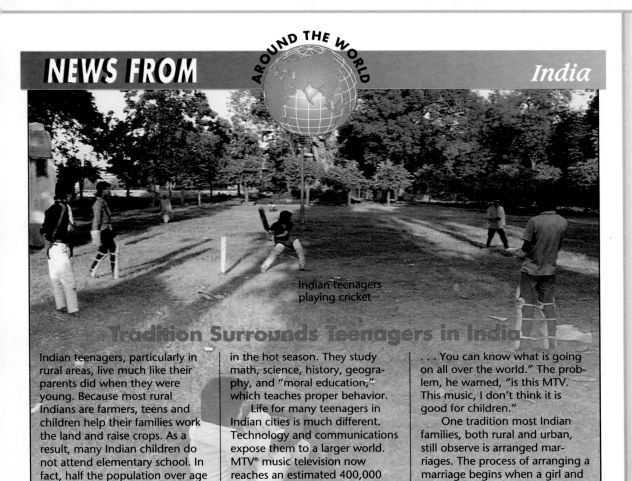

NEWS FROM

AROUND THE WORLD

India

Indian teenagers playing cricket

Tradition Surrounds Teenagers in India

Indian teenagers, particularly in rural areas, live much like their parents did when they were young. Because most rural Indians are farmers, teens and children help their families work the land and raise crops. As a result, many Indian children do not attend elementary school. In fact, half the population over age 15 is unable to read and write.

For those who are able to attend school in rural areas, conditions are poor. There are too few teachers, and materials and equipment are outdated. School often takes place under makeshift circumstances—on a porch, under a tree, in a village square. Children go to school from 10:00 A.M. to 4:00 P.M. in the cool season and from 7:00 A.M. to noon

in the hot season. They study math, science, history, geography, and "moral education," which teaches proper behavior.

Life for many teenagers in Indian cities is much different. Technology and communications expose them to a larger world. MTV® music television now reaches an estimated 400,000 homes in India. Eager Indian teenagers tune in to the channel's rap and rock music. "Most of my friends, they watch," says one teen. "In every house, they have their MTV."

Parents, however, often view such western influences as threats to traditional values. For example, one father says, "Satellite TV is very good, no doubt. There are so many programs you can get.

. . . You can know what is going on all over the world." The problem, he warned, "is this MTV. This music, I don't think it is good for children."

One tradition most Indian families, both rural and urban, still observe is arranged marriages. The process of arranging a marriage begins when a girl and boy are still in their early teens. The process can continue for years while parents exchange and analyze their children's horoscopes and arrange property settlements. Although many teenagers accept their parents' decisions, others growing up in the cities are beginning to choose their own spouses. Traditional ways of life in India, however, are slow to change.

Geographical Sidelight

For generations, upper- and middle-class Indians have hired dhobis, *washermen and washerwomen, to clean their clothes. The* dhobis, *traditionally of the untouchable class, wash the clothes by hand as they always have—by beating them on rocks along the banks of a river and allowing them to dry in the sun. Today, however,* dhobis *see their livelihoods dwindling as more and more middle-class families are buying washing machines.*

Government family planning programs have been more successful in urban areas than in rural areas. Migration to the cities, however, has raised the urban population to hold about one-fourth of the country's people. Millions of people from rural areas seek a better income in the cities and an escape from the caste system. Unfortunately, many move from rural poverty to urban poverty.

Urban Centers New Delhi (DEHL-ee), in northern India, is the capital of the country. Like Washington, D.C., this city was built to house the national government. Along with the neighboring old city of Delhi, New Delhi is the largest industrial area of northern India.

Bombay, located on the western coast, is India's largest city and a commercial and industrial

Text continues on page 582.

Themes in Geography

A Day in the Life

After students have read the feature, ask them to imagine that they live in a rural Indian village. Have them write an account of a typical day in their village. Accounts should describe the village and a person's home, occupation, family, and daily activities. Students might also include the villagers' plans for the future. Call on volunteers to read their accounts to the class. Students may want to include their accounts in their individual portfolios.

THEMES IN GEOGRAPHY

Indian Villages

Although India is gradually industrializing, the majority of its people are still peasant farmers. These farmers live in the hundreds of thousands of small villages scattered throughout the country.

Surrounded by farmland, each village consists of a number of small huts with one or two rooms. These huts cluster around a well or a stream, which forms the central point of the village. Each day, the women gather at the well with large clay pots for carrying water. Most do not have luxuries such as running water or electricity. And even in villages that do have electricity, many of the huts are not connected to the electricity supply. There are few shops in the villages.

Some villagers own their own plots of land, but they usually manage to grow only enough food to feed their own families. Most villagers, however, do not own land. Because

Small farming villages set the stage of daily life in rural India.

INTERACTION Most of the farming methods practiced in Indian villages today have been used for generations. Because the peasant farmers are very poor, they cannot afford modern farm equipment. They obtain their water from wells, using cattle to drive the pumps. Most crops, however, must depend on rainfall rather than irrigation. The great differences in rainfall throughout the country mean people must live off the crop that grows best in their area. **Why are there great differences in rainfall in India?**

580

Using Geography Themes

Have students find the two geography themes that are called out in the feature. *(place and human-environment interaction)* Then have students draw illustrations or create maps relating one of the other three geography themes (location, movement, region) to the material presented in the feature. Students' illustrations and maps should include extended captions explaining the relationship between the theme and the feature.

PLACE Family ties are very important in Indian communities. Marriages often are arranged by parents. After a couple marries, they usually live in the home of the husband's parents. In fact, grandparents, uncles, aunts, cousins, and other relatives may all live in one house. They combine their earnings, and all share in the work on the farm and in the house. **Can you name other countries in which marriages are still arranged?**

Geographical Sidelight

Cottage industries that produce traditional Indian handicrafts are important sources of rural employment in India. The export of these handicrafts also provides important foreign earnings. These handicrafts include hand-knotted carpets, art metalware, hand-printed textiles, and leatherware. The skills used to create many of these items have been passed down in families over the centuries.

Photo Caption Answer

Some areas of Europe, Asia, and Africa. Answers will vary.

India has such a large population, there is not enough land for everyone. Instead, many farmers must rent plots from landowners. It may take a village farmer a year to earn as much as an American farm laborer could earn in two weeks. Most Indian farmers make their money at markets in nearby towns, where they sell any extra produce they grow.

People living in neighboring villages may not be able to communicate with each other because India is divided into 15 states, each with its own language. Each of these different languages has more than 500 dialects. As a result, many villagers spend their entire lives without ever leaving their own village.

Most peasant farmers in India are illiterate. Their lives revolve around their land, and they have little time for learning to read and write. The government is trying to change this situation, however. Some villages now have schools,

which also serve as meeting places for the community. Adults and children alike can attend these schools.

The community schools play an important part in teaching villagers about new and better ways of farming. Many government officials know that the only way India can feed its rapidly increasing population is by using modern farming methods. They also are recognizing that education can help the people of India's villages improve their lives.

Closure

Remind students that India is a land of many contrasts. Ask students to make statements that reflect this fact. *(For example, "Ancient conflicts exist between Hindus and Muslims in India.")* Have students exchange statements and discuss them.

Extension

Ask interested students to research and prepare a report on the life and work of Mohandas Gandhi. Students may wish to provide a time line as background and then to concentrate their report on a particular aspect of his life or work. For example, students might discuss his concept of nonviolent protest or his identification with the poor and outcast. Ask students to present their reports orally to the class.

Geographical Sidelight

Textile production is an important economic activity in the city of Madras. A distinctive cloth called madras originated here. The cotton yarn used in madras is usually dyed with vegetable dyes that "bleed" during washing. The result is a soft blending of colors that distinguishes true madras from other fabrics.

Section 3 Review
ANSWERS

1. Languages: Hindi, English, Dravidian; religions: Hinduism, Islam, Christianity, Sikhism, Buddhism, animism

2. Rural: high birthrates; urban: migration from rural areas; much of rural India has five social class divisions, while urban India has a small number of wealthy people, a larger middle class, and a still larger poor population.

3. Rural: influences family size, diets, clothing, and social class divisions; urban: might determine where a person lives; country: religious tensions and movements have killed many people. Some want India to be a Hindu nation.

582

 Movie advertisements cover this small theater in **Madras**. India's large moviemaking industry is supported by millions of movie fans.

center. Actually, Bombay lies on a small island just off India's western coast. Several bridges link the island and the city to the mainland. Bombay's wide variety of industries includes India's large textile and movie industries. In fact, India is the number-one moviemaking nation in the world, and Bombay is sometimes called the "Hollywood of India." Its constant stream of movies offering romance and adventure delights large Indian audiences. Bombay is also a busy seaport, well situated for world trade.

Calcutta, located in eastern India on a branch of the Ganges River, is the other giant industrial and seaport city of India. Calcutta produces a wide range of goods, from tea and hides to steel and electronics. As thousands of people live in poverty in the streets, high-rise buildings tower above them, and a modern subway system runs beneath them. Most of the city's people are Hindu, though Muslims, Sikhs, and Christians make up about one-fifth of the population. Even

today, these groups tend to live in separate sections of the city.

Madras is southern India's leading industrial city and port. This busy shipping center on the country's southeastern coast also has automobile assembly plants, cotton mills, and iron, cement, and glass works.

SECTION REVIEW

1. List three languages spoken and four religions practiced by the people of India.
2. What factors are responsible for population growth in rural India? in urban India? What social class divisions exist in each of the two areas?
3. **Critical Thinking**   **Discuss** the influences of religion on life in India in rural areas, urban areas, and in the country as a whole.

Building a Vocabulary

1. Industries in which workers produce small items in homes and small workshops; handwoven carpets, metal tools, locks, jewelry, shoes, other leather goods
2. A social system in which people are divided by occupation; Brahmins (priests and intellectuals), Kshatriyas (warriors), Vaisyas (traders and merchants), Sudras (farmers and laborers); untouchables

Recalling and Reviewing

1. The Indian Ocean, the Bay of Bengal, and the Arabian Sea
2. They use cow dung for fuel because firewood is in short supply, possibly as a result of deforestation.
3. A large supply of low-cost labor; develop a more reliable infrastructure
4. New Delhi, Bombay, Calcutta, Madras

CHAPTER 47 CHECK

Palace of the Winds in **Jaipur**

Reviewing the Main Ideas

1. India borders several countries to the north and is surrounded by water to the west, east, and south. Most of India lies in the tropics and is strongly influenced by the monsoon. A major challenge for India is to clean up and protect its varied environment.
2. About three-fourths of India's people are involved in agriculture. Agricultural development programs are increasing crop yields. The country has a variety of mineral resources and industries. Today, small privately owned industries are the fastest-growing sector of the economy, and foreign investment is increasing.
3. India, a democratic republic, is one of the most culturally complex nations in the world, with many languages spoken and religions practiced. Religious tensions have plagued India for centuries. The differences between rural and urban life in India are striking.

Building a Vocabulary

1. What are cottage industries? What are some of the products they produce?
2. What is the caste system? What are Hinduism's four castes? What group is considered outside the caste system?

Recalling and Reviewing

1. What bodies of water does India border?
2. How is farmers' use of cow dung as a source of fuel related to India's deforestation?
3. What attracts foreign capital and industry to India? What might India do to further attract them?
4. What are four major cities in India?

Thinking Critically

1. India became a democratic republic after it became independent. In what way, however, did the country's government resemble a dynasty in the decades following independence? What do you think was responsible for this resemblance?
2. As you read in the chapter, many farmers in India work the land of large landowners and must give a share of the crops they grow to the owners. What Middle and South American system of agriculture does this resemble?

Learning About Your Local Geography

Individual Project

In India, rural houses are built of local materials such as straw, mud, and bamboo. From what materials are houses or other buildings made in your community? How do these materials reflect local geography in terms of resources found in or near your area? Draw some illustrations of buildings in your community. Write captions for your illustrations, describing the materials used in the buildings and where these materials might have originated.

Using the Five Themes of Geography

Write a paragraph discussing the theme of **human-environment interaction** in India. Focus on India's rivers. How have India's rivers affected human settlement, influenced religious beliefs, and served as resources? How has human activity affected the rivers?

Thinking Critically

1. For more than 40 years following independence, India's prime ministership almost appeared to be a hereditary office, with three generations of the same family holding it. Answers to the second part will vary but might include that the family was popular.
2. The hacienda system

Learning About Your Local Geography

Illustrations should show local buildings; captions should identify the building materials and their origins.

Using the Five Themes of Geography

Much of India's population depends on rivers for agriculture and hydroelectricity. Also, Hindus consider the Ganges to be sacred. Human activity has polluted the rivers; even the Ganges has been polluted by industrial wastes.

583

CHAPTER 48 The Indian Perimeter
Planning Guide

TEXTBOOK	TEACHING RESOURCES	OTHER AVAILABLE MATERIALS
Section		
1. Pakistan Pages 585–587	Reteaching Worksheet 48.1	*Study Guide for Spanish Speakers* *Lecture Notes,* Chapter 48
2. The Himalayan Kingdoms Pages 587–590	Reteaching Worksheet 48.2 Skills Worksheet 48 Challenge/Enrichment Worksheet 48	*Study Guide for Spanish Speakers* *Lecture Notes,* Chapter 48
3. Bangladesh Pages 590–591	Reteaching Worksheet 48.3	*Study Guide for Spanish Speakers* *Lecture Notes,* Chapter 48
4. Island Countries Pages 591–592	Reteaching Worksheet 48.4 Graphic Organizer 48 Critical Thinking Worksheet 48	*Study Guide for Spanish Speakers* *Lecture Notes,* Chapter 48
Chapter Review Page 593	Chapter 48 Test Chapter 48 Portfolio Assessment See also *Alternative Assessment Forms*	*Test Generator,* Chapter 48
Unit Review Pages 594–595	Unit 10 Test Unit 10 Portfolio Assessment See also *Alternative Assessment Forms*	*Test Generator,* Unit 10

CHAPTER RESOURCES

Lecture Notes, Chapter 48
Teaching Resources
    Chapter 48 Graphic Organizer
    Chapter 48 Skills Worksheet
    Chapter 48 Critical Thinking Worksheet
    Chapter 48 Challenge/Enrichment Worksheet

REVIEW AND ASSESSMENT

Chapter 48 Check, page 593
Chapter 48 Test, *Teaching Resources*
Chapter 48 Portfolio Assessment, *Teaching Resources*
Test Generator, Chapter 48
Alternative Assessment Forms

ADDITIONAL RESOURCES

Books for Teachers

Baxter, Craig. *Bangladesh: A New Nation in an Old Setting.* Boulder, CO: Westview Press, 1984.

Robinson, Francis. *The Cambridge Encyclopedia of India, Pakistan, Bangladesh, Sri Lanka, Nepal, Bhutan, and the Maldives.* New York: Cambridge University Press, 1989.

Taylor-Ide, Daniel. *Something Hidden Behind the Ranges: A Himalayan Quest.* San Francisco: Mercury House, 1995.

Books for Students

Ahluwalia, H. P. *Everest: Where the Snow Never Melts.* New Delhi: National Council of Educational Research and Training, 1987.

Foster, Leila Merrell. *Bhutan.* Chicago: Children's Press, 1989.

Nepal–In Pictures. Minneapolis, MN: Lerner Publications, 1989.

Pakistan–In Pictures. Minneapolis, MN: Lerner Publications, 1989.

The World Today *Videodisc Program*

A variety of still images, short videos, and activities are available for you to use as you teach this chapter. *Lecture Notes* include bar codes for specific images and videos.

Multimedia Resources

The Birth of Pakistan. (VHS, 20 minutes) Films for the Humanities & Sciences, P.O. Box 2053, Princeton, NJ 08543–2053.

Pakistan: Children of the Road. (VHS, 26 minutes) Films for the Humanities & Sciences, P.O. Box 2053, Princeton, NJ 08543–2053.

Picture Atlas of the World. (interactive maps, photo bank, animation, audio on CD-ROM) National Geographic Society, Washington, DC.

Return to Everest. (VHS/laserdisc, 60 minutes) National Geographic Video, National Geographic Society, Washington, DC 20036.

MEETING INDIVIDUAL NEEDS

Limited English Proficiency (LEP)

Write the title of Chapter 48 on the board. Explain the meaning of *perimeter*. Then help students understand the context of *The Indian Perimeter* by indicating on a wall map the locations of Pakistan, Nepal, Bhutan, Bangladesh, Sri Lanka, and the Maldives relative to India. Have students consider their own communities, counties, and states to identify the "perimeters" of each of these regions.

Learners Having Difficulty

Organize students into six groups, and assign each group one of the countries of the Indian perimeter. Tell the groups to complete a chart for their country with the following headings: Physical Geography, Economic Geography, Human Geography, challenging Issues. Call on volunteers to copy their group's chart on the board. Discuss similarities and differences among the countries.

Auditory Learners

Provide students with a list of statements that describe aspects of the physical or human geography of the Indian perimeter countries but that do not identify the country by name. As you read each statement aloud, call on volunteers to name the country described and then to indicate its location on a wall map.

Visual Learners

Instruct students to write new captions that reflect one or more of the Five Themes of Geography for the Chapter 48 photographs.

Tactile/Kinesthetic Learners

Organize students into small groups. Have students imagine that they work in the art department of a geographic magazine that is planning a special issue on a country of the Indian perimeter. Have each group design and create a cover that illustrates significant aspects of the country's geography.

Gifted Learners

Give a group of students the following scenario: They are members of a United Nations committee that has approval to allocate 100 million dollars in foreign aid to the six Indian perimeter nations. The committee must decide how to divide the money and justify the allocations. The group should present its recommendations to other students, who play the role of other representatives. Have them vote on the committee's recommendations.

MULTICULTURAL FOCUS

Two cultures that have coexisted on Sri Lanka for centuries, the Sinhalese and the Tamil, have maintained distinct languages, religions, and traditions. Sinhala, spoken by the Sinhalese majority, is in the Indo-Aryan group of Indo-European languages. Tamil is a Dravidian language, belonging to another language family.

The Sinhalese migrated from northern India more than 2,000 years ago. Buddhism and a sophisticated irrigation system were the foundations for a great Sinhalese civilization that flourished from the 100s to the 1200s A.D. The Tamils began arriving in Sri Lanka from southern India as early as the 300s B.C. The Sri Lanka Tamil are mainly Hindu, although many converted to Christianity during the colonial period.

Under British colonial rule, Tamil culture placed a strong emphasis on study in professional fields such as law and engineering. As a result, the minority Tamils made up a large share of the professional class. Many Sinhalese resented this, and after independence, Sinhalese politicians made Sinhala the only official language and set university and employment quotas. These acts sparked violent protests by Tamils, the first outbursts of the ongoing conflict.

Assign two groups of students to do additional research on each culture and its perspective on the current conflict. Have the groups work together to decide how to present (for example, role-playing or a panel discussion) these conflicting viewpoints to the class.

CHAPTER 48
THE INDIAN PERIMETER

Section 1: Pakistan
Section 2: The Himalayan Kingdoms
Section 3: Bangladesh
Section 4: Island Countries

OVERVIEW

In Chapter 48, students will explore the countries that surround India: Pakistan, Nepal, Bhutan, Bangladesh, Sri Lanka, and the Maldives.

Section 1 focuses on Pakistan, describing the physical, economic, political, and cultural characteristics of this Islamic republic.

Section 2 describes characteristics of the Himalayan kingdoms of Nepal and Bhutan.

Section 3 examines Bangladesh, highlighting its physical geography and the many challenges that face this nation.

Section 4 profiles the island nations of Sri Lanka and the Maldives, noting their physical and economic characteristics as well as concerns for the future.

About the Author

H.P.S. Ahluwalia grew up near the Himalayas. In 1965, he joined a team of Indian mountain climbers to scale Mount Everest. The quotation on this page, from *Everest: Where the Snow Never Melts*, describes his view from the summit.

Primary Source

Description of Change: excerpted
Rationale: excerpted to focus on main idea

CHAPTER 48

The Indian Perimeter

66 We took a long wheeling look from the highest point in the world. There was Makalu and Lhotse, Nuptse, and Kanchenjunga looming on the horizon and a host of other giants, now all far below us, in a maze of rock and snow, fluted ice-walls plunging to glaciers and glaciers plunging to the valleys below. We gazed north, towards the Tibetan plateau, and south, towards the plains of India. . . . The view was unforgettable. 99

Hari Pal Singh Ahluwalia

India borders Pakistan in the west and the Himalayan kingdoms of Nepal and Bhutan in the north. To the east is Bangladesh, and to the south are the island nations of Sri Lanka and the Maldives. These countries make up the Indian perimeter.

A majority religious group dominates each of the region's countries. Pakistan, Bangladesh, and the Maldives all have a Muslim majority. Sri Lanka and Bhutan are Buddhist, and Nepal is Hindu. Today, ethnic tension between majority and minority groups, along with economic development, challenge most of these countries.

Decorative buckle from **Pakistan**

GEOGRAPHY DICTIONARY

partition
cease-fire
storm surge

Mount Everest

Using the Quotation

Read the quotation on page 584 aloud to the class. Tell students that the author describes the view from the summit of Mount Everest, the highest mountain in the world. Refer them to the background photograph of Mount Everest. Tell students that Makalu, Lhotse, Nuptse, and Kanchenjunga are other peaks in the Himalayan range.

Using the Geography Dictionary

Ask students if they can guess the meaning of **partition.** Underline the word *part,* and explain that when something is partitioned it is divided into parts. Ask students where they may have heard the term **cease-fire** and in what context. *(Students are likely to be familiar with the term from television news reports or newspaper articles.)*

SECTION 1
OBJECTIVES

The objectives reflect the Focus questions on this page and are assessed in the review on page 587.

1. Describe each of Pakistan's land-form regions.

2. Examine the economic, political, and cultural characteristics of Pakistan.

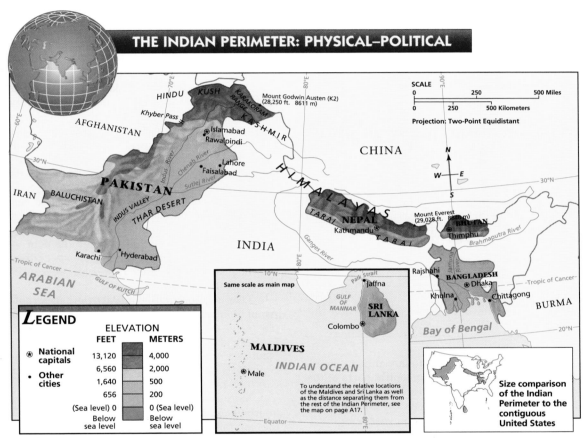

THE INDIAN PERIMETER: PHYSICAL–POLITICAL

Using the Physical–Political Map: The Indian Perimeter

Focus students' attention on the map on this page. Ask students to name all the mainland Indian perimeter countries. *(Pakistan, Nepal, Bhutan, Bangladesh)* Then ask students to locate the island nations that are part of the Indian perimeter. *(Sri Lanka, the Maldives)* Discuss the ways (geographically, politically, strategically, economically, culturally) in which India dominates this region.

COOPERATIVE LEARNING

Organize the class into groups of four or five. Assign each group a topic related to Pakistan (landform regions, irrigation, economy, cities, government, cultures) about which the group will prepare a presentation for the class. Groups should decide how best to present the topic—using maps, charts, and so on—and how to divide tasks within the group.

PAKISTAN

FOCUS

- *What are Pakistan's landform regions? What are the characteristics of each?*

- *What are the economic, political, and cultural characteristics of Pakistan?*

Pakistan is the second largest country in South Asia. It formed in 1947 when the British government **partitioned**, or divided, British India into Muslim and Hindu areas. Pakistan, with its Muslim majority, originally was divided into West Pakistan and East Pakistan. Then in 1971, East Pakistan became the independent country of Bangladesh. Pakistan today consists of the former West Pakistan.

Physical Geography Pakistan can be divided into three landform regions: the mountains

in the north, the deserts in the west and east, and the Indus River valley. (See the map above.)

The northern mountain region includes the glacier-covered Karakoram Range, which forms Pakistan's northern border with China. The mountains of this range are some of the highest and most rugged in the world. Mt. Godwin Austen is the world's second highest peak, soaring 28,250 feet (8,611 m). The mountain is also called K2 because it was the second peak measured in the Karakoram Range. This mountain region is the source of the Indus River system. It also holds most of the nation's remaining forests. Less than five percent of Pakistan is forested.

Deserts lie in the east and west of Pakistan. In the east, the Thar Desert separates Pakistan and India. In the west, a small strip of desert runs to the border with Iran. Rugged hills surround this arid land.

Pakistan's third landform region follows the Indus River, which flows from the Himalaya and Karakoram ranges south to the Arabian Sea. The

Motivator

Have students study the map on page 585, focusing on Pakistan. Ask: How might Pakistan be divided into regions, based on information given by the map? How might life in the northern part of the country differ from that in the south? How might the Indus Valley be different from the Thar Desert? Tell students that they will learn more about Pakistan in Section 1.

Analyzing Relationships

Have students complete a graphic organizer about Pakistan. Tell them to label the hubs Economy, Political Geography, and Culture/Population and to draw connecting lines between them. Direct them to write examples above each line specifying relationships between the hubs. *(For instance, between Political and Culture/Population—"Pakistan is an Islamic republic.")*

Closure

Organize students into pairs, and have them work together to answer the Focus questions on page 585.

ASSESSMENT

Assign the Section 1 Review questions on page 587.

Reteaching

Assign Section 1 of Reteaching Worksheet 48.

Chart Caption Answer

Maldives, Bhutan

In many traditional Punjab villages, the main entrance is through the *durwaza,* a long gateway that arches over the road. The *durwaza* is a center of social and economic activity. It is a popular meeting place as well as the main stop for traveling craftspeople and traders. Ask students to consider what, if any, place in their school or community serves as a kind of *durwaza. (Students may mention places such as a school entrance, a shopping mall, or a restaurant.)* Ask what purposes are served by having an agreed-upon public meeting place.

586

FACTS IN BRIEF	The Indian Perimeter		
COUNTRY POPULATION (1994)	LIFE EXPECTANCY (1994)	LITERACY RATE (1990)	PER CAPITA GDP (1993)
BANGLADESH 125,149,469	55, male 55, female	35%	$1,000
BHUTAN 716,380	51, male 50, female	15% (1989)	$700
MALDIVES 252,077	63, male 66, female	93% (1989)	$620 (1991)
NEPAL 21,041,527	52, male 53, female	26%	$1,000
PAKISTAN 128,855,965	57, male 58, female	35%	$1,900 (GNP)
SRI LANKA 18,129,850	69, male 75, female	88%	$3,000
UNITED STATES 260,713,585	73, male 79, female	97% (1991)	$24,700

Sources: *The World Factbook 1994,* Central Intelligence Agency; *The World Almanac and Book of Facts 1995*

Which two countries of the Indian perimeter have the lowest per capita GDP?

river's valley, a wide and flat alluvial plain, dominates the central and eastern central parts of the country. The valley was the birthplace of the Indus civilization and has been the site of irrigation farming for several thousand years.

Today, the Indus River valley is one of the world's largest irrigated regions. In fact, virtually all of Pakistan's agriculture and population are located along the river and its tributaries. Deforestation and overgrazing in the Indus River's upper drainage basin has led to severe flooding in the Indus Valley in recent years.

Economic Geography Pakistan is a poor country in which agriculture produces about half of the nation's exports. Cotton is the major export crop, while wheat and rice are the major food crops. Most of Pakistan's industries produce cotton textiles and food products.

Hydroelectric power from the Indus River is an important source of energy. Even so, the country must import oil. To reduce this dependence, Pakistan is developing its natural-gas deposits.

The Pakistan government is looking for other ways to improve the country's economy. For example, the government now promotes the privatization of inefficient state-owned industries. It also is trying to attract foreign investment to help increase export earnings. The nation's poor infrastructure, particularly its roads and communications systems, is improving as well.

Human Geography Pakistan is officially an Islamic republic based on Islamic law. In reality, though, power has often been in the hands of military governments since the country's independence. There have been a few brief periods under elected civilian governments, but military coups have usually put an end to these periods. Political corruption is a serious problem.

More than 95 percent of the country's people are Muslim. Punjabis form the largest ethnic group. The official language is Urdu, but English is spoken in the cities. The Baluchi of Baluchistan, a western desert region, speak their own language. Many other languages are spoken in the isolated northwest.

Like neighboring India, Pakistan faces rapid population growth. Its more than 120 million people may double in number by about 2025. Pakistan also has a large number of refugees. (See "Global Connections" on pages 588–589.) The country is struggling to provide schools, health care, and housing for this exploding population.

More than one-quarter of all Pakistanis live in urban areas. The country's major cities include the capital city of Islamabad (is-LAHM-uh-bahd) in the far north and Karachi, the nation's largest city and major seaport. Karachi lies near the Indus River delta on the Arabian Sea coast.

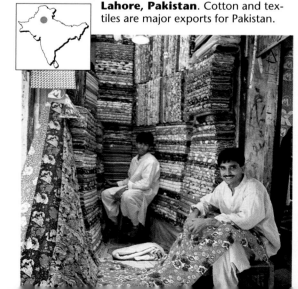

Merchants display fabrics for sale in their shop in **Lahore, Pakistan**. Cotton and textiles are major exports for Pakistan.

SECTION 2
OBJECTIVES

The objectives reflect the Focus questions on this page and are assessed in the review on page 590.

1. Identify physical, economic, cultural, and environmental characteristics of Nepal.

2. Compare and contrast Bhutan and Nepal.

TEACHING SECTION 2
Describing Nepal

Have students analyze Nepal's significant characteristics by completing an outline with four headings: Physical Geography, Economy, Culture, Environment. Instruct students to use the information on pages 587 and 590 to complete their outlines. Ask them to include at least two facts under each heading.

Guided Practice

Compare and contrast the characteristics of Nepal and Bhutan. First have students point out similarities. (*Both are landlocked nations with a mountainous terrain. Most people are subsistence farmers.*) Then have them point out differences. (*Nepal is primarily Hindu and has deforestation and environmental problems that are related to much tourism. Bhutan is primarily Buddhist; it has little tourism and fewer environmental problems.*)

Lahore, near the border with India, is Pakistan's cultural center and second largest city.

LOCATION **Kashmir** For nearly 50 years, Pakistan and India have fought over the mountainous Muslim region of Kashmir. Pakistan tried to take the region after the partition of British India in 1947, sparking a war with India. In 1949, the United Nations initiated a **cease-fire**, an agreement to halt active hostility, between Pakistan and India. Pakistan was given control over the northern portion of Kashmir, and India had control over the southern part. More fighting, however, broke out in 1965, 1971, and 1990. Kashmir remains a major source of conflict between the two countries.

SECTION REVIEW

1. What are Pakistan's three landform regions? Which region holds most of the country's people and agriculture?

2. How is Pakistan's government trying to improve the nation's economy? What types of governments has Pakistan had since independence? To what religion do most Pakistanis belong?

3. **Critical Thinking** **Agree or disagree** with the following: "Pakistan should be considered part of Southwest Asia rather than South Asia." Support your answer.

THE HIMALAYAN KINGDOMS

FOCUS

● *What are the physical, economic, cultural, and environmental characteristics of Nepal?*

● *How is Bhutan similar to Nepal? How is it different?*

REGION Sandwiched between India and China are the small landlocked kingdoms of Nepal and Bhutan. (See the map on page 585.) Both are extremely mountainous and depend on India for most of their trade and foreign aid.

Nepal Nepal rises steadily from south to north. In the south, near the border with India, lies a low, tropical plain called the Tarai. Just north of the Tarai is a region of hills. Farther north tower the ice-capped Himalayas. Mount Everest, the world's highest mountain, rises above Nepal's border with China.

Over 90 percent of Nepal's population is involved in agriculture, most of it subsistence. The Tarai is one of Nepal's most agriculturally productive regions. It supports typical tropical crops, such as jute, rice, and sugarcane.

In the hills north of the Tarai, farmers grow corn, millet, and wheat. This central region includes the Kathmandu (kat-man-DOO) Valley, where most Nepalese live. The capital city of Kathmandu lies here as well. At an elevation of 4,344 feet (1,324 m), the city's back door opens on the Himalayas. *Kathmandu* means "wooden temple" and refers to a Buddhist temple that was built in the area in the late sixteenth century. A temple in the city's central square is believed to be the original structure. Kathmandu is Nepal's only major city, and it is the center of the country's business and transportation.

For the most part, the Himalayas are not agriculturally productive. Some of the lower slopes can be used for grazing, and some herbs grown here are used in medicines in demand around the world. In addition, some wheat grows in the highlands. Farmers raise the grain along narrow terraces that conserve soil moisture and prevent erosion. Most of the higher mountain areas, however, are uninhabitable. Indeed, snow covers more than 15 percent of Nepal's total land area year-round.

Although many of the people in northern Nepal are Buddhists, most Nepalese live farther south and are Hindus. Nepali is the official language, but there are many local languages. The government is a constitutional monarchy in which the king has limited power. Democracy has been increasing, particularly since 1990 when the government legalized political parties.

With few mineral resources and little access to world trade routes, Nepal's economic development has been limited. Tourism is important to the country's economy. Visitors browsing in the streets of Kathmandu will find shops so small that shopkeepers can sit on the floors of their stores and reach all items for sale simply by moving their arms. Most tourists in Nepal,

Text continues on page 590.

Section 1 Review
ANSWERS

1. Northern mountains, western and eastern deserts, Indus River valley; Indus River valley

2. Privatization of industry, foreign investment, improvement of infrastructure; mostly military governments; Islam

3. Some students may agree because of Pakistan's Islamic government and peoples. Other students may say that Pakistan's colonial history is tied to India's, and thus it is part of South Asia.

GLOBAL PERSPECTIVES

Have students list benefits and drawbacks of international tourism to environmentally sensitive places in developing countries such as Nepal. (*benefits: economic gains, promoting cultural ties; drawbacks: ecological damage, emphasis on tourism at the expense of other types of development*)

Global Connections

Analyzing Immigration

Help students understand that the desire to move from one place to another to seek a better life has been constant throughout history. Ask students to name major reasons for immigration mentioned in this feature. *(war, economic hardship, religious or political persecution, environmental problems)* Then have students examine the pie graphs on the opposite page. Have them work with partners to research how specific events have contributed to the figures in the graphs. *(For example, students might link wars in recent years in El Salvador, Nicaragua, Vietnam, and Cambodia with the third graph.)*

Geographical Sidelight

The United Nations in 1948 issued the Universal Declaration of Human Rights, which states that "everyone has the right to seek and to enjoy in other countries asylum from persecution." Interpretation of what constitutes "persecution" however, has led to inconsistent responses by countries in providing asylum to individuals or groups.

GLOBAL CONNECTIONS

The Movement of People

In the 1500s, Ferdinand Magellan's ship took three years to sail around the world. Today, airplanes fly across entire continents in a matter of hours, and communication technology brings distant lands into contact. As news circles the earth, the world's people are moving from place to place in greater numbers than ever before.

Some agencies estimate that at least 100 million people live outside the countries in which they were born.

Immigration

Many people move to find employment or political freedom. Others seek a community of people who share their particular religion or culture. Fathers often send back money to their families to ease the economic hardship in their home countries, but women and children are moving in larger numbers as well.

In the 1960s, for the first time in its history, the United States received more immigrants from Middle and South America than from Europe. Most arrived in search of jobs, though some came to escape civil wars and political persecution. These immigrants brought distinct cultures with them and helped make cities such as Los Angeles and Miami rich in diversity.

Many western European nations also have received large numbers of immigrants in recent decades. Arabs have moved to France, Turks to Germany, and Pakistanis to Britain. These newcomers have introduced new food, languages, religions, and customs to western Europe. In the opposite direction, more than 1 million Jews have emigrated to Israel in recent years from the former Soviet Union. They seek not jobs but the opportunity to live without discrimination among other Jews.

The largest migrations, however, are taking place within and between developing countries. In 1993, the United Nations Population Fund reported that each year 20 to 30 million people move from rural to urban areas in developing countries. By the year 2000, 18 of the world's largest urban areas will be in non-Western countries such as India. Mexico City, Mexico, and São Paulo, Brazil, will lead the list. London and Paris will not make the top 20.

Refugees

Many people migrate because they are in danger of losing their lives. They are refugees, fleeing war-torn nations and governments that persecute them because of their political beliefs, religion, or ethnic background. In 1995, the world's refugees numbered as many as 23 million.

Recent refugees include the 5 million who fled from the 1990–1991 Persian Gulf War. Kuwaitis, along with foreign-born workers from Yemen, Egypt, Pakistan, Bangladesh, Sri Lanka, and the Philippines, escaped to Jordan. Then at the war's end, Palestinian workers in Kuwait fled or were deported. About 2 million Rwandans

Hundreds of thousands of refugees from Rwanda fled in 1994 to camps such as this one in **Zaire** to escape ethnic violence in their country. War and ethnic violence have been only two of the reasons for the mass movement of people throughout history.

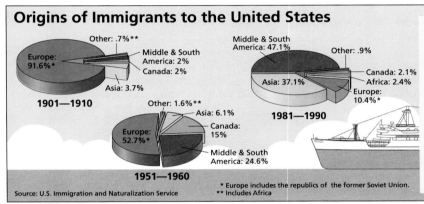

Origins of Immigrants to the United States

1901—1910
Europe: 91.6%*
Other: .7%**
Middle & South America: 2%
Canada: 2%
Asia: 3.7%

1951—1960
Europe: 52.7%*
Other: 1.6%**
Asia: 6.1%
Canada: 15%
Middle & South America: 24.6%

1981—1990
Middle & South America: 47.1%
Other: .9%
Canada: 2.1%
Africa: 2.4%
Europe: 10.4%*
Asia: 37.1%

Source: U.S. Immigration and Naturalization Service
* Europe includes the republics of the former Soviet Union.
** Includes Africa

These pie graphs break down by percentage the origins of immigrants to the United States during three periods of this century. Today, immigrants to the United States are more likely to be from Asia or Middle or South America than from anywhere else. In the decade after the turn of the century, 92 percent of immigrants were from Europe. Most Americans are immigrants or are decendants of immigrants.

fled ethnic fighting in their country in 1994. Many refugees, fearing for their lives, refuse to return home.

Some regions of the world are filled with refugees who have left their homes to escape a devastated environment. This is particularly true in parts of Africa, where severe drought and overgrazing have forced many to migrate in search of fertile soil. Often, such migration leads to a cycle of environmental damage as people crowd onto neighboring lands.

Immigration Policies

Because migration can influence a nation's economy, culture, and physical environment, migrants are not always made to feel welcome in their new country. People worry that newcomers will take jobs or introduce customs that threaten accepted ways of life. Such attitudes often produce social conflict.

For example, in the early 1990s, more than 1 million Germans from Eastern Europe and the former Soviet Union returned to Germany. At the same time, another 1 million non-Germans were seeking to immigrate to the country. Germans, however, were struggling with their own economic problems brought about by the unification of East Germany and West Germany. Some Germans resented immigrants, who were provided with food and housing by the government. Anti-foreigner groups attacked not only recent immigrants but also many Turks who had lived in Germany for decades.

Most nations have policies dealing with the admission of immigrants and refugees. The U.S. Congress sets the annual limit for immigrants and refugees allowed into the United States.

Americans remain divided on immigration. Some believe that the United States should have an open-door policy, extending a welcome to all who knock. Others, however, believe that the country should close its doors to all but a few immigrants.

Regardless of nations' policies, people continue to move, seeking better lives. As a result of this migration, cultures mix, languages and religions spread, areas gain and lose population, and the natural environment is sometimes altered.

The large number of Jewish immigrants from the former Soviet Union has forced the government of **Israel** to speed up the pace of home building.

▼ YOUR TURN

Imagine you are an immigrant who has just arrived in a new country. **Write** a few journal entries describing your experiences in your new home. **Explain** why you have moved and what you are hoping to find.

589

Closure

Have students use question-and-answer cards to quiz each other about Nepal and Bhutan.

ASSESSMENT

Assign the Section 2 Review questions on this page.

Reteaching

Assign Section 2 of Reteaching Worksheet 48.

SECTION 3
OBJECTIVES

The objectives reflect the Focus questions on this page and are assessed in the review on page 591.

1. Analyze Bangladesh's physical geography.

2. Identify challenges facing Bangladesh.

TEACHING SECTION 3
Analyzing Physical Geography

Ask students to write a short essay explaining how the physical geography of Bangladesh affects life there. Tell them to be sure to discuss natural disasters common to Bangladesh, including tropical cyclones and their accompanying storm surges.

Section 2 Review
ANSWERS

1. Agriculture in all three, tourism in the hills and mountains; many people in the northern mountains are Buddhists, but most live farther south and are Hindus.

2. Bhutan has a smaller population and has made efforts to preserve its forests. Buddhists, rather than Hindus, form a majority in Bhutan.

3. No. Nepal's has occurred because of tourists and a need for fuel. In Southeast Asia, it has occurred mainly because tropical hardwoods are valuable exports.

THEME
Place

Thimphu, the capital of Bhutan since 1962, developed around an ancient fortified Buddhist monastery, now the meeting hall of the National Assembly. Before 1962, the seat of government was wherever the king resided.

however, come not to shop but to hike in the mountains. The Sherpa, who live near Mount Everest, are skilled mountain climbers, and sometimes serve as guides.

Tourism has brought much-needed income to Nepal, but it has also damaged the country's environment. A constant stream of hikers tramps over the hillsides. These visitors' need for firewood has contributed to deforestation and soil erosion already begun by Nepal's growing population with its own fuel needs. Serious flood and landslide hazards now exist. To combat the problem, the government has restricted access to some of the more popular hiking routes.

Bhutan The tiny Buddhist kingdom of Bhutan lies east of Nepal. Until 1974, the Bhutan government promoted a policy of almost total isolation. Even today, the country remains somewhat cut off from the rest of the world. Bhutan has worked to preserve its forests, particularly on the steepest slopes. These policies and Bhutan's smaller population have led to fewer of the deforestation problems that plague Nepal.

Bhutan's culture is closely related to that of nearby Tibet. About 70 percent of the people are Buddhists. The rest are mainly Hindus from Nepal. Most of the people are subsistence farmers who grow rice and wheat in the few fertile valleys. Fruit and cardamom are cash crops, and the country sells lumber and hydroelectricity to India.

Even though Bhutan may appear poor by the standards of developed countries, most of its people are self-sufficient in food. The king believes that people's ability to provide food for themselves is more important than the development of an export economy.

This golden statue of Buddha is draped in colorful clothing and jewelry for a festival in **Kathmandu, Nepal**. Many Hindu and Buddhist temples are found throughout Nepal, although the vast majority of the people are Hindus.

2 SECTION REVIEW

1. What economic activities take place in each of Nepal's landform regions? How do the country's cultural characteristics change from north to south?

2. Why has Bhutan had fewer problems with deforestation than Nepal? How is Bhutan's cultural composition different from that of Nepal?

3. Critical Thinking **Decide** whether deforestation in Nepal has occurred for the same reasons it has occurred in Southeast Asia. Explain your answer.

3 BANGLADESH

FOCUS

● *What is the physical geography of Bangladesh?*

● *What challenges does the country face?*

LOCATION **Physical Geography** Bangladesh is surrounded by India on the north, west, and east. Its only other neighbor is Burma, with whom it shares a short border at its southeastern tip. To the south, Bangladesh has a low, marshy coastline on the Bay of Bengal. (See the map on page 585.)

The low flood plain and delta of the Ganges and Brahmaputra (called the Jamuna in Bangladesh) rivers make up most of the country. A complex network of more than 5,000 miles (8,045 km) of waterways make river transportation the most common form of travel. Except for a few hilly areas along the eastern border, most of Bangladesh has suffered deforestation.

The entire country has a humid-tropical climate and depends on the monsoon's seasonal rainfall. Temperatures average about 80°F (27°C) year-round, and rice and vegetables thrive throughout the year in the alluvial soils.

Natural disasters are common in Bangladesh. During the wet monsoon, between June and September, heavy monsoon rains flood the lower river areas and delta islands. Often, more than half the nation is flooded for several weeks each year.

Guided Practice

Point out to students that since independence Bangladesh has depended on foreign aid and imported food. Lead a class discussion about the connection between Bangladesh's huge, rapidly growing population and the challenges it faces in the future. Use cause-and-effect diagrams to show that problems such as poverty, unemployment, and pollution are made more difficult by overpopulation.

Closure

Have students create outlines of Section 3 organized by physical geography and human geography.

ASSESSMENT

Assign the Section 3 Review questions on this page.

Reteaching

Assign Section 3 of Reteaching Worksheet 48.

SECTION 4
OBJECTIVES

The objectives reflect the Focus questions on this page and are assessed in the review on page 592.

1. Identify physical and economic characteristics of Sri Lanka and the Maldives.
2. Discuss challenges facing Sri Lanka and the Maldives.

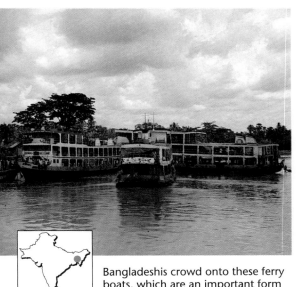

Bangladeshis crowd onto these ferry boats, which are an important form of transportation on the many rivers that cross the low-lying country.

The greatest danger, however, is from the high **storm surges** that accompany tropical cyclones in the Bay of Bengal. A storm surge is a rise in tide level that occurs as a storm approaches the coast, washing huge waves of seawater over the land. In 1991, more than 130,000 people in Bangladesh drowned when a 20-foot-high (6-m) storm surge from a tropical cyclone hit the low-lying coast. Outbreaks of deadly diseases, such as cholera, often follow such disasters. Carried by contaminated water supplies, the diseases are hard to control.

Human Geography The Bangladeshis are closely related to the West Bengalis of neighboring India. They speak the same language but differ in religion. Most Bangladeshis are Muslims, while many West Bengalis are Hindus.

The country's capital and largest city is Dhaka (DAK-uh). Most of the city's several million people are poor. Many live in the streets or in surrounding shantytowns without electricity, waste disposal, or a safe water supply. Most industry is located in polluted factory towns near Dhaka.

Rice and jute are the country's most important crops. Jute is the major export, though its value is declining because of competition from synthetic fibers. Bangladesh also has a textile industry that is growing in importance.

The major challenge facing Bangladesh is how to feed, educate, and employ its more than 125 million people. Over 80,000 children are born each week in Bangladesh, and the country has the eighth largest population in the world. There is little room for all these people. The country already has a population density of about 2,200 people per square mile (868 per sq. km). In contrast, the U.S. state of Iowa is about the same size as Bangladesh and has a population density of only about 50 people per square mile (19 per sq. km). For Iowa to have the same population density as Bangladesh, almost half of the entire U.S. population would have to crowd into the state. At its present rate of growth, Bangladesh's population will double by 2025.

SECTION REVIEW

1. What countries border Bangladesh? What climate does Bangladesh have?
2. What economic challenges must Bangladesh overcome?
3. **Critical Thinking** **Explain** how storm surges and other natural disasters can be deadly in more than one way. How might Bangladesh reduce the effects of these natural disasters?

ISLAND COUNTRIES

FOCUS

- *What are the physical and economic characteristics of Sri Lanka? of the Maldives?*
- *What challenges does each country face?*

LOCATION **Sri Lanka** Sri Lanka, formerly Ceylon, is a pear-shaped tropical island located just off the southern tip of India. It sits on the continental shelf of the Indian subcontinent. (See the map on page 585.) Sri Lanka has three major landform regions: the central mountains, the coastal plain, and the northern limestone plain.

The low mountains of the island's interior reach an elevation of 8,280 feet (2,524 m) and are

Geographical Sidelight

Bangladesh has been hit by about five dozen tropical cyclones since 1900. Little of the country is more than 30 feet (9 m) above sea level. Flooding in 1987, when other parts of South Asia were experiencing severe drought, left 2 million people homeless.

Section 3 Review
ANSWERS

1. India and Burma; humid tropical
2. Poverty, declining value of jute, growing population
3. Disasters cause deaths directly, by drowning, and indirectly, by the outbreaks of disease that follow the disasters. Answers to the second part will vary but might include making water supplies less subject to contamination.

Identifying the Island Countries

Ask students to write three sentences. Each one should describe a physical or economic characteristic of either Sri Lanka or the Maldives without using the country's name. As students read their sentences, have the class identify the country.

Guided Practice

Write the heading Island Nation Challenges on the board. Have volunteers list current and future challenges for Sri Lanka and the Maldives. After each challenge is listed, discuss how it might be met.

Closure

Ask students to draw connections between the geography of the island nations and the challenges they face.

ASSESSMENT

Assign the Section 4 Review questions on this page.

Reteaching

Assign Section 4 of Reteaching Worksheet 48.

Geographical Sidelight

The people of the Maldives call themselves Divehi, *which means "islander," and their country* Divehi Rājje, *meaning "kingdom."*

Section 4 Review

ANSWERS

1. Tea growing; fish products

2. The civil war between the Tamils and the government/ Sinhalese; tropical cyclones

3. Global warming may cause a rise in sea levels, which would devastate the Maldives since they are so close to sea level.

592

These women are picking tea leaves on a plantation in southern **Sri Lanka**. To learn how tea is harvested and processed for consumers, see the illustration on pages 512–513.

an important tea-growing region. In fact, Sri Lanka is the world's largest exporter of tea, and tea is the nation's leading source of income. The mountain area also supplies the nation with forest products and hydroelectricity. In addition, mines tap into deposits of precious gems and graphite. Graphite is a mineral used in a wide variety of products, including pencils.

The coastal plain along the south and west is Sri Lanka's most densely populated area. Coconuts and rubber are the major crops here. Most of the nation's rice and other food crops also are produced in this area. Colombo, the capital and largest city, lies on the western coastal plain.

Sri Lanka's third landform region is the northern limestone plain, which is wide but drier and less arable than the coastal plain. Irrigation projects have helped turn some of this land into a food-growing area.

The region's monsoons and Sri Lanka's location just north of the equator strongly influence the island's climate. The central mountains are the coolest and wettest part of the island, while the southwest portion of the island has a wet humid-tropical climate. The drier northeast part has a tropical-savanna climate.

The population of Sri Lanka is divided into two major culture groups. About three-quarters of the people are Sinhalese, who are Buddhist and

of Aryan background. They arrived in Sri Lanka from northern India more than 2,000 years ago. About one-fifth of the population is Tamil, who are Hindu. The Tamils have migrated from southern India over many centuries and live mostly in the northeast portion of the island.

During the British colonial era, the two groups lived together peacefully. Since independence in 1948, however, the Sinhalese have discriminated against the Tamils. Some Tamils initiated terrorist activities against the government in the 1980s. The democratic government granted some autonomy to the Tamils, but the Tamils demanded separation. Over the past decade, the violence has intensified. Since 1983 more than 35,000 people have died as a result of civil war. The war has drained the country economically and has prevented Sri Lanka from developing its industry and tourism.

LOCATION **The Maldives** The Maldives are a group of tropical islands in the Indian Ocean. They lie just north of the equator, about 400 miles (644 km) southwest of Sri Lanka. (See the map on page 585.) The nation's 1,200 tiny coral islands sit on top of an ancient submerged volcanic plateau. About 200 of the islands are inhabited, mainly by Muslim Sinhalese. The highest point on the islands is less than 10 feet (3 m) above sea level. Tropical cyclones threaten all of the islands.

Fish products account for almost half of the country's income. Coconuts and yams are its main food crops. Most other food must be imported. Tourism and a small clothing industry are a growing source of income.

4 SECTION REVIEW

1. What economic activity provides Sri Lanka's major source of income? What products provide much of the Maldives' income?

2. What is the major challenge facing Sri Lanka? What type of storm threatens the Maldives?

3. **Critical Thinking** **Predict** how the Maldives might be affected by global warming.

ANSWERS

Building a Vocabulary

1. To divide; India and Pakistan
2. A cease-fire is an agreement to halt active hostility. The 1949 cease-fire worked temporarily, but fighting eventually broke out again.
3. Storm surge

Recalling and Reviewing

1. Bangladesh and the Maldives
2. The Indus; the Ganges and the Brahmaputra
3. Nepal and Bhutan; Nepal
4. Agricultural products: tea, coconuts, rubber, rice; natural resources: forest products, hydroelectricity, gems, graphite

---

CHAPTER **48** CHECK

Muslims worship at this mosque in **Lahore, Pakistan**.

Reviewing the Main Ideas

1. Pakistan, officially an Islamic republic, is South Asia's second largest country. Agriculture produces about half of the nation's exports. The region of Kashmir remains a source of controversy between Pakistan and India.
2. Nepal and Bhutan are small landlocked kingdoms located between India and China. Over 90 percent of Nepal's population is involved in agriculture, but tourism is the country's largest industry. Bhutan is somewhat isolated from the rest of the world.
3. Bangladesh is one of the world's poorest and most densely populated countries. Most of Bangladesh has suffered deforestation, and natural disasters are common. The major challenge facing Bangladesh is how to feed, educate, and employ its many people.
4. Sri Lanka is a tropical island located off the southern tip of India. Its population is divided into two major culture groups, the Sinhalese and the Tamils. The Maldives are a group of tropical islands in the Indian Ocean.

Building a Vocabulary

1. Define *partition*. Into what two countries did the British partition British India in 1947?
2. What is a cease-fire? Was the 1949 cease-fire between Pakistan and India effective?
3. What term describes a rise in tide level that occurs as a storm approaches the coast?

Recalling and Reviewing

1. Which countries discussed in this chapter do not have a mountain region?
2. What is Pakistan's most important river? What two rivers are important to Bangladesh?
3. Which two countries of South Asia are monarchies? Which of these has had more of a problem with deforestation?
4. What are Sri Lanka's major agricultural products? natural resources?

Thinking Critically

1. Why do you think Pakistan was originally divided into West Pakistan and East Pakistan, when the two parts were separated by 1,000 miles (1,609 km) of Indian territory? Why do you think East Pakistan eventually separated from West Pakistan to become Bangladesh?
2. Three of the countries discussed in this chapter have borders with China. Do you think China takes a great strategic or economic interest in these countries? Why or why not?

Learning About Your Local Geography

Cooperative Project

With your group, research the population densities of various communities within your town or city, of towns and cities within your county, or of counties within your state. Make a bar graph showing the different population densities. Below the graph, suggest a few possible reasons for why the different areas have different population densities.

Using the Five Themes of Geography

Construct a table with six columns, one for each country discussed in this chapter. In each column, list some of that country's major **place** characteristics. When you have finished filling in the table, decide whether place characteristics alone support the grouping of the six countries together as one **region**. Next to your table, write a few sentences explaining your decision.

Thinking Critically

1. Both had majority Muslim populations. Except for religion, East Pakistan had little in common with West Pakistan, and the separation may have made government logistically difficult.
2. Possible answer: probably not, as high mountains separate the South Asian countries from China, and the countries have little to offer China economically

Learning About Your Local Geography

Graphs should show population densities of various places. Reasons for differences might include physical or economic characteristics.

Using the Five Themes of Geography

Tables should list place characteristics of each country discussed in the chapter. Sentences should say why the countries should or should not be grouped as a region.

593

Unit 10 Review

SKILLS

Point out to students that people in different regions not only adapt to their environments differently but may also change their environments in different ways. For example, suppose both the United States and India are constructing factories. Whereas the United States may be able to afford expensive pollution controls on its new factory, India may not be. As such, India's environment may be more likely to become polluted. Have students suggest other examples of how region influences the ways people change their environments.

Linking Physical and Human Geography

In the summer of 1993, parts of the United States and South Asia were both devastated by rains and floods. Two months of almost continuous rain poured down on the Midwestern United States. Rivers swelled, putting close to 20,000 square miles (51,800 sq. km) under water. Similarly, four weeks of floods flowed through India, Nepal, and Bangladesh, as a result of the heaviest monsoon rains in decades. In Bangladesh alone, waters covered more than 18,500 square miles (47,915 sq. km), one-third of the country's total area.

The consequences of these floods, however, were far from similar for the two regions. While the U.S. floods claimed only about 45 lives, some 4,200 people died in South Asia. In the United States, the floods forced some 74,000 people to leave their homes. In contrast, the South Asian floods left more than 7 million people homeless. Yet while the floods had much higher human costs in South Asia, economic costs were higher in the United States. Estimates of damage in South Asia were in the millions of dollars, but cost estimates for the U.S. floods were more than 12 billion dollars.

Why did the South Asian floods cause so many more deaths, and leave so many more people homeless, than the U.S. floods did? Why did the U.S. floods cost so much more money? Several factors account for these differences. In the United States, an extensive system of dams, levees, and reservoirs helped control the waters and saved lives. No such system exists in South Asia. Dams and levees are expensive to build, and South Asia does not have the funds available for such projects. Furthermore, the building of major dams to control South Asia's rivers would require

the consent of several countries, each with its own interests to protect.

Communication and transportation systems also played a role. South Asian residents did not receive the advance warnings that were available to their U.S. counterparts, nor did they have the benefit of evacuation plans. And when the floods did hit, local building materials became a factor. While in the United States most homes are built of materials such as brick or concrete, many structures in South Asia are made from mud and other substances that are easily washed away.

Another factor was population density. Much of the flooding in the United States took place on thinly populated farmland, so fewer people were affected. In South Asia, on the other hand, the river plains are some of the region's most densely settled areas. Differences in population density and land usage also explains, in part, the differences in economic costs. The U.S. farmlands drowned by the floods were mostly commercial farms worth hundreds of thousands of dollars. In contrast, much of the land flooded in South Asia was made up of small subsistence farms, whose primary worth was only to the families that lived on and worked them.

As you can see from this example, the effects of natural disasters are not the same the world over. Levels of economic development, political concerns, and a variety of other human factors all play a role in determining how people will respond to environmental challenges. In short, people have different ways of adapting to and changing their environments. Human geography and physical geography are firmly linked.

Practicing the Skill
ANSWERS

Summaries and presentations should discuss the effects of a type of natural disaster on two or more regions that have both experienced that type of disaster over the past decade.

PRACTICING THE SKILL

Research the effects of a different type of natural disaster—hurricane, drought, earthquake— on two or more regions that have experienced that type of disaster over the past decade. How did the event affect each region? What factors were responsible for any differences? Did ways in which people had changed their environments play a role in how the disaster affected them? Did either region's response to the disaster involve further change to the environment? Write a summary of your findings and present it to the class.

594

BUILDING YOUR PORTFOLIO

Decide whether students will work individually or in groups on the unit projects described below. Students should then choose which project they will work on. Provide students with copies of the Portfolio Assessment Worksheets in *Teaching Resources*. Chapters 46–48 worksheets and the Unit Portfolio Assessment Worksheet will guide students step-by-step to complete either project.

Teaching with Sketch Maps

Have students sketch a map of South Asia from memory. Tell them to include as many details as they can about place names, landforms, and boundaries of regions they studied in Unit 10. Students should compare this map with the sketch map they completed at the beginning of Unit 10 to note their improvement in identifying geographic features.

BUILDING YOUR PORTFOLIO

Individually or in a group, complete one of the following projects to show your understanding of the geography concepts involved.

A Cities of the Subcontinent

Select any city of South Asia. Cities you might select include (but are not limited to) New Delhi, Bombay, Calcutta, and Madras, in India; Islamabad, Karachi, and Lahore, in Pakistan; Kathmandu, in Nepal; Dhaka, in Bangladesh; and Colombo, in Sri Lanka. Imagine that you are a government official in the city you have chosen. Your job is to write a brochure profiling the city's history, population, economy, and anything else you think is significant about the city. To prepare your brochure, you will need to consider the following questions.

1. What are the city's origins? How did location and physical features play a role in the city's origins? How has the city grown over time, and why has it grown? Is it still growing today? For your brochure, draw a map of the city itself or of its location in the region. Include a caption explaining the city's origins and pattern of growth.
2. How large is the city's population? What is the economic and cultural make-up of its citizens? Do different groups live in different areas of the city? A chart might help your brochure present information on the city's population.
3. What major economic activities take place in the city? What forms of entertainment are available? What are the city's main sights? You might use illustrations to show this information in your brochure.

Organize your materials into the brochure. Present the brochure to other government officials (the rest of your class). Conclude with a few statements discussing challenges facing the city and predicting the city's future over the next 20 years.

B A Year in Rural India

Imagine that you are a farmer in rural India. You live on the outskirts of a village on a small piece of land that your family has worked for decades. The village has changed little in your lifetime.

One year, however, a large factory is built in your village by a company based in Bombay. The factory employs a large number of workers, all of whom have moved to your community. How does the arrival of this factory and its workers affect your life? Write a series of monthly journal entries discussing the ways in which your life changes. Your journal entries might address the following questions.

1. How does the introduction of the factory affect your village and its physical environment? What resources does the factory need? How does the factory receive supplies and send its products to market? Where do the factory workers live?
2. How does the introduction of the factory and the arrival of the factory workers affect your life economically? Particularly consider the relationship between the newly increased population and the amounts of food, other goods, and money available.
3. How does the introduction of the factory and its workers change your belief systems, if at all? Particularly consider beliefs regarding tradition, caste, and so on. Also consider matters of literacy and the status of women.

Share your journal entries with other villagers (the rest of your class). Include a discussion of how you have responded to the changes in your life caused by the introduction of the factory.

Project A:

Brochures should display information on cities' origins, histories, populations, and economies. Presentations should conclude with statements discussing challenges facing the city and predicting the city's future over the next 20 years.

Project B:

Journal entries might mention such items as increased pollution and noise as the factory and access roads are built; the introduction of a whole new group of people making demands on the community's resources and possibly having more money to pay for these resources than do the original villagers; the introduction of a whole new set of beliefs brought in by people affiliated with the new factory; and changes in the villagers' own beliefs necessitated by all of the other changes.

OVERVIEW

Chapter 49 • AUSTRALIA describes the country's landforms, climates, water and mineral resources, unique plant and animal life, agriculture, industry, and population distribution.

Chapter 50 • THE PACIFIC ISLANDS AND ANTARCTICA introduces the distinctive physical characteristics of the islands and island groups of the Pacific region, New Zealand, and Antarctica.

OBJECTIVES

1. Describe the landforms and climates in the Pacific World and Antarctica.

2. Identify the indigenous culture groups and their ways of life in Australia, New Zealand, and the South Pacific islands.

3. Explain how Australia has benefited from its natural resources.

4. Describe how New Zealand's economy is dependent on agriculture and foreign trade.

5. Identify the environmental challenges that confront the Pacific World and Antarctica.

6. Interpret special-purpose maps to analyze the relationships among climate, population patterns, and economic resources in the Pacific World.

7. Develop awareness of how events, immigration, politics, and shared goals create global connections among peoples and nations.

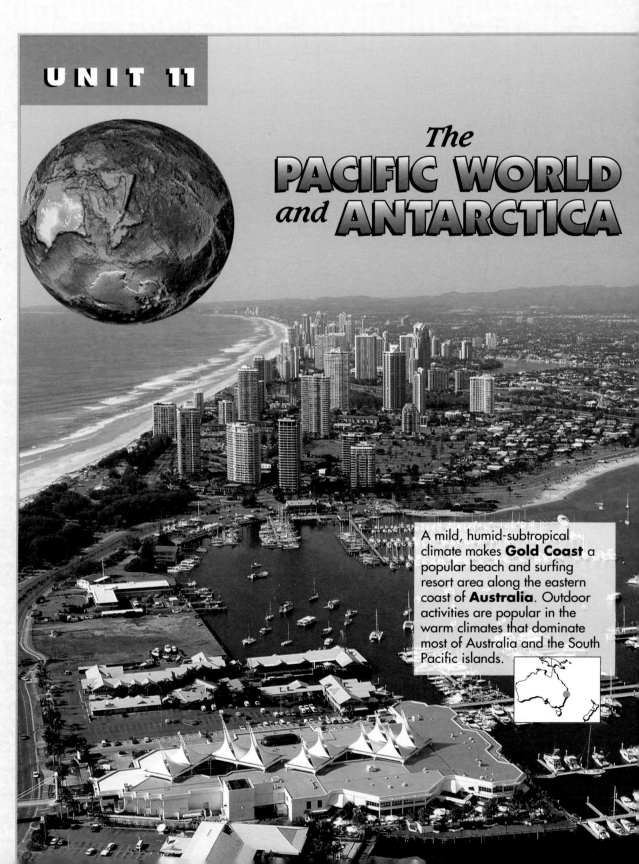

UNIT 11

The
PACIFIC WORLD
and ANTARCTICA

A mild, humid-subtropical climate makes **Gold Coast** a popular beach and surfing resort area along the eastern coast of **Australia**. Outdoor activities are popular in the warm climates that dominate most of Australia and the South Pacific islands.

TEACHING UNIT 11

Using the Illustrations

Tell students that the photograph on the opposite page illustrates an important population pattern of Australia. Refer them to the inset map near the photo caption that shows the location of the Gold Coast. Have them try to predict how population is distributed in Australia. *(Most of the population resides in urban areas on the coast.)* Then ask how the physical features of Pacific islands such as Bora-Bora (shown below) might influence their economies. *(Their scenic beauty attracts tourists.)*

Teaching with Sketch Maps

Have students draw a map of Australia and the islands in the South Pacific from memory. Tell them to label any bodies of water, physical features, island names, and cities they know. Students might add this map to their portfolios and compare it with another they will draw after they have completed Unit 11.

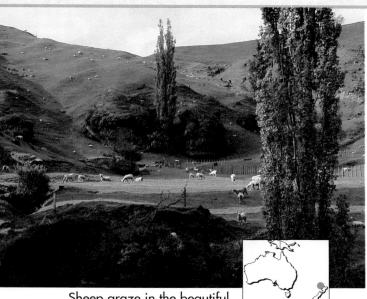

Sheep graze in the beautiful countryside south of **Auckland, New Zealand**. New Zealand is one of the world's major wool producers.

UNIT 11 CHAPTERS

49 Australia

50 The Pacific Islands and Antarctica

Clam-shell pendants from the **Solomon Islands**

Aboriginal bark-painting of the Big Dipper from **Australia**

THE WORLD TODAY VIDEODISC PROGRAM

The videodisc profiles Australia.

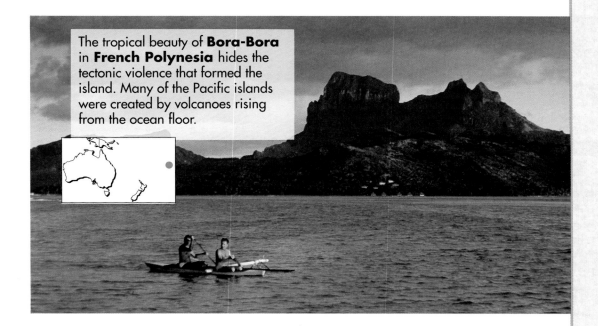

The tropical beauty of **Bora-Bora** in **French Polynesia** hides the tectonic violence that formed the island. Many of the Pacific islands were created by volcanoes rising from the ocean floor.

One World, Many Voices

LITERATURE

Point out that the words of this Maori poem are shown below in both the Maori language and English. Tell students that oral traditions are very important to Maori culture and that action songs, or waiata, are usually performed. Focus students' attention on the intricate Maori wood carving on this page and refer them to the author's reference to the "treasured arts of our ancestors."

More about Wiremu Kingi Kerekere

Wiremu Kingi Kerekere received an OBE, Order of the British Empire, from Queen Elizabeth II of England for his original compositions. He has also served as a consultant about Maori culture for Radio New Zealand.

ONE WORLD, MANY VOICES
LITERATURE

Maori stick dance, **New Zealand**

Wiremu Kingi Kerekere (1923–), of Maori heritage, is a poet and singer. Maori peoples are descendants of the original inhabitants of New Zealand. One Maori cultural tradition is the waiata, or action song, which is performed in a dance form. The poem below is a waiata. When Kerekere was only 13 years of age, he performed some of his original compositions on radio broadcasts. In recent years, Kerekere's work has been performed on television and by many theater, dance, and opera groups in New Zealand.

Primary Source

Description of Change: excerpted
Rationale: excerpted to focus on main idea

Interpreting Literature

Answers

1. Their cultural traditions expressed through art, customs, and stories
2. Honoring and understanding ancestral traditions and beliefs will help people in the modern world live together in harmony.

Taku Waiata

Horahia mai ki au nga taonga tupuna,
Nga kowhaiwhai, nga whakairo, nga
    pakiwaitara, nga waiata Maori tuturu
    o te ao tawhito,
Tuituia mai ki nga tikanga o toku ao o
    tenei ra,
Te ao tawhito, te ao hou, Maori, Pakeha,
Homai nga tohu tika, i runga i te aroha.
Tohatohaina tenei kupu ki te ao katoa
Puta ake i te po, ka ao, ka ao, ka marama.

Maori wood carving

Give to us the treasured arts of our
    ancestors,
The carvings, the designs and patterns,
    the stories and the traditional chants.
Let us of this generation, try to combine
    them with what talents we have in this
    changing world of today,
Give to us your greatest gift, 'aroha', love
    in its many connotations,
It will bring peace, goodwill, friendship
    among peoples,
And will take us from the world of darkness
    and ignorance, into the
    light of knowledge and understanding.

Interpreting Literature

1. What does the author admire about his Maori ancestors?
2. What does this poem say about how the past and the present are connected?

FOR THE RECORD

You might introduce the primary source by asking students if they have ever read about, witnessed in person, or seen on television a large brush fire or grass fire. *(Some students might recall the disastrous fires that destroyed tens of thousands of acres in Southern California in 1993.)* Discuss the effects of wind on such fires to explain why brush fires can be difficult to bring under control. After students have read this selection, use a wall map to point out the distance between Victoria and Tasmania, and the location of the Bass Strait.

More about J.H. Kerr

J.H. Kerr wrote about his impressions of Australia and his experiences as an immigrant. *Glimpses of Life in Victoria by a Resident* was published in Scotland in 1872.

FOR THE RECORD

J. H. Kerr was one of the many European settlers who came to Australia in the mid-1800s. Settlers such as Kerr from the British Isles confronted a very different climate from what they knew in their home countries. Australia's dry climate and heat sparked intense bushfires that could rage across large areas within a very short time. In the passage below, Kerr describes a particularly destructive fire that engulfed Victoria in southern Australia on February 6, 1851, later known as Black Thursday.

"I drew aside a curtain to peep at the weather, and found the glass of the window panes so heated that I was glad quickly to withdraw my hand. . . . The wind was still rising as the day wore on, moaning and whistling through the streets and hollows in fiery suffocating blasts, that seemed to issue from a furnace heated seven times. Clouds of dust and smoke obscured the sun, while the air was dense with dead leaves and other charred matter, whirled on high by the wind. . . . It was evident even from the town that there must be extensive bush-fires blazing not far distant, but the full extent of the damage which had been done was not suspected at the time. The storm attained its great height at about two o'clock p.m., when it became a perfect hurricane. Some idea may be formed of its fury, from the fact that singed leaves and other light matter were blown as far as to the adjoining colony of Tasmania, and fell in showers on the decks of ships in Bass's Straits. In the afternoon, as is usual after these burning gales, the wind suddenly veered round to the south, blowing back the dust and smoke once more, till at length its fury was exhausted and it sank to rest as night closed. Those who have not witnessed the violence of tempests in tropical climates cannot realize the force of those wild tumults of nature, in which it often seems as though 'Heaven and earth must pass away.' "

Primary Source
Description of Change: excerpted
Rationale: excerpted to focus on main idea

Analyzing Primary Sources

Answers

1. He noticed that a hot wind was blowing in bits of burned debris and that smoke obscured the sun.
2. The fire burned for the greater part of an afternoon, fueled by an intense wind that blew matter across the Bass Strait to Tasmania.

Analyzing Primary Sources

1. How did the author first realize that a fire was burning somewhere in Victoria?
2. What details in this passage indicate that the fire covered a large area?

CHAPTER 49 Australia
Planning Guide

TEXTBOOK	TEACHING RESOURCES	OTHER AVAILABLE MATERIALS
Section		
1. Physical Geography Pages 604–606	Reteaching Worksheet 49.1 Skills Worksheet 49	*REGIONS OF THE WORLD Map Transparencies:* The Pacific World and Antarctica *Creative Strategies for Teaching World Geography* LESSON 22 Building an Ecosystem *The World Today* Videodisc Program *Study Guide for Spanish Speakers* *Lecture Notes,* Chapter 49
2. Economic Geography Pages 606–607	Reteaching Worksheet 49.2 Challenge/Enrichment Worksheet 49	*Study Guide for Spanish Speakers* *Lecture Notes,* Chapter 49
3. Population Geography Pages 607–610	Reteaching Worksheet 49.3 Critical Thinking Worksheet 49 Graphic Organizer 49	*Study Guide for Spanish Speakers* *Lecture Notes,* Chapter 49 *Creative Strategies for Teaching World Geography* LESSON 23 Australian Slang
Chapter Review Page 611	Chapter 49 Test Chapter 49 Portfolio Assessment See also *Alternative Assessment Forms*	*Test Generator,* Chapter 49

CHAPTER RESOURCES

Lecture Notes, Chapter 49
Teaching Resources
    Chapter 49 Graphic Organizer
    Chapter 49 Skills Worksheet
    Chapter 49 Critical Thinking Worksheet
    Chapter 49 Challenge/Enrichment Worksheet

REVIEW AND ASSESSMENT

Chapter 49 Check, page 611
Chapter 49 Test, *Teaching Resources*
Chapter 49 Portfolio Assessment, *Teaching Resources*
Test Generator, Chapter 49
Alternative Assessment Forms

ADDITIONAL RESOURCES

Books for Teachers

The Illustrated Encyclopedia of New Zealand. Boston: G. K. Hall (division of Macmillan Publishing), 1990.

Kurian, George Thomas. *Australia and New Zealand.* New York: Facts on File, 1990.

Morris, Jan. *Sydney.* New York: Random House, 1992.

Pilger, John. *A Secret Country: The Hidden Australia.* New York: Knopf, 1992.

Books for Students

Australia. Alexandria, VA: Time-Life Books, 1986.

Bambrick, Susan, ed. *The Cambridge Encyclopedia of Australia.* New York: Cambridge University Press, 1994.

Shadbolt, Maurice. *Reader's Digest Guide to New Zealand.* New York: Reader's Digest Press, 1988. (Distributed by Random House)

Williams, Bruce, ed. *The Real World: Understanding Our World Through the New Geography.* Boston: Houghton Mifflin, 1991.

The World Today *Videodisc Program*

A variety of still images, short videos, and activities are available for you to use as you teach this chapter. *Lecture Notes* include bar codes for specific images and videos.

Multimedia Resources

Australia. (16mm film/VHS, 25 minutes) Physical Geography of the Continents series. National Geographic Society, Educational Services, Washington, DC 20036.

Australia's Aborigines. (VHS, 59 minutes) National Geographic Society, Educational Services, Washington, DC 20036.

Australia/New Zealand: Why Is the World Shrinking? (VHS, 15 minutes) Agency for Instructional Technology, Box A, Bloomington, IN 47402–0120.

The Discovery of Australia. (VHS, 50 minutes) Films for the Humanities & Sciences, P.O. Box 2053, Princeton, NJ 08543–2053.

Picture Atlas of the World. (interactive maps, photo bank, animation, audio on CD-ROM) National Geographic Society, Washington, DC.

MEETING INDIVIDUAL NEEDS

Limited English Proficiency (LEP)

Play a "Password" type of word game using key terms relating to the geography of Australia. Organize players into teams of two. If possible pair LEP students with students proficient in English. Two teams will compete with one another at a time. Reveal a term to one player on each team. These players will take turns giving one-word clues to their partners until one player guesses the password, scoring a point for that team. Declare a winner after a determined number of rounds.

Learners Having Difficulty

Refer students to the chapter summary in Reviewing the Main Ideas on page 611. On a sheet of paper, have students list the sentences separately to form a skeleton outline of Chapter 49. Students should review the chapter to fill in specific information, such as the names of Australia's landform regions and its resources.

Auditory Learners

After students have read Chapter 49, begin a class discussion about where students might wish to live or travel in Australia if they had the opportunity. Encourage them to specify any details about the country's physical geography that influenced their choice.

Visual Learners

Organize students into three groups. Assign each group the task of collecting or drawing pictures of Australian endemic species, alien species, or ocean life of the Great Barrier Reef. Ask students to write captions for each picture. Have a volunteer compile the pictures and captions into a class book titled "Australia—Its Unique Biogeography." Display the book in the classroom.

Tactile/Kinesthetic Learners

If possible, take a field trip to a nearby zoo, oceanarium, or conservatory. Arrange for a member of the staff to give students a tour of the facility, with a focus on the endemic animals, plants, or marine life of Australia. Have students prepare questions for the tour guide.

Gifted Learners

Have interested students conduct research in current magazine and newspaper articles on the changes and challenges facing Australia, such as rising immigration, urban smog, deforestation, and soil erosion. Have students form panel groups to discuss the changes and challenges and to propose solutions for them.

MULTICULTURAL FOCUS

Aboriginal peoples of Australia have traditionally linked their spiritual beliefs to the earth's physical features. Refer students to the photograph of Ayers Rock on page 611. Point out that this giant monolith, rising from the desert plains in the Northern Territory, is a sacred location to the Aborigines.

In 1988, the Australian government returned Ayers Rock to the Pitjantjatjara people, known as the "guardians" of the rock and the lands that surround it. The caves at the base of Ayers Rock are the site of many Aboriginal carvings and paintings.

The Aboriginal peoples of Australia believe that the world was created during a period called "the dreaming" or "dreamtime." Explain that this does not relate to the conventional meaning of the word as it relates to sleep, but rather a period when Aborigines' ancestors—some of whom they believe were supernatural beings—came from within the earth to shape its physical features and landforms and give them names. This and other Aboriginal beliefs are passed on from generation to generation.

Despite the many challenges facing them and the outside influences on their way of life, the Aboriginal peoples of Australia have managed to maintain a distinct culture.

CHAPTER 49

AUSTRALIA

Section 1: Physical Geography
Section 2: Economic Geography
Section 3: Population Geography

OVERVIEW

In this chapter, students will learn how Australia's physical geography and location have influenced its natural history, its economy, and the settlement patterns of its people.

Section 1 describes the landforms, climate, and water resources of Australia as well as the country's unique plant and animal life.

Section 2 focuses on Australia's agriculture and mineral wealth. It discusses Australia's major industries and explores the history of Australia's trade relationships.

Section 3 explains why Australia's population is largely urban. It discusses the issues all Australians face because of rapid population growth.

About the Author

David Attenborough (1926–) is a film producer and a prolific writer of natural history books. The quotation on this page is from *Life on Earth.* Attenborough, a British citizen, has made zoological and ethnographic filming expeditions to many parts of the world.

Primary Source

Description of Change: excerpted and bracketed
Rationale: excerpted to focus on main idea; bracketed to clarify meaning

CHAPTER 49

Australia

Aboriginal art from **Australia**

❝**T**he Great Barrier Reef swarms with life. The tides surging through the coral heads charge the water with oxygen and the tropical sun warms it and fills it with light. All the main kinds of sea animals seem to flourish here. . . . Purple eyes peer out from beneath shells; black sea urchins swivel their spines as they slowly perambulate [stroll] on needle tip; starfish of an intense blue spangle the sand; and patterned rosettes unfurl from holes in the smooth surface of the coral.❞

David Attenborough

Settled by Great Britain in 1788 as a prison colony, Australia today is a modern and prosperous democratic nation. It is the sixth largest country in the world, with a land area about the same size as the contiguous United States. Because it lies south of the equator, Australia is sometimes called the "Land Down Under."

The nation is made up of six states and two federal territories. The states include New South Wales, Victoria, Queensland, South Australia, Western Australia, and the island state of Tasmania. The two federal territories are the Northern Territory and the tiny Australian Capital Territory, in which Canberra, the national capital, is located. Australia also governs several small islands in the South Pacific and Indian oceans. The Australian population is concentrated in a few large coastal cities. The rest of the continent, known as the Outback, is dry, flat, and almost uninhabited.

GEOGRAPHY DICTIONARY

artesian well
biogeography
endemic species
marsupial
alien species

Great Barrier Reef off **Australia**

Using the Quotation

Point out that in the quotation on page 600 the author paints a picture of animal life in the Great Barrier Reef off the coast of Australia. Locate the Great Barrier Reef on a wall map for students. Ask volunteers to read aloud the descriptive phrases in the quotation. Have students choose one of the descriptive phrases and illustrate it. You might combine the illustrations in a bulletin board display.

Using the Geography Dictionary

Ask for a few volunteers to look up the entry for **artesian well** in different encyclopedias, and have them summarize for the class what they learn about these wells. Then have students find the terms **biogeography, endemic species,** and **marsupials** in the Glossary. Have students write a sentence using all three terms, and ask volunteers to write their sentences on the board.

MAPPING THE REGION

Using the Physical-Political Map

Direct students' attention to the map on this page. Have them use the scales of the main map and of the inset map to draw conclusions about the relative sizes of Australia and Antarctica. (*The scales are similar; Antarctica is larger than Australia.*) Then have students compare the physical features of the two continents. (*Antarctica is covered by an ice cap.*)

MAP TRANSPARENCIES

Overhead transparencies from the maps on pages 601–603 are available in *REGIONS OF THE WORLD Map Transparencies.*

CREATIVE STRATEGIES FOR TEACHING WORLD GEOGRAPHY

LESSON 22 Building an Ecosystem: The Great Barrier Reef

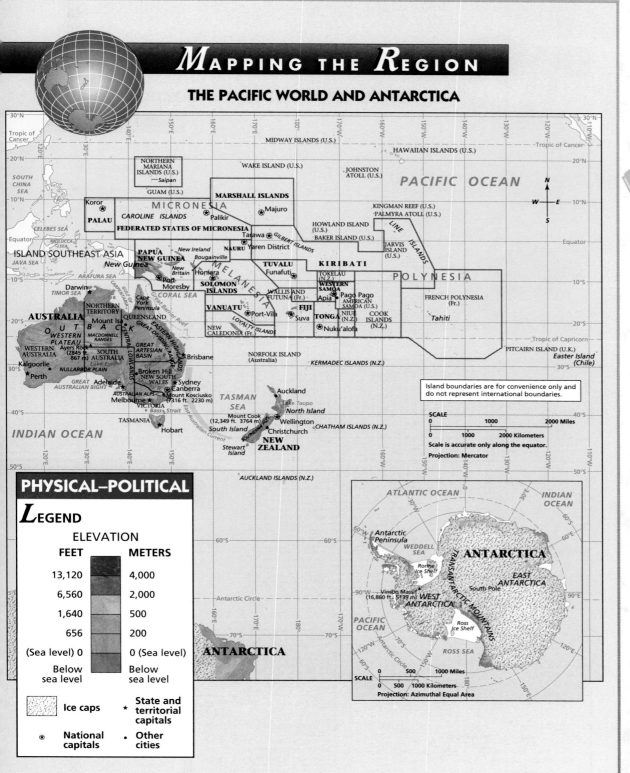

MAPPING THE REGION

THE PACIFIC WORLD AND ANTARCTICA

PHYSICAL–POLITICAL

LEGEND

ELEVATION

FEET		METERS
13,120		4,000
6,560		2,000
1,640		500
656		200
(Sea level) 0		0 (Sea level)
Below sea level		Below sea level

Ice caps

★ State and territorial capitals

⊛ National capitals

• Other cities

Island boundaries are for convenience only and do not represent international boundaries.

SCALE
0 — 1000 — 2000 Miles
0 — 1000 — 2000 Kilometers
Scale is accurate only along the equator.
Projection: Mercator

SCALE
0 — 500 — 1000 Miles
0 — 500 — 1000 Kilometers
Projection: Azimuthal Equal Area

MAPPING THE REGION

Using the Climate Map

Direct students' attention to the climate map of the Pacific World and Antarctica on this page. To help students understand the dominant climates in Australia and New Zealand, have them compare this climate map with the one of the United States and Canada on page 108. Have them identify areas of the United States that have similar climates to those found in Australia and New Zealand. (*For example, the southwestern United States has the same desert climate as west and central Australia.*) List students' responses on the board.

MAPPING THE REGION

Using the Population Map

Have students study the population map on this page and the climate map above it to draw conclusions about the link between climate and population density in Australia. (*The largest concentrations of people are in marine west coast, humid subtropical, and Mediterranean climates; the lowest concentrations are in the desert climate.*) Point out that only two Australian cities have populations of more than 2 million.

CLIMATE

Answers

1. Desert, steppe
2. Humid tropical, highland
3. Mediterranean

POPULATION

Answers

1. Sydney, Melbourne
2. Marine west coast, Mediterranean, humid subtropical
3. Mining, dairying, geothermal power generation, livestock raising, natural gas production (offshore), manufacturing

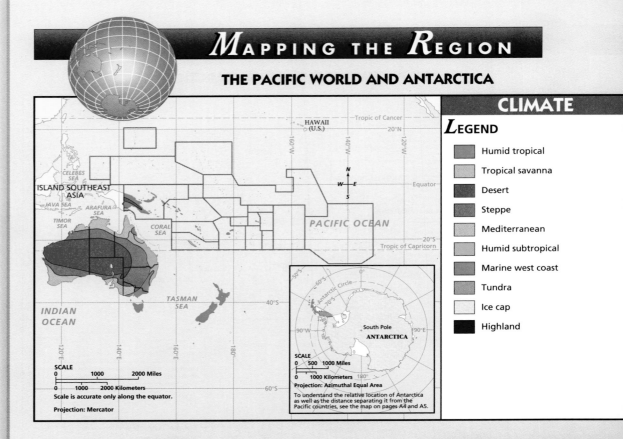

MAPPING THE REGION

THE PACIFIC WORLD AND ANTARCTICA

CLIMATE

LEGEND

- Humid tropical
- Tropical savanna
- Desert
- Steppe
- Mediterranean
- Humid subtropical
- Marine west coast
- Tundra
- Ice cap
- Highland

POPULATION

LEGEND

POPULATION DENSITY

Persons per sq. mile	Persons per sq. km
520	200
260	100
130	50
25	10
3	1
0	0

- ● Metropolitan areas with more than 2 million inhabitants
- ● Metropolitan areas with 1 million to 2 million inhabitants

MAPPING THE REGION

Using the Economy Map

Explain that Australia's Outback has been described as "a storehouse of mineral wealth." Have students study the map and the map key on this page and list the mineral resources found in Australia. Then have students consider the relationship between mineral wealth and economic prosperity: Do they automatically go hand in hand? Why or why not? You might wish to discuss the pros and cons of developing Antarctica's economic resources.

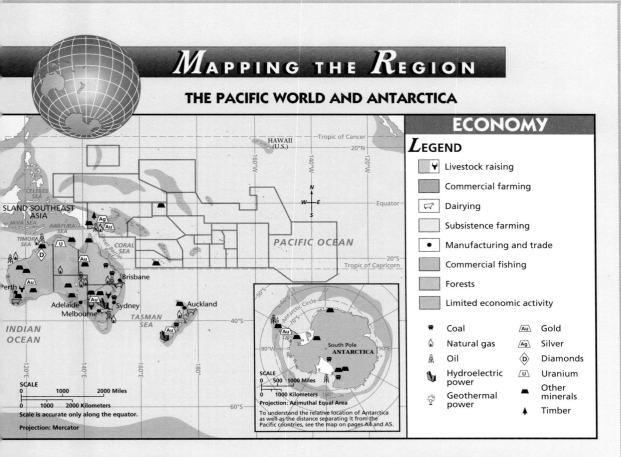

MAPPING THE REGION

THE PACIFIC WORLD AND ANTARCTICA

ECONOMY

LEGEND

- Livestock raising
- Commercial farming
- Dairying
- Subsistence farming
- Manufacturing and trade
- Commercial fishing
- Forests
- Limited economic activity

- Coal
- Natural gas
- Oil
- Hydroelectric power
- Geothermal power
- Au Gold
- Ag Silver
- Diamonds
- U Uranium
- Other minerals
- Timber

To understand the relative location of Antarctica as well as the distance separating it from the Pacific countries, see the map on pages A4 and A5.

SCALE 0 500 1000 Miles
0 1000 Kilometers
Projection: Azimuthal Equal Area

SCALE 0 1000 2000 Miles
0 1000 2000 Kilometers
Scale is accurate only along the equator.
Projection: Mercator

Compare this map to the physical–political map on page 601.

Compare this map to the climate map on page 602.

CLIMATE

◆ Most of Australia is generally warm, dry, and sunny. Most precipitation occurs in regions along the coasts.

◆ Most of the Pacific islands have tropical climates.

◆ Because of its oceanic and middle-latitude location, New Zealand is dominated by a marine-west-coast climate.

(?) Which two climates cover most of Australia?

(?) Which climate types are dominant in Papua New Guinea?

Compare this map to the physical–political map on page 601.

(?) In which climate region are Perth and Adelaide located?

POPULATION

◆ Australia's most populous region is in the southeast.

◆ Most of New Zealand's population lives on North Island.

◆ Except for scientists at research stations, Antarctica has no permanent human population.

(?) Which Australian metropolitan areas have populations of more than 2 million?

Compare this map to the climate map on page 602.

(?) Which climate regions in Australia have the densest populations?

Compare this map to the economic map above.

(?) What economic activities are found on New Zealand's most populous island?

ECONOMY

◆ Papua New Guinea has an abundance of natural resources, including minerals, which are the country's dominant exports.

◆ Because of its extensive dry climates, the vast majority of Australia's land is not used for farming.

◆ Although believed to be rich in natural resources, Antarctica is not open to mining.

(?) Name the manufacturing and trade centers shown in Australia.

(?) Where in Australia will you find commercial farming?

(?) Which natural resources can be found in Antarctica?

ECONOMY

Answers

1. Perth, Adelaide, Melbourne, Sydney, Brisbane
2. Primarily in the east and southeast and coastal southwest
3. Gold and various other minerals, coal, offshore oil deposits; thought to be rich in as yet undiscovered resources

603

The objectives reflect the Focus questions on this page and are assessed in the review on page 606.

1. Identify Australia's landforms, climates, and water resources.

2. Explain the significance of Australia's animal, plant, and ocean life.

TEACHING SECTION 1
Motivator

Write *Australia* on the board, and ask students to suggest the images that come to mind when they think of Australia. Write their ideas on the board. As students read the chapter, have them note which images are consistent with the photographs and information in the text.

Building a Vocabulary

Contrast the terms **endemic species** and **alien species**. Have students suggest why so many endemic lifeforms are found in Australia. Discuss the changes that occur when alien species are introduced to an area and the reasons why they sometimes flourish. *(no natural enemies to keep their populations in check)*

THE WORLD TODAY VIDEODISC PROGRAM

Australia: Life in the Outback

1A 22850 4:07

Play   Pause

Geographical Sidelight

Australia's main mountain system is called the Great Dividing Range because its waters on the east flow into the Pacific Ocean and those on the west flow into the Murray-Darling river system.

Size comparison of Australia to the contiguous United States

The Finke River winds its way through the **Northern Territory** of **Australia**. Australians call the less-populated interior of the country the Outback. The Outback is dominated by the dry desert and steppe climates.

PHYSICAL GEOGRAPHY

FOCUS

- *What are the landforms and climates of Australia? What water resources are found in Australia?*

- *How is Australia's animal, plant, and ocean life significant?*

REGION **Landforms** Australia is the world's smallest, flattest, and lowest continent. It can be divided into three major landform regions: the Western Plateau, the Central Lowlands, and the Eastern Highlands. (See the map on page 601.)

The Western Plateau covers more than one-half of the continent, including Western Australia, the Northern Territory, and most of South Australia. The treeless Nullarbor Plain, the flattest landform region on Earth, stretches along the southern edge of the plateau. Ayers Rock, a huge rock formation, lies just south of the Macdonnell Ranges in the western part of the region.

To the east of the Western Plateau are the Central Lowlands, which extend from the Gulf of Carpentaria in the north to the Great Australian Bight in the south. They include interior Queensland and New South Wales and adjoining areas of Victoria and South Australia. The area around dry Lake Eyre in South Australia is below sea level. The lake basin fills with water only about once every 10 years.

Australia's highest elevations are found in the Great Dividing Range of the Eastern Highlands. This range stretches from Cape York Peninsula in the north, along the country's east coast, to Victoria in the south. The mountains appear again on the island state of Tasmania. Mount Kosciusko (kahz-ee-ES-koe) is Australia's highest peak at 7,316 feet (2,230 m).

Climate Some geographers describe Australia as a desert with green edges. Indeed, most of the continent receives little rain, mainly for two reasons. First, the stable air of the subtropical high-pressure system prevents rain from reaching many areas. Second, the Great Dividing Range rises in the path of the southeast trade winds.

The range acts as a barrier blocking moist Pacific winds from reaching the leeward rain-shadow side of the range and the interior of the continent. Instead, moisture falls on the windward eastern slopes, forming a narrow band of humid climate regions on the east coast of the continent.

As a result, about half of Australia has a desert climate. Desert covers much of the continent's interior and stretches to the coast of Western Australia. Forming a semiarid rim around the desert is a steppe climate, dominated by scrub bushes.

On the northern coast of the Northern Territory and the Cape York Peninsula of Queensland is a tropical-savanna climate. The monsoon influences this hot region, causing distinct wet and dry seasons. A half year of drought and several months of flooding are typical.

A small area of humid-tropical climate and rain forests lies in the far northeast corner of coastal Queensland. Meanwhile, most of coastal Queensland and New South Wales has a humid-subtropical climate. Farther south, the cooler, marine-west-coast climate prevails. Here, winter storms pushed by the westerly winds bring rain and sometimes snow to the mountain areas. The marine-west-coast climate extends along coastal Victoria, northward into the mountains of New South Wales, and southward over all of the island of Tasmania. The windward western side of Tasmania supports lush temperate forests.

A Mediterranean climate bathes the southwest coast of Western Australia near Perth and the coast of South Australia near Adelaide. Winters here are mild and rainy, and summers are long, dry, and sunny.

Water Resources Almost all of Australia is subject to drought. These long dry spells can lead to severe water shortages, brushfires, and dust storms. Severe tropical cyclones (hurricanes) also occasionally threaten the northern and northeastern coasts.

Australia has only one major river system, the Murray-Darling. It flows from the Great Dividing Range westward across the dry, steppe country of southeastern Australia. In contrast, the smaller rivers along the eastern coast flow into the sea within a short distance of their origins. Large water projects have tried to move water through tunnels from the wet eastern slopes to the dry interior slopes.

Fortunately, groundwater is abundant in Australia. The largest source of underground well water lies in the Great Artesian Basin of interior Queensland. Here, **artesian wells** dot the area. Artesian wells are those in which water rises toward the surface without being pumped. This water originates as rain falling on the wet mountains to the east. The water enters an aquifer and flows westward under the desert. Much of this well water, however, is of poor quality and can be used only to water sheep.

Biogeography The study of the geographic distribution of plants and animals is called **biogeography**. Australia's island location has helped create a biogeography here that is much different from that in other parts of the world. Because of Australia's isolation from other continents, many **endemic species** developed. Endemic species are plants or animals that originate only in a particular geographic region. Most of Australia's plants and animals are found nowhere else on Earth.

Australia is especially well known for its **marsupials**, such as the kangaroo and the koala. Marsupials are mammals that carry their young in pouches. There are almost 50 varieties of kangaroos in Australia, from small mouselike ones to some that are larger than humans. The koala, on the other hand, is a tree-dwelling vegetarian known for its furry, bearlike appearance. Australia also is home to the platypus and spiny anteater, the only egg-laying mammals in the world.

The endemic plants of Australia are unusual in that they are dominated by a single plant variety, the eucalyptus (yoo-kuh-LIP-tus). Australia has more than 500 species of eucalyptus, and more than 90 percent of its trees are some variety of this plant.

FACTS IN BRIEF	Australia			
COUNTRY POPULATION (1994)	**LIFE EXPECTANCY** (1994)	**LITERACY RATE**	**PER CAPITA GDP** (1993)	
AUSTRALIA 18,077,419	74, male 81, female	99% (1992)	$19,100	
UNITED STATES 260,713,585	73, male 79, female	97% (1991)	$24,700	

Sources: *The World Factbook 1994*, Central Intelligence Agency; *The World Almanac and Book of Facts 1995*

The life expectancy rates of Australians are slightly higher than those for residents of the United States.

SECTION 2
OBJECTIVES

The objectives reflect the Focus questions on this page and are assessed in the review on page 607.

1. List Australia's agricultural and mineral resources.
2. Discuss global forces affecting industry and trade in Australia.

TEACHING SECTION 2
Motivator

Refer students to the Chapter 14 discussion of the Wheat Belt in the United States on page 152. Point out that Australia also has a wheat belt, west of the Great Dividing Range in Victoria and New South Wales. Tell students that wheat is Australia's major crop, and locate Australia's wheat belt on a wall map.

Guided Practice

On the board, create a chart with the following headings: Mining, Ranching, Agriculture, Services, Tourism. Have students work in pairs to identify at least two facts about Australia in each category. Then have volunteers take turns coming to the board and filling in facts. Discuss which economic activities are likely to be important to Australia in 10 years, 50 years, and 100 years.

Section 1 Review
ANSWERS

1. Much of inland Australia has a desert climate. Coastal areas tend to be wetter and therefore have greener vegetation.

2. Endemic animals include kangaroos, koalas, platypuses, and spiny anteaters. The Great Barrier Reef is the world's largest coral reef and supports a complex ecosystem.

3. Possible answer: if a population of a certain kind of animal gets too large, the animals may interfere with vegetation and other animals in the area. Examples from Australia include rabbits and kangaroos.

Many plants and animals in Australia, however, are **alien species**. Alien species are those introduced to a region by humans. The Aborigines, the original inhabitants of Australia, brought the dingo, a dog from Asia. Europeans later introduced sheep, rabbits, foxes, water buffalo, birds, and hundreds of plants. Due to a lack of native predators, the number of rabbits in Australia exploded. They overran the continent, destroying crops and grasslands. In 1950, a disease introduced by humans killed most of the rabbits. The rabbit population, however, is increasing once again.

Managing Australia's unique wildlife is a source of concern and controversy to the country's people. Some Australian animals, such as the koala, may become endangered, mainly due to loss of habitat. Other animals, especially kangaroos, exist in great numbers and are hunted to protect sheep-grazing lands. Some oppose killing any of the animals.

The koala lives in forested areas of eastern **Australia** and is protected by law. Eucalyptus leaves are its favorite food.

The Great Barrier Reef In the Coral Sea, off the northeastern coast of Queensland, lies the Great Barrier Reef. The world's largest coral reef, it stretches more than 1,200 miles (1,931 km). The reef first formed along the edges of the Australian continental shelf. These shallow waters, swept by the warm East Australian Current, are ideal for the formation of coral reefs. Then, in the years since the last ice age, the coral continued to grow upward with rising sea levels. Eventually, numerous coral sand islands built up on the reef.

Except for the tropical rain forests, no ecosystem on Earth is as complex as that of the Great Barrier Reef. It is home to the world's greatest variety of ocean life, including fish, shellfish, and birds. Many people are concerned that tourism and offshore oil drilling could damage the reef. To protect it, the government has made most of the Great Barrier Reef a national park.

SECTION REVIEW

1. Why do some geographers describe Australia as a desert with green edges?
2. What are two of Australia's endemic animals? What is significant about the Great Barrier Reef?
3. **Critical Thinking** **Explain** how problems can arise when the population of a certain kind of animal becomes very large. Use two examples from the section.

ECONOMIC GEOGRAPHY

FOCUS

● *What agricultural and mineral resources does Australia have?*

● *How have Australia's industry and trade been influenced by global forces?*

Agriculture Australia produces about one-third of the world's wool supply. The country exports nearly all its wool, mainly to Japan and China. In the wetter eastern portions of Australia, sheep are raised for mutton and lamb, which are meat products. Southwest Asia imports much of this meat as well as many live sheep.

Australia is also the leading beef exporter in the world. It exports most of its beef to the United States and Japan. Beef cattle and sheep are usually raised on large ranches, called stations, in the nation's dry interior. Most dairy farms are located in wetter regions, such as Victoria and Tasmania, and near large urban areas.

Because of its dry climates, only about 6 percent of Australia's land is used for farming. Despite the limited amount of arable land and good water, Australia is self-sufficient in food and is a major food exporter as well. Its small population, efficient farmers, and the many different climate regions contribute to this success.

Wheat, the nation's major crop, covers more than half of Australia's cropland. The wheat belt stretches across the plains west of the Great Dividing Range in Victoria and New South Wales. Some wheat is grown, however, in all of the mainland Australian states. The country is one of the world's largest wheat exporters.

Closure

As you write their responses on the board, call on students to name the successful industries in Australia and to explain the reasons for the success of these industries. Then ask students to name some global forces affecting Australian trade and industry.

ASSESSMENT

Assign the Section 2 Review questions on this page.

Reteaching

Assign Section 2 of Reteaching Worksheet 49.

Extension

Have interested students imagine that they are executives of a U.S. high-tech, mining, or food-processing company. Have students research and write a brief report outlining why opening an office in Australia might be a good move for the company.

The objectives reflect the Focus questions on this page and are assessed in the review on page 610.

1. Examine population distribution and composition patterns in Australia.
2. Discuss environmental challenges facing Australia.

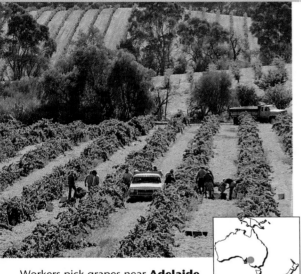

Workers pick grapes near **Adelaide, South Australia**, where a Mediterranean climate supports many vineyards.

Australia's other major export crops are sugar, rice, and cotton, which are grown mainly in coastal New South Wales and Queensland. Also from the Queensland coast come bananas, pineapples, and other tropical fruits. Citrus fruits, olives, and wine grapes grow well in the humid-subtropical and Mediterranean climate regions. Crops such as corn, apples, peaches, and pears grow in the marine-west-coast climate region on the southeastern coast and in Tasmania.

Mineral Resources Australia's Outback is a storehouse of mineral wealth. In fact, Australia is a world leader in iron-ore, bauxite, and coal exports and in diamond and opal processing. The country also has vast uranium reserves and deposits of nickel, gold, lead, zinc, and copper.

Australia produces about 70 percent of its needed oil. The country's major oil fields lie in the Bass Strait, between Tasmania and Victoria, on the offshore shelf areas of Western Australia, and in southern Queensland. Huge natural-gas reserves also have been discovered on the continental shelf off the northwest coast. Most of the country's coal mining is in eastern Australia.

Industries Although it has a small consumer population and a small labor force, Australia is an industrialized nation. Many industries, such as iron, steel, and food processing, are based on rich natural resources. Most of Australia's industries, however, are now in decline due to for-

eign competition. As a result, service industries are becoming the country's major employer. For example, Australia has a large transportation industry to maintain its vast network of roads and operate its railroads and airlines. The tourism and entertainment industries also are growing rapidly.

Asian-Pacific Connections With the formation of the European Community, Australia lost some of its close trade ties with the United Kingdom and other European countries. Since then, Australia has looked north to its Asian neighbors for international trade. Asian countries make up Australia's largest market. Japan in particular is a major customer for Australia's crops, meat, and minerals. The United States also is an important trade partner.

SECTION REVIEW

1. List three of Australia's most important agricultural exports and three of its major mineral resources.
2. How have foreign competition and the formation of the European Community affected Australia's industry and trade?
3. **Critical Thinking** **Determine** what factors have contributed to Australia's economic success.

POPULATION GEOGRAPHY

FOCUS

- *How is population distributed in Australia? What groups of people have settled in Australia?*

- *What environmental challenges does Australia face?*

Urban Geography Although Australians are often portrayed as rugged Outback settlers, more than 85 percent of all Australians live in urban areas. Each of Australia's six states is dominated by one large nodal capital city and seaport, with suburbs and transportation networks fanning out

Text continues on page 610.

Section 2 Review
ANSWERS

1. Agricultural exports: wool, mutton, lamb, beef, wheat, sugar, rice, cotton; mineral resources: iron ore, bauxite, coal, diamonds, opals, uranium, nickel, gold, lead, zinc, copper, oil, natural gas

2. Foreign competition has caused a decline of some of Australia's industries. The formation of the European Union caused Australia to lose its close trade ties with European countries and to form new ties with Asian and other countries.

3. Factors include Australia's self-sufficiency in food (due to its small population, efficient farmers, and many climate regions), its vast mineral resources, and its service, transportation, tourism, and entertainment industries.

607

Cities of the World

Living Down Under

Have students imagine that they are "Sydneysiders." Assign students occupations (for example, businessperson, construction worker, sailing instructor, university professor). Have them use classroom and library resources to find detailed maps of the city. Ask them to describe the geography of their workday by answering the following questions: Where do they work? *(Students' choices should be based on their assigned occupations.)* Where do they live in the city and why? *(Students may choose locations near the water or on a hill overlooking* the harbor or near their work.) How do they get to work? *(Students might list the roads or ferries they would take.)* Have volunteers "walk" the class through the geography of their workdays.

CITIES OF THE WORLD

Sydney

*N*o longer a lonely British outpost, Sydney and its harbor open Australia to the Pacific Ocean and world trade.

Unlike other major cities of the world, Sydney did not begin as a bustling port, an industrial or transportation center, or a mining town. Instead, the British established Sydney as a prison colony in 1788. Australia's then-isolated location seemed an ideal place to house the empire's convicts.

For their survival, the first prisoners and their guards depended on supplies shipped from Great Britain. Soon, the small settlement on the shore of Sydney Cove became an important port for the entire continent and for the South Pacific. It also served as a base for explorers of Australia's interior.

Modern Sydney

In the past two centuries, Sydney has grown from a small, isolated prison colony to a city with a population of more than 3.5 million. The city covers 670 square miles (1,735 sq. km) and is Australia's largest urban area and leading seaport. Its residents are called Sydneysiders.

Sydney's growth is due in large part to its excellent harbor, which is one of the deepest in the world. The two main bays of the harbor area are Port Jackson and Middle Harbor. Dozens of fingerlike peninsulas creep into the harbor area, giving the impression that the city is built on a series of islands. The communities on each peninsula have their own distinct character.

Sydney grew rapidly in its early years. The colony first spread inland from Sydney Cove, then west and east along the harbor. Later, settlements sprang up across the harbor on the north shore. The original plan of the city was a regular grid pattern. As the city spread, however, this pattern changed.

The main roads into the interior of Australia followed the open areas between the hills, making travel easy for animal-drawn vehicles. When suburbs later grew around these roads, they followed the same winding pattern. The irregular coastline of the harbor's bays also adds to Sydney's unusual layout.

Also unusual is the fact that a high percentage of people in the city own their own house and garden. When Sydney first developed, large areas of land were inexpensive, and many people could afford to own large amounts of land. Land prices have risen dramatically, however. Today, a shortage of space, especially in the central business district, has resulted in many towering apartment complexes and high-rise buildings springing up.

Living Down Under

Growth brings many challenges to a city the size of Sydney. The irregular layout of the streets and the large

The Parramatta River, dotted with sailboats, flows into the harbor of **Sydney**. A mild, humid-subtropical climate helps make outdoor activities popular for Sydney residents and tourists.

Using Geography Themes

To emphasize the themes of place and location, first organize the class into two groups. Have students in one group find photographs in magazines and newspapers or sketch drawings of Sydney that show its human and physical place characteristics. Group members should assemble a collage of the photographs and sketches and write a caption summarizing Sydney's place characteristics.

Students in the second group should work together to draw three maps: one of Australia, one of the South Pacific, and one of the Pacific Rim. Each map should highlight Sydney's location. Captions should explain how Sydney's location within each of the three regions affects its human and place characteristics. Have both groups present their materials and display them in the classroom. Then have students find at least five images in the first group's collage that support the statements in the second group's maps about the effects of Sydney's location.

SYDNEY
- Major highways
- Minor highways
- Railroads
- City limits
- Parks and forests
- Airport

0 — 3 Miles
0 — 4 Kilometers

AUSTRALIA
Sydney

The metropolitan area of **Sydney** is divided by the Parramatta River and the harbor. Port Jackson and Middle Harbor provide Sydney's busy shipping industry access to the Tasman Sea and the Pacific Ocean beyond.

The Sydney Opera House, a famous city landmark, is located near the Sydney Harbor Bridge. The Australians use the British spelling of *harbour*.

Geographical Sidelight

Jorn Utzon, a Danish architect, won an international competition in 1955 for the design of the Sydney Opera House. Utzon submitted a rough design only. Details about how to actually build the large vaulted, curved ceilings of the opera house required several years and the combined efforts of engineers, mathematicians, and other architects. Construction took 14 years and $103 million, nearly 15 times the original estimated cost. Utzon ultimately disagreed with a governmental committee about the execution of his design, and he never returned to Sydney to see the finished structure. The Opera House complex seats 5,200 and is also home to other artistic companies of Australia devoted to theater and dance.

number of commuters to the city's center each day have resulted in growing traffic congestion. People who live on the mainly residential north shore must drive to work over the Harbor Bridge to the commercial areas of the city in the south and west. Ferries ease some of this congestion, but automobiles remain residents' favorite form of transportation.

Sydney, host of the 2000 Olympics, is filled with opportunities for entertainment and recreation. City dwellers enjoy the Opera House; sports such as tennis, golf, and horse racing; and all water activities, especially sailing. The coastline provides many sandy beaches for swimming and surfing. These attractions, combined with the city's excellent harbor, business opportunities, and location on the Pacific Rim make Sydney an increasingly popular world city.

The famous Sydney Opera House, completed in 1973, overlooks the city's harbor. The Opera House, with its distinctive white concrete shells that resemble sails, hosts concerts, opera, and theater.

TEACHING SECTION 3

Motivator

Write the names of Australia's five largest cities on the board (Sydney, Melbourne, Brisbane, Perth, Adelaide). Have a volunteer locate the cities on a wall map. Ask: Why do most Australians live in these cities? (*They are located near coasts.*) Tell students that they will learn more about Australia's population in Section 3.

Thinking Critically

Population   Point out to students that social scientists who study migration from one country to another refer to "push factors," or situations that cause people to move away from one area, and "pull factors," conditions that attract people to a new location. Based on their reading of the chapter, have students suggest possible pull factors for immigration to Australia. (*developed economy, economic opportunities, open spaces, warm climates*)

Closure

Review with students the issues facing Australia today in terms of population and environment.

ASSESSMENT

Assign the Section 3 Review questions on this page.

Reteaching

Assign Section 3 of Reteaching Worksheet 49.

MULTICULTURAL PERSPECTIVES

Point out that Australia's economy is increasingly linked to Asian countries. Some Australian schools have begun teaching Asian languages, history, and literature. Discuss how these educational practices might benefit Australians. (*They might promote more understanding between Australians and Asian immigrants in Australia and prepare Australians for trade with Asia.*)

Section 3 Review
ANSWERS

1. The southeastern corner; increasing number of Asians
2. They have caused increased smog, mining, damming of rivers, and destruction of forests.
3. Both peoples originated in Asia. Large numbers of both groups died from diseases introduced by Europeans and lost rights when Europeans settled around them. The ways of life of both groups have changed dramatically. Today, problems related to poverty plague some members of both groups.

610

into the hinterland. In some cases, the hinterland is an entire state.

The southeastern corner of Australia is the country's heartland, containing the nation's three largest cities and most of its population. Sydney, Australia's oldest and largest city, is the capital of the most populous state, New South Wales. (See "Cities of the World" on pages 608–609.) Melbourne, the capital of Victoria, is Australia's second largest city. It is a busy seaport and a commercial and industrial center. Located inland, about halfway between Sydney and Melbourne, is Canberra, the nation's capital.

Brisbane, the capital of Queensland, is Australia's third largest city. Its economy is based mainly on agriculture and tourism. Near Brisbane are the Gold Coast, known as "surfer's paradise," and the Great Barrier Reef.

The continent's other nodal capital cities are Perth in Western Australia, Adelaide in South Australia, and Darwin in the Northern Territory. Tasmania, the smallest and only island state, has Hobart as its capital city.

The Aborigines   The first inhabitants of Australia were the Aborigines. Originally a nomadic hunting-and-gathering people, they migrated to Australia from Southeast Asia about 50,000 years ago. When the first European settlers arrived during the 1700s, about 300,000 Aborigines lived in Australia. Diseases brought by Europeans greatly reduced this number. Today, Aborigines make up only about one percent of the population, and their nomadic way of life has disappeared.

An Australian Aborigine artist produces a bark painting. Aboriginal art forms, including cave paintings, have been practiced for centuries.

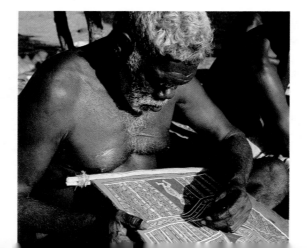

In recent years, the Aborigines have gained back some of their rights that were lost when Europeans settled the continent. They now have full citizenship, and the federal government has returned some of the land in the Northern Territory to them. Yet most Aborigines live in urban areas in poverty. Substandard housing, high crime rates, alcoholism, and low health standards plague the Aborigine people. The life expectancy of Aborigines is 20 years less than that of other Australians.

Issues   Australia's ethnic composition is changing rapidly. Like the United States and Canada, Australia has always been a land of immigration. British and Irish colonial settlers were later joined by other Europeans, especially from Greece and Italy. Today, an increasing number of Australia's immigrants are from Asia, particularly Vietnam, Malaysia, China, and Cambodia. Now more than one-fourth of all Australians were born outside the country.

Some Australians are opposed to immigration growth, which over the past 40 years has helped double Australia's population as well as change its composition. Others, however, believe that these new citizens contribute to the nation's economic growth and diversity.

Rapid population growth and economic development, however, have led to environmental concerns among Australians. Although Australia's cities generally are clean, smog hangs over many urban centers. In addition, mining, damming of rivers, and forest destruction all threaten Australia's environment. Salinity and wind erosion of topsoil have become serious problems as well, especially during drought years.

SECTION REVIEW

1. What is the most densely populated part of Australia? How is Australia's ethnic composition changing?
2. How have population growth and economic development affected Australia's environment?
3. **Critical Thinking**   **Compare** the experiences of the Aborigines in Australia with the experiences of Native Americans in the Western Hemisphere.

CHAPTER 49 CHECK
ANSWERS

Building a Vocabulary

1. A well in which water rises to the surface without being pumped; in the Great Artesian Basin of interior Queensland; an aquifer under the desert whose water originates as rain falling on the eastern mountains
2. The study of the geographic distribution of plants and animals; because Australia is isolated from other continents
3. A species that originated only in a particular geographic region; the eucalyptus
4. Marsupials carry their young in pouches; the kangaroo and the koala

Recalling and Reviewing

1. It is a range in Australia's Eastern Highlands. It blocks moist winds from reaching the interior of the continent, so the continent's east coast is humid but Australia's interior is dry.
2. Iron, steel, food processing, services, transportation, tourism, entertainment
3. Sydney, Melbourne, Brisbane

Thinking Critically

1. Possible answer: its location away from other continents made it difficult for prisoners to escape.
2. Ranching and wheat farming are important to the economies of all three regions.

Learning About Your Local Geography

Posters should be about local wildlife and the protection of endangered species.

Using the Five Themes of Geography

Paragraphs should mention that Australia's location away from other continents led to the development of many endemic plants and animals. Effects of human-environment interaction include the decrease of the koala population because of loss of habitat, the hunting of kangaroos to protect grazing lands, and the destruction of crops and grasslands that occurred when rabbits were introduced.

CHAPTER 49 CHECK

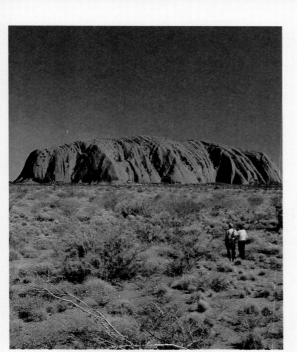

Ayers Rock, **Northern Territory, Australia**

Building a Vocabulary

1. What is an artesian well? In what part of Australia are artesian wells found, and what is the source of their water?
2. Define *biogeography*. Why is Australia's biogeography distinct?
3. What is an endemic species? Give an example of a plant that is an endemic species in Australia.
4. What distinguishes marsupials from other mammals? Name two marsupials that live in Australia.

Recalling and Reviewing

1. What is the Great Dividing Range, and what effect does it have on Australia's climates?
2. What are Australia's major industries?
3. What are the country's three largest cities?

Thinking Critically

1. Australia was originally settled as a prison colony by Great Britain in 1788. Why might the British have chosen Australia for a prison colony?
2. What does Australia's agriculture have in common with that of Argentina and of the Great Plains region of the United States?

Learning About Your Local Geography

Cooperative Project
Contact a local wildlife organization to see what plants and animals are endemic species in your area and whether any of them are endangered. Make a poster showing the local wildlife and explaining why it is important to protect endangered species.

Using the Five Themes of Geography
Write a paragraph discussing the effects of **location** and **human-environment interaction** on Australia's biogeographical **place** characteristics.

Reviewing the Main Ideas

1. Australia is the world's smallest continent. It can be divided into three major landform regions. Although Australia has many climate types, most of the country is warm, sunny, and subject to drought. Because of Australia's isolation from other continents, many of its animals and plants are unique.
2. Australia is rich in agricultural and mineral resources. It also is an industrialized nation, although some industries are now in decline. The countries of Asia are Australia's major trade partners.
3. More than 85 percent of all Australians live in urban areas, particularly in the southeastern corner of the country. The Aborigines were Australia's first inhabitants but today make up only about one percent of the population. Many of Australia's immigrants now come from Asia. Population growth and economic development have presented some environmental challenges.

CHAPTER 50 The Pacific Islands and Antarctica

Planning Guide

TEXTBOOK	TEACHING RESOURCES	OTHER AVAILABLE MATERIALS
Section **1.** Islands of the Pacific Pages 613–615	Reteaching Worksheet 50.1 Critical Thinking Worksheet 50	*The World Today* Videodisc Program *Study Guide for Spanish Speakers* *Lecture Notes,* Chapter 50
2. New Zealand Pages 615–617	Reteaching Worksheet 50.2 Graphic Organizer 50	*Study Guide for Spanish Speakers* *Lecture Notes,* Chapter 50
3. Antarctica Pages 617–620	Reteaching Worksheet 50.3 Skills Worksheet 50 Challenge/Enrichment Worksheet 50	*Study Guide for Spanish Speakers* *Lecture Notes,* Chapter 50
Chapter Review Page 621	Chapter 50 Test Chapter 50 Portfolio Assessment See also *Alternative Assessment Forms*	*Test Generator,* Chapter 50
Unit Review Page 622–623	Unit 11 Test Unit 11 Portfolio Assessment See also *Alternative Assessment Forms*	*Test Generator,* Unit 11

CHAPTER RESOURCES

Lecture Notes, Chapter 50
Teaching Resources
    Chapter 50 Graphic Organizer
    Chapter 50 Skills Worksheet
    Chapter 50 Critical Thinking Worksheet
    Chapter 50 Challenge/Enrichment Worksheet

REVIEW AND ASSESSMENT

Chapter 50 Check, page 621
Chapter 50 Test, *Teaching Resources*
Chapter 50 Portfolio Assessment, *Teaching Resources*
Test Generator, Chapter 50
Alternative Assessment Forms

ADDITIONAL RESOURCES

Books for Teachers

Doubilet, David. *Pacific: An Undersea Journey.* New York: Little, Brown, & Co., 1992.

Linnekin, Jocelyn, and Lin Poyer. *Cultural Identity and Ethnicity in the Pacific.* Honolulu: University of Hawaii Press, 1990.

Pyne, Stephen. *The Ice: A Journey to Antarctica.* Iowa City: University of Iowa Press, 1986.

Smith, Bernard. *Imagining the Pacific: In the Wake of the Cook Voyages.* New Haven, CT: Yale University Press, 1992.

Books for Students

Marke, Julie. *Living on Islands.* New York: Franklin Watts, 1987.

Marshall, Bruce, ed. *The Real World: Understanding Our World Through the New Geography.* Boston: Houghton Mifflin, 1991.

Theroux, Paul. *The Happy Isles of Oceania: Paddling the Pacific.* New York: Putnam, 1992.

Wright, Ronald. *On Fiji Islands.* New York: Viking, 1986.

The World Today *Videodisc Program*

A variety of still images, short videos, and activities are available for you to use as you teach this chapter. *Lecture Notes* include bar codes for specific images and videos.

Multimedia Resources

Antarctica. (16mm film/VHS; 25 minutes) Physical Geography of the Continents series. National Geographic Society, Educational Services, Washington, DC 20036.

Antarctica: The Last Continent. (Multimedia CD-ROM) Cambrix Publishing, 6269 Variel Avenue, Suite B, Woodland Hills, CA 91367.

Antarctic Wildlife Adventure. (VHS, 59 minutes) National Geographic Society, Educational Services, Washington, DC 20036.

Picture Atlas of the World. (interactive maps, photo bank, animation, audio on CD-ROM) National Geographic Society, Washington, DC.

MEETING INDIVIDUAL NEEDS

Limited English Proficiency (LEP)

Help students appreciate the vast area that the Pacific islands cover by asking volunteers to describe the latitudes and longitudes of major islands, as you write the information on the board. Ask students to determine the distance in degrees from the northernmost Pacific island to the southernmost, and from the eastern edge of the region to the western edge.

Learners Having Difficulty

Assign each student a subsection of Chapter 50. Ask students to write sentences about the main ideas of the subsection. You may wish to have students read their sentences aloud as you write them on the board or on an overhead projector in the form of a chapter summary. Then have students work in pairs to ask each other questions based on the summary.

Auditory Learners

You might obtain from your local library one of Richard E. Byrd's books about his Antarctic explorations and then read a selection aloud to the class. A high-interest example is *Alone,* published in 1938. It describes Byrd's solitary experience at an Antarctic base camp for one year.

Visual Learners

Organize students into three groups, and assign each group a section of Chapter 50. Have group members study the section to create a time line of important dates and events in the history of the region. Combine the three time lines into one showing the history of the Pacific islands and Antarctica.

Tactile/Kinesthetic Learners

Have students construct three-dimensional models of one of the regions or places discussed in this chapter. For example, students might use a deep bowl filled with water and other materials to represent the Pacific island region, mud and hot water to show geothermal activity on New Zealand's North Island, or a plate of irregularly-shaped ice to represent Antarctica.

Gifted Learners

Have interested students investigate the latest scientific information discovered by researchers in Antarctica. Ask students to present their findings to the class in an oral report.

MULTICULTURAL FOCUS

Point out to students that the geographic seclusion of many Pacific ethnic groups allows anthropologists to study how a society's cultural practices may change because of outside influences.

The Kim-Yal, who live in an isolated village high in the central mountains of Irian Jaya on the island of New Guinea, provide an example of this. The Kim-Yal are among the last known makers of stone tools. Carefully grinding and polishing blue-gray lava rocks, these crafters create tools sharp enough to cut through a tree as thick as a telephone pole in five to 10 minutes.

The Kim-Yal have traditionally traded these axes for goods from other parts of New Guinea. However, they have recently begun to use metal axes manufactured elsewhere and brought into their region. As a result, very few of the younger members of the group are learning to make stone tools anymore. Anthropologists believe that the ability to make the axes will be lost in two or three generations.

Elicit comments from students regarding how the experience of the Kim-Yal reflects positive and negative aspects of new technologies replacing older ones, or handcrafted tools being replaced by mass-produced ones. *(The newer axes might provide convenience to the Kim-Yal, but the stone tools and the ability to make them are linked to their cultural traditions.)*

OVERVIEW

In this chapter, students will learn about the unique physical characteristics of the islands and island groups of the Pacific region, New Zealand, and Antarctica.

Section 1 describes the significant geographic features, historical development, and economic life of the major Pacific island groups.

Section 2 focuses on the physical features, history, and culture of New Zealand. It explains the need for good trade relations between New Zealand and its trading partners to support the country's economy.

Section 3 describes the harsh environment and scientific importance of Antarctica, where the only human population is made up of researchers and scientists from many countries.

About the Author

John Gunther (1901–1970) was an American journalist and newspaper correspondent in Europe during World War II. He is well known for his numerous political and travel books. The quotation on this page is from *Inside Australia and New Zealand.*

Primary Source

Description of Change: excerpted and bracketed
Rationale: excerpted to focus on main idea; bracketed to clarify meaning

---

CHAPTER 50

The Pacific Islands and Antarctica

❝ The North Island [of New Zealand] is all juts and spits and gulfs and bays. A protuberance [bulge] reaches north like a strand of taffy pulled away from the bulk of the island; no spot on . . . this peninsula is more than twenty miles from the sea. . . . All is fairly flat here for a hundred miles to the south of Auckland—then the country rises to volcanically active elevations of between three and nine thousand feet, and they dominate the rest of the island. This country is prone to convulse, rumble, boil, hiss, gurgle, seethe, tremble, and blow up. ❞

John Gunther

GEOGRAPHY DICTIONARY

Oceania

atoll

Exclusive Economic Zone (EEZ)

polar desert

katabatic wind

Story board from **Papua New Guinea**

Covering more than one-third of the earth's surface, the Pacific Ocean is the largest geographic feature on Earth. It contains more than 25,000 islands, more than are in all the other oceans and seas combined. Because most of the Pacific region is water, it is sometimes called **Oceania**.

Some of the Pacific islands are little more than piles of jagged rock or coral reefs. Others cover thousands of square miles. Likewise, some islands are heavily populated, while others remain uninhabited. The first peoples arrived in the Pacific thousands of years ago from Southeast Asia. They were experienced navigators and eventually settled even the most isolated island groups. Today, modern influences have greatly changed the islanders' traditional ways of life.

South of the Pacific, and the rest of the world, lies the ice-covered continent of Antarctica. Its only human inhabitants are researchers studying this vast, frozen land.

Northern **Papua New Guinea**

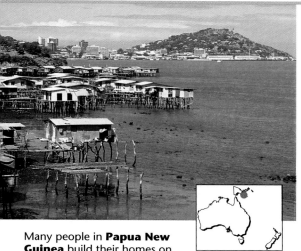

Many people in **Papua New Guinea** build their homes on stilts, like these houses near **Port Moresby**. The elevated houses are cooler and are better protected from rising water.

ISLANDS OF THE PACIFIC

FOCUS

● *What is the physical and historical geography of the Pacific islands?*

● *What are the major islands and island groups of the Pacific? What are their characteristics?*

REGION **Physical Geography** The islands of the Pacific can be divided into three island regions: Melanesia, Micronesia, and Polynesia. All of these groups contain high islands and low islands. The high islands have fresh water, good soils, and forest resources, and can thus support large human populations. Most high islands were formed by volcanoes rising from the deep sea. Tahiti, Hawaii, and Guam are examples of volcanic high islands. In the South Pacific, however, lie some large high islands, formed not from ocean volcanoes, but from sections of ancient continents. Examples of high continental islands are New Guinea, New Caledonia, and New Zealand.

The low islands, most of which are just above sea level, are made of coral. Because these islands have few resources, limited fresh water, thin soils, and few trees, they support only small

human populations. Many small coral islands form **atolls**. An atoll is a ring of small islands built up on a coral reef surrounding a shallow lagoon.

Most of the Pacific islands lie in the tropics and have a humid-tropical climate, with rainfall all year. Some have a tropical-savanna climate, with most rainfall occurring during the Southern Hemisphere's summer.

MOVEMENT **Historical Geography** The first humans reached the Pacific islands about 50,000 years ago. Probably the first to arrive were the people who settled the large South Pacific islands of Melanesia. Thousands of years later, the Polynesian culture developed around the

FACTS IN BRIEF **The Pacific Islands**

COUNTRY POPULATION (1994)	LIFE EXPECTANCY (1994)	LITERACY RATE (1990)	PER CAPITA GDP (1993)
FIJI 764,382	63, male 68, female	85%	$4,000
KIRIBATI 77,853	53, male 56, female	90% (1985)	$525 (1990)
MARSHALL ISLANDS 54,031	62, male 65, female	86%	$1,500 (1992)
MICRONESIA 120,347	66, male 70, female	90% (1991)	$1,500 (1989) (GNP)
NAURU 10,019	64, male 69, female	99% (1988)	$10,000 (1989) (GNP)
NEW ZEALAND 3,388,737	73, male 80, female	99% (1991)	$15,700
PALAU 16,366	69, male 73, female	92% (1980)	$2,260 (1986)
PAPUA NEW GUINEA 4,196,806	56, male 57, female	52%	$2,000
SOLOMON ISLANDS 385,811	68, male 73, female	60% (1989)	$2,500 (1991)
TONGA 104,778	66, male 70, female	99% (1988)	$2,000
TUVALU 9,831	62, male 64, female	96%	$700 (1990) (GNP)
VANUATU 169,776	58, male 61, female	90%	$1,050 (1990)
WESTERN SAMOA 204,447	66, male 70, female	90% (1989)	$2,000 (1992)
UNITED STATES 260,713,585	73, male 79, female	97% (1991)	$24,700

Sources: *The World Factbook 1994*, Central Intelligence Agency; *The World Almanac and Book of Facts 1995*

Which is the most populous country of the region?

613

Motivator

Have students refer to page 612 and review the quotation. Point out that 75 percent of New Zealand's population lives on North Island. Ask students to name areas of the United States that are subject to many of the same convulsions, rumbles, and trembling, yet are heavily populated. *(California, Hawaii)* Discuss why people might want to live in these areas.

Guided Practice

Ask students to give examples of the power sources, agricultural products, and industrial products found in New Zealand, and write the examples on the board. Then work with students to create a poster titled "The Economic Geography of New Zealand." Have students design the poster, illustrate the examples they included on the board, and write captions for each illustration.

Closure

Review the main features of New Zealand's physical, economic, and human geography. Ask students to say, based on the information in this section, whether they would like to live in or visit New Zealand and to explain why.

COOPERATIVE LEARNING

Organize the class into groups of four or five. Ask each group to prepare a television documentary about New Zealand. The documentary should present the physical features, ways of making a living, and cultural activities found there. Tasks might be divided into doing background research and organizing, writing, illustrating, and narrating the documentary. Have groups present their documentaries to the class.

THEME
Region

New Zealand devotes a major part of its economic assistance resources to projects in the Pacific island countries. New Zealand belongs to a group called the South Pacific Forum, which promotes regional economic cooperation.

NEWS FROM AROUND THE WORLD — *New Zealand*

New Zealand Teenagers Feel Global Influences

Thousands of miles and 11 time zones separate New Zealand and Great Britain. Yet, as a member of the British Commonwealth, New Zealand—particularly its schools—is steeped in British tradition. Morning tea is served promptly at 10 a.m. each school day to teachers and administrators, while students entertain themselves nearby. For 15 to 30 minutes, teachers share news, enjoy tea or coffee, and gear up for the day ahead.

Also borrowed from British tradition are school uniforms, like those shown above. New Zealanders believe that uniforms encourage students to take education seriously. They also maintain that uniform clothing lessens the distinction between rich and poor. For the most part, uniforms are worn proudly, and students are discouraged from any behavior that might "disgrace the school uniform."

Although most of the population is of British descent, New Zealand is a nation of diverse peoples. In recent years, increased immigration to New Zealand has meant a blending of world peoples and a sharing of new traditions. So that students of all backgrounds feel welcome, a class might be greeted each morning in several languages: "namaste" in Hindi, "magandáng umága" in Tagalog, "zǎo shàng hǎo" in Chinese, "kia ora" in Maori, "talofa" in Samoan, or "good morning" in English.

---

islands, North Island and South Island. (See the map on page 601.)

Mountains and hills cover almost three-fourths of New Zealand. On South Island, the spectacular glaciated Southern Alps rise from the Tasman Sea. Mount Cook, New Zealand's highest peak, is found here. North Island, meanwhile, has four volcanic peaks separated by a volcanic plateau. Most of the rest of the island is rolling hills and plains. New Zealand's location on the Pacific Ring of Fire makes the nation prone to earthquakes and volcanic eruptions. Geothermal activity on North Island causes hot springs, geysers, and steam vents.

Because of its oceanic, middle-latitude location, New Zealand has a mild marine-west-coast climate. The warmest area is on the northern peninsula of North Island. The coolest area is Stewart Island in the far south.

New Zealand also lies in the path of the westerly winds. As the westerlies rise over the Southern Alps, rain and snow fall on the windward side of the range. Some areas on the west coast receive up to 320 inches (813 cm) of rain per year. Dry air descending on the leeward side of the mountains results in some areas receiving only 14 inches (36 cm) of rain per year.

As in Australia, the biogeography of New Zealand is distinct. The country's endemic animals include several species of bats and flightless birds. The most famous flightless bird is the kiwi (KEE-wee), the country's national symbol. The many alien species, including deer, sheep, and opossums, have greatly altered the ecology of the islands.

MOVEMENT **Historical Geography** The first people to reach New Zealand were the Moa hunters, who arrived about 1,000 years ago. They were named after the huge, flightless birds they hunted for food. The Moas eventually died out, perhaps because they hunted the moa birds to extinction.

The Maoris (MAU-reez), a Polynesian people, settled the islands sometime between the tenth

ASSESSMENT

Assign the Section 2 Review questions on this page.

Reteaching

Assign Section 2 of Reteaching Worksheet 50.

Extension

Have interested students research and report on Maori art such as wood carvings. Students should indicate the role of the environment in creating the artwork. Encourage students to include examples of Maori art with their reports.

SECTION 3
OBJECTIVES

The objectives reflect the Focus questions on this page and are assessed in the review on page 620.
1. Describe Antarctica's physical geography.
2. Discuss human activity in Antarctica.

and fifteenth centuries. Although they built an advanced civilization, fierce wars broke out among various Maori groups.

The first European to sight New Zealand was the Dutch explorer Abel Tasman in 1642. Then in 1769, Captain James Cook claimed the islands for Great Britain. The first British settlements were established in the 1840s. Although the British settlers and the Maoris signed a treaty in 1840, fighting between them continued until the early 1870s. Today, the Maoris form about 10 percent of the population.

New Zealand is a parliamentary democracy and, like Australia, is a member of the British Commonwealth of Nations. It has had self-government since 1852.

Economic Geography New Zealand is a major exporter of lamb, beef, butter, cheese, and wool. Its mild climate also allows a great variety of crops, including grains, vegetables, and fruit. New Zealand grows enough food to supply most of the nation's needs. Its leading export crops are apples and kiwifruit.

The country must import oil, but it has large quantities of coal and offshore natural-gas deposits. Hydroelectric power and geothermal energy, however, supply more than 80 percent of the country's electricity. New Zealand is not rich in mineral resources, but the country does have some iron and gold deposits.

New Zealand's industries include tourism, processed foods, minerals, and wood and paper products. The country also produces steel, aluminum, boats, and clothing. Most manufactured goods, however, must be imported.

Before the formation of the European Union, New Zealand's exporting was largely dependent on British markets. Today, Japan, Australia, and the United States, as well as the EU, are its major trade partners. New Zealand seeks to maintain good relations with all countries because of its dependence on foreign trade.

Population About 75 percent of New Zealand's population lives on North Island, where most industry, agriculture, and the largest cities are found. In fact, more than 85 percent of New Zealanders live in urban areas. The city of Auckland has the nation's largest population, most industry, and main seaport. Wellington, the capital, is New Zealand's second largest city.

New Zealanders, sometimes called "Kiwis," generally enjoy a high standard of living. In return for high taxes, the government provides an extensive social-welfare system.

 SECTION REVIEW

1. What climate type dominates New Zealand?
2. What are New Zealand's major exports? Who are the Maoris?
3. **Critical Thinking** **Explain** why most of New Zealand's population, industry, and agriculture are located on North Island.

 ANTARCTICA

FOCUS

● *What is the physical geography of Antarctica?*

● *What human activity takes place in Antarctica?*

LOCATION **The Frozen Continent** At the southernmost part of the world lies Antarctica, Earth's coldest, driest, highest, and windiest continent. It remains the planet's last frontier, a place where harsh physical geography reigns over human activity.

Antarctica's total land area is about 5.5 million square miles (14.3 million sq. km), or about 10 percent of the world's land. Yet less than 5 percent of the continent is free of ice. For six months of the year, Antarctica is hidden from the sun and remains in total darkness. During its short summer, the sun never sets.

The Antarctic ice sheet contains more than 90 percent of the world's ice and averages over one mile (1.6 km) in depth. In some areas, the ice buries entire mountain ranges. The weight of the ice sheet causes it to flow slowly off the continent. When it reaches the sea, huge icebergs then break away and drift northward into the ocean. Some of these icebergs are larger than the state of Rhode Island.

The towering peaks of the Transantarctic Mountains divide the continent into East

Text continues on page 620.

Section 2 Review
ANSWERS

1. Marine west coast
2. Lamb, beef, butter, cheese, wool, apples, kiwifruit; a Polynesian people who settled the islands before the Europeans
3. North Island's landforms are better suited to population, industry, and agriculture. Whereas South Island is mostly mountainous, much of North Island has rolling hills and plains. Also, geothermal activity on North Island provides a source of power.

Planet Watch

Expressing Points of View

Lead a class discussion on the following questions: Should individual nations be allowed to claim and control Antarctica? How would these claims be verified and enforced? Or instead, should Antarctica be designated a world park on behalf of all the earth's people? How would Antarctica's "neutrality" be upheld? List students' arguments for and against these two proposals on the board.

····· Geographical
Sidelight

Antarctica also offers many opportunities for astronomical and space research. NASA sometimes uses the continent as a training ground for astronauts and as a laboratory for developing new technologies. Dark, 24-hour nights during the winter, frigid temperatures, and the high altitude combine to provide advantages for using optical telescopes to study the heavens.

PLANET WATCH

Preserving Antarctica

In 1912, a search party combing the frozen lands of Antarctica found the body of explorer Robert Falcon Scott. Scott had written in his diary, "Great God! This is an awful place." His dream of being the first person to reach the South Pole had been thwarted. Scott's expedition reached the pole only to find the tent and flag of Norwegian explorer Roald Amundsen, who had arrived five weeks earlier than Scott.

Nearly a century later, Antarctica's modern-day explorers are research scientists. For the most part, Antarctica has become a successful experiment in international scientific research and cooperation.

Scientists believe that understanding Antarctica is an important key to understanding our planet.

A Scientific Laboratory

A wasteland to some, Antarctica is a scientific laboratory to many. The continent is of central importance to understanding our planet. For example, the icy waters bordering Antarctica

move north, cooling warmer waters. This movement helps explain ocean currents, clouds, and weather patterns. The world's climate is also affected by Antarctic sea ice. The ice acts as a shield, keeping the earth cool by reflecting rather than absorbing the sun's heat energy.

The continent's ice, meanwhile, gives clues to the past. Buried within the layers of ice are gas bubbles and pollutants that are a "record" of Earth's air. Scientists have been able to compare, for example, traces of atmospheric gas trapped in Antarctic ice with atmospheric gas of today. From this comparison, they have learned that the use of fossil fuels has raised the amount of carbon dioxide in the air to the highest levels in human history.

Eyes on Antarctica

Antarctica is not owned by any one nation. Although a few countries claim sectors of Antarctica, none of these claims is recognized. In 1959, the Antarctic Treaty established Antarctica as "a continent for science and peace." Signed by 12 nations and later agreed to by 26 additional nations, the treaty banned military activity in the region and made Antarctica a nuclear-free zone. It also promoted free access to scientific research on the continent and banned further territorial claims by nations.

The Antarctic Treaty, however, did not cover mining rights on the continent. So when Antarctica's mineral riches were discovered, some nations began viewing the continent as a source of untapped wealth. Geographers, scientists, and environmentalists also took notice. They feared that if mining took place on Antarctica, the continent's environment would suffer. In particular, if a practical, inexpensive method of

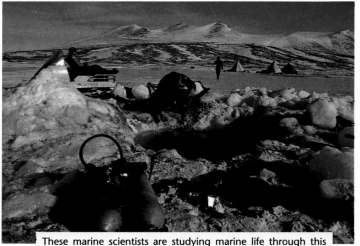

These marine scientists are studying marine life through this six-foot hole cut through the ice around **Antarctica**. A rich variety of marine life inhabits the water under Antarctica's ice.

extracting the offshore oil were found, Antarctica's coastline and rich marine life would be threatened by oil spills.

At the same time, evidence of environmental neglect at the continent's research stations appeared. Environmentalists voiced concern about scientists' careless disposal of trash and sewage. A U.S. Coast Guard captain who worked on icebreakers described pollution at McMurdo, a U.S. scientific base. "Trash was just rolled down the hill. . . . One of the jobs of the icebreakers was to break up the ice where the trash was and push it out to sea." Concern for the Antarctic waters increased when, in 1989, an Argentine supply ship ran aground along Antarctica's rocky shore, spilling 170,000 gallons of oil.

Antarctica

Seven countries have claimed parts of Antarctica. The United States and Russia, which do not recognize these claims, are among the nations that have research stations on the continent.

Protecting Antarctica

In 1991, 39 nations abiding by the original Antarctic Treaty signed a new agreement called the Madrid Protocol. Unlike the original treaty of nearly 40 years ago, the new agreement is designed to protect the continent. It forbids most activities on Antarctica that do not have a scientific purpose. It also bans mining and drilling on the continent by the signing nations and sets limits on tourism. Pollution concerns are addressed specifically.

The treaty will remain in effect for 50 years. At that time, the agreement can be changed if two-thirds of the signing nations agree. Then, it becomes the responsibility of another generation to preserve Antarctica.

Penguins and elephant seals share this island off the coast of **Antarctica**. Most of Antarctica's land animals live along the edges of the continent and on surrounding islands.

▼ YOUR TURN

Imagine you are helping develop the Madrid Protocol. **Write** a short description of the provisions you want included in the treaty and why you think they are important.

619

TEACHING SECTION 3

Motivator

Give students a reference of a mile they can relate to, such as the distance from a familiar point A to a familiar point B. Then ask them to imagine an ice sheet deeper than a mile covering a continent. Have a volunteer point out Antarctica on a globe or a map, and ask students to suggest why the continent is covered with ice.

Guided Practice

Have students assume the role of a scientist at a research station in Antarctica. Ask them to write diary entries describing their impressions of the physical features of Antarctica and the difficulties involved in working there. Call on volunteers to read their entries.

Closure

Lead a discussion on how the human activity that occurs in Antarctica relates to its physical geography.

ASSESSMENT

Assign the Section 3 Review questions on this page.

Reteaching

Assign Section 3 of Reteaching Worksheet 50.

Geographical Sidelight

Although the massive ice sheet covering Antarctica is more than a mile thick, it moves and changes. Several forces act on the ice sheet to cause shifting and even melting. These forces include calving, or breaking off, of ice chunks into the sea, global warming, and volcanic eruptions occurring under the ice.

An iceberg in the bitter cold climate of **Antarctica** is a haven for these penguins. Except for researchers who temporarily live on the continent, only penguins and other wildlife make up Antarctica's population.

Section 3 Review

ANSWERS

1. More than 95 percent of the continent is covered with ice, and the Antarctic ice sheet contains more than 90 percent of the world's ice and averages over a mile thick.

2. Scientific research

3. No. The Antarctic Convergence Zone is where the Antarctic waters meet the middle-latitude waters, forming the boundary of the Antarctic region. It does not involve economic control of marine resources, as do Exclusive Economic Zones.

620

Antarctica and West Antarctica, a few dry coastal valleys, and the Antarctic Peninsula. The highest peak on the continent is Vinson Massif at 16,860 feet (5,139 m). Except for a few insects, this frozen continent supports no land animals and has never had a native human population. Only tundra vegetation survives in the rare ice-free regions.

Antarctica's polar ice-cap climate is severe. Because of its high average elevation and because more than 95 percent of the land is ice covered, Antarctica is much colder than even the Arctic region. For most of the year, temperatures remain bitterly cold, dropping to -120°F (-84°C) or lower. The Russian research station in East Antarctica recorded the world's record low temperature of -128.6°F (-89.2°C). Most of Antarctica is a **polar desert** and receives less precipitation than the Sahara in Africa. Here, however, there is almost no evaporation and melting. As a result, ice has built up over thousands of years.

Air flowing off this icy land makes Antarctica the windiest place on Earth. **Katabatic winds** gusting more than 100 miles per hour (161 kph) race along Antarctica's coast. A katabatic wind is a rapid flow of dense cold air down the slopes of high mountains or ice caps.

Antarctic Waters The "Southern Ocean" isolates Antarctica from the other continents. The Southern Ocean is really the far southern reaches of the Atlantic, Pacific, and Indian oceans. It has a rich and distinct ecosystem, with a large population of penguins, seals, and whales. This ocean life depends on a small shrimplike animal that forms

the basis of the Antarctic food chain. Without these tiny shelled creatures, the Antarctic waters would have little or no ocean life.

During winter in the Southern Hemisphere, the Southern Ocean freezes over with about five feet (1.5 m) of drifting sea ice. During the summer, the sea ice melts, and research and supply ships can reach some sections of the Antarctic coast. The stormy waters around Antarctica are the world's most severe, with high winds and huge waves. The world's largest ocean current, the West Wind Drift, flows clockwise around the continent. Where the cold Antarctic waters meet the warmer middle-latitude waters is the Antarctic Convergence Zone. This is at about 50° south latitude and forms the true boundary of the Antarctic region.

INTERACTION **Research** British explorer James Cook was the first person known to reach Antarctic waters. His sighting of icebergs in 1772 confirmed that a vast continent existed. It was not until 1911, however, that the first recorded human expedition reached the South Pole. Today, the only people found in Antarctica are researchers working in scientific research stations, which are located at various points on the ice cap.

The Antarctic environment holds clues to the ice ages, global warming, the ozone layer, meteorites (found in the ice), and marine ecosystems. Indeed, research here has provided new and fascinating information on our planet and solar system. Minerals have been found on Antarctica, and oil is thought to be there as well. Mining of these minerals is prohibited. (See "Planet Watch" on pages 618–619.) Researchers, however, continue to mine Antarctica for answers to questions about the planet's past, present, and future.

SECTION REVIEW

1. Why is Antarctica called "the frozen continent"?
2. In what activity are most of the people in Antarctica involved?
3. **Critical Thinking** **Determine** whether the Antarctic Convergence Zone is an Exclusive Economic Zone. Explain your reasoning.

Building a Vocabulary

1. Because most of the region is ocean, rather than land
2. A ring of small islands built on a coral reef around a shallow lagoon; no, because atolls are made of coral
3. Katabatic wind
4. No, polar deserts receive little precipitation but are covered with ice.

Recalling and Reviewing

1. New Zealand has a marine-west-coast climate, whereas most of the other Pacific islands have a humid-tropical or tropical-savanna climate. New Zealand's higher latitude accounts for the difference.
2. Australia, Japan, and the United States
3. Preserving traditional ways of life and distinct cultures in the face of rapid acculturation
4. The continent itself supports no land animals except for a few insects; tundra is the only vegetation it supports, and only in a few ice-free regions. The Southern Ocean, however, supports much marine life, including penguins, seals, and whales.

CHAPTER 50 CHECK

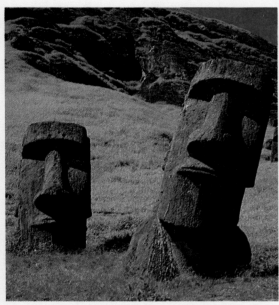

Carved hundreds of years ago, some 600 stone statues called *moai* (MOE-ie) still watch over isolated **Easter Island**.

Reviewing the Main Ideas

1. There are more than 25,000 islands in the Pacific Ocean. Some are small and uninhabited, while others are large and populated. The islands are divided into three regions: Melanesia, Micronesia, and Polynesia. All the island cultures have been influenced by Asian and European explorers and settlers.
2. New Zealand is made up of two large continental islands: North Island and South Island. New Zealand exports many agricultural products but must import oil and most manufactured goods. Auckland is New Zealand's largest city.
3. Ice covers almost all of Antarctica. The Southern Ocean supports life, but the only human inhabitants of the continent are scientific researchers.

Building a Vocabulary

1. Why is the Pacific region sometimes called Oceania?
2. Define *atoll*. Are the soils of atolls agriculturally productive?

3. What is the term for a rapid flow of dense cold air down high mountain slopes or ice caps?
4. Is a polar desert dry? Why or why not?

Recalling and Reviewing

1. How does New Zealand's climate differ from the climates of most of the other Pacific islands? What accounts for the difference?
2. What countries are the major trading partners of the Pacific islands and New Zealand?
3. What cultural challenges do the Pacific islands face?
4. Describe Antarctica's biogeography.

Thinking Critically

1. Both cultural diffusion and cultural isolation characterize the Pacific islands. Explain how both are possible in the same region and how geographic factors are involved.
2. Compare and contrast the economic geography of New Zealand and Australia.

Learning About Your Local Geography

Cooperative Project
World War II had a great impact on the Pacific islands. To what extent did World War II affect your own community? Interview an individual who has lived in your community long enough to remember the war and its local effects. After the interview, write a summary of what you have learned.

Using the Five Themes of Geography

What forms of **human-environment interaction** are currently possible in Antarctica, given its physical **place** characteristics? Make a list of adaptations humans might make to enable increased **movement** to and settlement in the region. Do you think that modification of the environment is possible? In your opinion, would such modification be desirable?

Thinking Critically

1. Answers will vary, but students should mention the ocean as both an isolating factor and one that has prompted trade, migration, and the diffusion of cultures.
2. Both countries are major exporters of wool, lamb, and beef. Both grow many crops, have some industries, and have had to find new trading partners after the formation of the European Union. Both countries have coal and natural gas, but Australia has more oil and other mineral resources.

Learning About Your Local Geography

Summaries should discuss effects of World War II on students' local communities, as researched through interviews. A possible focus might be local economic activities.

Using the Five Themes of Geography

Lists of adaptations might mention domed or underground structures.

621

Unit 11 Review

SKILLS

Use "The Foods We Eat" as an example of global linkages. Call on students to find out what foods they have eaten over the past few days. Write each food on the board. Then, with the class, write a possible region or country of origin for each food listed. For example, bananas might have come from Middle or South America; rice might have come from East or Southeast Asia. Point out that even the very basic things we do for survival—such as eat—are made possible by global connections.

UNIT 11
REVIEW ✦ SKILLS

Developing Global Awareness

Some say all the world's people are connected by a long chain of acquaintances—that everyone knows someone who knows someone else, and so on. While the existence of such a chain would be difficult to prove, the idea that everyone on Earth is connected in some way is a reality in modern times. Things that happen on the other side of the planet can affect you, and things that you do in your community can have worldwide impact. An understanding of such relationships is called global awareness.

You might think that the Pacific islands, spread throughout the vast Pacific Ocean, have somehow managed to escape global importance. But nothing could be further from the truth. Despite the water that physically separates the islands from each other and from the continents, they are connected to the rest of the world. Many people from Europe, Asia, and the Americas have settled on the islands, and the societies there are a blend of cultures. Also, while the islands depend on income from tourism, they must adapt to tourists' expectations that the islands are a sort of paradise. This often means further cultural change.

Furthermore, due to their location between Asia and the Americas, the islands have global strategic value. Several of the islands have been used as military bases by larger nations. Some Pacific islands were the scene of intense fighting between the United States and Japan during World War II. And because of the islands' relative isolation and small populations, some of them have been used as sites for weapons testing. As a result of this testing, certain islands are now uninhabitable.

As you read or watch the news, think about the global meaning of the events being discussed. How is the United States involved politically, economically, militarily, environmentally, or culturally in events that take place throughout the world? How are other countries involved in U.S. affairs? How do events in the opposite hemisphere affect you directly or indirectly, and how do your actions affect people in the opposite hemisphere or even just on the opposite side of the country? With the pace of communication growing ever faster, the world's regions become linked in more ways, and almost everything we do can have a global effect.

Practicing the Skill
ANSWERS

1. Paragraphs should analyze the global significance of a recent news event. Students should name countries involved both directly and indirectly and should discuss how the outcomes of the event will affect the rest of the world.

2. Check to see that students have completed logs explaining how daily activities have global impact.

PRACTICING THE SKILL

1. Analyze the global significance of a news event you have read or heard about. What countries are involved directly? What countries are involved indirectly? What effects will the outcomes of the event have on the rest of the world? Write a paragraph answering these questions.

2. Keep a log of your activities during one day. Identify from your log activities that have global impact, however minor. Make notes next to those selected entries, explaining their global implications.

622

BUILDING YOUR PORTFOLIO

Decide whether students will work individually or in groups on the unit projects described below. Students should then choose which project they will work on. Provide students with copies of the Portfolio Assessment Worksheets in *Teaching Resources*. Chapters 49–50 worksheets and the Unit Portfolio Assessment Worksheet will guide students step-by-step to complete either project.

Teaching with Sketch Maps

Have students sketch a map of the Pacific World and Antarctica from memory. Tell them to include as many details as they can about place names, landforms, and boundaries of regions they studied in Unit 11. Students should compare this map with the sketch map they completed at the beginning of Unit 11 to note their improvement in identifying geographic features.

BUILDING YOUR PORTFOLIO

Individually or in a group, complete one of the following projects to show your understanding of the geography concepts involved.

A Biogeography

As you know, biogeography is the study of the geographic distribution of plants and animals. From this unit or from outside sources, identify and select any animal or plant species found in Australia, the Pacific islands, Antarctica, the Pacific Ocean, or the Southern Ocean. Once you have made your selection, use library resources to research the animal or plant's geographic origins, its unique or unusual features, its role in and interaction with its environment, and its interaction with humans. Design a poster to display your findings. You will need to consider the following questions as you prepare your poster.

1. How did the animal or plant come to its present location? Is it endemic to the area, or was it introduced from another region? What unique or unusual features does the species have? How do these features relate to geography? Draw some pictures of the species for your poster. Next to your illustrations, write about the geographic origins and unusual features of the plant or animal.
2. How does the plant or animal interact with its environment, including humans? Where is the plant or animal in the food chain? Do many other species depend on it for survival, or is it near the top of the food chain? Has the plant or animal's presence changed the environment? Do humans hunt or use the plant or animal? If so, for what purpose? Do any human activities endanger the plant or animal? Create diagrams, graphic organizers, or flowcharts showing the plant or animal's patterns of interaction with its environment and with humans. Place these materials on your poster.

Present your poster to the rest of your class.

B Antarctica and Outer Space

You have read about how Antarctica is not a country or territory but is peacefully shared by many different countries. Antarctica is viewed by some as a model of a much larger mostly unexplored area—outer space. Using library resources, collect information on the Antarctic Treaty. Then, using what you have learned about the Antarctic Treaty, write a treaty outlining how Earth's nations might cooperate in their efforts to explore outer space, use its resources, and learn about any life-forms discovered in space in the future. To complete this project, you will need to do the following.

1. Make an outline of the provisions of the Antarctic Treaty. How is the treaty divided? What issues does the treaty address? How does the treaty provide for potential problems in dealing with Antarctica?
2. Based on your outline of the Antarctic Treaty, write a treaty for how nations should approach the exploration and use of outer space. Should space be divided into territories, each belonging to a country, or should it be shared by many countries, as Antarctica is? If valuable resources are discovered in space, what rules will govern their use? Do the environments of other planets and of space itself need protection? What policy should govern encounters with life-forms in space, should such encounters occur?

Present your outline of the Antarctic Treaty and your own treaty for outer space to your class. Discuss the extent to which the two treaties are similar, and explain why there are differences.

Project A:

Students' posters might focus on the kangaroo, koala, platypus, spiny anteater, eucalyptus, dingo, sheep, rabbit, kiwi, penguin, or other animal or plant species found in the Pacific World or Antarctica. Posters should include illustrations of the species, explanations of its geographic origins, descriptions of its unusual features, and diagrams, graphic organizers, or flowcharts showing its patterns of interaction with its environment, including humans.

Project B:

Outlines of the Antarctic Treaty might indicate that the treaty consists of several articles dealing with usage, international cooperation, exchanges of information and people, territorial claims, environmental protection, areas of application, inspections, dispute settlement, and review of the treaty itself. Students' treaties for outer space may have similar elements; students may also add other items they think are important. Presentations should compare and contrast the two treaties.

$\mathbf{A}$PPENDIX

Metric conversion table 624
Statistical Charts 624
The United States and Canada 624
Middle and South America 624
Europe 626
Russia and Northern Eurasia......................629
Southwest Asia 629
Africa 630
East and Southeast Asia 634
South Asia 635
The Pacific World..635
The United States (by state) 636

Metric Conversion Table

	If you have	multiply by	to get
LENGTH	miles	1.609	kilometers
	kilometers	.62	miles
	feet	.3048	meters
	meters	39.37, then divide by 12	feet
	inches	2.54	centimeters
	centimeters	.39	inches
WEIGHT	pounds	.454	kilograms
	kilograms	2.2046	pounds
AREA	acres	.405	hectares
	hectares	2.47	acres
	square miles	2.59	square kilometers
	square kilometers	.3861	square miles
CAPACITY	quarts	.946	liters
	liters	1.057	quarts
TEMPERATURE	degrees Fahrenheit	subtract 32, then multiply by $5/9$	degrees Celsius
	degrees Celsius	multiply by $9/5$, then add 32	degrees Fahrenheit

Country ★ Capital	Population (1994)	Land Area (1994)	Principal Languages
THE UNITED STATES AND CANADA			
United States ★ Washington, D.C.	260,713,585	3,618,765 sq. mi. 9,372,610 sq. km	English, Spanish, Native American languages, other languages of the world
Canada ★ Ottawa	28,113,997	3,851,788 sq. mi. 9,976,140 sq. km	English, French
MIDDLE AND SOUTH AMERICA			
Antigua and Barbuda ★ St. John's	64,762	170 sq. mi. 440 sq. km	English, local dialects
Argentina ★ Buenos Aires	33,912,994	1,068,296 sq. mi. 2,766,890 sq. km	Spanish, English, Italian, German, French

624

Country ★ Capital	Population (1994)	Land Area (1994)	Principal Languages
Bahamas ★ Nassau	273,055	5,382 sq. mi. 13,940 sq. km	English, Creole
Barbados ★ Bridgetown	255,827	166 sq. mi. 430 sq. km	English
Belize ★ Belmopan	208,949	8,865 sq. mi. 22,960 sq. km	English, Spanish, Maya, Garifuna (Carib)
Bolivia ★ La Paz ★ Sucre	7,719,445	424,162 sq. mi. 1,098,580 sq. km	Spanish, Quechua, Aymara,
Brazil ★ Brasília	158,739,257	3,286,470 sq. mi. 8,511,965 sq. km	Portuguese, Spanish, English, French
Chile ★ Santiago	13,950,557	292,258 sq. mi. 756,950 sq. km	Spanish
Colombia ★ Bogotá	35,577,556	439,733 sq. mi. 1,138,910 sq. km	Spanish
Costa Rica ★ San José	3,342,154	19,730 sq. mi. 51,100 sq. km	Spanish, English
Cuba ★ Havana	11,064,344	42,803 sq. mi. 110,860 sq. km	Spanish
Dominica ★ Roseau	87,696	290 sq. mi. 750 sq. km	English, French patois
Dominican Republic ★ Santo Domingo	7,826,075	18,815 sq. mi. 48,730 sq. km	Spanish
Ecuador ★ Quito	10,677,067	109,483 sq. mi. 283,560 sq. km	Spanish, Quechua and other Native American languages
El Salvador ★ San Salvador	5,552,511	8,124 sq. mi. 21,040 sq. km	Spanish, Nahuatl
Grenada ★ St. George's	94,109	131 sq. mi. 340 sq. km	English, French patois
Guatemala ★ Guatemala City	10,721,387	42,042 sq. mi. 108,890 sq. km	Spanish, Native American dialects
Guyana ★ Georgetown	729,425	83,000 sq. mi. 214,970 sq. km	English, Native American dialects
Haiti ★ Port-au-Prince	6,491,450	10,714 sq. mi. 27,750 sq. km	French, Creole

Country ★ Capital	Population (1994)	Land Area (1994)	Principal Languages
Honduras ★ Tegucigalpa	5,314,794	43,278 sq. mi. 112,090 sq. km	Spanish, Native American dialects
Jamaica ★ Kingston	2,555,064	4,243 sq. mi. 10,990 sq. km	English, Creole
Mexico ★ Mexico City	92,202,199	761,602 sq. mi. 1,972,550 sq. km	Spanish, Mayan dialects
Nicaragua ★ Managua	4,096,689	49,998 sq. mi. 129,494 sq. km	Spanish, English, Native American languages
Panama ★ Panama City	2,630,000	30,193 sq. mi. 78,200 sq. km	Spanish, English
Paraguay ★ Asunción	5,213,772	157,046 sq. mi. 406,750 sq. km	Spanish, Guaraní
Peru ★ Lima	23,650,671	496,223 sq. mi. 1,285,220 sq. km	Spanish, Quechua, Aymara
St. Kitts and Nevis ★ Basseterre	40,671	104 sq. mi. 269 sq. km	English
St. Lucia ★ Castries	145,090	239 sq. mi. 620 sq. km	English, French patois
St. Vincent and the Grenadines ★ Kingstown	115,437	131 sq. mi. 340 sq. km	English, French patois
Suriname ★ Paramaribo	422,840	63,039 sq. mi. 163,270 sq. km	Dutch, English, Surinamese, Hindi, Javanese
Trinidad and Tobago ★ Port-of-Spain	1,328,282	1,981 sq. mi. 5,130 sq. km	English, Hindi, French, Spanish
Uruguay ★ Montevideo	3,198,910	68,039 sq. mi. 176,220 sq. km	Spanish
Venezuela ★ Caracas	20,562,405	352,143 sq. mi. 912,050 sq. km	Spanish, Native American dialects

EUROPE

Country ★ Capital	Population (1994)	Land Area (1994)	Principal Languages
Albania ★ Tiranë	3,374,085	11,100 sq. mi. 28,750 sq. km	Albanian, Greek
Andorra ★ Andorra la Vella	63,930	174 sq. mi. 450 sq. km	Catalan, French, Castilian

Country ★ Capital	Population (1994)	Land Area (1994)	Principal Languages
Austria ★ Vienna	7,954,974	32,374 sq. mi. 83,850 sq. km	German
Belgium ★ Brussels	10,062,836	11,780 sq. mi. 30,510 sq. km	Flemish (Dutch), French, German
Bosnia and Herzegovina ★ Sarajevo	4,651,485	19,781 sq. mi. 51,233 sq. km	Serbo-Croatian
Bulgaria ★ Sofia	8,799,986	42,822 sq. mi. 110,910 sq. km	Bulgarian, Turkish
Croatia ★ Zagreb	4,697,614	21,829 sq. mi. 56,538 sq. km	Serbo-Croatian
Czech Republic ★ Prague	10,408,280	30,387 sq. mi. 78,703 sq. km	Czech, Slovak
Denmark ★ Copenhagen	5,187,821	16,629 sq. mi. 43,070 sq. km	Danish, Faroese, Greenlandic, German
Estonia ★ Tallinn	1,616,882	17,413 sq. mi. 45,100 sq. km	Estonian, Latvian, Lithuanian, Russian
Finland ★ Helsinki	5,068,931	130,127 sq. mi. 337,030 sq. km	Finnish, Swedish, Lapp, Russian
France ★ Paris	57,840,445	211,208 sq. mi. 547,030 sq. km	French, regional dialects and languages
Germany ★ Berlin	81,087,506	137,803 sq. mi. 356,910 sq. km	German
Greece ★ Athens	10,564,630	50,942 sq. mi. 131,940 sq. km	Greek, English, French
Hungary ★ Budapest	10,319,113	35,919 sq. mi. 93,030 sq. km	Hungarian
Iceland ★ Reykjavik	263,599	39,768 sq. mi. 103,000 sq. km	Icelandic
Ireland ★ Dublin	3,539,296	27,135 sq. mi. 70,280 sq. km	English, Irish (Gaelic)
Italy ★ Rome	58,138,394	116,305 sq. mi. 301,230 sq. km	Italian, German, French, Slovene
Latvia ★ Riga	2,749,211	24,749 sq. mi. 64,100 sq. km	Lettish, Lithuanian, Russian

Country ★ Capital	Population (1994)	Land Area (1994)	Principal Languages
Liechtenstein ★ Vaduz	30,281	62 sq. mi. 160 sq. km	German, Alemannic dialect
Lithuania ★ Vilnius	3,848,389	25,174 sq. mi. 65,200 sq. km	Lithuanian, Polish, Russian
Luxembourg ★ Luxembourg	401,900	998 sq. mi. 2,586 sq. km	Luxembourgisch, German, French, English
Macedonia ★ Skopje	2,213,785	9,781 sq. mi. 25,333 sq. km	Macedonian, Albanian, Turkish, Serbo-Croatian
Malta ★ Valletta	366,767	124 sq. mi. 320 sq. km	Maltese, English
Monaco ★ Monaco	31,278	.73 sq. mi. 1.9 sq. km	French, English, Italian, Monegasque
Netherlands ★ Amsterdam ★ The Hague	15,367,928	14,413 sq. mi. 37,330 sq. km	Dutch
Norway ★ Oslo	4,314,604	125,181 sq. mi. 324,220 sq. km	Norwegian, Lapp, Finnish
Poland ★ Warsaw	38,654,561	120,726 sq. mi. 312,680 sq. km	Polish
Portugal ★ Lisbon	10,524,210	35,552 sq. mi. 92,080 sq. km	Portuguese
Romania ★ Bucharest	23,181,415	91,699 sq. mi. 237,500 sq. km	Romanian, Hungarian, German
San Marino ★ San Marino	24,091	23 sq. mi. 60 sq. km	Italian
Slovakia ★ Bratislava	5,403,505	18,859 sq. mi. 48,845 sq. km	Slovak, Hungarian
Slovenia ★ Ljubljana	1,972,227	7,836 sq. mi. 20,296 sq. km	Slovenian, Serbo-Croatian
Spain ★ Madrid	39,302,665	194,884 sq. mi. 504,750 sq. km	Castilian Spanish, Catalan, Galician, Basque
Sweden ★ Stockholm	8,778,461	173,731 sq. mi. 449,964 sq. km	Swedish, Lapp, Finnish
Switzerland ★ Bern	7,040,119	15,942 sq. mi. 41,290 sq. km	German, French, Italian, Romansh

Country ★ Capital	Population (1994)	Land Area (1994)	Principal Languages
United Kingdom ★ London	58,135,110	94,525 sq. mi. 244,820 sq. km	English, Welsh, Scottish Gaelic
Vatican City ★ Vatican City	821	0.169 sq. mi. 0.44 sq. km	Italian, Latin, other languages of the world
Yugoslavia (Serbia and Montenegro) ★ Belgrade	10,759,897	39,517 sq. mi. 102,350 sq. km	Serbo-Croatian, Albanian

RUSSIA AND NORTHERN EURASIA

Country ★ Capital	Population (1994)	Land Area (1994)	Principal Languages
Armenia ★ Yerevan	3,521,517	11,506 sq. mi. 29,800 sq. km	Armenian, Russian
Azerbaijan ★ Baki	7,684,456	33,436 sq. mi. 86,600 sq. km	Azeri, Russian, Armenian
Belarus ★ Minsk	10,404,862	80,154 sq. mi. 207,600 sq. km	Belorussian, Russian
Georgia ★ T'bilisi	5,681,025	26,911 sq. mi. 69,700 sq. km	Georgian, Russian, Armenian, Azeri
Kazakhstan ★ Almaty	17,267,554	1,049,150 sq. mi. 2,717,300 sq. km	Kazakh, Russian
Kyrgyzstan ★ Bishkek	4,698,108	76,641 sq. mi. 198,500 sq. km	Kyrgyz, Russian
Moldova ★ Chişinău	4,473,033	13,012 sq. mi. 33,700 sq. km	Moldovan, Romanian, Russian, Turkish dialect
Russia ★ Moscow	149,608,953	6,592,735 sq. mi. 17,075,200 sq. km	Russian
Tajikistan ★ Dushanbe	5,995,469	55,251 sq. mi. 143,100 sq. km	Tajik, Russian
Turkmenistan ★ Ashgabat	3,995,122	188,455 sq. mi. 488,100 sq. km	Turkmen, Russian, Uzbek
Ukraine ★ Kyyiv	51,846,958	233,089 sq. mi. 603,700 sq. km	Ukrainian, Russian, Romanian, Polish, Hungarian
Uzbekistan ★ Toshkent	22,608,866	172,741 sq. mi. 447,400 sq. km	Uzbek, Russian, Tajik

SOUTHWEST ASIA

Country ★ Capital	Population (1994)	Land Area (1994)	Principal Languages
Afghanistan ★ Kabul	16,903,400	250,000 sq. mi. 647,500 sq. km	Pashtu, Afghan Persian, Uzbek, Turkmen

Country ★ Capital	Population (1994)	Land Area (1994)	Principal Languages
Bahrain ★ Manama	585,683	239 sq. mi. 620 sq. km	Arabic, English, Farsi, Urdu
Cyprus ★ Nicosia	730,084	3,571 sq. mi. 9,250 sq. km	Greek, Turkish, English
Iran ★ Tehran	65,615,474	636,293 sq. mi. 1,648,000 sq. km	Persian and Persian dialects, Turkic dialects, Kurdish, Luri, Baloch, Arabic, Turkish
Iraq ★ Baghdad	19,889,666	168,753 sq. mi. 437,072 sq. km	Arabic, Kurdish, Assyrian, Armenian
Israel ★ Jerusalem	5,050,850	8,019 sq. mi. 20,770 sq. km	Hebrew, Arabic, English
Jordan ★ Amman	3,961,194	34,445 sq. mi. 89,213 sq. km	Arabic, English
Kuwait ★ Kuwait City	1,819,322	6,880 sq. mi. 17,820 sq. km	Arabic, English
Lebanon ★ Beirut	3,620,395	4,015 sq. mi. 10,400 sq. km	Arabic, French, Armenian, English
Oman ★ Muscat	1,701,470	82,031 sq. mi. 212,460 sq. km	Arabic, English, Balochi, Urdu, Indian dialects
Qatar ★ Doha	512,779	4,447 sq. mi. 11,000 sq. km	Arabic, English
Saudi Arabia ★ Riyadh	18,196,783	756,981 sq. mi. 1,960,582 sq. km	Arabic
Syria ★ Damascus	14,886,672	71,498 sq. mi. 185,180 sq. km	Arabic, Kurdish, Armenian, Aramaic, Circassian, French
Turkey ★ Ankara	62,153,898	301,382 sq. mi. 780,580 sq. km	Turkish, Kurdish, Arabic
United Arab Emirates ★ Abu Dhabi	2,791,141	29,182 sq. mi. 75,581 sq. km	Arabic, Persian, English, Hindi, Urdu
Yemen ★ Sanaa	11,105,202	203,849 sq. mi. 527,970 sq. km	Arabic

AFRICA

Algeria ★ Algiers	27,895,068	919,590 sq. mi. 2,381,740 sq. km	Arabic, French, Berber dialects

630

Country ★ Capital	Population (1994)	Land Area (1994)	Principal Languages
Angola ★ Luanda	9,803,576	481,351 sq. mi. 1,246,700 sq. km	Portuguese, Bantu, other African languages
Benin ★ Porto-Novo	5,341,710	43,483 sq. mi. 112,620 sq. km	French, Fon, Yoruba, other local languages
Botswana ★ Gaborone	1,359,352	231,803 sq. mi. 600,370 sq. km	English, Setswana
Burkina Faso ★ Ouagadougou	10,134,661	105,869 sq. mi. 274,200 sq. km	French, local languages
Burundi ★ Bujumbura	6,124,747	10,745 sq. mi. 27,830 sq. km	Kirundi, French, Swahili
Cameroon ★ Yaoundé	13,132,191	183,567 sq. mi. 475,440 sq. km	English, French, African languages
Cape Verde ★ Praia	423,120	1,556 sq. mi. 4,030 sq. km	Portuguese, Crioulo
Central African Republic ★ Bangui	3,142,182	240,533 sq. mi. 622,980 sq. km	French, Sangho, Arabic, Hunsa, Swahili
Chad ★ N'Djamena	5,466,771	495,752 sq. mi. 1,284,000 sq. km	French, Arabic, Sara, Sango, other dialects
Comoros ★ Moroni	530,136	838 sq. mi. 2,170 sq. km	Arabic, French, Comoran
Congo ★ Brazzaville	2,446,902	132,046 sq. mi. 342,000 sq. km	French, Lingala, Kikongo, other African languages
Côte d'Ivoire ★ Abidjan	14,295,501	124,502 sq. mi. 322,460 sq. km	French, Dioula, native dialects
Djibouti ★ Djibouti	412,599	8,494 sq. mi. 22,000 sq. km	French, Arabic, Somali, Afar
Egypt ★ Cairo	60,765,028	386,660 sq. mi. 1,001,450 sq. km	Arabic, English, French
Equatorial Guinea ★ Malabo	409,550	10,830 sq. mi. 28,050 sq. km	Spanish, pidgin English, Fang, Bubi, Ibo
Eritrea ★ Asmara	3,782,543	46,842 sq. mi. 121,320 sq. km	Tigre and Kunama, Tigre, Nora Bana, Arabic, other dialects
Ethiopia ★ Addis Ababa	54,927,108	435,184 sq. mi. 1,127,127 sq. km	Amharic, Tigrinya, Orominga, Guaraginga, Somali, Arabic, English

Country ★ Capital	Population (1994)	Land Area (1994)	Principal Languages
Gabon ★ Libreville	1,139,006	103,347 sq. mi. 267,670 sq. km	French, Fang, Myene, Bateke, Bapounou/Eschira, Bandjabi
Gambia ★ Banjul	959,300	4,363 sq. mi. 11,300 sq. km	English, Mandinka, Wolof, Fula, other local dialects
Ghana ★ Accra	17,225,185	92,100 sq. mi. 238,540 sq. km	English, Akan, Moshi-Dagomba, Ewe, Ga
Guinea ★ Conakry	6,391,536	94,927 sq. mi. 245,860 sq. km	French, local languages
Guinea-Bissau ★ Bissau	1,098,231	13,946 sq. mi. 36,120 sq. km	Portuguese, Criolo, African languages
Kenya ★ Nairobi	28,240,658	224,961 sq. mi. 582,650 sq. km	English, Swahili, local languages
Lesotho ★ Maseru	1,944,493	11,718 sq. mi. 30,350 sq. km	Sesotho, English, Zulu, Xhosa
Liberia ★ Monrovia	2,972,766	43,000 sq. mi. 111,370 sq. km	English, Niger-Congo languages
Libya ★ Tripoli	5,057,392	679,358 sq. mi. 1,759,540 sq. km	Arabic, Italian, English
Madagascar ★ Antananarivo	13,427,758	226,656 sq. mi. 587,040 sq. km	French, Malagasy
Malawi ★ Lilongwe	9,732,409	45,745 sq. mi. 118,480 sq. km	English, Chichewa, other regional languages
Mali ★ Bamako	9,112,950	478,764 sq. mi. 1,240,000 sq. km	French, Bambara, African languages
Mauritania ★ Nouakchott	2,192,777	397,953 sq. mi. 1,030,700 sq. km	Hasaniya Arabic, Pular, Soninke, Wolof
Mauritius ★ Port Louis	1,116,923	718 sq. mi. 1,860 sq. km	English, Creole, French, Hindi, Urdu, Hakka, Bojpoori
Morocco ★ Rabat	28,558,635	172,413 sq. mi. 446,550 sq. km	Arabic, Berber dialects, French
Mozambique ★ Maputo	17,346,280	309,494 sq. mi. 801,590 sq. km	Portuguese, many local dialects
Namibia ★ Windhoek	1,595,567	318,694 sq. mi. 825,418 sq. km	English, Afrikaans, German, local languages

632

Country ★ Capital	Population (1994)	Land Area (1994)	Principal Languages
Niger ★ Niamey	8,971,605	489,189 sq. mi. 1,267,000 sq. km	French, Hausa, Djerma
Nigeria ★ Abuja	98,091,097	356,668 sq. mi. 923,770 sq. km	English, Hausa, Yoruba, Ibo, Fulani
Rwanda ★ Kigali	8,373,963	10,170 sq. mi. 26,340 sq. km	Kinyarwanda, French, Kiswahili
São Tomé and Príncipe ★ São Tomé	136,780	371 sq. mi. 960 sq. km	Portuguese
Senegal ★ Dakar	8,730,508	75,749 sq. mi. 196,190 sq. km	French, Wolof, Pulaar, Diola, Mandingo
Seychelles ★ Victoria	72,113	176 sq. mi. 455 sq. km	English, French, Creole
Sierra Leone ★ Freetown	4,630,037	27,699 sq. mi. 71,740 sq. km	English, Mende, Temne, Krio
Somalia ★ Mogadishu	6,666,873	246,201 sq. mi. 637,660 sq. km	Somali, Arabic, Italian, English
South Africa ★ Pretoria	43,930,631	471,008 sq. mi. 1,219,912 sq. km	Afrikaans, English, Ndebele, Pedi, Sotho, Swati, Tsonga, Tswana, Venda, Xhosa, Zulu
Sudan ★ Khartoum	29,419,798	967,493 sq. mi. 2,505,810 sq. km	Arabic, Nubian, Ta Bedawie, various other dialects, English
Swaziland ★ Mbabane	936,369	6,703 sq. mi. 17,360 sq. km	English, siSwati
Tanzania ★ Dar es Salaam	27,985,660	364,899 sq. mi. 945,090 sq. km	Swahili, English, local languages
Togo ★ Lomé	4,255,090	21,927 sq. mi. 56,790 sq. km	French, Ewe, Mina, Dagomba, Kabye
Tunisia ★ Tūnis	8,726,562	63,170 sq. mi. 163,610 sq. km	Arabic, French
Uganda ★ Kampala	19,121,934	91,135 sq. mi. 236,040 sq. km	English, Luganda, Swahili, other Bantu and Nilotic languages
Zaire ★ Kinshasa	42,684,091	905,563 sq. mi. 2,345,410 sq. km	French, Lingala, Swahili, Kingwana, Kikongo, Tshiluba

Country ★ Capital	Population (1994)	Land Area (1994)	Principal Languages
Zambia ★ Lusaka	9,188,190	290,583 sq. mi. 752,610 sq. km	English, local languages
Zimbabwe ★ Harare	10,975,078	150,803 sq. mi. 390,580 sq. km	English, Shona, Sindebele

EAST AND SOUTHEAST ASIA

Country ★ Capital	Population (1994)	Land Area (1994)	Principal Languages
Brunei ★ Bandar Seri Begawan	284,653	2,228 sq. mi. 5,770 sq. km	Malay, English, Chinese
Burma (Myanmar) ★ Rangoon (Yangon)	44,277,014	261,969 sq. mi. 678,500 sq. km	Burmese, local languages
Cambodia ★ Phnom Penh	10,264,628	69,900 sq. mi. 181,040 sq. km	Khmer, French
China ★ Beijing	1,190,431,106	3,705,386 sq. mi. 9,596,960 sq. km	Mandarin Chinese, other Chinese dialects
Indonesia ★ Jakarta	200,409,741	741,096 sq. mi. 1,919,440 sq. km	Bahasa Indonesia, English, Dutch, Javanese and other local dialects
Japan ★ Tokyo	125,106,937	145,882 sq. mi. 377,835 sq. km	Japanese
Laos ★ Vientiane	4,701,654	91,428 sq. mi. 236,800 sq. km	Lao, French, English
Malaysia ★ Kuala Lumpur	19,283,157	127,316 sq. mi. 329,750 sq. km	Malay, English, Chinese dialects, Tamil, local dialects
Mongolia ★ Ulaanbaatar	2,429,762	604,247 sq. mi. 1,565,000 sq. km	Khalkha Mongol, Russian, Chinese, Turkic languages
North Korea ★ P'yŏngyang	23,066,573	46,540 sq. mi. 120,540 sq. km	Korean
Philippines ★ Manila	69,808,930	115,830 sq. mi. 300,000 sq. km	Pilipino, English
Singapore ★ Singapore	2,859,142	244 sq. mi. 633 sq. km	Chinese, Malay, Tamil, English
South Korea ★ Seoul	45,082,880	38,023 sq. mi. 98,480 sq. km	Korean, English
Taiwan ★ T'aipei	21,298,930	13,892 sq. mi. 35,980 sq. km	Mandarin Chinese, Taiwanese (Min), Hakka dialects

Country ★ Capital	Population (1994)	Land Area (1994)	Principal Languages
Thailand ★ Bangkok	59,510,471	198,455 sq. mi. 514,000 sq. km	Thai, English, regional dialects
Vietnam ★ Hanoi	73,103,898	127,243 sq. mi. 329,560 sq. km	Vietnamese, French, Chinese, English, Khmer, local languages

SOUTH ASIA

Country ★ Capital	Population (1994)	Land Area (1994)	Principal Languages
Bangladesh ★ Dhaka	125,149,469	55,598 sq. mi. 144,000 sq. km	Bangla, English
Bhutan ★ Thimphu	716,380	18,147 sq. mi. 47,000 sq. km	Dzongkha, Tibetan dialects, Nepalese dialects
India ★ New Delhi	919,903,056	1,269,338 sq. mi. 3,287,590 sq. km	Hindi, English, Hindustani, local languages
Maldives ★ Male	252,077	116 sq. mi. 300 sq. km	Divehi, English
Nepal ★ Kathmandu	21,041,527	54,363 sq. mi. 140,800 sq. km	Nepali, local dialects
Pakistan ★ Islamabad	128,855,965	310,401 sq. mi. 803,940 sq. km	English, Punjabi, Sindhi, Pashtu, Balochi, Urdu
Sri Lanka ★ Colombo	18,129,850	25,332 sq. mi. 65,610 sq. km	Sinhala, Tamil, English

THE PACIFIC WORLD

Country ★ Capital	Population (1994)	Land Area (1994)	Principal Languages
Australia ★ Canberra	18,077,419	2,967,893 sq. mi. 7,686,850 sq. km	English, local languages
Fiji ★ Suva	764,382	7,054 sq. mi. 18,270 sq. km	English, Fijian, Hindustani
Kiribati ★ Tarawa	77,853	277 sq. mi. 717 sq. km	English, Gilbertese
Marshall Islands ★ Majuro	54,031	70 sq. mi. 181 sq. km	English, Marshallese dialects, Japanese
Micronesia, Federated States of ★ Palikir	120,347	271 sq. mi. 702 sq. km	English, Trukese, Pohnpeian, Yapese, Kosrean
Nauru ★ Yaren District	10,019	8 sq. mi. 21 sq. km	Nauruan, English

Country ★ Capital	Population (1994)	Land Area (1994)	Principal Languages
New Zealand ★ Wellington	3,388,737	103,737 sq. mi. 268,680 sq. km	English, Maori
Palau ★ Koror	16,366	177 sq. mi. 458 sq. km	English, Palauan, Japanese, Sonsorolese, other local languages
Papua New Guinea ★ Port Moresby	4,196,806	178,259 sq. mi. 461,690 sq. km	local languages, English, pidgin English, Motu
Solomon Islands ★ Honiara	385,811	10,985 sq. mi. 28,450 sq. km	local languages, Melanesian pidgin, English
Tonga ★ Nuku'alofa	104,778	289 sq. mi. 748 sq. km	Tongan, English
Tuvalu ★ Funafuti	9,831	10 sq. mi. 26 sq. km	Tuvaluan, English
Vanuatu ★ Port-Vila	169,776	5,699 sq. mi. 14,760 sq. km	English, French, pidgin
Western Samoa ★ Apia	204,447	1,104 sq. mi. 2,860 sq. km	Samoan (Polynesian), English

■ THE UNITED STATES

State	Capital	Date entered Union
Alabama	Montgomery	1819
Alaska	Juneau	1959
Arizona	Phoenix	1912
Arkansas	Little Rock	1836
California	Sacramento	1850
Colorado	Denver	1876
Connecticut	Hartford	1788
Delaware	Dover	1787
Florida	Tallahassee	1845
Georgia	Atlanta	1788
Hawaii	Honolulu	1959
Idaho	Boise	1890
Illinois	Springfield	1818
Indiana	Indianapolis	1816
Iowa	Des Moines	1846
Kansas	Topeka	1861
Kentucky	Frankfort	1792
Louisiana	Baton Rouge	1812
Maine	Augusta	1820
Maryland	Annapolis	1788
Massachusetts	Boston	1788
Michigan	Lansing	1837
Minnesota	St. Paul	1858
Mississippi	Jackson	1817
Missouri	Jefferson City	1821

State	Capital	Date entered Union
Montana	Helena	1889
Nebraska	Lincoln	1867
Nevada	Carson City	1864
New Hampshire	Concord	1788
New Jersey	Trenton	1787
New Mexico	Santa Fe	1912
New York	Albany	1788
North Carolina	Raleigh	1789
North Dakota	Bismarck	1889
Ohio	Columbus	1803
Oklahoma	Oklahoma City	1907
Oregon	Salem	1859
Pennsylvania	Harrisburg	1787
Rhode Island	Providence	1790
South Carolina	Columbia	1788
South Dakota	Pierre	1889
Tennessee	Nashville	1796
Texas	Austin	1845
Utah	Salt Lake City	1896
Vermont	Montpelier	1791
Virginia	Richmond	1788
Washington	Olympia	1889
West Virginia	Charleston	1863
Wisconsin	Madison	1848
Wyoming	Cheyenne	1890

GLOSSARY

Pronunciation Key This key will help you understand the pronunciations of unfamiliar words given in the textbook.

Key Word	Example	Key Word	Example
map	Africa (AF-ri-kuh)	book	Cooktown (KOOHCK-town)
day	Bombay (bahm-BAY)	student	Peru (puh-ROO)
date	Maine (MANE)	sun	Kentucky (ken-TUHCK-ee)
air	Canary (kuh-NAIR-ee)	deep	Armenia (ahr-MEE-nee-uh)
father	Baja (BAH-HAH)	tip	Britain (BRIT-en)
law	Bengal (ben-GAWL)	pie	Thailand (TIE-land)
wet	Genoa (JEHN-uh-wuh)	tide	Palestine (PAL-uh-stine)
boy	Detroit (di-TROIT)	top	Bosporus (BAHS-puh-ruhs)
now	Moscow (MAHS-kow)	toe	Ohio (oe-HIE-oe)
moon	Rangoon (ran-GOON)	note	Roanoke (ROE-uh-noke)

A

abdicate to resign or give up power, 329

absolute location the exact spot on Earth where something is found, often stated in latitude and longitude, 3

abyssal plain a large, flat area on the ocean floor, 47

acculturation a process by which one culture changes a great deal through its meeting with another culture, 69

acid rain a type of polluted rain, produced when acids from smokestacks combine with water vapor, that can harm lakes, forests, and human health, 92

agribusiness farming on large modern farms that often are owned by big corporations, 160

air pressure the measurement of the force exerted by air, 20

alien species a species of plant or animal that has been introduced to a region by humans from a different region, 606

alliance an agreement between countries to support one another against enemies, 265

alluvial fan a fan-shaped deposit of sediment located at the base of a mountain, 49

altiplano the broad highland plain between the eastern and western ridges of the Andes in Peru and Bolivia, 247

animism a religious belief that honors the memories of ancestors and holds that waters, animals, trees, mountains, and other objects in nature have spirits, 427

annex to formally join, as a territory, to one's own country or territory, 521

Antarctic Circle the line of latitude, located 66½° south of the equator, south of which all areas have 24 hours of daylight during the December solstice and 24 hours of darkness during the June solstice, 15

anthracite coal hard coal, 125

apartheid the South African government policy of separation of races; the policy began to disappear in the late 1980s, 479

aquaculture farming in seas and ponds, 500

aqueduct a canal for carrying water, 155

aquifer a rock layer through which groundwater flows easily, 39

arable suitable for growing crops, 294

archipelago a large group of islands, 540

Arctic Circle the line of latitude, located 66½° north of the equator, north of which all areas have 24 hours of darkness during the December solstice and 24 hours of daylight during the June solstice, 15

artesian well a well in which water rises toward the surface without being pumped, 605

atlas an organized collection of maps in one book, S4

atmosphere the layer of gases that surrounds the earth, 16

atoll a ring of small islands built up on a coral reef surrounding a shallow lagoon, 613

autarky economic self-sufficiency, 330

autonomy the right to self-government, 300

B

badlands lands eroded by wind and water into small gullies stripped of vegetation and soil, 151

balance of power a political stability that results when members of a group, such as countries of a region, have equal levels of strength, 265

balance of trade the relationship between the value of a country's exports and its imports, 304

barrier island a long, narrow, sandy island separated from the mainland by a shallow lagoon or wetland, 133

basin a low area of land, generally surrounded by mountains, 113

basin irrigation the type of irrigation in which farmers divide their land into shallow basins by building walls of earth to trap floodwaters, 433

bayou a small, sluggish stream that cuts through the low and swampy land of a delta, 135

Bedouin a migrating herder of Southwest Asia, 384

bilingual having the ability to speak two languages, 138

biogeography the study of the geographic distribution of plants and animals, 605

biome a plant and animal community that covers a very large land area, 58

biosphere the part of the earth in which all of the planet's plant and animal life exists, 16

biotechnology the application of biology to industrial processes, 123

birthrate the number of births per 1,000 people in a given year, 83

bituminous coal soft coal, 125

bog   soft ground that is soaked with water, *276*

borough   an administrative unit of a city, *127*

break-of-bulk center   a place where goods are moved from one mode of transportation to another, *127*

Buddhism   a religion, practiced primarily in Asia, whose followers seek enlightenment through meditation, *503*

buffer state   a small country that separates two larger, competing countries, *234*

C

cacao   a small tree on which the cocoa bean grows, *198*

caldera   a large depression formed after a major eruption and collapse of a volcanic mountain, *164*

calico   a fine fabric woven from cotton and printed with colorful patterns, *567*

calypso   a type of music that originated in Trinidad and Tobago, *219*

canton   a largely self-governing state within a country, as in Switzerland, *291*

capital   a source of wealth used to make more wealth, *438*

capitalism   an economic system, based on free enterprise, in which resources, industries, and businesses are owned by private individuals, *81*

cardamom   a spice popular in many of the foods of Southwest Asia, *215*

cartel   a group of business organizations that agree to limit supplies of a product and thereby control prices, *249*

cartography   the branch of geography that studies maps and mapmaking, *8*

cash crop   a crop produced primarily for direct sale in a market, rather than for other purposes such as consumption by the farmer, *204*

cassava   a crop whose fleshy roots are nutritious, *443*

caste system   the social system developed in ancient India which divides society into major classes, or castes, according to occupation, *578*

cease-fire   an agreement to halt active hostility, *587*

center pivot irrigation   the type of irrigation that uses sprinkler systems mounted on huge rotating wheels, *153*

channel   a natural or artificial course for running water; to straighten a river by carving a deep lane for ships, *143*

chinook   a strong, dry wind that blows from the Rocky Mountains onto the Great Plains, *152*

Christianity   the religion founded on the teachings of Jesus Christ, *380*

city-state   a self-governing city and its surrounding area, *286*

clan   a group of people descended from a common ancestral family, *364*

client state   a country that is politically, economically, or militarily dependent on a more powerful country, *331*

climate   weather conditions in an area averaged over a long period of time, *18*

climax community   the stable plant group that is best suited to use the resources of a particular area and forms the last stage of plant succession, *55*

coalition government   a government in which several political parties join together to run the country, *304*

coke   coking coal, used in blast furnaces to purify iron ore for making steel, *125*

command economy   an economy in which the government determines wages, the kinds and amounts of goods produced, and the prices of goods, *81*

commercial agriculture   the type of agriculture in which people grow crops for sale, usually on large farms using modern technology, *72*

commonwealth   (1) a self-governing political unit, *223* (2) an association of self-governing states with similar backgrounds and united by a common loyalty, *275*

commune   a large cooperative group, *509*

communism   an economic and political system in which the government owns or controls almost all of the facilities and resources necessary to make products, *82*

compass rose   a directional indicator that has arrows pointing north, south, east, and west, *S6*

complementary region   a region particularly suited to certain activities, such as industry or mining, so that when combined with a region of different strengths, such as agriculture, the regions benefit each other, *312*

condensation   the process by which water vapor changes from a gas into liquid droplets, *25*

confederation   a group of states joined together for a common purpose, *291*

Confucianism   a religion or code of ethics, practiced primarily in East Asia, that centers around family, social relationships, and duty, *503*

coniferous forest   a middle-latitude forest whose plants remain green year-round, *58*

constitutional monarchy   a country that has a reigning king or queen (a monarch), as well as a parliament that serves as the lawmaking branch of government, *273*

contiguous   connecting or bordering; the 48 contiguous U.S. states border each other as a single unit, *113*

continent   one of seven large landmasses on Earth, *S2*

Continental Divide   the crest of the Rocky Mountains that divides the rivers of North America into those that flow east and those that flow west, *154*

continental shelf   the part of the sea floor that slopes gently down from the continents; the most shallow part of the ocean, *39*

cooperative   an organization owned by and run for the mutual benefit of its members, *222*

copra   the dried meat of the coconut, from which coconut oil is obtained, *485*

coral reef   a ridge found in warm, tropical water close to shore, made of a rocky limestone material formed by the skeletons of tiny marine animals, *56*

cork   a bark stripped from the trunk of the Mediterranean oak tree, *299*

Corn Belt   the area of the Midwestern United States, stretching mainly from Ohio to Iowa, in which raising corn is a major economic activity, *143*

cosmopolitan   having many foreign influences, *290*

Cossack   one of a group of people who controlled the northern Eurasian steppe region beginning in the 1500s, *353*

cottage industry   an industry in which workers make small consumer items in homes and small workshops, *577*

coup   an overthrow of an existing government by another political or military group, *248*

crop rotation   a method of preserving soil fertility in which different types of crops are grown on the same piece of land over time, *87*

culture   the features of a society's way of life, passed down from generation to generation through teaching, example, and imitation, *67*

culture hearth   a region where many important new ideas and developments originate, *72*

culture region   an area with many shared culture traits, *67*

culture trait   an activity or behavior repeatedly practiced by people in a given culture, *67*

czar   the title for the rulers of the Russian Empire (pre-Soviet era); literally, "emperor," *329*

D

Dairy Belt the area of the Midwestern United States, located north of the Corn Belt, in which dairy farming is a major economic activity, *144*

death rate the number of deaths per 1,000 people in a given year, *83*

deciduous forest a middle-latitude forest whose plants lose their leaves during a certain season each year, *58*

defoliant a chemical that makes plants' leaves drop off, usually killing the plants, *535*

deforestation the clearing or destruction of forests, *88*

degree a unit of measurement of temperature or of distance on the earth's surface, *53*

delta the landform at the mouth of a river created by deposits of sediment, *49*

demilitarized zone (DMZ) a buffer zone into which military forces may not enter, *528*

democratic government a government in which everyone in the society has a voice, *75*

demography the study that emphasizes statistics to look at human population distribution, population density, and trends in population, *82*

deport to send out of a country, *175*

desalinization a process by which saltwater is changed into fresh water, *89*

desertification a process by which desertlike conditions are created by a loss of plant cover and soil due to human activity, *87*

developed country a wealthy country that has a good educational system, widely available health care, and many manufacturing and service industries, *80*

developing country a poorer country in which many people live by subsistence farming and in which there are few manufacturing and service industries, *81*

dialect a regional variety of a major language, *300*

diffusion a process by which an innovation or other culture trait spreads from one culture region to another, *68*

dike a wall built to keep water out of a low-lying area, *290*

directional indicator an element of a map that shows which directions are north, south, east, and west, *S6*

diversify to produce a variety of crops or goods rather than just one, *135*

doldrums calm areas centered along the equator, *21*

domestication the growing of a plant or taming of an animal by a people for their own use, *72*

dominion a territory or sphere of influence, *175*

double cropping an agricultural practice, enabled by a long growing season, in which two crops are grown in one year on the same land, *511*

dredge to dig or clean out a river to make it deeper, *143*

drip irrigation a process by which pipes placed near plant roots drip only as much water as the plants need, *382*

drought a long, dry period with little or no rain, *4*

dryland agriculture agriculture that depends on rainfall rather than irrigation to water crops, *361*

dynasty a government ruled by a family whose power lasts through several generations, *507*

E

earthquake the shock waves, or vibrations, caused by movement along a fault, *45*

earth system the interactions of elements on and around the earth, *16*

economic association an organization formed to break down trade barriers among member nations, *267*

ecosystem an interdependent community of plants and animals combined with its physical environment, *54*

edge city a suburb with large employment and commercial centers, *116*

ejido a communal farm, in Mexico, on which farmers work land together or work individual plots, *205*

elevation height on the earth's surface above or below sea level, *26*

El Niño a weather pattern, involving a warming of part of the Pacific Ocean, that occurs every five to ten years and causes a decrease in fish and an increase in flooding in some areas, *248*

embargo a legal restriction against the movement of freight, *480*

emigrant someone who moves out of a country, *84*

endemic species a plant or animal that originates only in a particular region, *605*

entrepot an intermediate port used to store goods before they are shipped to their final destinations, *544*

equator the imaginary line of latitude that lies halfway between the North and South poles and circles the globe, *S2*

equinox one of two times each year when the earth's polar axis is not pointed toward or away from the sun but is at a 90° angle to the sun, *15*

erg a large area covered by sand, *431*

erosion the wearing away of land by water, wind, waves, ice, and other agents, *42*

escarpment a steep slope capped by a plateau, *421*

estuary a semi-enclosed coastal body of water where seawater and fresh water mix, *37*

ethnic group a population that shares a common cultural background, *68*

ethnocentrism a form of cultural bias in which people use their own culture and experiences as a point of reference from which to make statements about other cultures, *552*

Eurasia the world's largest landmass, made up of the continents of Europe and Asia, *258*

European Union (EU) an economic association, made up of several European countries, whose ultimate goal is for Europe to be a unified economic region, *267*

evaporation the process by which water is changed from a liquid to a gas, *25*

evapotranspiration a measure of the total loss of water from the land to the atmosphere through evaporation and transpiration, *37*

exclave a part of a country that is separated from the rest of the country and surrounded by foreign territory, *473*

Exclusive Economic Zone (EEZ) an area off a country's coast, whose resources are claimed and controlled by that country, *615*

exile a person who has been forced to leave his or her country, *480*

exotic river a river that flows from a humid environment across a desert into a sea, *378*

export economy an economy in which most manufactured products are made for export, *527*

extinct no longer in existence, *60*

F

Fall Line the line in the United States along which the Coastal Plain mees the Piedmont, marked by waterfalls and rapids indicating the elevation change, *111*

fallow unused, as cropland, *400*

famine an extreme shortage of food, *266*

fault a feature of the earth's crust created by the breaking and movement of rock layers below the surface, *45*

favela a slum area near a Brazilian city, *237*

fjord a narrow, deep inlet of the sea between high, rocky cliffs created by glaciers, *277*

floodplain a landform of level ground built by sediment deposited by a river or stream, *49*

foehn a warm, dry wind in Europe that blows down the slopes of the Alps, *294*

fold a feature of the earth's crust created by the bending of rock layers, *45*

food chain a series of organisms in which energy is passed along through living things, *53*

foreign aid loans or gifts of money from one country to another, *81*

fossil fuel a fuel that may have been formed from the remains of prehistoric plants and animals, *93*

fossil water groundwater that is not replaced by rainfall, *379*

free enterprise an economic principle under which prices are determined mostly through competition and people are free to choose what and when to sell and buy, *81*

free port a port in which unloaded goods are subject to few or no taxes, *438*

front a zone formed by the meeting of two air masses with very different temperatures and amounts of moisture, *21*

fundamentalism a movement that stresses the strict following of basic traditional principles, *75*

furrow a narrow trench made by a plow, *379*

G

geothermal energy a renewable energy source generated by the escape of underground heat to the surface, *95*

geyser a hot spring that shoots hot water and steam into the air, *280*

ghetto a part of a city where one or more minority groups are concentrated because of economic pressure and social discrimination, *116*

ghost town a town that once flourished but is deserted today, usually because of the depletion of a nearby resource such as gold or silver, *154*

glacier a thick mass of ice that moves slowly across Earth's surface, *43*

glasnost the policy of open discussion promoted in the Soviet Union shortly before its breakup, *331*

globe a scale model of Earth, especially useful for looking at the entire earth or at large areas of its surface, *S2*

granite a speckled, hard, crystalline rock formed deep in the earth's crust, *122*

great-circle route the shortest route between any two points on the earth, *S5*

greenhouse effect (1) the warming of the earth created by the trapping of the sun's energy in Earth's atmosphere, *19* (2) the warming of the earth caused by the buildup of carbon dioxide in the lower atmosphere, possibly as a result of human industrial activity, *93*

Green Revolution the name given to the widespread introduction of new high-yield grains, *566*

grid the pattern of lines that circle the earth in east-west and north-south directions, *S2*

gross domestic product (GDP) the value of goods and services produced within a country in a year, not including income earned outside that country, *80*

gross national product (GNP) the value of goods and

services produced by a country in a year, both inside and outside the country, *80*

groundwater fresh water found beneath the earth's surface in tiny spaces between soil and rock grains, *39*

growth pole a metropolitan area given official help to strengthen its economic development, *238*

guerrilla a person who fights as a member of an armed band and takes part in irregular warfare, such as harassment and raids, *229*

Gulf Stream an Atlantic Ocean current that moves warm water north along the U.S. East Coast, *114*

gum arabic a product that comes from the sap of acacia trees and is used in some medicines and candies, *455*

H

habitation fog a fog caused by fumes and smoke from cities, *336*

hacienda a land system, introduced by the Spanish to Middle and South America, in which large family-owned estates (haciendas) are worked by peasants in return for small plots on which to grow their own crops, *198*

headwaters the first and smallest streams to form from the runoff of precipitation flowing down the slopes of hills and mountains, *37*

heavy industry industry that focuses on manufacturing based on metals such as steel, *342*

hemisphere half of the earth, such as the Northern or Southern hemispheres or the Eastern and Western hemispheres, *S3*

Hinduism the major religion of India and Nepal, *564*

hinterland a region beyond the center of an area, *176*

Holocaust the name given to the persecution and killing of millions of Jews by Nazi Germany during World War II, *390*

homogeneous of the same kind, *355*

hot spot an area where molten material from the earth's interior rises through the crust, *169*

human-environment interaction one of the Five Themes of Geography; deals with how people adapt to and change their environment, *5*

humidity the amount of water vapor in the air, *25*

humus an important ingredient of soil, made of decayed plant or animal matter, *55*

hurricane a tropical storm that brings violent winds, heavy rain, and high seas; also called a typhoon, *25*

hydroelectricity electricity produced by harnessing the power of running water, *35*

hydrologic cycle the circulation of water among parts of the hydrosphere, *36*

hydrosphere the physical system made up of all of Earth's water, *16*

I

icebreaker a ship that can break up the ice of frozen waterways, *264*

immigrant someone who moves into a country, *84*

imperialism the policy of gaining control over territory outside a nation, *265*

indigenous native to a certain area, *188*

indigo a plant used to make a blue dye, *196*

industrialization the process in which manufacturing based on machine power becomes widespread in an area, *35*

inflation the rise in prices caused by a decrease in the value of a nation's currency, *212*

infrastructure facilities and structures, such as roads,

640

bridges, and power plants, that are needed to build industries and move goods in and out of a country, *80*

innovation   a new idea that is accepted into a culture, *68*

intensive agriculture   agriculture that uses much human labor, *532*

irrigation   the watering of land through pipes, ditches, or canals, *35*

Islam   the religion founded by Muhammad, *380*

island   a landmass smaller than a continent and completely surrounded by water, *S2*

isthmus   a narrow neck of land that acts as a bridge to connect two larger bodies of land, *188*

J

Judaism   the religion of the Jews, which appeared in Southwest Asia between 3500 B.C. and 600 B.C., *380*

jute   a plant fiber used to produce coarse fabrics, such as burlap and twine, *565*

K

kampong   a traditional village in island Southeast Asia; a slum area around Jakarta, Indonesia, *546*

katabatic wind   a rapid flow of dense cold air down the slopes of high mountains or ice caps, *620*

kibbutz   an Israeli collective farm, *390*

L

labor intensive   requiring a large human work force, *198*

landform   a shape on the earth's surface, *46*

landlocked   completely surrounded by land, without access to an ocean, *242*

land reclamation   a method used after mining to prevent erosion and return the land to its original contours, *153*

land reform   the changing of a land ownership system; usually the breaking up of large landholdings to allow more people to own their own land, *200*

Landsat   a satellite that collects information about Earth's surface and the condition of Earth's environment, *23*

latifundia   a land system, introduced by Europeans to South America, in which tenant farmers work part of an estate in exchange for food; the estates themselves, *250*

latitude   a measure of distance north or south of the equator, *S2*

lava   melted liquid rock, or magma, from within the earth that spills out on the earth's surface, *45*

leaching   the process by which nutrients are washed down out of the topsoil by rainfall, *56*

legacy   something received from a previous era, *311*

legend   an element of a map that explains what the symbols on the map represent; also called a key, *S6*

levee   a dirt ridge along a river that hinders flooding, *135*

light industry   industry that focuses on producing lightweight goods, such as clothing, *342*

lingua franca   a common language spoken and understood by the majority of people in an area, although other languages may be spoken at home, *445*

literacy   the ability to read or write, *80*

lithosphere   the solid surface of the planet that forms the continents and the ocean floor, *16*

llanos   the plains of southern Colombia and Venezuela, *193*

location   one of the Five Themes of Geography; deals with where places and regions are located; defined in terms of absolute and relative location, *3*

loch   a long, deep lake carved by a glacier, *273*

lock   a part of a waterway enclosed by gates, used to raise or lower boats from one water level to another, *135*

loess   dust-sized soil particles deposited by wind, *287*

longitude   a measure of distance east or west of the prime meridian, *S2*

M

map   a flat diagram of all or part of the earth's surface, *S4*

map projection   one of a variety of ways of presenting a round earth on a flat map, *S4*

maquiladora   a foreign company's factory in Mexico, usually near the U.S. border, in which products are assembled for export, *208*

market economy   an economy in which consumers determine what is to be bought or sold by buying or not buying certain goods and services, *81*

marsupial   a mammal, such as a kangaroo, that carries its young in a pouch, *605*

martial law   military rule, *515*

Megalopolis   the giant urban area that stretches along the U.S. eastern seaboard, *120*

mental map   a map that we create and see in our minds to help make sense of the world, *S9*

merengue   the national music and dance of the Dominican Republic, *219*

meridian   a line of longitude, *S3*

Meseta   the rocky, treeless plateau that covers much of Spain, *297*

mestizo   a person, usually from Middle or South America, with both Native American and European ancestors, *196*

meteorology   the field of geography that specializes in weather and weather forecasting, *22*

microstate   a very small country, *258*

mid-ocean ridge   a chain of mountains on the ocean floor formed when molten rock rises from below the seafloor to fill openings created by plate movement, *47*

millet   a grain crop that is resistant to drought, *443*

minaret   the tower of a mosque, *395*

minifundio   a small farm in Middle or South America formed when a large estate is broken up, *250*

mistral   a strong, cool wind in Europe that blows from the Alps toward the Mediterranean coast, following the valley of the Rhone River, *284*

mixed forest   an area where deciduous and coniferous forests blend, *59*

model   a plan or pattern used to represent something, such as a city, *182*

monopoly   a business that has no competitors and thus can control an industry's supply of goods and prices, *212*

monotheism   the belief in one god, *381*

monsoon   the wind that blows from the same direction for months at a definite season of the year, *27*

moraine   a ridge of rocks, gravel, and sand deposited along the margins of a glacier or ice sheet, *122*

mosque   an Islamic place of worship, *360*

movement   one of the Five Themes of Geography; deals with the movement of people, goods, and ideas, *6*

mulatto   a person, usually from Middle or South America, with both African and European ancestors, *196*

multicultural region   an area in which people of many cultures have settled, *161*

multilingual   having the ability to speak three or more languages, *267*

multinational company   a business with activities in many countries, *81*

muskeg a forested swamp that is frozen except during the summer, *179*

Muslim a person who follows Islam, *380*

N

nationalism feelings of pride in one's country, *74*

nationalize to take over by the government, such as to assume government control of an industry, *274*

natural boundary a physical feature, such as a mountain range or a river, that determines a political border, *262*

navigable able to be sailed by large ships, such as a wide and deep river, *263*

neutral choosing not to take sides in international conflicts, *279*

newly industrialized country (NIC) a nation that is undergoing or has recently undergone rapid industrialization and economic growth, *501*

newsprint an inexpensive paper used mainly for printing newspapers, *174*

nodal region a central area distinguished by the movement or activities that take place in and around it, *101*

nonrenewable resource a resource, such as oil or gold, that cannot be replaced by natural processes or is replaced extremely slowly, *86*

O

oasis a site in the desert where water can be found in springs or wells or under dry riverbeds, *359*

ocean the largest part of the earth's water surface, often divided into four smaller oceans, *S2*

Oceania the ocean region containing the Pacific islands, *612*

orographic effect an effect of mountains on climate that produces moisture on the windward side and dryness on the leeward side, *26*

ozone a gas formed from an interaction between oxygen and sunlight; the ozone layer is a region in the earth's upper atmosphere that protects life beneath it by filtering out dangerous ultraviolet solar radiation, *92*

P

paddy (padi) a water-covered field usually used for rice cultivation, *500*

pampas the wide grassy plains of Argentina and neighboring countries in South America, *193*

panhandle a narrow arm of land attached to a larger region, *165*

parallel a line of latitude, *S2*

páramo the elevation zone of the Andes just above the tree line, *228*

parliament a lawmaking body, *175*

partition to divide an area, *585*

pastoralism an agricultural way of life that involves raising livestock and usually includes some farming, *443*

peat a substance made of decayed vegetable matter, usually mosses, that can be used for fuel, *276*

peninsula a landform bordered by water on three sides, *122*

perennial irrigation the type of irrigation in which dams store floodwater and release it year-round to irrigate crops, *433*

perestroika the policy of economic reform and restructuring promoted in the U.S.S.R. just before its breakup, *331*

perimeter outer boundary, *405*

periodic market an open-air trading market held regularly at a crossroads or in a town, *466*

permafrost water below a tundra surface that remains frozen throughout the year, *32*

petrochemical a chemical produced from petroleum; some chemical fertilizers, pesticides, food additives, explosives, and medicines are petrochemicals, *93*

photochemical smog a type of air pollution produced by the interaction of exhaust gases with sunlight, *92*

photosynthesis the process by which a plant converts sunlight into chemical energy, *53*

pictogram a written language character that is a simple picture of the object or idea it represents, *502*

piedmont an area of plains and low hills at or near the foot of a mountain region; in the United States, the Piedmont is a rolling plateau region inland from the Coastal Plain, east and south of the Appalachian Mountains, *111*

pilgrimage a journey, often for religious purposes, *404*

place one of the Five Themes of Geography; deals with the physical and human characteristics of a place, *4*

plain a nearly flat area of land, *46*

planet a spherical object that orbits around a star, *11*

plantain a type of banana, *223*

plantation a large farm that concentrates on one major crop, such as cotton or coffee, *115*

plant community a group of plants that live and grow together, *53*

plant succession the process by which one group of plants replaces another in a community, *55*

plateau an elevated flatland, *46*

plate tectonics theory that the earth's crust is divided into several rigid, slow-moving plates *46*

plaza a public square in the center of a town that serves as a marketplace and gathering place, *298*

polar desert a high-latitude region that receives very little precipitation, such as Antarctica, *620*

polar region a cold area near the North or South pole, *14*

polar wind the type of wind that dominates the high latitudes and brings cold conditions into the middle latitudes, usually blowing from the east, *21*

polder a lowland area that has been drained of water, *290*

portage a low land area across which boats and their cargoes can be carried between bodies of water, *175*

potash a mineral that is an important raw material for manufacturing fertilizers, *178*

prairie a region of tall grassland and fertile soil, *59*

precipitation condensed droplets of water vapor that fall as rain, snow, sleet, or hail, *25*

prevailing wind a wind that usually blows from the same direction year-round, *21*

primary industry an economic activity, such as agriculture, forestry, or mining, that makes direct use of natural resources or raw materials, *79*

primate city a city that ranks first in a nation in terms of population and economy, *284*

prime meridian the imaginary line of longitude that runs through Greenwich, England, from the North Pole to the South Pole, *S3*

privatization the selling of government-owned businesses or lands to private owners, *212*

puppet government a government that is controlled by outside forces, *509*

Q

quaternary industry an economic activity involving the movement and processing of information, *79*

quota a limit on the amount of a particular good that can be imported, 76

Qur'an (Koran) the book of Islam's sacred writings, 381

R

rain-shadow effect the forming of a desert on the leeward side of mountains due to a lack of moisture, 27

reforestation a process by which forests are renewed through the planting of seeds or young trees, 87

refugee a person who flees his or her own country to another, often for economic or political reasons, 223

reg a gravel-covered plain from which wind has blown the sand and dust away, 432

reggae a type of music that originated in Jamaica, 219

region one of the Five Themes of Geography, 7; an area with more common characteristics than differences, 100

regionalism the political and emotional support for one's region before support for one's country, 180

regional specialization a plan in which each country of a region specializes in products it makes or grows best, 311

relative location the position of a place in relation to other places, 3

relief the difference in elevation between the top and bottom of a landform, 49

remote sensor a device attached to a satellite that collects information about Earth, 22

Renaissance the era from the 1300s to the 1600s, during which a renewed interest in learning spread throughout Europe; literally, "rebirth," 301

renewable resource a resource, such as solar energy, soil, water, or forests, that can be replaced by the earth's natural processes, 86

reunification the process of reuniting, or rejoining, into one unit, 266

revolution the orbit of an object in space; the orbit of the earth around the sun, which takes one Earth year, 13

rift a long, deep valley usually with mountainous slopes on both sides, 421

rock weathering the process that breaks up rocks and causes them to decay, 43

rotation the act of turning around an axis; one complete spin of Earth on its axis, which takes 24 hours, 12

S

Sahel a region of bush savannas and grasslands along the southern edge of Africa's Sahara, 422

sanction an economic or political measure, such as an embargo, used by one or more nations to force another nation to cease an illegal or immoral act, 480

sanctuary an area in which endangered species are protected, 572

sand dune a hill of wind-deposited sand, 44

satellite a body that orbits a larger body, 11

savanna a tropical grassland with scattered trees and shrubs, 59

scale an element of a map that indicates distances between points on the map, 56

seaboard a land area near the ocean, 115

secondary industry an economic activity that takes goods made in primary industries and processes them into products that are useful to consumers, 79

second-growth forest a forest consisting of trees covering an area after the original forest has been removed, 123

secularism the indifference to religion, 381

sediment small particles of mud, sand, or gravel created by rock weathering, 43

selvas the dense rain forest vegetation of South America's Amazon basin, 245

serf a peasant bound to the land, 340

shantytown a poor settlement, often near a city, of small, makeshift shelters, 199

shelterbelt a row of trees along a field that blocks the wind and protects the soil, 152

shifting agriculture a type of agriculture in which farmers move to new areas every few years, 532

Shinto a Japanese religion whose followers believe that gods inhabit natural objects and that these gods should be worshiped in community rituals, 504

Sikhism a religion formed from the blending of Muslim and Hindu teachings, 565

silt sediment coarser than clay but finer than sand, 432

sisal a strong fiber used to make rope and twine, 458

site the place where something is located, 3

situation the position of a place in relation to other places, 3

slash-and-burn farming the type of agriculture in which forests are cut and burned to clear land for planting, 231

smelter a factory that processes mineral ores, 342

socialism an economic system in which the government owns and controls the means of producing goods, 278

soil exhaustion the loss in soil nutrients that results from always planting the same crop in a particular area, 236

soil horizon a distinct layer of soil formed over a long period of time, 55

soil salinization salt buildup in the soil often caused by evaporation of irrigation water, 87

solar system the sun and the nine planets and all other objects that revolve around the sun, 11

solstice one of two times each year when the earth's polar axis points at its greatest angle toward or away from the sun, 14

sorghum a grain crop that is resistant to drought, 443

sovereignty authority to rule, 550

soviet the local governing council of each republic of the Soviet Union during the Soviet era, 330

spatial interaction the movement of people, goods, and ideas, 6

Special Economic Zone (SEZ) an area designed to attract foreign companies and investment, 512

staple a main food crop in an area, 443

steppe short-grass vegetation; a steppe climate is often found between deserts and more humid climates, 30

stereotype a generalization about a group of people, 552

storm surge a rise in tide level that occurs as a storm approaches the coast, washing huge waves of seawater over the land, 591

strip mining the type of mining in which soil and rock are stripped away by large machines to get at coal or other minerals buried just beneath the surface, 153

subcontinent a large landmass that is smaller than a continent, 558

subduction the process in which a tectonic plate collides with and dives under another tectonic plate, 47

subregion a region that is part of a larger region, 100

subsidy financial support given by a government to an individual or industry within a country, 521

subsistence agriculture the type of agriculture in which people grow food on small farms mostly for their own use, 72

sultanate   a country that is ruled by a Muslim monarch (sultan), *406*

superpower   a huge, powerful nation, *331*

suq   a marketplace in parts of Southwest Asia and North Africa, *395*

sustained yield use   a method of forest preservation in which people can take products from a forest without destroying the trees, *88*

Swahili   a language used in East Africa that developed as a trade language, *454*

T

taiga   the huge forest region, located in the Northern Hemisphere just south of the tundra, of tall, evergreen needle-leaf trees, *337*

tannin   a substance, obtained from the quebracho tree, used in preparing leather, *240*

Taoism   a religion practiced mainly in East Asia whose main idea is to live a simple life close to nature, *503*

tariff   a tax placed on an import or export, *76*

telecommunications   electronically transmitted communication, *80*

temperature   the measurement of heat, *19*

tenant farmer   a farmer who rents a plot of land on which to grow crops, *135*

terrorism   the use of violence as a means of political force, *248*

tertiary industry   an economic activity that provides services to primary and secondary industries and to consumers; also called a service industry, *79*

textile   any of a number of products made from cloth, *121*

theocracy   a country governed by religious law, *403*

tierra caliente   the elevation zone of the Andes that is closest to sea level, *227*

tierra fría   the elevation zone of the Andes just below the tree line, *228*

tierra helada   the highest elevation zone, permanently covered with snow, of the Andes, *228*

tierra templada   the elevation zone of the Andes, just above the *tierra caliente,* which has mild climates, *228*

topographic   relating to surface features; a topographic map shows a region's surface features, *8*

toponym   a place-name, *156*

tornado   a small, twisting, highly destructive storm that usually forms along fronts in the middle latitudes, *26*

totalitarian government   a government in which only one or a few people decide what is best for everyone in the society, *75*

township and range system   the system, started by the U.S. government in the late 1780s, by which the land northwest of the Ohio River was divided into sections and offered to people who agreed to farm them, *143*

trade deficit   the gap that occurs when the value of a nation's imports exceeds that of its exports, *118*

trade surplus   the economic condition in which a country exports more than it imports, *522*

trade wind   an east-to-west wind that blows from the subtropical high-pressure zone toward the equatorial low-pressure zone, *21*

transition zone   an area that contains characteristics from more than one region as a result of these regions overlapping with each other, *101*

transpiration   the process by which plants give off water vapor through their leaves, *37*

tree line   the elevation, on mountain slopes, above which trees cannot grow, *154*

trench   a long, deep valley on the ocean floor caused by subduction, *47*

tributary   a stream or river that flows into a larger stream or river, *37*

tropical cyclone   a hurricane-like storm that usually forms in low latitudes, *564*

Tropic of Cancer   the line of latitude, located 23$\frac{1}{2}$° north of the equator, struck by the sun's most direct rays of energy during the June solstice, *15*

Tropic of Capricorn   the line of latitude, located 23$\frac{1}{2}$° south of the equator, struck by the sun's most direct rays of energy during the December solstice, *15*

tropics   warm areas in the low latitudes, *14*

truck farm   a farm whose products are trucked to nearby city markets, *125*

trust territory   a region that is placed under the control of another nation until it can govern itself, *472*

tsunami   a large, potentially destructive sea wave caused by tectonic activity, *520*

typhoon   a tropical storm that brings violent winds, heavy rain, and high seas; also called a hurricane, *25*

U

uniform region   an area distinguished by one or more common characteristics, *101*

uninhabitable   unable to support human life and settlements, *279*

urban agglomeration   a densely inhabited and contiguous region surrounding a central city, *523*

urbanization   a growth in the proportion of people living in towns and cities, *72*

V

veld   a term used for a vegetation and landform region of South Africa, *477*

volcano   the opening in the earth's crust through which lava flows; the surface feature created by the lava, *45*

voodoo   a religion, common in Haiti, whose followers believe that spirits of good and evil play an important part in daily life, *223*

W

water table   the top of an underground zone that is saturated with groundwater, *39*

watershed   an area of land drained by a river and its tributaries, *37*

weather   the condition of the atmosphere at a given place and time, *18*

westerly   a west-to-east wind that dominates the middle latitudes, blowing from the subtropical high-pressure zone to the subpolar low-pressure zone, *21*

wetlands   productive land areas that are flooded for at least part of the year, *38*

Wheat Belt   the area of the United States, located in the Great Plains, in which wheat farming is a major economic activity, *152*

work ethic   a belief that work is good in and of itself, *522*

Z

Zionism   a movement, started in Europe in the 1800s, that believed a Jewish state should be created in Palestine and Jews from around the world should move there, *390*

zonal   running in east-west bands of latitude, or zones; often used to describe climate patterns, *441*

INDEX

(def.) indicates definition   g indicates graphic   m indicates map   p indicates photograph

A

Abadan, Iran, 402, p 402
abdicate, (def.) 329
Abidjan, Côte d'Ivoire, 450
Aborigines, 69, p 600, 606, 610, p 610
absolute location, (def.) 3
Abuja, Nigeria, 449, 450
abyssal plains, (def.) 47
Accra, Ghana, 450
acculturation, (def.) 69
acid rain, (def.) 92, 123
Acquired Immune Deficiency Syndrome (AIDS), 84, g 84, 423
Adamawa Mountains, 465
Addis Ababa, Ethiopia, 461
Adelaide, Australia, 605, p 607, 610
Aden, Yemen, 406
Adirondack Mountains, 125
Adriatic Sea, 300
Aegean Sea, 304, 306, 389
Afghanistan, 364, g 382, m 399, g 400, 403, p 403, p 407
Africa: artifacts, p 412, p 416, p 430, p 440, p 452, p 464, p 476; climate, m 418, 422–23; contact with Europe, 424–25; culture regions, 68; early civilizations, 423–24; economy, m 420, 428, p 428; European colonization, 425, m 425; forest zone, 442; historical geography, m 423, 423–26; human geography, 426–27; independence, 425–26; landforms, 421–22; language, 426, g 427; natural challenges, 423; physical geography, m 417, 421–23; population, m 419; religion, 426–27; resources, 427–28; Sahara, 4–5, 30, 45; subregions, 100. See also specific subregions and countries
African National Congress (ANC), 480
Afrikaans, 478
Afro-Asiatic languages, g 427
Agra, India, p 555, p 570
agribusiness, (def.) 160, 197
agriculture: commercial, 72, 197, 204, 379; dryland, 361; hacienda system, 198; intensive, 532–33; irrigation, p 117, 152–53, 155, g 382, 382, 401, 485; plantation, 115, 197–98, 204; shifting, 532; slash-and-burn farming, 231, 532; soil conservation, 87; subsistence, 72
Agua Prieta, Mexico, p 208
Ahaggar Mountains, 431
Ahluwalia, Hari Pal Singh, 584
Ahmadabad, India, p 567
AIDS. See Acquired Immune Deficiency Syndrome
air pollution: 92–93, p 92; global air pollution, 92–93
air pressure: (def.) 20; and wind, g 20; zones, 20–21, g 21
Alabama, m 133, 135, g 136

Alaska: m 159, g 160; climate, 114–15, 168; human geography, 168–69; issues, 169; landforms, 113, 165, p 168, 168, p 171; mountains, 113; oil reserves, 94; resources 168; wetlands, p 38
Alaska Range, 113, 165
Albania, m 309, g 310, 313
Albanian language, g 267, 310
Albany, NY, 126
Alberta, Canada, p 43, 177, 178, 180
Aleutian Islands, 113, 165, 169
Aleuts, 168–69
Alexander the Great, 358
Alexandria, Egypt, 424, 433, 434
Algeria, m 431, p 432, g 433, 437, p 437
alien species, (def.) 606
Allegheny Plateau, 112
Allegheny River, 126
alliances, (def.) 265
alluvial fans, g 46, (def.) 49
Almaty, Kazakhstan, 361
Alpine countries, m 283, g 285, 291, 294. See also Austria; Switzerland
Alps, p 100, p 255, 262, 283, 291, 300, 301
altiplano, (def.) 247, 249
Amazon River, 37, 193, 194, 238
Amazon River basin, 193, 235
Amharic, 462
Amman, Jordan, 396
Amritsar, India, p 577
Amsterdam, Netherlands, 291, 292, p 292
Amu Darya, 359, 360
Anatolia, 387, 388
ANC. See African National Congress
Andean Group, 200
Andes Mountains, p 30, p 185, 193, 195, p 226, 227–28, g 228, 240, p 244, 245, 247
Andorra, m 297, g 298, 306
Anegada, 224
Angara River, 336
Angel Falls, p 184, 230
Anglo-Boer War, 479
Angola, m 465, g 470, 473, p 473
animism, (def.) 427
Ankara, Turkey, 388
annex, (def.) 521
Antarctica, p 17, 617–620, p 618–20, m 619
Antarctic Circle, g 14, (def.) 15
anthracite coal, (def.) 125
Antigua and Barbuda, m 215, g 216, 224
Antofagasta, Chile, 249
ANWR. See Arctic National Wildlife Refuge
apartheid, (def.) 479, 479–80
Apennines, 262, 300, 301
Appalachian Mountains, 111–12, 117, 121, 124, 133–34, 145, 173
`Aqaba, Jordan, 396
aquaculture, (def.) 500
aqueducts, (def.) 155

aquifers, (def.) 39
Aquino, Corazon, 550
Arabian Gulf. See Persian Gulf
Arabian perimeter, 405–406. See also individual countries
Arabic, m 411, g 427
Arab-Israeli conflict, 381, 384, 391–92, 394, 395, 396
arable, (def.) 294
Aral Sea, 359, 360, p 360
archipelago, (def.) 540
Arctic Circle, g 14, (def.) 15
Arctic National Wildlife Refuge (ANWR), 168
Arctic Ocean, 39, 168, 169, 332, 334, 335
Ardennes, 262, 283
Argentina: g 236, p 240; climate, 193; desert biome, p 59; economic geography, 240–41; oil reserves, 197; physical geography, m 235, 240; political geography, 241
Arica, Chile, 249
Arizona: g 152; agriculture, p 117; cities, 155–56; climate, 155; Grand Canyon, p 44, p 102, 155; mining, 155; Native Americans, 156; physical geography, m 151, 155
Arkansas, m 133, 134, 135, g 136
Arkansas River, 134
Armenia, m 351, g 352, 355–56, p 356
Arno River, 300, 301
Aroostook Valley, 122
artesian wells, (def.) 605
artifacts. See specific regions and countries
Aruba, 224
ASEAN. See Association of Southeast Asian Nations
Ashgabat, Turkmenistan, 364
Asia. See Central Asia; East and Southeast Asia; Northern Eurasia; South Asia
Association of Southeast Asian Nations (ASEAN), 533
Asunción, Paraguay, p 242
Aswān, Egypt, p 439
Aswān High Dam, 89, 433
Atacama Desert, 194, 249
Athens, Greece, p 271, 306, p 307
Atlanta, GA, 137
Atlantic Ocean, S2, 39, m 48
Atlantic South America, 234–42, m 235. See also specific countries
atlases, (def.) S4
Atlas Mountains, p 421, 421, 438
atmosphere: (def.) 16, g 16; atmospheric effects, 25–27; composition, 18; elevation and temperature, 26; global energy systems and, 19–24; orographic effect, 26–27, g 26; storms, p 25, 25–26; water and, 25
atolls, (def.) 613
Attenborough, David, 42, 52, 600
Atwood, Margaret, 172

Auckland, New Zealand, p 597, 617
Austin, TX, p 119, 137
Australia: m 601, g 605; Aborigines, 69, 610, p 610; agriculture, 606–607; artifact, p 600; Asian-Pacific connections, 607; biogeography, 605–606, p 606; climate, m 602, 604–605; coal reserves, 93; economic geography, m 603, 606–607; Great Barrier Reef, p 600, 606; industries, 607; issues, 610; landforms, 604; mineral resources, 607; population and urban geography, 607–10; water resources, 605
Australian Capital Territory, 600
Austria, g 83, m 283, g 285, 294
autarky, (def.) 330
automobile industry: 117–18, 146; solar power and, 128–29, p 128–29
autonomy, (def.) 300
Ayers Rock, 604, p 611
Aymara, g 196
Azerbaijan, m 351, g 352, 356
Aztec Empire, 195, m 195

B

Bab al-Mandab, 406, 462
badlands, (def.) 151
Baghdad, Iraq, 399–400
Bahamas, m 215, g 216, 219, 224
Bahasa Indonesia language, 547
Bahrain, m 399, g 400, 405–06
Baja California, 203–204
Baku, Azerbaijan, 356
balance of power, (def.) 265
balance of trade, (def.) 304, 435
Balearic Islands, 297
Bali, Indonesia, p 491
Balkan Peninsula, 309, m 309, 313, 387
Baltic countries, 264, g 318, 318–19, p 318–19, m 319
Baltic languages, g 267
Baltic Sea, 263, 336
Baltimore, MD, p 126, 130
Bandar Seri Begawan, Brunei, p 542
Bangkok, Thailand, 537, p 537
Bangladesh, p 51, 82, 88, p 563, m 585, g 586, 590–91, p 591
Banneker, Benjamin, 130
Bantu language, 454, 459, 464, 478
Baraboo, WI, p 140
Barbados, m 215, g 216, 219, 224
Barcelona, Spain, 299
bar graphs, S10, g S10
barrier islands, (def.) 133
basins, (def.) 113
basin irrigation, (def.) 433
Basin and Range region, 155
Basques, 270, 300
Baton Rouge, LA, 136
Batu, 328
Bavaria, 287
Bavarian Plateau, 262
Bay of Bengal, 39, 571–72
bayous, (def.) 135
bays, 39
Bechuanaland. See Botswana
Bedouins, (def.) 384, 404
Beijing, China, p 490, 499, 513–14, p 513, p 517
Beira, Mozambique, 485

Beirut, Lebanon, 395–96
Belarus, m 351, g 352, 354–55, p 354
Belém, Brazil, 238
Belgian Congo. See Zaire
Belgium, m 283, g 285, 290, p 290
Belize, m 215, 216, g 216
Belo Horizonte, Brazil, 237
Belorussian language, g 267
Benelux countries, m 283, g 285, 290–91. See also Belgium; Luxembourg; Netherlands
Benghazi, Libya, 436
Benin, m 441, g 445, 450
Bering Sea, 168, 169
Berlin, Germany, 287, 288–89, p 288–89, m 289
Berlin Wall, 71
Bernard, Maine, p 123
Bern, Switzerland, 294
Bhutan, 562, m 585, g 586, 590
Biafra, 449
bilingual, (def.) 138
Bingham, Hiram, 186, 187, p 187
biogeography, (def.) 605–606
biomes: (def.) 58; barren regions, m 57, 60; desert biome, m 57, 60; forest biome, m 57, p 58, 58–59; grassland biome, m 57, 59–60; people and, 60; savanna biome, m 57, 59; tundra biome, m 57, 60; world plant regions, m 57, p 58–59, 58–60
biosphere, (def.) 16, g 16, 52, 60
biotechnology, (def.) 123
Birmingham, AL, 136
birthrate, (def.) 83
bituminous coal, (def.) 125
Black Forest, Germany, 262, p 264, 286
Black Sea, p 315, 336
blizzards, p 25
Bloemfontein, Orange Free State, 479
Blue Ridge Mountains, 111, 133
Bobo-Dioulasso, Burkina Faso, p 413
bogs, (def.) 276
Bogotá, Colombia, 229, p 229
Bohemian Forest, 286
Bohemian Highlands, 262
Bolivia, m 245, g 246, 248–49
Bombay, India, p 27, 579, 582
Bonaire, 224
Bond, Ruskin, 34
Bonifacio, France, p 285
Bora-Bora, p 597
boreal forests, m 57, 59
Borneo, p 17, 541
boroughs, (def.) 127
Bosnia and Herzegovina, m 309, g 310, 313, 314
Bosporus, 263, 387
Boston, MA, p 122, 123, 127, p 131
Botswana, 421, m 477, g 484
Bousnian, Mongi, 66
Brahmaputra River, 562, 572, 590
Brahmins, 578
Brasília, Brazil, 237
Bratislava, Slovakia, 312
Brazil: g 236, p 238; Amazon region, 238; Campo Cerrado, 237; Carnival, p 234; historical geography, 236; issues, 238–39; minerals, 197; mountains, 193; Northeast region, 236–37; oil reserves,

197; physical geography, m 235, 235–36; population density, 82; rain forests, 88, p 88; regions of, 236–38; Southeast region, 237
Brazzaville, Congo, 471
break-of-bulk centers, (def.) 127
Brisbane, Australia, 610
Britain. See United Kingdom
British Columbia, 174, 179, 180
British Empire: 265, 275; in South Asia, 565–66, 567
British Guiana. See Guyana
British Honduras. See Belize
British Isles. See United Kingdom; Ireland
British North America, 175
Brooks Range, 165
Brunei, m 541, 542–43, g 542, p 542
Brussels, Belgium, 290, p 290
Bryce Canyon, p 43, 155
Bucharest, Romania, 312
Budapest, Hungary, p 255, 312
Buddhism, p 100, 362, (def.) 503, 514, 520, 532, 537, 565, 576, 590, 592
Buenos Aires, Argentina, 241, p 241
Buffalo, NY, 126
buffer states, (def.) 234
Bukhoro, Uzbekistan, m 363, p 363
Bulgaria, m 309, g 310, 313, p 315
Bulgarian language, g 267
Bunker Hill, 122
Burkina Faso, p 413, m 441, g 445, 446, p 446
Burma (Myanmar), m 531, g 532, 537–38, p 538
Burundi, 459–60, g 459
Buryat, 338
Byelorussia. See Belarus

C

Cabinda, Angola, 473, p 473
Cabora Bassa, Mozambique, 478
Cabot, John, 175
cacao: (def.) 198; production, g 448–49
cacti, p 53, 54, 60
Cairo, Egypt, p 435, p 9, 424, 434
Calcutta, India, p 569, 582
calderas, (def.) 164
Calgary, Alberta, 179
calicoes, (def.) 567
California: g 160; agriculture, p 72, 160; cities, 161–63; climate, 160; Death Valley, 49, 113, 159; earthquakes, 159–60; industry, 161; irrigation, 35; issues, 161; landforms, 159–60; physical geography, 159, m 159; population density, 82; resources, 160–61; San Andreas Fault, 48, 159–60, p 160; solar energy, 95, p 95; Yosemite National Park, m 3, p 4, 5–6, p 6. See also specific cities
calypso, (def.) 219
Cambodia, 504, m 531, g 532, 532, 534, 535–36, p 539
Cameroon, m 465, g 470, p 472, 471–72, p 475
Canada: g 174; artifacts, p 172; border with United States, p 106; British in, 175; climate, m 108, 174; Dominion of Canada, 175–76; economy, m 110, 180; French in, 175; French-speaking people in, 177, 178, 180; historical geography,

175–76; Hudson Bay, 39; hydroelectricity, 95, p 95, 174; issues, 180; landforms, 173–74; parks, p 172; physical geography, m 107, m 173, 173–74; physiographic regions, m 112; population, 96, m 109, g 239; provinces, 176–79; regionalism, 180; resources, 174; Rocky Mountains, p 43, 154; teenagers, 178. *See also* specific cities and provinces
Canadian Interior Plains, 173
Canadian North, 179
Canadian Shield, 173
Canary Islands, 297
Canberra, Australia, 600, 610
Cancún, Mexico, 209
Cantabrian Mountains, 297
cantons, (def.) 291
Cape Cod, 122
Cape of Good Hope, 478
Cape Province, South Africa, p 422, p 478
Cape Town, South Africa, 478, 479
Cape Verde, m 441, g 445, 447–48
capital, (def.) 438
capitalism, (def.) 81
Capri, Italy, 304
Caracas, Venezuela, 231
carbon dioxide, 93
cardamom, (def.) 215
Caribbean islands: 214; human geography 219, 222; island countries, 222–24; physical geography, m 215, 219. *See also* specific islands
Caribbean Sea, 39
Caribbean South America, 226–32, m 227. *See also* specific countries
Carpathian Mountains, 262, 309, 351
Carson National Forest, p 34
Cartagena, Colombia, p 197, 229
cartels, (def.) 249
Carthage, Tunisia, 424
Cartier, Jacques, 175
cartograms, g 239
cartography, (def.) 8
Casablanca, Morocco, 438
Cascade Mountains, 44, 113, 164
case studies. *See* environment; Five Themes of Geography; global connections
cash crops, (def.) 204
Caspian Sea, 38, 336, 355, 356, 361
cassava, (def.) 443
caste system, (def.) 578
Castro, Fidel, 222
Catherine the Great, 329
Catholicism. *See* Roman Catholicism
Catskill Mountains, 111, 125
Caucasus Mountains, 337, 387
Caucasus region, 355–56
Cave of Machpelah, 386
cease-fires, (def.) 587
Celebes. *See* Sulawesi
Celtic languages, g 267
center pivot irrigation, (def.) 153
Central Africa: g 470; climate, 465; landforms, 462; Northern Central Africa, 471–72; physical geography, 465, m 465; Southern Central Africa, 473–74; ways of life, 465–66
Central African Republic, m 465, g 470, 471
Central America: 214–19, m 215, g 216;

culture regions, 68; formation, 193· as isthmus region, 215. *See also* specific countries
Central American Common Market, 200
Central Asia: g 364; economic geography, 360; human geography, 360–61; issues, 361; landforms and climate, 359; language, m 369; physical geography, 359–60, m 359; as region in transition, 368–69, p 368–69; Silk Road, 360, 362–63, p 362–63, m 363; water resources, 359–60
Central Valley, CA, 113, 159, 160
Ceylon, 566. *See also* Sri Lanka
Chaco, 240, 242
Chad, m 441, g 445, 446
Chad Basin, 421
Chang River (Yangtze), 499, 511
channel, (def.) 143
Chao Phraya, 531
Charlemagne, 265
charts, S12–S13, g S12–S13
Chechnya, Russia, 347
Chekhov, Anton, 322, p 322
Chelyabinsk, Russia, 342
Chen, Yuan-tsung, 492, p 492
Cheng Ho, 493
Cherbourg, France, 284
Chernobyl, 94
Chesapeake Bay, p 37, 37, 124, 130
Chiang Kai-shek, 509
Chiapas, Mexico, 212
Chicago, IL, 116, 147–48, p 147
Chichén Itzá, p 202
Chile, 193, 194, 197, 244, p 244, m 245, g 246, 249–50, p 250, p 251
China: g 509, g 568; agriculture, p 510, 511–12, g 512–13; artifacts, p 491, p 506, p 508; coal reserves, 93; communism, 504, 509–11; cultural development, 507–508; historical geography, 504, 507–11, g 508; under Mao, 509–10; Northeastern China, 514; Northern China, 512–14; outside influences, 508–09; physical geography, m 507; political turmoil, 509; population, 116, g 568; population density, 82; regions, 511–15; Southern China, 511–12; Western China, 514–15
Chinese language, p 502, 502–503, 616
chinooks, (def.) 152
chlorofluorocarbons (CFCs), 93
Choptank River, p 37
Christianity, (def.) 380, 381, 395, 503, 576
Cincinnati, OH, p 115, 146
cities. *See* urban areas and urbanization; specific cities
citizenship, 8
city-states, (def.) 286
Ciudad Juárez, Mexico, 205, p 206
civil wars, 115, 217, 229, 313–14, 396, 403, 449, 462, 470, 473, 485, 592
Civil War, U.S., 115, 567
clans, (def.) 364
Clark, William, 105
Cleopatra, 390
Cleveland, OH, 148
client states, (def.) 331
climate: (def.) 18; dry, m 28–29, 30–31, p 31; global, 27–32, m 28–29; highland, m 28–29, 32; high-latitude,

m 28–29, 32; low-latitude, 27, m 28–29, 30; middle-latitude, m 28–29, 31. *See also* specific continents, regions, and countries
climate maps, S8
climax community, (def.) 55
climographs, S11, g S11, g 194
coal. *See* anthracite coal, bituminous coal
coalition governments, (def.) 304
Coastal Plain, U.S., 111, 124, 133, 134
Coast Ranges, U.S., 113, 159, 164
coke, (def.) 125
Cold War, 265–66, 288–89, 331
Colombia: p 19, 197, 227–29, m 227, p 229, g 229, p 233
Colón, Panama, 219
colonialism, 115, 195–96, 380, 425–26, 504, 565–66
Colorado: g 152; agriculture, 152; mining, 154; Native Americans, 156; oil production, 153; parks, 155; physical geography, m 151, 155; Rocky Mountains, p 150
Colorado Plateau, 155
Colorado River, p 44, 117, 155, p 157
Columbia Basin, 164
Columbia Plateau, 155
Columbia River, 95, 117, 164, 165
Columbus, Christopher, 195
Columbus, OH, 146
COMECON. *See* Council for Mutual Economic Assistance
command economies, (def.) 81–82
commercial agriculture, (def.) 72
commonwealth, (def.) 223, 275
communes, (def.) 509
communism, (def.) 82, 222, 310, 313, 330, 473, 509–11, 534–35
Comoros, m 477, g 484, 486, m 486
compass roses, (def.) S6
complementary regions, (def.) 312
Conakry, Guinea, 447
condensation, (def.) 25
confederations, (def.) 291
Confucianism, (def.) 503
Congo, m 465, g 470, 471
conic projections, S4, g S4
coniferous forests, (def.) 58
Connecticut, m 121, g 124
Connecticut Valley, 112
Constantinople, 380. *See also* Istanbul, Turkey
constitutional monarchy, (def.) 273
contiguous, (def.) 113
Continental Divide, (def.) 154
continental shelves, (def.) 39, g 39
continents, (def.) S2, m S2. *See also* specific continents
Contras, 217
Cook, James, 614, 620
cooperation. *See* economic cooperation; international cooperation
cooperatives, (def.) 222
copra, (def.) 485
coral reefs, (def.) 56, p 56, p 61
Coral Sea, 39
Córdoba, Argentina, 241
Corfu, m 297, 306
cork, (def.) 299, p 299
Corn Belt, U.S., 143–44, g 144–45
corn production, g 144–45

Corsica, 285, p 285
Cortés, Hernán, 195, 196
cosmopolitan, (def.) 290
Cossacks, (def.) 353
Costa of Ecuador, 245, 246
Costa Rica, 68, m 215, g 216, 217–18
Côte d'Ivoire, p 426, m 441, 442, g 445, 450, p 451
cottage industries, (def.) 577
Coulee Dam, WA, p 35
Council for Mutual Economic Assistance (COMECON), 310–11
coups, (def.) 248
Cozumel, Mexico, 209
Crater Lake, 38, 164, p 164
Creole language, 223
Crete, m 297, 305
Crimea, m 351, 353–54
Croatia, m 309, g 310, 313, 314
Cronon, William, 140
crop rotation, (def.) 87
Cuba, 193, m 215, g 216, 219, p 222, 222–23
cultural bias, 552
cultural diversity. See ethnic diversity
cultural geography. See human geography
culture: (def.) 67, 67–69; change and, 68–69; economics and, 75–76; nationalism and, 74; politics and, 75; religion and, 74–75, m 74; traditional versus modern values, 75; world culture, 70–71; world events and, 74–76. See also specific regions and countries
culture hearths, 72, m 73
culture regions, (def.) 67–68
culture traits, (def.) 67, p 67
Cumberland Plateau, 112, 133
Curaçao, 224
Cushite speakers, 454
Cuyahoga River, 148
cylindrical projections, S4, g S4
Cyprus, 306, m 387, g 388, 389
Cyrillic alphabet, p 319
czars, (def.) 329
Czech language, g 267
Czechoslovakia, 266, 312
Czech Republic, m 309, g 310, 312

D

Dahomey, 450
daily life. See teenagers; specific regions and countries
Dairy Belt, U.S., 144–45
dairy farming, 122, p 140, 144–45, 164
Dakar, Senegal, 447, p 447
Dalai Lama, 86, 514, p 514
Dallas-Forth Worth area, TX, 138
Damascus, Syria, 395, p 395
dams, 35, 89, 95, 153, 155, 197, 336, 433
Danish language, g 267
Danube River, p 255, 264, 286, 294, 309
Dardanelles, 387
Dar es Salaam, Tanzania, 459
Darwin, Australia, 610
Dayton, OH, 146
Dead Sea, 38, 379, p 385, 389, 390
Dean, Captain Harry, 415
death rate, (def.) 83
Death Valley, CA, p 40, 49, 113, 159
deciduous forests, (def.) 58

defoliants, (def.) 535
deforestation, (def.) 88, p 88, 456–57, p 456
de Cervantes, Miguel, 256, p 256
de Gramont, Sanche, 440
degrees, (def.) S3
Delaware, p 37, m 121, g 124, 125
Delaware Bay, p 37, 127
Delaware River, 124, 127, 130
Delmarva Peninsula, p 37
deltas, (def.) 49
demilitarized zone (DMZ), 528
democratic governments, (def.) 75
demography, (def.) 82
Denali National Park, p 168
Deng Xiaoping, 510
Denmark, m 273, g 274, 279
Denver, CO, 153
Depoe Bay, OR, p 41
deport, (def.) 175
desalinization, (def.) 89
desalinization plants, p 40
desert biome, m 57, p 59, 60
desertification, (def.) 87
deserts: desert biome, m 57, 60; as dry climates, g 28, 30–31, p 31; plant community of, p 53, 53–54; polar, 620; rainshadow effect and, 27. See also specific deserts
Detroit, MI, 148
Detroit River, 142
developed countries: (def.) 80–81; economics, 80–82; politics, 81–82; population, 84. See also specific countries
developing countries: (def.) 81; economics, 80–82; politics, 82; population, 84. See also specific countries
Dhaka, Bangladesh, p 563, 591
dialects, (def.) 300
diffusion, (def.) 68
dikes, (def.) 290, 292–93, p 293
Dinaric Alps, 309
Diop, Birago, 414, p 414
directional indicators, (def.) S6
District of Columbia. See Washington, DC
diversify, (def.) 135
diversity. See ethnic diversity
Djibouti, m 453, g 459, 462
DMZ. See demilitarized zone
doldrums, (def.) 21
Dolomites, p 255
domestication, (def.) 72
Dominica, m 215, g 216, 224
Dominican Republic, m 215, g 216, 219, 223
Dominion of Canada, 175–76
dominions, (def.) 175
Donets Basin, 352
Donets'k, Ukraine, 352
double cropping, (def.) 511
Douglas, William O., 373, p 373
Douro River, 298, 299
Drakensberg Escarpment, 421, 477, p 478
Dravidian language, 576
dredge, (def.) 143
drip irrigation, (def.) 382
droughts, (def.) 4, 89, 114, 151–52, 382, 423, 441–42, 445, 446
dry climates, g 28, m 28–29, 30–31, p 31
dryland agriculture, (def.) 361
Dubai, United Arab Emirates, p 381

Dublin, Ireland, p 276, 277
Duluth, MN, p 119
Durban, South Africa, 422
Dutch language, g 196, g 267, 290
dynasties, (def.) 507–508

E

Earth: atmosphere and climates, 18–32; earth system, 16, g 16; forces beneath crust, g 45, 45–46; interior, g 45, 45; landforms, 42–50; as planet, g 11; plants and soil, 53–54, g 56, 53–56; revolution, 13; rotation, 12–13; tilt, g 13, 13; as water planet, 34–40
earthquakes, (def.) 45, 46, 48, 159–60, p 162–63, 193, 211, 519–20
earth system, (def.) 16, g 16
East Africa: g 459; climate, 453–54; economic geography, 455; ethnic heritage, 454–55; Horn of Africa, 461–62; human geography, 454–55; landforms, 453; physical geography, m 453, 453–54; savanna, 59
East and Southeast Asia: g 494; agriculture, 500; artifacts, p 491, p 494, p 506, p 518, p 521, p 530, p 540; climate, m 496, 499; colonial history, 504; fishing, 500; forests, 500; human geography, 502–504; industrialization, 501–502; landforms, 498–99; language, 502–503; mineral and energy resources, 500–501; political geography, 504; physical geography, m 495; population, m 496; post–World War II, 504; rainfall, 499; religion, 503–504. See also specific subregions and countries
East Asia. See East and Southeast Asia
Easter Island, p 621
Eastern Europe: g 310; artifacts, p 308; as cultural crossroads, 310; economic geography, 310–11; physical geography, m 309, 309; as region in conflict, 313–14; Soviet control in, 310
Eastern Mediterranean, 386–96, m 387. See also specific countries
Eastern Orthodox Christianity, 310, 312, 353, 354, 380
Eastern Rift Valley, 421–22, 453, p 454
East Germany, 265, 266, 287. See also Germany
East Pakistan, 566. See also Bangladesh
Ebro River, 298
economic associations, (def.) 267
economic cooperation: in Eastern Europe, 310–11; in Europe, 267; in North America, 206–207; in Middle and South America, 200; in Pacific Rim, 166–67; in Southeast Asia, 533
economic development. See developed countries; developing countries
economic geography, 79. See also specific regions and countries
economic indicators, 79–80
economic maps, S9, m S9
economics: choices in economic development, 81; in developed and developing countries, 80–82; global, 79–82; linked with geography, 366; politics and, 81–82
ecosystems, (def.) 54

Ecuador, p 193, 194, 245–47, m 245, g 246
edge cities, 116
Edmonton, Alberta, 179
Edwards Plateau, 134
EEZ. *See* Exclusive Economic Zone
EFTA. *See* European Free Trade Association
Egypt, 82, 394, 421, m 431, g 433, 433–35, p 434–35, p 439
ejidos, (def.) 205, 212
El-Baz, Farouk, 430
Elbe River, 286
Elburz Mountains, 401
El Djouf, 421
electricity, 94, 95
elevation, (def.) 26
elevation maps, m 50
elevation profiles, g 50
El Niño, (def.) 248
El Paso, TX, p 206
El Salvador: m 215, g 216, 217; culture regions, 68
embargoes, (def.) 480
emigrants, (def.) 84
endangered animals, 572, 574–75
endemic species, (def.) 605
energy resources. *See also* global energy systems; specific types of energy resources
England. *See* United Kingdom
English language, g 196, g 267, 503, 550, 576, 578
entrepots, (def.) 544
environment: air pollution, 6, 92–93, p 92, 130; rain forests, 220–21; energy resources, 93–95; forests, 87–88, p 88; insecticides, 548–49; organizations committed to protection of, 91; population and, 96; renewable versus nonrenewable resources, 86, 94–95; soil, 87; students' strategies for solving environmental problems, 90–91; water, 382–83; water pollution, 146; water quantity and quality, 89, 92; wildlife and, 574–75. *See also* human-environment interaction
equator, (def.) S2, g S3
Equatorial Guinea, m 465, g 470, 472
equinoxes, g 14, (def.) 15
ergs, (def.) 431
Erie Canal, 126
Eritrea, m 453, g 459 462
erosion, (def.) 42, 43–46, p 43, p 44, 49
escarpments, (def.) 421
Eskimo. *See* Inuit
Estonia, 318, g 318, m 319
Estonian language, g 267, 318
estuaries, (def.) 37
Ethiopia, 95, 422, 425, m 453, g 459, 461–62
ethnic diversity, 68, 161. *See also* specific regions and countries
ethnic groups, (def.) 68. *See also* specific countries
ethnocentrism, (def.) 552
EU. *See* European Union
Euphrates River, 379, 399
Eurasia, (def.) 258
Europe: agriculture, 264; artifacts, p 255, p 257, p 258, p 272, p 282, p 296, p 308; climate, m 260, 262–63;

colonization of Africa, 424–25, m 425; economic cooperation, 267; economy, m 261, 267; exchange students from, 305; historical geography, 265–66; human geography, 266–67, 269; issues, 270; landforms, 262; languages, 267, g 267; minerals and energy, 264; natural resources, 263–64; physical geography, m 259, 262–63; population changes, 266–67; population, 96, m 260; as region in transition 318–19; religion, 267; tourism, 268; water resources, 263–64; ways of life, 270. *See also* Northern Eurasia; Russia; specific subregions and countries
European Free Trade Association (EFTA), 267, 279
European Union (EU), (def.) 267, 268
Evansville, IN, 146
evaporation, (def.) 25
evapotranspiration, (def.) 37
Everglades, 38, 133, 137, 138, p 139
exclaves, (def.) 473
Exclusive Economic Zone (EEZ), 615
exiles, (def.) 480
exotic rivers, (def.) 378, 384
export economies, (def.) 527
extinct, (def.) 60

F

factories. *See* industries and industrialization
Fall Line, U.S., (def.) 111, 133
fallow, (def.) 400
famines, (def.) 266
FAO. *See* Food and Agriculture Organization
farming. *See* agriculture
faults, (def.) 45, g 45, 46, 48
favelas, (def.) 237
Federated States of Micronesia, m 601, g 613, 614
Fertile Crescent, 379, 399
Fiji, p 269, g 613, 614
Finke River, p 604
Finland, m 273, g 274, 280, 318
Finnish language, g 267, 318
Finno-Ugric languages, g 267
Five Themes of Geography: 7; case studies, 206–207, 292–93, 468–69, 580–81; human-environment interaction, 5; location, 3; movement, 6; as organizers, 7; place, 4; region, 7. *See also* human-environment interaction; location; movement; place; region
fjords, (def.) 277–78, p 277
flat-plane projections, S5, g S5
flood control and floods, 89, 135, 293, 594. *See also* monsoons
floodplains, (def.) 49
Florence, Italy, 301
Florida: m 133; g 136; agriculture, 135, 136; climate, 134; Everglades, 38, 133, 138, p 139; hurricanes, 134; resources, 136
Florida Keys, 137
flowcharts, S13, g S13
foehn, (def.) 294
folds, (def.) 45, g 45, 46
Food and Agriculture Organization (FAO), 548

food chains, (def.) 53
food webs, g 53
Ford, Henry, 117–18
foreign aid, 81
forest biome, m 57, p 58, 58–59
forests: quebracho forests, 240, 242; rain forests, p 17, p 30, m 57, 58, 88, p 88, p 199, 220–21, p 220–21, 235, 422, 533; as resource, 87–88, p 88; second-growth forests, 123
Formosa. *See* Taiwan
Fort Duquesne, 126
Forten, Charlotte, 132
Fort Pitt, 126
fossil fuels, (def.) 93, 274
fossil water, (def.) 379
Fouta Djallon, 421
France: g 285; culture, 67; economic geography, 284; issues, 285; Mont Saint Michel, p 12; nuclear power plants, 94; physical geography, 283–84, m 283. *See also* Paris, France
Frankfurt, Germany, 287
free enterprise, (def.) 81
free ports, (def.) 438
French Guiana, 226, m 227, 232, p 232
French language, 177, g 196, g 267, 284, m 294, 447, 503
French Polynesia, p 597
French Riviera, 283
Fried, Robert, 234
fronts, (def.) 21
fundamentalism, (def.) 75, 381, 403, 435, 437
furrows, (def.) 379
al-Fuzay, Khalil, 372

G

Gabon, m 465, g 470, 472
Galápagos Islands, 246–47
galaxies, 10
Galeano, Eduardo, 186, p 186
Galveston, TX, 134
Gambia, m 441, g 445, 447
Gambia River, 447
Gandhi, Indira, 577
Gandhi, Mohandas, 566, 578
Gandhi, Rajiv, 577
Ganges River, 562, 571, 590
Garonne River, 283
GATT. *See* General Agreement on Tariffs and Trade
gauchos, 240
Gaza Strip, m 387, 394
GDP. *See* gross domestic product
General Agreement on Tariffs and Trade (GATT), 118
Geneva, Switzerland, 294
Genghis Khan, 328, 516
geographic information systems (GISs), 23
geographic method, 252
geography: (def.) 2; branches, 7–8; economic geography, 79; fine art and, 316; history and, 408; human geography, 7, 67–69, 72–76; models used, 182–83, g 182; photographs as geographical evidence, 488; physical geography, 8; population geography, 82–84; as profession, 8; urban geography, 7
Georges Bank, 123

Georgia, Eurasia, p 350, m 351, g 352, 355, p 357
Georgia, U.S.: m 133, g 136; agriculture, 135, 136; Appalachians, 133; industry, 136; Okefenokee Swamp, 133; Piedmont, 133
geothermal energy, (def.) 95
Germanic languages, g 267
German language, g 267, m 294
Germany: g 285; Berlin Wall, 71; castles, p 286; cathedral, p 270; division, 265, 288–89; East Germany, 265, 266, 287; economic geography, 286–87; Holocaust, 390–91; issues, 287; Northwestern Germany, 287; nuclear energy, 94; physical geography, m 283, 286; population, 266; reunification, 266, 287, 289; Rhineland, 287; Ruhr, 287; Southern Germany, 287; World Wars and, 265, 288, 390–91
geysers, p 279, (def.) 280
Gezira Scheme, 460
Ghana, m 441, g 445, 450
ghettos, (def.) 116–17
ghost towns, (def.) 154
Gibraltar, m 297, 306
GISs. See geographic information systems
Giza, Egypt, p 435
glaciers, (def.) 43–44, p 43, p 171
Glasgow, Scotland, 274
glasnost, (def.) 331
Glen Canyon Dam, 155
global air pollution, 92–93
global awareness, 622
global climates, 27–32, m 28–29
global connections: migration, 588–89; tourism, 268–69; trade, 166–67, 362–63, m 363, 544–45, m 545
global economics: choices in economic development, 81; developed and developing countries, 80–82; economic geography and, 79; economic indicators, 79–80; politics and, 81–82
global energy systems: air pressure and wind, g 20–21, 20–21; global wind belts, 21–24; greenhouse effect, g 19, 19; ocean circulation, 24
global issues. See entries beginning with "global" and "international"
global population density, 82
global population growth, 82–84, g 83, 96
global warming, 93. See also greenhouse effect
global wind belts, 21–24
globes, S2–S3, g S3
GMT. See Greenwich mean time
GNP. See gross national product
Gobi, 30, 338, 499, 516
Godavari River, 572
GOES. See Geostationary Operational Environmental Satellite
Gogol, Nikolay, 350
Golan Heights, 394
Gold Coast, Australia, p 596, 610
Gold Coast. See Ghana
Gondwanaland, 48
Gorbachev, Mikhail, 331, 339
government, 75, 82, 403. See also specific countries and types of government
Graham, Peter, 316, p 316
Grand Canyon, p 44, p 102, 155

Grand Coulee Dam, p 35
granite, (def.) 122
graphic organizers, S13, g S13
graphs, S10–S11, g S10–S11
grassland biome, m 57, p 59, 59–60
Great Barrier Reef, p 600, 606, 610
Great Basin, U.S., 30, 113, 155
Great Britain. See United Kingdom
great-circle routes, (def.) S5
Great Dividing Range, 604, 606
Greater Antilles, 193, 219
Greater Khingan, 498
Great Lakes, 38, 112, 113, 117, 141–42, m 142, 145, 146, 173, 175
Great Plains: 112; buffalo herds, 60; climate, 114, 151–52; and end of open range, 152; landforms, 151; Ogallala Aquifer, 39; resources and industry, 153; Wheat Belt, 152
Great Salt Lake, 38
Great Serpent Mound, p 115
Great Smoky Mountains, p 111, 133, 137
Great Wall of China, p 490, 507
Grebenshikov, Boris, 70
Greece, m 297, g 298, 304–306, 388–89. See also Cyprus
Greek language, g 267, m 369
Green Belt Movement, 456–57
greenhouse effect, (def.) 19, g 19
Greenland: m 259, 279; 44, 60; ice cap, p 30; ice sheets, 44
Green Mountains, 121
Green Revolution, (def.) 566
Greenwich mean time (GMT), S14
Grenada, m 215, g 216, 224
grids, (def.) S2, g S3
gross domestic product (GDP), (def.) 80
gross national product (GNP), (def.) 80
groundwater: g 38, 38–39, (def.) 39; pollution, 92
growth poles, (def.) 238
Guadalajara, Mexico, 205
Guadalcanal: contour map, m 50; elevation map, m 50; elevation profile, g 50
Guadalquivir River, 298
Guadiana River, 298
Guam, 615
Guanajuato, Mexico, p 205, p 209
Guangzhou, China, p 510
Guaraní, g 196, 242
Guatemala: 215–16, m 215, g 216; culture regions, 68
Guatemala City, Guatemala, 215
guerillas, (def.) 229
Guiana Highlands, 193, 230, 231
Guinea, m 441, g 445, 446, 447
Guinea-Bissau, m 441, g 445, 447
Gulf of Alaska, 39
Gulf of `Aqaba, 390, 396
Gulf of Bothnia, 280
Gulf of Guinea, 422
Gulf of Mexico, 39, 135, 136
Gulf States, 405–406
Gulf Stream, (def.) 114, 122
gum arabic, (def.) 455
Gunther, John, 612
Guyana, 226, m 227, g 229, 231–32

H

habitation fogs, (def.) 336

hacienda system, (def.) 198, 205
The Hague, Netherlands, 291
Haifa, Israel, 394
Haiti, 88, m 215, g 216, 223
Half Moon Bay, CA, p 158
Halifax, Nova Scotia, 176
Hamburg, Germany, p 263, 287
Han Chinese, 514
Han dynasty, 507, g 508
Harare, Zimbabwe, p 412
Havana, Cuba, 222, p 222
Havel, Vaclav, 308
Hawaii: m 159, g 160; climate, 114, 170; cultural geography, 170; economic geography, 170; ethnic diversity, 170, p 170; Hawaii Volcanoes National Park, p 42; island environment, 169–70; issues, 170; volcanoes, 113, 169
Hayslip, Le Ly, 530
headwaters, (def.) 37
heavy industry, (def.) 342
Hebrides Islands, m 273, 274
Helsinki, Finland, 280, 318
hemispheres, (def.) S3
highland climates, m 28–29, 32
high-latitude climates, m 28–29, 32
High Plains, U.S., 134, 152
highways: U.S. interstate highway system, 147; Pacific Coast Highway, p 158
Hilton Head, SC, 137
Himalayan kingdoms. See Nepal and Bhutan
Himalayas, 48, 49, p 269, 498, 562, 571
Hindi language, 576, 616
Hinduism, (def.) 564, 576
Hindu Kush range, 403, p 403
hinterlands, (def.) 176
Hispaniola, 193, m 215, 219
HIV. See Human Immunodeficiency Virus
Hobart, Australia, 610
Ho Chi Minh City, Vietnam, p 534, 535
Hokkaido, 519
Hollywood, CA, 161
Holocaust, (def.) 390–91
Homer, 18
homogeneous, (def.) 355
Honduras, m 215, 216, g 216
Hong Kong, p 491, 501
Hong River, 531
Honolulu, HI, 169, 170
Honshu, 519, p 522
Hoover Dam, 155, p 157
Hopi, p 103, 156
hot spots, (def.) 169
Houston, Sam, 138
Houston, TX, 136, 138
Huang He, 89, 499, 513
Hudson Bay, 39
Hudson River, 117, 124, 130
Hudson River Valley, 112, 126
human-environment interaction, (def.) 5; agriculture and, 72; biosphere and people, 60; water quality, 89–90; water quantity, 89. See also specific regions and countries
human geography: (def.) 7; agriculture and, 72, p 72; biomes and, 60; city life, 72–73; culture and, 67–69; economics and, 75–76; industrialization, 73; issues, 76; landforms and, 49–50, linked with physical geography, 594; nationalism,

74; politics, 75; religion, 74–75, m 74; traditional versus modern values, 75
Human Immunodeficiency Virus (HIV), 84. *See also* Acquired Immune Deficiency Syndrome
humidity, (def.) 25
humus, 55, g 56
Hungarian language, g 267, 310
Hungary, 94, 266, m 309, g 310, p 311, 312
hunting and gathering, 69
Huntington, Henry, 166
hurricanes, 22, p 22, (def.) 25, p 25, 134, 138
Huxley, Elspeth, 452
hydroelectricity, (def.) 35
hydrologic cycle, g 36, 36–37
hydrosphere, (def.) 16, g 16, 36

I

Ibadan, Nigeria, 449
Iberian Peninsula, 297, m 297
Ibo, 449
icebergs, p 41
icebreakers, (def.) 264, p 337
ice-cap climates, 32, 60
Iceland, 47, p 49, 95, m 273, g 274, p 279, 279–80
Icelandic language, g 267
ice sheets, 44
Idaho, m 151, g 152, 154, 155
Idrisi, Abu Abdallah Mohammed, 257
Ife, Nigeria, 440
Illinois: m 141, g 143; agriculture, 144; cities, 147–48; mining, 145; Ohio River valley, 146. *See also* specific cities
Illinois River, 143
immigrants, (def.) 84
immigration, 116, 588–89, g 589. *See also* movement
imperialism, (def.) 265. *See also* colonialism
Incan Empire, 195, m 195
India: agriculture, 572–73; caste system, 578; city life, 578–79, 582; climate, 572; colonialism, 565; economic geography, 572–73; environmental issues, 572; geothermal energy, 95; historical geography, 564–66; human geography, 564–66, independence, 565–66; industrialization, 567–68; industry, 573, 576; landforms, 571–72; language, 576; mineral and energy resources, 573; monsoon, p 27, 570; political geography, 576–77; population, 116, 568, g 568, 570; population density, 82; religion, 564, 576; rural life, 577, 580–81, p 580; teenagers, 579; temples, p 577; wildlife, 574–75, p 574, m 575
Indiana, m 141, g 143, 145, 146
Indian Ocean, S2, 39, 406, 421
Indian perimeter, 584–93, m 585, g 586. *See also* specific countries
indigenous, (def.) 188
indigo, (def.) 196
Indo-European languages, g 267
Indonesia, p 67, 88, 95, p 491, p 540, m 541, g 542, 543–47, p 546, p 548, 549
Indus River, 562, 585–86
industries and industrialization: (def.) 35,

73; air pollution and, 92; primary industries, 79; quaternary industries, 79; secondary industries, 79; tertiary industries, 79. *See also* specific regions and countries
inflation, (def.) 212
information age, p 65, 76, p 76
infrastructure, (def.) 80
Inner Mongolia, 516
innovation, (def.) 68
inset maps, S7
Integrated Pest Management (IPM), 548–49
intensive agriculture, (def.) 532
Interior Plains, U.S., 112, m 112, 134, 141
Interior Southwest Asia, 398–407, m 399, g 400. *See also* specific countries
Interior West, U.S., U.S.: g 152; agriculture, 152–53, 155; climate, 114, 151–52, 154, 155; economic geography, 152–55; landforms, 151, 154, 155; physical geography, m 151, 151–52, 154, 155; resources, 153, 154, 155
Intermountain West, U.S.: agriculture, 155; cities, 155–56; human geography, 156; physical geography, m 151, 155; resources, 155
international agencies. *See* specific agencies, such as European Union (EU) and United Nations (UN)
international cooperation, 75. *See also* economic cooperation; United Nations
international date line, m S14, S15
international issues. *See* headings beginning with "global" and "international"
international trade: balance of trade, 304; between Canada and United States, 180; in Pacific Rim, 166–67, m 167; shipping routes, m 545; Silk Road, 362–63, p 362–63, m 363; United States and, 118
interstate highway system, U.S., 147
Inuit (Eskimo), 69, 168–69, 179, 279
Ionian Sea, 304
Iowa, m 141, 143–44, g 143
IPM. *See* Integrated Pest Management
Iquique, Chile, 249
Iquitos, Peru, 247
IRA. *See* Irish Republican Army
Iran, 381, g 382, 384, m 399, g 400, 401–403, p 401
Iranic languages, m 369
Iraq, 379, 381, m 399, 399–401, g 400, p 400
Ireland: m 273, g 274; climate, p 32; Northern Ireland, 275; Republic of Ireland, 276–77
Irian Jaya, m 541, 543, 546
Irish Gaelic language, g 267
Irish Republican Army (IRA), 275
Iron Curtain, 310
Irrawaddy River, 531, 537
irrigation: (def.) 35, 382–83; basin, 433; center pivot, 153; drip, 382; perennial, 433. *See also* agriculture
Ise Bay, 523
Isfahan, Iran, 402
Islam, 67, 310, 360, (def.) 380–81, 388, 402, 408, 435, 437, 438, 443, 586. *See also* Muslims
Islamabad, Pakistan, 586

islands, (def.) S2
island Southeast Asia, 540–51, m 541, g 542
Israel: g 388; agriculture, 390; cities, 392–93, 394; historical geography, 390–91, 394; issues, 381, 382–83, 384, 394; physical geography, m 387, 389; PLO accord, 381, 394; population, 394; resources and industry, 390;
Istanbul, Turkey, 380, 388, p 389, p 410
isthmuses, (def.) 188
Isthmus of Corinth, 305
Isthmus of Panama, 188
Isthmus Region, Central America, 215, m 215
Isthmus of Tehuantepec, 203
Italian language, g 267, m 294
Italy, 95, p 100, m 297, g 298, 300–304
Ivan III, 329
Ivan IV, 329
Izmir, Turkey, p 371

J

Jaipur, India, p 573, p 583
Jakarta, Indonesia, p 546
Jamaica, 193, p 201, 214, p 214, m 215, g 216, 219, 222, 224
Jamestown, VA, 115
Jamuna River, 590
Japan: 518, m 519, g 520; agriculture, 521; automobile industry, 146; birthrate, 96; climate, 520; culture, 67; earthquake, 519–20; geothermal energy, 95; historical geography, 504, 508, 509, 520–21; industry, 501–502, 522; international influence, 522–23; landforms, 519–20; language, 503; nuclear energy, 94; religion, 503, 504, 520; resources, 521–22; urbanization, 523, 524–25; way of life, 523, 526
Japanese Alps, 519
Java, m 541, 543, 552
Jefferson, Thomas, 105, p 105
Jenkins, Peter and Barbara, 150
Jerusalem, Israel, 391, 392–93, p 392–93, m 393, 394
Johannesburg, South Africa, 479, 481, 482–83, p 482–83, m 482
Jordan, p 374, 382, m 387, g 388, 396, p 396
Jordan River, 382, 389, 396
Jost Van Dyke, 224
Judaism, 67, (def.) 380
Jura Mountains, 262, 283, 291
Juruá River, p 20
jute, (def.) 565, 591

K

Kabul, Afghanistan, 403
Kalahari Basin, 421
Kalahari Desert, 69, 422, 484
Kalahari River, 421
Kalimantan, m 541, 543
Kaliningrad, 319
Kamchatka Peninsula, 336, 340
Kampala, Uganda, 460
kampongs, (def.) 546
Kano, Nigeria, 449
Kansas, 148, 151, m 151, g 152

Kansas City, KS, 148
Kansas City, MO, 148
Kansas River, 148
Kanuri language, g 427
Kaohsiung, Taiwan, 516
Karachi, Pakistan, 586
Karaganda. *See* Qaraghandy
Karakoram Range, 585
Kara Kum Canal, 360
Kara Sea, p 337
Karenga, Maulana, p 71
Kariba Dam, 478
Kashmir, p 555, m 571, 587
Kasindi, Zaire, 468–69, p 468–69, m 469
katabatic winds, (def.) 620
Kathmandu, Nepal, 587
Kauai, HI, m 159, 169, p 169, 170
Kazakhstan, 342, 358, m 359, 361, 364,
    g 364, p 365
Kazantzakis, Nikos, 272
Kentucky, p 103, m 133, 134, g 136
Kenya, g 83, 95, 96, p 413, 425, 453,
    m 453, 454, 455–58, p 455–57, p 463
Kerekere, Wiremu Kingi, 598
Kerr, J. H., 599
Khabarovsk, Russia, 343
Khartoum, Sudan, 460
Khmer, 532
Khmer Rouge, 535–36
Khoi (Hottentots), 478, 484
Khoisan languages, g 427
Khomeini, Ayatollah, 403
Khyber Pass, 403, p 407
kibbutzim, (def.) 390
Kiev, 328, 329. *See also* Kyyiv, Ukraine
Kikuyu, 454, 458
Kilimanjaro, 26, p 413, 453, 458
Kincaid, Jamaica, 214
Kinshasa, Zaire, 470
Kiribati, m 601, g 613, 614
Kiruna, Sweden, 278
Kitwe, Zambia, p 473
Kjølen Mountains, 262, 277, 278
klongs, 537
Kluane National Park, Canada, p 172
Kobe, Japan, 523
Komsomol, 348
Komsomol'sk-na-Amure, Russia, 343
Kopet Mountains, 359, 401
Koran. *See* Qur'an
Korea, m 519, 526–28. *See also* North
    Korea; South Korea
Korean language, 503
Korean War, 527
Kourou, French Guiana, 232, p 232
Krishna River, 572
Kuala Lumpur, Malaysia, 541
Kuna, 220–21, p 220–21
Kunlun range, 498
Kurds, 384, 388, 588–89
Kuril Islands, 346, 519, m 519
Kursk, Russia, p 321
Kush, kingdom of, 424
Kuwait, m 399, g 400, 400–401, 405
Kuzbas, 343
Kuznetsk Basin, 340, 343
Kwanzaa, p 71
Kyoto, Japan, 523
Kyrgyzstan, m 359, 364, g 364
Kyushu, 519, m 519, 520
Kyyiv, Ukraine, 351, 352. *See also* Kiev

L

labor intensive, (def.) 198
Labrador, 176
Labrador Current, 122
Lagos, Nigeria, 449, 450
Lahore, Pakistan, p 586, p 593
Lake Baykal, 336, 338–39, p 338
Lake Chad, 421, 446
Lake Erie, 126, 142, 146
Lake Huron, 142
Lake Malawi (Lake Nyasa), 421, 474
Lake Maracaibo, 230
Lake Michigan, 142, 147
Lake Nyasa, 421, 474
Lake Ontario, 142
Lake of the Ozarks, p 146
Lake Poopó, 249
lakes: S2; formation, 38; pollution, 92. *See*
    also specific lakes
Lake St. Clair, 142
Lake Superior, 142, 145
Lake Tanganyika, 458
Lake Titicaca, 249, p 249
Lake Victoria, 460
Lanai, HI, m 159, 169
landforms: (def.) 46; classification, 49;
    development, 49–50; Earth's interior
    and, g 45, 45; forces shaping the land,
    43–46; glaciers and, 43–44, p 43; peo-
    ple and, 49–50; plate tectonics, 46–48,
    g 47; primary and secondary, 49; rock
    and, 43, p 43; types, g 46; water and,
    p 43, 43; wind and, p 44, 44–45
landlocked, (def.) 242
land reclamation, (def.) 153
land reform, (def.) 200. *See also* agriculture
land use. *See* agriculture; mining; urbaniza-
    tion
Landsat satellites, (def.) 23, p 23
language. *See* culture; specific languages,
    regions, and countries
language groups, g 196, g 267, m 294,
    m 369, m 411, g 427
Laos, 504, m 531, g 532, 534, 535
La Paz, Bolivia, 249
Lapland, 280
Las Vegas, NV, 155
latifundia system, (def.) 250
Latin-American Integration Association, 200
Latin America. *See* Middle and South
    America; specific countries
latitude: (def.) S2–S3, g S3; and solar ener-
    gy, 14
Latvia, 318, p 318, g 318, m 319
Latvian language, g 267
Laurasia, 48
lava, (def.) 45
leaching, (def.) 56, g 56
Lebanon, 381, m 387, g 388, 395–96
legacies, (def.) 311
legends, (def.) S6, S7, m S7
Le Havre, France, 284
Lena River, 335
L'Enfant, Pierre, 130
Leningrad, Russia. *See* St. Petersburg, Russia
Lenin, V. I., 329–30
Lesotho, m 477, g 484, 486, m 486
Lesser Antilles, 193, m 215, 219
levees, (def.) 135
Lewis, Meriwether, 105

Lhasa, Tibet, p 514
Liberia, 425, m 441, g 445, 448
Libreville, Gabon, 472
Libya, m 431, g 433, 435–36
Liechtenstein, m 283, g 285, 306
light industry, (def.) 342. *See also* industry
    and industrialization
lightning, 26
Lima, Peru, 248
Limpopo River, 485
line graphs, S10, g S10
lingua franca, (def.) 445
Lisbon, Portugal, 299
literacy, (def.) 80
literature, 104, 186, 256, 322, 372, 414,
    492, 556, 598. *See also* primary sources
lithosphere, (def.) 16, g 16
Lithuania, 318, g 318, m 319
Lithuanian language, g 267
Liverpool, England, 274
Ljubljana, Slovenia, 314
llanos, (def.) 193, 227, 230
location: absolute versus relative location,
    3; elevation and temperature, 26. *See*
    also absolute location, relative location
lochs, (def.) 273
locks, (def.) 135, g 142, p 149
loess, (def.) 287, 513
Loire River, 283
London, England, 274, 275, p 281
Longfellow Mountains, 121
Long Island, 124, 125
longitude, (def.) S2, S3, g S3
Los Angeles, CA: m 163; earthquakes, 48,
    p 162–63; innovation, 69; as metropol-
    itan area, 116; as multicultural area,
    161, 162–63, p 163, 166; population
    growth, 161, 162, 163; smog, 92
Louisiana: m 133, g 136; agriculture, 135;
    cities, 137–38; fishing industry, 136;
    hurricanes, 134; Mississippi Delta, 133,
    135; oil production, 136; oil reserves,
    94; resources, 136
Louisiana Purchase, 105
low-latitude climates, 27, m 28–29, 30
Lubumbashi, Zaire, 470
Luxembourg, m 283, g 285, 291
Lyons, France, 285

M

Maastricht Treaty, 267
Maathai, Wangari, 456–57, p 457
Macao, m 507, 512
Macedonia, m 309, g 310, 314
Macedonian language, g 267
Machu Picchu, p 185, 186, 247
Madagascar, p 33, 426, m 477, g 484, 486
Madrid, Spain, 299
Magnitogorsk, Russia, 342
Ma Huan, 493, 552
Maine, 121, m 121, 122, p 123, g 124
mainland Southeast Asia: g 532; agricul-
    ture, 532–33; climate, 531–32; physical
    geography, 531, m 531; regional coop-
    eration, 533; resources, 533; settlement,
    532. *See also* specific countries
Makarakomburu mountain, g 50, m 50
Malagasy language, g 427, 486
Malawi, m 465, g 470, 474
Malayo-Polynesian languages, g 427

Malaysia, 541–42, m 541, g 542, p 551
Maldives, m 585, g 586, 592
Mali, 4, m 441, g 445, 446
malls. *See* shopping malls
Malta, m 297, g 298, 306
Managua, Nicaragua, 217
Manaus, Brazil, 238
Manchu dynasty, 508, g 508
Mandela, Nelson, 480, p 480
Manitoba, Canada, m 173, 177–78, 180
manufacturing. *See* industries and
    industrialization
Maoris, 598, p 598, 616–17
Mao Zedong, 509–10
mapmaking, S4–S5, g S4–S5
map projections, (def.) S4, g S4–S5, S5
maps: (def.) S4; elements, S6–S7, g S6–S7;
    historical examples, m 1; inset maps, S7;
    mental, S9; time-zone maps, S14–S15,
    m S14–S15; types, S8–S9, m S8–S9
Maputo, Mozambique, 485
maquiladoras, 206–207, (def.) 208, p 208
Marble Canyon, p 44
Marcos, Ferdinand, 550
Mariana Trench, 39
marine ecoystems, p 56, 56
Maritime Provinces, Canada, 176, 180
markets, (def.) 81
market economies, (def.) 81
Marseilles, France, 284–85
Marshall Islands, m 601, g 613, 614
marsupials, (def.) 605
Martha's Vineyard, 122
martial law, (def.) 515
Martinique, 224
Marx, Karl, 329–30
Mary, Turkmenistan, m 363
Maryland, p 37, m 121, 124, g 124, 125.
    See also specific cities
Masai, 454
Massachusetts, m 121, 122, 123, g 124.
    See also specific cities
Massif Central, 262, 283
mass production, 117–18
Maui, HI, m 159, 169
Mau Mau, 458
Mauritania, 4, m 441, 445–46, g 445
Mauritius, m 477, g 484, 486, m 486
Mayas, p 188, 195, m 195, 209
Mayle, Peter, 282
Mecca, Saudi Arabia, p 371, 380, 404, 408
Mediterranean Sea, 39, 263, 296, 304
Megalopolis: (def.) 120, 127, 130
Meiji Restoration, 520
Meir, Golda, 386
Mekong Delta, 531, 535
Mekong River, 531, 535, 536
Melanesia, m 601, 613, 614
Melbourne, Australia, 610
mental maps, (def.) S9
Mercator projections, S4, g S4
merengue, (def.) 219
Mérida, Mexico, 209
meridians, (def.) S3, g S3
Mesabi Range, 146
Mesa Verde, 155
Meseta, (def.) 297, 299
Mesopotamia, 72, 379–80, 399, m 399
mestizos, (def.) 196, 231, 242
meteorology, (def.) 8
Meuse River, 290

Mexico: g 204; agriculture, 204, p 205;
    Central Interior region, 208–209; cli-
    mate and vegetation, 204; culture
    regions, 208–209; economic geography,
    204–205, 208; formation, 193; future,
    212; Greater Mexico City region, 208;
    industrial growth, 205, 206–208;
    landforms, 203–204; land reform, 200,
    205; minerals, 197; northern border
    region, 205, 206–208, p 206–208,
    m 207, 209; oil production, 209; oil
    reserves, 197; physical geography,
    203–204, m 203; politics, p 212;
    social and economic issues, 212;
    Southern Mexico, 209; Yucatán
    Peninsula, p 188, 203, 209. *See also*
    Mexico City, Mexico
Mexico City, Mexico: m 211; challenges
    facing, 210–11; as culture region, 208;
    earthquakes, 211; as industrial center,
    205; history, 210; landforms, 203; land-
    marks, p 210–11, 210, p 213; pollution,
    92, 210–11
Miami, FL, 137
Michigan: m 141, g 143; cities, 148; Dairy
    Belt, 144; forests, 145; mining, 145. *See
    also* specific cities
Micronesia, m 601, 613, g 613, 614, 615
microstates, (def.) 258
Middle Ages, 265
Mid-Atlantic Ridge, m 48
Middle Atlantic states: m 121; agriculture,
    125; climate, 124–25; forests, 125;
    industry, 126; issues, 126; landforms,
    124; population, 124; resources, 125;
    urban growth, 126. *See also* specific
    states
Middle East. *See* Southwest Asia
middle-income countries, 81, 311
middle-latitude climates, m 28–29, 31
Middle and South America: 188; agricul-
    ture, 196, 197–98, p 201; Amazon
    River, 37; artifacts, p 188, p 202, p 214,
    p 226, p 234, p 244, p 247; climate,
    m 190, 193–94, g 194; economic geog-
    raphy, m 192, 197–98; economic issues,
    200; European colonization, 196; histori-
    cal geography, 195–97, m 195; hunter-
    gatherers, 69; independence, 196–97;
    industry, 198; landforms, 193; resources,
    197; physical geography, m 189,
    193–94; political issues, 199–200; popu-
    lation, m 191, 199, g 239; rain forests,
    p 30, 220–21; regional cooperation,
    200; regional issues, 199–200; river sys-
    tems, 194; settlement patterns, 195–96,
    m 195. *See also* specific countries
mid-ocean ridge, (def.) 47, m 48
Midway, m 601, 615
Midwestern United States: g 143; agricul-
    ture, 143–45; cities, 145, 146, 147–48;
    climate, 114, 141; Corn Belt, 143–44;
    g 144–45; Dairy Belt, 144–45; forests,
    145; industry, 145–46; interstate high-
    ways,147; issues, 146; landforms, 141;
    mining, 145–46; physical geography,
    m 141, 141–43; resources, 145, 146
migration, 84, 223, 588–89, g 589. *See also*
    movement; refugees
Mijas, Spain, p 296
Milan, Italy, p 300, 301

Milford Sound, New Zealand, p 51
Mili Atoll, Marshall Islands, p 61
Milky Way, 10
millet, (def.) 443
Millet, Jean-Francois, p 316
minarets, (def.) 395, p 406
minifundios, (def.) 250
Ming dynasty, 508, g 508
mining: land reclamation, 153; strip, 153
Minneapolis, MN, 148
Minnesota, 141, m 141, g 143, 144, 145,
    146, 148. *See also* specific cities
Minnesota River, 148
Minsk, Belarus, 354, p 354
Mississippi, m 133, 135, 136, g 136
Mississippi Delta, 133, 135, 136
Mississippi River, 37, 112, 117, 134–35,
    143, 146, 148
Mississippi Valley, 133, 136
Missouri: m 141, g 143; agriculture,
    143–44; cities, 148; Lake of the Ozarks,
    p 146; Ozark Plateau, 112, 134; Ozarks,
    145. *See also* specific cities
Missouri River, 37, 112, 134, 143, 148
Missouri River Valley, 141
mistral, (def.) 284
Mitre Peak, New Zealand, p 51
Mittelland, Switzerland, 291, 294
Mitumba Mountains, 465
mixed forests, (def.) 59
Mobile, AL, 135
models, use of in geography, 182, g 182
Mohawk River Valley, 126
Moldova, 318, g 318, 319, p 319, m 319
Molokai, HI, m 159, 169
Momaday, N. Scott, 2
Monaco, m 283, g 285, 306
Mongolia, 336, m 507, g 509, 516
Mongols: 328–29, 346, 516; dynasty in
    China (Yuan) 508, g 508
Monongahela River, 126
monopolies, (def.) 212
monotheism, (def.) 381
monsoons, (def.) 27, 499, 537, 563, p 563,
    570, 590, 592
Montana, m 151, 152, g 152, 154
Mont Blanc, 262
Monterrey, Mexico, 205, 209
Montevideo, Uruguay, 241
Montreal, Canada, p 103, 175, 177, 180
Mont Saint Michel, France, p 12
moon: in solar system, 11; tides and, g 12,
    40
Moors, 299
moraine, (def.) 122
Mormons, 116, 156
Morocco, p 410, p 421, m 431, g 433,
    437–38, p 438
Morris, Jan, 296
Moscow, Russia, p 70, p 320, 329, 330,
    332, p 332, 340, 341–42, p 342
mosques, (def.) 360, p 365, p 381, p 410,
    p 542
mountains: folds and faults in creation,
    g 45, 46; glaciers, p 43, 44; orographic
    effect and, 26–27, g 26; plate boundaries
    and formation, g 47, 47–48; tempera-
    ture and, 26. *See also* specific mountains
Mount Cameroon, 471–72
Mount Cook, 616
Mount El'brus, 355

Mount Elgon, 453
Mount Etna, 304
Mount Everest, p 269, 562, p 584
Mount Fuji, 519
Mount Godwin Austen, 585
Mount Hood, 113
Mount Kenya, 453
Mount Kosciusko, 604
Mount McKinley, 113, 165, p 168
Mount Orizaba, 203
Mount Pinatubo, 540
Mount Rainier, 113
Mount St. Helens, 113, p 113
Mount Shasta, 113
Mount Vesuvius, 304
Mount Washington, 121
movement: cultural change and, 68–69;
    Earth's rotation, 12–13; global wind
    belts, 21, 24; hunting and gathering,
    69; hydrologic cycle, g 36, 36–37;
    immigration, 588–89, g 589; ocean cir-
    culation, 24; plate boundaries, 46–48,
    g 47; refugees, 588–89; spatial interac-
    tion, 6; wind, p 44, 44–45
Mozambique, m 477, g 484, 485–86
muezzin, p 381, 395
Muhammad, 380–81
Muir, John, 158
mulattos, (def.) 196
multicultural region, (def.) 161
multilingual, (def.) 267
multinational companies, (def.) 81
Munich, Germany, 287
Murmansk, Russia, p 92
Murray-Darling river system, 605
music, 70–71, p 138, 219, p 225, p 238,
    p 318, p 353, p 475, p 547
muskegs, (def.) 179
Muslims, 116, 310, (def.) 380–81, 388,
    403, 443, 541, 566, 576, 586, 591. See
    also Islam
Myanmar. See Burma

N

NAFTA. See North American Free-Trade
    Agreement
Nagasaki, Japan, 520, 521
Nagorno-Karabakh, 356
Nagoya, Japan, 523
Nahuatl, g 196
Naipaul, V. S., 557, p 557
Nairobi, Kenya, 455, 458, p 458
Nama language, g 427
Namib Desert, p 44, 422
Namibia, m 477, 484, g 484
Nance, John, 540
Nanticoke River, p 37
Napa Valley, 160
Narayan, R. K., 556, p 556
Narmada River, 572, 573
Nashville, TN, 137
Natal, 479
nationalism, (def.) 74
nationalize, (def.) 274
National Park Service (U.S.), 5
Native Americans: artifacts, p 103, p 172;
    in Canada, 175; as early inhabitants of
    North America, 115, p 115; Great
    Serpent Mound, p 115; in
    Intermountain West, 156; languages,

g 196; in Middle and South America,
    195–96, m 195, 220–21, p 220–21,
    238, p 242, 246; in Midwestern United
    States, 140; in Pacific United States,
    164; ways of life, 152. See also specific
    groups
NATO. See North Atlantic Treaty
    Organization
natural boundaries, (def.) 262
natural resources. See resources
Nauru, m 601, g 613, 614
Navajo, 156
navigable, (def.) 263
Nazi Germany, 391
N'Djamena, Chad, 446
Nebraska, 39, 144, 151, m 151, 152, g 152
Nehru, Jawaharlal, 576–77
Nepal, p 554, 562, m 585, g 586, 587, 590
Neruda, Pablo, 244
Netherlands, m 283, g 285, 290–91,
    292–93, p 291–93
neutral, (def.) 279
Nevada, m 151, g 152, 155, p 157
New Brunswick, m 173, 176
New Delhi, India, p 83, 579
New England: m 121; climate, 122; fishing,
    122–23; industry, 123; issues, 123; land-
    forms, 121–22; resources, 122–23. See
    also specific states
Newfoundland, m 173, 176, p 176, 180
New France, 175
New Hampshire, 111, 121, m 121, g 124
New Jersey, m 121, 124, g 124, 125
newly industrialized countries (NICs),
    (def.) 501
New Mexico: g 152; Carson National
    Forest, p 34; climate, 155; mining, 154,
    155; Native Americans, 156; natural gas,
    94; oil production, 153; physical geog-
    raphy, m 151, 155
New Orleans, LA, 137–38, p 138
New South Wales, 600, m 601, 604, 605,
    606, 607, 610
newsprint, (def.) 174
New York City, NY: p 78, p 120; communi-
    cations, p 65; description, 127; landfills,
    p 90; as large metropolitan area, 116,
    120; people and traffic, p 64; satellite
    image of, p 23
New York state, m 121, g 124, 125, 126
New Zealand: p 51, 95, 96, p 598, g 613,
    p 615; economic geography, 617; his-
    torical geography, 616–17; physical
    geography, m 601, 615–16; population,
    617
Niagara Falls, 142
Niagara River, 142
Niamey, Niger, m 3, 446
Nicaragua: m 215, g 216, 217; culture, 68;
    geothermal energy, 95
Nice, France, p 31
Nicosia, Cyprus, 389
NICs. See newly industrialized countries
Niger, m 3, 4, m 441, g 445, 446
Niger-Congo language, g 427, 454
Nigeria, 68, 82, 426, m 441, 442, g 445,
    448–50
Niger River, 421, 449, p 449
Nile Delta, 434, p 434
Nile River, 89, 421, 432, 433, 434, 454
Nile River valley, 432, p 432

Nilo-Saharan languages, g 427, 454
nodal regions, (def.) 101
nonrenewable resources, (def.) 86
Normandy, France, p 266
North Africa: g 433; physical geography,
    431–32, m 431; as region in transition,
    410–11, m 411; Sahara, 4–5, 30, 45,
    421, 422, 431–32. See also specific
    countries
North America: m 107–10; grasslands
    biome, p 59; lowest point, 113, 159;
    population density, 82. See also Canada;
    Mexico; United States
North American Free-Trade Agreement
    (NAFTA), 118, 207, 250
North Atlantic Treaty Organization (NATO),
    265–66
North Carolina: m 133, g 136;
    Appalachians, 133; Blue Ridge Parkway,
    p 2; industry, 136; mountains, 111,
    p 111; Piedmont, 133
North Dakota, 151, m 151, 152, g 152
Northeastern United States: g 124; agricul-
    ture, 122, 125; climate, 113, 122,
    124–25; industry, 123, 126; issues, 123,
    126, 130; landforms, 121–22, 124;
    physical geography, m 121; resources,
    122–23, 125. See also specific subre-
    gions and states
Northern Central Africa, 471–72
Northern Eurasia: climate, m 326; collapse
    of Soviet Union, 331; economic geogra-
    phy, m 327, 332; environmental issues,
    332; human geography, 331–32; physi-
    cal geography, m 325; population,
    m 326. See also Russia; Soviet Union
Northern Europe, 272–81, m 273, g 274.
    See also specific countries
Northern Ireland, 275. See also Ireland
Northern Mariana Islands, 615
Northern Rhodesia. See Zambia
Northern Territory, Australia, 600, m 601,
    604, p 604, 605, p 611
North Korea, 504, m 519, g 520, 526–28
North Pole, 13, 14
North Sea, 263, 277, 291
North Star, 13
Northwest Territories, Canada, 179, 180
Norway, 95, m 273, g 274, 277–78, p 277
Norwegian language, g 267
Nouakchott, Mauritania, 445
Nova Scotia, m 173, 175, 176
nuclear energy, 94, p 94, p 97
Nullarbor Plain, 604
Nunavut, m 173, 179

O

Oahu, HI, m 159, 169, 170, p 170
oases, (def.) 359–60
OAS. See Organization of American States
OAU. See Organization of African Unity
Ob River system, 335
Oceania, (def.) 612
oceans: (def.) S2, characteristics, 39–40;
    division of world ocean, S2; circulation,
    24, 40; pollution, 89, 92; temperature,
    40. See also specific oceans
Ogallala Aquifer, 39, 153
Ohio, m 141, 143, g 143, 145, 146, 148
Ohio River, 37, 112, 117, 134, 135, 143, 146

Ohio River Valley, 141, 146
oil deposits and production, 93–94. *See also* specific regions and countries
Ojike, Mbonu, 464
Okavango River, 484
Okefenokee Swamp, 133
Okinawa, 519, m 519
Oklahoma, 134, 151, m 151, 152, g 152
Olduvai Gorge, 458
Olmecs, 195, m 195
Olympic Mountains, 164
Olympic National Forest, p 165
Oman, m 399, g 400, 405, 406
Ontario, Canada, 177, 178, 180
Ōoka, Makoto, 518
OPEC. *See* Organization of Petroleum Exporting Countries
Oporto, Portugal, 299
Orange Free State, 477, 479
Orange River, 421, 477, 478
Oregon: m 159, g 160; agriculture, 164; cities, 165; climate, 164; Crater Lake, 38, p 164; Depoe Bay, p 41; issues, 165; landforms, 164; resources and industry, 164–65
Organization of African Unity (OAU), 462
Organization of American States (OAS), 200
Organization of Petroleum Exporting Countries (OPEC), 379
Oriente, Ecuador, 245
Oriente, Peru, 247
Orinoco River, 194, 230
Orlando, FL, 137
orographic effect, (def.) 26–7, g 26
Orontes River, 395
Osaka, Japan, 523, p 523
Oslo, Norway, 278
Osman, 380
Ottawa, Canada, p 174, p 175, 177, 180
Ottoman Empire, 380, 408
Ouagadougou, Burkina Faso, 446, p 446
Outer Mongolia, 516
Ozark Plateau, 112, m 133, 134, 141, 145
ozone layer, (def.) 92–93, g 92

P

Pacific islands, 613–15, 622. *See also* specific island countries
Pacific Ocean, S2, 39. *See also* entries beginning with "Pacific"
Pacific plate, 46, m 47
Pacific Rim, 166–67, m 167
Pacific Ring of Fire, 113, 498, 547
Pacific South America, 244–50, m 245, g 246. *See also* specific countries
Pacific states, U.S.: g 160; agriculture, 160, 164, 170; climate, 114, 160, 164, 168, 170; industry, 161, 165; landforms, 113, 159–60, 164, 165, 168, 169; Pacific Rim trade, 166–67; physical geography, m 159, 159–60, 164–65, 168, 169–70; resources, 160–61, 164, 165, 168. *See also* specifc states
paddies (padis), (def.) 500
Pakistan, 96, p 407, 566, 585–87, m 585, g 586, p 586
Palau, g 613, 614
Palestine, 381, 390, 391
Palestinian Liberation Organization (PLO), 394

Pamirs, 359
pampas, (def.) 193, 240
Panama, 68, 200, m 215, g 216, 218–19, p 219, p 220–21, m 221
Panama Canal, 219, p 219
Panama City, Panama, 219
Pangaea, 48, m 48
panhandles, (def.) 165
Papua New Guinea, m 601, p 612, g 613, p 613, 614
Paraguay, m 235, g 236, 242, p 242
Paraguay River, 242
parallels, (def.) S2–S3, g S3
Paramaribo, Suriname, 232
páramo, (def.) 228, g 228
Paraná River system, 194
Paris, France, p 254, p 268, 284, p 284, p 295
parks: in Canada, p 172; in United States, 5–6, 154, 155, p 168
parliament, (def.) 175
partition, (def.) 585
pastoralism, (def.) 443
Patagonia, 240
Pathans, 403
Paton, Alan, 476
Paz, Octavio, 202
peat, (def.) 276, 277
Peloponnesus, 305
peninsulas, (def.) 122. *See also* specific peninsulas
Pennsylvania, m 121, g 124, 125, p 125, 126, 130
Penn, William, 130
perennial irrigation, (def.) 433
perestroika, (def.) 331
perimeters, (def.) 405
periodic markets, (def.) 466, 468–69, p 468
permafrost, (def.) 32
Persian Gulf, 39, 379
Persian Gulf War, 400–401, 588–89
Perth, Australia, 605, 610
Peru: m 245, g 246; Andes highlands, p 30, 247; climate, 194; coastal region, 248; minerals, 197; Oriente, 247; political geography, 248
pesticides, 548–49
Peter the Great, 329, 344
petrochemicals, (def.) 93
Petrograd, Russia. *See* St. Petersburg, Russia
petroleum. *See* oil deposits and production
Philadelphia, PA, 127, 130
Philippines: 504, 540, g 542, p 547, p 550; economic geography, 550; historical geography, 550; physical geography, m 541, 547; political geography, 550; population 547, 550. *See also* Spratly Islands
Phoenix, AZ, 155–56, p 156
phosphates, 136, 438
photochemical smog, (def.) 92
photographs, interpreting, 488
photosynthesis, (def.) 53
physical geography: (def.) 8; linked with human geography, 594. *See also* specific continents, regions, and countries
physical–political maps: S6–S7, m S6–S7, S8
pictograms, (def.) 502–503, p 502
piedmont, (def.) 111
Piedmont, U.S., 111, 124, 133

pie graphs, S10, g S10
pilgrimages, (def.) 404
Pilipino language, 550
Pindus Mountains, 306
Piraeus, Greece, 306
Pittsburgh, PA, 126
place: (def.) 4; human characteristics, 4; physical characteristics, 4
plains, (def.) 46
planets, (def.) 11, g 11
plantains, (def.) 223
plantation agriculture, 115, 198, 204
plantations, (def.) 115
plant communities, (def.) 53–54, p 53
plants: biomes and, m 57, p 58–59, 58–60; climax community, 55; forest succession after a fire, g 54, 54–55; growth, 53–54; marine ecosystems, p 56, 56; photosynthesis, 53; plant succession, g 54, 54–55
plant succession, g 54, 54–55
Plateau of Mexico, 203, 204
plateaus, (def.) 46
plate boundaries, 46–48, g 47
plate tectonics, (def.) 46, 46–48, m 47, 159–60, 193, 203–204, 398, 562
plazas, (def.) 298
PLO. *See* Palestinian Liberation Organization
Ploiesti oil fields, 312
Pocono Mountains, 111
Pointe-Noire, Congo, 471
points of view, analysis of, 98–99
Poland, m 309, g 310, 311–12, p 312, 319
polar deserts, (def.) 620
polar ice-cap climate. *See* ice-cap climate
polar regions, (def.) 14
polar winds, (def.) 21
polders, (def.) 290, 292–93, m 293, p 293
Polish language, g 267
politics: culture and, 75; economics and, 81–82. *See also* specific countries
pollution: air, 92, 130, 161, 163, p 163, 301, p 303; water, 89, 92, 264
Polo, Marco, 358, 363
Polynesia, 268, m 601, 613, g 614, 615
Pompeii, Italy, 304
Pontic Mountains, 387
popular culture, 70–71
population: global density, 82; growth, 82–84, g 83, 96; resources and, 96
population geography, 82–84. *See also* human geography
population maps, S9, m S9
population pyramids, S11, g S11, g 83
portages, (def.) 175
Port-au-Prince, Haiti, 223
Port-Gentil, Gabon, 472
Portland, OR, 165, 166
Portugal, 297–300, m 297, g 298
Portuguese language, 188, g 196, 236, 241, g 267
potash, (def.) 178
Potomac River, 124
poultry farming, 122, 136
Po Valley, 300, 301
Prague, Czech Republic, p 308, 312
prairies, (def.) 59
Prairie Provinces, Canada, 177–79
precipitation: (def.) 25; global climates and, 27–32, m 28–29; groundwater and, 39
Pretoria, South Africa, 479
prevailing winds, (def.) 21

prewriting, S17
primary industries, (def.) 79
primary landforms, 49
primary sources, (def.) S17, 2, 10, 18, 34,
    42, 52, 66, 78, 86, 98, 105, 120, 132,
    140, 150, 158, 172, 187, 202, 214,
    226, 234, 244, 257, 272, 282, 296,
    308, 323, 334, 350, 358, 373, 386,
    398, 415, 430, 440, 452, 464, 476,
    493, 506, 518, 530, 540, 557, 570,
    584, 599, 600, 612. See also literature
primate cities, (def.) 284
prime meridian, (def.) S3, g S3
Prince Edward Island, m 173, 176
Priština, Serbia, p 314
privatization, (def.) 212
Promontory Point, Utah, 166
proofreading, S17
Protestants, 116, 267
Prudhoe Bay, 168
Puebla, Mexico, p 65, 205
Puerto Rico, 193, m 215, g 216, 219, 223,
    p 223
Puget Sound, 37, 164
Puget Sound Lowland, 113
Punjabis, 586
puppet governments, (def.) 509
P'yŏngyang, North Korea, 527
Pyrenees, 262, 283, 298

Q

al-Qaddafi, Muammar, 436
qanats, p 382, 401, p 401
Qaraghandy, 361
Qatar, m 399, g 400, 405–406
Qattara Depression, 432
Qin Ling, 498
Qom, Iran, 402
quaternary industries, (def.) 79
Quebec, Canada, p 103, m 173, 175,
    176–77, p 177, 178, 180, p 181
quebracho forests, 240, 242
Quechua, g 196, 246
Queensland, Australia, 600, m 601, 604,
    605, 606, 607, 610
Quiché, g 196
Quito, Ecuador, p 193, 246, p 246
quotas, (def.) 76
Qur'an, (def.) 381

R

Rabat, Morocco, 438
rain. See precipitation
rain forests, p 17, p 30, m 57, 58, 88, p 88,
    p 199, 220–21, 235, 422, 533, p 533
rain-shadow effect, (def.) 27
Raleigh, NC, 137
Rangoon, Burma, 537, p 538
recycling, 90–91
Red River, 134
Red Sea, 398, 406
reforestation, (def.) 87–88
refugees, (def.) 223, p 396, 588–89
reggae, (def.) 219
region: (def.) 7, 100, m 101; boundaries,
    100–101; characteristics, 100; comple-
    mentary, 312; nodal, 101; subregions,
    100; types, 101; uniform, 101
regional cooperation: in Eastern Europe,

311; in Europe, 267; in Middle and
    South America, 200; in North America,
    206–207; in Southeast Asia, 533
regionalism, (def.) 180
regional specialization, (def.) 311
regions in transition: Baltic countries,
    318–19, m 319; Central Asia, 368–69,
    m 369; defining, 100–101; Europe,
    318–19; North Africa, 410–11, m 411
regs, (def.) 432
relative location, (def.) 3
relief, (def.) 49
religion. See culture; specific religions,
    regions, and countries
remote sensors, (def.) 22
Renaissance, (def.) 301
renewable resources, (def.) 86, 94–95, 129
Republic of South Africa, m 477, 479–81,
    g 484
resources: energy, 93–95, p 94–95; forests,
    87–88, p 88; nonrenewable, 86; popula-
    tion and, 96; renewable, 86, 94–95,
    129; soil, 87; water quantity and quality,
    89, 92. See also specific regions, coun-
    tries, and types of resources
reunification: (def.) 266, in Germany, 287;
    in Korea, 527–28; in Vietnam, 534–35
revolution, of Earth, (def.) 13
Rhineland, 287
Rhine River, p 262, 264, 286, 291
Rhode Island, m 121, g 124
Rhone River, 283, 291
rifts, (def.) 421–22
Ring of Fire. See Pacific Ring of Fire
Rihani, Ameen, 398
Rio de Janeiro, Brazil, p 234, 236, 237,
    p 243
Río de la Plata, 194, 240, 241
Rio Grande, 135
Río Negro, 194
rivers: dredging and channeling, 143; exot-
    ic, 378, 384; hydroelectricity, 94–95;
    pollution, 89, 92. See also specific rivers
Riyadh, Saudi Arabia, 404
roads. See highways
Robinson projections, S5
rock weathering, (def.) 43, p 43
Rocky Mountains: 112–13, 117, 134,
    p 150; in Canada, p 43, 178–79; eco-
    nomic geography, 154; glaciers, p 43,
    44; physical geography, m 151, 154
Roman alphabet, p 319
Roman Catholicism, 116, 267, 275, 276,
    300, 310, 318, 353, 354, 380, 550,
    p 550
Roman Empire, 265, 299, 301, 392
Romance languages (Latin or Italic), g 267
Romania, m 309, g 310, 312–13
Romanian language, g 267, 310
Romansh language, m 294
Rome, Italy, 301, p 301
Rosario, Argentina, 241
Rose, Wendy, 104, p 104
rotation, of Earth, (def.) 12–13
Rotterdam, Netherlands, 291
Rub` al-Khali, 404
Ruhr, 287
Russia: agriculture, 340–41; artifacts, p 321,
    p 324, p 334; climate, m 326, 336–37;
    coal reserves, 93; diverse heritage,
    328–29, g 347; economic and industrial

regions, 341–43, 346; economy, m 327,
    366; fall of Russian Empire, 329; forests,
    p 334; hydroelectricity, 95; issues,
    347–48; Lake Baykal, 336, 338–39, p 338;
    landforms, 335–36; Moscow region,
    341–42; physical geography, m 325,
    m 335, 335–37; political geography,
    346–48; population, 96, m 326;
    resources, 337, 340; Russian Empire,
    318, 328–29, m 328; Russian Far East
    region, 343, 346; Russian Federation,
    346–47; St. Petersburg region, 342,
    344–45, p 344–45, m 345; Siberia,
    343; teenagers, 348; Urals region, 342;
    vegetation, 337; Volga region, 342.
    See also Soviet Union
Russian Empire, 318, 328–29, m 328
Russian Far East, 343, 346
Russian Federation, 346–47. See also Russia
Russian language, g 267
Rwanda, m 453, 459–60, g 459
Ryukyu Islands, 519, m 519

S

Sadik, Nafis, 78
Safi, Morocco, p 438
Sahara, 4–5, 30, 45, 421, 422, 431–32,
    441–43, 445
Sahel, 4–5, p 4–7, (def.) 422, 441–43, 445
Sahel countries, 445–46
Saigon, Vietnam, 534–35. See also Ho Chi
    Minh City
St. Clair River, 142
St. Croix, m 215, 224
St. John, m 215, 224
St. Kitts and Nevis, m 215, g 216, 224
St. Lawrence Lowlands, 173
St. Lawrence River, 142, m 142, 175
St. Lawrence Seaway, 142, m 142
St. Louis, MO, 148
St. Lucia, m 215, g 216, 224, p 225
St. Marys River, 142
St. Paul, MN, 148
St. Petersburg, FL, 137
St. Petersburg, Russia, p 341, 342, 344–45,
    p 344–45, m 345, p 349
St. Thomas, m 215, 224
St. Vincent and the Grenadines, m 215,
    g 216, 224
Salonika, Greece, 306
Salt Lake City, UT, 156
Salton Sea, p 72
Samarqand, Uzbekistan, p 362, m 363
samurai, 520, p 521
San (Bushmen), 69, 426, 478
Sanaa, Yemen, 406, p 406
San Andreas Fault, 48, 159–60, p 160
San Antonio, TX, 138
San Blas Mountains, 220
sanctions, (def.) 480
sanctuaries, (def.) 572
sand dunes, (def.) 44–45, p 44
Sand Hills, 151
San Diego, CA, 161, 206, 209
Sandinistas, 217
San Francisco Bay Area, CA, 48, 161, p 167
San Juan, Puerto Rico, p 223
San Marino, m 297, g 298, 306
Sanskrit, 564
Santa Cruz, Bolivia, 249

São Paulo, Brazil, 236, 237, p 237
São Tomé and Príncipe, m 465, g 470, 472
Sapi, 440
Sarajevo, Bosnia and Herzegovina, 314
sari, p 578
Saskatchewan, Canada, 177, 178, p 179
Saskatoon, Saskatchewan, p 179
satellites: (def.) 11, 22–23; Landsat, 23, p 23; weather, p 22, 22–23
Saudi Arabia: p 40, 94, p 101, p 370, p 371, p 384, 400, g 400; economic geography, 404–405; global role, 405; physical geography, m 399, 404; population, 404; water resources, 382
Sault Ste. Marie, MI, p 149
savannas, (def.) 59
savanna biome, m 57, p 58, 59
scales, (def.) S6
Scandinavia. See Nordic countries
Schelde River, 290
Scotland. See United Kingdom
Scott, Robert Falcon, 618
Scottish Gaelic language, g 267
seaboards, (def.) 115
Sea of Galilee, 390
Sea of Japan, 519
seas, S2. See also specific seas
seasons, g 14–15, 14–15
Seattle, WA, p 114, 165, 166
seawater, 40
secondary industries, (def.) 79
secondary sources, S17
second-growth forests, (def.) 123
secularism, (def.) 381
sediment, (def.) 43
Seine River, 283, 284
selvas, (def.) 245
Semitic languages, m 369
Senegal, p 429, m 441, g 445, 447, p 447
Seoul, South Korea, 527, p 527
Serbia and Montenegro (Yugoslavia), m 309, g 310, 313, 314
Serbo-Croatian language, g 267
Serengeti Plain, 422, 458
serfs, (def.) 340
Serra dos Carajás, Brazil, 238
Seth, Vikram, 506
Seward, William H., 168
Seychelles, m 477, g 484, 486
SEZs. See Special Economic Zones
Shannon River, 276
shantytowns, (def.) 199
shelterbelts, (def.) 152
Shenandoah Valley, 112
Sheridan, WY, p 153
Shi`a, 381, 402, 403
shifting agriculture, (def.) 532
Shikoku, p 498, 519
Shilka River, p 336
Shinto, p 503, (def.) 504
shipping. See international trade
shoguns, 520
shopping malls, 3–4, p 5
Siberia, p 32, 335, p 336, 337, 340, 343
Sicily, m 297, 300, 304
Sierra Leone, 440, m 441, g 445, 448
Sierra Madre Occidental, 203
Sierra Madre Oriental, 203
Sierra Morena, Spain, 298
Sierra Nevada, Spain, 298
Sierra Nevada, U.S., 44, 113, 159, 160

Sikhism, (def.) 565
Sikhs, 565, 576, 577
Silk Road, 360, 362–63, m 363, 508
silt, (def.) 432
Simia, India, p 565
Simon, Paul, 71
Sinai Peninsula, m 387, 394
Singapore, 96, 501, m 541, 542, g 542, 544–45, p 544–45, m 545
Sinhalese, 592
sisal, (def.) 458
sites, (def.) 3
situations, (def.) 3
sketch maps, 62
Skills Handbook: charts, S12–S13, g S12–S13; diagrams, S10–S11, g S10–S11; elements of maps, S6–S7, m S6–S7; globe, S2–S3, m S2–S3; graphs, S10–S11, g S10–S11; mapmaking, S4–S5, g S4–S5; tables, S12–S13, g S12–S13; thinking critically, S16; time-zone maps, S14–S15, m S14–S15; types of maps, S8–S9, m S8–S9; writing skills, S16–S17
Skopje, Macedonia, 314
slash-and-burn farming, (def.) 231, 532
slavery, 115, 135, 447–48
Slavic languages, g 267, 310
Slovakia, p 67, m 309, g 310, 312
Slovak language, g 267
Slovene language, g 267
Slovenia, m 309, g 310, 313, 314
smelters, (def.) 342
smog, 92, p 162–63, 210–11
Snake River, 155
socialism, (def.) 278–79
Sofia, Bulgaria, 313
soil: as resource, 87; exhaustion, 236; formation, 55; horizons, 55–56, g 56; salinization, 87
solar energy: as resource, 95; factors controlling amount falling on Earth, 12–13, g 13; for homes, p 129; hydrologic cycle and, 36; latitude and, 14; seasons and, g 14–15, 14–15; solar-powered automobiles, 128–29, p 128–29
solar system, 11, g 11
Solomon Islands, m 601, g 613
solstices, (def.) 14, g 14
Solzhenitsyn, Aleksandr, 334
Somalia, m 453, g 459, 462
Somalis, 425–26
Soo Locks, p 149
sorghum, (def.) 443
South Africa, Republic of, p 478, 479–81, p 480, 482–83
South America. See Middle and South America; specific countries
South Asia: agriculture, 566–67; artifacts, p 555, p 558, p 570, p 584; climate, m 560, 563–64; cultural influences, 564–65; economic geography, m 561, 566–68; floods, 594; historical geography, 564–66; island countries, 591–92; landforms, 562–63; monsoon, 563–64; physical geography, m 559, 562–64; population, m 560, 568, g 568; resources and industry, 567–68; as subcontinent, 558; tectonic origins, 562. See also specific countries
South Australia, 600, m 601, 604, 605, p 607, 610

South Carolina, 133, m 133, 136, g 136
South China Sea, m 495, 499, 511, 512, 547, 550
South Dakota, 144, 151, m 151, g 152
Southeast Asia. See East and Southeast Asia; specific subregions and countries
Southern Africa: m 477, g 484; climate, 477–78; historical geography, 478–79; landforms, 477
Southern Central Africa, 473–74
Southern Europe 296–307, m 297, g 298. See also specific countries
Southern Ocean, 620
Southern United States: g 136; agriculture, 135–36; cities, 137–38; climate, 114, 134; economic geography, 135–38; industry, 136–37; issues, 138; landforms, 133–34; physical geography, m 133, 133–35; resources, 136. See also specific states
South Island, New Zealand, p 615
South Korea, 94, 501, g 501, 504, m 519, g 520, 527–28, p 529
South Pole, 13, 620
Southwest Asia: agriculture, 378–79; artifacts, p 371, p 374, p 386, p 398; climate, m 376; culture and conflict, 381, 384; economy, m 377; historical geography, 379–80; issues, 384; language, m 369; oil production, p 101; physical geography, m 375, 378; population, m 376; as region in transition, 368–69, m 369, 410–11, m 411; religion, 380–81, 408; resources and industry, 379; water resources, 382–83. See also specific countries
sovereignty, (def.) 550
soviets, (def.) 330
Soviet Union: Chernobyl, 94; Cold War and, 331; collapse, 222, 318, 324, 331, 516; daily life, 330–31; economic geography, 330; industry, 341–42; principles underlying, 329–30; world culture and, 70–71, p 70. See also Russia
Spain, 297–300, m 297, g 298
Spanish–American War, 550
Spanish language, 188, g 196, 241, g 267
spatial interaction, (def.) 6
Special Economic Zones (SEZs), 512
Spratly Islands, m 541, 550
Sri Lanka, 75, 566, m 585, g 586, 591–92
Stalin, Joseph, 330
staples, (def.) 443
stars, 10
steppe, g 28, (def.) 30–31, 59–60
steppe region, p 321, 322, 328, 337
stereotypes, (def.) 552
Stockholm, Sweden, p 278, 279
storms, p 25, 25–26
storm surges, (def.) 591
Strait of Gibraltar, 296
Strait of Messina, 304
strip mining, (def.) 153
subcontinents, (def.) 558
subduction, (def.) 47
subregions, 100
subsidies, (def.) 521
subsistence agriculture, (def.) 72
suburbanization, 116
Sudan, m 453, g 459, 460–61, p 460
Sudd swamps, 421

Suez Canal, 394, 430, p 430, m 545
Sukuma, 454
Sulawesi, m 541, 543
sultanates, (def.) 406
Sumatra, p 540, m 541, 543
sun: energy, 12–15, g 13–14, p 15; seasons and, g 14–15, 14, p 15; in solar system, g 11, 11–12; tides and, g 12, 40
Sunbelt, 116
Sunni, 381
Sun Yat-sen, 509
Sung dynasty, 508
superpowers, (def.) 331
suqs, (def.) 395
surface water, p 37–38, 37–38
Suriname, p 30, 196, 226, m 227, g 229, 232, p 232
Surtsey, p 49, 49
Susquehanna River, 124
sustained yield use, (def.) 88
Swahili, g 427, (def.) 454
Swaziland, m 477, g 484, 486
Sweden: m 273, g 274, 278–79, p 278; nuclear energy, 94
Swedish language, g 267
Switzerland: p 282, m 283, g 285, 291, 294; languages, m 294
Sydney, Australia, 608–609, 610, p 608–609, m 609
Syr Darya, 359, 360
Syria, 383, m 387, g 388, 394–95, p 395

T

tables, S12–S13, g S12–S13
Tabriz, Iran, 402
Tagalog, 550, 616
Tagus River, 298
Tahiti, m 601, 615
taiga, p 334, (def.) 337
T'aipei, Taiwan, 516
Taiwan, m 507, g 509, 515–16
Tajikistan, m 359, 364, g 364, p 368–69
Tajiks, 360, 364
Taj Mahal, p 555
Taklimakan Desert, 30, 499
Tallinn, Estonia, 318
Tamils, 592
Tampa, FL, 137
T'ang dynasty, 508
Tangier, Morocco, 438
tannin, (def.) 240, 242
Tanzania, p 67, p 413, 422, m 425, m 453, 454, 458–59, g 459
Taoism, (def.) 503
TAPS. *See* Trans-Alaska Pipeline System
tariffs, (def.) 76
Tarim Basin, 499
Tashkent. *See* Toshkent, Uzbekistan
Tasmania, 600, m 601, 604, 605, 606, 607, 610
Taurus Mountains, 387
T'bilisi, Georgia, 355
tea production, g 512–13
technology, information age and, p 65, p 76
tectonic activity, (def.) 46
teenagers: in Canada, 178; in Costa Rica, 218; environmental action and, 90–91, p 90, p 91; European exchange students, 305; in India, 579; in Israel, 391;

in Mexico, p 205; in New Zealand, 616; in Niger, 444; in Russia, 348; in South Korea, 528; in Tour de Sol, 128
Tehran, Iran, 402
Tel Aviv, Israel, 394
telecommunications, (def.) 80
Television InfraRed Observation Satellite (TIROS), 22
temperature: (def.) 19; of Earth's interior, g 45; elevation and, 26; global climates and, 27–32, m 28–29; of oceans, 40
tenant farmers, (def.) 135
Tennessee, m 133, 134, g 136, 137
Tennessee River, 117, 134, 135
Tennessee Valley Authority (TVA), 135
Tenn-Tom Waterway, 135
Tenochtitlán, 195, 210
terrorism, (def.) 248
tertiary industries, (def.) 79
Texas: m 133, g 136; agriculture, 135, 136, 152; cities, 138; climate, p 31, 134; fishing industry, 136; Great Plains, 133, 134; industry, 136; Interior Plains, 134; irrigation, 35; landforms, 134; natural gas, 94; natural regions, 134, m 134; Ogallala Aquifer, 39; oil production, 136; oil reserves, 94; resources, 136. *See also* specific cities
textiles, (def.) 121
Thailand, 96, p 100, m 531, g 532, p 533, 536–37
Thames River, 274, 275
Thar Desert, 563–64, 572, 585, m 585
themes. *See* Five Themes of Geography; human-environment interaction; location; movement; place; region
theocracies, (def.) 403
Theroux, Paul, 226
Thessaly, plain of, 306
thinking critically, S16
thunderstorms, 26, p 33
Tian Shan, 359, 498
Tiber River, 300, 301
Tibet, m 507, 514
tides, g 12, 40
tierra caliente, (def.) 227, g 228
tierra fría, (def.) 228, g 228
tierra helada, (def.) 228, g 228
tierra templada, (def.) 228, g 228
Tigre, 462
Tigris River, 379, 399
Tijuana, Mexico, 205, 209
Timbuktu, Mali, 424
time lines, S12, g S12
time-zone maps, S14–S15, m S14–S15
Timor, 543
Tirane, Albania, 313
TIROS. *See* Television InfraRed Observation Satellite
Tito, Josef Broz, 313
Togo, m 441, g 445, 450
Tokyo, Japan: 523, 524–25, p 524–25, m 525; world culture and, p 70
Tombigbee River, 135
Tonga, m 601, g 613, 614
topographic maps, (def.) 8
toponyms, (def.) 156
topsoil. *See* humus
tornadoes, (def.) 26
Toronto, Ontario, 175, 177, 180

Tortola, 224
Toshkent, Uzbekistan, 364
totalitarian governments, (def.) 75, 330
Tour de Sol, 128–29
tourism, 5–6, 268–69
township and range system, (def.) 143
trade. *See* international trade
trade deficits, (def.) 118
trade surpluses, (def.) 522–23
trade winds, (def.) 21
Trans-Alaska Pipeline System (TAPS), 168, p 168
transition regions. *See* regions in transition
transition zones, (def.) 101
transpiration, (def.) 37
Transvaal, 477, p 478, 479
tree lines, (def.) 154
trenches, (def.) 47
tributaries, (def.) 37
Trinidad and Tobago, 214, m 215, g 216, 219, 224
Tripoli, Libya, 436
tropical cyclones, (def.) 564
tropical rain forests. *See* rain forests
Tropic of Cancer, g 14, (def.) 15
Tropic of Capricorn, g 14, (def.) 15
tropics, (def.) 14
truck farms, (def.) 125
trust territories, (def.) 472
tsunami, (def.) 520
Tswana, 479, 484
Tuareg, 444
Tucson, AZ, p 114, 155
tundra, p 32, 32, m 57, p 59, 60
Tunisia, 424, m 431, g 433, 436–37
Tūnis, Tunisia, 436, p 436
Turin, Italy, 301
Turkic languages, 360, 368, m 369
Turkey: p 371, 380, g 388, p 389; economic geography, 388; modern society, 388–89; physical geography, m 387, 387–88; urban geography, 388. *See also* Cyprus
Turkmenistan, p 358, m 359, 364, g 364
Tuvalu, m 601, g 613, 614
TVA. *See* Tennessee Valley Authority
Twin Cities, MN, 148
typhoons, (def.) 25
Tyrrhenian Sea, 304

U

Uganda, m 453, g 459, 460, p 460
Ukraine: p 321, p 333, g 352, p 353; Crimea, 353–54; economic geography, 352–53; human geography, 353; physical geography, 351, m 351; resources, 351–52
Ukrainian language, g 267
Ulaanbaatar, Mongolia, 516
Uniate Church, 353
uniform regions, (def.) 101
uninhabitable, (def.) 279
Union of Soviet Socialist Republics (USSR). *See* Soviet Union
United Arab Emirates, p 381, m 399, g 400, 405–406
United Kingdom: g 274; economic geography, 274; global role, 275; physical geography, m 273, 273–74. *See also* British Empire; London, England